Mind on Statistics

Chapter 8

Chapter 9

Chapter 10

Chapter 11

Chapter 12

Chapter 13

Chapter 14

Chapter 15

Chapter 16

Chapter 17

Supplemental Topic 1

Supplemental Topic 2

Supplemental Topic 3

Supplemental Topic 4

Supplemental Topic 5

Mind on Statistics

Fourth Edition

Jessica M. Utts

University of California, Irvine

Robert F. Heckard

Pennsylvania State University

BROOKS/COLE
CENGAGE Learning™

Australia • Brazil • Japan • Korea • Mexico • Singapore • Spain • United Kingdom • United States

BROOKS/COLE
CENGAGE Learning™

***Mind on Statistics,* Fourth Edition**
Jessica M. Utts and Robert F. Heckard

Publisher: Richard Stratton
Senior Sponsoring Editor: Molly Taylor
Senior Development Editor: Jay Campbell
Associate Editor: Dan Seibert
Senior Editorial Assistant: Shaylin Walsh
Media Editor: Andrew Coppola
Marketing Manager: Ashley Pickering
Marketing Communications Manager: Mary Anne Payumo
Content Project Manager: Cathy Labresh Brooks
Art Director: Linda Helcher
Rights Acquisitions Specialists: Amanda Groszko, Katie Huha
Print Buyer: Diane Gibbons
Production Service: Graphic World Inc.
Text Designer: Itzhack Shelomi
Cover Designer: Rokusek Design
Compositor: Graphic World Inc.

Library of Congress Control Number: 2010931695

ISBN-13: 978-0-538-73348-9

ISBN-10: 0-538-73348-9

Brooks/Cole
20 Channel Center Street
Boston, MA 02210
USA

Printed in the United States of America
1 2 3 4 5 6 7 14 13 12 11 10

To Bill Harkness—energetic, generous, and innovative educator, guide, and friend—who launched our careers in statistics and continues to share his vision.

and

To our students, from whom we continue to learn, and who teach us how to be better teachers.

Brief Contents

Contents

Instructors: the Supplemental Topics are available on the book companion website on CourseMate, or print copies may be custom published.

Preface

A Challenge

Before you continue, think about how you would answer the question in the first bullet, and read the statement in the second bullet. We will return to them a little later in this preface.

- What do you *really know* is true, and how do you know it?
- The diameter of the moon is about 2160 miles.

What Is Statistics, and Who Should Care?

Because people are curious about many things, chances are that your interests include topics to which statistics has made a useful contribution. As written in Chapter 17, "information developed through the use of statistics has enhanced our understanding of how life works, helped us learn about each other, allowed control over some societal issues, and helped individuals make informed decisions. There is almost no area of knowledge that has not been advanced by statistical studies."

Statistical methods have contributed to our understanding of health, psychology, ecology, politics, music, lifestyle choices, business, commerce, and dozens of other topics. A quick look through this book, especially Chapters 1 and 17, should convince you of this. Watch for the influences of statistics in your daily life as you learn this material.

How Is This Book Different? Two Basic Premises of Learning

We wrote this book because we were tired of being told that what statisticians do is boring and difficult. We think statistics is useful and not difficult to learn, and yet the majority of college graduates we've met seemed to have had a negative experience taking a statistics class in college. We hope this book will help to overcome these misguided stereotypes.

Let's return to the two bullets at the beginning of this preface. Without looking, do you remember the diameter of the moon? Unless you already had a pretty good idea or have an excellent memory for numbers, you probably don't remember. One premise of this book is that **new material is much easier to learn and remember if it is related to something interesting or previously known**. The diameter of the moon is about the same as the air distance between Atlanta and Los Angeles, San Francisco and Chicago, London and Cairo, or Moscow and Madrid. Picture the moon sitting between any of those pairs of cities, and you are not likely to forget the size of the moon again. Throughout this book, new material is presented in the context of interesting and useful examples. The first and last chapters (1 and 17) are exclusively devoted to examples and case studies, which illustrate the wisdom that can be generated through statistical studies.

Now answer the question asked in the first bullet: What do you *really know* is true, and how do you know it? If you are like most people, you know because it's something you have experienced or verified for yourself. It is not likely to be something you were told or heard in a lecture. The second premise of this book is that **new material is easier to learn if you actively ask questions and answer them for yourself**. *Mind on Statistics* is designed to help you learn statistical ideas by actively thinking about them. Throughout most of the chapters there are boxes titled *Thought Questions.* Thinking about the questions in those boxes will help you to discover and verify important ideas for yourself. We encourage you to think and question, rather than simply read and listen.

New to This Edition

- New Case Studies and Examples were written for the new edition. All data in examples, case studies, and exercises also have been updated to the latest information available.
- The exercise sets have been reorganized and revised significantly.
  - Almost all odd-numbered exercises are answered in the back of the book. For all critical concepts and methods, there are both odd-numbered exercises with answers given and even-numbered exercises with answers not given.
  - The number of exercises in the book has increased. Also, in response to reviewers, the book contains a significant number of new drill exercises.
- The order of Chapters 3 through 6 has been modified. Third edition Chapters 5 and 6 and Chapters 3 and 4 have been switched. The chapters on sampling and research studies now appear in Chapters 5 and 6, respectively, and the chapters on relationships between quantitative and categorical variables now appear in Chapters 3 and 4, respectively. All of the descriptive statistics chapters (Chapters 2 to 4) are now grouped together.
- The number of In Summary boxes has been increased, and the boxes are placed more consistently throughout the chapters. A listing of the In Summary boxes now appears at the end of the chapter.
- The language has been tightened and simplified whenever possible.
- A discussion and Case Study addressing the dangers of multiple testing have been added to Chapters 1 and 13.

Text Features

Chapters 9 to 13, which contain the core material on sampling distributions and statistical inference, are organized in a modular, more flexible format. There are six modules for each of the topics: sampling distributions, confidence intervals, and hypothesis testing. The first module presents an introduction and the remaining five modules each deal with a specific parameter, such as one mean, one proportion, or the difference in two means. Chapter 9 covers sampling distributions, Chapters 10 and 11 cover confidence intervals, and Chapters 12 and 13 cover hypothesis testing.

This structure emphasizes the similarity among the inference procedures for the five parameters discussed. It allows instructors to illustrate that each procedure covered is a specific instance of the same process. We recognize that instructors have different preferences for the order in which to cover inference topics. For instance, some prefer to first cover all topics about proportions and then cover all topics about means. Others prefer to first cover everything about confidence intervals and then cover everything about hypothesis testing. **With the modular format, instructors can cover these topics in the order they prefer.**

To aid in the navigation through these modular chapters, the book contains **color-coded, labeled tabs that correspond to the introductory and parameter modules**. The table below, also found in Chapter 9, lays out the color-coding system as well as the flexibility of these new chapters. In addition, the table is a useful course planning tool.

Organization of Chapters 9 to 13

Parameter	Chapter 9: Sampling Distributions (SD)	Chapter 10: Confidence Intervals (CI)	Chapter 11: Confidence Intervals (CI)	Chapter 12: Hypothesis Tests (HT)	Chapter 13: Hypothesis Tests (HT)
0. Introductory	**SD Module 0** Overview of sampling distributions	**CI Module 0** Overview of confidence intervals		**HT Module 0** Overview of hypothesis testing	
1. Population Proportion (p)	**SD Module 1** SD for one sample proportion	**CI Module 1** CI for one population proportion		**HT Module 1** HT for one population proportion	
2. Difference in two population proportions ($p_1 - p_2$)	**SD Module 2** SD for difference in two sample proportions	**CI Module 2** CI for difference in two population proportions		**HT Module 2** HT for difference in two population proportions	
3. Population mean (μ)	**SD Module 3** SD for one sample mean		**CI Module 3** CI for one population mean		**HT Module 3** HT for one population mean
4. Population mean of paired differences (μ_d)	**SD Module 4** SD for sample mean of paired differences		**CI Module 4** CI for population mean of paired differences		**HT Module 4** HT for population mean of paired differences
5. Difference in two population means ($\mu_1 - \mu_2$)	**SD Module 5** SD for difference in two sample means		**CI Module 5** CI for difference in two population means		**HT Module 5** HT for difference in two population means

To add to the flexibility of topic coverage, Supplemental Topics 1 to 5 on discrete random variables, nonparametric tests, multiple regression, two-way ANOVA, and ethics are now available for use in both print and electronic formats. Instructors, please contact your sales representative to find out how these chapters can be custom published for your course.

Student Resources: Tools for Learning

There are a number of tools provided in this book and beyond to enhance your learning of statistics.

Tools for Conceptual Understanding

Updated! **Thought Question** boxes appear throughout each chapter to encourage active thinking and questioning about statistical ideas. *Hints* are provided at the bottom of the page to help you develop this skill.

THOUGHT QUESTION 2.4 Redo the bar graph in Figure 2.4 using counts instead of percentages. The necessary data are given in Table 2.3. Would the comparison of frequency of myopia across the categories of lighting be as easy to make using the bar graph with counts? Generalize your conclusion to provide guidance about what should be done in similar situations.*

***HINT:** Which graph makes it easier to compare the percentage with myopia for the three groups? What could be learned from the graph of counts that isn't apparent from the graph of percentages?

Updated! **Skillbuilder Applet** sections provide opportunities for in-class or independent hands-on exploration of key statistical concepts. The applets that accompany this feature can be found on the book's companion website.

SKILLBUILDER APPLET

5.7 Random Sampling in Action

It should be obvious by now that each time you take a random sample from a population, you are likely to get a different group of individuals, even though the population remains fixed. The **Sampling** applet at the website that accompanies this text allows us to watch what happens when we repeatedly take random samples from the same population. Figure 5.6 illustrates how the applet appears when it starts. Each stick figure represents one member of a population of 100 college students: 55 females (blue) and 45 males (red). Their heights have been recorded as well, and if you look carefully, you will see that the little stick figures are all of slightly different heights. The mean height for the population is 68 inches.

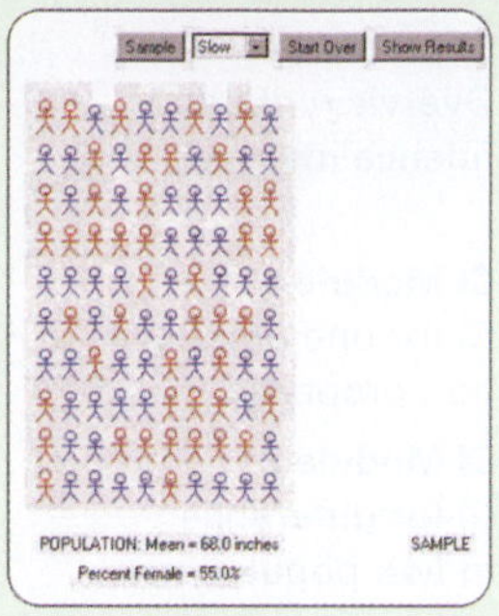

Figure 5.6 The sampling applet starting point

Updated! **Technical Notes** boxes provide additional technical discussion of key concepts.

TECHNICAL NOTE **A Philosophical Issue about Probability**

There is some debate about how to represent probability when an outcome has been determined but is unknown, such as if you have flipped a coin but not looked at it. Technically, any particular outcome has either happened or not. If it has happened, its probability of happening is 1; if it hasn't, its probability of happening is 0. In statistics, an example of this type of situation is the construction of a 95% confidence interval, which was introduced in Chapter 5 and which we will study in detail in Chapters 10 and 11. Before the sample is chosen, a probability statement makes sense. The probability is .95 that a sample will be selected for which the computed 95% confidence interval covers the truth. After the sample has been chosen, "the die is cast." Either the computed confidence interval covers the truth or it doesn't, although we may never know which is the case. That's why we say that we have *95% confidence* that a computed interval is correct, rather than saying that *the probability* that it is correct is .95.

Investigating Real-Life Questions

Updated! Relevant **Examples** form the basis for discussion in each chapter and walk you through real-life uses of statistical concepts.

Example 3.15 **Hours of Sleep and Hours of Study** Figure 3.14 displays, for a sample of 116 college students, the relationship between the reported hours of sleep during the previous 24 hours and the reported hours of study during the same period. The correlation value for this scatterplot is $r = -0.36$, indicating a negative association that is not particularly strong. On average, the hours of sleep decrease as hours of study increase, but there is substantial variation in the hours of sleep for any specific hours of study.

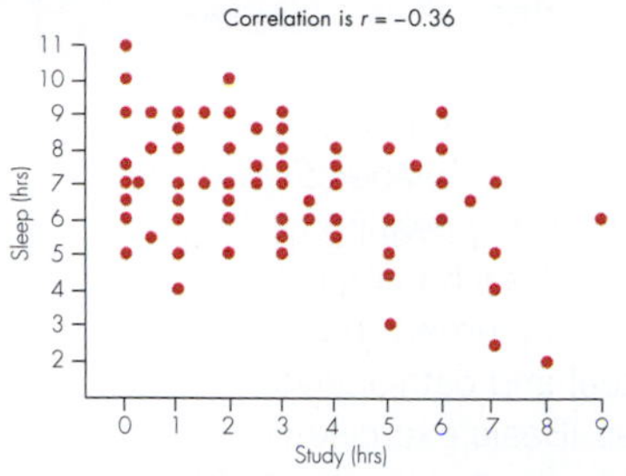

Figure 3.14 Hours of study and hours of sleep *(Source: Class data collected by one of the authors.)*

Updated! **Case Studies** apply statistical ideas to intriguing news stories. As the Case Studies are developed, they model the statistical reasoning process.

CASE STUDY 5.2 No Opinion of Your Own? Let Politics Decide

This is an excellent example of how people will respond to survey questions when they do not know about the issues, and how the wording of questions can influence responses. In 1995, the *Washington Post* decided to expand on a 1978 poll taken in Cincinnati, Ohio, in which people were asked whether they "favored or opposed repealing the 1975 Public Affairs Act" (Morin, 1995, p. 36). There was no such act, but about one-third of the respondents expressed an opinion about it.

In February 1995, the *Washington Post* added this fictitious question to its weekly poll of 1000 randomly selected respondents: "Some people say the 1975 Public Affairs Act should be repealed. Do you agree or disagree that it should be repealed?" Almost half (43%) of the sample expressed an opinion, with 24% agreeing that it should be repealed and 19% disagreeing!

The *Post* then tried another trick that produced even more disturbing results. This time, they polled two separate groups of 500 randomly selected adults. The first group was asked: "President Clinton [a Democrat] said that the 1975 Public Affairs Act should be repealed. Do you agree or disagree?" The second group was asked: "The Republicans in Congress said that the 1975 Public Affairs Act should be repealed. Do you agree or disagree?" Respondents were also asked about their party affiliation.

Overall, 53% of the respondents expressed an opinion about repealing this fictional act! The results by party affiliation were striking: For the Clinton version, 36% of the Democrats but only 16% of the Republicans agreed that the act should be repealed. For the "Republicans in Congress" version, 36% of the Republicans but only 19% of the Democrats agreed that the act should be repealed.

Updated! Original **Journal Articles** for many of the Examples and Case Studies can be found on the companion website on CourseMate, http://www.cengage.com/statistics/Utts4e. By reading the original, you are given the opportunity to learn much more about how the research was conducted, what statistical methods were used, and what conclusions the original researchers drew.

Example 2.2 **Lighting the Way to Nearsightedness** A survey of 479 children found that those who had slept with a nightlight or in a fully lit room before the age of 2 had a higher incidence of nearsightedness (myopia) later in childhood (*Sacramento Bee*, May 13, 1999, pp. A1, A18). The raw data for each child consisted of two categorical variables, each with three categories. Table 2.3 gives the categories and the number of children falling into each combination of them. The table also gives percentages (relative frequencies) falling into each eyesight category, where percentages are computed within each nighttime lighting category. For example, among the 172 children who slept in darkness, about 90% (155/172 = .90) had no myopia.

Read the original source on the companion website, http://www.cengage.com/statistics/Utts4e.

Getting Practice

***Updated!* Basic Exercises,** comprising 25% of all exercises found in the text, focus on practice and review. These exercises, found under the header *Skillbuilder Exercises* and appearing at the beginning of each exercise section, complement the conceptual and data-analysis exercises. Basic exercises give you ample practice for these key concepts.

Relevant conceptual and data analysis **Exercises** have been added and updated throughout the text. All exercises are found at the end of each chapter, with corresponding exercise sets written for each section and chapter. You will find well more than 1500 exercises, allowing for ample opportunity to practice key concepts.

Exercises

◆ Indicates that the dataset is available on the companion website (http://www.cengage.com/statistics/Utts4e) but is not required to solve the exercise.

Bold-numbered *exercises have answers in the back of the text.*

Note: Many of these exercises will be repeated in later chapters in which the relevant material is covered in more detail.

Skillbuilder Exercises

1.1 Refer to the data and five-number summaries given in Case Study 1.1. Give a numerical value for each of the following.

a. The fastest speed driven by anyone in the class.
b. The slowest of the "fastest speeds" driven by a male.
c. The speed for which one-fourth of the women had driven at that speed or faster.
d. The proportion of females who had driven 89 mph or faster.
e. The number of females who had driven 89 mph or faster.

1.2 A five-number summary for the heights in inches of the women who participated in the survey in Case Study 1.1 is as shown.

	Female Heights (inches)	
Median	65	
Quartiles	63.5	67.5
Extremes	59	71

a. What is the median height for these women?
b. What is the range of heights—that is, the difference in heights between the shortest woman and the tallest woman?
c. What is the interval of heights containing the shortest one-fourth of the women?
d. What is the interval of heights containing the middle one-half of the women?

1.5 Refer to Case Study 1.3, in which teens were asked about their dating behavior.

a. What population is represented by the random sample of 602 teens?
b. What population is represented by the 496 teens in the sample who had dated?

1.6 Using Case Study 1.6 as an example, explain the difference between a population and a sample.

1.7 A CBS News poll taken in December 2009, asked a random sample of 1048 adults in the United States, "In general, do you think the education most children are getting today in public schools is better, is about the same, or is worse than the education you received?" About 34% said "Better," 24% said "About the same," and 38% said "Worse." (The remaining 4% were unsure.)

a. What is the population for this survey?
b. What is the approximate margin of error for this survey?
c. Provide an interval that is 95% certain to cover the true percentage of U.S. adults in December 2009 who would have answered "Better" to this question if asked.

1.8 A telephone survey of 2000 Canadians conducted March 20–30, 2001, found that "Overall, about half of Canadians in the poll say the right number of immigrants are coming into the country and that immigration has a positive effect on Canadian communities. Only 16 percent view it as a negative impact while one third said it had no impact at all" (*The Ottawa Citizen*, August 17, 2001, p. A6.).

a. What is the population for this survey?
b. How many people were in the sample used for this survey?
c. What is the approximate margin of error for this survey?
d. Provide an interval of numbers that is 95% certain to cover the true percentage of Canadians who view immigration as having a negative impact.

Section 3.2

Skillbuilder Exercises

3.13 Suppose that a regression equation for the relationship between y = weight (pounds) and x = height (inches) for men aged 18 to 29 years old is

Average weight = −250 + 6 (Height)

a. Estimate the average weight for men in this age group who are 70 inches tall.
b. What is the slope of the regression line for average weight and height? Write a sentence that interprets this slope in

General Section Exercises

3.18 ◆ The average August temperatures (y) and geographic latitudes (x) of 20 cities in the United States were given in the table for Exercise 3.9. (The data are part of the **temperature** dataset on the website for this book.) The regression equation for these data is

$$\hat{y} = 113.6 - 1.01x$$

a. What is the slope of the line? Interpret the slope in terms of how the mean August temperature is affected by a change in latitude.

◆ Dataset available but not required **Bold-numbered** exercises answered in the back

Answers to Selected Odd-Numbered Exercises, indicated by bold numbers in the Exercise sections, have final answers or partial solutions found in the back of the text for checking your answers and guiding your thinking on similar exercises. Most odd-numbered exercises have answers in back of the book.

Answers to Selected Odd-Numbered Exercises

*The following are partial or complete answers to the exercises numbered in **bold** in the text.*

Chapter 1

1.1 a. 150 mph. b. 55 mph. c. 95 mph. d. 1/2. e. 51.
1.3 a. .00043. b. .00043. c. Rate is based on past data; risk uses past data to predict an individual's likelihood of developing cervical cancer.
1.5 a. All teens in the U.S. at the time the poll was taken. b. All teens in the U.S. who had dated at the time the poll was taken.
1.7 a. All adults in the U.S. at the time the poll was taken. b. $\frac{1}{\sqrt{1048}}$ = .031 or 3.1%. c. 30.9% to 37.1%.
1.9 a. 400.
1.11 a. Self-selected or volunteer sample. b. No; readers with strong opinions will respond.
1.13 a. Randomized experiment. b. Observational study. c. Observational study.
1.15 Answers will vary, but one possibility is general level of activity.
1.17 How large the difference in weight loss was for the two groups.
1.19 How many different relationships were examined.

Chapter 2

2.1 a. 4. b. State in the United States. c. n = 50.
2.3 a. Whole population. b. Sample.
2.5 a. Population parameter. b. Sample statistic. c. Sample statistic.
2.7 Sex and self-reported fastest ever driven speed. b. Students in a statistics class. c. Answer depends on whether interest is in this class only or in a larger group represented by this class.
2.9 Population summary if we restrict interest to fiscal year 1998. Sample summary if 1998 value is used to represent errors in other years.
2.11 a. Categorical. b. Quantitative. c. Quantitative. d. Categorical.
2.13 a. Categorical. b. Ordinal. c. Quantitative
2.15 a. Explanatory is score on the final exam; response is final course grade. b. Explanatory is gender; response is opinion about the death penalty.
2.17 a. Not continuous. b. Continuous. c. Continuous.

Technology for Developing Concepts and Analyzing Data

To access additional course materials and companion resources, please visit www.cengagebrain.com. At the CengageBrain.com home page, search for the ISBN of your title (from the back cover of your book) using the search box at the top of the page. This will take you to the product page where free companion resources can be found.

Aplia™ (ISBN: 0-538-73726-3) is an online interactive learning solution that improves comprehension and outcomes by increasing student effort and engagement. Founded by a professor to enhance his own courses, Aplia provides automatically graded assignments with detailed, immediate explanations on every question and innovative teaching materials. Aplia's easy-to-use system has been used by more than 1,000,000 students at over 1800 institutions.

Mind on Statistics includes **Statistics CourseMate**, a complement to the textbook. Statistics CourseMate includes:

- An interactive eBook with highlighting, note taking, and search capabilities
- Interactive teaching and learning tools including:
  - Online quizzes
  - Conceptual applets
  - Flashcards
  - Videos
  - and more
- An online activities manual
- Step-by-Step technology manuals for TI-84 Plus calculators, Microsoft® Excel®, Minitab®, SPSS®, JMP, and R
- Downloadable datasets (in ASCII as well as the native file formats for each software and calculator model covered by the Step-by-Step manuals)
- Original journal articles for select Examples and Case Studies, where you can learn much more about how the research was conducted, what statistical methods were used, and what conclusions the original researchers drew
- *Engagement Tracker*, a first-of-its-kind tool that monitors student engagement in the course

Go to login.cengage.com to access these resources, and look for this icon, which denotes a resource available within CourseMate.

Step-by-Step technology manuals, written specifically for *Mind on Statistics*, Fourth Edition, walk you through the statistical software and graphing calculator—step by step. You will find manuals for:

- **TI-84 Calculators**, written by Roger E. Davis, Pennsylvania College of Technology
- **Microsoft Excel**, written by Tom Mason, University of St. Thomas
- **Minitab**, written by Edith Seier and Robert M. Price, East Tennessee State University (student CD bundle available)
- **SPSS®**, written by Brenda K. Gunderson and Kirsten T. Namesnik at the University of Michigan at Ann Arbor (student CD bundle available)
- **JMP®**, written by Jerry Reiter and Christine Kohnen, Duke University
- **R**, written by Mark A. Rizzardi, Humboldt State University

Note: These technology manuals are available in electronic formats. Instructors, contact your sales representative to find out how these manuals can be custom published for your course.

Updated! **Minitab, Excel, TI-84, and SPSS Tips** in the text offer key details on the use of technology.

MINITAB TIP **Computing a Chi-Square Test for a Two-Way Table**

- If the raw data are stored in columns of the worksheet, use **Stat > Tables > Cross Tabulation and Chi-Square**. Specify a categorical variable in the "For rows" box and a second categorical variable in the "For columns" box. Then click the ***Chi-Square*** button and select "Chi-Square analysis."
- If the data are already summarized into counts, enter the table of counts (excluding totals) into columns of the worksheet, and then use **Stat > Tables > Chi-Square Test (Table in Worksheet)**. In the dialog box, specify the columns that contain the counts.

EXCEL TIP The *p*-value can also be computed by using Microsoft Excel. The function CHIDIST(x,df) provides the *p*-value, where x is the value of the chi-square statistic and df is a number called "degrees of freedom," which will be explained later in this book. The formula for df is (# of rows − 1)(# of columns − 1). For instance, corresponding to the information in Example 4.13, df = (2 − 1)(2 − 1) = 1, and the *p*-value is CHIDIST(7.659,1) = .005649, or about .006 as given by Minitab.

Tools for Review

Key Terms at the end of each chapter, organized by section, can be used as a "quick-finder" and as a review tool.

Key Terms

Section 3.1
scatterplot, 70
explanatory variable, 70
response variable, 70
dependent variable, 70
y variable, 70
x variable, 70
positive association, 70, 71
linear relationship, 70, 71, 74
negative association, 71
nonlinear relationship, 71
curvilinear relationship, 71
outliers in regression, 73

Section 3.2
regression analysis, 74
regression equation, 74, 75, 76
prediction, 74
regression line, 74, 75, 76
simple linear regression, 75
slope of a straight line, 74, 77
intercept of a straight line, 74, 77
y-intercept, 74
predicted *y* ($\hat{y}$), 76
estimated *y*, 76
predicted value, 76
deterministic relationship, 77
statistical relationship, 77
prediction error, 79
residual, 79
least squares, 80
least squares line, 80
least squares regression, 80
sum of squared errors (SSE), 80, 87

Section 3.3
correlation, 82
Pearson product moment correlation, 82
correlation coefficient, 82
squared correlation (r^2), 85
proportion of variation explained by *x*, 85
sum of squares total (SSTO), 87
sum of squares due to regression (SSR), 87

Section 3.4
extrapolation, 89
interpolation, 89
influential observations, 90

Section 3.5
causation versus correlation, 94–95

Updated! **In Summary** boxes serve as a useful study tool, appearing at appropriate points to enhance key concepts and calculations. More In Summary boxes have been added for this edition.

IN SUMMARY **Bell-Shaped Distributions and Standard Deviation**

- The standard deviation measures the variability among data values.
- The formula for **sample standard deviation** is $s = \sqrt{\dfrac{\Sigma(x_i - \bar{x})^2}{n-1}}$
- For bell-shaped data, about 68% of the data values fall within 1 standard deviation of the mean either way, about 95% fall within 2 standard deviations of the mean either way, and about 99.7% fall within 3 standard deviations either way.
- A **standardized score**, also called a ***z*-score**, measures how far a value is from the mean in terms of standard deviations.

In Summary Boxes

Basic Data Concepts, 17
Types of Variables and Roles for Variables, 20
Bar Graphs for Categorical Variables, 25
Using Visual Displays to Identify Interesting Features of Quantitative Data, 25
Numerical Summaries of Quantitative Variables, 45
Possible Reasons for Outliers and Reasonable Actions, 46
Bell-Shaped Distributions and Standard Deviation, 52

Tools for Active Learning

To access additional course materials and companion resources, please visit www.cengagebrain.com. At the CengageBrain.com home page, search for the ISBN of your title (from the back cover of your book) using the search box at the top of the page. This will take you to the product page where free companion resources can be found.

The **Student Solutions Manual** (ISBN 0-538-73604-6), prepared by Jessica M. Utts and Robert F. Heckard, provides worked-out solutions to most of the odd-numbered problems in the text.

The online **Activities Manual**, written by Jessica M. Utts and Robert F. Heckard, includes a variety of activities for students to explore individually or in teams. These activities guide students through key features of the text, help them understand statistical concepts, provide hands-on data collection and interpretation team-work, include exercises with tips incorporated for solution strategies, and provide bonus dataset activities. Information can be found on the companion website on CourseMate.

Instructor Resources: Tools for Assessment

Solution Builder: This online instructor database offers complete worked solutions to all exercises in the text, allowing you to create customized, secure solutions printouts (in PDF format) matched exactly to the problems you assign in class. Access available via www.cengage.com/solutionbuilder or the PowerLecture CD (see PowerLecture description).

ExamView® testing software allows instructors to quickly create, deliver, and customize tests for class in print and online formats and features automatic grading. Included is a test bank with hundreds of questions customized directly to the text, with all questions also provided in PDF and Microsoft Word formats for instructors who opt not to use the software component. The test questions, written by Brenda K. Gunderson and Kirsten T. Namesnik at the University of Michigan at Ann Arbor, are grouped by section and are a combination of multiple-choice and free-response questions. ExamView is available within the PowerLecture CD (see PowerLecture description).

The PowerLecture™ CD-ROM (ISBN: 0-538-73609-7) provides the instructor with dynamic media tools for teaching. Create, deliver, and customize tests (both print and online) in minutes with ExamView computerized testing featuring algorithmic equations. Easily build solution sets for homework or exams using Solution Builder's online solutions manual. Microsoft PowerPoint lecture slides and figures from the book are also included on this CD-ROM.

JoinIn™ on **TurningPoint®** is also available on the PowerLecture and offers instructors text-specific content for electronic response systems. You can transform your classroom and assess students' progress with instant in-class quizzes and polls. TurningPoint software lets you pose book-specific questions and display students' answers seamlessly within Microsoft PowerPoint lecture slides, in conjunction with a choice of "clicker" hardware. Enhance how your students interact with you, your lecture, and each other.

A Note to Instructors

The entire *Mind on Statistics* learning package has been informed by the recommendations put forth by the ASA/MAA Joint Curriculum Committee and the GAISE (Guidelines for Assessment and Instruction in Statistics Education) College Report, for which Jessica Utts was one of the authors. Each of the pedagogical features and ancillaries listed in the section entitled "Student Resources: Tools for Expanded Learning" and "Instructor Resources: Tools for Assessment" has been categorized by suggested use to provide you with options for designing a course that best fits the needs of your students.

In addition to these tools you will also have access to CourseMate, containing the items mentioned previously as well as

- Course outlines and syllabi
- Class projects
- Suggested discussions for the Thought Questions located throughout the text
- Supplemental topic solutions
- List of applications and methods

Acknowledgments

We thank William Harkness, Professor of Statistics at Penn State University, for continued support and feedback throughout our careers and during the writing of this book, and for his remarkable dedication to undergraduate statistics education. Preliminary editions of *Mind on Statistics*, the basis for this text, were used at Penn State; the University of California, Davis; and Texas A & M University, and we thank the many students who provided comments and suggestions on those and on subsequent editions. Thanks to Dr. Melvin Morse (Valley Children's Clinic and University of Washington) for suggesting the title for Chapter 17 and to Deb Niemeier, University of California, Davis, for suggesting that we add a supplemental chapter on Ethics (available on the companion website on CourseMate). We are indebted to Neal Rogness, Grand Valley State University, for help with the SPSS Tips, and Larry Schroeder and Darrell Clevidence, Carl Sandburg College, for help with the TI-84 Tips. At Penn State, Dave Hunter, Steve Arnold, and Tom Hettmansperger have provided many helpful insights. At the University of California, Davis, Rodney Wong has provided insights as well as material for some exercises and the test bank. We extend special thanks to Phyllis Curtiss, Grand Valley State University, and Brian Kotz, Montgomery College, each of whom provided hundreds of valuable suggestions for improving this edition of the book.

For providing datasets used in the book and available at the companion website on CourseMate, we thank Susan Jelsing, as well as William Harkness and Laura Simon from Penn State University.

The following reviewers offered valuable suggestions for this and previous editions:

Patricia M. Buchanan, Penn State University
Elizabeth Clarkson, Wichita State University
Ian Clough, University of Cincinnati–Clermont College
Patti B. Collings, Brigham Young University
James Curl, Modesto Junior College
Boris Djokic, Keiser University
Wade Ellis, West Valley College
Linda Ernst, Mt. Hood Community College
Anda Gadidov, Kennesaw State University

Joan Garfield, University of Minnesota
Jonathan Graham, University of Montana
Jay Gregg, Colorado State University
Donnie Hallstone, Green River Community College
Donald Harden, Georgia State University
Rosemary Hirschfelder, University Sound
Sue Holt, Cabrillo Community College
Mark Johnson, University of Central Florida
Tom Johnson, North Carolina University
Yevgeniya Kleyman, University of Michigan
Andre Mack, Austin Community College
Jean-Marie Magnier, Springfield Technical Community College
Suman Majumdar, University of Connecticut
D'Arcy Mays, Virginia Commonwealth
Megan Meece, University of Florida
Jack Osborn Morse Jr., University of Georgia
Emily Murphree, Miami University
Mary Murphy, Texas A & M University
Helen Noble, San Diego State University
Thomas Nygren, Ohio State University
Thomas J. Pfaff, Ithaca College
Nancy Pfenning, University of Pittsburgh
Jennifer Lewis Priestley, Kennesaw State University
Lawrence Ries, University of Missouri
David Robinson, St. Cloud State University
Neal Rogness, Grand Valley State University
Kelly Sakkinen, Lansing Community College
Heather Sasinouska, Clemson University
Kirk Steinhorst, University of Idaho
Robert Talbert, Franklin College
Gwen Terwilliger, University of Toledo
Ruth Trygstad, Salt Lake Community College
Robert Alan Wolf, University of San Francisco

Our sincere appreciation and gratitude also goes to Molly Taylor, Cathy Brooks, Jay Campbell, and the staff at Cengage Learning, as well as Mike Ederer of Graphic World Publishing Services, who oversaw the development and production of the fourth edition. We also wish to thank Carolyn Crockett and Danielle Derbenti, without whom this book could not have been written, and Martha Emry who kept us on track throughout the editing and production of the first three editions of the book. Finally, for their support, patience, and numerous prepared dinners, we thank our families and friends, especially Candace Heckard, Molly Heckard, Wes Johnson, Claudia Utts-Smith, and Dennis Smith.

Jessica M. Utts
Robert R. Heckard

1

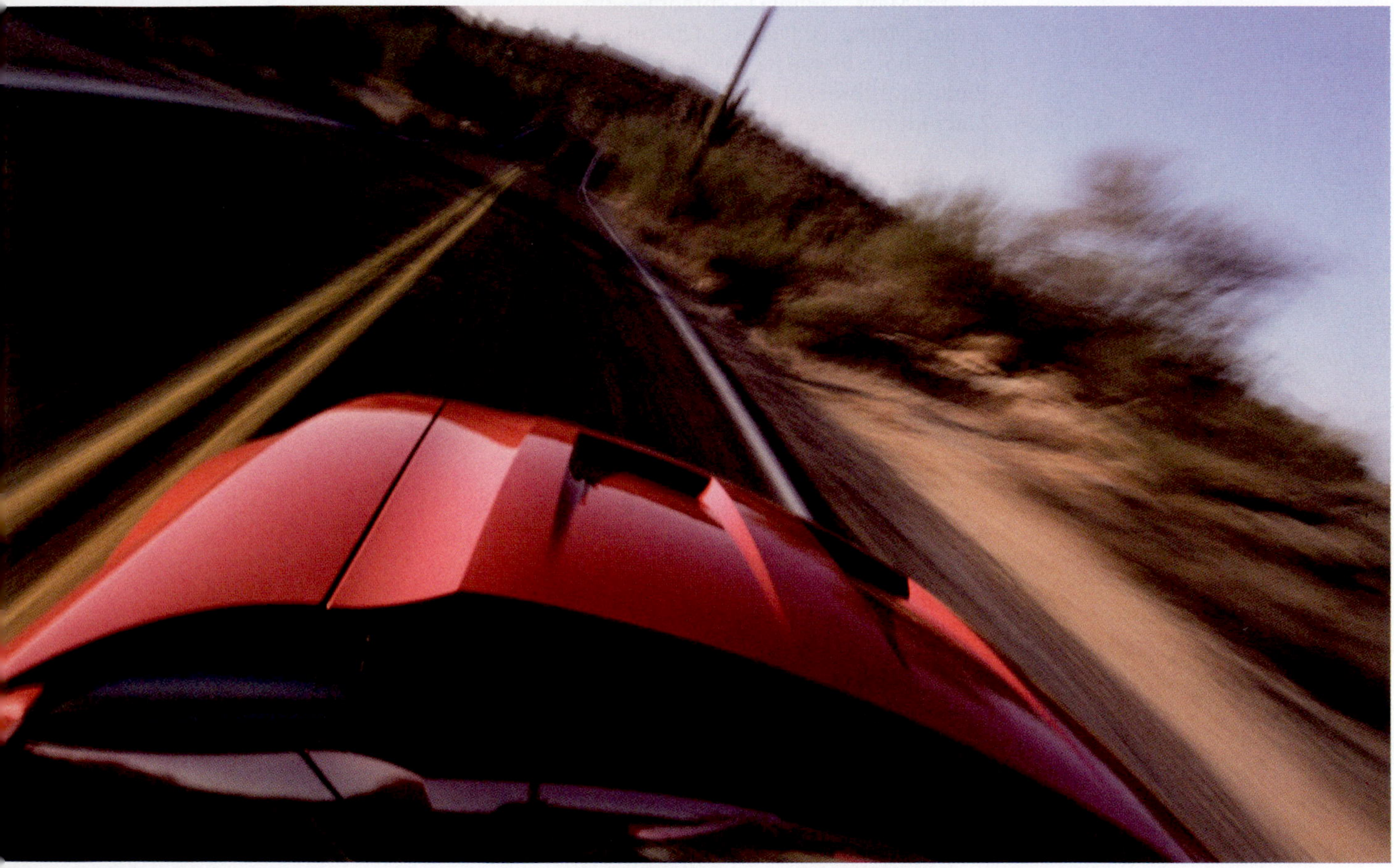

Zoomstock/Masterfile

Is a male or a female more likely to be behind the wheel of this speeding car?

See Case Study 1.1 *(p. 2)*

Statistics Success Stories and Cautionary Tales

The eight stories in this chapter are meant to bring life to the term ***statistics.*** After reading these stories, if you think the subject of statistics is lifeless or gruesome, check your pulse!

Let's face it. You're a busy person. Why should you spend your time learning about statistics? In this chapter, we give eight examples of situations in which statistics provided either enlightenment or misinformation. After reading these examples, we hope you will agree that learning about statistics may be interesting and useful.

Each of the stories in this chapter illustrates one or more concepts that will be developed throughout the book. These concepts are given as "the moral of the story" after a case is presented. Definitions of some terms used in the story also are provided following each case. By the time you read all of these stories, you already will have an overview of what statistics is all about.

1.1 What Is Statistics?

When you hear the word *statistics* you probably think of lifeless or gruesome numbers, such as the population of your state or the number of violent crimes committed in your city last year. The word *statistics,* however, actually is used to mean two different things. The better-known definition is that statistics are numbers measured for some purpose. A more complete definition, and the one that forms the substance of this book, is the following:

DEFINITION **Statistics** is a collection of procedures and principles for gathering data and analyzing information to help people make decisions when faced with uncertainty.

The stories in this chapter are meant to bring life to this definition. After reading them, if you think the subject of statistics is lifeless or gruesome, check your pulse!

1.2 Eight Statistical Stories with Morals

The best way to gain an understanding of some of the ideas and methods used in statistical studies is to see them in action. Each of the eight stories presented in this section includes interesting lessons about how to extract information from data. The methods and ideas will be expanded throughout the book, but these stories will give you an excellent overview of why it is useful to study statistics. To help you understand some basic statistical principles, each case study is accompanied by a "moral of the story" and by some definitions. All of the ideas and definitions will be discussed in greater detail in subsequent chapters.

CASE STUDY 1.1 Who Are Those Speedy Drivers?

A survey taken in a large statistics class at Penn State University contained the question "What's the fastest you have ever driven a car? ____ mph." The *data* provided by the 87 males and 102 females who responded are listed here.

Males: 110 109 90 140 105 150 120 110 110 90 115 95 145 140 110 105 85 95 100 115 124 95 100 125 140 85 120 115 105 125 102 85 120 110 120 115 94 125 80 85 140 120 92 130 125 110 90 110 110 95 95 110 105 80 100 110 130 105 105 120 90 100 105 100 120 100 100 80 100 120 105 60 125 120 100 115 95 110 101 80 112 120 110 115 125 55 90

Females: 80 75 83 80 100 100 90 75 95 85 90 85 90 90 120 85 100 120 75 85 80 70 85 110 85 75 105 95 75 70 90 70 82 85 100 90 75 90 110 80 80 110 110 95 75 130 95 110 110 80 90 105 90 110 75 100 90 110 85 90 80 80 85 50 80 100 80 80 80 95 100 90 100 95 80 80 50 88 90 90 85 70 90 30 85 85 87 85 90 85 75 90 102 80 100 95 110 80 95 90 80 90

From these numbers, can you tell which sex tends to have driven faster and by how much? Notice how difficult it is to make sense of the *data* when you are simply presented with a list. Even if the numbers had been presented in numerical order, it would be difficult to compare the two groups.

Your first lesson in statistics is how to formulate a simple summary of a long list of numbers. The **dotplot** shown in Figure 1.1 helps us see the pattern in the data. In the plot, each dot represents the response of an individual student. We can see that the men tend to claim a higher "fastest ever driven" speed than do the women.

The graph shows us a lot, and calculating some statistics that summarize the data will provide additional insight. There are a variety of ways to do so, but for this example, we examine a **five-number summary** of the data for males and females. The five numbers are the lowest value; the cut-off points for one-fourth, one-half, and three-fourths of the data; and the highest value. The three middle values of the summary (the cutoff points for one-fourth, one-half, and three-fourths of the data) are called the *lower quartile, median,* and *upper quartile,* respectively. Five-number summaries can be represented like this:

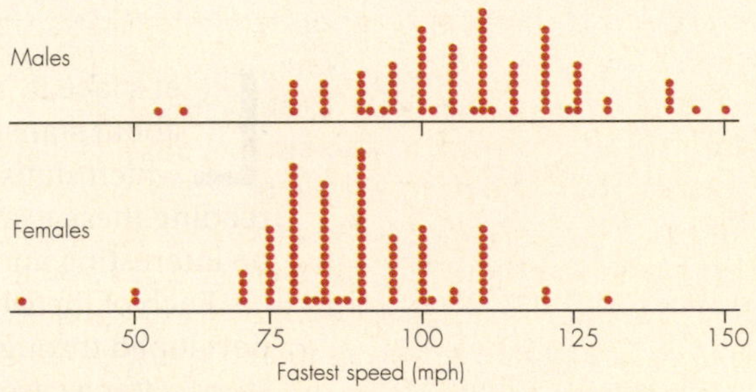

Figure 1.1 Responses to "What's the fastest you've ever driven?"

	Males (87 Students)		Females (102 Students)	
Median	110		89	
Quartiles	95	120	80	95
Extremes	55	150	30	130

Some interesting facts become immediately obvious from these summaries. By looking at the medians, you see that half of the men have driven 110 miles per hour or more, whereas the halfway point for the women is only 89 miles per hour. In fact, three-fourths of the men have driven 95 miles per hour or more, but only one-fourth of the women have done so. These facts were not at all obvious from the original lists of numbers.

Moral of the Story: *Simple summaries of data can tell an interesting story and are easier to digest than long lists.*

Definitions: Data is a plural word referring to numbers or nonnumerical labels (such as male/female) collected from a set of entities (people, cities, and so on). The **median** of a numerical list of data is the value in the middle when the numbers are put in order. For an even number of entities, the median is the average of the middle two values. The **lower quartile** and **upper quartile** are (roughly) the medians of the lower and upper halves of the data.

CASE STUDY 1.2 Safety in the Skies?

If you fly often, you may have been relieved to see the *New York Times* headline on October 1, 2007, proclaiming "Fatal airline crashes drop 65%" (Wald, 2007). And you may have been dismayed if you had seen an earlier headline in *USA Today* that read, "Planes get closer in midair as traffic control errors rise" (Levin, 1999). The details were even more disturbing: "Errors by air traffic controllers climbed from 746 in fiscal 1997 to 878 in fiscal 1998, an 18% increase."

So, are the risks of a fatal airline crash or an air traffic control error something that should be a major concern for airline passengers? Don't cancel your next vacation yet. A look at the statistics indicates that the news is actually pretty good! The low risk becomes obvious when we are told the *base rate* or *baseline risk* for these problems. According to the *New York Times* article, "the drop in the accident rate [from 1997 to 2007] will be about 65%, to one fatal accident in about 4.5 million departures, from 1 in nearly 2 million in 1997." And according to the 1999 *USA Today* story, "The errors per million flights handled by controllers climbed from 4.8 to 5.5." So the *rate* of fatal accidents changed from about 1 in 2 million departures in 1997 to 1 in 4.5 million departures in 2007, and the ominous rise in air traffic controller errors in 1998 still led to a very low rate of only 5.5 errors per million flights.

Fortunately, the rates for these problems were provided in both stories. This is not always the case in news reports of changes in rates or risk. For instance, an article may say that the risk of a certain type of cancer is doubled if you eat a certain unhealthful food. But what good is that information unless you know the actual risk? Doubling your chance of getting cancer from 1 in a million to 2 in a million is trivial, but doubling your chance from 1 in 50 to 2 in 50 is not.

Moral of the Story: *When you read about the change in the rate or risk of occurrence of something, make sure you also find out the base rate or baseline risk.*

Definitions: The **rate** at which something occurs is simply the number of times it occurs per number of opportunities for it to occur. In fiscal year 1998, the rate of air traffic controller errors was 5.5 per million flights. The **risk** of a bad outcome in the future can be estimated by using the past rate for that outcome, if it is assumed the future will be like the past. Based on recent data, the estimated risk of a fatal accident for any given flight is 1 in 4.5 million, which is 1/4,500,000 or about .00000022. The **base rate** or **baseline risk** is the rate or risk at a beginning time period or under specific conditions. For instance, the base rate of fatal airline crashes from which the 65% decrease for 2007 was calculated was about 1 crash per 2 million flights, for fiscal year 1997.

CASE STUDY 1.3 Did Anyone Ask Whom You've Been Dating?

"According to a new *USA Today*/Gallup Poll of teenagers across the country, 57% of teens who go out on dates say they've been out with someone of another race or ethnic group" (Peterson, 1997). That's over half of the dating teenagers, so of course it was natural for the headline in the *Sacramento Bee* to read, "Interracial dates common among today's teenagers." The article contained other information as well, such as "In most cases, parents aren't a major obstacle. Sixty-four percent of teens say their parents don't mind that they date interracially, or wouldn't mind if they did."

There are millions of teenagers in the United States whose experiences are being reflected in this story. How could the polltakers manage to ask so many teenagers these questions? The answer is that they didn't. The article states that "the results of the new poll of 602 teens, conducted Oct. 13–20, reflect the ubiquity of interracial dating today." They asked only 602 teens? Could such a small sample possibly tell us anything about the millions of teenagers in the United States? The answer is "yes" if those teens constituted a *random sample* from the *population* of interest.

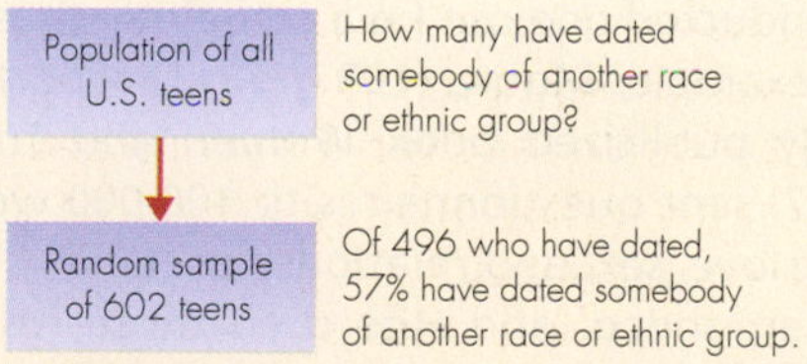

Figure 1.2 Population and sample for the survey.

The featured statistic of the article is that "57 percent of teens who go out on dates say they've been out with someone of another race or ethnic group." Only 496 of the 602 teens in the poll said that they date, so the 57% value is actually a percentage based on 496 responses. In other words, the pollsters were using information from only 496 teenagers to estimate something about all teenagers who date. Figure 1.2 illustrates this situation.

How accurate could this *sample survey* possibly be? The answer may surprise you. The results of this *poll* are accurate to within a *margin of error* of about 4.5%.

(continued)

As surprising as it may seem, the true percentage of all dating teens in the United States who date interracially is reasonably likely to be within 4.5% of the reported percentage that's based only on the 496 teens asked! We'll be conservative and round the 4.5% margin of error up to 5%. At the time the poll was taken, the percentage of all dating teenagers in the United States that would say they had dated someone of another race or ethnic group was likely to be in the range 57% ± 5%, or between 52% and 62%. (The symbol ± is read "plus and minus" and means that the value on the right should be added to and subtracted from the value on the left to create an interval.)

Polls and *sample surveys* are frequently used to assess public opinion and to estimate population characteristics such as the percent of teens who have dated interracially or the proportion of voters who plan to vote for a certain candidate. Many sophisticated methods have been developed that allow pollsters to gain the information they need from a very small number of individuals. The trick is to know how to select those individuals. In Chapter 5, we examine a number of other strategies that are used to ensure that sample surveys provide reliable information about populations.

Moral of the Story: *A representative sample of only a few thousand, or perhaps even a few hundred, can give reasonably accurate information about a population of many millions.*

Definitions: A **population** is a collection of all individuals about which information is desired. The "individuals" are usually people but could also be schools, cities, pet dogs, agricultural fields, and so on. A **random sample** is a subset of the population selected so that every individual has a specified probability of being part of the sample. In a **poll** or **sample survey,** the investigators gather opinions or other information from each individual included in the sample. The **margin of error** for a properly conducted survey is a number that is added to and subtracted from the sample information to produce an interval that is 95% certain to contain the true value for the population. In the most common types of sample surveys, the margin of error is approximately equal to 1 divided by the square root of the number of individuals in the sample.

Hence, a sample of 496 teenagers who have dated produces a margin of error of about $1/\sqrt{496} = .045$, or about 4.5%. In some polls the margin of error is called the **margin of sampling error** to distinguish it from other sources of errors and biases that can distort the results. The next Case Study illustrates a common source of bias that can occur in surveys, discussed more fully in Chapter 5.

CASE STUDY 1.4 Who Are Those Angry Women?

A well-conducted survey can be very informative, but a poorly conducted one can be a complete disaster. As an extreme example, Moore (1997, p. 11) reports that for her highly publicized book *Women and Love,* Shere Hite (1987) sent questionnaires to 100,000 women asking about love, sex, and relationships. Only 4.5% of the women responded, and Hite used those responses to write her book. As Moore notes, "The women who responded were fed up with men and eager to fight them. For example, 91% of those who were divorced said that they had initiated the divorce. The anger of women toward men became the theme of the book." Do you think that women who were angry with men would be likely to answer questions about love relationships in the same way as the general population of women?

The Hite sample exemplifies one of the most common problems with surveys: The sample data may not represent the population. Extensive *nonparticipation (nonresponse)* from a random sample, or the use of a *self-selected* (i.e., *all volunteer*) *sample,* will probably produce biased results. Those who voluntarily respond to surveys tend to care about the issue and therefore have stronger and different opinions than those who do not respond.

Moral of the Story: *An unrepresentative sample, even a large one, tells you almost nothing about the population.*

Definitions: Nonparticipation bias (also called **nonresponse bias**) can occur when many people who are selected for the sample either do not respond at all or do not respond to some of the key survey questions. This may occur even when an appropriate random sample is selected and contacted. The survey is then based on a nonrepresentative sample, usually those who feel strongly about the issues. Some surveys don't even attempt to contact a random sample but instead ask anyone who wishes to respond to do so. Magazines, television stations, and Internet websites routinely conduct this kind of poll, and those who respond are called a **self-selected sample** or a **volunteer sample.** In most cases, this kind of sample tells you nothing about the larger population at all; it tells you only about those who responded.

CASE STUDY 1.5 Does Prayer Lower Blood Pressure?

Read the original source on the companion website, http://www.cengage.com/statistics/Utts4e.

News headlines are notorious for making one of the most common mistakes in the interpretation of statistical studies: jumping to unwarranted conclusions. A headline in *USA Today* read, "Prayer can lower blood pressure" (Davis, 1998). The story that followed continued the possible fallacy it began by stating, "Attending religious services lowers blood pressure more than tuning into religious TV or radio, a new study says." The words "attending religious services lowers blood pressure" imply a direct cause-and-effect relationship. This is a strong statement, but it is not justified by the research project described in the article.

The article was based on an *observational study* conducted by the U.S. National Institutes of Health, which followed 2391 people aged 65 or older for 6 years. The article described one of the study's principal findings: "People who attended a religious service once a week and prayed or studied the Bible once a day were 40% less likely to have high blood pressure than those who don't go to church every week and prayed and studied the Bible less" (Davis, 1998). So the researchers did observe a relationship, but it's a mistake to think that this justifies the conclusion that prayer actually *causes* lower blood pressure.

When groups are compared in an observational study, the groups usually differ in many important ways that may contribute to the observed relationship. In this example, people who attended church and prayed regularly may have been less likely than the others to smoke or to drink alcohol. These could affect the results because smoking and alcohol use are both believed to affect blood pressure. The regular church attendees may have had a better social network, a factor that could lead to reduced stress, which in turn could reduce blood pressure. People who were generally somewhat ill may not have been as willing or able to go out to church. We're sure you can think of other possibilities for *confounding variables* that may have contributed to the observed relationship between prayer and lower blood pressure.

Moral of the Story: *Cause-and-effect conclusions cannot generally be made on the basis of an observational study.*

Definitions: An **observational study** is one in which participants are merely observed and measured. Comparisons based on observational studies are comparisons of naturally occurring groups. A **variable** is a characteristic that differs from one individual to the next. It may be numerical, such as blood pressure, or it may be categorical, such as whether or not someone attends church regularly. A **confounding variable** is a variable that is not the main concern of the study but may be partially responsible for the observed results.

(Source: International Journal of Psychiatry in Medicine *by Koenig, H.G., L.K. George, J.C. Hays, and D.B. Larson. [See p. 701 for complete credit.])*

CASE STUDY 1.6 Does Aspirin Reduce Heart Attack Rates?

Read the original source on the companion website, http://www.cengage.com/statistics/Utts4e.

In 1988, the Steering Committee of the Physicians' Health Study Research Group released the results of a 5-year *randomized experiment* conducted using 22,071 male physicians between the ages of 40 and 84. The purpose of the experiment was to determine whether or not taking aspirin reduces the risk of a heart attack. The physicians had been *randomly assigned* to one of the two *treatment* groups. One group took an ordinary aspirin tablet every other day, while the other group took a *placebo*. None of the physicians knew whether he was taking the actual aspirin or the placebo.

The results, shown in Table 1.1, support the conclusion that taking aspirin does indeed help to reduce the risk of having a heart attack. The rate of heart attacks in the group taking aspirin was only about half the rate of heart attacks in the placebo group. In the aspirin group, there were 9.42 heart attacks per 1000 participating doctors, while in the placebo group, there were 17.13 heart attacks per 1000 participants.

Table 1.1 The Effect of Aspirin on Heart Attacks

Treatment	Heart Attacks	Doctors in Group	Attacks per 1000 Doctors
Aspirin	104	11,037	9.42
Placebo	189	11,034	17.13

Because the men in this experiment were randomly assigned to the two conditions, other important risk factors such as age, amount of exercise, and dietary habits should have been similar for the two groups. The only important difference between the two groups should have been whether they took aspirin or a placebo. This makes it possible to conclude that taking aspirin actually *caused* the lower rate of heart attacks for that group. In a later chapter, you will learn how to determine that the difference seen in this sample is *statistically significant*. In other words, the observed sample difference probably reflects a true difference within the population.

To what population does the conclusion of this study apply? The participants were all male physicians, so the

(continued)

conclusion that aspirin reduces the risk of a heart attack may not hold for the general population of men. No women were included, so the conclusion may not apply to women at all. More recent evidence, however, has provided additional support for the benefit of aspirin in broader populations.

Moral of the Story: *Unlike with observational studies, cause-and-effect conclusions can generally be made on the basis of randomized experiments.*

Definitions: A **randomized experiment** is a study in which treatments are randomly assigned to participants. A **treatment** is a specific regimen or procedure assigned to participants by the experimenter. A **random assignment** is one in which each participant has a specified probability of being assigned to each treatment. A **placebo** is a pill or treatment designed to look just like the active treatment but with no active ingredients. A **statistically significant** relationship or difference is one that is large enough to be unlikely to have occurred in the sample if there was no relationship or difference in the population.

(Source: International Journal of Psychiatry in Medicine *by Koenig, H.G., L.K. George, J.C. Hays, and D.B. Larson. [See p. 701 for complete credit.])*

CASE STUDY 1.7 Does the Internet Increase Loneliness and Depression?

It was big news. Researchers at Carnegie Mellon University had found that "greater use of the Internet was associated with declines in participants' communication with family members in the household, declines in size of their social circle, and increases in their depression and loneliness" (Kraut et al., 1998, p. 1017). An article in the *New York Times* reporting on this study was titled "Sad, lonely world discovered in cyberspace" (Harmon, 1998). The study included 169 individuals in 73 households in Pittsburgh, Pennsylvania, who were given free computers and Internet service in 1995, when the Internet was still relatively new. The participants answered a series of questions at the beginning of the study and either 1 or 2 years later, measuring social contacts, stress, loneliness, and depression. The *New York Times* reported:

> In the first concentrated study of the social and psychological effects of Internet use at home, researchers at Carnegie Mellon University have found that people who spend even a few hours a week online have higher levels of depression and loneliness than they would if they used the computer network less frequently. . . . it raises troubling questions about the nature of "virtual" communication and the disembodied relationships that are often formed in cyberspace.
>
> *(Source: "Sad, Lonely World Discovered in Cyberspace," by A. Harmon, New York Times, August 30, 1998, p. A3. Reprinted with permission of the New York Times Company.)*

Given these dire reports, one would think that using the Internet for a few hours a week is devastating to one's mental health. But a closer look at the findings reveals that the changes were actually quite small, though statistically significant. Internet use averaged 2.43 hours per week for participants. The number of people in the participants' "local social network" decreased from an average of 23.94 people to an average of 22.90 people, hardly a noticeable loss. On a scale from 1 to 5, self-reported loneliness decreased from an average of 1.99 to 1.89 (lower scores indicate greater loneliness). And on a scale from 0 to 3, self-reported depression dropped from an average of .73 to an average of .62 (lower scores indicate higher depression).

The *New York Times* did report the magnitude of some of the changes, noting for instance that "one hour a week on the Internet was associated, on average, with an increase of .03, or 1% on the depression scale." But the attention the research received masked the fact that the impact of Internet use on depression, loneliness, and social contact was actually quite small, and thus may not have been of much practical significance.

As a follow-up to this study, in July 2001, *USA Today* (Elias, 2001) reported that in continued research, the bad effects had mostly disappeared. The article, titled "Web use not always a downer: Study disputes link to depression," began with the statement "Using the Internet at home doesn't make people more depressed and lonely after all." However, the article noted that the lead researcher, Robert Kraut of Carnegie Mellon University, believes that the earlier findings were correct but that "the Net has become a more social place since the study began in 1995." His explanation for the change in findings is that "either the Internet has changed, or people have learned to use it more constructively, or both." Research on this topic continues to develop. A study released in February 2010 (Morrison and Gore, 2010) identified 18 "Internet addicted" individuals out of 1319 study participants. They found that the Internet addicts scored in the "moderately-to-severely depressed range" on a test called the Beck Depression Inventory, while an equivalent group of non-addicts scored "firmly in the non-depressed range." As the authors point out, it is not clear whether Internet use causes depression, depression causes more Internet use, or some other factors lead to abnormal scores in both for some people.

Moral of the Story: *A statistically significant finding does not necessarily have practical significance or importance. When a study reports a statistically significant finding, find out the magnitude of the relationship or difference. A secondary moral to this story is that the implied direction of cause and effect may be wrong. In this case, it could be that people who were more lonely and depressed were more prone to using the Internet. And remember that, as the follow-up research makes clear, "truth" doesn't necessarily remain fixed across time. Any study should be viewed in the context of society at the time it was done.*

CASE STUDY 1.8 Did Your Mother's Breakfast Determine Your Sex?

Read the original source on the companion website, http://www.cengage.com/statistics/Utts4e.

You've probably heard that "you are what you eat," but did it ever occur to you that you might be who you are because of what your mother ate? A study published in 2008 by the British Royal Society seemed to find just that. The researchers reported that mothers who ate breakfast cereal prior to conception were more likely to have boys than mothers who did not (Mathews et al., 2008). But 9 months later, just enough time for the potential increased cereal sales to have produced a plethora of little baby boys, another study was published that dashed cold milk on the original claim (Young et al., 2009). The dispute was based on something statisticians call *multiple testing,* which can lead to erroneous findings of statistical significance. The authors of the original study had asked 740 women about 133 different foods they might have eaten just before getting pregnant. They found that 59% of the women who consumed breakfast cereal daily gave birth to a boy, compared to only 43% of the women who rarely or never ate cereal (http://www.cbsnews.com/stories/2008/04/22/health/webmd/main4036102.shtml). The result was highly statistically significant, but almost none of the other foods tested showed a statistically significant difference in the ratio of male to female births.

As previously discussed, statistical significance is how statisticians assess whether a difference found in a sample, in this case of 740 women, is large enough to conclude that the difference is likely to represent more than just chance. But sometimes what looks like a statistically significant difference is actually a *false positive*—a difference that looks like it wasn't due to chance when it really was. The more differences that are tested, the more likely it is that one of them will be a false positive. The criticism by Young et al. was based on this idea. When 133 food items that in fact do not affect the sex of a baby are all tested, it is likely that at least one of them will show up as a false positive, showing a big enough difference in the proportion of male to female births to be statistically significant when in fact the difference is due to chance.

The authors of the original study defended their work (Mathews et al., 2009). They noted that they only tested the individual food items after an initial test based on total pre-conception calorie consumption showed a difference in male and female births. They found that 56% of the mothers in the top third of calorie consumption had boys, compared with only 45% of the mothers in the bottom third of calorie consumption. That was one of only two initial tests they did; the other had to do with vitamin intake. With only two tests, it is unlikely that either of them would be a false positive. Unfortunately the media found the cereal connection to be the most interesting result in the study, and that's what received overwhelming publicity. The best way to resolve the debate, as in most areas of science, is to ask the same questions in a new study and see if the results are consistent. The authors of the original study have stated their intention to do that.

Moral of the Story: *When you read about a study that found a relationship or difference, try to find out how many different things were tested. The more tests that are done, the more likely it is that a statistically significant difference is a false positive that can be explained by chance. You should be especially wary if dozens of things are tested and only one or two of them are statistically significant.*

Definitions: Multiple testing or **multiple comparisons** in statistics refers to the fact that researchers often test many different hypotheses in the same study. This practice may result in statistically significant findings by mistake, called **false positive** results. Sometimes this practice is called **data snooping** because researchers snoop around in their data until they find something interesting to report.

1.3 The Common Elements in the Eight Stories

The eight stories were meant to bring life to our definition of statistics. Let's consider that definition again:

> **Statistics** is a collection of procedures and principles for gathering data and analyzing information to help people make decisions when faced with uncertainty.

Think back over the stories. In each of them, *data are used to make a judgment about a situation.* This common theme is what statistics is all about. The stories should also help you realize that you can be misled by the use of data, and learning to recognize how that happens is one of the themes of this book.

The Discovery of Knowledge

Each story illustrates part of the process of discovery of new knowledge, for which statistical methods can be very useful. The basic steps in this process are as follows:

1. *Asking the right question(s)*
2. *Collecting useful data,* which includes deciding how much is needed
3. *Summarizing and analyzing data,* with the goal of answering the questions
4. *Making decisions and generalizations* based on the observed data
5. *Turning the data and subsequent decisions into new knowledge*

We'll explore these five steps throughout the book, concluding with a chapter on "Turning Information into Wisdom." We're confident that your active participation in this exploration will benefit you in your everyday life and in your future professional career.

In a practical sense, almost all decisions in life are based on knowledge obtained by gathering and assimilating data. Sometimes the data are quantitative, as when an instructor must decide what grades to give on the basis of a collection of homework and exam scores. Sometimes the information is more qualitative and the process of assimilating it is informal, such as when you decide what you are going to wear to a party. In either case, the principles in this book will help you to understand how to be a better decision maker.

THOUGHT QUESTION 1.1 Think about a decision that you recently had to make. What "data" did you use to help you make the decision? Did you have as much information as you would have liked? If you could freely use them, how would you use the principles in this chapter to help you gain more useful information?*

***HINT:** As an example, how did you decide to live where you are living? What additional data, if any, would have been helpful?

IN SUMMARY Some Important Statistical Principles

The "moral of the story" items for the case studies presented in this chapter give a good overview of many of the important ideas covered in this book. Here is a summary:

- Simple summaries of data can tell an interesting story and are easier to digest than long lists.
- When you read about the change in the rate or risk of occurrence of something, make sure you also find out the base rate or baseline risk.
- A representative sample of only a few thousand, or perhaps even a few hundred, can give reasonably accurate information about a population of many millions.
- An unrepresentative sample, even a large one, tells you almost nothing about the population.
- Cause-and-effect conclusions cannot generally be made on the basis of an observational study.
- Unlike with observational studies, cause-and-effect conclusions can generally be made on the basis of randomized experiments.
- A statistically significant finding does not necessarily have practical significance or importance. When a study reports a statistically significant finding, find out the magnitude of the relationship or difference.
- When you read about a study that found a relationship or difference, try to find out how many different things were tested. The more tests that are done, the more likely it is that a statistically significant difference is a false positive that can be explained by chance.

Key Terms

Every term in this chapter is discussed more extensively in later chapters, so don't worry if you don't understand all of the terminology that has been introduced here. The following list indicates the page number(s) where the important terms in this chapter are introduced and defined.

Section 1.1
statistics, 1

Case Study 1.1
dotplot, 2
five-number summary, 2
data, 2
median, 2
lower quartile, 2
upper quartile, 2

Case Study 1.2
rate, 3
risk, 3
base rate, 3
baseline risk, 3

Case Study 1.3
population, 3, 4
random sample, 3, 4
poll, 3, 4
sample survey, 3, 4
margin of error, 3, 4
(margin of) sampling error, 4

Case Study 1.4
nonparticipation bias, 4
nonresponse bias, 4
self-selected sample, 4
volunteer sample, 4

Case Study 1.5
observational study, 5
variable, 5
confounding variable, 5

Case Studies 1.6 and 1.7
randomized experiment, 5, 6
treatment, 5, 6
random assignment, 5, 6
placebo, 5, 6
statistically significant, 5, 6, 7
practical significance, 6, 7
practical importance, 6, 7

Case Study 1.8
multiple testing, 7
multiple comparisons, 7
false positive, 7
data snooping, 7

Exercises

◆ Denotes that the dataset is available on the companion website, http://www.cengage.com/statistics/Utts4e, but is not required to solve the exercise.

Bold *exercises have answers in the back of the text.*

Note: Many of these exercises will be repeated in later chapters in which the relevant material is covered in more detail.

Skillbuilder Exercises

1.1 Refer to the data and five-number summaries given in Case Study 1.1. Give a numerical value for each of the following.

a. The fastest speed driven by anyone in the class.

b. The slowest of the "fastest speeds" driven by a male.

c. The speed for which one-fourth of the women had driven at that speed or faster.

d. The proportion of females who had driven 89 mph or faster.

e. The number of females who had driven 89 mph or faster.

1.2 A five-number summary for the heights in inches of the women who participated in the survey in Case Study 1.1 is as shown.

	Female Heights (inches)	
Median	65	
Quartiles	63.5	67.5
Extremes	59	71

a. What is the median height for these women?

b. What is the range of heights—that is, the difference in heights between the shortest woman and the tallest woman?

c. What is the interval of heights containing the shortest one-fourth of the women?

d. What is the interval of heights containing the middle one-half of the women?

1.3 In recent years, Vietnamese American women have had the highest rate of cervical cancer in the country. Suppose that among 200,000 Vietnamese American women, 86 developed cervical cancer in the past year.

a. Calculate the rate of cervical cancer for these women.

b. What is the estimated risk of developing cervical cancer for Vietnamese American women in the next year?

c. Explain the conceptual difference between the rate and the risk, in the context of this example.

1.4 The risk of getting lung cancer at some point in one's life for men who have never smoked is about 13 in 1000. The risk for men who smoke is just over 13 times the risk for nonsmokers. (Source: Villenueve and Lau, 1994)

a. What is the base rate for lung cancer in men over a lifetime?

b. What is the approximate lifetime risk of getting lung cancer for men who smoke?

1.5 Refer to Case Study 1.3, in which teens were asked about their dating behavior.

a. What population is represented by the random sample of 602 teens?

b. What population is represented by the 496 teens in the sample who had dated?

1.6 Using Case Study 1.6 as an example, explain the difference between a population and a sample.

1.7 A CBS News poll taken in December 2009, asked a random sample of 1048 adults in the United States, "In general, do you think the education most children are getting today in public schools is better, is about the same, or is worse than the education you received?" About 34% said "Better," 24% said "About the same," and 38% said "Worse." (The remaining 4% were unsure.)

a. What is the population for this survey?

b. What is the approximate margin of error for this survey?

c. Provide an interval that is 95% certain to cover the true percentage of U.S. adults in December 2009 who would have answered "Better" to this question if asked.

1.8 A telephone survey of 2000 Canadians conducted March 20–30, 2001, found that "Overall, about half of Canadians in the poll say the right number of immigrants are coming into the country and that immigration has a positive effect on Canadian communities. Only 16 percent view it as a negative impact while one third said it had no impact at all" (*The Ottawa Citizen*, August 17, 2001, p. A6.).

a. What is the population for this survey?

b. How many people were in the sample used for this survey?

c. What is the approximate margin of error for this survey?

d. Provide an interval of numbers that is 95% certain to cover the true percentage of Canadians who view immigration as having a negative impact.

1.9 In Case Study 1.3, the margin of error for the sample of 496 teenagers was about 4.5%. How many teenagers should be in the sample to produce an approximate margin of error of .05 or 5%?

1.10 About how many people would need to be in a random sample from a large population to produce an approximate margin of error of .30 or 30%?

1.11 A popular Sunday newspaper magazine often includes a yes-or-no survey question such as "Do you think there is too much violence on television?" or "Do you think parents should use physical discipline?" Readers are asked to mail their answers to the magazine, and the results are reported in a subsequent issue.

a. What is this type of sample called?

b. Do you think the results of these polls represent the opinions of all readers of the magazine? Explain.

1.12 A proposed study design is to leave 100 questionnaires by the checkout line in a student cafeteria. The questionnaire can be picked up by any student and returned to the cashier. Explain why this volunteer sample is a poor study design.

1.13 For each of the examples given here, decide whether the study was an observational study or a randomized experiment.

a. A group of students enrolled in an introductory statistics course were randomly assigned to take either a web-based course or a traditional lecture course. The two methods were compared by giving the same final examination in both courses.

b. A group of smokers and a group of nonsmokers who visited a particular clinic were asked to come in for a physical exam every 5 years for the rest of their lives to monitor and compare their health status.

c. CEOs of major corporations were compared with other employees of the corporations to see if the CEOs were more likely to have been the first child born in their families than were the other employees.

1.14 For each of the studies described, explain whether the study was an observational study or a randomized experiment.

a. A group of 100 students was randomly divided, with 50 assigned to receive vitamin C and the remaining 50 to receive a placebo, to determine whether or not vitamin C helps to prevent colds.

b. A random sample of patients who received a hip transplant operation at Stanford University Hospital during 2000 to 2010 will be followed for 10 years after their operation to determine the success (or failure) of the transplant.

c. Volunteers with high blood pressure were randomly divided into two groups. One group was taught to practice meditation and the other group was given a low-fat diet. After 8 weeks, reduction in blood pressure was compared for the two groups.

1.15 Read Case Study 1.5. Give an example of a confounding variable that might explain why elderly people who attended religious services might have lower blood pressure than those who did not. Do not use one of the variables already mentioned in the Case Study.

1.16 Suppose that an observational study showed that students who got at least 7 hours of sleep performed better on exams than students who got less than 7 hours of sleep. Which of the following are possible confounding variables, and which are not? Explain why in each case.

a. Number of courses the student took that term.

b. Weight of the student.

c. Number of hours the student spent partying in a typical week.

1.17 A randomized experiment was done in which overweight men were randomly assigned to either exercise or go on a diet for a year. At the end of the study there was a statistically significant difference in average weight loss for the two groups. What additional information would you need in order to determine if the difference in average weight loss had *practical* importance?

1.18 Explain the distinction between statistical significance and practical significance. Can the result of a study be statistically significant but not practically significant?

1.19 A (hypothetical) study of what people do in their spare time found that people born under the astrological sign of Aries were significantly more likely to be regular swimmers than people born under other signs. What additional information would you want to know to help you determine if this result is a false positive?

1.20 Explain what is meant by a "false positive" in the context of conclusions in statistical studies.

Chapter Exercises

1.21 Refer to Case Study 1.6, in which the relationship between aspirin and heart attack rates was examined. Using the results of this experiment, what do you think is the base rate of heart attacks for men like the ones in this study? Explain.

1.22 Students in a statistics class at Penn State were asked, "About how many minutes do you typically exercise in a week?" Responses from the *women* in the class were

60, 240, 0, 360, 450, 200, 100, 70, 240, 0, 60, 360, 180, 300, 0, 270

Responses from the *men* in the class were

180, 300, 60, 480, 0, 90, 300, 14, 600, 360, 120, 0, 240

a. Compare the women to the men using a dotplot. What does your plot show you about the difference between the men and the women?

b. For each sex, determine the median response.

c. Do you think there's a "significant" difference between the weekly amount that men and women exercise? Explain.

1.23 Refer to Exercise 1.22.

a. Create a five-number summary for the men's responses. Show how you found your answer.

b. Use your five-number summary to describe in words the exercise behavior of this group of male students.

1.24 Refer to Exercise 1.22.

a. Create a five-number summary for the women's responses. Show how you found your answer.

b. Use your five-number summary to describe in words the exercise behavior of this group of female students.

1.25 An article in the magazine *Science* (Service, 1994) discussed a study comparing the health of 6000 vegetarians and a similar number of their friends and relatives who were not vegetarians. The vegetarians had a 28% lower death rate from heart attacks and a 39% lower death rate from cancer, even after the researchers accounted for differences in smoking, weight, and social class. In other words, the reported percentages were the remaining differences after adjusting for differences in death rates due to those factors.

a. Is this an observational study or a randomized experiment? Explain.

b. On the basis of this information, can we conclude that a vegetarian diet causes lower death rates from heart attacks and cancer? Explain.

c. Give an example of a potential confounding variable and explain what it means to say that it is a confounding variable.

1.26 Refer to Exercise 1.25, comparing vegetarians and nonvegetarians for two causes of death. Were base rates given for the two causes of death? If so, what were they? If not, explain what a base rate would be for this study.

1.27 An article in the *Sacramento Bee* (March 8, 1984, p. A1) reported on a study finding that "men who drank 500 ounces or more of beer a month (about 16 ounces a day) were three times more likely to develop cancer of the rectum than nondrinkers." In other words, the rate of cancer in the beer-drinking group was three times that of the non–beer drinkers in this study. What important numerical information is missing from this report?

1.28 Dr. Richard Hurt and his colleagues (Hurt et al., 1994) randomly assigned volunteers wanting to quit smoking to wear either a nicotine patch or a placebo patch to determine whether wearing a nicotine patch improves the chance of quitting. After 8 weeks of use, 46% of those wearing the nicotine patch but only 20% of those wearing the placebo patch had quit smoking.

a. Was this a randomized experiment or an observational study?
b. The difference in the percentage of participants who quit (20% versus 46%) was statistically significant. What conclusion can be made on the basis of this study?
c. Why was it advisable to assign some of the participants to wear a placebo patch?

1.29 Refer to the study in Exercise 1.28, in which there was a statistically significant difference in the percentage of smokers who quit using a nicotine patch and a placebo patch. Now read the two cautions in the "moral of the story" for Case Study 1.7. Discuss each of them in the context of this study.

1.30 Refer to the study in Exercises 1.28 and 1.29, comparing the percentage of smokers who quit using a nicotine patch and a placebo patch. Refer to the definition of statistics given on page 1, and explain how it applies to this study.

1.31 Case Study 1.6 reported that the use of aspirin reduces the risk of heart attack and that the relationship was found to be "statistically significant." Does either of the cautions in the "moral of the story" for Case Study 1.7 apply to this result? Explain.

1.32 A random sample of 1001 University of California faculty members taken in December 1995 was asked, "Do you favor or oppose using race, religion, sex, color, ethnicity, or national origin as a criterion for admission to the University of California?" (Roper Center, 1996). Fifty-two percent responded "favor."

a. What is the population for this survey?
b. What is the approximate margin of error for the survey?
c. Based on the results of the survey, could it be concluded that a majority (over 50%) of *all* University of California faculty members favor using these criteria? Explain.

1.33 The Roper Organization conducted a poll in 1992 (Roper, 1992) in which one of the questions asked was whether or not the respondent had ever seen a ghost. Of the 1525 people in the 18- to 29-year-old age group, 212 said "yes."

a. What is the approximate margin of error that accompanies this result?
b. What is the interval that is 95% certain to contain the actual proportion of people in this age group who have seen a ghost?

1.34 Refer to Exercise 1.33. What is the risk of someone in this age group having seen a ghost?

1.35 Refer to Exercise 1.33. The Roper Organization selected a random sample of adults in the United States for this poll. Suppose listeners to a late-night radio talk show were asked to call and report whether or not they had ever seen a ghost.

a. What is this type of sample called?
b. Do you think the proportion reporting that they had seen a ghost for the radio poll would be higher or lower than the proportion for the Roper poll? Explain.

1.36 The CNN website sometimes has a small box called "Quick vote" that contains a question about an interesting topic in the news that day. For example, one question in February 2010 asked "Should the U.S. military let gays and lesbians serve openly?" Visitors to the website are invited to click their response and to view the results. When the results are displayed they contain the message "This is not a scientific poll."

a. What type of sample is obtained in this Quick vote?
b. What do you think is meant by the message that "This is not a scientific poll?"

1.37 Explain what is meant by "data snooping."

1.38 A headline in a major newspaper read, "Breast-fed youth found to do better in school."

a. Do you think this statement was based on an observational study or a randomized experiment? Explain.
b. Given your answer in part (a), which of these two alternative headlines do you think would be preferable: "Breast-feeding leads to better school performance" or "Link found between breast-feeding and school performance"? Explain.

1.39 In this chapter, you learned that cause and effect can be concluded from randomized experiments but generally not from observational studies. Why don't researchers simply conduct all studies as randomized experiments rather than observational studies?

1.40 Why was the study described in Case Study 1.5 conducted as an observational study instead of an experiment?

1.41 Give an example of a question you would like to have answered, such as whether or not eating chocolate helps to prevent depression. Then explain how a randomized experiment or an observational study could be done to study this question.

1.42 Suppose you were to read the following news story: "Researchers compared a new drug to a placebo for treating high blood pressure, and it seemed to work. But the researchers were concerned because they found that significantly more people got headaches when taking the new drug than when taking the placebo. Headaches were the only problem out of the 20 possible side effects the researchers tested."

a. Do you think the research used an observational study or a randomized experiment? Explain.

b. Do you think the researchers are justified in thinking the new drug would cause more headaches in the population than the placebo would? Explain.

1.43 Refer to Case Study 1.5. Explain what mistakes were made in the implementation of steps 4 and 5 of "The Discovery of Knowledge" when *USA Today* reported the results of this study.

1.44 Refer to Case Study 1.6. Go through the five steps listed under "The Discovery of Knowledge" in Section 1.3, and show how each step was addressed in this study.

2

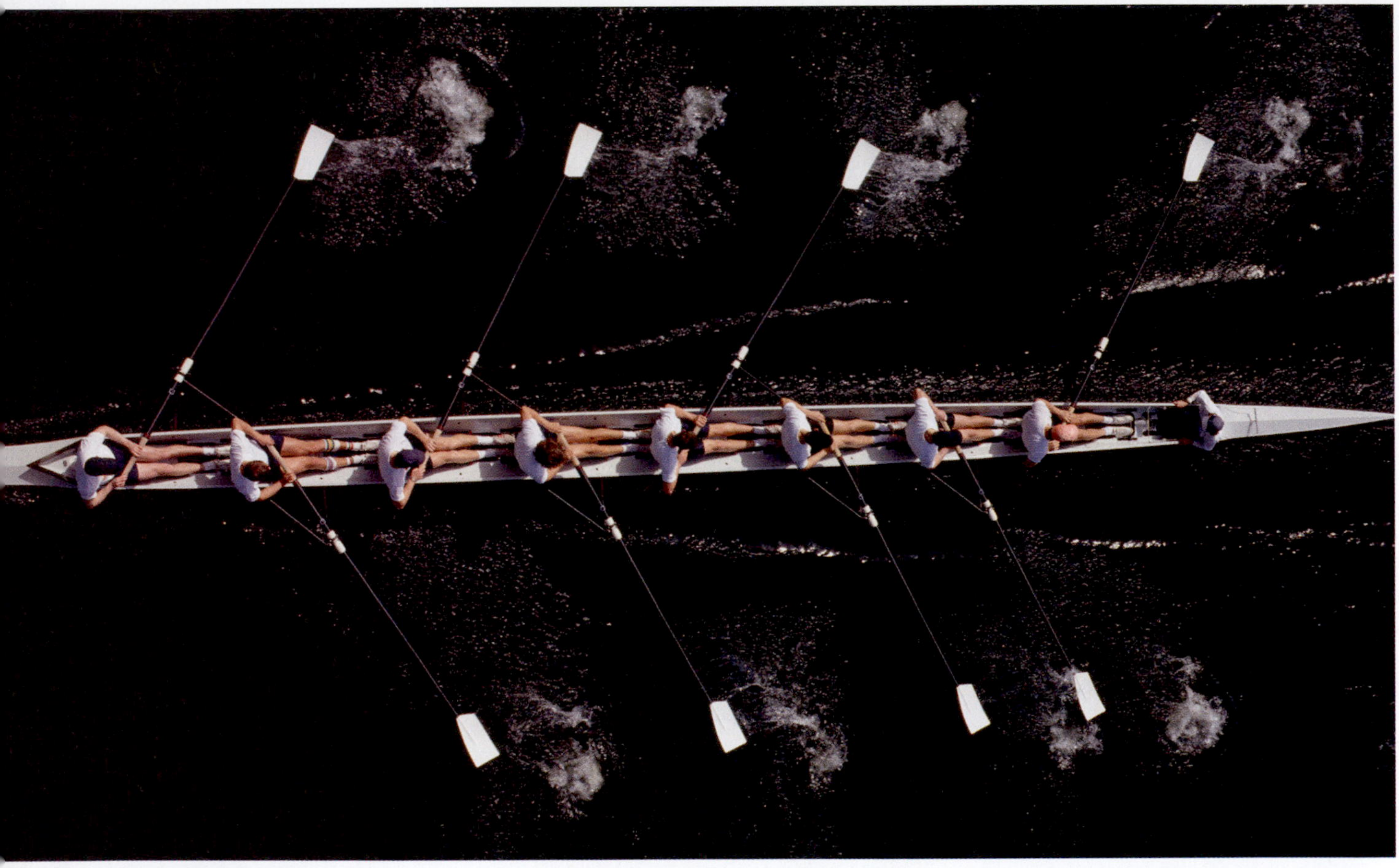

Why might the average weight of the men in this boat be lower than you think?

See Example 2.16 *(p. 46)*

Turning Data into Information

Looking at a long list of numbers is about the same as looking at a scrambled set of letters. To get information from numerical data, you have to organize it in ways that allow you to answer questions of concern to you.

In Case Study 1.1, we analyzed the responses that 189 college students gave to the question "What's the fastest you've ever driven a car?" The "moral of the story" for that case study was that *simple summaries of data can tell an interesting story and are easier to digest than long lists.* In this chapter, you will learn how to create simple summaries and pictures from various kinds of raw data.

2.1 Raw Data

Raw data is a term used for numbers and category labels that have been collected but have not yet been processed in any way. For example, here is a list of questions asked in a large statistics class and the "raw data" given by one of the students:

1. What is your sex (m = male, f = female)?	Raw data: m
2. How many hours did you sleep last night?	Raw data: 5 hours
3. Randomly pick a letter—*S* or *Q*.	Raw data: *S*
4. What is your height in inches?	Raw data: 67 inches
5. Randomly pick a number between 1 and 10.	Raw data: 3
6. What's the fastest you've ever driven a car (mph)?	Raw data: 110 mph
7. What is your right handspan in centimeters?	Raw data: 21.5 cm
8. What is your left handspan in centimeters?	Raw data: 21.5 cm

For questions 7 and 8, a centimeter ruler was provided on the survey form, and handspan was defined as the distance covered on the ruler by a stretched hand from the tip of the thumb to the tip of the small finger. For question 3, about one-half of the students saw the choice of letters in reverse order, so their question was "Randomly pick a letter — *Q* or *S*." This was done to learn whether or not students might be more likely to pick the first choice offered, regardless of whether it was the *S* or the *Q*. If you do Exercise 2.35 at the end of this chapter, you will learn the result. You may be wondering why question 5 was asked. Your curiosity will be satisfied as you keep reading this chapter.

Datasets, Observations, and Variables

A **variable** is a characteristic that can differ from one individual to the next. Students in the statistics class provided raw data for eight variables: sex, hours of sleep, choice of a letter, height, choice of a number, fastest speed ever driven, right handspan, and left handspan. The instructor imposed a ninth variable: the order of listing *S* and *Q* in question 3.

An **observational unit** is a single individual entity, a person for instance, in a study. More simply, an individual entity may be called an **observation**. The word *observation* might also be used to describe the value of a single measurement, such as height = 67 inches. The **sample size** for a study is the total number of observational units. The letter n is used to represent the sample size; one hundred and ninety students participated in the class survey, so the sample size is $n = 190$.

A **dataset** is the complete set of raw data, for all observational units and variables, in a survey or experiment. When statistical software or a spreadsheet program will be used to summarize the raw data, the dataset typically is organized so that each row of the dataset gives the data for one observational unit and each column gives the raw data for a particular variable. For the statistics class survey, Figure 2.1 shows the first five rows of a dataset created for the Minitab statistical software program. These five rows give the raw data for five of the 190 students in the total dataset. Note that each column label gives a clue to which variable is in the column. (T after a column number indicates text.) The final column indicates the order of presenting the letters in question 3.

pennstate1.mtw ***

→	C1-T	C2	C3-T	C4	C5	C6	C7	C8	C9-T
	Sex	HrsSleep	SQpick	Height	RandNumb	Fastest	RtSpan	LftSpan	Form
1	M	5.00	S	67.000	3	110	21.50	21.50	SorQ
2	M	7.00	S	75.000	9	109	22.50	22.50	SorQ
3	M	6.00	S	73.000	7	90	23.50	24.00	SorQ
4	F	7.50	S	64.000	8	80	20.00	21.00	SorQ
5	F	7.00	S	63.000	7	75	19.00	19.00	SorQ

Figure 2.1 Minitab worksheet with dataset

Data from Samples and Populations

Researchers often use sample data to make inferences about the larger population represented by the data. Occasionally, in a **census,** data are collected from all members of a population.

- **Sample data** have been collected when measurements have been taken from a subset of a population.
- **Population data** have been collected when all individuals in a population have been measured.

Sometimes the reason for collecting the data creates this distinction. For instance, data measured in a statistics class are sample data when we use them to represent a larger collection of students, but are population data if we use them to describe only that class.

It is generally important to determine whether raw data are sample data or population data. However, most of the descriptive methods for summarizing data explained in this chapter are the same for both sample and population data. Therefore, in this chapter we will only distinguish between sample and population data when the notation differs for the two situations. We will begin emphasizing the distinction between samples and populations in Chapter 9.

Parameters and Statistics

The generic names used for summary measures from sample and population data also differ. A summary measure computed from sample data is called a **statistic**, while a summary measure using data for an entire population is called a **parameter**. This distinction is often overlooked when we are interested only in numerical summaries for either a sample or a population. In that case, the summary numbers are simply called **descriptive statistics** for either a sample or a population.

2.1 Exercises are on page 55.

IN SUMMARY Basic Data Concepts

- An **observational unit**, or **observation,** is an individual entity in a study. An individual measurement is also called an *observation.*
- A **variable** is a characteristic that may differ among individuals.
- **Sample data** are collected from a subset of a larger population.
- **Population data** are collected when all individuals in a population are measured.
- A **statistic** is a summary measure of sample data.
- A **parameter** is a summary measure of population data.

THOUGHT QUESTION 2.1 There were almost 200 students who answered the survey questions shown on page 15. Formulate four interesting questions that you would like to answer using the data from these students. What kind of summary information would help you answer your questions?*

2.2 Types of Variables

We learned in the previous section that a **variable** is a characteristic that differs from one individual to the next. A variable may be a *categorical* characteristic, such as a person's sex, or a *numerical* characteristic, such as hours of sleep last night. To determine what type of summary might provide meaningful information, you first have to recognize which type of variable you want to summarize.

For a **categorical variable,** the raw data consist of group or category names that don't necessarily have any logical ordering. Each individual falls into one and only one category. For a categorical variable, the most fundamental summaries are how many individuals and what percent of the group fall into each category.

The term **ordinal variable** may be used to describe the data when a categorical variable has ordered categories. For example, suppose that you are asked to rate your driving skills compared to the skills of other drivers, using the codes 1 = better than average, 2 = average, and 3 = worse than average. The response is an ordinal variable because the response categories are ordered perceptions of driving skills.

Following are a few examples of categorical variables and their possible categories. The final variable in the list, the rating of a teacher, is ordinal because the response categories convey an ordering.

Categorical Variable	Possible Categories
Dominant hand	Left-handed, Right-handed, Ambidextrous
Regular church attendance	Yes, No
Opinion about marijuana legalization	Yes, No, Not sure
Eye color	Brown, Blue, Green, Hazel, Other
Teacher Rating	Scale of 1 to 7, 1 = Poor, 7 = Excellent

***HINT:** An example is, "What was the average amount of sleep for these students?" Case Study 1.1 could be utilized to generate another example.

For a **quantitative variable**, the raw data are either numerical measurements or counts taken on each individual. All individuals can be meaningfully ordered according to these values, and averaging and other arithmetic operations make sense for these data. A few examples of quantitative variables follow:

Quantitative Variable	Possible Responses
Height	Measured height in inches
Weight	Measured weight in pounds
Amount of sleep last night	Self-reported sleep in hours
Classes missed last week	Count of missed classes
Number of siblings	Count of brothers and sisters

Not all numbers fit the definition of a quantitative variable. For instance, Social Security numbers or student identification numbers may carry some information (such as region of the country where the Social Security number was obtained), but it is not generally meaningful to put them into numerical order or to determine the average Social Security number.

Measurement variable and **numerical variable** are synonyms for a quantitative variable. The term **continuous variable** can also be used for quantitative data when every value within some interval is a possible response. For example, height is a continuous quantitative variable because any height within a particular range is possible. The limitations of measuring tapes, however, don't allow us to measure heights accurately enough to find that a person's actual height is 66.5382617 inches. Even if we could measure that accurately, we would usually prefer to round such a height to 66.5 inches. The distinction between quantitative variables that are continuous and those that are not will be expanded in Chapter 8 when we study probability distributions.

A variable type can depend on how something is measured. For instance, household income is a numerical value with two digits after the decimal place, and if it is recorded this way, it is a quantitative variable. Researchers often collect household income data using ordered categories, however, such as 1 = less than $25,000, 2 = $25,000 to $49,999, 3 = $50,000 to $74,999, and so on. With categories like these, household income becomes a categorical variable or, more specifically, an ordinal variable. In some situations, household income could be categorized very broadly, as when it is used to determine whether or not someone qualifies for a loan. In that case, income may be either "high enough" to qualify for the loan or "not high enough."

Raw data for quantitative and categorical variables are summarized differently. It makes sense, for example, to calculate the average number of hours of sleep last night for the members of a group, but it doesn't make sense to calculate the average gender (male, female) for the group. For gender data, it makes more sense to determine the number and proportion of the group who are male and the number and proportion who are female. Usually ordinal variables are summarized using the same methods used for categorical variables, although occasionally they are summarized as quantitative variables.

TECHNICAL NOTE

Summarizing Ordinal Variables

The way we summarize an ordinal variable can depend on the purpose. As an example, consider the self-rating of driver skill with 1 = better than average, 2 = average, and 3 = worse than average. We can summarize the responses in the way we usually do for a categorical variable, which is to find the number and percentage of the sample who responded in each category. It could also be informative to treat the variable as quantitative data and find the average of the responses to see whether or not it is close to 2, which it should be if all respondents give an honest appraisal of their abilities. On the other hand, it would not make sense to talk about "average household income" using the numerical codes attached to broad income categories like the ones shown earlier on this page.

THOUGHT QUESTION 2.2 Review the data collected in the statistics class, listed in Section 2.1, and identify a type for each variable. The only one that is ambiguous is question 5. That question asks for a numerical response, but as we will see later in this chapter, it is more interesting to summarize the responses as if they are categorical.*

Asking the Right Questions

As with most situations in life, the information you get when you summarize a dataset depends on how careful you are about asking for what you want. Here are some examples of the types of questions that are most commonly of interest for different kinds of variables and combinations of variables.

One Categorical Variable

Example: What percentage of college students favors the legalization of marijuana, and what percentage of college students opposes legalization of marijuana?

Opinion about the legalization of marijuana is a categorical variable with two possible response categories (favor or oppose). For one categorical variable, it is useful to ask what percentage of individuals falls into each category.

Two Categorical Variables

Example: In Case Study 1.6, the researchers asked if the likelihood of a male physician having a heart attack depends on whether he has been taking aspirin or taking a placebo.

The two categorical variables here are whether or not a physician had a heart attack and whether a physician took aspirin or a placebo. For two categorical variables, we ask if there is a relationship between the two variables. Does the chance of falling into a particular category for one variable depend on which category an individual is in for the other variable?

One Quantitative Variable

Example: What is the average body temperature for adults, and how much variability is there in body temperature measurements?

Body temperature is a quantitative variable. To summarize one quantitative variable, we typically ask about the interesting summary measures, such as the average or the range of values.

One Categorical and One Quantitative Variable

Example: Do men and women drive at the same "fastest speeds" on average?

We are considering how a quantitative variable (fastest ever driven) is related to a categorical variable (sex). A general question about this type of situation is whether the quantitative measurements are similar across the categories or they differ. This question could be approached by examining whether or not the average measurement (such as average fastest speed ever driven) is different for the two categories (men and women). We might also ask whether or not the range of measurements is different across the categories.

Two Quantitative Variables

Example: Does average body temperature change as people age?

Age and body temperature, the two variables in this example, are both quantitative variables. A question we ask about two quantitative variables is whether they are related so that when measurements are high (low) on one variable the measurements for the other variable also tend to be high (low).

***HINT:** For each variable, consider whether the raw data are meaningful quantities or category names.

Explanatory and Response Variables

Three of the questions just listed were about the relationship between two variables. In these instances, we identify one variable as the **explanatory variable** and the other variable as the **response variable**. The value of the *explanatory variable* is thought to partially explain the value of the *response variable* for an individual. For example, in the relationship between smoking and lung cancer, whether or not an individual smokes is the explanatory variable, and whether or not he or she develops lung cancer is the response variable. If we note that people with higher education levels generally have higher incomes, education level is the explanatory variable and income is the response variable.

The identification of one variable as "explanatory" and the other as "response" does not imply that there is a *causal* relationship. It simply implies that knowledge of the value of the explanatory variable may help provide knowledge about the value of the response variable for an individual.

2.2 Exercises are on pages 55–56.

IN SUMMARY Types of Variables and Roles for Variables

- A **categorical variable** is a variable for which the raw data are group or category names that don't necessarily have a logical ordering. Examples include eye color and country of residence.
- An **ordinal variable** is a categorical variable for which the categories have a logical ordering or ranking. Examples include highest educational degree earned and T-shirt size (S, M, L, XL).
- A **quantitative variable** is a variable for which the raw data are numerical measurements or counts taken on each individual observation. Examples include height and number of siblings.
- In a relationship between two variables, regardless of type, an **explanatory variable** is one that partially explains the value of a **response variable** for an individual.

2.3 Summarizing One or Two Categorical Variables

Numerical Summaries

To summarize a categorical variable, first count how many individuals fall into each possible category. Percentages usually are more informative than counts, so the second step is to calculate the percentage in each category. These two easy steps can also be used to summarize a combination of two categorical variables.

Example 2.1 **Seatbelt Use by Twelfth-Graders** One question asked in a 2003 nationwide survey of American high school students was, "How often do you wear a seatbelt when driving a car?" The biennial survey, organized by the U.S. Centers for Disease Control and Prevention, is conducted as part of a federal program called the Youth Risk Behavior Surveillance System. Possible answers for the seatbelt question were Always, Most times, Sometimes, Rarely, and Never. Respondents could also say that they didn't drive.

Table 2.1 summarizes responses given by twelfth-grade students who said that they drive. The total sample size for the table is $n = 3042$ students. Note that a majority, $1686/3042 = .554$, or 55.4%, said that they always wear a seatbelt when driving, while just $115/3042 = .038$, or 3.8%, said that they never wear a seatbelt. To find the percentage who either rarely or never wears a seatbelt, we sum the percentages in the Rarely and Never categories. This is 8.2% + 3.8% = 12%.

Table 2.1 Seatbelt Use by Twelfth-Graders When Driving

Response	Count	Percent
Always	1686	55.4%
Most times	578	19.0%
Sometimes	414	13.6%
Rarely	249	8.2%
Never	115	3.8%
Total	3042	100%

Source: Centers for Disease Control and Prevention, http://www.cdc.gov/HealthyYouth/yrbs/index.htm.

One stereotype about males and females is that males are more likely to engage in risky behaviors than females are. Are females more likely to say that they always wear a seatbelt? Are males more likely to say they rarely or never wear a seatbelt? Table 2.2 summarizes seatbelt use for twelfth-grade males and females in the sample. Percentages are given within each sex. Among females, 915 out of 1467 = 62.4% said that they always wear a seatbelt compared to 771 out of 1575 = 49.0% of the males. Males were more likely than females to rarely or never use seatbelts. Adding the percentages for Rarely and Never gives 10.5% + 5.7% = 16.2% for the males and 5.7% + 1.7% = 7.4% for the females.

Do these sample data provide enough information for us to infer that sex and seatbelt use are related variables in the larger population of all U.S. twelfth-grade drivers? We will learn how to answer this type of question in Chapters 4 and 15.

Table 2.2 Sex and Seatbelt Use by Twelfth-Graders When Driving

	Always	Most Times	Sometimes	Rarely	Never	Total
Female	915 (62.4%)	276 (18.8%)	167 (11.4%)	84 (5.7%)	25 (1.7%)	1467 (100%)
Male	771 (49.0%)	302 (19.2%)	247 (15.7%)	165 (10.5%)	90 (5.7%)	1575 (100%)

Source: http://www.cdc.gov/HealthyYouth/yrbs/index.htm.

Frequency and Relative Frequency

In general, the **distribution** of a variable describes how often the possible responses occur.

- A **frequency distribution** for a categorical variable is a listing of all categories along with their frequencies (counts).
- A **relative frequency distribution** is a listing of all categories along with their relative frequencies (given as proportions or percentages, for example).

It is commonplace to give the frequency and relative frequency distributions together, as was done in Table 2.1.

Example 2.2 Read the original source on the companion website, http://www.cengage.com/statistics/Utts4e.

Lighting the Way to Nearsightedness A survey of 479 children found that those who had slept with a nightlight or in a fully lit room before the age of 2 had a higher incidence of nearsightedness (myopia) later in childhood (*Sacramento Bee*, May 13, 1999, pp. A1, A18). The raw data for each child consisted of two categorical variables, each with three categories. Table 2.3 gives the categories and the number of children falling into each combination of them. The table also gives percentages (relative frequencies) falling into each eyesight category, where percentages are computed within each nighttime lighting category. For example, among the 172 children who slept in darkness, about 90% (155/172 = .90) had no myopia.

Table 2.3 Nighttime Lighting in Infancy and Eyesight

Slept with:	No Myopia	Myopia	High Myopia	Total
Darkness	155 (90%)	15 (9%)	2 (1%)	172 (100%)
Nightlight	153 (66%)	72 (31%)	7 (3%)	232 (100%)
Full Light	34 (45%)	36 (48%)	5 (7%)	75 (100%)
Total	342 (71%)	123 (26%)	14 (3%)	479 (100%)

(Source: From Nature *1999, Vol. 399, pp. 113–114. [See p. 701 for complete credit.])*

The pattern in Table 2.3 is striking. As the amount of sleeptime light increases, the incidence of myopia also increases. However, this study does not prove that sleeping with light actually *caused* myopia in more children. There are other possible explanations. For example, myopia has a genetic component, so those children whose parents have myopia are more likely to suffer from it themselves. Maybe nearsighted parents are more likely to provide light while their children are sleeping.

THOUGHT QUESTION 2.3 Can you think of possible explanations for the observed relationship between use of nightlights and myopia, other than direct cause and effect? What additional information might help to provide an explanation?*

Explanatory and Response Variables for Categorical Variables

In many summaries of two categorical variables, we can identify one variable as an explanatory variable and the other as a response variable (**outcome variable**). For instance, in Example 2.1, sex (male, female) was an explanatory variable and how often a student wears a seatbelt when driving was the response variable. In Example 2.2, the amount of sleeptime lighting was an explanatory variable and the degree of myopia was the response variable.

In both Tables 2.2 and 2.3, the explanatory variable categories defined the rows and the response variable categories defined the columns. Tables often are formed this way, although not always. When they are, row percentages are more informative than column percentages. In Tables 2.2 and 2.3, percentages were given across rows. For instance, Table 2.3 shows that 90% of children who slept in darkness did not have myopia but only 45% of those who had slept in full light did not have myopia.

No matter how the table is constructed, determine whether one variable is an explanatory variable and the other is a response variable. Within each explanatory variable category we are interested in the percentage falling into each response variable category.

***HINT:** Reread Example 2.2, in which one possible explanation is mentioned. What data would we need to investigate the possible explanation mentioned there?

MINITAB TIP **Numerically Describing One or Two Categorical Variables**

- To determine how many and what percentage fall into the categories of a single categorical variable, use **Stat > Tables > Tally Individual Variables.** In the dialog box, specify a column containing the raw data for a categorical variable. Click on any desired options for counts and percentages under "Display."
- To create a two-way table for two categorical variables, use **Stat > Tables > Crosstabulation and Chi-Square.** Specify a categorical variable in the "For rows" box and another categorical variable in the "For columns" box. Select any desired percentages (row, column, and/or total) under "Display."

SPSS TIP **Numerically Describing One or Two Categorical Variables**

- To create a frequency table for one categorical variable, use **Analyze > Descriptive Statistics > Frequencies.**
- To create a two-way table for two categorical variables, use **Analyze > Descriptive Statistics > Crosstabs.** Use the ***Cells*** button to request row and/or column percentages.

Visual Summaries for Categorical Variables

Two simple visual summaries are used for categorical data:

- **Pie charts** are useful for summarizing a single categorical variable if there are not too many categories.
- **Bar graphs** are useful for summarizing one or two categorical variables and are particularly useful for making comparisons when there are two categorical variables.

Both of these simple graphical displays are easy to construct and interpret, as the following examples demonstrate.

Example 2.3 **Humans Are Not Good Randomizers** Question 5 in the class survey described in Section 2.1 asked students to "Randomly pick a number between 1 and 10." The pie chart shown in Figure 2.2 illustrates that the responses are not even close to being evenly distributed across the numbers. Note that almost 30% of the students chose 7, while only just over 1% chose the number 1.

Figure 2.3 illustrates the same results with a bar graph. This bar graph shows the frequencies of responses on the vertical axis and the possible response categories on the horizontal axis. The display makes it obvious that the number of students who chose 7 was more than double that of the next most popular choice. We also see that very few students chose either 1 or 10.

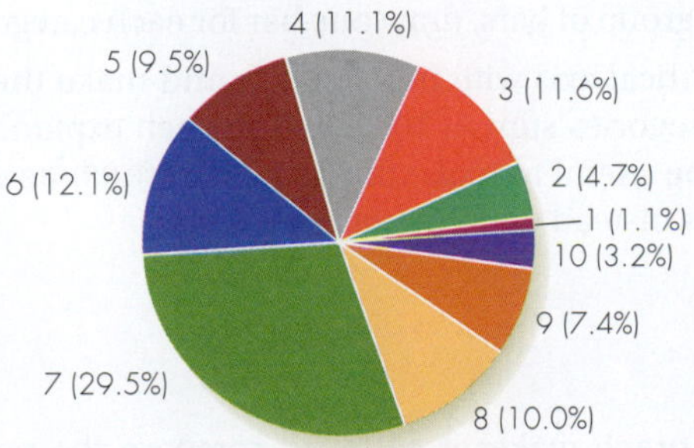

Figure 2.2 Pie chart of numbers picked

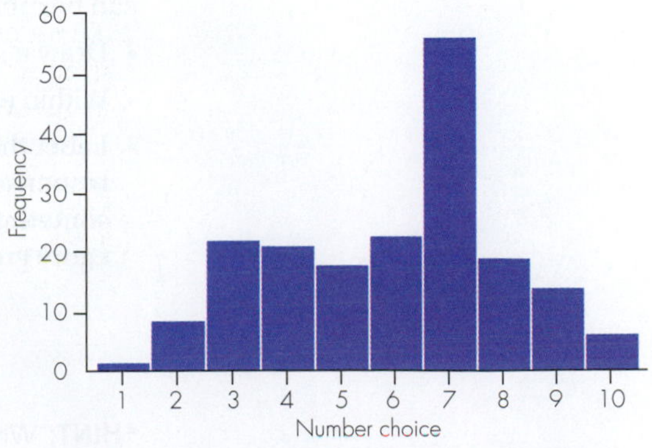

Figure 2.3 Bar graph of numbers picked

Example 2.4 **Revisiting Example 2.2: Nightlight and Nearsightedness** Figure 2.4 illustrates the data presented in Example 2.2 with a bar chart showing, for each lighting group, the percentage that ultimately had each level of myopia. This bar chart differs from the one in Figure 2.3 in two respects. First, it is used to present data for two categorical variables instead of just one. Second, the vertical axis represents percentages instead of counts, with the percentages for myopia status computed separately within each lighting category. Within each sleeptime lighting category, the percentages add to 100%.

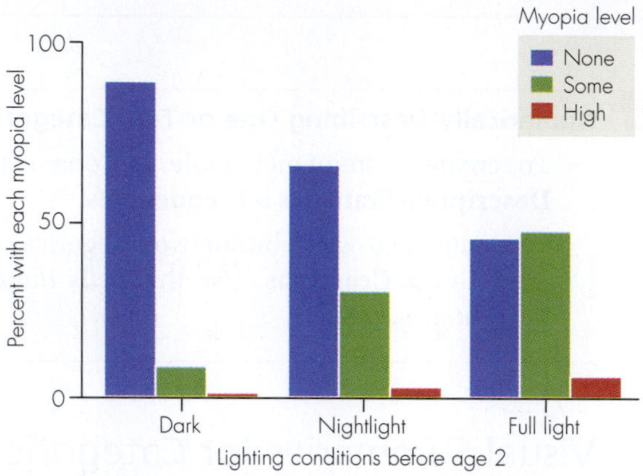

Figure 2.4 Bar chart for myopia and nighttime lighting in infancy

THOUGHT QUESTION 2.4 Redo the bar graph in Figure 2.4 using counts instead of percentages. The necessary data are given in Table 2.3. Would the comparison of frequency of myopia across the categories of lighting be as easy to make using the bar graph with counts? Generalize your conclusion to provide guidance about what should be done in similar situations.*

IN SUMMARY Bar Graphs for Categorical Variables

In a bar graph for *one categorical variable*, you can choose one of the following to display as the height of a bar for each category, indicated by labeling the vertical axis:

- Frequency or count
- Relative frequency = number in category/overall number
- Percentage = relative frequency × 100%

In a bar graph for *two categorical variables*, if an explanatory and response variable can be identified, it is most common to:

- Draw a separate group of bars for each category of the explanatory variable.
- Within each group of bars, draw one bar for each category of the response variable.
- Label the vertical axis with percentages and make the heights of the bars for the response categories sum to 100% within each explanatory category group. It can sometimes be useful to make the heights of the bars equal the counts in the category groups instead of percentages.

***HINT:** Which graph makes it easier to compare the percentage with myopia for the three groups? What could be learned from the graph of counts that isn't apparent from the graph of percentages?

MINITAB TIP **Graphically Describing One or Two Categorical Variables**

- To draw a *bar graph,* use **Graph > Bar Chart.** In the resulting display, select ***Simple*** to graph one variable or select ***Cluster*** to graph the relationship between two variables. Then, in the "Categorical variables" box, specify the column(s) containing the raw data for the variable(s). To graph percentages rather than counts, use the Bar Chart Options button.
- To draw a *pie chart,* use **Graph > Pie Chart.** Use the ***Multiple Graphs*** button to create separate pie charts for subgroups within the dataset.

2.3 Exercises are on pages 57–58.

2.4 Exploring Features of Quantitative Data with Pictures

Looking at a long, disorganized list of data values is about the same as looking at a scrambled set of letters. To begin finding the information in quantitative data, we have to organize it using visual displays and numerical summaries. In this section, we focus on interpreting the main features of quantitative variables. More specific details will be given in the following sections.

Table 2.4 displays the raw data for the right handspan measurements (in centimeters) made in the student survey described in Section 2.1. The measurements are listed separately for males and females but are not organized in any other way. Imagine that you know a female whose stretched right handspan is 20.5 cm. Can you see how she compares to the other females in Table 2.4? That will probably be hard because the list of data values is disorganized.

In Case Study 1.1, we graphed the "fastest ever driven" responses with a simple **dotplot**. We also summarized the data using a **five-number summary**, which consists of the median, the quartiles (roughly, the medians of the lower and upper halves of the data), and the extremes (low, high). Let's use those methods to organize the handspan data in Table 2.4.

Table 2.4 Stretched Right Handspans (centimeters) of 190 College Students

Males (87 students):
21.5, 22.5, 23.5, 23, 24.5, 23, 26, 23, 21.5, 21.5, 24.5, 23.5, 22, 23.5, 22, 22, 24.5, 23, 22.5, 19.5, 22.5, 22, 23, 22.5, 20.5, 21.5, 23, 22.5, 21.5, 25, 24, 21.5, 21.5, 18, 20, 22, 24, 22, 23, 22, 22, 23, 22.5, 25.5, 24, 23.5, 21, 25.5, 23, 22.5, 24, 21.5, 22, 22.5, 23, 18.5, 21, 24, 23.5, 24.5, 23, 22, 23, 23, 24, 24.5, 20.5, 24, 22, 23, 21, 22.5, 21.5, 24.5, 22, 22, 21, 23, 22.5, 24, 22.5, 23, 23, 23, 21.5, 19, 21.5

Females (103 students):
20, 19, 20.5, 20.5, 20.25, 20, 18, 20.5, 22, 20, 21.5, 17, 16, 22, 22, 20, 20, 20, 20, 21.7, 22, 20, 21, 21, 19, 21, 20.25, 21, 22, 18, 20, 21, 19, 22.5, 21, 20, 19, 21, 20.5, 21, 22, 20, 20, 18, 21, 22.5, 22.5, 19, 19, 19, 22.5, 20, 13, 20, 22.5, 19.5, 18.5, 19, 17.5, 18, 21, 19.5, 20, 19, 21.5, 18, 19, 19.5, 20, 22.5, 21, 18, 22, 18.5, 19, 22, 17, 12.5, 18, 20.5, 19, 20, 21, 19, 19, 21, 18.5, 19, 21.5, 21.5, 23, 23.25, 20, 18.8, 21, 21, 20, 20.5, 20, 19.5, 21, 21, 20

Example 2.5 **Right Handspans** In Figure 2.5, each dot represents the handspan of an individual student, with the value of the measurement shown along the horizontal axis. From this dotplot, we learn that a majority of the females had handspans between 19 and 21 cm and a good number of the males had handspans between 21.5 and 23 cm. We also see that there were two females with unusually small handspans compared to those of the other females.

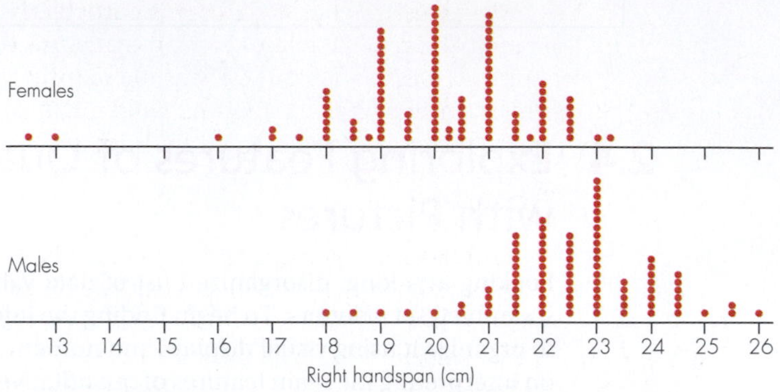

Figure 2.5 Stretched right handspans (in centimeters) of college students

Here are five-number summaries for the male and females handspan measurements given in Table 2.4 and graphed in Figure 2.5:

	Males (87 Students)		Females (103 Students)	
Median	22.5		20.0	
Quartiles	21.5	23.5	19.0	21.0
Extremes	18.0	26.0	12.5	23.25

Remember that the five-number summary approximately divides the dataset into quarters. For example, about 25% of the female handspan measurements are between 12.5 and 19.0 cm, about 25% are between 19.0 and 20.0 cm, about 25% are between 20.0 and 21.0 cm, and about 25% are between 21.0 and 23.25 cm. The five-number summary, along with the dotplot, gives us a good idea of where our imagined female with the 20.5-cm handspan fits into the distribution of handspans for females. She is in the third quarter of the data, slightly above the median (the middle value).

Summary Features of Quantitative Variables

The **distribution** of a quantitative variable is the overall pattern of how often the possible values occur. For most quantitative variables, three summary characteristics of the overall distribution of the data tend to be of the most interest. These are the **location** (center, average), the **spread** (variability), and the **shape** of the data. We also will be interested in whether or not there are any **outliers**—individual values that are unusual compared to the bulk of the other values—in the data.

Location (Center, Average)

The first concept for summarizing a quantitative variable is the idea of the "center" of the distribution of values, also called the *location* of the data. What is a typical or average value? The **median**, approximately the middle value in the data, is one estimate of location. The **mean**, which is the usual arithmetic average, is another. Details about how to compute these are given in Section 2.5.

Spread (Variability)

The **variability** among the individual measurements is an important feature of any dataset. How spread out are the values? Are all values about the same? Are most of them together but with a few that are unusually high or low?

In a five-number summary, we can assess the amount of spread (variability) in the data by looking at the difference between the two extremes (called the *range*) and the difference between the two quartiles (called the *interquartile range*). Later in this chapter, you will learn about the *standard deviation,* another important measure of variability.

An assessment of variability is particularly important in interpreting data. For instance, to know whether or not the amount of rainfall during a year at a location is unusual, we have to know about the natural variation in annual rainfall amounts. To determine whether a 1-year-old child might be growing abnormally, we need to know about the natural variation in the heights of 1-year-old children.

Shape

A third feature to consider is the shape of how the values of a quantitative variable are distributed. Using appropriate visual displays, we can address questions about *shape* such as the following: Are most of the values clumped in the middle, with values tailing off at each end (like the handspan measurements shown in Figure 2.5)? Are most of the values clumped together on one end (either high or low), with the remaining few values stretching relatively far toward the other end? We will discuss *shape* more completely later in this section, on page 33.

Outliers

We will also want to consider whether or not any individual values are outliers. There is no precise definition for an outlier, but in general, an outlier is a data point that is not consistent with the bulk of the data. For a single variable, an outlier is a value that is unusually high or low. When two variables are considered, an outlier is an unusual combination of values. For instance, in Example 2.5 about handspans, a female with a handspan of 24.5 cm would be an outlier because this handspan is well past the largest of the measurements made by the 103 females. A male with a handspan of 24.5 cm, however, is not an outlier because this measurement is clearly within the normal range for males.

The extreme values, low and high, in a dataset do not automatically qualify as outliers. To qualify as an outlier a data value must be unusually low or high compared to the rest of the data. We will describe a method for identifying outliers in Section 2.5.

Example 2.6 **Annual Compensation for Highest Paid CEOs in the United States** Figure 2.6 is a dotplot of the paid compensation (in millions of $) for the 50 highest-paid CEOs in 2008 for companies on *Fortune Magazine*'s list of Top 500 companies in the United States. Somewhat vague indications of *location* and *spread* are shown on the figure. The median compensation for these 50 CEOs was about $35.6 million, and that's approximately where "location" is indicated on Figure 2.6. Overall, the data spread from $24.3 million to $557 million, although the value at $557 million looks to be an outlier, a data value inconsistent with the bulk of the data. By the way, this astounding amount was paid to Lawrence J. Ellison, CEO of Oracle. The *shape* of the dataset is that most values are clumped on the lower end of the scale with the remaining values stretching relatively far toward the high end (called a skewed shape).

The data are given in the **ceodata08** dataset on the companion website, http://www.cengage.com/statistics/Utts4e.

(*Source:* http://www.forbes.com/lists/2009/12/best-boss-09_CEO-Compensation_CompTotDisp.html)

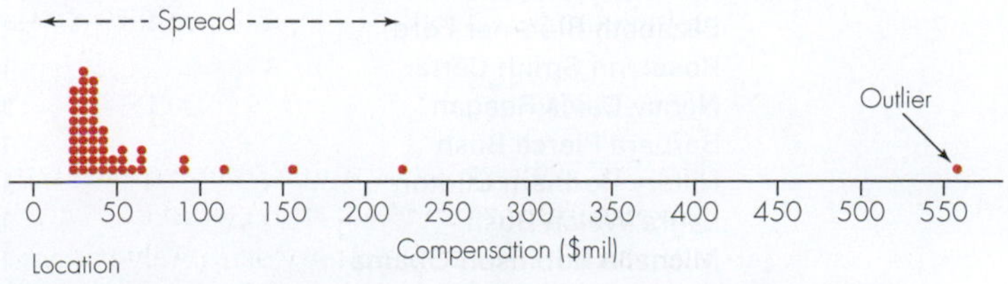

Figure 2.6 Dotplot of CEO compensation in 2008

The next example illustrates that sometimes the extreme points are the most interesting features of a dataset, even if they might not be outliers.

Example 2.7 **Ages of Death of U.S. First Ladies** Much has been written about ages of U.S. presidents when elected and at death, but what about their wives? Do these women tend to live short lives or long lives? Table 2.5 lists the approximate ages at death for first ladies of the United States as listed at the White House website. It is not completely accurate to label all of these women "first ladies" if the strict definition is "the wife of a president while in office." For example, Harriet Lane served socially as "first lady" to President James Buchanan, but he was unmarried and she was his niece.

Table 2.5 The First Ladies of the United States of America

Name	Born–Died	Age at Death
Martha Dandridge Custis Washington	1731–1802	71
Abigail Smith Adams	1744–1818	74
Martha Wayles Skelton Jefferson	1748–1782	34
Dolley Payne Todd Madison	1768–1849	81
Elizabeth Kortright Monroe	1768–1830	62
Louisa Catherine Johnson Adams	1775–1852	77
Rachel Donelson Jackson	1767–1828	61
Hannah Hoes Van Buren	1783–1819	36
Anna Tuthill Symmes Harrison	1775–1864	89
Letitia Christian Tyler	1790–1842	52
Julia Gardiner Tyler	1820–1889	69
Sarah Childress Polk	1803–1891	88
Margaret Mackall Smith Taylor	1788–1852	64
Abigail Powers Fillmore	1798–1853	55
Jane Means Appleton Pierce	1806–1863	57
Harriet Lane	1830–1903	73
Mary Todd Lincoln	1818–1882	64
Eliza McCardle Johnson	1810–1876	66
Julia Dent Grant	1826–1902	76
Lucy Ware Webb Hayes	1831–1889	58
Lucretia Rudolph Garfield	1832–1918	86
Ellen Lewis Herndon Arthur	1837–1880	43
Frances Folsom Cleveland	1864–1947	83
Caroline Lavinia Scott Harrison	1832–1892	60
Ida Saxton McKinley	1847–1907	60
Edith Kermit Carow Roosevelt	1861–1948	87
Helen Herron Taft	1861–1943	82
Ellen Louise Axson Wilson	1860–1914	54
Edith Bolling Galt Wilson	1872–1961	89
Florence Kling Harding	1860–1924	64
Grace Anna Goodhue Coolidge	1879–1957	78
Lou Henry Hoover	1874–1944	70
Anna Eleanor Roosevelt Roosevelt	1884–1962	78
Elizabeth Virginia Wallace Truman	1885–1982	97
Mamie Geneva Doud Eisenhower	1896–1979	83
Jacqueline Lee Bouvier Kennedy Onassis	1929–1994	65
Claudia Taylor Johnson	1912-2007	95
Patricia Ryan Nixon	1912–1993	81
Elizabeth Bloomer Ford	1918–	
Rosalynn Smith Carter	1927–	
Nancy Davis Reagan	1923–	
Barbara Pierce Bush	1925–	
Hillary Rodham Clinton	1947–	
Laura Welch Bush	1946–	
Michelle Robinson Obama	1964–	

Sources: http://www.whitehouse.gov/WH/glimpse/firstladies/html/firstladies.html, and http://www.firstladies.org/biographies/.

A few of the women listed died before their husband's term in office. Nonetheless, we will use the data as provided by the White House and summarize the ages at death for these women. Following is a five-number summary for these ages:

	First Ladies' Ages at Death	
Median	70.5	
Quartiles	60	82
Extremes	34	97

If you are at all interested in history, this summary will make you curious about the extreme points. Who died at 34? Who lived to be 97? The extremes are more interesting features of this dataset than is the summary of ages in the middle, which tend to match what we would expect for ages at death. From Table 2.5, you can see that Thomas Jefferson's wife, Martha, died in 1782 at age 34, almost 20 years before he entered office. He reportedly was devastated, and he never remarried, although historians believe that he may have had other children in his relationship with Sally Hemings. At the other extreme, Bess Truman died in 1982 at age 97; her husband, Harry, preceded her in death by 10 years, but he too lived a long life—he died at age 88.

Should we attach the label "outlier" to either of the most extreme points in the list of ages at death for the first ladies? To study this issue, we have to examine all of the data to see whether or not the two extremes clearly stand apart from the other values. If you look over Table 2.5, you may be able to form an opinion about whether Martha Jefferson and Bess Truman should be called outliers by comparing them to the other first ladies. Making sense of a list of numbers, however, is difficult. The most effective way to look for outliers is to graph the data, which we will learn more about in the remainder of this section.

Pictures of Quantitative Data

Three similar types of pictures are used to represent quantitative variables, all of which are valuable for assessing location, spread, shape, and outliers. **Histograms** are similar to bar graphs and can be used for any number of data values, although they are not particularly informative when the sample size is small. **Stem-and-leaf plots** and **dotplots** present all individual values, so for very large datasets, they are more cumbersome than histograms.

Figures 2.7 to 2.9 illustrate a histogram, a stem-and-leaf plot, and a dotplot, respectively, for the females' right handspans displayed in Table 2.4. Figure 2.9 is merely a portion of the dotplot shown previously in Figure 2.5 on page 26. Examine the three figures. Note that if the stem-and-leaf plot were turned on its side, all three pictures would look similar. Each picture shows the *distribution* of the data — the pattern of how often the various measurements occurred.

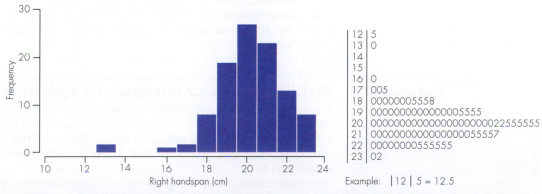

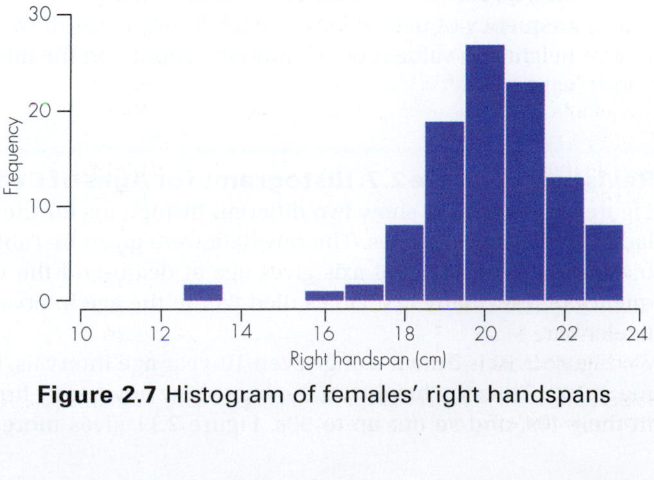

Figure 2.7 Histogram of females' right handspans

```
12 | 5
13 | 0
14 |
15 |
16 | 0
17 | 005
18 | 00000005558
19 | 0000000000000005555
20 | 00000000000000000000022555555
21 | 0000000000000000055557
22 | 00000000555555
23 | 02
```

Example: | 12 | 5 = 12.5

Figure 2.8 Stem-and-leaf plot of females' right handspans

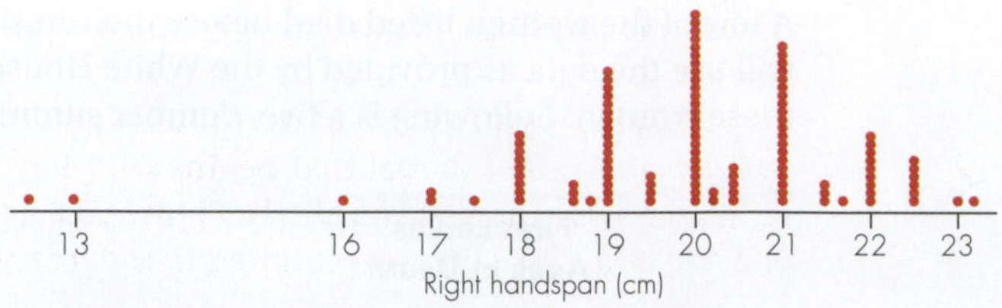

Figure 2.9 Dotplot of females' right handspans

A fourth kind of picture, called a **boxplot** or **box-and-whisker plot**, displays the information given in a five-number summary. It is especially useful for comparing two or more groups and for identifying outliers. We will examine boxplots at the end of this section.

Interpreting Histograms, Stem-and-Leaf Plots, and Dotplots

Each of these pictures is useful for assessing the location, spread, and shape of a distribution, and each is also useful for detecting outliers. For the data presented in Figures 2.7 to 2.9, note that the values are *centered* at about 20 cm, which we learned in Example 2.5 is indeed the median value. There are two possible *outlier* values that are low in comparison to the bulk of the data. These are identifiable in the stem-and-leaf plot as 12.5 and 13.0 cm, but are evident in the other two pictures as well. Except for those values, the handspans have a *range* of about 7 cm, extending from about 16 to 23 cm. They tend to be clumped around 20 and taper off toward 16 and 23.

There are many computer programs that can be used to create these pictures. Figures 2.7, 2.8, and 2.9, for instance, are slight modifications of pictures created using Minitab. We will go through the steps for creating each type of picture by hand, but keep in mind that statistical software such as Minitab automates most of the process.

Creating a Histogram

A histogram is a bar chart of a quantitative variable that shows how many values are in various intervals of the data. The steps in creating a histogram are as follows:

Step 1: Decide how many *equally spaced* intervals to use for the horizontal axis. The experience of many researchers is that somewhere between 6 and 15 intervals is a good number for displaying the bulk of the data, although occasionally more may be needed to accommodate outliers. Use intervals that make the range of each interval convenient.

Step 2: Decide whether to use *frequencies* or *relative frequencies* on the vertical axis. A frequency is the actual number of observations in an interval. A relative frequency is either the proportion or the percent in an interval.

Step 3: Draw the appropriate number of equally spaced intervals on the horizontal axis; be sure to cover the entire data range. Determine the frequency or relative frequency of data values in each interval and draw a bar with corresponding height. If a value is on a boundary, count it in the interval that begins with that value.

Example 2.8 **Revisting Example 2.7: Histograms for Ages of Death of U.S. First Ladies**

Figures 2.10 and 2.11 show two different histograms for the ages of death for the first ladies of the United States. The raw data were given in Table 2.5 on page 28. In each histogram, the horizontal axis gives age at death and the vertical axis gives the frequency of how many first ladies died within the age interval represented by any particular bar.

Figure 2.10 is drawn using seven 10-year age intervals, beginning at 30 and ending at 100. Thus the heights of the bars show how many first ladies died in their 30s, in their 40s, and so on, up to 90s. Figure 2.11 gives more detail by using fourteen

5-year age intervals, beginning with the interval 30 to 35 and ending with the interval 95 to 100.

A first lady with an age of death falling on the boundary between two age intervals is counted in the interval that begins with her age. For instance, there were two first ladies who died at 60 years old. In Figure 2.10, they are counted in the interval 60 to 70, and in Figure 2.11 they are counted in the interval 60 to 65.

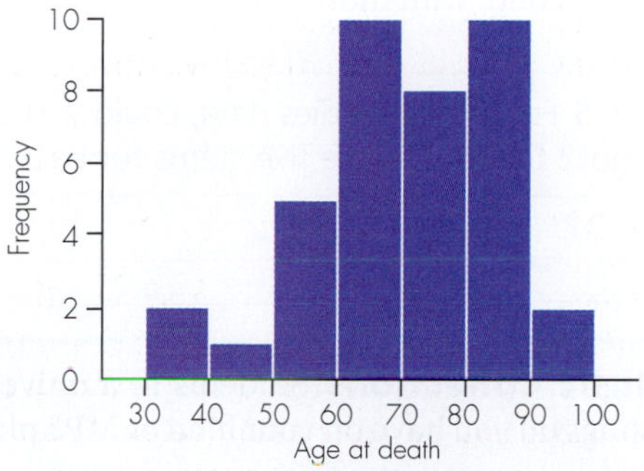

Figure 2.10 Histogram of ages of death of U.S. first ladies using seven 10-year intervals

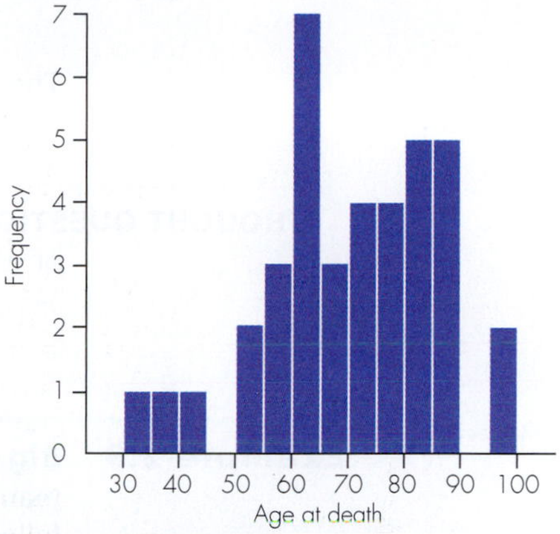

Figure 2.11 Histogram of ages of death of U.S. first ladies using fourteen 5-year intervals

Creating a Dotplot

To create a dotplot, the first step is to draw a number line (horizontal axis) that covers the range from the smallest to the largest data value. Then, for each observation, place a dot above the number line located at the observation's data value. When there are multiple observations with the same value, the dots are stacked vertically as in Figure 2.9.

Creating a Stem-and-Leaf Plot

A stem-and-leaf plot is created much like a histogram, except that every individual data value is shown. This plot is a quick way to summarize small datasets and is also useful for ordering the data from lowest to highest. A row in the plot starts with a "stem" and each stem gives the first part of a data value. A value within a row is called a "leaf" and it gives information about the last part of a data value.

To simplify the work, data values sometimes are truncated or rounded off. To truncate a value, simply drop digits. The number 23.58 is truncated to 23.5, but it is rounded off to 23.6. In Figure 2.8, the handspan values are truncated to one digit after the decimal point. The largest handspan is 23.25 cm, which is truncated to 23.2. In the figure, this value is shown on the "23" stem along with the value of 23.0 cm. These two handspans are displayed as |23|02. In other words, this stem has two leaves, each representing a different individual, one with a handspan of about 23.0 and the other with a handspan of 23.2.

Here are the steps required to create a stem-and-leaf plot:

Step 1: Determine the stem values. Remember that the "stem" contains all but the last of the displayed digits of a number. As with histograms, it is reasonable to have between 6 and 15 stems where each stem defines an interval of values. The number of stems should be sufficient to allow us to see the shape of the data. The stems should define equally spaced intervals.

Step 2: For each individual, attach a "leaf" to the appropriate stem. A "leaf" is the last part of the displayed digits of a number. It is standard, but not mandatory, to put the leaves in increasing order at each stem value.

Note: There will be more than one way to define equally spaced stems. For example, the ages at death of first ladies range from 34 to 97. We could have stem values representing the decades (3, 4, . . . , 9) for a total of seven stems, *or* we could allow two stem values for each decade, for a total of 14 stem values. With two stems for each decade, the first instance of each stem value would receive leaves of 0 to 4, and the second would receive leaves of 5 to 9. So the two deaths in the 30s, at ages 34 and 36, could be represented in two different ways:

|3|46 or |3|4
|3|6

The first method, with fewer stem values, is generally preferable for small datasets, while the second, with more stem values, is preferable for larger datasets.

THOUGHT QUESTION 2.5 For the first ladies data, could you use three stems for each decade? Why or why not? Could you use five stems for each decade? Why or why not?*

Example 2.9 **Big Music Collections** Students in a university statistics class were asked, "How many songs do you have on your iPod or MP3 player?" Responses from $n = 24$ students follow:

2510, 500, 500, 1300, 687, 600, 500, 2600, 30, 900, 800, 0,
750, 1500, 1500, 2400, 800, 2017, 1150, 5000, 4000, 1250, 1700, 3305

Here is a stem-and-leaf plot of the data, as drawn by Minitab:

Reported songs on iPod or MP3 player for $n = 24$ students, stem unit = 1000s, leaf unit = 100s.

```
0 | 00
0 | 555667889
1 | 123
1 | 557
2 | 04
2 | 56
3 | 3
3 |
4 | 0
4 |
5 | 0
```

Source: Class data collected by Robert Heckard in 2010

For the plot, Minitab considered all values to be four-digit numbers and used the first digit, the 1000s value, for stem (row) labels. The software used the second digit, the 100s value, as a leaf value and then truncated the last two digits. Three examples of this procedure follow:

- 2510 has a stem label of 2 and a leaf value of 5.
- 500 written as a four-digit number is 0500 giving a stem label of 0 and a leaf value of 5.
- 30 can be written as 0030 so its stem value is 0 and its leaf value is also 0.

Note that there are two stems for each 1000s possibility. The first is for the leaf values 0, 1, 2, 3, 4 and the second is for leaf values 5, 6, 7, 8, 9. These data have a shape called "skewed to the right," which we will define following this example.

***HINT:** Is the number of possibilities for the second digit of age evenly divisible by 3?

MINITAB TIP **Drawing a Histogram, Dotplot, or Stem-and-Leaf Plot**

- To draw a *histogram* of a quantitative variable, use **Graph > Histogram.** In the resulting display, select ***Simple.*** In the "Graph Variables" box, specify the column containing the raw data for the variable. To graph percents rather than counts, use the ***Scale*** button and then select the "Y-Scale Type" tab.
- To draw a *dotplot* of a quantitative variable, use **Graph > Dotplot.** Select ***Simple*** under "One Y" to create a dotplot showing the entire sample, as in Figure 2.9. Select ***With Groups*** under "One Y" to compare subgroups within the sample, as in Figure 2.5.
- To draw a *stem-and-leaf plot* of a quantitative variable, use **Graph > Stem-and-Leaf.**

Describing Shape

The *shape* of a dataset is usually described as either **symmetric**, meaning that it is similar on both sides of the center, or **skewed**, meaning that the values are more spread out on one side of the center than on the other. If it is **skewed to the right**, the higher values (toward the right on a number line) are more spread out than the lower values. If it is **skewed to the left**, the lower values (toward the left on a number line) are more spread out than the higher values. Figure 2.12 illustrates the distinction between the two different directions of skewness.

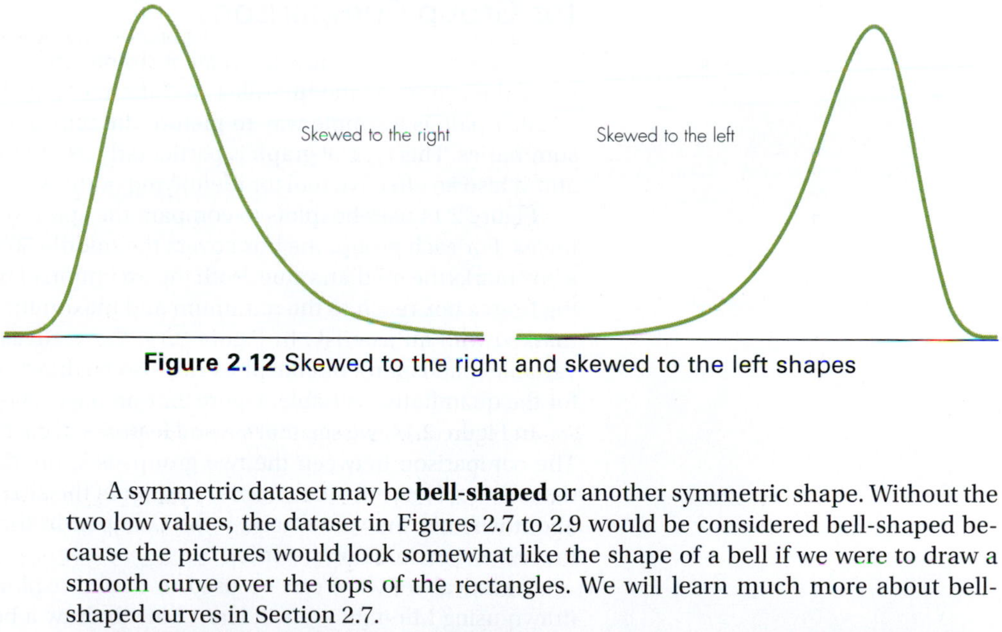

Figure 2.12 Skewed to the right and skewed to the left shapes

A symmetric dataset may be **bell-shaped** or another symmetric shape. Without the two low values, the dataset in Figures 2.7 to 2.9 would be considered bell-shaped because the pictures would look somewhat like the shape of a bell if we were to draw a smooth curve over the tops of the rectangles. We will learn much more about bell-shaped curves in Section 2.7.

The **mode** of a dataset is the most frequent value. The shape is called **unimodal** if there is a single prominent peak in a histogram, stemplot, or dotplot, as in Figures 2.7 to 2.9. The shape is called **bimodal** if there are two prominent peaks in the distribution. Figure 2.13 (on the next page) shows a histogram with a bimodal shape. The data are eruption durations (minutes) for $n = 230$ eruptions of the Old Faithful geyser in Yellowstone Park. There is a prominent peak around 2.0 minutes and another prominent peak around 4.5 minutes.

Some datasets will be described best with a combination of these terms, w oth- ers will require only one of them. The description "bell-shaped" already c information that the shape is symmetric and that it is unimodal. On the knowing that a shape is symmetric or skewed does not tell us about th modes. The dataset could be unimodal, bimodal, or neither. For instan might be described as bimodal and slightly skewed to the left, since th the data is on the right.

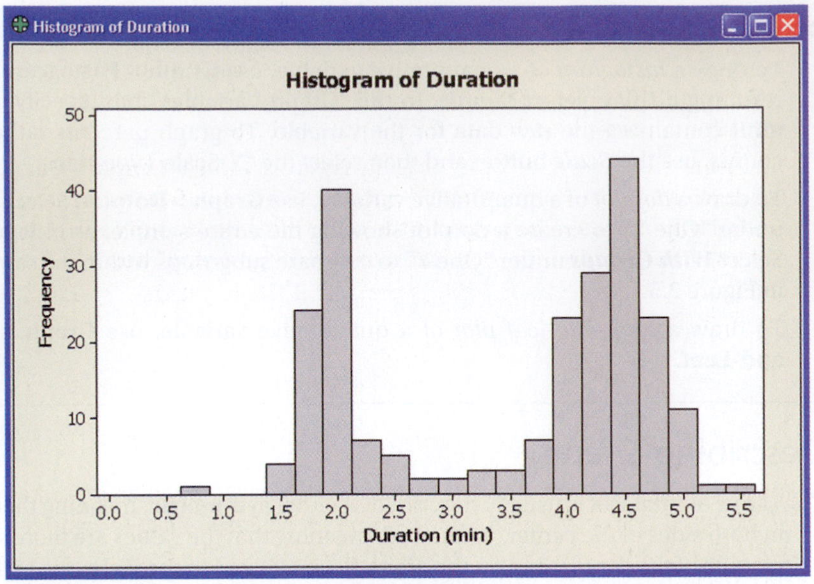

Figure 2.13 Histogram of eruption durations (minutes) for Old Faithful geyser *(Source: Hand et al., 1994.)*

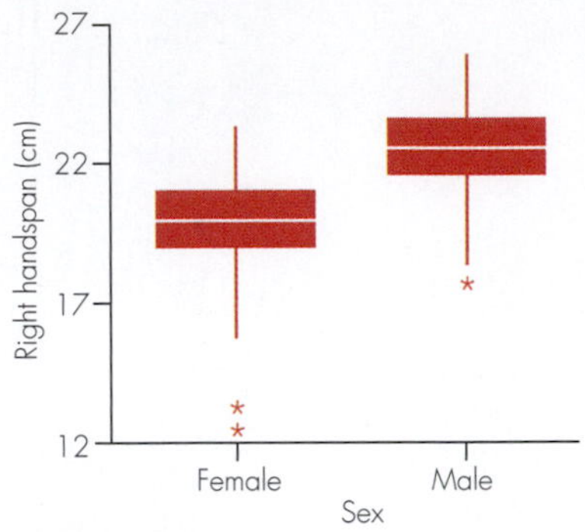

Figure 2.14 Boxplots for right handspans of men and women

Boxplots: Picturing Location and Spread for Group Comparisons

We have already seen how to create a simple and informative five-number summary using the extremes, the quartiles, and the median. A boxplot, also called a box-and-whisker plot, is a simple way to picture the information in one or more five-number summaries. This type of graph is particularly useful for comparing two or more groups and is also an effective tool for identifying outliers.

Figure 2.14 uses boxplots to compare the spans of the right hands of males and females. For each group, the box covers the middle 50% of the data, and the line within a box marks the median value. With the exception of possible outliers, the lines extending from a box reach to the minimum and maximum data values. Possible outliers are marked with an asterisk. In Figure 2.14, the vertical axis is used for the quantitative variable (handspan), but boxplots can also be drawn so that the horizontal axis is used for the quantitative variable. Figure 2.21 on page 44 is an example.

In Figure 2.14, we see that several features of each group are immediately obvious. The comparison between the two groups is simplified as well. The only feature of a dataset that is not obvious from a boxplot is the shape, although there is information about whether the values tend to be clumped in the middle or tend to stretch more toward one extreme or the other.

Statistical software can be used to draw boxplots. Figure 2.14, for instance, was drawn using Minitab. We will learn how to draw a boxplot by hand in Section 2.5, in which we cover the details of determining numerical summaries such as the median and quartiles and the steps for determining possible outliers.

THOUGHT QUESTION 2.6 Using the boxplots in Figure 2.14, what can you say about the respective handspans for males and females? Are there any surprises, or do you see what you would expect?*

***HINT:** Is there a difference in the location of the data for the two graphs? Is either group more spread out than the other?

MINITAB TIP **Drawing a Boxplot**

- To draw a *boxplot* of a quantitative variable, use **Graph > Boxplot**. In the resulting display, select ***Simple*** under "One Y." In the "Graph Variables" box, specify the column containing the raw data for the variable.
- To create a comparative boxplot as in Figure 2.14, use **Graph > Boxplot** and select ***With Groups*** under "One Y." Specify the quantitative response variable in the "Graph Variables" box; specify the categorical explanatory variable that creates the groups in the "Categorical variables for grouping" box.

Strengths and Weaknesses of the Four Visual Displays

Histograms, stem-and-leaf plots, dotplots, and boxplots organize quantitative data in ways that let us begin to find the information in a dataset. As to the question of which type of display is the best, there is no unique answer. The answer depends on what feature of the data may be of interest and, to a certain degree, on the sample size. Let's consider some strengths and weaknesses of each type of plot.

Histograms

Strengths: A histogram is an excellent tool for judging the shape of a dataset with moderate or large sample sizes. There is flexibility in choosing the number as well as the width of the intervals for the display. Between 6 and 15 intervals usually gives a good picture of the shape.

Weaknesses: With a small sample, a histogram may not "fill in" sufficiently well to show the shape of the data. With either too few intervals or too many, we may not see the true shape of the data.

Figures 2.15 (below) and 2.16 (on the next page) are histograms for the same dataset, the self-reported average hours of studying per week for 3179 college students. Fifteen intervals were used in Figure 2.15, whereas only four intervals were used in Figure 2.16. With 15 intervals we get a more detailed look at shape and we also learn from Figure 2.15 that the peak occurs between 5 and 15 hours, something we can't see in Figure 2.16 with fewer intervals.

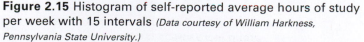

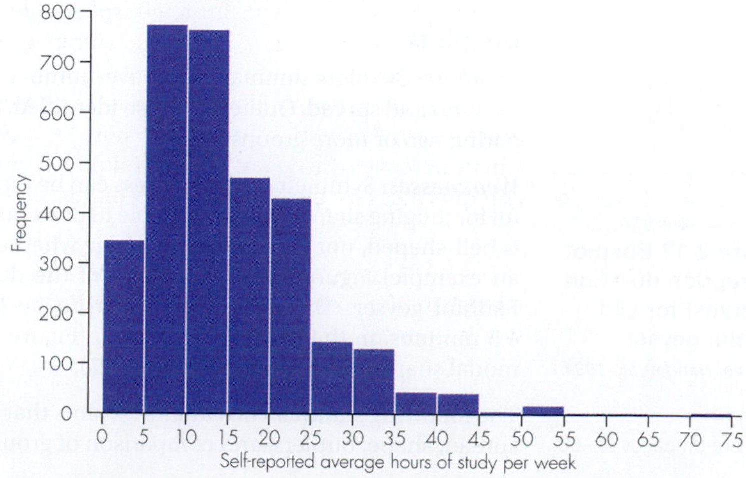

Figure 2.15 Histogram of self-reported average hours of study per week with 15 intervals *(Data courtesy of William Harkness, Pennsylvania State University.)*

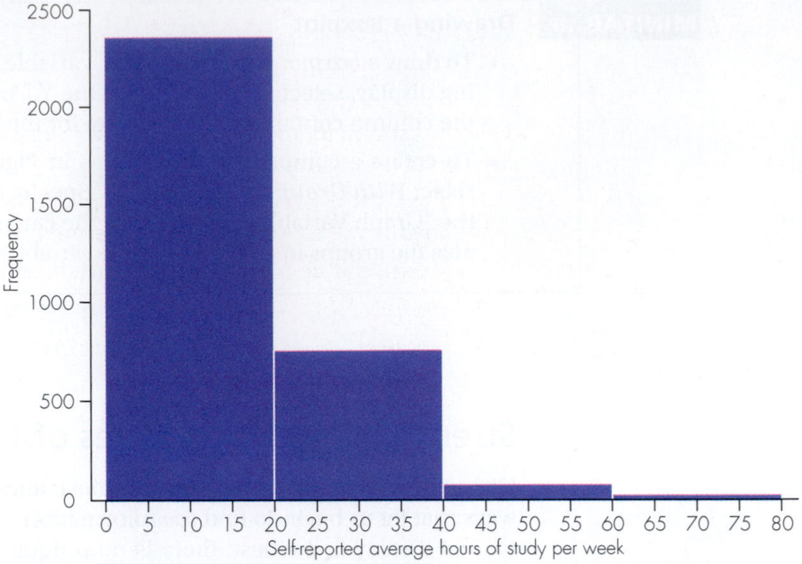

Figure 2.16 Histogram of self-reported average hours of study per week with four intervals

Stem-and-Leaf Plots

Strengths: A stem-and-leaf plot is an excellent tool for sorting data, and with a sufficient sample size, it can be used to judge shape. Somewhere between 6 and 15 stems (intervals) is usually suitable for judging shape.

Weaknesses: With a large sample size, a stem-and-leaf plot may be too cluttered because the display shows all individual data points. Also, compared to a histogram, we are restricted in choices for "intervals," which in a stem-and-leaf plot are determined by the rules for allocating leaf values to stems.

Dotplots

Strengths: We see all individual data values and a dotplot is easy to create. With moderate or small sample sizes, comparative dotplots are useful for comparing two or more groups, as in Figure 2.5 on page 26.

Weaknesses: A dotplot will be too cluttered with a large sample size. For large sample sizes, we can judge shape better by summarizing the data into intervals as we do when creating histograms and stem-and-leaf plots.

Boxplots

Stengths: Boxplots summarize the five-number summary, so they give a direct look at location and spread. Outliers are also identified. Boxplots are an excellent tool for comparing two or more groups.

Weaknesses: Symmetry and skewness can be judged, but boxplots are not entirely useful for judging shape. It is not possible to use a boxplot to judge whether or not a dataset is bell-shaped, nor is it possible to judge whether or not a dataset may be bimodal. As an example, Figure 2.17 is a boxplot of the duration times of eruptions of the Old Faithful geyser. These data were seen to be bimodal, with peaks around 2.0 and 4.5 minutes, in the histogram shown in Figure 2.13 on page 34. We don't see the bimodal shape in the boxplot (Figure 2.17).

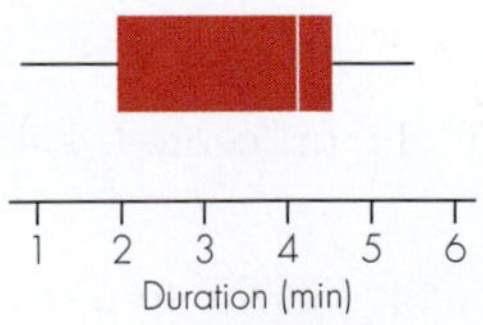

Figure 2.17 Boxplot of eruption duration (minutes) for Old Faithful geyser *(Source: Hand et al., 1994.)*

The following summary box identifies plots that may work well for examining location, spread, shape, outliers, and comparison of groups.

2.4 Exercises are on pages 58–60.

IN SUMMARY

Using Visual Displays to Identify Interesting Features of Quantitative Data

Here is a summary of how histograms, stem-and-leaf-plots, dotplots, and boxplots are useful for various purposes:

- To illustrate ***location*** and ***spread,*** any of the pictures work well. A boxplot marks the median, so it may be best for identifying location.
- To illustrate ***shape***, histograms and stem-and-leaf plots are best.
- To see ***individual values***, use stem-and-leaf plots and dotplots.
- To ***sort the values***, use stem-and-leaf plots. This can be a useful first step for creating a five-number summary.
- To ***compare groups***, use side-by-side boxplots, or create versions of one of the other pictures using the same scale for each group.
- To ***identify outliers*** using the standard definition, use a boxplot (see p. 43). Any of the other pictures will enable you to identify possible outliers.

2.5 Numerical Summaries of Quantitative Variables

In this section, we learn how to compute numerical summaries of these features for quantitative data. Recall from Section 2.1 that the numbers in a dataset are called *raw data*. To write formulas for some of the summaries in this section, we need notation for the raw data.

FORMULA

Notation for Raw Data

n = the number of individuals in a dataset

$x_1, x_2, x_3, \ldots, x_n$ represent the individual raw data values

Example: A dataset consists of handspan values in centimeters for six females; the values are 21, 19, 20, 20, 22, and 19. Then,

$n = 6$

$x_1 = 21$, $x_2 = 19$, $x_3 = 20$, $x_4 = 20$, $x_5 = 22$, and $x_6 = 19$

Describing the Location of a Dataset

The word *location* is used as a synonym for the "middle" or "center" of a dataset. There are two common ways to describe this feature.

- The **mean** is the usual numerical average, calculated as the sum of the data values divided by the number of values. It is nearly universal to represent the mean of a sample with the symbol $\bar{x}$, read as "x-bar."
- The **median** of a sample is the middle data value for an odd number of observations, after the sample has been ordered from smallest to largest. It is the average of the middle two values, in an ordered sample, for an even number of observations. We will use the letter M to represent the median of a sample.

FORMULA **Determining the Mean and Median**

The Mean

The symbol $\bar{x}$ is nearly always used to represent the mean of a sample. Notation for calculating the sample mean of a list of values is

$$\bar{x} = \frac{\Sigma x_i}{n}$$

The capital Greek letter sigma, written as Σ, is the universal symbol meaning "add up whatever follows." Therefore, for a dataset with individual values $x_1, x_2, x_3, \ldots, x_n$, the notation Σx_i is the same as saying "add together all the values."

The Median

It would require more notation than is convenient to write a formula for the median, so we simply write the rule:

- If n is odd, the median M is the middle of the ordered values. Find M by counting $(n + 1)/2$ up from the bottom or down from the top of the ordered list.
- If n is even, the median M is the average of the middle two of the ordered values. Find M by averaging the values that are $(n/2)$ and $(n/2) + 1$ from the top or the bottom of the ordered list.

Note: If you are determining a median "by hand," your first step should be to put the data in order from lowest to highest.

The first of the next two examples illustrates how to find the median when the sample size is odd; the second shows how to find the median when the sample size is even. The second example also demonstrates how skewness can cause the values of the mean and median to differ.

Example 2.10 **Median and Mean Quiz Scores** Suppose that scores on a quiz for $n = 7$ students in a class are

91 79 60 94 89 93 86

To find the median score, first write the data in order from smallest to largest. The ordered scores are

60 79 86 **89** 91 93 94

Because the number of data values is odd ($n = 7$), the median is the middle value of the ordered scores. The median, $M = 89$, is underlined and in bold typeface in the ordered list. To find the mean, add the seven data values, and then divide that sum by $n = 7$. The sum of the seven scores is 592, so the mean is $\bar{x} = 592/7 = 84.57$. The low score of 60 pulls the mean down slightly in comparison to the median.

Example 2.11 **Example 2.9 Revisited: Median and Mean Number of Songs on Student iPods or MP3 Players** In Example 2.9 on page 32, we gave data for the number of songs $n = 24$ college students said they have on their iPod or MP3 player. The ordered list of data values is

0	30	500	500	500	600	687	750
800	800	900	1150	1250	1300	1500	1500
1700	2017	2400	2501	2600	3305	4000	5000

The sample size is even ($n = 24$), so the median M is the average of the middle two values in the ordered data. These will be the 12th- and 13th-highest values in the dataset, so the median is

$$M = (1150 + 1250)/2 = 1200.$$

Half of the students had fewer than 1200 songs on their iPod or MP3 player and the other half had more than 1200.

The mean, the sum of the values divided by the sample size, is $\bar{x} = 36290/24 = 1512.08$. The general skew to the right in the data (toward big values) causes the mean to be greater than the median.

It is not necessarily correct to say that the mean and median measure the "typical" value in a dataset. This is true only when data values are clumped within a central region. In some situations, values near the mean and median may be relatively rare. The histogram in Figure 2.13 (p. 34) summarizing data for the durations of Old Faithful eruption shows an example. The mean duration is about 3.5 minutes, a relatively rare duration as most values in the bimodal distribution fall near either 4.5 minutes or 2 minutes.

Statements such as "the value is way above normal" also do not have a clear meaning unless accompanied by a description of the overall variability in a dataset. This point is made in the next example.

Example 2.12 **Will "Normal" Rainfall Get Rid of Those Odors?** A company (that will remain unnamed) located near Davis, California, was having an odor problem in its wastewater facility. Blame it on the weather:

> [According to a company official] "Last year's severe odor problems were due in part to the 'extreme weather conditions' created in the Woodland area by El Niño." She said Woodland saw 170 to 180 percent of its normal rainfall. "Excessive rain means the water in the holding ponds takes longer to exit for irrigation, giving it more time to develop an odor." (Amy Goldwitz, *The Davis Enterprise,* March 4, 1998, p. A1.)

This wording is typical of weather-related stories in which it is often remarked that rainfall is vastly "above normal" or "below normal." In fact, these stories occur so frequently that one wonders if there is *ever* a normal year. The annual rainfall (inches) for Davis, California, for 47 years leading up to and including the year in question is shown in Table 2.6. A histogram of annual rainfall amounts is shown in Figure 2.18.

What is the "normal" annual rainfall in Davis? The mean in this case is 18.69 inches, and the median is 16.72 inches. For the year under discussion in the article, 1997, rainfall was 29.69 inches; hence the comment that it was "170 to 180 percent of normal." But 29.69 inches of rain is within the range of rainfall values over this 47-year period, and more rain occurred in 4 of the other 46 years. The next time you hear about weather conditions that are not "normal," pay attention to whether they are truly outliers or just different from the *average* value. In this case, the company's excuse for the problem just doesn't hold water.

Table 2.6 Annual Rainfall for Davis, California (July 1–June 30), in Inches

Year	Rainfall	Year	Rainfall	Year	Rainfall	Year	Rainfall
1951	20.66	1963	11.2	1975	6.14	1987	16.3
1952	16.72	1964	18.56	1976	7.69	1988	11.38
1953	13.51	1965	11.41	1977	27.69	1989	15.79
1954	14.1	1966	27.64	1978	17.25	1990	13.84
1955	25.37	1967	11.49	1979	25.06	1991	17.46
1956	12.05	1968	24.67	1980	12.03	1992	29.84
1957	28.74	1969	17.04	1981	31.29	1993	11.86
1958	10.98	1970	16.34	1982	37.42	1994	31.22
1959	12.55	1971	8.6	1983	16.67	1995	24.5
1960	12.75	1972	27.69	1984	15.74	1996	19.52
1961	14.99	1973	20.87	1985	27.47	1997	29.69
1962	27.1	1974	16.88	1986	10.81		

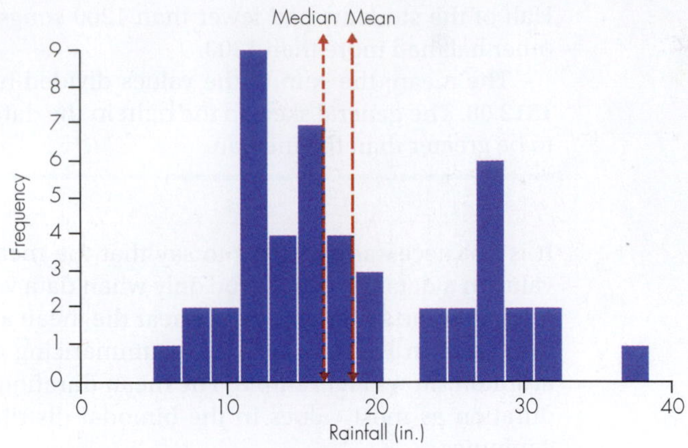

Figure 2.18 Annual rainfall in Davis, California

The Influence of Outliers on the Mean and Median

Outliers have a larger influence on the mean than on the median. Outliers at the high end will increase the mean, while outliers at the low end will decrease it. For instance, suppose that we use the ages of death for a person's grandparents and great grandparents as an indicator of a "typical" lifespan, and these ages at death are 76, 78, 80, 82, and 84. Then the median and the mean age of death would both be 80. But now suppose that the youngest age was 46 instead of 76. The median would *still* be 80 years old. But the mean would be

$$\bar{x} = \frac{46 + 78 + 80 + 82 + 84}{5} = 74 \text{ years}$$

Remember this when you hear statistics about "average life expectancy." Those values are calculated on the basis of averaging the anticipated age at death for all babies born in a given time period. The majority of individuals will live to be older than this "average," but those who die in infancy are the outliers that pull down the average. Datasets that have large outliers at the high end generally have higher means than medians. Examples of data that might have large outliers include annual incomes of executives and sales prices of homes for a large area.

The Influence of Shape on the Mean and Median

In a perfectly symmetric dataset, the mean and the median are equal; but in a skewed dataset, they differ. When the data are skewed to the left, the mean will tend to be smaller than the median. When the data are skewed to the right, with extreme high values, the mean will tend to be larger than the median. Example 2.9, about the number of iPod or MP3 songs that students have, illustrates an extreme case with data strongly skewed to the right, and the mean of about 1512 songs is much larger than the median of 1200.

Figures 2.19 and 2.20 show two more examples of how shape affects the relative sizes of the mean and median. Figure 2.19, a histogram of hours of sleep the previous night for $n = 173$ college students, is more or less symmetric. The mean of the sample data is $\bar{x} = 6.94$ hours, and the median is nearly the same at $M = 7$ hours. Having similar values for the mean and median is characteristic of a symmetric dataset. Figure 2.20 is a histogram of self-reported hours spent per week using a computer for the same 173 students who were used for Figure 2.19. The histogram in Figure 2.20 is skewed to the right. The mean weekly time of 14.3 hours is decidedly greater than the median time of 10 hours. There is an outlier at 84 hours, but deleting it decreases the mean only to 13.9 hours, which is still clearly greater than the median.

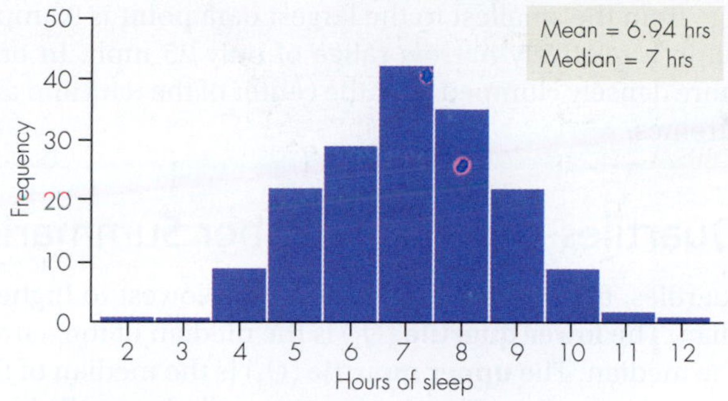

Figure 2.19 A symmetric shape—hours of sleep the previous night for $n = 173$ college students

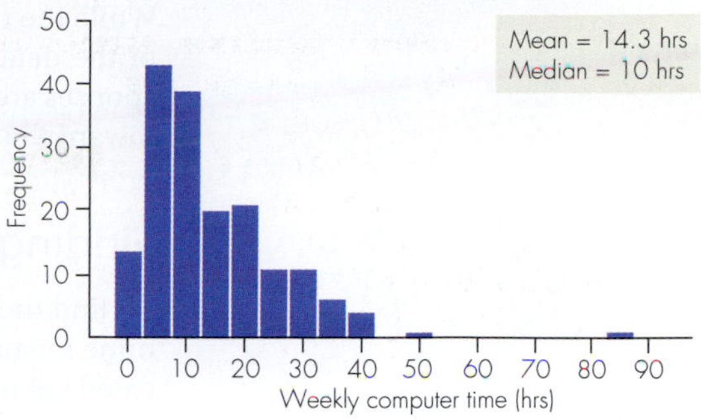

Figure 2.20 Data skewed to the right—weekly hours using the computer for $n = 173$ college students

Describing Spread: Range and Interquartile Range

Three summary measures that describe the *spread* or *variability* of a dataset are

- **Range** = high value − low value
- **Interquartile range** = upper quartile − lower quartile. The notation **IQR** is often used to represent the interquartile range.
- **Standard deviation**

The standard deviation is easiest to interpret in the context of bell-shaped data, so we postpone a description of it until Section 2.7, in which bell-shaped distributions are discussed. In this section, we describe how to calculate the range and the interquartile range.

Before going into detail about how to compute these measures of spread, let's revisit part of Case Study 1.1. This example shows how informative it is to reduce a dataset to a few simple summary values, providing information about location, spread, and outliers.

Example 2.13 **Range and Interquartile Range for Fastest Speeds Ever Driven** In Case Study 1.1, we summarized responses to the question, "What's the fastest you've ever driven a car?" The five-number summary for the 87 males surveyed is as follows:

	Males (87 Students)	
Median	110	
Quartiles	95	120
Extremes	55	150

This summary provides substantial information about the location, spread, and possible outliers. Remember that the median, 110 mph in this case, measures the *center* or *location* of the data. The other four numbers in the five-number summary can be used to describe how *spread out* or *variable* the responses are.

- The two *extremes* describe the spread over 100% of the data. Here, the responses are spread from 55 mph to 150 mph.
- The two *quartiles* describe the spread over approximately the middle 50% of the data. About 50% of the men gave responses between 95 and 120 mph.

Given the values of the extremes and the quartiles, it's simple to calculate the range and the interquartile range (IQR). For the fastest speed reportedly driven by males, values for these two measures of variability are

- Range = high − low = 150 − 55 = 95 mph
- IQR = upper quartile − lower quartile = 120 − 95 = 25 mph

While the range from the smallest to the largest data point is 95 mph, the middle 50% of the data fall in a relatively narrow range of only 25 mph. In other words, the responses are more densely clumped near the center of the data and are more spread out toward the extremes.

Finding Quartiles and Five-Number Summaries

To find the quartiles, first put the data values from lowest to highest and then determine the median. The **lower quartile** (Q_1) is the median of the data values that are located below the median. The **upper quartile** (Q_3) is the median of the data values that are located above the median. These values are called **quartiles** because, along with the median and the extremes, they divide the ordered data approximately into quarters.

FORMULA

Finding Quartiles

Q_1 = **lower quartile** = median of lower half of the ordered data values
Q_3 = **upper quartile** = median of upper half of the ordered data values

Note: The Minitab program uses a different procedure. Its quartile estimates may differ slightly from those determined by the procedures described here. Other software packages may do this as well.

Example 2.14 **Fastest Driving Speeds for Men** Here are the 87 males' responses to the question about how fast they have driven a car, as given in Case Study 1.1, except now the data are in numerical order. To make them easier to count, the data are arranged in rows of ten numbers:

55	60	80	80	80	80	85	85	85	85
90	90	90	90	90	92	94	95	95	95
95	***95***	95	100	100	100	100	100	100	100
100	100	101	102	105	105	105	105	105	105
105	105	109	**110**	110	110	110	110	110	110
110	110	110	110	110	112	115	115	115	115
115	115	120	120	120	***120***	120	120	120	120
120	120	124	125	125	125	125	125	125	130
130	140	140	140	140	145	150			

The median is the middle value in an ordered list, so for 87 values, the median is the $(87 + 1)/2 = 88/2 = 44$th value in the ordered list. The 44th value is 110, and this value is shown in bold in the data list. Note that several observations equal 110. We ignore these ties when finding the median and simply count (in this case) to the 44th location, regardless of ties.

There are 43 locations for data on either side of the median. To find the *quartiles,* simply find the median of each set of 43 values. Again, we don't worry about tied values. The lower quartile is at the $(43 + 1)/2 = 22$nd location from the bottom of the data, and the upper quartile is at the 22nd location from the top. These values are in bold and italics in the data list; $Q_1 = 95$ and $Q_3 = 120$.

The five-number summary divides the data into intervals with approximately equal numbers of values but it does not necessarily divide the data into equally wide intervals. For example, the lowest one-fourth of the males had responses ranging over the 40-mph interval from 55 mph to 95 mph, while the next one-fourth had responses ranging over only a 15-mph interval, from 95 to 110. Similarly, the third one-fourth had responses in only a 10-mph interval (110 to 120), while the top one-fourth had responses in a 30-mph interval (120 to 150). It is common to see the majority of values clumped in the middle and the remainder tapering off into a wider range.

THOUGHT QUESTION 2.7 A **resistant statistic** is a numerical summary of the data that is "resistant" to the influence of outliers. In other words, an outlier is not likely to have a major influence on its numerical value. Two of the summary measures from the list ***mean, median, range,*** and ***interquartile range*** are ***resistant,*** while the other two are not. Explain which two are resistant and which two are not.*

How to Draw a Boxplot and Identify Outliers

Now that you know how to find a five-number summary, you can draw a boxplot.

Step 1: Label either a vertical axis or a horizontal axis with numbers from the minimum to the maximum of the data.

Step 2: Draw a box with the lower end of the box at the lower quartile (denoted as Q_1) and the upper end at the upper quartile (Q_3).

Step 3: Draw a line through the box at the median.

Step 4: Calculate IQR = $Q_3 - Q_1$.

Step 5: Draw a line that extends from the lower quartile end of the box to the smallest data value not smaller than the value of ($Q_1 - 1.5 \times$ IQR). Also, draw a line that extends from the upper quartile end of the box to the largest data value that is not greater than the value of ($Q_3 + 1.5 \times$ IQR).

Step 6: Mark the location of any data points smaller than ($Q_1 - 1.5 \times$ IQR) or larger than ($Q3 + 1.5 \times$ IQR) with an asterisk. In other words, values more than one and a half IQRs beyond the quartiles are considered to be *outliers.*

The following example illustrates the process of finding five-number summaries, identifying outliers, and presenting the information in a boxplot. Note that boxplots can be displayed vertically, as in Figure 2.14, or horizontally, as in Figure 2.21.

Example 2.15 **Example 2.9 Revisited: Five-Number Summary and Outlier Detection for Songs on iPod or MP3** For the example about number of songs on students' iPods or MP3 players started in Example 2.9 on page 32 and continued in Example 2.11 on page 38, the ordered list of $n = 24$ data values follows:

0	30	500	500	500	600	687	750	800	800	900	1150
1250	1300	1500	1500	1700	2017	2400	2501	2600	3305	4000	5000

Previously we found that the median is $M = 1200$. The lower quartile is the median of the lower half of the data, the 12 values that are smaller than 1200. The upper quartile is the median of the upper half of the data, the 12 values larger than 1200. Calculations are:

$Q_1 = (600 + 687)/2 = 643.5$, median of the first row of ordered values above
$Q_3 = (2017 + 2400)/2 = 2208.5$, median of the second row of ordered values above

The resulting five-number summary for these data follows:

	Number of Songs	
Median		1200	
Quartiles	643.5	2208.5
Extremes	0	5000

***HINT:** Which of the statistics incorporate all data values at the extremes in the calculations?

A boxplot for these data is given in Figure 2.21. We see that the data are skewed to the right as the display stretches further to the right of the median than to the left. The value 5000 is marked as an outlier. Here, IQR = 2208.5 − 643.5 = 1565. On the high side of the data, a value is marked as an outlier when it is greater than $Q_3 + (1.5 \times IQR)$. For this example, that boundary is 4556, so the value 5000 is marked as an outlier because it is larger.

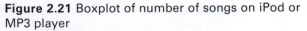

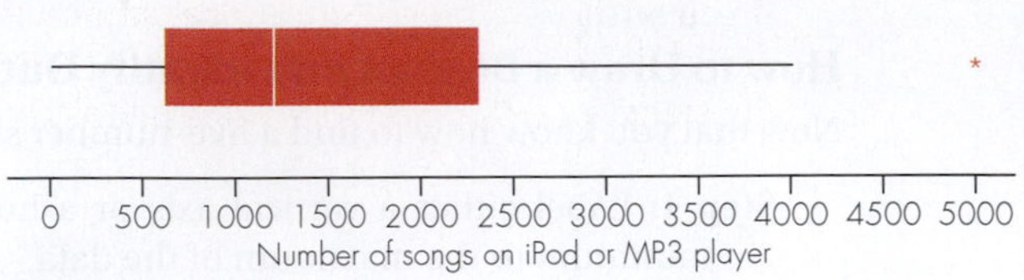

Figure 2.21 Boxplot of number of songs on iPod or MP3 player

Percentiles

2.5 Exercises are on pages 60–62.

The quartiles and the median are special cases of **percentiles** for a dataset. In general, the kth *percentile* is a number that has k% of the data values at or below it and $(100 - k)$% of the data values at or above it. The lower quartile, median, and upper quartile are also the 25th percentile, 50th percentile, and 75th percentile, respectively. If you are told that you scored at the 90th percentile on a standardized test (such as the SAT), it indicates that 90% of the scores were at or below your score, while 10% were at or above your score.

EXCEL TIP

Suppose the dataset has been stored in a range of cells, which we represent by the word *list* in what follows. For instance, if the dataset is in column A, rows 1 to 30, then *list* is A1:A30. You can also "list" the actual numerical values themselves, rather than the range of cells containing them. All of these commands are part of the "statistical functions" provided by Excel. You can insert them directly into a cell by preceding the command with either the symbol @ or =. Some values have multiple options. For instance, there are many commands that give the minimum value.

Average(list) = mean

Quartile(list, 0) = *Min(list)* = minimum value

Quartile(list, 1) = lower quartile

Quartile(list, 2) = *Median(list)* = median

Quartile(list, 3) = upper quartile

Quartile(list, 4) = *Max(list)* = maximum value

Small(list,k) gives the kth-smallest value, for example, *small(list,1)* = minimum

Large(list,k) gives the kth-largest value, for example, *large(list,1)* = maximum

Percentile(list,p) gives the kth percentile, where $p = k/100$. In other words, you must express the desired percentile as a proportion rather than a percent. For instance, to find the 90th percentile, use *percentile(list,.9)*.

Count(list) = n, the number of values in the dataset.

Note: As with most computer programs, Excel uses a more precise algorithm to find the upper and lower quartiles than the one we recommend using if you are finding them "by hand," so your values may differ slightly.

MINITAB TIP **Numerical Summaries of a Quantitative Variable**

- To determine summary statistics for a quantitative variable, use **Stat > Basic Statistics > Display Descriptive Statistics.** In the dialog box, specify one or more columns containing the raw data for a quantitative variable(s). Use the ***Statistics*** button to select or deselect summary statistics that will be displayed. The ***Graphs*** button provides options for several different graphs.
- If you wish to compare numerical summaries of a quantitative variable across categories (for example, to compare the handspans of men and women), specify the categorical variable that defines the groups in the "By Variables" box.

IN SUMMARY **Numerical Summaries of Quantitative Variables**

- The **median** and the **mean** summarize the location (center) of a dataset.
- **IQR**, **Range**, and **standard deviation** summarize the spread (variability) of a dataset.
- For nearly symmetric data, the mean and median are nearly equal. For data skewed to the right, the mean will tend to be larger than the median. For data skewed to the left, the mean will tend to be smaller than the median.
- In a boxplot, data points smaller than ($Q_1 - 1.5 \times \text{IQR}$) or larger than ($Q_3 + 1.5 \times$ IQR are marked as **outliers**.

2.6 How to Handle Outliers

Outliers need special attention because they can have a big influence on conclusions drawn from a dataset and because they can lead to erroneous conclusions if they are not treated appropriately. Outliers can also cause complications in some statistical analysis procedures, as you will learn throughout this book. As a result, some researchers wrongly discard them rather than treating them as legitimate data. Outliers should never be discarded without justification. The first step in deciding what to do with outliers is determining why they exist so that appropriate action can be taken. Let's consider three possible reasons for outliers and what action to take in each case.

The Outlier Is a Legitimate Data Value and Represents Natural Variability for the Group and Variable(s) Measured

The characterization of natural variability is one of the most important themes in statistics and data analysis. We should not discard legitimate values that inherently occur, unless the goal is to study only a partial range of the possible values. In the handspan data given in Table 2.4 and Figure 2.5, the two smallest female handspan measurements, 12.5 cm and 13.0 cm, were well below the other female measurements. If they are legitimate measurements, they provide important information about the spectrum of possibilities for female handspans, and they should be retained in the dataset. Discarding them would result in an erroneous depiction of female handspans, with a measure of variability that is too low and a mean that is too high.

A Mistake Was Made While Taking a Measurement or Entering It into the Computer

Faulty measuring equipment, unclear instructions on a survey, or typing errors when data are entered into a computer cause many of the outliers that occur in datasets. For example, a stopwatch with a dying battery might give the wrong value for the time needed to do a task. Or the handspan measurement for a woman whose span is 20 cm would be recorded as 8 if she misunderstood the instructions and reported the measurement in inches instead of centimeters. A typing mistake could cause a height of 68 inches to be recorded as 86 inches. Fortunately, outliers caused by these kinds of problems are often easy to identify. If possible, the values for outliers caused by mistakes should be corrected and retained. If it is not possible to correct them, they should be discarded.

The Individual in Question Belongs to a Different Group Than the Bulk of Individuals Measured

A group sometimes includes a few individuals that are different from the others in an important way. For instance, a college class might include mostly traditional-age college students (perhaps aged 18 to 22), and a few returning students who are much older. Measurements for the older students are likely to be outliers for any variables related to age, such as dollar value of assets owned or length of longest romantic relationship.

If we know that outliers are individuals that are different from the others for a specific reason, our reason for studying the data should be considered in deciding whether to discard them or not. For example, in measuring assets owned by college students, we should not include the older, returning students if we want to study the assets owned by traditional-age students. If we are interested in the value of assets owned by all college students, we should retain all measurements. In that case, the relationship between age and assets may also be of interest.

Example 2.16 **Tiny Boatmen** Here are the weights (in pounds) of 18 men who were on the crew teams at Oxford and Cambridge universities (*The Independent,* March 31, 1992; also Hand et al., 1994, p. 337):

Cambridge: 188.5, 183.0, 194.5, 185.0, 214.0, 203.5, 186.0, 178.5, 109.0

Oxford: 186.0, 184.5, 204.0, 184.5, 195.5, 202.5, 174.0, 183.0, 109.5

Read over the list. Do you notice anything unusual? The last weight given in each list is very different from the others. In fact, those two men were the coxswains for their teams, while the other men were the rowers. What is the mean weight for the crew team members? If all members are included, it is 181 lb. If only the rowers are included, it is 190 lb. Different questions, different answers.

2.6 Exercises are on page 62.

IN SUMMARY **Possible Reasons for Outliers and Reasonable Actions**

- *The outlier is a legitimate data value and represents natural variability for the group and variable(s) measured.* Values may not be discarded in this case — they provide important information about location and spread.
- *A mistake was made while taking a measurement or entering it into the computer.* If this can be verified, the values should be discarded or corrected.
- *The individual in question belongs to a different group than the bulk of individuals measured.* Values may be discarded if a summary is desired and reported for the majority group only.

2.7 Bell-Shaped Distributions and Standard Deviations

Nature seems to follow a predictable pattern for many kinds of measurements. Most individuals are clumped around the center, and the greater the distance that a value is from the center, the fewer individuals have that value. Except for the two outliers at the lower end, that pattern is evident in the females' right handspan measurements in Figures 2.7 to 2.9. If we were to draw a smooth curve connecting the tops of the bars on a histogram with this shape, the smooth curve would resemble the shape of a symmetric bell.

Numerical variables that follow this pattern are said to follow a **bell-shaped curve**, or to be "bell-shaped." A special case of this distribution of measurements is so common it is also called a **normal distribution** or **normal curve**. There is a precise mathematical formula for this smooth curve, which we will study in more depth in Chapter 8, but in this chapter we will limit ourselves to a few convenient descriptive features for data of this type. Most variables with a bell shape do not fit the mathematical formula for a normal distribution exactly, but they come close enough that the results in this section can be applied to them to provide useful information.

Example 2.17 **The Shape of British Women's Heights** A representative sample of 199 married British couples, taken in 1980, provided information on five variables: height of each spouse (in millimeters), age of each spouse, and husband's age at the time they were married (Hand et al., 1994). Figure 2.22 displays a histogram of the wives' heights, with a normal curve superimposed. The particular normal curve shown in Figure 2.22 was generated using Minitab statistical software, and of all the possible curves of this type, it was chosen because it is the best match for the histogram. The mean height for these women is 1602 millimeters, and the median at 1600 millimeters is very close. Although it is difficult to tell precisely in Figure 2.22, the normal curve is centered at the mean of 1602. For bell-shaped curves, there is a useful measure of spread called the *standard deviation,* which we describe next.

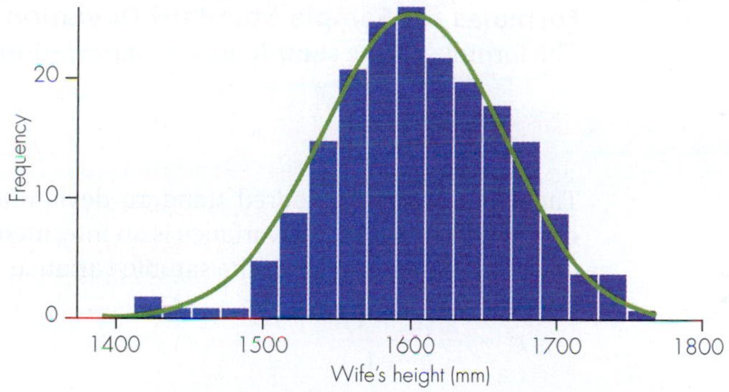

Figure 2.22 Histogram of wife's height and normal curve

MINITAB TIP **Superimposing a Normal Curve onto a Histogram**

- To draw a normal curve onto the histogram of a quantitative variable, use **Graph** > **Histogram**. In the resulting display, select ***With Fit***. In the "Graph Variables" box, specify the column containing the raw data for the variable.

Describing Spread with Standard Deviation

Because normal curves are so common in nature, a whole set of descriptive features has been developed that apply mostly to variables with that shape. In fact, two summary features uniquely determine a normal curve, so if you know those two summary numbers, you can draw the curve precisely. The first summary number is the mean, and the bell shape is centered on that number. The second summary number is called the **standard deviation**, and it is a measure of the spread of the values. The symbol s is used to represent the standard deviation of a sample.

The Concept of Standard Deviation

You can think of the standard deviation as *roughly the average distance that values fall from the mean.* Put another way, it measures variability by summarizing how far individual data values are from the mean. Consider, for instance, the standard deviations for the following two sets of numbers, both with a mean of 100:

Set	Numbers	Mean	Standard Deviation
1	100, 100, 100, 100, 100	100	0
2	90, 90, 100, 110, 110	100	10

In the first set of numbers, all values equal the mean value, so there is no variability or spread at all. For this set, the standard deviation is 0, as will *always* be the case for such a set of numbers. In the second set of numbers, one number equals the mean, while the other four numbers are each 10 points away from the mean, so the average distance away from the mean is close to 10, the standard deviation for this set of data.

Calculating the Standard Deviation

The formula for calculating the standard deviation is a bit more involved than the conceptual interpretation that we just discussed. This is the first instance of a summary measure that differs based on whether the data represent a sample or an entire population. The version given here is appropriate when the dataset is considered to represent a sample from a larger population. This distinction will become clear later in the book.

FORMULA **Formulas for Sample Standard Deviation and Variance**

The formula for the **sample standard deviation** is

$$s = \sqrt{\frac{\Sigma(x_i - \bar{x})^2}{n - 1}}$$

The value of s^2, the squared standard deviation, is called the **sample variance**. In descriptive statistics, the variance is an intermediate step in calculating the standard deviation. The formula for the sample variance is

$$s^2 = \frac{\Sigma(x_i - \bar{x})^2}{n - 1}$$

In practice, statistical software such as Minitab or a spreadsheet program such as Excel typically is used to find the standard deviation for a dataset. For situations in which you have to calculate the standard deviation by hand, here is a step-by-step guide to the steps involved:

Step 1: Calculate $\bar{x}$, the sample mean.

Step 2: For each observation, calculate the difference between the data value and the mean.

Step 3: Square each difference calculated in Step 2.

Step 4: Sum the squared differences calculated in Step 3, and then divide this sum by $n - 1$. The answer for this step is called the sample variance.

Step 5: Take the square root of the sample variance calculated in Step 4.

Example 2.18 **Calculating a Standard Deviation** Calculate the standard deviation of the four pulse rates 62, 68, 74, and 76.

Step 1: The sample mean is

$$\bar{x} = \frac{62 + 68 + 74 + 76}{4} = 70$$

Steps 2 and 3: For each observation, calculate the difference between the data value and the mean. Then square this difference. The results of these two steps are shown here:

Data Value	Step 2 Value − Mean	Step 3 $(\text{Value} - \text{Mean})^2$
62	$62 - 70 = -8$	$(-8)^2 = 64$
68	$68 - 70 = -2$	$(-2)^2 = 4$
74	$74 - 70 = 4$	$4^2 = 16$
76	$76 - 70 = 6$	$6^2 = 36$

Step 4: The sum of Step 3 quantities is $64 + 4 + 16 + 36 = 120$. Divide this sum by $n - 1 = 4 - 1$ to get the variance:

$$s^2 = \frac{120}{4 - 1} = \frac{120}{3} = 40$$

Step 5: Take the square root of the variance computed in Step 4:

$$s = \sqrt{40} = 6.3$$

TECHNICAL NOTE

Population Mean and Standard Deviation

For reasons that will become clear later in this book, datasets are commonly treated as if they represent a sample from a larger population. However, in situations in which the dataset includes measurements for an entire population, the notations for the mean and standard deviation are different, and the formula for the standard deviation is also slightly different. A **population mean** is represented by the Greek letter μ ("mu"), and a **population standard deviation** is represented by the Greek letter σ ("sigma"). The formula for the population standard deviation is

$$\sigma = \sqrt{\frac{\sum(x_i - \mu)^2}{N}}$$

Note that one difference between this formula and the sample version is that the denominator is now N instead of $n - 1$. The letter N represents the number of measurements in the population. Also, the appropriate notation for the mean of the population is used.

EXCEL TIP

The **Excel** commands for the standard deviation and variance follow:

Stdev(list) = sample standard deviation
Stdevp(list) = population standard deviation
Var(list) = sample variance
Varp(list) = population variance

Interpreting the Standard Deviation for Bell-Shaped Curves: The Empirical Rule

Once you know the mean and standard deviation for a bell-shaped curve, you can also determine the approximate proportion of the data that will fall into any specified interval. We will learn much more about how to do this in Chapter 8, but for now, here are some useful benchmarks.

DEFINITION The **Empirical Rule** states that for any bell-shaped curve, approximately

- 68% of the values fall within 1 standard deviation of the mean in either direction
- 95% of the values fall within 2 standard deviations of the mean in either direction
- 99.7% of the values fall within 3 standard deviations of the mean in either direction

Note: A small percentage, 0.3%, falls farther than 3 standard deviations from the mean.

Combining the Empirical Rule with knowledge that bell-shaped variables are symmetric allows the "tail" ranges to be specified as well. The first statement of the Empirical Rule implies that about 16% of the values fall more than 1 standard deviation *below* the mean and 16% fall more than 1 standard deviation *above* the mean. Similarly, about 2.5% fall more than 2 standard deviations below the mean and 2.5% fall more than 2 standard deviations above the mean.

Example 2.19 **Example 2.17 Revisited: Women's Heights and the Empirical Rule** The mean for the 199 British women's heights is 1602 millimeters, and the standard deviation is 62.4 millimeters. Figure 2.23 illustrates how the Empirical Rule would apply if the distribution exactly followed a normal curve.

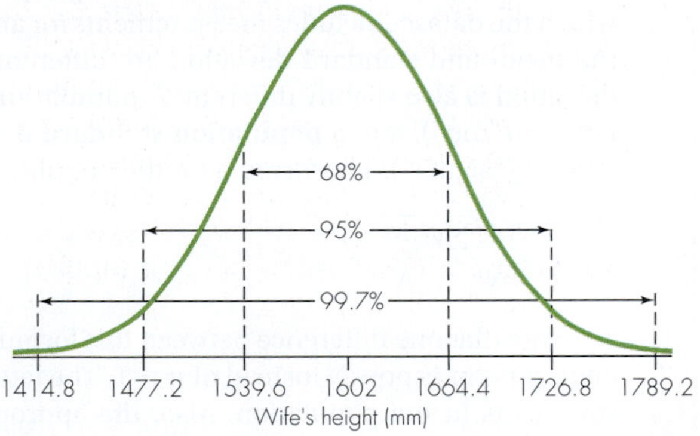

Figure 2.23 The Empirical Rule applied to British women's heights

For instance, about 68% of the 199 heights would fall into the range 1602 ± 62.4, or 1539.6 to 1664.4 mm. (The symbol "±" is read "plus or minus" and indicates that you form an interval by first subtracting and then adding the value that follows the symbol from the value that precedes it.) About 95% of the heights would fall into the interval 1602 ± (2 × 62.4), or 1477.2 to 1726.8 mm. And about 99.7% of the heights would be in the interval 1602 ± (3 × 62.4), or 1414.8 to 1789.2 mm. In fact, these intervals work well for the actual data. Here is a summary of how well the Empirical Rule compares with the actual numbers and percents of heights falling within 1, 2, and 3 standard deviations (s.d.) of the mean:

Interval	Numerical Interval	Empirical Rule % and Number	Actual Number	Actual Percent
Mean ± 1 s.d.	1539.6 to 1664.4	68% of 199 = 135	140	140/199 or 70%
Mean ± 2 s.d.	1477.2 to 1726.8	95% of 199 = 189	189	189/199 or 95%
Mean ± 3 s.d.	1414.8 to 1789.2	99.7% of 199 = 198	198	198/199 or 99.5%

Note that the women's heights, although not perfectly bell-shaped, follow the Empirical Rule quite well.

The Empirical Rule, the Standard Deviation, and the Range

The Empirical Rule implies that the range from the minimum to the maximum data values equals about 4 to 6 standard deviations. For relatively large samples, you can get a rough idea of the value of the standard deviation by dividing the range of the data values by 6. In other words, the standard deviation can be approximated as

$$s \approx \frac{\text{Range}}{6}$$

This approximation works reliably only for bell-shaped data with a sample size of about 200 or more observations. It does not work reliably for skewed data or for smaller sample sizes.

In Example 2.17, about British women's heights, the data are approximately bell-shaped. The sample size of $n = 199$ is large enough to use the formula to estimate the standard deviation. The minimum height was 1410 mm, while the maximum was 1760, for a range of $1760 - 1410 = 350$ mm. Therefore, a reasonable guess for the standard deviation is Range/6 = 350/6 = 58.3 mm. This is indeed close to the actual standard deviation of 62.4 mm.

Standardized *z*-Scores

The standard deviation is also useful as a "yardstick" for measuring how far an individual value falls from the mean. Suppose you were told that scores on your last statistics exam were bell-shaped (they often are) and that your test score was 2 standard deviations above the mean for your class. Without even knowing your score or the class mean score, you would know that only about 2.5% of the students had scores exceeding yours. From the Empirical Rule, we know that the scores for about 95% of the class are within 2 standard deviations of the mean. Of the remaining 5% of the scores, about half, or 2.5%, will be more than 2 standard deviations above the mean.

The **standardized score** or ***z*-score** is a useful measure of the relative value of any observation in a dataset. The formula for this score is simple:

$$z = \frac{\text{Observed value} - \text{Mean}}{\text{Standard deviation}}$$

Note that a *z*-score is simply the distance between the observed value and the mean, measured in terms of number of standard deviations. Data values below the mean have negative *z*-scores, and values above the mean have positive *z*-scores.

As an example, suppose that the mean resting pulse rate for adult men is 70 beats per minute, the standard deviation is 8 beats per minute, and we calculate the standardized score for a resting pulse rate of 80 beats per minute. The calculation is

$$z = \frac{80 - 70}{8} = 1.25$$

The value $z = 1.25$ indicates that a pulse rate of 80 is 1.25 standard deviations above the mean pulse rate for adult men.

DEFINITION The **Empirical Rule** for bell-shaped data can be restated for standardized scores as follows:

- About 68% of values have *z*-scores between −1 and +1.
- About 95% of values have *z*-scores between −2 and +2.
- About 99.7% of values have *z*-scores between −3 and +3.

Figure 2.24 illustrates this version of the Empirical Rule.

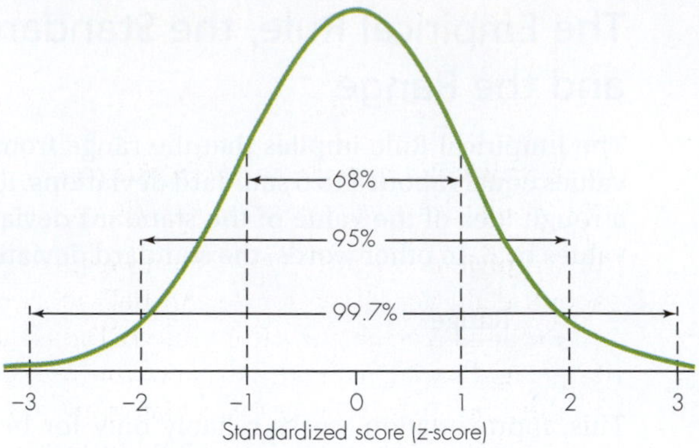

Figure 2.24 The Empirical Rule applied to standardized scores (*z*-scores)

Many computer programs and calculators will find the approximate proportion of a bell-shaped variable falling below any *z*-score you specify. For instance, the function NORMSDIST(z) in Excel does this. Remember that a common special case of a bell-shaped distribution is called the "normal distribution," and it is this special case that is used by Excel. As an example, NORMSDIST(−1) = .158655, or 15.8655%, corresponding to the information from the Empirical Rule that about 16% of values fall more than 1 standard deviation below the mean. In Chapter 8, you will learn a more precise interpretation for *z*-scores.

2.7 Exercises are on pages 62–64.

THOUGHT QUESTION 2.8 Why do you think measurements with a bell-shaped distribution are so common in nature? For example, why do you think women's heights are distributed in this way rather than, for instance, being equally spread out from about 5 feet tall to 6 feet tall?*

TI-84 TIP **Numerical Summaries of a Quantitative Variable**

- First, store the data values into a list, say L1.
- Press [STAT]. Scroll horizontally to **CALC**, and then scroll vertically to **1:1-Var Stats** and press [ENTER]. Assuming that the data are in list L1, complete the expression as **1-Var Stats L1**, followed by [ENTER]. The display will show the mean, standard deviation (both sample and population), the five-number summary, the sample size, the sum of *x*-values, and the sum of x^2 values.

IN SUMMARY Bell-Shaped Distributions and Standard Deviation

- The standard deviation measures the variability among data values.
- The formula for **sample standard deviation** is $s = \sqrt{\dfrac{\Sigma(x_i - \bar{x})^2}{n - 1}}$
- For bell-shaped data, about 68% of the data values fall within 1 standard deviation of the mean either way, about 95% fall within 2 standard deviations of the mean either way, and about 99.7% fall within 3 standard deviations either way.
- A **standardized score**, also called a ***z*-score**, measures how far a value is from the mean in terms of standard deviations.

***HINT:** What factors contribute to a person's adult height? Considering these factors, why would it be more likely that heights are close to the mean than far from the mean?

SKILLBUILDER APPLET

2.8 The Empirical Rule in Action

The **Empirical** *applet described in this section is available on the companion website, http://www.cengage.com/statistics/Utts4e.*

The **Empirical** applet can be used to explore how well the Empirical Rule works for each of eight variables, some with bell-shaped distributions and some with skewed distributions. The data are from the **UCDavis1** and **pennstate1** datasets. A description of the eight variables is at the top of the web page that includes the applet. For each variable, the applet will display a histogram along with information about the intervals ***mean*** **± *s*** and ***mean*** **± *2s***. When the Empirical Rule applies, these two intervals should include about 68% and 95% of the sample, respectively.

What to Do

Open the **Empirical** applet. Figure 2.25 shows the initial applet display, a summary of the hours of sleep the previous night for $n = 173$ students in a UC Davis statistics class. A histogram of the hours of sleep data is displayed with superimposed vertical lines indicating the intervals ***mean*** **± *s*** and ***mean*** **± *2s***. Note that the histogram has approximately a bell shape. Below the histogram, we see that the sample mean = 6.935 hours and the standard deviation is $s = 1.705$. Note also that the interval ***mean*** **± *s*** = (5.23, 8.64) contains 114 of the 173 data values, which is 65.9%. The interval ***mean*** **± *2s*** = (3.525, 10.35) contains 166/173 = 95.95% of the data values. These percentages are consistent with the Empirical Rule—not surprising since the distribution is approximately bell-shaped.

Now click on ***TV Hours***, the second variable in the menu at the left of the applet display. Figure 2.26 displays the result, a summary of self-reported weekly hours of watching television for the same 173 students in the hours of sleep example. Note that the distribution is skewed to the right, and there is an extreme outlier at 100 hours, so the Empirical Rule won't work well. Here, the interval ***mean*** **± *s***, given as −1.484 to 19.26 hours, contains about 89% of the dataset, much more than the (approximate) 68% that would be in this interval if the Empirical Rule applied. Another difficulty is that the lower value of the interval is negative, an impossible value for weekly hours of watching television. This also is the case for the interval ***mean*** **± *2s*** = (−11.86, 29.63).

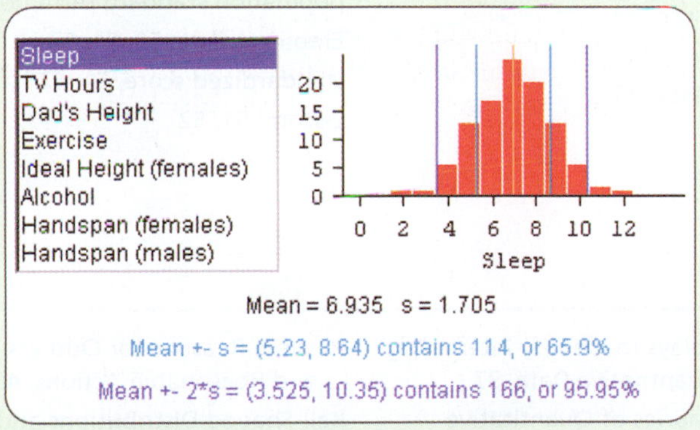

Figure 2.25 The Empirical applet display of hours of sleep reported by $n = 173$ students

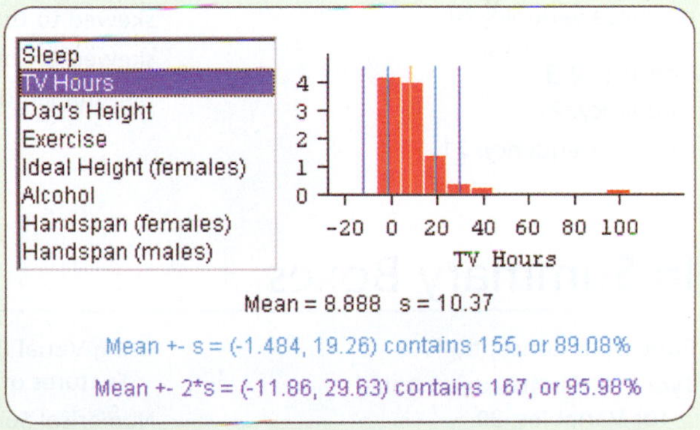

Figure 2.26 The Empirical applet display of weekly hours watching television reported by $n = 173$ students

Click on each of the other six variables to further explore the connection between the shape of the histogram and the applicability of the Empirical Rule. For each variable, judge the shape of the distribution and take note of the percents in the two intervals given. Which of the variables are well described by the Empirical Rule? Typically, what is the approximate shape of the distributions of these variables? Which variables are not well described by the Empirical Rule? What shape do the distributions of these variables typically have?

(continued)

Lessons Learned

You'll see that the Empirical Rule works well when the distribution is more or less bell-shaped. But when the distribution is skewed or an extreme outlier is present, you will see that the interval ***mean* ± *s*** tends to include noticeably more than 68% of the dataset, and the boundary values for the interval ***mean* ± 2*s*** may not make sense for the variable of interest.

Section 2.8 Exercises are on page 64.

Key Terms

In Summary Boxes

Exercises

◆ Denotes that the dataset is available on the companion website, http://www.cengage.com/statistics/Utts4e, but is not required to solve the exercise.

Bold exercises have answers in the back of the text.

Section 2.1

Skillbuilder Exercises

2.1 A sociologist assembles a dataset consisting of the poverty rate, per capita income, serious crime rate, and teen birth rate for the 50 states of the United States.

a. How many variables are in this dataset?
b. What is an observational unit in this dataset?
c. What is the sample size for the dataset?

2.2 Suppose that in a national survey of 620 randomly selected adults, each person is asked how important religion is to him or her (very, fairly, not very), and whether the person favors or opposes stricter regulation of what can be broadcast on network television.

a. How many variables are measured in this survey?
b. What is an observational unit in this study?
c. What is the sample size for this survey?

2.3 In each situation, explain whether it would be more appropriate to treat the observed data as a sample from a larger population or as data from the whole population.

a. An instructor surveys all the students in her class to determine whether students would prefer a take-home exam or an in-class exam.
b. The Gallup Organization polls 1000 individuals to estimate the percent of American adults who approve of the President's job performance.

2.4 In each situation, explain whether it would be more appropriate to treat the observed dataset as sample data or as population data.

a. A historian summarizes the ages at death for all past presidents of the United States.
b. A nutritionist wants to determine which of two weight-loss programs is more effective. He assigns 25 volunteers to each program and records each participant's weight loss after 2 months.

2.5 For each of the following statistical summaries, explain whether it is a population parameter or a sample statistic.

a. In the 2000 census of the United States, it was determined that the average household size was 2.59 persons per household (http://www.census.gov).
b. In an *ABC News poll* completed in June 2004, 36% of $n = 500$ persons surveyed said that they supported replacing the portrait of Alexander Hamilton on the U.S. $10 bill with a portrait of Ronald Reagan (www.pollingreport.com/news.htm).
c. To estimate average normal body temperature of all adults, a doctor measures the temperatures of 100 healthy adults. The average temperature for that group is 98.2 degrees Fahrenheit.

2.6 For each of the following statistical summaries, explain whether it is a population parameter or a sample statistic.

a. A highway safety researcher wants to estimate the average distance at which all drivers can read a highway sign at night. She measures the distance for a sample of 50 drivers; the average distance for these drivers is 495 feet.
b. The average score on the final exam is 76.8 for $n = 83$ students in a statistics class. The instructor is only interested in describing the performance of this particular class.
c. Case Study 1.3 (p. 3) reported that in a Gallup poll, 57% of $n = 496$ teens who date said they have been out with someone of another race or ethnic group.

General Section Exercises

2.7 Case Study 1.1 (p. 2) was about the fastest speeds that students in a statistics class claimed they have ever driven.

a. What variables are described in Case Study 1.1?
b. What are the observational units in the study?
c. Explain whether you think it would be more appropriate to treat the data as sample data or as population data.

2.8 Read Case Study 1.5 (p. 5) about prayer and blood pressure.

a. What was the sample size for the observational study conducted by the National Institutes of Health?
b. Describe the observational units in this study.
c. Describe two variables that the researchers related to each other in Case Study 1.5.
d. Explain whether you think the researchers treated the observed data as sample data or as population data.

2.9 Case Study 1.2 (p. 3) gave the information that the rate of errors made by air traffic controllers in the United States during fiscal year 1998 was 5.5 errors per million flights. Discuss whether this summary value is a population summary (a parameter) or a sample summary (a statistic).

2.10 Read Case Study 1.6 (p. 5) about aspirin and heart attack rates.

a. What two variables are measured on each individual in Case Study 1.6?
b. Describe the observational units in this study.
c. What was the sample size for the study?
d. Explain whether you think the researchers treated the observed data as sample data or as population data.

Section 2.2

Skillbuilder Exercises

2.11 For each of the following variables, indicate whether the variable is categorical or quantitative.

a. Importance of religion to respondent (very, somewhat, or not very important).
b. Hours of sleep last night.
c. Weights of adult women, measured in pounds.
d. Favorite color for an automobile.

2.12 For each of the following characteristics of an individual, indicate whether the variable is categorical or quantitative.

a. Length of forearm from elbow to wrist (in centimeters).
b. Whether or not the person has ever been the victim of a crime.
c. Number of music CDs owned.
d. Feeling about own weight (overweight, about right, underweight).

2.13 For each of the following, indicate whether the variable is ordinal or not. If the variable is not ordinal, indicate its variable type.

a. Opinion about a new tax law (favor or oppose).
b. Letter grade in a statistics course (A, B, and so on).
c. Heights of men (in inches).

2.14 For each of the following quantitative variables, explain whether the variable is continuous or not.

a. Body weight (in pounds).
b. Number of text messages a person sends in a day.
c. Number of coins presently in someone's pockets and/or purse.

2.15 For each pair of variables, specify which variable is the explanatory variable and which is the response variable in the relationship between them.

a. Score on the final exam and final course grade in a psychology course.
b. Opinion about the death penalty (favor or oppose), and sex (male or female).

2.16 For each of the following, indicate whether the variable is ordinal or not. If the variable is not ordinal, indicate its variable type.

a. Whether or not the person believes in love at first sight.
b. Student rating of teacher effectiveness on a 7-point scale where 1 = not at all effective and 7 = extremely effective.
c. Number of text messages received in a day.

2.17 For each of the following quantitative variables, explain whether the variable is continuous or not.

a. Number of classes a student misses in a week.
b. Head circumference (in centimeters).
c. Time it takes students to walk from their dorm to a classroom.

2.18 For each pair of variables, specify which variable is the explanatory variable and which is the response variable in the relationship between them.

a. Amount a person walks or runs per day and performance on a test of lung function.
b. Feeling about importance of religion and age of respondent.

General Section Exercises

2.19 For each of the following situations reported in the news, specify what variable(s) were measured on each individual and whether they are best described as categorical, ordinal, or quantitative.

a. A *Los Angeles Times* survey found that 60% of the 1515 adult Californians polled supported a recent state law banning smoking in bars (*Sacramento Bee*, May 28, 1998, p. A3).
b. According to the *College Board News* (December 1998, p. 1), "Students using either one of two major coaching programs [for the SAT] were likely to experience an average gain of 5 to 19 points on verbal and 5 to 38 points on math."

2.20 According to the *Associated Press* (June 19, 1998), "Smokers are twice as likely as lifetime nonsmokers to develop Alzheimer's disease and other forms of dementia . . . [according to a study that] followed 6,870 men and women ages 55 and older." For this situation, specify what variables were measured on each individual and whether they are best described as categorical, ordinal, or quantitative.

2.21 A physiologist records the pulse rates of 30 men and 30 women.

a. Specify the two variables measured in this situation.
b. For each variable, explain whether it is categorical or quantitative.
c. Using the examples under the "Asking the Right Questions" heading in Section 2.2 (p. 19) as a guide, write a question that would be helpful for comparing the pulse rates of men and women. What summary information would be useful for making this comparison?

2.22 Give an example of an ordinal variable that is likely to be treated as a categorical variable because numerical summaries like the average would not make much sense.

2.23 Give an example of an ordinal variable for which a numerical summary like the average would make sense.

2.24 Find an example of a study that uses statistics in a magazine, newspaper, or website. Determine what variables were measured, and, for each variable, determine its type. Which of the questions listed under "Asking the Right Questions" (p. 19) were addressed in this study? Describe the question(s) in the context of the study, and then explain what answer was found.

2.25 To answer the following questions, researchers would measure two variables for each individual unit in the study. In each case, specify the two variables, and the variable type for each. Then, specify which is the explanatory variable and which is the response variable.

a. Is the average IQ of left-handed people higher than the average IQ of right-handed people?
b. For married couples, is there a relationship between owning a pet and whether or not they get divorced?

2.26 To answer the following questions, researchers would measure two variables for each individual unit in the study. In each case, specify the two variables, and the variable type for each. Then, specify which is the explanatory variable and which is the response variable.

a. For college students, is there a relationship between grade point average (GPA) and average number of hours spent studying each week?
b. Individuals in the United States fall into one of a small number of tax brackets based on level of income. Is there a relationship between a person's tax bracket and the percent of income donated to charities?

Section 2.3

Skillbuilder Exercises

2.27 Table 2.1 (p. 21) summarized frequency of seatbelt use while driving for twelfth-grade participants in the 2003 Youth Risk Behavior Surveillance System (YRBSS) survey. In 2001, YRBSS survey students were asked the same question. For the 2001 survey, a summary of responses given by 2530 students in the twelfth grade who said that they drive follows.

Wears Seatbelt	Frequency
Never	105
Rarely	248
Sometimes	286
Most times	464
Always	1427

Source: http://www.cdc.gov/nccdphp/dash/yrbs.

a. What percent of the twelfth-grade students who drive said that they always wear a seatbelt when driving?

b. What percent of the twelfth-grade students who drive said that they do not always wear a seatbelt when they drive?

c. Find the percentage in each of the five response categories.

d. Draw a bar graph of the percentages found in part (c).

2.28 In the 2008 General Social Survey, participants were asked, "Would you say that you are very happy, pretty happy, or not too happy?" The results were that 599 people said very happy, 1100 people said pretty happy, and 316 people said not too happy (*Data source*: http://sda.berkeley.edu).

a. Write the frequency and relative frequency table for these data. Use Table 2.1 on p. 21 for guidance.

b. Draw a bar graph of the percentages (or proportions) found for the relative frequency distribution in part (a).

c. What percent of respondents said either "very happy" or "pretty happy?"

2.29 In a survey done in 2010, students in a statistics class were asked, "How do you prefer to use your cell phone—to talk or to text?" Of the 106 women who responded, 22 women said to talk and 84 said to text. Of the 83 men who responded, 34 men said to talk and 49 men said to text.

a. Summarize the observed counts by making a table similar to Table 2.2 on p. 21.

b. For the women, calculate the percent that said "to talk" and the percent that said "to text."

c. For the men, calculate the percent that said "to talk" and the percent that said "to text."

d. Write one or two sentences that summarize the difference between males and females in their preferred use of a cell phone.

2.30 Refer to Exercise 2.27. Students also were asked what grades they usually get in school. For twelfth-grade students who responded to this question and the question about how often they wear seatbelts when driving, a summary of frequency counts for combinations of responses to the two questions is as follows:

	Usual School Grades			
Wears Seatbelt	A and B	C	D and F	Total
Never	52	32	18	102
Rarely	128	93	22	243
Sometimes	166	104	8	278
Most times	298	128	24	450
Always	1056	300	41	1397
Total	1700	657	113	2470

a. The total number of students in the table is 2470. What percentage of these 2470 students said that they usually get A's and B's in school?

b. What percentage of the 1700 students who said that they usually get A's and B's said that they always wear a seatbelt when driving?

c. What percentage of the 657 students who said that they usually get C's said that they always wear a seatbelt when driving?

d. What percentage of the 113 students who said that they usually get D's and F's said that they always wear a seatbelt when driving?

2.31 For each of the following situations, which is the explanatory variable and which is the response variable?

a. The two variables are whether or not someone smoked and whether or not the person developed Alzheimer's disease.

b. The two variables are whether or not somebody voted in the last election and the person's political party (Democrat, Republican, Independent, or Other).

c. The two variables are income level and whether or not the person has ever been subjected to a tax audit.

2.32 In 2006 the age distribution for mothers in the United States who had a first child that year was as follows (Martin et al., p. 31):

Under 20	20–24	25–29	30–34	35 and Over
20.9%	30.6%	24.7%	15.7%	8.1%

a. Draw a bar graph to represent the data.

b. Draw a pie chart to represent the data.

c. Explain which picture—bar graph or pie chart—you think is more informative.

2.33 ◆ A sample of college students was asked how they felt about their weight. Of the 143 women in the sample who responded, 38 women said that they felt overweight, 99 felt that their weight was about right, and 6 felt that they were underweight. Of the 78 men in the sample, 18 men felt that they were overweight, 35 felt that their weight was about right, and 25 felt that they were underweight (*Data source:* **pennstate3** dataset on the companion website).

a. In the relationship between feelings about weight and sex, which variable is the explanatory variable and which is the response variable?

b. Summarize the observed counts by creating a table similar to Table 2.3 (p. 22).

c. For the 143 women, find the percentage responding in each category for how they felt about their weight.

d. For the 78 men, find the percentage responding in each category for how they felt about their weight.

e. Using the percentages found in parts (c) and (d), summarize how the women and men differed in how they felt about their weight.

2.34 Refer to Exercise 2.33 concerning feelings about weight. To compare the men and women, draw a bar graph of the percents found in parts (c) and (d). Use Figure 2.4 (p. 24) for guidance.

General Section Exercises

2.35 In the sample survey described in Section 2.1, there were 92 students who responded to "Randomly pick a letter — *S* or *Q*." Of these 92 students, 61 picked *S* and 31 picked *Q*. The order of the letter choices was reversed for another 98 students who responded to "Randomly pick a letter — *Q* or *S*." Of these 98 students, 45 picked *S* and 53 picked *Q*.

a. Construct a two-way table of counts summarizing the relationship between the letter listed first in the survey question and the letter picked by the student.

b. For the 92 students who saw *S* listed first in the question, determine the percents who picked *S* and *Q*.

c. For the 98 students who had *Q* listed first, determine the percents who picked *S* and *Q*.

d. Draw a bar chart of the percents found in parts (b) and (c) to show the relationship between the letter listed first and the letter picked.

e. Explain whether or not you think the letter listed first in the question affected the choice of letter.

2.36 Refer to Exercise 2.35.

a. Reconstruct the table using the two categorical variables "letter listed first (*S* or *Q*)" and "ordering of letter chosen (listed first or second)."

b. Draw an appropriate picture to accompany your numerical summary.

c. Explain whether you think the variables used in Exercise 2.35 or the variables used in this exercise were more appropriate for illustrating the point of this dataset.

2.37 *This is the same as Exercise 1.1.* The five-number summaries of the fastest ever driven data given in Case Study 1.1 (page 2) were as follows:

	Males (87 students)		Female (102 Students)	
Median	110		89	
Quartiles	95	120	80	95
Extremes	55	150	30	130

Give a numerical value for each of the following:

a. The fastest speed driven by anyone in the class.

b. The slowest of the "fastest speeds" driven by a male.

c. The speed for which one-fourth of the women had driven at that speed or faster.

d. The proportion of females who had driven 89 mph or faster.

e. The number of females who had driven 89 mph or faster.

2.38 Refer to the five-number summaries given in Exercise 2.37.

a. Using the appropriate summary value, compare the *location* of the fastest ever driven response for males to the location for females.

b. Explain whether the *spread* is greater for one sex than the other or whether it is about the same.

2.39 In an experiment, one female and one male restaurant server drew happy faces on the checks of randomly chosen dining parties. The figure for this exercise is a dotplot comparing tip percentages for the female (n = 22 checks) to the tip percentages for the male (n = 23 checks).

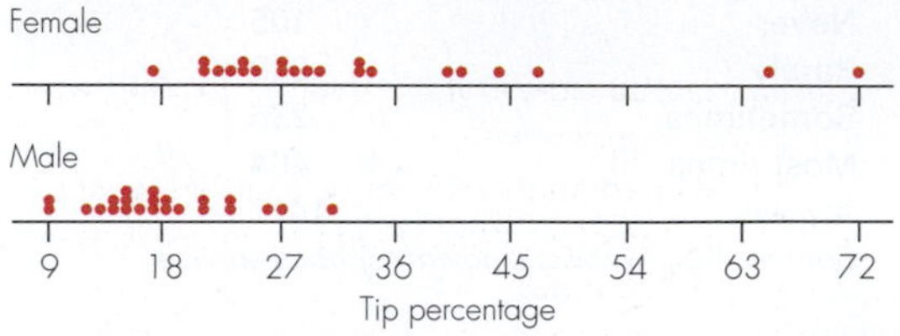

a. Compare the two servers with respect to the approximate centers (locations) of their tip percentages.

b. Compare the two servers with respect to the variation (spread) among tip percentages.

c. Explain whether you think there are any outliers in the dataset or not. If you think there are outliers, give their approximate values.

Section 2.4

Skillbuilder Exercises

2.40 *This is the same as Exercise 1.2.* A five-number summary for the heights in inches of the women who participated in the survey described in Section 2.1 follows:

	Female Heights (inches)	
Median	65	
Quartiles	63.5	67.5
Extremes	59	71

a. What is the median height for these women?

b. What is the range of heights, that is, the difference in heights between the shortest and the tallest women?

c. What is the interval of heights containing the shortest one-fourth of the women?

d. What is the interval of heights containing the middle one-half of the women?

2.41 Refer to Exercise 2.40.

a. Give a value from the five-number summary that characterizes the *location* of the data.

b. Describe the spread of the data using values from the five-number summary.

2.42 ◆ The figure for this exercise is a histogram summarizing the responses given by 137 college women to a question asking how many ear pierces they have (*Data source:* **pennstate2** dataset on the companion website).

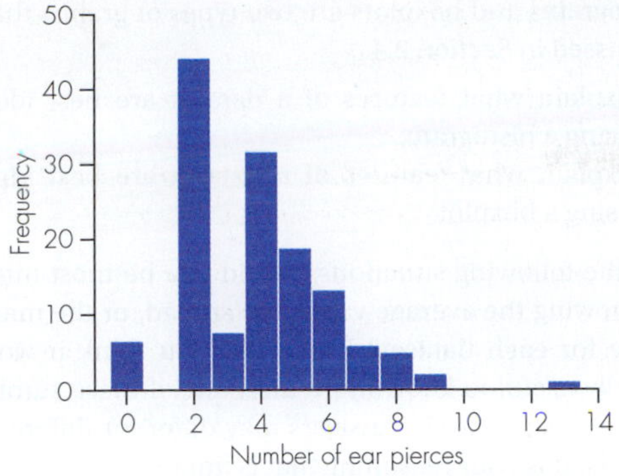

a. Describe the shape of the dataset. Explain whether it is symmetric or skewed.
b. Are there any outliers? For any outlier, give a value for the number of ear pierces, and explain why you think the value is an outlier.
c. What number of ear pierces was the most frequently reported value? Roughly, how many women said they have this number of ear pierces?
d. Roughly, how many women said they have four ear pierces?

2.43 ◆ The figure for this exercise is a histogram summarizing the responses given by 116 college students to a question asking how much they had slept the previous night (*Data source:* **sleepstudy** dataset on the companion website).

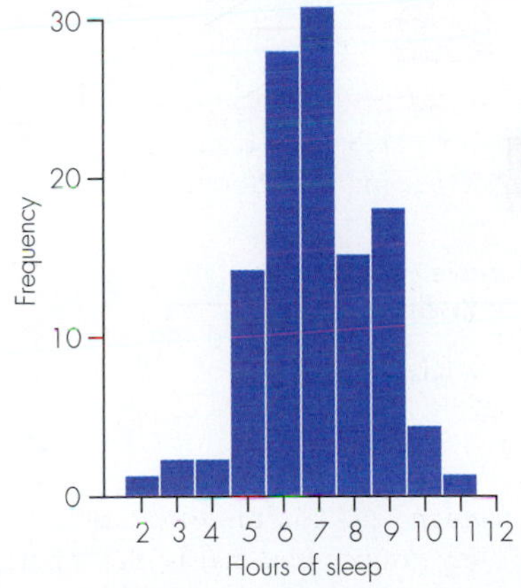

a. Describe the shape of the dataset. Explain whether it is symmetric or skewed.
b. Are there any outliers? For any outlier, give an approximate value for the amount of sleep, and explain why you think the value is an outlier.
c. What was the most frequently reported value (approximately) for the amount of sleep the previous night?
d. Roughly, how many students said that they slept 8 hours the previous night?

2.44 Hand et al. (1994, p. 148) provide data on the number of words in each of 600 randomly selected sentences from the book *Shorter History of England* by G. K. Chesterton. They summarized the data as follows:

Number of Words	Frequency	Number of Words	Frequency
1–5	3	31–35	68
6–10	27	36–40	41
11–15	71	41–45	28
16–20	113	46–50	18
21–25	107	51–55	12
26–30	109	56–60	3

a. Create a histogram for the number of words in the 600 randomly selected sentences.
b. Provide a summary of the dataset based on your histogram.
c. Explain why you could not create a stem-and-leaf plot for this dataset.
d. Count the number of words in the first 20 sentences in Chapter 1 of this book (not including headings), and create a histogram of sentence lengths. Compare the sentence lengths to those in the *Shorter History of England.*

2.45 ◆ The following stem-and-leaf plot is for the mean August temperatures (Fahrenheit) in 20 U.S. cities. The "stem" (row label) gives the first digit of a temperature, while the "leaf" gives the second digit (*Data source:* **temperature** dataset on the companion website).

6	44
6	89
7	01124
7	56667
8	1223
8	5
9	2

a. Describe the shape of the dataset. Is it skewed or is it symmetric?
b. What is the highest temperature in the dataset?
c. What is the lowest temperature in the dataset?
d. What percent of the 20 cities have a mean August temperature in the 80s?

2.46 ◆ About how many music CDs do you own? Responses to this question for 24 students in a senior-level statistics course in 1999 follow:

220, 20, 50, 450, 300, 30, 20, 50, 200, 35, 25, 50, 250, 100, 0, 100, 20, 13, 200, 2, 125, 150, 90, 60

The data are also given in the **musiccds** dataset on the companion website.

a. Draw a stem-and-leaf plot of these data.
b. Draw a histogram of these data.
c. Characterize the shape of the data

General Section Exercises

2.47 A set of exam scores is as follows:

75, 84, 68, 95, 87, 93, 56, 87, 83, 82, 80, 62, 91, 84, 75

a. Draw a stem-and-leaf plot of the scores.
b. Draw a dotplot of the scores.

2.48 Cholesterol levels for $n = 20$ individuals follow:

196 212 200 242 206 178 184 198 160 182
198 182 222 198 188 166 204 178 164 230

a. Draw a histogram of these data. Make the bars cover intervals of cholesterol that are 10 wide beginning at 155 (155 to 165, 165 to 175, and so on).
b. Create a stem-and-leaf plot of the data.
c. Are there any notable outliers in the data?
d. How would you describe the shape of the data?

2.49 ◆ Annual rainfall for Davis, California, for 1951 to 1997 is given in Table 2.6 in Section 2.5 and in the **rainfall** dataset on the companion website. A histogram is shown in Figure 2.18 (p. 40).

a. Create a stem-and-leaf plot for the rainfall data, rounded (not truncated) to the nearest inch.
b. Create a dotplot for the rainfall data, rounded to the nearest inch.
c. Describe the shape of the rainfall data.

2.50 ◆ Case Study 1.1 (p. 2) presented data on the fastest speed that men and women had driven a car, and dotplots were shown for each sex. Data for the men are also in the **pennstate1M** dataset on the companion website.

a. Create a stem-and-leaf plot for the male speeds.
b. Create a histogram for the male speeds.
c. Compare the pictures created in (a) and (b) and the dotplot in Case Study 1.1. Comment on which is more informative, if any of them are, and comment on any other differences that you think are important.
d. How would you describe the shape of this dataset?

2.51 Does a stem-and-leaf plot provide sufficient information to determine whether or not a dataset contains an outlier? Explain.

2.52 ◆ Here are the ages, arranged in order, for the 50 highest-paid CEOs on the Fortune 500 list of top companies in the United States (*Data source*: http://www.forbes.com/lists/2009/12/best-boss-09_CEO-Compensation_CompTotDisp.html). These data are part of the **ceodata08** dataset on the companion website.

42, 47, 48, 49, 49, 50, 50, 50, 50, 51, 51, 51, 52, 54, 54, 55, 55, 55, 55, 56, 57, 57, 57, 57, 57, 58, 58, 59, 59, 60, 60, 60, 61, 61, 62, 62, 62, 62, 62, 63, 63, 64, 64, 64, 64, 67, 67, 69, 74, 78

a. Create a histogram for these ages.
b. Create a stem-and-leaf plot for these ages.
c. Create a dotplot for these ages.
d. Describe the shape of this dataset.
e. Are there any outliers in this dataset?
f. In general, do you think that outliers would be more likely to occur in the salaries of heads of companies or in the ages of heads of companies? Explain.

2.53 About 75% of the students in a class score between 80 and 100 on a quiz. The other 25% of the students have scores spread out between 35 and 79. Characterize the shape of the distribution of quiz scores. Explain.

2.54 Construct an example and sketch a histogram for a measurement that you think would be bimodal.

2.55 Histograms and boxplots are two types of graphs that were discussed in Section 2.4.

a. Explain what features of a dataset are best identified using a histogram.
b. Explain what features of a dataset are best identified using a boxplot.

2.56 For the following situations, would you be most interested in knowing the average value, the spread, or the maximum value for each dataset? Explain. If you think it would be equally useful to know more than one of these summaries, explain that as well. (Answers may differ for different individuals. It is your reasoning that is important.)

a. A dataset with the annual salaries for all employees in a large company that has offered you a job.
b. You need to decide from which of two statistics instructors you will take a class. You have two datasets, with previous final exam scores given by each of the two instructors.
c. A dataset with ages at death for 20 of your relatives who died of natural causes.

2.57 The figure for this exercise is a boxplot comparing tip percentages for a male and a female restaurant server, each of whom drew happy faces on the checks of randomly selected dining parties. A dotplot of the data was given as the figure for Exercise 2.39. Discuss the ways in which the tip percentages for the two servers differed.

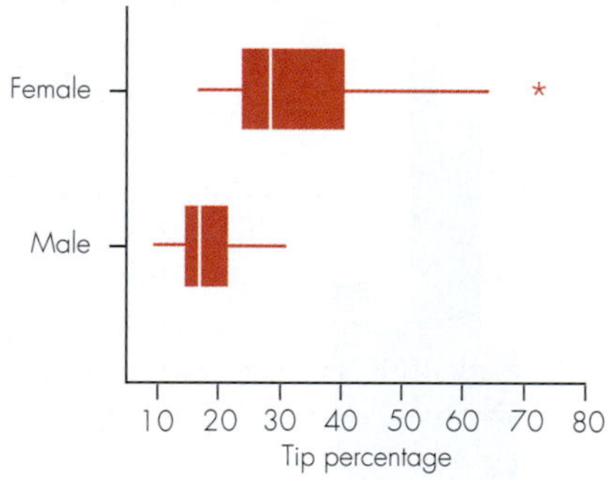

Section 2.5

Skillbuilder Exercises

2.58 The heights (in inches) of seven adult men are 73, 68, 67, 70, 74, 72, and 69.

a. Find the median height for this list.
b. Find the value of the mean height.

2.59 Find the mean and the median for each list of values:

a. 64, 68, 72, 76, 80, 86
b. 10, 6, 2, 7, 100
c. 30, 10, 40, 30

2.60 Refer to part (b) of Exercise 2.59. Explain why there is such a large difference between the mean and median values.

2.61 ◆ Sixty-three college men were asked what they thought was their ideal weight. A five-number summary of the responses (in pounds) follows:

Median	175	
Quartiles	155	190
Extremes	123	225

Data source: **idealwtmen** dataset on the companion website.

a. Find the value of the range for these data.
b. Find the value of the interquartile range (IQR).
c. About what percent of the men gave a response that falls in the interval 155 to 190 pounds?

2.62 ◆ Students in a statistics class wrote as many letters of the alphabet as they could in 15 seconds using their nondominant hand. The figure for this exercise is a boxplot that compares the number of letters written by males and females in the sample (*Data source:* **letters** dataset on the companion website).

a. What is the median number of letters written by females?
b. What is the median for males?
c. Explain whether the interquartile range is larger for males or for females.
d. Find the value of the range for males.
e. Find the value of the range for females.

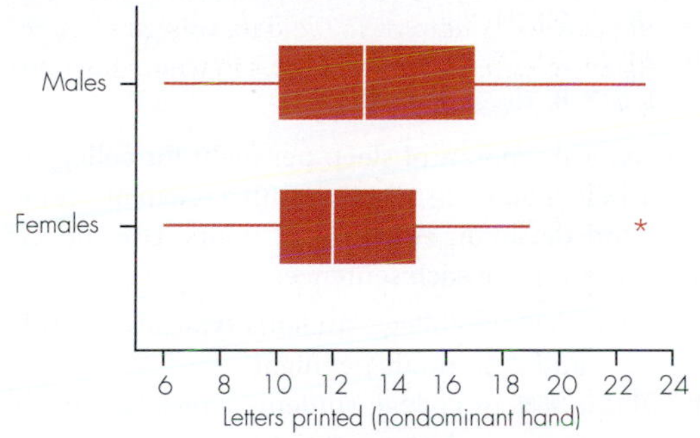

2.63 The weights (in pounds) for nine men on the Cambridge crew team were as follows (*The Independent,* March 31, 1992; also Hand et al., 1994, p. 337):

188.5, 183.0, 194.5, 185.0, 214.0, 203.5, 186.0, 178.5, 109.0

The nine men are comprised of eight rowers and a coxswain, a person who does not row but gives orders to the rowers about the rowing tempo.

a. Find a five-number summary for these data.
b. Identify whether or not any data points would qualify to be marked as an outlier on a boxplot. If there are outliers, specify the values.
c. Which individual do you think is the coxswain?

2.64 A set of eight systolic blood pressures follows:

110, 123, 132, 150, 127, 118, 102, 122

a. Find the median value for the dataset.
b. Find the values of the lower and upper quartiles.
c. Find the value of the interquartile range (IQR).
d. Identify any outliers in the dataset. Use the criterion that a value is an outlier if it is either more than 1.5 × IQR above Q_3 or more than 1.5 × IQR below Q_1.
e. Draw a boxplot of the dataset.

General Section Exercises

2.65 ◆ Create side-by-side boxplots for the "fastest ever driven a car" described in Case Study 1.1 (p. 2): one for males and one for females. Compare the two sexes based on the boxplots. (Five-number summaries are given in Exercise 2.37 and in Case Study 1.1. The raw data are on page 2 and in the **pennstate1** dataset on the companion website.)

2.66 *This is the same as parts (a) and (b) of Exercise 1.22.* Students in a statistics class were asked, "About how many minutes do you typically exercise in a week?" Responses from the *women* in the class were

60, 240, 0, 360, 450, 200, 100, 70, 240, 0, 60, 360, 180, 300, 0, 270

Responses from the *men* in the class were

180, 300, 60, 480, 0, 90, 300, 14, 600, 360, 120, 0, 240

a. Compare the women to the men using a dotplot. What does your plot show you about the difference between the men and the women?
b. For each sex, determine the median response.

2.67 *Parts (a) and (b) are the same as Exercise 1.23.* Refer to Exercise 2.66, which gives exercise times for men and women in a class.

a. Create a five-number summary for the men's responses. Show how you found your answer.
b. Use your five-number summary to describe in words the exercise behavior of this group of students.
c. Draw a boxplot of the men's responses.

2.68 ◆ Describe the data on first ladies' ages at death given in Table 2.5 (p. 28) and also in the dataset **firstladies** on the companion website. Compute whatever numerical summaries you think are appropriate, and then write a narrative summary based on the computed information. Include pictures if appropriate.

2.69 ◆ Refer to Exercise 2.68. Repeat that exercise to describe the rainfall data given in Table 2.6 (p. 39) and also in the dataset **rainfall** on the companion website.

2.70 Refer to Example 2.12, Table 2.6, and Figure 2.18 (p. 40) for the rainfall data. Specify whether the shape is skewed to the left or to the right, and explain whether or not the relationship between the mean and the median (which one is higher) is what you typically expect for data with that shape.

2.71 ◆ Create a five-number summary for the rainfall data in Example 2.12, Table 2.6 (p. 39). Write a few sentences describing the dataset. The data are in the dataset **rainfall** on the companion website.

2.72 The football team at the school of one of the authors won 4 of 11 games it played during the 2004 college football season. Point differences between teams in the 11 games were

+38, −14, +24, −13, −9, −7, −2, −11, −7, +4, +24

A positive difference indicates that the author's school won the game, and a negative difference indicates that the author's school lost.

a. Find the value of the mean point difference and the value of the median point difference for the 11 games.
b. Explain which of the two summary values found in part (a) is a better summary of the team's season.

2.73 Refer to the sentence-length dataset in Exercise 2.44. Note that you cannot compute exact summary values. Provide as much information as you can about the median, interquartile range, and range for the sample of sentence lengths from the *Shorter History of England*.

2.74 In an experiment conducted by one of this book's authors, 19 students were asked to estimate (in millions) the population of Canada, which was about 30 million at that time. Before they made their estimates, ten of the students (Group 1) were told that the population of the United States was about 290 million at that time. Nine of the students (Group 2) were told that the population of Australia was roughly 20 million at that time. The estimates for the population of Canada given by the students in each group were:

Group 1: 2, 30, 35, 70, 100, 120, 135, 150, 190, 200

Group 2: 8, 12, 16, 29, 35, 40, 45, 46, 95

a. Find the five-number summary for Group 1.
b. Find the five-number summary for Group 2.
c. Compare the values of the range for the two groups.
d. Draw a boxplot that compares the two groups. Refer to Figure 2.14 (p. 34) for guidance.

2.75 ◆ Create a five-number summary for the ages of the 50 highest-paid CEOs of Fortune top 500 companies listed in Exercise 2.52 and in the **ceodata08** dataset on the companion website. Write a few sentences describing the dataset.

2.76 ◆ Create a boxplot for the ages of the 50 highest-paid CEOs of Fortune top 500 companies listed in Exercise 2.52 and in the **ceodata08** dataset on the companion website.

2.77 ◆ Find the mean and median for the ages of CEOs of the 50 highest-paid CEOs of Fortune top 500 companies listed in Exercise 2.52 and in the **ceodata08** dataset on the companion website. Is the relationship between them what you would typically expect for data with the shape of this dataset? Explain.

Section 2.6

Skillbuilder Exercises

2.78 In the data discussed in Section 2.1, one student reported having slept 16 hours the previous night.

a. What additional information do you need to determine whether or not this value is an outlier?
b. If you do determine that the data value of 16 hours of sleep is an outlier, what additional information do you need to decide whether or not to discard it before summarizing the data?

2.79 A male whose height is 78 inches might be considered to be an outlier among males in a statistics class but not among males who are professional basketball players. Give another example in which the same measurement taken on the same individual would be considered to be an outlier in one dataset but not in another dataset.

2.80 One of the authors of this book (one male, one female) has a right handspan measurement of 23.5 cm. Would you consider this value to be an outlier? What additional information do you need to make a decision?

General Section Exercises

2.81 Refer to the rainfall data given in Table 2.6 (p. 39). Discuss whether or not there are any outliers, and, if so, whether to discard them or not.

2.82 Give an example, not given in Section 2.6, of a situation in which a measurement is an outlier because the individual belongs to a different group than the bulk of individuals measured. Be specific about the variable measured and the way in which the individual may differ from the others who were measured.

2.83 In a statistics class survey, students reported their heights in inches. The instructor entered the data into a computer file and used statistical software to find separate five-number summaries for men and women in the class. In the five-number summary for men, the minimum height was 17 inches, an obvious outlier. What is a possible reason for this outlier, and what should the instructor do about it?

Section 2.7

Skillbuilder Exercises

2.84 Find the mean and standard deviation for each set of values:

a. 18, 19, 20, 21, 22
b. 20, 20, 20, 20, 20
c. 1, 5, 7, 8, 79

2.85 The typical amount of sleep per night for college students has a bell-shaped distribution with a mean of 7 hours and a standard deviation equal to 1.7 hours. Use the Empirical Rule to complete each sentence:

a. About 68% of college students typically sleep between ____ and ____ hours per night.
b. About 95% of college students typically sleep between ____ and ____ hours per night.
c. About 99.7% of college students typically sleep between ____ and ____ hours per night.

2.86 Suppose that the distribution of speeds at an interstate highway location is bell-shaped with a mean of 71 mph and a standard deviation of 5 mph. Use the Empirical Rule to complete each sentence:

a. About 68% of vehicles at this location travel between ____ and ____ mph.
b. About 95% of vehicles at this location travel between ____ and ____ mph.
c. About 99.7% of vehicles at this location travel between ____ and ____ mph.

2.87 Find the mean and standard deviation for each set of values:

a. 22, 27, 30, 21
b. 25, 35, 40, 20

2.88 Suppose that the mean weight for men 18 to 24 years old is 170 pounds, and the standard deviation is 20 pounds. In each part, find the value of the standardized score (z-score) for the given weight:

a. 200 pounds.
b. 140 pounds.
c. 170 pounds.
d. 230 pounds.

2.89 Refer to Exercise 2.85 about hours of sleep per night for college students. Draw a picture of the distribution. Indicate the locations of the three intervals found in Exercise 2.85. Use Figure 2.24 on p. 52 for guidance.

2.90 Both of the following lists of $n = 8$ data values have a mean of 20:

List 1: 10, 10, 10, 10, 30, 30, 30, 30

List 2: 10, 15, 19, 20, 20, 21, 25, 30

Draw a dotplot comparing the two data lists and explain how the plot shows that the standard deviation for List 1 is greater than the standard deviation for List 2.

2.91 Suppose that the amount spent on textbooks in a semester for college students has a mean of \$350 and a standard deviation of \$100. In each part, find the value of the standardized score (z-score) for the given amount spent on textbooks.

a. \$300
b. \$460
c. \$650
d. \$210

2.92 The data for Exercise 2.64 was this set of systolic blood pressures:

110, 123, 132, 150, 127, 118, 102, 122

a. Find the mean and standard deviation for these data.
b. What is the variance for these data?

2.93 Write a set of seven numbers with a mean of 50 and a standard deviation of 0. Is there more than one possible set of numbers? Explain.

2.94 If you learn that your score on an exam was 80 and the mean was 70, would you be more satisfied if the standard deviation was 5 or if it was 15? Explain.

General Section Exercises

2.95 The scores on the final exam in a course with a large number of students have approximately a bell-shaped distribution. The mean score was 70, the highest score was 98, and the lowest score was 41.

a. Find the value of the range for the exam scores.
b. Refer to part (a). Use the value of the range to estimate the value of the standard deviation.

2.96 Scores on the Stanford-Binet IQ test have a bell-shaped distribution with mean = 100 and standard deviation = 16.

a. Use the Empirical Rule to specify the intervals into which 68%, 95%, and 99.7% of Stanford-Binet IQ scores fall.
b. Draw a picture similar to Figure 2.24 (p. 52) illustrating the intervals found in part (a).
c. What is the variance of Stanford-Binet IQ scores?

2.97 The mean for the women's right handspans is about 20 cm, with a standard deviation of about 1.8 cm. Using the stem-and-leaf plot in Figure 2.8 (p. 29), determine how well this set of measurements fits with the Empirical Rule.

2.98 Refer to the women's right handspan data. As you can see in Figures 2.7 to 2.9, there are two apparent outliers at 12.5 cm and 13.0 cm.

a. If these values are removed, do you think the mean will increase, decrease, or remain the same? What about the standard deviation? Explain.
b. With the outliers removed, the mean and standard deviation for the remaining 101 values are 20.2 cm and 1.45 cm, respectively. The range is 7.25, from 16.0 to 23.25. Determine whether or not the Empirical Rule for mean ± 3 standard deviations appears to hold for these values.
c. Refer to Figures 2.7 to 2.9. Based on those figures, do you think the Empirical Rule should hold when the outliers are removed? How about when the outliers have not been removed? Explain.
d. Is there any justification for removing the two outliers? Explain.

2.99 Head circumferences of adult males have a bell-shaped distribution with a mean of 56 cm and a standard deviation of 2 cm.

a. Explain whether or not it would be unusual for an adult male to have a 52-cm head circumference.
b. Explain whether or not it would be unusual for an adult male to have a 62-cm head circumference.

2.100 Refer to Exercise 2.99. What is the variance of head circumferences of adult males?

2.101 Suppose verbal SAT scores for students admitted to a university are bell-shaped with a mean of 540 and a standard deviation of 50.

a. Draw a picture of the distribution of these verbal SAT scores, indicating the cutoff points for the middle 68%, 95%, and 99.7% of the scores.
b. What is the variance of verbal SAT scores for students admitted to the university?

2.102 Exercise 2.99 gave the information that head circumferences of adult males have a bell-shaped distribution with mean = 56 cm and standard deviation = 2 cm.

a. What is the head circumference such that only 2.5% of adult males have a smaller head circumference?
b. What is the head circumference such that only 2.5% of adult males have a larger head circumference?
c. What is the head circumference such that only 16% of adult males have a larger head circumference?

2.103 Exercise 2.101 gave the information that the verbal SAT scores for students admitted to a university had a bell-shaped distribution with mean = 540 and standard deviation = 50.

a. What is the verbal SAT score such that only 16% of admitted students had a higher score?
b. What is the verbal SAT score such that only 2.5% of admitted students had a higher score?
c. What is the verbal SAT score such that only 16% of admitted students had a lower score?

2.104 Using a computer or calculator that provides proportions falling below a specified z-score, determine the approximate proportion for each of the following situations. In each case, assume the values are approximately bell-shaped.

a. The proportion of SAT scores falling below 450 for a group with a mean of 500 and a standard deviation of 100.

b. The proportion of boys with heights below 36.5 inches for a group with mean height of 34 inches and standard deviation of 1 inch.
c. The proportion of a large class that scored below you on a test for which the mean was 75, the standard deviation was 8, and your score was 79.
d. The proportion of a large class that scored below you on a test for which the mean was 75, the standard deviation was 4, and your score was 79.

2.105 Can a categorical variable have a bell-shaped distribution? Explain.

2.106 Remember that a resistant statistic is a numerical summary whose value is not unduly influenced by an outlier of any magnitude. Is the standard deviation a resistant statistic? Justify your answer by giving an example of a small dataset, and then adding a very large outlier and noting how the standard deviation is affected.

Section 2.8: Skillbuilder Applet Exercises

For these exercises, use the **Empirical** *applet described in Section 2.8 and available on the companion website,* http://www.cengage.com/statistics/Utts4e.

2.107 Examine the results given by the applet for each of the eight variables.

a. Among the eight variables, which variables are best described by the Empirical Rule, and which are not well described by the Empirical Rule?
b. Generally, what is the shape of the histogram for the variables that are well described by the Empirical Rule? What is the shape of the histogram for the variables that are not well described by the Empirical Rule?

2.108 The parts of this exercise concern the variable ***Dad's Height***, which is data for father's height as reported by $n = 167$ college students.

a. What is the shape of the histogram given for the variable ***Dad's Height***? Are there any outliers?
b. Refer to part (a). Based on the shape of the histogram, explain whether or not the Empirical Rule will apply.
c. What numerical values are given by the applet for the interval ***mean ± s***? What percentage of the data values are in this interval? Compare this percentage to the percentage that would be expected if the Empirical Rule applies.
d. What numerical values are given by the applet for the interval ***mean ± 2s***? What percentage of the data values are in this interval? Compare this percentage to the percentage that would be expected if the Empirical Rule applies.

2.109 The parts of this exercise concern the variable ***Alcohol***, which is data for number of alcoholic beverages consumed in a typical week as reported by $n = 167$ college students.

a. What is the shape of the histogram given for ***Alcohol***? Are there any outliers?
b. Refer to part (a). Based on the shape of the histogram, explain whether or not the Empirical Rule will apply.
c. What numerical values are given by the applet for the interval ***mean ± s***? What percentage of the data values are in this interval? Compare this percentage to the percentage that would be expected if the Empirical Rule applies.
d. What numerical values are given by the applet for the interval ***mean ± 2s***? Explain why this interval is not a good description of possible values for the variable ***Alcohol***.

2.110 The parts of this exercise concern the variable ***Ideal Height***, which is the respondent's desired height as reported by $n = 149$ college women.

a. Use information given by the applet to explain whether or not the Empirical Rule applies for ***Ideal Height***.
b. Assuming the Empirical Rule applies, give numerical values for the interval that will contain about 99.7% of the data values.
c. Compare the interval found in part (b) to the histogram of the data. Explain whether or not the interval is a reasonable description of the data.

2.111 The data for each of the following variables includes one or more outliers. In each case, identify the outlier(s). Then explain whether or not the Empirical Rule would apply to the remaining data if the outlier(s) were removed from the dataset.

a. ***TV Hours***.
b. ***Handspan (females)***.
c. ***Dad's Height***.

Chapter Exercises

2.112 Do Thought Question 2.4 on p. 24.

2.113 ◆ A question in the 2002 General Social Survey (GSS) conducted by the National Opinion Research Center asked participants how long they spend on e-mail each week. A summary of responses (hours) for $n = 1881$ respondents follows. (The data are in the dataset **GSS-02** on the companion website.)

Mean	StDev	Minimum	Q1	Median	Q3	Maximum
4.14	7.235	0	0	2	5	70

a. Explain how the summary statistics show us that at least 25% of the respondents said that they do not use e-mail.
b. What is the interval that contains the lower 50% of the responses?
c. What is the interval that contains the upper 50% of the responses?
d. Explain whether or not the maximum value, 70 hours, would be marked as an outlier on a boxplot.
e. Calculate Range/6 and compare the answer to the value of the standard deviation. What feature(s) of the data do you think causes the values to differ?
f. Compare the mean to the median. What feature(s) of the data do you think causes the values to differ?

2.114 For each of the following situations, would you prefer your value to be average, a low outlier, or a high outlier? Explain.

a. Number of children you have.
b. Your annual salary.
c. Gas mileage for your car.
d. Crime rate in the city or town where you live.

2.115 Specify the type (categorical, ordinal, quantitative) for each of the following variables recorded in a survey of telephone usage in student households:

a. Telephone exchange (first three numbers after area code).
b. Number of telephones in the household.
c. Dollar amount of last month's phone bill.
d. Long-distance phone company used.

2.116 In the same survey for which wives' heights are given in Example 2.19, husbands' heights were also recorded. A five-number summary of husbands' heights (mm) follows:

Husbands' Heights (n = 199)		
Median	1725	
Quartiles	1691	1774
Extremes	1559	1949

a. Construct a boxplot for the husbands' heights.
b. Use the range to approximate the standard deviation for these heights.
c. What assumption did you need to make in part (b) to make the approximation appropriate?
d. The mean and standard deviation for these heights are 1732.5 mm and 68.8 mm. Use the Empirical Rule to construct an interval that should cover 99.7% of the data, and compare your interval to the extremes. Does the interval cover both extremes?

2.117 Can a variable be *both* of the following types? If so, give an example.

a. An explanatory variable and a categorical variable.
b. A continuous variable and an ordinal variable.
c. A quantitative variable and a response variable.
d. A bell-shaped variable and a response variable.

2.118 Reach into your wallet, pocket, or wherever you can find at least ten coins, and sort all of the coins you have by type.

a. Count how many of each kind of coin you have (pennies, nickels, and so on, or the equivalent for your country). Draw a pie chart illustrating the distribution of your coins.
b. In part (a), "kind of coin" was the variable of interest for each coin. Is that a categorical, ordinal, or quantitative variable?
c. Now consider the total monetary value of all of your coins as a single data value. What type of variable is it?
d. Suppose you had similar data for all of the students in your statistics class. Write a question for which the variable "kind of coin" is the variable of interest for answering the question. Then write a question for which "total monetary value of the coins" is the variable of interest for answering the question.

2.119 For each of the following datasets, explain whether you would expect the mean or the median of the observations to be higher:

a. In a rural farming community, for each household the number of children is measured.
b. For all households in a large city, yearly household income is measured.
c. For all students in a high school (not just those who were employed), income earned in a job outside the home in the past month is measured.
d. For the coins in someone's pocket that has one-third pennies, one-third nickels, and one-third quarters, the monetary value of each coin is recorded.

2.120 Look around your living space or current surroundings, and find a categorical variable for which there are at least three categories and for which you can collect at least 20 observations (example: color of the shirts in your closet). Collect the data.

a. Draw a pie chart for your data.
b. Draw a bar graph for your data.
c. Is one of the pictures more informative than the other? Explain. (Your answer may depend on the variable you chose.)

2.121 A sample of $n = 500$ individuals is asked how many hours they typically spend using a computer in a week. The mean response is $\bar{x} = 8.3$ hours, and the standard deviation is $s = 7.2$ hours. Find values for the interval $\bar{x} \pm 2s$, and explain why the result is evidence that the distribution of weekly hours spent using the computer is not bell-shaped.

2.122 Look around your living space or current surroundings and find a quantitative variable for which you can collect at least 20 observations (examples: monetary amounts of the last 25 bills you received or your last 20 scores on tests and homework assignments). List the data with your response.

a. Create a five-number summary.
b. Draw a boxplot.
c. Draw your choice of a histogram, stem-and-leaf plot, or dotplot.
d. Refer to your picture in part (c) and comment on the shape and presence or absence of outliers.
e. Compute the mean and compare it to the median. Explain whether or not the relationship between them is what you would expect based on the information you discussed in part (d).

2.123 **a.** Would the first ladies' ages at death data in Table 2.5 (p. 28) be considered a population of measurements or a sample from some larger population? Explain.
b. Find the appropriate standard deviation (sample or population) for the "ages at death" data in Table 2.5 (p. 28).

2.124 Each of the following quotes is taken from an article titled, "Education seems to help in selecting husbands" (*Sacramento Bee*, December 4, 1998, p. A21), which reported on new data in the *Statistical Abstract of the United States.* Draw an appropriate graph to represent each situation.

a. "The data show that 3.8 percent of women who didn't complete high school had four or more husbands. For high school graduates, the share with four or more partners drops to 3 percent. Among those who attended college 2 percent had four or more husbands, and that fell to 1 percent for those with college degrees."
b. "From 1997 on, 5.5 percent of children lived with their grandparents, a share that has been rising steadily. It was only 3.6 percent in 1980, and by 1990 it was 4.9 percent."

c. "The center said 20.1 percent of Americans took part in some regular activity—21.5 percent of men and 18.9 percent of women."

2.125 In each case, specify which of the two variables is the explanatory variable and which is the response variable. If it is ambiguous, explain why.

a. Is there a relationship between the amount of beer people drink and their systolic blood pressure?
b. Is there a relationship between calories of protein consumed per day and incidence of colon cancer?

2.126 a. If a data value has a z-score of 0, the value equals one of the summary measures discussed in this chapter. Which summary measure is that?
b. Verify that a data value having a z-score of 1.0 is equal to the mean plus 1 standard deviation.

2.127 The data for 103 women's right handspans are shown in Figures 2.7 to 2.9 (pp. 29–30), and a five-number summary is given in Example 2.5 (p. 26).

a. Examine Figures 2.7 to 2.9 and comment on whether or not the Empirical Rule should hold.
b. The mean and standard deviation for these measurements are 20.0 cm and 1.8 cm, respectively. Determine whether or not the range of the data (found from the five-number summary) is about what would be expected using the Empirical Rule.

2.128 For each of the following two sets of data, explain which one is likely to have a larger standard deviation:

a. Set 1: Heights of the children in a kindergarten class.
Set 2: Heights of all of the children in an elementary school.
b. Set 1: Systolic blood pressure for a single individual taken daily for 30 days.
Set 2: Systolic blood pressure for 30 people who visit a health clinic in 1 day.
c. Set 1: SAT scores (which range from 200 to 800) for the students in an honors class.
Set 2: Final examination scores (which range from 0 to 100) for all of the students in the English classes at a high school.

2.129 ◆ Refer to Exercise 2.52, in which the ages for the highest-paid 50 CEOs of America's top 500 companies were given. These data are in the **ceodata08** dataset for this book.

a. Find the mean and standard deviation for these ages.
b. Recall that the range should be equivalent to 4 to 6 standard deviations for bell-shaped data. Determine whether or not that relationship holds for these ages.
c. Find the z-scores for the ages of the youngest and oldest CEO. Are they about what you would expect? Explain.

2.130 For a bell-shaped dataset with a large number of values, approximately what z-score would correspond to a data value equaling each of the following?

a. The median.
b. The lowest value.
c. The highest value.
d. The mean.

Exercises 2.131 to 2.134 each describe one or two variables and the individuals for whom they were measured. For each exercise, state an interesting research question about the situation. Use the examples under the "Asking the Right Questions" heading (p. 19) in Section 2.2 as a guide.

2.131 Individuals are all of the kindergarten children in a school district.
One variable: Adult(s) with whom the child lives (both parents, mother only, father only, one or both grandparents, other).

2.132 Individuals are all mathematics majors at a college.
Two variables: Grade point average and hours spent studying last week.

2.133 Individuals are a representative sample of adults in a large city.
Two variables: Ounces of coffee consumed per day and marital status (currently married or not).

2.134 Individuals are a representative sample of college students.
Two variables: Male or female and whether the person dreams in color (yes or no).

2.135 ◆ Exercise 2.46 gave the following data values for the number of CDs owned by 24 students in a statistics class (in 1999). (*Data source:* **musiccds** dataset on the companion website.)

220, 20, 50, 450, 300, 30, 20, 50, 200, 35, 25, 50, 250, 100, 0, 100, 20, 13, 200, 2, 125, 150, 90, 60

a. Find the five-number summary for these data.
b. Determine whether or not any data values would be marked as outliers on a boxplot.

2.136 In each case, specify which of the two variables is the explanatory variable and which is the response variable. If it is ambiguous, explain why.

a. Is there a relationship between eye color and whether or not corrective lenses are needed by age 18?
b. For women who are HIV-positive when they get pregnant, is there a relationship between whether or not the HIV is transmitted to the infant and the length of time the woman had been infected before getting pregnant?

2.137 Exercise 2.63 gave the following weights (in pounds) for nine men on the Cambridge crew team:

188.5, 183.0, 194.5, 185.0, 214.0, 203.5, 186.0, 178.5, 109.0

Draw a boxplot of these data.

2.138 Refer to Exercise 2.40, which gives a five-number summary of heights for college women. Draw a boxplot displaying the information in this five-number summary.

2.139 The interquartile range and the standard deviation are two different measures of spread. Which measure do you think is more affected by outliers? Explain.

2.140 Explain why women's heights are likely to have a bell shape but their ages at marriage do not.

◆ Dataset available but not required **Bold** exercises answered in the back

Dataset Exercises

Datasets required to solve these exercises are available on the companion website, http://www.cengage.com/statistics/Utts4e.

2.141 Use the **oldfaithful** dataset on the companion website; it gives data for $n = 299$ eruptions of the Old Faithful geyser.

a. The variable ***TimeToNext*** is the time until the next eruption after the present eruption. Draw a histogram of this variable. Describe the shape of the histogram.

b. Draw a boxplot of the ***TimeToNext*** variable. What important feature of the data cannot be seen in the boxplot that is seen in the histogram?

2.142 The data for this exercise are in the **GSS-08** dataset on the companion website. The variable ***gunlaw*** is whether a respondent favors or opposes stronger gun control laws.

a. Determine the percentage of respondents who favor stronger gun control laws and the percentage of respondents who oppose stronger gun control laws. (*Note:* Not all survey participants were asked the question about gun laws, so the sample size for ***gunlaw*** is smaller than the overall sample size.)

b. Draw a graphical summary of the ***gunlaw*** variable.

c. Create a two-way table of counts that shows the relationship between the variable ***sex*** (female, male) and opinion about stronger gun control laws. From looking at this table of counts, are you able to judge whether or not the two variables are related? Briefly explain.

d. What percentage of females favors stronger gun control laws? What percentage of males favors stronger gun control laws?

e. Based on the percentages found in part (d), do you think that sex and opinion about gun control are related? Briefly explain.

2.143 Use the **pennstate1** dataset on the companion website for this exercise. The data for the variable ***HrsSleep*** are responses by $n = 190$ students to the question, "How many hours did you sleep last night?"

a. Draw a histogram of the data for the ***HrsSleep*** variable. Describe the shape of this histogram, and comment on any other interesting features of the data.

b. Determine the five-number summary for these data.

c. What is the range of the data? What is the interquartile range?

2.144 Use the **pennstate2** dataset on the companion website for this exercise. The variable ***CDs*** is the approximate number of music CDs owned by a student.

a. Draw a stem-and-leaf plot for the ***CDs*** variable.

b. Draw a histogram for the ***CDs*** variable.

c. Draw a dotplot for the ***CDs*** variable.

d. Describe the shape of the data for the ***CDs*** variable, and comment on any other interesting features of the data.

e. Calculate the mean number and the median number of CDs. Compare these two values.

f. For these data, do you think the mean or the median is a better description of the location of the data? Briefly explain.

2.145 For this exercise, use the **GSS-08** dataset on the companion website. The variable ***cappun*** is the respondent's opinion about the death penalty for persons convicted of murder, and the variable ***polparty*** is the respondent's political party preference (Democrat, Republican, Independent, Other).

a. In this dataset, what percentage favors the death penalty? What percentage opposes it?

b. Create a table that displays the relationship between political party and opinion about the death penalty. Calculate an appropriate set of percentages for describing the relationship.

c. Are the variables ***polparty*** and ***cappun*** related? Explain.

2.146 Use the **cholest** dataset on the companion website for this exercise. The dataset contains cholesterol levels for 30 "control" patients and 28 heart attack patients at a medical facility. For the heart attack patients, cholesterol levels were measured 2 days, 4 days, and 14 days after the heart attack.

a. Calculate the mean, the standard deviation, and the five-number summary for the control patients.

b. Calculate the mean, the standard deviation, and the five-number summary for the heart attack patients' cholesterol levels 2 days after their attacks.

c. Generally, which group has the higher cholesterol levels? How much difference is there in the *location* of the cholesterol levels of the two groups?

d. Which group of measurements has a larger *spread*? Compare the groups with regard to all three measures of spread introduced in Sections 2.6 and 2.7.

e. Compare the control patients and the heart attack patients using a comparative dotplot (as in Case Study 1.1 on p. 2). Briefly explain what this plot indicates about the difference between the two groups.

2.147 Use the **pennstate1** dataset on the companion website for this exercise.

a. Draw a histogram of the ***height*** variable.

b. What is the shape of this histogram? Why do you think it is not a bell shape?

c. Draw a boxplot of the height variable.

d. Which graph, the histogram or the boxplot, is more informative about this dataset? Briefly explain.

2.148 Use the **GSS-08** dataset on the companion website. The variable ***degree*** indicates the highest educational degree achieved by a respondent.

a. Is the ***degree*** variable quantitative, categorical, or ordinal? Explain.

b. Determine the number and percentage falling into each degree category.

c. What percentage of the sample has a degree that is beyond a high school degree?

d. The variable ***tvhours*** is the self-reported number of hours of watching television in a typical day. Find the mean number of television-watching hours for each of the five degree groups.

e. Is there a relationship between self-reported hours of watching television and educational degree? Explain.

f. Draw any visual summary of the variable ***tvhours*** (for the whole sample). What are the interesting features of your graph?

3

TMI/Alamy

Does the driver's age affect the view?

See Example 3.2 *(p. 71)*

Relationships Between Quantitative Variables

A *statistical relationship* is different from a *deterministic relationship,* for which the value of one variable can be determined exactly from the value of the other variable. In a statistical relationship, there is variation from the average pattern. Our ability to predict what happens for an individual depends on the amount of natural variability from that pattern.

The description and confirmation of relationships between variables are so important in research that entire courses are devoted to the topic. You have already seen several examples that involved a potential relationship. For instance, in the observational study in Example 2.2, the investigators wanted to know whether incidence of myopia was related to how much light people slept with when they were infants. In the randomized experiment in Case Study 1.6, the researchers wanted to know whether aspirin consumption and risk of heart attack were related.

In this chapter, we will learn how to describe the relationship between two *quantitative variables.* Remember (from Chapter 2) that the terms *quantitative variable* and *measurement variable* are synonyms for data that can be recorded as numerical values and then meaningfully ordered according to those values. The relationship between weight and height is an example of a relationship between two quantitative variables.

The questions that we ask about the relationship between two variables often concern specific numerical features of the association. For example, we may want to know how much weight will increase on average for each 1-inch increase in height. Or we may want to estimate what the college grade point average will be for a student whose high school grade point average was 3.5.

We will use three tools to describe, picture, and quantify the relationship between two quantitative variables:

- **Scatterplot**, a two-dimensional graph of data values.
- **Correlation**, a statistic that measures the *strength and direction* of a linear relationship between two quantitative variables.
- **Regression equation**, an equation that describes the average relationship between a quantitative response variable and an explanatory variable.

THOUGHT QUESTION 3.1 For adults, there is a *positive association* between weight and height. For used cars, there is a *negative association* between the age of the car and the selling price. Explain what it means for two variables to have a positive association. Explain what it means when two variables have a negative association. What is an example of two variables that would have *no association*?*

***HINT:** Average weight increases as height increases. The selling price decreases as a car's age increases. Use these patterns to define positive and negative association more generally.

3.1 Looking for Patterns with Scatterplots

A **scatterplot** is a two-dimensional graph of the measurements for two numerical variables. A point on the graph represents the combination of measurements for an individual observation. The vertical axis, which is called the *y axis,* is used to locate the value of one of the variables. The horizontal axis, called the *x axis,* is used to locate the value of the other variable.

As we learned in Chapter 2, when looking at relationships, we can often identify one of the variables as an **explanatory variable** that may explain or cause differences in the **response variable**. The term **dependent variable** is used as a synonym for *response variable* because the value for the response variable may *depend* on the value for the explanatory variable. In a scatterplot, the response variable is plotted on the vertical axis (the *y* axis), so it may also be called the ***y* variable**. The explanatory variable is plotted along the horizontal axis (the *x* axis) and may be called the ***x* variable**.

Questions to Ask About a Scatterplot

- What is the *average* pattern? Does it look like a straight line, or is it curved?
- What is the direction of the pattern?
- How much do individual points vary from the average pattern?
- Are there any unusual data points?

Example 3.1 **Height and Handspan** Table 3.1 displays the first 12 observations of a dataset that includes the heights (inches) and fully stretched handspans (cm) of 167 college students. The data values for all 167 students are the raw data for studying the connection between height and handspan. Imagine how difficult it would be to see the pattern in the data if all 167 observations were shown in Table 3.1. Even when we just look at the data for 12 students, it takes a while to confirm that there does seem to be a tendency for taller people to have larger handspans.

Figure 3.1 is a scatterplot that displays the handspan and height measurements for all 167 students. The handspan measurements are plotted along the vertical axis (*y*), and the height measurements are plotted along the horizontal axis (*x*). Each point represents the two measurements for an individual.

We see that taller people tend to have greater handspan measurements than shorter people. When two variables tend to increase together, as they do in Figure 3.1, we say that they have a **positive association**. Another noteworthy characteristic of the graph is that we can describe the general pattern of this relationship with a straight line. In other words, the handspan and height measurements may have a **linear relationship**.

Table 3.1 Handspans and Height

Height (in)	Span (cm)
71	23.5
69	22.0
66	18.5
64	20.5
71	21.0
72	24.0
67	19.5
65	20.5
76	24.5
67	20.0
70	23.0
62	17.0

and so on, for $n = 167$ observations.

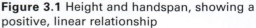

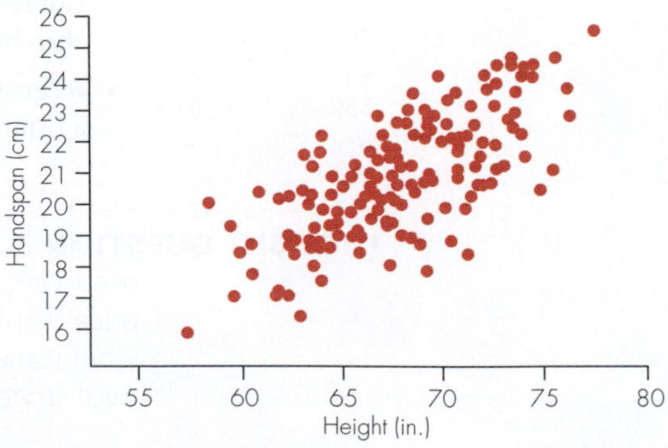

Figure 3.1 Height and handspan, showing a positive, linear relationship

DEFINITION

- Two variables have a **positive association** when the values of one variable tend to increase as the values of the other variable increase.
- Two variables have a **negative association** when the values of one variable tend to decrease as the values of the other variable increase.
- Two variables have a **linear relationship** when the pattern of their relationship resembles a straight line.

Example 3.2 **Driver Age and the Maximum Legibility Distance of Highway Signs**

In a study of the legibility and visibility of highway signs, a Pennsylvania research firm determined the maximum distance at which each of 30 drivers could read a newly designed sign. The 30 participants in the study ranged in age from 18 to 82 years old. The government agency that funded the research hoped to improve highway safety for older drivers and wanted to examine the relationship between age and the sign legibility distance.

Table 3.2 lists the data and Figure 3.2 shows a scatterplot of the ages and distances. The sign legibility distance is the response variable, so that variable is plotted on the *y* axis (the vertical axis). The maximum reading distance tends to decrease as age increases, so there is a negative association between distance and age. This is not a surprising result. As a person gets older, his or her eyesight tends to get worse, so we would expect the distances to decrease with age.

The researchers collected the data to determine numerical estimates for two questions about the relationship:

- How much does the distance decrease when age is increased?
- For drivers of any specific age, what is the average distance at which the sign can be read?

We'll examine these questions in the next section. For now, we simply point out that the pattern in the graph looks *linear*, so a straight-line equation that links distance to age will help us to answer these questions.

Table 3.2 Data Values for Example 3.2

Age	Distance (ft)	Age	Distance (ft)	Age	Distance (ft)
18	510	37	420	68	300
20	590	41	460	70	390
22	560	46	450	71	320
23	460	53	460	73	280
25	490	55	420	74	420
27	560	63	350	75	460
28	510	65	420	77	360
29	460	66	300	79	310
32	410	67	410	82	360

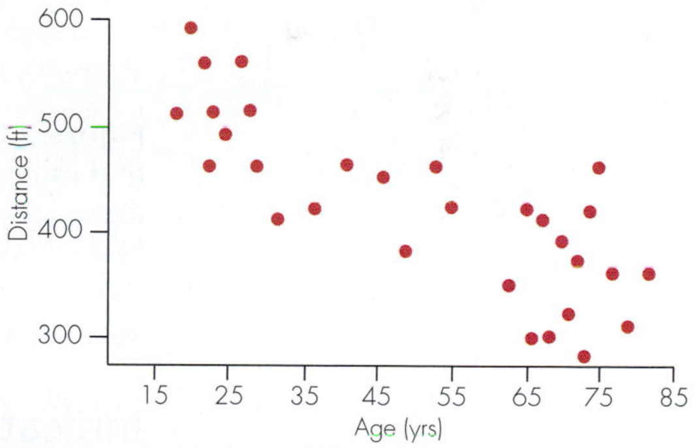

Figure 3.2 Driver age and the maximum distance at which a highway sign was read

(Source: Adapted from data collected by Last Resource, Inc., Bellefonte, PA.)

Curvilinear Patterns

A linear pattern is common, but it is not the only type of relationship. Sometimes, a curve describes the pattern of a scatterplot better than a line does, and when that's the case, the relationship is called **nonlinear** or **curvilinear**.

Example 3.3 **The Development of Musical Preferences** Will you always like the music that you like now? If you are about 20 years old, the likely answer is "yes," according to research reported in the *Journal of Consumer Research* (Holbrook and Schindler, 1989). The researchers concluded that we tend to acquire our popular music preferences during late adolescence and early adulthood.

Read the original source on the companion website, http://www.cengage.com/statistics/Utts4e.

In the study, 108 participants from 16 to 86 years old listened to 28 hit songs that had been on *Billboard*'s Top 10 list for popular music sometime between 1932 and 1986. Respondents rated the 28 songs on a 10-point scale, with 1 corresponding to "I dislike it a lot" and 10 corresponding to "I like it a lot." Each individual's ratings were then adjusted so that the mean rating for each participant was 0. On this adjusted rating scale, a positive score indicates a rating that was above average for a participant, whereas a negative score indicates a below-average rating.

For each of the 108 participants $\times$ 28 songs, a "song-specific age" was calculated representing how old the participant was when that song was popular. If the song was popular before the person was born, the song-specific age was negative. For example, the youngest participant in the study was born in 1971, so the song from 1932 was popular 39 years before that person was born, for a song-specific age of -39. The oldest participant was born in 1901, so the song from 1986 was popular 85 years after that person was born, for a song-specific age of $+85$. These were the two extremes, so the song-specific ages range from -39 to $+85$.

Figure 3.3 shows the relationship between the average adjusted song ratings and the song-specific ages. There are 124 points in the scatter plot, one for each song-specific age from -39 to $+85$. The overall pattern in Figure 3.3 looks somewhat like an inverted U, and the highest preference ratings occur when song-specific ages are in the late teens and early twenties. A straight line does not describe the overall pattern, so the association is called nonlinear or curvilinear.

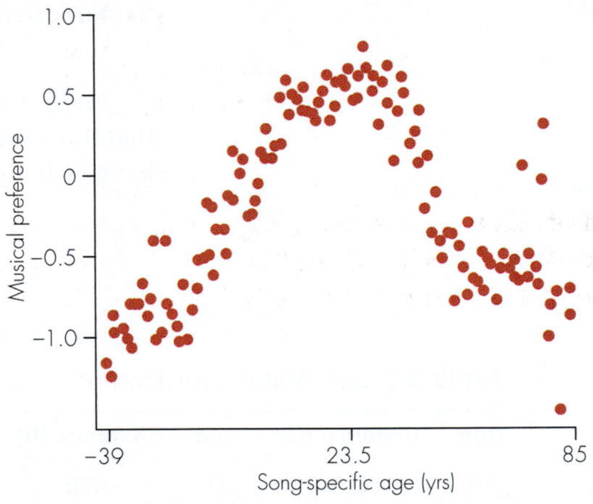

Figure 3.3 Song-specific age and music-preference score *(Source:* The Journal of Consumer Research, *Vol. 16 (1), pp. 119–124. [See p. 701 for complete credit.])*

Indicating Groups within the Data on Scatterplots

When we examine the connection between height and handspan in Example 3.1, you may wonder whether we should be concerned about the sex of the student. Both height and handspan tend to be greater for men than for women, so we should consider the possibility that sex differences might be completely responsible for the observed relationship.

It's easy to indicate subgroups on a scatterplot. We just use different symbols or different colors to represent the different groups. Figure 3.4 is the same as Figure 3.1 except that now different symbols are used for males and females. Note that the positive association between handspan and height appears to hold within each sex. For both men and women, handspan tends to increase as height increases. It's not always the case that the pattern in each subgroup is consistent with the pattern in the whole group. Later in this chapter, we will see that when we combine subgroups inappropriately, the relationship for the combined group can misrepresent the relationship that we see in each subgroup.

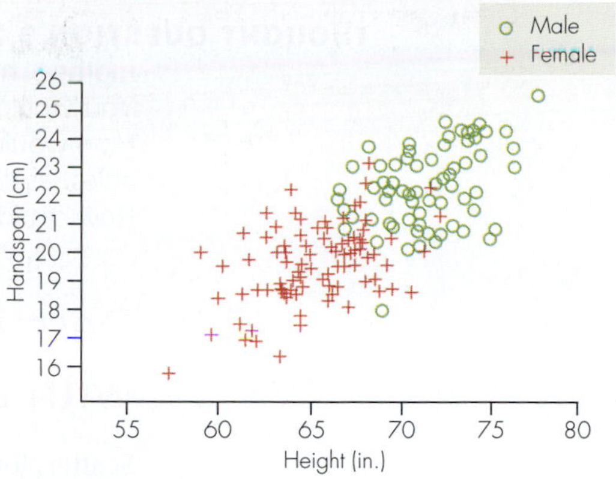

Figure 3.4 Height and handspan by sex

Look for Outliers

Outliers can have a big impact when we quantify a relationship, as we will see in more detail later in this chapter. When we consider two variables, an **outlier** is a point that has an unusual combination of data values. For instance, a man 6′3″ tall who weighs 130 pounds would probably be an outlier in a scatterplot of weights and heights because this is an unusual *combination* of weight and height measurements. As we learned in Chapter 2, outliers can occur because there are unusual and interesting data points, or they may occur because mistakes were made when the data were recorded or entered into the computer.

Example 3.4 **Heights and Foot Lengths of College Women** Figure 3.5 shows the relationship between foot length (cm) and height (inches) for a sample of 41 college women. The two shortest women have much greater foot lengths than we would predict on the basis of the rest of the data. On the other side of the height scale, the tallest woman (74 inches) has a much shorter foot length than we might expect. Fortunately, the students submitted their measurements on a paper form, and a look at those forms revealed that the heights of the three "unusual" women were incorrectly entered into the computer. The woman who appears to be 74 inches tall is actually 64 inches tall. The women with heights of 55 inches and 57 inches on the plot were actually 65 and 67 inches tall.

3.1 Exercises are on pages 99–101.

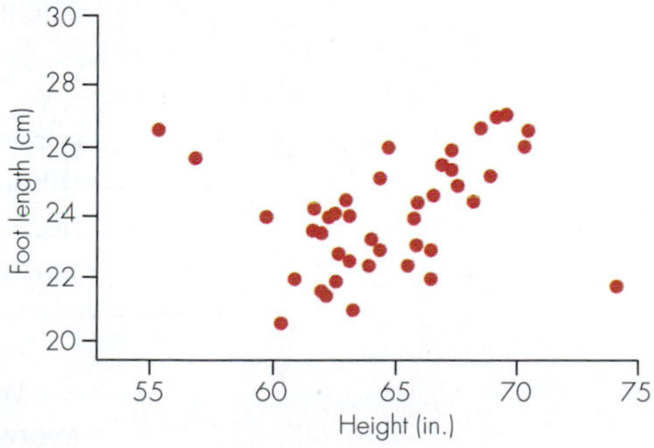

Figure 3.5 Outliers in the relationship between the height and foot length of women *(Data source: Collected in class by one of the authors.)*

MINITAB TIP **Graphing the Relationship between Two Quantitative Variables**

- To draw a scatterplot, use **Graph > Scatterplot**, and then select ***Simple***. In the dialog box, specify the columns containing the raw data for Y and X.
- To mark different subgroups with different symbols, select ***With Groups*** rather than ***Simple***. Specify the Y and X variables; then in the box labeled "Categorical variables for grouping," specify the column with the categorical variable that defines the subgroups.

THOUGHT QUESTION 3.2 Suppose you were to make a scatterplot of adult daughters' heights versus mothers' heights by collecting data on both variables from several of your female friends. You would now like to predict how tall your infant niece will be when she grows up. How would you use your scatterplot to help you make this prediction? What other variables, aside from her mother's height, might be useful for improving your prediction? How could you use these variables in conjunction with the mother's height?*

3.2 Describing Linear Patterns with a Regression Line

Scatterplots show us a lot about a relationship, but we often want specific numerical descriptions of how the response and explanatory variables are related. Imagine, for example, that we are examining the weights and heights of a sample of college women. We might want to know what the increase in average weight is for each 1-inch increase in height. Or we might want to estimate the average weight for women with a specific height, such as 5'10".

Regression analysis is the area of statistics that is used to examine the relationship between a quantitative response variable and one or more explanatory variables. A key element of regression analysis is the estimation of a **regression equation** that describes how, on average, the response variable is related to the explanatory variables. This regression equation can be used to answer the types of questions that we just asked about the weights and heights of college women.

There are many types of relationships and many types of regression equations. *The simplest kind of relationship between two variables is a straight line,* and that is the only type we will discuss here. Straight-line relationships, also called **linear relationships**, occur frequently in practice, so a straight line is a useful and important type of regression equation. Before we use a straight-line regression model, we should always examine a scatterplot to verify that the pattern actually is linear. We remind you of the music preference and age example, in which a straight line definitely does not describe the pattern of the data.

The straight line that best describes the linear relationship between two quantitative variables is called the **regression line**. Let's review the equation for a straight line relating y and x.

FORMULA

Equation for a Straight Line

The equation for a straight line relating y and x is

$$y = b_0 + b_1x$$

where b_0 is the "y-intercept" (sometimes just called "intercept") and b_1 is the slope. When $x = 0$, y is equal to the **y-intercept**. The letter y represents the vertical direction, and x represents the horizontal direction. The **slope** tells us how much the y variable changes for each increase of one unit in the x variable.

We can use available data pairs (x,y) to create a regression equation describing the average relationship between x and y, but the equation also can be used in the future to **predict** values of a response variable when we only know the values for the explanatory variable. For instance, it might be useful for colleges to have an equation for the connection between verbal SAT score and college grade point average (GPA). They could create the equation using students that have been through college already, and then use that equation to predict the potential GPAs of future students, based on their verbal SAT scores. Some colleges actually do this kind of prediction to help them de-

***HINT:** Perhaps use the approximate average of daughters' heights for mothers who are about the same height as your niece's mother.

cide whom to admit, but they use a collection of variables to predict GPA. The prediction equation for GPA usually includes high school GPA, high school rank, verbal and math SAT scores, and possibly other factors such as a rating of the student's high school or the quality of an application essay.

DEFINITION A **regression line** is a straight line that describes how values of a quantitative response variable (y) are related, on average, to values of a quantitative explanatory variable (x). The equation for the line is called the **regression equation.** A regression line is used for two purposes:

- To *estimate the average value* of y at any specified value of x
- To *predict the unknown value* of y for an individual, given that individual's x value

The term **simple linear regression** refers to methods used to analyze straight-line relationships.

Example 3.5 **Describing Height and Handspan with a Regression Line** In Figure 3.1 (p. 70), we saw that the relationship between handspan and height has a straight-line pattern. Figure 3.6 displays the same scatterplot as Figure 3.1, but now a regression line is shown that describes the average relationship between the two variables. We used statistical software (Minitab) to find the "best" line for this set of measurements. We will discuss the criterion for "best" later. For now, let's focus on what the line tells us about the data.

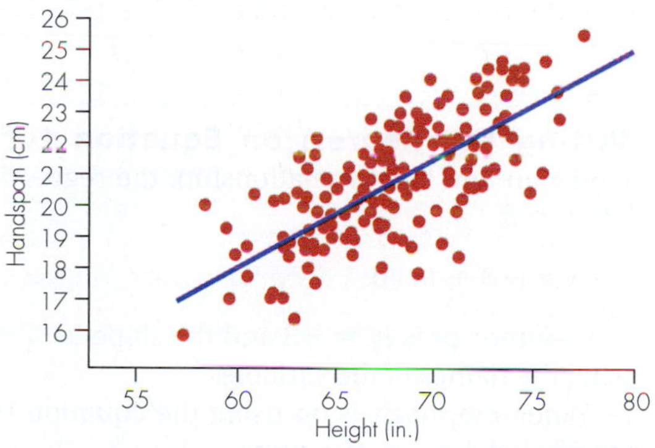

Figure 3.6 Regression line describing height and handspan

The regression line drawn through the scatterplot describes how average handspan is linked to height. For example, when the height is 60 inches, the vertical position of the line is at about 18 centimeters (cm). To see this, locate 60 inches along the horizontal axis (x axis), look up to the line, and then read the corresponding value on the vertical axis to determine the handspan value. The result is that we can estimate that people who are 60 inches tall have an *average* handspan of about 18 cm (roughly 7 inches; 1 inch = 2.5 cm). We can also use the line to *predict* the handspan for an individual whose height is known. For instance, someone who is 60 inches tall is *predicted* to have a handspan of about 18 cm.

Let's use the line to estimate the average handspan for people who are 70 inches tall. Using the regression line, we see that the handspan value corresponding to a height of 70 inches is somewhere between 21 and 22 cm, perhaps about 21.5 cm (roughly 8.5 inches). So when height is increased from 60 inches to 70 inches, average handspan increases from about 18 cm to about 21.5 cm.

The average handspan increased by 3.5 cm (about 1.5 inches) when the height was increased by 10 inches. This is a rate of $3.5/10 = 0.35$ cm per 1-inch increase in height, which is the slope of the line. For each 1-inch difference in height, there is about a 0.35-cm average difference in handspan.

The Equation for the Regression Line

Remember that for a linear regression relationship the points do not all fall exactly on the line. Therefore, we need to distinguish between the actual value of y for an individual and the value that would be predicted if the individual fell exactly on the line. We do this by defining the **predicted y,** denoted by $\hat{y}$, to be the value that falls exactly on the line for a given value of x. The **regression equation** for a **regression line** describes the relationship between x and the **predicted values** of y. It is written as:

$$\hat{y} = b_0 + b_1 x$$

- $\hat{y}$ is spoken as "y-hat," and it is also referred to as **predicted** y or **estimated** y.
- b_0 is the *intercept* of the straight line. The intercept is the value of $\hat{y}$ when $x = 0$.
- b_1 is the *slope* of the straight line. The slope tells us how much of an increase (or decrease) there is for the predicted or average value of the y variable when the x variable increases by one unit. The sign of the slope tells us whether $\hat{y}$ increases or decreases when x increases.

In any given situation, the sample is used to determine numbers that replace b_0 and b_1. They are based on a method called *least squares estimation* that is described in detail on page 80. Although formulas are provided on page 81, we generally will use statistical software to find these values.

One important note is that the relationship between x and y is not interchangeable when we use least squares estimation. If we reverse the roles and allow y to be the explanatory variable and x to be the response variable, the equation above cannot be used to describe the relationship. We would have to use statistical software (or formulas provided on page 81) to find the new relationship.

Example 3.6 **Writing the Regression Equation for Height and Handspan** For the handspan and height relationship, the regression equation determined by statistical software is

$$\hat{y} = -3 + 0.35x$$

The y-intercept is $b_0 = -3$ and the slope is $b_1 = 0.35$. We also can write the equation using the names of the variables.

When emphasis is on using the equation to estimate the average handspans for specific heights, we may write

Average handspan = −3 + 0.35 (Height)

When emphasis is on using the equation to predict an individual handspan, we might instead write

Predicted handspan = −3 + 0.35 (Height)

As examples, let's use the regression equation to estimate the average handspans for some specific heights.

For height = 60, average handspan = −3 + 0.35(60) = −3 + 21 = 18 cm

For height = 67, average handspan = −3 + 0.35(67) = −3 + 23.45 = 20.45 cm

For height = 70, average handspan = −3 + 0.35(70) = −3 + 24.5 = 21.5 cm

The handspan values just calculated for heights of 60, 67, and 70 inches can also be used to predict the handspans of any individuals with those specific heights.

Interpreting the *y*-Intercept and the Slope

In the handspan and height equation, the *y*-intercept value is $b_0 = -3$. This would be the estimated or predicted handspan for someone whose height (x) is 0 inches. Of course that has no meaning in this context, as will often be the case when interpreting the *y*-intercept. It will have a meaningful interpretation only in situations for which $x = 0$ is a reasonable value for x.

The value $b_1 = 0.35$ multiplies the height. This value is the *slope* of the straight line that links handspan and height. In general, the slope of a line measures how much the *y* variable changes per each one-unit increase in the value of the *x* variable. Consistent with our previous estimates, the slope in this example tells us that handspan increases by 0.35 cm, on average, for each increase of 1 inch in height. We can use the slope to estimate the average difference in handspan for any difference in height. If we consider two heights that differ by 7 inches, our estimate of the difference in handspans would be $7 \times 0.35 = 2.45$ cm, or approximately 1 inch.

Statistical Relationships versus Deterministic Relationships

As we have already noted, in most regression relationships the points do not all fall exactly on the line, so knowing the value of *x* does not allow us to precisely determine the value of *y*. In a **deterministic relationship**, if we know the value of one variable, we can exactly determine the value of the other variable. For example, the relationship between the volume and weight of water is deterministic. Every pint of water weighs 1.04 pounds, so we can determine exactly the weight of any number of pints of water. If we had a bucket that weighed 3 pounds and we filled it with *x* pints of water, we could determine the exact value of *y*, the weight of the filled bucket. It would be $y = 3 + 1.04x$.

In a **statistical relationship**, there is variation from the average pattern. You can see from Figure 3.6 that the regression line does not predict exactly what will happen for each individual. Most individuals do not have a handspan exactly equal to $-3 + 0.35$ (Height), the handspan that would be predicted from the regression equation.

Our ability to predict what happens for an individual depends on the amount of natural variability from the overall pattern. If most measurements are close to the regression line, we may be able to accurately predict what will happen for an individual. When there is substantial variation from the line, we will not be able to accurately predict what will happen for an individual. In Section 3.3 we will learn one way to quantify how much natural variation there is in a linear relationship.

IN SUMMARY Interpreting a Regression Line

- $\hat{y}$ estimates the average *y* for a specific value of *x*. It also can be used as a prediction of the value of *y* for an individual with a specific value of *x*.
- The **slope** of the line estimates the average or predicted increase in *y* for each one-unit increase in *x*.
- The **intercept** (or **y-intercept**) of the line is the value of $\hat{y}$ when $x = 0$. Note that interpreting the intercept in the context of statistical data makes sense only if $x = 0$ is included in the range of observed *x* values.

Example 3.7 **Regression for Driver Age and the Maximum Legibility Distance of Highway Signs** Example 3.2 (p. 71) described a study in which researchers measured the maximum distance at which an automobile driver could read a highway sign. Thirty drivers participated. The regression line $\hat{y} = 577 - 3x$ describes how the maximum sign legibility distance (the y variable) is related to driver age (the x variable). Statistical software was used to calculate this equation and to create the graph shown in Figure 3.7. Earlier, we asked these two questions about distance and age:

- How much does the distance decrease when age is increased?
- For drivers of any specific age, what is the average distance at which the sign can be read?

The slope of the equation can be used to answer the first question. Remember that the slope is the number that multiplies the x variable and the sign of the slope indicates the direction of the association. Here, the slope of -3 tells us that, on average, the legibility distance decreases 3 feet when age increases by 1 year. This information can be used to estimate the average change in distance for any difference in ages. For an age *increase* of 30 years, the estimated *decrease* in legibility distance is 90 feet because the slope is -3 feet per year.

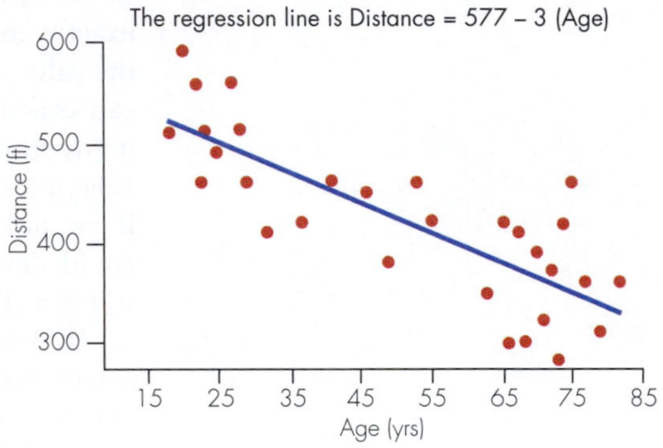

Figure 3.7 Regression line for driver age and sign legibility distance

The question about estimating the average legibility distances for a specific age is answered by using the specific age as the x value in the regression equation. To emphasize this use of the regression line, we write it as follows:

Average distance $= 577 - 3$ (Age)

Here are the results for three different ages:

Age	Average Distance
20	$577 - 3(20) = 517$ feet
50	$577 - 3(50) = 427$ feet
80	$577 - 3(80) = 337$ feet

The equation can also be used to predict the distance measurement for an *individual* driver with a specific age. To emphasize this use of the regression line, we write the equation as follows:

Predicted distance $= 577 - 3$ (Age)

For example, we can predict that the legibility distance for a 20-year-old will be 517 feet. For a 50-year-old the predicted legibility distance is 427 feet and for an 80-year-old it is 337 feet.

MINITAB TIP **Finding the Regression Line or Scatterplot with Regression Line**

- To find a simple regression equation, use **Stat > Regression > Regression**. In the dialog box, specify the column containing the raw data for the response variable (Y) as the "Response," and specify the column containing the data for the explanatory variable (X) as a "Predictor."
- To find a regression line and also have Minitab draw this line onto a scatterplot of the data, use **Stat > Regression > Fitted Line Plot**. Specify the response variable (Y) and the predictor (X) in the dialog box.

EXCEL TIP **Finding the Regression Equation**

It is easiest to explain using an example. To find the regression equation for the 13 *height* and *handspan* values shown in Table 3.1 on page 70, enter the data as shown below.

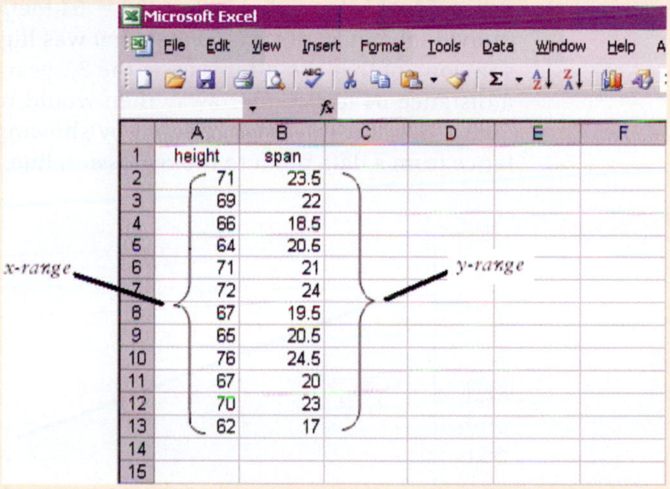

	A	B	C	D	E	F
1	height	span				
2	71	23.5				
3	69	22				
4	66	18.5				
5	64	20.5				
6	71	21				
7	72	24				
8	67	19.5				
9	65	20.5				
10	76	24.5				
11	67	20				
12	70	23				
13	62	17				
14						
15						

Then the slope and intercept are found as follows:

INTERCEPT(B2:B13, A2:A13) = −14.0128
SLOPE(B2:B13, A2:A13) = 0.514822

Note that for the intercept and slope, the *y* range must be listed first.

Prediction Errors and Residuals

Although we generally use the prediction capability of regression for future situations when only *x* is known and we want to predict *y*, we can check on how well the line works by predicting *y* for the cases in our dataset, and comparing the *predicted values* ($\hat{y}$) to the *observed y values*. To do this for any particular point, we plug the observed *x* value into the regression equation and compute $\hat{y}$. The **prediction error** for an observation is the difference between the observed *y* value and the predicted value $\hat{y}$; the formula is *error* $= (y - \hat{y})$. The terminology "error" is somewhat misleading, since the amount by which an individual differs from the line is usually due to natural variation rather than "errors" in the measurements. A more neutral term for the difference $(y - \hat{y})$ is that it is the **residual** for that individual. In Chapter 14, we will learn that the standard deviation of the residuals for a dataset is a useful measure of the "typical" difference between actual and predicted values of the *y* variable.

Example 3.8 **Prediction Errors for the Highway Sign Data** Examples 3.2 and 3.7 described a study in which y = maximum distance at which a person can read a highway sign was related to x = age. The regression equation for these data is $\hat{y} = 577 - 3x$. To calculate $\hat{y}$ for an individual, substitute his or her age for x in the equation. For individuals in the sample, an observed value of y is available, and the residual $(y - \hat{y})$ can then be found. For the first three individuals shown in Table 3.2, the *residuals,* or *prediction errors,* are calculated as follows:

x = Age	y = Distance	$\hat{y} = 577 - 3x$	Residual = $y - \hat{y}$
18	510	577 − 3(18) = 523	510 − 523 = −13
20	590	577 − 3(20) = 517	590 − 517 = 73
22	560	577 − 3(22) = 511	560 − 511 = 49

This process could be carried out for any of the 30 observations in the dataset. The seventh individual in Table 3.2, for instance, has age = 27 years and distance = 560 feet. The predicted distance for this person is $\hat{y} = 577 - 3(27) = 496$ feet, so the residual is $(y - \hat{y}) = 560 - 496 = 64$ feet. A positive residual indicates that the individual had an observed value that was higher than what would be predicted for someone of that age. In this case, the 27-year-old in the study could see the sign at a distance 64 feet farther away than would be predicted for someone of that age. Figure 3.8 illustrates this residual by showing that the residual is the vertical distance from a data point to the regression line.

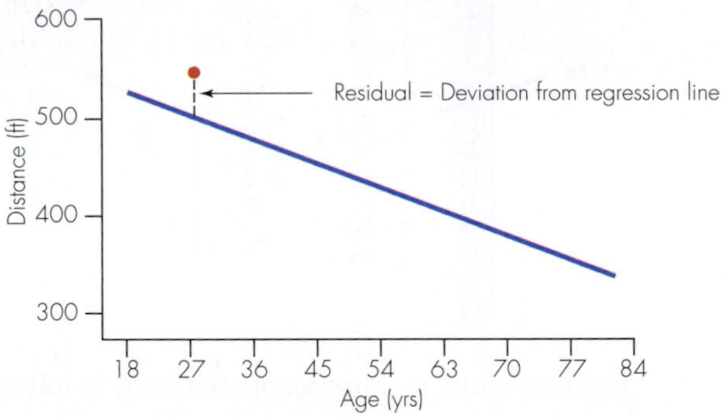

Figure 3.8 Residual from regression line for 27-year-old who saw sign at a distance of 560 feet. The residual, also called the prediction error, is the difference between observed y = 560 feet and $\hat{y}$ = 496 feet.

The Least Squares Estimation Criterion

A mathematical criterion called **least squares** is nearly always the basis for estimating the equation of a regression line. The term *least squares* is a shortened version of "least sum of squared errors." A **least squares line** or **least squares regression line** has the property that the sum of squared differences between the observed values of y and the predicted values $\hat{y}$ is smaller for that line than it is for any other line. Put more simply, the least squares line minimizes the sum of the squared prediction errors (squared residuals) for the observed data set. The notation SSE, which stands for **sum of squared errors,** is used to represent the sum of squared prediction errors. The least squares line (the regression line) has a smaller SSE than any other regression line that might be used to predict the response variable.

There is a mathematical solution that produces general formulas for computing the slope and intercept of the least squares line. These formulas are used by all statistical software, spreadsheet programs, and statistical calculators. To be complete, we include the formulas. In practice, however, regression analysis is done using a computer, so we don't include an example showing how to calculate the slope and intercept for the least squares line "by hand."

FORMULA

Formulas for the Slope and Intercept of the Least Squares Line

b_1 is the slope and b_0 is the y-intercept:

$$b_1 = \frac{\sum_i (x_i - \bar{x})(y_i - \bar{y})}{\sum_i (x_i - \bar{x})^2}$$

$$b_0 = \bar{y} - b_1\bar{x}$$

x_i represents the x measurement for the ith observation.

y_i represents the y measurement for the ith observation.

$\bar{x}$ represents the mean of the x measurements.

$\bar{y}$ represents the mean of the y measurements.

Example 3.9 **Calculating the Sum of Squared Errors** Suppose that x = score on exam 1 in a course and y = score on exam 2 and that the first two rows in Table 3.3 (shown below) give x values and y values for $n = 6$ students. For these data, the least squares regression line is $\hat{y} = 20 + 0.8x$ (found using Minitab). Values of $\hat{y}$ for all observations are given in the third row of Table 3.3, and the fourth row gives the corresponding values of the prediction errors $(y - \hat{y})$. For instance, $x = 70$ and $y = 75$ for the first observation shown in Table 3.3, so $\hat{y} = 20 + 0.8(70) = 76$ and $(y - \hat{y}) = 75 - 76 = -1$. The sum of the squared prediction errors for the regression line is

$$\begin{aligned} \text{SSE} &= (-1)^2 + (2)^2 + (-4)^2 + (2)^2 + (2)^2 + (-1)^2 \\ &= 1 + 4 + 16 + 4 + 4 + 1 = 30 \end{aligned}$$

Table 3.3 Values of x, y, $\hat{y}$, and $(y - \hat{y})$ for Example 3.9

x = Exam 1 score	70	75	80	80	85	90
y = Exam 2 score	75	82	80	86	90	91
$\hat{y} = 20 + 0.8x$	76	80	84	84	88	92
$(y - \hat{y})$	−1	2	−4	2	2	−1

The line $\hat{y} = 20 + 0.8x$ is the least squares line, so any other line will have a sum of squared errors greater than 30. As an example, if the line $\hat{y} = 4 + x$ were used to predict the values of y, the sum of squared values of $(y - \hat{y})$ would be

$$(75 - 74)^2 + (82 - 79)^2 + (80 - 84)^2 + (86 - 84)^2 + (90 - 89)^2 + (91 - 94)^2$$
$= 40$, which is obviously greater than 30.

Why Regression Is Called Regression

You may wonder why the word *regression* is used to describe the study of statistical relationships. Most of the vocabulary used by statisticians has at least some connection to the common usage of the words, but this doesn't seem to be true for *regression*. The statistical use of the word *regression* dates back to Francis Galton, who studied heredity in the late 1800s. (See Stigler, 1986 or 1989, for a detailed historical account.) One of Galton's interests was whether or not a man's height as an adult could be predicted by his parents' heights. He discovered that it could, but the relationship was such that very tall parents tended to have children who were shorter than they were and very short parents tended to have children who were taller than themselves. He initially described this phenomenon by saying that there was "reversion to mediocrity" but later changed the terminology to "regression to mediocrity." Thereafter, the technique of determining such relationships was called *regression*.

3.2 Exercises are on pages 101–102.

THOUGHT QUESTION 3.3 Suppose the statistics community is having a contest to rename *regression* to something more descriptive of what it actually does. What would you suggest as a name for the entire procedure? As a name for the regression line?*

3.3 Measuring Strength and Direction with Correlation

The linear pattern is so common that a statistic was created to characterize this type of relationship. The statistical **correlation** between two quantitative variables is a number that *indicates the strength and the direction of a straight-line relationship.*

- The *strength* of the relationship is determined by the *closeness of the points to a straight line.*
- The *direction* is determined by whether one variable generally increases or generally decreases when the other variable increases.

As used in statistics, the meaning of the word *correlation* is much more specific than it is in everyday life. A statistical correlation describes only *linear relationships.* Whenever a correlation is calculated, a straight line (the regression line) is used as the frame of reference. When the pattern is nonlinear, as it was for the music preference data shown in Figure 3.3, a correlation is not an appropriate way to measure the strength of the relationship.

Correlation is represented by the letter r. Sometimes this measure is called the **Pearson product moment correlation** or the **correlation coefficient**. Unlike for the calculation of a regression equation, it doesn't matter which of the two variables is called the x variable and which is called the y variable. The value of the correlation is the same either way. For instance, the correlation between height and foot length is the same regardless of whether you use height as the y variable or use foot length as the y variable. Another useful feature of the correlation coefficient is that its value doesn't change when the measurement units are changed for either or both of the variables. For instance, the correlation between weight and height is the same whether the measurements are in pounds and inches or in kilograms and centimeters. (Of course, this assumes the data values aren't rounded off to the nearest whole number after they are converted to a new scale of measurement.)

The formula for calculating the correlation coefficient looks complicated, although it can be described rather simply in terms of standardized scores (introduced in Section 2.7). Approximately, the correlation value is the average product of standardized scores for variables x and y. Calculating a correlation value by hand, however, generally involves much labor, so all statistical software programs and many calculators provide a way to easily calculate this statistic. In this section, we focus on how to interpret the correlation coefficient rather than how to calculate it.

FORMULA **A Formula for Correlation**

$$r = \frac{1}{n-1}\sum_i \left(\frac{x_i - \bar{x}}{s_x}\right)\left(\frac{y_i - \bar{y}}{s_y}\right)$$

n is the sample size.

x_i is the x measurement for the ith observation.

$\bar{x}$ is the mean of the x measurements.

s_x is the standard deviation of the x measurements.

y_i is the y measurement for the ith observation.

$\bar{y}$ is the mean of the y measurements.

s_y is the standard deviation of the y measurements.

***HINT:** Two purposes for using regression are given in the definition box on page 75.

Interpreting the Correlation Coefficient

Some specific features of the correlation coefficient are as follows:

- Correlation coefficients are always between −1 and +1.
- The *magnitude* of the correlation indicates the strength of the relationship, which is the overall closeness of the points to a straight line. The *sign* of the correlation does not matter when assessing the strength of the linear relationship.
- A correlation of either −1 or +1 indicates that there is a perfect linear relationship and all data points fall on the same straight line.
- The *sign* of the correlation indicates the direction of the relationship. A *positive* correlation indicates that the two variables tend to increase together (a positive association). A *negative* correlation indicates that when one variable increases, the other is likely to decrease (a negative association).
- A correlation of 0 indicates that the best straight line through the data is exactly horizontal, so knowing the value of x does not change the predicted value of y.

The following examples illustrate these features.

Example 3.10 **The Correlation Between Handspan and Height** In Example 3.1 we saw that the relationship between handspan and height appears to be linear, so a correlation is useful for characterizing the strength of the relationship. For these data, the correlation is $r = 0.74$, a value that indicates a somewhat strong *positive* relationship. Figure 3.9 (which is the same as Figure 3.1) shows us that average handspan definitely increases when height increases, but within any specific height there is some natural variation among individual handspans.

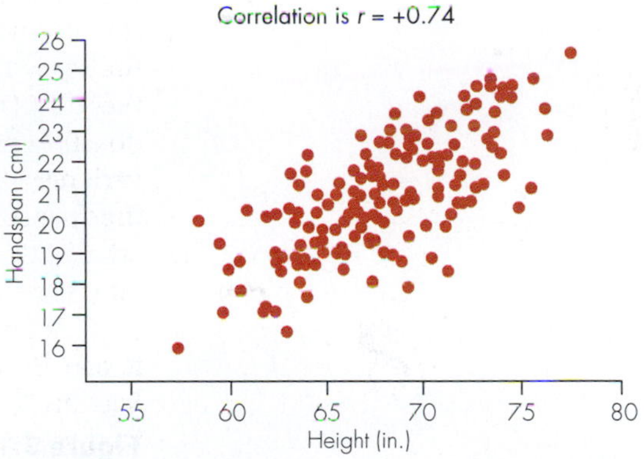

Figure 3.9 Height and handspan

Example 3.11 **The Correlation Between Age and Sign Legibility Distance** For the data shown in Figure 3.10 (which is the same as Figure 3.2) relating driver age and sign legibility distance, the correlation is $r = -0.8$. This value indicates a somewhat strong *negative* association between the variables.

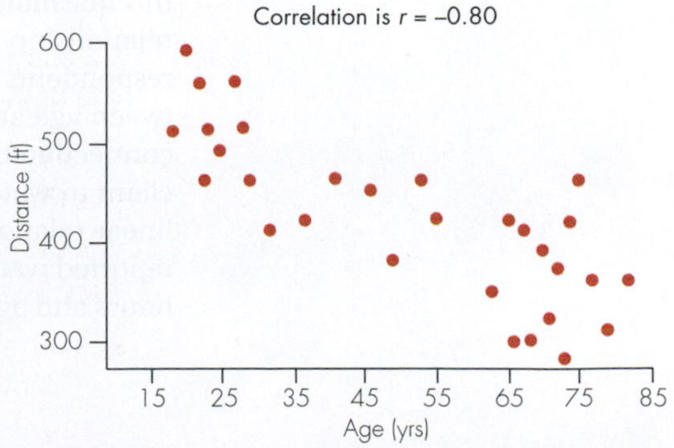

Figure 3.10 Driver age and the maximum distance at which a highway sign was read *(Source: Adapted from data collected by Last Resource, Inc., Bellefonte, PA.)*

Example 3.12 **Left and Right Handspans** If you know the span of a person's right hand, do you think you could accurately estimate his or her left handspan? Figure 3.11 displays the relationship between the right and left handspans (in cm) of the 190 college students in the dataset of Chapter 2. In the plot, the points nearly fall into a straight line. The correlation coefficient for this strong positive association is 0.95.

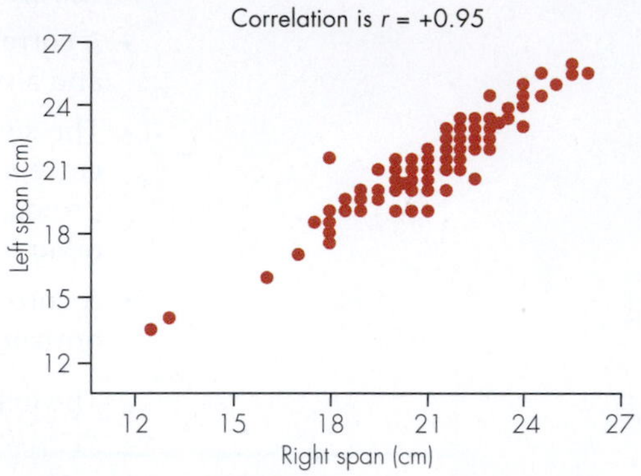

Figure 3.11 Right handspan and left handspan

Example 3.13 **Verbal SAT and GPA** The scatterplot in Figure 3.12 shows the grade point averages (GPAs) and verbal SAT scores for a sample of 100 students at a university in the northeastern United States. The correlation for the data in the scatterplot is $r = 0.485$, a value that indicates only a moderately strong relationship.

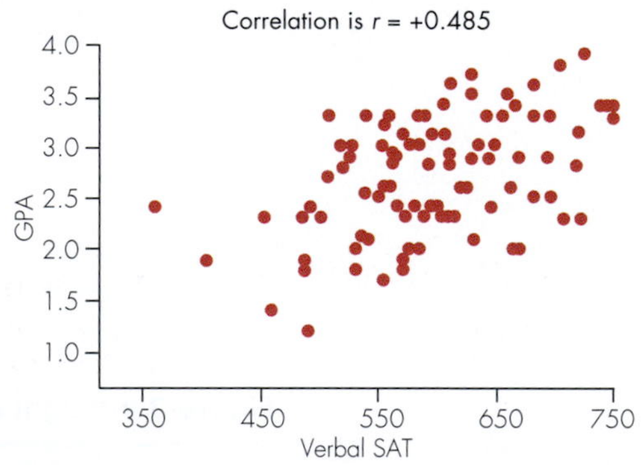

Figure 3.12 Verbal SAT and grade point average

Example 3.14 **Age and Hours of Television Watching per Day** On a typical day, how many hours do you spend watching television? The National Opinion Research Center asks this question in its General Social Survey. In Figure 3.13 (on the next page), we see the relationship between respondent age and hours of daily television viewing for 1299 respondents in the 2008 survey. There does not seem to be much of a relationship between age and television hours, and the correlation of only 0.136 confirms this weak connection between the variables. We also see some odd responses. Four respondents claim to watch television 24 hours per day! Given these outliers, it is not clear that a linear relationship is appropriate at all. But even if we restrict the data to those who reported watching 14 hours or less of television a day the correlation between viewing hours and age remains weak, at 0.171.

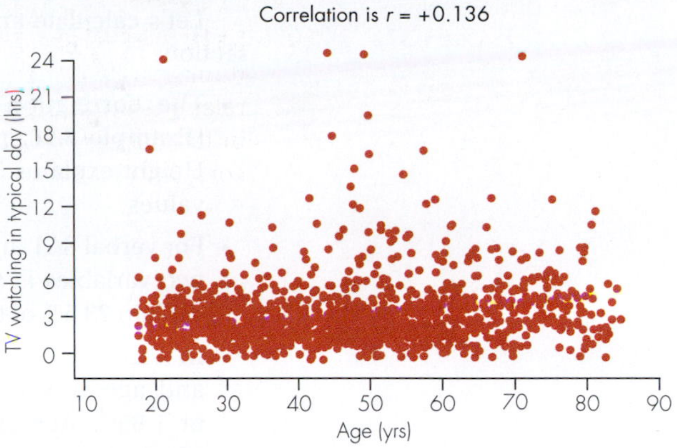

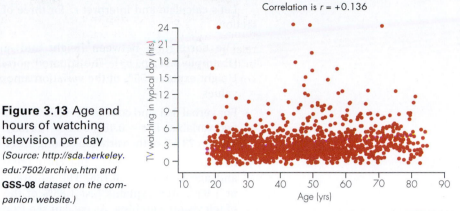

Figure 3.13 Age and hours of watching television per day *(Source: http://sda.berkeley.edu:7502/archive.htm and* ***GSS-08*** *dataset on the companion website.)*

Example 3.15 **Hours of Sleep and Hours of Study** Figure 3.14 displays, for a sample of 116 college students, the relationship between the reported hours of sleep during the previous 24 hours and the reported hours of study during the same period. The correlation value for this scatterplot is $r = -0.36$, indicating a negative association that is not particularly strong. On average, the hours of sleep decrease as hours of study increase, but there is substantial variation in the hours of sleep for any specific hours of study.

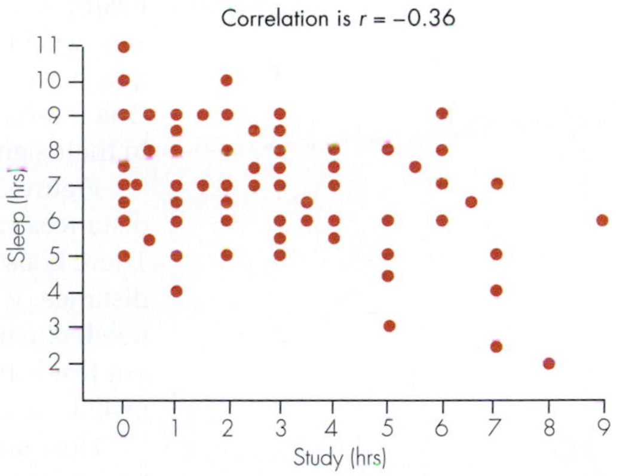

Figure 3.14 Hours of study and hours of sleep *(Source: Class data collected by one of the authors.)*

Interpreting the Squared Correlation, r^2

The squared value of the correlation can also be used to describe the strength of a linear relationship. A **squared correlation, r^2**, always has a value between 0 and 1, although some computer programs will express its value as a percent between 0 and 100%. By squaring the correlation, we retain information about the strength of the relationship, but we lose information about the direction.

The phrase "**proportion of variation explained by *x***" is sometimes used in conjunction with the squared correlation, r^2. For example, if a correlation has the value $r = 0.5$, the squared correlation is $r^2 = (0.5)^2 = 0.25$, or 25%, and a researcher may write that the explanatory variable explains 25% of the variation among observed values of the response variable. This interpretation stems from the use of the least squares line as a prediction tool, and will be explained in detail in Example 3.16 below.

Let's calculate and interpret r^2 for three of the examples given previously in this section.

- The correlation between height and stretched right handspan is $r = 0.74$ (Example 3.10, p. 83). The squared correlation is $r^2 = (0.74)^2 = 0.55$, or 55%. Height explains 55% of the variation among observed stretched right handspan values.
- For verbal SAT and college GPA (Example 3.13, p. 84), the correlation between the two variables is $r = 0.485$, so $r^2 = (0.485)^2 = 0.235$, or 23.5%. Verbal SAT scores explain 23.5% of the variation among observed GPAs.
- In Example 3.14 (p. 84), the correlation between television watching hours and age is only $r = 0.136$. The squared correlation is $r^2 = (0.136)^2 = 0.0185$, or 1.85%. Age explains just 1.85% of the variation among observed amounts of television watching. As we can see from Figure 3.13, knowing a person's age does not help us much in predicting how much television the person watches per day.

Formula for r^2

Although r^2 is the correlation coefficient squared, it can be described using a completely different rationale. This explanation will illustrate why it is sometimes called the proportion of the variation in the y's "explained" by knowing x. Let's illustrate the rationale and the corresponding formula with an example.

Example 3.16 **How Much Variability in Vision Is Explained by Age?** We learned in Example 3.7 that the distance at which drivers can see a highway sign decreases with age, and in Example 3.11 we saw that the relationship was fairly strong, with a correlation between age and sign legibility distance of -0.8. That means $r^2 = 0.64$, or 64%. One interpretation of this is that knowing drivers' ages "explains" 64% of the variability in their sign-reading distances. What does this interpretation mean?

Figure 3.15 (on the next page) shows a scatterplot of the ages and sign-reading distances for the 30 drivers in the data set, with the regression line illustrated in blue. There is also a black line shown at a distance value of 423.3 feet. That value is the mean distance, $\bar{y}$, for the 30 drivers. If we had no additional information about a driver, but needed to predict sign-reading distance, the best we could do is use this mean value as our prediction. In other words, our regression line would be a flat line with y-intercept of 423.3 and slope of 0.

How much better are we able to predict sign-reading distance with the benefit of knowing someone's age? Figure 3.15 illustrates the answer to this question for two of the drivers in the sample. One of the drivers was 20 years old and could see the sign at a distance of 590 feet. That's a full 166.7 feet more than we would have predicted if we relied only on the mean distance of 423.3 feet as our prediction. In fact, knowing that the driver was 20 years old, we would have predicted the distance to be 517 feet, the point on the regression line for $x = 20$. Comparing the two predictions of 423.3 feet (without knowledge of age) and 517 feet (with knowledge of age) shows that we were partially able to explain why this driver could see so much farther than the average. In fact, of the total error of 166.7 feet in our naïve prediction, knowing that the person was 20 years old helped to explain $(\hat{y} - \bar{y}) = (517 - 423.3) = 93.7$ feet. However, the residual $(y - \hat{y}) = (590 - 517) = 73$ feet remains unexplained even when we know that the driver was 20 years old.

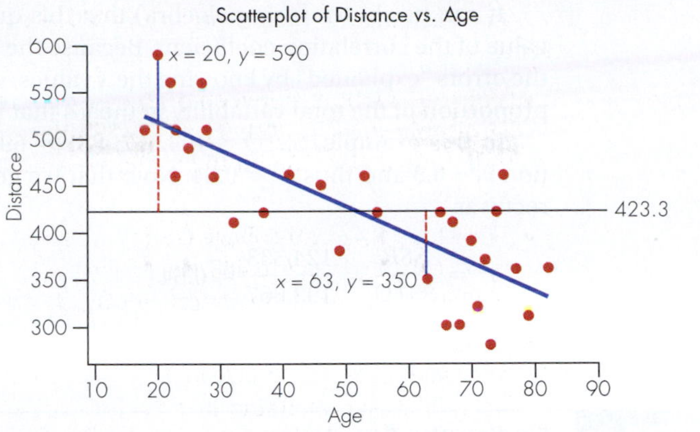

Figure 3.15 Illustration of explained error (dashed red lines) and unexplained error (solid blue lines) for interpreting r^2

We can do a similar calculation for each driver. We identify the "unexplained" residual, shown by the vertical blue line segment in Figure 3.15, as the *prediction error* for that driver. The distance between the mean and the regression line, illustrated by the dashed red line segments in Figure 3.15, is *error explained by regression.* The sum of these two sources of error is called the *total error* for that individual.

For the two individuals shown in Figure 3.15, the total error is as follows:

Age, Distance	**Total Error $y - \bar{y}$**	**Unexplained Residual $y - \hat{y}$**	**Explained by Regression $\hat{y} - \bar{y}$**
20, 590	590 − 423.3 = 166.7	590 − 517 = 73	517 − 423.3 = 93.7
63, 350	350 − 423.3 = −73.3	350 − 388 = −38	388 − 423.3 = −35.5

To summarize these errors over the entire sample, we square them to get rid of negative signs, and then sum them over all of the individuals. We define the results as follows:

- Total errors: The sum of squared differences between observed y values and the sample mean $\bar{y}$ is called the *total variation* in y or **sum of squares total** and is denoted by **SSTO**.
- Unexplained residuals (prediction errors): The sum of squared differences between observed y values and the predicted values based on the regression line is called the sum of squared errors and is denoted by *SSE*. This is the same SSE the least squares estimation criterion minimizes, as explained on page 80.
- Errors explained by regression: The sum of squared differences between the sample mean and the predicted values is called **sum of squares due to regression** and is denoted by **SSR**.

Although it is not immediately obvious, through algebra you can verify that SSTO = SSE + SSR. The squared correlation, r^2, can be calculated using SSTO, SSR and SSE as follows:

$$r^2 = \frac{\text{SSTO} - \text{SSE}}{\text{SSTO}} = \frac{\text{SSR}}{\text{SSTO}}$$

It can be shown (using algebra) that this quantity is exactly equal to the squared value of the correlation coefficient. Because the numerator is the sum of the portion of the errors "explained" by knowing the *x* values, we have illustrated that r^2 is indeed the proportion of the total variability in the *y*'s that is explained by the *x*'s.

In this example, SSTO = 193,667, SSE = 69,334 and SSR = 124,333. The correlation is −0.8 and thus $r^2 = 0.64$. Note that we get the same value if we compute r^2 directly as

$$r^2 = \frac{SSR}{SSTO} = \frac{124{,}333}{193{,}667} = 0.64.$$

MINITAB TIP **Finding the Correlation**

To calculate a correlation coefficient, use **Stat > BasicStatistics > Correlation**. Specify two or more columns as variables.

Reading Computer Results for Regression

Many statistical computer packages are available that will do all of the regression calculations for you. Figure 3.16 illustrates the basic results of using the statistical package Minitab for the data in Figure 3.11 (p. 84). The explanatory variable is right handspan, and the response variable is left handspan.

```
Session

The regression equation is
LftSpan = 1.46 + 0.938 RtSpan


Predictor      Coef   SE Coef      T      P
Constant     1.4635    0.4792   3.05  0.003
RtSpan      0.93830   0.02252  41.67  0.000

S = 0.638615    R-Sq = 90.2%    R-Sq(adj) = 90.2%

Analysis of Variance

Source           DF      SS      MS        F      P
Regression        1  708.15  708.15  1736.38  0.000
Residual Error  188   76.67    0.41
```

Figure 3.16 Minitab regression results for the data shown in Figure 3.11

When we revisit regression in Chapter 14, the computer results will become clearer. For now, you should be able to recognize the following features:

- Regression equation: $\hat{y} = b_0 + b_1x = 1.46 + 0.938x$, where x = right handspan
- Slope: $b_1 = 0.938$, expanded in another part of the display to 0.93830
- Intercept: $b_0 = 1.46$, expanded in another part of the display to 1.4635
- $r^2 = 0.902$, or 90.2%
- SSR = 708.15
- SSE = 76.67

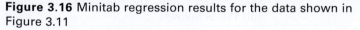

The value of the correlation coefficient r can be found using the value of "R-Sq" along with the sign of the slope. The calculation is $r = \sqrt{r^2} = \sqrt{0.902} = 0.95$. We know the correlation coefficient has a positive sign because the value of the slope is positive ($b_1 = 0.938$).

3.3 Exercises are on pages 102–103.

THOUGHT QUESTION 3.4 Sometimes the main purpose of a regression analysis is to determine the nature of the relationship between two variables, and sometimes the main purpose is to use the equation in the future to predict a y value when the x value is known. Explain which purpose is likely to be the main reason for a regression analysis between

x = percent fat consumed in diet, y = blood pressure

x = SAT score, y = college grade point average

x = height at age 4, y = height at age 21

x = hours of sleep per night, y = score on IQ test*

3.4 Regression and Correlation Difficulties and Disasters

Each of the following actions will cause misleading regression and correlation results:

- Extrapolating too far beyond the observed range of x values
- Allowing outliers to overly influence the results
- Combining groups inappropriately
- Using correlation and a straight-line equation to describe curvilinear data

Extrapolation

It is risky to use a regression equation to predict values outside the range of the observed data, a process called **extrapolation.** There is no guarantee that the relationship will continue beyond the range for which we have observed data. Suppose that a sample of adult men is used to estimate a regression equation that relates weight to height and the equation of the line is Weight = $-180 + 5$ (Height). This equation should work well in the range of heights that we see in adult men, but it will not describe the weights of children. If we use the equation to estimate the weight of a boy who is 36 inches tall, the answer is $-180 + 5(36) = 0$ pounds. The straight-line equation developed for adult men doesn't accurately describe the connection between the weights and heights of children.

Extrapolation also is an issue when regression methods are used to predict future values of a y variable when the x variable is related to calendar time. For instance, a straight line describes the relationship between y = winning time in Olympic women's 100-meter backstroke swim and x = Olympic year. This straight line could be used to predict the winning time in the near future, but it should not be used to predict the time in the year 3000.

It is quite acceptable to use a regression equation for **interpolation,** in which y values are estimated or predicted for new values of x that were not in the original dataset, but are in the range of values covered by the x's in the dataset. For example, if you are between the ages of 18 and 82 you could use the equation found in Example 3.6 to predict how far away you would be able to read the highway sign used in that study. Your age might not be the same as one of the drivers in the study, but the equation can be used for any age in the range of x values used to obtain it, which is 18 to 82 years old.

***HINT:** Colleges use SAT scores to predict GPAs of applicants, so predicting a future y value is more likely in that instance.

The Influence of Outliers

Earlier in this chapter we learned that an outlier in regression is a data point that has an unusual *combination* of *x* and *y* values. These outliers can have an impact on correlation and regression results. This is particularly true for small samples. In Example 3.4, we learned that sometimes outliers occur because mistakes are made when the data are recorded or are entered into the computer. In these cases, we may be able to make the necessary corrections. When outliers are legitimate data, we have to carefully consider their effect on the analysis. We may exclude outliers from some analyses, but we shouldn't forget about them. As in everyday life, the unusual data is often the most interesting data.

Outliers with extreme *x* values have the most *influence* on correlation and regression and are called **influential observations**. Depending on whether these points line up with the rest of the data, they can either deflate or inflate a correlation. An outlier at an extreme *x* value can also have a big effect on the slope of the regression line.

Example 3.17 **Height and Foot Length of College Women** Figure 3.5 (p. 73) displayed a scatterplot of the foot lengths and heights of 41 college women. We saw three outliers in that plot, all of which occurred because heights were incorrectly entered into the computer. If we do not correct these mistakes, the correlation between the foot lengths and heights in Figure 3.5 is only $r = 0.28$. For the corrected data set, the correlation is $r = 0.69$, a markedly higher value. The outliers also have a big effect on the equation of the least squares line.

- For the uncorrected dataset, foot length $= 15.4 + 0.13$ (Height)
- For the corrected dataset, foot length $= -3.2 + 0.42$ (Height)

The slope of the correct line is more than three times the size of the slope of the line for the incorrect data. This is a big difference. For instance, let's consider a 12-inch difference in heights. The correct estimate of the associated difference in average foot lengths is $12 \times 0.42 \approx 5$ cm (about 2 inches). If we use the incorrect data, the estimated difference in average foot lengths is only $12 \times 0.13 \approx 1.6$ cm (about 5/8 of an inch).

Example 3.18 **Earthquakes in the Continental United States** Table 3.4 lists the major earthquakes that occurred in the continental United States between 1850 and 2009. These include all earthquakes that were at least magnitude 7.0 and/or in which there were at least 20 fatalities. The correlation between deaths and magnitude for these 14 earthquakes is 0.26, showing a somewhat weak positive association. It implies that, on average, higher death tolls accompany stronger earthquakes, which seems logical. However, if you examine the scatterplot of the data shown in Figure 3.17, you will notice that the positive correlation is entirely due to the famous San Francisco earthquake of 1906. In fact, for the remaining earthquakes, the trend is actually *reversed*, as the scatterplot in Figure 3.18 shows. Without the 1906 quake, the correlation for these 13 earthquakes is strongly *negative*, at -0.824, indicating that fewer deaths are associated with *greater* magnitudes.

Table 3.4 Major Earthquakes in Continental United States, 1850–2009

Date	Location	Deaths	Magnitude
January 9, 1857	Fort Tejon, CA	1	7.9
October 21, 1868	Hayward, CA	30	6.8
March 26, 1872	Owens Valley, CA	27	7.4
August 31, 1886	Charleston, SC	60	6.6
April 18–19, 1906	San Francisco, CA	503	7.8
March 10, 1933	Long Beach, CA	115	6.2
May 19, 1940	Imperial Valley, CA	9	7.1
April 13, 1949	Puget Sound, WA	8	7.1
July 21, 1952	Kern County, CA	12	7.3
August 18, 1959	Hebgen Lake, MT	28	7.3
February 9, 1971	San Fernando Valley, CA	65	6.6
October 17, 1989	Loma Prieta, CA	62	6.9
June 28, 1992	Landers, CA	3	7.3
January 17, 1994	Northridge, CA	60	6.7

Source: U.S. Geological Survey, http://earthquake.usgs.gov/earthquakes/states/historical.php.

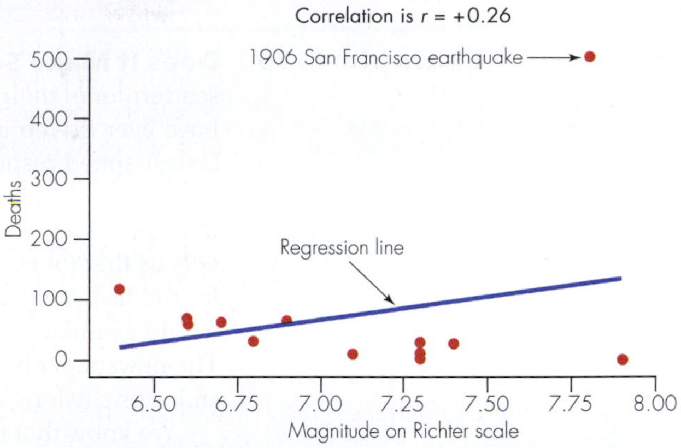

Figure 3.17 Earthquake magnitude and deaths

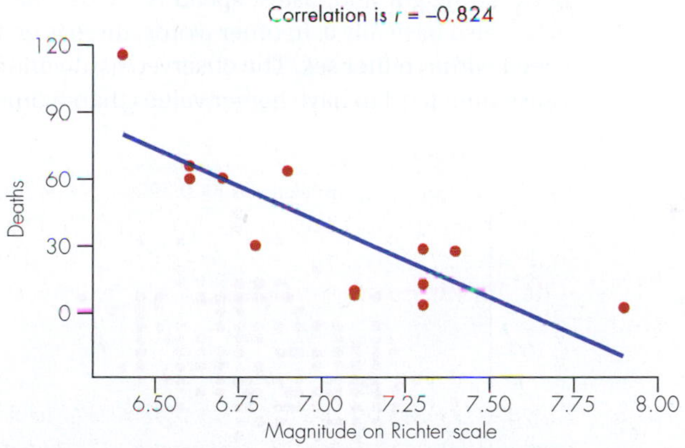

Figure 3.18 Earthquake magnitude and deaths, without 1906 San Francisco earthquake

Clearly, trying to interpret the correlation between magnitude and death toll for this small group of earthquakes is a misuse of statistics. The San Francisco earthquake in 1906 occurred before earthquake building codes were enforced. Many of the largest earthquakes occurred in very sparsely populated areas.

Inappropriately Combining Groups

The next example demonstrates that combining two distinctly different groups may cause illegitimate results.

Example 3.19 **Does It Make Sense? Height and Lead Feet** For a sample of college students, a scatterplot of their heights and their responses to the question "What is the fastest you have ever driven a car?" is displayed in Figure 3.19. Height is the *x* variable, and the fastest-speed response is the *y* variable. We see that the fastest-speed response tends to increase as height increases. The correlation is +0.39, and the least squares line that describes the average pattern is $\hat{y} = -20 + 1.7x$. The slope of the equation tells us that for every 1-inch increase in height, there is an average increase of 1.7 mph for the fastest-speed response. This means that for a 12-inch difference in heights, we would estimate the difference in the fastest-speed response to be $12 \times 1.7 \approx 20$ mph. The newspaper headline might read "Height and Lead Foot Go Together." Is this a sensible conclusion? Why might these results be misleading?

We know that men tend to be taller than women. Those of you with good memories may recall Case Study 1.1, which indicated that men tend to claim a higher fastest speed than women do. These sex differences could be causing the positive association that we see in Figure 3.19. One way to examine this possibility is to look separately at men and women. Figure 3.20 shows two scatterplots, one for each group. For men, the correlation between height and fastest speed is −0.01—basically 0. For women, the correlation is 0.04—also basically 0. In other words, there is no relationship between height and fastest speed within either sex. The observed association in the combined data occurs only because men tend to have higher values than women do for both variables.

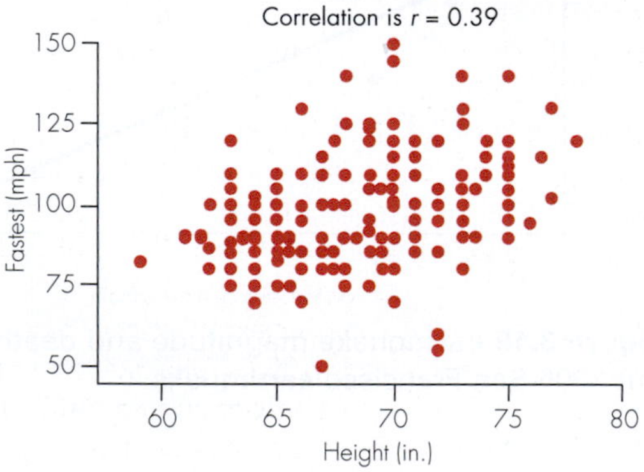

Figure 3.19 Height and the fastest that college students have ever driven *(Data source: Class data collected in 1998 by one of the authors.)*

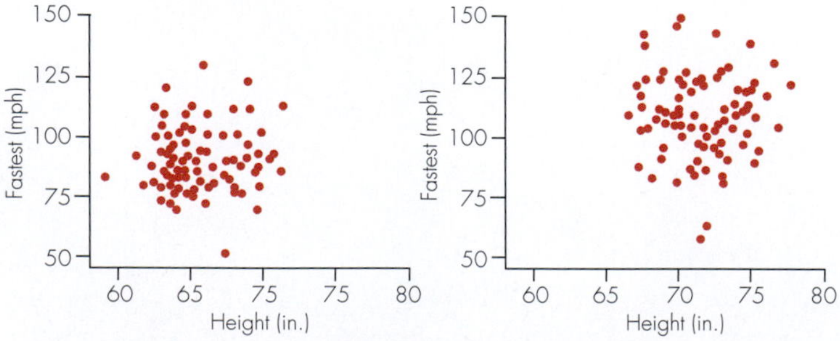

Figure 3.20 Fastest speed and height separately for men and women

The height and fastest-speed example demonstrates a common mistake that can lead to an illegitimate correlation, which is combining two or more groups when the groups should be considered separately. In Chapter 6 and Exercise 3.89, we will see an example that describes a relationship between the cost of a book and the number of pages in the book. Surprisingly, there seems to be a negative relationship. As the number of pages increases, the cost tends to decrease. The data include hardcover and softcover books, and this explains the unexpected negative association. Hardcover books generally have fewer pages than softcover books, but the cost of hardcover books is higher because they are more likely to be limited distribution technical books and textbooks. When the book types are considered separately, the association between pages and cost is positive for each type.

Sometimes a similar association exists whether groups are combined or examined separately. In Figure 3.4, we saw that the positive association between handspan and height holds for each sex. The correlation for the combined group is 0.74, but within each sex, the correlation is about 0.6. Combining males and females does inflate the correlation somewhat, but sex differences do not completely account for the observed association.

Curvilinear Data

The next example shows us that it is important to look at a scatterplot before we calculate a regression line. When the data are curvilinear, predictions based on a straight line are likely to be inaccurate.

Example 3.20

Table 3.5 U.S. Population (millions) in Census Years since 1790

Year	Pop.	Year	Pop
1790	3.9	1900	76.2
1800	5.3	1910	92.2
1810	7.2	1920	106.0
1820	9.6	1930	123.2
1830	12.9	1940	132.2
1840	17.1	1950	151.3
1850	23.2	1960	179.3
1860	31.4	1970	203.3
1870	38.6	1980	226.5
1880	50.2	1990	248.7
1890	63.0	2000	281.4

Source: U.S. Census Bureau, http://www.census.gov/population/www/censusdata/files/table-16.pdf, for 1790 to 1990.

Does It Make Sense? U.S. Population Predictions Table 3.5 lists the population of the United States (in millions) for each census year between 1790 and 2000. There is, of course, a positive association between y = population size and x = year, because the population size has been steadily increasing through the years. The correlation between population size and year is $r = +0.96$, indicating a very strong relationship. The least squares line for these data has the equation $\hat{y} = -2348 + 1.289$ (Year). If we use this equation to predict the population in 2030, our estimate is about 269 million. Does this estimate make sense? In February 2010, the U.S. population was estimated to be 308.7 million, already notably higher than our linear regression prediction of the population 20 years later. You can see what the estimated population is at the time you are reading this at http://www.census.gov/main/www/popclock.html.

Your first thought may be that we should not extrapolate a prediction to 2030 because that's too far past the end of the data. That's a good thought, but the extrapolation issue is not the biggest problem here. Why does the regression line produce such a poor estimate for 2030? The reason is that the pattern of population growth is actually *curved,* so a straight-line equation isn't the right type of equation to use. Figure 3.21 shows the situation. The least squares line is shown, as well as a curve that is a much better fit for the actual data pattern. Using the curve, the estimate of the population for the year 2030 is 363.5 million. The U.S. Census Bureau projects a population size of 363.58 million in the year 2030, essentially the same as the prediction based on the curve in Figure 3.21 (http://www.census.gov/population/www/projections/usinterimproj/natprojtab02a.pdf).

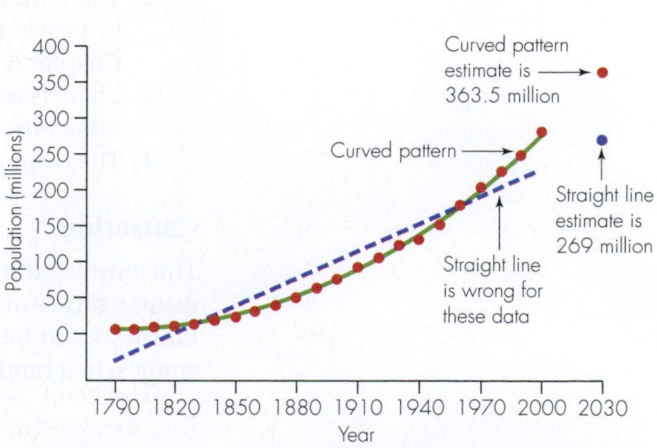

Figure 3.21 Estimating the U.S. population in 2030 with a line or a curve

The U.S. population example illustrates that we can make big mistakes if we use a straight line to describe curved data. We should describe a curvilinear pattern with an equation for a curve. This is easier said than done. There are many different types of equations that describe curves, and it is hard to judge which type we should use just by looking at the scatterplot.

The music-preference and age relationship in Example 3.3 illustrates another way in which correlation can be misleading when the pattern is curvilinear. For the data shown in Figure 3.3, the correlation will be around 0. This value could make us believe that there is no relationship, but in fact there is. Remember that the frame of reference for calculating a correlation is a straight line. For the inverted U pattern in Figure 3.3, the best straight line will be nearly horizontal, so the correlation will be around 0. Clearly, a straight line should not be used to describe those data, so the correlation value is meaningless.

3.4 Exercises are on pages 104–105.

THOUGHT QUESTION 3.5 Sketch a scatterplot with an outlier that would inflate the correlation between the two variables. Sketch a scatterplot with an outlier that would deflate the correlation between the two variables.*

3.5 Correlation Does Not Prove Causation

The saying "correlation does not imply causation" is used so frequently that you may already have encountered it in everyday life or in another academic course. It is easy to construct silly, obvious examples of observed associations that don't have a causal connection. For example, there would be a positive correlation between weekly flu medication sales and weekly coat sales for an area with extreme seasons because values of both variables would tend to be large in the winter and small in the summer.

In most situations, the explanation for an observed relationship is not as obvious as it is in the flu medication and coat sales example. Suppose, for example, that the finding in an observational study is that people who use vitamin supplements get fewer colds than do people who don't use vitamin supplements. One possible explanation is that the use of vitamin supplements causes a reduced risk of a cold. It is easy, however, to think of other explanations for the observed association. Perhaps those who use supplements also sleep more and it is the sleep difference that is causing the difference in the frequency of colds. Or perhaps the extra glass of water required to take the vitamins makes a difference.

Interpretations of an Observed Association

There are at least four possible interpretations of an observed association between an explanatory variable and a response variable.

1. There is causation. The explanatory variable is causing a change in the response variable.
2. There may be causation, but confounding factors make this causation difficult to prove. (Confounding variables are defined on p. 5 and in more detail in Chapter 6.)
3. There is no causation. The observed association can be explained by how one or more other variables affect both the explanatory and response variables.
4. The response variable is causing a change in the explanatory variable.

Causation

The most legitimate way to establish a causal connection statistically is to collect data using a randomized experiment. We learned in Chapter 1 that cause-and-effect relationships can be inferred from randomized experiments but not from observational studies. In a randomized experiment, there is random assignment of the experimental

***HINT:** Examples 3.4 and 3.18 might be helpful.

units to specific values of the explanatory variable. Because the treatments are randomly assigned to the units, the values of confounding variables should approximately even out across treatment groups. This reduces the chances that an observed association is due to confounding variables, even those confounding variables that we have neglected to measure. We will learn more about this topic in Chapter 6.

Confounding Factors Make Causation Difficult to Prove

The data from an observational study, in the absence of any other evidence, simply cannot be used to establish causation. It is nearly impossible to separate the effect of confounding variables from the effect of the explanatory variable. For example, even if we observe that smokers tend to have higher blood pressure than nonsmokers, we cannot definitively say that smoking causes high blood pressure. There may be alternative explanations. Perhaps smokers are more stressed than nonsmokers and this causes the higher blood pressure.

Other Variables May Explain the Association Between the Explanatory and Response Variables

The association between height and fastest speed in Example 3.19 (p. 92) had this interpretation. Sex differences in both height and the fastest speed ever driven explained the observed association. As another example of an observed association that does not imply causation, recall the association between verbal SAT scores and college GPAs exhibited in Figure 3.12 (p. 84). Almost certainly, higher SAT scores do not directly cause higher grades in college. However, the causes responsible for verbal SAT being high (or low) may be the same as those responsible for college GPA being high (or low). Those causes might include things such as intelligence, motivation, and ability to perform well on tests.

A common situation that may produce a misleading association between two variables is when they are both changing over time. For example, suppose we were to measure x = number of sodas sold in the year and y = number of divorces in the year for years from 1950 to 2010. We would almost certainly see a strong relationship because the population has increased substantially over those years, and thus the numbers for both variables have steadily increased. It is more appropriate to use rates such as number of divorces per 1000 marriages when comparing data across time.

A whimsical situation for which strong correlations can be found is when the x and y variables both are related to the size of the unit for which they are observed. For instance, if we were to measure the number of teachers and the number of bars for cities in California, we would find a strong correlation. This is because large cities have many more teachers and bars than smaller cities. Again, it would make more sense to measure these characteristics on a *per capita* basis.

The Response Variable Is Causing a Change in the Explanatory Variable

Sometimes the causal connection is the opposite of what might be expected or claimed. For instance, suppose that an observational study finds, for men and women over 60 years old, that regular church attendance is associated with better health. Should we conclude that church attendance causes better health? An alternative explanation is that the causation may be in the opposite direction. Healthy people are more able to attend church, so good health may increase the likelihood of church attendance.

3.5 Exercises are on pages 105–106.

THOUGHT QUESTION 3.6 An article in the *Centre Daily Times* (April 19, 1997, p. 8A) included data from the United States and several European countries that indicated a negative correlation between the cost of cigarettes and annual per capita cigarette consumption. Does this result mean that if the United States increased its cigarette tax to increase the price of cigarettes, the result would be that people would smoke less? What are some other explanations for the negative correlation between cigarette price and annual cigarette consumption?*

***HINT:** Interpretation 4 (p. 94) might provide one explanation.

SKILLBUILDER APPLET

3.6 Exploring Correlation

The **Correlation** *applet described in this section is available on the companion website,* http://www.cengage.com/statistics/Utts4e.

The **Correlation** applet on the website accompanying this book can be used to explore how the correlation coefficient, *r*, is related to the strength and direction of the relationship between two quantitative variables. Remember that the strength of a linear relationship is measured by the absolute value of the correlation coefficient. The sign of the correlation value indicates whether the two variables have a positive association or a negative association.

What Happens

Your goal is to create a scatterplot so that the correlation value for the points on the graph is close to a "goal" value. You place points onto a graph by using the mouse to click on locations in a graph. The applet recalculates and displays the correlation value after each point is added. The applet will declare "Goal Reached!" when the correlation is within ± 0.05 of the goal value after 15 or more points have been added to the plot.

What to Do

Open the **Correlation** applet. You will see three different scatterplot regions, each with a different goal value for the correlation. Figure 3.22 shows the first of these regions before any points have been added. For that region, the goal is $r = 0.5$, so your task is to create a graph with at least 15 data points for which the correlation is anywhere between 0.45 and 0.55.

Begin adding points to the graph by using the mouse to click on desired locations. An example of how the applet might look after five data points have been added is displayed in Figure 3.23. Note that for the five points shown in the figure, the correlation is $r = 0.6087$. Continue adding points until the goal is reached. You can delete points by clicking on the ***Delete*** radio button and then clicking on the point you wish to delete. The ***Clear!*** button can be used to remove all points.

The correlation goals for the second and third scatterplots on the applet page are $r = -0.8$ and $r = 0$, respectively. Create scatterplots that achieve those targets. In each instance, first try achieving the correlation by making your *x* values range from about 10 to 90 and not including any outliers. Remember that at least 15 points should be placed on a graph. Figures 3.10, 3.12, and 3.20 in this chapter provide models for the three target correlations of -0.8, 0.5 and 0, respectively.

After reaching your goals, clear the plots and then explore how an outlier affects a correlation value. For instance, try reaching the $r = 0.5$ goal for the first scatterplot by adding 14 points for which the correlation is above 0.8, and then adding an outlier that brings the correlation down to about 0.5. Or, where the goal is $r = -0.8$, put 14 points in the upper left portion of the graph so that the correlation for those points is between -0.2 and $+0.2$, and then add an outlier that makes the correlation become about -0.8.

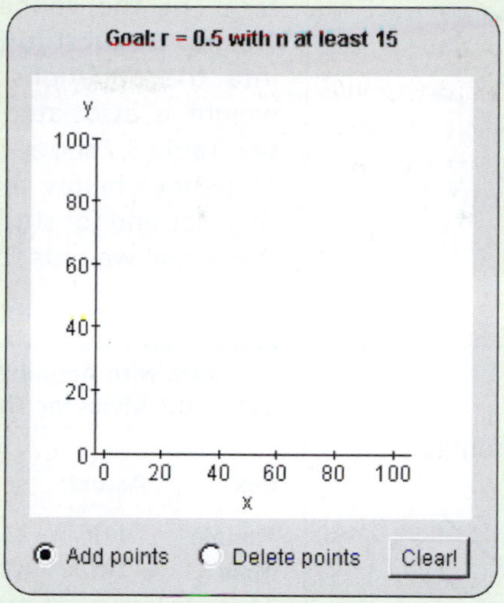

Figure 3.22 Starting point for the **Correlation** applet. The goal is $r = 0.5$, and at least 15 points should be placed on the graph.

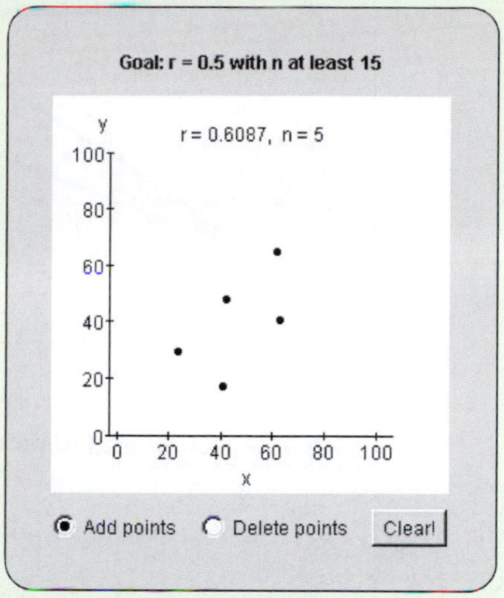

Figure 3.23 Appearance of **Correlation** applet after five points have been added. For these five points, the correlation is $r = 0.6087$, a value shown at the top of the scatterplot.

Lessons Learned

The three target correlations exemplify how correlation is related to the strength and direction of a relationship. The algebraic sign of the correlation value gives the direction of the relationship. For linear relationships with no outliers present in the data, the absolute value of the correlation measures the strength of a relationship, which has to do with the overall closeness of points to a line. By using the applet to put outliers on the plots, you can see that an outlier may either increase or decrease a correlation.

3.6 Exercises are on pages 106–109.

CASE STUDY 3.1 A Weighty Issue

In a large statistics class, students (119 females and 63 males) were asked to report their actual and ideal weights. It is well known that males and females differ with regard to actual weights and their views of their weight, so the two groups should be separated for the analysis. Table 3.6 displays the mean actual and ideal weights for men and women. For women, the mean ideal is 10.7 pounds less than the mean actual, while for men, the mean ideal is only about 2.5 pounds less than mean actual.

We can use a scatterplot and regression to learn more about the connection between actual and ideal weight. Figure 3.24 (next page) shows a scatterplot of the two variables for the females, and Figure 3.25 is the same plot for the males. Each point represents one student (or multiple students with the same values), whose ideal weight can be read on the vertical axis and actual weight can be read on the horizontal axis.

If everyone had responded that his or her ideal weight was the same as his or her actual weight, all points would fall on a line with the following equation (shown in blue in Figures 3.24 and 3.25 on the next page):

Ideal = Actual

That line is drawn in each figure. Most of the women fall below that line, indicating that their ideal weight is *below* their actual weight. The situation is not as clear for the men, but a pattern is still evident. The majority of men weighing less than 175 pounds would prefer to weigh the same as or more than they do, and they fall on or above the line. The majority of men weighing over 175 pounds fall on or below the line and would prefer to weigh the same or less than they do.

Table 3.6 Mean Actual and Ideal Weights by Sex (in pounds)

	Actual	Ideal	Difference
Females ($n = 119$)	132.8	122.1	10.7
Males ($n = 63$)	176.1	173.6	2.5

(continued)

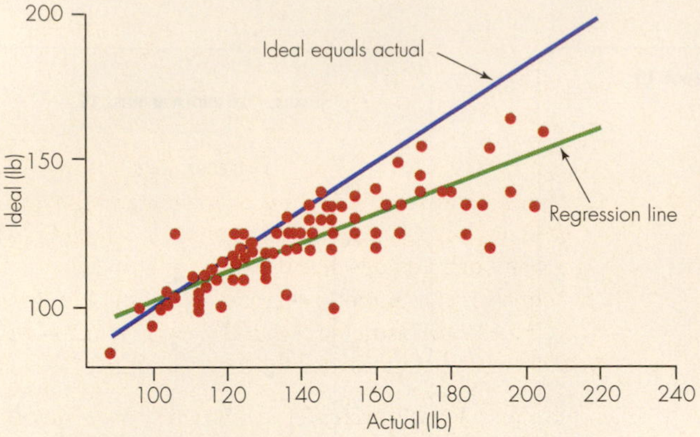

Figure 3.24 Actual and ideal weight: females

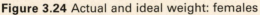

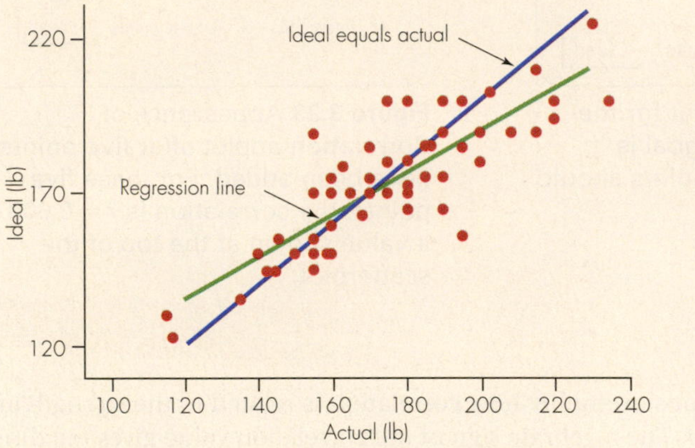

Figure 3.25 Actual and ideal weight: males

The least squares regression line is also shown on each scatterplot. The approximate regression equations follow:

Women: Average ideal = 44 + 0.6 Actual
Men: Average ideal = 53 + 0.7 Actual

The regression equations tell us the "average pattern" of the connection between actual and ideal weight. By substituting some different actual weights into the equations, we can explore how the ideal weight is associated with the actual weight for each sex. Table 3.7 shows regression calculations for students 15 pounds below the mean of the actual weights for their sex and for students 15 pounds above the mean of the actual weights for their sex.

Table 3.7 Regression Estimates of Ideal Weight

Students with Actual Weight 15 Pounds Below the Mean for Their Sex:

Sex	Actual	Ideal Based on Regression	Average Preference
Female	118	$44 + 0.6(118) \approx 115$	Lose 3 pounds
Male	161	$53 + 0.7(161) \approx 166$	Gain 5 pounds

Students with Actual Weight 15 Pounds Above the Mean for Their Sex:

Sex	Actual	Ideal Based on Regression	Average Preference
Female	148	$44 + 0.6(148) \approx 133$	Lose 15 pounds
Male	191	$53 + 0.7(191) \approx 187$	Lose 4 pounds

The results in Table 3.7 reveal interesting sex differences. For instance, consider women who weigh 118 pounds, which is about 15 pounds less than the mean weight for women. On average, their ideal weight is about 115 pounds, which is 3 pounds less than their actual weight. On the other hand, men who weigh 15 pounds less than the mean for men, on average would like to gain about 5 pounds. Women who weigh 15 pounds more than the mean for women would like to lose about 15 pounds. Men who weigh 15 pounds more than the mean for men would like to lose only about 4 pounds.

Key Terms

In Summary Box

Interpreting a Regression Line, 77

Exercises

◆ Denotes that the dataset is available on the companion website, http://www.cengage.com/statistics/Utts4e, but is not required to solve the exercise.

Bold exercises have answers in the back of the text.

Section 3.1

Skillbuilder Exercises

3.1 For each of the following pairs of variables, is there likely to be a positive association, a negative association, or no association? Briefly explain your reasoning.

a. Amount of alcohol consumed and performance on a test of coordination, where a high score represents better coordination.

b. Height and grade point average for college students.

c. Weight of a car and average number of miles it can go on a gallon of gas.

3.2 For each of the following pairs of variables, is there likely to be a positive association, a negative association, or no association? Briefly explain your reasoning.

a. Miles of running per week and time for a 5-kilometer run.

b. Forearm length and foot length.

c. Grade level and height for children in grades 1 through 10.

3.3 ◆ The figure for this exercise is a scatterplot of y = average math SAT score in 1998 versus x = percent of graduating seniors who took the test that year for the 50 states and the District of Columbia. The data are from the **sats98** dataset on the companion website.

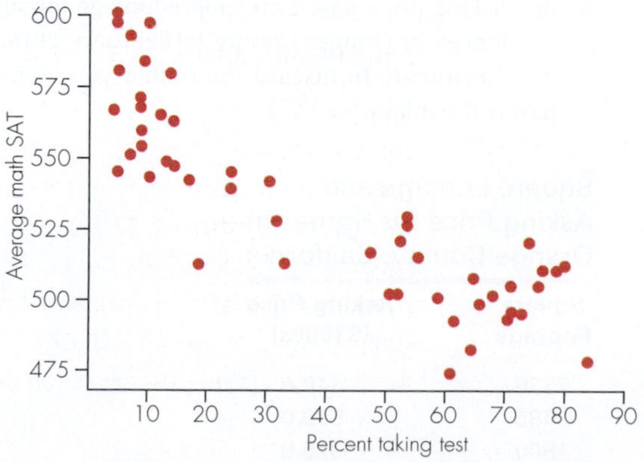

a. Does the plot show a positive association, a negative association, or no association between the two variables? Explain.

b. Explain whether you think the pattern of the plot is linear or curvilinear.

c. About what was the highest average math SAT for the 50 states and District of Columbia? Approximately, what percent of graduates took the test in that state?

d. About what was the lowest average math SAT for the 50 states and District of Columbia? Approximately what percent of graduates took the test in that state?

3.4 ◆ The figure for this exercise is a scatterplot of y = head circumference (cm) versus x = height (inches) for the 30 females in the **physical** dataset on the companion website.

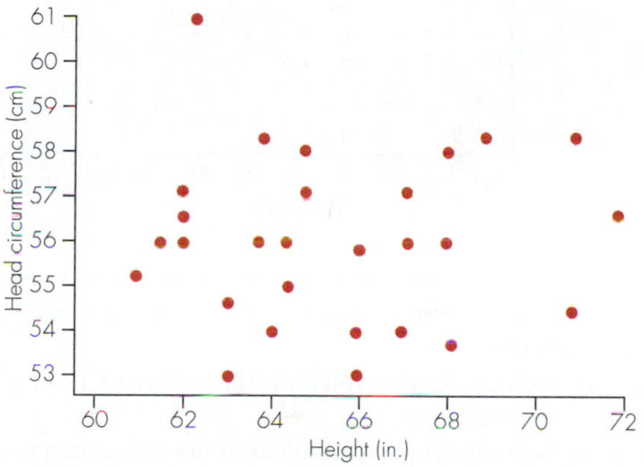

a. Does the plot show a positive association, a negative association, or no association between the two variables? Explain.

b. One data point appears to be an outlier. What are the approximate values of height and head circumference for that point?

3.5 Identify whether a scatterplot would or would not be an appropriate visual summary of the relationship between the following variables. In each case, explain your reasoning.

a. Blood pressure and age.

b. Region of country and opinion about stronger gun control laws.

3.6 Identify whether a scatterplot would or would not be an appropriate visual summary of the relationship between the following variables. In each case, explain your reasoning.

a. Verbal SAT score and math SAT score.

b. Handspan and sex (male or female).

General Section Exercises

3.7 The following table shows the relationship between the speed of a car (mph) and the average stopping distance (feet) after the brakes are applied:

Speed (mph)	0	10	20	30	40	50	60	70
Distance (ft)	0	20	50	95	150	220	300	400

Source: Defensive Driving: Managing Time and Space, *American Automobile Association, Pamphlet #3389, 1991.*

a. In the relationship between these two variables, which is the response variable (y) and which is the explanatory variable (x)?

b. Draw a scatterplot of the data. Characterize the relationship between stopping distance and speed.

3.8 ◆ The figure for this exercise is a scatterplot of y = pulse rate after marching in place for 1 minute versus x = resting pulse rate measured before marching in place. (The data are in the **pulsemarch** dataset on the companion website.)

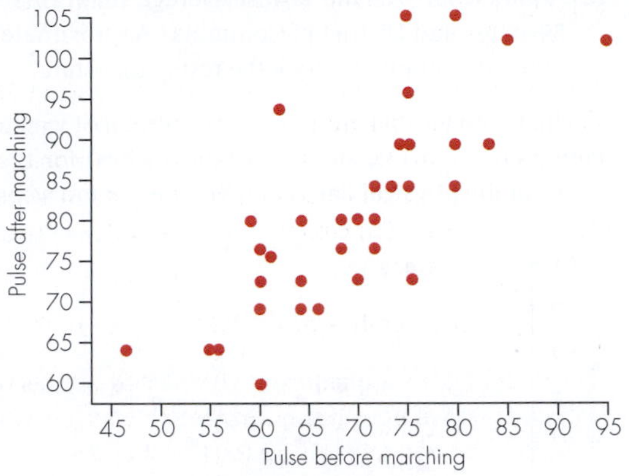

a. Does the plot show a positive association, a negative association, or no association between the two variables? Explain.

b. Explain whether you think the pattern of the plot is linear or curvilinear.

c. According to MayoClinic.com normal resting pulse rates for adults range from about 60 to 100, and for well-conditioned athletes they range from about 40 to 60. Using this information, explain whether there are any outliers in the scatterplot that are obvious mistakes. If there are outliers, describe where they are located on the plot.

3.9 ◆ The data in the following table are the geographic latitudes and the average August and January temperatures (Fahrenheit) for 20 cities in the United States. The cities are listed in geographic order from south to north. (These data are part of the **temperature** dataset on the companion website.)

Latitude and Mean Temperatures in Aug. and Jan.

City	Latitude	August Temperature	January Temperature
Miami, FL	26	83	67
Houston, TX	30	82	50
Mobile, AL	31	82	50
Phoenix, AZ	33	92	43
Dallas, TX	33	85	54
Los Angeles, CA	34	75	58
Memphis, TN	35	81	40
Norfolk, VA	37	77	39
San Francisco, CA	38	64	49
Baltimore, MD	39	76	32
Kansas City, MO	39	76	28
Washington, DC	39	74	31
Pittsburgh, PA	40	71	25
Cleveland, OH	41	70	25
New York, NY	41	76	32
Boston, MA	42	72	29
Syracuse, NY	43	68	22
Minneapolis, MN	45	71	12
Portland, OR	46	69	40
Duluth, MN	47	64	7

Data source: The World Almanac and Book of Facts, *1999, pp. 220, 456. Reprinted by permission.*

a. Draw a scatterplot of y = August temperature versus x = latitude.

b. Is the pattern linear or curvilinear? What is the direction of the association?

c. Are there any cities that appear to be outliers because they don't fit the pattern of the rest of the data? If so, which city or cities are they?

3.10 ◆ Refer to the latitude and temperature data in the table presented in Exercise 3.9, which also appear in the **temperature** dataset on the companion website.

a. Draw a scatterplot of y = January temperature versus x = latitude.

b. Is the pattern linear or curved?

c. Is the direction of the association positive or negative? Is this direction what you would expect for these data? Explain.

d. Are there any cities that appear to be outliers because they don't fit the pattern of the rest of the data? If so, which city or cities are they?

3.11 The data in the following table show the square footage and asking price (in thousands of dollars) for nine homes for sale in Orange County, California in February 2010. Orange County has a mixture of residential areas, including suburban neighborhoods and exclusive beachfront properties.

a. In the relationship between square footage and asking price, which is the response variable (y) and which is the explanatory (x) variable?

b. Draw a scatterplot of the data.

c. There is an obvious outlier in the data. Refer to the reasons for outliers described in Section 2.6, and explain which one of the reasons is the mostly likely cause of the outlier in this situation.

d. If you wanted to establish a regression equation to predict asking price based on square footage for suburban residences in Orange County in February 2010, would it be legitimate to discard the outlier you identified in part (c)? Explain.

Square Footage and Asking Price for Homes in Orange County, California

Square Footage	Asking Price ($1000s)
2336	448.0
2485	500.0
1800	325.0
1300	499.0
2700	589.9
1881	745.0
2100	574.9
2200	569.0
5500	1600.0

3.12 ◆ The following table shows sex, height (inches), and mid-parent height (inches) for a sample of 18 college students. The variable mid-parent height is the average of mother's height and father's height. (These data are in the dataset **UCDchap3** on the companion website; they are sampled from the larger dataset **UCDavis2**.)

Sex, Height, and Mid-Parent Height for 18 College Students

Sex	Height	Mid-Parent Height
M	71	64.0
F	60	63.5
F	66	67.0
M	70	64.5
F	65	65.5
F	66	69.5
M	74	72.5
F	67	67.5
F	63	65.5
M	67	64.0
F	69	70.0
M	65	63.0
M	72	69.0
M	68	67.0
F	63	63.0
F	61	63.0
M	74	69.5
F	65	67.5

a. In the relationship between height and mid-parent height, which variable is the response variable (y) and which is the explanatory variable (x)?

b. Draw a scatterplot of the data for the y and x variables defined in part (a). Use different symbols for males and females.

c. Briefly interpret the scatterplot. Does the association appear to be linear? What are the differences between the males and females? Which points, if any, are outliers?

d. Calculate the difference between height and mid-parent height for each student, and draw a scatterplot of y = difference versus x = mid-parent height. Use different symbols for males and females. What does this graph reveal about the connection between height and mid-parent height?

Section 3.2

Skillbuilder Exercises

3.13 Suppose that a regression equation for the relationship between y = weight (pounds) and x = height (inches) for men aged 18 to 29 years old is

$$\text{Average weight} = -250 + 6\,(\text{Height})$$

a. Estimate the average weight for men in this age group who are 70 inches tall.

b. What is the slope of the regression line for average weight and height? Write a sentence that interprets this slope in terms of how much average weight changes when height is increased by 1 inch.

3.14 Refer to Exercise 3.13 in which a regression equation is given that relates average weight and height for men in the 18- to 29-year-old age group.

a. Suppose a man in this age group is 72 inches tall. Use the regression equation given in the previous exercise to predict the weight of this man.

b. Suppose this man, who is 72 inches tall, weighs 190 pounds. Calculate the residual (prediction error) for this individual.

3.15 ◆ Refer to the scatterplot for Exercise 3.3 showing the relationship between the average math SAT score and the percentage of high school graduates taking the test for the 50 states and District of Columbia. (The data are from the **sats98** dataset on the companion website.) The regression line for these data is

$$\text{Average math} = 575 - 1.11\,(\text{Percent took})$$

a. The slope of the equation is −1.11. Interpret this value in the context of how average math SAT changes when the percent of graduates taking the test changes.

b. In Missouri, only 8% of graduates took the SAT test. What is the predicted average math SAT score for Missouri?

c. In 1998, the average math SAT score for Missouri was 573. What is the residual (prediction error) for Missouri?

3.16 A school cafeteria has a salad bar that is priced based on weight, with salads costing 30 cents an ounce. Students fill a container that weighs 8 ounces when it is empty. Define x to be the weight of the filled container (in ounces) and y to be the price the student is charged (in dollars). The relationship is $y = -2.40 + 0.30x$.

a. Is the relationship between x and y a deterministic relationship or a statistical relationship? Explain.

b. Explain what it would mean if $x = 8$.

c. Does the y-intercept have a useful interpretation in this example? Explain.

d. Explain the meaning of the slope of 0.30 in this situation.

e. If the container plus ingredients weighs 20 ounces, how much does the salad cost the student?

3.17 The equation for converting a temperature from x = degrees Celsius to y = degrees Fahrenheit is $y = 32 + 1.8x$. Does this equation describe a statistical relationship or a deterministic relationship? Briefly explain your answer.

General Section Exercises

3.18 ◆ The average August temperatures (y) and geographic latitudes (x) of 20 cities in the United States were given in the table for Exercise 3.9. (The data are part of the **temperature** dataset on the companion website.) The regression equation for these data is

$$\hat{y} = 113.6 - 1.01x$$

a. What is the slope of the line? Interpret the slope in terms of how the mean August temperature is affected by a change in latitude.

b. Estimate the mean August temperature for a city with latitude of 32.

c. San Francisco has a mean August temperature of 64, and its latitude is 38. Use the regression equation to estimate the mean August temperature in San Francisco, and then calculate the prediction error (residual) for San Francisco.

3.19 A regression equation for y = handspan (cm) and x = height (inches) was discussed in Section 3.2. If the roles of the variables are reversed and only women are considered, the regression equation is

$$\text{Average height} = 51.1 + 0.7\,(\text{Handspan}).$$

a. Interpret the slope of 0.7 in terms of how height changes as handspan increases.

b. What is the estimated average height of women with a handspan of 20 cm?

c. Molly has a handspan of 20 cm and is 66.5 inches tall. What is the prediction error (residual) for Molly?

3.20 Imagine a regression line that relates y = average systolic blood pressure to x = age. The average blood pressure for people 30 years old is 120, while for those 50 years old the average is 130.

a. What is the slope of the regression line?

b. What is the estimated average systolic blood pressure for people who are 34 years old?

3.21 Iman (1994) reports that for professional golfers, a regression equation relating x = putting distance (in feet) and y = success rate (in percent) based on observations of distances ranging from 5 feet to 15 feet is

$$\text{Success rate} = 76.5 - 3.95\,(\text{Distance})$$

a. What percentage of success would you expect for these professional golfers if the putting distance was 10 feet?

b. Explain what the slope of 3.95 means in terms of how success changes with distance.

3.22 ◆ The figure for Exercise 3.8 is a scatterplot of pulse rate after marching in place for 1 minute (y) versus resting pulse rate measured before marching (x) for n = 63 individuals. (The data are in the **pulsemarch** dataset on the companion website.) The regression equation for these data is

$$\text{Pulse after marching} = 17.8 + 0.894\,(\text{Resting pulse})$$

a. What is the slope of this equation? Write a sentence that interprets this slope in the context of this situation.

b. Predict the pulse rate after marching for somebody with a resting pulse rate of 50 beats per minute.

c. Predict the pulse rate after marching for somebody with a resting pulse rate of 90 beats per minute.

d. Use the results of parts (b) and (c) to draw the regression line. Clearly label the axes of your graph.

3.23 ◆ Refer to Exercise 3.22.

a. Predict the pulse rate after marching for somebody with a resting pulse rate of 70.

b. Suppose the pulse rate after marching is 76 for somebody whose resting pulse rate is 70. What is the residual (prediction error) for this individual?

3.24 ◆ The average January temperatures (y) and geographic latitudes (x) of 20 cities in the United States were given in the table for Exercise 3.9. (The data are part of the **temperature** dataset on the companion website.) The regression equation for these data is

$$\hat{y} = 1.26 - 2.34x$$

a. What is the slope of the line? Interpret the slope in terms of how mean January temperature is related to change in latitude.

b. Pittsburgh, Pennsylvania, has a latitude of 40, and Boston, Massachusetts, has a latitude of 42. Use the slope to predict the difference in expected average January temperatures for these two cities. Compare your answer to the actual difference in average January temperature for these two cities using the data shown in the table for Exercise 3.9.

c. Predict the average January temperature for a city with latitude 33.

d. Refer to part (c). Identify the two cities in the table that have a latitude of 33 and compute the residual (prediction error) for each of these cities. Discuss the meaning of these two residuals in the context of this example, identifying whether each city is warmer or cooler than predicted.

3.25 The data for this exercise are as follows:

x	1	2	3	4
y	4	10	14	16

a. Determine the sum of squared errors (SSE) for each of the following two lines:

Line 1: $\hat{y} = 3 + 3x$

Line 2: $\hat{y} = 1 + 4x$

b. By the least squares criterion, which of the two lines is better for these data? Why is it better?

3.26 The least squares regression equation for the data in the following table is $\hat{y} = 5 + 2x$.

x	4	4	7	10	10
y	15	11	19	21	29

a. Calculate the value of $\hat{y}$ for each data point.

b. Calculate the sum of squared errors for this equation.

Section 3.3

Skillbuilder Exercises

3.27 Which of the numbers 0, 0.25, −1.7, −0.5, and 2.5 could not be values of a correlation coefficient? In each case, explain why.

3.28 Remember that r^2 can be expressed as a proportion or as a percent. (When written as a percent, the percent sign will always be included.)

a. Explain which of the following could not be a value for r^2: 0, −0.25, 0.3, 1.0, 1.7, 25%, −50%, +200%.

b. Refer to the values in part (a). Which one of the legitimate values for r^2 represents the strongest relationship between x and y?

3.29 For $n = 188$ students, the correlation between y = fastest speed ever driven and x = number randomly picked between 1 and 10 is about $r = 0$. Describe what this correlation indicates about the association between the fastest speed driven and picking a number between 1 and 10.

3.30 Suppose the value of r^2 is 100% for the relationship between two variables.

a. What is indicated about the strength of the relationship?

b. What are the two possible values for the correlation coefficient for the two variables?

3.31 ◆ For 19 female bears, the correlation between x = length of the bear (inches) and y = chest girth (inches) is $r = 0.82$. (*Data source:* **bears-female** dataset on the companion website.)

a. Describe how chest girth will change when length is increased.

b. Assuming that there are no outliers and the relationship is linear, explain what the correlation indicates about the strength of the relationship.

c. If the measurements were made in centimeters rather than inches, what would be the value of the correlation coefficient?

3.32 Which implies a stronger linear relationship: a correlation of +0.4 or a correlation of −0.6? Briefly explain.

3.33 In Figure 3.11 (p. 84), we observed that the correlation between the left and right handspans of college students was 0.95. The handspans were measured in centimeters. What would be the correlation if the handspans were converted to inches? Explain.

General Section Exercises

3.34 Explain how two variables can have a perfect curved relationship yet have zero correlation. Draw a picture of a set of data meeting those criteria.

3.35 Suppose two variables have a deterministic linear relationship with a positive association. What is the value of the correlation between them?

3.36 Sketch a scatterplot showing data for which the correlation is $r = -1$.

3.37 The figure for this exercise (see next figure) shows four graphs. Assume that all four graphs have the same numerical scales for the two axes. Which graph shows the strongest relationship between the two variables? Which graph shows the weakest?

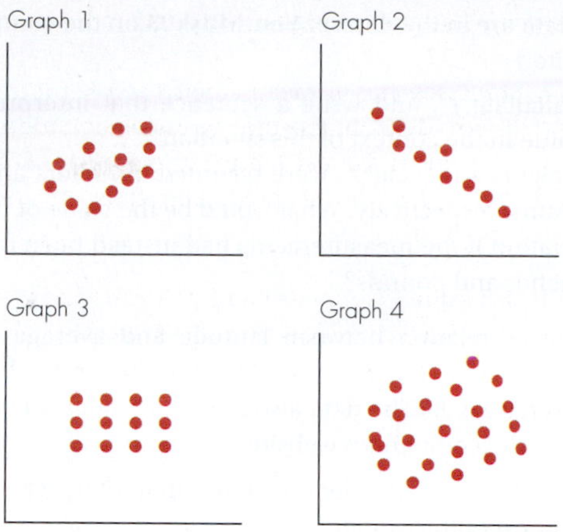

3.38 Refer to the figure for the previous exercises. In scrambled order, correlation values for these four graphs are −0.9, 0, +0.3, and +0.6. Match these correlation values to the graphs.

3.39 In the 1996 General Social Survey, the correlation between respondent age and hours of daily television viewing for $n = 1913$ respondents was $r = +0.12$. Using this value, characterize the nature of the relationship between age and hours of television watching in 1996.

3.40 ◆ The dataset **bodytemp** on the companion website gives age in years and body temperature in degrees Fahrenheit for 100 blood donors ranging in age from 17 to 84 years old. A scatterplot showed a linear relationship with a correlation between age and body temperature of −0.313. Using this value, characterize the relationship between age and body temperature.

3.41 For each pair of variables, identify whether the pair is likely to have a positive correlation, a negative correlation, or no correlation. Briefly indicate your reasoning.

a. Hours of television watched per day and grade point average for college students.

b. Number of liquor stores and number of ministers in Pennsylvania cities.

c. Performance on a strength test and age for people between 40 and 80 years old.

3.42 For each pair of variables, identify whether the pair is likely to have a positive correlation, a negative correlation, or no correlation. Briefly indicate your reasoning.

a. Verbal skills and age for children under 12 years old.

b. Height of husband and height of wife.

c. Number of dogs and number of fire hydrants for cities in New York State.

d. Average number of bicycles per household and average January temperature for cities in the United States.

3.43 ◆ The correlation between height and weight is $r = 0.40$ for 12th-grade male respondents ($n = 1501$) in a survey done in 2003 by the U.S. Centers for Disease Control and Prevention as part of the Youth Risk Behavior Surveillance System. (The raw data are in the dataset **YouthRisk03** on the companion website.)

a. Calculate r^2, and write a sentence that interprets this value in the context of this situation.

b. Heights and weights were recorded in meters and kilograms, respectively. What would be the value of the correlation if the measurements had instead been made in inches and pounds?

3.44 ◆ The correlation between latitude and average August temperature is −0.78 for the 20 cities shown in the table for Exercise 3.9. (The data also are in the dataset **temperature** on the companion website.)

a. Calculate r^2 and write a sentence that interprets it in the context of this situation.

b. If temperature were to be converted to Centigrade (without rounding off) what would be the value of the correlation between latitude and temperature?

3.45 Calculate r^2 for Example 3.15 in this chapter (about hours of sleep and hours of study) in which the correlation is −0.36. Write a sentence that interprets this value.

3.46 ◆ Refer to Exercise 3.12 and the table for Exercise 3.12 in which heights and mid-parent heights are given for 18 college students (*Data source:* **UCDchap3** dataset on the website for this book). Draw a scatterplot for the data, using different symbols for males and females as instructed in part (b) of Exercise 3.12. Based on the scatterplot, would you say that the correlation between height and mid-parent height is higher for the females in the sample or for the males? Or are the correlation values about the same for males and females? Explain your reasoning.

3.47 In a regression analysis, the total sum of squares (SSTO) is 800, and the error sum of squares (SSE) is 200. What is the value for r^2?

3.48 ◆ The data in the table for Exercise 3.12 show the heights and average of parents' heights, called "mid-parent height," for each of 18 college students. The data are also in the file **UCDchap3** where the mother's and father's heights are provided for each student as well. Student's height can be predicted using any of the three possible explanatory variables of mother's height, father's height, and mid-parent height. The following table shows the values of SSR, SSE, and SSTO for the three possible explanatory variables when used in a regression relationship with y = student's height as the response variable.

Explanatory variable	SSR	SSE	SSTO
Mother's height	84	200	284
Father's height	88	196	284
Mid-parent height	122.4	161.6	284

a. Explain why SSTO has the same value no matter which explanatory variable is used.

b. Without doing any calculations, use the information provided to explain which explanatory variable "explains" the most variability in students' heights.

c. Calculate r^2 for the model using father's height as the explanatory variable and write a sentence that interprets this value.

3.49 Suppose you know that the slope of a regression line is $b_1 = +3.5$. Based on this value, explain what you know and do not know about the strength and direction of the relationship between the two variables.

3.50 The average January temperatures (y) and geographic latitudes (x) of 20 cities in the United States were given in the table for Exercise 3.9. The regression equation for these data was given in Exercise 3.24 as $\hat{y} = 1.26 - 2.34x$. The value of r^2 for this relationship is 73.3%. What is the correlation between average January temperature and latitude for these 20 cities?

Section 3.4

Skillbuilder Exercises

3.51 An article in the *Sacramento Bee* (May 29, 1998, p. A17) noted, "Americans are just too fat, researchers say, with 54 percent of all adults heavier than is healthy. If the trend continues, experts say that within a few generations virtually every U.S. adult will be overweight." This prediction is based on extrapolation, which assumes that the current rate of increase will continue indefinitely. Is that a reasonable assumption? Do you agree with the prediction? Explain.

3.52 ◆ The **physical** dataset on the companion website gives heights (inches) and head circumferences (cm) for a sample of college students. For females only, the correlation between the two variables is 0.05, while for males only, the correlation is 0.19. For the combined sample of males and females, however, the correlation is 0.42. Explain why the correlation in the combined sample is higher than the correlations in the separate samples of males and females. Refer to Example 3.19 for guidance.

3.53 Sketch a scatterplot in which the presence of an outlier decreases the observed correlation between the response and explanatory variables. Indicate on your plot which point is the outlier.

3.54 Sketch a scatterplot in which the correlation without an outlier is negative, but the correlation when the outlier is added is positive. Indicate on your plot which point is the outlier.

General Section Exercises

3.55 Refer back to Exercise 3.7 about stopping distance and vehicle speed. The least squares line for these data is

$$\text{Average distance} = -44.2 + 5.7\,(\text{Speed})$$

a. Use this equation to estimate the average stopping distance when the speed is 80 miles per hour. Do you think this is an accurate estimate? Explain.

b. Draw a scatterplot of the data, as instructed in Exercise 3.7(b). Use the scatterplot to estimate the average stopping distance for a speed of 80 mph.

c. Do you think the data on stopping distance and vehicle speed shown in Exercise 3.7 describe the relationship between these two variables for all situations? What are some other variables that should be considered when the relationship between stopping distance and vehicle speed is analyzed?

3.56 When a correlation value is reported in research journals, there often is not an accompanying scatterplot. Explain why reported correlation values should be supported with either a scatterplot or a description of the scatterplot.

3.57 A memorization test is given to ten women and ten men. The researchers find a negative correlation between scores on the test and height. Explain which of the reasons listed at the beginning of Section 3.4 for misleading correlations might explain this finding. Sketch a scatterplot for the relationship between the variables that is consistent with your explanation.

3.58 The data in the table for Exercise 3.11 gave the square footage and asking price for nine homes for sale in Orange County, California in February 2010. The house with a square footage of 5500 is an obvious outlier. The value of r^2 for the relationship between y = asking price and x = square footage for all nine homes is 82.6%. If the outlier were to be removed, do you think the value of r^2 would increase or decrease? Explain your reasoning, using the interpretation of r^2 as "the proportion of variability in y explained by knowing x" as a guide. (*Hint:* A scatterplot may help.)

3.59 Give an example of a prediction that is an extrapolation. Do not give an example that is already in this chapter.

3.60 ◆ The data in the following table come from a time when the United States had a maximum speed limit of 55 miles per hour in all states. An issue of some concern at that time was whether lower speed limits reduce the highway death rate. (These data are called **speedlimit** on the companion website.)

Highway Death Rates and Speed Limits

Country	Death Rate (per 100 million vehicle miles)	Speed Limit (in miles per hour)
Norway	3.0	55
United States	3.3	55
Finland	3.4	55
Britain	3.5	70
Denmark	4.1	55
Canada	4.3	60
Japan	4.7	55
Australia	4.9	60
Netherlands	5.1	60
Italy	6.1	75

Source: D. J. Rivkin, "Fifty-five mph speed limit is no safety guarantee," New York Times *(letters to the editor), November 25, 1986, p. 26.*

a. In the relationship between death rate and speed limit, which variable is the response variable and which is the explanatory variable?

b. Plot the data in the table, and discuss the result. Does there appear to be an association? Are there any outliers? If so, what is their influence on the correlation?

3.61 In Exercise 3.21, a regression equation relating x = putting distance (feet) to y = success rate (in percent) for professional golfers was given as

$$\text{Success rate} = 76.5 - 3.95\,(\text{Distance})$$

The equation was based on observations of distances ranging from 5 feet to 15 feet.

a. Use the equation to predict success rate for a distance of 2 feet and for a distance of 20 feet.

b. The original data included values beyond those used to determine the regression equation (5 feet to 15 feet). At a distance of 2 feet, the observed success rate was 93.3%, and at a distance of 20 feet, 15.8% of observed putts were successful. Compare your results in part (a) to the observed success rates for distances of 2 feet and 20 feet. Utilize your results from part (a) to explain why it is not a good idea to use a regression equation to predict information beyond the range of values used to determine the equation.

c. Draw a graph of what you think the relationship between putting distance and success rate would look like for the entire range from 2 feet to 20 feet.

3.62 ◆ The table for Exercise 3.9 gave the average August temperature (y) and geographic latitude (x) for 20 cities in the United States. (The data are part of the **temperature** dataset on the companion website.) Exercise 3.18 gave the information that the regression equation relating these two variables is

$$\hat{y} = 113.6 - 1.01x$$

a. The latitude at the equator is 0. Using the regression equation, estimate the average August temperature at the equator.

b. Explain why we should not use this equation to estimate average August temperature at the equator.

Section 3.5

Skillbuilder Exercises

3.63 Explain why a strong correlation would be found between weekly sales of firewood and weekly sales of cough drops over a 1-year period.

3.64 Based on the data for the past 50 years in the United States, there is a strong correlation between yearly beer sales and yearly per capita income. Would you interpret this to mean that increasing a person's income will cause him or her to drink more beer? Explain.

3.65 ◆ The **pennstate2** dataset on the companion website includes heights and the total number of ear pierces for each person in a sample of college students. The correlation between the two variables is −0.495. What third variable may explain this observed correlation? Explain how that third variable could create the negative correlation.

General Section Exercises

3.66 Suppose a positive relationship had been found between each of the following sets of variables. For each set, discuss possible reasons why the connection may not be causal. Refer to the list of possible reasons for an observed association in Section 3.5.

a. Number of deaths from automobiles and soft drink sales for each year from 1950 to 2010.

b. Amount of daily walking and quality of health for men over 65 years old.

3.67 Suppose the indicated relationship has been found between each of the following sets of variables. For each set, discuss possible reasons why the connection may not be causal. Refer to the list of possible reasons for an observed association in Section 3.5.

a. A negative relationship between average number of cigarettes smoked per day and age of death.

b. A positive relationship between number of ski accidents and average wait time for the ski lift for each day during one winter at a ski resort.

3.68 Suppose that in an observational study, it is observed that the risk of heart disease increases as the amount of dietary fat consumed increases. Write a paragraph discussing why this result does not imply that diets high in fat cause heart disease.

3.69 Give an example of a situation in which it would be reasonable to conclude that an explanatory variable causes changes in a response variable.

3.70 Suppose a medical researcher finds a negative correlation between amount of weekly walking and the incidence of heart disease for people over 50 years old; in other words, people who walked more had a lower incidence of heart disease. One possible explanation for this observed association is that increased walking reduces the risk of heart disease. What are some other possible explanations?

3.71 Give an example not given elsewhere in this chapter of two variables that are likely to be correlated because they are both changing over time.

3.72 It is said that a higher proportion of drivers of red cars are given tickets for traffic violations than drivers of any other car color. Does this mean that if you drive a red car rather than a car of some other color, it will cause you to get more tickets for traffic violations? Explain.

3.73 Researchers have shown that there is a positive correlation between average fat intake and the breast cancer rate across countries. In other words, countries with higher fat intake tend to have higher breast cancer rates. Does this correlation prove that dietary fat is a contributing cause of breast cancer? Explain.

3.74 Example 2.2 (p. 22) described an observational study in which it was found that children who slept with a night-light or in a fully lit room before the age of 2 were more likely to be nearsighted than children who slept in darkness. Does this mean that sleeping with a light on as an infant causes nearsightedness? What are some other possible explanations?

Section 3.6: Skillbuilder Applet Exercises

For these exercises, use the **Correlation** *applet described in Section 3.6 and available on the companion website, http://www.cengage.com/statistics/Utts4e. In each exercise, you are asked to sketch a facsimile of a graph you create with the applet. Alternatively, you might use "Print Screen" on your keyboard to copy the screen image, and then paste it to a word-processing document.*

3.75 Using the applet, create a plot for the target correlation $r = +0.5$. Don't include any outliers. Sketch an approximate facsimile of your resulting graph.

3.76 Using the applet, create a plot for the target correlation $r = -0.8$. Don't include any outliers. Sketch an approximate facsimile of your resulting graph.

3.77 Using the applet, create a plot for the target correlation $r = 0$. Don't include any outliers. Sketch an approximate facsimile of your resulting graph.

3.78 Using the applet, create a plot for the target correlation $r = +0.5$ in which one point is an outlier that decreases the correlation. Make the plot such that if the outlier were removed, the correlation for the remaining points would be greater than $r = 0.7$. Sketch an approximate facsimile of your resulting graph.

3.79 Using the applet, create a plot for the target correlation $r = -0.8$ in which one point is an outlier that inflates the correlation. Make the plot such that if the outlier were removed, the correlation for the remaining points would be between −0.2 and +0.2. Sketch an approximate facsimile of your resulting graph. *Hint:* Start by putting points in the upper left corner of the plot.

3.80 Using the applet with the target correlation $r = 0$, make a plot that has a curvilinear pattern for which the correlation is 0. Sketch an approximate facsimile of your resulting graph.

Chapter Exercises

3.81 ◆ The dataset **bodytemp** on the companion website gives age in years and body temperature in degrees Fahrenheit for 100 blood donors ranging in age from 17 to 84 years old. The regression equation is $\hat{y} = 98.6 - 0.0138x$.

a. In the regression relationship shown, which variable is the response variable (y) and which is the explanatory variable (x)?

b. What is the predicted body temperature for someone who is 50 years old?

c. One of the donors was 50 years old and had a body temperature of 97.6°. What is the residual for this person? Explain what the residual tells you about this person's body temperature in comparison to the average body temperature for someone his age.

3.82 ◆ Refer to Exercise 3.81 in which the regression relationship between age in years and body temperature in degrees Fahrenheit is given as $\hat{y} = 98.6 - 0.0138x$, based on data from 100 blood donors ranging in age from 17 to 84 years old.

a. What is the y-intercept for this relationship? Does it have a useful meaning in this situation? Explain.
b. Give the value of the slope, and interpret what in means in this situation.
c. When you are 40 years older than you are today, how is your body temperature predicted to differ from what it is now?
d. Is it reasonable to use this regression equation to predict the body temperature for someone who is 100 years old? Explain.
e. Is it reasonable to use this regression equation to predict the body temperature for someone who is 30 years old? Explain.

3.83 ◆ The regression relationship for y = student height and x = father's height for the 10 female students listed in the table for Exercise 3.12 and in the dataset **UCDchap3** is

$$\hat{y} = 19.42 + 0.658x$$

a. Give the value of the y-intercept. Does it have a meaningful interpretation in this situation? Explain.
b. Give the value of the slope and interpret what it means in this situation.
c. Use the equation to predict the height of a female student whose father is 70 inches tall.
d. One student was 67 inches tall and her father was 70 inches tall. Find the residual for this student and explain what it tells you about this student in relation to other female students whose fathers are 70 inches tall.
e. Would it make sense to use this regression equation to predict the height of a male student whose father is 70 inches tall? Explain.

3.84 The regression line relating verbal SAT scores and college GPA for the data exhibited in Figure 3.12 is

$$\text{Average GPA} = 0.539 + 0.00362\ (\text{Verbal SAT})$$

a. Estimate the average GPA for those with verbal SAT scores of 600.
b. Explain what the slope of 0.00362 represents in terms of the relationship between GPA and SAT.
c. For two students whose verbal SAT scores differ by 100 points, what is the estimated difference in college GPAs?
d. Explain whether the intercept has any useful interpretation in the relationship between GPA and verbal SAT score. Keep in mind that the lowest possible verbal SAT score is 200.

3.85 Refer to Case Study 3.1, in which regression equations are given for males and females relating ideal weight to actual weight. The equations are

$$\text{Women: Ideal} = 44 + 0.6\ (\text{Actual})$$
$$\text{Men: Ideal} = 53 + 0.7\ (\text{Actual})$$

a. Predict the ideal weight for a man who weighs 140 pounds and for a woman who weighs 140 pounds. Compare the results.
b. Do the intercepts have logical physical interpretations in the context of this example? Explain.
c. Do the slopes have logical interpretations in the context of this example? Explain.

3.86 ◆ The heights (inches) and foot lengths (cm) of 33 college men are shown in the following table. (These data are in the dataset **heightfoot** on the companion website.)

Height (in) and Foot Length (cm) for 33 College Students

Student	Height	Foot Length
1	66.5	27.0
2	73.5	29.0
3	70.0	25.5
4	71.0	27.9
5	73.0	27.0
6	71.0	26.0
7	71.0	29.0
8	69.5	27.0
9	73.0	29.0
10	71.0	27.0
11	69.0	29.0
12	69.0	27.2
13	73.0	29.0
14	75.0	29.0
15	73.0	27.2
16	72.0	27.5
17	69.0	25.0
18	68.0	25.0
19	72.5	28.0
20	78.0	31.5
21	79.0	30.0
22	71.0	28.0
23	74.0	29.0
24	66.0	25.5
25	71.0	26.7
26	71.0	29.0
27	71.0	28.0
28	84.0	27.0
29	77.0	29.0
30	72.0	28.0
31	70.0	26.0
32	76.0	30.0
33	68.0	27.0

Data source: William Harkness.

a. Draw a scatterplot with y = foot length (cm) and x = height (inches). Does the relationship appear to be linear? Are there any outliers? If so, do you think the outliers are legitimate data values?
b. Use statistical software or a calculator to calculate the correlation between height and foot length. If heights

were converted to centimeters, what would be the correlation between height and foot length?

c. If there are any outliers, remove them and recalculate the correlation. Describe how the correlation changed from part (b).

3.87 ◆ Refer to Exercise 3.86 about y = foot length and x = height. (*Data source:* the **heightfoot** dataset on the companion website.) If the person who reportedly is 84 inches tall is excluded, the regression equation for the remaining 32 men is $\hat{y} = 0.25 + 0.384x$.

a. How much does average foot length increase for each 1-inch increase in height?

b. Predict the difference in the foot lengths of men whose heights differ by 10 inches.

c. Suppose Max is 70 inches tall and has a foot length of 28.5 cm. On the basis of the regression equation, what is the predicted foot length for Max? What is the value of the prediction error (residual) for Max?

3.88 The winning time in the Olympic men's 500-meter speed skating race over the years 1924 to 2006 can be described by the following regression equation:

$$\text{Winning time} = 272.63 - 0.1184\,(\text{Year})$$

Note: Beginning with the 1998 Olympics each competitor skated twice and the average of the two times defined the winner. In this analysis the data used for the relevant years is the average of the two times for the winner (*Source: http://www.infoplease.com/ipsa/A0758122.html*).

a. Is the correlation between winning time and year positive or negative? Explain how you know, and explain what that means in the context of this situation.

b. In 2010, the actual winning time for the gold medal was 34.91 seconds. Use the regression equation to predict the winning time for 2010, and compare the prediction to what actually happened.

c. Explain what the slope of −0.1184 indicates in terms of how winning times change from one set of Olympic games to the next. Olympic games occur every 4 years.

d. Why should we not use this regression equation to predict the winning time for the men's 500-meter speed skating race in the 2080 Winter Olympics?

3.89 ◆ The following table lists the number of pages and the price for 18 books, sorted in order of increasing number of pages. Ten of the books are hardcover and eight are softcover. (These data are in the dataset **ProfBooks** on the companion website.)

a. Draw a scatterplot of y = price versus x = pages. Use different symbols for hardcover and softcover books.

b. For all 18 books, determine the correlation between price and pages.

c. Separate the books by type. Determine the correlation between price and pages for hardcover books only. Determine the correlation between price and pages for softcover books only.

d. Which of the reasons listed in Section 3.4 for misleading correlations is illustrated in this exercise?

Pages versus Price for Books

Pages	Price	Type
200	35.00	H
256	47.50	H
305	29.95	H
370	64.95	H
384	74.95	H
436	15.99	S
480	35.00	H
545	69.95	H
565	19.95	S
601	79.95	H
612	50.00	H
639	60.00	H
641	14.95	S
673	24.99	S
747	21.95	S
833	29.95	S
877	20.00	S
907	25.00	S

H = hardcover; S = softcover.

3.90 ◆ U.S. Census Bureau estimates of the average number of persons per household in the United States for census years between 1850 and 2000 are shown in the following table. (These data are in the file **perhouse** on the companion website.)

Persons per Household in the United States

Year	Per Household
1850	5.55
1860	5.28
1870	5.09
1880	5.04
1890	4.93
1900	4.76
1910	4.54
1920	4.34
1930	4.11
1940	3.67
1950	3.37
1960	3.35
1970	3.14
1980	2.76
1990	2.63
2000	2.59

Data source: The World Almanac and Book of Facts, *1999, p. 383, and U.S. Bureau of the Census.*

a. Draw a scatterplot for the relationship between persons per household and year. Is the relationship linear or curvilinear? Is the association between persons per household and year positive or negative?

b. On your scatterplot, add a line that you believe fits the data pattern. Extend this line to the year 2010. On the basis of this line, estimate the number of persons per household in the United States in the year 2010.

3.91 ◆ Refer to Exercise 3.90 about the trend in number of persons per household.

a. Using statistical software, determine the least squares line for these data. Use the equation of this line to estimate the number of persons per household in the year 2010 (*Data source:* **perhouse** dataset on the website for this book).

b. What is the slope of the line? Interpret the slope in the context of these variables.

c. Based on the regression line, what would be the predicted persons per household in the year 2200? Realistically, what is the lowest possible value of the persons per household number? How does the estimate for 2200 compare to this value?

d. Part (c) illustrates that the observed pattern can't possibly continue in the same manner forever. Sketch the pattern for the trend in persons per household that you think might occur between now and the year 2200.

3.92 ◆ For a statistics class project at a large northeastern university, a student examined the relationship between the following two variables:

x = body weight (in pounds)

y = time to chug a 12-ounce beverage (in seconds)

We'll leave it to you to imagine the beverage. The student collected data from 13 individuals, and those data are in the following table. (This dataset is named **chugtime** on the website for this book.)

Body Weight (pounds) and Chug Time (seconds) for 13 College Students

Person	Weight	Chug Time
1	153	5.6
2	169	6.1
3	178	3.3
4	198	3.4
5	128	8.2
6	183	3.5
7	177	6.1
8	210	3.1
9	243	4.0
10	208	3.2
11	157	6.3
12	163	6.9
13	158	6.7

Data source: William Harkness.

a. Draw a scatterplot of the measurements. Characterize the relationship between chug time and body weight.

b. The heaviest person appears to be an outlier. Do you think that observation is a legitimate observation, or do you think an error was made in recording or entering the data?

c. Outliers should not be thrown out unless there's a good reason, but there are several reasons why it may be legitimate to conduct an analysis without them (for instance, see part (e)). Delete the data point for the heaviest person, and determine a regression line for the remainder of the data.

d. Use the regression line from part (c) to estimate the chug time for an individual who weighs 250 pounds. Do you think this time could be achieved by anybody?

e. Sometimes the relationship between two variables is linear for a limited range of x values and then changes to a different line or curve. Using this idea, draw a sketch that illustrates what you think the actual relationship between weight and chug time might be for the range of weights from 100 to 300 pounds.

f. Discuss plausible reasons why the heaviest person appears to be an outlier with regard to his combination of weight and chug-time measurements.

3.93 Give an example of a situation not mentioned elsewhere in this chapter in which two variables have no causal connection but are highly correlated because they are both related to a third variable. Explain what the third variable is.

3.94 Measure the heights and weights of ten friends of the same sex.

a. Draw a scatterplot of the data, with weight on the vertical axis and height on the horizontal axis. Draw a line onto the scatterplot that you believe describes the average pattern. On the basis of two points on this line, estimate the slope of the relationship between weight and height.

b. Using statistical software, compute the least squares line, and compare the slope to your estimated slope from part (a).

3.95 The following is from Thought Question 3.4 on page 89. Sometimes the main purpose of a regression analysis is to determine the nature of the relationship between two variables, and sometimes the main purpose is to use the equation in the future to predict a y value when the x value is known. Explain which purpose is likely to be the main reason for a regression analysis between

a. x = percent fat consumed in diet, y = blood pressure

b. x = SAT score, y = college grade point average

c. x = height at age 4, y = height at age 21

d. x = hours of sleep per night, y = score on IQ test

Dataset Exercises

Datasets required to solve these exercises are available on the companion website, http://www.cengage.com/statistics/Utts4e.

3.96 Use the dataset **ceodata08** on the companion website for this exercise, which gives the ages (***Age***) and salaries

(***Salary***) for the 50 highest-paid CEOs on the Fortune 500 list of top companies in the United States (*Data source: http://www.forbes.com/lists/2009/12/best-boss-09_CEO-Compensation_CompTotDisp.html*).

a. In the relationship between age and salary, which is the response variable and which is the explanatory variable?
b. Plot ***Salary*** versus ***Age***. Are there any obvious outliers in the plot?
c. Use your plot from part (b) to discuss whether linear regression is appropriate for predicting CEO salaries from age for the top Fortune 500 companies.

3.97 Use the dataset **poverty** on the companion website; it includes teenage mother birth rates and poverty rates for the 50 states and the District of Columbia. The variable ***PovPct*** is the percent of a state's population in 2000 living in households with incomes below the federally defined poverty level. The variable ***Brth15to17*** is the birth rate for females 15 to 17 years old in 2002, calculated as births per 1000 persons in this age group.

a. Plot ***Brth15to17*** (y) versus ***PovPct*** (x). Describe the direction and strength of the relationship, and comment on whether there are any outliers.
b. Determine the equation of the regression line relating y = ***Brth15to17*** to x = ***PovPct***. Write the equation.
c. What is the value of the slope of the equation? Write a sentence that interprets the slope in the context of these variables.
d. Based on the equation, what is the estimated birth rate for females 15 to 17 years old in a state with a poverty rate of 15%?

3.98 Use the dataset **oldfaithful** on the companion website; it gives data for $n = 299$ eruptions of the Old Faithful geyser. The variable ***Duration*** is the duration (minutes) of an eruption, and the variable ***TimeNext*** is the time interval (minutes) until the next eruption.

a. Plot ***TimeNext*** (y) versus ***Duration*** (x). Describe the direction and strength of the relationship, and comment on whether there are any outliers.
b. Determine the equation of the regression line relating y = ***TimeNext*** to x = ***Duration***. Write the equation.
c. What is the value of the slope of the equation? Write a sentence that interprets the slope in the context of these variables.
d. Estimate the interval of time until the next eruption following one that lasts 4 minutes.

3.99 Use the dataset **cholesterol** on the companion website. For $n = 28$ heart attack patients, the variables ***2-Day*** and ***4-Day*** are cholesterol levels measured 2 days and 4 days, respectively, after the attacks.

a. Plot ***4-Day*** (y) versus ***2-Day*** (x). Describe the direction and strength of the relationship, and comment on whether there are any outliers.
b. Determine the equation of the regression line relating y = ***4-Day*** to x = ***2-Day***. Write the equation.
c. What is the value of the slope of the equation? Write a sentence that interprets the slope in the context of these variables.
d. Separately estimate cholesterol levels four days after the attack for patients with ***2-Day*** values of 200, 250, and 300.
e. Utilizing the results of part (d), describe how cholesterol levels are predicted to change in the time from 2 days to 4 days after their heart attacks.

3.100 Use the dataset **sats98** on the companion website for this exercise. The variable ***Verbal*** contains the average scores on the verbal SAT in 1998 for the 50 states and the District of Columbia. ***PctTook*** is the percent of high school graduates, in each state, who took the SAT that year.

a. Make a scatterplot showing the connection between average verbal SAT (y) and the percent of graduates who took the SAT in a state (x). Describe the relationship between these two variables.
b. Compute the least squares regression line for the relationship between these two variables. Write a sentence that interprets the slope of this equation in a way that could be understood by people who don't know very much about statistics.
c. Based on the appearance of the scatterplot, do you think that a straight line is an appropriate mathematical model for the connection between ***Verbal*** and ***PctTook***? Why or why not?
d. Explain why the intercept of the equation computed in part (b) would not have a sensible interpretation for these two variables.

3.101 Use the **sats98** dataset on the companion website.

a. Plot the relationship between average verbal (***Verbal***) and average math (***Math***) SAT scores in the 50 states. Describe the characteristics of the relationship.
b. What states are outliers? In what specific way are they outliers?

3.102 The dataset **bodytemp** on the companion website includes sex, age, and body temperature for 100 blood donors who ranged in age from 17 to 84.

a. Create a scatter plot of body temperature (y) and age (x) using different symbols for men and women. Is there an obvious difference in the relationship between age and body temperature for men and women? Explain.
b. Find the regression equations relating body temperature and age for the men and the women separately. Compare the two equations. Do they indicate that the relationship between the two variables is similar for men and for women? Explain.

3.103 Use the dataset **idealwtmen** on the companion website. It contains data for the men used for Case Study 3.1. The variable ***diff*** is the difference between actual and ideal weights and was computed as ***diff*** = ***actual*** − ***ideal***.

a. Plot ***diff*** (y) versus ***actual*** (x, actual weight). Does the relationship appear to be linear, or is it curvilinear?
b. Compute the equation of the regression line for the relationship between ***diff*** and ***actual***. Estimate the average difference for men who weigh 150 pounds. On average, do 150-pound men want to weigh more or less than they actually do?
c. Repeat part (b) for men who weigh 200 pounds.

d. What is the value of r^2 for the relationship between ***diff*** and ***actual***?

3.104 In 1993, *Forbes Magazine* identified what it considered to be America's 60 best small companies, and published the ages and salaries of their CEOs. The data are in the dataset **ceodata** on the companion website. The annual salaries (in thousands of dollars) for 59 of these CEOs are in the dataset along with the ages.

a. Plot ***Salary*** versus ***Age***.

b. Compute the correlation coefficient and r^2.

c. Characterize the relationship between annual salary and age. What is the pattern of the relationship? How strong is the association?

3.105 Use the dataset **UCDwomht** on the companion website. For a sample of college women, the variable ***height*** is student's height (in inches), and the variable ***midparent*** is the average height of the student's parents (in inches) as reported by the student.

a. Compute the regression equation for predicting a student's height from the average of her parents' heights.

b. Use the regression equation to predict the height for a college woman with parents who have an average height of 68 inches.

c. Use the regression equation to predict the height of a college woman whose mother is 62 inches tall and whose father is 70 inches tall.

d. What other summaries of the data should be done to determine the strength of the relationship between ***height*** and ***midparent*** height?

3.106 Use the dataset **temperature** on the companion website. A portion of this dataset was presented in Exercise 3.9, in which the relationship between mean August temperature and geographic latitude was analyzed. For predicting mean April temperature (***AprTemp***), which of these two variables in the dataset is a stronger predictor: geographic latitude (***latitude***) or mean January temperature (***JanTemp***)? Support your answer with relevant statistics and plots.

4

Joe Muszynski/Digital Light Source/Photolibrary

Is this woman typical of her age group?

See Example 4.1 *(p. 115)*

Relationships Between Categorical Variables

Whenever risk statistics are reported, there is a risk that they are misreported. Journalists often present risk data in a way that produces the best story rather than in a way that provides the best information. Very commonly, news reports either don't contain or don't emphasize the information that you need to understand risk.

This chapter is about the analysis of the relationship between two categorical variables, so let's begin by recalling the meaning of the term *categorical variable.* The raw data from categorical variables consist of group or category names that don't necessarily have any ordering. Eye color and hair color, for instance, are categorical variables.

We can also use the methods of this chapter to examine *ordinal* variables. Ordinal variables can be thought of as categorical variables for which the categories have a natural ordering. For example, a researcher might define categories for quantitative variables such as age, income, or years of education.

Although there are many questions that we can and will ask about two categorical variables, the principal question that we ask in most cases is, "Is there a relationship between the two variables such that the category into which individuals fall for one variable seems to depend on the category they are in for the other variable?"

THOUGHT QUESTION 4.1 Hair color and eye color are related characteristics. What exactly does it mean to say that these two variables are related? Suppose that you know the hair colors and the eye colors of 200 individuals. How would you assess whether the two variables are related for those individuals?*

4.1 Displaying Relationships Between Categorical Variables

We have already encountered several examples of the type of problem we will study in this chapter. In Chapter 2, for instance, we described a study of 479 children that found that children who slept either with a nightlight or in a fully lit room before the age of 2 had a higher incidence of myopia (nearsightedness) later in childhood. Table 4.1, which is also Table 2.3 in Chapter 2, is used to explore this relationship.

Table 4.1 Nighttime Lighting in Infancy and Eyesight

Slept with:	No Myopia	Myopia	High Myopia	Total
Darkness	155 (90%)	15 (9%)	2 (1%)	172
Nightlight	153 (66%)	72 (31%)	7 (3%)	232
Full Light	34 (45%)	36 (48%)	5 (7%)	75
Total	342 (71%)	123 (26%)	14 (3%)	479

***HINT:** Read the sentence just before Thought Question 4.1.

The table displays the number of children in each combination of the categories of two categorical variables: the sleep-time lighting condition and the child's eyesight classification at the time of the study. The table also gives row percentages—the percentages of children within each row who fall into the different eyesight categories. For example, 90% (155/172) of those who slept in darkness had no myopia, but only 45% (34/75) of those who slept in full light had no myopia. Figure 4.1, which is also Figure 2.4 in Chapter 2, displays the percentages of children falling into the different eyesight categories for each category of sleep-time lighting.

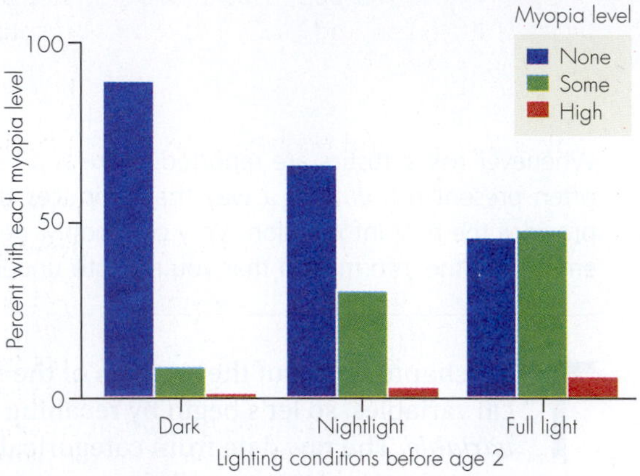

Figure 4.1 Bar chart for myopia and nighttime lighting in infancy

From these percentages, we learn that the incidence of myopia and high myopia increases when the amount of sleep-time light increases. This result does not prove that sleeping with light *causes* myopia, but we can say that for some reason, the characteristics of eyesight and sleep-time lighting are *associated* characteristics in this sample. Later in this chapter, in Example 4.9, we will see that a third variable, parents' eyesight, may explain this association.

Displays like Table 4.1 are called **contingency tables** because they cover all contingencies for the combinations of the two variables. Because the categories of two variables are used to create the table, a contingency table is also called a **two-way table**. Each row category and column category combination in the table is called a **cell**.

The first step in analyzing the relationship between two categorical variables is to count how many observations fall into each cell of the contingency table. It's difficult, however, to look at a table of counts and make useful judgments about a relationship. Usually, we need to consider the **conditional percentages** within either the rows or the columns of the table.

There are two types of conditional percentages that we can compute for a contingency table. **Row percentages**, like the percentages in the myopia table, are the percentages within a given row of a contingency table. Row percentages are based on the total number of observations in the row. **Column percentages** are the percentages within a given column of a contingency table. Column percentages are based on the total number of observations in the column.

In some cases, one variable can be designated as the *explanatory variable* and the other variable as the *response variable*. As we learned in Chapter 2, in these situations it is customary to define the rows using the categories of the explanatory variable and the columns using the categories of the response variable. When this is done, the row percentages can be used to examine the relationship because they tell us what percentage *responded* in each possible way. We can then see whether individuals responded equivalently for each category of the explanatory variable. For instance, we can see whether the myopia response was the same for different lighting conditions.

Example 4.1 **Age and Main News Source** Where do you get most of your information about current news events? This question was asked in the 2008 General Social Survey, a national survey of randomly selected Americans. Possible answers included television, Internet, and newspapers, as well as other possibilities such as radio, family, and friends.

Table 4.2 is a two-way contingency table that summarizes the results by age group. The table gives counts along with the percentages within each age group that picked the various news sources. For instance, within the 18- to 29-year-old age group, 41.6% (109/262) said television, 35.1% (92/262) said Internet, 9.5% (25/262) said newspapers, and 13.7% (36/262) said something other than television, Internet, or newspapers.

Table 4.2 Age and Main Source of Information for Current News

	Main News Source				
Age Group	**TV**	**Internet**	**Newspapers**	**Other**	**Total**
18–29	109 (41.6%)	92 (35.1%)	25 (9.5%)	36 (13.7%)	262 (100%)*
30–49	272 (46.9%)	157 (27.1%)	88 (15.2%)	63 (10.9%)	580 (100%)*
≥50	345 (54.6%)	59 (9.3%)	165 (26.1%)	63 (10.0%)	632 (100%)*
Total	726 (49.3%)	308 (20.9%)	278 (18.9%)	162 (11.0%)	1474 (100%)

Data Source: http://sda.berkeley.edu/archive.htm, accessed March 9, 2010.
** Row percentages may not add exactly to 100% due to rounding.*

A comparison of the row percentages for the three age groups shows that age group and main news source are related for this sample. The percentages for both television and newspapers increase as age increases, while the percentages for Internet and other sources both decrease as age increases. The percentage using the Internet as their main source of news information is strikingly different for the youngest and oldest age groups, 35.1% for 18- to 29-year-olds compared to only 9.3% for respondents aged 50 and older. The survey results are bad news for the future of newspapers. Only 9.5% of the 18- to 29-year-olds use newspapers as their main source of information about current news.

Figure 4.2 graphs the row percentages using a bar chart, as described in Chapter 2. We can see clearly that the variables are related. The increasing percentage for television and the decreasing percentage for Internet are easy to see as you move from the youngest age group to the oldest, as is the increasing percentage for newspapers.

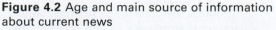

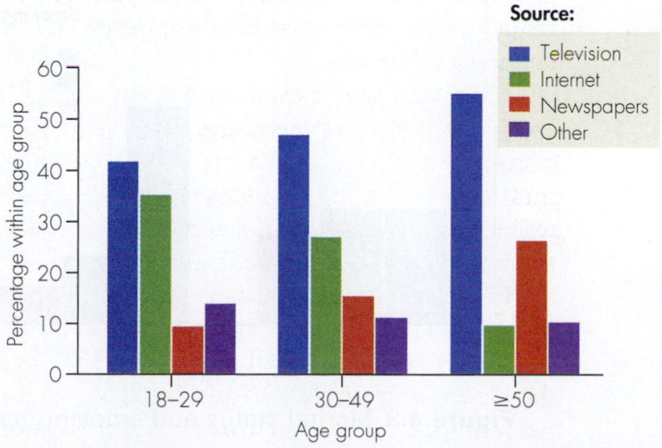

Figure 4.2 Age and main source of information about current news

In the previous example, we viewed age group as an explanatory variable and the main source of news information as a response variable. Because the explanatory variable, age group, defined the rows of the table, we used row percentages to examine the relationship. This allowed us to compare the age groups to see if they had different favored news sources. In the next example, it may be interesting to describe the data using both row percentages and column percentages.

Example 4.2 Read the original source on the companion website, http://www.cengage.com/statistics/Utts4e.

Smoking and Divorce Australian researchers surveyed 1498 married couples with children in each of 3 consecutive years to learn what factors may lead to marital separation or divorce (Butterworth et al., 2008). During the study period from 2001 to 2003, 114 of the 1498 couples separated or divorced. The researchers identified several differences between the couples who had and had not separated. Among them was a difference in smoking rates for the separated and intact couples.

Table 4.3 summarizes counts and row percentages for the association between smoking habits of the couples (neither partner smoked, one smoked but not both, or both smoked) and marital status at the end of the study period (separated or not separated). The row percentages give the separation rates for each of the three smoking habits categories. Note in Table 4.3 that the percentage who had separated is higher in couples where either one or both partners smoked (12.4% and 16.4%, respectively) compared to couples where neither partner smoked (4.2%).

Table 4.3 Smoking and Marital Status: Counts and Row Percentages

	Marital Status		
Smoking	**Separated**	**Not Separated**	**Total**
Neither smoked	41 (4.2%)	931 (95.8%)	972 (100%)
One smoked	41 (12.4%)	290 (87.6%)	331 (100%)
Both smoked	32 (16.4%)	163 (83.6%)	195 (100%)
Total	114 (7.6%)	1384 (92.4%)	1498 (100%)

The column percentages for these data give a comparison of the smoking habits of the separated couples and the couples that did not separate. Those percentages are given in Table 4.4 (next page) and plotted in the bar chart shown in Figure 4.3. We see a large difference between the smoking habits of the separated and not separated couples. For instance, the percentage where neither partner smoked was only 36.0% for separated couples compared to 67.3% for couples who did not separate.

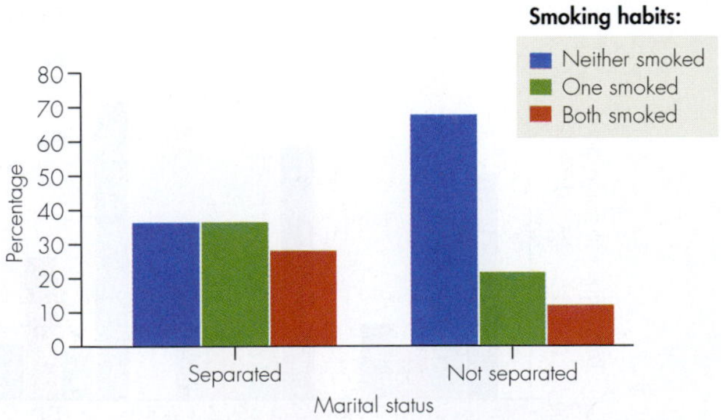

Figure 4.3 Marital status and smoking habits

Although the association between smoking habits and the likelihood of marital separation is strong, we cannot conclude that smoking causes divorce. This was an observational study, so there may be confounding factors. Perhaps higher smoking rates are associated with other factors that the researcher found to increase the likelihood of separation, such as financial difficulties, general dissatisfaction with life, and having parents who divorced.

Table 4.4 Smoking and Marital Status: Column Percentages

	Marital Status	
Smoking	**Separated**	**Not Separated**
Neither smoked	36.0% (41/114)	67.3% (931/1384)
One smoked	36.0% (41/114)	21.0% (290/1384)
Both smoked	28.1% (32/114)	11.8% (163/1384)
Total	100%*	100%*

** Column percents do not exactly add to 100% due to rounding.*

The next example describes a situation in which there is a weak relationship between the two categorical variables that form the two-way table. The distribution of row percentages is nearly the same for the two rows of the two-way table.

Example 4.3 **Gender and Rating of Quality of Public Education** In a poll conducted by *CBS News* between December 17 and 22, 2009, 1048 American adults were surveyed about various issues. The respondents were selected using random-digit dialing methods (explained in detail in Chapter 5) to reach both landline and cell phones. One of the questions was, "How would you grade the U.S. on the quality of public schools in this country?"

Table 4.5 gives the percentage distributions of responses for the 438 men and 594 women who gave grades. Note that the pattern of responses is only slightly different for the two sexes. As a result, it is difficult to say whether sex and the grade assigned to the quality of education are related variables or not. If a relationship does exist, it is a weak one.

Table 4.5 Sex and Grade Assigned to Quality of Public Education

	A	**B**	**C**	**D or F**	**Total**
Men	5.0%	21.2%	31.3%	42.5%	100%
Women	5.1%	26.4%	33.8%	34.7%	100%

Source: http://www.cbsnews.com/sections/opinion/polls/main500160.shtml, accessed March 9, 2010.

4.1 Exercises are on pages 134–137.

IN SUMMARY

Row Percentages, Column Percentages, and Relationships in Two-Way Tables

- A **row percentage** uses a total row count as the basis for computing a percentage. It is the percentage of the observations within a particular row category that are in a specified category of the column variable.
- A **column percentage** uses a total column count as the basis for computing a percentage. It is the percentage of observations within a particular column category that are in a specified category of the row variable.
- There is a **relationship** between the categorical variables that form a two-way table if two or more rows have different distributions of row percentages. Equivalently, there is a relationship if two or more columns have different distributions of column percentages.

4.2 Risk, Relative Risk, and Misleading Statistics about Risk

When a particular outcome is undesirable, researchers and journalists may describe the *risk* of that outcome. The **risk** that a randomly selected individual within a group falls into the undesirable category is simply the proportion in that category.

$$\text{Risk} = \frac{\text{Number in category}}{\text{Total number in group}}$$

It is commonplace to express risk as a percentage rather than as a proportion. Suppose, for instance, that within a group of 200 individuals, asthma affects 24 people. In this group, the *risk* of asthma is $24/200 = .12$, or 12%.

Relative Risk

We often want to know how the risk of an outcome relates to an explanatory variable. One statistic that is used for this purpose is **relative risk,** which is the ratio of the risks in two different categories of an explanatory variable.

$$\text{Relative risk} = \frac{\text{Risk in category 1}}{\text{Risk in category 2}}$$

Relative risk describes the risk in one group as a multiple of the risk in another group. For example, suppose that a researcher states that for those who drive while under the influence of alcohol, the relative risk of an automobile accident is 15. This means that the risk of an accident for those who drive under the influence is 15 times the risk for those who don't drive under the influence.

Some features of relative risk are as follows:

- When two risks are the same, the relative risk is 1.
- When the category in the numerator has the higher risk, the relative risk is greater than 1. When the category in the numerator has the lower risk, the relative risk is less than 1.
- The risk in the denominator of the ratio is often, although not always, a **baseline risk**, which is the risk for a category in which no additional treatment or behavior is present.

THOUGHT QUESTION 4.2 Based on one study in Section 4.1, the relative risk of developing any myopia later in childhood is 5.5 for babies sleeping in full light compared with babies sleeping in darkness. Restate this information in a sentence that the public would understand.*

Percent Increase or Decrease in Risk

An increase or decrease in risk might be presented as a percent change instead of a multiple. The **percent increase** (or decrease) **in risk** when comparing two groups can be calculated as follows:

$$\text{Percent increase in risk} = \frac{\text{Difference in risks}}{\text{Baseline risk}} \times 100\%$$

Equivalently, when the relative risk has already been determined, the percent increase in risk can be calculated by using the following relationship:

$$\text{Percent increase in risk} = (\text{relative risk} - 1) \times 100\%$$

When a risk is smaller than the baseline risk, the relative risk is less than 1, and the percent "increase" will actually be negative. In this situation, the term *percent decrease* should be used to describe the percent change in the risks.

***HINT:** See the interpretation of relative risk in Example 4.4.

Example 4.4 **Sex and the Risk of Childhood Asthma** The National Health Interview Survey is an annual national health survey of people living in about 35,000 randomly selected households in the United States. On the basis of the data collected in the 2006 survey, the Centers for Disease Control and Prevention estimated that 15.7% of boys and 11.2% of girls under the age of 18 had at some time been diagnosed with asthma (Bloom and Cohen, 2007, p. 59).

The relative risk of ever having asthma for boys compared to girls is the ratio of the risks for the two sexes.

$$\text{Relative risk} = \frac{\text{Risk for boys}}{\text{Risk for girls}} = \frac{15.7\%}{11.2\%} = 1.40$$

Thus the risk of asthma for boys is 1.40 times the risk for girls. Because we already know the relative risk is 1.40, we can calculate the percent increase in risk as follows:

$$\text{Percent increase in risk} = (1.40 - 1) \times 100\% = 40\%$$

The interpretation of this percent increase in risk is that boys under 18 have a risk of asthma that is 40% higher than the risk of asthma for girls. An equivalent calculation of the percent increase in risk is

$$\text{Percent increase in risks} = \frac{\text{Difference in risks}}{\text{Baseline risk}} \times 100\%$$

$$= \frac{15.7 - 11.2}{11.2} \times 100\% = 40\%$$

By the way, the direction of the relative risk for males and females is reversed in adulthood. Adult women are more likely to have asthma than adult men.

Odds and Odds Ratio

DEFINITION The **odds** of an event compare the chance that the event happens to the chance that it does not. Odds are typically expressed using a phrase with the structure "*a* to *b*," so a ratio is implied but not actually computed.

As an example, suppose there is a 60% chance that it will rain tomorrow, so the chance that it will not rain is 40%. The odds that it will rain tomorrow can be expressed as 60 to 40, a comparison of the chance that it rains to the chance that it does not. Dividing both sides of the odds of rain by 20 lets us express these odds as 3 to 2.

DEFINITION An **odds ratio** compares the odds of an event for two different categories. It is calculated as the ratio:

$$\text{Odds ratio} = \frac{\text{Odds in category 1}}{\text{Odds in category 2}}$$

In the calculation of an odds ratio, the odds values for the two categories being compared are computed as ratios.

Some features of odds ratios are as follows:

- When the odds are the same in two categories, the odds ratio is 1.
- When the odds of an event are higher in the category in the numerator, the odds ratio is greater than 1.
- When the odds of an event are lower in the category in the numerator, the odds ratio is less than 1.
- The value stays the same if the roles of the response and explanatory variables are reversed.

Example 4.5 **Example 4.4 Revisited: Odds Ratio for Sex and Childhood Asthma**

Odds compare the chance that an event happens to the chance that the event does not happen. In Example 4.4, the risk of ever having had asthma was given as 15.7% for boys under 18 years old and 11.2% for girls under 18.

- For boys, 100% − 15.7% = 84.3% is the chance of not ever having had asthma, so the odds of asthma for boys are 15.7 to 84.3, about 1 to 5.37 if we divide both values by 15.7.
- For girls, 100% − 11.2% = 88.8% is the chance of not having had asthma, so the odds of asthma for girls are 11.2 to 88.8, about 1 to 7.93 if we divide both values by 11.2.

The odds ratio that compares the odds of ever having asthma for boys compared to girls is

$$\text{Odds ratio} = \frac{\text{Odds of asthma for boys}}{\text{Odds of asthma for girls}} = \frac{(15.7/84.3)}{(11.2/88.8)} = 1.48$$

The interpretation is that the odds of ever having asthma for boys are 1.48 times the odds for girls.

IN SUMMARY Statistics on Risk, Relative Risk, Odds, and Odds Ratios

Let's summarize the various ways in which risk, relative risk, odds, and odds ratios are constructed from a two-way contingency table. These measures are usually employed when there is a definable explanatory and response variable and when there is a baseline condition, so we present them using those distinctions. In situations in which those distinctions cannot be made, simply be clear about which condition is in the numerator and which is in the denominator.

	Response Variable		
Explanatory Variable	**Response 1**	**Response 2**	**Total**
Category of Interest	A_1	A_2	T_A
Baseline Category	B_1	B_2	T_B

- Risk (of response 1) for category of interest = A_1/T_A
- Odds (of response 1 to response 2) for category of interest = A_1 to A_2
- Relative risk = $\dfrac{A_1/T_A}{B_1/T_B}$
- Odds ratio = $\dfrac{A_1/A_2}{B_1/B_2}$

Misleading Statistics about Risk

THOUGHT QUESTION 4.3 Suppose a newspaper article claims that drinking coffee doubles your risk of developing a certain disease. Assume that the statistic was based on legitimate, well-conducted research. What additional information would you want about the risk before deciding whether or not to quit drinking coffee?*

Whenever risk statistics are reported, there is a risk that they are misreported. Unfortunately, journalists often present risk data in a way that produces the best story rather than in a way that provides the best information. Very commonly, news reports either don't contain or don't emphasize the information that you need to understand risk.

***HINT:** Would your decision be the same if you knew that the disease was rare in non-coffee drinkers of your age as it would be if you knew that it was common?

You should always ask the following questions when you encounter statistics about risk:

1. What are the actual risks? What is the baseline risk?
2. What is the population for which the reported risk or relative risk applies?
3. What is the time period for this risk?

News reports sometimes give counts of how often a "bad" event has happened without giving information about how many people have been exposed to the risk. Without information on the amount of exposure, we don't know the actual risk and can't properly interpret the news report. The next two examples show that consideration of the amount of exposure to a risk is important to understanding the risk.

Example 4.6 **The Risk of a Shark Attack** A terrifying risk of going to the beach, especially where the water is warm, is the possibility of being attacked by a shark. Media stories about shark attacks on humans often include counts of how many attacks have occurred within the past year or so, and those numbers can sound scary. For instance, the website of the Florida Museum of Natural History Ichthyology Department reports that from 1959 to 2008, there were 936 shark attacks in the United States, resulting in 25 fatalities. That may sound like a lot, but when estimated beach attendance is taken into account, we see that the risk of being attacked by a shark is extremely low. At the Florida Museum website, the risk of a shark attack in the year 2000 is estimated to have been only 1 in 11.5 million beach visits. In contrast, from 1959 to 2008 the relative risk of being killed by lightning in coastal areas, compared with being killed by a shark, was 77.2 (http://www.flmnh.ufl.edu/Fish/sharks/attacks/relarisk.htm). And even that risk was extremely low, with an average of 38.6 lightning strike deaths per year.

Example 4.7 **Case Study 1.2 Revisited: Disaster in the Skies?** Case Study 1.2 described a *USA Today* article that told us, "Errors by air traffic controllers climbed from 746 in fiscal 1997 to 878 in fiscal 1998, an 18% increase." To airplane travelers, this may sound frightening, but a look at the risk of controller error per flight should ease their fear. In 1998, there were only 5.5 errors per million flights compared to 4.8 errors per million flights in 1997. The risk of controller error did increase, but the actual risk is extremely small.

In the next example, we see that a reported risk value may be meaningless if we don't know what specific population is being described or the time period of exposure to the risk.

Example 4.8 **Dietary Fat and Breast Cancer** "Italian scientists report that a diet rich in animal protein and fat—cheeseburgers, french fries, and ice cream, for example—increases a woman's risk of breast cancer threefold," according to *Prevention Magazine*'s *Giant Book of Health Facts* (1991, p. 122). The statement attributed to the Italian scientists is nearly useless information for at least two reasons:

1. We don't how the data were collected, so we don't know what population these women represent.
2. We don't know the ages of the women studied, so we don't know the baseline rate of breast cancer for these women.

Age is a critical factor. A frequently stated statistic about breast cancer is that about one in eight women will develop breast cancer, but this is actually an accumulated lifetime risk to the age of 90, so a woman who dies before she reaches age 90 does not have a 1 in 8 risk. According to researchers at the National Cancer Institute (Horner et al., 2009), the accumulated lifetime risk of developing breast cancer by certain ages for women presently 30 years old is as follows:

by age 40: 1 in 227
by age 50: 1 in 54
by age 60: 1 in 24
by age 90: 1 in 8.2

The annual risk of developing breast cancer is only about 1 in 3700 for women in their early 30s (Fletcher et al., 1993, p. 1644). If the Italian study was done on young women, the threefold increase in risk represents a small increase. Unfortunately, *Prevention Magazine's Giant Book of Health Facts* did not even give enough information to lead us to the original research report, so it is impossible to intelligently evaluate the claim.

4.2 Exercises are on pages 137–138.

CASE STUDY 4.1 Is Smoking More Dangerous for Women?

"Higher heart risk in women smokers" was the headline of an April 3, 1998, article at the Yahoo!® Health news website. The article, from the Reuters news agency, described Danish research that was interpreted as evidence that smoking affects the risk of a heart attack more for women than for men. Here is part of the article:

> Women who smoke have a greater than 50% higher risk of a heart attack than male smokers according to a study from Denmark. The researchers suggest that this difference may be related to the interaction of tobacco smoke and the female hormone, estrogen.
>
> "Women may be more sensitive than men to some of the harmful effects of smoking," the team writes in the April 4th edition of the *British Medical Journal.*
>
> Analyzing data from nearly 25,000 Danish men and women, Dr. Eva Prescott of the University of Copenhagen and colleagues report that women who smoke have a 2.24 relative risk of myocardial infarction, or heart attack, compared with nonsmokers. This is significantly higher than the relative risk of male smokers compared with nonsmokers.

Unfortunately, the only information that we are given, for each sex, is the *relative risk* of a heart attack for smokers. What's missing from the news article is any mention of the estimated risk of a heart attack for any group of interest. A look at those risks makes the research interpretation debatable.

The *British Medical Journal* article (Prescott, Hippe, Schnor, and Vestbo, 1998) that originally reported the research provided more complete information about the risks. That information is displayed in Table 4.6. Based on the data in the table, do you agree with the conclusion stated in the Web article?

Table 4.6 Smoking, Gender, and the Risk of a Heart Attack

	Sample Size	Heart Attacks	Risk of Heart Attack
Men			
Smokers	8490	902	10.62% (902/8490)
Nonsmokers	4701	349	7.42% (349/4701)
Women			
Smokers	6461	380	5.88% (380/6461)
Nonsmokers	5011	132	2.63% (132/5011)

The researchers focused on relative risk, which in this case is the ratio of the risks of a heart attack for smokers compared to nonsmokers. The relative risks for women and men are calculated as follows:

- For women who smoke, the relative risk of a heart attack is 5.88%/2.63% = 2.24.
- For men who smoke, the relative risk of a heart attack is 10.62%/7.42% = 1.43.

The first sentence of the news article says, "Women who smoke have a greater than 50% higher risk of a heart attack than male smokers..." That is an incorrect statement. In Table 4.6, we can see that women who smoke have a lower risk of a heart attack than males (5.88% risk for women compared to 10.62% risk for males). The researchers actually compared relative risks, not risks; the relative risk of a heart attack for women smokers, 2.24, is about 50% higher than the relative risk for male smokers, 1.43.

In a subsequent issue of the *British Medical Journal,* several letter writers argued that the correct interpretation of these data is as follows:

- For both smokers and nonsmokers, men have a higher risk of heart attack than women do. (See the risks in Table 4.6.)
- For men, the difference in the risks of a heart attack for smokers versus non-smokers is 10.62% − 7.42% = 3.20%. For women, the difference in the risks of a heart attack for smokers versus non-smokers is 5.88% − 2.63% = 3.25%. In other words, the additive effect of smoking, the amount that smoking adds to the risk of a heart attack, is almost the same for the two sexes.

The researchers' response to these letters was that they believed that it was valid to consider the multiplicative effect of smoking on risk. Dr. Prescott and her colleagues believe it is important that smoking multiplies the risk of a heart attack by 2.24 for women but by only 1.43 for men.

The moral of the story here is that a *relative risk* is affected by the size of the *baseline risk,* so it is important to know the baseline risk. For example, if a risk increases to 3% from a baseline risk of 1%, the relative risk is 3. If a risk increases to 22% from a baseline risk of 20%, the relative risk is only 22/20 = 1.1, although the difference in risks is still 2%. Is an increase in risk from 1% to 3% more serious than an increase from 20% to 22%? That may depend on the situation.

THOUGHT QUESTION 4.4 If you were a frequent beer drinker and were worried about getting colon cancer, would it be more informative to you to know the *risk* of colon cancer for frequent beer drinkers or the *relative risk* of colon cancer for frequent beer drinkers compared to nondrinkers? Which of those statistics would likely be of more interest to the media? Explain your responses.*

4.3 The Effect of a Third Variable and Simpson's Paradox

In the first three chapters, you have seen several examples in which a confounding or lurking variable may have affected the relationship between an explanatory variable and a response variable. An observational study is one in which the researchers only measure or observe all variables and do not assign any treatments or conditions. In such studies, a confounding variable might explain an apparent relationship between an explanatory and response variable or, in some instances, it can mask a relationship. Whenever the data are the product of an observational study, you should carefully consider the possibility that a third variable may affect an observed relationship.

Example 4.9 **Example 2.2 Revisited: Sleep-Time Lighting, Child Vision, and Parents' Vision** Table 4.1 and Example 2.2 gave data showing that children who slept with some type of light in their room before the age of 2 had a higher incidence of nearsightedness later in childhood than children who slept in complete darkness. In Example 2.2, we mentioned the possibility that parental vision could be a confounding factor that affects the child's vision and the type of nighttime lighting used in the child's room. Two studies done after the original research found this to be true. Both studies found that nearsighted parents were more likely to use a nightlight in their child's room than parents with good vision (Zadnik et al., 2000, and Gwiazda et al., 2000). Nearsightedness is genetic, so nearsighted children are more likely to have nearsighted parents who, in turn, are more likely to use some form of nighttime lighting in their child's bedroom.

Simpson's Paradox

Occasionally, the effect of a confounding factor is strong enough to produce a paradox known as **Simpson's paradox**. The paradox is that the direction of a relationship or difference is reversed within subgroups compared to the direction of the relationship within the whole group. This is caused by a combination of differing sample sizes within the subgroups and differing magnitudes of response variable summaries. The next two examples illustrate Simpson's paradox.

Example 4.10 **U.S. Unemployment in 2009 and 1982** The United States experienced economic recessions in 1982 and in 2009. Which year had the higher unemployment rate? In October 2009, the national unemployment rate was 10.2%, slightly lower than the 10.8% unemployment rate in November and December of 1982. Paradoxically, the October 2009 unemployment rate was higher than the November/December 1982 rate for workers in every educational subgroup—for college graduates, for those with some college, for high school graduates, and for high school dropouts.

How could the overall unemployment rate have been lower in 2009 when it was higher for every educational subgroup? The answer is that the potential workforce in 2009 had a higher proportion of college graduates and workers with some college, and these groups tend to have lower unemployment rates. Thus the overall national rate was weighted more heavily with more educated workers in 2009 than in 1982 and that caused the overall rate to be lower (Source: Tuna, *Wall Street Journal*, December 2, 2009).

***HINT:** Suppose the relative risk is 3.0. Does that mean the same to you as an individual if the disease is rare as it does if it is common?

Example 4.11 **Blood Pressure and Oral Contraceptive Use** This is a hypothetical example of an observational study done to examine the association between oral contraceptive use and blood pressure. Although the data are hypothetical, they are similar to data from several actual studies of the same problem. Suppose that 2400 women are categorized according to whether or not they use oral contraceptives and whether or not they have high blood pressure. The results shown in Table 4.7 indicate that the percentage with high blood pressure is about the same among oral contraceptive users as it is for nonusers. In fact, a slightly higher percentage of the nonusers have high blood pressure.

We're certain that you can think of many factors that affect blood pressure. If the users and nonusers of oral contraceptives differ with respect to one of these factors, that factor confounds the results in Table 4.7. Age is a critical factor. The users of oral contraceptives tend to be younger than the nonusers. This is important because blood pressure increases with age.

Table 4.7 Percentage with High Blood Pressure for Users and Nonusers of Oral Contraceptives

	Sample Size	Number with High Blood Pressure	% with High Blood Pressure
Use Oral Contraceptives	800	64	64 of 800 = 8.0%
Don't Use Oral Contraceptives	1600	136	136 of 1600 = 8.5%

One way to control for the effect of age is to create separate contingency tables for women in different age groups. Table 4.8 divides the data from our hypothetical study into two age groups. In each age group, the percentage with high blood pressure is higher for the users than for the nonusers. This reverses the direction of the relationship that we see when the age factor is not considered. In Table 4.7, not controlling for age differences masked the true nature of the relationship. Note that the older women were less likely to be oral contraceptive users but more likely to have high blood pressure.

Table 4.8 Controlling for the Effect of Age

	Age 18–34		Age 35–49	
	Sample Size	*n* and % with High Blood Pressure	Sample Size	*n* and % with High Blood Pressure
Use Oral Contraceptives	600	36 (6%)	200	28 (14%)
Don't Use Oral Contraceptives	400	16 (4%)	1200	120 (10%)

4.3 Exercises are on pages 138–139.

IN SUMMARY Confounding Variables and Simpson's Paradox

- **Confounding variables** may explain or partially explain an observed relationship when the data have been gathered in an observational study.
- **Simpson's paradox** occurs when the direction of a relationship is reversed within subgroups compared to the direction of the relationship within the total group.

Section 4.4 can be studied after Chapters 10 through 13 have been covered with no loss in the continuity of the material.

4.4 Assessing the Statistical Significance of a 2 × 2 Table

Observed data often are a sample from a larger population. When this is the case, the purpose typically is to use the sample information to make generalizations about the population. Because we are using sample evidence to *infer* something about the population, rather than being certain by measuring the entire population, the statistical methods that are used are called **inferential statistics**.

This section introduces a special case of **hypothesis testing**, one of the two most common statistical procedures making up inferential statistics. The other common inference procedure is the use of a *confidence interval* to estimate a population value, a method that will be introduced in Chapter 5. Hypothesis testing and confidence intervals will be covered in detail in Chapters 10 through 16, but the brief introduction here and in Chapter 5 will help you to become familiar with some basic concepts and prepare you for those chapters.

When we have sample data for two categorical variables, the question of interest is whether the two variables are related in the population represented by the sample. Specifically, we wish to assess whether the relationship observed in the sample is strong enough to allow us to infer that the relationship also holds in the population. The research question asked is: On the basis of the sample data, can we infer that the two variables are related in the population?

Statistical Significance

In Chapter 1 (p. 5), the term *statistically significant* was defined as follows:

> A **statistically significant relationship** or difference is one that is large enough to be unlikely to have occurred in the observed sample if there is no relationship or difference in the population.

When we call an observed relationship *statistically significant*, we are inferring that a relationship exists in the population. We make this inference by establishing that if there were really no relationship in the population, it would be unlikely that we would have observed such a strong relationship in the sample.

Example 4.12 **Case Study 1.6 Revisited: Aspirin and the Risk of a Heart Attack** Case Study 1.6 described the Physicians' Health Study, in which more than 22,000 male physicians were randomized to take either aspirin or a placebo daily over a 5-year period. This was done to see if taking aspirin daily would reduce the risk of a heart attack. Results are shown in Table 4.9, which shows a reduced heart attack risk in the aspirin group. Can we infer that daily use of aspirin will reduce the risk of a heart attack in the larger population represented by the sample? The rate of heart attacks was 9.42 per 1000 physicians (or 0.942%) in the aspirin group compared to 17.13 heart attacks per 1000 physicians (1.713%) in the placebo group. The sample difference was found to be *statistically significant*, which means that a sample difference this large would be unlikely to occur if there is actually no difference within the larger population. As a result, it can be inferred that the observed difference reflects an actual difference within the population.

Table 4.9 Results from the Physicians' Health Study

The Effect of Aspirin on Heart Attacks

Treatment	Heart Attacks	Doctors in Group	Attacks per 1,000 Doctors
Aspirin	104	11,037	9.42
Placebo	189	11,034	17.13

THOUGHT QUESTION 4.5 A random sample includes 110 women and 90 men. Of the women, approximately 9% are left-handed, while approximately 11% of the men are left-handed. Based on these observed data, do you think there is a relationship between sex and handedness in the population represented by this sample? Why or why not?*

***HINT:** Consider *how many* men and women in the samples are left-handed. Now consider how the percentages would change if one more or one fewer had been left-handed.

The Five Steps to Determining Statistical Significance

Five steps are required in any hypothesis testing situation. These steps lead to a decision about whether the observed result is statistically significant. In this section, we describe how these steps are implemented for a 2×2 contingency table. In Chapter 15, you will learn how to implement them for a contingency table of any size, and in Chapters 12 and 13 you will learn how they apply to various research questions.

The five steps, as they apply to contingency tables, are as follows:

Step 1: Determine the null and alternative hypotheses, two possible inferences about the population. For two categorical variables, the null hypothesis is that the variables are not related and the alternative hypothesis is the two variables are related.

Step 2: Summarize the data into an appropriate test statistic after first verifying that necessary data conditions are met. For two categorical variables, the test statistic is called a chi-square statistic.

Step 3: Find the *p*-value, the probability that the test statistic would be as extreme as it is, or more so, calculated assuming the null hypothesis is true.

Step 4: Decide whether the result is statistically significant based on the *p*-value. The conventional rule is that a result is statistically significant when the *p*-value is less than or equal to .05.

Step 5: Report the conclusion in the context of the situation.

Step 1: Null and Alternative Hypotheses

We are considering whether the sample data permit us to infer that two variables are related in the population. Said another way, we are assessing two possible hypotheses about the population. In statistical language, the possibilities are called the **null hypothesis** and the **alternative hypothesis**, and they can be stated as follows:

Null hypothesis: The two variables are not related in the population.

Alternative hypothesis: The two variables are related in the population.

The two hypotheses just given are specific to the context of two-way tables. In Chapters 12 and 13, you will learn how to write null and alternative hypotheses for a broad range of research questions.

Step 2: The Chi-Square Statistic

A statistic known as the **chi-square statistic** (χ^2) is used to examine the association between two categorical variables. To use this statistic, the sample should be representative of the population. The value of this statistic is sensitive to the strength of the relationship between the two variables. A relatively large value indicates a strong relationship in the sample. We will describe how to compute the chi-square statistic later in this section.

Step 3: The *p*-Value of the Chi-Square Test

A relatively large value of the chi-square statistic reflects a strong relationship in the sample, but how large should the value be for us to be able to declare statistical significance? In practice, this question is transformed into a different but equivalent question: If there is actually no relationship in the population, what is the likelihood that the chi-square statistic could be as large as it is or even larger? The answer to this question is called the *p-value* of the test, and it is used to decide whether the sample relationship is statistically significant.

DEFINITION The ***p*-value** for a chi-square test is computed by assuming that the null hypothesis is true and then determining the likelihood that the chi-square statistic would be as large as the one observed, or even larger. It answers the question "How likely is it that a relationship of the magnitude observed, or one even stronger, would occur in the sample if there is no relationship in the population?"

The p-value will be part of the computer output provided by almost any statistical software. If you are computing the chi-square statistic "by hand" instead of using statistical software, you can use Excel to find the p-value; see the Excel Tip on page 128. Alternatively, you can determine statistical significance using a rule about the value of the chi-square statistic for a 2×2 table. The decision process is as follows.

Steps 4 and 5: Making and Reporting a Decision

A relatively large chi-square statistic and subsequently a relatively small p-value provide evidence that a real relationship exists in the population. To determine whether to decide in favor of the alternative hypothesis that the two variables are related in the population, we will use the rule that most researchers commonly use:

- When the p-value is less than or equal to .05 (5%), we say that the observed relationship in the sample is statistically significant. In this event, we reject the null hypothesis in favor of the alternative hypothesis and infer that the two variables are related in the population. For a 2×2 contingency table, an equivalent rule is that we declare statistical significance when the chi-square statistic value is greater than or equal to 3.84.
- When the p-value is greater than .05 (5%), we say that the observed relationship is not statistically significant. In that event, we cannot reject the null hypothesis. For a 2×2 contingency table, an equivalent rule is that we cannot declare statistical significance when the chi-square statistic value is less than 3.84.

Example 4.13 **Sex and Opinion about Banning Cell Phone Use while Driving** In a survey done in 2010, students in a statistics class were asked whether they favored or opposed banning the use of cell phones while driving. Table 4.10 summarizes the responses by sex. Among the females in the class, 64.8% (68/105) favored banning cell phone use by drivers. Among the males, only 44.6% (37/83) favored a ban. The difference between these percentages indicates that sex and opinion are related in the sample.

Table 4.10 Sex and Opinion about Banning Cell Phone Use While Driving

	Ban Cell Phone Use by Drivers?		
Sex	**Oppose**	**Favor**	**Total**
Female	37 (35.2%)	68 (64.8%)	105 (100%)
Male	46 (55.4%)	37 (44.6%)	83 (100%)
Total	83 (44.15%)	105 (55.85%)	188 (100%)

Source: Robert Heckard.

On the basis of the relationship observed in the sample, can we generalize that sex and opinion about banning cell phone use by drivers are related in the larger population represented by this sample of students? We answer this question by carrying out a hypothesis test.

In any hypothesis test, the first step is to write the null and alternative hypotheses. For this example, they are:

Null hypothesis: Sex and opinion about banning cell phone use by drivers are not related.

Alternative hypothesis: Sex and opinion about banning cell phone use by drivers are related.

Steps 2 and 3 are to compute the chi-square statistic and determine the p-value. For these data, the Minitab computer program gives the result that the chi-square value is 7.659 and the p-value is .006. Steps 4 and 5 are to make and report the decision. Because the p-value is less than .05, we can say that the relationship is statistically significant. Equivalently, we can declare statistical significance because the chi-square value is greater than 3.84. The null hypothesis can be rejected in favor of the alternative. We can infer that sex and opinion about banning cell phone use while driving are related variables in the population represented by these students.

MINITAB TIP **Computing a Chi-Square Test for a Two-Way Table**

- If the raw data are stored in columns of the worksheet, use **Stat > Tables > Cross Tabulation and Chi-Square**. Specify a categorical variable in the "For rows" box and a second categorical variable in the "For columns" box. Then click the ***Chi-Square*** button and select "Chi-Square analysis."
- If the data are already summarized into counts, enter the table of counts (excluding totals) into columns of the worksheet, and then use **Stat > Tables > Chi-Square Test (Table in Worksheet)**. In the dialog box, specify the columns that contain the counts.

EXCEL TIP The p-value can also be computed by using Microsoft Excel. The function CHIDIST(x,df) provides the p-value, where x is the value of the chi-square statistic and df is a number called "degrees of freedom," which will be explained later in this book. The formula for df is (# of rows − 1)(# of columns − 1). For instance, corresponding to the information in Example 4.13, df = (2 − 1)(2 − 1) = 1, and the p-value is CHIDIST(7.659,1) = .005649, or about .006 as given by Minitab.

Expected Counts: What to Expect When the Null Hypothesis Is True

A chi-square statistic measures the overall difference between the **observed counts**, the actual counts in the data, and a set of **expected counts**, which are the hypothetical counts that would occur if the two variables are not related. Thus the first step in the calculation of a chi-square statistic is to determine a table of expected counts.

FORMULA **Calculating Expected Counts**

The expected count for each cell can be calculated as

$$\frac{\text{Row total} \times \text{Column total}}{\text{Total } n \text{ for table}}$$

Do not round off the expected counts to whole numbers. It is standard practice to retain at least two digits after the decimal point.

A table of expected counts has the following two properties:

1. There is no association between the two variables. For instance, there are no differences between the conditional percentages in the rows (or columns) of a table of expected counts.
2. The expected counts in each row and column sum to the same totals as the observed numbers.

The second property can be utilized when computing the expected counts. For a 2×2 table, it is only necessary to use the formula for expected counts for one of the cells. The other three expected counts can be found using subtractions that make the sums in each row and column match the corresponding sums for the observed counts.

TECHNICAL NOTE **Verifying Conditions for the Test**

A sufficiently large, representative sample is necessary to use the method in this section. Use the chi-square statistic only when at least three of the four expected counts are 5 or more and no expected counts are less than 1.

Calculating the Chi-Square Statistic

The specific way in which the chi-square statistic (χ^2) measures the difference between the observed and expected counts is relatively simple:

First, for each cell in the table, we compute

$$\frac{(\text{Observed count} - \text{Expected count})^2}{\text{Expected count}}$$

Then we total these quantities over all cells of the table:

$$\chi^2 = \text{Sum of } \frac{(\text{Observed count} - \text{Expected count})^2}{\text{Expected count}}$$

The first of the following two examples demonstrates the calculation of expected counts and the chi-square statistic. The second carries out all five steps of the hypothesis testing procedure.

Example 4.14 **Example 4.13 Revisited: Expected Counts and Chi-Square Statistic for Sex and Opinion about Banning Cell Phone Use while Driving** The observed data for Example 4.13 are given in the following table.

	Ban Cell Phone Use by Drivers?		
Sex	Oppose	Favor	Total
Female	37	68	105
Male	46	37	83
Total	83	105	188

Table 4.11 gives the expected counts for these data. For the "Female, Oppose" cell, the calculation of the expected count is

$$\frac{\text{Row total} \times \text{Column total}}{\text{Total } n \text{ for table}} = \frac{105 \times 83}{188} = 46.36$$

The other expected counts can be determined using subtractions that make the row and column totals match the corresponding totals for the observed data. For example, there were 105 females in the survey, so the expected count for females favoring a ban is $105 - 46.36 = 58.64$. Expected counts for the males can be found by subtracting the expected counts for the females from the observed column totals.

The row percentages for the expected counts are the same for the two sexes. For instance, based on expected counts, the percent opposed to a ban is 44.15% both for females (46.36/105) and for males (36.64/83). In other words, the expected counts are counts that indicate no relationship between the two variables.

Note that the expected counts are not whole numbers, so they are not actual counts that we would expect to observe in a particular study. Instead, they are averages of what we would expect to observe over many studies if in fact there were no relationship between the two variables in the population. When using them to compute a chi-square statistic, do *not* round the expected counts to whole numbers.

Table 4.11 Expected Counts for Sex and Opinion about Banning Cell Phone Use While Driving

	Ban Cell Phone Use by Drivers?		
Sex	**Oppose**	**Favor**	**Total**
Female	$\frac{105 \times 83}{188} = 46.36$	$105 - 46.36 = 58.64$	105
Male	$83 - 46.36 = 36.64$	$105 - 58.64 = 46.36$	83
Total	83	105	188

The value of the chi-square statistic is computed as follows:

$$\chi^2 = \text{Sum of } \frac{(\text{Observed count} - \text{Expected count})^2}{\text{Expected count}}$$

$$= \frac{(37 - 46.36)^2}{46.36} + \frac{(68 - 58.64)^2}{58.64} + \frac{(46 - 36.64)^2}{36.64} + \frac{(37 - 46.36)^2}{46.36} = 7.66$$

Example 4.15 **Breast Cancer Risk Stops Hormone Replacement Therapy Study** On July 17, 2002, the *Journal of the American Medical Association* published the results of a study that affected the lives of millions of women. The study was the first large randomized experiment to test the effects of combined estrogen and progestin, the most commonly prescribed hormone replacement therapy for postmenopausal women. The big news was that "On May 31, 2002, after a mean of 5.2 years of follow-up, the data and safety monitoring board recommended stopping the trial of estrogen plus progestin vs placebo because the test statistic for invasive breast cancer exceeded the stopping boundary for this adverse effect and the global index statistic supported risks exceeding benefits" (Writing Group for the Women's Health Initiative Investigators, 2002, p. 321).

The study measured several medical outcomes, but the two that received the most attention were breast cancer and coronary heart disease. Table 4.12 shows the results for breast cancer (results for coronary heart disease are given in Exercise 4.58).

Table 4.12 Results of a Randomized Experiment Comparing Hormone Therapy and Placebo

	Invasive Breast Cancer?			
	Yes	**No**	**Total**	**Risk of Breast Cancer**
Hormones	166	8,340	8,506	.0195
Placebo	124	7,978	8,102	.0153
Total	290	16,318	16,608	.0175

Within the sample, there is a higher risk for breast cancer in the group that used hormones than in the placebo group. Is there a statistically significant effect of the hormone treatment on the risk of breast cancer? Here are the five steps of a hypothesis test:

Step 1: The Hypotheses

Null hypothesis: Occurrence of breast cancer is unrelated to hormone therapy.

Alternative hypothesis: Occurrence of breast cancer is related to hormone therapy.

Step 2: The Conditions and Test Statistic

The expected count for the "Hormones, Yes" cell is [(8506)(290)]/16,608 = 148.53. The other expected counts can be found similarly, or they can be found by subtraction because the row and column totals are the same as they are for the observed counts. For instance, the expected count for the "Placebo, Yes" cell is 290 − 148.53 = 141.47. The sample size condition is met, since all expected counts exceed 5. The chi-square statistic is

$$= \frac{(166 - 148.53)^2}{148.53} + \frac{(8340 - 8357.47)^2}{8357.47} + \frac{(124 - 141.47)^2}{141.47} + \frac{(7978 - 7960.53)^2}{7960.53} = 4.288$$

Steps 3, 4, and 5: The *p*-Value and Decision in Context

Using Excel, the *p*-value is found as CHIDIST(4.288,1) = .038382, or about .04. Using the standard convention, the result is statistically significant because $.04 < .05$. In the context of the problem, this means that there is a statistically significant relationship between hormone therapy and occurrence of breast cancer *in the population similar to the women in this experiment*. Because the data are from a randomized experiment, it can be concluded that hormone therapy *causes* a change in the incidence of breast cancer. It was for this reason that the study was terminated early.

Note that the relative risk of breast cancer for the women in the study is .0195/.0153 = 1.27, so that women who took hormones had an increased risk of about 27%. The baseline risk was .0153, and the difference in risk for the two groups was .0042. This represents about four additional cases of breast cancer per 1000 women taking hormones.

Factors That Affect Statistical Significance

In general, whether we can infer that an observed relationship represents a relationship in the population depends on two factors:

- The strength of the observed relationship
- How many people were studied

For 2×2 tables, the strength of the relationship can be measured by the difference between the two categories of the row variable with respect to the percentages falling into a category of the column variable. For instance, in Example 4.12, we saw that 64.8% of the females favored a ban on the use of cell phones while driving, while only 44.6% of the males favored a ban. The difference between these two percentages indicates the strength of the relationship between the two variables.

The sample size of a study also affects the significance of the results. Imagine, for example, that a psychiatric researcher compares the effectiveness of two treatments for depression by conducting a randomized experiment in which only ten patients receive each treatment. The results are that after 3 months, eight of the ten patients using treatment A show improvement, but among those who used treatment B, only five of ten patients show improvement. Although the difference between the percentages is large (80%–50%), the study is too small to safely infer that treatment A would be better in the larger population of patients. The chi-square statistic and *p*-value are 1.978 and .16, respectively. On the other hand, if 100 patients were in each treatment group and the same percentages were observed (80% and 50%), the chi-square statistic would be 19.78 and the *p*-value would 8.7×10^{-8} (nearly 0). With the larger sample size, we would almost certainly believe that treatment A was better than treatment B.

In the following example, you will see that with a large sample size, a relatively small arithmetic difference can be statistically significant. Remember that "statistical significance" means that we are convinced that a real relationship exists in the larger population represented by the sample. It makes sense that if we have a very large sample and it is exhibiting a relationship, we would be more convinced that a relationship exists in the population than we would be if a very small sample exhibited that same relationship.

Example 4.16 **Case Study 1.6 Revisited: Aspirin and Heart Attacks** In Case Study 1.6, the possible categories for the explanatory and response variables were as follows:

Explanatory variable = took placebo or took aspirin
Response variable = heart attack or no heart attack

```
Expected counts are printed below observed counts

                Heart Attack?
             Yes          No        Total
Aspirin      104       10933        11037
          146.52    10890.48

Placebo      189       10845        11034
          146.48    10887.52

Total        293       21778        22071

Chi-Sq = 12.339 + 0.166 +
         12.343 + 0.166 = 25.014
DF = 1, P-Value = 0.000
```

Read the original source on the companion website, http://www.cengage.com/statistics/Utts4e.

Figure 4.4 Minitab output for aspirin study (Example 4.16)

Figure 4.4 shows the results of asking the Minitab program to compute the chi-square statistic. Because the *p*-value, .000, is less than .05, we can reject the null hypothesis of no relationship and declare statistical significance. Even further, the *p*-value of .000, which is considerably less than .05, tells us that there is almost no chance that we could observe such a strong relationship if there is really no difference between the effects of aspirin and placebo on heart attack risk for male physicians. Excel provides the *p*-value to more decimal places, and gives 5.7×10^{-7}.

In the placebo group, 189/11,034 = 1.71% had a heart attack during the study period, while in the aspirin group, 104/11,037 = 0.94% had a heart attack. The difference has practical importance because the percent decrease in risk for those taking aspirin is (1.71 − 0.94)/1.71 = 45%. However, the arithmetic difference in the percentages who experienced a heart attack is 1.71% − 0.94% = 0.77%, or less than 1%. If a smaller study had been done, the researchers may not have been able to declare statistical significance, but since they used a sample with over 22,000 participants, they were able to detect this important relationship.

Practical Versus Statistical Significance

While statistical significance is an informative label, it does not necessarily mean that the relationship between the two variables has **practical significance**. Practical significance speaks to the magnitude of the relationship and whether or not that magnitude is important. A table based on a very large number of observations will have little trouble achieving the status of statistical significance even if the relationship between the two variables in the population is only minor. When interpreting a statistically significant result, find out the magnitude of the relationship or difference so that you can consider whether or not it has practical importance.

Interpreting a Nonsignificant Result

When we cannot claim statistical significance, we have to be careful about how we state the conclusion. The correct interpretation of a **nonsignificant result** is that the sample results are not strong enough to safely conclude that there is a relationship in the population. A *p*-value that is too large for us to declare significance means that the observed relationship would not be an unusual occurrence if the null hypothesis were true. This is not the same as saying that the null hypothesis is true.

In the following case study, the difference between the relevant conditional percentages is larger than the difference we saw in the aspirin and heart attacks example (Example 4.16). The result of the chi-square test, however, will be that we cannot declare statistical significance. You will see that the sample size for the case study is much smaller than the sample size for the aspirin study. Remember that the sample size affects our ability to declare significance.

4.4 Exercises are on pages 139–141.

SPSS TIP **Computing a Chi-Square Test for a Two-Way Table**

- Use **Analyze > Descriptive Statistics > Crosstabs**. Enter variable names for "Row(s)" and "Column(s)" and then click ***Statistics*** to select "Chi-Square."
- Use the ***Cells*** button to request expected counts and/or conditional percentages of interest.

CASE STUDY 4.2 Drinking, Driving, and the Supreme Court

In the early 1970s, a young man challenged an Oklahoma state law that prohibited the sale of 3.2% beer to males under 21 but allowed its sale to females in the same age group. The case (*Craig v. Boren*, 429 U.S. 190, 1976) was ultimately heard by the U.S. Supreme Court.

Laws are allowed to use gender-based differences as long as they "serve important governmental objectives" and "are substantially related to the achievement of these objectives" (Gastwirth, 1988, p. 524). The defense argued that traffic safety was an important governmental objective and that data clearly show that young males are more likely to have alcohol-related accidents than young females.

The Supreme Court examined evidence from a "random roadside survey" that measured information on age, sex, and whether or not the driver had been drinking alcohol in the previous 2 hours. Although the survey was called a "random" survey of drivers, it probably was not. In roadside surveys, police tend to stop all drivers at certain locations at the time of the survey. This procedure does not really provide a random sampling of drivers in an area, but we'll treat it as though it does. Table 4.13 gives the results of the roadside survey for the drivers under 20 years of age.

Table 4.13 Results of Roadside Survey for Young Drivers

	Drank Alcohol in Last Two Hours?			
	Yes	**No**	**Total**	**Percent Who Drank**
Males	77	404	481	16.0%
Females	16	122	138	11.6%
Total	93	526	619	15.0%

Source: Gastwirth, 1988, p. 526.

Note that the percentage of young men who had been drinking alcohol is slightly higher than the percentage of young women. The difference is 16% − 11.6% = 4.4%. However, we cannot rule out chance as a reasonable explanation for this difference. In other words, if there really is no difference between the percentages of young male and female drivers in the population who drink and drive, we could possibly see a difference as large as the one observed in a sample of this size.

In Figure 4.5, we present the results of asking the Minitab program to compute the chi-square statistic for this example. The chi-square summary statistic is 1.637, and the *p*-value for this statistic is .201. This *p*-value tells us that if there is really no association in the population (the null hypothesis), there is about a 20% chance that the sample would have a chi-square statistic as large as 1.637 or larger. In other words, the observed relationship could easily have occurred even if there is no relationship in the population represented by the sample.

```
Expected counts are printed below observed counts

            Drank in Last 2 Hours?
              Yes        No       Total
Males          77       404         481
            72.27    408.73

Females        16       122         138
            20.73    117.27

  Total        93       526         619

Chi-Sq = 0.310 + 0.055 +
         1.081 + 0.191 = 1.637
DF = 1, P-Value = 0.201
```

Figure 4.5 Minitab output for Case Study 4.2

The Supreme Court overturned the law, concluding that "the showing offered by the appellees does not satisfy us that sex represents a legitimate, accurate proxy for the regulation of drinking and driving" (Gastwirth, 1988, p. 527). Based on the chi-square analysis, you can see why the Supreme Court was reluctant to conclude that the difference in the sample represented sufficient evidence for a real difference in the population.

IN SUMMARY Testing Hypotheses about Two Categorical Variables

- A **null hypothesis** about two categorical variables is that the two variables are not related in the population represented by the sample.
- An **alternative hypothesis** about two categorical variables is that the two variables are related in the population represented by the sample.
- A **chi-square** test is used to decide between the null and alternative hypotheses.
- The ***p*-value** for a chi-square test is the probability that the value of the chi-square statistic would be as large as it is, or larger, if the null hypothesis were true.
- A standard decision rule is to decide in favor of the alternative hypothesis when the *p*-value is .05 or less. This result is called a **statistically significant** result.

Key Terms

In Summary Boxes

Exercises

◆ Denotes that the dataset is available on the companion website, http://www.cengage.com/statistics/Utts4e, but is not required to solve the exercise.

Bold exercises have answers in the back of the text.

Section 4.1

Skillbuilder Exercises

4.1 ◆ The following table shows data for grades usually achieved in school and how often the respondent puts on sunscreen when going out in the sun for more than 1 hour. Respondents are 12th-grade participants in the 2003 Youth Risk Behavior Surveillance System survey. The survey, sponsored by the U.S. Centers for Disease Control and Prevention, is a national survey of high school students. (Raw data are in the **YouthRisk03** dataset on the companion website.)

Grades and Sunscreen Use for Twelfth Graders

	Sunscreen Use			
Grade	**Never or Rarely**	**Sometimes**	**Always or Most Times**	**Total**
A's and B's	1322	450	285	2057
C's	568	83	47	698
D's and F's	85	15	3	103
Total	1975	548	335	2858

a. Among students who usually get A's and B's in school, what percentage never or rarely uses sunscreen when going out in the sun for more than one hour? Explain whether this value is a row percentage or a column percentage.

b. Among students who sometimes wear sunscreen, what percentage usually gets C's in school? Explain whether this value is a row percentage or a column percentage.

c. What percentage of the overall sample usually gets A's and B's in school and also uses sunscreen always or most times when going out in the sun for more than one hour?

d. Determine a complete table of row percentages.

e. Briefly explain whether the percentages found in part (d) indicate that there is a relationship between grades usually achieved and frequency of sunscreen use.

4.2 The following two-way table of counts summarizes whether respondents smoked or not and whether they had ever divorced or not for persons in the 1991–1993 General Social Surveys who had ever been married.

	Ever Divorced?		
Smoke?	**Yes**	**No**	**Total**
Yes	238	247	485
No	374	810	1184
Total	612	1057	1669

Data source: http://sda.berkeley.edu/archive.htm.

a. Among those who smoked, what percentage has ever been divorced? Is this value a row percentage or a column percentage?

b. Among those who have ever been divorced, what percentage smoked? Is this value a row percentage or a column percentage?

c. What percentage of the overall sample did not smoke and had not ever been divorced?

d. Determine a complete set of row percentages.

e. Briefly explain whether the percentages found in part (d) indicate that there is an observed relationship between smoking habits and whether persons ever married had ever divorced or not.

4.3 Each fall, auditions for the band and orchestra are held at a large university. Last fall, the numbers of males and females in each class who auditioned follow:

Class	Female	Male	Total
Freshman	170	100	270
Sophomore	50	50	100
Junior	60	20	80
Senior	20	30	50
Total	300	200	500

a. Calculate the row percentage for freshman females and explain what it means.

b. Calculate the column percentage for freshman females and explain what it means.

c. Which class had the highest percentage of female applicants? Support your answer with numbers.

d. Which sex had a higher percentage of sophomore applicants? Support your answer with numbers.

4.4 The following two-way table of counts summarizes data for age group and frequency of reading newspapers for respondents in the 2008 General Social Survey.

	Frequency of Reading Newspapers				
Age Group	**Every Day**	**A Few Times a Week**	**Once a Week**	**Less Than Once a Week**	**Total**
18–29	45	68	38	83	234
30–49	118	125	100	175	518
≥50	260	99	69	126	554
Total	423	292	207	384	1306

Data source: http://sda.berkeley.edu/archive.htm.

a. Determine a table of row percentages for these data.

b. Briefly explain whether the percentages found in part (a) indicate that age and frequency of reading newspapers are related in the sample. Briefly describe any differences among the age groups.

4.5 For each pair of variables, indicate whether or not a two-way table would be appropriate for summarizing the relationship. In each case, briefly explain why or why not.

a. Political party (Republican, Democrat, etc.) and opinion about a new gun control law.

b. Weight (pounds) and height (inches).

4.6 For each pair of variables, indicate whether or not a two-way table would be appropriate for summarizing the relationship. In each case, briefly explain why or why not.

a. Age group (under 20, 21–29, etc.) and rating of a song on 1 to 5 scale (1 = hate it, 5 = love it).

b. Gender and opinion about capital punishment.

c. Head circumference (centimeters) and gender.

4.7 Suppose a study on the relationship between gender and political party included 200 men and 200 women and found 180 Democrats and 220 Republicans. Is that information sufficient for you to construct a contingency table for the study? If so, construct the table. If not, explain why not.

General Section Exercises

4.8 In the 2008 General Social Survey, religious preference and opinion about when premarital sex might be wrong were among the measured variables. The contingency table of counts for these variables is as follows:

Religious Preference and Opinion about Premarital Sex

	When Is Premarital Sex Wrong?				
Religion	**Always**	**Almost Always**	**Sometimes**	**Never**	**Total**
Protestant	221	54	98	288	661
Catholic	45	17	54	179	295
Jewish	2	1	8	18	29
None	15	10	32	164	221
Other	20	7	12	41	80
Total	303	89	204	690	1286

Source: http://sda.berkeley.edu/archive.htm.

a. For each religious preference category, determine the percentage of respondents who think premarital sex is always wrong.

b. Do the percentages computed in part (a) indicate that there is a relationship between the two variables? Briefly explain why or why not.

4.9 Students in a class were asked whether they preferred an in-class or a take-home final exam and were then categorized as to whether or not they had received an A on the in-class midterm. Of the 25 A students, 10 preferred a take-home exam, while of the 50 non-A students, 30 preferred a take-home exam.

a. Display the data in a contingency table.

b. In the relationship between grade on the midterm and opinion about type of final, which variable is the response variable and which is the explanatory variable?

c. Determine an appropriate set of conditional percentages for determining whether there is a relationship between grade on the midterm and opinion about the type of final. Based on these percentages, does it appear that there is a relationship? Why or why not?

4.10 Do grumpy old men have a greater risk of having coronary heart disease than men who aren't so grumpy? Harvard Medical School researchers examined this question in a prospective observational study reported in the November 1994 issue of *Circulation* (Kawachi et al., 1994). For 7 years, the researchers studied men between the ages of 46 and 90. All study participants completed a survey of anger symptoms at the beginning of the study period. Among 199 men who had no anger symptoms, there were 8 cases of coronary heart disease. Among 559 men who had the most anger symptoms, there were 59 cases of coronary heart disease.

a. Construct a contingency table for the relationship between degree of anger and the incidence of heart disease.

b. Among those with no anger symptoms, what percentage had coronary heart disease?

c. Among those with the most anger symptoms, what percentage had coronary heart disease?

d. Draw a bar graph of these data. Based on this graph, does there appear to be an association between anger and the risk of coronary heart disease? Explain.

4.11 In a study done in England, Voss and Mulligan (2000) collected data on height (short or not) and whether or not the student had ever been bullied in school for 209 secondary school students.

A student was categorized as short if he or she was below the third percentile for height on school entry, but the researchers intentionally sampled in a way that made short students constitute almost half of the sample.

The following table displays a contingency table of the data.

Height and Bullying in School

	Ever Bullied?		
Height	**Yes**	**No**	**Total**
Short	42	50	92
Not Short	30	87	117
Total	72	137	209

a. Among students in the "short" category, what percentage have ever been bullied?

b. Among students in the "not short" category, what percentage have ever been bullied?

c. Is there a relationship between height and the likelihood of having been bullied? Briefly justify your answer.

4.12 In a class survey, statistics students at a university were asked, "Regarding your weight, do you think you are: About right? Overweight? Underweight?" The following table displays the results by sex:

Sex and Perception of Weight

	Perception of Weight			
Sex	**About Right**	**Overweight**	**Underweight**	**Total**
Female	87	39	3	129
Male	64	3	16	83
Total	151	42	19	212

Source: The authors.

a. Write a sentence that explains what would be measured by the row percentages for this table. Make your answer specific to this situation.

b. Determine the row percentages.

c. Draw a bar graph of the row percentages.

d. Briefly describe how males and females differ in their perceptions of weight.

e. An important objective of statistics is the use of sample information to make generalizations about a larger population. What population do you think is represented by this sample?

4.13 In a 1997 poll conducted by the *Los Angeles Times*, 1218 southern California residents were surveyed about their health and fitness habits. One of the questions was, "What is the most important reason why you try to take care of your body: Is it mostly because you want to be attractive to others, or mostly because you want to keep healthy, or mostly because it helps your self-confidence, or what?" The following table gives the percentage distribution of responses for men and for women.

	Healthy	**Self-Confidence**	**Attractive**	**Don't Know**	**Total**
Men	76%	16%	7%	1%	100%
Women	74%	20%	4%	2%	100%

Source: www.latimes.com, poll archives, study #401.

Briefly discuss whether gender and reason for taking care of your body are related variables in this sample or not.

4.14 Anton and Edward often play a game together, so they decide to see whether or not who goes first affects who wins. They keep track of 50 games, with each going first 25 times. Of the 25 times Anton went first, he won 15 times. Of the 25 times Edward went first, he won 12 times. In constructing a contingency table for these results, the two variables are "Who went first?" and "Did the person who went first win?"

a. What are the categories for each of the two variables?

b. If the explanatory variable is used to define the rows of the contingency table, which variable would be the row variable?

c. Construct a contingency table for the results.
d. Overall, what percent of the games were won by the person who went first?
e. Is there an advantage to going first? Explain.

Section 4.2

Skillbuilder Exercises

4.15 For each of the following measures, give the value that implies no difference between the two groups being compared.

a. Relative risk.
b. Odds ratio.
c. Percent increase in risk.

4.16 *This is a modification of Exercise 1.4.* The risk of getting lung cancer at some point in one's life for men who have never smoked is about 13 in 1000. The risk for men who smoke is just over 13 times the risk for nonsmokers (Source: Villeneeve and Lau, 1994).

a. Expressed as a percentage, what is the lifetime risk of lung cancer in men who never smoked?
b. What is the approximate lifetime risk of getting lung cancer for men who smoke?
c. In the introduction to this exercise it was stated that "The risk for men who smoke is just over 13 times the risk for non-smokers." Using the terminology of this chapter, what name applies to the statistic in that statement (e.g., odds, risk, relative risk)?

For Exercises 4.17 and 4.18: A study is done to compare side effects for two different drugs used to treat a medical condition. One hundred people are given each drug. Results are as shown in the following table:

	Headache?		Nausea?	
	Yes	**No**	**Yes**	**No**
Drug 1	10	90	15	85
Drug 2	5	95	5	95

4.17 For the headache side effect, compute each of the following.

a. The risk of experiencing a headache for each drug (separately).
b. The relative risk of a headache for Drug 1 compared to Drug 2.
c. The percent increase in the risk of a headache for Drug 1 compared to Drug 2.
d. The odds ratio for comparing the odds of a headache for Drug 1 compared to Drug 2.

4.18 For nausea, compute each of the following.

a. The risk of experiencing nausea for each drug (separately).
b. The relative risk of nausea for Drug 1 compared to Drug 2.
c. The percent increase in the risk of nausea for Drug 1 compared to Drug 2.
d. The odds ratio for comparing the odds of nausea for Drug 1 compared to Drug 2.

4.19 If the baseline risk of a certain disease for nonsmokers is 1% and the relative risk of the disease is 5 for smokers compared to nonsmokers, what is the risk of the disease for smokers?

4.20 a. For a relative risk of 2.1, what is the percent increase in risk?
b. For a percent increase in risk of 40%, what is the relative risk?

4.21 **a.** For a relative risk of 1.53, what is the percent increase in risk?
b. For a percent increase in risk of 140%, what is the relative risk?

General Section Exercises

4.22 *Science News* (February 25, 1995, p. 124) reported a study of 232 people aged 55 or over who had had heart surgery. The patients were asked whether or not their religious beliefs give them feelings of strength and comfort and whether or not they regularly participate in social activities. Of those who said yes to both, about 1 in 50 died within 6 months after their operation. Of those who said no to both, about 1 in 5 died within 6 months after their operation. What is the relative risk of death (within 6 months) for the two groups? Write your answer in a sentence or two that would be understood by someone with no training in statistics.

4.23 The relative risk of contracting a certain coronary disease is 2.0 for male smokers compared to male nonsmokers and 3.0 for female smokers compared to female nonsmokers. Is this enough information to determine whether male smokers or female smokers are more likely to contract the disease? If so, make that determination. If not, explain what additional information would be needed to make that determination.

4.24 Using the terminology of this chapter, what name applies to each of the boldface numbers in the following quotes (e.g., odds, risk, relative risk)?

a. "Fontham found increased risks of lung cancer with increasing exposure to secondhand smoke, whether it took place at home, at work, or in a social setting. A spouse's smoking alone produced an overall **30** percent increase in lung-cancer risk" (*Consumer Reports*, January 1995, p. 28).
b. "What they found was that women who smoked had a risk [of getting lung cancer] **27.9** times as great as nonsmoking women; in contrast, the risk for men who smoked regularly was only **9.6** times greater than that for male nonsmokers" (Taubes, 1993, p. 1375).
c. "**One student in five** reports abandoning safe-sex practices when drunk" (*Newsweek*, December 19, 1994, p. 73).

4.25 Exercise 4.10 described a study in which men were classified according to how many anger symptoms they exhibit and whether they have coronary heart disease or not. Among 559 men with the most anger symptoms, 59 had coronary heart disease. Among 199 men with no anger symptoms, 8 had coronary heart disease.

a. For those with the most anger symptoms, what is the relative risk of heart disease (compared to those with no anger symptoms)?

b. For those with the most anger symptoms, what is the percent increase in the risk of heart disease?

c. Calculate the odds ratio that compares the odds of having heart disease for the men with the most anger symptoms to the odds for men with no anger symptoms.

4.26 Exercise 4.11 gave data on height and whether or not a student had ever been bullied for 209 secondary school students in England. Among 92 short students, 42 had been bullied. Among 117 students who were not short, 30 had been bullied.

a. For each height category, calculate the risk of having been bullied.

b. What is the relative risk for short students of having been bullied? Write a sentence that interprets this relative risk.

c. What is the increased risk of having been bullied for short students? Write a sentence that interprets this increased risk.

d. Calculate the odds ratio that compares the odds of having been bullied for the short students to the odds for students who are not short. Write a sentence that interprets this ratio.

4.27 The Roper Organization (1992) conducted a study as part of a larger survey to ascertain the number of U.S. adults who had experienced phenomena such as seeing a ghost, "feeling as if you left your body," and seeing a UFO. A representative sample of adults (18 and over) in the continental United States were interviewed in their homes during July, August, and September 1991. The results when respondents were asked about seeing a ghost are shown in the following table:

	Reportedly Has Seen a Ghost		
	Yes	No	Total
Aged 18 to 29	212	1313	1525
Aged 30 or over	465	3912	4377
Total	677	5225	5902

Data source: The Roper Organization (1992), Unusual Personal Experiences, Las Vegas: Bigelow Holding Corp., p. 35.

a. In each age group, find the percentage who reported seeing a ghost.

b. What is the relative risk of reportedly seeing a ghost for adults under 30 compared to adults 30 and over? Write your answer in a sentence that could be understood by someone who knows nothing about statistics.

c. Repeat part (b) using increased risk instead of relative risk.

d. What are the odds of reportedly seeing a ghost to not seeing one in the older group?

4.28 Suppose a newspaper article states that drinking three or more cups of coffee per day doubles the risk of gall bladder cancer. Before giving up coffee, what questions should be asked by a person who drinks this much coffee?

4.29 Suppose that you read in your hometown newspaper that there were 80 home burglaries in your town in the past year compared to only 65 in 1990. Explain why this might not mean that your home is more at risk for a burglary now than it was in 1990.

4.30 Discuss how you might estimate your risk of being injured falling down stairs in the next year.

4.31 The teacher of a class you're taking tells the class that he thinks students who often skip classes have twice the risk of failing the class compared to students who regularly attend class. To better understand his claim, what question(s) would be reasonable to ask the professor?

Section 4.3

Skillbuilder Exercises

4.32 Case Study 1.5 (p. 5) was called "Does Prayer Lower Blood Pressure?" One of the results quoted in that study was, "People who attended a religious service once a week and prayed or studied the Bible once a day were 40% less likely to have high blood pressure than those who don't go to church every week and prayed and studied the Bible less."

a. What is the explanatory variable in this study?

b. What is the response variable in this study?

c. Give an example of a third variable that might at least partially account for the observed relationship.

4.33 Refer to Exercise 4.32, in which one or more third variables are at least partially likely to account for the observed relationship between religious activities and reduced incidence of high blood pressure. Is this an example of Simpson's paradox? Explain.

4.34 This exercise presents a real example of Simpson's paradox (Wagner, 1982). The total income and total taxes paid in each of five income categories are given for 2 years, 1974 and 1978, in the following table:

	1974		1978	
Adjusted Gross Income	**Income**	**Tax**	**Income**	**Tax**
Under $5,000	41,651,643	2,244,467	19,879,622	689,318
$5,000 to $9,999	146,400,740	13,646,348	122,853,315	8,819,461
$10,000 to $14,999	192,688,922	21,449,597	171,858,024	17,155,758
$15,000 to $99,999	470,010,790	75,038,230	865,037,814	137,860,951
$100,000 or more	29,427,152	11,311,672	62,806,159	24,051,698
Total	880,179,247	123,690,314	1,242,434,934	188,577,186

a. The "tax rate" for any income category is calculated as tax divided by income. Calculate the tax rates for each income bracket in each year.

b. Calculate the overall tax rate for each of the 2 years.

c. Did the tax rates within each income bracket increase or decrease from 1974 to 1978?

d. Did the overall tax rate increase or decrease from 1974 to 1978?

e. Compare your results in parts (c) and (d). Explain them in terms of Simpson's paradox.

General Section Exercises

4.35 In a 1997 Marist College Institute for Public Opinion survey of 995 randomly selected Americans, 31% of the men and 12% of the women surveyed said they have dozed off while driving (Source: www.mipo.marist.edu). Think of a third variable that might at least partially explain the observed relationship between sex of the driver and dozing when driving. Briefly explain.

4.36 Suppose two hospitals are willing to participate in an experiment to test a new treatment, and both hospitals agree to include 1100 patients in the study. Because the researchers who are conducting the experiment are on the staff of hospital A, they decide to perform the majority of cases with the new procedure. They randomly assign 1000 patients to the new treatment, with the remaining 100 receiving the standard treatment. Hospital B, which is a bit reluctant to try something new on too many patients, agrees to randomly assign 100 patients to the new treatment, leaving 1000 to receive the standard treatment. The following table displays the results:

Survival Rates for Standard and New Treatments at Two Hospitals

	Hospital A			Hospital B		
Treatment	**Survive**	**Die**	**Total**	**Survive**	**Die**	**Total**
Standard	5	95	100	500	500	1000
New	100	900	1000	95	5	100
Total	105	995	1100	595	505	1100

a. Which treatment was more successful in hospital A? Justify your answer with relevant percents.
b. Which treatment was more successful in hospital B? Justify your answer with relevant percents.
c. Combine the data from the two hospitals into a single contingency table that shows the relationship between treatment and outcome. Which treatment has the higher survival rate in this combined table? Justify your answer.
d. Explain how this exercise is an example of Simpson's paradox.

4.37 A well-known example of Simpson's paradox, published by Bickel, Hammel, and O'Connell (1975), examined admission rates for men and women who had applied to graduate programs at the University of California at Berkeley. The actual data for specific programs is confidential, so we are using similar, hypothetical numbers. For simplicity, we assume there are only two graduate programs. The figures for acceptance to each program are shown in the following table:

	Program A		Program B	
	Admit	**Deny**	**Admit**	**Deny**
Men	400	250	50	300
Women	50	25	125	300

a. Combine the data for the two programs into a single table. What percentage of all men who applied was admitted? What percentage of all women who applied was admitted? Which sex was more successful in the admissions process?
b. What percentage of the men who applied to program A did program A admit? What percentage of the women who applied to program A did program A admit? Repeat the question for program B. Which sex was more successful in getting admitted to program A? Into program B?
c. Explain how this problem is an example of Simpson's paradox. Provide a potential explanation for the observed figures by guessing what type of programs A and B might have been.

4.38 A researcher observes that, compared to students who do not procrastinate, students who admit to frequent procrastination are more likely to miss class due to illness. Does this mean that procrastinating increases illness? What is another explanation?

4.39 The success rates of two treatments (A and B) for clinical depression are being compared. The research team included five doctors, and the participants were 200 patients with depression. The doctors were supposed to randomly assign treatments to patients, but two doctors didn't do this. Instead, they often assigned more severely depressed patients to use treatment A and less severely depressed patients to use treatment B.

a. Three variables measured by the investigators were the outcome of treatment (successful or not), method of treatment (A or B), and a rating of the initial severity of the depression (mild or severe). Which of these variables is the response variable in this investigation?
b. The doctors were surprised by the final result, which was that treatment B had a higher success rate than treatment A. Based on results attained by other investigators, they had expected the opposite to occur. Explain how the actions of the two doctors who did not always randomly assign treatments may have caused an unfair comparison of the two treatments.
c. Taking Example 4.11 (p. 124) into consideration, describe how the researchers might control for the effect of the initial severity of patient depression when they analyze the relationship between treatment method and outcome.

Section 4.4

Skillbuilder Exercises

4.40 Researchers studied a random sample of North Carolina high school students who participated in interscholastic athletics to learn about the risk of lower-extremity injuries (anywhere between hip and toe) for interscholastic athletes (Yang et al., 2005). Of 999 participants in girls' soccer, 74 experienced lower-extremity injuries. Of 1667 participants in boys' soccer, 153 experienced lower-extremity injuries.

a. Write null and alternative hypotheses about sex and the risk of a lower-extremity injury while playing interscholastic soccer.
b. For these data, the value of the chi-square statistic is 2.51, and the p-value for the chi-square test is .113. Based on these results, state a conclusion about the two variables in this situation and explain how you came to this conclusion.

c. For each sex separately, calculate the percent of participants who had a lower-extremity injury. Explain how the difference between these percentages is consistent with the conclusion you stated in part (b).

4.41 Refer to Exercise 4.40.

a. Write a two-way table of observed counts for sex and whether a participant in soccer had a lower-extremity injury or not.

b. Determine a two-way table of expected counts for these data.

c. Show calculations verifying that the value of the chi-square statistic is 2.51.

4.42 In a national survey, $n = 1000$ randomly selected adults are asked how important religion is in their own lives (very, fairly, or not very) and whether they approve or disapprove of same-sex marriage. Write null and alternative hypotheses about the relationship between the two variables in this situation. Make your hypothesis statements specific to this situation.

4.43 ◆ The 2 × 2 contingency table that follows shows data for sex and opinion on the death penalty for respondents in the 2008 General Social Survey in which a random sample of adults in the United States was surveyed. (Raw data are in the **GSS-08** dataset on the companion website.)

Opinion on Death Penalty

	Oppose	Favor	All
Male	254	631	885
Female	385	632	1017
All	639	1263	1902

a. Write null and alternative hypotheses about the relationship between the two variables in this problem.

b. For this table, the value of the chi-square statistic is 17.78, and to three decimal places, the corresponding p-value is .000. Based on these results, state a conclusion about the two variables, and explain why you came to this conclusion.

c. For each sex separately, calculate the percentage who were opposed to the death penalty. Explain whether these percentages are evidence that opinion about the death penalty is related to sex.

4.44 Refer to Exercise 4.43.

a. Calculate a table of expected counts for the data in Exercise 4.43.

b. Show calculations verifying that the value of the chi-square statistic is 17.78.

4.45 In a national survey, $n = 1500$ randomly selected adults are asked if they favor or oppose a ban on texting while driving and if they have personally texted while driving during the previous month. Write null and alternative hypotheses about the relationship between the two variables in this situation. Make your hypothesis statements specific to this situation.

4.46 For each of the following results, explain what conclusion can be made about the null hypothesis that there is no relationship between two variables that form a 2 × 2 contingency table.

a. p-value = .001.
b. p-value = .101.
c. p-value = .900.
d. Chi-square statistic = 4.01.

4.47 For each of the following results, explain what conclusion can be made about the null hypothesis that there is no relationship between two variables that form a 2 × 2 contingency table.

a. p-value = .049.
b. p-value = .755.
c. p-value = .027
d. Chi-square statistic = 2.98.

4.48 A sample of 90 men and 110 women is asked about their handedness. A two-way table of observed counts follows:

	Left-handed	Right-handed	Total
Men	11	79	90
Women	9	101	110
Total	20	180	200

a. Calculate a table of expected counts for these data.

b. Calculate the value of the chi-square statistic for testing the null hypothesis of no relationship between sex and handedness.

General Section Exercises

4.49 If a relationship is statistically significant, does that guarantee that it also has practical significance? Explain.

4.50 If a relationship has practical significance, does it guarantee that statistical significance will be achieved in every study that examines it? Explain.

4.51 Explain whether each of the following is possible.

a. A relationship exists in the observed sample but not in the population from which the sample was drawn.

b. A relationship does not exist in the observed sample but does exist in the population from which the sample was drawn.

c. A relationship does not exist in the observed sample, but an analysis of the sample shows that there is a statistically significant relationship, so it is inferred that there is a relationship in the population.

4.52 The table for Exercise 4.27 gave data for the relationship between age group and whether or not a person reports ever having seen a ghost.

a. Write null and alternative hypotheses about the possible relationship between the two variables.

b. The Minitab output for a chi-square test of the relationship follows. Based on this output, what can be concluded about the relationship? What is the basis for this conclusion?

```
Expected counts are printed below observed counts

              Yes        No     Total
18-29         212      1313      1525
           174.93   1350.07

30+           465      3912      4377
           502.07   3874.93

  Total       677      5225      5902

Chi-Sq = 7.857 + 1.018 +
         2.737 + 0.355 = 11.967
DF = 1, P-Value = 0.001
```

c. Note in the Minitab output that the p-value is .001. Write a sentence that interprets this number.

4.53 Refer to the Minitab output for Exercise 4.52. Show how the first term of the chi-square statistic, 7.857 (corresponding to the "18–29, Yes" cell), was computed.

4.54 Refer to the Minitab output for Exercise 4.52.

a. Demonstrate how the *expected count* was computed for the "age 18–29, yes" cell of the table.
b. Verify that the expected count for the "30+, yes" cell can be determined by subtracting the answer for part (a) from the total count for the "yes" column of the table.
c. In each row of the table, express the *expected count* for yes as a percent of the total count for that row. How do these two percentages compare to each other?

4.55 Imagine that 50 men and 50 women are asked, "Do you favor or oppose capital punishment for those convicted of murder?" In the observed data, 38/50 = .76, or 76% of the men favor capital punishment compared to 32/50 = .64, or 64% of the women.

a. Write null and alternative hypotheses about the possible relationship between sex and opinion about capital punishment.
b. A chi-square test of the statistical significance of the relationship has a p-value of .19. Is this evidence that there is a relationship? What is the justification for this conclusion?
c. Calculate the value of the chi-square statistic.
d. In part (b), the p-value was given as .19. Write a sentence that interprets this number.

4.56 Refer to Exercise 4.55.

a. Suppose that 500 men and 500 women had been surveyed, rather than 50 of each sex as in Exercise 4.55. Further suppose that the proportions in favor of capital punishment remained the same, at 380/500 or 76% of the men and 320/500 or 64% of the women. Calculate the chi-square statistic in this case.
b. The p-value for the chi-square statistic in part (a) is .000035. Is this evidence that there is a relationship between sex and opinion about capital punishment?
c. Explain the reason for the discrepancy between the result in Exercise 4.55 part (b) and the result in part (b) of this exercise.

4.57 Considering the effect of sample size on the chi-square test, explain why a finding that a relationship is "not statistically significant" should not be interpreted as absolute proof that there is no relationship in the population.

4.58 In Example 4.15 (p. 130), a statistically significant relationship was found between hormone therapy and invasive breast cancer. In the same study, the following observed counts were found for death from coronary heart disease (CHD). Carry out the five steps leading to a conclusion about whether there is a statistically significant relationship between hormone therapy and death from CHD.

	Death from CHD?		
	Yes	**No**	**Total**
Hormones	33	8,473	8,506
Placebo	26	8,076	8,102
Total	59	16,549	16,608

4.59 ◆ The following 2 × 2 contingency table shows data for sex and opinion about the legalization of marijuana for respondents in the 2008 General Social Survey in which a random sample of U.S. adults was surveyed. (Raw data are in the **GSS-08** dataset on the companion website.)

	Opinion on Marijuana		
	Legalize	**Don't Legalize**	**All**
Male	241	315	556
Female	255	436	691
All	496	751	1247

a. Write null and alternative hypotheses about the relationship between the two variables in this problem.
b. For this table, the value of the chi-square statistic is 5.34, and the corresponding p-value is .02. Based on these results, state a conclusion about the two variables and explain how you came to this conclusion.
c. For each sex separately, calculate the percentage who were opposed to the legalization of marijuana. Explain why these percentages are evidence that the two variables in this exercise are related.
d. Verify that the value of the chi-square statistic is 5.34.

Chapter Exercises

4.60 Wechsler and Kuo (2000) used data from the 1999 College Alcohol Study to examine the relationship between student alcohol use and student definitions of binge drinking. In the study, approximately 14,000 college students from 119 schools answered questions about drinking habits. Using the definition that binge drinking is five consecutive drinks for men and four consecutive drinks for women, the researchers categorized students into four "type of drinker" categories on the basis of their answers to questions about personal alcohol use. Students in the survey were also asked how many drinks in a row they thought constituted binge drinking for men and for women. The following is a contingency table showing the relationship between the student definitions of binge drinking for a man and student alcohol use:

Type of Drinker and Personal Definition of Binge Drinking

	Definition of Binge Drinking for Men (Drinks in a Row)				
Type of Drinker	≤4	5	6	≥7	*Total*
Abstainer	814	659	484	702	2,659
Non-Binge	714	1,129	997	2,223	5,063
Occasional Binge	231	471	480	1,780	2,962
Frequent Binge	179	442	367	2,147	3,135
Total	1,938	2,701	2,328	6,852	13,819

a. Among all students, what percentage defined binge drinking for men as being seven or more drinks in a row?
b. Write a sentence describing what would be measured by row percentages in the table. Make your answer specific to these variables.
c. Write a sentence describing what would be measured by column percentages in this table. Make your answer specific to these variables.
d. Calculate the row percentages in the table. In a few sentences, describe the relationship between type of drinker and the personal definition of binge drinking.

4.61 The following table contains data on ear piercings and tattoos for a sample of 1375 college women. The ear-pierce response is the total number of ear piercings for a woman, and this has been categorized in the rows of the table.

Ear Piercings and Tattoos, 1375 College Women

Piercings	No Tattoo	Have Tattoo
2 or less	498	40
3 or 4	374	58
5 or 6	202	77
7 or more	73	53

Data source: One of the authors.

a. For each ear-piercing category, determine the percentage with a tattoo. On the basis of these percentages, would you say there is a relationship between these two variables? Explain.
b. Draw a bar graph of the column percentages. What does the graph show us about the relationship between having a tattoo and the number of ear piercings?
c. Calculate the percentage with at least five ear piercings for women who have a tattoo and separately for women who don't have a tattoo.
d. What percentage of the total sample has a tattoo?
e. What percentage of the total have two or fewer ear piercings and also do not have a tattoo?

4.62 Researchers asked women who were pregnant with planned pregnancies how long it took them to get pregnant (Baird and Wilcox, 1985; see also Weiden and Gladen, 1986). Length of time to pregnancy was measured according to the number of cycles between stopping birth control and getting pregnant. Women were also categorized on whether or not they smoked, with smoking defined as having at least one cigarette per day for at least the first cycle during which they were trying to get pregnant. The following table summarizes the observed counts.

Time to Pregnancy for Smokers and Nonsmokers

	Pregnancy Occurred After	
	First Cycle	Two or More Cycles
Smoker	29	71
Nonsmoker	198	288

a. Among smokers, what percentage were pregnant after the first cycle?
b. Among nonsmokers, what percentage were pregnant after the first cycle?
c. Draw a bar graph that can be used to examine the relationship between smoking and how long it took to get pregnant.
d. Among those who were pregnant after the first cycle, what percentage smoked?
e. Among those who took two or more cycles to become pregnant, what percentage smoked?
f. Do you think there is a relationship between smoking and how long it takes to get pregnant? Explain.
g. Can these data be used as evidence of a causal connection between smoking and how long it takes to get pregnant? Explain.

4.63 In a study in Berlin, reported by Kohlmeier et al. (1992) and by Hand et al. (1994), researchers asked 239 lung-cancer patients and 429 controls (matched to the lung-cancer cases by age and sex) whether or not they had kept a pet bird during adulthood. Of the 239 lung-cancer cases, 98 said yes. Of the 429 controls, 101 said yes.

a. Construct a contingency table for the data.
b. Compute the risk of lung cancer for bird owners and for those who had never kept a bird.
c. Can the risks computed in part (b) be used as baseline risks for the populations of those who have and have not owned birds? Explain.
d. What is the relative risk of lung cancer for bird owners?
e. What additional information about the risk of lung cancer would you want before you made a decision about whether or not to own a pet bird?
f. Can these data be used to establish a causal connection between owning a bird and the risk of lung cancer? Explain.
g. For any study of lung cancer, the effect of smoking should be considered. How would you determine whether smoking might be a confounding variable in this study?

4.64 Compute the chi-square statistic, and assess the statistical significance for the relationship between bird ownership and lung cancer, based on the data in Exercise 4.63. State a conclusion about the relationship.

4.65 Compute the chi-square statistic, and assess the statistical significance for the relationship between smoking and time to pregnancy in Exercise 4.62. State a conclusion about the relationship.

4.66 Pagano and Gauvreau (1993, p. 133) reported data for women participating in the first National Health and Nutrition Examination Survey (Carter et al., 1989). The explanatory variable was whether or not the woman gave birth to her first child at the age of 25 or older, and the outcome variable was whether or not she developed breast cancer. Observed counts are in the following table.

Age at Birth of First Child and Breast Cancer

Age at First Child	Breast Cancer	No Breast Cancer	Total
25 or Older	31	1597	1628
Under 25	65	4475	4540
Total	96	6072	6168

Source: Pagano and Gauvreau (1993).

a. Calculate the risk of breast cancer for women who were under 25 years old when they had their first child.
b. Calculate the risk of breast cancer for women who were 25 or older when they had their first child.
c. Calculate the relative risk of breast cancer for women who were 25 or older when they had their first child compared with women who were under 25. Write a sentence that interprets this relative risk in the context of this problem.
d. What is the percent increase in the risk of breast cancer for women who were 25 or older when they had their first child?
e. Can these data be used as evidence of a causal connection between age at first child and the risk of breast cancer? Explain.

4.67 In Example 4.15 (p. 130) a statistically significant relationship was found between hormone therapy and invasive breast cancer. There were 166 cases of invasive breast cancer out of 8506 women taking hormones and 124 cases of invasive breast cancer out of 8102 women taking placebo.

a. What is the relative risk of invasive breast cancer when taking hormones compared with taking placebo?
b. What is the baseline risk of invasive breast cancer, assuming that the placebo group is representative of the relevant population of women?
c. Explain why it is important to report the baseline risk found in part (b) when reporting the relative risk found in part (a).

4.68 Example 4.15 (p. 130), was about the relationship between hormone therapy and invasive breast cancer. One of the results reported in the paper was, "Absolute excess risks per 10,000 person-years attributable to estrogen plus progestin were . . . 8 more invasive breast cancers" (Writing Group for the Women's Health Initiative Investigators, 2002, p. 321). In other words, for every 10,000 "person-years" of women taking the hormones instead of a placebo, there would be eight additional cases of invasive breast cancer. Discuss practical versus statistical significance in this situation.

4.69 Exercise 4.10 concerned the relationship between anger and the risk of coronary heart disease. Some computer output for a chi-square test is shown below. What do the results indicate about the relationship? Explain.

Expected counts are printed below observed counts

	Heart Disease Yes	Heart Disease No	
No Anger	8	191	199
	17.59	181.41	
Most Anger	59	500	559
	49.41	509.57	
Total	67	691	758

Chi-Sq = 5.228 + .507 + 1.861 + .180 = 7.777
DF = 1, P-Value = .0005

4.70 Refer to the data in Exercise 4.69, showing the frequency of coronary heart disease for men with no anger and men with the most anger. Show how the numbers in the following statement were calculated: The odds of remaining free of heart disease versus getting heart disease are about 24 to 1 for men with no anger, whereas those odds are only about 8.5 to 1 for men with the most anger.

4.71 Refer to Exercises 4.69 and 4.70. Find the odds ratio for remaining free of heart disease for men with no anger compared with men with the most anger. Give the result in a sentence that someone with no training in statistics would understand.

4.72 "Saliva test predicts labor onset" was the headline of a Reuters Health story that appeared May 23, 2000, at the Yahoo! Health News website. The story described a medical test called SalEst. A positive SalEst test indicates an elevated estrogen level, and this knowledge may help to predict how soon a pregnant woman will give birth because estrogen level increases in the 2 to 3 weeks before delivery. Researchers employed by the manufacturers of SalEst tested 642 women who had been pregnant for 39 weeks. Among the 615 women who delivered before the 42nd week of the pregnancy, about 59% had a positive SalEst test. Among the 27 women who delivered in the 42nd week or later, 33% had a positive SalEst test.

a. Construct a contingency table of counts for the relationship between the SalEst test result and the time of delivery.
b. Which of the two variables that were used to make the contingency table is the explanatory variable, and which is the response variable?

c. Among all 642 women, what percentage delivered before 42 weeks?
d. The article states that if a test result is positive, there is a 98% chance that the woman will deliver before 42 weeks. Based on the contingency table, justify this claim.
e. Among those who had a negative test, what percentage delivered before 42 weeks?
f. Would you advise a pregnant woman to spend $90 for the test? Explain why or why not.

4.73 In Exercise 4.36, data were given for admissions to two graduate programs for men and women. The data are given again here for use in this exercise, with the combined data presented as well.

	Program A		Program B		Combined	
	Admit	Deny	Admit	Deny	Admit	Deny
Men	400	250	50	300	450	550
Women	50	25	125	300	175	325

a. Give the odds of being admitted versus being denied to program A separately for the men and women who applied to that program. Then give the odds ratio for women compared to men. Write a sentence explaining the odds ratio in words.
b. Give the odds of being admitted versus being denied for the combined programs separately for men and women. Then give the odds ratio for women compared to men. Write a sentence explaining the odds ratio in words.
c. Using the results from parts (a) and (b), write one sentence that sounds as if the university favors female applicants and another sentence that sounds as if the university favors male applicants.

4.74 According to a study on partner abuse reported by the *Sacramento Bee* (July 14, 2000, p. A6, Associated Press), 25% of women with male partners had been assaulted by their current or a former partner, whereas 11% of women with female partners had been assaulted. What is the relative risk of assault for women with male partners compared to women with female partners? Write your answer in a sentence that would be understood by someone with no training in statistics.

Dataset Exercises

Datasets required to solve these exercises are available on the companion website, http://www.cengage.com/statistics/Utts4e.

4.75 For this exercise, use the **UCDavis2** dataset on the companion website. The variable ***WtFeel*** contains UC Davis student responses to the question "Do you think you are: Underweight? About right? Overweight?"

a. Create a contingency table for the relationship between ***Sex*** (female or male) and perception of weight (***WtFeel***).
b. For each sex category, calculate the conditional percentages for the perception of weight variable. Of the men, what percentage thinks they are underweight? Of the women, what percentage thinks they are underweight?
c. In Section 4.5, the chi-square test was used only for 2×2 tables, but it can be used for larger tables as well. Use software to do a chi-square test of statistical significance for the relationship between sex and perception of weight. What is the *p*-value of the test? On the basis of this *p*-value, what conclusion can be reached about the relationship?

4.76 For this exercise, use the **UCDavis2** dataset on the companion website. The variable ***Cheat*** contains answers to a question about whether or not the respondent would tell the instructor if he or she saw another student cheating on an examination.

a. Create a two-way table for the relationship between ***Sex*** (male or female) and the variable ***Cheat***. Of all students, what percentage said that they would tell the instructor about witnessing cheating?
b. Of the females, what percentage said that they would tell the instructor about witnessing cheating? What percentage of the males said that they would do this?
c. Is the observed relationship between sex and response to the cheating question statistically significant? Justify your answer by reporting the results of a chi-square test.
d. For this question about cheating, what population do you think is represented by the students in the dataset?

4.77 For this exercise, use the **GSS-08** dataset on the companion website. The variable ***gunlaw*** has responses to a question about whether or not the respondent would favor a law requiring a police permit to be obtained before the purchase of any gun. The variable ***owngun*** indicates whether or not there are any guns in the respondent's home (including garages and sheds).

a. Write null and alternative hypotheses about the possible relationship between ***owngun*** and ***gunlaw***.
b. Create a two-way table that summarizes the relationship between ***owngun*** and ***gunlaw***. What percentage of all respondents had a gun in their home?
c. Among those with a gun in their home, what percentage favors a gun permit law? Among those with no gun in their home, what percentage favors a gun permit law? On the basis of the difference between these two percentages, would you say that there is a relationship between the two variables? Why or why not?
d. What is the *p*-value of a chi-square test of statistical significance for the relationship between ***owngun*** and ***gunlaw***? Is the observed relationship statistically significant?
e. Explain what it means to say that an observed relationship is statistically significant.

4.78 For this exercise, use the **GSS-08** dataset on the companion website. Write a paragraph that describes the relationship between ***race*** (respondent's race) and ***cappun*** (opinion about capital punishment). Include a two-way table of counts and a relevant set of conditional percentages.

4.79 For this exercise, use the **Student0405** dataset on the companion website. The variable ***Seat*** gives student responses to where they prefer to sit in a classroom (Front, Middle,

Back) and the variable ***ReligImp*** indicates how important students feel religion is in their lives (Very, Fairly, Somewhat).

a. Create a two-way table that summarizes the relationship between ***Seat*** and ***ReligImp***.

b. For each seat area, determine the percentages that fall into each category of the religious importance variable.

c. Write a few sentences that describe the observed relationship between the two variables in this exercise.

4.80 For this exercise use the **Student2010** dataset on the companion website. The variable ***UseCell*** gives student responses to a question that asked students how they mainly used a cell phone (to talk, to text).

a. Create a two-way table that summarizes the relationship between ***Sex*** and ***UseCell***.

b. Fill in the following table with row percentages.

	UseCell		
Sex	**To Talk**	**To Text**	**Total**
Female			
Male			

c. Explain why the percentages found in part (b) are evidence that ***Sex*** and ***UseCell*** are related in the sample.

d. Write null and alternative hypotheses about the possible relationship between ***Sex*** and ***UseCell***.

e. What is the p-value of a chi-square test of statistical significance for the relationship between ***Sex*** and ***UseCell***? Is the observed relationship statistically significant?

5

Space/NASA Sites

How many people would like to fly to the moon?

See Example 5.4 *(p. 152)*

Sampling: Surveys and How to Ask Questions

Here is a fact that may astound you. If you use proper methods to sample 1500 people from a population of millions and millions, you can almost certainly gauge the percentage of the entire population who have a certain trait or opinion to within 3%. The tricky part is that you have to use a proper sampling method.

People are curious about each other. When a newsworthy event occurs, people want to know what others think. In the midst of election campaigning, people want to know which candidates are in the lead. We devour news stories about the behavior and opinions of others. For this reason, news organizations such as *USA Today* and CNN often conduct sample surveys. Many recent surveys can be found on the Web. In this chapter, we learn how to conduct surveys, how to make sure they are representative, and what can go wrong.

5.1 Collecting and Using Sample Data Wisely

Remember that there are two major categories of statistical techniques that can be applied to data. The first is **descriptive statistics**, in which we use numerical and graphical summaries to characterize a dataset or describe a relationship between two variables. We learned many ways to describe variables in Chapter 2, and we learned how to describe relationships between two variables in Chapters 3 and 4.

The second important category of statistical techniques is **inferential statistics**, in which we use sample data to make conclusions about a broader range of individuals than just those who are observed. In Chapter 4 we learned about *hypothesis testing*, in which we used a sample to *infer* whether or not a real relationship existed in a population. In Case Study 1.6 about aspirin use and the risk of heart disease, the data from a sample of 22,071 physicians was used to *infer* that taking aspirin helps to prevent heart attacks for all men who are similar to the participants. In this chapter we will introduce *confidence intervals*, which are another important method of making an inference about a population based on a sample. Confidence intervals and hypothesis tests are covered in much more detail in Chapters 10 to 13.

In Chapters 5 and 6, you will learn how to collect representative data. In these chapters, you will learn that the data collection method used affects the extent to which sample data can be used to make inferences about a larger population. Descriptive summaries such as the mean and the standard deviation, as well as graphical techniques, can be used whether the data are from a sample or from an entire population, but inferential methods can be used only when the data in hand are from a *representative* sample for the question being asked about a larger population. When you use inferential methods, a key concept is that you have to think about *both* the source of the data *and* the question(s) of interest. A dataset may contain representative information for some questions but not for others.

DEFINITION **The Fundamental Rule for Using Data for Inference** is that available data can be used to make inferences about a much larger group *if the data can be considered to be representative with regard to the question(s) of interest.*

Example 5.1 **Do First Ladies Represent Other Women?** If we want to know how old American women live to be, it would not make sense to use the ages of death given in Chapter 2 for the first ladies. Even if you ignore the fact that most of the first ladies did not live in modern times with modern health care, you can probably think of many reasons why those women do not represent the full population of American women. In fact, there is probably no larger group that is represented by these women. Past first ladies are not even likely to represent future first ladies on the question of age at death because medical, social, and political conditions keep changing in ways that may affect their health.

Example 5.2 **Do Penn State Students Represent Other College Students?** One dataset in Chapter 2 was collected at Penn State University. Do Penn State students represent all college students? This probably depends on the question. If we want to know the average handspan of female college students in the United States, we could convincingly argue that the data collected in statistics classes at Penn State University are indeed representative. We could even argue that the Penn State women represent all females in the same age group, not just college students. On the other hand, Penn State students may not represent all college students when questioned about how fast they have ever driven a car. Penn State is in rural Pennsylvania, where there are many country roads and little traffic. Students in larger cities, even nearby ones such as New York or Philadelphia, may not have access to the same open spaces and may not have had the opportunity to drive a car as fast as students in rural areas. Many students in larger cities may not even drive cars at all.

These examples show that it may require knowledge of the subject matter to determine whether a sample represents the larger group for the question of interest. There are, however, a small number of common strategies that researchers use to improve the likelihood that the data they collect can be used to make inferences about a larger population. We will examine some of those strategies in the remainder of this chapter and the next. Also, we will discuss the strengths of those strategies and some problems that may arise when they are used.

Populations, Samples, and Simple Random Samples

In most statistical studies, the objective is to use a small group of units to make an inference about a larger group. The larger group of units about which inferences are to be made is called the **population**. The smaller group of units actually measured is called the **sample**. Sometimes measurements *are* taken on the whole group of interest, in which case *these measurements comprise a* ***census*** *of the whole population.* Occasionally, you will see someone make the mistake of trying to use census data to make inferences to some hypothetical "larger group" when there isn't one.

DEFINITION

- A **population** is the entire group of units about which inferences will be made.
- A **sample** is the group of units that are actually measured or surveyed.
- A **census** is taken when every unit in the population is measured or surveyed.

Remember the fundamental rule for making valid inferences about the group represented by the sample for which the data were measured: *The data must be representative of the larger group with respect to the question(s) of interest.* The principal way to guarantee that sample data represents a larger population is to use a **simple random sample** from the population.

DEFINITION With a **simple random sample**, every conceivable group of units of the required size from the population has the same chance to be the selected sample.

An ideal data collection method is to obtain a simple random sample of the population of interest or to collect sample data by using one of the more complex random sampling methods described later in this chapter. In some research studies, however, random sampling is not possible for practical or ethical reasons (or both). For instance, suppose researchers want to study the effect of using marijuana to reduce pain in cancer patients. It would be neither practical nor ethical to select a random sample of all cancer patients to participate. Instead, the researchers would use volunteers who want to take part and hope that those volunteers represent the larger population of all cancer patients for the question of interest. The use of volunteers will be discussed more fully in Chapter 6 when we cover randomized experiments.

Simple random samples and related sampling methods *are* typically used for one type of statistical study: sample surveys or polls. Remember from Chapter 1 that in a *sample survey,* the investigators gather opinions or other information from each individual included in the sample. Because this gathering of information is usually not time-consuming or invasive, it is often both practical and ethical to contact a large random sample from the population of interest. Throughout this chapter, we will learn more about how to select simple random samples and how to conduct sample surveys.

In a **sample survey**, a subgroup of a large population is questioned on a set of topics. The investigator simply asks the participants to answer some questions. There is no manipulation of a respondent's behavior in this type of research. The results from a sample survey are used as if they represent the larger population, which they will if the sample is chosen correctly and if those who are selected to participate cooperate in responding.

Advantages of a Sample Survey over a Census

When a Census Isn't Possible

Suppose you need a laboratory test to find out whether or not your blood has too high a concentration of a certain substance. Would you prefer that the lab measure the entire population of your blood, or would you prefer to give the lab a sample? Similarly, suppose that a manufacturer of firecrackers wants to know what percentage of its products are duds. The company would not make much of a profit if it tested them all, but it could get a reasonable estimate of the desired percentage by testing a properly selected sample. As these examples illustrate, there are situations in which measurements destroy the units being tested and therefore a census is not feasible.

Speed

Another advantage of a sample survey over a census is speed. For instance, it takes several years to plan and attempt to execute a census of the entire population of the United States, something that is done every 10 years. Getting figures such as monthly unemployment rates would be impossible with a census; they would be quite out of date by the time they were released. It is much faster to collect a sample than a census if the population is large.

Accuracy

A final advantage to a sample survey is that you can devote your resources to getting the most accurate information possible from the sample you have selected. It is easier to train a small group of interviewers than a large one, and it is easier to track down a small group of nonrespondents than the larger one that would inevitably result from trying to conduct a census. For the census done in the United States every 10 years, residents receive a form in the mail, which they are asked to fill in and return. But typically only about two-thirds of households complete the form. For households that don't respond or who don't have regular mail service, the government must send someone to collect the information. It is estimated that it cost the U.S. government about $2.5 billion to perform that task for the 2010 census.

Bias: How Surveys Can Go Wrong

The purpose of most surveys is to use a characteristic of the sample (such as the proportion with a certain opinion) as an estimate of the corresponding characteristic of the population from which the sample was selected. While it is unlikely that the sample value will equal the population value precisely, the goal of a good survey is to get an **unbiased** sample value. Survey results are **biased** if the method used to obtain those results would produce values that are either consistently too high or consistently too low. When the fundamental rule for inference does not apply, it's likely that the results will be biased.

There are three common types of bias that might occur in surveys:

- **Selection bias** occurs if the method for selecting the participants produces a sample that does not represent the population of interest. For instance, if shoppers at a mall are surveyed to determine attitudes about raising the sales tax, the results are not likely to represent all area residents.
- **Nonparticipation bias (nonresponse bias)** occurs when a representative sample is chosen for a survey but some of the selected individuals do not participate, either because they cannot be contacted or because they do not respond. For instance, a survey conducted by phoning people at home in the evening would omit people who work during that time. Their opinions would not be represented, so if their opinions were much different, the sample results would be biased.
- **Biased response** or **response bias** occurs when participants provide incorrect information. The way questions are worded, the way the interviewer behaves, as well as many other factors might lead an individual to answer inaccurately or untruthfully. For instance, surveys about socially unacceptable behavior such as heavy smoking or drinking, abusive behavior, and so on must be worded and conducted carefully to minimize the possibility of response bias.

5.1 Exercises are on pages 176–177.

IN SUMMARY Three Types of Bias in Surveys

Type of Bias	Cause of Bias
Selection bias	Using a poor method for selecting whom to ask to participate.
Nonparticipation (nonresponse) bias	Those who were selected to participate choose not to do so or can't be reached.
Response bias (biased response)	Participants' answers to questions don't truly reflect how they feel or behave.

THOUGHT QUESTION 5.1 The terms *nonresponse bias* and *response bias* sound very similar. Do they refer to two aspects of the same kind of problem? Explain.*

5.2 Margin of Error, Confidence Intervals, and Sample Size

Surprisingly, a sample does not have to include a large portion of the population to be reasonably accurate. National polling organizations such as the Gallup Organization often question only slightly over 1000 individuals. How can a survey of such a small number of individuals possibly obtain accurate information about the opinions of millions?

Here is some information that may astound you. If you use commonly accepted methods to sample 1600 people from an entire population of millions and millions, you can

***HINT:** In each case, think about whether the bias has to do with *who* responds or with *how* they respond.

fairly certainly gauge the percentage of the entire population who have a certain trait or opinion to within 2.5%. Even more amazing is the fact that this result does not depend on how big the population is. It depends only on how many are in the sample! A sample of 1600 would do about equally well for a population of 4 billion as it would for a population of 10 million. Of course, the tricky part is that you have to use a proper sampling method.

You can see why researchers are content to rely on public opinion polls rather than trying to ask everyone for their opinions. It is much cheaper to ask 1600 people than several million, especially when you can get an answer that is almost as accurate. It also takes much less time to conduct a sample survey than it does to conduct a census, in which everyone in the population is measured. And because fewer interviewers are needed for a survey than for a census, there may be better quality control of interviewing procedures when only a small sample of individuals is measured.

Margin of Error: The Accuracy of Sample Surveys

Sample surveys are often used to estimate the *proportion* or *percentage* of people who have a certain trait or opinion. For example, during election years polls based on a sample of about a thousand likely voters are used to predict the voting behavior of an entire population of millions of voters. Newspapers and magazines routinely survey only 1000 to 2000 people to determine public opinion on current topics of interest. As we have said, if properly conducted, these surveys are amazingly accurate.

When a survey is used to find a proportion based on a sample of only a few thousand individuals, the obvious question is how close the sample proportion comes to the true population proportion. The **margin of error**, sometimes called the **margin of sampling error** to distinguish it from errors caused by bias, is a measure of the accuracy of a sample proportion as an estimate of the population proportion. It provides an upper limit on the difference between the sample proportion and the population proportion that holds for at least 95% of simple random samples of a specific size. In other words, the difference between the sample proportion and the unknown value of the population proportion is less than the margin of error at least 95% of the time, or in at least 19 of every 20 situations. A conservative estimate of the margin of error for a sample proportion is calculated as $1/\sqrt{n}$, where n is the sample size. To express results in terms of percentages instead of proportions, simply multiply everything by 100.

DEFINITION The **conservative margin of error** for a sample proportion is calculated by using the formula $1/\sqrt{n}$, where n represents the **sample size**, the number of people in the sample. The amount by which the sample proportion differs from the true population proportion is less than this quantity in at least 95% of all random samples. Survey results reported in the media are usually expressed as percentages. The conservative margin of error for a sample percentage is

$$\frac{1}{\sqrt{n}} \times 100\%$$

In Chapter 10, we will learn a more complicated and precise formula for the margin of error that depends on the actual sample proportion. It will never give an answer larger than the formula given here, so we call this formula a *conservative* margin of error. For simplicity, many media reports use the conservative formula. For example, with a sample of 1600 people, we will usually get an estimate that is accurate to within $1/\sqrt{1600} = 1/40 = .025$, i.e., to within 2.5% of the true population percentage.

Confidence Intervals

You might see results such as, "Fifty-five percent of respondents support the President's economic plan. The margin of error for this survey is plus or minus 2.5 percentage points." This means that it is almost certain that between 52.5% and 57.5% of the entire population supports the plan. In other words, add and subtract the margin of error to the sample value (55% in this example) to create an interval. If you were to follow this method every time you read the results of a properly conducted survey, the interval would miss covering the true population percentage only about 1 in 20 times (5%) and would cover the truth the remaining 95% of the time.

A **confidence interval**, also called an **interval estimate**, is an interval of values that *estimates* an unknown population value. In the preceding paragraph, the percentage of *all* U.S. adults favoring the President's economic policy is the unknown value we wish to estimate. The confidence interval estimate of this value is 52.5% to 57.5%, calculated as *sample percentage* ± *margin of error*.

The word *confidence* refers to the fraction or percentage of random samples for which a confidence interval procedure works, by which we mean that the sample gives an interval that includes the unknown value of a population parameter. For the procedure described here, the confidence is at least 95%. This means that intervals calculated using this formula will cover the population proportion for at least 95% of all properly selected random samples.

DEFINITION **95% Confidence Interval for a Population Proportion:** For about 95% of properly conducted sample surveys, the interval

$$\textit{sample proportion} - \frac{1}{\sqrt{n}} \text{ to } \textit{sample proportion} + \frac{1}{\sqrt{n}}$$

will contain the actual population proportion. This interval is called an **approximate 95% confidence interval** for the population proportion. In Chapter 10, a more precise formula will be given.

Another way of writing the approximate 95% confidence interval is

$$\textit{sample proportion} \pm \frac{1}{\sqrt{n}}$$

To report the results in percentages instead of proportions, multiply everything by 100%.

Here are two examples of polls, with the margin of error and an approximate 95% confidence interval for each one. In each case, the poll is based on a nationwide random sample of Americans aged 18 and older.

Example 5.3 **The Importance of Religion for Adult Americans** For a Gallup poll conducted December 11–13, 2009, a random sample of $n = 1025$ adult Americans was asked, "How important would you say religion is in your own life: very important, fairly important, or not very important?" (http://www.pollingreport.com/religion.htm, March 2, 2010). The percentages that selected each response were as follows:

Very important: 56%
Fairly important: 25%
Not very important: 19%

$$\text{Conservative margin of error: } \frac{1}{\sqrt{1025}} = .03 \text{ or } .03 \times 100\% = 3\%$$

Approximate 95% confidence intervals for the proportion and percentage of *all* adult Americans who would say religion is very important are

Proportion: .56 ± .03 or [.56 − .03 to .56 + .03] or .53 to .59
Percentage: 56% ± 3% or [56% −3% to 56% +3%] or 53% to 59%

Example 5.4 **Do You Want to Fly to the Moon?** An *ABC News/Good Morning America* poll conducted by phone January 25–29, 2008, posed the following question to a random sample of $n = 1019$ adult Americans: "If you had a chance in your lifetime to travel in outer space, would you do so, or not?" (http://abcnews.go.com/images/PollingUnit/1059a2SpaceTravel.pdf, accessed March 2, 2010). The percentages that selected each response were as follows:

Would: 39%
Would not: 60%
Unsure: 1%

$$\text{Conservative margin of error: } \frac{1}{\sqrt{1019}} = .03 \text{ or } .03 \times 100\% = 3\%$$

Approximate 95% confidence intervals for the proportion and percentage of *all* adult Americans who would say they would travel in outer space:

Proportion: .39 ± .03 or [.39 − .03 to .39 + .03] or .36 to .42
Percentage: 39% ± 3% or [39% −3% to 39% +3%] or 36% to 42%

Most news reports of polls include mention of the size of the sample and the margin of error. The news story accompanying this poll reported that, "This ABC News poll was conducted by telephone January 25–29, 2008, among a random national sample of 1019 adults. The results have a 3-point error margin" (Langer, 2008).

Interpreting the Confidence Intervals in Examples 5.3 and 5.4

In both examples of surveys, an approximate 95% confidence interval was found. There is no way to know whether one, both, or neither of these intervals actually covers the population value of interest. For instance, the interval from 53% to 59% may or may not contain the percentage of adult Americans who considered religion to be very important in their lives in December 2009. But in the long run, this procedure will produce intervals that cover the unknown population values about 95% of the time, as long as the procedure is used with properly conducted surveys. This long-run performance is usually expressed after an interval is computed by saying that we are 95% confident that the population value is covered by the interval. We will learn more about how to interpret a confidence interval and the accompanying confidence level (such as 95%) in Chapter 10.

Choosing a Sample Size for a Survey

When surveys are planned, the **choice of a sample size** is an important issue. One commonly used strategy is to use a sample size that provides a **desired margin of error** for a 95% confidence interval. Table 5.1 displays the margin of error for several different sample sizes. As we will learn in Chapter 10, the exact margin of error formula involves the sample proportion found after the sample has been measured. The margin of error calculations in Table 5.1 were done by using the "conservative" formula $1/\sqrt{n}$. This is commonly done by polling organizations because, before the sample is observed, there is no way to know what sample proportion to use in the more exact margin of error formula. With a table like Table 5.1, researchers can pick a sample size that provides suitable accuracy for any sample proportion, within the constraints of the time and money available for the survey.

Table 5.1 Relationship between Sample Size and Margin of Error for 95% Confidence

Sample Size n	Margin of Error $1/\sqrt{n}$
100	.10 (10%)
400	.05 (5%)
625	.04 (4%)
1,000	.032 (3.2%)
1,600	.025 (2.5%)
2,500	.02 (2%)
10,000	.01 (1%)

Two important features of Table 5.1 are

1. When the sample size is *increased,* the margin of error *decreases.*
2. When a large sample size is made even larger, the improvement in accuracy is relatively small. For example, when the sample size is increased from 2500 to 10,000, the margin of error decreases only from 2% to 1%. In general, cutting the margin of error in half requires a fourfold increase in sample size.

Polling organizations determine a sample size that is accurate enough for their purposes and is also economical. Many national surveys use a sample size of about 1000, which, as you can see from Table 5.1, makes the margin of error roughly 3%. This is a reasonable degree of accuracy for most questions asked in these surveys.

Some federal government surveys utilize much larger sample sizes, sometimes as large as $n = 120{,}000$, to make accurate estimates of quantities such as the unemployment rate. Also, when researchers want to make accurate estimates for subgroups within the population, they have to use a very large overall sample size. For instance, to get information from approximately 1000 African-American women in the 18- to 29-year-old age group, a random sample of 120,000 Americans might be necessary.

The Effect of Population Size

You might wonder how the number of people in the population affects the accuracy of a survey. The surprising answer is that for most sample surveys, the number of people in the population has almost no influence on the accuracy of sample estimates. The margin of error for a sample size of 1000 is about 3% whether the number of people in the population is 30,000 or 200 million.

The formulas for margin of error in this chapter were derived by assuming that the number of units in the population is essentially infinite. In practice, as long as the population is at least ten times as large as the sample, we can ignore the specific size of the population. For small populations, a "finite population correction" is used. We will not discuss it in this book, but you can consult any book on survey sampling for details.

5.2 Exercises are on pages 178–179.

IN SUMMARY Conservative Margin of Error and Confidence Intervals

- Sample size for a survey = n
- Conservative margin of error for proportions is $\frac{1}{\sqrt{n}}$
- Conservative margin of error for percents is $\frac{1}{\sqrt{n}} \times 100\%$.

An approximate 95% confidence interval for a population proportion or percent is:

sample proportion or percent ± conservative margin of error

where the symbol ± means to add and subtract. In other words, use the value on the right to create an interval by subtracting it from and adding it to the value on the left.

If a desired margin of error for a confidence interval for a population proportion is m.e., the required sample size is $n = \frac{1}{(m.e.)^2}$. For example, for a desired margin of error of .05, $n = \frac{1}{(.05)^2} = 400$.

THOUGHT QUESTION 5.2 Suppose that a survey of 400 students at your school is conducted to assess student opinion about a new academic honesty policy. Based on Table 5.1, about what will be the margin of error for the poll? How many students attend your school? Given this figure, do you think the values in Table 5.1 should be used to estimate the margin of error for a survey of students at your school? Explain.*

5.3 Choosing a Simple Random Sample

The ability of a relatively small sample to accurately reflect a huge population does not happen haphazardly. It happens only if proper sampling methods are used. A **probability sampling plan** is one in which everyone in the population has a specified probability to be selected for the sample.

The most basic probability sampling plan is to use a **simple random sample**. Remember that with a simple random sample, every conceivable group of units of the required size has the same chance of being the selected sample. In this section, we discuss how to choose a simple random sample. A variety of other probability sampling methods will be discussed in Section 5.4.

***HINT:** The sample size is $n = 400$. Is the number of students at your school more than ten times the sample size?

Choosing a simple random sample is somewhat like choosing the winning numbers in many state lotteries. For instance, in the Pennsylvania Cash 5 lottery game, five numbers are randomly selected from the choices 1, 2, . . . , 43. Every possible set of five numbers is equally likely to be the winning set. There are actually 962,598 different possible sets, which is why the odds of any specific individual guessing the winning set are so small!

Similarly, the chances of any *particular* group of units getting selected to be the random sample from a large population is quite small, but whatever group is selected is likely to be *representative.* For instance, in a simple random sample of 1000 people in your state or country, it is extremely unlikely that you and your next-door neighbor would both be selected. But it is extremely likely that someone in the sample will be representative of each of you, having similar opinions to yours.

By the way, if there were 100 million people in a population, the number of different possible samples of 1000 individuals would be incomprehensively large. It would take about one and a half pages of this book to show the value. Approximately, the number of possible different samples of 1000 individuals selected from a population of 100 million is 247 followed by 5430 zeros.

To actually produce a simple random sample, you need only two things. First, you need a list of the units in the population. For instance, in drawing the winning lottery numbers, the list of units is the numbers 1, 2, . . . , 43. In selecting a simple random sample of students from your school, the list of units is all students in the school (which the registrar can usually produce).

Second, you need a source of **random numbers**. Computer programs such as Minitab and Excel or the right calculator can be used to generate random numbers. Random numbers can also be found in tables called **tables of random digits**, which are (exceedingly boring!) tables that contain lists of integers from 0 to 9 in random sequence. They are generated using a method equivalent to writing the numbers from 0 to 9 on slips of paper, mixing them well, drawing one and recording the number, returning it to the pile, drawing another, and so on. Typing "Table of random digits" into an Internet search engine should provide examples of such tables. If the population isn't very large, physical methods can be used, such as in lotteries in which the numbers are written on small, hollow plastic balls and five of them are physically selected.

Suppose your school has 5000 students and you want to select a random sample of 10 of them for a focus group. You could number the students from 1 to 5000, and then ask a computer program such as Minitab to randomly select 10 of them. Table 5.2 illustrates three samples of 10 individuals each, resulting from asking Minitab to draw a random sample of 10 from a population of 5000 people. The samples are generated by the equivalent of writing the numbers from 1 to 5000 on slips of paper, mixing them well, and then choosing 10 of them. Using the computer is a lot less work!

Table 5.2 Three Samples of $n = 10$ Chosen from a Population of 5000 Individuals

Sample										
1	309	450	1117	1529	1822	2389	3193	3535	3855	4056
2	71	545	576	780	1011	2370	3500	4238	4566	4991
3	278	370	703	858	2405	2423	2726	4050	4068	4795

MINITAB TIP

Picking a Random Sample

- To create a column of ID numbers, use **Calc>Make Patterned Data>Simple Set of Numbers**. In the dialog box, specify a column for storing the ID numbers, and specify the first and last possible ID number for the population.
- To sample values from a column, use **Calc>Random Data>Sample from Columns**. In the dialog box, specify how many items (rows) will be selected from a particular column, and specify a column where the sample will be stored.

Note: Items can be randomly selected from a column of names or data values, so it may not be necessary to assign ID numbers to the units in the population in order to select a sample.

Example 5.5 **Choosing a Random Sample of Colleges in the United States** The University of Texas, Austin website (http://www.utexas.edu/world/univ) provides a comprehensive list of regionally accredited four-year colleges and universities in the United States. As of March 2010, there were 2042 schools listed. Suppose a company wants to select a random sample of 20 of these schools at which to conduct an advertising campaign. The schools could be numbered from 1 to 2042, and 20 numbers from that list could be drawn to select the sample. Equivalently, using computer software such as Minitab, the schools could be entered into one column and the software could be instructed to draw a random sample of 20 of them. The list below illustrates the results of asking Minitab to do that.

Carnegie Mellon University (PA)
Frostburg State University (MD)
Northern Illinois University (IL)
Corcoran College of Art & Design (DC)
Green Mountain College (VT)
Baker University (KS)
State University of New York at Stony Brook (NY)
Clark University (MA)
Universidad Interamericana de Puerto Rico (PR), San Germán Campus
Coppin State University (MD)
Pennsylvania State University–Wilkes-Barre (PA)
State University of New York–College at Oneonta (NY)
Emmanuel School of Religion (TN)
North Carolina State University (NC)
New Jersey City University (NJ)
Rutgers University–Camden (NJ)
University of Medicine & Dentistry of New Jersey (NJ)
Bucknell University (PA)
Howard University (DC)
Colorado Technical University (CO)

Although your school may not be in the sample, if you scan the list you probably will find a school that is similar to yours (if you are attending a four-year college or university). You can see that there are private and public schools listed, large and small, general and specialized, from a variety of locations.

Example 5.6 **Representing the Heights of British Women** In Example 2.17 in Chapter 2, we examined data on the heights of 199 British women. Suppose you had a list of these 199 women and wanted to choose 10 of them to test-drive a sporty, but small, automobile model and give their opinions about its comfort. The heights of the women in the sample should be representative of the range of heights in the larger group. You would not want your sample of 10 to include only short women or only tall women. Here's how you could choose a simple random sample:

1. Assign an ID number from 001 to 199 to each woman.
2. Use a table of random digits, a computer, or a calculator to randomly select 10 numbers between 001 and 199, and sample the heights of the women with those ID numbers.

We used the statistical package Minitab to choose a simple random sample of heights, and then used a table of random digits to choose another one. The samples are listed along with their sample means and the list of 10 random numbers between 001 and 199 that generated them (with leading zeros dropped). In Chapter 2, the heights were given in millimeters, but here they have been converted to inches.

- Sample 1 (Minitab)
  ID numbers of the women selected: 176, 10, 1, 40, 85, 162, 46, 69, 77, 154
  Heights: 60.6, 63.4, 62.6, 65.7, 69.3, 68.7, 61.8, 64.6, 60.8, 59.9; mean = 63.7 inches
- Sample 2 (Table of random digits)
  ID numbers of the women selected: 41, 93, 167, 33, 157, 131, 110, 180, 185, 196
  Heights: 59.4, 66.5, 63.8, 62.6, 65.0, 60.2, 67.3, 59.8, 67.7, 61.8; mean = 63.4 inches

As you can see, each sample is different, but each should be representative of the entire group of 199 women. Within each sample, the range of heights is from about 60 inches to about 69 inches. The sample means are both close to the mean height for the larger group, which was given in Chapter 2 as 1602 mm, or about 63 inches.

5.3 Exercises are on pages 179–180.

5.4 Other Sampling Methods

For large populations, it may not be practical to take a simple random sample because it may be difficult to get a numbered list of the units. For instance, if a polling organization wanted to take a simple random sample of all voters or all adults in a country or region, the organization would need to get a numbered list of them, which is simply an impossible task in most cases. Instead, it relies on more complicated sampling methods, all of which are good substitutes for simple random sampling in most situations. In fact, they often have advantages over simple random sampling. For instance, one of these other methods, stratified random sampling, can be used to increase the chance that the sample represents important subgroups within the population by specifying how large a sample to take in each subgroup separately.

To visually illustrate the various sampling plans discussed in this chapter, let's suppose that a college administration would like to survey a sample of students living in dormitories. The college has undergraduate and graduate dormitories. The undergraduate dormitories have 3 floors each with 12 rooms per floor. The graduate dormitories have 5 floors each with 8 rooms per floor.

Figure 5.1 illustrates a simple random sample of 30 rooms in the dormitories. Any collection of 30 rooms has an equal chance of being the selected sample. Note that for the sample illustrated, there are 12 undergraduate rooms and 18 graduate rooms in the sample.

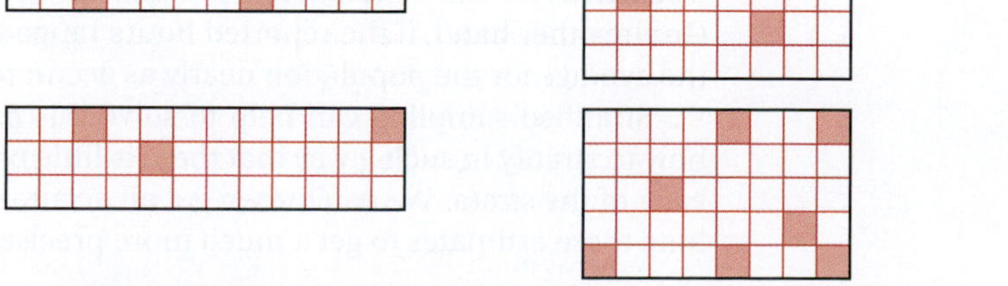

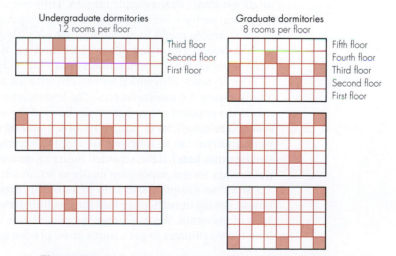

Figure 5.1 A simple random sample of 30 dorm rooms

Stratified Random Sampling

A **stratified random sample** is collected by first dividing the population of units into **strata** (subgroups of the population) and then taking a simple random sample from each one. The strata are subgroups that seem important to represent properly in the sample and might differ with regard to values of the response variable(s) measured.

For example, public opinion pollsters often take separate random samples from each region (strata) of the country so that they can spot regional differences as well as improve the likelihood that all regions are properly represented in the overall national sample. Or political pollsters may separately sample from each political party to compare opinions by party and to be sure that each party is properly represented.

Figure 5.2 illustrates a stratified sample for the college survey. There are two strata: the undergraduate and graduate dorms. A random sample of 15 rooms is taken from each of the two strata. Each collection of 15 undergraduate rooms has an equal chance of being the selected sample for the undergraduate dorms, and each collection of 15 graduate rooms has an equal chance of being the selected sample for the graduate dorms. But the total of 15 rooms to be sampled within each stratum (undergraduate or graduate) is fixed before the sample is selected.

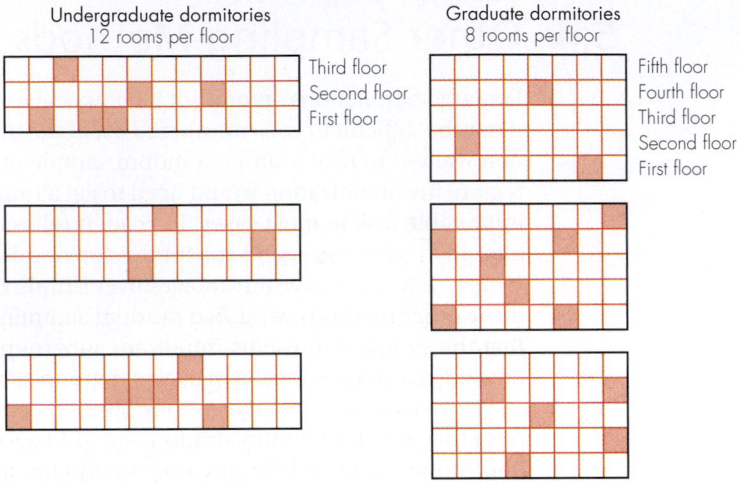

Figure 5.2 A stratified sample of 15 undergraduate and 15 graduate dorm rooms

A principal benefit of stratified sampling is that it can be used to improve the chance that the selected sample properly represents important subgroups in the population. It also is used to create more accurate estimates of population values than we might get from using a simple random sampling method.

So far, we have been focusing on measuring categorical variables, such as opinions or traits people might have. Surveys are also used to measure quantitative variables, such as hours of studying per week or number of cigarettes smoked per day. We are often interested in the population average for such measurements.

The accuracy with which we can estimate the average depends on the natural variability among the measurements. The less variable they are, the more precisely we can assess the population average on the basis of the sample values. For instance, if everyone in a relatively large sample reports that they studied between 20 and 25 hours a week, then we can be relatively sure that the average in the population is close to that. On the other hand, if the reported hours range from 2 to 70, then we cannot pinpoint the average for the population nearly as accurately.

Stratified sampling can help to solve this problem. Suppose we could figure out how to stratify in such a way that there is little natural variability in the answers within each of the strata. We could then get an accurate estimate for each stratum and combine these estimates to get a much more precise answer for the overall group.

Cluster Sampling

In **cluster sampling**, the population is divided into **clusters** (subgroups), but rather than sampling within each cluster, we select a random sample of clusters and include only members of these selected clusters in the sample. After clusters are selected, usually all members of a cluster are included in the sample. Occasionally a subsample (instead of all members) is taken from each of the selected clusters. A cluster typically comprises units that are physically close to each other in some way, such as the students living on one floor of a college dormitory, all individuals listed on a single page of a telephone directory, or all passengers on a particular airplane flight.

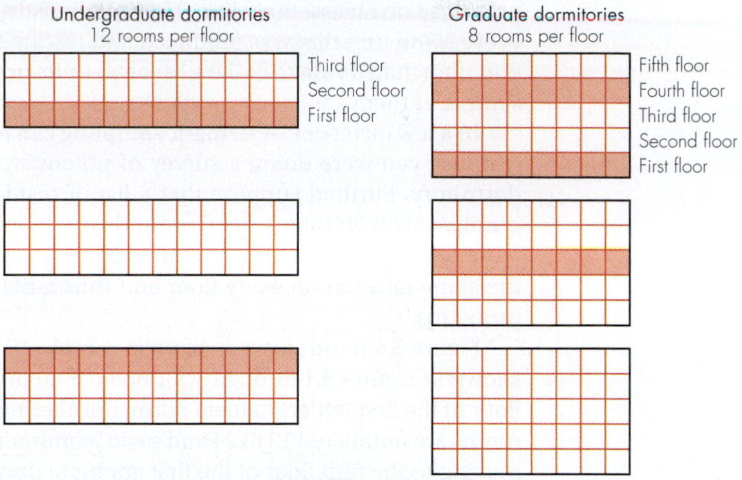

Figure 5.3 A cluster sample in which five floors (clusters) are randomly selected

Figure 5.3 illustrates a cluster sampling plan for the college survey. Each floor of each dormitory is a cluster in this particular plan. A random sample of five floors is selected from the 24 floors of the three undergraduate and three graduate dorms. Any collection of five floors has an equal chance of being the selected sample of floors. Once the five floors have been selected, all students on the five selected floors are surveyed. This plan is efficient logistically because the data collection team will have to visit only five different dormitory floors to collect data.

Cluster sampling is often confused with stratified sampling, but it is a radically different concept and can be much easier to accomplish. In most applications of stratified sampling, the population is divided into a few *large* strata, such as regions of the country, and a small subset is then *randomly sampled from each of the strata.* In most applications of cluster sampling, the population is divided into *small* clusters (such as city blocks), a large number of clusters are randomly sampled, and either everyone or a sample is measured in those clusters selected. In stratified sampling, all subgroups (strata) are represented in the sample. In cluster sampling, some clusters within the population are included in the sample, and others are not.

One obvious advantage of cluster sampling is that you need only a list of clusters instead of a list of all individual units. City blocks are commonly used as clusters in surveys that require door-to-door interviews. To measure customer satisfaction, airlines sometimes randomly sample a set of flights, and then distribute a survey to everyone on those flights. Each flight is a cluster. It is clearly much easier for the airline to choose a random sample of flights than it would be to identify and locate a random sample of individual passengers to whom to distribute surveys.

If cluster sampling is used, the analysis must proceed differently because there may be similarities among the members of the clusters that must be taken into account. Numerous books are available that describe proper analysis methods based on which sampling plan was employed. (See, for example, *Sampling: Design and Analysis, 2nd ed.*, by Sharon Lohr, Brooks/Cole, Cengage Learning, 2010.)

Systematic Sampling

Suppose you want to select a sample of 100 individuals from a list of 5000 names. That means you would want to select 1 out of every 50 people on the list. An idea that would work in most cases is to choose every 50th name on the list.

If you did so, you would be using a **systematic sampling plan**. With this plan, you divide the list into as many consecutive segments as you need, randomly choose a starting point in the first segment, and then sample at that same point in each segment.

In our example, you would randomly choose a starting point in the first 50 names, and then sample every 50th name after that. When you were finished, you would have selected one person from each of 100 segments, equally spaced throughout the list.

As another example, suppose that a company wants to assess the quality of items produced on an assembly line. They might use a systematic sampling plan by selecting every 200th unit that was produced and testing its quality. This would be a much simpler plan than trying to make a list of all units produced and selecting a simple random sample of them.

In a few instances, systematic sampling can lead to a biased sample. As an example, suppose you were doing a survey of potential noise problems in a high-rise college dormitory. Further, suppose that a list of residents was provided, arranged by room number, with 20 rooms per floor and two people per room. If you were to take a systematic sample of, say, every 40th person on the list, you would get people who lived in the same location on every floor and thus a biased sampling of opinions about noise problems.

Figure 5.4 illustrates a systematic sample of the college dormitory rooms that were shown in Figures 5.1 through 5.3. Consider an ordered list of rooms starting with the top floor of the first undergraduate dorm, numbering those rooms 1 to 12. The second-floor rooms are numbered 13 to 24 and so on, continuing with the undergraduate dorms, then moving to the fifth floor of the first graduate dorm. In the figure, room 3 was randomly selected as a starting point chosen from the first 11 rooms. After that, every eleventh room is sampled. Note that there is some pattern to the rooms selected, which may or may not lead to biased results, depending on the questions asked in the survey.

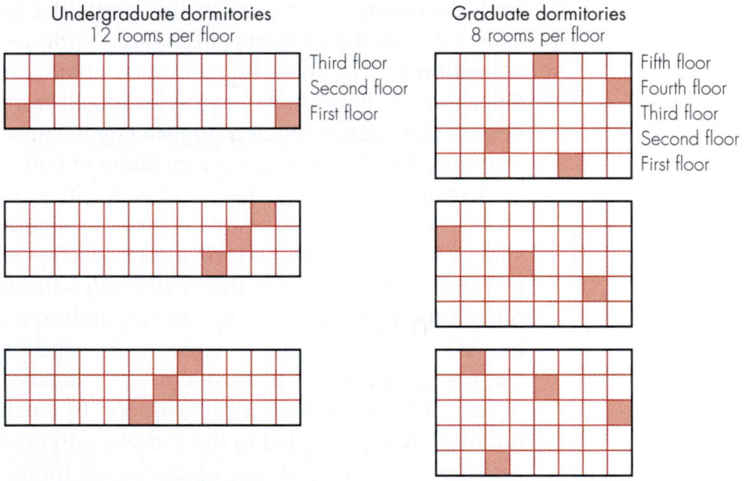

Figure 5.4 A systematic sample of every 11th room, randomly starting with the third room

Random-Digit Dialing

It is essentially impossible to take a simple random sample of a large population such as the adult population of the United States. Most national polling organizations and many government surveys in the United States now use a sampling method called **random-digit dialing**. Until recently, the technique consisted of a sophisticated method of randomly sampling landline telephone numbers and calling them. This method results in a sample that approximates a simple random sample of all households in the United States that have landline telephones. In recent years cell phones have been added because many households no longer have a landline.

The method proceeds as follows. First, the polling organization makes a list of all possible telephone *exchanges*, where the exchange consists of the area code and the next three digits. Then a computer is used to generate a sample of exchanges. The pollsters can approximate the proportion of all households in the country that have a specific exchange, and this proportion governs the chance that the exchange is sampled. Next, the same method is used to randomly sample *banks* within each exchange, where a bank consists of the next two numbers. Phone companies assign numbers using banks, so certain banks are mainly assigned to businesses, certain ones are held for future neighborhoods, and so on. Finally, to complete the number, the computer randomly generates the last two digits from 00 to 99.

Once a phone number has been determined, the pollsters should make multiple attempts to reach someone at that household. Sometimes they will ask to speak to a male because females are more likely to answer the phone and would thus be over-represented if the researchers always survey the first individual who answers the phone.

One of the complications of including cell phones is that geographic data may not be accurate. When people move they often keep their original cell phone numbers, so knowing someone's area code no longer ensures that the pollsters know where they live. Younger adults are more likely to live in households that have no landline, so pollsters must take care to ensure that they are adequately sampled.

Example 5.7 **An *ABC News* Poll on Parental Permissiveness** Most polling agencies now include news stories highlighting interesting results of recent polls on their websites. A brief description of how the poll was conducted is usually included at the end of the news story. For example, an *ABC News/Good Morning America Weekend* poll surveyed parents to find out what they thought their teenage children should be permitted to do. A news story at the website abcnews.go.com/PollingUnit reported many results of the poll, such as the following:

> "Question: Mom, Dad, can I have a credit card? Answer: Nope. Nor, for most parents, can a teenager have a glass of beer at a family event, attend an unsupervised party, or, if a girl, stay out past 11 p.m. So much for rampant permissiveness. Instead this ABC News/Good Morning America Weekend poll finds significant parental constraints on children's activities. Four in 10, for instance, rule out social networking websites and unsupervised use of the Internet. A quarter forbid the pre-technology sport of hanging out at the mall. And for those who do see these and other activities as appropriate for teens, the starting age is the mid-teens, generally 15 or 16 years old" (Langer and Mokrzycki, 2010).

A description of how the poll was conducted was given in the last paragraph of the story, as follows:

> "METHODOLOGY—This ABC News poll was conducted by telephone Feb. 3–7, 2010, among a random national sample of 779 parents (landline and cell-phone respondents). Results for the full sample have a 4-point error margin. Sampling, data collection and tabulation by Social Science Research Solutions at ICR-International Communications Research of Media, Pa."

Multistage Sampling

Many large surveys use a combination of the methods we have discussed. In a **multistage sampling plan**, the researchers may combine methods to sample successively smaller portions of the population to reach an individual unit. The survey designers might stratify the population by region of the country, then stratify by urban, suburban, and rural, and then choose a random sample of communities within those strata. They would then divide those communities into city blocks or fixed areas, as clusters, and sample some of those. Everyone on the block or within the fixed area may then be sampled.

Example 5.8 **The Current Population Survey** Unemployment rates in the United States are estimated each month based on interviews with a small percentage of households in the country. They are part of the Current Population Survey conducted by the Bureau of Labor Statistics. It is important for these households to represent the entire U.S. population while also making sure that unemployment rates can be estimated for each state.

The sample of households is chosen in three stages. In the first stage, the country is divided into slightly over 2000 "primary sampling units" (PSUs), which are defined based on a combination of population and size. In most cases, one or two adjoining counties constitute a PSU. Within each state, strata are created such that PSUs within the same strata are likely to have similar unemployment rates. Then one PSU is randomly selected from each of the strata.

In the second stage, "ultimate sampling units" (USUs) are randomly selected from each PSU that was chosen in the first stage. USUs are small groups of housing units (usually four addresses) in the same location, so that an interviewer can visit them all in a relatively short time. The groups of housing units from which the USUs are formed are determined from the census that takes place every 10 years, updated in between with new housing units. Multiple USUs are selected in each strata. In most cases the sampling stops there, and each household in the USU is visited. Households remain in the sample for 4 months, rotate out for 8 months, and then are surveyed again for another 4 months.

The third stage of sampling is necessary only when the USU selected contains too many households. In that case a systematic sampling scheme is used to randomly choose which of the households will be part of the sample.

Note that this is a multistage sample. Households are organized into PSUs based on location, which are then grouped into *strata* based on demographics. Within each of the strata, a representative PSU is selected. Within that PSU, multiple *clusters* of housing units (USUs) are randomly selected. Every household in the selected clusters is visited. In all, about 72,000 housing units are visited. From interviews with those households, monthly unemployment rates are determined. For more information, see http://www.census.gov/prod/2006pubs/tp-66.pdf.

5.4 Exercises are on pages 180–181.

THOUGHT QUESTION 5.3 In Section 5.1, the Fundamental Rule for Using Data for Inference stated that "available data can be used to make inferences about a much larger group *if the data can be considered to be representative with regard to the question(s) of interest.*" Read the description of how the *ABC News* poll in Example 5.7 was conducted. Do you think the results of the poll can be extended to a larger group than the 779 parents in the sample? If so, to what group can the results be extended and why?*

5.5 Difficulties and Disasters in Sampling

In theory, designing a good sampling plan is easy and straightforward. However, the real world rarely cooperates with well-designed plans, and it can be difficult to collect a proper sample. Difficulties that can occur in practice need to be considered when you evaluate a study. If a proper sampling plan is not implemented, the results may be misleading and inaccurate. Some problems can occur even when a sampling plan has been well designed. Here is a list of possible problems that might cause either selection bias or nonparticipation (nonresponse) bias:

- Using the wrong sampling frame
- Not reaching the individuals selected
- Nonresponse or nonparticipation
- Self-selected sample
- Convenience or haphazard sample

Using the Wrong Sampling Frame

DEFINITION The **sampling frame** is the list of units from which the sample is selected. This list may or may not be the same as the list of all units in the desired "target" population.

***HINT:** Read the section on random-digit dialing in this chapter.

Using the wrong sampling frame is one way to create *selection bias.* Suppose that our target population for a survey is the general adult population in a community and that we select individuals by randomly sampling the community's paper telephone directory. The sampling frame consists of people who are listed in the directory, but this excludes individuals with unlisted home numbers (e.g., many physicians and teachers), those who use a cell phone as their only telephone, and people who cannot afford a telephone. So the sampling frame doesn't entirely cover the target population of all adults. This is an example of **undercoverage**, in which the sampling frame does not cover the entire population of interest. Figure 5.5 illustrates this situation. A partial solution is to use random-digit dialing, but this still excludes those who don't have a landline telephone unless cell phones are included.

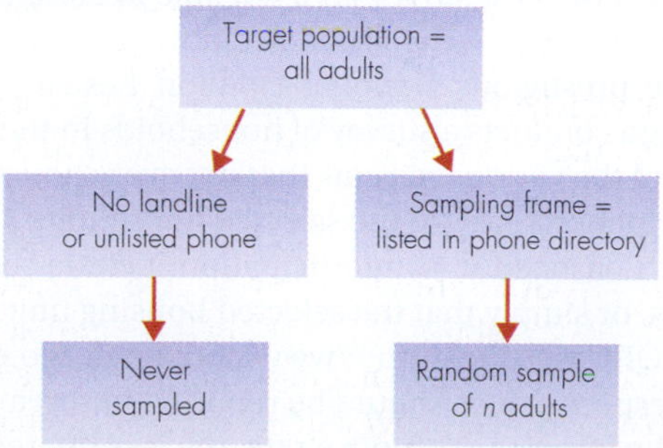

Figure 5.5 The sampling frame doesn't cover the target population, leading to selection bias

As another example of using the wrong sampling frame, suppose we sample a list of registered voters to predict election outcomes. This sampling frame will include individuals who are likely to vote (the target population) but will also include individuals who are not likely to vote. We might be able to partially solve this problem by having interviewers ascertain the voting history of the people they contact by asking where they vote and then continuing only if the person knows the answer.

Not Reaching the Individuals Selected

Even if a sample of units is selected properly, researchers might not reach the desired units. For example, *Consumer Reports* magazine mails a lengthy survey to its subscribers to obtain information on the reliability of various products. If you received such a survey and had a close friend who had been having trouble with a highly rated automobile, you may very well have decided to pass the questionnaire on to your friend. That way, he would get to register his complaints about the car, but *Consumer Reports* would not have reached the intended recipient.

Telephone surveys tend to reach a disproportionate number of women because they are more likely to answer the phone. Researchers sometimes ask to speak to the oldest adult male at home to try to counter that problem. Surveys are also likely to have trouble contacting people who work long hours and are rarely home and those who tend to travel extensively. Some people screen their calls or simply refuse to cooperate even when they do answer the phone.

In recent years, there has been pressure on news organizations to produce surveys of public opinions quickly. When a controversial story breaks, people want to know how others feel about it. This pressure results in what *Wall Street Journal* reporter Cynthia Crossen calls "**quickie polls**." As she notes, these are "most likely to be wrong because questions are hastily drawn and poorly pretested, and it is almost impossible to get a random sample in one night" (Crossen, 1994, p. 102). Even if the computer can randomly generate phone numbers for the sample, there will be groups of people who are not likely to be home that night, and they may have different opinions from those who are likely to be home. Most responsible reports about polls include information about the dates during which the poll was conducted. For instance, the poll in Example 5.7 was conducted from Wednesday to Sunday in the first week of February. Because the target population was parents of teenagers, including weekdays and weekends during a time period when most schools are in session should have helped minimize selection bias.

Nonresponse, Volunteer Response, and Nonparticipation

Another problem with polls occurs when a sample is selected properly, but the individuals cannot be contacted or refuse to participate. When only some members of a selected sample choose to participate, these **volunteer responses** are not likely to represent the entire sample. This leads to *nonresponse* or *nonparticipation bias*. It is important, once a sample has been selected, that those individuals are the ones who are actually measured. It is better to put resources into getting a smaller sample than to get one that has been biased by moving on to the next person on the list when someone is initially unavailable.

Even the best surveys are not able to contact everyone on the list of selected units, and not everyone who is contacted will respond. The General Social Survey (GSS), run by the prestigious National Opinion Research Center (NORC) at the University of Chicago conducts a survey of households in the United States every 2 years. They keep track of the various reasons that not every household selected for the survey responds. Participation rates for the selected households have been roughly 70% to 75% in recent years. Reasons for nonparticipation include refusal to participate, language barriers, illness, or simply that the selected housing unit is vacant. More details can be found at the NORC website, http://www.norc.uchicago.edu/projects/gensoc.asp.

Response rates should be reported in research summaries. As a reader, remember that the lower the response rate, the less the results can be generalized to the population as a whole. Responding to a survey (or not) is voluntary, and those who respond are likely to have stronger opinions than those who do not respond. As we mentioned earlier, this type of **nonresponse bias** or **nonparticipation bias** can lead to systematically overestimating or underestimating the truth about a population.

With mail and email surveys, it may be possible to compare those who respond immediately with those who need a second prodding; and in telephone surveys, you could compare those who respond on the first try with those who require numerous call-backs. If those groups differ on the measurement of interest, then those who were never reached are probably different as well.

In a mail survey, researchers should not just accept that those who did not respond the first time can't be cajoled into responding. Often, sending a reminder with a brightly colored stamp or following up with a personal phone call will produce the desired effect. Surveys that simply use those who respond voluntarily are sure to be biased in favor of those with strong opinions or with time on their hands.

THOUGHT QUESTION 5.4 Suppose you want to know how students at your school feel about the computer services that are offered. You are able to obtain the list of e-mail addresses for all students who are taking statistics classes, so you send a survey to a simple random sample of 100 of those students and 65 respond. Using the difficulties discussed so far in this section, explain to whom you could extend the results of your survey and why.*

Example 5.9 **Which Scientists Trashed the Public?** According to a poll taken among scientists and reported in the prestigious journal *Science* (Mervis, 1998), scientists don't have much faith in either the public or the media. The article reported that based on the results of a "recent survey of 1400 professionals" in science and in journalism, 82% of scientists "strongly or somewhat agree" with the statement, "The U.S. public is gullible and believes in miracle cures or easy solutions." Eighty percent agreed that "the public doesn't understand the importance of federal funding for research." Eighty-two percent also trashed the media, agreeing with the statement, "The media do not understand statistics well enough to explain new findings." It isn't until the end of the article that we learn who responded: "The study reported a 34% response rate among scientists, and the typical respondent was a white, male physical scientist over the age of 50

*HINT: What type of bias is represented and how is it likely to affect the results?

doing basic research." Remember that those who feel strongly about the issues in a survey are the most likely to respond. With only about a third of those contacted responding, it is inappropriate to generalize these findings and conclude that most scientists have so little faith in the public and the media. This is especially true since we were told that the respondents represented only a narrow subset of scientists.

Not Using a Probability Sample

Basing a sample survey on a **self-selected sample** (also called a **volunteer sample**) or a **convenience sample** is usually so problematic that the results cannot be extended to anyone beyond the sample.

Self-Selected Sample

While relying on *volunteer response* from a properly selected random sample presents somewhat of a difficulty (i.e., nonresponse bias), relying on a self-selected sample consisting of only those who pick themselves to participate (i.e., a self-selected or *volunteer sample*) is usually a complete waste of time. When a television or radio station, magazine, or Internet site presents a survey and asks any readers or viewers who are interested to respond, the results reflect the opinions of only those listeners or readers who decide to take the trouble to respond. And people who weren't watching or listening to these stations or who weren't reading the magazine or website that presented the survey would not even know the poll existed.

As we noted earlier, those who have a strong opinion about the question are more likely to respond than those who do not have a strong opinion. Thus, the self-selected responding group is simply not representative of any larger group. Most media outlets now acknowledge that such polls are "unscientific" when they report the results, but most readers are not likely to understand how misleading the results can be. The next example illustrates the contradiction that can result between a scientific poll and one relying solely on a self-selected sample.

Example 5.10 **A Meaningless Poll** On February 18, 1993, shortly after Bill Clinton became president of the United States, a television station in Sacramento, California, asked viewers to respond to the question, "Do you support the President's economic plan?" The next day the results of a properly conducted study that asked the same question were published in the newspaper. Here are the results from these two different surveys:

	Television Poll	Survey
Yes (support plan)	42%	75%
No (don't support plan)	58%	18%
Not sure	0%	7%

As you can see, those who were dissatisfied with the president's plan were much more likely to respond to the television poll than those who supported it, and no one who was "Not sure" called the television station, since they were not invited to do so. Trying to extend the television poll results to the general population is misleading. It is irresponsible to publicize such studies without a warning that they result from an unscientific survey and are not representative of general public opinion. You should never interpret such polls as anything other than a count of who bothered to respond.

Convenience or Haphazard Sample

Another problematic sampling technique for surveys is to use the most convenient group available or to decide haphazardly on the spot whom to sample. It is usually the case that these types of samples break the Fundamental Rule of Using Data for Inference given in Section 5.1. The responses from a **convenience sample** or **haphazard sample** rarely represent any larger population for the question of interest.

Example 5.11 **Haphazard Sampling** A few years ago, the student newspaper at a California university announced as a front-page headline, "Students ignorant, survey says." The article explained that a "random survey" indicated that American students were less aware of current events than international students were.

The article quoted the undergraduate researchers, who were international students themselves, as saying "the students were randomly sampled on the quad." The quad is a wide expanse of lawn where students relax, eat lunch, and so on. There is simply no proper way to collect a random sample of students by selecting them in an area like that. In such situations, the researchers are likely to approach people who they think will support the results they intended for their survey. Or they are likely to approach friendly-looking people who look as though they will easily cooperate. This haphazard sample cannot be expected to be representative at all.

You have seen the proper way to collect a sample and have been warned about the many difficulties and dangers inherent in the process. Here is a famous example that helped researchers learn some of these pitfalls.

CASE STUDY 5.1 The Infamous *Literary Digest* Poll of 1936

Before the presidential election of 1936, a contest between Democratic incumbent Franklin Delano Roosevelt and Republican Alf Landon, the magazine *Literary Digest* had been extremely successful in predicting the results in U.S. presidential elections. But 1936 turned out to be the year of its downfall, when it predicted a 3-to-2 victory for Landon. To add insult to injury, young pollster George Gallup, who had just founded the American Institute of Public Opinion in 1935, not only correctly predicted Roosevelt as the winner of the election, but also predicted that the *Literary Digest* would get it wrong. He did this before they had even conducted their poll! And Gallup surveyed only 50,000 people, while the *Literary Digest* sent questionnaires to 10 million people (Freedman et al., 1991, p. 307).

The *Literary Digest* made two classic mistakes. First, the lists of people to whom they mailed the 10 million questionnaires were taken from magazine subscribers, car owners, telephone directories, and, in a few instances, lists of registered voters. In 1936, those who owned telephones or cars or subscribed to magazines were more likely to be wealthy individuals who were not happy with the Democratic incumbent.

Despite what many accounts of this famous story conclude, the bias produced by the more affluent list was not likely to have been as severe as the second problem (Bryson, 1976). *The main problem was nonresponse bias caused by volunteer response.* The *Literary Digest* received 2.3 million responses, a response rate of only 23%. Those who felt strongly about the outcome of the election were most likely to respond. That included a majority of those who wanted a change: the Landon supporters. Those who were happy with the incumbent were less likely to bother to respond.

Gallup, on the other hand, knew the value of random sampling. He not only was able to predict the election, but also predicted what the results of the *Literary Digest* poll would be to within 1%. How was he able to predict the results of the *Literary Digest* survey? According to Freedman et al. (1991, p. 308), "he just chose 3,000 people at random from the same lists the *Digest* was going to use, and mailed them all a postcard asking them how they planned to vote."

Case Study 5.1 illustrates the beauty of random sampling and the folly of trying to base conclusions on nonrandom and biased samples. The *Literary Digest* went bankrupt the following year, so it never had a chance to revise its methods. The organization founded by George Gallup has flourished, although it has made a few sampling blunders as well. (See Exercise 5.61.)

5.5 Exercises are on page 181.

5.6 How to Ask Survey Questions

You may be surprised at how much the answers to questions can change on the basis of simple changes in wording. Here is one example. Loftus and Palmer (1974, quoted in Plous, 1993, p. 32) showed college students films of an automobile accident, after which they asked the students a series of questions. One group was asked the question, "About how fast were the cars going when they contacted each other?" The average response was 31.8 miles per hour. Another group was asked, "About how fast were the cars going when they collided with each other?" In that group, the average response was 40.8 miles per hour. Simply changing from the word *contacted* to the word *collided* increased the estimates of speed by 9 miles per hour, or 28%, even though the respondents had witnessed the same film.

The wording and presentation of questions can significantly influence the results of a survey. The seven pitfalls discussed in this section are all possible sources of response bias in surveys.

Possible Sources of Response Bias in Surveys

Many pitfalls leading to response bias can be encountered when asking questions in a survey or experiment. Here are seven of them, each of which will be discussed in turn:

- Deliberate bias in questions
- Unintentional bias in questions
- Desire of respondents to please
- Asking the uninformed
- Unnecessary complexity
- Ordering of questions
- Confidentiality and anonymity concerns

Deliberate Bias in Questions

Sometimes, if a survey is being conducted to support a certain cause, questions are deliberately worded in a biased manner (**deliberate bias**). Be careful about survey questions that begin with phrases such as "Do you agree that . . . ?" Most people want to be agreeable and will be inclined to answer "yes" unless they have strong feelings the other way.

For example, suppose that an anti-abortion group and a pro-choice group each wanted to conduct a survey in which it would find the best possible agreement with its position. Here are two questions that would each produce an estimate of the proportion of people who think abortion should be completely illegal. The questions are almost certain to produce different estimates:

1. Do you agree that abortion, which many consider to be the murder of innocent beings, should be outlawed?
2. Do you agree that there are circumstances under which abortion should be legal to protect the rights and health of the mother?

The wording of a question should not indicate a desired answer. For instance, the Gallup Poll has been asking opinions about abortion for many years, with this question: "Do you think abortions should be legal under any circumstances, legal only under certain circumstances, or illegal in all circumstances?" As another example, a November 2009 Associated Press–Stanford University poll on the environment asked: "In general, would you favor or oppose building more nuclear power plants at this time?" In case you are curious, opinions were evenly split with 49% favoring, 48% opposing, and 3% unsure (http://www.ap-gfkpoll.com/pdf/AP-Stanford_University_Environment_Poll_Topline.pdf).

Unintentional Bias in Questions

Sometimes questions are worded in such a way that the meaning is misinterpreted by a large percentage of the respondents (**unintentional bias**). For example, if you were to ask people whether they use drugs, you would need to specify whether you mean prescription drugs, illegal drugs, over-the-counter drugs, or common substances such as caffeine. If you were to ask people to recall the most important date in their life, you would need to clarify whether you meant the most important calendar date or the most important social engagement with a potential partner. (It is unlikely that anyone would mistake the question as being about a "date" as in the shriveled fruit, but you can see that the same word can have multiple meanings.)

Example 5.12 **Laid Off or Fired?** Plewes (1994) described a number of changes implemented at the beginning of 1994 to the Current Population Survey (CPS), which is the monthly survey (described in Example 5.8) used to determine government unemployment figures and other government statistics in the United States. One change involved clarification and expansion of questions about being laid off. Plewes explained that people were answering that they had been laid off when in reality they had been fired: "The CPS concept of 'on layoff' differs from the everyday usage by respondents. CPS defines 'on layoff' as embodying an expectation of recall to the job; common usage considers it a euphemism for 'fired'" (p. 38).

Desire of Respondents to Please

Most survey respondents have a desire to please the person who is asking the question. They tend to understate their responses about undesirable social habits and opinions and vice versa. For example, in recent years, estimates of the prevalence of cigarette smoking based on surveys do not match those based on cigarette sales. Either people are not being completely truthful or lots of cigarettes are ending up in the garbage.

Example 5.13 **Most Voters Don't Lie, but Some Liars Don't Vote** Abelson, Loftus, and Greenwald (1992) discuss how the National Election Surveys always find that more people claim to vote than actually did. They explain that researchers have tried to make it more socially acceptable to admit to not voting by adding a preamble to questions about voting. The preamble reads, "In talking to people about elections, we often find that a lot of people were not able to vote because they weren't registered, they were sick, or they just didn't have time. How about you—did you vote in the elections this November?" (p. 139). In a survey 6 to 7 months after the November 1986 election, the authors of the report surveyed people for whom they knew whether they had actually voted. The question started with the preamble and finished with "did you vote in the 1986 elections for United States Congress?" (p. 141). They found that of the 211 respondents who had actually voted, 203 (96%) accurately reported that they did. But of the 98 respondents who had *not* voted, 39 (40%) claimed that they *did* vote.

Asking the Uninformed

See Case Study 5.2 (p. 173) for an extended example of how people respond even when they have no knowledge of the topic.

People do not like to admit that they don't know what you are talking about when you ask them a question. Crossen (1994, p. 24) gives an example: "When the American Jewish Committee studied Americans' attitudes toward various ethnic groups, almost 30% of the respondents had an opinion about the fictional Wisians, rating them in social standing above a half-dozen other real groups, including Mexicans, Vietnamese and African blacks."

Political pollsters, who are interested in surveying only those who will actually vote, learned long ago that it is useless to simply ask people whether they plan to vote. Most of them will say "yes." Instead, they ask questions to establish a history of voting, such as "Where did you go to vote in the last election?"

Unnecessary Complexity

If questions are to be understood, they must be kept simple. A question such as "Shouldn't former drug dealers not be allowed to work in hospitals after they are released from prison?" is sure to lead to confusion. Does a "yes" answer mean they should or should not be allowed to work in hospitals? It would take a few readings to figure that out.

Another way in which a question can be unnecessarily complex is by actually asking more than one question at once. An example would be a question such as "Do you support the president's health care plan, since it would ensure that all Americans receive health coverage?" If you agree with the idea that all Americans should receive health coverage but disagree with the remainder of the plan, do you answer "yes" or "no"? Or what if you support the president's plan, but not for that reason?

Example 5.14 **Why Weren't You at Work Last Week?** In the work leading to the revision of the Current Population Survey discussed in Examples 5.8 and 5.12, researchers discovered that some questions were too complex to be understood by many respondents. For instance, as Plewes (1994, p. 38) described, one such question on the old version was, "Did you have a job from which you were temporarily absent or on layoff LAST WEEK?" In the revised version of the questionnaire, people are first asked whether they were absent in the last week, and if they were, further questions about temporary absence are asked. A new and separate series of questions asks about layoffs.

Ordering of Questions

If one question requires respondents to think about something that they may not have otherwise considered, then the order in which questions are presented can change the results. For example, suppose a survey were to ask, "To what extent do you think teenagers today worry about peer pressure related to drinking alcohol?" and then ask, "Name the top five pressures you think face teenagers today." It is quite likely that respondents would use the idea they had just been given and name peer pressure related to drinking alcohol as one of the five choices.

Example 5.15 **Is Happiness Related to Dating?** Clark and Schober (1992, p. 41) report on a survey that asked the following two questions:

1. How happy are you with life in general?
2. How often do you normally go out on a date? (about ___ times a month)

When the questions were asked in this order, there was almost no relationship between the two answers. But when Question 2 was asked first, the answers were highly related. Clark and Schober speculate that in that case, respondents consequently interpreted Question 1 to mean "Now, considering what you just told me about dating, how happy are you with life in general?"

Confidentiality and Anonymity Concerns

People will often answer questions differently based on the degree to which they believe they are anonymous. Because researchers often need to perform follow-up surveys, it is easier to try to ensure confidentiality than anonymity. In ensuring **confidentiality**, the researcher promises not to release identifying information about respondents. In an **anonymous** survey, the researcher does not know the identity of the respondents.

Questions on issues such as sexual behavior and income are particularly difficult because people consider those to be private matters. A variety of techniques have been developed to help ensure confidentiality, but surveys on such issues are hard to conduct accurately.

The problems discussed in this section are well known to professional pollsters, and most reputable polling agencies go to great lengths to overcome them to the extent possible. The survey questions in Examples 5.3 and 5.4 in Section 5.2 are excellent illustrations of how questions should be worded. For instance, the question in Example 5.3, "How important would you say religion is in your own life: very important, fairly important, or not very important?" is worded so as not to suggest which answer the interviewer would prefer. Here are two examples that illustrate attempts to overcome bias but also illustrate how difficult that can be.

Example 5.16 **When Will Adolescent Males Report Risky Behavior?** It is particularly difficult to measure information on behavior that is sensitive, risky, or illegal. In an article in *Science*, Turner et al. (1998) describe the use of audio, computer-assisted self-interviews as part of the 1995 National Survey of Adolescent Males (aged 15 to 19). The respondents were randomly assigned to answer the survey using the traditional paper form ($n = 368$) or using a laptop computer ($n = 1361$). The paper questionnaire was filled out and sealed in an envelope to return to the surveyor but was accompanied by identifying code numbers. The computer method included

Read the original source on the companion website, http://www.cengage.com/statistics/Utts4e.

listening to questions through headphones, and then recording the answers on a laptop computer. The authors believed that this method would allow respondents to feel that their responses were more private. In particular, respondents who could not read would need to have the paper version read to them, whereas the audio-computer method allowed complete self-administration of the survey.

Results indicated that the additional perception of privacy increased the reported incidence of certain behaviors while decreasing the reporting of others. The behaviors with increased reported incidence for the audio-computer version tended to be less socially acceptable, while those with higher reported incidence for the paper version tended to be more socially acceptable (for the adolescent males in this survey). Here are some examples reported in the article:

Question:	**Estimated Prevalence per 100 Males in Population**	
Have you ever . . .	*Paper Version*	*Audio-Computer Version*
Had sex with a prostitute	0.7	2.5
Had intercourse with a female	68.1	63.9
Made a girl pregnant	7.9	6.5
Had sex with another male	1.5	5.5
Taken street drugs using a needle	1.4	5.2

Example 5.17 **Politics Is All in the Wording** Crossen (1994, p. 112) described three very different sets of responses to a question about whether the president of the United States should be able to veto specific items in the federal budget. The first set came from a questionnaire published in *TV Guide* in which founder of the Reform Party and presidential candidate Ross Perot asked, "Should the president have the line item veto to eliminate waste?" Respondents were volunteers who mailed in the survey. Of those who responded, 97% said "yes." The second set of responses came from asking a properly selected random sample the same question, and 71% said "yes." But clearly, the question is written to elicit a "yes" answer. When the question was reworded to read, "Should the president have the line item veto, or not?" and a proper random sample was queried, only 57% answered yes.

Cohn and Cope (2001, pp. 145–146) report that Perot also asked the question, "Should laws be passed to eliminate all possibilities of special interests giving huge sums of money to candidates?" Again the "yes" responses were an overwhelming 99%. When a properly selected random sample was asked the same, biased version of the question, 80% said "yes." But when a properly selected random sample was asked, "Should laws be passed to prohibit interested groups from contributing to campaigns, or do groups have a right to contribute to the candidate they support?" only 40% said that such laws should be passed.

THOUGHT QUESTION 5.5 You have now learned that survey results have to be interpreted in the context of *who* responded and *to what questions* they responded. When you read the results of a survey, for which of these two areas do you think it would be easier for you to recognize and assess possible biases? Why?*

Be Sure You Understand What Was Measured

Sometimes words mean different things to different people. When you read about survey results, you should get a precise definition of what was actually asked or measured. Here are two examples that illustrate that even common terminology may mean different things to different people.

***HINT:** What information would you need in each case, and what information is more likely to be included in a description of the survey?

Example 5.18 **Teenage Sex** A letter to advice columnist Ann Landers stated, "According to a report from the University of California at San Francisco . . . sexual activity among adolescents is on the rise. There is no indication that this trend is slowing down or reversing itself." The letter went on to explain that these results were based on a national survey (*Davis Enterprise*, February 19, 1990, p. B-4). On the same day, in the same newspaper, an article titled, "Survey: Americans Conservative with Sex" reported that "teenage boys are not living up to their reputations. [A study by the Urban Institute in Washington] found that adolescents seem to be having sex less often, with fewer girls and at a later age than teenagers did a decade ago" (*Ibid.*, p. A-9).

Here we have two apparently conflicting reports on adolescent sexuality, both reported on the same day in the same newspaper. One indicated that teenage sex was on the rise, while the other indicated that it was on the decline. Although neither report specified exactly what was measured, the letter to Ann Landers proceeded to note that "national statistics show the average age of first intercourse is 17.2 for females and 16.5 for males." The article stating that adolescent sex was on the decline measured it in terms of *frequency.* It was based on interviews with 1880 boys between the ages of 15 and 19, in which "the boys said they had had six sex partners, compared with seven a decade earlier. They reported having had sex an average of three times during the previous month, compared with almost five times in the earlier survey."

Thus, it is not enough to know that both surveys were measuring adolescent or teenage sexual behavior. In one case, the author was, at least partially, discussing the *age* of first intercourse, whereas in the other case, the author was discussing the *frequency* of intercourse.

Example 5.19 **The Unemployed** Ask people whether they know anyone who is unemployed and, invariably, they will say that they do. Most people, however, don't realize that to be officially unemployed and included in the unemployment statistics given by the U.S. government, you must meet very stringent criteria. To be classified as unemployed, someone either must be waiting for a call back to a job from which they have been laid off, waiting to report to a new job within 30 days, or must have "engaged in any specific job seeking activity within the past 4 weeks such as registering at a public or private employment office, meeting with prospective employers, checking with friends or relatives, placing or answering advertisements, writing letters of application, or being on a union or professional register" (http://www.census.gov/apsd/techdoc/cps/mar97/glossary.html). The government excludes people who are discouraged enough that they have not sought a job in the past 4 weeks. If you knew someone like that, you would undoubtedly think of him or her as unemployed. But they are not included in the official statistics. You can see that the true number of people who are not working is higher than the government statistics would lead you to believe.

Some Concepts Are Hard to Precisely Define

Sometimes it is not the language but the concept itself that is ill-defined. For example, there is still not universal agreement on what should be measured with intelligence tests. IQ tests were developed at the beginning of the 20th century to determine the mental level of schoolchildren. The intelligence quotient (IQ) of a child was found by dividing the child's "mental level" by his or her chronological age. The "mental level" was determined by comparing the child's performance on the test with that of a large group of "normal" children to find the age group the individual's performance matched. Thus, if an 8-year-old child performed as well on the test as a "normal" group of 10-year-old children, he or she would have an IQ of $100 \times (10/8) = 125$.

IQ tests have been expanded and refined since the early days, but they continue to be surrounded by controversy. One reason is that it is very difficult to define what is meant by *intelligence.* It is hard to measure something if you can't even agree on what it is you are trying to measure. If you are interested in knowing more about these tests and the surrounding controversies, you can find numerous books on the subject. Anastasi (1988) provides a detailed discussion of a large variety of psychological tests, including IQ tests.

Measuring Attitudes and Emotions

Similar problems exist with trying to measure attitudes and emotions such as self-esteem and happiness. The most common method for trying to measure such things is to have respondents read statements and determine the extent to which they agree with the statement. For example, a test for measuring happiness might ask respondents to indicate their level of agreement, from "strongly disagree" to "strongly agree," with statements such as "I generally feel optimistic when I get up in the morning." To produce agreement on what is meant by characteristics such as "introversion," psychologists have developed standardized tests that claim to measure those attributes.

Open or Closed Questions: Should Choices Be Given?

An **open question** is one in which respondents are allowed to answer in their own words; a **closed question** is one in which respondents are given a list of alternatives from which to choose their answer. Usually, the latter form offers a choice of "other" in which the respondent is allowed to fill in the blank.

Problems with Closed Questions

To show the limitations of closed questions, Schuman and Scott (1987) asked about "the most important problem facing this country today." Half of the sample, 171 people, were given this as an open question. The most common responses were

Unemployment (17%)
General economic problems (17%)
Threat of nuclear war (12%)
Foreign affairs (10%)

In other words, one of these four choices was volunteered by over half of the respondents.

The other half of the sample was given the question as a closed question, with the four specific choices and the percentages that chose them being

The energy shortage (5.6%)
The quality of public schools (32.0%)
Legalized abortion (8.4%)
Pollution (14.0%)

These choices combined were mentioned by only 2.4% of respondents in the open question survey, yet they were selected by 60% of respondents in the closed question, where they were the only specific choices given. Don't think that respondents had no choice; in addition to the list of four, they were told, "If you prefer, you may name a different problem as most important." On the basis of the closed-form questionnaire, policymakers would have been seriously misled about what is important to the public. It is possible to avoid this kind of astounding discrepancy. If closed questions are preferred, they first should be presented as open questions to a test sample before the real survey, and the most common responses should then be included in the list of choices for the closed question. This kind of exercise is usually done as part of what's called a **pilot survey**, in which various aspects of a study design can be tried before it's too late to change them.

Problems with Open Questions

The biggest problem with open questions is that the results can be difficult to summarize. If a survey includes thousands of respondents, it can be a major chore to categorize their responses. Another problem, found by Schuman and Scott (1987), is that the wording of the question might unintentionally exclude answers that would have been appealing had they been included in a list of choices (such as in a closed question). To test this, they asked 347 people the following question in open form: Name one or two of the most important national or world event(s) or change(s) during the past 50 years. The most common choices and the percentages that mentioned them were

World War II (14.1%)
Exploration of space (6.9%)
Assassination of John F. Kennedy (4.6%)
The Vietnam war (10.1%)
Don't know (10.6%)
All other responses (53.7%)

The same question was then repeated in closed form to a new group of 354 people. Five choices were given: the preceding ones and "invention of the computer." Of the 354 respondents, the percentages that selected each choice were

World War II (22.9%)
Exploration of space (15.8%)
Assassination of John F. Kennedy (11.6%)
The Vietnam war (14.1%)
Invention of the computer (29.9%)
Don't know (.3%)
All other responses (5.4%)

The most frequent response was "invention of the computer," which had been mentioned by only 1.4% of respondents in the open question. Clearly, the wording of the question led respondents to focus on "events" rather than "changes," and the invention of the computer did not readily come to mind. When it was presented as an option, however, people realized that it was indeed one of the most important events or changes during the past 50 years.

In summary, there are advantages and disadvantages to both approaches. One compromise is to ask a small test sample to list the first several answers that come to mind, and then use the most common of those. These could be supplemented with additional answers, like "invention of the computer," that may not readily come to mind.

Remember that as the reader, you have an important role in interpreting the results. You should always be informed as to whether questions were asked in open or closed form; if the latter, you should be told what the choices were. You should also be told whether "don't know" or "no opinion" was offered as a choice in either case.

5.6 Exercises are on pages 182–183.

CASE STUDY 5.2 No Opinion of Your Own? Let Politics Decide

This is an excellent example of how people will respond to survey questions when they do not know about the issues, and how the wording of questions can influence responses. In 1995, the *Washington Post* decided to expand on a 1978 poll taken in Cincinnati, Ohio, in which people were asked whether they "favored or opposed repealing the 1975 Public Affairs Act" (Morin, 1995, p. 36). There was no such act, but about one-third of the respondents expressed an opinion about it.

In February 1995, the *Washington Post* added this fictitious question to its weekly poll of 1000 randomly selected respondents: "Some people say the 1975 Public Affairs Act should be repealed. Do you agree or disagree that it should be repealed?" Almost half (43%) of the sample expressed an opinion, with 24% agreeing that it should be repealed and 19% disagreeing!

The *Post* then tried another trick that produced even more disturbing results. This time, they polled two separate groups of 500 randomly selected adults. The first group was asked: "President Clinton [a Democrat] said that the 1975 Public Affairs Act should be repealed. Do you agree or disagree?" The second group was asked: "The Republicans in Congress said that the 1975 Public Affairs Act should be repealed. Do you agree or disagree?" Respondents were also asked about their party affiliation.

Overall, 53% of the respondents expressed an opinion about repealing this fictional act! The results by party affiliation were striking: For the Clinton version, 36% of the Democrats but only 16% of the Republicans agreed that the act should be repealed. For the "Republicans in Congress" version, 36% of the Republicans but only 19% of the Democrats agreed that the act should be repealed.

SKILLBUILDER APPLET

5.7 Random Sampling in Action

The Sampling applet described in this section is available on the companion website, http://www.cengage.com/statistics/Utts4e.

It should be obvious by now that each time you take a random sample from a population, you are likely to get a different group of individuals, even though the population remains fixed. The **Sampling** applet at the website that accompanies this text allows us to watch what happens when we repeatedly take random samples from the same population. Figure 5.6 illustrates how the applet appears when it starts. Each stick figure represents one member of a population of 100 college students: 55 females (blue) and 45 males (red). Their heights have been recorded as well, and if you look carefully, you will see that the little stick figures are all of slightly different heights. The mean height for the population is 68 inches.

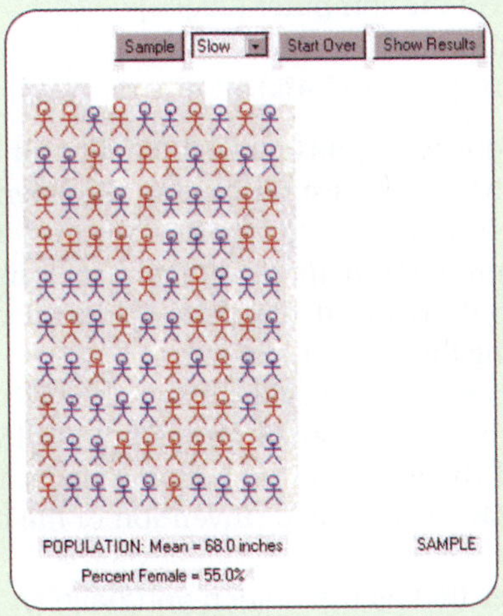

Figure 5.6 The sampling applet starting point

If we repeatedly take simple random samples of 10 people from this population, how much do you think the proportion of females in the samples will vary? Will it almost always be 50% or 60%, given that it must be a multiple of 10 and the population is 55% female? Will the sample ever consist of all males or all females? Pick one of the stick figures to represent yourself. How often do you think you would end up in the sample? Will the mean height of the sample always be close to the mean height of all 100 students (68 inches)? If the mean height of your sample were to be 70 inches, would that indicate a possible problem in the selection of the sample, or is such a large value to be expected in some samples? In Chapters 7 to 9, you will learn the mathematical answers to these questions, but playing with the applet can give you a good intuitive idea of what happens.

What Happens

The applet always chooses a simple random sample of 10 individuals when you press the button labeled ***Sample*** (in the upper left corner). You can control the speed of the sampling to be slow, fast, or batch. With slow and fast sampling, you get to watch each individual move from the population into the sample, which is illustrated to the right of the population. With batch sampling, the whole sample appears as soon as you press the ***Sample*** button. Figures 5.7 and 5.8 illustrate two examples of the results after pressing the ***Sample*** button. Note that in Figure 5.7, only three females were selected, so 30% of the sample is female. Not surprisingly, given that the sample has a higher percentage of men than the population, the mean height for the sample (69.6 inches) is higher than the 68-inch mean height for the population. The reverse happens in Figure 5.8, in which females make up 60% of the sample and the mean height is 67.2 inches.

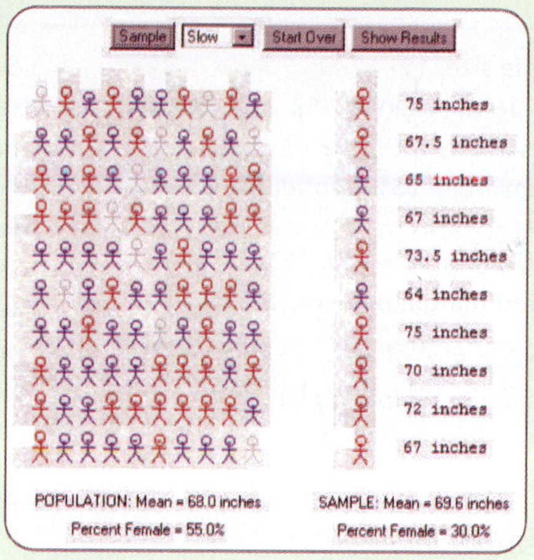

Figure 5.7 Sampling applet results

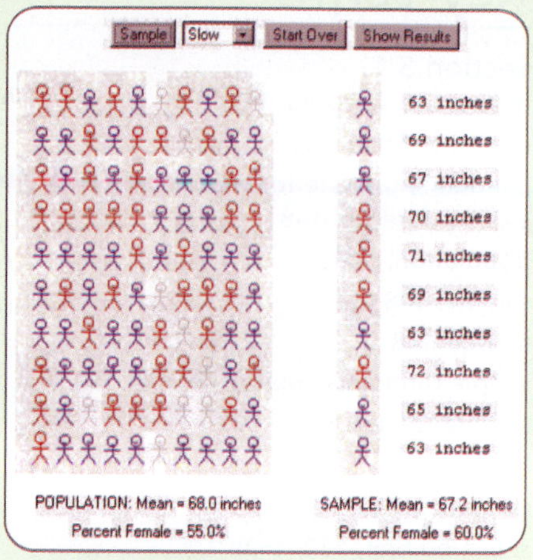

Figure 5.8 Additional sampling applet results

What to Do

Open the ***Sampling*** applet at the book's website, and then repeatedly press the ***Sample*** button to choose many samples. In the option box to the right of the ***Sample*** button you can decide whether you want to select a sample slowly so that you can carefully watch each person being taken from the population and placed in the sample, fast (you can still watch), or in a batch. This decision won't change the results. Do *not* press the ***Start Over*** button.

Note how the percentage of females changes for each sample and how the mean height changes. Pick one of the stick figures to represent yourself, and notice how often you make it into the sample. At any time, you can see a summary of the samples you have selected so far by pushing the button labeled ***Show Results***. When you press the ***Start Over*** button, all past results are cleared.

Take dozens of samples, and then press ***Show Results***. Are the percent female values and the mean height values for the samples more spread out or less spread out than you would have anticipated? The number of females must be 0 or 1 or 2, and so on, up to 10, so the "percent female" can only be one of 11 possibilities: 0%, 10%, 20%, 30%, and so on, up to 100%. Count how often each of the 11 possibilities occurred in your repeated sampling. Does the result surprise you, or is it what you expected? Draw a histogram of the means from the samples you chose. What do you notice? Do sample mean values far away from the true population mean of 68 inches occur as often as sample mean values close to 68 inches? Would you suspect a problem with the sample if a sample had a mean of 70 inches? How about 72 inches?

Note for Minitab Users: The results in the "Show Results" window can be copied and then pasted into a Minitab worksheet. Minitab can then be used to analyze the results. For instance, **Graph > Histogram** could be used to draw a histogram of the means of the chosen samples.

Lessons Learned

This applet lets you see simple random sampling in action. Much of the remaining material in this book follows from the idea that although samples vary, the amount by which they vary over the long run can be quantified. Remember that a simple random sample is supposed to be representative of the population. However, it should be obvious to you that with only ten individuals per sample, both the percentage of females and the mean height vary considerably from sample to sample. In fact, the conservative margin of error for a sample of $n = 10$ is over 30%! Therefore, you should expect to see a wide range in the percentage of females across samples.

5.7 Exercises are on page 183.

Key Terms

In Summary Boxes

Exercises

◆ Denotes that the dataset is available on the companion website, http://www.cengage.com/statistics/Utts4e, but is not required to solve the exercise.

Bold *exercises have answers in the back of the text.*

Section 5.1

Skillbuilder Exercises

5.1 According to the Fundamental Rule for Using Data for Inference, when can available data be used to make inferences about a much larger group?

5.2 For each of the following examples from the text, explain whether there would be more interest in descriptive statistics (for the acquired dataset only) or in inferential statistics (extending results to a larger population).

a. Ages at death of first ladies.
b. The relationship between nicotine patch use and cessation of smoking.
c. Annual rainfall data for Davis, California.

5.3 In each case, briefly discuss whether you think the available sample data can be used to make an inference about the larger population.

a. The heights of women in a psychology class will be used to estimate the average height of all women at a college.
b. Parents of children attending a daycare facility are surveyed about whether the state government should increase funding to daycare providers to provide lower rates for low-income families. The data are used to estimate statewide support for increased state funding of daycare facilities.

5.4 The median income for a random sample of households in a school district was found to be $56,300. From this information, the school board decided that the median income for all households in the district was probably between about $52,000 and $60,000. Describe the difference between descriptive statistics and inferential statistics in the context of this example.

5.5 A political scientist surveys 400 voters randomly selected from the list of all registered voters in a community. The purpose is to estimate the proportion of registered voters who will vote in an upcoming election.

a. What is the population of interest in this survey?
b. What is the sample in this survey?

5.6 To estimate the percentage of households in the United States that use a DVD player, a researcher surveys a randomly selected sample of 500 households and asks about DVD player usage.

a. What is the population of interest in this survey?
b. What is the sample in this survey?

5.7 Briefly explain what it means to say that a survey method produces selection bias.

5.8 Briefly explain what it means to say that a survey method produces nonparticipation bias.

5.9 Give an example of a situation in which a sample must be used because a census is not possible.

5.10 In a class of 20 students (John, Maria, Inez, Bill, etc.), a simple random sample of two people will be selected from the class. Use the definition of a simple random sample to compare the chance that John and Maria will be the group selected with the chance that Maria and Bill will be the group selected.

5.11 Briefly explain the difference between a sample survey and a census.

5.12 A friend has recommended the work of a musician who has recorded five CDs, each containing ten selections. You decide to visit a music store and listen to a simple random sample of four songs. Explain how you could select the four songs.

5.13 Refer to Exercise 5.12. Suppose that you randomly chose four of the CDs, and then randomly chose one song from each one. Would this be considered a simple random sample of the musician's songs? Explain.

General Section Exercises

5.14 In each of the following situations, indicate whether the potential bias is a selection bias, a nonparticipation bias, or a response bias.

a. A survey question asked of unmarried men was, "What is the most important feature you consider when deciding whether to date somebody?" The results were found to depend on whether the interviewer was male or female.
b. In a study of women's opinions about community issues, investigators randomly selected a sample of households and interviewed a woman from each selected household. When no woman was present in a selected household, a next-door neighbor was interviewed instead. The survey was done during daytime hours, so working women might have been disproportionately missed.
c. A telephone survey of 500 residences was conducted. People refused to talk to the interviewer in 200 of the residences.

5.15 For each definition, identify the correct term for the type of bias being defined. Possible answers are *selection bias, nonparticipation bias,* and *response bias.*

a. Participants respond differently from how they truly feel.
b. The method for selecting the participants produces a sample that does not represent the population of interest.
c. A representative sample is chosen, but a subset of the sample cannot be contacted or does not respond.

5.16 Refer to the three types of bias given in Section 5.1. Which type of bias do you think would be introduced in each of the following situations? Explain.

a. In a college town, college students are hired to conduct door-to-door interviews to determine whether city residents think there should be a law forbidding loud music at parties.
b. A magazine sends a survey to a random sample of its subscribers asking them whether they would like the frequency of publication reduced from biweekly to monthly or would prefer that it remain the same.
c. A random sample of registered voters is contacted by phone and asked whether or not they are going to vote in the upcoming presidential election.

5.17 Refer to the three types of bias given in Section 5.1. Which type of bias do you think would be introduced in each of the following situations? Explain.

a. A list of registered automobile owners is used to select a random sample for a survey about whether people think homeowners should pay a surtax to support public parks.
b. A survey is mailed to a random sample of residents in a city asking whether or not they think the current mayor is doing an acceptable job.

5.18 For each of the following situations, explain whether or not the Fundamental Rule for Using Data for Inference holds.

a. *Available Data:* Opinions on whether or not the legal drinking age should be lowered to 19 years old, collected from a random sample of 1000 adults in the state.

Research Question: Do a majority of adults in the state support lowering the drinking age to 19?

b. *Available Data:* Opinions on whether or not the legal drinking age should be lowered to 19 years old, collected from a random sample of parents of high school students in the state.

Research Question: Do a majority of adults in the state support lowering the drinking age to 19?

5.19 For each of the following situations, explain whether or not the Fundamental Rule for Using Data for Inference holds.

a. *Available Data:* Salaries for a random sample of male and female professional basketball players.

Research Question: Are women paid less than men who are in equivalent jobs?

b. *Available Data:* Pulse rates for smokers and nonsmokers in a large statistics class at a major university.

Research Question: Do college-age smokers have higher pulse rates than college-age nonsmokers?

5.20 The U.S. government gathers numerous statistics based on random samples, but every 10 years, it conducts a census of the U.S. population. What can it learn from a census that cannot be learned from a sample?

5.21 Give an example of a survey situation that is likely to produce nonparticipation bias.

5.22 Give an example of a survey situation that is likely to produce each of the following types of bias:

a. Response bias.
b. Selection bias.

Section 5.2

Skillbuilder Exercises

5.23 A television rating agency produces the ratings of television shows by asking a random sample of about 5000 households with TV sets what shows they watch. The rating for a show is the proportion (or percentage) of all televisions in the sample that were tuned in to the show. Calculate a conservative margin of error for this survey as a proportion and as a percentage.

5.24 A survey is planned to estimate the proportion of voters in a community who plan to vote for Candidate Y. Calculate the conservative margin of error, as a proportion and as a percentage, for each of the following possible sample sizes:

a. $n = 100$.
b. $n = 400$.
c. $n = 900$.

5.25 In a random sample of 90 students at a university, 72 students (80% or .80 of the sample) say that they use a laptop computer.

a. Calculate the conservative margin of error for the survey.
b. Compute an approximate 95% confidence interval for the population *proportion* that uses a laptop computer.

5.26 In a CNN/*Time* poll conducted December 17–18, 1998, a sample of $n = 1031$ adults in the United States was asked, "Do you think the police should or should not be allowed to collect DNA information from suspected criminals, similar to how they take fingerprints?" Of those sampled, 66% answered "should" (http://www.pollingreport.com/crime.htm).

a. Calculate the conservative margin of error for the survey, as a percentage.
b. Compute an approximate 95% confidence interval for the percentage of all American adults who think police should be allowed to collect DNA information from suspected criminals.

5.27 *This is also Exercise 1.8c.* For a survey based on 2000 adults, what is the approximate margin of error?

5.28 *This is also Exercises 1.9 and 1.10.* What sample size produces each of the following as the approximate margin of error?

a. Margin of error = .05 or 5%.
b. Margin of error = .30 or 30%.

General Section Exercises

5.29 In an *ABC News* poll conducted between January 21 and 26, 2000, a random sample of $n = 1006$ adult Americans was asked, "Compared to buying things by mail order or in a store, do you think that buying things over the Internet poses more of a threat to your personal privacy, less of a threat, or about the same?" (http://www.pollingreport.com/computer.htm, September 17, 2002). The percentages that selected each response were

More of a threat: 40%
Less of a threat: 7%
About the same: 47%
No opinion: 6%

a. What is the margin of error for this poll? Give your answer as both a proportion and a percentage.
b. What is a conservative 95% confidence interval for the population proportion that would have responded, "More of a threat?" Give your answer for the population percentage as well.
c. Write a sentence or two interpreting the interval that you found in part (b).
d. Based on the interval you found in part (b), could you conclude that in January 2000 fewer than half of all adult Americans perceived Internet shopping as posing more of a threat than buying things by mail order or in a store? Explain.

5.30 A *CBS News* poll conducted between December 17 and 22, 2009, asked a random sample of $n = 563$ married adult Americans, "Would you say that your marriage with your spouse is better, worse or about the same as your parents' marriage?" (http://www.pollingreport.com/life.htm, March 9, 2010). The percentages that selected each response follow:

Better: 55%
Worse: 3%
About the same: 41%
Unsure: 1%

a. What is the margin of error for this poll? Give your answer as both a proportion and a percentage.
b. What is a conservative 95% confidence interval for the population proportion that would have responded "Better?" Give your answer for the population percentage as well.
c. Write a sentence or two interpreting the interval that you found in part (b).
d. Based on the interval you found in part (b), could you conclude that in December 2009 more than half of all married adult Americans thought that their own marriage was better than their parents' marriage? Explain.

5.31 In the year leading up to a big election, there are many polls conducted to estimate what percent of the population is likely to vote for each candidate. The true percent can change during the year, so the polls are not necessarily estimating the same thing every time. Suppose 100 such polls are done and each one is used to find a 95% confidence interval for the percent of the population who plan to vote for a certain candidate at the time of the poll.

a. About how many of the 100 confidence intervals are likely to cover the true percent?
b. Is there any way to know which of the polls cover the true percent and which ones don't?

5.32 Suppose a national polling agency conducted 100 polls in a year, using proper random sampling, and reported a 95% confidence interval for each poll. About how many of those confidence intervals would we expect *not* to cover the true population value?

5.33 An Internet report on a 1999 Gallup poll (http://www.gallup.com/poll/index.asp, August 2, 1999) included the following statement: "The results below are based on telephone interviews with a randomly selected national sample of 1021 adults, 18 years and older, conducted July 22–25, 1999. For results based on this sample, one can say with 95 percent confidence that the maximum error attributable to sampling and other random effects is plus or minus 3 percentage points." What is the margin of error for this survey? Give two separate ways in which you know this is the answer.

5.34 The question about the importance of religion in Example 5.3 was asked in a survey of elementary statistics students at Penn State University from 2007 to 2010. The results were as follows:

Very important: 730 = 21.4%
Fairly important: 1512 = 44.3%
Not very important: 1171 = 34.3%
Total: 3417 = 100%

Because these students were not a random sample from a larger population, and because statistics students may not represent other students on questions of religion, treat these 3417 students as a population. Therefore, we know that for the population of elementary statistics students at Penn State from 2007 to 2010, 21.4% felt that religion was very important in their lives.

a. Based on the results in Example 5.3, can we conclude that this percentage differs from the percentage of the population of all adult Americans who felt that religion was very important in their lives in December 2009? Explain, taking into account the margin of error for the survey in Example 5.3.
b. Based on the preceding results and those in Example 5.3, answer the question in part (a) for the percentage who felt religion was fairly important and for the percentage who felt it was not very important.
c. Write a short news article comparing the two surveys as if you were writing it for the student newspaper.
d. Suppose we believe that the Penn State sample is representative of all students in the academic programs that require the elementary statistics course, and thus treat it as a sample from a larger population. Calculate an approximate 95% confidence interval for the population proportion that would say that religion is very important in their lives.

5.35 Refer to Exercise 5.33. One of the questions asked was, "Do you think there will or will not come a time when Israel and the Arab nations will be able to settle their differences and live in peace?" The choices and percentage choosing them were "Yes, will be a time" (49%), "No, will not" (47%), "No opinion" (4%).

a. Give the interval of values that is likely to cover the true proportion of the population who would answer, "Yes, will be a time."
b. Give the interval of values that is likely to cover the true proportion of the population who would answer, "No, will not."
c. Compare your answers in parts (a) and (b). Is there a clear majority of the population for either opinion? Explain.

5.36 A soft drink company has a new recipe for one of its products, and it would like to know if a majority of consumers of its drinks prefer the new recipe over the old recipe. The company plans to ask a representative sample of the population to taste both drinks. The company will then measure the percent that prefer the new recipe.

a. What would be the approximate margin of error for this percent if the company asks 600 consumers?
b. How many consumers would need to be in the sample to provide a margin of error of about 5%?

5.37 Suppose that a researcher is designing a survey to estimate the proportion of adults in your state who oppose a proposed law that requires all automobile passengers to wear a seat belt.

a. What would be the approximate margin of error if the researcher randomly sampled 400 adults?
b. What sample size would be needed to provide a margin of error of about 2%?

5.38 Explain whether the width of a confidence interval would increase, decrease, or remain the same as a result of each of the following changes:

a. Increase the sample size from 1000 to 2000.
b. Decrease the sample size from 1000 to 500.
c. Increase the population size from 10 million to 20 million.

5.39 In which of the following three samples will the margin of error be the smallest? Explain. Assume that each sample is a random sample.

Sample A: sample of $n = 1000$ from a population of 10 million.

Sample B: sample of $n = 2500$ from a population of 200 million.

Sample C: sample of $n = 400$ from a population of 50,000.

5.40 An epidemiologist plans a survey to estimate the prevalence of Alzheimer's disease in the population of adults aged 65 years or older. The desired margin of error for a 95% confidence interval is to be no more than 3%. What sample size is needed?

5.41 A college dean plans a student survey to estimate the percentage of currently enrolled students who plan to take classes during the next summer session. The desired margin of error for a 95% confidence interval is to be at most 5%. What is the necessary sample size?

Section 5.3

Skillbuilder Exercises

5.42 Samples can be chosen using a probability sampling plan and they can be chosen by selecting a simple random sample. One of these methods is a special case of the other. Explain which one is a special case of the other.

5.43 Define each of the following terms:

a. Probability sampling plan.
b. Simple random sample.

5.44 In Example 2.3 of Chapter 2 we learned that when students are asked to "randomly pick a number between 1 and 10" they are much more likely to choose 7 than any other number, and much less likely to choose 1 than any other number. Explain how you could actually "randomly pick a number between 1 and 10." Your method should be such that all 10 numbers have the same chance of being selected.

General Section Exercises

5.45 A radio station has a contest each day for a week in which the DJ randomly selects one birthday (month and day, not year) and announces it on the air. The first person with that birthday who calls the station wins a prize. A new birthday is selected for each of the 7 days, and the station doesn't mind if the same birthday is selected more than once over the course of the contest.

a. What is the list of units from which a selection will be made each day?

b. If the station did not have access to appropriate computer software or a table of random digits, explain how they could randomly choose a birthday each day.

5.46 There are 8000 items in a population, and these items are labeled by using four-digit numbers ranging from 0000 to 7999. Use the following stream of random digits to select four items from the population. Explain how you determined your answer.

76429 69730 23395 12694 43387

5.47 A lottery game is played by choosing six whole numbers between 1 and 49. The grand prize is won if all six numbers chosen match the winning numbers drawn. We can think of choosing the winning numbers as the same thing as choosing a simple random sample of size 6.

a. What is the list of units from which the sample will be chosen?

b. Use a computer random number generator or invent your own random mechanism to draw six numbers. Explain your process, and write down the six numbers chosen.

c. George plays this game weekly. Assume that the winning numbers are drawn fairly, so that any number has the same chance of occurrence. Would George have a better chance of winning if he chose the same numbers every week, if he chose different numbers every week, or does it matter?

5.48 ◆ The right handspan measurements (in cm) for 103 female college students in Table 2.4 in Chapter 2 are given again here and in the **pennstate1F** dataset on the companion website.

Females (103 Students): 20, 19, 20.5, 20.5, 20.25, 20, 18, 20.5, 22, 20, 21.5, 17, 16, 22, 22, 20, 20, 20, 20, 21.7, 22, 20, 21, 21, 19, 21, 20.25, 21, 22, 18, 20, 21, 19, 22.5, 21, 20, 19, 21, 20.5, 21, 22, 20, 20, 18, 21, 22.5, 22.5, 19, 19, 19, 22.5, 20, 13, 20, 22.5, 19.5, 18.5, 19, 17.5, 18, 21, 19.5, 20, 19, 21.5, 18, 19, 19.5, 20, 22.5, 21, 18, 22, 18.5, 19, 22, 17, 12.5, 18, 20.5, 19, 20, 21, 19, 19, 21, 18.5, 19, 21.5, 21.5, 23, 23.25, 20, 18.8, 21, 21, 20, 20.5, 20, 19.5, 21, 21, 20.

Suppose a random sample of 5 handspans is to be drawn from this list.

a. How many units are in the list from which the sample will be taken?

b. What is the handspan value for unit #10 in this list?

c. In Example 5.6 on page 156, two random samples of women were chosen from a list of 199 of them. The first sample consisted of women who were numbered 176, 10, 1, 40, 85, 162, 46, 69, 77, and 154 in the list of 199 units. Use this same list of random numbers to choose a random sample of 5 of the right handspans in this exercise, ignoring random numbers that are not relevant for this list of units. Write down the 5 handspan values in the sample you chose.

5.49 Refer to Exercise 5.48, which lists the right handspan measurements for 103 female college students.

a. Draw three simple random samples of ten measurements each from this dataset (an individual can be in more than one of your samples). Explain how you chose the samples and list the ten handspan measurements in each sample.

b. Find the median for each of your three samples. Compare them to the median for the full dataset (20.0 cm).

c. Find the mean for each of your three samples. Compare them to the mean for the full dataset (also 20.0 cm).

d. There were 21 women whose recorded handspan measurements were 20.0 cm. Would it be possible to have a random sample of ten measurements with a mean of 20.0 cm and a standard deviation of 0 cm? Explain.

Section 5.4

Skillbuilder Exercises

5.50 In a factory producing television sets, every 100th set produced is inspected. Is the collection of sets inspected a simple random sample, a stratified random sample, a cluster sample, or a systematic sample?

5.51 A class of 200 students is numbered from 1 to 200, and a table of random digits is used to choose 60 students from the class. Is the group of students selected a simple random sample, a stratified random sample, a cluster sample, or a systematic sample?

5.52 In each part, identify whether the sample is a stratified random sample or a cluster sample.

a. A class of 200 students is seated in 10 rows of 20 students per row. Three students are randomly selected from every row.

b. An airline company randomly chooses one flight from a list of all international flights taking place that day. All passengers on that selected flight are asked to fill out a survey on meal satisfaction.

General Section Exercises

5.53 Suppose a state has 10 universities, 25 four-year colleges, and 50 community colleges, each of which offer multiple sections of an introductory statistics class each year. Researchers want to conduct a survey of students taking introductory statistics in the state. Explain a method for collecting each of the following types of samples:

a. A stratified sample.

b. A cluster sample.

c. A simple random sample.

5.54 Refer to Exercise 5.53. Give one advantage of each sampling method in the context of the problem.

5.55 Is a sample that is found by using random-digit dialing more like a stratified sample or a cluster sample? Explain.

◆ Dataset available but not required **Bold** exercises answered in the back

5.56 Find an example of a survey routinely conducted by the U.S. government. (The Internet is a good source; for instance, try http://www.fedstats.gov.) Explain how the survey is conducted, and identify the type(s) of sampling used.

Section 5.5

Skillbuilder Exercises

5.57 A local government wants to determine whether taxpayers support increasing local taxes to provide more public funding to schools. They randomly select 500 schoolchildren from a list of all children enrolled in local schools and then survey the parents of these children about possible tax increases.

a. What is the population of interest for the local government?

b. What sampling frame did the government use for the survey?

c. Explain why the problem called "using the wrong sampling frame" might lead to a biased estimate of taxpayer support for increasing taxes.

5.58 A group of biologists wants to estimate the abundance of barrel cactus in a desert. They divide the desert into a grid of 100 rectangular areas but exclude 10 of those areas because they are difficult to access. The biologists then measure the density of cactus in a randomly selected sample of 40 of the 90 accessible areas.

a. What was the sampling frame in this study, and how did it differ from the population of interest?

b. Explain why the problem called "using the wrong sampling frame" might lead to a biased estimate of the abundance of cactus in the desert.

5.59 In each part, indicate whether the sample should be called a self-selected sample or a convenience sample.

a. To assess passenger satisfaction, an airline distributed questionnaires to 100 passengers in the airline's frequent flyer lounge. All 100 individuals responded, and 95 respondents said that they had a high degree of satisfaction with the airline.

b. A magazine contains a survey about sexual behavior. Readers are asked to mail in their answers, and $n = 2000$ readers do so.

5.60 In each part, indicate whether the sample should be called a self-selected sample or a convenience sample.

a. A political scientist surveys the 80 people in a class he teaches to evaluate student political views.

b. A soft drink company wants to know which of two of their drinks consumers prefer. They set up a table at a mall and people who are passing by can stop and participate in a taste test if they wish. The sample consists of those who agree to participate.

General Section Exercises

5.61 Despite his success in 1936, George Gallup failed miserably in trying to predict the winner of the 1948 U.S. presidential election. His organization, as well as two others, predicted that Thomas Dewey would beat incumbent Harry Truman. All three used what is called **quota sampling**. The interviewers were told to find a certain number, or quota, of each of several types of people. For example, they might have been told to interview six women under age 40, one of whom was African American and the other five of whom were Caucasian. Imagine that you are one of their interviewers, trying to follow these instructions. Whom would you ask? Now explain why you think these polls failed to predict the true winner and why quota sampling is not a good method.

5.62 Explain why the main problem with the *Literary Digest* poll is described as "volunteer response" and not "volunteer sample."

5.63 Gastwirth (1988, p. 507) describes a court case in which Bristol Myers was ordered by the Federal Trade Commission to stop advertising that "twice as many dentists use Ipana as any other dentifrice" and that more dentists recommended it than any other dentifrice. Bristol Myers had based its claim on a survey of 10,000 randomly selected dentists from a list of 66,000 subscribers to two dental magazines. They received 1983 responses, with 621 saying they used Ipana and only 258 reporting that they used the second most popular brand. As for the recommendations, 461 respondents recommended Ipana, compared with 195 for the second most popular choice.

a. Specify the sampling frame for this survey, and explain whether or not you think the problem of "using the wrong sampling frame" was a difficulty here, based on what Bristol Myers was trying to conclude.

b. Of the remaining four "Difficulties and Disasters in Sampling" listed in Section 5.5 (other than "using the wrong sampling frame"), which do you think was the most serious in this case? Explain.

c. What could Bristol Myers have done to improve the validity of the results after it had mailed the 10,000 surveys and received 1983 of them back? Assume that it kept track of who had responded and who had not.

5.64 Find an example of a poll based on a self-selected sample. Report the wording of the questions and the number who chose each response. Comment on whether or not you think the results can be extended to any population.

5.65 *This is also Exercise 1.11.* A popular Sunday newspaper magazine often includes a yes-or-no survey question such as "Do you think there is too much violence on television?" or "Do you think parents should use physical discipline?" Readers are asked to phone their answers to the magazine, and the results are reported in a subsequent issue.

a. What is this type of sample called?

b. Do you think the results of these polls represent the opinions of all readers of the magazine? Explain.

5.66 *This is also Exercise 1.12.* A proposed study design is to leave 100 questionnaires by the checkout line in a student cafeteria. The questionnaire can be picked up by any student and returned to the cashier. Explain why this volunteer sample is a poor study design.

5.67 *This is a modification of Exercise 1.35.* Suppose listeners to a late-night radio talk show were asked to call and report whether or not they had ever seen a ghost.

a. What is this type of sample called?

b. Do you think the proportion reporting that they had seen a ghost for the radio poll would be higher or lower than the proportion for a poll done using a random sample of adults in the United States? Explain.

Section 5.6

Skillbuilder Exercises

5.68 A survey question will be asked to determine whether people think smoking should be banned at all airports.

a. Write a version of the question that is as neutral and unbiased as possible.
b. Write a version of the question that is likely to get people to respond that smoking should be forbidden at all airports.
c. Write a version of the question that is likely to get people to respond that smoking should be allowed at selected locations in airports.

5.69 An example of an unnecessarily complex survey question is "Shouldn't former drug dealers not be allowed to work in hospitals after they are released from prison?" Restate this question so that it is clearer.

5.70 In a planned survey about movie-going, two questions that will be asked follow:

- How many times per month do you go to the movies?
- Do you consider yourself to be well-informed about recent movies or not?

The questions could be asked in either order. Briefly explain how you think the question order for asking them might affect how people answer them.

5.71 Medical tests, such as those for detecting HIV, sometimes concern such sensitive information that people do not want to give their names when they take the test. In some instances, a person taking such a medical test is given a number or code that he or she can use later to learn the test result by phone. Explain whether this procedure is an example of anonymous testing or confidential testing.

5.72 In presidential election years in the United States, a Gallup poll is conducted in which the first survey question asks which presidential candidate the voter prefers. Subsequent questions concern other political, social, and election issues. Explain why this question order might be the best order for estimating how people may vote in the presidential election.

General Section Exercises

5.73 Refer to Example 5.16, "When Will Adolescent Males Report Risky Behavior?" Explain which two of the seven "Possible Sources of Response Bias in Surveys" are illustrated by this example.

5.74 An Internet poll sponsored by a site called About.com asked Internet users to pick one of two choices in response to the question, "Should jurors opposed to gun control laws refuse to convict defendants even if they have clearly broken gun laws?" The two choices and the number and percentage choosing them were

- Yes, that's an effective way to defeat unjust laws (16,864, 23%).
- No, that undermines the legal system (55,519, 77%).

Discuss this poll, including whether or not you think the results are representative of all adults and whether you think the wording is appropriate.

5.75 Give an example of two survey questions for which you think the results would be substantially different depending on which order they were asked.

5.76 A Gallup poll that was released on July 9, 1999, included a series of questions about possible religious activities in public schools. The poll was based on telephone interviews with a randomly selected sample of 1016 U.S. adults conducted June 25–27, 1999.

The questions, asked in random order, included the following:

- Do you favor or oppose teaching creationism ALONG WITH evolution in public schools? Favor—68%, Oppose—29%, No opinion—3%.
- Do you favor or oppose teaching creationism INSTEAD OF evolution in public schools? Favor—40%, Oppose—55%, No opinion—5%.

a. What is the margin of error for this poll?
b. Write a news story of a few sentences describing the results of the survey, being truthful but biasing the story in favor of teaching creationism.
c. Repeat part (b), but bias the story in favor of not teaching creationism.
d. Repeat part (b), but write an unbiased story.
e. Explain which of the pitfalls in Section 5.6 is illustrated by this example.

5.77 Explain which of three methods—a door-to-door interview, a telephone interview, or a mail survey—would be *most* likely to suffer from each of the following problems:

a. Bias due to desire to please the interviewer.
b. Volunteer response.
c. Bias due to perceived lack of confidentiality.

5.78 Explain which of three methods—a door-to-door interview, a telephone interview, or a mail survey—would be *least* likely to suffer from each of the following problems:

a. Bias due to desire to please the interviewer.
b. Volunteer response.
c. Bias due to perceived lack of confidentiality.

5.79 An *NBC News/Wall Street Journal* poll conducted at the end of the 20th century (September 9–12, 1999) asked a random sample of $n = 1010$ adult Americans, "Which one of the following do you consider to be the best American movie of the 20th century?" The choices given and the percentages who chose them were: *Gone With the Wind* 28%, *Schindler's List* 18%, *Titanic* 11%, *Star Wars* 11%, *Casablanca* 8%, *The Godfather* 7%, *Citizen Kane* 6%, *The Graduate* 1%, and Not sure 4%. In addition, 2% volunteered the answer "all of them," and 4% volunteered an answer not listed (http://www.pollingreport.com/hollywoo.htm, September 29, 2002).

a. Was this asked as an open-form or closed-form question?
b. Do you think the form that was used was appropriate for trying to ascertain what movies adult Americans thought were the best of the 20th century? Explain.
c. Give one advantage and one disadvantage each for open-form and closed-form questions in this situation.

5.80 An advertiser of a certain brand of aspirin (let's call it Brand B) claims that it is the preferred painkiller for headaches, on the basis of the results of a survey of headache sufferers. But further investigation reveals that the choices given to respondents were Tylenol, Extra-Strength Tylenol, Brand B aspirin, and Advil.

a. Is this an open- or closed-form question? Explain.
b. Comment on the choices given to respondents.
c. Comment on the advertiser's claim.
d. What choices do you think should have been given? Explain.

5.81 Rock singer Elvis Presley died on August 16, 1977, and his life received substantial publicity as the 25th anniversary of his death approached on August 16, 2002. On August 7–11, 2002, an *ABC News* poll asked a random sample of $n = 1023$ adult Americans, "Who do you think is the greatest rock 'n' roll star of all time?" There were 128 different stars volunteered by respondents. The top responses and the percentages they received were Elvis Presley 38%, Jimi Hendrix 4%, John Lennon 2%, Mick Jagger 2%, Bruce Springsteen 2%, Paul McCartney 2%, Eric Clapton 2%, Michael Jackson 2%, Other 25%, None 7%, No opinion 15% (http://abcnews.go.com/sections/us/DailyNews/elvis_poll.html, September 29, 2002).

a. Was this an open- or closed-form question?
b. Do you think the percentage who responded "Elvis Presley" was influenced by the timing of the poll? Explain.
c. Do you think the percentage who responded "Elvis Presley" would have been higher, lower, or about the same if the question had given specific choices?

5.82 Refer to Exercise 5.81. Another question that was asked in the poll was "Do you consider yourself a fan of Elvis Presley, or not?" The responses were almost equally split, with 49% saying "yes" and 51% saying "no" (http://www.pollingreport.com/music.htm, September 29, 2002). Do you think the order in which this question and the one in Exercise 5.81 were asked would influence the results to either question? Explain.

Section 5.7: Skillbuilder Applet Exercises

For these exercises, use the **Sampling** *applet described in Section 5.7 and available on the companion website,* http://www.cengage.com/statistics/Utts4e.

5.83 Choose one stick figure to represent yourself, and identify which one it was in your answers to this exercise by giving the row number (1 to 10) and column number (1 to 10) where you reside.

a. Suppose you are in a class of 100 individuals represented by these stick figures. Your teacher randomly selects 10 people each class period and asks them to answer a question about the reading assignment. What are your chances of being in the sample each day?
b. Repeat the sample selection process 20 times, each time noting whether you made it into the sample. How many times did you make it into the sample? Does that surprise you, or is it about what you expected?

5.84 Repeat the sampling process 50 times without pressing the ***Start Over*** button. Use ***Show Results*** to display the results of the 50 samples. Cut and paste the results into a computer program capable of creating a histogram, or write them down.

a. Display a histogram of the 50 means of the chosen samples.
b. Describe the histogram in part (a), including shape, center, and spread.
c. Display a histogram of the 50 values of "percent female."
d. Describe the histogram in part (c).

5.85 Repeat the sampling process 20 times without pressing the ***Start Over*** button. Use ***Show Results*** to display the results of the 20 samples. Write down the results, or cut and paste them into a computer program.

a. What is the median of the 20 sample means?
b. What is the range of the 20 sample means?
c. Display a box-plot of the 20 sample means.
d. We know that the population has a mean height of 68 inches. Based on your results in parts (a), (b), and (c), discuss how useful the mean of a sample of size 10 is for estimating the population mean in this situation.

5.86 Repeat the sampling process 20 times without pressing the ***Start Over*** button. Use ***Show Results*** to display the results of the 20 samples. Write down the results, or cut and paste the results into a computer program. Find the mean of the 20 sample means. This value represents the mean of all 200 sample values. Is this combined mean a better estimate of the population mean of 68 than most of the individual sample means? Is that what you would expect, or not? Explain.

5.87 Refer to Figure 5.7, showing the results of taking a simple random sample of ten stick figures from Figure 5.6. Label the stick figures in Figure 5.6 from 00 to 99, going across rows. Suppose a table of random digits had been used to select the sample in Figure 5.7. Write down a stream of random digits that would have led to the selection of the sample shown.

5.88 Refer to Exercise 5.87 and write down a stream of random digits that would have led to the selection of the sample shown in Figure 5.8.

5.89 Generate one random sample, and print the display showing the applet and your result. Show what stream of numbers from a table of random digits would have produced your sample, numbering the stick figures from 00 to 99, starting at the top and going across rows.

5.90 The **Sampling** applet described in Section 5.7 drew a simple random sample of 10 individuals from a population of 45 men and 55 women and measured their heights. The average height for the sample could be used as an estimate of the average height in the population. Another possibility for estimating the average height would be to use stratified sampling, in which men and women would be the two strata, then estimate the heights of men and women separately and use the knowledge that 45% of the population is male to combine the two estimates. Do you think stratified sampling would be a more accurate or less accurate method than simple random sampling for estimating the mean height in the population? Explain.

Chapter Exercises

5.91 *This is also Exercise 1.7.* A *CBS News* poll taken in December 2009 asked a random sample of 1048 adults in the United States, "In general, do you think the education most children are getting today in public schools is better, is about the same, or is worse than the education you received?" About 34% said "Better," 24% said "About the same," and 38% said "Worse." (The remaining 4% were unsure.)

a. What is the population for this survey?
b. What is the approximate margin of error for this survey?
c. Provide an interval that is 95% certain to cover the true percentage of U.S. adults in December 2009 who would have answered "Better" to this question if asked.

5.92 *This is also Exercise 1.36.* The CNN website sometimes has a small box called "Quick vote" that contains a question about an interesting topic in the news that day. For example, one question in February 2010 asked, "Should the U.S. military let gays and lesbians serve openly?" Visitors to the website are invited to click their response and to view the results. When the results are displayed, they contain the message, "This is not a scientific poll."

a. What type of sample is obtained in this "Quick vote?"
b. What do you think is meant by the message that "This is not a scientific poll?"

5.93 Suppose a community group wants to convince city officials to put more trash containers on the city streets. They decide to conduct a survey, but rather than a scientifically valid result, they want their results to show that as many citizens as possible want additional trash containers. Give an example of how they could use each of the following to their advantage:

a. Self-selected sample.
b. Deliberate bias.
c. Ordering of questions.
d. Desire to please.

5.94 Refer to Example 5.17, "Politics Is All in the Wording." Explain which one of the seven "Possible Sources of Response Bias in Surveys" is illustrated by this example.

5.95 Refer to Example 5.17, "Politics Is All in the Wording." Explain which one of the "Difficulties and Disasters" listed in Section 5.5 is illustrated by this example.

5.96 Suppose you wanted to estimate the proportion of adults who write with their left hands and decide to watch a sample of n people signing credit card receipts at a mall.

a. Do you think this sample would be representative for the question of interest? Explain.
b. Assuming that the sample was representative, how many people would you have to include to estimate the proportion with a margin of error of 5%?
c. Suppose you observed 500 people and 60 of them signed with their left hand. What interval of values would you give as being likely to contain the true proportion of people in the larger population who would sign with their left hand?
d. Is the sampling plan that was used a probability sampling plan? Explain.

5.97 A Pew Research Center for the People & the Press and Pew Forum on Religion & Public Life survey conducted by Princeton Survey Research Associates between February 25 and March 10, 2002, asked a random sample of $n = 2002$ adult Americans, "Do you favor or oppose scientific experimentation on the cloning of human beings?" (http://www.pollingreport.com/science.htm, September 17, 2002). The responses and the percentages that selected them were as follows: Favor (17%), Oppose (77%), Don't know or Refused to answer (6%).

a. What is the approximate margin of error for this survey?
b. Find approximate 95% confidence intervals for the proportion and the percentage of adult Americans who would say they favor this experimentation.

5.98 Refer to Exercise 5.97, in which a survey question was, "Do you favor or oppose scientific experimentation on the cloning of human beings?" Suppose an agency that advocated cloning of human beings wanted to conduct a survey that would show support for their cause. Explain how it might use each of the following to do so:

a. Selection bias.
b. Deliberate bias.
c. Desire to please.
d. Ordering of questions.

5.99 Suppose that a survey reported that 55% of respondents favored gun control, with a margin of error of ± 3 percentage points.

a. What was the approximate size of the sample?
b. What is an approximate 95% confidence interval for the percentage of the corresponding population who favor gun control?
c. Based on the material in this chapter, write a statement interpreting the interval you found in part (b) that could be understood by someone with no training in statistics. (In Chapter 10, you will learn a more precise way to interpret this type of interval.)

5.100 The faculty senate at a large university wanted to know what proportion of the students thought that a foreign language should be required for everyone. The statistics department offered to cooperate in conducting a survey, and a simple random sample of 500 students was selected from all students enrolled in statistics classes. A survey form was sent by e-mail to these 500 students. For parts (a) to (c) discuss the extent to which the specified type of bias would be likely to occur in this survey:

a. Selection bias.
b. Nonresponse bias.
c. Response bias.
d. Which of these three types of bias do you think would be the most serious? Explain.

5.101 Refer to Exercise 5.100.

a. What is the population of interest to the faculty senate?

b. What is the sampling frame?

c. What is the sample?

d. Is the sample representative of the population of interest? Explain.

5.102 A large medical professional organization with membership consisting of doctors, nurses, and other medical employees wanted to know how its members felt about HMOs (health maintenance organizations). Name the type of sampling plan they used in each of the following scenarios:

a. They randomly selected 500 members from each of the lists of all doctors, all nurses, and all other employees and surveyed those 1500 members.

b. They randomly selected ten cities from all cities in which the members lived, and then surveyed all members in those cities.

c. They randomly chose a starting point from the first 50 names in an alphabetical list of members, and then chose every 50th member in the list starting at that point.

5.103 An article in the *Sacramento Bee* was headlined, "Drop found in risky behavior among teens over last decade" (June 7, 2000, p. A6; reprinted from the *Los Angeles Times*). The article reported that "the findings show that for the general population [not a specific ethnic group], the share of teens abstaining from all 10 risky behaviors jumped from 20 percent to 25 percent [from the beginning to the end of the 1990s]." The article did not say who was measured but was obviously based on *samples* taken at each end of the decade. What additional information would you need to know to determine whether or not this drop represents a real drop in the percentage of the *population* of teens abstaining? Explain.

5.104 One evening around 6 P.M., authorities in a major metropolitan area received hundreds of phone calls from people reporting that they had just seen an unidentified flying object (UFO). A local television station would like to report the story on their 10 P.M. news and wants to include an estimate of the proportion of area residents who witnessed the UFO. Explain how each of the following methods of obtaining this information would be biased, if at all:

a. A "quickie poll" taken between 7 P.M. and 9 P.M. that evening, using random-digit dialing for residents of the area.

b. During commercial breaks between 7 P.M. and 9 P.M. that evening, providing two numbers for viewers to phone: one if they had seen the UFO and one if they had not.

c. A door-to-door survey taken between 7 P.M. and 9 P.M., based on a multistage sampling plan, in which 10 city blocks (or equivalent area) were randomly selected and interviewers knocked on the doors of all homes in those areas and surveyed those who answered the door.

d. An e-mail survey sent at 7 P.M. to all e-mail addresses for local residents served by a local Internet service provider in the area, using responses received by 9 P.M.

5.105 Refer to Exercise 5.104. Acknowledging that some bias is inevitable when information is required on such short notice, which of the four methods would you recommend that the television station use to get the information it desires? Explain.

5.106 Refer to the *Literary Digest* poll in Case Study 5.1. Discuss the extent to which each of the following types of bias played a role in producing the disastrous results, if at all.

a. Selection bias.

b. Nonparticipation bias.

c. Response bias.

5.107 Refer to Exercises 5.104 and 5.105. Suppose that a local newspaper conducted a survey during the next week, based on random-digit dialing of 1400 residents of the area, and found that 20% of them reported having seen a UFO on the evening in question.

a. What is the margin of error for the survey?

b. Give an interval of values that probably covers the true percentage of the population that saw the UFO.

5.108 ◆ For this exercise, use the data in Case Study 1.1 (p. 2) or in the file **pennstate1** on the companion website.

a. Select a random sample of ten males and a random sample of ten females, and write down their answers to the question "What's the fastest you've ever driven a car?"

b. Compare the males' and females' responses, using appropriate numerical and/or graphical summary information.

c. Recall from Chapters 1 and 2 that there was an obvious difference between male and female responses for the set of all students. Is this difference obvious in your samples? Explain.

d. It is true in statistics that the larger the actual difference between two groups is in the population, the smaller the sample sizes needed to detect the difference. Discuss your answer to part (c) in the context of this statement.

5.109 Refer to the survey described in Example 5.9, "Which Scientists Trashed the Public?" Explain which type of bias (selection bias, nonparticipation bias, or biased response) was the most problematic in that survey.

5.110 An ABCNews.com poll conducted by TNS Intersearch June 13–17, 2001, posed the following question to a random sample of $n = 1024$ adult Americans: "Scientists can change the genes in some food crops and farm animals to make them grow faster or bigger and be more resistant to bugs, weeds and disease. Do you think this genetically modified food, also known as bio-engineered food, is or is not safe to eat?" (http://www.pollingreport.com/food.htm, September 17, 2002). The percentages that selected each response were as follows: Is safe (35%), Is not safe (52%), and No opinion (13%).

a. What is the approximate margin of error for this survey?
b. Find approximate 95% confidence intervals for the proportion and the percentage of adult Americans who would have said in June 2001 that they think genetically modified food is *not* safe.
c. Note that a majority of the sample (52%) thought that genetically modified food is not safe. Based on the confidence interval you found in part (b), can you also conclude that a majority of the *population* at that time thought that genetically modified food was not safe? Explain.

5.111 Suppose an (unscrupulous!) organization wanted to conduct a survey in which the results supported its position that drinking coffee in public should be illegal except in designated coffee bars. Explain how the organization could use each of the following sources of bias to help produce the results it wants:

a. Selection bias.
b. Nonresponse bias.
c. Response bias.

5.112 Find an example of a poll conducted by a reputable scientific polling organization. Write a few paragraphs describing how the poll was conducted, what was asked, and the results. Explain whether or not any of the problems discussed in Sections 5.5 and 5.6 were likely to have biased the results of the poll.

5.113 Find an example of a poll that would be considered unscientific because it does not represent a population other than those who responded. Write a few paragraphs describing how the poll was conducted, what was asked, and the results. Explain which of the problems discussed in Sections 5.5 and 5.6 were likely to have biased the results of the poll.

Exercises 5.114 to 5.117 are based on a report of a Field poll printed in the Sacramento Bee *on July 7, 2000 (Dan Smith, p. A4), with the headline,* "Ratings Up for Boxer, Feinstein in New Poll." *The results were reported with a graphical display containing the information that "results are based on a statewide survey of 750 adults conducted June 9–18. The sample includes 642 voters considered likely to vote in the Nov. 7 general election. The margin of error is 3.8 percentage points."*

5.114 Read the headline that accompanied the article, which stated that ratings were "up" for both senators. For the June 2000 poll, 57% approved of Feinstein's performance, compared with 56% and 53%, respectively, for similar polls taken in February 2000 and October 1999. Presumably, the headline refers to all registered voters in the state, not just those in the samples. Considering the margin of error accompanying each poll, do you agree with the headline? Explain.

5.115 The text of the article said that the 750 adults surveyed were actually 750 *registered voters.* Based on the information about who was surveyed, if the poll results are to be used to predict the proportion of voters who were likely to vote for Senator Feinstein in November, should the results of all 750 voters be used? Explain using the Fundamental Rule for Using Data for Inference in your explanation.

5.116 The report stated that the margin of error was 3.8 percentage points. This value is based on the 642 voters who were considered likely to vote and uses the precise formula you will learn in Chapter 10.

a. Compute the conservative margin of error for this poll based on those likely to vote, and compare it to the precise margin of error presented.
b. For this sample, 57% reported that they approve of Dianne Feinstein's job performance. What is an approximate 95% confidence interval for the percent of all likely voters who approve of her performance?

5.117 The article also reported that "among GOP voters, 39 percent give Feinstein high marks." Would the margin of error accompanying this result be larger, smaller, or the same as the margin of error for the entire sample? Explain.

Exercises 5.118 to 5.120 refer to a survey of University of California faculty members on affirmative action policies for the university (Roper Center, 1996). The survey was based on telephone interviews with a random sample of 1001 faculty members conducted in December 1995.

5.118 One of the questions asked was, "Do you favor or oppose using race, religion, sex, color, ethnicity, or national origin as a criterion for admission to the University of California?" Of the 804 male respondents, 47% said "Favor," and of the 197 female respondents, 73% said "Favor."

a. Find the conservative margin of error for the males, and use it to compute an approximate 95% confidence interval for the percent of males in the population who favored this policy at the time of the poll.
b. Repeat the analysis in part (a) for females.
c. Compare the approximate 95% confidence intervals you found in parts (a) and (b), and comment on whether you think male and female faculty members in the population differed on this question.

5.119 Refer to Exercise 5.118. Of the 166 arts and humanities faculty members, 66% favored the policy, while of the 229 engineering, mathematical, and physical sciences faculty, 38% favored it.

a. Find the conservative margin of error for the arts and humanities faculty, and use it to compute an approximate 95% confidence interval for the percent of them in the population who favored this policy.
b. Repeat the analysis in part (a) for the engineering and sciences faculty.
c. Compare the approximate 95% confidence intervals you found in parts (a) and (b), and comment on whether you

think faculty members in these disciplines in the population differed on this question at the time of the poll.

5.120 One question asked was as follows: "The term *affirmative action* has different meanings to different people. Please tell me which statement best describes what you mean by the term: First, affirmative action means granting preferences to women and certain racial and ethnic groups. Second, affirmative action means promoting equal opportunities for all individuals without regard to their race, sex, or ethnicity." The options "Neither" or "Both" were not given and were recorded only if the respondent offered them. The first statement was chosen by 37%, the second one by 43%, "Neither" by 13%, and "Both" by 2%. The remainder said, "Don't know." The organization that commissioned the poll, the California Association of Scholars, issued a press release that said in reference to this question, "These findings demolish the claim that the faculty at UC wants preferential policies." Refer to the "Possible Sources of Response Bias in Surveys" in Section 5.6, and discuss them in the context of this question and resulting press release.

6

© Brooklyn Production/Corbis

Does prayer lower blood pressure?

See Example 6.1 *(p. 191)*

Gathering Useful Data for Examining Relationships

Probably the biggest misinterpretation made when reporting studies is to imply that a *cause-and-effect* relationship can be concluded on the basis of an observational study. Groups that naturally differ for the explanatory variable of interest are almost certain to differ in other ways, any of which may contribute to differences in the responses.

In this chapter, we learn about ways to collect data in order to examine relationships between variables. We have already seen several examples that involved possible links between variables. Case Study 1.6 described a study that demonstrated a link between taking an aspirin a day and a decreased risk of heart attacks for men. In Chapter 2, Example 2.1 was about the connection between sex and seat belt use for 12th-grade students. Example 2.2 was about a possible connection between the use of night-lights in infancy and nearsightedness. In Chapter 3, we saw many examples of relationships between two quantitative variables, and in Chapter 4, we saw many examples of relationships between two categorical variables.

In many of these cases, we want to know whether a **cause-and-effect relationship** exists. That is, we want to know whether changing the value of one variable *causes* changes in another variable. We will learn in this chapter that the way in which a study is conducted affects our ability to infer that a cause-and-effect relationship exists.

6.1 Speaking the Language of Research Studies

Although there are a number of different strategies for collecting meaningful data, there is common terminology used in most of them. Statisticians tend to borrow words from common usage and apply a slightly different meaning, so be sure that you are familiar with the special usage of a word in a statistical context.

Types of Research Studies

Two basic types of statistical research studies are conducted to detect relationships between variables:

- Observational studies
- Experiments

In an **observational study**, the researchers simply observe or question the participants about opinions, behaviors, or outcomes. Participants are not asked to do anything differently. For example, Case Study 1.5 described an observational study in which blood pressure and frequency of certain types of religious activity (such as prayer and church attendance) were measured. The goal was to determine whether people with higher frequency of religious activity had lower blood pressure. Researchers simply measured blood pressure and frequency of religious activity. They did not ask participants to change how often they prayed or went to religious services or to change any other aspect of their lives.

In an **experiment**, researchers manipulate something and measure the effect of the manipulation on some outcome of interest. **Randomized experiments** are experiments in which the participants are *randomly assigned* to participate in one condition or another. The different "conditions" are called **treatments**.

As an example of a randomized experiment, Case Study 1.6 described a randomized experiment in which physicians were randomly assigned to take an aspirin or a placebo every other day. The goal was to determine whether regular intake of aspirin resulted in lower risk of heart attacks than that experienced by the group taking only a placebo. The two possible treatments were taking aspirin or taking a placebo. The random assignment should have ensured that the groups were similar in all other respects, so any difference in heart attack rates could actually be attributed to the difference in the pill they took (aspirin or placebo).

DEFINITION In an **observational study** researchers only observe or measure the participants and do not assign any treatments or conditions.

In a **randomized experiment** participants are randomly assigned to participate in one condition or another.

Observational Study or Randomized Experiment?

A major theme of this chapter is that a randomized experiment provides stronger evidence of a cause-and-effect relationship than an observational study does. In many situations, however, there are practical and ethical issues that force researchers to conduct an observational study rather than an experiment. For instance, in an investigation of the effect of oral contraceptive usage on blood pressure, it would be unreasonable to randomly assign some women to use oral contraceptives and others to use a placebo. Also, some variables, such as handedness, are inherent traits that cannot be randomly assigned. A study done to find out whether handedness affects the risk of a workplace accident, for example, would have to be done as an observational study.

Sometimes researchers can choose between the strategies for collecting data and must weigh the advantages and disadvantages of each strategy. For instance, suppose researchers want to study whether people lose more weight by exercising or by limiting their fat intake. They could conduct an *observational study*, in which they question people about their exercise and eating habits, as well as their history of weight loss, and compare people who exercise with those who eat limited fat. Or they could conduct a *randomized experiment* in which they randomly assign people to either exercise or eat a specific diet and then measure and compare weight loss. As we will see later in this chapter, each type of study has advantages and disadvantages.

THOUGHT QUESTION 6.1 For many randomized experiments, researchers recruit volunteers who agree to accept whichever treatment is randomly assigned to them. Why do you think this strategy cannot always be used, thus requiring observational studies to be used instead?*

Who Is Measured: Units, Subjects, Participants

Researchers use a variety of terms to describe who is measured. As discussed in Chapter 5, the generic term **unit** is used to indicate a single individual or object being measured. In experiments, the most basic entity (person, plant, and so on) to which different treatments can be assigned is called an **experimental unit**. For instance, in Case Study 1.6, each physician was an experimental unit, because each one was randomly assigned to take aspirin or a placebo. When the experimental units are people, they often are called **subjects**. In both experiments and observational studies, the subjects may also be called **participants**.

***HINT:** See the discussion under "Types of Research Studies" on pages 189–190.

Explanatory and Response Variables

Most experiments and observational studies are conducted because researchers want to learn about the relationship between two or more variables. For instance, in the observational study in Case Study 1.5, the researchers wanted to know whether religious activity and blood pressure were related. In the randomized experiment in Case Study 1.6, they wanted to know if aspirin consumption and risk of heart attack were related.

We defined the roles that variables play in a relationship in Chapter 2. You may recall that an **explanatory variable** is one that may explain or may cause differences in a **response variable** (sometimes called an **outcome variable**). For example, in Case Study 1.5, frequency of religious activity is the explanatory variable and blood pressure is the response variable. We distinguished between explanatory variables and response variables in Chapter 2 when creating summaries for categorical variables. The data for a response variable or an explanatory variable may be either categorical or quantitative, as we learned in Chapters 3 and 4.

The term **dependent variable** is sometimes used for the response variable because the values may be thought to *depend* on the values of the explanatory variable(s). In contrast, and completely separate from the common usage of the word *independent,* explanatory variables are sometimes called **independent variables**.

Confounding Variables—Measured or Not

Most relationships are complex, and it is unlikely that any explanatory variable is the direct and sole explanation for the values of the response variable. In many studies, particularly in observational studies, *confounding* variables may be present. For example, suppose that an observational study finds that people who take at least 500 mg of vitamin C every day get fewer colds than other people do. But perhaps people who take vitamins also eat more healthful foods, and the better diet reduces the risk of a cold. Or perhaps the extra glass of water that is used to take the vitamin pill helps to prevent colds.

DEFINITION A **confounding variable** is a variable that both *affects the response variable* and also *is related to the explanatory variable.* The effect of a confounding variable on the response variable cannot be separated from the effect of the explanatory variable.

Confounding might cause the appearance of a nonexistent meaningful relationship between the explanatory and response variables, or it can even mask an actual relationship. The term **lurking variable** is sometimes used to describe a potential confounding variable that is not measured and is not considered in the interpretation of a study. Whether the researchers measure confounding variables or not, such variables are an inherent property of the nature of the relationship between explanatory and response variables. When interpreting the results of an observational study, always consider the possibility that confounding variables may exist.

Confounding variables are less likely to be a problem in the interpretation of randomized experiments. Because the treatments are randomly assigned to the units, the values of confounding variables should approximately even out across treatment groups. For example, the group of physicians that was randomly assigned to take aspirin and the group of physicians randomly assigned to take a placebo in Case Study 1.6 should have had similar compositions in terms of other health-related variables. Therefore, health-related variables such as diet and exercise would not be confounded with the explanatory variable, which was whether a physician took aspirin or a placebo.

Example 6.1 **Case Study 1.5 Revisited: What Confounding Variables Lurk behind Lower Blood Pressure?** In Case Study 1.5, people who attended church regularly had lower blood pressure than people who stayed home and watched religious services on television. But it may be that those who attended church had lower blood pressure as a result of a strong social support network rather than from attending church regularly. If this is the case, "amount of social support" is a *confounding* variable. Note that two conditions must be met for "amount of social support" to qualify as a confounding variable:

- Amount of social support (confounding variable) *affects* blood pressure (the response variable).
- Amount of social support (confounding variable) *is related to* attending church regularly (the explanatory variable) because the same people who attend church regularly are likely to have a strong social support network.

Health status is another possible confounding variable, since those who are not well may not be able to attend church regularly. There are numerous other possibilities, such as age and attitude toward life, that could affect the observed relationship between religious activity and blood pressure.

THOUGHT QUESTION 6.2 Choose a possible confounding variable for the situation in Example 6.1, other than the ones mentioned in the example, and explain how it meets the two conditions necessary to qualify as a confounding variable.*

Example 6.2 **The Fewer the Pages, the More Valuable the Book?** If you peruse the bookshelves of a typical college professor, you will find a variety of books, ranging from textbooks to esoteric technical publications to paperback novels. To determine whether there is a relationship between the price of a book and the number of pages it contains, a college professor recorded the number of pages and price for 18 books on one shelf.

The numbers are shown in Table 6.1. The books are ordered from fewest to most number of pages. Do prices increase accordingly so that books with the greatest number of pages cost the most? No, the trend appears to be the reverse: Many of the books with fewer pages seem to be more expensive than many of the books with a larger number of pages! In fact, the correlation is −0.31, which indicates a somewhat weak negative association.

Table 6.1 Pages versus Price for the Books on a Professor's Shelf

Book #	Pages	Price	Book #	Pages	Price
1	200	35.00	10	601	79.95
2	256	47.50	11	612	50.00
3	305	29.95	12	639	60.00
4	370	64.95	13	641	14.95
5	384	74.50	14	673	24.99
6	436	15.99	15	747	21.95
7	480	35.00	16	833	29.95
8	545	69.95	17	877	20.00
9	565	19.95	18	907	25.00

Why does the relationship seem to be the opposite of what we might expect? The answer is that there is a confounding variable: the type of book. The books on the professor's shelf consisted of paperback books, which tend to be long but inexpensive, and hardcover technical books, which tend to be shorter but expensive. As is illustrated in Figure 6.1, when we look at the relationship between price (the response variable) and pages (the explanatory variable) for each type of book sepa-

***HINT:** Think of something that would probably be different for people who attend church compared to those who don't and that might affect blood pressure.

rately, we see that the price does tend to increase with number of pages, especially for the technical books. Note that type of book (confounding variable) not only *affects* price (response variable) but it also *is related to* number of pages (explanatory variable) because hardcover technical books tend to have fewer pages than paperback books.

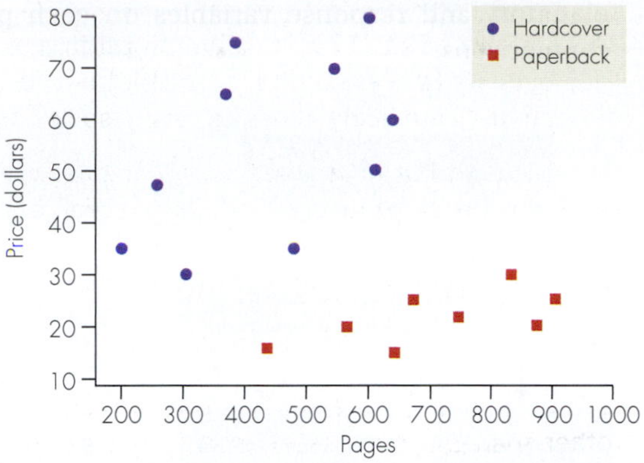

Figure 6.1 Price versus pages for two types of books

As Examples 6.1 and 6.2 illustrate, it is unlikely that any explanatory variable is the direct and sole explanation for the values of the response variable. There are almost always confounding variables, particularly in observational studies. These confounding variables might be measured and accounted for in the analysis of the data, or they could be unmeasured lurking variables. In either case, always think about the possible effect of confounding variables when you consider the results of statistical studies. Confounding variables can be especially problematic in interpreting the results of observational studies. Randomized experiments are designed to help control the influence of confounding variables.

CASE STUDY 6.1 Lead Exposure and Bad Teeth

The following article appeared on the *USA Today* website (http://www.usatoday.com/life/news/health/default.htm). As you read the article, identify the explanatory and response variables and any possible lurking or confounding variables mentioned.

Lead exposure linked to bad teeth in children Children exposed to lead are more likely to suffer tooth decay, and vitamin C might help lower blood lead levels, say two studies out Thursday. In the first of the reports in the *Journal of the American Medical Association,* researchers calculate that lead exposure could account for tooth decay in 2.7 million children. "Other people may debate that, but that's our position," says head researcher Mark Moss of the University of Rochester (N.Y.) School of Medicine and Dentistry. Prior studies showed that lead exposure can depress a child's IQ. "There are a lot worse things that lead can do to you than hurt your teeth," Moss says. He notes, however, that one of the key questions in dentistry is why low-income people experience more tooth decay than higher-income people. "This study suggests lead might be one of the reasons," he says.

The study involved 24,901 children ages 2 and older. It showed that the greater the child's exposure to lead, the more decayed or missing teeth. "The risk of getting tooth decay increased as the amount of lead went up," Moss says.

Thomas Matte of the Centers for Disease Control and Prevention says that nearly 1 million U.S. children younger than age 6 still have too much lead in their blood, particularly poor children living in older housing. An estimated 57 million residences still have lead paint on the walls, Matte says. He cautions in an editorial, however, that the problems with tooth decay might be associated with something other than lead, perhaps a sugar-rich diet or poor access to fluoridated water. "Those are good points," Moss says. "But I doubt it. We controlled for income level, the proportion of diet due to carbohydrates, calcium in the diet and the number of days since the last dental visit. We still saw this consistent pattern between dental decay and blood lead level—whether they were baby teeth or adult teeth."

In the second study, Joel Simon and his colleagues at the University of California at San Francisco studied 19,578 people who had no history of excess lead exposure. They found that the higher a person's intake of vitamin C, the lower his blood lead level (Steve Sternberg, *USA Today*, June 23, 1999, accessed online).

(Source: "Lead Exposure Linked to Bad Teeth in Children," by S. Sternberg, USA Today, *June 23, 1999. Internet. Reprinted with permission.)*

Discussion of Case Study 6.1

The *USA Today* article is based on two *observational studies,* the first of which is described in the most detail.

Figure 6.2 illustrates the process that the researchers used in the first study. Because it is an observational study, there is no randomization step in the process. The steps for this study include identifying the participants, measuring the explanatory and response variables on each participant, and making appropriate comparisons.

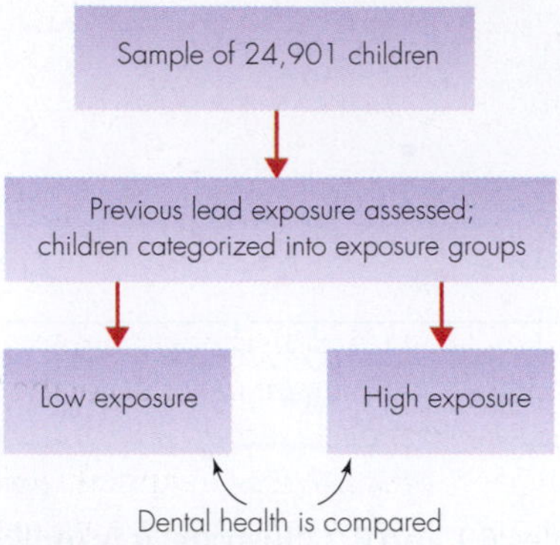

Figure 6.2 An observational study in Case Study 6.1: Researchers assess prior lead exposure and compare dental health of low- and high-exposure groups

The researchers are interested in studying the relationship between exposure to lead and tooth decay/loss. The *explanatory variable* is level of exposure to lead, and the *response variable* is the extent to which the child has missing or decayed teeth. The article mentions a number of possible *confounding variables.* Remember that confounding variables are variables whose effect on the response variable cannot be separated from the effect of the explanatory variable. In this example, these are factors that may differ for children exposed to high and low amounts of lead and that may also affect tooth decay. The researchers measured some of these and tried to account for their impact, including "income level, the proportion of diet due to carbohydrates, calcium in the diet and the number of days since the last dental visit." They did not measure or account for other possible *lurking variables* that could also be confounding variables, such as amount of fluoride in the water and amount of general health care available to these children. These variables may differ for children exposed to high and low amounts of lead.

6.1 Exercises are on pages 210–212.

THOUGHT QUESTION 6.3 For most randomized experiments, such as medical studies comparing a new treatment with a placebo, it is unrealistic to recruit a simple random sample of people to participate. Why is this the case? What can be done instead to make sure that the Fundamental Rule for Using Data for Inference (see Section 5.1) is not violated?*

***HINT:** As an example, in a study to test a new medication to treat people with athlete's foot, would it be possible to identify the population of those afflicted, get a random sample, and get their cooperation? How might a representative sample be found instead?

IN SUMMARY The Language of Research Studies

- **Observational studies** and **randomized experiments** are the main two types of research studies done to examine how an explanatory variable is related to a response variable.
- The researchers only observe or measure all the variables in an observational study, whereas they randomly assign experimental units to categories of an explanatory variable in a randomized experiment.
- **Confounding variables** are likely to affect the response variable in observational studies, and this effect cannot be separated from the effect of the explanatory variable on the response. Experiments are done to reduce the risk of confounding.

6.2 Designing a Good Experiment

An **experiment** measures the effect of manipulating the environment of the participants in some way. The idea is to measure the effect of the feature being manipulated, the explanatory variable, on the response variable. With human participants, the manipulation may include receiving a drug or medical treatment, going through a training program, agreeing to a special diet, and so on. Experiments are also done on other kinds of experimental units, such as when different growing conditions are compared for their effect on plant yield or when different paints are applied on highways to find out which paints last longer.

Randomized experiments are important because they often allow us to determine whether there is a cause-and-effect relationship between two variables. In a **randomized experiment**, participants usually are randomly assigned either to be in a specific treatment group or possibly to be in a control group. The benefit of the random assignment is that it should make the groups approximately the same in all respects except for the explanatory variable, which is purposely manipulated. Response variable differences among the groups, if large enough to rule out natural chance variability, can then be attributed to the manipulation of the explanatory variable.

CASE STUDY 6.2 Kids and Weight Lifting

Read the original source on the companion website, http://www.cengage.com/statistics/Utts4e.

Is weight training good for children? If so, is it better for them to lift heavy weights for a few repetitions or to lift moderate weights a larger number of times? Researchers at the University of Massachusetts set out to answer this question with a randomized experiment using 43 young volunteers between the ages of 5.2 and 11.8 years old. The children, recruited from a YMCA after-school program, were randomly assigned to one of three groups. Group 1 performed 6 to 8 repetitions with a heavy load, Group 2 performed 13 to 15 repetitions with a moderate load, and Group 3 was a control group.

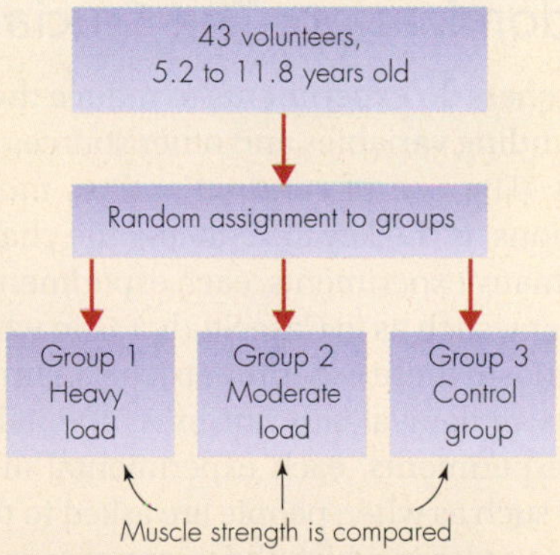

Figure 6.3 The randomized experiment in Case Study 6.2: Children were randomly assigned to one of three different treatment groups.

Figure 6.3 illustrates the process used in this study. The steps for this randomized experiment include recruiting volunteers to participate, randomly assigning participants to the three groups, carrying out the weight-lifting treatments (the explanatory variable), and comparing muscular strength and endurance (the response variables) for the three groups.

The exercises were performed twice a week for 8 weeks. According to the research report, the higher-repetition, moderate-load group (Group 2) showed the best results. Here is a sample of the results:

> Leg extension strength significantly increased in both exercise groups compared with that in the control subjects. Increases of 31.0% and 40.9%, respectively, for the low repetition–heavy load and high repetition–moderate load groups were observed (Faigenbaum et al., 1999, p. e5).

Because the children were randomly assigned to the exercise groups, they should have been similar, on average, before the weight-training intervention. Therefore, it is reasonable to conclude that the different training programs actually *caused* the difference in results.

Who Participates in Randomized Experiments?

Researchers often recruit **volunteers** to participate in randomized experiments. These volunteers agree to receive whichever treatment is assigned to them, and they often don't know which treatment it is until the experiment is concluded. For example, volunteers in a study of the effectiveness of a new vaccine may be randomly assigned to receive either the vaccine or a placebo. Because the volunteers sign an "informed consent" to participate, it is deemed ethically acceptable to ask them to receive whichever treatment is assigned to them.

Sometimes the participants are passive volunteers, such as when all patients who are treated at a particular medical facility are asked to sign a consent form agreeing to participate in a study. Often researchers recruit volunteers through the newspaper. For instance, a "Well-Being" feature page in *The Sacramento Bee* had an article titled "Volunteers for Study." It said:

> People who suffer from anxiety disorders are needed for a study being conducted by the University of California, Davis, Department of Psychiatry. Researchers are examining the effectiveness of a new medication on moderate-to-severe symptoms of anxiety. The study requires volunteers who are at least 18 and in generally good health and who are not taking oral medications for their anxiety. The study lasts 14 weeks and requires routine visits to the Department of Psychiatry. All study procedures, visits and medications are free. To enroll, call. . . (*Sacramento Bee*, July 7, 1999, p. G5).

When researchers recruit volunteers for studies, they may be able to extend the results to a larger population, but this is not always the case. For instance, investigators often recruit volunteers by offering a small payment or free medical care as compensation for participation. Those who respond are more likely to be from lower socioeconomic backgrounds, and the results may not apply to the general population.

Remember the Fundamental Rule for Using Data for Inference given in Section 5.1: Available data can be used to make inferences about a much larger group *if the data can be considered to be representative with regard to the question(s) of interest.* In many randomized experiments, the question of interest is whether the treatment and control groups differ on the outcome variable(s). Common sense should enable you to figure out whether the volunteers in the study are representative of a larger population for that question.

Randomization: The Crucial Element

Researchers do experiments to reduce the likelihood that the results will be affected by confounding variables and other sources of bias that often are present in observational studies. The use of **randomization**, meaning random assignment to treatments or conditions, is the key to reducing the chance of confounding.

In many experiments, each experimental unit receives or participates in only a single treatment, such as in Case Study 1.6, in which the physicians were given either aspirin or a placebo. In these experiments, the treatments are randomly assigned to the experimental units, or equivalently, the experimental units are assigned to the treatment groups. In other experiments, each experimental unit receives or participates in all of the treatments, such as when people are asked to taste and compare three brands of soda. In this case, the order in which the treatments are presented must be randomized.

Randomizing the Type of Treatment

Randomly assigning the treatments to the experimental units keeps the researchers from making assignments that are favorable to their hypotheses and also helps to protect against hidden or unknown biases and confounding variables. For example, suppose that in the experiment in Case Study 1.6, approximately the first 11,000 physicians who enrolled were given aspirin, and the remaining physicians were given placebos. It could be that the healthier, more energetic physicians enrolled first, thus giving aspirin an unfair advantage.

In statistics, *random* is not synonymous with *haphazard*, despite what your thesaurus might say. As we have seen, random assignments may not be possible or ethical under some circumstances. But in situations in which randomization is feasible, it is usually not difficult to accomplish. It can be done with a table of random digits, with a computer, or even, if done carefully, by physical means such as flipping a coin or drawing numbers from a hat. The important features of proper randomization are that the chances of being assigned to each condition are the same for each participant, and that randomness is the only treatment assignment criterion.

Randomizing the Order of Treatments

In some experiments, all treatments are applied to each unit. In that case, randomization should be used to determine the *order* in which they are applied. For example, suppose that an experiment is conducted to determine the extent to which drinking alcohol or smoking marijuana impairs driving ability. Because drivers are all different, it makes sense to test the same drivers under all three conditions (alcohol, marijuana, and sober) rather than using different drivers for each condition. But if everyone were tested under alcohol first, then under marijuana, and then sober, the performances could improve on the second and third times just from having learned something about the test.

A better method would be to randomly assign some drivers to each of the possible orderings of the treatments, so that the learning effect would average out over the three treatments. Note that it is important that the assignments be made *randomly*. If we let the experimenter decide which drivers to assign to which ordering, assignments could be made to give an unfair advantage to one of the treatments.

Example 6.3 **Revisiting Case Study 6.2: Randomly Assigning Children to Weight-Lifting Groups** In Case Study 6.2, 43 children were randomly assigned to one of three treatment groups. Children in Group 1 performed weight-lifting repetitions with a heavy load, Group 2 performed more repetitions but with a moderate load, and Group 3 was a control group that did not lift weights. There were 15 children assigned to Group 1, 16 to Group 2, and 12 to Group 3.

Suppose that we are asked to randomly assign children to the treatment groups. How could we carry out this randomization? One way would be to use a computer program such as Minitab or Excel. In one method, we would first number the participating children from 1 to 43. Then, we would enter the numbers 1 to 43 into a column of a Minitab worksheet or Excel spreadsheet. In either program, it's relatively easy to permute these numbers (1 to 43) into a random order. In the randomly permuted list of child numbers, we would assign children with the first 15 numbers appearing in the list to be in Group 1, the children with the next 16 numbers to be in Group 2, and the remaining 12 to be in Group 3.

As an example, the following is a random permutation of the numbers 1 to 43 created in Minitab:

21 37 28 31 12 4 16 24 30 17 19 6 10 34 20 | 13 23 1 39 2 33 5
15 3 43 26 7 36 40 38 22 | 42 29 8 14 9 27 32 18 41 25 11 35

Group 1 would be composed of the children with the first 15 numbers in the list: 21, 37, 28, ..., 20 (just before the vertical bar in the first row). Group 2 would be made up of the children with the next 15 numbers in the list: 13, 23, 1, ...,22. Group 3 would comprise children with the final 12 numbers: 42, 29, 8, ..., 35.

Other methods of making the random assignments would also work. For instance, you could write the names of the children on identical slips of paper, put them in a bag, and mix them well. You would then reach in and select 15 of them for Group 1, another 16 for Group 2, and the remaining 12 would constitute the control group. A random digit table could also be used to create the random assignments.

MINITAB TIP **Randomly Permuting a Column**

- Use **Calc > Random Data > Sample From Columns**. In the dialog box, specify the number of rows in the column to be permuted, the column to be permuted, and designate a new column for storage of the permuted list.
- To display the permuted column, use **Data > Display Data**.

Replication

Replication in an experiment refers to the idea that more than one experimental unit is assigned to each treatment condition. This is a crucial component of a well-designed experiment. The number of experimental units per treatment should be large enough to provide suitably accurate estimates of response differences among the treatments. With sample sizes that are too small, it is difficult to rule out natural chance variation as the reason for any observed differences.

For example, in Case Study 1.6 it would have been meaningless to assign one physician to take aspirin and one to take a placebo. Even if the placebo recipient had a heart attack and the aspirin recipient did not, we would not be able to make any meaningful conclusion. Much of the benefit of statistical methods over anecdotes follows from the fact that replication allows us to separate real differences and relationships from natural chance variability.

More generally, **replication in science** refers to the principle that a single experiment rarely provides sufficient evidence for anything, so it is important to have independent researchers try to reproduce findings. Any individual experiment could have unseen flaws, so a variety of *replications* of similar findings by different researchers provides more evidence of a real phenomenon than does a single experiment. In Chapter 13, we will learn that experimental results don't have to be exactly the same to provide sufficient evidence of a real phenomenon. What is important is that the results are similar enough that any discrepancies can be attributed to known factors and chance variability.

Control Groups, Placebos, Blinding, and Double Dummy

Control Groups

To determine whether a drug, weight-lifting routine, meditation technique, and so on has an effect, we need to know what would have happened to the response variable if the treatment had not been applied. To find that out, experimenters create **control groups,** which are treated identically in all respects except that they don't receive the active treatment. Occasionally, a control group will receive a standard or existing treatment against which a new treatment is to be compared.

Placebos

A special kind of control group is often used in studies of the effectiveness of drugs. A substantial body of research shows that people respond not only to active drugs but also to **placebos**. A placebo looks like the real drug but has no active ingredients. Placebos can be amazingly effective; studies have shown that they can help up to 62% of headache sufferers, 58% of those suffering from seasickness, and 39% of those with postoperative wound pain.

Because the **placebo effect** is so strong, drug research is generally conducted by randomly assigning half of the volunteers to receive the drug and the other half to receive a placebo, without telling any of the volunteers which they are receiving. The placebo looks just like the real thing, so the participants will not be able to distinguish between it and the actual drug and therefore will not be influenced by belief biases.

Blinding

It isn't only the patient who can be affected by knowing whether he or she has received an active drug or not. If the researcher measuring the patients' responses were to know their group assignment, the measurements could be taken in a biased fashion.

To avoid these biases, **blinding** is used. A **single-blind** experiment is one in which the participants do not know which treatment they have been assigned. An experiment would also be called *single-blind* if the participants knew the treatments but the researcher was kept blind. Good experiments use double-blind procedures whenever possible. A **double-blind** experiment is one in which neither the participant nor the researcher taking the measurements knows who had which treatment.

Although double-blind experiments are preferable, they are not always possible. For example, to test the effect of meditation on blood pressure, the subjects would obviously know whether they were in the meditation group or the control group. Only a single-blind study is possible, one in which the person measuring the blood pressures would not know who was in which group.

Double Dummy

In an experiment to compare two treatments, it sometimes is impossible to mask which treatment is which, so each group receives one active treatment and one placebo treatment. For instance, suppose researchers wanted to compare the effects of nicotine patches to nicotine gum for helping people quit smoking. Obviously, participants would know if they were wearing a patch or chewing gum. So participants are asked to both wear a patch and to chew gum. One of the two will have the active ingredient (nicotine), but the other will be a placebo. In other words, the treatments would be

Group One: Nicotine patch, placebo gum

Group Two: Placebo patch, nicotine gum

This method of giving each participant two "treatments" to ensure that the experiment is blind is called **double dummy**. Sometimes a third group is used and given placebos of both methods to compare the results from each treatment method to what would happen with no active treatment at all. In our example, a third group would receive a placebo patch and placebo gum.

Case Study 6.3 describes a well-designed experiment that was done to assess the effectiveness of a nicotine patch as an aid to quitting smoking. The researchers used randomization, a placebo group, and a double-blind method.

CASE STUDY 6.3 Quitting Smoking with Nicotine Patches

There is no longer any doubt that smoking cigarettes is hazardous to your health and the health of those around you. Yet for someone who is addicted to smoking, quitting is no simple matter. One promising technique is to apply a patch to the skin that dispenses nicotine into the blood. In fact, these nicotine patches have become one of the most frequently prescribed medications in the United States.

To test the effectiveness of these patches on the cessation of smoking, Dr. Richard Hurt and his colleagues (1994) recruited 240 smokers at Mayo Clinics in Rochester, Minnesota; Jacksonville, Florida; and Scottsdale, Arizona. Volunteers were required to be between the ages of 20 and 65, have an expired carbon monoxide level of 10 ppm or greater (showing that they were indeed smokers), be in good health, have a history of smoking at least 20 cigarettes per day for the past year, and be motivated to quit.

Volunteers were randomly assigned to receive either 22-mg nicotine patches or placebo patches for 8 weeks. They were also provided with an intervention program recommended by the National Cancer Institute, in which they received counseling before, during, and for many months after the 8-week period of wearing the patches.

After the 8-week period of patch use, almost half (46%) of the nicotine group had quit smoking, while only one-fifth (20%) of the placebo group had done so. Having quit was defined as "self-reported abstinence (not even a puff) since the last visit and an expired air carbon monoxide level of 8 ppm or less" (p. 596). After a year, rates in both groups had declined, but the group that had received the nicotine patch still had a higher percentage who had successfully quit than did the placebo group, 27.5% versus 14.2%.

The study was double-blind, so neither the participants nor the nurses taking the measurements knew who had received the active nicotine patches. The study was funded by a grant from Lederle Laboratories and was published in the *Journal of the American Medical Association.* Because this was a well-designed randomized experiment, it is likely that the nicotine patches actually *caused* the additional percentage who had quit in the nicotine patch group compared with the control group.

Matched-Pair Designs and Blocking

In some experiments, researchers match people on traits that are likely to be related to the outcome, such as age, IQ, or weight. They then randomly assign each of the treatments to one member of each matched pair or grouping. For example, in a study comparing chemotherapy to surgery to treat cancer, patients might be matched by sex, age, and level of severity of the illness prior to treatment assignment. One from each pair would then be randomly chosen to receive the chemotherapy, and the other would receive surgery. (Of course, such a study would be ethically feasible only if there were no prior knowledge that one treatment was superior to the other. Patients in such cases are always required to sign an informed consent.)

Matched-Pair Designs

A **matched-pair design** is an experimental design in which individuals are first matched on important characteristics and then each member of a pair receives a different treatment. An important feature of these designs is that randomization is used to assign the order of the two treatments. A special case of matched pairs occurs when the same individual is measured twice under different conditions, or before and after an intervention.

Block Designs

The matched-pair design is a special case of a **block design**. In a block design, experimental units are first divided into homogeneous groups called **blocks**, and each treatment is randomly assigned to one or more units within each block. This somewhat peculiar terminology results from the fact that these ideas were first used in agricultural experiments, in which the experimental units were plots of land that had been subdivided into homogeneous blocks.

Blocks and matched pairs are used to control and reduce known sources of variability in the response variable. Blocking should be used when such sources are known and can be determined for each experimental unit. For example, if you wanted to study the relationship between exercise and weight loss, it would make sense to separate people based on whether they were initially underweight, overweight, or about right. Initial weight would be used to create matched pairs or blocks. Within each matched pair or block, individuals would be randomly assigned to different exercise programs.

Example 6.4 **Blocked Experiment for Comparing Memorization Methods** Imagine an experiment done to compare two methods for increasing memorization skills. Age affects memorization ability so the researchers might block participants into age groups (young, middle-aged, elderly) and then randomly assign the two methods to half of the individuals in each age group. This ensures the same overall age distribution for participants using the two memory methods. Comparison of the two treatments could also be done separately in each age group to examine whether age affects the magnitude of difference between the two methods. Figure 6.4 illustrates the steps that would be taken in such an experiment.

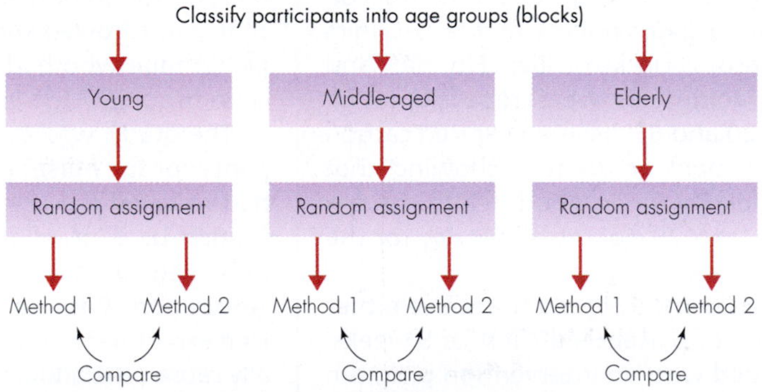

Figure 6.4 A blocked experiment to compare two memorization methods. Within each of three age groups (blocks), individuals are randomly allocated to the two methods

THOUGHT QUESTION 6.4 Students are sometimes confused by the reasons for blocking and for randomization. One method is used to control *known* sources of variability among the experimental units, and the other is used to control *unknown* sources of variability. Explain which is which, and provide examples illustrating these ideas.*

***HINT:** Consider the example in Figure 6.4. Two reasons for differing ability to memorize might be age and individual brain structure. Which one can be measured and used to assign people to blocks?

Repeated Measures

In a **repeated measures design** each experimental unit receives all treatments, ideally in a random order. This is an easy and efficient way to control for variation among individuals. We encountered this idea when we discussed how to compare driving ability under the influence of alcohol and marijuana and when sober. Each participating driver could be tested in all three conditions. The order of the three conditions should be randomized for each participant. For three conditions, there are six possible orderings, and it would be desirable to assign equal numbers of drivers to each possible ordering. This would prevent any one of the conditions from having an advantage due to when it was used compared to the other conditions.

Repeated measures designs are a special case of block designs, or with only two treatments, a special case of a matched-pairs design. Each individual can be thought of as a block, and the different time periods can be thought of as the experimental units. All treatments are assigned to each block (individual), in random order, with one treatment per time period. In this type of situation, the term *repeated measures design* is commonly used in the social sciences, instead of *block design*.

Design Terminology and Examples

An experiment is called a **completely randomized design** when treatments are randomly assigned to experimental units without using matched pairs or blocks. When matched pairs are used the experiment has a **matched-pair design,** and when blocks are used the experiment has a **randomized block design**.

Note that there is some overlap in the terminology used for these designs because a matched-pair design could also be considered to be a randomized block design, with each pair representing one block. Therefore, matched-pair designs are a subset of randomized block designs, in which there are only two treatments, each assigned only once within each block.

To help you distinguish these three types of experimental design, here are two examples of experiments that might be done by using them and how each design would be accomplished.

Experiment 1: The Effect of Caffeine on Swimming Speed

Suppose a researcher wants to know whether taking caffeine an hour before swimming affects the time it takes swimmers to complete a 1-mile swim and that 50 volunteers are available for the study.

- *Completely randomized design.* Randomly assign 25 volunteers to take a caffeine pill and the remaining 25 to take a placebo, have them all swim a mile, and compare the times for the two groups in the analysis.
- *Matched-pair design.* There are two ways in which this could be done.
  - *Method 1:* Have each of the 50 swimmers take both treatments, with adequate time, perhaps a week, between them. In week 1, randomly assign swimmers to take either a caffeine pill or a placebo. Have them swim a mile and record the time. The next week, give each swimmer the treatment he or she did not have the previous week, have them swim a mile, and record the time. For each swimmer, take the difference in the two times, and use those measurements for the analysis. (This is also called a repeated measures design.)
  - *Method 2:* Create 25 pairs of swimmers based on speed. Before assigning treatments, measure how long it takes each one to swim a mile. Match the two fastest swimmers as one pair, the next two as the second pair, and so on. A week later, randomly assign one swimmer in each pair to take a caffeine pill and the other to take a placebo. Measure their times to swim a mile. For each pair, take the difference in the two times, and use those measurements for the analysis.
- *Randomized block design.* This would be most appropriate if there were different types of swimmers. For instance, suppose that 10 of the volunteers were males who swim competitively, 20 were males who don't swim competitively, 8 were females who swim competitively, and the remaining 12 were females who don't swim competitively. Then each of these four groups would constitute a block, and half of

each block would be randomly assigned to take a caffeine pill while the other half would take a placebo. Each person would swim a mile. The times for caffeine versus placebo would be compared within each block, as well as across the four blocks combined.

Experiment 2: Advertising Strategies for Selling Coffee near Colleges

Suppose that a national chain of cafes wants to compare three methods for increasing sales at their cafes near college campuses. Each method involves placing a weekly advertisement in the campus newspaper. With Method 1, a coupon for a free cup of coffee is included in the ad. With Method 2, a coupon for a free cookie with any coffee drink purchase is included in the ad. The third method is to place the ad without any coupons. They place the ads for one full semester and measure how much sales change from the same semester the previous year.

- *Completely randomized design.* For this design, they would randomly choose one-third of the campuses in the study to receive each of the treatments, and the analysis would compare the change in sales for the three treatment groups.
- *Matched-pair design.* This design would not be possible if they want to compare all three treatments.
- *Randomized block design.* Decide what known factors contribute to differing sales across campuses. Group or "block" the campuses on the basis of those factors. For instance, they might categorize the campuses on the basis of size of the student body (small or large) and population of the town or city (small, medium, large) for a total of six blocks. Once the campuses have been blocked, randomly assign one-third of the campuses within each block to receive each treatment. Compare change in sales within each block, and then across blocks. Other blocking variables might include initial sales, distance from the campus, and whether or not there is a competing cafe in the neighborhood.

6.2 Exercises are on pages 212–213.

TECHNICAL NOTE

The Number of Units per Block

Often, although not always, the number of experimental units in a block will equal the number of treatments so that each treatment is assigned only once in each block. That allows as many sources of known variability as possible to be controlled. For the example of the effect of caffeine on swimming speed, there are two treatments, so blocks of size 2 (matched pairs) would be created. This could be accomplished by matching people on known variables such as sex, initial swim speed, usual caffeine consumption, age, and so on.

IN SUMMARY

Designing a Good Experiment

- Always use **randomization** to assign conditions or treatments to experimental units, or to assign the order of using treatments when units receive more than one treatment.
- **Replications**, the sample sizes per treatment, should be large enough to obtain suitably accurate estimates of treatment differences.
- **Matched-pairs** and **blocking** can be used to control known sources of variation in the response variable.
- **Control groups** and **placebo** groups are helpful for determining the effect of a treatment because these groups are dealt with in exactly the same way as the treatment group(s) except that they don't receive the active treatment.

6.3 Designing a Good Observational Study

For the purpose of establishing cause-and-effect links, observational studies start with a distinct *disadvantage* in comparison to randomized experiments. In observational studies, the researchers observe, but do not control, the explanatory variables. However, these studies have the possible *advantage* that participants are measured in their natural setting. The following case study, about a possible association between baldness and heart disease, is an example of a well-designed observational study.

CASE STUDY 6.4 Baldness and Heart Attacks

On March 8, 1993, *Newsweek* announced, "A really bad hair day: Researchers link baldness and heart attacks" (p. 62). The article reported that "men with typical male pattern baldness . . . are anywhere from 30 to 300 percent more likely to suffer a heart attack than men with little or no hair loss at all." (Male pattern baldness is the type affecting the crown or vertex and not the same thing as a receding hairline; it affects approximately one-third of middle-aged men.)

The report was based on an observational study conducted by researchers at Boston University School of Medicine. They compared 665 men who had been admitted to the hospital with their first heart attack to 772 men in the same age group (21 to 54 years old) who had been admitted to the same hospitals for other reasons. There were 35 hospitals involved, all in eastern Massachusetts and Rhode Island. Lesko, Rosenberg, and Shapiro reported the full results in the *Journal of the American Medical Association* (1993).

The study found that the percentage of men who showed some degree of male-pattern baldness was substantially higher for those who had had a heart attack (42%) than for those who had not (34%). Furthermore, when they used sophisticated statistical tests to ask the question in the reverse direction, they found that there was an increased risk of heart attack for men with any degree of male-pattern baldness. The analysis methods included adjustments for age and other heart attack risk factors. The increase in risk was more severe with increasing severity of baldness, after adjusting for age and other risk factors.

The authors of the study speculated that a third variable, perhaps a male hormone, might simultaneously increase the risk of heart attacks and the propensity for baldness. With an observational study such as this one, scientists can establish a connection, and they can then look for causal mechanisms in future investigations.

A later report using data gathered in the Harvard Physicians' Health Study provided additional evidence of an association between male-pattern baldness and heart attack risk (Lotufo et al., 2000). Baldness patterns and coronary heart disease incidence were recorded for the 22,071 physicians who participated in the study described in Case Study 1.6 about aspirin use and heart attack rates. After controlling for factors such as age, treatment assignment (aspirin or placebo), smoking habits, cholesterol levels, and hypertension, the researchers found an increased risk of heart disease for men with baldness.

Types of Observational Studies

Retrospective and Prospective Studies

Observational studies can be classified according to the relevant time frame of the data. In a **retrospective study**, the data are from the past. Participants may be asked to recall past events, or the researchers may use information that already has been recorded about the participants (medical records, for instance). In a **prospective study**, researchers follow participants into the future and record relevant events and variables. The prospective approach generally is a better procedure because people often do not remember past events accurately. A possible difficulty with a prospective study, however, is that participants may change their behavior because they know that it is being recorded.

Case-Control Studies

In a **case-control study**, "cases" who have a particular attribute or condition are compared to "controls" who do not. A distinguishing feature of a case-control study is that individuals are sampled within categories of the response variable (cases and controls). The idea is to compare the cases and controls to see how they differ on explanatory variables of interest. In medical settings, the cases usually are individuals who have been diagnosed with a particular disease. Researchers then identify a group of controls who are as similar as possible to the cases except that they don't have the disease. To achieve this similarity, researchers often use patients who have been hospitalized for other causes as the controls.

The first study described in Case Study 6.4 about baldness and heart disease is an example of a case–control study. In this example, men who had been admitted to the hospital with a heart attack were the cases, and men who had been admitted for other reasons were the controls. The cases and controls were compared on an explanatory variable of interest, which in Case Study 6.4 was the degree of male-pattern baldness. In this example, the response variable is whether a participant is a case or a control (heart attack or not). Figure 6.5 illustrates the steps that were taken in this study.

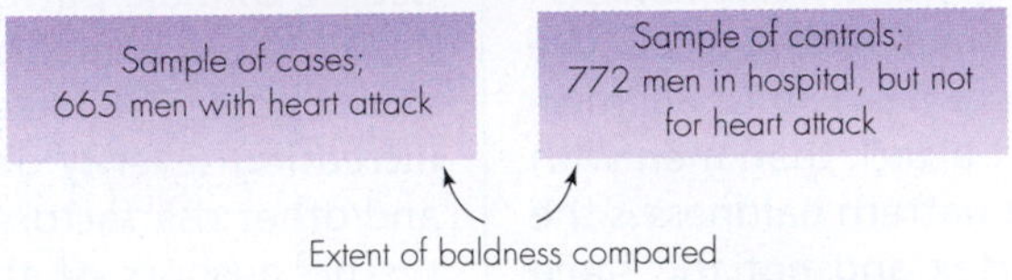

Figure 6.5 The case–control design in Case Study 6.4: Samples of male heart attack patients (cases) and other male hospital patients (controls) were compared for extent of baldness

Sometimes cases are matched with controls on an individual basis. That type of design is similar to a matched-pair experimental design. The analysis proceeds by first comparing the pair, and then summarizing over all pairs. Unlike a matched-pair experiment, the researcher does not randomly assign treatments within pairs; no treatments are given in a case–control study. For example, to identify whether left-handed people die at a younger age than right-handed people, researchers might match each left-handed case with a right-handed sibling as a control and compare their ages at death. Handedness obviously cannot be randomly assigned to the two individuals, so confounding factors (such as the fact that most tools are made for right-handed people) might be responsible for any observed differences.

Advantages of Case–Control Studies

Case-control studies have become increasingly popular in medical research, and with good reason. They do not suffer from the ethical considerations that are inherent in the random assignment of potentially harmful or beneficial treatments. In addition, case-control studies can be efficient in terms of time, money, and the inclusion of enough people with the disease.

Efficiency—In an exercise given in Chapter 4, researchers were interested in whether or not owning a pet bird is related to the incidence of lung cancer. Imagine trying to design an experiment to study this problem. You would randomly assign people to either own a bird or not and then wait to see how many in each group contracted lung cancer. The problem is that you would have to wait a long time, and even then, you would observe very few cases of lung cancer in either group. In the end, you may not have enough cases for a valid comparison.

A case-control study, in contrast, would identify a large group of people who had just been diagnosed with lung cancer (response variable), and then ask them whether or not they had owned a pet bird (explanatory variable). A similar control group (without lung cancer) would be identified and asked the same question. A comparison would then be made between the proportion of cases (lung cancer patients) who had birds and the proportion of controls who had birds.

Reducing Potential Confounding Variables—Another advantage of case-control studies over other observational studies is that the controls are chosen to try to reduce potential confounding variables. For example, in Case Study 6.4, suppose that it was simply the case that balding men were less healthy than other men and were therefore more likely to get sick in some way. An observational study that recorded only whether or not someone had baldness and whether or not he had experienced a heart attack would not be able to control for that fact. By using other hospitalized patients as the controls, the researchers were able to partially account for general health as a potential confounding factor.

You can see that careful thought is needed to choose controls that reduce potential confounding factors and do not introduce new ones. For example, suppose we want to know whether heavy exercise induces heart attacks, and as cases, we use people who were recently admitted to the hospital for a heart attack. We would certainly not want to use other newly admitted patients as controls. People who are sick enough to enter the hospital (for anything other than sudden emergencies) would probably not have been recently engaging in heavy exercise. When you read the results of a case-control study, you should pay attention to how the controls were selected.

6.3 Exercises are on pages 213–214.

IN SUMMARY Some Special Types of Observational Studies

- A **retrospective study** uses data from the past.
- A **prospective study** follows participants into the future and uses data from that future time.
- A **case-control study** compares individuals with a condition (cases) and individuals without the condition (controls) to determine whether cases and controls differ with respect to explanatory variables of interest.

6.4 Difficulties and Disasters in Experiments and Observational Studies

A number of complications can arise if researchers are not careful when they conduct and report experiments and observational studies. Some difficulties described here apply only to observational studies, some apply only to experiments, and some apply to both. We first list the possible problems and then describe each one in detail.

- Confounding variables and the implication of causation in observational studies
- Extending results inappropriately
- Interacting variables
- Hawthorne and experimenter effects
- Ecological validity and generalizability
- Using the past as a source of data

Confounding Variables and the Implication of Causation in Observational Studies

Probably the most common mistake made in media reports of research studies is to imply that a *cause-and-effect* relationship can be concluded from an observational study. When a link between the explanatory and response variables is demonstrated, it is tempting to assume that the differences in the explanatory variable *caused* the differences in the response variable. A conclusion of cause and effect may be justified when a randomized experiment is done but not when the data are from an observational study. With an observational study, it is difficult, perhaps impossible, to separate the effects of confounding variables from the effects of the main explanatory variable(s) of interest. Consequently, we are not able to know the extent to which a specific explanatory variable causes changes in the response variable.

DEFINITION The **Rule for Concluding Cause and Effect** is that cause-and-effect relationships can be inferred from randomized experiments but not from observational studies.

A partial solution for researchers who are carrying out an observational study is to measure possible confounding variables and take them into account in the analysis. For instance, if researchers are studying the connection between oral contraceptive use and blood pressure in women, they should measure and take into account other factors that are known to affect blood pressure, such as age, diet, and smoking habits. Another partial solution can be achieved when a case–control study is conducted by choosing the controls to be as similar as possible to the cases. There also may be other considerations that can help you to determine whether a causal link is plausible, such as a reasonable explanation of why cause and effect could occur in the particular situation. One part of the solution is up to you as a consumer of information: Don't be misled into thinking that a causal relationship exists when it may not.

Example 6.5 **Will Preventing Artery Clog Prevent Memory Loss?** An observational study based on a random sample of over 6000 older people (average age 70 when the study began) found that 70% of the participants did not lose cognitive functioning over time. One important finding, however, was that "those who have diabetes or high levels of atherosclerosis in combination with a gene for Alzheimer's disease are eight times more likely to show a decline in cognitive function." This finding led one of the researchers to comment that "that has implications for prevention, which is good news. If we can prevent atherosclerosis, we can prevent memory loss over time, and we know how to do that with behavior changes—low-fat diets, weight control, exercise, not smoking, and drug treatments" (Source: "Study: Age Doesn't Sap Memory," *Sacramento Bee*, Kathryn Doré Perkins, July 7, 1999, pp. A1 and A10).

But note that this conclusion assumes that the memory loss is *caused* by atherosclerosis. This conclusion is not justified on the basis of an observational study. Perhaps certain people have a genetic predisposition to atherosclerosis and to memory loss, in which case preventing the former condition with lifestyle changes will not reduce the memory loss. Or perhaps people who suffer from depression are more likely to eat poorly and not exercise due to their depression, so they have atherosclerosis and they also suffer from memory loss due to their depression. There are simply too many possible confounding factors that could be related to both atherosclerosis and memory loss to conclude that preventing atherosclerosis would prevent memory loss.

Extending Results Inappropriately

In interpreting both randomized experiments and observational studies, keep in mind the Fundamental Rule for Using Data for Inference: Available data can be used to make inferences about a much larger group *if the data can be considered to be representative with regard to the question(s) of interest.* Some observational studies are based on random samples, in which case the results can readily be extended to the population from which the sample was drawn. Example 6.5 is one such example; the study on aging and memory loss was part of the large Cardiovascular Health Study, sponsored by the National Heart, Lung and Blood Institute, part of the National Institutes of Health.

Many observational studies and almost all randomized experiments use convenience samples or volunteers. It is up to you to determine whether the results can be extended to any larger group for the question of interest. For example, the weight-lifting experiment in Case Study 6.2 used 11 girls and 32 boys who volunteered to participate in the study. There is no reason to suspect that these volunteers would differ from other children in terms of the differential effects of no training, moderate lifting, and heavy lifting measured in the study. Thus, although the sample consisted solely of volunteers, the results can probably be extended to other children in this age group.

Interacting Variables

Sometimes a second explanatory variable *interacts* with the principal explanatory variable in its relationship with the response variable. This can be a problem if the results of a study are reported without describing the effects of an **interacting variable**. Although the reported results may be true on average, they will differ in specific subgroups when interaction is present.

DEFINITION An **interacting variable** is an explanatory variable that affects the magnitude of the relationship between another explanatory variable and a response variable. The interaction is said to be between the two explanatory variables.

In Case Study 6.3, there was an interaction between the treatment (nicotine or placebo patch) and whether or not there were other smokers at home. The researchers measured and reported this interaction, which is illustrated in Figure 6.6. When there were no other smokers in the participant's home, the eight-week success rate was actually 58% for the nicotine patch users compared to 20% for the placebo patch group for a 38% difference between the nicotine and placebo patch groups. When there were other smokers in the home, the eight-week success rate for the nicotine patch was only 31%, but the success rate for the placebo users remained at about 20%, giving only an 11% difference between the two treatment categories. In other words, the difference between the nicotine and placebo groups was affected by whether or not other smokers were present in the home. Therefore, it would be misleading to merely report that 46% of the nicotine recipients had quit without also providing the information about how the results were affected by the presence of another smoker in the home.

Interaction has to do with the combined effect of two explanatory variables on a response, and it can occur whether the two interacting explanatory variables are related or not. In Case Study 6.3, for instance, treatments were randomly assigned so there was no correlation between treatment assignment and whether other smokers were present in the home. Because it was not related to treatment assignments, the presence (or absence) of other smokers in the home was not a confounding variable. A confounding variable, by definition, is related to both the primary explanatory variable and the response variable.

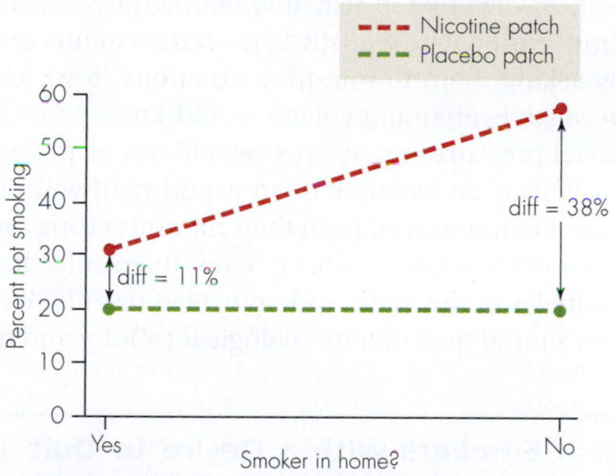

Figure 6.6 Interaction in Case Study 6.3: The difference between the nicotine and placebo patches is greater when there are no smokers in the home than when there are smokers in the home

Hawthorne and Experimenter Effects

We have already discussed the fact that a placebo can have a very strong effect on experimental outcomes because the power of suggestion is somehow able to affect the result. A similar difficulty is that participants in an experiment respond differently than they otherwise would, just because they are in the experiment. This is called the **Hawthorne effect** because it was first detected in the late 1920s in studying factory workers at the Western Electric Company's Hawthorne plant in Cicero, Illinois. (The phrase was not actually coined until much later; see French, 1953.)

The Hawthorne effect is a common problem in medical research. Many treatments have been observed to have a higher success rate in clinical trials than they do in actual practice. This may occur because patients and researchers are highly motivated to correctly carry out a treatment protocol in a clinical trial.

Related to these effects are numerous ways in which the experimenter can bias the results. These **experimenter effects** include recording the data erroneously to match the desired outcome, treating subjects differently on the basis of which condition they are receiving, and subtly making the subjects aware of the desired outcome. Most of these problems can be overcome by using double-blind designs and by including a placebo group or a control group that gets identical handling except for the active part of the treatment. Other problems, such as incorrect data recording, should be addressed by allowing data to be automatically entered into a computer when they are collected, if possible. Depending on the experiment, there may still be subtle ways in which experimenter effects can sneak into the results. You should be aware of these possibilities when you read the results of a study.

Example 6.6 **Dull Rats** In a classic experiment that was designed to test whether the experimenter's expectations could really influence the results, Rosenthal and Fode (1963) deliberately conned 12 experimenters. They gave each one five rats that had been taught to run a maze. They told six of the experimenters that the rats had been bred to do well (i.e., they were "maze bright") and told the other six that their rats were "maze dull" and should not be expected to do well. Sure enough, the experimenters who had been told that they had bright rats found learning rates far superior to those found by the experimenters who had been told they had dull rats. Hundreds of other studies have since confirmed the experimenter effect.

Ecological Validity and Generalizability

Suppose you want to compare three assertiveness training methods to find out which is most effective in teaching people how to say "no" to unwanted requests for their time. Would it be realistic to give participants the training and then measure the results by asking them to role-play situations in which they should say "no"? Probably not, because everyone involved would know that they were only role-playing. The usual social pressures to say "yes" would not be present.

This is an example of an experiment with little **ecological validity**. The variables have been removed from their natural setting and are measured in the laboratory or in some other artificial setting. Thus, the results do not accurately reflect the impact of the variables in the real world or in everyday life. Whenever you read the results of a study, you should question its ecological validity and its generalizability.

Example 6.7 **Real Smokers with a Desire to Quit** The researchers in Case Study 6.3, the nicotine patch study, did many things to help ensure ecological validity and generalizability. First, they used a standard intervention program available from and recommended by the National Cancer Institute instead of inventing their own, so that other physicians could follow the same program. Next, they used participants at three different locations around the country rather than in one community only, and they involved a wide range of ages (20 to 65). They included individuals who lived in households with other smokers as well as those who did not. Finally, they recorded numerous other variables (sex, ethnicity, education, marital status, psychological health, and so on) and checked to make sure these were not related to the response variable or the patch assignment.

Using the Past as a Source of Data

Retrospective observational studies can be particularly unreliable because they ask people to recall past behavior. Some studies, in which the response variable is whether or not someone has died, can be even worse because they rely on the memories of relatives and friends rather than the actual participants to measure explanatory and other variables. Retrospective studies also suffer from the fact that variables that confounded things in the past may no longer be similar to those that would currently be confounding variables, and researchers may not think to measure them (see Example 6.8).

If at all possible, prospective studies should be used. That is not always possible. For example, researchers who first considered the potential causes of AIDS or toxic-shock syndrome had to start with those who were ill and try to find common factors from their pasts. If possible, retrospective studies should use authoritative sources such as medical records rather than relying on memory.

Example 6.8 **Do Left-Handers Die Young?** Many years ago, a highly publicized study pronounced that left-handed people did not live as long as right-handed people (Coren and Halpern, 1991). In one part of the study, the researchers sent letters to next of kin for a random sample of recently deceased individuals, asking which hand the deceased had used for writing, drawing, and throwing a ball. The researchers found that the average age of death for those who had been left-handed was 66, while for those who had been right-handed, it was 75.

What the researchers failed to take into account was that in the early part of the 20th century, many children were forced to write with their right hands, even if their natural inclination was to be left-handed. Therefore, people who died in their 70s and 80s during the time of this study were more likely to be right-handed than were those who died in their 50s and 60s. The confounding factor of how long ago one learned to write was not taken into account. A better study would be a prospective one, following current left- and right-handers to see which group survived longer, but you can imagine the practical difficulties of conducting such a study. The participants could well outlive the researchers.

6.4 Exercises are on pages 214–215.

IN SUMMARY **Difficulties in Experiments and Observational Studies**

- **Confounding** is likely to affect the results of observational studies.
- Cause-and-effect conclusions are possible for randomized experiments but not for observational studies.
- The results of a research study should only be extended to the population that is represented by the participants in the study.
- **Interacting variables** may make the magnitude of a relationship be different for different subgroups.
- The **Hawthorne effect**, **experimenter effect**, lack of **ecological validity**, and relying on memories of past behavior can all bias results.

Key Terms

Introduction
cause-and-effect relationship, 189

Section 6.1
observational study, 189, 190, 195
experiment, 190,195
randomized experiment, 190,195
treatments, 190
unit, 190
experimental units, 190
subjects, 190
participants, 190
explanatory variable, 191
response variable, 191
outcome variable, 191
dependent variable, 191
independent variable, 191
confounding variable, 191
lurking variable, 191

Section 6.2
volunteers, 196
randomization, 196
replication, 198
replication in science, 198
control groups, 198
placebo, 198
placebo effect, 198
blinding, 198
single-blind, 198
double-blind, 198
double dummy, 199
matched-pair design, 200, 201
block design, 200
blocks, 200
repeated measures design, 201
completely randomized design, 201
randomized block design, 201

Section 6.3
retrospective study, 203, 205
prospective study, 203, 205
case–control study, 203, 205

Section 6.4
Rule for Concluding Cause and Effect, 206
interacting variable, 207, 209
Hawthorne effect, 208
experimenter effect, 208
ecological validity, 208

Exercises
effect modifier, 215

In Summary Boxes

The Language of Research Studies, 195
Designing a Good Experiment, 202
Some Special Types of Observational Studies, 205
Difficulties in Experiments and Observational Studies, 209

Exercises

◆ Denotes that the dataset is available on the companion website, http://www.cengage.com/statistics/Utts4e, but is not required to solve the exercise.

Bold exercises have answers in the back of the text.

Section 6.1

Skillbuilder Exercises

6.1 In each situation, indicate whether you think an observational study or a randomized experiment would be used. Explain why in each case.

a. A teacher wants to compare the grade point averages of female students who are in sororities to the grade point averages of female students who are not in sororities.

b. A doctor wants to compare the effectiveness of two allergy medications.

c. A social psychologist wants to determine whether restaurant servers will get better tips if they introduce themselves by name to the people they serve.

d. A child psychologist wants to find out whether there is an association between the amount of television 6-year-old children watch and how often they bully other children in school.

6.2 In each part, identify the response variable and the explanatory variable in the relationship between the two given variables.

a. Daughter's height and mother's height.

b. Age and weight for children between the ages of 3 and 10 years old.

c. Opinion about the death penalty for persons convicted of murder and political party membership (Democrat, Republican, and so on).

d. Smoking habits and performance on a test of lung function.

6.3 Refer to Exercise 6.1. In each part, specify the explanatory variable and the response variable.

6.4 Suppose that a statistics teacher wants to know whether the number of hours students spend studying in a group affects the final course grade. In each part, explain whether the research method described is a randomized experiment or an observational study.

a. Each student keeps a log of the hours he or she spends studying in a group and reports the total after the course is completed.

b. Students are randomly assigned to study groups. The teacher tells each group how often to meet. This varies from 1 hour the day before each exam to 2 hours per week.

c. Students voluntarily join groups based on how often the groups will meet. The groups are designated as meeting weekly, meeting only before exams, or meeting whenever enough members feel that it is necessary.

6.5 Remember that a confounding variable affects the response variable and is related to the explanatory variable. For each of the following situations, explain how the given confounding variable meets these criteria.

a. *Response variable* = Math skills of children aged 6 to 12 years old; *explanatory variable* = shoe size; *confounding variable* = age.

b. *Response variable* = Whether or not a student gets a cold during a school term; *explanatory variable* = whether or not the student procrastinates in doing assignments; *confounding variable* = number of hours per week the student spends socializing.

6.6 *This is also Exercise 1.16.* Suppose that an observational study showed that students who got at least seven hours of sleep performed better on exams than students who didn't. Which of the following are possible confounding variables, and which are not? Explain why in each case.

a. Number of courses the student took that term.

b. Weight of the student.

c. Number of hours the student spent partying in a typical week.

6.7 *This is also Exercise 1.13.* For each of the examples given here, decide whether the study was an observational study or a randomized experiment.

a. A group of students who were enrolled in an introductory statistics course were randomly assigned to take a web-based course or to take a traditional lecture course. The two methods were compared by giving the same final examination in both courses.

b. A group of smokers and a group of nonsmokers who visited a particular clinic were asked to come in for a physical exam every 5 years for the rest of their lives to monitor and compare their health status.

c. CEOs of major corporations were compared with other employees of the corporations to see if the CEOs were more likely to be the first child born in their families than were the other employees.

6.8 Remember that a confounding variable affects the response variable and is related to the explanatory variable. For each of the following situations, explain how the given confounding variable meets these criteria.

a. *Response variable* = Child's IQ at age 10; *explanatory variable* = whether mother smokes or not; *confounding variable* = mother's educational level.

b. *Response variable* = Number of missing or decayed teeth a child has; *explanatory variable* = amount of past exposure to lead; *confounding variable* = number of visits to a dentist the child has made in the past 3 years.

6.9 Specify what an individual unit is in each of the following studies. Then specify what two variables were measured on each unit.

a. A study finds that college students who often procrastinate tend to be sick more often than students who do not procrastinate.

b. A study finds that sport utility vehicles (SUVs) made by one car manufacturer tend to be more heavily damaged in a crash test than SUVs made by a second car manufacturer.

6.10 *This is also Exercise 1.14.* For each of the studies described, explain whether the study was an observational study or a randomized experiment.

a. A group of 100 students was randomly divided, with 50 assigned to receive vitamin C and the remaining 50 to receive a placebo, to determine whether vitamin C helps to prevent colds.

b. A random sample of patients who received a hip transplant operation at Stanford University Hospital during 2000 to 2010 will be followed for 10 years after their operation to determine the success (or failure) of the transplant.

c. Volunteers with high blood pressure were randomly divided into two groups. One group was taught to practice meditation and the other group was given a low-fat diet. After 8 weeks, reduction in blood pressure was compared for the two groups.

General Section Exercises

6.11 For the research scenarios in parts (a) and (b), explain whether a randomized experiment could be used.

a. To study the relationship between long-term practice of meditation and blood pressure.

b. To determine whether a special training program improves scores on a standard college admissions test.

c. For each of parts (a) and (b) of this exercise, state the explanatory variable and the response variable.

d. For each of parts (a) and (b) of this exercise, give an explanation of a possible confounding variable that could be present if an observational study were to be done.

6.12 For the research scenarios in parts (a) and (b), explain whether a randomized experiment could be used.

a. To compare two programs to reduce the number of commuters who drive to work: providing discount coupons for the bus, or providing shuttle service for people who need to run errands during the workday.

b. To study the relationship between age and opinion about whether marijuana should be legal.

c. For each of parts (a) and (b) of this exercise, state the explanatory variable and the response variable.

d. For each of parts (a) and (b) of this exercise, give an explanation of a possible confounding variable that could be present if an observational study were done.

6.13 A news article by Reuters on October 6, 1998, reported:

> In a new study, Dr. Matti Uhari and colleagues at the University of Oulu in Finland randomly gave 857 healthy children in daycare centers xylitol in syrup, gum, or a lozenge form, or a placebo gum or syrup in five doses per day for 3 months. According to a report in the October issue of the journal *Pediatrics* [Vol. 102, pp. 879–884, 971–972, 974–975], the incidence of ear infections was reduced by 40% in children given xylitol chewing gum, 30% in those given syrup and 20% in those given lozenges when compared to children given a placebo.

a. Is this an observational study or a randomized experiment? Explain.

b. What are the explanatory and response variables?

c. Are confounding variables likely to be a problem in this study? Explain.

6.14 A news article on the *Reuters Health* website, December 18, 1998, reported that:

> A study of over 1200 people over age 65 showed that the "owls"—those who go to sleep after 11 p.m. and rise after 8 a.m.—tend to be as healthy and intelligent as "larks"—those who go to bed before 11 p.m. and rise before 8 a.m.—according to a report in the December 19/26 issue of the *British Medical Journal* [Vol. 317, pp. 1675–1677]. The researchers also found that owls tend to have higher average income than the early birds.

a. Do you think this is based on an observational study or a randomized experiment? Explain.
b. It is mentioned that the "owls" have higher average income than the "larks" (early birds). Explain whether or not income is a possible confounding variable in this study.
c. Give an example of a possible confounding variable that is not mentioned.
d. Draw a figure similar to Figure 6.2 (p. 194), illustrating the steps in this study.

6.15 Case Study 1.5 in Chapter 1 was about the association between blood pressure and frequency of participating in religious activities. Why was this study conducted as an observational study instead of an experiment?

Section 6.2

Skillbuilder Exercises

6.16 An experiment will be done to determine whether college students learn information better when listening to rap music, when listening to classical music, or when not listening to music. In the study, each participant will read a chapter from a history textbook and then take a test about the material. Ninety students will participate.

a. Briefly describe how a randomized experiment would be performed in this situation.
b. For the experiment you described in part (a), draw a figure similar to Figure 6.3 (p. 195) illustrating the steps of the study.
c. Describe the replication in this study.

6.17 Suppose that we want to determine whether people who want to lose weight would lose more in three months by dieting or by exercising. Sixty people have volunteered to participate. Briefly describe how a randomized experiment would be done in this situation.

6.18 Two weight-training regimens designed to improve arm strength will be compared. The response variable will be the improvement in arm strength at the end of six weeks of training. Forty individuals—20 males and 20 females—will participate in the study.

a. Describe how a completely randomized experiment would be done in this situation.
b. Describe how a matched-pair experiment could be done in this situation. Be specific about what constitutes a matched pair and how treatments would be assigned.
c. Describe how a block experiment could be done in this situation. Be specific about how blocks would be formed and how weight-training methods would be assigned.

6.19 Twenty students agreed to participate in a study on colds. Ten were randomly assigned to receive vitamin C, and the remaining 10 received a tablet that looked and tasted like vitamin C but in fact contained only sugar and flavoring. The students did not know whether they were taking vitamin C or not, but the investigators did. The students were followed for 2 months to see who came down with a cold and who didn't. Explain whether each of the following terms applies to this randomized experiment.

a. Placebo.
b. Single-blind.
c. Double-blind.
d. Matched pairs.
e. Repeated measures.
f. Replication.

6.20 Case Study 6.3 (p. 199) was about an experiment done to determine whether nicotine patches helped smokers to quit smoking. For that experiment, draw a figure similar to Figure 6.3 (p. 195) illustrating the steps of the study.

General Section Exercises

6.21 Twenty grocery stores will participate in an experiment to compare the effectiveness of two methods for displaying a product. The response variable will be the number of items of the product sold during a 1-week period.

a. Describe a completely randomized design for this experiment.
b. Describe a matched-pairs design for this experiment.

6.22 Researchers want to design a study to compare topical cream (applied to the skin) and a drug taken in capsule form for the treatment of a certain type of skin rash. Sixty volunteers who have the rash agree to participate in the study.

a. If each volunteer is given only one treatment, can the study be double-blind? Explain.
b. If each volunteer is given only one treatment, can the study be single-blind? Explain.
c. Could a double-blind, double-dummy study be done? If not, explain why. If so, explain how it would be done.

6.23 Twenty volunteers aged 40 to 65 years old will participate in an experiment to compare two methods of memorizing information. The age variation in the participants concerns the researchers because memorization skills decrease with age.

a. Describe how the study could be done using a matched-pair design.
b. Explain why a matched-pair design is a good idea in this situation.

6.24 A school district has 20 elementary schools, and each school has 12 classes that can be used for a study.

a. Using schools as blocks, describe a randomized block design to compare three teaching methods.
b. Explain why schools should be used as blocks.

6.25 One hundred volunteers agree to participate in an experiment. There are four treatments, each of which will be assigned to 25 participants. Using Example 6.3 (p. 197) for guidance, explain how you could use statistical software such as Minitab to randomly assign participants to treatments.

6.26 Refer to Case Study 6.3 (p. 199), "Quitting Smoking with Nicotine Patches," which used a completely randomized design. Explain how the experiment could have been done by using a randomized block design instead.

6.27 Explain whether each of the following experiments was single-blind, double-blind, or neither.

a. An electric company wanted to know whether residential customers would use less electricity during peak hours if they were charged more during those hours. One hundred customers were randomly selected to receive the special time-of-day rates, and they were matched with 100 customers who were similar in terms of past electrical use and household size. Special meters were installed for all 200 customers, and at the end of three months, the usage during peak hours was compared for each pair. The technician who read the meters did not know who had which rate plan.

b. To test an herbal treatment for depression, 100 volunteers who suffered from mild depression were randomly divided into two groups. Each person was given a month's supply of tea bags. For one group, the tea contained the herb mixed with a spice tea; for the other group, the bags contained only the spice tea. Participants were not told which type of tea they had and were asked to drink one cup of the tea per day for a month. At the end of the month, a psychologist evaluated them to determine whether their mood had improved. The psychologist did not know who had the tea with the herbal ingredient added.

c. To compare three packaging designs for a new food, a manufacturer randomly selected three grocery stores in each of 50 cities. The three designs were randomly assigned to the three stores in each city, and sales were compared for a 2-month period.

6.28 A fast-food chain sells its burgers alone or as part of a "value meal" that includes fries and a drink. They know that some customers are health-conscious. They want to do an experiment to determine whether the proportion of customers choosing the meal would increase if they offered baby carrots with the meal as an alternative to fries. They will compare three treatments: (1) status quo, (2) offering carrots as an alternative but with no advertising, and (3) offering carrots as an alternative and advertising this option in the local area. They have restaurants in three types of areas: cities, suburban areas, and along major highways.

a. Explain how they would conduct this experiment using a completely randomized design.

b. Explain how they would conduct this experiment using a randomized block design.

c. Explain why it makes sense to use the type of area in which a restaurant is located as a block determinant in this experiment.

d. Explain why a matched-pair design could not be used for this experiment.

6.29 Refer to Exercise 6.27. In each case, designate the explanatory variable and the response variable.

6.30 Refer to Exercise 6.27. In each case, explain whether the experiment was a matched-pair design, a block design, or neither.

6.31 Refer to Exercise 6.27. In each case, explain whether a control group was used and whether a placebo treatment was used.

6.32 Echinacea is an herb that may help to prevent colds and flu. Explain how researchers could conduct a double-blind, double-dummy experiment to compare echinacea tea with chewable vitamin C pills for effectiveness in preventing colds.

Section 6.3

Skillbuilder Exercises

6.33 A teacher wants to determine how class attendance during the term affects grades on the final exam.

a. Describe how the data would be collected in a prospective observational study.

b. Describe how the data would be collected in a retrospective observational study.

c. Refer to parts (a) and (b). Give one disadvantage for each method of collecting the data in this situation.

6.34 Suppose you want to know whether men or women spend more time talking on a cell phone during a week's time.

a. Describe how you would collect the data in a retrospective observational study.

b. Describe how you would collect the data in a prospective observational study.

c. Refer to parts (a) and (b). Give one disadvantage for each method of collecting the data in this situation.

6.35 Students whose undergraduate major was economics were followed for 10 years after graduation to study the number of different jobs they took. Explain whether each of the following terms applies to this observational study.

a. Prospective study.

b. Retrospective study.

c. Case-control study.

6.36 Students who had meningitis were matched with students without meningitis using sex, undergraduate (or graduate) status, and college. The students' recent activities were examined to discover risk factors for meningitis. Explain whether each of the following terms applies to this observational study.

a. Prospective study.

b. Retrospective study.

c. Case-control study.

6.37 A medical researcher wants to know how lifetime sun exposure affects the risk of skin cancer for persons over 50 years old. Describe how data might be collected for a retrospective case-control study.

General Section Exercises

6.38 Suppose researchers were interested in determining the relationship, if any, between the use of cellular telephones and the incidence of brain cancer. Would it be better to use a randomized experiment, a case-control study, or an observational study that did not use cases and controls? Explain.

6.39 A study will be done to determine the relationship, if any, between where a college student lives (on-campus or off-campus) and grade point average. Would it be better to use a randomized experiment or an observational study? Explain.

6.40 Find an example of an observational study in the news. Determine whether or not it was a case-control study and whether it was prospective or retrospective. Specify the explanatory and response variables, and explain whether confounding variables were likely to be a major problem in interpreting the results. Be sure to include the news article with your response.

6.41 A story at ABCNews.com ("Pet Contact" by Rita Rubin, March 17, 1998) reported that Karen Allen, a researcher at the University of Buffalo, found that couples who own cats or dogs have more satisfying marriages and are less stressed out than those who don't own pets. Pet owners also have more contact with each other and with other people. Allen compared 50 couples who owned either cats or dogs with 50 pet-free couples. The volunteers completed a standard questionnaire assessing their relationship and how attached they were to their pets. They also kept track of their social contacts over a 2-week period. To see how the couples responded to stress, Allen monitored their heart rates and blood pressure while they discussed sore subjects. Pet-owning couples started out with lower blood pressure readings than the others, and their numbers didn't rise as much when they argued.

a. Was this a case-control study, a randomized experiment, or an observational study that was not case-controlled?
b. What are the explanatory and response variables in this study?
c. Give an example of a possible confounding variable.
d. Draw a figure illustrating the steps in this study, similar to Figure 6.2 (p. 194).

6.42 Reuters (June 24, 1997) reported on a study published in the *Journal of the American Medical Association* (1997, vol. 277, pp. 1940–1944) in which researchers recruited 276 volunteers (aged 18 to 55) and used nose drops to infect them with a cold virus. The volunteers were then quarantined and observed to see whether they came down with a cold or infection. They also answered questions about their social lives, including the number of different types of contacts they had (family, work, community, religious groups, and so on). Reuters reported that "those with one to three types of social contacts were 4.2 times as likely to come down with cold symptoms and signs of infection compared to those with six or more contacts."

a. Explain why this is an observational study, not a randomized experiment.
b. What were the explanatory and response variables of interest to the researchers?
c. The researchers in this study took an unusual step to help reduce the impact of confounding variables. Explain what they did and why it might have helped to reduce the impact of confounding variables.

Section 6.4

Skillbuilder Exercises

6.43 Explain why it is preferable, whenever possible, to conduct a randomized experiment rather than an observational study.

6.44 *This is also Exercise 1.38.* A headline in a major newspaper read, "Breast-Fed Youth Found to Do Better in School."

a. Do you think this statement was based on an observational study or a randomized experiment? Explain.
b. Given your answer in part (a), which of these two alternative headlines do you think would be preferable: "Breast-Feeding Leads to Better School Performance" or "Link Found Between Breast-Feeding and School Performance"? Explain.

6.45 *This is also Exercise 1.25.* An article in *Science* magazine (Service, 1994) discussed a study comparing the health of 6000 vegetarians and a similar number of their friends and relatives who were not vegetarians. The vegetarians had a 28% lower death rate from heart attacks and a 39% lower death rate from cancer, even after the researchers accounted for differences in smoking, weight, and social class. In other words, the reported percentages were the differences remaining after adjusting for differences in death rates due to those factors.

a. Is this an observational study or a randomized experiment? Explain.
b. On the basis of this information, can we conclude that a vegetarian diet causes lower death rates from heart attacks and cancer? Explain.
c. Give an example of a potential confounding variable in this situation, and explain what it means to say that it is a confounding variable.

6.46 For each of the following, explain whether the study summary describes an interacting variable or a confounding variable.

a. A comparison of the mean grade point averages of male and female students finds more difference for fraternity and sorority members than for students who are not fraternity and sorority members.
b. A study finds that wine drinkers tend to have better health than beer drinkers but also finds that wine drinkers tend to smoke less than beer drinkers.
c. It is found that drawing a happy face on a restaurant bill increases the mean tip percentage for female servers but does not increase the mean tip percentage for male servers.

6.47 Researchers have found that women who take oral contraceptives (birth control pills) are at higher risk of having a heart attack or stroke and that the risk is substantially higher if a woman smokes. In investigating the relationship between taking oral contraceptives (the explanatory variable) and having a heart attack or stroke (the response variable), would smoking be called a confounding variable or an interacting variable? Explain.

6.48 A new medication is observed to cause a weight loss in women but a weight gain in men. Identify what the response variable, explanatory variable, and interacting variable are in this situation.

6.49 In a study on worker productivity done in 1927 in Cicero, Illinois, it was observed that productivity improved when the lighting was increased. However, it was observed that productivity also improved when the lighting was decreased. What term is used to describe the fact that productivity increased simply because the subjects knew they were in an experiment?

General Section Exercises

6.50 An investigator believes that a new medication will help to reduce anxiety and that a placebo will not. In a single-blind study (patients are blind), she finds that the group taking the new medication had reduced anxiety while the placebo group did not. Explain how the experimenter effect may have played a role in the results.

6.51 An experiment on elephants in captivity was done to find out whether breeding would occur more often when the elephants were kept together in a herd or when a male was isolated with a female. Explain how ecological validity affects the ability to generalize these results to wild elephants.

6.52 In a sample of 500 college students aged 18 to 21, it was found that offering discount coupons for restaurant meals substantially increased the likelihood that the students would eat at those restaurants. Explain how "extending results inappropriately" may be a problem in the interpretation of this study.

6.53 A retrospective study of 517 veterans who never smoked was done to determine whether there was an association between lung disease and exposure to workplace gases, dust, and fumes (Clawson, *Stanford Report*, November 1, 2000). In this study, subjects were asked whether or not they remembered being exposed to gases, dust, or fumes in the workplace. A concern expressed by the investigator was that "patients who have histories of lung diseases may be more likely to recall a history of exposure—whether accurate or not—than subjects reporting good health." Discuss whether each of the following is likely to be a problem in this study.

a. Relying on memory.
b. Confounding variable.
c. Hawthorne effect.

6.54 Refer to Exercise 6.41, describing a study about pet ownership and marriage. Explain whether each of the following is likely to be a problem for that study:

a. Confounding variables and the implication of causation in observational studies.
b. Hawthorne and experimenter effects.
c. Ecological validity and generalizability.

6.55 Refer to Exercise 6.27, in which three randomized experiments are described. For each of the experiments described, pick one of the "difficulties and disasters" described in Section 6.4, and explain how it might be a problem in that experiment.

6.56 Is the experimenter effect more likely to be a problem in a study that is double-blind, single-blind, or not blind at all? Explain.

6.57 Pick the two "difficulties and disasters" that are most likely to be a problem in the following study, and explain why each would be a problem: A researcher at a large medical clinic wants to determine whether a high-fat diet in childhood is more likely to result in heart disease in later life. As a case-control study, she gives a questionnaire about childhood diet to 500 patients with heart disease and 500 patients without heart disease matched by age and sex.

6.58 Refer to Exercise 6.57, describing a case-control study relating childhood diet and heart disease. Comment on the extent to which each of the following is likely to be a problem.

a. Confounding variables and the implication of causation in observational studies.
b. Extending results inappropriately.
c. Interacting variables.
d. Hawthorne and experimenter effects.
e. Ecological validity and generalizability.
f. Relying on memory or secondhand sources.

6.59 A categorical interacting variable defines subgroups for which the effect of the explanatory variable on the outcome variable differs. For instance, as explained in the text, for the nicotine patch experiment described in Case Study 6.3, one interacting variable was whether or not there were other smokers at home. When the difference in subgroups is thought to hold for the population, the subgroup variable is sometimes called an **effect modifier**. Explain why this term makes sense, using the context of the smokers at home in the nicotine experiment. For instance, what "effect" is being modified?

6.60 Pick the two "difficulties and disasters" that are most likely to be a problem in each of the following studies, and explain why they would be a problem.

a. To compare marketing methods, a marketing professor randomly divides a large class into three groups and randomly assigns each group to watch one of three television commercials. The students are then asked to rate how likely they would be to buy the advertised product.
b. Volunteers are recruited through a newspaper article to participate in a study of a new vaccine for hepatitis C, which is transmitted by contact with infected blood and is a particular problem for intravenous drug users. The article specifically requests both intravenous drug users and medical workers to volunteer, assuring them confidentiality and offering them free medical care for the life of the study. Volunteers are then randomly assigned to receive either a placebo or the vaccine, and a year later they are tested to see whether they have the disease.

Chapter Exercises

Exercises 6.61 to 6.68 refer to a study published in the Journal of the American Medical Association *and reported in the* Sacramento Bee *on January 26, 2000 ("Study Ties Hormone Replacement, Cancer" by Shari Roan, p. A1). A sample of 46,355 postmenopausal women were studied for 15 years and asked about their use of hormone therapy and whether or not they had breast cancer; 2082 developed the disease. The article reported that "for each year of combined [estrogen and progestin] therapy, a woman's risk of breast cancer was found to increase by 8 percent compared to a 1 percent increase in women taking estrogen alone."*

6.61 Do you think this research was an observational study or a randomized experiment? Explain.

6.62 Discuss whether this research is the following:

a. A case-control study.
b. A retrospective study or a prospective study.

6.63* What are the explanatory and response variables for this study? Are they quantitative or categorical? If they are categorical, what are the categories?

6.64* To whom are the women taking combined therapy and estrogen alone being compared when the increased risks of 8% and 1% are computed?

6.65* Near the end of the *Sacramento Bee* article, it was noted that "the NCI researchers analyzed their results by the woman's weight. . . There was no increased breast cancer risk in heavier women." Which of the "Difficulties and Disasters" in Section 6.4 is represented by this statement? Explain.

6.66* Assuming that the women and their physicians made the decision about which treatment to pursue (combined hormones, estrogen alone, or no hormones), discuss the possibility of confounding variables in this study.

6.67* One drawback of the study quoted in the *Sacramento Bee* was that "the women in the study may not resemble women today, because many women now take lower doses of progestin than was used during the follow-up years of the study (1979–95)." Which of the "difficulties and disasters" in Section 6.4 is represented by this statement? Explain.

6.68* Another drawback of the study quoted in the *Sacramento Bee* was that "people who develop breast cancer are much more likely to remember whether they took hormones compared to women who don't develop the disease." Which of the "difficulties and disasters" in Section 6.4 is represented by this statement? Explain.

Exercises 6.69 to 6.73 refer to a study described in a Sacramento Bee *article titled "Much Ado Over Those Bad Hairdos" (January 26, 2000, p. A21). Here is part of the article:*

> Researchers surveyed 60 men and 60 women from 17 to 30, most of them Yale students. They were separated into three groups. One group was questioned about times in their lives when they had bad hair. The second group was told to think about bad product packaging, like leaky containers, to get them in a negative mind-set. The third group was not asked to think about anything negative. The groups were tested for self-esteem and self-judgment, and the bad hair group scored lower than the others.

6.69 Was this an observational study or a randomized experiment? Explain.

6.70 What were the explanatory and response variables for this study? Were they categorical variables or quantitative variables?

6.71 Discuss whether this study could have been blind or double-blind, and whether you think it was blind or double-blind.

6.72 Discuss each of the following "difficulties and disasters" in the context of this research.

a. Ecological validity.
b. Extending results inappropriately.
c. Experimenter effect.

6.73 The article also reported that "contrary to popular belief men's self-esteem may take a greater licking than women's when their hair just won't behave. Men were likely to feel less smart and less capable when their hair stuck out, was badly cut or otherwise mussed." In the context of this quote, is male/female a confounding variable, an interacting variable, or neither?

6.74 Refer to Example 6.3 in which children were randomly assigned to three treatment groups. Suppose now that 60 children will be randomly assigned to these three groups. Describe how you could carry out the random assignments.

6.75 In this chapter, you learned that cause and effect can be concluded from randomized experiments but generally not from observational studies. Why don't researchers simply conduct all studies as randomized experiments rather than observational studies?

6.76 Refer to the study reported in Example 2.2, relating the use of nightlights in childhood and the incidence of subsequent myopia.

a. Was the research based on an observational study or a randomized experiment? Explain.
b. Which one of the "difficulties and disasters" in Section 6.4 do you think is most likely to be a problem in this research? Explain.

6.77 Refer to Case Study 6.2 (p. 195), "Kids and Weight Lifting." Did the experiment described there use a completely randomized design, a matched-pair design, or a randomized block design? Explain.

6.78 Find an example of an observational study in the news. Answer the following questions about it. Be sure to include the news article with your response.

a. What are the explanatory and response variables, or is this distinction not possible?
b. Briefly describe how the study was done. For instance, was it a case-control study, a prospective or retrospective study?
c. Discuss possible confounding variables. For instance, were any possible confounding variables included in the study and news report? Can you think of possible confounding variables that were not mentioned?
d. Pick one of the "Difficulties and Disasters" in Section 6.4 and discuss how it applies to the study.

6.79 Is it possible for each of the following to be used in the same study (on the same units)? Explain or give an example of such a study.

a. A placebo and a double-blind procedure.
b. A matched-pair design and a retrospective study.
c. A case-control study and random assignment of treatments.

6.80 Find an example of a randomized experiment in the news. Answer the following questions about it. Be sure to include the news article with your response.

a. What are the explanatory and response variables? What relationship was found, if any?
b. What treatments were assigned? Was a control group or placebo used?
c. Was the study a matched-pair design, a block design, or neither? Explain.
d. Was the study single-blind, double-blind, or neither? Explain.

*See scenario on previous page.

6.81 Specify what an individual "unit" is in each of the following studies. Then specify what two variables were measured on each unit.

a. A study found that tomato plants raised in full sunlight produced more tomatoes than did tomato plants raised in partial shade.

b. A study found that gas mileage was higher for automobiles when the tires were inflated to their maximum possible pressure than when the tires were underinflated.

c. Ten randomly selected classrooms in a school district were assigned to have a morning fruit snack break, and another ten classrooms of the same grade level in similar schools were measured as a control group. The number of children who performed better than average on standardized tests was measured for each classroom.

6.82 Refer to Exercise 6.81 and answer these questions.

a. For each study, on the basis of the information given, is it clearly a randomized experiment? If not, explain what additional information would make it clear that the study was a randomized experiment rather than an observational study.

b. For each study, discuss whether or not matched pairs were used. If it is not clear, explain what additional information would make it clear.

c. Discuss whether any of the studies would suffer from each of the following: the Hawthorne effect, interacting variables, ecological validity.

6.83 Refer to the study in Example 6.8 that compares ages at death for left-handed and right-handed people. Draw a figure similar to one of Figures 6.2 through 6.5 that illustrates the steps for Example 6.8.

6.84 Refer to Case Study 1.6, in which physicians were randomly assigned to take aspirin or a placebo, and heart attack rates were compared. Draw a figure similar to one of Figures 6.2 through 6.5 that illustrates the steps for Case Study 1.6.

6.85 Explain whether it is possible for a variable to be both a confounding variable and an interacting variable.

6.86 A study was done (fictional) to compare the proportion of children who developed myopia after sleeping with and without a nightlight. The study found that the results differed based on whether at least one parent suffered from myopia by age 20. The percents of children suffering from myopia were as follows:

	Parent(s) Had Myopia	Parent(s) Did Not Have Myopia
Slept with:	**% with Myopia**	**% with Myopia**
No Light	10%	10%
Some Light	50%	30%

a. Identify the explanatory variable, the response variable, and an interacting variable.

b. Draw a figure similar to Figure 6.6 (p. 207), illustrating the interaction in this study.

c. Write a few sentences that would be understood by someone with no training in statistics explaining the concept of interaction in this study.

6.87 Refer to Example 6.6 (p. 208), "Dull Rats," which was done by using a completely randomized design.

a. What were the treatments in this experiment?

b. Were the experimental units the 60 individual rats or the 12 individual experimenters? Explain.

c. Explain how the experiment could have been done using a matched-pair design instead.

6.88 Give an example of an observational study, and explain the difference between a confounding variable and a lurking variable in the context of your example.

6.89 Explain whether a variable can be:

a. Both a confounding variable and a lurking variable.

b. Both a response variable and a confounding variable.

c. Both an explanatory variable and a dependent variable.

6.90 Refer to Exercise 6.27, in which three randomized experiments are described. In each case, explain the extent to which you think the results from the sample in the experiment could be extended to a larger population.

6.91 Explain why confounding variables are more of a problem in observational studies than in randomized experiments. Give an example.

6.92 Refer to the first observational study described in Case Study 6.1 (p. 193). In the study, a link was found between tooth decay and exposure to lead for 24,901 children. Give an example of a possible confounding variable in addition to those described in the text. Explain why it could be a confounding variable.

6.93 In an experiment done at an English university, 64 students held their hands in ice water for as long as they could while repeating a swear word of their choice. The same students also held their hands in ice water for as long as they could while repeating a neutral word. For each condition, the researchers measured pain tolerance, the length of time of submersion, and physiological responses such as heart rate. The result was that the students had better pain response while swearing (Stephens et al., 2009).

a. Explain how this experiment fits the definition of a repeated measures design.

b. What randomization should be used in this experiment?

c. What is the replication in this experiment?

7

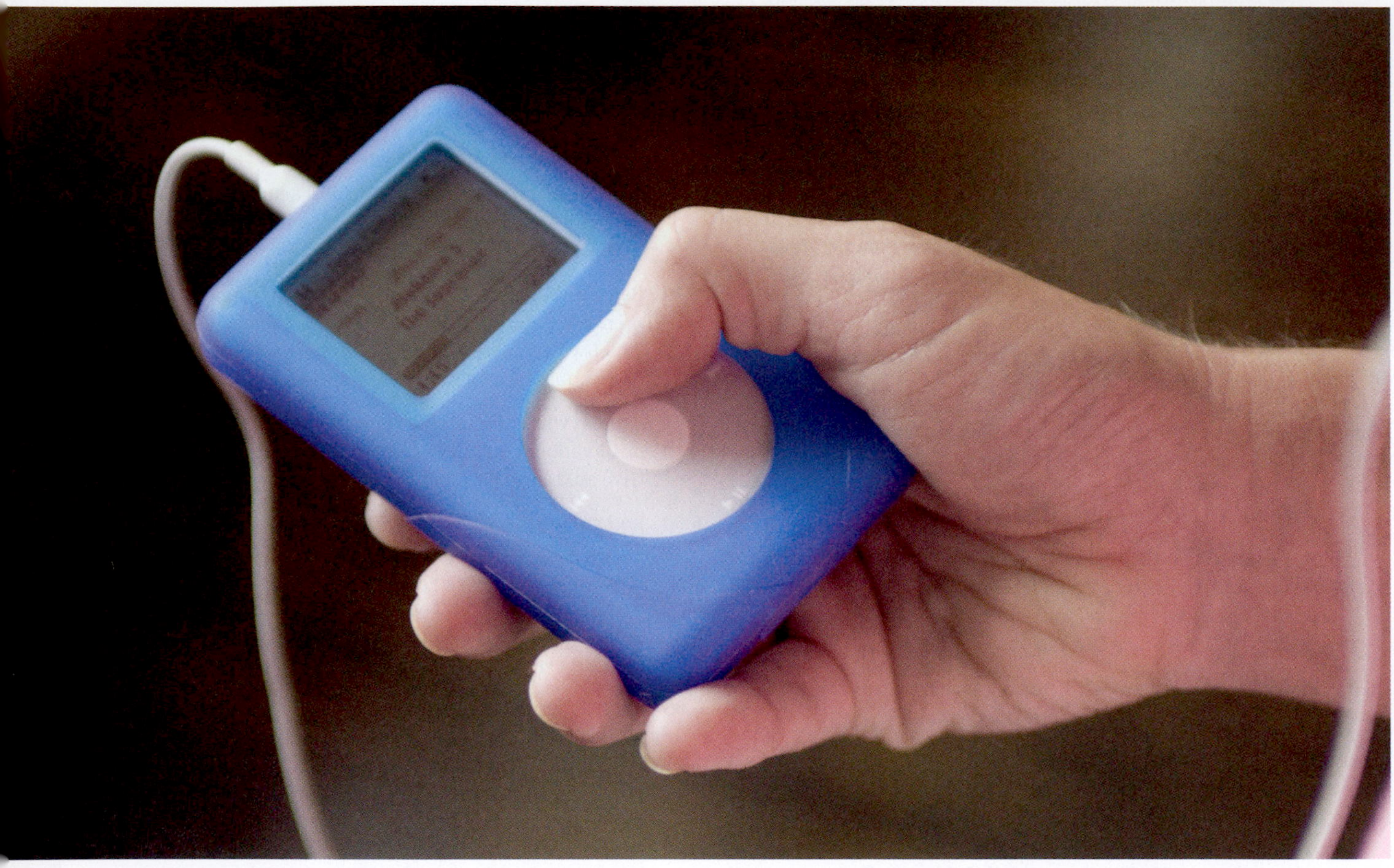

Raymond Forbes/age fotostock/Photo Library

Does the shuffle function really mix the song order randomly?

See Case Study 7.2 *(p. 251)*

Probability

We rarely stop to think about the precise meaning of the word *probability*. When we speak of the probability that we will win a lottery based on buying a single ticket, are we using the word in the same way that we do when we speak about the probability that we will buy a new house in the next 5 years?

Statistical methods are used to evaluate information in uncertain situations and probability plays a key role in that process. Remember our definition of statistics from Chapter 1: *Statistics is a collection of procedures and principles for gathering data and analyzing information to help people make decisions when faced with uncertainty.* Decisions like whether to buy a lottery ticket, whether to buy an extended warranty on a computer, or which of two courses to take are examples of decisions that you may have to make that involve uncertainty and the evaluation of probabilities.

Probability calculations are a key element of *statistical inference,* in which we use sample information to make conclusions about a larger population. For example, consider Case Study 1.6 in which 22,071 physicians were randomly assigned to take either aspirin or a placebo. In the aspirin group, there were 9.42 heart attacks per 1000 participating doctors, whereas in the placebo group there were 17.13 heart attacks per 1000 participants. It can be determined that the probability is only .00001 that the observed difference between heart attack rates would be so large if in truth there was no difference between taking aspirin and taking placebo. This is strong evidence that the observed difference did not occur just by chance. From this, we conclude that taking aspirin does reduce the risk of a heart attack.

7.1 Random Circumstances

The next case study is hypothetical but contains the kinds of situations that people encounter every day involving probability. We will use the elements of this story throughout the chapter to illustrate the concepts and calculations necessary to understand the role that probability plays in our lives.

CASE STUDY 7.1 A Hypothetical Story: Alicia Has a Bad Day

Last week, Alicia went to her physician for a routine medical exam. This morning, her physician phoned to tell her that one of her tests came back positive, indicating that she may have a disease that we will simply call D. Thinking there must be some mistake, Alicia inquired about the accuracy of the test. The physician told her that the test is 95% accurate as to whether someone has disease D or not. In other words, when someone has D, the test detects it 95% of the time. When someone does not have D, the test is correctly negative 95% of the time.

Therefore, according to the physician, even though only 1 out of 1000 women of Alicia's age actually has D, the test is a pretty good indicator that Alicia may have the disease. Alicia doesn't know it yet, but her physician is wrong to imply that it's likely that Alicia has the disease. Actually, her chance of having the disease is small, even given the positive test result. Later in this chapter you will discover why this is true.

Alicia had planned to spend the morning studying for her afternoon statistics class. At the beginning of each class, her professor randomly selects three different students to answer questions about the material. There are 50 students in the class, so Alicia reasons that she is not likely to be selected. Rather than studying, she uses the time to search the Web for information about disease D.

At the statistics class that afternoon, the 50 student names are written on slips of paper and put into a bag. One name is drawn (without replacement) for each of the three questions. Alicia twice breathes a sigh of relief. She is not picked to answer either of the first two questions. But probability is not in Alicia's favor on this day: She gets picked to answer the third question.

Random Circumstances in Alicia's Day

A **random circumstance** is one in which the outcome is unpredictable. In many cases, *the outcome is not determined until we observe it.* It was not predetermined that Alicia would be selected to answer one of the questions in class. This happened when the professor drew her name out of the bag. In other cases, *the outcome is already determined, but our knowledge of it is uncertain.* Alicia either has disease D or not, but she and her physician don't know which possibility is true.

One lesson of this chapter is that the probabilities associated with random circumstances sometimes depend on other random circumstances. Alicia's test results were positive. The probability that Alicia would have a positive test depends on another random circumstance, which is whether she actually has disease D or not. In her statistics class, Alicia was not selected to answer questions 1 or 2. So at the time of the drawing of a name for the third question, the probability that she would be selected was 1/48 because there were only 48 names left in the bag instead of the original 50. If she had been selected to answer either question 1 or question 2, her probability of being selected to answer question 3 would have been 0.

Here is a list of the random circumstances in Alicia's story and the possible outcomes for each of them:

Random Circumstance 1: Disease status

- Alicia has D.
- Alicia does not have D.

Random Circumstance 2: Test result

- Test is positive.
- Test is negative.

Random Circumstance 3: First student's name is drawn

- Alicia is selected.
- Alicia is not selected.

Random Circumstance 4: Second student's name is drawn

- Alicia is selected.
- Alicia is not selected.

Random Circumstance 5: Third student's name is drawn

- Alicia is selected.
- Alicia is not selected.

Assigning Probabilities to the Outcomes of a Random Circumstance

We said that the probability was 1/48 that Alicia would be picked for the third question given that she had not been picked for either of the first two. Note that this probability is expressed as a fraction. A probability is a value between 0 and 1 and is written either as a fraction or as a decimal fraction. From a purely mathematical point of view, a **probability** simply is a number between 0 and 1 that is assigned to a possible outcome of a random circumstance. Additionally, for the complete set of distinct possible outcomes of a random circumstance, the total of the assigned probabilities must equal 1.

In practice, of course, we should assign probabilities to outcomes in a meaningful way. A probability should provide information about how likely it is that a particular outcome will be the result of a random circumstance. In the next section, we discuss two different ways to assign and interpret probabilities. One of these ways, the relative frequency definition of probability, forms the foundation for the statistical inference methods that we will examine in later chapters.

7.1 Exercises are on page 252.

THOUGHT QUESTION 7.1 Based on your understanding of probability and random events, assign probabilities to the two possible outcomes for Random Circumstance 3.*

THOUGHT QUESTION 7.2 At the beginning of Alicia's day, the outcomes of the five random circumstances listed were uncertain to her. Which of them were uncertain because *the outcome was not yet determined* and which were uncertain because of Alicia's *lack of knowledge of the outcome*?**

7.2 Interpretations of Probability

The word *probability* is so common that in all probability, you will encounter it today in everyday language. But we rarely stop to think about the precise meaning of the word. When we speak of the probability that we will win a lottery based on buying a single ticket, are we using the word in the same way that we do when we speak about the probability that we will buy a new house in the next 5 years? We can quantify the chance of winning the lottery exactly, but our assessment of the chance that we will buy a new house is based on our personal and subjective belief about how life will evolve for us.

The conceptual difference illustrated by these two different situations creates two different interpretations of what is meant by the word *probability*. In most situations in statistics, a probability is assigned to a possible outcome on the basis of what will or has happened over the long run of repeatedly observing a random circumstance. There are situations, however, in which a probability may be assigned based on the expert assessment of an individual.

The Relative Frequency Interpretation of Probability

The relative frequency interpretation of probability applies to situations in which we can envision repeatedly observing the results of a random circumstance. For example, it is easy to imagine flipping a coin over and over again and counting the number of heads and tails. It makes sense to interpret the probability that the coin lands with heads up to be the relative frequency, over the long run, with which it lands with heads up. When we say that the probability of flipping heads is 1/2, we can interpret this to mean that about half of a large number of flips will result in heads. Here are some other situations to which the relative frequency interpretation of probability can be applied:

- Buying lottery tickets regularly and observing how often you win.
- Drawing a student's name out of a hat and seeing how often a particular student is selected.

*HINT: See Example 7.3 on page 223.
**HINT: Only two of them were already determined.

- Commuting to work daily and observing how often a certain traffic signal is red when we encounter it.
- Surveying many adults and determining what proportion smokes.
- Observing births and noting how often the baby is a female.

DEFINITION For situations that we can imagine repeating many times, we define the **probability** of a specific outcome as *the proportion of times it would occur over the long run.* This also is called the **relative frequency** of that particular outcome.

Note the emphasis on what happens *in the long run*. We cannot accurately assess the probability of a particular outcome by observing it only a few times. For example, consider a family with five children in which only one child is a boy. We should not take this as evidence that the probability of having a boy is only 1/5. To more accurately assess the probability that a baby is a boy, we have to observe many births.

Example 7.1 **Probability of Male versus Female Births** According to the U.S. Centers for Disease Control and Prevention, the long-run relative frequency of males born in the United States is about .512. In other words, over the long run, 512 male babies and 488 female babies are born per 1000 births (source: http://www.cdc.gov/nchs/data/nvsr/nvsr53/nvsr53_20.pdf).

Suppose that we were to tally births and the relative frequency of male births in a certain city for the next year. Table 7.1, which was generated with a computer simulation, shows what we might observe. In the simulation, the chance was .512 that any individual birth was a boy, and the average number of births per week was 24. Note how the *proportion* or *relative frequency* of male births is relatively far from .512 over the first few weeks but then settles down to around .512 in the long run. After 52 weeks, the *relative frequency* of boys was 639/1237 = .517. If we used only the data from the first week to estimate the probability of a male birth, our estimate would have been far from the actual value. We need to look at a large number of observations to accurately estimate a probability.

Table 7.1 Relative Frequency of Male Births over Time

Weeks of Watching	Total Births	Total Boys	Proportion of Boys
1	30	19	.633
4	116	68	.586
13	317	172	.543
26	623	383	.615
39	919	483	.526
52	1237	639	.517

Determining the Relative Frequency Probability of an Outcome

There are two ways to determine a **relative frequency probability**. The first method involves *making an assumption about the physical world.* The second method involves *making a direct observation of how often something happens.*

Method 1: Make an Assumption about the Physical World

Sometimes, it is reasonable to assign probabilities to possible outcomes based on what we think about physical realities. We generally assume, for example, that coins are manufactured in such a way that they are equally likely to land with heads or tails up when flipped. Therefore, we conclude that the probability of a flipped coin showing heads up is 1/2 or .5.

Example 7.2 **A Simple Lottery** A lottery game that is run by many states in the United States is one in which players choose a three-digit number between 000 and 999. A player wins if his or her three-digit number is chosen. If we assume that the physical mechanism that is used to draw the winning number gives each possibility an equal chance, we can determine the probability of winning. There are 1000 possible three-digit numbers (000, 001, 002, . . . , 999), so the probability that the state picks the player's number is 1/1000. In the long run, a player should win about 1 out of 1000 times. Note that this does not mean that a player will win *exactly* once in every thousand plays.

Example 7.3 **The Probability That Alicia Has to Answer a Question** In Case Study 7.1, there were 50 student names in the bag when a student was selected to answer the first question. If we assume that the names have been well mixed in the bag, then each student is equally likely to be selected. Therefore, the probability that Alicia would be selected to answer the first question was 1/50 or .02.

Method 2: Observe the Relative Frequency

Another way to determine the probability of a particular outcome is by *observing the relative frequency of the outcome over many, many repetitions of the situation.* We illustrated this method with Example 7.1 when we observed the relative frequency of male births in a given city over a year. If we consider many births, we can get a very accurate figure for the probability that a birth will be a male. As we mentioned, the relative frequency of male births in the United States has been consistently close to .512. In 2002, for example, there were 4,021,726 live births in the United States, of which 2,057,979 were males. Assuming that the probability does not change in the future, the probability that a live birth will result in a male is about 2,057,979/4,021,726 = .5117 (source: http://www.cdc.gov/nchs/data/nvsr/nvsr53/nvsr53_20.pdf).

Example 7.4 **The Probability of Lost Luggage** According to the Bureau of Transportation Statistics, 3.91 per thousand passengers on U.S. airline carriers in 2009 temporarily or permanently lost their luggage. This number is based on data collected over the long run (a full year) and is found by dividing the number of passengers who lost their luggage by the total number of passengers. Another way to state this fact is to say that the probability is 3.91/1000, about 1/256, or about .004, that a randomly selected passenger on a U.S. carrier in 2009 would lose luggage (source: http://www.bts.gov/press_releases/2010/dot027_10/pdf/dot027_10.pdf).

Proportions and Percentages as Probabilities

Relative frequency probabilities are often derived from the proportion of individuals who have a certain characteristic, or from the proportion of the time a certain outcome occurs in a random circumstance. The relative frequency probabilities are therefore numbers between 0 and 1. It is commonplace, however, to be somewhat loose in how relative frequencies are expressed. For instance, here are four ways to express the relative frequency of lost luggage:

- The proportion of passengers who lose their luggage is 1/256 or about .004.
- About 0.4% of passengers lose their luggage.
- The probability that a randomly selected passenger will lose his or her luggage is about .004.
- The probability that you will lose your luggage is about .004.

This last statement is not exactly correct, because *your* probability of lost luggage depends on a number of factors, including whether you check any luggage and how late you arrive at the airport for your flight. The probability of .004 really applies only to the long-run relative frequency across all passengers, or the probability that a *randomly selected* passenger is one of the unlucky ones.

Estimating Probabilities from Observed Categorical Data

When appropriate data are available, the long-run relative frequency interpretation of probability can be used to estimate the probabilities of outcomes and the probabilities of combinations of outcomes for categorical variables. Assuming that the data are representative, the probability of a particular outcome is estimated to be the relative frequency (proportion) with which that outcome was observed. As we learned in Case Study 1.3 and in Chapter 5, such estimates are not precise, and an approximate margin of error for the estimated probability is calculated as $1/\sqrt{n}$. We will learn a more precise way to calculate the margin of error in Chapter 10.

Example 7.5 **Night-lights and Myopia Revisited** In an example discussed in Chapters 2 and 4, we investigated the relationship between nighttime lighting in early childhood and the incidence of myopia later in childhood. The results are shown again in Table 7.2.

Table 7.2 Nighttime Lighting in Infancy and Eyesight

Slept with:	No Myopia	Myopia	High Myopia	Total
Darkness	155 (90%)	15 (9%)	2 (1%)	172
Night-light	153 (66%)	72 (31%)	7 (3%)	232
Full Light	34 (45%)	36 (48%)	5 (7%)	75
Total	342 (71%)	123 (26%)	14 (3%)	479

Assuming that these data are representative of a larger population, what is the approximate probability that someone from that population who sleeps with a night-light in early childhood will develop some degree of myopia? There were 232 infants who slept with a night-light. Of those, 72 developed myopia and 7 developed high myopia, for a total of 79 who developed some degree of myopia. Therefore, the relative frequency of some myopia among night-light users is $(72 + 7)/232 = 79/232 = .34$. From this, we can conclude that the approximate probability of developing some myopia, given that an infant slept with a night-light, is .34. This estimate is based on a sample of 232 people, so there is a margin of error of about $1/\sqrt{232} = .066$.

The Personal Probability Interpretation

The relative frequency interpretation of probability clearly is limited to repeatable conditions. Yet uncertainty is a characteristic of most events, whether or not they are repeatable under similar conditions. We need an interpretation of probability that can be applied to situations even if they will never happen again.

If you decide to drive downtown this Saturday afternoon, will you be able to find a good parking space? Should a movie studio release its potential new hit movie before Christmas, when many others are released, or wait until January when it might have a better chance to be the top box-office attraction? If the United States enters into a trade alliance with a particular country, will that cause problems in relations with another country? Will you have a better time if you go to Cancun or to Florida for your next spring break?

These are unique situations, not likely to be repeated. They require people to make decisions based on an assessment of how the future will evolve. Each of us could assign a personal probability to these events based on our own knowledge and experiences, and we could use that probability to help us with our decision. Different people may not agree on what the probability of an event happening is, but nobody would be considered wrong.

DEFINITION We define the **personal probability** of an event to be the degree to which a given individual believes that the event will happen. Sometimes, the term **subjective probability** is used because the degree of belief may be different for each individual.

There are very few restrictions on personal probabilities. They must fall between 0 and 1 (or, if expressed as a percent, between 0% and 100%). They must also fit together in certain ways if they are to be **coherent**. By *coherent,* we mean that your personal probability of one event doesn't contradict your personal probability of another. For example, if you thought that the probability of finding a parking space downtown on Saturday afternoon was .20, then to be coherent, you must also believe that the probability of not finding one is .80. We will explore some of these logical rules later in this chapter.

How We Use Personal Probabilities

People routinely base decisions on personal probabilities. This is why committee decisions are often so difficult. For example, suppose a committee is trying to decide which candidate to hire for a job. Each member of the committee has a different assessment of the candidates, and they may disagree on the probability that a particular candidate would best fit the job. We are all familiar with the problem that juries sometimes have when trying to agree on someone's guilt or innocence. Each member of the jury has his or her own personal probability of guilt and innocence. One of the benefits of committee or jury deliberations is that they may help members to reach some consensus in their personal probabilities.

Personal probabilities often take relative frequencies of similar events into account. For example, the late astronomer Carl Sagan believed that the probability of a major asteroid hitting Earth soon is high enough to be of concern. "The probability that the Earth will be hit by a civilization-threatening small world in the next century is a little less than one in a thousand" (Arraf, 1994, p. 4, quoting Sagan). To arrive at that probability, Sagan obviously could not use the long-run frequency definition of probability. He would have to use his own knowledge of astronomy combined with past asteroid behavior.

Note that unlike relative frequency probabilities, personal probabilities assigned to unique events are not equivalent to proportions or percentages. For instance, it does not make sense to say that major asteroids will hit approximately 1 out of 1000 Earths in the next century.

TECHNICAL NOTE

A Philosophical Issue about Probability

There is some debate about how to represent probability when an outcome has been determined but is unknown, such as if you have flipped a coin but not looked at it. Technically, any particular outcome has either happened or not. If it has happened, its probability of happening is 1; if it hasn't, its probability of happening is 0. In statistics, an example of this type of situation is the construction of a 95% confidence interval, which was introduced in Chapter 5 and which we will study in detail in Chapters 10 and 11. Before the sample is chosen, a probability statement makes sense. The probability is .95 that a sample will be selected for which the computed 95% confidence interval covers the truth. After the sample has been chosen, "the die is cast." Either the computed confidence interval covers the truth or it doesn't, although we may never know which is the case. That's why we say that we have *95% confidence* that a computed interval is correct, rather than saying that *the probability* that it is correct is .95.

7.2 Exercises are on pages 252–253.

IN SUMMARY Interpretations of Probability

The *relative frequency probability* of an outcome is the proportion of times the outcome would occur over the long run. Relative frequency probabilities can be determined by either of these methods:

- Making an assumption about the physical world and using it to define relative frequencies
- Observing relative frequencies of outcomes over many repetitions of the same situation or measuring a representative sample and observing relative frequencies of possible outcomes

The *personal probability* of an outcome is the degree to which a given individual believes it will happen. Sometimes data from similar events in the past and other knowledge are incorporated when determining personal probabilities.

THOUGHT QUESTION 7.3 You are about to enroll in a course for which you know that 20% of the students will receive As. Do you think that the probability that *you* will receive an A in the class is .20? Do you think the probability that a randomly selected student in the class will receive an A is .20? Explain the difference in these two probabilities, using the distinction between relative frequency probability and personal probability in your explanation.*

7.3 Probability Definitions and Relationships

In this section, we cover the fundamental definitions for how outcomes of one or more random circumstances may be related to each other. We also give some elementary rules for calculating probabilities. In the next section, we will cover more detailed rules for calculating probabilities.

The first set of definitions has to do with the possible outcomes of one random circumstance. A specific possible outcome is called a **simple event**, and the collection of all possible outcomes is called the **sample space** for the random circumstance. The term **event** is used to describe any collection of one or more possible outcomes, and the term **compound event** is sometimes used if an event includes at least two simple events.

DEFINITION

- A **simple event** is a unique possible outcome of a random circumstance.
- The **sample space** for a random circumstance is the collection of all simple events.
- A **compound event** is an event that includes two or more simple events.
- An **event** is any collection of one or more simple events in the sample space; events can be simple events or compound events. Events are often written using capital letters A, B, C, and so on.

Example 7.6 **Days per Week of Drinking Alcohol** Suppose that we randomly sample college students and ask each student how many days he or she drinks alcohol in a typical week. The *simple events* in the *sample space* of possible outcomes for each student are 0 days, 1 day, 2 days, and so on, up to 7 days. These simple events will not be equally likely. We may want to learn the probability of the *event* that a randomly selected student drinks alcohol on 4 or more days in a typical week. The *event* "4 or more" is comprised of the *simple events* {4 days, 5 days, 6 days, and 7 days}.

***HINT:** Look at the preceding IN SUMMARY box. Which of the methods described applies in each case?

Assigning Probabilities to Simple Events

The notation $P(A)$ is used to denote the probability that event A occurs. A probability value is assigned to each *simple event* (unique outcome) using either relative frequency or personal probabilities. The relative frequency method is used most often.

Two Conditions for Valid Probabilities

The assigned probabilities for the possible outcomes of a random circumstance must meet the following two conditions:

1. Each probability is between 0 and 1.
2. The sum of the probabilities over all possible simple events is 1. In other words, the total probability for all possible outcomes of a random circumstance is equal to 1.

Probabilities for Equally Likely Simple Events

If there are k equally likely simple events in the sample space, then the probability is $1/k$ for each one. For instance, suppose that a fair six-sided die is rolled and the number that lands face up is observed. Then the sample space is {1, 2, 3, 4, 5, 6}, and the probability is 1/6 for each outcome. As another example, suppose that two fair coins are tossed and the outcome of interest is the combination of results for the first two tosses. There are four equally likely events, {HH, HT, TH, TT}, where H = heads and T = tails, so the probability is 1/4 for each outcome.

Example 7.7 **Probabilities for Some Lottery Events** Example 7.2 (p. 223) described a lottery game in which players choose a three-digit number between 000 and 999.

Random Circumstance: A three-digit winning lottery number is selected.
Sample Space: {000, 001, 002, 003, . . . , 997, 998, 999}. There are 1000 simple events.
Probabilities for Simple Event: The probability that any specific three-digit number is a winner is 1/1000. We assume that all three-digit numbers are equally likely.

The assumption that all simple events are equally likely can be used to find probabilities for events that include more than one simple event.

- Let event A = the last digit is a 9. Note that event A is comprised of the set of all simple events that end with a 9. We can write this as A = {009, 019, . . . , 999}. Because one out of every 10 three-digit numbers is included in this set, $P(A) = 1/10$.
- Let event B = the three digits are all the same. This event includes 10 simple events, which are {000, 111, 222, 333, 444, 555, 666, 777, 888, 999}. Because event B contains 10 of the 1000 equally likely simple events, $P(B) = 10/1000$, or 1/100.

Complementary Events

An event either happens or it doesn't. A randomly selected person either has blue eyes or does not have blue eyes. Alicia either has a disease or does not. "Opposite" events are called **complementary events**.

DEFINITION One event is the **complement** of another event if the two events do not contain any of the same simple events *and* together they cover the entire sample space. For an event A, the notation A^C represents the complement of A.

Because the total probability for a sample space must be equal to 1, the probabilities of complementary events must sum to 1. In symbols, $P(A) + P(A^C) = 1$. As a result, $P(A^C) = 1 - P(A)$. In words, the probability that an event does not happen is equal to one minus the probability that it does.

Example 7.8 **The Probability of Not Winning the Lottery** If A is the event that a player buying a single three-digit lottery ticket wins, then A^C is the event that he or she does not win. As we have seen in Example 7.2, $P(A) = 1/1000$, so it makes sense that $P(A^C) = 1 - 1/1000 = 999/1000$.

Mutually Exclusive Events

While complementary events cover the entire sample space, we sometimes are interested in two events with distinctly separate simple events that may not cover all possible outcomes in the sample space. For example, we might want to know the probability that a randomly selected man is unusually short or tall; for instance, that he is either at least 6 inches shorter than the mean height or is at least 6 inches taller than the mean height. The events "at least 6 inches shorter than the mean" and "at least 6 inches taller than the mean" are separate events that cannot occur simultaneously, so they are called *mutually exclusive* or *disjoint events.* They do not cover the entire sample space (leaving out the event that the man is within 6 inches of the mean), so they are not complementary events.

DEFINITION Two events are **mutually exclusive** if they do not contain any of the same simple events (outcomes). Equivalent terminology is that the two events are **disjoint**.

Example 7.9 **Mutually Exclusive Events for Lottery Numbers** Here are two mutually exclusive (disjoint) events when a winning lottery number is drawn from the set 000 to 999.

A = All three digits are the same.
B = The first and last digits are different.

Note that events A and B together do not cover all possible three-digit numbers, so they are not complementary events.

When A and B are mutually exclusive events, the probability that either event A or event B occurs is found by adding probabilities for the two events. In symbols, $P(\text{either A or B}) = P(A) + P(B)$ for mutually exclusive events. For example, the probability that a randomly selected person has either blue eyes or green eyes is equal to the probability of having blue eyes plus the probability of having green eyes.

Independent and Dependent Events

The next two definitions have to do with whether the fact that one event will take place (or has taken place) affects the probability that a second event could happen.

DEFINITION Two events are **independent** of each other if knowing that one will occur (or has occurred) does not change the probability that the other occurs.

Two events are **dependent** if knowing that one will occur (or has occurred) changes the probability that the other occurs.

The definitions can apply *either* to events *within the same random circumstance* or to events *from two separate random circumstances.* As an example of two dependent events within the same random circumstance, suppose that a random digit from 0 to 9 is selected. If we know that the randomly selected digit is an *even number,* our assignment of a probability to the event that *the number is 4* changes from 1/10 to 1/5. Often, we examine questions about the dependence of the outcomes of two separate random circumstances. For instance, does the knowledge of whether a randomly selected person is male or female (random circumstance 1) affect the probability that he or she is opposed to abortion (random circumstance 2)? If the answer is yes, the outcomes of the two random circumstances are dependent events.

Example 7.10 **Winning a Free Lunch** Some restaurants display a glass bowl in which customers deposit their business cards, and a drawing is held once a week for a free lunch. The bowl is emptied after each week's drawing. Suppose that you and your friend Vanessa each deposit a card in 2 consecutive weeks. Define

Event A = You win in Week 1.
Event B = Vanessa wins in Week 1.
Event C = Vanessa wins in Week 2.

Events A and B are *dependent* events. If you win in Week 1 (event A), then Vanessa cannot win in Week 1 (event B cannot happen). In this case, the probability of event B is 0. If you do not win in Week 1 (A does not happen), it is possible that Vanessa could win in Week 1, so the probability of event B is greater than 0. Note that events A and B refer to the same random circumstance: the drawing in Week 1.

Events A and C are *independent*. In general, knowing who wins in Week 1 gives no information about who wins in Week 2. In particular, knowing whether you win in the first week (event A) gives no information about the probability that Vanessa wins in the second week (event C). Note that events A and C refer to two *different* random circumstances: the drawings in separate weeks.

In Example 7.10, the events from two different random circumstances were independent. This is not always the case. The next example illustrates that events from two separate random circumstances may be dependent.

Example 7.11 **The Probability That Alicia Has to Answer a Question** In Case Study 7.1 (p. 220), Alicia was worried that her statistics teacher might call on her to answer a question during class. Define

Event A = Alicia is selected to answer question 1.
Event B = Alicia is selected to answer question 2.

In Example 7.3, we determined that $P(\text{A}) = 1/50$. Are A and B independent? Note that they refer to different random circumstances, but the probability that B occurs is affected by knowing whether A occurred. If Alicia answered question 1, her name is already out of the bag, so the probability that B occurs is 0. If event A does not occur, there are 49 names left in the bag including Alicia's, so the probability that B occurs is 1/49. Therefore, knowing whether A occurred changes the probability that B occurs.

Conditional Probabilities

Suppose that two events are dependent. From the definition of dependent events, it follows that knowing that one event will occur (or has occurred) *does* change the probability that the other occurs. Therefore, we need to distinguish between two probabilities:

- $P(\text{B})$, the unconditional probability that the event B occurs
- $P(\text{B}|\text{A})$, read "probability of B given A," the conditional probability that the event B occurs *given that we know A has occurred or will occur*

DEFINITION The **conditional probability** of the event B, given that the event A has occurred or will occur, is the long-run relative frequency with which event B occurs when circumstances are such that A has occurred or will occur. This probability is written as $\boldsymbol{P(\text{B}|\text{A})}$.

Example 7.12 **Probability That a Teenager Gambles Differs for Boys and Girls** Based on a survey of nearly all of the ninth- and twelfth-graders in Minnesota public schools in 1998, a total of 78,564 students, Stinchfield (2001) reported proportions admitting that they had gambled at least once a week during the previous year. He found that these proportions differed considerably for males and females, as well as for ninth- and twelfth-graders. One result he found was that 22.9% of the ninth-grade boys but only 4.5% of the ninth-grade girls admitted gambling at least weekly. Assuming these figures are representative of all Minnesota ninth-graders, we can write these as conditional probabilities for ninth-grade Minnesota students:

Read the original source on the companion website, http://www.cengage.com/statistics/Utts4e.

$P(\text{student is a weekly gambler} \mid \text{student is a boy}) = .229$
$P(\text{student is a weekly gambler} \mid \text{student is a girl}) = .045$

The vertical line | denotes the words *given that.* The probability that the student is a weekly gambler, *given that* the student is a boy, is .229. The probability that the student is a weekly gambler, *given that* the student is a girl, is .045.

Note the dependence between the outcomes for the random circumstance "weekly gambling habit" and the outcomes for the random circumstance "sex." Knowledge of a student's sex changes the probability that he or she is a weekly gambler.

7.3 Exercises are on pages 253–254.

THOUGHT QUESTION 7.4 Remember that there were 50 students in Alicia's statistics class and that student names were not put back in the bag after being selected. Define the events A = Alicia is selected to answer question 1 and B = Alicia is selected to answer question 2. Describe each of these four conditional probabilities in words and determine a value for each.

$$P(B|A),\ P(B^C|A),\ P(B|A^C),\ P(B^C|A^C)$$

Now, based on your answers, can you formulate a general rule about the value of $P(B|A) + P(B^C|A)$?*

IN SUMMARY Key Probability Definitions and Notation

- A **simple event** is a unique possible outcome of a random circumstance.
- The **sample space** for a random circumstance is the collection of all simple events.
- A **compound event** is an event that includes two or more simple events.
- An **event** is any collection of one or more simple events in the sample space; events can be simple events or compound events.
- Events are often written using capital letters A, B, C, and so on, and their probabilities are written as $P(A)$, $P(B)$, $P(C)$, and so on.
- One event is the **complement** of another event if the two events do not contain any of the same simple events *and* together they cover the entire sample space. For an event A, the notation A^C represents the complement of A.
- Two events are **mutually exclusive** if they do not contain any of the same simple events (outcomes). Equivalent terminology is that the two events are **disjoint**.
- Two events are **independent** of each other if knowing that one will occur (or has occurred) does not change the probability that the other occurs.
- Two events are **dependent** if knowing that one will occur (or has occurred) changes the probability that the other occurs.
- The **conditional probability of the event B, given that the event A has occurred or will occur**, is the long-run relative frequency with which event B occurs when circumstances are such that A has occurred or will occur. It is written as $\mathbf{P(B|A)}$.

***HINT:** For $P(B|A)$ and $P(B|A^C)$, see Example 7.11.

7.4 Basic Rules for Finding Probabilities

In this section, we cover four rules that can be used to find probabilities for complicated events. We introduced parts of these rules in Section 7.3, and now we give more details. When stating these rules, we use the letters A and B to represent events, and we use $P(A)$ and $P(B)$ for their respective probabilities. Each rule is followed by examples that translate the formulas into words. Common words that signal the use of the rule are shown in *italics* in the examples.

Probability That an Event Does Not Occur (Rule 1)

DEFINITION **Rule 1 (for "not the event"):** To find the probability of A^C, the complement of A, use $P(A^C) = 1 - P(A)$. $P(A^C)$ is the probability that the event A will *not* occur.

Rule 1 follows directly from the observation that $P(A) + P(A^C) = 1$, which was stated as an immediate consequence after the definition of A^C. Therefore, if $P(A)$ is known, it is easy to find the probability that A does not occur. Simply subtract $P(A)$ from 1. For instance, in Example 7.8, it was easy to determine that the probability of *not* winning the lottery was .999, once we knew that the probability of winning was .001.

Rule 1 is simple but surprisingly useful. When you are confronted with finding the probability of an event, always ask yourself whether it would be easier to find the probability of its complement and then subtract that from 1.

Example 7.13 **Probability That a Stranger Does *Not* Share Your Birth Date** Assuming that all 365 birth dates are equally likely, the probability that the next stranger you meet will share your birthday is 1/365. Therefore, the probability that the stranger will *not* share your birthday is $1 - (1/365) = (364/365) = .9973$. This does not hold if you were born on February 29, but if that's you, you are used to ingenuity when it comes to birthdays!

Probability That One or Both of Two Events Happen (Rule 2)

DEFINITION **Rule 2 (addition rule for "either/or/both"):** To find the probability that *either* A *or* B or both happen:

Rule 2a (general): $P(A \text{ or } B) = P(A) + P(B) - P(A \text{ and } B)$

Rule 2b (for mutually exclusive events): *If* A and B are *mutually exclusive* events, $P(A \text{ or } B) = P(A) + P(B)$

Rule 2b is a special case of Rule 2a because when A and B are mutually exclusive events, $P(A \text{ and } B) = 0$. In other words, events A and B cannot both happen at once.

To find the probability that either of two *mutually exclusive* events will be the outcome of a random circumstance, we add the probabilities for the two events. For instance, the probability that a randomly selected integer between 1 and 10 is either 3 or 7 is $.1 + .1 = .2$. When the two events are not mutually exclusive, we modify the sum of the probabilities in the manner described in Rule 2a and illustrated in Example 7.14.

Example 7.14 **Roommate Compatibility** Brett is off to college. There are 1000 additional new male students, and one of them will be randomly assigned to share Brett's dorm room. He is hoping that it won't be someone who likes to party or who snores. The 2×2 table for the 1000 students follows:

	Snores	Doesn't Snore	Total
Likes to Party	150	100	250
Doesn't Like to Party	200	550	750
Total	350	650	1000

What is the probability that Brett will be disappointed and get a roommate who *either* likes to party or snores or both? Relevant events and their probabilities can be found directly from the frequencies in the table, as follows:

$$A = \text{likes to party} \quad P(A) = \frac{250}{1000} = .25$$

$$B = \text{snores} \quad P(B) = \frac{350}{1000} = .35$$

$$A \text{ and } B = \text{likes to party } \textit{and} \text{ snores} \quad P(A \text{ and } B) = \frac{150}{1000} = .15$$

Using Rule 2a, $P(\text{A or B}) = P(\text{A}) + P(\text{B}) - P(\text{A and B}) = .25 + .35 - .15 = .45$. The probability is .45 that Brett will be assigned a roommate who either likes to party, or snores, or both. Note that the probability is therefore $1 - .45 = .55$ that he will get a roommate who is acceptable, using Rule 1. We could also determine this probability from the table of counts because there are 550 suitable roommates out of 1000 possibilities.

We can use Example 7.14 to illustrate why Rule 2a makes sense. To find $P(\text{A or B})$, we need to include all possible outcomes involved in either event. In Brett's case, 250 (.25) of the potential roommates like to party, and 350 (.35) snore, for a total of 600 (.6). But notice that the 150 potential roommates who like to party *and* who snore have been counted twice. So we need to subtract them once, resulting in $250 + 350 - 150 = 450$ individuals out of the 1000 total who have one or both traits. So Rule 2a tells you to add the probabilities for all of the outcomes that are part of event A plus the probabilities for all of the outcomes that are part of event B, then to subtract the probabilities for all of the outcomes that have been counted twice, $P(\text{A and B})$.

Example 7.15 **Probability of *Either* Two Boys *or* Two Girls in Two Births** What is the probability that a woman who has two children has *either* two girls *or* two boys? The question asks about two *mutually exclusive* events because with only two children, she can't have two girls *and* two boys. Rule 2b applies to this situation. We'll use the information in Example 7.1 that the probability of a boy is .512 and the probability of a girl is .488, and we'll also take a sneak peek at Rule 3, to be introduced next.

Event A = two girls. From Rule 3b (next), $P(\text{A}) = (.488)(.488) = .2381$.
Event B = two boys. From Rule 3b (next), $P(\text{B}) = (.512)(.512) = .2621$.

Therefore, the probability that the woman has *either* two girls *or* two boys is

$$P(\text{A } \textit{or} \text{ B}) = P(\text{A}) + P(\text{B}) = .2381 + .2621 = .5002$$

Probability That Two or More Events Occur Together (Rule 3)

Rule 2 allowed us to answer "either one or both" questions about events. But what if we want to restrict consideration to the combination of *both* events only? In that case, we want to find $P(\text{A and B})$. For instance, if you flip a nickel and a dime, what is the probability that they *both* land heads up? You probably know intuitively that the answer is 1/4 or could be convinced of that by noting that the four possibilities, HH, HT, TH, and TT, are equally likely. A more general way of arriving at the answer is to notice that 1/4 is the result of multiplying $1/2 \times 1/2$, the two individual probabilities of heads. This general method works for finding $P(\text{A and B})$ as long as A and B are independent events.

If A and B are dependent events, then finding the probability that they both occur requires that we know more than their individual probabilities. Sometimes, we can find $P(\text{A and B})$ directly as a relative frequency. In Example 7.14, the probability that Brett's roommate will be someone who likes to party *and* who snores can be found directly from the table of counts as $150/1000 = .15$. If we can't find $P(\text{A and B})$ directly, then we need to know either $P(\text{A}|\text{B})$ and $P(\text{B})$ or $P(\text{B}|\text{A})$ and $P(\text{A})$.

DEFINITION **Rule 3 (multiplication rule for "and"):** To find the probability that two events A *and* B both occur simultaneously or in a sequence:

Rule 3a (general): $P(\text{A and B}) = P(\text{A})P(\text{B}|\text{A}) = P(\text{B})P(\text{A}|\text{B})$

Rule 3b (for independent events): *If* A and B are *independent* events, $P(\text{A and B}) = P(\text{A})P(\text{B})$

Extension of Rule 3b to more than two independent events: For several independent events, $P(\text{A}_1 \text{ and } \text{A}_2 \ldots \text{ and } \text{A}_n) = P(\text{A}_1)P(\text{A}_2) \ldots P(\text{A}_n)$

Rule 3b is a special case of Rule 3a because when A and B are independent, $P(\text{B}|\text{A})$ is equal to $P(\text{B})$.

To find the probability that two or more independent events occur together, multiply the probabilities of the events. For instance, the probability is $(.5)(.5)(.5) = .125$ that the next three people of your sex you see all are taller than the median height for your sex. (Remember that the median is the value for which half of a group has a greater value and half has a smaller value, so the probability is .5 that a randomly selected person is taller than the median. We are assuming that nobody's height is at the exact median, and that the next three people you see are equivalent to a random selection.)

Example 7.16

Read the original source on the companion website, http://www.cengage.com/statistics/Utts4e.

Probability That a Randomly Selected Ninth-Grader Is a Male *and* a Weekly Gambler In a 1998 survey of most of the ninth-grade students in Minnesota, 22.9% of the boys and 4.5% of the girls admitted that they had gambled at least once a week during the previous year (Stinchfield, 2001). The population consisted of 50.9% girls and 49.1% boys. Assuming these students represent all ninth-grade Minnesota teens, what is the probability that a randomly selected student will be a male *who also* gambles at least weekly?

Event A = A male is selected.
Event B = A weekly gambler is selected.

Events A and B are dependent, so we use Rule 3a (general multiplication rule) to determine the probability that A and B occur together. Here is what we know:

$P(\text{A}) = .491$ (probability that a male is selected)
$P(\text{B}|\text{A}) = .229$ (probability that a gambler is selected, given that a male is selected)

The answer to our question is therefore

$$P(\text{male } \textit{and} \text{ weekly gambler}) = P(\text{A } \textit{and} \text{ B}) = P(\text{A})P(\text{B}|\text{A}) = (.491)(.229) = .1124$$

About 11% of all ninth-grade teenagers are males *and* weekly gamblers.

Example 7.17

Probability That Two Strangers Both Share Your Birth Month Assuming that birth months are equally likely, what is the probability that the next two unrelated strangers you meet *both* share your birth month? Because the strangers are unrelated, we assume independence between the two events that have to do with them sharing your birth month. Define

Event A = first stranger shares your birth month: $P(A) = 1/12$.
Event B = second stranger shares your birth month: $P(B) = 1/12$.

The events are independent, so Rule 3b applies. The probability that *both* individuals share your birth month is the probability that A occurs *and* B occurs:

$$P(\text{A and B}) = P(\text{A})P(\text{B}) = \frac{1}{12} \times \frac{1}{12} = \frac{1}{144} = .007$$

The Extension of Rule 3b allows this result to be generalized. For instance, the probability that four unrelated strangers *all* share your birth month is $(1/12)^4$.

Determining a Conditional Probability (Rule 4)

DEFINITION

Rule 4 (conditional probability): This rule is simply an algebraic restatement of Rule 3a, but it is sometimes useful to apply the following form of that rule, which gives the probability that B occurs *given* that A has occurred or will occur:

$$P(\text{B}|\text{A}) = \frac{P(\text{A and B})}{P(\text{A})}$$

The assignment of letters to events A and B is arbitrary, so it is also true that

$$P(\text{A}|\text{B}) = \frac{P(\text{A and B})}{P(\text{B})}$$

Often, a conditional probability is given to us directly. In Example 7.12, for instance, it was reported that the estimated conditional probability is .045 that a ninth-grader is a weekly gambler if the student is a female. In other cases, the physical situation can make it easy to determine a conditional probability. For instance, in drawing 2 cards from an ordinary deck of 52 cards, the probability that the second card drawn is red, *given* that the first card drawn was red, is 25/51 because 25 of the remaining 51 cards are red. Occasionally, it is useful to use Rule 4, an algebraic restatement of the multiplication rule (Rule 3a), to calculate a conditional probability.

Example 7.18 **Probability That Alicia Is Picked for the First Question Given That She Is Picked to Answer a Question** This example continues Case Study 7.1 (p. 220). Here is an example for which it is easy to see the logical answer. If *we know* that Alicia is picked to answer one of the questions, what is the probability that it was the first question? Once we know that she had to answer one of the three questions, the conditional probability that it was the first one should be 1/3. All three questions would be equally likely to be the question for which Alicia was picked.

Let's use Rule 4 to verify this logic. Define

A = Alicia is called on to answer the first question; $P(A) = 1/50$.
B = Alicia is called on to answer any one of the questions; $P(B) = 3/50$, since there are 50 students and 3 students are chosen to answer questions.

Note that the event that A and B *both* occur is synonymous with the event A occurring, since A is a subset of B. In other words, $P(\text{A and B}) = P(\text{A})$ when A is a subset of B. Therefore, $P(\text{A and B}) = 1/50$. Applying Rule 4, we have

$$P(\text{A}|\text{B}) = \frac{P(\text{A and B})}{P(\text{B})} = \frac{1/50}{3/50} = \frac{1}{3}$$

Example 7.19 **The Probability of Guilt and Innocence Given a DNA Match** This example is modified from one given by Paulos (1995, p. 72). Suppose that there has been a crime and it is known that the criminal is a person within a population of 6,000,000. Further, suppose it is known that in this population only about one person in a million has a DNA type that matches the DNA found at the crime scene, so let's assume there are six people in the population with this DNA type. Someone in custody has this DNA type. Let's call him Dan. *We know* Dan's DNA matches, but what is the probability that he is actually innocent? In the absence of any other evidence beyond the DNA match, the probability of guilt is only 1/6 because the criminal is one of the six individuals who have this DNA type in the population. Therefore, the probability of innocence is 5/6.

We now know the answer, but to illustrate Rule 4, we will look at how it can be used to determine this probability. Imagine Dan was randomly selected from the population. Define

Event A = Dan's DNA matches the DNA at the crime scene.
Event B = Dan is innocent of the crime.

We want to determine

$$P(\text{B}|\text{A}) = \frac{P(\text{A and B})}{P(\text{A})}$$

The event "A and B" is the event that Dan is innocent *and* his DNA matches. There are five such people in the population, so $P(\text{A and B}) = 5/6{,}000{,}000$. The probability of a DNA match is $P(\text{A}) = 6/6{,}000{,}000$, and

$$P(\text{B}|\text{A}) = \frac{P(\text{A and B})}{P(\text{A})} = \frac{5/6{,}000{,}000}{6/6{,}000{,}000} = \frac{5}{6}$$

If you were on the jury, it would be important to realize that without additional evidence, the probability that Dan is *innocent* is 5/6, even though his DNA matches. The prosecutor surely would emphasize a different conditional probability—specifically, the very small probability that an innocent person's DNA type would match the crime scene DNA (remember, there are 5,999,999 innocent people). Do not confuse these two different conditional probabilities.

You may be thinking that it is preposterous that the police would just happen to arrest someone whose DNA matches the DNA found at the crime scene. But, just as with fingerprints, there are now databases of DNA matches, so it is quite reasonable that the police would find someone to arrest who has a DNA match, even if that person had nothing to do with the crime.

THOUGHT QUESTION 7.5 Continuing the DNA example, verify that the conditional probability

$$P(\text{DNA match} \mid \text{innocent person}) = \frac{5}{5{,}999{,}999} = .00000083$$

Then provide an explanation that would be understood by a jury for the distinction between the two statements:

- The probability that a person who has a DNA match is innocent is 5/6.
- The probability that a person who is innocent has a DNA match is .00000083.*

***HINT:** It might be helpful to construct a 2 × 2 table showing where all 6 million people fall, where row categories are "innocent or not," and column categories are "DNA match (yes, no)."

IN SUMMARY

Independent and Mutually Exclusive Events and Probability Rules

Students sometimes confuse the definitions of independent and mutually exclusive events. Remember these two concepts:

- When two events are *mutually exclusive* and one happens, the other event cannot happen simultaneously, so it has probability 0.
- When two events are *independent* and one happens, the probability of the other event is unaffected.

Probabilities under these two situations follow these rules:

When Events Are:	P(A *or* B) is:	P(A *and* B) Is:	P(A|B) Is:
Mutually exclusive	$P(A) + P(B)$	0	0
Independent	$P(A) + P(B) - P(A)P(B)$	$P(A)P(B)$	$P(A)$
Any	$P(A) + P(B) - P(A \text{ and } B)$	$P(A)P(B\|A)$	$\dfrac{P(A \text{ and } B)}{P(B)}$

Sampling with and without Replacement

Suppose that a classroom has 30 students in it and 3 of the students are left-handed. If the teacher randomly picks one student, the probability that a left-handed student is selected is 3/30, or 1/10. The probability that a right-handed student is selected is 27/30, or 9/10. Now suppose that the teacher randomly picks a student again. What is the probability that another left-handed student is chosen? The answer is that it depends on whether the first student selected is eligible to be selected again.

DEFINITION

- A **sample** is drawn **with replacement** if individuals are returned to the eligible pool for each selection.
- A **sample** is drawn **without replacement** if sampled individuals are not eligible for subsequent selection.

To determine the probability of a sequence of events, we use the multiplication rule. If sampling is done *with replacement*, the Extension of Rule 3b holds. If sampling is done *without replacement*, probability calculations are more complicated because the probabilities of the possible outcomes at any specific time in the sequence are *conditional* on previous outcomes.

Example 7.20 **Choosing Left-Handed Students** A class of 30 children includes 27 who are right-handed and 3 who are left-handed. The teacher plans to randomly choose one child on Monday and one on Tuesday to help demonstrate a science experiment. The teacher, who is left-handed herself, enjoys it when a left-handed assistant is chosen. She wonders whether there is a higher probability of choosing left-handed students on both Monday and Tuesday if she samples with replacement or without replacement. Here is how you would find the probability that a left-handed student is drawn on both days, when sampling with and without replacement:

Sampled with replacement:

$$P(\text{Left and Left}) = \left(\frac{3}{30}\right)\left(\frac{3}{30}\right) = \frac{9}{900} = \frac{1}{100}$$

Sampled without replacement:

$$P(\text{Left and Left}) = \left(\frac{3}{30}\right)\left(\frac{2}{29}\right) = \frac{6}{870} = \frac{1}{145}$$

When sampling without replacement the probability that the second student is left-handed, given that the first student was left-handed, is a conditional probability. It is found by noticing that there are 29 students remaining, of whom 2 are left-handed. Note that there is a higher probability of choosing a left-handed student both days if sampling is *with* replacement. This makes sense because the conditional probability of choosing a left-handed student Tuesday is higher when the left-handed student chosen Monday remains in the pool of possible choices.

If a sample is drawn from a very large population, the distinction between sampling with and without replacement becomes unimportant. For example, suppose a city has 300,000 people of whom 30,000 are left-handed. If two people are sampled, the probability that they are both left-handed is $1/100 = .01$ if they are drawn with replacement and .0099997 if they are drawn without replacement. In most polls, individuals are drawn without replacement, but the analysis of the results is done as if they were drawn with replacement. The consequences of making this simplifying assumption are negligible.

7.4 Exercises are on pages 254–256.

IN SUMMARY Probability Rules 1 through 4

Rule 1 (for "not the event"): To find the probability of A^C, the complement of A, use $P(A^C) = 1 - P(A)$

Rule 2 (addition rule for "either/or/both"): To find the probability that *either* A *or* B or both happen:

- **Rule 2a (general):** $P(\text{A or B}) = P(\text{A}) + P(\text{B}) - P(\text{A and B})$
- **Rule 2b (for mutually exclusive events):** *If* A and B are *mutually exclusive* events: $P(\text{A or B}) = P(\text{A}) + P(\text{B})$

Rule 3 (multiplication rule for "and"): To find the probability that two events, A *and* B, both occur simultaneously or in a sequence:

- **Rule 3a (general):** $P(\text{A and B}) = P(\text{A})P(\text{B}|\text{A}) = P(\text{B})P(\text{A}|\text{B})$
- **Rule 3b (for independent events):** *If* A and B are *independent* events, $P(\text{A and B}) = P(\text{A})P(\text{B})$
- **Extension of Rule 3b to more than two independent events:** For several independent events, $P(\text{A}_1 \text{ and A}_2 \ldots \text{ and A}_n) = P(\text{A}_1)P(\text{A}_2) \ldots P(\text{A}_n)$

Rule 4 (conditional probability): To find the probability that B occurs given that A has occurred or will occur:

$$P(\text{B}|\text{A}) = \frac{P(\text{A and B})}{P(\text{A})}$$

The assignment of letters to events A and B is arbitrary, so it is also true that

$$P(\text{A}|\text{B}) = \frac{P(\text{A and B})}{P(\text{B})}$$

7.5 Finding Complicated Probabilities

Much as we build complicated sentences using basic words and simple rules, we can compute complicated probabilities using the basic rules for probability relationships. Finding probabilities of complicated events is somewhat like trying to drive from one city to another. There often are many different routes that get you there. Some routes are faster than others, but all will get you to your destination. When you attempt to solve a probability problem, remember that you may take a different approach than this book or your professor or friend. The important thing is that you all get to the same answer eventually. The next two examples illustrate an easy and a difficult way to solve each of the problems.

Example 7.21 **Winning the Lottery** Suppose that you have purchased a lottery ticket with the number 956. The winning number will be a three-digit number between 000 and 999. What is the probability that you will win? Let the event A = winning number is 956. Here are two methods for finding $P(A)$:

Method 1: With the physical assumption that all 1000 possibilities are equally likely, it is clear that $P(A) = 1/1000$.

Method 2: Define three events:

- B_1 = First digit drawn is 9
- B_2 = Second digit drawn is 5
- B_3 = Third digit drawn is 6

Event A occurs if and only if all three of these events occur. With the physical assumption that each of the ten digits (0, 1, . . . , 9) is equally likely on each draw, $P(B_1) = P(B_2) = P(B_3) = 1/10$. Since these events are all independent, we can apply the Extension of Rule 3b to find $P(A) = (1/10)^3 = 1/1000$.

Both of these methods are effective in determining the correct answer, but the first method is faster.

Example 7.22 **Prizes in Cereal Boxes** A particular brand of cereal contains a prize in each box. There are four possible prizes, and any box is equally likely to contain each of the four prizes. What is the probability that you will receive two different prizes if you purchase two boxes? Here are the known facts:

- The probability of receiving each prize is 1/4 for each box.
- The prizes you receive in two boxes are independent of each other.

There are a number of equally correct ways to solve this problem. The simplest method is to recognize that no matter what you receive in the first box, the probability of receiving a different prize in the second box is 3/4. Therefore, the probability of receiving two different prizes is 3/4. Here is a more detailed method: Label the four prizes as a, b, c, d and define an outcome as the pair of prizes from the first and second box. Since outcomes are independent, the probability of any pair is found from Rule 3b to be $(1/4)(1/4) = 1/16$. The sample space is

{aa, ab, ac, ad, ba, bb, bc, bd, ca, cb, cc, cd, da, db, dc, dd}

There are 12 disjoint ways to get two different prizes, so if you reapply Rule 2b, the probability of getting two different prizes is $12(1/16) = 12/16 = 3/4$.

None and at Least One

A common situation in probability is when we want to know the probability that a particular outcome never happens when a random circumstance is repeated several times, with independent outcomes from one time to the next. A similar problem in this type of situation is to find the probability that a particular outcome happens at least once. The following example shows how we can use the probability rules from Section 7.4 to solve these kinds of problems.

Example 7.23 **Will Shaun's Friends Be There for Him?** Shaun has been studying all night for an exam and is afraid he will oversleep and miss the exam. He sends messages to three friends asking them to call him around 11 A.M. and wake him up. Suppose that the probability that each friend will call is .7 and is independent for the three friends. What is the probability that none of them will call? What is the probability that at least one of them will call?

Note that "at least one" is the complement of "none" because either no friends call, or at least one of them calls, and these two outcomes are mutually exclusive. Therefore, $P(\text{none call}) + P(\text{at least one calls}) = 1$.

Finding $P(\text{none call})$ is easier, because there is only one way that can happen. Here is how the probability rules are used to find it:

For each friend, use Rule 1 to find $P(\text{friend does not call}) = 1 - P(\text{friend calls}) = 1 - .7 = .3$.

Now use the extension of Rule 3b:

$P(\text{none call}) = P(\text{friend 1 does not call } \textit{and} \text{ friend 2 does not call } \textit{and} \text{ friend 3 does not call}) = (.3)(.3)(.3) = .027$.

Therefore, $P(\text{at least one friend calls}) = 1 - P(\text{none call}) = 1 - .027 = .973$.

Bayes' Rule

Another common situation in probability occurs when we know conditional probabilities in one direction and need to find them in the other direction. For example, we may know the probability that a medical test gives a positive result, conditional on having the disease and conditional on not having the disease. If your test result is positive, what is of most interest to you is the conditional probability that you have the disease, given that the test result is positive.

If you know $P(B|A)$ but are trying to find $P(A|B)$, you can use Rule 3a to find the two pieces in $P(B) = P(A \text{ and } B) + P(A^C \text{ and } B)$, then use Rule 4 to find $P(A|B)$. We can represent this with a formula, called **Bayes' Rule**.

FORMULA **Bayes' Rule states:**

$$P(A|B) = \frac{P(A \text{ and } B)}{P(B|A)P(A) + P(B|A^C)P(A^C)}$$

This rule looks complicated, but note that the denominator is simply $P(B)$ broken into two parts, $P(A \text{ and } B) + P(A^C \text{ and } B)$, each written using Rule 3a.

Example 7.24 **Optimism for Alicia—She Is Probably Healthy** In Case Study 7.1 (p. 220), Alicia learned that she had a positive result for a diagnostic test of a certain disease. If you were in Alicia's position, you would be most interested in finding the probability of having the disease *given* that the test was positive. Let's use Bayes' Rule to find that probability. Define the following events for her disease status:

A = Alicia has the disease
A^C = Alicia does not have the disease
$P(A) = 1/1000 = .001$ (from her physician's knowledge)
$P(A^C) = 1 - .001 = .999$

Define the following events for her test results:

B = test is positive
B^C = test is negative

For these events, we only know conditional probabilities, from her physician:

$P(B|A) = .95$ (positive test given disease)
$P(B^C|A) = .05$ (negative test given disease)
$P(B|A^C) = .05$ (positive test given no disease)
$P(B^C|A^C) = .95$ (negative test given no disease)

Let's find the denominator for Bayes' Rule first. The denominator provides $P(B)$:

$P(B) = P(B|A)P(A) + P(B|A^C)P(A^C) = (.95)(.001) + (.05)(.999) = .00095 + .04995 = .0509.$

Using Bayes' Rule:

$$P(A|B) = \frac{P(\text{A and B})}{P(B)} \frac{.00095}{.0509} = .019$$

Note what this tells Alicia! Even though her test was positive the probability that she *actually has disease D is only .019!* There is less than a 2% chance that Alicia has the disease, even though her test was positive. The fact that so many more people are disease-free than diseased means that the 5% of disease-free people who falsely test positive far outweigh the 95% of those with the disease who legitimately test positive.

Two-Way Tables and Tree Diagrams—Two Useful Tools

Probability questions such as the one in Example 7.24 are much easier to solve if we use visual tools. There are two useful tools for illustrating combinations of events and answering questions about them. The two tools, constructing a hypothetical two-way table and constructing a tree diagram, are particularly useful when conditional probabilities are already known in one direction, such as $P(A|B)$. The tools can then be used to find $P(\text{A and B})$, $P(B|A)$, and other probabilities of interest.

Constructing a Two-Way Table: The Hypothetical Hundred Thousand

When conditional or joint probabilities are known for two events, it is sometimes useful to construct a hypothetical two-way table of outcomes for a round number of individuals, and 100,000 is a good number because usually it results in whole numbers of people in each cell. Therefore, we will call this method the **hypothetical hundred thousand**. Let's illustrate the idea with an example.

Example 7.25

Read the original source on the companion website, http://www.cengage.com/statistics/Utts4e.

Two-Way Table for Teens and Gambling In Example 7.12, a study was reported in which ninth-grade Minnesota teens were asked whether they had gambled at least once a week during the past year. The sample consisted of 49.1% boys and 50.9% girls. The proportion of boys who had gambled weekly was .229, while the proportion of girls who had done so was only .045. If we were to construct a table for a hypothetical group of 100,000 teens reflecting these proportions, there would be (.491)(100,000) = 49,100 boys and 50,900 girls. Of the 49,100 boys, (.229 × 49,100) or 11,244 would be weekly gamblers, and of the 50,900 girls, .045, or 2291 girls, would be weekly gamblers. The hypothetical hundred thousand table can be constructed by starting with these numbers and filling in the rest by subtraction and addition. The resulting table follows.

	Weekly Gambler	Not Weekly Gambler	Total
Boy	11,244	37,856	49,100
Girl	2,291	48,609	50,900
Total	13,535	86,465	100,000

It's easy to find various probabilities from this table. Here are some examples:

$P(\text{boy and weekly gambler}) = 11{,}244/100{,}000 = .11244$, so about 11% of ninth graders are boys who gamble weekly.

P(boy | weekly gambler) = 11,244/13,535 = .8307, so about 83% of weekly gamblers are boys.

P(weekly gambler) = 13,535/100,000 = .13535, so about 13.5% of ninth graders are weekly gamblers.

Tree Diagrams

Another useful tool for combinations of events is a tree diagram. A **tree diagram** is a schematic representation of the sequence of events and their probabilities, including conditional probabilities based on previous events for events that happen sequentially. It is easiest to illustrate by example.

Example 7.26 **Alicia's Possible Fates** Let's consider what could have happened with Alicia's medical test and the probabilities associated with those options. Two random circumstances are involved. The first circumstance is whether she has the disease or not (*D* or *not D*), and the second circumstance is whether the test result is positive or negative (*positive test* or *negative test*). We have already specified the probabilities of each of the pieces, and using Rule 3a, we can find the probabilities of the various combinations, such as the probability that Alicia has the disease but the test result is negative.

Figure 7.1 illustrates these probabilities in a tree diagram. The initial set of branches shows the two possibilities and associated probabilities for her disease status. The next set of branches shows the two possibilities for the outcome of the test and associated *conditional probabilities* depending on whether she has the disease or not. The final probability shown for each combination of branches is found by multiplying the probabilities on the branches, applying Rule 3a. Thus, for instance, the probability that Alicia has the disease *and* has a positive test is .00095.

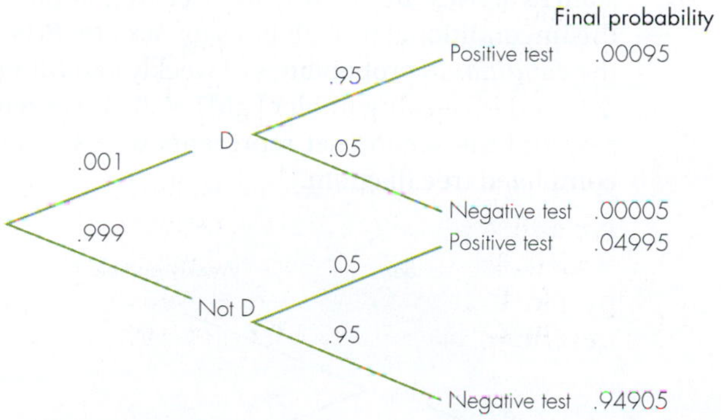

Figure 7.1 Tree diagram for Alicia's medical test

Here are the general steps for creating a tree diagram:

Step 1: Determine the first random circumstance in the sequence, and create the first set of branches to illustrate possible outcomes for it. Create one branch for each outcome, and write the associated probability on the branch. The first set of branches is the only one that contains *unconditional* probabilities.

Step 2: Determine the next random circumstance, and append branches for the possible outcomes to each of the branches in Step 1. Write the associated *conditional probabilities* on the branches, where the outcome for Step 2 is *conditional* on the branch taken in Step 1.

Step 3: Continue this process for as many steps (and sets of branches) as necessary.

Step 4: To determine the probability of following any particular sequence of branches, multiply the probabilities on those branches. This is an application of Rule 3a.

Step 5: To determine the probability of any collection of sequences of branches, add the individual probabilities for those sequences, as found in Step 4. This is an application of Rule 2b.

Example 7.27 **The Probability That Alicia Has a Positive Test** Note that Alicia's test is positive if she follows branch D and then branch "positive test" *or* if she follows branch "not D" and then branch "positive test." The overall probability that the test is positive can be found by adding the probabilities from these two sets of branches. Hence,

$$P(\text{positive test}) = (.001)(.95) + (.999)(.05) = .00095 + .04995 = .0509$$

In other words, for all women like Alicia who are tested, including the .001 of them with the disease and the .999 of them without the disease, a positive test will occur with probability .0509. This is equivalent to saying that .0509, or 5.09%, of these women will have a positive test.

It is always the case that the first set of branches in a tree diagram displays probabilities for the outcomes of one random circumstance, then the second set of branches displays *conditional* probabilities for the outcomes of another random circumstance, *given* what happened on the first one. In general, the first set of branches corresponds to the circumstance that occurs first chronologically, if the events have a time order. If they don't have a time order, then think about what probabilities are known. You must know the unconditional probabilities of outcomes for the events on the first set of branches.

Example 7.28 **Tree Diagram for Teens and Gambling** Let's construct a tree diagram using the information provided in Example 7.12 about the gambling behavior of ninth-grade Minnesota teenagers. For a randomly selected ninth grader, the two random circumstances are sex (boy or girl) and weekly gambler (yes or no). For this study we know that the unconditional probabilities for sex are: $P(\text{boy}) = .491$ and $P(\text{girl}) = .509$. We know the *conditional* probabilities of weekly gambling, *given* sex: $P(\text{weekly gambler} \mid \text{boy}) = .229$ and $P(\text{weekly gambler} \mid \text{girl}) = .045$. Therefore, the first set of branches represents sex, and the second set represents weekly gambler or not. Figure 7.2 illustrates the completed tree diagram.

Read the original source on the companion website, http://www.cengage.com/statistics/Utts4e.

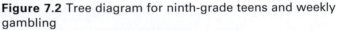

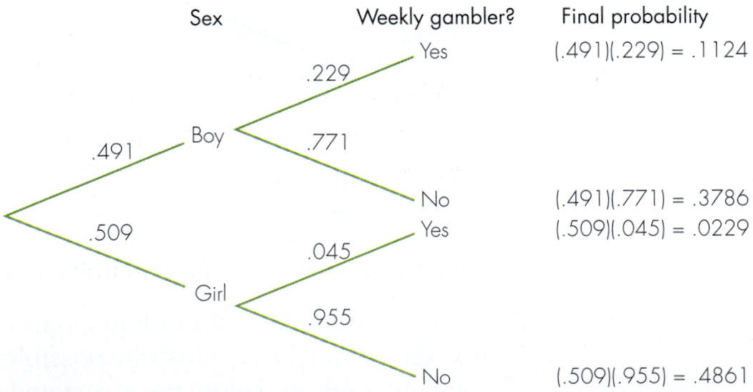

Figure 7.2 Tree diagram for ninth-grade teens and weekly gambling

Once the tree diagram is completed, we can easily find probabilities for any combination of sex *and* weekly gambling by multiplying the probabilities across branches. For instance,

$$P(\text{boy and weekly gambler}) = (.491)(.229) = .1124$$
$$P(\text{girl and not weekly gambler}) = (.509)(.955) = .4861$$

We can find other probabilities by adding outcomes from two sets of branches. For instance,

$$P(\text{weekly gambler}) = P(\text{boy and weekly gambler}) + P(\text{girl and weekly gambler}) = (.1124) + (.0229) = .1353$$

Finally, we can find conditional probabilities in the opposite direction to those known by finding probabilities such as those we just found and then applying Rule 4. For instance,

7.5 Exercises are on pages 256–257.

$$P(\text{boy} \mid \text{weekly gambler}) = \frac{P(\text{boy and weekly gambler})}{P(\text{weekly gambler})} = \frac{.1124}{.1353} = .8307$$

TECHNICAL NOTE **Probabilities, Proportions, and Percentages**

In many of the examples in this chapter, we have discussed probabilities associated with a "randomly selected individual" from a population. These are equivalent to percentages in the population with specific characteristics, but multiplying percentages doesn't work. For instance, in Example 7.28, 49.1%, or .491, of the teens were boys, and of those, 22.9%, or .229, were weekly gamblers. To find what *percentage* of the teens were boys *and* weekly gamblers, it would not work to multiply (49.1%)(22.9%) = 1124.39%. Instead, convert everything to proportions, multiply, then convert back to a percentage by multiplying by 100%: (.491)(.229) = .1124, or .1124 × 100% = 11.24% of the teens are boys who gamble weekly.

THOUGHT QUESTION 7.6 Explain why the tree diagram in Figure 7.1 (p. 241) displayed disease status first and test results second rather than the other way around.*

7.6 Using Simulation to Estimate Probabilities

Some probabilities are so difficult or time consuming to calculate that it is easier to **simulate** the situation repeatedly by using a computer or calculator and observe the relative frequency of the event of interest. If you simulate the random circumstance n times and the outcome of interest occurs in x out of those n times, then the estimated probability for the outcome of interest is x/n. This is an estimate of the long-run relative frequency with which the outcome would occur in real life.

Example 7.29 **Getting All the Prizes** Refer to the situation in Example 7.22 (p. 238). Suppose that cereal boxes each contain one of four prizes and any box is equally likely to contain any of the four prizes. You would like to collect all four prizes. If you buy six boxes of cereal, how likely is it that you will get all four prizes?

It would be quite time consuming to solve this problem using the rules presented in the previous sections. It is much simpler to simulate the situation using a computer or calculator. Table 7.3 shows 50 simulations using Minitab computer software. In each case, six digits were drawn, each equally likely to be 1, 2, 3, or 4. The outcomes that contain all four digits are highlighted in boldface.

Table 7.3 50 Simulations of Prizes from Cereal Boxes

112142	443222	323324	321223	**332314**
224123	342324	**232413**	121123	422244
244412	**342431**	121333	**244132**	434224
234323	**144332**	121142	**432121**	144313
443313	211141	**421342**	441211	111134
113432	433424	**312314**	241114	313411
214443	222422	144441	**213141**	312232
312341	411124	111422	312213	**323314**
143124	111131	**441233**	121223	424433
213341	114333	**243311**	442244	**214133**

*__HINT:__ See the paragraph just before Example 7.28 on page 242.

Counting the outcomes in bold reveals that 19 out of 50, or .38, of these simulations satisfy the condition that all four prizes are collected. In fact, this simulation had particularly good results. There are $4^6 = .4096$ equally likely possible outcomes for listing six digits randomly chosen from 1, 2, 3, 4, and 1560 of those outcomes include all four digits. So the actual probability of getting all four prizes is $1560/4096 = .3809$.

In general, determining the accuracy of simulation for estimating a probability is similar to determining the margin of error when estimating a population proportion from a sample survey. The probability estimated by simulation will be no further off than $1/\sqrt{n}$ most of the time, where n is the number of simulated repetitions of the situation.

This information about the accuracy of the simulation approach for estimating probabilities should make it clear that you need to repeat the simulated situation a large number of times to obtain accurate results. Therefore, it is most convenient if you can use a computer to generate the simulated outcomes and also to count how many of them satisfy the event of interest.

Example 7.30 **Finding Gifted ESP Participants** An ESP test is conducted by randomly selecting one of five video clips and playing it in one building while a participant in another building tries to describe what is playing. Later, the participant is shown the five video clips and is asked to determine which one best matches the description he or she had given. By chance, the participant would get this correct with probability 1/5. Individual participants are each tested eight times, with five new video clips each time. They are identified as "gifted" if they guess correctly at least five times out of the eight tries. Suppose that people actually do have some ESP and can guess correctly with probability .30 (instead of the .20 expected by chance). What is the probability that a participant will be identified as "gifted"?

In Chapter 8, you will learn how to solve this kind of problem, but we can simulate the answer using a random number generator that produces the digits 0, 1, 2, ... , 9 with equal likelihood. Many calculators and computers will simulate these digits. Here are the steps needed for one "repetition":

- Each "guess" is simulated with a digit, equally likely to be 0 to 9.
- For each participant, we simulate eight "guesses" resulting in a string of eight digits.
- If a digit is 7, 8, or 9, we count that guess as "correct," so $P(\text{correct}) = 3/10 = .3$, as required in the problem. If the digit is 0 to 6, the guess is "incorrect." (There is nothing special about 7, 8, 9; we could have used any three digits.)
- If there are five or more "correct" guesses (digits 7, 8, 9), we count that as "gifted."

The entire process is repeated many times, and the proportion of times that the process results in a "gifted" participant provides an estimate of the desired probability. Let's use Minitab to simulate the experiment for 1000 participants, each making eight guesses.

Simulation Process and Results Using Minitab

1. Generate 1000 rows of eight columns, using random integers from 0 to 9. Each row represents the eight guesses for one participant.
2. Use the "code" feature to recode all of the values, (7,8,9)→1, (0,1,2,3,4,5,6)→0.
3. For each row (representing a participant) you now have eight column entries, and each is either 1 (correct) or 0 (incorrect). Sum the eight column entries; the sum is the number of "correct" guesses. Enter these 1000 sums into a new column.
4. The sums are now in a column with 1000 rows. Tally this column to find out how many of the participants had five or more "correct guesses."
5. The results of doing this were as follows:

Number Correct	0	1	2	3	4	5	6	7	8
Frequency	64	187	299	271	119	44	14	2	0

Note that there were 44 + 14 + 2 + 0 = 60 individuals who got five or more correct, so the probability of finding a "gifted" participant each time is about 60/1000 = .06. In other words, if everyone is equally talented and each guess is correct with probability .3, then there will be five or more correct guesses out of eight tries with probability .06, or about 6% of the time.

7.6 Exercises are on pages 257–258.

THOUGHT QUESTION 7.7 Does using simulation to estimate probabilities rely on the relative frequency or the personal probability interpretation of probability? Explain. Whichever one you chose, could simulation be used to find probabilities for the other interpretation? Explain.*

7.7 Flawed Intuitive Judgments about Probability

People have poor intuition about probability assessments. In Case Study 7.1, Alicia's physician made a common mistake by informing her that the medical test was a good indicator of her disease status because it was 95% accurate as to whether she had the disease or not. As we learned in Example 7.24, the probability that she actually has the disease is small, about 2%. In this section, we explore this phenomenon, called *confusion of the inverse*, and some other ways in which intuitive probability assessments can be seriously flawed.

Confusion of the Inverse

Eddy (1982) posed the following scenario to 100 physicians:

> One of your patients has a lump in her breast. You are almost certain that it is benign; in fact you would say there is only a 1% chance that it is malignant. But just to be sure, you have the patient undergo a mammogram, a breast x-ray designed to detect cancer.
>
> You know from the medical literature that mammograms are 80% accurate for malignant lumps and 90% accurate for benign lumps. In other words, if the lump is truly malignant, the test results will say that it is malignant 80% of the time and will falsely say it is benign 20% of the time. If the lump is truly benign, the test results will say so 90% of the time and will falsely declare that it is malignant only 10% of the time.
>
> Sadly, the mammogram for your patient is returned with the news that the lump is malignant. What are the chances that it is truly malignant?

Most of the physicians to whom Eddy posed this question thought the probability that the lump was truly malignant was about 75% or .75. In truth, given the probabilities described in the scenario, *the probability is only .075*. The physicians' estimates were 10 times too high! When Eddy asked the physicians how they arrived at their answers, he realized they were confusing the answer to the actual question with the answer to a different question. "When asked about this, the erring physicians usually report that they assumed that the probability of cancer given that the patient has a positive x-ray was approximately equal to the probability of a positive x-ray in a patient with cancer" (1982, p. 254).

Robyn Dawes has called this phenomenon **confusion of the inverse** (Plous, 1993, p. 132). The physicians were confusing the conditional probability of cancer *given a positive x-ray* with the inverse, the conditional probability of a positive x-ray *given that the patient has cancer!*

It is not difficult to see that the correct answer is indeed .075. Let's construct a hypothetical table of 100,000 women who fit this scenario. In other words, these are women who would present themselves to the physician with a lump for which the

***HINT:** Could you tell a computer what proportions to use if you and your friend do not agree on what to use?

probability that it was malignant seemed to be about 1%. Thus, of the 100,000 women, about 1%, or 1000 of them, would have a malignant lump. The remaining 99%, or 99,000, would have a benign lump.

Further, given that the test was 80% accurate for malignant lumps, it would show a malignancy for 800 of the 1000 women who actually had one. Given that it was 90% accurate for the 99,000 women with benign lumps, it would show benign for 90%, or 89,100 of them, and malignant for the remaining 10%, or 9900 of them. Table 7.4 shows how the 100,000 women would fall into these possible categories.

Table 7.4 Breakdown of Actual Status versus Test Status for a Rare Disease

	Test Says Malignant	Test Says Benign	Total
Actually Malignant	800	200	1,000
Actually Benign	9,900	89,100	99,000
Total	10,700	89,300	100,000

Let's return to the question of interest. Our patient has just received a positive test for malignancy. Given that her test showed malignancy, what is the actual probability that her lump is malignant? Of the 100,000 women, 10,700 of them would have an x-ray showing malignancy. But of those 10,700 women, only 800 of them actually have a malignant lump! Thus, given that the test showed a malignancy, the probability of malignancy is just $800/10,700 = 8/107 = .075$.

Sadly, many physicians are guilty of confusion of the inverse. Remember, in a situation in which the *base rate* for a disease is very low and the test for the disease is less than perfect, there will be a relatively high probability that a positive test result is a false positive. If you ever find yourself in a situation similar to the one just described, you may wish to construct a table like the one above.

To determine the probability of a positive test result being accurate, you need only three pieces of information:

1. What the base rate or probability that you are likely to have the disease is, without any knowledge of your test results.
2. What the **sensitivity** of the test is, which is the proportion of people who correctly test positive when they actually have the disease.
3. What the **specificity** of the test is, which is the proportion of people who correctly test negative when they don't have the disease.

Note that items 2 and 3 are measures of the accuracy of the test. They do not measure the probability that people have the disease when they test positive or the probability that they do not have the disease when they test negative. Those probabilities, which are obviously the ones of interest to the patient, can be computed by constructing a table similar to Table 7.4 or by the other methods shown earlier in this chapter.

Specific People versus Random Individuals

According to the Federal Aviation Administration (http://www.ntsb.gov/aviation/Table6.htm), between 1989 and 2008, there were 64 fatal airline accidents for regularly scheduled flights on U.S. carriers. During that same time period, there were 227,166,913 flight departures. That means that the relative frequency of fatal accidents was about one accident per 3.5 million departures. Based on these kinds of statistics, you will sometimes hear statements like, "The probability that you will be in a fatal plane crash is 1 in 3.5 million" or "The chance that your marriage will end in divorce is 50%."

Do these probability statements really apply to you personally? In an attempt to personalize the information, reporters express probability statements in terms of individuals when they actually apply to the aggregate. Obviously, if you never fly, the probability that you will be in a fatal plane crash is 0. Here are two equivalent, correct ways to restate the aggregate statistics about fatal plane crashes:

- In the long run, about 1 flight departure out of every 3.5 million ends in a fatal crash.

- The probability that a randomly selected flight departure ends in a fatal crash is about 1/3,500,000.

If you have had a terrific marriage for 30 years, it is not likely to end in divorce now. Your probability of a divorce surely is less than the 50% figure often reported in the media. Here are two correct ways to express aggregate statistics about divorce:

- In the long run, about 50% of marriages end in divorce.
- At the beginning of a randomly selected marriage, the probability that it will end in divorce is about .50.

Note the emphasis on "random selection" in the second version of each statement. Sometimes the phrase *randomly selected* is omitted, but it should always be understood that randomness is part of the communication.

Coincidences

When one of the authors of this book was in college in upstate New York, she visited Disney World in Florida during summer break. While there, she ran into three people she knew from her college, none of whom were there together. A few years later, she visited the top of the Empire State Building in New York City and ran into two friends (who were there together) and two additional unrelated acquaintances. Years later, when traveling from London to Stockholm, she ran into a friend at the airport in London while waiting for the flight. The friend not only turned out to be taking the same flight but had been assigned an adjacent seat.

These events are all examples of what would commonly be called coincidences. They are certainly surprising, but are they improbable? Most people think that coincidences have low probabilities of occurring, but we shall see that our intuition can be quite misleading regarding such phenomena. We will adopt the definition of **coincidence** proposed by Persi Diaconis and Fred Mosteller (1989, p. 853):

DEFINITION A **coincidence** is a surprising concurrence of events, perceived as meaningfully related, with no apparent causal connection.

The mathematically sophisticated reader may wish to consult the article by Diaconis and Mosteller, in which they provide some instructions on how to compute probabilities for coincidences. For our purposes, we will need nothing more sophisticated than the simple probability rules we encountered earlier in this chapter.

Following are some examples of coincidences that at first glance seem highly improbable.

Example 7.31 **Two George D. Brysons** "My next-door neighbor, Mr. George D. Bryson, was making a business trip some years ago from St. Louis to New York. Since this involved weekend travel and he was in no hurry . . . and since his train went through Louisville, he asked the conductor, after he had boarded the train, whether he might have a stopover in Louisville.

"This was possible, and on arrival at Louisville he inquired at the station for the leading hotel. He accordingly went to the Brown Hotel and registered. And then, just as a lark, he stepped up to the mail desk and asked if there was any mail for him. The girl calmly handed him a letter addressed to Mr. George D. Bryson, Room 307, that being the number of the room to which he had just been assigned. It turned out that the preceding resident of Room 307 was another George D. Bryson" (Weaver, 1963, pp. 282–283).

Example 7.32 **Identical Cars and Matching Keys** Plous (1993, p. 154) reprinted an Associated Press news story describing a coincidence in which a man named Richard Baker and his wife were shopping on April Fool's Day at a Wisconsin shopping center. Mr. Baker went out to get their car, a 1978 maroon Concord, and drove it around to pick up his wife. After driving for a short while, they noticed items in the car that did not look familiar. They checked the license plate, and sure enough, they had someone else's car. When they drove back to the shopping center (to find the police waiting for them), they discovered that the owner of the car they were driving was a Mr. Thomas Baker, no relation to Richard Baker. Thus, both Mr. Bakers were at the same shopping center at the same time, with identical cars and with matching keys. The police estimated the odds as "a million to one."

Example 7.33 **Winning the Lottery Twice** Moore (1997, p. 330) reported on a *New York Times* story of February 14, 1986, about Evelyn Marie Adams who won the New Jersey lottery twice in a short time period. Her winnings were \$3.9 million the first time and \$1.5 million the second time. When Ms. Adams won for the second time, the *New York Times* claimed that the odds of one person winning the top prize twice were about 1 in 17 trillion. Then, in May 1988, Robert Humphries won a second Pennsylvania lottery, bringing his total winnings to \$6.8 million.

Most people think that the events just described are exceedingly improbable, and they are. *What is not improbable is that someone, somewhere, someday will experience those events or something similar.*

When we examine the probability of what appears to be a startling coincidence, we ask the wrong question. For example, the figure quoted by the *New York Times* of 1 in 17 trillion is the probability that a *specific* individual who plays the New Jersey State Lottery exactly twice will win both times (Diaconis and Mosteller, 1989, p. 859). However, millions of people play the lottery every day, and it is not surprising that someone, somewhere, someday would win twice.

In fact, Purdue professors Stephen Samuels and George McCabe (cited in Diaconis and Mosteller, 1989, p. 859) calculated those odds to be practically a sure thing. They calculated that there was at least a 1 in 30 chance of a double winner in a 4-month period and better than even odds that there would be a double winner in a 7-year period somewhere in the United States. And they used conservative assumptions about how many tickets past winners had purchased.

When you experience a coincidence, remember that there are almost 7 billion people in the world and over 308 million in the United States. If something has a 1 in a million probability of occurring to each individual on a given day, it will occur to an average of over 308 people in the United States *each day* and to almost 7000 people in the world *each day*. Of course, probabilities of specific events depend on individual circumstances, but we hope you can see that it is not unlikely that something surprising will happen quite often.

Example 7.34 **Sharing the Same Birthday** Here is a famous example that you can use to test your intuition about surprising events. How many people would need to be gathered together to be at least 50% sure that two of them share the same birthday (month and day, not necessarily year)? Most people provide answers that are much higher than the correct answer, which is that only 23 people are needed.

There are several reasons why people have trouble with this problem. If your answer was somewhere close to 183, or half the number of birthdays, then you may have confused the question with another one, such as the probability that someone in the group has *your* birthday or that two people have a specific date as their birthday.

It is not difficult to see how to calculate the appropriate probability using our probability rules. Note that the only way to avoid two people having the same birthday is if all 23 people have different birthdays. To find that probability, we simply use the rule that applies to the word *and* (Rule 3), thus multiplying probabilities. The probability that the first three people have different birthdays is the probability that the second person does not share a birthday with the first, which is 364/365 (ignoring February 29), and the third person does not share a birthday with either of the first two, which is 363/365. (Two dates were already taken.)

Continuing this line of reasoning, the probability that none of the 23 people share a birthday is

$$\frac{(364)(363)}{(365)(365)} \cdots \frac{(343)}{(365)} = .493$$

The probability that at least two people share a birthday is the probability of the complement, or $1 - .493 = .507$.

If you find it difficult to imagine that this could be correct, picture it this way. Imagine each of the 23 people shaking hands with the other 22 people and asking them about their birthday. There would be 253 handshakes and birthday conversations. Surely, there is a relatively high probability that at least one of those pairs would discover a common birthday.

By the way, the probability of a shared birthday in a group of ten people is already better than one in nine, at .117. (There would be 45 handshakes.) With only 50 people, it is almost certain, with a probability of .97. (There would be 1225 handshakes.)

The technique used in Example 7.34 can be used to illustrate why some coincidences are not at all unlikely. Consider the situation described in Example 7.31, in which two George D. Brysons occupied the same hotel room in succession. It is not at all unlikely that occasionally, two successive occupants of a hotel room somewhere, sometime will have the same name. To see that this is so, consider the probability of the complement—that no matter who occupies a hotel room, the next occupant will have a different name. Even ignoring common names, if we add these probabilities across many thousands of hotel rooms and many days, weeks, and years, it is extremely unlikely that successive occupants would always have different names.

Example 7.35 **Unusual Hands in Card Games** As a final example of unlikely coincidences, consider a card game such as bridge, in which a standard 52-card deck is dealt to four players, so they each receive 13 cards. Any specific set of 13 cards is equally likely, each with a probability of about 1 in 635 billion. You would probably not be surprised to get a mixed hand, say, the 4, 7, and 10 of hearts; 3, 8, 9, and jack of spades; 2 and queen of diamonds; and 6, 10, jack, and ace of clubs. Yet that specific hand is just as unlikely as getting all 13 hearts! The point is that any very specific event, surprising or not, has extremely low probability, but there are many, many surprising events, and their combined probability is quite high.

Magicians sometimes exploit the fact that many small probabilities add up to one large probability by doing a trick in which they don't tell you what to expect in advance. They set it up so that *something* surprising is almost sure to happen. When it does, you are likely to focus on the probability of *that particular outcome*, rather than realizing that a multitude of other outcomes would have also been surprising and that one of them was likely to happen.

To summarize, most coincidences seem improbable only if we ask for the probability of that specific event occurring at that time, to us. If, instead, we ask the probability of it occurring some time, to someone, the probability can become quite large. Further, because of the multitude of experiences we each have every day, it is not surprising that some of them may appear to be improbable. That specific event is undoubtedly improbable. What is not improbable is that something "unusual" will happen to each of us once in awhile.

The Gambler's Fallacy

Another common misperception about random events is that they should be self-correcting. Another way to state this is that people think the long-run frequency of an event should apply even in the short run. This misperception has classically been called the **gambler's fallacy**. Researchers William Gehring and Adrian Willoughby (2000) have discovered that this fallacy and related decision making while gambling are reflected in activity in the brain, in "a medial-frontal region in or near the anterior cingulate cortex" (pp. 2279).

The gambler's fallacy can lead to poor decision making, especially if applied to gambling. For example, people tend to believe that a string of good luck will follow a string of bad luck in a casino. Unfortunately, independent chance events have no such "memory". Making ten bad gambles in a row doesn't change the probability that the next gamble will also be bad.

Note that the gambler's fallacy primarily applies to independent events. The gambler's fallacy may not apply to situations in which knowledge of one outcome affects probabilities of the next. For instance, in card games using a single deck, knowledge of what cards have already been played provides information about what cards are likely to be played next, and gamblers routinely make use of that information. If you normally receive lots of mail but have received none for two days, you would probably assess that you are likely to receive more than usual the next day, and you probably would be correct.

A somewhat different but related misconception is what Tversky and Kahneman (1982) call the belief in the **law of small numbers**, "according to which [people believe that] even small samples are highly representative of the populations from which they are drawn" (p. 7). They report that "in considering tosses of a coin for heads and tails, for example, people regard the sequence HTHTTH to be more likely than the sequence HHHTTT, which does not appear random, and also more likely than the sequence HHHHTH, which does not represent the fairness of the coin" (p. 7). Remember that any specific sequence of heads and tails is just as likely as any other sequence if the coin is fair, so the idea that the first sequence is more likely is a misperception.

7.7 Exercises are on pages 258–259.

THOUGHT QUESTION 7.8 If you wanted to pretend to be psychic, you could do a "cold reading" on someone you do not know. Suppose that you are doing this for a 25-year-old woman. You make statements such as the following:

- I see that you are thinking of two men, one with dark hair and the other one with slightly lighter hair or complexion. Do you know who I mean?
- I see a friend who is important to you but who has disappointed you recently.
- I see that there is some distance between you and your mother that bothers you.

Using the material in this section, explain why this would often work to convince people that you are psychic.*

***HINT:** Think about all of the possible ways in which these statements could be true for you.

CASE STUDY 7.2 Doin' the iPod Random Shuffle

The ability to play a collection of songs in a random order is a popular feature of portable digital music players. As an example, the first generation Apple iPod Shuffle player had 512 megabytes of memory and could store about 120 songs. Newer players with larger memory can store and randomly order thousands of songs. When the shuffle function is used, the stored songs are played in a random order.

We mention the iPod because there has been much grumbling, particularly on the Internet, that its shuffle might not be random. Some users complain that a song might be played within the first hour in two or three consecutive random shuffles. A similar complaint is that there are clusters of songs by the same group or musician within the first hour or so of play. *Newsweek* magazine's technology writer, Steven Levy (31 January 2005), wrote, "From the day I first loaded up my first iPod, it was as if the little devil liked to play favorites. It had a particular fondness for Steely Dan, whose songs always seemed to pop up two or three times in the first hour of play. Other songs seemed to be exiled to a forgotten corner" (p. 10). Conspiracy theorists even accuse Apple of playing favorites, giving certain musicians a better chance to have their songs played early in the shuffle.

One fundamental cause of the complaints may be a dislike of the consequences of randomness. People may be expecting a more systematic reordering than randomness often provides. For instance, there seems to be a belief that two songs by the same musician or group should not land near each other in the reordering. In a random shuffle, every possible reordering of the items is equally likely. (With a 120-song list, the number of possible shuffles is about 669 followed by 196 zeros.) Among all possible reorderings, some will have a few songs by the same musician being played nearly consecutively. This is analogous to occasionally getting several heads in a row when flipping coins or finding three aces within the first few cards of a randomly shuffled deck of playing cards.

For illustrative purposes, suppose that an iPod Shuffle user has stored 10 albums with 12 songs per album. What is the probability that two or more songs from the same album will be among the first four songs played in a random shuffle? We can find this by first determining the probability that all of the first four songs of the shuffle are from different albums and then subtracting that probability from 1. Define events:

A = first song is anything (120 choices)
B = second song is from a different album than the first song (108 of the remaining 119 songs)
C = third song is from a different album than the first two songs (96 of the remaining 118 songs)
D = fourth song is from a different album than the first three songs (84 of the remaining 117 songs)

The probability that all four songs are from different albums is

$$P(\text{A and B and C and D}) = \frac{120}{120} \times \frac{108}{119} \times \frac{96}{118} \times \frac{84}{117} = .53$$

Thus, the probability that all four songs are *not* from different albums, meaning that at least two songs are from the same album, is found using Rule 1:

P(at least two of first four are from the same album) $= 1 - .53 = .47$

At least two of the first four songs will be from the same album about half of the time. If we extend the calculation to the first five songs in the random order, we can use the same method that was just used to learn that the probability is .671 that at least two songs will be from the same album. In fact, at least three of the first five songs in the random shuffle will be from the same album about 10% of the time.

We have set up a specific scenario to do these calculations, but the basic moral of the story remains the same for any situation in which we randomly select or order items. We too often imagine that we see patterns in random sequences and we tend to expect more structure than randomness provides. For example, we may expect a random shuffle to play one song by every musician before playing a second song by any musician. If this does not happen, we think there is a nonrandom pattern.

Key Terms

In Summary Boxes

Exercises

◆ Denotes that the dataset is available on the companion website, http://www.cengage.com/statistics/Utts4e, but is not required to solve the problem.

Bold exercises have answers in the back of the text.

Section 7.1

Skillbuilder Exercises

7.1 According to a U.S. Department of Transportation website (http://www.bts.gov/press_releases/2010/dot045_10/html/dot045_10.html), 78.7% of domestic flights flown by the top 18 U.S. airlines in January 2010 arrived on time. Represent this in terms of a random circumstance and an associated probability.

7.2 Answer Thought Question 7.1 on page 221.

7.3 Jan is a member of a class with 20 students that meets daily. Each day for a week (Monday to Friday), a student in Jan's class is randomly selected to explain how to solve a homework problem. Once a student has been selected, he or she is not selected again that week. If Jan was not one of the four students selected earlier in the week, what is the probability that she will be picked on Friday? Explain how you found your answer.

General Section Exercises

7.4 Find information on a random circumstance in the news. Identify the circumstance and possible outcomes, and assign probabilities to the outcomes. Explain how you determined the probabilities.

7.5 Identify three random circumstances in the following story, and give the possible outcomes for each of them:

It was Robin's birthday and she knew she was going to have a good day. She was driving to work, and when she turned on the radio, her favorite song was playing. Besides, the traffic light at the main intersection she crossed to get to work was green when she arrived, something that seemed to happen less than once a week. When she arrived at work, rather than having to search as she usually did, she found an empty parking space right in front of the building.

7.6 Answer Thought Question 7.2 on page 221.

7.7 Give an example of a random circumstance in which:

a. The outcome is not determined until we observe it.
b. The outcome is already determined, but our knowledge of it is uncertain.

Section 7.2

Skillbuilder Exercises

7.8 Is each of the following values a legitimate probability value? Explain any "no" answers.

a. .50
b. .00
c. 1.00
d. 1.25
e. −.25

7.9 Suppose that you live in a city that has 125,000 households with telephones and a polling organization randomly selects 1000 of them to phone for a survey. What is the probability that your household will be selected?

7.10 A car dealer has noticed that 1 out of 25 new-car buyers returns the car for warranty work within the first month.

a. Write a sentence expressing this fact as a proportion.
b. Write a sentence expressing this fact as a percent.
c. Write a sentence expressing this fact as a probability.

7.11 Which interpretation of probability (relative frequency or personal) applies to each of the following situations? If it's the relative frequency interpretation, explain which of the methods listed in the "In Summary" box at the end of Section 7.2 applies.

a. If a spoon is tossed 10,000 times and lands with the rounded head face up 3000 of those times, we would say that the probability of the rounded head landing face up for that spoon is about .30.
b. In a debate with you, a friend says that she thinks there is a 50:50 chance that God exists.
c. Based on data from the 1991 to 1993 General Social Survey, the probability was about 612/1669 = .367 in those years that a randomly selected adult who had ever been married had divorced.

General Section Exercises

7.12 Explain which interpretation of probability (relative frequency or personal) applies to each of these statements and how you think the probability was determined.

a. According to Krantz (1992, p. 161), the probability that a randomly selected American will be injured by lightning in a given year is 1/685,000.
b. According to my neighbor, the probability that the tomato plants she planted last month will actually survive to produce fruit is only 1/2.
c. According to the nursery where my neighbor purchased her tomato plants, if a plant is properly cared for, the probability that it will produce tomatoes is .99.
d. The probability that a husband will outlive his wife (for U.S. couples) is 3/10 (Krantz, 1992, p. 163).

7.13 Refer to Exercise 7.5. Suppose that Robin wants to find the probability associated with the outcomes in the random circumstances contained in the story. Identify one of the circumstances, and explain how she could determine the probabilities associated with its outcomes.

7.14 Casino games often use a fair die that has six sides with 1 to 6 dots on them. When the die is tossed or rolled, each of the six sides is equally likely to come out on top. Using the physical assumption that the die is fair, determine the probability of each of the following outcomes for the number of dots showing on top after a single roll of a die:

a. Six dots.
b. One or two dots.
c. An even number of dots.

7.15 A computer solitaire game uses a standard 52-card deck and randomly shuffles the cards for play. Theoretically, it should be possible to find optimal strategies for playing and then to compute the probability of winning based on the best strategy. Not only would this be an extremely complicated problem, but also the probability of winning for any particular person depends on that person's skill. Explain how an individual could determine his or her probability of winning this game. (*Hint:* It might take a while to determine the probability!)

7.16 Give an example of a situation for which a probability statement makes sense but for which the relative frequency interpretation could not apply, such as the probability given by Carl Sagan for an asteroid hitting Earth.

7.17 Alicia's statistics class meets 50 times during the semester, and each time it meets, the probability that she will be called on to answer the first question is 1/50. Does this mean that Alicia will be called on to answer the first question exactly once during the semester? Explain.

7.18 Every day, John buys a lottery ticket with the number 777 for the lottery described in Example 7.2. He has played 999 times and has never won. He reasons that since tomorrow will be his 1000th time and the probability of winning is 1/1000, he will have to win tomorrow. Explain whether John's reasoning is correct.

7.19 Refer to Example 7.5 (p. 224). What is the probability that a randomly selected child who slept in darkness would develop some degree of myopia?

Section 7.3

Skillbuilder Exercises

7.20 A student wants to send a bouquet of roses to her mother for Mother's Day. She can afford to buy only two types of roses and decides to randomly pick two different varieties from the following four choices: Blue Bell, Yellow Success, Sahara, and Aphrodite. Label these varieties B, Y, S, A.

a. Make a list of the six simple events in the sample space.
b. Assuming that all outcomes are equally likely, what is the probability that she will pick Sahara and Aphrodite?

7.21 Suppose that we randomly select a student and record how many days in the previous week the student exercised for at least half an hour. Make a list of the simple events in the sample space.

7.22 Refer to Exercise 7.21 in which the number of days a randomly selected student exercised in the previous week is recorded. List the simple events that make up each of these events:

a. The student exercised on at least 5 days.
b. The student exercised on at most 2 days.
c. The student did not exercise at all.

7.23 Remember that the event A^C is the complement of the event A.

a. Are A and A^C mutually exclusive? Explain.
b. Are A and A^C independent? Explain.

7.24 A penny and a nickel are each tossed once. Explain whether the outcomes for the two coins are:

a. Independent events.
b. Complementary events.
c. Mutually exclusive events.

7.25 Suppose that A, B, and C are all disjoint possible outcomes for the same random circumstance. Explain whether each of the following sets of probabilities is possible.

a. $P(A) = 1/3, P(B) = 1/3, P(C) = 1/3$.
b. $P(A) = 1/2, P(B) = 1/2, P(C) = 1/4$.
c. $P(A) = 1/4, P(B) = 1/4, P(C) = 1/4$.

7.26 Suppose that events A and B are mutually exclusive with $P(A) = 1/2$ and $P(B) = 1/3$.

a. Are A and B independent events? Explain how you know.
b. Are A and B complementary events? Explain how you know.

General Section Exercises

7.27 Refer to Example 7.10 (p. 229), in which you and your friend Vanessa enter a drawing for a free lunch in Week 1 and again in Week 2. Events defined were A = you win in Week 1, B = Vanessa wins in Week 1, C = Vanessa wins in Week 2.

a. Are events B and C independent? Explain.
b. Suppose that after Week 1, the cards that were not drawn as the winning card are retained in the bowl for the drawing in Week 2. Are events B and C independent? Explain.

7.28 Jill and Laura have lunch together. They flip a coin to decide who pays for lunch and then flip a coin again to decide who pays the tip. Define a possible outcome to be who pays for lunch and who pays the tip, in order—for example, "Jill, Jill."

a. List the simple events in the sample space.
b. Are the simple events equally likely? If not, why not?
c. What is the probability that Laura will have to pay for lunch and leave the tip?

7.29 When a fair die is tossed, each of the six sides (numbers 1 to 6) is equally likely to land face up. Two fair dice, one red and one green, are tossed. Explain whether the following pairs of events are mutually exclusive, independent, both, or neither:

a. A = red die is a 3; B = red die is a 6.
b. A = red die is a 3; B = green die is a 6.

7.30 Refer to Exercise 7.29, in which a red die and a green die are each tossed once. Explain whether the following pairs of events are mutually exclusive, independent, both, or neither:

a. A = red die and green die sum to 4; B = red die is a 3.
b. A = red die and green die sum to 4; B = red die is a 4.

7.31 Use the information given in Case Study 7.1 and the "physical assumption" method of assigning probabilities to argue that on any given day, the probability that Alicia has to answer one of the three questions is 3/50.

7.32 When 190 students were asked to pick a number from 1 to 10, the number of students selecting each number were as follows:

Number	1	2	3	4	5	6	7	8	9	10	Total
Frequency	2	9	22	21	18	23	56	19	14	6	190

a. What is the approximate probability that someone asked to pick a number from 1 to 10 will pick the number 3?
b. What is the approximate probability that someone asked to pick a number from 1 to 10 will pick one of the two extremes, 1 or 10?
c. What is the approximate probability that someone asked to pick a number from 1 to 10 will pick an odd number?

7.33 Refer to Exercise 7.32. Suppose that a student is asked to choose a number from 1 to 10. Define event A to be that the student chooses the number 5, and event B to be that the student chooses an even number.

a. What is $P(A \text{ and } B)$?
b. Are events A and B independent? Explain how you know.
c. Are events A and B mutually exclusive? Explain how you know.

7.34 According to Krantz (1992, p. 102), "[In America] women between the ages of 20 and 24 have a 90 percent chance of being fertile while women between 40 and 44 have only a 37 percent chance of bearing children [i.e., being fertile]." Define appropriate events, and write these statements as conditional probabilities, using appropriate notation.

7.35 Refer to Exercise 7.34. Suppose that an American woman is randomly selected. Are her age and her fertility status independent? Explain.

7.36 Refer to Exercise 7.34. Which method of finding probabilities do you think was used to find the "90 percent chance" and "37 percent chance"? Explain.

7.37 Refer to Case Study 7.1. Define C_1, C_2, and C_3 to be the events that Alicia is called on to answer questions 1, 2, and 3, respectively.

a. Based on the physical situation used to select students, what is the (unconditional) probability of each of these events? Explain.
b. What is the *conditional* probability of C_3, *given* that C_1 occurred?
c. Are C_1 and C_3 independent events? Explain.

Section 7.4

Skillbuilder Exercises

7.38 Two fair dice are rolled. The event "A = getting the same number on both dice" has probability = 1/6.

a. Describe in words what the event A^C is.
b. What is the probability of A^C?

7.39 A fair coin is tossed three times. The event "A = getting all heads" has probability = 1/8.

a. Describe in words what the event A^C is.
b. What is the probability of A^C?

7.40 Two fair coins are tossed. Define

A = Getting a head on the first coin

B = Getting a head on the second coin

A and B = Getting a head on both the first and second coins

A or B = Getting a head on the first coin, or the second coin, or both coins

a. Find P(A) = the probability of A.
b. Find P(B) = the probability of B.
c. Using the multiplication rule (Rule 3b), find P(A and B).
d. Using the addition rule (Rule 2a), find P(A or B).

7.41 Suppose that people are equally likely to have been born on any day of the week. You meet a new friend, and ask her on what day of the week she was born.

a. List the simple events in the sample space.
b. What is the probability that she was born on a weekend (Saturday or Sunday)?

7.42 Refer to Exercise 7.41. Now suppose you meet two new friends independently, and ask each of them on what day of the week they were born.

a. What is the probability that the first friend was born on a Friday?
b. What is the probability that both friends were born on a Friday?
c. What is the probability that both friends were born on the same day of the week?

7.43 A popular lottery game is one in which three digits from 0 to 9 are chosen, so the winning number can be any of the 1000 numbers from 000 to 999. Define:

A = the first digit is odd

B = the first digit is even

C = the second digit is odd

a. State two of these events that are mutually exclusive.
b. State two of these events that are independent.
c. Are any two of these events complements? If so, which two events are they?

7.44 Refer to Exercise 7.43. Find the following probabilities:

a. P(A)
b. P(B)
c. P(A and B)
d. P(A or B)
e. P(A and C)
f. P(A or C)

7.45 Julie is taking English and history. Suppose that at the outset of the term, her probabilities for getting As are

P(grade of A in English class) = .70.

P(grade of A in history class) = .60.

P(grade of A in both English and history classes) = .50.

a. Are the events "grade of A in English class" and "grade of A in history class" independent? Explain how you know.
b. Use the addition rule for two events (Rule 2) to find the probability that Julie will get at least one A between her English and history classes.

7.46 In a recent election, 55% of the voters were Republicans, and 45% were not. Of the Republicans, 80% voted for Candidate X, and of the non-Republicans, 10% voted for Candidate X. Consider a randomly selected voter. Define

A = Voter is Republican.

B = Voted for Candidate X.

a. Write values for P(A), $P(A^C)$, P(B|A), and $P(B|A^C)$.
b. Find P(A and B), and write in words what outcome it represents.
c. Find $P(A^C$ and B), and write in words what outcome it represents.
d. Using the results in parts (b) and (c), find P(B). (*Hint:* The events in parts (b) and (c) cover all of the ways in which B can happen.)
e. Use the result in part (d) to state what percent of the vote Candidate X received.

7.47 In each situation, explain whether the selection is made with replacement or without replacement.

a. The three digits in the lottery in Example 7.2 (p. 223).
b. The three students selected to answer questions in Case Study 7.1 (p. 220).
c. Five people selected for extra security screening while boarding a particular flight.

7.48 In each situation, explain whether the selection is made with replacement or without replacement.

a. The two football teams selected to play in the Rose Bowl in a given year.
b. The cars stopped by the police for speeding in five consecutive mornings on the same stretch of highway.
c. The winning lottery numbers for 2 consecutive days of the same daily lottery game.

General Section Exercises

7.49 In Example 7.17 (p. 233), we found the probability that both of two unrelated strangers share your birth month. In this exercise, we find the probability that at least one of the two strangers shares your birth month. Assume that all 12 months are equally likely.

a. What is the probability that the first stranger does *not* share your birth month?
b. What is the probability that the second stranger does *not* share your birth month?
c. What is the probability that *neither* of them shares your birth month?
d. Use the result from part (c) to find the probability that at least one of them shares your birth month.

7.50 Refer to Exercise 7.49. In this exercise, another method is used for finding the probability that at least one of two unrelated strangers shares your birth month.

a. What is the probability that the first stranger shares your birth month?
b. What is the probability that the second stranger shares your birth month?
c. What is the probability that both of them share your birth month?
d. Use Rule 2a ("either A or B or both") to find the probability that at least one of the strangers shares your birth month.

7.51 In Example 7.15, we found that the probability that a woman with two children either has two girls or two boys is .5002. What is the probability that she has one child of each sex?

7.52 Harold and Maude plan to take a cruise together, but they live in separate cities. The cruise departs from Miami, and they each book a flight to arrive in Miami an hour before they need to be on the ship. Their travel planner explains that Harold's flight has an 80% chance of making it on time for him to get to the ship and that Maude's flight has a 90% chance of making it on time.

a. How do you think the travel planner determined these probabilities?

b. Assuming that the probabilities quoted are long-run relative frequencies, do you think whether Harold's plane is on time is independent of whether Maude's plane is on time on the particular day they travel? Explain. (*Hint:* Consider reasons why planes are delayed.)

c. Whether realistic or not, assuming that the probabilities are independent, what is the probability that Harold and Maude will both arrive on time?

d. What is the probability that one of the two will be cruising alone?

7.53 A raffle is held in a club in which 10 of the 40 members are good friends with the president. The president draws two winners.

a. If the two winners are drawn *with replacement*, what is the probability that a friend of the president wins each time?

b. If the two winners are drawn *without replacement*, what is the probability that a friend of the president wins each time?

c. If the two winners are drawn *with replacement*, what is the probability that neither winner is a friend of the president?

d. If the two winners are drawn *without replacement*, what is the probability that neither winner is a friend of the president?

7.54 A robbery has been committed in an isolated town. Witnesses all agree that the criminal was driving a red pickup truck and had blond hair. Evidence at the scene indicates that the criminal also smoked cigarettes. Police determine that 1/50 of the vehicles in town are red pickup trucks, 30% of the residents smoke, and 20% of the residents have blond hair. The next day they notice a red pickup truck whose driver has blond hair and is smoking. They arrest the driver for the robbery.

a. Is it reasonable to assume that whether someone drives a red pickup truck, smokes cigarettes, and has blond hair are all independent traits? Explain.

b. Assuming that the traits listed are all independent, what proportion of the vehicle owners in town are red pickup truck owners who smoke and have blond hair?

c. Continuing to assume the traits listed are independent, if there are 10,000 vehicle owners in town, how many of them fit the description of the criminal?

d. Assuming that there is no other evidence, what is the probability that the driver arrested by the police is innocent of the robbery?

e. The prosecuting attorney argues that the probability that someone would own a red pickup truck and smoke cigarettes and have blond hair is very small, so the person arrested by the police must be guilty. Do you agree with this reasoning? Explain.

Section 7.5

Skillbuilder Exercises

7.55 A public library carries 50 magazines, each of which focuses on either news or sports. Thirty of the magazines focus on news and the remaining 20 focus on sports. Among the 30 news magazines, 20 include international news and 10 include national, state, or local news only. Among the 20 sports magazines, 5 focus on international sports, and the remaining 15 focus on national, state, or local sports.

a. Create a table illustrating these numbers, with type of magazine (news, sports) as the row variable.

b. A customer randomly picks a magazine from the shelf and notices that it is a news magazine. Given that it is a news magazine, what is the probability that it includes international news?

c. What proportion of the magazines include international news or international sports?

7.56 In a large general education class, 60% (.6) are science majors and 40% (.4) are liberal arts majors. Twenty percent (.2) of the science majors are seniors, while 30% (.3) of the liberal arts majors are seniors.

a. If there are 100 students in the class, how many of them are science majors?

b. If there are 100 students in the class, how many of them are science majors and seniors?

c. Create a hypothetical table of 100 students with major (science, liberal arts) as the row variable and class (senior, nonsenior) as the column variable, illustrating the proportions given in the exercise.

d. Use the table created in part (c) to determine what percent of the class are seniors.

7.57 Refer to Exercise 7.56.

a. Create a tree diagram for this situation.

b. Use the tree diagram in part (a) to determine what percentage of the class are seniors.

7.58 In a computer store, 30% (.3) of the computers in stock are laptops and 70% (.7) are desktops. Five percent (.05) of the laptops are on sale, while 10% (.1) of the desktops are on sale. Use a tree diagram to determine the percentage of the computers in the store that are on sale.

7.59 In an Italian breakfast café, a waitress has observed that 80% of her customers order coffee and 25% of her customers order both biscotti and coffee. Define

A = a randomly selected customer orders coffee.

B = a randomly selected customer orders biscotti.

a. Express the waitress's observations as probability statements involving events A and B.
b. What is the conditional probability that a randomly selected customer would order a biscotti, given that he or she orders coffee?
c. What is the conditional probability that a customer would not order a biscotti, given that he or she orders coffee?

7.60 Two students each use a random number generator to pick an integer between 1 and 7. What is the probability that they pick the same number?

General Section Exercises

7.61 A standard poker deck of cards contains 52 cards of which four are aces. Suppose that two cards are drawn sequentially, so that one random circumstance is the result of the first card drawn and a second random circumstance is the result of a second card drawn. Find the probability that the first card is an ace and the second card is not an ace.

7.62 Suppose that a magnet high school includes grades 11 and 12, with half of the students in each grade. Half of the senior class and 30% of the junior class are taking calculus. Create a hypothetical table of 100 students for this situation. Use grade (junior, senior) as the row variable.

7.63 Refer to Exercise 7.62. Suppose that a calculus student is randomly selected to accompany the math teachers to a conference. What is the probability that the student is a junior?

7.64 An airline has noticed that 40% of its customers who buy tickets don't take advantage of advance-purchase fares and the remaining 60% do. The no-show rate for those who don't have advance-purchase fares is 30%, while for those who do have them, it is only 5%.

a. Create a tree diagram for this situation.
b. Create a "hypothetical hundred thousand" table for this situation.
c. Using the tree diagram from part (a) or the table from part (b), find the percent of customers who are no-shows.
d. Given that a customer is a no-show, what is the probability that he or she had an advance-purchase fare?
e. Refer to the answer you found in part (d). Write a sentence presenting this result that someone with no training in statistics would understand.

7.65 Refer to the part of Example 7.20 (pp. 236–237) in which two students are drawn without replacement from a class of 30 students in which three are left-handed. Draw a tree diagram to illustrate that the probability of selecting two left-handed students is 1/145.

7.66 In a test for ESP, a picture is randomly selected from four possible choices and hidden away. Participants are asked to describe the hidden picture, which is unknown to anyone in the environment. They are later shown the four possible choices and asked to identify the one they thought was the hidden target picture. Because the correct answer had been randomly selected from the four choices before the experiment began, the probability of guessing correctly by chance alone is 1/4 or .25. Each time the experiment is repeated, four new pictures are used.

a. Researchers are screening people for potentially gifted ESP participants for a later experiment. They will accept only people who get three correct answers in three repetitions of the screening experiment. What is the probability that someone who is just guessing will be accepted for the later experiment?
b. Refer to part (a). Suppose that an individual has some ESP ability and can correctly guess with probability .40 each time. What is the probability that such a person will be accepted for the later experiment?
c. Refer to parts (a) and (b). Suppose that half of the people in the population being tested are just guessing and have no ESP ability, while the other half can actually guess correctly with probability .40 each time. Given that someone is accepted for the upcoming experiment, what is the probability that the person actually has ESP ability?

Section 7.6

Skillbuilder Exercises

7.67 Five fair dice were tossed, and the sum of the resulting tosses was recorded. This process was repeated 10,000 times using a computer simulation. The number of times the sum of the five tosses equaled 27 was 45. What is the estimated probability that the sum of the five dice will be 27?

7.68 The observed risk of an accident per month at a busy intersection without any stoplights was 1%. The potential benefit of adding a stoplight was studied by using a computer simulation modeling the typical traffic flow for a month at that intersection. In 10,000 repetitions, a total of 50 simulated accidents occurred.

a. What is the estimated probability of an accident in a month at the intersection with a stoplight added?
b. Would adding the stoplight be a good idea? Explain.

7.69 Refer to Example 7.29 (p. 243) and use the results given in Table 7.3 to estimate probabilities of the following outcomes:

a. Prize 4 is received at least three times.
b. Prize 4 is received at least three times *given* that the full collection of all four prizes was not received.
c. Prize 4 is received at least three times *and* the full collection of all four prizes was received.

General Section Exercises

7.70 Refer to the Minitab simulation results given on pages 244–245 for Example 7.30. What is the estimated probability that a participant would guess four or more correctly?

7.71 Refer to Example 7.30 (p. 244). Explain how you would change the simulation procedure if the assumption was that everyone was randomly guessing, so that the probability of a correct guess was .20 each time.

7.72 Refer to Example 7.30 (p. 244). Suppose that the probability of a correct guess each time is .40.

a. Explain how you would simulate this situation.
b. Carry out the simulation, and estimate the probability that a participant will be identified as gifted.

7.73 Suppose that in a state lottery game players choose three digits, each from the set 0 to 9, as they do in many state lottery games. But for this game, what counts is the *sum* of the three chosen digits. The state selects a winning sum from the possibilities 0 $(0 + 0 + 0)$ to 27 $(9 + 9 + 9)$ by randomly choosing three digits and using their sum. For instance, if the three digits 483 are selected, then a winning ticket is any ticket in which the three digits sum to $4 + 8 + 3 = 15$. As an example, a ticket with 195 would win because $1 + 9 + 5 = 15$.

a. Using the physical assumption that all 1000 choices (000, 001, ... , 999) are equally likely, find the probability that three digits chosen at random will sum to 15, so that the winning sum is 15.
b. Simulate playing this game 1000 times, randomly selecting the three digits each time. Based on the simulation, find the approximate probability that the winning sum is 15.
c. Based on information about how people respond when asked to pick random digits, do you think the proportion of people who would actually pick three digits summing to 15 is close to the probabilities you found in parts (a) and (b)? Explain.

7.74 Janice has noticed that on her drive to work, there are several things that can slow her down. First, she hits a red light with probability .3. If she hits the red light, she also has to stop for the commuter train with probability .4, but if she doesn't hit the red light, she has to stop with only probability .2.

a. Find the probability that Janice has to stop for both the red light and the train.
b. Find the probability that Janice has to stop for the red light *given* that she has to stop for the commuter train.
c. Explain how you would conduct a simulation to estimate the probabilities in parts (a) and (b).
d. Conduct the simulation described in part (c). Use at least 50 repetitions.
e. Compare the answers that you computed in parts (a) and (b) to the estimated probabilities from your simulation in part (d).

Section 7.7

Skillbuilder Exercises

7.75 In tossing a fair coin 10 times, if the first nine tosses resulted in all tails, will the chance be greater than .5 that the tenth toss will turn up heads? Explain.

7.76 The Pap smear is a screening test to detect cervical cancer. Estimate the sensitivity and specificity of the test if a study of 200 women with cervical cancer resulted in 160 testing positive, and in another 200 women without cervical cancer, 4 tested positive.

7.77 Which of the following sequences resulting from tossing a fair coin five times is most likely: HHHHH, HTHHT, or HHHTT? Explain your answer.

7.78 Using material from Section 7.7, explain what is wrong with the following statement: "The probability that you will win the million dollar lottery is about the same as the probability that you will give birth to quintuplets."

General Section Exercises

7.79 The University of California at Berkeley's *Wellness Encyclopedia* (1991) contains the following statement in its discussion of HIV testing: "In a high-risk population, virtually all people who test positive will truly be infected, but among people at low risk the false positives will outnumber the true positives. Thus, for every infected person correctly identified in a low-risk population, an estimated ten noncarriers [of HIV] will test positive" (p. 360). Suppose that you have a friend who is part of this low-risk population but who has just tested positive.

a. Using the numbers in the statement above, what is the probability that your friend actually carries the virus?
b. Understandably, your friend is upset and doesn't believe that the probability of being infected with HIV isn't really near 1. After all, the test is accurate, and it came out positive! Explain to your friend how the quoted statement can be true even though the test is very accurate both for people with HIV and for people who don't carry it. If it's easier, you can make up numbers to put in a table to support your argument.

7.80 A rare disease occurs in about 1 out of 1000 people who are similar to you. A test for the disease has sensitivity of 95% and specificity of 90%.

a. Create a hypothetical one hundred thousand table illustrating this situation, where the row categories are disease (yes, no) and the column categories are test results (positive, negative).
b. Draw a tree diagram for this situation.
c. Using the table from part (a) or the tree diagram from part (b), find the proportion of people who will test positive for the disease.
d. Find the probability that you actually have the disease given that your test results are positive.

7.81 Suppose that there are 30 people in your statistics class and you are divided into 15 teams of 2 students each. You happen to mention that your birthday was last week, upon which you discover that your teammate's mother has the same birthday you have (month and day, not necessarily year). Assume that the probability is 1/365 for any given day.

a. What is the probability that your teammate's mother would have the same birthday as yours?
b. Suppose that your teammate has two siblings and two parents, for a family of size five. What is the probability that at least one of your teammate's family has the same birthday as yours, assuming their birth dates are all independent?
c. Is the fact that your teammate's mother has the same birthday as yours a surprising coincidence? Explain.

7.82 Give an example of a situation in which the gambler's fallacy would not apply because the events are not independent.

7.83 A friend has three boys and would like to have a girl. She explains to you that the probability that her next baby will be a girl is very high because the law of averages says that she should have half of each and she already has three boys. Is she correct? Explain.

7.84 A friend, quite upset, calls you because she had a dream that a building had been bombed and she was helping to search for survivors. The next day, a terrorist bombed an embassy building in another country. Your friend is convinced that her dream was a warning and that she should have told someone in authority. Comment on this situation.

7.85 Tomorrow morning when you first arise, pick a three-digit number (anything from 000 to 999). You can choose randomly or simply decide what number you want to use. As you go through the day, note whether you encounter that number. It could be in a book, on a license plate, in a newspaper or magazine, on a digital clock when you happen to glance at it, part of the serial number on a bill, and so on. Write down all of the instances in which you encounter your chosen three-digit number. Are you surprised at how many or how few times you encountered the number? Explain.

7.86 Give an example of a coincidence that has occurred in your life. Using the material from this chapter, try to approximate the probability of exactly that event happening. Discuss whether the answer convinces you that something very odd happened to you.

7.87 Suppose that you are seated next to a stranger on an airplane and you start discussing various topics such as where you were born (what state or country), what your favorite movie of all time is, your spouse's occupation, and so on. For simplicity, assume that the probability that your details match for any given topic is 1/50 and is independent from one topic to the next. If you discuss 10 topics, how surprising would it be to find that you match on at least one of them?

7.88 Do you think that all coincidences can be explained by random events? Explain why or why not, using probability as the focus of your explanation. (There is no correct answer; your reasoning is what counts.)

Chapter Exercises

7.89 Answer Thought Question 7.7 on page 245.

7.90 Answer Thought Question 7.8 on page 250.

7.91 A psychological test identifies people as being one of eight types. For instance, Type 1 is "Rationalist" and applies to 15% of men and 8% of women. Type 2 is "Teacher" and applies to 12% of men and 14% of women. Each person fits one and only one type.

a. What is the probability that a randomly selected male is either a "Rationalist" or a "Teacher"?

b. What is the probability that a randomly selected female is not a "Teacher"?

c. Suppose that college roommates have a particularly hard time getting along with each other if they are both "Rationalists." A college randomly assigns roommates of the same sex. What proportion of male roommate pairs will have this problem?

d. Refer to part (c). What proportion of female roommate pairs will have this problem?

e. Using your answers to parts (c) and (d) and assuming that half of college roommate pairs are male and half are female, what proportion of all roommate pairs will both be "Rationalists"?

7.92 Refer to Exercise 7.91. A psychologist has noticed that "Teachers" and "Rationalists" get along particularly well with each other, and she thinks that they tend to marry each other. One of her colleagues disagrees and thinks that the types of spouses are independent of each other.

a. If the types are independent, what is the probability that a randomly selected opposite-sex married couple would consist of one "Rationalist" and one "Teacher"?

b. In surveys of thousands of randomly selected married couples, the psychologist has found that about 5% of them have one "Rationalist" and one "Teacher." Does this contradict her colleague's theory that the types of spouses are independent of each other? Explain.

7.93 New spark plugs have just been installed in a small airplane with a four-cylinder engine. For each spark plug, the probability that it is defective and will fail during its first 20 minutes of flight is 1/10,000, independent of the other spark plugs.

a. For any given spark plug, what is the probability that it will *not* fail during the first 20 minutes of flight?

b. What is the probability that none of the four spark plugs will fail during the first 20 minutes of flight?

c. What is the probability that at least one of the spark plugs will fail?

7.94 In Chapter 4, we learned that when there is no relationship between two categorical variables in a population, a "statistically significant" relationship will appear in 5% of the samples from that population, over the long run. Suppose that two researchers independently conduct studies to see whether there is a relationship between drinking coffee (regularly, sometimes, never) and having migraine headaches (frequently, occasionally, never).

a. If there really is no relationship in the population, what is the probability that the first researcher finds a statistically significant relationship?

b. If there really is no relationship in the population, what is the probability that both researchers find a statistically significant relationship?

7.95 A professor has noticed that even though attendance is not a component of the grade for his class, students who attend regularly obtain better grades. In fact, 40% of those who attend regularly receive As in the class, while only 10% of those who do not attend regularly receive As. About 70% of students attend class regularly. Find the following percentages:

a. The percentage who receive As *given* that they attend class regularly.

b. The percentage who receive As *given* that they do not attend class regularly.

c. The overall percentage who receive As.

7.96 Refer to Exercise 7.95. Draw a tree diagram, and use it to find the overall percent who receive As.

7.97 Refer to Exercises 7.95 and 7.96. Construct a hypothetical hundred thousand table for this situation.

7.98 Refer to Exercises 7.95, 7.96, and 7.97. Given that a randomly chosen student receives an A grade, what is the probability that he or she attended class regularly?

7.99 Recall from Chapter 2 that the median of a dataset is the value with at least half of the observations at or above it and at least half of the observations at or below it. Suppose that four individuals are randomly selected *with replacement* from a large class and asked how many hours they studied last week. Assume that the number of students in the class is even and that nobody has a response exactly equal to the median.

a. What is the probability that a particular individual will give a response that is above the median for the class?

b. Use the Extension of Rule 3b to find the probability that all four individuals will give a response that is above the median for the class.

c. If the individuals had been selected without replacement, would the Extension of Rule 3b still be applicable? Explain.

7.100 About one-third of all adults in the United States have type O+ blood. If three randomly selected adults donate blood, find the probability of each of the following events.

a. All three are type O+.

b. None of them is type O+.

c. Two out of the three are type O+.

7.101 The 2007 Working Group on California Earthquake Probabilities reported that "the likelihood of [an earthquake in California] of magnitude 7.5 or greater in the next 30 years is 46%." The report noted that "[t]he final forecast is a sophisticated integration of scientific fact and expert opinion" (source: http://pubs.usgs.gov/fs/2008/3027/fs2008-3027.pdf). Explain whether the figure of 46% is based on personal probability, long-run relative frequency, physical assumptions about the world, or some combination.

7.102 Recall that the Empirical Rule in Chapter 2 stated that for bell-shaped distributions, about 68% of the values fall within one standard deviation of the mean. The heights of women at a large university are approximately bell-shaped, with a mean of 65 inches and standard deviation of 2.5 inches. Use this information to answer the following questions:

a. What is the probability that a randomly selected woman from this university is 67.5 inches or taller?

b. What is the probability that two randomly selected women from this university are both 62.5 inches or shorter?

c. What is the probability that of two randomly selected women, one is 62.5 inches or shorter and the other is 62.5 inches or taller?

d. What is the probability that two randomly selected women are both 65 inches or taller?

7.103 According to the U.S. Census Bureau, "only 35 percent of the foreign-born people in the United States in 1997 were naturalized citizens, compared with 64 percent in 1970" (*Sacramento Bee*, October 15, 1999, p. A1).

a. What is the probability that two randomly selected foreign-born people in the United States in 1997 were both naturalized citizens?

b. Refer to part (a) of this exercise. Suppose that a married couple was randomly selected instead of two separate individuals. Do you think the probability that they were both naturalized citizens would be the same as the probability in part (a)? Explain.

c. Do you think the probabilities reported by the Census Bureau are relative frequency probabilities or personal probabilities? Explain.

d. A student writing a report about these statistics wrote, "If you had lived in the United States in 1970 and were foreign-born, the probability that you would be a naturalized citizen would have been .64. But by 1997, if you still lived in the United States, the probability that you would be a naturalized citizen would be only .35." Rewrite this sentence in a way that conveys the information correctly.

The following scenario applies to the remaining exercises. In Exercise 4.2, we presented a two-way table of counts for a random sample of people who had ever been married, demonstrating the proportions who smoked and who had ever been divorced. The numbers are shown again in the following table.

Smoking and Divorce, GSS Surveys, 1991–1993

	Ever Divorced?		
Smoke?	***Yes***	***No***	***Total***
Yes	238	247	485
No	374	810	1184
Total	612	1057	1669

Data source: SDA archive at U.C. Berkeley website, http://csa.berkeley.edu/archive.htm

Because this survey was based on a random sample in the United States in the early 1990s, the data should be representative of the adult population who had ever been married at that time. Answer these questions for a randomly selected member of that population.

7.104 Draw a tree diagram illustrating this situation, where the first set of branches represents smoking status and the second set represents "ever divorced."

7.105 What is the approximate probability that the person smoked?

7.106 What is the approximate probability that the person had ever been divorced?

7.107 Given that the person had been divorced, what is the probability that he or she smoked?

7.108 Given that the person smoked, what is the probability that he or she had been divorced?

7.109 Suppose that two people were randomly selected from that population, without replacement. What is the probability that one of them smoked but the other one did not (in either order)?

7.110 Suppose that two people were randomly selected from that population, without replacement. What is the probability that they had both been divorced?

7.111 Suppose that two people were randomly selected from that population. Given that one of them had been divorced, what is the probability that the other one had also been divorced?

8

Digital Vision/Jupiter Images

What is random about this event?

See Example 8.1 *(p. 264)*

Random Variables

Dogs come in a variety of breeds, sizes, and temperaments, but all dogs fit certain patterns on which veterinarians can rely when treating nearly any type of dog. Similarly, situations involving uncertainty and probability fall into certain broad classes, and we can use the same set of rules and principles for all situations within a class.

The numerical outcome of a random circumstance is called a *random variable.* In this chapter, we'll learn how to characterize the pattern of the distribution of the values that a random variable may have, and we'll learn how to use the pattern to find probabilities.

Patterns make life easier to understand and decisions easier to make. For instance, dogs come in a variety of breeds, sizes, and temperaments, but all dogs fit certain patterns on which veterinarians can rely when treating nearly any type of dog. If a veterinarian had to learn a different pattern for treating every breed, it might be nearly impossible for any individual to learn enough to be able to treat dogs in general.

Similarly, situations involving uncertainty and probability fall into certain broad classes, and we can use the same set of rules and principles for all situations within a class. We encountered this idea in Chapter 2 when we learned the Empirical Rule for bell-shaped distributions. All variables that follow a bell-shaped distribution have the same pattern, which is that about 68% of the values fall within 1 standard deviation of the mean, about 95% of the values fall within 2 standard deviations of the mean, and so on. This pattern holds whether the variable is heights of adult males, handspans of college-age females, or SAT scores of high school seniors.

8.1 What Is a Random Variable?

We usually assign a numerical value to a possible outcome of a random circumstance. As examples, we might count how many people in a random sample have type O blood, how many times we win when we play a lottery game every day for a month, or how much weight we will lose if we use a diet plan. Numerical characteristics like these are called *random variables.*

DEFINITION A **random variable** assigns a number to each outcome of a random circumstance, or, equivalently, a random variable assigns a number to each unit in a population.

Random variables are classified into two broad classes, and within each broad class, there are many specific families of random variables. A **family of random variables** consists of all random variables for which the same formula would be used to find probabilities. In considering random variables, the first step is to identify how the random variable fits into any known family. Then, the formulas for that family can be used. This will be easier than having to create rules for the random variable using the basic probability rules covered in Chapter 7.

Classes of Random Variables

The two different broad classes of random variables follow:

1. *Discrete random variables*
2. *Continuous random variables*

DEFINITION A **discrete random variable** can take one of a countable list of distinct values.

A **continuous random variable** can take any value in an interval or collection of intervals.

An example of a discrete random variable is the number of people with type O blood in a sample of ten individuals. The possible values are 0, 1, … , 10, a list of distinct values. An example of a continuous random variable is height for adult women. With accurate measurement to any number of decimal places, any height is possible within the range of possibilities for heights. Between any two heights, there always are other possible heights, so possible heights fall on an infinite continuum.

TECHNICAL NOTE **Discrete or Continuous?**

Sometimes a random variable fits the technical definition of a discrete random variable, but it is more convenient to treat it as a continuous random variable. Examples include incomes, prices, and exam scores. Sometimes continuous random variables are rounded off to whole units, giving the appearance of a discrete random variable, such as age in years or pulse rate to the nearest beat. In most of these situations, the number of possible values is large, and we are more interested in probabilities concerning intervals than specific values, so the methods for continuous random variables will be used.

Example 8.1 **Random Variables at an Outdoor Graduation or Wedding** Suppose you are participating in a major outdoor event, such as a graduation or wedding ceremony. Several random factors will determine how enjoyable the event will be, such as the temperature and the number of airplanes that fly overhead during the important speeches. In this context, *temperature* is a *continuous random variable* because it can take on any value in an interval. We often round off continuous random variables to the nearest whole number, as with temperature in degrees, but conceptually the value can be anything in an interval. In contrast, the *number of airplanes* that fly overhead during the event is a *discrete random variable.* The value can be 0, 1, 2, 3, and so on, but cannot be anything in between the integers, such as 2.5 airplanes.

A discrete random variable often has a finite number of outcomes, but an infinite number of outcomes is also possible. A countable list of values is such that the values can be counted one at a time, even if the counting never finishes. Example 8.2 illustrates that a discrete random variable can have an infinite number of possible values.

Example 8.2 **It's Possible to Toss Forever** Consider repeatedly tossing a fair coin, and define the random variable as follows:

$X =$ number of tosses until the first head occurs

Note that theoretically, X could equal any integer value from 1 to infinity because the coin could continue to come up tails indefinitely. The number of tosses that are required to get the first head is not likely to be large, but any number of flips is a possible outcome.

Families within the Two Broad Classes

There are specific "families" of random variables within each broad class. Variables within a family share the same structure and general rules for finding probabilities. Many computer software programs (such as Minitab and Excel) and calculators are programmed to provide probabilities for some of the more commonly encountered families.

The next example describes three seemingly different questions that all can be answered by using the same set of rules. We'll learn these rules in Section 8.4, where we will cover the family of *binomial random variables* for counting how often a particular event happens in a number of independent tries.

Example 8.3 **Probability an Event Occurs Three Times in Three Tries** As an example of a family of random variables, think about what the following three questions have in common:

- What is the probability that three tosses of a fair coin will result in three heads?
- Assuming that boys and girls are equally likely each time (which we learned in Chapter 7 is not quite true), what is the probability that three births will result in three girls?
- Assuming the probability is 1/2 that a randomly selected individual will be taller than the median height of a population, what is the probability that three randomly selected individuals will all be taller than the median?

These three questions all have the same answer, which is 1/8, found by using the multiplication rule in Chapter 7. Furthermore, if we modify each question to ask about the probability of getting one outcome of one type and two of the other, such as one boy and two girls, the answers will again be the same. (It will be 3/8.)

Probabilities for the number of heads in three tosses, girls in three births, and individuals falling above the median in three random selections are all found by using the same process. In each case, a random variable can be defined as X = number of times the "outcome of interest" occurs in three independent tries. It doesn't matter whether the "outcome of interest" is a head on a coin toss, a girl in a birth, or a randomly selected male taller than the median. The pattern for finding probabilities will be the same.

Probabilities for Discrete and Continuous Variables

We consider discrete and continuous random variables separately because probabilities are specified differently for them.

- For *discrete random variables,* we can find probabilities for exact outcomes.
- For *continuous random variables,* we cannot find probabilities for exact outcomes. Instead, we are limited to finding probabilities for intervals of values.

For example, we can find the probability that exactly two of the next three births at a hospital will be girls, because the number of girls is a discrete random variable. But we cannot find the probability that a newborn's weight is exactly 7 pounds, because weight is a continuous random variable. We could, however, find the probability that a newborn's weight is in an interval such as 6.9 to 7.1 pounds.

Example 8.4 **Waiting on Standby** Suppose you are on the standby list to board an airplane and you are first on the list. A discrete random variable of interest to you is the number of standby passengers who will be allowed to board the plane. You would be interested in the probability that the number is at least one. This could be found by adding the probabilities that exactly one standby gets on, exactly two standbys get on, and so on.

Alternatively, it could be found by subtracting the probability that no standbys get on from 1, because the total probability for all outcomes must equal 1.

You might also wonder about the probability that the flying time (a continuous random variable) will equal the time specified in the flight schedule. The probability that the flying time *exactly equals* that amount, to the exact second, would be essentially zero. In a practical sense, you might really be wondering about the probability that the flight time is within an interval of times close to the scheduled time, or the probability that the flight time will be less than the scheduled time. Generally, it doesn't make practical sense to talk about probabilities of exact values for continuous random variables. Moreover, there are mathematical difficulties that prevent us from being able to specify probabilities for all possible exact outcomes over an infinite continuum.

8.1 Exercises are on page 302.

THOUGHT QUESTION 8.1 If you know that the number of possible values a random variable can have is finite, do you know whether the random variable is discrete or continuous? Answer the same question for a random variable that can have an infinite number of possible values.*

IN SUMMARY Discrete and Continuous Random Variables

- A **discrete random variable** can take one of a countable list of distinct values. For discrete random variables we can find probabilities for exact outcomes.
- A **continuous random variable** can take any value in an interval or collection of intervals. For continuous variables we are limited to finding probabilities for intervals of values.

8.2 Discrete Random Variables

In this section, we learn how to organize and use probabilities for the possible values of a discrete random variable. Keep in mind that a discrete random variable has a countable list of distinct possibilities for its value and that we can specify a probability for each separate outcome value.

Probability Notation for a Discrete Random Variable

We will use the following notation to specify a probability for a possible outcome of a discrete random variable:

- X = the random variable.
- k = a specified number the discrete random variable could assume.
- $P(X = k)$ is the probability that X equals k.

As an example, suppose that we are interested in the probability that there are two girls in the next three births at a hospital and we find that the probability that this happens is 3/8. In this case, the random variable X = number of girls in the next three births, $k = 2$, and $P(X = 2) = 3/8$.

The Probability Distribution of a Discrete Random Variable

For a discrete random variable, we may either create a table or use a formula to give probabilities for each possible value. The correspondence between values and probabilities is called the **probability distribution function (pdf)** for the variable.

***HINT:** An interval has an infinite number of values in it. Also, see Example 8.2 (p. 264).

DEFINITION The **probability distribution function (pdf)** for a discrete random variable X is a table or rule that assigns probabilities to the possible values of the random variable X.

Note that the word *function* is used to mean either a table of values and probabilities or a formula (rule) that assigns probabilities to values.

Example 8.5 **Probability Distribution Function for Number of Courses** Suppose that 35% of the students at a college are currently taking four courses, 45% are taking five courses, and the remaining 20% are taking six courses. Let $X =$ number of courses that a randomly selected student is taking. Possible values are 4, 5, and 6, with probabilities .35, .45, and .20, respectively. The following table shows the probability distribution function for X by listing each value along with its probability.

k **= number of courses**	4	5	6
P(X = k)	.35	.45	.20

Note that the sum of probabilities over the three possible values of X is equal to 1 (calculated as $.35 + .45 + .20$). Also, each probability is between 0 and 1.

Conditions for Probabilities for Discrete Random Variables

Two conditions must always apply to the probabilities for discrete random variables:

Condition 1: The sum of the probabilities over all possible values of a discrete random variable must equal 1. Stated mathematically,

$$\Sigma_k P(X = k) = 1$$

Condition 2: The probability of any specific outcome for a discrete random variable must be between 0 and 1. Stated mathematically,

$$0 \le P(X = k) \le 1 \text{ for any value } k$$

Using the Sample Space to Find Probabilities for Discrete Random Variables

A random variable X gives a value to each simple event in the sample space of unique possible outcomes of a random circumstance. The probability for any particular value of X can be found by adding the probabilities for all of the simple events that have that value of X. The basic steps are as follows:

Step 1: List all simple events in the sample space.

Step 2: Identify the value of the random variable X for each simple event.

Step 3: Find the probability for each simple event (often simple events are equally likely).

Step 4: To find $P(X = k)$, the probability that $X = k$, add the probabilities for all simple events where $X = k$.

Example 8.6 **Probability Distribution Function for Number of Girls** Suppose that a family has three children and that the probability of a girl is 1/2 for each birth.[1] Define the random variable $X =$ number of girls among the three children. Possible values for X are 0, 1, 2, 3. What are the probabilities for these values?

The following chart helps to organize the solution. All eight possible arrangements of sexes for three children are listed (B = boy, G = girl), the number of girls is counted for each arrangement, and we note that each simple event has probability 1/8. (The eight simple events are equally likely.)

[1] In Chapter 7, we learned that *P*(girl) is about .488, but for simplicity here, we use .5.

Simple Event	BBB	BBG	BGB	GBB	BGG	GBG	GGB	GGG
X = No. of Girls	0	1	1	1	2	2	2	3
Probability	1/8	1/8	1/8	1/8	1/8	1/8	1/8	1/8

We can find the probability for each value of X by adding the probabilities of the simple events that have that value. The probabilities for the possible values of X are as follows:

$P(X = 0) = 1/8$ (BBB is the only event with $X = 0$)
$P(X = 1) = 3/8$ (added probabilities for the three simple events with $X = 1$)
$P(X = 2) = 3/8$ (added probabilities for the simple events with $X = 2$)
$P(X = 3) = 1/8$ (GGG is the only event with $X = 3$)

This set of probabilities make up the probability distribution (pdf) for X = number of girls. Note that these probabilities sum to 1 over all possible values of X and that all probabilities are between 0 and 1.

Graphing the Probability Distribution Function

It is often useful to represent a probability distribution function with a picture similar to a histogram.

- The possible outcome values are placed on the horizontal axis, and their probabilities are placed on the vertical axis.
- A bar is drawn centered on each possible value, with the height of the bar equal to the probability for that value.

The bars are scaled so that their total area is 1. Then the probability for any combination of values can be found as the area of the bars (rectangles) for those values. This method will be displayed in the next example.

Example 8.7 **Example 8.6 Revisited: Graph of pdf for Number of Girls** In Example 8.6, we determined the probability distribution function (pdf) for the number of girls in three births. Figure 8.1 displays the pdf. Note that the bar over each value forms a rectangle that has a width of 1 and a height equal to the probability for that value. Therefore, the area of each bar is equivalent to the probability for that value. For instance, $P(X = 2) = 3/8$ coincides with the area of the rectangle centered on $X = 2$, a rectangle that has height = 3/8 and width = 1. Later in this chapter, you will learn that equating probability to an area is a crucial and necessary idea when working with continuous variables. For discrete variables, however, we generally can use more direct ways to find the probability that X takes a particular value, so for discrete variables, we usually don't have to find the area of a rectangle to find a probability.

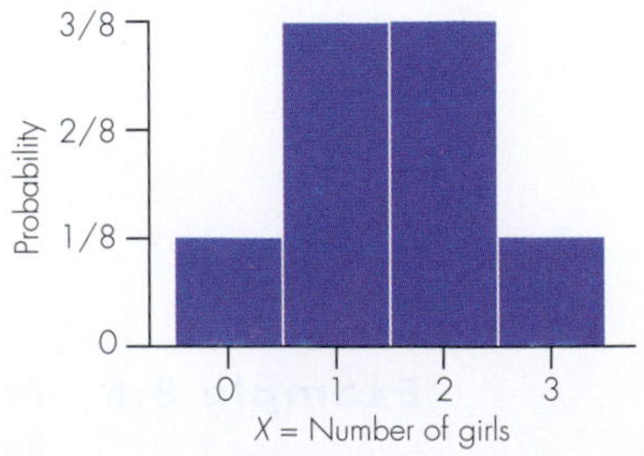

Figure 8.1 The pdf for X = number of girls in three births

The Cumulative Distribution Function of a Discrete Random Variable

A **cumulative probability** is the probability that the value of a random variable is *less than or equal to* a specific value. A few examples follow:

- Probability of getting *6 or fewer* answers right when guessing at answers to ten true-false questions
- Probability that there are *2 or fewer* televisions in a randomly selected household
- Probability that *3 or fewer* individuals have type O+ blood in a random sample of $n = 10$ people

The notation for a cumulative probability is $P(X \leq k)$, where X represents the random variable and k is a specific value.

DEFINITION The **cumulative distribution function (cdf)** for a random variable X is a table or rule that provides the probabilities $P(X \leq k)$ for any real number k. Generally, the term **cumulative probability** refers to the probability that X is less than or equal to a particular value.

For a discrete random variable, the cumulative probability $P(X \leq k)$ is the sum of probabilities for all values of X less than or equal to k.

Example 8.8 **Example 8.6 Revisited: Cumulative Distribution for the Number of Girls** In Example 8.6, we found that the pdf for $X =$ number of girls among three children in a family is as follows:

k	0	1	2	3
$P(X = k)$	1/8	3/8	3/8	1/8

For each specific value of X, the cumulative probability is the sum of probabilities for all values less than or equal to that value. The cumulative distribution function (cdf) for X is calculated as follows:

k	0	1	2	3
$P(X \leq k)$	1/8	1/8 + 3/8 = 4/8	1/8 + 3/8 + 3/8 = 7/8	1

Note that the cumulative probability for $X = 3$ must equal 1 because all possible values of X are less than or equal to 3.

Finding Probabilities for Complex Events Based on Random Variables

The probability rules in Chapter 7 all apply to events defined for random variables and can be used to find probabilities of complicated events.

Example 8.9 **Example 8.6 Revisited: A Mixture of Children** What is the probability that a family with three children will have at least one child of each sex? In Example 8.6, we defined the random variable $X =$ number of girls in three children. If there is at least one child of each sex, then either the family has one girl and two boys ($X = 1$) or two girls and one boy ($X = 2$). In Example 8.6, we found that the probability is 3/8 for each of these two possibilities. The probability can be found as

$$P(X = 1 \text{ or } X = 2) = P(X = 1) + P(X = 2) = 3/8 + 3/8 = 6/8 = 3/4$$

Note that we can simply add these probabilities because the events $X = 1$ and $X = 2$ are mutually exclusive events. A family with three children cannot have both exactly one girl ($X = 1$) and exactly two girls ($X = 2$).

THOUGHT QUESTION 8.2 In Example 8.9, we added the probabilities of $X = 1$ and $X = 2$ because they were mutually exclusive events. For any discrete random variable X, is it always true that $X = k$ and $X = m$ are mutually exclusive events, where k and m represent two values that X can have? Explain.*

Example 8.10 **Probabilities for Sum of Two Dice** Many gambling games use dice in which the number of dots rolled is equally likely to be 1, 2, 3, 4, 5, or 6. For instance, craps is a game in which two dice are rolled and the sum of the dots on the two dice determines whether the gambler wins, loses, or continues rolling the dice.

Let X = sum of two fair dice. Here is the sample space for the two dots and the corresponding values of the random variable X:

Dots	X	Dots	X	Dots	X	Dots	X	Dots	X	Dots	X
1,1	2	2,1	3	3,1	4	4,1	5	5,1	6	6,1	7
1,2	3	2,2	4	3,2	5	4,2	6	5,2	7	6,2	8
1,3	4	2,3	5	3,3	6	4,3	7	5,3	8	6,3	9
1,4	5	2,4	6	3,4	7	4,4	8	5,4	9	6,4	10
1,5	6	2,5	7	3,5	8	4,5	9	5,5	10	6,5	11
1,6	7	2,6	8	3,6	9	4,6	10	5,6	11	6,6	12

There are 36 equally likely simple events in the sample space, so each simple event has probability 1/36. To construct the probability distribution (pdf) for X, count the number of simple events resulting in each value of X and divide by 36. The probability distribution for the sum of two dice is as follows:

k	2	3	4	5	6	7	8	9	10	11	12
P(X = k)	1/36	2/36	3/36	4/36	5/36	6/36	5/36	4/36	3/36	2/36	1/36

Figure 8.2 displays the pdf with the rectangles for $X = 4$, 5, and 6 shaded. We can find $P(4 \le X \le 6)$ by adding the areas of the rectangles over the interval or by adding the probabilities for all individual values in that range. Here, $P(4 \le X \le 6) = 3/36 + 4/36 + 5/36 = 12/36 = 1/3$.

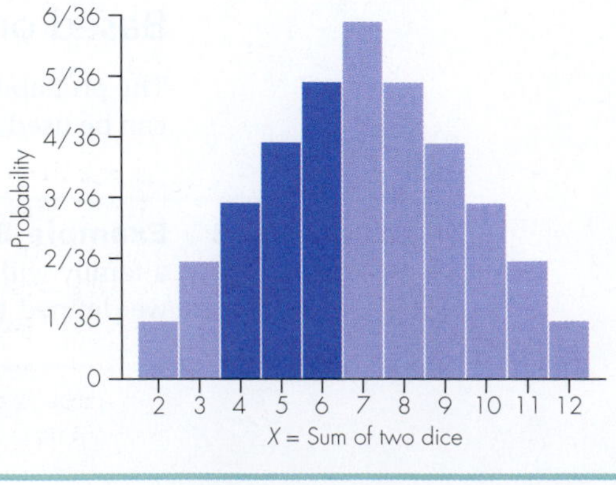

Figure 8.2 Probability distribution function for sum of two dice showing $P(4 \le X \le 6)$

8.2 Exercises are on pages 302–303.

***HINT:** See Step 4 on page 267. Are there simple events for which $X = k$ and $X = m$ at the same time?

THOUGHT QUESTION 8.3 Refer to the probability distribution for the sum of two dice. What is the value of $P(X \le 4)$? What does this probability measure? Explain what is measured by the value of $1 - P(X \le 4)$.*

IN SUMMARY Probabilities for Discrete Random Variables

- The **probability distribution function** for a discrete random variable is a table or a rule that gives probabilities for all possible outcomes of the variable.
- A **cumulative probability** is the probability that a variable X is less than or equal to a particular value.
- A **cumulative distribution function** is a table or rule that gives $P(X \le k)$ for any real number k.

8.3 Expectations for Random Variables

If we know probabilities for all possible values of a random variable, we can determine the mean outcome over the long run. This long-run average is called the *expected value* of the random variable. Synonymously, it is the **mean value of the random variable**.

As an example, suppose you play a lottery game once every week. For one play, the random variable that interests you is the amount that you win or lose. You won't be able to accurately predict the value of this random variable in advance for any given week, but as we show in this section, you will be able to determine the average amount that you can expect to win or lose per play over many weeks. This long-run average amount won or lost per play is the expected value, or mean value, of the random variable that describes possible outcomes for a single play.

DEFINITION The **expected value** of a random variable X is the mean value of the variable in the sample space or population of possible outcomes. *Expected value* can also be interpreted as the mean value that would be obtained from an infinite number of observations of the random variable.

Calculating Expected Value

The calculation of expected value, or mean value, for a discrete random variable is

Expected value = Sum of "value × probability"

In this formula, *value* is a possible numerical outcome for the random variable, and *probability* is the probability of that outcome. We compute "*value × probability*" separately for each possible outcome and then add these quantities to find the expected value.

Example 8.11 **Gambling Losses** Suppose that in a gambling game, the probability of winning \$2 is .3 and the probability of losing \$1 is .7. Let X = amount a player "gains" on a single play. The following table gives the probability distribution for X by listing possible values along with their probabilities. In the table, a loss of \$1 is written as a negative "gain."

X = amount gained	\$2	−\$1
Probability	.3	.7

*HINT: What possible values of k are included if $X \le 4$?

What is the mean amount a player will "gain" per bet over many plays? The calculation is

$$\begin{aligned}\text{Expected value} &= \text{Sum of "value} \times \text{probability"} \\ &= (\$2)(.3) + (-\$1)(.7) \\ &= -\$0.10\end{aligned}$$

Over a large number of plays, a player can expect to lose an *average* of 10 cents per play. Note that the expected value in this example is not a possible outcome of a single play.

Notation and Formula for Expected Value

The notation $E(X)$ represents the mean or expected value of a random variable X. The Greek letter μ ("mu") can also be used to represent the mean or expected value. In other words, $\mu = E(X)$.

FORMULA

Calculating Expected Value for a Discrete Random Variable

If X is a discrete random variable with possible values $x_1, x_2, x_3, \ldots$, occurring with probabilities $p_1, p_2, p_3, \ldots$, then the **expected value** of X is calculated as

$$E(X) = \sum x_i p_i$$

The formula says to calculate the sum of "value times probability."

Example 8.12 **California Decco Lottery Game** Decco is one of the many games played in the California lottery over the years. In this game, a player chooses four cards, one each from the four suits in a regular deck of playing cards. For example, the player might choose "4 of hearts," "3 of clubs," "10 of diamonds," and "jack of spades." The state draws a winning card from each suit. When one or more of the player's choices matches the winning cards drawn, the player wins a prize, the prize amount depending on the number of matches.

We will define the random variable X to be a player's net gain for any single play. Table 8.1 lists the prizes, net gains, and probabilities for all possible numbers of matches, including 0 matches. It costs \$1.00 to play the game, so the net gain for any prize is \$1.00 less than the prize awarded. Note that we count a free ticket as an even trade because it would cost \$1.00 to buy a new one, but see the technical note following this example for discussion of whether or not a free ticket actually is an even trade.

Table 8.1 Probabilities for the Decco Game

Number of Matches	Prize	X = Net Gain	Probability
4	\$5,000	\$4,999	.000035
3	\$50	\$49	.00168
2	\$5	\$4	.0303
1	Free ticket	0	.242
0	None	−\$1	.726

Over the long run, how much would you win or lose per ticket in this game? The expected value of X, denoted as $E(X)$, is the answer, and the calculation is

$$\begin{aligned}E(X) &= (\$4{,}999 \times .000035) + (\$49 \times .00168) + (\$4 \times .0303) + (\$0 \times .242) + (-\$1 \times .726) \\ &= -\$0.35\end{aligned}$$

This tells us that over many repetitions of the game, players will *lose* an *average* of 35 cents per play. From the perspective of the Lottery Commission, this means that it gains about 35 cents and pays out about 65 cents for each \$1 ticket it sells for this game.

TECHNICAL NOTE **Free Tickets Are Not Worth the Price**

The situation is actually worse than presented in Example 8.12. Many lottery games give a free ticket as a prize, and you may think the value of that ticket is equal to its cost, usually $1.00. It's not, unless you can sell the ticket to someone else for that price. The real value of the ticket is the expected win for one play of the game. In the computation in Example 8.12, for simplicity, we used a "net gain" of 0 when the prize was a free ticket, but because it is actually worth less than that, the average loss per play is even more than $0.35. We will continue to use an expected loss of $0.35 for simplicity in discussions of the Decco game.

Note that the expected value in Example 8.12 is not one of the possible outcomes, so the phrase *expected value* is a bit misleading. In other words, the expected value may not be a value that is ever expected on a single random outcome. Instead, it is the average over the long run.

THOUGHT QUESTION 8.4 Suppose the probability of winning in a gambling game is .001, and when a player wins, his or her net gain is $999. When a player loses, the net amount lost is $1 (the cost to play). Is this game fair? Why or why not? How would you define a fair game? Does the number of times the game is played affect your view of whether the game is fair or not? Explain.*

Standard Deviation for a Discrete Random Variable

We first encountered the idea of standard deviation in Chapter 2, where we learned how to compute the standard deviation as a measure of spread for a quantitative variable and then used it in the Empirical Rule. Similarly, the **standard deviation of a discrete random variable** quantifies how spread out the possible values of a discrete random variable might be, weighted by how likely each value is to occur. As with the standard deviation of a set of measurements, the standard deviation of a random variable is roughly the average distance the random variable falls from its mean, or expected value, over the long run.

FORMULA **Variance and Standard Deviation of a Discrete Random Variable**

If X is a random variable with possible values of $x_1, x_2, x_3, \ldots$, occurring with probabilities $p_1, p_2, p_3, \ldots$, and with expected value $E(X) = \mu$, then,

Variance of X $= V(X) = \sigma^2 = \sum (x_i - \mu)^2 p_i$

Standard deviation of X = square root of $V(X) = \sigma = \sqrt{\sum (x_i - \mu)^2 p_i}$

The sum is taken over all possible values of the random variable X.

Example 8.13 **Stability or Excitement—Same Mean, Different Standard Deviations**

Suppose you decide to invest $100 in a scheme that you hope will make some money. You have two choices of investment plans, and you must decide which one to choose. For each $100 invested under the two plans, the possible net gains 1 year later and their probabilities are as shown:

Plan 1		Plan 2	
X = Net Gain	Probability	Y = Net Gain	Probability
$5,000	.001	$20	.3
$1,000	.005	$10	.2
$0	.994	$4	.5

***HINT:** A game is fair if the expected value of the gain is the same as the cost to play.

Examine the possible gains and associated probabilities. Which plan would you choose? Let's calculate the expected value for each plan. The expected value will tell us what the average long-run gain per investment is for a particular plan.

For Plan 1,

$$E(X) = \$5{,}000 \times (.001) + \$1{,}000 \times (.005) + \$0 \times (.994) = \$10.00$$

For Plan 2,

$$E(Y) = \$20 \times (.3) + \$10 \times (.2) + \$4 \times (.5) = \$10.00$$

In other words, the expected net gain is the same for the two plans.

What is different about the two plans is the variability in the possible net gains. Under the first plan, you could make a lot of money but with a low probability. Most likely, you will make nothing at all. Under the second plan, you will be sure to make a small amount, but there is no possibility of a big gain. This difference in the variability in possible amounts under the two plans can be quantified by finding the standard deviations.

The calculations of the standard deviations are displayed in Table 8.2. Summing the third column under each plan provides the variances. Taking the square root of a variance provides the standard deviation of the net gain for a plan.

Table 8.2 Standard Deviation Calculations for Example 8.13

Plan 1			Plan 2		
$(X-\mu)^2$	p	$(X-\mu)^2p$	$(Y-\mu)^2$	p	$(Y-\mu)^2p$
$(\$5000 - \$10)^2 = \$24{,}900{,}100$	.001	\$24,900.10	$(\$20 - \$10)^2 = \$100.00$	.3	\$30.00
$(\$1000 - \$10)^2 = \$980{,}100$	.005	\$4,900.50	$(\$10 - \$10)^2 = \$0.00$	.2	\$0.00
$(\$0 - \$10)^2 = \$100$	.994	\$99.40	$(\$4 - \$10)^2 = \$36.00$	.5	\$18.00

For Plan 1,

$$V(X) = \$29{,}900.00 \text{ and } \sigma = \sqrt{29{,}900.00} = \$172.92$$

For Plan 2,

$$V(Y) = \$48.00 \text{ and } \sigma = \sqrt{48.00} = \$6.93$$

These values demonstrate that the possible outcomes for Plan 1 are much more variable than the possible outcomes for Plan 2. If you wanted to invest cautiously, you would choose Plan 2, but if you wanted to have the chance to gain a large amount of money, you would choose Plan 1.

Expected Value (Mean) and Standard Deviation for a Population

When we have a population of measurements, we can think of each individual measurement as the value of a random variable, and we can calculate the mean and standard deviation in either of two ways: (1) We can create a probability distribution for the measurements and use the definitions of expected value and standard deviation in this section, or (2) we can find the mean and standard deviation by using the individual measurements directly in the formulas of Chapter 2. The two methods will give the same answers.

Example 8.14 **Mean Hours of Study for the Class Yesterday** A class of $N = 200$ students is asked, "To the nearest hour, how many hours did you study yesterday?" The instructor is interested only in the responses of the students in his class, so the 200 students in the class constitute a population. (As in Chapter 2, we use N to represent a population size.) The probability distribution of the responses is shown in Table 8.3.

Table 8.3 Hours of Study Yesterday for a Population of College Students

X = Hours of Study	Number of Students	Probability
0	16	16/200 = .08
1	28	28/200 = .14
2	50	50/200 = .25
3	36	36/200 = .18
4	26	26/200 = .13
5	22	22/200 = .11
6	14	14/200 = .07
7	6	6/200 = .03
8	2	2/200 = .01
Total	200	1.00

What is the mean number of hours studied the previous day in this population? The answer can be found by using the formula $E(X) = \Sigma x_i p_i$, and the calculation is

$$E(X) = (0 \times .08) + (1 \times .14) + (2 \times .25) + (3 \times .18) + (4 \times .13) + (5 \times .11) + (6 \times .07) + (7 \times .03) + (8 \times .01) = 2.96 \text{ hours}$$

This mean can also be calculated in the usual way that we average 200 individual values, which is that we can total the 200 individual values and divide that total by $N = 200$.

8.3 Exercises are on pages 303–304.

FORMULA

Mean and Standard Deviation for a Population

Suppose a population has N individuals and a measurement X is of interest. Define:

k_i = value of X for individual i.

$x_1, x_2, x_3, \ldots$ as the distinct possible values for the measurement X.

$p_1, p_2, p_3, \ldots$ as the proportions of the population with the values $x_1, x_2, x_3, \ldots$

Then, the **population mean** and **population standard deviation** are

$$E(X) = \frac{1}{N}\Sigma k_i = \Sigma x_i p_i$$

$$\text{Standard deviation of } X = \sigma = \sqrt{\frac{\Sigma(k_i - \mu)^2}{N}} = \sqrt{\Sigma(x_i - \mu)^2 p_i}$$

IN SUMMARY Expected Value

- The **expected value** of a random variable X is the mean value of the variable in the sample space or population of all possible outcomes.
- Expected value can be interpreted as the mean value that would be obtained from an infinite number of observations of the random variable.
- Expected value is computed as $E(X) = \Sigma x_i p_i$, the sum of "value × outcome" where the sum is over all possible outcomes.

8.4 Binomial Random Variables

In this section, we consider an important family of *discrete* random variables called *binomial random variables*. Certain conditions must be met for a variable to fall into this family, but the basic idea is that a binomial random variable is a count of how many times an event occurs (or does not occur) in a particular number of independent observations or trials that make up a random circumstance.

Binomial Experiments and Binomial Random Variables

The number of heads in three tosses of a fair coin, the number of girls in six independent births, and the number of men who are six feet tall or taller in a random sample of ten adult men from a large population are all examples of binomial random variables. A **binomial random variable** is defined as X = number of successes in the n trials of a binomial experiment.

DEFINITION A **binomial experiment** is defined by the following conditions:

1. There are n "trials," where n is specified in advance and is not a random value.
2. There are two possible outcomes on each trial. The outcomes are called "success" and "failure" and denoted S and F.
3. The outcomes are independent from one trial to the next.
4. The probability of a "success" remains the same from one trial to the next, and this probability is denoted by p. The probability of a "failure" is $1 - p$ for every trial.

Each row in Table 8.4 contains an example of a binomial experiment and a binomial random variable. For each example, go through the conditions for a binomial experiment and make sure you understand why those conditions apply to that example.

Table 8.4 Examples of Binomial Experiments and Binomial Random Variables

Random Circumstance	Random Variable	Success	Failure	n	p
(1) Toss three fair coins	X = number of heads	Head	Tail	3	1/2
(2) Roll a die eight times	X = number of 4s and 6s	4, 6	1, 2, 3, 5	8	2/6 = 1/3
(3) Randomly sample 1000 U.S. adults	X = number who have seen a UFO	Seen UFO	Have not seen UFO	1000	Proportion of all adults who have seen a UFO
(4) Roll two dice once	X = number of times sum is 7	Sum is 7	Sum is not 7	1	6/36 = 1/6

Additional Features of Binomial Random Variables

The examples in Table 8.4 illustrate a few subtle features of binomial random variables. Keep these points in mind when trying to determine whether or not a random variable fits the binomial description:

- *There may be more than two possible simple events for each trial,* but the random variable counts how many times a particular subset of the possibilities occurs. For example, a single die can display either 1, 2, 3, 4, 5, or 6 dots, but a success can be defined by a particular set of them (such as 4 and 6), while a failure is anything else. Look at the second example in Table 8.4.
- *Sample surveys can produce a binomial random variable* when we count how many individuals in the sample have a particular opinion or trait (see the third example in Table 8.4). A "trial" is one sampled individual. A "success" is that the individual has the opinion or trait. The probability of "success" is the proportion of the population that has the opinion or trait. If a random sample is taken without replacement from a large population, the conditions of a binomial experiment are considered to be met, although the probability of a "success" actually changes very slightly from one trial to the next as each sampled individual is removed from eligibility.

- *Sampling without replacement from a small population does not produce a binomial random variable.* Suppose a class consists of ten boys and ten girls. Five children are randomly selected to be in a play. X = the number of girls selected. Note that X is *not* a binomial random variable because the probability that a girl is selected each time depends on who is already in the sample and who is left and is not the same for all trials. X is an example of a *hypergeometric random variable*, discussed in Supplemental Topic 1 at the companion website for this book, http://www.cengage.com/statistics/Utts4e.
- *Any individual random circumstance can be treated as a binomial experiment* with $n = 1$ and p = probability of a particular outcome. In this case, the value of the random variable X is either 0 or 1, and it may also be called a **Bernoulli random variable**. The fourth example in Table 8.4 illustrates such a variable.

THOUGHT QUESTION 8.5 The word *binomial* is from the Latin *bi* "two," and *nomen* "name." Explain why the word *binomial* is appropriate for a binomial random variable.*

Finding Probabilities for Binomial Random Variables

For a binomial random variable, the probabilities for the possible values of X are given by the formula

$$P(X = k) = \frac{n!}{k!(n-k)!}p^k(1-p)^{n-k} \quad \text{for } k = 0, 1, 2, \ldots, n$$

The formula for $P(X = k)$ is made up of two parts.

1. The first part, $n!/[k!(n-k)!]$, gives the number of simple events in the sample space (consisting of all possible listings of successes and failures in n trials) that have exactly k successes. The notation $n!$ is read "n-factorial," and it is the product of the integers from 1 to n. For instance, $3! = 1 \times 2 \times 3 = 6$. By convention, $0! = 1$.
2. The second part, $p^k(1-p)^{n-k}$, gives the probability for each of the simple events for which $X = k$. It follows from the extension of Rule 3b in Chapter 7. If $X = k$, there are k successes and $(n - k)$ failures. Multiply p for each success and $(1 - p)$ for each failure to get $p^k(1-p)^{n-k}$.

Remember that a *probability distribution function* for a discrete random variable is a table or rule that assigns probabilities to the possible values of the random variable X. It is synonymous to say that a random variable X is a *binomial random variable* and to say that X has a **binomial distribution**. In both cases, the *probability distribution function* for X is the formula just given for $P(X = k)$.

Example 8.15 **Probability of Two Wins in Three Plays** Suppose the probability that you win a game is .2 for each play and plays of the game are independent of one another. Let X = number of wins in three plays. What is $P(X = 2)$, the probability that you win exactly twice in three plays? For this problem, X is a binomial random variable with $n = 3$ and success probability $p = .2$. Also, $1 - p = .8$ and $k = 2$ in the formula for $P(X = k)$.

- There are $\frac{3!}{2!(3-2)!} = \frac{6}{2(1)} = 3$ simple events that produce $X = 2$. The three sequences with two wins are WWL, WLW, and LWW.
- For each of these simple events, the probability is $p^k(1-p)^{n-k} = (.2)^2(.8)^1 = .032$. So the probability of exactly two wins is $P(X = 2) = 3(.032) = .096$.

The complete probability distribution function (pdf) for X = number of wins in three plays is graphed in Figure 8.3. See if you can verify that $P(X = 1) = .384$.

*HINT: See the second criterion in the definition of a binomial experiment.

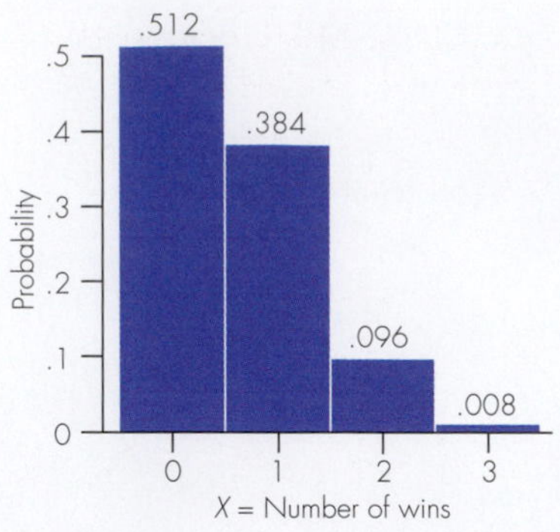

Figure 8.3 Probability distribution function for binomial random variable with $n = 3$ and $p = .2$

EXCEL TIP **Using Excel to Find Binomial Probabilities**

Fortunately, binomial probabilities are available in many computer software programs and calculators. To find binomial probabilities using Excel:

- $P(X = k)$ is calculated with the command BINOMDIST(k,n,p,false). The "false" indicates that we want the probability for exactly k successes.
- $P(X \leq k)$ is calculated with the command BINOMDIST(k,n,p,true). The "true" indicates that the desired probability is "cumulative" and that we want the probability of k or fewer successes.

Example 8.16 **Excel Calculations for Number of Girls in Ten Births** Let X = number of girls in ten births, and assume that $p = .488$ is the probability that any birth is a girl. As was noted in Chapter 7, this value of p is based on birth records in the United States. We can use Excel to find that the probability of exactly seven girls in ten births is = BINOMDIST(7,10,.488,false) = .106. The probability of seven or fewer girls is = BINOMDIST(7,10,.488,true) = .953.

MINITAB TIP **Finding Binomial Probabilities**

- Use **Calc>Probability Distributions>Binomial.**
- In the dialog box, select "Probability" or "Cumulative Probability" depending on whether you want $P(X = k)$ or $P(X \leq k)$. Specify the number of trials and the probability of success. Click on "Input constant" and fill in the corresponding box with the value of k.

***Note* 1:** To find probabilities for several values of k at once, first store the values in a column of the worksheet, and then specify that column as "Input column."

***Note* 2**: In Version 15 and above, **Graph > Probability Distribution Plot > View Probability** can be used to find cumulative binomial probabilities as well as probabilities for other types of intervals, and to see those probabilities on a plot of the distribution.

Example 8.17 **Guessing Your Way to a Passing Score** You have been busy lately, so busy that you're surprised to learn when you arrive at today's statistics class that a 15-question true–false quiz is on the agenda. The quiz is about readings that you haven't done, so you are forced to guess at every question. You will pass the quiz if you get ten or more correct answers, so you wonder about $P(X \geq 10)$, where X = number of correct answers.

X is a binomial random variable with $n = 15$ trials, and $p = .5$ is the success probability for any question. Figure 8.4 shows the probability distribution for X with darker shading for the region of ten or more correct guesses. Getting ten or more questions right is the complement of getting nine or fewer right, so

Cumulative Distribution Function	
Binomial with n = 15 and p = .500000	
x	P(X <= x)
9.0	0.8491

$$P(X \geq 10) = 1 - P(X \leq 9)$$

Minitab output showing the value of the cumulative probability $P(X \leq 9)$ is shown in the margin.

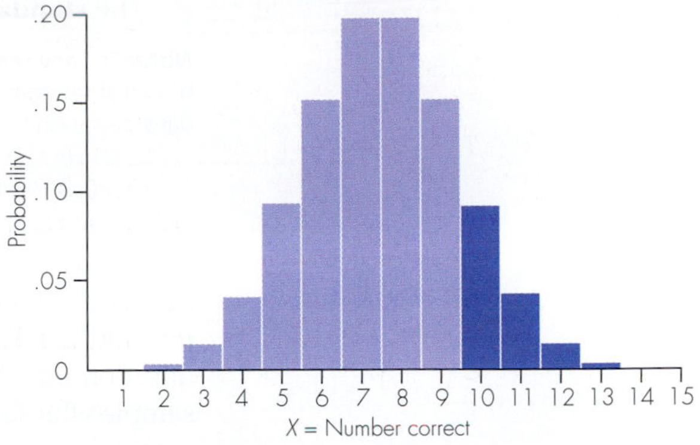

Figure 8.4 Probabilities for number correct when guessing on a 15-question true–false quiz

The probability that you pass (guess ten or more correct) is

$$P(X \geq 10) = 1 - .8491 = .1509$$

You've got a chance to pass (but the probability that you don't pass is much higher!).

TI-84 TIP

Calculating Binomial Probabilities

- To begin, press [2nd] [VARS] to reach the **Distribution** menu.
- To find $P(X = k)$, the probability of exactly k successes, use **0:binompdf(**. Enter numerical values for n, p, k (separated by commas) and enter the closing parenthesis. For example, **binompdf(**15, .5, 9) gives the probability of *exactly* $k = 9$ successes in a binomial experiment with $n = 15$ and $p = .5$.
- To find $P(X \leq k)$, the cumulative probability of k or fewer successes, use **A:binomcdf(**. Enter numerical values for n, p, k (separated by commas) and enter the closing parenthesis. For example, **binomcdf(**15, .5, 9) gives the probability of 9 *or fewer* successes in a binomial experiment with $n = 15$ and $p = .5$.

Expected Value (Mean) and Standard Deviation for a Binomial Random Variable

The **mean value for a binomial** random variable is $\mu = np$, where n = number of trials and p = probability of success. We don't have to use the formula from Section 8.3 for expected value. The simpler formula $\mu = np$ applies to all binomial random variables.

If you flip a fair coin 100 times, for instance, how many heads would you expect to result, on average? The answer is $100 \times 1/2 = 50$ heads, which is the number of flips times the probability of heads for any single flip. Or if you guess the answer for every question on a multiple-choice test with 25 questions, each with five choices, you could expect to get about $25 \times 1/5 = 5$ right answers.

There also is a simplified formula for the standard deviation of a binomial random variable. The formulas for the mean (expected value) and standard deviation of a binomial random variable were derived by using algebra and the formulas that were introduced earlier in this chapter for expected value, standard deviation, and $P(X = k)$ when X is a binomial random variable. You won't have to know how to do these derivations, but the results are useful for later applications in this textbook, so remember where to find them.

FORMULA **Mean, Standard Deviation, and Variance of a Binomial Random Variable**

For a *binomial random variable* X based on n trials with success probability p,

The **mean** is $\mu = E(X) = np$.

The **standard deviation** is $\sigma = \sqrt{np(1-p)}$.

Note: For any random variable, the standard deviation is the square root of the **variance**. For a binomial random variable, the variance is $V(X) = np(1-p)$, which leads to the formula for standard deviation.

Example 8.18

Is There Extraterrestrial Life? Suppose that 50% of a large population would say "yes" if they were asked, "Do you believe there is extraterrestrial life?" A sample of $n = 100$ is taken from this population, and each person in the sample is asked the question about belief in extraterrestrial life. The random variable X = number in the sample who say "yes" is a binomial random variable with $n = 100$ and $p = .5$.

- The mean of X is $\mu = np = 100(.5) = 50$.
- The standard deviation is $\sqrt{np(1-p)} = \sqrt{100(.5)(1-.5)} = 5$.

In repeated samples of size 100 from this population, on average 50 people would say "yes" to the question. The amount by which that number would differ from sample to sample is represented by the standard deviation of 5. We will discuss this kind of "sample variability" in much more detail in Chapter 9.

8.4 Exercises are on pages 304–305.

THOUGHT QUESTION 8.6 In Example 8.17, we determined the probability that you could guess your way to a passing score on a quiz with 15 true–false questions. If you did guess at each of 15 true–false questions, what is the expected value of X = number of correct answers? Is the expected value a possible score on the quiz? What exactly does the expected value tell us?*

CASE STUDY 8.1 Does Caffeine Enhance the Taste of Cola?

Soft drink manufacturers claim that caffeine enhances the flavor of cola drinks and that this is why they make caffeinated colas. Researchers from Johns Hopkins University disputed this claim in a study reported in the *Archives of Family Medicine* (Griffiths and Vernotica, 2000). The researchers conducted a taste test to find out if people can tell by tasting whether a cola contains caffeine or not.

Each participant tried 20 times to identify which of two cola drinks was "sample A" and which was "sample B." Unknown to the participants, one sample (A or B) contained caffeine and one did not. The particular sample (A or B) with caffeine was the same for all 20 trials. On each trial, a participant tasted drinks from two unlabeled cups and guessed which was sample A. After every trial, the participant was told whether he or she was correct. To learn about the tastes of the two samples, participants did five practice trials before the 20 trials that counted in the results.

Suppose that a participant cannot ever detect the difference between the cola samples and randomly guesses on every trial. In that case, X = number of correct guesses is a binomial random variable with $n = 20$ trials and $p = .5$ (the probability of correctly guessing just by luck).

The investigators called a participant's flavor detection performance "significant" when the individual made 15 or more correct identifications in 20 trials. The probability is only .0207 (about 1 in 50) that somebody who guesses every time could do this well. The calculation of this probability is

$$P(X \geq 15) = 1 - P(X \leq 14) = 1 - .9793 = .0207.$$

The binomial probability of $P(X \leq 14)$ could be found using Minitab, Excel, or a calculator such as a TI-84.

Twenty-five individuals participated, and only two made 15 or more correct identifications. This isn't much better than what would be expected if everybody ran-

(continued)

***HINT:** Use the formula for $E(X)$ for a binomial random variable.

domly guessed for all trials. As we already determined, the probability of 15 or more correct when randomly guessing is .0207, which is about 1 in 50. If all 25 participants guessed, we would expect 25(.0207) = .5175, or about 0 or 1 participants to fall into the "significant" category.

The researchers interpreted their results to mean that people generally can't taste whether a cola contains caffeine. Soft drink manufacturers objected to several aspects of the study, including the sample sizes for trials and participants and the way in which the researchers prepared the caffeinated cola samples.

IN SUMMARY Binomial Random Variables

- A **binomial random variable** is X = number of successes in n independent trials of a random circumstance in which p = probability of a success is the same in each trial.
- The probability distribution function for a binomial random variable is called a **binomial distribution**.
- The **mean value for a binomial** random variable is $\mu = np$.
- The **standard deviation for a binomial** random variable is $\sigma = \sqrt{np(1-p)}$.

8.5 Continuous Random Variables

We learned in Section 8.1 that a *continuous random variable* is one for which the outcome can be any value in an interval or collection of intervals. In practice, all measurements are rounded to a specified number of decimal places, so we may not be able to accurately observe all possible outcomes of a continuous variable. For example, the limitations of weighing scales keep us from observing that a weight may actually be 128.3671345993 pounds. Generally, however, we call a random variable a continuous random variable if there are a large number of observable outcomes covering an interval or set of intervals.

By definition, the probability is 0 that a continuous random variable exactly equals any specified value. For a continuous random variable, we are only able to find the probability that X falls between two values. In other words, unlike discrete random variables, continuous random variables do not have probability distribution functions specifying the exact probabilities of specified values. Instead, a continuous random variable has a probability *density* function, which is used to find the probability that the random variable falls into a specified interval of values.

DEFINITION The **probability density function** for a continuous random variable X is a curve such that the area under the curve over an interval equals the probability that X is in that interval. In other words, the probability that X is between the values a and b is the area under the density curve over the interval between the values a and b.

Notation for Probability in an Interval

- The two endpoints of an interval are represented using the letters a and b.
- The interval of values of X that falls between a and b, including the two endpoints, is written $a \leq X \leq b$.
- The probability that X has a value between a and b is written $P(a \leq X \leq b)$.

For example, the probability that a randomly selected 3-year-old weighs between 31 and 33 pounds would be written $P(31 \leq X \leq 33)$. The random variable X = weight of a 3-year-old, $a = 31$ and $b = 33$.

Example 8.19 **Time Spent Waiting for the Bus** A bus arrives at a bus stop every 10 minutes. If a person arrives at the bus stop at a random time, how long will he or she have to wait for the next bus? Define the random variable X = waiting time until the next bus arrives. The value of X could be any value between 0 and 10 minutes, and X is a continuous random variable. (In practice, the limitations of watches would force us to round off the exact time.) Figure 8.5 shows the probability density function for the waiting time. Possible waiting times are along the horizontal axis, and the vertical axis is a density scale. The height of the "curve" is .1 for all X between 0 and 10, so the total area between 0 and 10 minutes is $(10)(.1) = 1$.

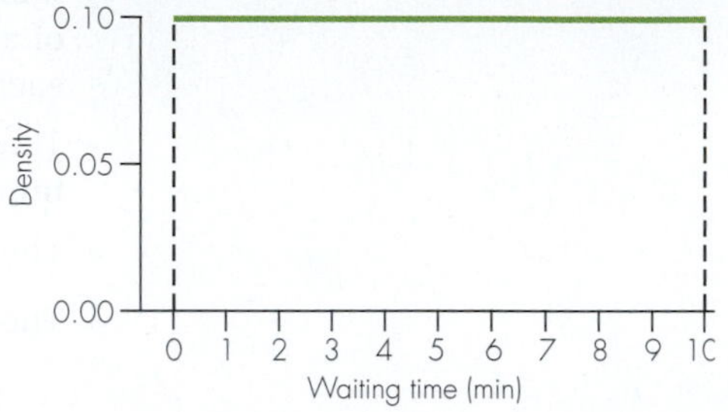

Figure 8.5 A uniform probability density function for Example 8.19

The density function shown in Figure 8.5 is a flat line that covers the interval of possible values for X. There is a "uniformity" to this density curve in that every interval with the same width has the same probability. A random variable with this property is called a **uniform random variable** and is the simplest example of a continuous random variable. For this type of variable, the probability for an interval equals the area of a rectangle.

Example 8.20 **Example 8.19 Revisited: Probability That the Waiting Time Is 5 to 7 Minutes** In Example 8.19, suppose we wanted to find the probability that the waiting time X was in the interval from 5 to 7 minutes. The general principle for any continuous random variable is that the probability $P(a \le X \le b)$ is the area under the density curve over the interval from a to b. In this example, the area under the density curve is the area of a rectangle that has width $= 7 - 5 = 2$ minutes and height $= .1$. This area is $(2)(.1) = .2$, which is the probability that the waiting time is between 5 and 7 minutes. In Figure 8.6, the shaded area represents the desired probability.

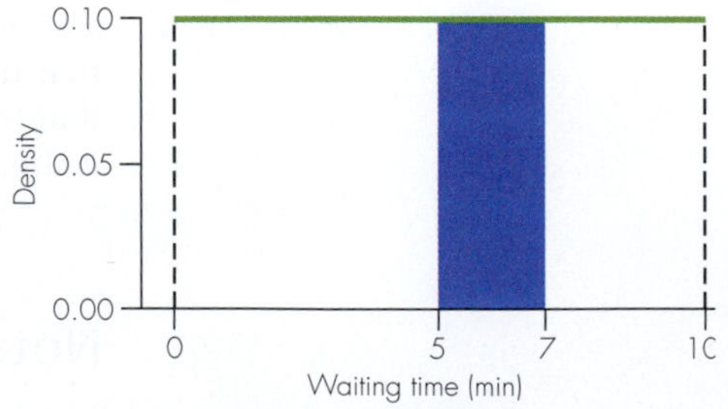

Figure 8.6 The probability that waiting time is between 5 and 7 minutes

Theoretically, the use of calculus is needed to find an area under a density curve (unless it's a simple rectangle or other simple shape, as in Examples 8.19 and 8.20). In practice, however, tables of appropriate probabilities usually are available. Computer software and graphing calculators also can be used to automate the calculations for the most commonly encountered models.

8.5 Exercises are on page 306.

THOUGHT QUESTION 8.7 The total area under the probability density function over the entire range of values the random variable X can possibly take is the same for all continuous random variables. What is that total area? What probability does it represent?

Suppose the random variable is the amount of rainfall in Davis, California, for a randomly selected year and can range from 0 inches to 50 inches. What is the area under the appropriate density curve over the range from 0 to 50 inches? What probability does it represent?*

IN SUMMARY Probabilities for Continuous Random Variables

- For a continuous random variable, we are only able to find probabilities for intervals of values.
- A **probability density function** (pdf) for a continuous random variable is a curve such that the area under the curve over an interval equals the probability that X is in that interval.

8.6 Normal Random Variables

The most commonly encountered type of continuous random variable is the **normal random variable**, which has a specific form of a bell-shaped probability density curve called a **normal curve**. A normal random variable is also said to have a **normal distribution**. Any normal random variable is completely characterized by specifying values for its mean μ and standard deviation σ.

Nature provides numerous examples of measurements that follow a normal curve. The fact that so many different kinds of measurements follow a normal curve is not surprising. On many attributes, the majority of people are somewhat close to average, and as you move further from the average, either above or below, there are fewer people with such values.

Features of Normal Curves and Normal Random Variables

As with any continuous random variable, the probability that a normal random variable falls into a specified interval is equivalent to an area under its density curve. Also, $P(X = k) = 0$, meaning that the probability is 0 that a normal random variable X exactly equals any specified value, a property of any continuous random variable.

Some features shared by all normal curves and normal random variables (X) are as follows:

1. The normal curve is symmetric and bell-shaped (but not all symmetric bell-shaped density curves are normal curves).
2. $P(X \leq \mu) = P(X \geq \mu) = .5$, meaning that there are equal probabilities for a measurement being less than the mean and greater than the mean. This happens because the curve is symmetric about the mean.
3. $P(X \leq \mu - d) = P(X \geq \mu + d)$ for any positive number d. This means that the probability that X is more than d units below the mean equals the probability that X is more than d units above the mean.
4. The Empirical Rule holds:
   - $P(\mu - \sigma \leq X \leq \mu + \sigma) \approx .68$
   - $P(\mu - 2\sigma \leq X \leq \mu + 2\sigma) \approx .95$
   - $P(\mu - 3\sigma \leq X \leq \mu + 3\sigma) \approx .997$

***HINT:** For X = Davis rainfall, as described, what is $P(0 \leq X \leq 50)$?

Example 8.21 **College Women's Heights** The distribution of the heights of college women can be described reasonably well by a normal curve with mean $\mu = 65$ inches (5 feet 5 inches) and standard deviation $\sigma = 2.7$ inches. A normal curve with these characteristics is shown in Figure 8.7. Heights are shown along the horizontal axis, with tick marks given at the mean and at 1, 2, and 3 standard deviations above and below the mean. Note that half (.5) of the area is above the mean of 65.0 and half is below it.

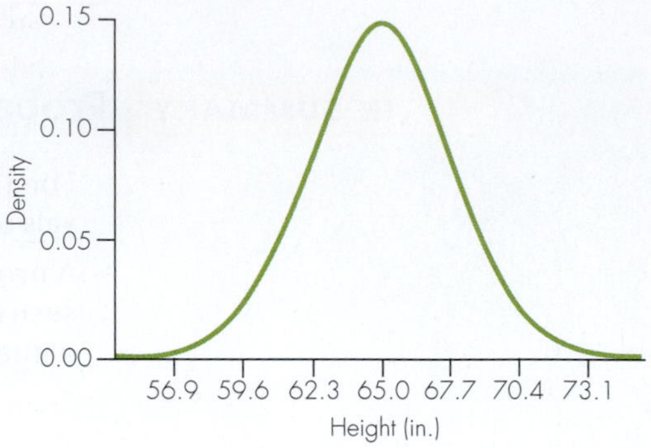

Figure 8.7 Normal curve for heights of college women

Most of the normal curve is over the interval from 56.9 to 73.1 inches. This is consistent with the Empirical Rule that we encountered in Chapter 2. One part of the Empirical Rule is that for a bell-shaped curve, about 99.7% of the values fall within 3 standard deviations of the mean in either direction. Here, that interval is $65 \pm 3 \times 2.7$, which is 65 ± 8.1 inches, or 56.9 and 73.1 inches. The Empirical Rule is generally used as an approximate rule for describing sample data with a bell-shaped curve, but the features of that rule are exact characteristics of a normal curve model (except for minor rounding of the percent values).

THOUGHT QUESTION 8.8 Which of the following measurements do you think are likely to have a normal distribution: heights of college men, incomes of 40-year-old women, pulse rates of college athletes? Explain your reasoning for each variable. For those variables that are likely to be normally distributed, give approximate values for the mean and standard deviation.*

Useful Probability Relationships for Normal Random Variables

Recall that a cumulative probability is the probability that a random variable is less than or equal to a specified value. Figure 8.8 illustrates this type of probability for a normal random variable. Note that the probability is the area under the normal curve to the left of a specified value of X.

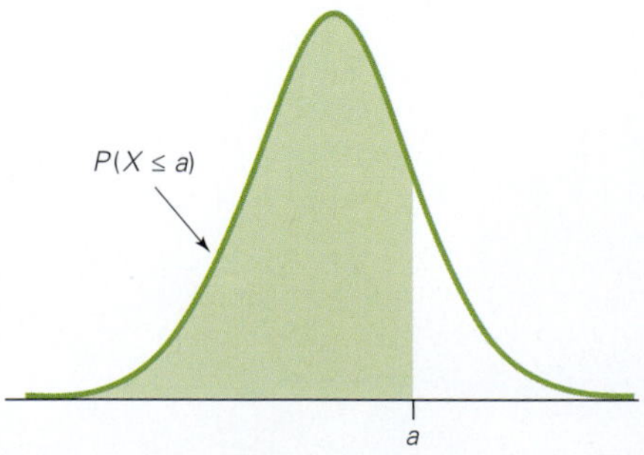

Figure 8.8 $P(X \le a)$, the probability that X is less than or equal to the value a

***HINT:** Which of these variables might have large outliers or a skewed distribution?

Software such as Minitab and Excel and some calculators can be used to find cumulative probabilities for a normal random variable. Cumulative probabilities can also be found using a table. We'll explain that method later in this section.

We don't always ask questions for which the answer is a cumulative probability. The following three rules are useful for using cumulative probabilities to find other types of probabilities for a normal random variable X:

Rule 1: $P(X > a) = 1 - P(X \leq a)$

Rule 2: $P(a < X < b) = P(X \leq b) - P(X \leq a)$

Rule 3: $P(X > \mu + d) = P(X < \mu - d)$

These relationships are illustrated in Figures 8.9, 8.10, and 8.11.

- Rule 1, shown in Figure 8.9, can be used to find the probability that X is *greater than* the value a.
- Rule 2, shown in Figure 8.10, can be used to find the probability that X is *between* the values a and b.
- Rule 3, shown in Figure 8.11, is a *symmetry* property. The probability that X is more than d units above the mean equals the probability that X is more than d units below the mean.

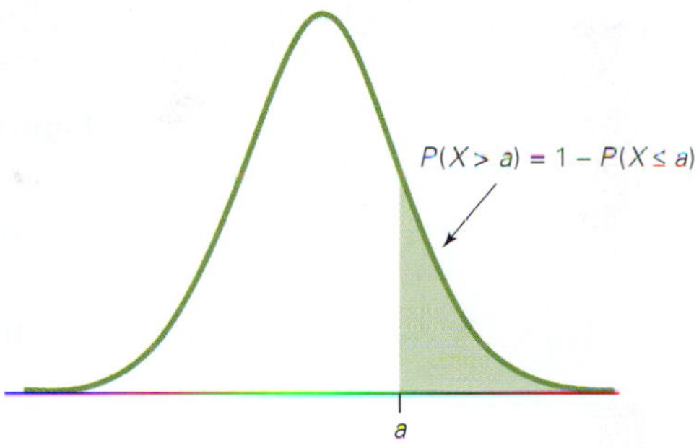

Figure 8.9 $P(X > a)$, the probability that X is greater than the value a

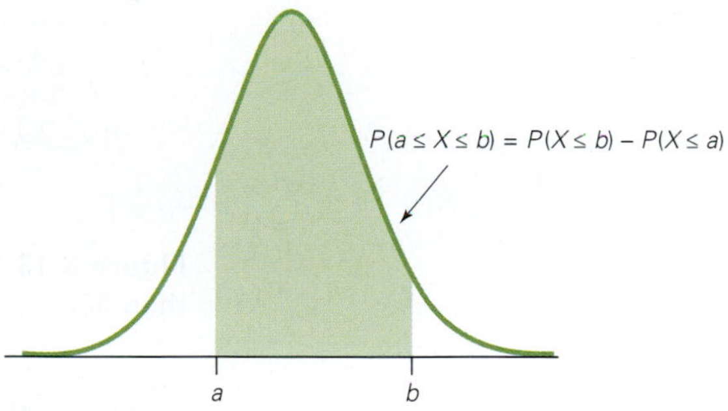

Figure 8.10 $P(a \leq X \leq b)$, the probability that X is between the values a and b

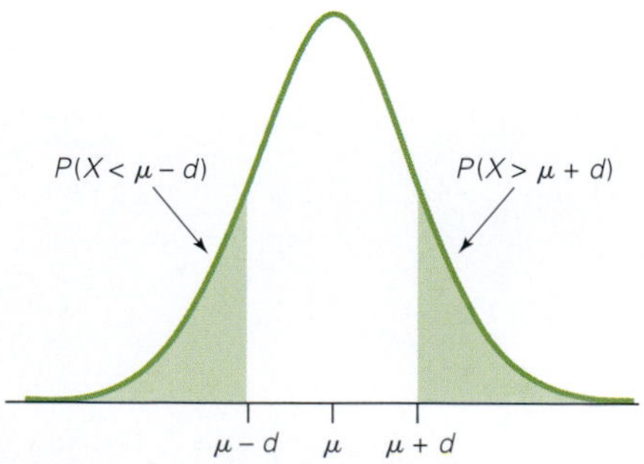

Figure 8.11 Symmetry property of a normal random variable

Generally, in solving normal curve problems, it helps to draw a picture of the problem, highlight the desired area, and then figure out how the desired area relates to the rules just given.

The following example illustrates how to use the rules just given to find probabilities for various types of intervals.

Example 8.22 **Probabilities for Math SAT Scores** Suppose that for the SAT test given to prospective college students in the United States, scores on the math section have a normal distribution with mean $\mu = 515$ and standard deviation $\sigma = 100$. Let's consider some questions that illustrate probability relationships.

Question 1: What is the probability that a randomly selected test-taker had a score less than or equal to 600? Said another way, what is the cumulative probability for a score of 600?

Answer: The answer, .8023, is the area under the normal curve to the left of 600. Note that the graph is centered at 515, the mean value. Figure 8.12 shows the solution, found using the **Graph > Probability Distribution Plot** feature available in Versions 15 and 16 of Minitab. Any method for finding cumulative probabilities for a normal random variable will give virtually the same answer.

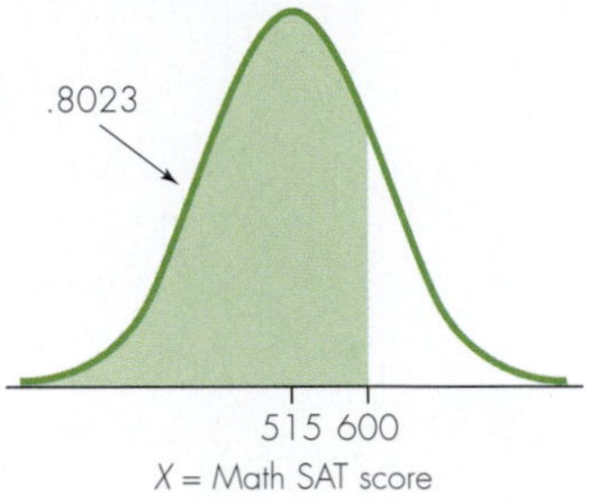

Figure 8.12 Probability of a score less than 600

Question 2: What is the probability that a randomly selected test-taker scored higher than 600?

Answer: A score "higher than 600" is the complement (opposite) of "less than or equal to 600" so the answer is $1 - .8023 = .1977$. Figure 8.13 illustrates the solution.

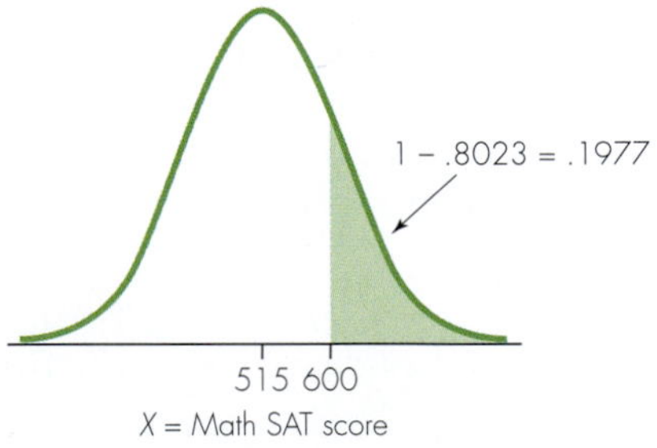

Figure 8.13 Probability of a score greater than 600

Question 3: What is the probability that a randomly selected test-taker scored between 515 and 600?

Answer: The answer is .3023, found as the difference between the cumulative probabilities for scores of 600 and 515. The cumulative probability for 600 is .8023, the answer to question 1. The cumulative probability for 515 is .5000, because 515 is the mean and the mean divides the normal curve into two equal parts. The difference is $.8023 - .5000 = .3023$. Figure 8.14 illustrates the solution.

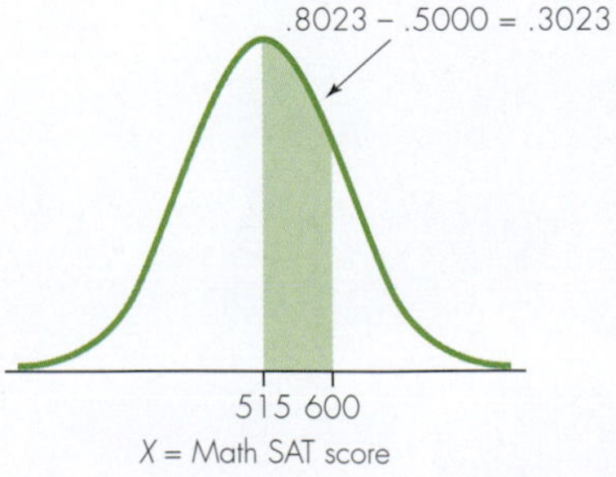

Figure 8.14 Probability of a score between 515 and 600

Question 4: What is the probability that a randomly selected test-taker's score was more than 85 points from the mean in either direction?

Answer: The answer is .3954, the sum of probabilities in two symmetric tails of a normal curve. Any score greater than 600 is more than 85 points above the mean. As the answer to question 2, we found that the probability is .1977 for a score greater than 600. Any score less than 430 is more than 85 points below the mean. By symmetry, the probability below 430 equals the probability above 600. Thus, the total probability for the two intervals is .1977 + .1977 = .3954. Figure 8.15 illustrates the solution.

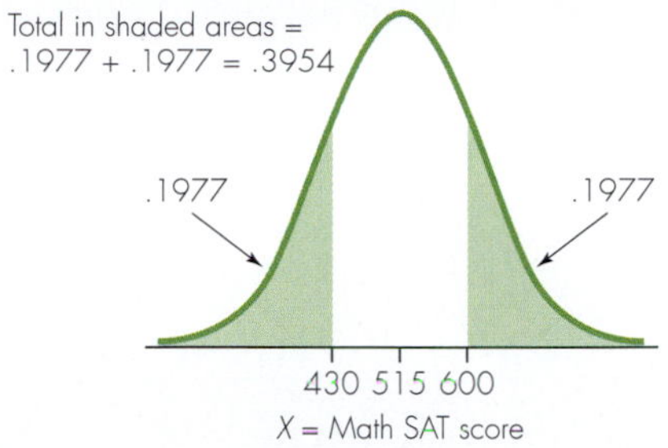

Figure 8.15 Probability of being more than 85 points from the mean in either direction

THOUGHT QUESTION 8.9 Use the answer to Question 3 in Example 8.22 to find the probability of being outside the interval 515 to 600. Then, use the answer to Question 4 in that example to find the probability of being between 430 and 600. In figures such as Figures 8.14 and 8.15, what is a general rule for finding the "non-shaded" area once you have the shaded area?*

Using a Table to Find Probabilities for a Normal Random Variable

If you have computer software that will calculate cumulative probabilities for normal random variables, you won't need to use the tools that we describe next to compute probabilities. However, you may need them when we discuss *statistical inference* in later chapters.

A normal random variable with mean $\mu = 0$ and standard deviation $\sigma = 1$ is said to be a **standard normal random variable** and to have a **standard normal distribution**. When we convert values for any normal random variable to *z*-scores, it is equivalent to converting the random variable of interest to a standard normal random variable. We use the letter *Z* to represent a standard normal random variable. An important fact about normal random variables is that any probability problem about a normal random variable can be converted to a problem about a standard normal random variable.

Table A.1, "Standard Normal Probabilities," in this book's Appendix (and also inside the back cover) is a table of cumulative probabilities for a standard normal random variable. In other words, we can use Table A.1 to find $P(Z \leq z^*)$, the probability that a standard normal random variable (Z) is less than a specific value (written as z^*).

Reading Table A.1 in the Appendix

As an example, the following (incomplete) row from Table A.1 shows that $P(Z \leq 1.82) = .9656$.

z	.00	.01	.02	.03	.04	.05	.06	.07	.08	.09
1.8	⟶	⟶	.9656							

The first column, with the heading *z*, gives the digit before the decimal place and the first decimal digit of a value of *z*. The column headings give the second decimal digit for *z*. Cumulative probabilities are given in the body of the table.

***HINT:** How are probabilities of complementary events related?

As a second example of reading Table A.1, the following row of the table gives the information that $P(Z \leq -2.59) = .0048$.

z	.00	.01	.02	.03	.04	.05	.06	.07	.08	.09
−2.5	→									.0048

Here are two other examples that you should verify by looking in Table A.1 to be sure you understand how to read the table:

$P(Z \leq 1.31) = .9049$

$P(Z \leq -2.00) = .0228$

Technical Detail: $P(Z = z^*) = 0$, so it is equivalent to write $P(Z \leq z^*)$ or $P(Z < z^*)$.

Probabilities for Extreme z-Scores

At the bottom of both pages of the table, there is a section titled "In the Extreme" where cumulative probabilities are given for selected "extreme" *z*-scores. For instance, from that section of the table, you can learn that $P(Z \leq -4.75) = .000001$ (one in a million!).

Calculating a Standardized Score

We learned in Chapter 2 that a **standardized score**, also called a ***z*-score**, is the distance between a specified value and the mean, measured in number of standard deviations. We repeat the definition here using notation for random variables.

FORMULA

Calculating a *z*-Score

The formula for converting any value *x* to a ***z*-score** is

$$z = \frac{\text{Value} - \text{Mean}}{\text{Standard deviation}} = \frac{x - \mu}{\sigma}$$

where μ and σ are the mean and standard deviation, respectively, for the random variable *X*.

A *z*-score measures the number of standard deviations that a value falls from the mean.

Example 8.23 **Example 8.21 Revisited: *z*-Score for a Height of 62 Inches** Assume that for the population of college women, mean height is 65 inches, and the standard deviation of heights is 2.7 inches. The *z*-score corresponding to a height of 62 inches is

$$z = \frac{\text{Value} - \text{Mean}}{\text{Standard deviation}} = \frac{62 - 65}{2.7} = \frac{-3}{2.7} = -1.11$$

This *z*-score tells us that 62 inches is 1.11 standard deviations below the mean height for this population.

Finding a Cumulative Probability $P(X \leq a)$ for Any Normal Random Variable

Suppose *X* is a normal random variable with mean μ and standard deviation σ. There are two steps in finding $P(X \leq a)$, the probability that *X* is less than or equal to the value *a*. These steps are not necessary if you have a computer or calculator that can determine $P(X \leq a)$ directly.

Step 1: Calculate a *z*-score for the value *a*; call it z^*.

Step 2: Use a table (such as Table A.1), calculator, or computer to find $P(Z \leq z^*)$, the cumulative probability for z^*. This equals the desired probability $P(X \leq a)$.

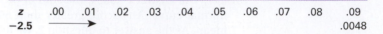

Example 8.24 **Example 8.21 Revisited: Probability That Height Is Less Than 62 Inches** Assume that the heights of college women have a normal distribution with mean $\mu = 65$ inches and standard deviation $\sigma = 2.7$ inches. What is the probability that a randomly selected college woman is 62 inches or shorter? This is the same as asking what *proportion* of college women are 62 inches or shorter.

Step 1: $z^* = \dfrac{62 - 65}{2.7} = -1.11$

Step 2: $P(X \leq 62) = P(Z \leq -1.11) = .1335$ (found by using Table A.1)

In other words, about 13% of college women are 62 inches or shorter. Figure 8.16 illustrates the area of interest under the normal curve for college women's heights.

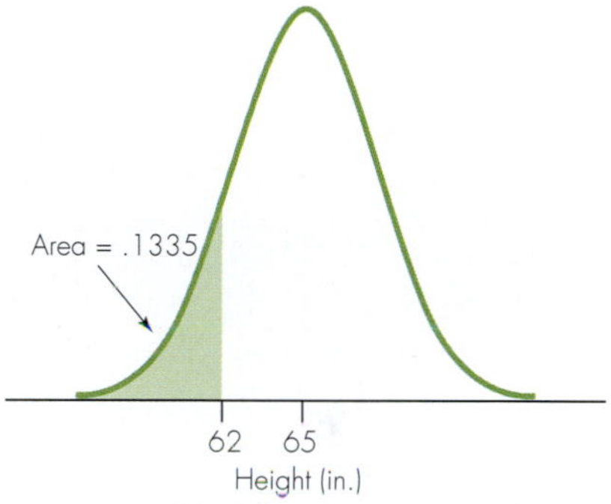

Figure 8.16 College women's heights and $P(\text{height} \leq 62 \text{ inches})$

Example 8.25 **Example 8.22 Revisited: Using Table A.1 to Find Probabilities for Math SAT Scores** In Example 8.22, we answered four probability questions about scores on the math section of the SAT test. We assumed that the scores have a normal distribution with mean $\mu = 515$ and standard deviation $\sigma = 100$. Now, we'll answer the same questions using z-scores and Table A.1. The answers will remain the same as they were in Example 8.22.

Question 1: What is the probability that a randomly selected test-taker had a score less than or equal to 600? Said another way, what is the cumulative probability for a score of 600?

Answer: The standardized score for 600, is $z^* = \dfrac{600 - 515}{100} = 0.85$. In Table A.1, the cumulative probability for this standardized score is .8023.

Question 2: What is the probability that a randomly selected test-taker scored higher than 600?

Answer: We subtract the cumulative probability for 600 from 1. The answer is $1 - .8023 = .1977$. This can be written as

$$P(X > 600) = P(Z > 0.85) = 1 - P(Z \leq 0.85) = 1 - .8023 = .1977.$$

Question 3: What is the probability that a randomly selected test-taker scored between 515 and 600?

Answer: The first step is to calculate standardized scores for the two scores defining the interval. We already know that for 600, $z^* = 0.85$. For 515, $z^* = \dfrac{515 - 515}{100} = 0$. We use Table A.1 to find cumulative probabilities for each standardized score and then take the difference between the two cumulative values. The solution is

$$P(515 \leq X \leq 600) = P(0 \leq Z \leq 0.85)$$
$$= P(Z \leq 0.85) - P(Z \leq 0) = .8023 - .5000 = .3023$$

Question 4: What is the probability that a randomly selected test-taker's score was more than 85 points from the mean in either direction?

Answer: A score that's more than 85 points below the mean is less than 430. For a score of 430, $z^* = \frac{430 - 515}{100} = -0.85$. In Table A.1, we find $P(Z \leq -0.85) = .1977$. By symmetry, the probability of a score more than 85 above the mean is also .1977. The combined probability of being more than 85 points from the mean in either direction is $.1977 + .1977 = .3954$, the answer to this question.

TI-84 TIP **Calculating Normal Curve Probabilities**

- To begin, press [2nd] [VARS] to reach the Distribution menu.
- Use **2:normalcdf** to find the probability that a normal random variable has a value in a specified interval.
- Enter numerical values (separated by commas) to complete the expression **normalcdf**(*lower value, upper value, mean, standard deviation*). For example, **normalcdf** (60, 70, 62, 5) gives the probability that a normal random variable with $\mu = 62$ and $\sigma = 5$ has a value between 60 and 70.

Note: To find a cumulative probability using **normalcdf**, make *lower value* be an extreme negative value, such as −10000000. To find the probability greater than a specified value, make *upper value* be an extreme positive value, such as 10000000.

Finding Percentiles

In some problems, we want to know what value of a variable has a given percentile ranking. For example, we may want to know what pulse rate is the 25th percentile of pulse rates for men. Note that the word **percentile** refers to the *value* of a variable. The **percentile rank** corresponds to the *cumulative probability* (the area to the left under the density curve) for that value.

Suppose that the 25th percentile of pulse rates for adult males is 64 beats per minute. This means that 25% of men have a pulse rate below 64. The *percentile* is 64 beats per minute (a value of the variable), and the *percentile rank* is 25% or .25 (a cumulative probability). Figure 8.17 illustrates this concept.

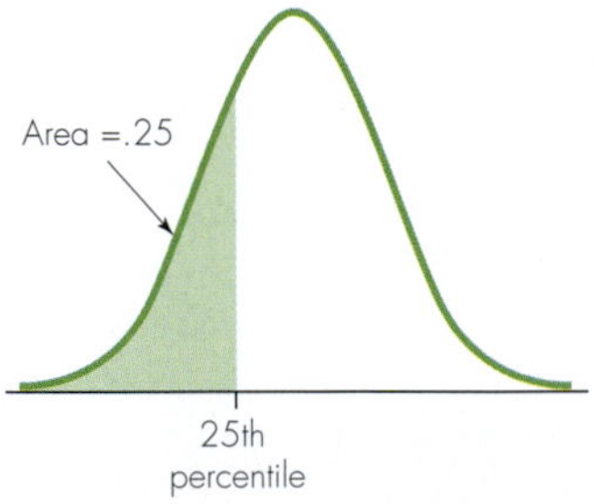

Figure 8.17 The 25th percentile for a random variable with a normal distribution

Finding a Percentile or a Value with Specified Cumulative Probability

Remember that the goal is to find the value of the variable that has a given cumulative probability. You may be able to use software or a calculator to directly determine the solution. Using Table A.1 to find a percentile for a normal random variable, there are two steps:

Step 1: Find the value z^* that has the specified cumulative probability. This does not involve calculation; in the body of the table find the specified cumulative probability (or the closest value to it) and read the z-score from row and column headings.

Step 2: Compute $x = z^* \sigma + \mu$. This is the desired percentile.

Example 8.26 **The 75th Percentile of Systolic Blood Pressures** Suppose that the blood pressures of men aged 18 to 29 years old have a normal distribution with mean $\mu = 120$ and standard deviation $\sigma = 10$. What value of blood pressure is the 75th percentile for this population?

Figure 8.18 illustrates the problem. We want to find the blood pressure for which the cumulative probability (area to the left under the density curve) is .75. With Table A.1, the steps are as follows:

Step 1: Search for .75 among the cumulative probabilities in the body of Table A.1 and then look at the row and column labels to read z^*. The approximate z-score for the 75th percentile is $z^* = 0.67$.

Step 2: $x = z^*\sigma + \mu = (0.67)(10) + 120 = 126.7$

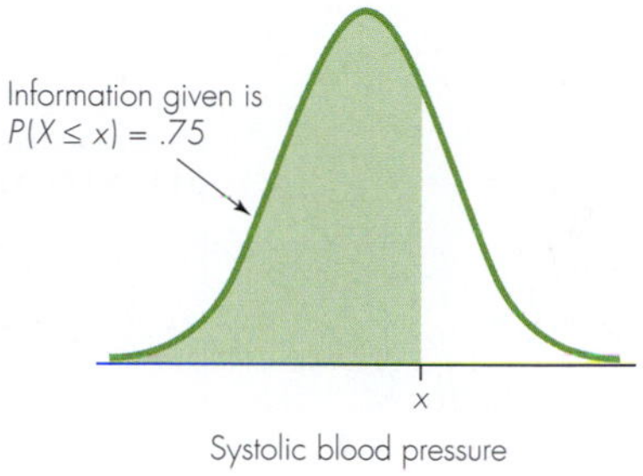

Figure 8.18 The 75th percentile for systolic blood pressure

Using this method, we determine that the 75th percentile of blood pressures for men aged 18 to 29 years old is 126.7, or approximately 127.

8.6 Exercises are on pages 306–307.

MINITAB TIP **Calculating Normal Curve Probabilities and Percentiles**

- In Version 15 and above, use **Graph>Probability Distribution Plot>View Probability**.
- Specify the distribution to be "Normal" and enter the values of the mean and standard deviation.
- To find a probability, select the "Shaded Area" tab, click "X value," select the type of region of interest and enter the value(s) of X.
- To find a percentile, select the "Shaded Area" tab, click "Probability," select a "Left Tail" region and enter the given value of the cumulative probability.

Note: In any version of Minitab, you can use **Calc>Probability Distribution >Normal** to determine either cumulative probabilities or percentiles (as inverse cumulative probabilities).

IN SUMMARY Normal Random Variables

- A **normal random variable** has a symmetric, bell-shaped distribution referred to as the normal distribution.
- The symbol μ is used to denote the mean of a normal random variable and σ is used to denote the standard deviation.
- The **standard normal distribution** has mean $\mu = 0$ and standard deviation $\sigma = 1$. The standard normal distribution is the distribution of **standardized scores** for a normal random variable.
- Any probability question about a normal random variable can be answered by solving the equivalent question about standardized scores for the values in question.

8.7 Approximating Binomial Distribution Probabilities

One useful application of the normal distribution is that it can be used to approximate probabilities for some other types of random variables. In this section, we learn how to use the normal distribution to approximate cumulative probabilities for binomial random variables. This approximation is useful for computational reasons, and as we will learn in Chapter 9, it has important conceptual uses as well.

When X has a binomial distribution with a large number of trials, the binomial probability formula is difficult to use because the factorial expressions in the formula become very large. The work required to find binomial probabilities when n is large may be enormous, and even some computer programs and calculators may not be able to do the task. Fortunately, the normal distribution can be used to approximate probabilities for a binomial random variable when this situation occurs.

DEFINITION The **normal approximation to the binomial distribution** is based on the following result, derived from mathematics. If X is a binomial random variable based on n trials with success probability p, and n is sufficiently large, then X is also approximately a normal random variable. For this normal random variable,

$$\text{Mean} = \mu = np$$

$$\text{Standard deviation} = \sigma = \sqrt{np(1-p)}$$

Conditions: The approximation works well when both np and $n(1 - p)$ are at least 10.

Example 8.27 **The Number of Heads in 60 Flips of a Coin** Figure 8.19 displays the probability distribution function of a binomial random variable based on $n = 60$ trials and success probability $p = .5$. This distribution could, for example, describe $X =$ the number of heads observed when you flip a coin 60 times. The bell-shaped pattern of the distribution is clear, and a normal curve could be used to approximate this distribution because both np and $n(1 - p)$ are greater than 10.

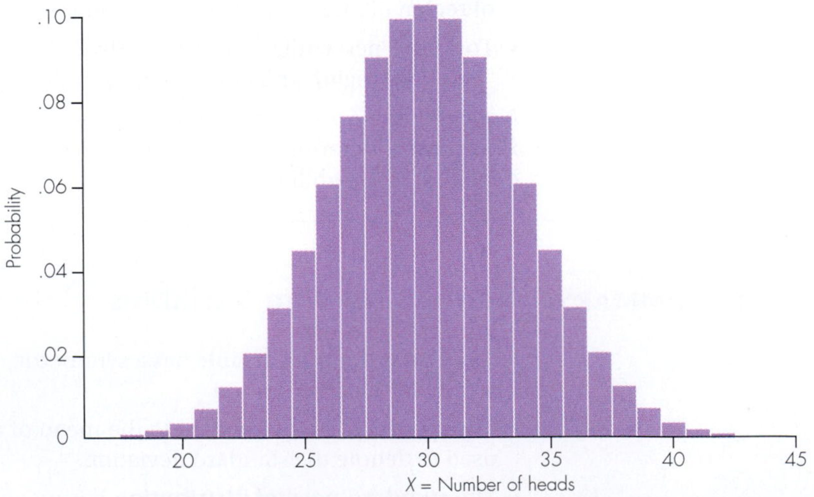

Figure 8.19 Binomial distribution for $n = 60$ and $p = .5$

Recall that 99.7%, or nearly all, of a normal curve falls within 3 standard deviations of the mean. For the binomial random variable displayed in Figure 8.19, $\mu = np = (60)(.5) = 30$, and $\sigma = \sqrt{np(1-p)} = \sqrt{60(.5)(1-.5)} = 3.873$. The interval $\mu \pm 3\sigma$ is 30 ± 11.62, which is roughly 18 to 42. Note how this interval fits with the binomial distribution shown in Figure 8.19. The probability is small that X would take a value outside this interval.

Example 8.28 Normal Approximation to Binomial Distribution with $n = 300$ and $p = .3$ Consider a binomial distribution with $n = 300$ and $p = .3$. For instance, this might be the distribution for X = number with a particular genetic trait in a sample of $n = 300$ persons randomly sampled from a population in which 30% has the trait. For this binomial random variable,

$$\mu = np = 300(.3) = 90$$
$$\sigma = \sqrt{np(1-p)} = \sqrt{300(.3)(1-.7)} = 7.937$$

Figure 8.20 shows a normal distribution with mean $\mu = 90$ and standard deviation $\sigma = 7.937$ superimposed over a binomial distribution with $n = 300$ and $p = .3$. In this situation, the normal curve closely approximates the binomial distribution.

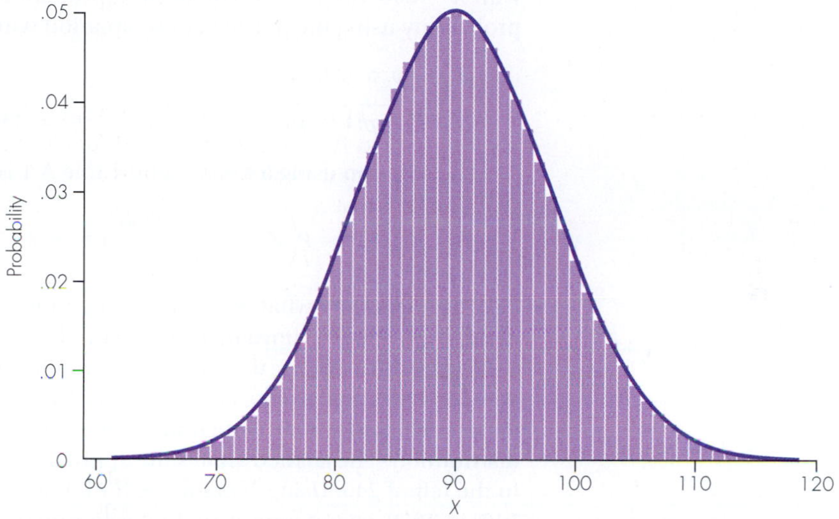

Figure 8.20 Normal approximation for binomial distribution with $n = 300$ and $p = .3$.

Approximating Cumulative Probabilities for a Binomial Random Variable

To approximate probabilities for a binomial random variable, you may be able to use statistical software or a calculator to determine probabilities for a normal random variable with mean $\mu = np$ and standard deviation $\sigma = \sqrt{np(1-p)}$. If you use standardized scores and Table A.1, the normal curve approximation can be used to find cumulative probabilities for binomial random variables, as follows:

$$P(X \le k) = P(Z \le z^*)$$

where

$$z^* = \frac{k - np}{\sqrt{np(1-p)}}$$

In other words, to find a cumulative probability, do the following:

1. Calculate a z-score for the value of interest, k.
2. Use the standard normal curve to determine the cumulative probability for the z-score found in Step 1.

More complex probabilities can be found as well, using the same relationships as for any normal random variable X.

It is important to note that you should *not* use the normal curve approximation if your statistical software or calculator can determine a binomial probability exactly. In this regard, the capabilities of different software programs and calculators vary. For the next example, for instance, Minitab versions 13 and higher can give an exact answer using the binomial distribution, but older versions cannot.

Example 8.29 **Political Woes** A politician is convinced that at least 50% of her constituents favor a woman's right to choose abortion, but is concerned because the latest poll of 500 adults in her district found that slightly fewer than half, only 240, or 48% of the sample, supported this position. If indeed 50% of adults in the district support a woman's right to choose abortion, what is the probability that a random sample of 500 adults would find 240 or fewer of them holding this position?

If X = number in sample who support a woman's right to choose abortion and if 50% of all adults in the district actually support it, then X is a binomial random variable with $n = 500$ and $p = .5$. The desired probability is $P(X \leq 240)$. We can estimate this probability using the normal approximation with

$$\mu = np = 500(.5) = 250$$

$$\sigma = \sqrt{np(1-p)} = \sqrt{500(.5)(1-.5)} = 11.18$$

The solution using a z-score and Table A.1 is

$$P(X \leq 240) \approx P\left(Z \leq \frac{240 - 250}{11.18}\right) = P(Z \leq -0.89) = .1867$$

Thus, it is somewhat possible that a minority of 48% or less of a sample would favor the right to choose abortion, even if 50% of adults in the district favor it.

Figure 8.21 shows the approximate solution. A normal distribution with mean $\mu = 250$ and standard deviation $\sigma = 11.18$ is superimposed on a binomial distribution for $n = 500$ and $p = .5$. Note that the normal curve closely approximates the binomial distribution. The shaded area is the approximate answer, .1867, the normal curve area to the left of 240. Using Minitab we find that the exact binomial probability for $P(X \leq 240)$ is .1978, so our approximation is reasonably close to the exact answer.

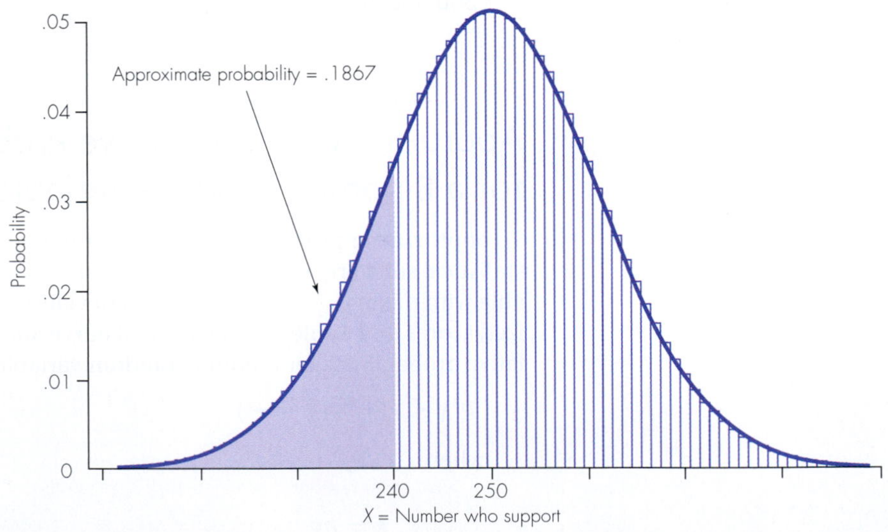

Figure 8.21 Approximate solution for binomial $P(X \leq 240)$

Example 8.30 **Guessing and Passing a True–False Test** Suppose you need 70%, or 21 questions, correct to pass a 30-question true–false test on which you are just guessing at every answer. We wish to find $P(X \geq 21)$ for a binomial random variable with $n = 30$ and $p = .5$. The number of correct answers is discrete, so the complement of this event is $P(X \leq 20)$. For a binomial random variable with $n = 30$ and $p = .5$, we have

$$\mu = np = 30(.5) = 15$$

$$\sigma = \sqrt{np(1-p)} = \sqrt{30(.5)(1-.5)} = 2.739$$

The approximate solution is

$$P(X \geq 21) = 1 - P(X \leq 20) \approx 1 - P\left(Z \leq \frac{20 - np}{\sqrt{np(1-p)}}\right) = 1 - P(Z \leq 1.83)$$

$$= 1 - .9664 = .0336.$$

Using Minitab, we find the exact probability that $X \geq 21$ is $1 - .9786 = .0214$.

Figure 8.22 illustrates both the approximate solution and the exact binomial solution for this situation. In the exact binomial solution, note that $P(X \geq 21)$ corresponds to the areas of the rectangles centered above 21, 22, ... , 30, of which only the ones above 21, 22, and 23 are visible. After that, the probabilities, and thus the heights of the rectangles, are nearly 0. For instance, $P(X = 24) = .0006$.

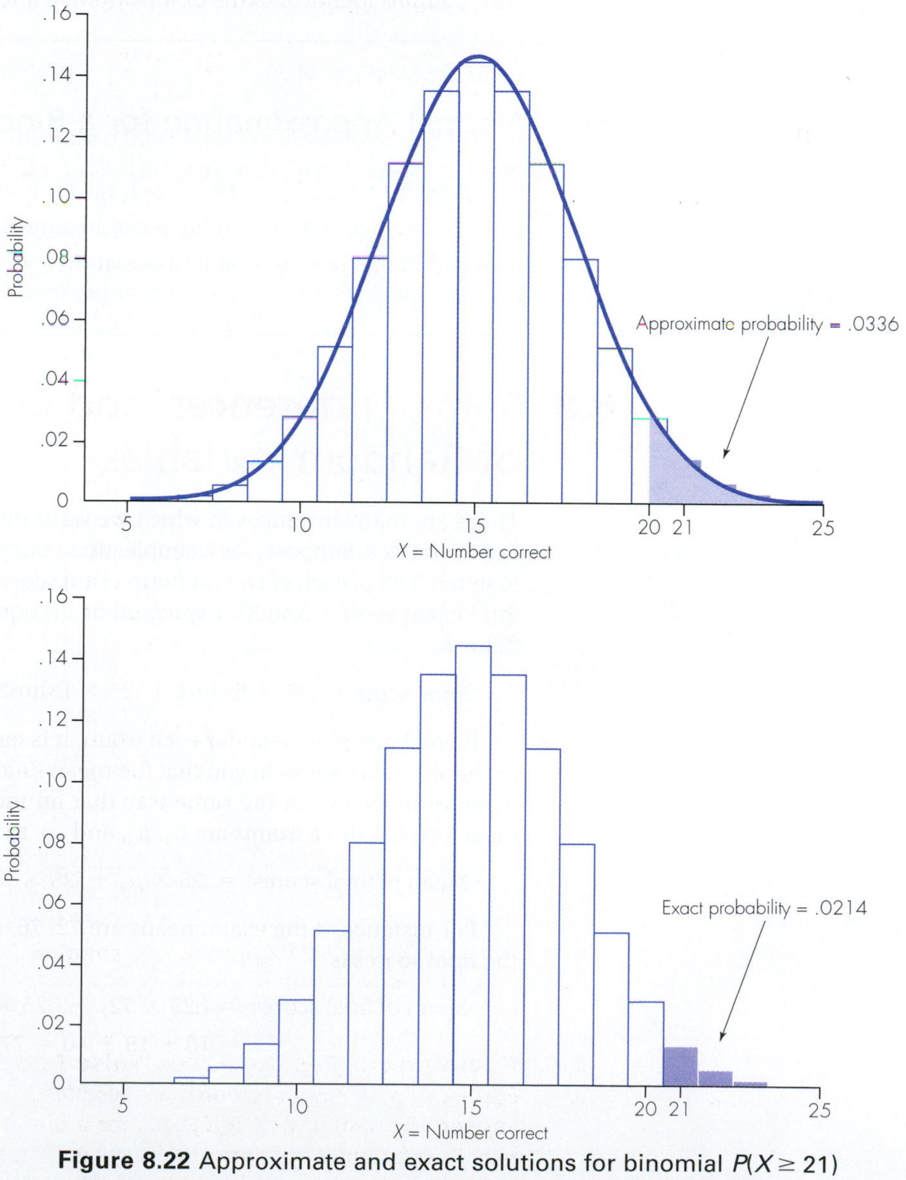

8.7 Exercises are on pages 307–308.

Figure 8.22 Approximate and exact solutions for binomial $P(X \geq 21)$

TECHNICAL NOTE

Continuity Correction

Note in Figure 8.22 for Example 8.30 that the rectangle centered on $X = 20$ actually ends at 20.5, not at 20. But in the normal approximation to the binomial calculation, we found the area under a normal curve to the left of 20 when we computed $1 - P(X \leq 20)$. In doing so, we omitted half of the rectangle centered at 20. To make the approximation more accurate, sometimes a **continuity correction** is used by adding or subtracting .5, based on which rectangles are desired. For Example 8.30, we could find the normal curve probability to the left of 20.5 to approximate $P(X \leq 20)$ for the binomial variable. The solution for Example 8.30 then becomes

$$1 - P(X \leq 20.5) \approx 1 - P\left(Z \leq \frac{20.5 - 15}{2.739}\right) = 1 - P(Z \leq 2.01) = .0222$$

which is much closer to the exact binomial probability of .0224 than the approximation of .0336 that was found without the continuity correction.

Similarly, in Example 8.29, we could use the continuity correction when we find $P(X \leq 240)$ so that we would find $P(X \leq 240.5)$ instead. This gives an answer of .1977, almost identical to the exact binomial answer of .1978.

IN SUMMARY

Normal Approximation for a Binomial Random Variable

When both np and $n(1 - p)$ are at least 10, a binomial random variable based on n trials with success probability p can be approximated by a normal random variable with mean $\mu = np$ standard deviation $\sigma = \sqrt{np(1 - p)}$.

8.8 Sums, Differences, and Combinations of Random Variables

There are many instances in which we want information about combinations of random variables. Suppose, for example, that your final score in a class is found by adding together 25% of each of two midterm exam scores (Exam1 and Exam2) and 50% of your final exam score (Exam3). Expressed as an equation, the final score is calculated as follows:

$$\text{Final score} = .25 \times \text{Exam1} + .25 \times \text{Exam2} + .50 \times \text{Exam3}$$

If you know the mean for each exam, it is easy to calculate the mean final score. It probably makes sense to you that the mean final score can be found by combining the mean exam scores in the same way that an individual's scores are combined. If the means on the three exams are μ_1, μ_2, and μ_3, the formula for the mean final score is

$$\text{Mean of final scores} = .25 \times \mu_1 + .25 \times \mu_2 + .50 \times \mu_3$$

For instance, if the exam means are 72, 76, and 80, respectively, then the mean of the final scores is

$$\begin{aligned}\text{Mean of final scores} &= (.25 \times 72) + (.25 \times 76) + (.50 \times 80) \\ &= 18 + 19 + 40 = 77\end{aligned}$$

We are assuming that the class constitutes the entire population of interest, so the means are represented as population means (μ) instead of sample means ($\bar{x}$). The formula for calculating the mean final score, however, would hold for sample means as well.

Linear Combinations of Random Variables

The combination of exam scores that we just considered is an example of a **linear combination of random variables**. In a linear combination, we add (or subtract) variables, but some of the combined variables may be multiplied by a numerical value, as occurred in calculating the final score. For two random variables X and Y, the most commonly encountered linear combinations are the **sum** $X + Y$ and the **difference** $X - Y$.

DEFINITION A **linear combination of random variables**, $X, Y, \ldots$ is a combination of the form

$$L = aX + bY + \cdots$$

where a, b, and so on are numbers that could be positive or negative. The most commonly encountered linear combinations of two variables are

$$\text{Sum} = X + Y$$
$$\text{Difference} = X - Y$$

Mean and Standard Deviation for Linear Combinations

A number of rules tell us about the statistical properties of a linear combination of random variables. The most basic rule is for the mean. This rule applies to any set of random variables, whether they are discrete or continuous, independent of each other or not, and regardless of their distributions. The only assumption is that the variables that are combined all have finite means.

FORMULA **Mean of a Linear Combination of Random Variables (Including Sums and Differences)**

If $X, Y, \ldots$ are random variables, $a, b, \ldots$ are numbers, either positive or negative, and

$$L = aX + bY + \cdots$$

The mean of L is

$$\text{Mean}(L) = a\text{Mean}(X) + b\text{Mean}(Y) + \cdots$$

In particular,

$$\text{Mean}(X + Y) = \text{Mean}(X) + \text{Mean}(Y)$$
$$\text{Mean}(X - Y) = \text{Mean}(X) - \text{Mean}(Y)$$

Rules for the variance and standard deviation of a linear combination are complicated, and we will only consider the specific situation in which the random variables are independent. Two random variables are **statistically independent** if the probability for any event associated with one random variable is not altered by whether or not any particular event for the other random variable has happened. In a more practical sense, it is usually safe to say that *two random variables are independent if there is no physical connection between the two variables or if there is no apparent reason why the value of one variable should influence the value of the other.*

FORMULA **Variance and Standard Deviation of a Linear Combination of Independent Random Variables**

If X, Y, etc. are independent random variables, a, b, and so on are numbers, and

$$L = aX + bY + \ldots$$

then the *variance* and *standard deviation* of L are

$$\text{Variance}(L) = a^2\,\text{Variance}(X) + b^2\,\text{Variance}(Y) + \cdots$$
$$\text{Standard deviation of } L = \sqrt{\text{Variance}(L)}$$

In particular,

$$\text{Variance}(X + Y) = \text{Variance}(X) + \text{Variance}(Y)$$
$$\text{Variance}(X - Y) = \text{Variance}(X) + \text{Variance}(Y)$$

Keep in mind that the standard deviation is the square root of the variance. Note also that the variance of the *difference* between two independent random variables is the *sum* of the variances because $b = -1$ and thus $b^2 = +1$.

Simply knowing the mean and standard deviation of a combination of random variables is of little practical use in most cases. To find probabilities associated with various outcomes, we also need to know the distribution.

Combining Independent Normal Random Variables

Some families of random variables keep the same shape when linear combinations of variables within the class are formed. This occurs for normal curve distributions. Any linear combination of normally distributed variables also has a normal distribution, but we consider only independent variables and focus on rules for the *sum* and for the *difference* of two such variables.

FORMULA **Linear Combinations of Independent Normal Random Variables**

If X, Y, etc. are independent, normally distributed random variables and a, b, ... are numbers, either positive or negative, then the random variable $L = aX + bY + \ldots$ is normally distributed. In particular,

$X + Y$ is normally distributed with mean $\mu_x + \mu_y$ and standard deviation $\sqrt{\sigma_X^2 + \sigma_Y^2}$.

$X - Y$ is normally distributed with mean $\mu_x - \mu_y$ and standard deviation $\sqrt{\sigma_X^2 + \sigma_Y^2}$.

Example 8.31 **Will Meg Miss Her Flight?** Meg travels often and lately has become lax about giving herself enough time to get to the airport. She leaves home 45 minutes before the last call for her flight will occur. Her travel time from her front door to the airport parking lot is normally distributed with a mean of 25 minutes and a standard deviation of 3 minutes. From the parking lot, she must then take a shuttle bus to the terminal and go through security screening. The mean time for doing this is 15 minutes, and the standard deviation is 2 minutes; this time also is normally distributed. The driving time and the airport time are independent of each other. What is the probability that Meg will miss her flight because her total time for getting to the plane is more than 45 minutes?

Let's define some random variables:

X = driving time; normally distributed with $\mu_x = 25$ minutes and $\sigma_x = 3$ minutes
Y = airport time; normally distributed with $\mu_y = 15$ minutes and $\sigma_y = 2$ minutes
$T = X + Y$ = total time.

The random variable T has a normal distribution because it is a sum of two independent, normally distributed random variables. For T, the mean is the sum of the means for X and Y, which is $\mu = 25 + 15 = 40$ minutes, and the standard deviation is $\sigma = \sqrt{\sigma_X^2 + \sigma_Y^2} = \sqrt{3^2 + 2^2} = \sqrt{9 + 4} = \sqrt{13} = 3.6$. Using Table A.1 in the Appendix (or statistical software or certain calculators), we find that the probability that the total time T exceeds 45 minutes is

$$P(T > 45) = P\left(Z > \frac{45 - 40}{3.6}\right)$$

$$= P(Z > 1.39) = 1 - P(Z < 1.39)$$

$$= 1 - .9177 = .0823$$

Meg should leave sooner. If she continues this behavior, she will miss her flight about 8% of the time.

Example 8.32 **Can Alison Ever Win?** Alison and her sister Julie each swim a mile every day. Alison's times are normally distributed with mean = 37 minutes and standard deviation = 1 minute. Julie is faster but less consistent than Alison, and her times are normally distributed with mean = 33 minutes and standard deviation = 2 minutes. On any day, their times are independent of each other. Can Alison ever beat Julie?

Let's define three random variables for a given day:

X = Alison's time; normally distributed with $\mu_x = 37$ minutes and $\sigma_x = 1$ minute
Y = Julie's time; normally distributed with $\mu_Y = 33$ minutes and $\sigma_Y = 2$ minutes
$D = X - Y$ = Difference in their times.

The random variable D is normally distributed because it is the difference of two normally distributed variables. For D, the difference between the sisters' times, the mean is $\mu = 37 - 33 = 4$ minutes and the standard deviation is $\sigma = \sqrt{\sigma_X^2 + \sigma_Y^2} = \sqrt{1^2 + 2^2} = \sqrt{5} = 2.236$. Alison beats Julie when $X < Y$ or, equivalently, when $D = X - Y < 0$. The probability that $D < 0$ is

$$P(D < 0) = P\left(Z < \frac{0 - 4}{2.236}\right) = P(Z < -1.79) = .0367$$

Alison swims a faster time about 3.7% of the time, which is about 1 day per month.

THOUGHT QUESTION 8.10 In Example 8.32, the time it takes Alison to swim a mile on any given day was assumed to be *independent* of the time that it takes Julie to swim a mile on that same day. Refer to the explanation of when two random variables are independent, given earlier in this section. Explain whether or not it is reasonable to assume that Alison's and Julie's times on the same day are independent in each of the following scenarios:

1. They swim at different times, and the one who swims first does not report her time to the other until after they are both done.
2. They swim side by side in two lanes of a pool.
3. They swim at different times, but the one who swims second knows the time it took the other to swim the mile that day.*

***HINT:** In each case, is there any apparent reason why the value of one should influence the value of the other?

Combining Independent Binomial Random Variables

A linear combination of binomial random variables is generally not a binomial variable, but there is one situation where it is. When each independent binomial variable has the same success probability, the sum has a binomial distribution. Put another way, independent binomial experiments with the same success probability can be combined and analyzed as a single binomial experiment. Imagine that you flip a coin five times while a friend flips a coin ten times. Probability questions about the total number of heads flipped by you and your friend can be answered by using the binomial distribution for $n = 5 + 10 = 15$ trials and $p = .5$ (assuming that you both flipped fairly!).

Adding Binomial Random Variables with the Same Success Probability

If X, Y, ... are independent binomial random variables with n_X, n_Y, $\cdots$ trials and all have the same success probability p, then the sum $X + Y + \cdots$ is a binomial random variable with $n = n_X + n_Y + \cdots$ and success probability p.

Note: If the success probabilities differ, the sum is not a binomial random variable.

Example 8.33 **Donations Add Up** A volunteer organization is having a fundraising drive. Three volunteers agree to call potential donors. In past drives, about 20% of those who were called agreed to donate, independent of who called them. If the three volunteers make 10, 12, and 18 calls, respectively, what is the probability that they get at least ten donors?

Let's define a random variable for each volunteer:

X = number of donors secured by volunteer 1; binomial $n = 10$, $p = .2$.
Y = number of donors secured by volunteer 2; binomial $n = 12$, $p = .2$.
W = number of donors secured by volunteer 3; binomial $n = 18$, $p = .2$.

The total number of donors, $T = X + Y + W$, is a binomial random variable with $n = 10 + 12 + 18 = 40$ trials and $p = .2$. Note that either there will be ten or more donors or there will be nine or fewer, so $P(T \geq 10) = 1 - P(T \leq 9) = 1 - .7318$ (from Minitab) $= .2682$. So in about 27% of the cases in this type of fundraising drive, there will be at least ten donors when 40 people are called.

Example 8.34 **Strategies for Studying When You Are Out of Time** You have a two-part multiple-choice test tomorrow, with 10 questions on one part and 10 on the other; all questions have four choices. You don't have time to study well. From past experience you know that if you study all of the material in the time you have, you can narrow all 20 questions down to two choices and get them correct with probability .5. If you study only the first set of material, you will get the questions right for that set with probability .8 but have to guess completely on the other set, which means $p = .25$ for those 10 questions. You need to get 13 or more questions right to pass the test. What should you do?

Strategy 1: Study all material. The score for the test has a binomial distribution with $n = 20$ and $p = .5$. The probability that you pass is $P(\text{Score} \geq 13) = 1 - P(\text{Score} \leq 12)$, which is $1 - .8684$ (from Minitab) $= .1316$.

The mean and standard deviation of the score will be useful for comparing the two strategies:

$\text{Mean(Score)} = np = (20)(.5) = 10$ questions
$\text{Variance(Score)} = np(1 - p) = (20)(.5)(1 - .5) = 5$
Standard deviation $= 2.24$ questions

Strategy 2: Study only set one. Score $= X + Y$, where X = number correct on the first part and Y = number correct on the second part. First, determine the mean and variance for each of X and Y:

X is a binomial with $n = 10$, $p = .8$
$\text{Mean}(X) = np = 10(.8) = 8$
$\text{Variance}(X) = np(1 - p) = 10(.8)(.2) = 1.6$

Y is a binomial with $n = 10$, $p = .25$
$\text{Mean}(Y) = np = 10(.25) = 2.5$
$\text{Variance}(Y) = np(1 - p) = 10(.25)(.75) = 1.875$

With these values, we can determine the mean and standard deviation of Score $= X + Y$.

Mean(Score) $=$ Mean(X) $+$ Mean(Y) $= 8 + 2.5 = 10.5$ questions
Variance(Score) $=$ Variance(X) $+$ Variance(Y) $= 1.6 + 1.875 = 3.475$
Standard deviation $= \sqrt{3.475} = 1.86$ questions

Strategy 1 has a lower mean (10) than Strategy 2 (10.5), so in the long run, Strategy 2 will be slightly better. But you are taking only a single exam, so an effective long-run strategy is not your concern. Strategy 1 has a higher standard deviation (2.24) than Strategy 2 (1.86), so perhaps the chance of a score well above the mean is greater for Strategy 1. This turns out not to be the case. For Strategy 1, the passing score of 13 is $(13 - 10)/2.24 = 1.34$ standard deviations above the mean. For Strategy 2, a score of 13 is $(13 - 10.5)/1.86 = 1.34$ standard deviations above the mean. It does not appear to matter which strategy you use if your only concern is the probability of passing the test.

When the success probabilities are unequal (as in Strategy 2), it is cumbersome to find probabilities for a sum of binomial random variables with a small number of trials. For Strategy 2, calculating the probability of passing involves finding probabilities for every possible combination of X and Y for which the sum is at least 13.

This is a case in which simulation may be an effective way to estimate the probability. With the Minitab package, Strategy 2 was simulated by randomly generating values for X and Y and adding them together. This was done 1000 times, and a passing grade of 13 or more was achieved in 130 of the 1000 repetitions. The estimated probability that you pass the test using Strategy 2 is $130/1000 = .130$, almost identical to the probability of .1316 for Strategy 1. Neither strategy offers you much hope!

8.8 Exercises are on pages 308–309.

Key Terms

Section 8.1
random variable, 263
family of random variables, 263, 265
discrete random variable, 264, 266
continuous random variable, 264, 266

Section 8.2
probability distribution function (pdf), 266–267, 271
cumulative probability, 269, 271
cumulative distribution function (cdf), 269, 271

Section 8.3
mean value of a discrete random variable, 271
expected value, 271, 272, 275
standard deviation of a discrete random variable, 273
variance of a discrete random variable, 273
mean for a population, 274–275
standard deviation for a population, 274–275

Section 8.4
binomial random variable, 275, 281
binomial experiment, 276
Bernoulli random variable, 277
binomial distribution, 277, 281
mean for binomial, 279, 280, 281
standard deviation for binomial, 280, 281
variance for binomial, 280

Section 8.5
probability density function, 281, 283
uniform random variable, 282

Section 8.6
normal random variable, 283, 291
normal curve, 283
normal distribution, 283
standard normal random variable, 287
standard normal distribution, 287, 291
standardized score, 288, 291
z-score, 288
percentile, 290
percentile rank, 291

Section 8.7
normal approximation to the binomial distribution, 292, 296
continuity correction, 296

Section 8.8
linear combination of random variables, 297
sum of random variables, 297, 298
difference of random variables, 297, 298
statistically independent, 297
independent random variables, 297

In Summary Boxes

Discrete and Continuous Random Variables, 266
Probabilities for Discrete Random Variables, 271
Expected Value, 275
Binomial Random Variables, 281
Probabilities for Continuous Random Variables, 283
Normal Random Variables, 291
Normal Approximation for a Binomial Random Variable, 296

Exercises

◆ Denotes that the dataset is available on the companion website, http://www.cengage.com/statistics/Utts4e, but is not required to solve the exercise.

Bold exercises have answers in the back of the text.

Section 8.1

Skillbuilder Exercises

8.1 Decide whether each of the following characteristics of a television news broadcast is a continuous or discrete random variable:

a. Number of commercials shown (e.g., five commercials).
b. Length of the first commercial shown (e.g., 15 seconds).
c. Whether there were any fatal car accidents reported (0 = there were no fatal accidents reported, 1 = there was at least one fatal accident reported).
d. Whether rain was forecast for the next day (0 = no rain forecast, 1 = rain forecast).

8.2 A book is randomly chosen from a library shelf. For each of the following characteristics of the book, decide whether the characteristic is a continuous or a discrete random variable:

a. Weight of the book (e.g., 2.3 pounds).
b. Number of chapters in the book (e.g., 10 chapters).
c. Width of the book (e.g., 8 inches).
d. Type of book (0 = hardback or 1 = paperback).
e. Number of typographical errors in the book (e.g., 4 errors).

8.3 For each characteristic, explain whether the random variable is continuous or discrete.

a. The number of left-handed individuals in a sample of 100 people.
b. Time taken to complete an exam for students in a class.
c. Vehicle speeds at a highway location.
d. The number of accidents reported last year at a highway location.

8.4 For each characteristic, explain whether the random variable is continuous or discrete.

a. Time to read a news article on the Internet.
b. Number of losing instant lottery tickets purchased before buying a winning ticket.
c. Body weights of 8-year-old children.
d. Number of people with brown eyes in a sample of 30 people.

General Section Exercises

8.5 Refer to Example 8.1 (p. 264), the scheduling of an outdoor event. Give an example of another continuous random variable (in addition to temperature) and another discrete random variable (in addition to number of planes flying overhead) that would influence the enjoyment of the event. Give the sample space for those random variables.

8.6 Suppose you regularly play a lottery game. Give an example of a discrete random variable in this context.

8.7 Explain the difference between how probabilities may be represented for discrete and continuous random variables.

8.8 Answer Thought Question 8.1 on page 266.

Section 8.2

Skillbuilder Exercises

8.9 What is the missing value, represented by the question mark (?) in each of the following probability distribution functions?

a.

k	1	4	5	7
$P(X = k)$	1/6	1/6	1/6	?

b.

k	100	200	300	400	500
$P(X = k)$	.1	.2	.3	.3	?

8.10 A professor gives a weekly quiz with varying numbers of questions and uses a randomization device to decide how many questions to include. Let the random variable X = the number of questions on an upcoming quiz. The probability distribution for X is given in the following table, but one probability is missing. What is the probability that $X = 30$?

No. of Questions, X	10	15	20	25	30
Probability	.05	.20	.50	.15	?

8.11 For a large population, probabilities for X = the number of meals eaten yesterday are:

Meals, X	1	2	3	4
Probability	.10	.32	.56	.02

a. Write a table that gives the cumulative probability distribution for X.
b. What is the value of $P(X \leq 2)$, the probability that a randomly selected person ate two or fewer meals yesterday?

8.12 Answer Thought Question 8.3 on page 271.

8.13 For a fair coin tossed three times, the eight possible simple events are HHH, HHT, HTH, THH, HTT, THT, TTH, and TTT. Let X = number of heads. Find the probability distribution for X by writing a table showing each possible value of X along with the probability that the value occurs.

8.14 Suppose the probability distribution for X = number of jobs held during the past year for students at a school is as follows:

No. of Jobs, X	0	1	2	3	4
Probability	.14	.37	.29	.15	.05

a. Find $P(X \leq 2)$, the probability that a randomly selected student held two or fewer jobs during the past year.
b. Find the probability that a randomly selected student held either one or two jobs during the past year.
c. Find $P(X > 0)$, the probability that a randomly selected student held at least one job during the past year.
d. Create a table that gives the cumulative probability distribution for X.

8.15 The following table gives the probability distribution for X = number of classes skipped yesterday by students at a college.

No. of Classes skipped, X	0	1	2	3	4
Probability	.73	.16	.06	.03	.02

a. What is the probability that a randomly selected student skipped either two or three classes yesterday?
b. What is the probability that a randomly selected student skipped at least one class yesterday?
c. Create a table that gives the cumulative distribution function for X.

8.16 The following table gives the probability distribution for X = number of wins in 3 plays of a game for which the chance of winning each game = .2, and plays are independent.

X = Number of wins	0	1	2	3
Probability	.512	.384	.096	.008

a. Find $P(X \leq 1)$, the probability of winning one or fewer games in three plays.
b. Find $P(X > 1)$, the probability of winning more than one game in three plays.
c. Create a table that gives the cumulative probability distribution for X.

General Section Exercises

8.17 Refer to Example 8.10 (p. 270), in which the probability distribution is given for the sum of two fair dice. Use the distribution to find the probability that the sum is even.

8.18 A kindergarten class has three left-handed children and seven right-handed children. Two children are selected without replacement for a shoe-tying lesson. Let X = the number who are left-handed.

a. Write the simple events in the sample space. For instance, one simple event is RL, indicating that the first child is right-handed and the second one is left-handed.
b. Find the probability for each of the simple events in the sample space. (*Hint:* A tree diagram may help you solve this.)
c. Find the probability distribution function for X.
d. Draw a picture of the probability distribution function for X.

8.19 Let the random variable X = number of phone calls you will get in the next 24 hours. Suppose the possible values for X are 0, 1, 2, or 3, and their probabilities are .1, .1, .3, and .5, respectively. For instance, the probability that you will receive no calls is .1.

a. Verify that the "Conditions for Probabilities for Discrete Random Variables" given on page 267 are met.
b. Draw a picture of the probability distribution function.
c. Write a table that gives the cumulative probability distribution for X.

8.20 Consider three tosses of a fair coin. Write the sample space, and then find the probability distribution function for each of the following random variables:

a. X = number of tails.
b. Y = the difference between the number of heads and the number of tails.
c. Z = the sum of the number of heads and the number of tails.

8.21 A woman decides to have children until she has her first girl or until she has four children, whichever comes first. Let X = number of children she has. For simplicity, assume that the probability of a girl is .5 for each birth.

a. Write the simple events in the sample space. Use B for boy and G for girl. For instance, one simple event is BG, because the woman would stop having children once she has a girl.
b. Find the probability for each of the simple events in the sample space.
c. Find the probability distribution function for X.
d. Draw a picture of the probability distribution function for X.

8.22 Refer to Example 8.10 (p. 270). Find the cumulative distribution function (cdf) for the sum of two fair dice.

8.23 Explain which would be of more interest in each of the following situations, the probability distribution function or the cumulative probability distribution function for X. If you think both would be of interest, explain why.

a. A politician wants to know how her constituents feel about a proposed new law. She hires a survey firm to take a random sample of voters in her district, and X = number in the sample who oppose the new law.
b. Each time someone is exposed to a virus, he or she could become infected with it. Let X = number of exposures a person has prior to becoming infected. For instance, if X = 0, the person became infected on the first exposure; if X = 1, the person became infected on the second exposure, and so on.

8.24 Explain which would be of more interest in each of the following situations—the probability distribution function or the cumulative probability distribution function for X. If you think both would be of interest, explain why.

a. A pharmaceutical company wants to show that its new drug is effective for lowering blood pressure. The drug is administered to a sample of 800 patients with high blood pressure, and X = the number for whom the drug reduces blood pressure.
b. A caterer who serves dinner at wedding receptions is told in advance the number of people who have said they will attend; X is the number of people who actually attend the dinner.

Section 8.3

Skillbuilder Exercises

8.25 Suppose that in a gambling game, the probability of winning \$4 is .3 and the probability of losing \$2 is .7.

a. Write a table that gives the probability distribution for X = amount won in a single play. Losing \$2 can be expressed as "winning" −\$2.
b. Calculate the expected value of X = amount won on a single play.
c. Write a sentence that interprets the expected value of X in this situation.

8.26 The Brann family wants to be financially prepared to have children. A financial advisor informs them that on the basis of families with similar characteristics, the probability distribution for the random variable X = number of children they might have is as follows:

No. of Children	0	1	2	3
Probability	.05	.60	.30	.05

a. What is the expected value of X?
b. Explain what "expected value" means in this situation.
c. Is $E(X)$ a possible outcome for the number of children that the Brann family will have?

8.27 The probability that Mary will win a game is .01, so the probability that she will not win is .99. If Mary wins, she will be given \$100; if she loses, she must pay \$5. If X = amount of money Mary wins (or loses), what is the expected value of X?

8.28 Exercise 8.11 gave the following distribution for X = the number of meals eaten yesterday by individuals in a large population:

Meals, X	1	2	3	4
Probability	.10	.32	.56	.02

a. Calculate the expected value of X.
b. Write a sentence that interprets the expected value of X in this situation.

8.29 Exercise 8.16 gave the following table for the probability distribution of X = number of wins in three independent plays of a game for which the chance of winning each game = .2.

X = Number of wins	0	1	2	3
Probability	.512	.384	.096	.008

a. What is the expected value of X?
b. Write a sentence that interprets the expected value of X in this situation.

8.30 A random variable X has the following probability distribution:

X	−1	0	1
Probability	.25	.50	.25

a. Calculate the mean of X.
b. Calculate the variance of X.
c. Calculate the standard deviation of X.

8.31 A random variable X has the following probability distribution:

X	−2	0	2
Probability	.25	.50	.25

a. Calculate the mean of X.
b. Calculate the variance of X.
c. Calculate the standard deviation of X.

General Section Exercises

8.32 Find the expected value for the sum of two fair dice. The probability distribution function was found in Example 8.10 (p. 270).

8.33 Find the expected value for the number of girls in a family with three children, assuming that boys and girls are equally likely. Use the probability distribution function given in Example 8.6 (p. 267).

8.34 Find the standard deviation for the sum of two fair dice. The probability distribution was found in Example 8.10 (p. 270).

8.35 Find the standard deviation for the net gain in the California Decco lottery game in Example 8.12 (p. 272).

8.36 Suppose the probability that you get an A in any class you take is .3 and the probability that you get a B is .7. To construct a grade point average, an A is worth 4.0 and a B is worth 3.0.

a. What is the expected value for your grade point average?
b. Would you expect to have this grade point average for each quarter or semester of your school career? Explain.

8.37 An insurance company expects 10% of its policyholders to collect claims of \$500 this year and the remaining 90% to collect no claims. What is the expected value for the amount they will pay out in claims per person?

8.38 Refer to Exercise 8.37. If the insurance company wants to make a net profit of \$10 per policyholder for the year, how much should it charge each person for insurance? Ignore administrative and other costs.

8.39 In 2004, 70% of children in the United States were living with both parents, 23% were living with mother only, 3% were living with father only, and 4% were not living with either parent (http://www.census.gov/prod/2008pubs/p70-114.pdf, p. 4).

a. What is the expected value for the *number* of parents that a randomly selected child was living with in 2004?
b. Does the concept of expected value have a meaningful interpretation for this example? Explain.

8.40 Suppose you have to cross a train track on your commute. The probability that you will have to wait for a train is 1/5, or .20. If you don't have to wait, the commute takes 15 minutes, but if you have to wait, it takes 20 minutes.

a. What is the expected value of the time it takes you to commute?
b. Is the expected value ever equal to the actual commute time? Explain.

Section 8.4

Skillbuilder Exercises

8.41 A fair coin is flipped 200 times. The random variable X = number of heads out of the 200 tosses.

a. Is X a binomial random variable? If so, specify n and p. If not, explain why not.
b. What is the expected value of X?

8.42 For each of the examples below, decide if X is a binomial random variable. If so, specify n and p. If not, explain why not.

a. X = number of heads from flipping the same coin ten times, where the probability of a head = 1/2.
b. X = number of heads from flipping two coins five times each, where the probability of a head for one coin = 1/2 and the probability of a head for the other coin = 1/4.
c. X = number of cities in which it will rain tomorrow among five neighboring cities located within 10 miles of each other.
d. X = number of children who will get the flu this winter in a kindergarten class with 20 children.

8.43 For each of the following binomial random variables, specify n and p:

a. A fair die is rolled 30 times. X = number of times a 6 is rolled.
b. A company puts a game card in each box of cereal and 1/100 of them are winners. You buy ten boxes of cereal, and X = number of times you win.
c. Jack likes to play computer solitaire and wins about 30% of the time. X = number of games he wins out of his next 20 games.

8.44 Refer to Exercise 8.43. Find μ in each case.

8.45 Assuming that X is a binomial random variable with $n = 10$ and $p = .20$, find the probability for each of the following values of X:

a. $X = 5$.
b. $X = 2$.
c. $X = 1$.
d. $X = 9$.

8.46 Find the mean and standard deviation for a binomial random variable X with each of the following values of n and p:

a. $n = 10, p = .50$.
b. $n = 1, p = .40$.
c. $n = 100, p = .90$.
d. $n = 30, p = .01$.

8.47 Find the expected value and standard deviation for a binomial random variable with each of the following values of n and p:

a. $n = 10, p = 1/2$.
b. $n = 100, p = 1/4$.
c. $n = 2500, p = 1/5$.
d. $n = 1, p = 1/10$.
e. $n = 30, p = .4$.

8.48 Assuming that X is a binomial random variable with $n = 4$ and $p = .7$, find the probability for each of the following values of X.

a. $X = 2$.
b. $X = 0$.
c. $X = 1$.

General Section Exercises

8.49 Explain which of the conditions for a binomial experiment is *not* met for each of the following random variables:

a. A football team plays 12 games in its regular season. X = number of games the team wins.
b. A woman buys a lottery ticket every week for which the probability of winning anything at all is 1/10. She continues to buy them until she has won three times. X = the number of tickets she buys.
c. A poker hand consists of 5 cards drawn from a standard deck of 52 cards. X = the number of aces in the hand.

8.50 A computer chess game and a human chess champion are evenly matched. They play ten games. Find probabilities for the following events.

a. They each win five games.
b. The computer wins seven games.
c. The human chess champion wins at least seven games.

8.51 Suppose that in a very large population, 10% of individuals are left-handed. Five individuals will be sampled. X = number of left-handed individuals among these five people. Find probabilities for the following events.

a. Exactly one of the five individuals is left-handed.
b. At most one of the five individuals is left-handed.
c. Two or more of the five individuals is left-handed.

8.52 In an ESP test, a "participant" tries to draw a hidden "target" photograph that is unknown to anyone in the room. After the drawing attempt, the participant is shown four choices and asked to determine which one had been the real target. The real target is randomly selected from the four choices in advance, so the probability of a correct match by chance is 1/4. The test is repeated ten times, and four new photographs are used each time.

a. Go through the conditions for a binomial experiment, and explain how this situation fits each one of them, assuming that the participant is just guessing each time.
b. Let X = number of correct choices in the ten tests. If the participant is just guessing, is X a binomial random variable? If not, explain why not. If so, specify n and p.
c. If the participant is just guessing, find $P(X = 6)$.
d. Suppose the participant actually has some psychic ability and can get each answer correct with probability .5 instead of .25. Find $P(X = 6)$.
e. Compare the answers in parts (c) and (d). If the participant actually selects six of the ten answers correctly, would you believe that he or she was just guessing or that he or she was using some psychic ability? Explain your answer. (Note that there is no correct answer here; your reasoning is what counts.)

8.53 For each of the following situations, assume that X is a binomial random variable with the specified n and p, and find the requested probability.

a. $n = 10, p = .5, P(X = 4)$.
b. $n = 10, p = .3, P(X \geq 4)$.

8.54 For each of the following situations, assume that X is a binomial random variable with the specified n and p, and find the requested probability.

a. $n = 10, p = .3, P(X \leq 3)$.
b. $n = 5, p = .1, P(X = 0)$.
c. $n = 5, p = .1, P(X \geq 1)$.

8.55 For each of the following scenarios, write the desired probability in a format such as $P(X = 10)$ and specify n and p. Do not actually compute the desired probability. If you cannot specify a numerical value for p, write in words what it represents.

a. A random sample of 1000 adults is drawn from the United States. $X =$ the number in the sample who are living with a partner but are not married. The desired probability is the probability that at least 1/4 of the sample fits this description.
b. A pharmaceutical company claims that 20% of those taking its new allergy medication will experience drowsiness. To test this claim, it randomly assigns 500 people to take the new medication and measures $X =$ the number who experience drowsiness. The desired probability is the probability that 110 or more people in the sample experience drowsiness if the claim is true.
c. A student has not studied the material for a 20-question true–false test and simply guesses on each question. A passing grade is 70%; the desired probability is the probability that the student passes the test.

8.56 A computer is used to generate $n = 6$ random integers, each between 0 and 9 (with replacement). $X =$ the number of even numbers among the six numbers picked. Find probabilities for the following events.

a. Three even numbers are picked.
b. Four or fewer even numbers are picked.
c. More than four even numbers are picked.
d. No even numbers are picked.

Section 8.5

Skillbuilder Exercises

8.57 Suppose that X is a uniform random variable where X is between 0 and 10.

a. What is the probability that X will be between 0 and 3?
b. What is the probability that X will be between 4 and 8?
c. What is the probability that X will be between 5 and 10?

8.58 Suppose that the time students wait for a bus can be described by a uniform random variable X, where X is between 0 minutes and 60 minutes.

a. What is the probability that a student will wait between 0 and 30 minutes for the next bus?
b. What is the probability that a student will have to wait at least 30 minutes for the next bus?

General Section Exercises

8.59 A game is played with a spinner on a circle, like the minute hand on a clock. The circle is marked evenly from 0 to 100, so, for example, the 3:00 position corresponds to 25, the 6:00 position to 50, and so on. The player spins the spinner, and the resulting number is the number of seconds he or she is given to solve a word puzzle. Let $X =$ amount of time the player is given to solve the puzzle.

a. Explain how you know that X is a uniform random variable.
b. Write down the density function for X. Be sure to specify the range of values for which it holds.
c. Find $P(X \leq 15 \text{ seconds})$.
d. Find $P(X \geq 40 \text{ seconds})$.
e. Draw a picture of the density function of X, and use it to illustrate the probabilities that you found in parts (c) and (d).
f. What is the expected value of X? This was not covered in the text, so explain your reasoning.

8.60 Draw the density curve corresponding to each of the following random variables, and then shade the area corresponding to the desired probability. You *do not* need to compute the probability.

a. X is a uniform random variable from 10 to 20, $P(10 \leq X \leq 13)$.
b. X is a uniform random variable from 0 to 20, $P(X \leq 5)$.
c. X is a uniform random variable from 0 to 8, $P(X \leq 4)$.

8.61 Give an example of a uniform random variable that might occur in your daily life.

8.62 Find the probabilities specified in each of the parts of Exercise 8.60.

Section 8.6

Skillbuilder Exercises

8.63 In each situation below, calculate the standardized score (or z-score) for the value x:

a. Mean $\mu = 0$, standard deviation $\sigma = 1$, value $x = 1.5$.
b. Mean $\mu = 10$, standard deviation $\sigma = 6$, value $x = 4$.
c. Mean $\mu = 10$, standard deviation $\sigma = 5$, value $x = 0$.
d. Mean $\mu = -10$, standard deviation $\sigma = 15$, value $x = -25$.

8.64 In each situation below, calculate the standardized score (or z-score) for the value x:

a. Mean $\mu = 65$, standard deviation $\sigma = 4$, value $x = 70$.
b. Mean $\mu = 120$, standard deviation $\sigma = 10$, value $x = 115$.
c. Mean $\mu = 72$, standard deviation $\sigma = 8$, value $x = 82$.
d. Mean $\mu = 72$, standard deviation $\sigma = 8$, value $x = 62$.

8.65 For each value of z^*, find the cumulative probability $P(Z \leq z^*)$:

a. $z^* = 0$.
b. $z^* = -0.35$.
c. $z^* = 0.35$.

8.66 For each value of z^*, find the cumulative probability $P(Z \leq z^*)$:

a. $z^* = 1.96$.
b. $z^* = -2.33$.
c. $z^* = 2.58$.
d. $z^* = 1.65$.

8.67 Weights (X) of men in a certain age group have a normal distribution with mean $\mu = 180$ pounds and standard deviation $\sigma = 20$ pounds. Find each of the following probabilities:

a. $P(X \leq 200)$ = probability the weight of a randomly selected man is less than or equal to 200 pounds.
b. $P(X \leq 165)$ = probability the weight of a randomly selected man is less than or equal to 165 pounds.
c. $P(X > 165)$ = probability the weight of a randomly selected man is more than 165 pounds.

8.68 Draw the density curve corresponding to each of the following normal random variables, and then shade the area corresponding to the desired probability. You *do not* need to compute the probability.

a. X is a normal random variable with $\mu = 75$ and $\sigma = 5$, $P(70 \leq X \leq 85)$.
b. X is a normal random variable with $\mu = 15$ and $\sigma = 10$, $P(5 \leq X \leq 20)$.
c. X is a normal random variable with $\mu = 100$ and $\sigma = 15$, $P(X > 115)$.

8.69 Find the following probabilities for a standard normal random variable Z:

a. $P(Z \leq -1.4)$.
b. $P(Z \leq 1.4)$.
c. $P(-1.4 \leq Z \leq 1.4)$.
d. $P(Z \geq 1.4)$.

8.70 Find the following probabilities for a standard normal random variable Z:

a. $P(Z \leq -3.72)$.
b. $P(-3.72 \leq Z \leq 3.72)$.
c. $P(Z \geq 1.5)$.
d. $P(Z > -0.67)$

General Section Exercises

8.71 Find the following probabilities for X = pulse rates of women, for which the mean is 75 and the standard deviation is 8. Assume a normal distribution.

a. $P(X \leq 71)$.
b. $P(X \geq 85)$.
c. $P(59 \leq X \leq 95)$.

8.72 Find the following probabilities for Verbal SAT test scores X, for which the mean is 500 and the standard deviation is 100. Assume that SAT scores are described by a normal curve.

a. $P(X \leq 500)$.
b. $P(X \leq 650)$.
c. $P(X \geq 700)$.
d. $P(500 \leq X \leq 700)$.

8.73 Refer to Exercise 8.71 about the pulse rates of women. What is the 10th percentile of women's pulse rates?

8.74 Suppose that the amounts that students at a college spend on textbooks this semester have a normal curve with mean $\mu = \$360$ and standard deviation $\sigma = \$120$.

a. What is the probability that a randomly selected student spends less than \$450 on textbooks this semester?
b. What is the probability that a randomly selected student spends more than \$240 on textbooks this semester?
c. What is the probability that a randomly selected student spends between \$240 and \$450 on textbooks this semester?

8.75 Suppose the yearly rainfall totals for a city in northern California follow a normal distribution, with a mean of 18 inches and a standard deviation of 6 inches. For a randomly selected year, what is the probability that total rainfall will be in each of the following intervals?

a. Less than 10 inches.
b. Greater than 30 inches.
c. Between 15 and 21 inches.
d. Greater than 35 inches.

8.76 Refer to Exercise 8.75. Suppose that in a given year, the total rainfall is only 6 inches. You work for the local newspaper, and your editor has asked you to write a story about the terrible drought the town is suffering and how abnormal the situation is. Write a few sentences that you could use to explain the statistical facts to your readers (and your editor). Be sure to comment on whether or not you agree that the situation is terribly abnormal.

8.77 Find the value z^* that satisfies each of the following probabilities for a standard normal random variable Z:

a. $P(Z \leq z^*) = .025$.
b. $P(Z \leq z^*) = .975$.
c. $P(-z^* \leq Z \leq z^*) = .95$.

8.78 Heights for college women are normally distributed with mean = 65 inches and standard deviation = 2.7 inches. Find the proportion of college women whose heights fall into the following ranges:

a. Between 62 inches and 65 inches.
b. Between 60 inches and 70 inches.
c. Less than 70 inches.
d. Greater than 60 inches.
e. Either less than 60 inches *or* greater than 70 inches.

8.79 Refer to Exercise 8.78. Find the height such that about 25% of college women are shorter than that height. What is the percentile ranking for that height?

8.80 Refer to Exercises 8.78 and 8.79. Find the height such that about 10% of college women are taller than that height. What is the percentile ranking for that height?

Section 8.7

Skillbuilder Exercises

8.81 Suppose that $p = .512$ is the proportion of one-child families in which the child is a boy.

a. For a random sample of $n = 50$ one-child families, use the normal approximation to the binomial distribution to estimate the probability that there will be 20 or fewer families with one boy.

b. Repeat part (a) using the continuity correction described in the Technical Note on page 296. To do this, find the probability that the number of families with one boy will be 20.5 or less.

c. Use statistical software, Excel, or a graphing calculator to find the exact binomial probability that the number of families with one boy will be 20 or less.

d. Which of the results in parts (a), (b), and (c) is the most accurate value and which is the least accurate value for the probability that the number of families with one boy will be 20 or less?

8.82 Suppose that a fair coin is flipped $n = 100$ times.

a. Calculate the mean and standard deviation of X = number of heads. (*Hint:* The variable X has a binomial distribution.)

b. Use the normal approximation to the binomial distribution to estimate the probability that the number of heads is greater than or equal to 60.

c. Repeat part (b) using the continuity correction described in the Technical Note on page 296. To do this, find the probability that the number of heads is greater than or equal to 59.5.

d. Use statistical software, Excel, or a graphing calculator to find the exact binomial probability that the number of heads will be greater than or equal to 60.

e. Which of the results in parts (b), (c), and (d) is the most accurate value and which is the least accurate value for the probability that the number of heads is greater than or equal to 60?

General Section Exercises

8.83 A random sample of 1000 eligible voters is drawn, and X = number who actually voted in the last election. It is known that 60% of all eligible voters did vote.

a. Use the normal approximation to the binomial distribution to approximate $P(X \leq 620)$, the probability that 620 or fewer individuals in the sample actually voted.

b. Use statistical software, Excel, or a graphing calculator to find the exact binomial probability that 620 or fewer individuals in the sample actually voted.

8.84 To be eligible for a certain job, women must be at least 62 inches tall, and 87% of women meet this criterion. In a random sample of 2000 women, X = number who qualify for the job (based on height).

a. Use the normal curve to approximate $P(X \leq 1700)$, the probability that 1700 or fewer women have the necessary height to qualify for the job.

b. Use statistical software, Excel, or a graphing calculator to find the exact binomial probability that 1700 or fewer women in the sample are tall enough to qualify for the job.

8.85 An allergy medication is successful for 70% of all patients who use it. The medication is given to 200 patients. Let X = number of successful treatments.

a. What are the mean and standard deviation of X?

b. Use the normal curve to approximate the probability that X is 128 or less.

8.86 In a test for extrasensory perception, the participant repeatedly guesses the suit of a card randomly sampled with replacement from an ordinary deck of cards. There are four suits, all equally likely. The participant guesses 100 times, and X = number of correct guesses.

a. Explain why X is a binomial random variable, and specify n and p, assuming that the participant is just guessing.

b. Find the mean and standard deviation for X if the participant is just guessing.

c. Suppose that the participant guesses correctly 33 times. Find the approximate probability of guessing this well or better by chance.

d. Suppose that the participant guesses correctly 50 times. Find the approximate probability of guessing this well or better by chance. In that circumstance, would you be convinced that the participant was doing something other than just guessing? Explain.

Section 8.8

Skillbuilder Exercises

8.87 Suppose the length of time a person takes to use an ATM machine is normally distributed with mean $\mu = 100$ seconds and standard deviation $\sigma = 10$ seconds. There are $n = 4$ people ahead of Jackson in a line of people waiting to use the machine. He is concerned about T = total time the four people will take to use the machine.

a. What is the mean value of T = total time for the four people ahead of Jackson?

b. Assuming that the times for the four people are independent of each other, determine the standard deviation of T.

c. Jackson hopes the total time T is less than or equal to 360 seconds (6 minutes). Find $P(T < 360)$, the probability that the total waiting time is less than 360 seconds.

8.88 Suppose the heights of adult males in a population have a normal distribution with mean $\mu = 70$ inches and standard deviation $\sigma = 2.8$ inches. Two unrelated men will be randomly sampled. Let X = height of the first man and Y = height of the second man.

a. Consider $D = X - Y$, the difference between the heights of the two men. What type of distribution will the variable D have?

b. What is the mean value for the distribution of D?

c. Assuming independence between the two men, find the standard deviation of D.

d. Determine the probability that the first man is more than 3 inches taller than the second man. That is, find $P(D > 3)$.

e. Find the probability that one of the men is at least 3 inches taller than the other. That is, find the probability that either $D > 3$ or $D < -3$.

8.89 Isabelle and Taylor work together on a group quiz that has 15 multiple-choice questions, each with four choices for the possible answer. Unfortunately, neither had time to study, so they decide to randomly guess at all answers. Isabelle guesses answers for the first 7 questions, and Taylor guesses for the other 8 questions.

a. For any single question on the quiz, what is p = chance of a correct guess?

b. Let X = number of correct guesses that Isabelle makes. What are the values of n and p for the binomial distribution that describes X?

c. Let Y = number of correct guesses that Taylor makes. What are the values of n and p for the binomial distribution that describes Y?

d. Consider $X + Y$ = total correct guesses that Isabelle and Taylor make. What are the values of n and p for the binomial distribution that describes $X + Y$?

8.90 The variable X has a normal distribution with mean $\mu = 75$ and standard deviation $\sigma = 6$. The variable Y has a normal distribution with mean $\mu = 70$ and standard deviation $\sigma = 8$. Variables X and Y are independent.

a. Find the mean and standard deviation of the sum $X + Y$.

b. Find the mean and standard deviation of the difference $X - Y$.

General Section Exercises

8.91 Give the mean, variance, and standard deviation of the sum $X + Y$ in each of the following cases. You can assume X and Y are independent. For any situation(s) covered in Section 8.8, name what distribution the sum $X + Y$ has.

a. X is a binomial random variable with $n = 10$, $p = .5$. Y is a binomial random variable with $n = 20$, $p = .4$.

b. X is a normal random variable with $\mu = 100$ and $\sigma = 15$. Y is a normal random variable with $\mu = 50$ and $\sigma = 10$.

c. X is a normal random variable with $\mu = 100$ and $\sigma = 15$. Y is a binomial random variable with $n = 200$, $p = .25$.

8.92 Ethan has a midterm in his statistics class, which starts 10 minutes after the scheduled end of his biology class. The biology teacher rarely ends class on time though, and the amount of time he is overtime is approximately normally distributed, with a mean of 2 minutes and a standard deviation of 1/2 minute. The time it takes Ethan to get from one class to another is also normal, with a mean of 6 minutes and a standard deviation of 1 minute.

a. Is it reasonable to assume that the two times are independent? Explain.

b. Assuming that the two times are independent, what is the probability that Ethan will be late for his exam?

8.93 Joe performs remarkably well on remote viewing ESP tests, which require the "viewer" to draw a picture, and then determine which of four possible photos was the intended "target." Because the target is randomly selected from among the four choices, the probability of a correct match by chance is 1/4, or .25. Joe participates in three experiments with $n = 10$, 20, and 50 trials, respectively. In the three experiments, his numbers correct are 4, 8, and 20, respectively, for a total of 32 correct out of 80 trials.

a. For each experiment, what is the expected number correct if Joe is just guessing?

b. Over all 80 trials, what is the expected number correct? Explain how you found your answer.

c. For each experiment separately, find the approximate probability that someone would get as many correct as Joe did, or more, if the person was just guessing.

d. Out of the 80 trials overall, find the approximate probability that someone would get as many correct as Joe did (32), or more, if the person was just guessing.

e. Compare your answers in parts (c) and (d). If Joe were trying to convince someone of his ability, would it be better to show the three experiments separately, or would it be better to show the combined data? Explain.

8.94 Charles, Julia, and Alex are in grades 4, 3, and 2, respectively, and are representing their school at a spelling bee. The school's team score is the sum of the number of words the individual students spell correctly out of 50 words each. Different words are given for each grade level. From practicing at school, it is known that the probability of spelling each word correctly is .9 for Charles and .8 for the younger two, Julia and Alex.

a. Find the mean and standard deviation for the number of correct words for each child.

b. Find the mean and standard deviation for the team score.

c. Assume that the individual scores are independent. Does the team score have a binomial distribution? Explain.

d. Although not obvious from the material in this chapter, the team score would be approximately normal with the mean and standard deviation you were asked to find in part (b). If last year's team score was 131, what is the approximate probability that this year's team scores as well or better?

8.95 Annmarie is trying to decide whether to take the train or the bus into the city. The train takes longer but is more predictable than the bus because there are no traffic delays. Train times are approximately normally distributed, with mean = 60 minutes and standard deviation = 2 minutes, while bus times are approximately normally distributed, with mean = 50 minutes and standard deviation = 8 minutes. The bus and train times are independent of each other. Find the probability that the train is faster on any given day.

Chapter Exercises

8.96 Do you think that a normal curve would be a good approximation to the distribution of the ages of all individuals in the world? Briefly explain your answer.

8.97 A histogram is drawn of the weights of all students in a class of 100 men and 100 women. Would this histogram have a bell shape? Briefly explain your answer.

8.98 Suppose that 10% of a population is left-handed. What is the probability that in a sample of $n = 10$ individuals, 3 or more individuals are left-handed?

8.99 In the casino game of roulette, a gambler can bet on which of 38 numbers will be the result when the roulette wheel is spun. On a $2 bet, a gambler gains $70 if he or she picks the right number but loses the $2 otherwise.

a. Let X = amount gained or lost on a $2 bet on a roulette number. Write out the probability distribution of X.

b. Calculate $E(X)$, the mean value of X. What does this value indicate about the advantage that a casino has over roulette players?

8.100 Suppose that a college determines the following distribution for X = number of courses taken by full-time students this semester:

k	3	4	5	6	7
P (X = k)	.07	.14	.52	.25	.02

a. Write out the cumulative distribution function of X.

b. What is the probability that a randomly selected full-time student is taking five or fewer courses this semester?

c. What is the probability that a randomly selected full-time student is taking more than five courses?

8.101 Refer to Exercise 8.100. What is the mean number of courses taken by full-time students?

8.102 Burt pays $30 a year for towing insurance. He thinks that the probability that he will need to have his car towed once in the next year is 1/10 and the probability that he will have to have it towed more than once is zero. It will cost $100 if his car is towed if he doesn't have insurance but will cost nothing if he does have insurance. Let X = Burt's cost next year for towing and/or insurance.

a. If Burt buys the insurance, X has only one possible value. What is it?

b. If Burt doesn't buy the insurance, X has two possible values. List the two values and their probabilities.

c. Refer to parts (a) and (b). In each case, find $E(X)$. Using these two values for $E(X)$, explain whether or not Burt should buy the insurance.

8.103 Explain which of the conditions for a binomial experiment is *not* met for each of the following random variables:

a. A ten-question quiz has five true–false questions and five multiple-choice questions. Each multiple-choice question has four possible choices for the answer. A student randomly picks an answer for every question. X = number of answers that are correct.

b. Four students are randomly picked without replacement from a class of ten women and ten men. X = number of women among the four selected students.

8.104 Use the direct formulas for expected value and standard deviation to verify that the mean and standard deviation for a binomial random variable with $n = 2$ and $p = .5$ are $\mu = 1$ and $\sigma = .7071$, respectively. (*Hint:* The only possible values of X are 0, 1, 2. Find their probabilities and use the formulas at the end of Section 8.3.)

8.105 The standard medical treatment for a certain disease is successful in 60% of all cases.

a. The treatment is given to $n = 200$ patients. What is the probability that the treatment is successful for 70% or more of these 200 patients? (*Hint*: 70% of $n = 200$ is 140 patients.)

b. The treatment is given to only $n = 20$ patients. What is the probability that the treatment is successful for 70% or more of these 20 patients?

8.106 Assume that right handspan measurements for college women are approximately normally distributed, with a mean of 20 cm and a standard deviation of 1.8 cm, and that for men, they are normally distributed, with a mean 22.5 cm and a standard deviation of 1.5 cm. Measure your own right handspan.

a. Assuming that you fit with the population of college students of your sex, what is the z-score corresponding to your right handspan measurement? (Be sure to give your sex and right handspan measurement with your answer.)

b. What proportion of students of your sex have right handspan measurements smaller than yours?

8.107 Kim and her sister Karen each plan to have four children. Assume that the probability of a girl is .50, independent across births.

a. If X = number of girls Kim will have, what is the distribution of X?

b. If Y = number of girls Karen will have, what is the distribution of Y?

c. What is the distribution for T, the total number of girls Kim and Karen have?

8.108 Shaun (3 years old) and Patrick (4 years old) are each allowed to pick one book for bedtime stories, and the parent on duty reads the two books sequentially. The times for reading (single) books in the collection from which they select are approximately normally distributed, with a mean of 5 minutes and a standard deviation of 2 minutes.

a. Does it make sense to assume that the times for the two books are independent? Explain.

b. On about what proportion of nights will story time exceed 15 minutes, assuming that the times for the two books are independent?

8.109 Suppose that mean body temperature for adults has a normal distribution with mean $\mu = 98.2$ degrees Fahrenheit and standard deviation $\sigma = 0.8$ degrees.

a. What is the probability that a randomly selected adult has a body temperature less than or equal to 98.6 degrees?

b. What is the probability that a randomly selected adult has a body temperature greater than 99.2 degrees?

c. What is the probability that a randomly selected adult has a body temperature between 97 and 98 degrees?

8.110 Use the rule referenced from Chapter 7 to verify that the relationship is correct for a standard normal random variable Z:

a. Use Rule 1 of Chapter 7 to verify that $P(Z > a) = 1 - P(Z \leq a)$.

b. Use Rule 2b of Chapter 7 to verify that $P(Z \leq b) = P(Z \leq a) + P(a \leq Z \leq b)$. Use that result to show that $P(a \leq Z \leq b) = P(Z \leq b) - P(Z \leq a)$.

8.111 You will need a computer program that simulates a large number of binomial random variables at once for this exercise. Refer to Example 8.34 (p. 300), "Strategies for Studying When You Are out of Time." Simulate what would happen if you used Strategy 2, in which the total score was the sum of two binomial random variables, one with $n = 10$ and $p = .8$, the other with $n = 10$ and $p = .2$. Run 1000 simulations, and determine the proportion of them for which you would have received a passing grade (13 or more). Explain what you did.

8.112 Answer Thought Question 8.4 on page 273.

8.113 The vehicle speeds at a particular interstate location can be described by a normal curve. The mean speed is 67 mph, and the standard deviation is 6 mph. What is the proportion of vehicle speeds at this location that are faster than 75 mph?

8.114 New spark plugs have just been installed in a small airplane with a four-cylinder engine. There is one spark plug per cylinder, so four spark plugs have been installed. For each spark plug, the probability that it is defective and will fail during its first 20 minutes of flight is 1/10,000, independent of the other spark plugs.

a. For each spark plug, what is the probability that it will not fail?

b. If one spark plug fails, the plane will shake and not climb higher, but it can be landed safely. What is the probability that this happens (as it did for one of the authors of this book)?

c. If two or more spark plugs fail, the plane will crash. What is the probability that this happens?

9

Richard Lord/PhotoEdit

Will her overeating lead to excessive weight gain during her freshman year?

See Example 9.1 *(p. 316)*

Understanding Sampling Distributions: Statistics as Random Variables

This chapter introduces the reasoning that allows researchers to make conclusions about entire populations using relatively small samples of individuals. The secret to understanding how things work is to understand what kind of dissimilarity we should expect to see among different samples from the same population.

One of the main purposes of statistical methods is to enable us to satisfy our curiosity about all kinds of things. Most of the time, the questions we want to answer extend beyond the sample data we are able to collect to questions about larger populations. Therefore, we need to know how to make use of sample data to answer questions about a bigger collection of individuals than the ones we have measured.

This chapter is an introduction to the reasoning that allows researchers to reach conclusions about an entire population on the basis of a relatively small sample of individuals. The basic idea is that we must work backward, from a sample to a population. We start with a question about a population such as "How many teenagers are infected with HIV?", "At what average age do left-handed people die?", or "What is the average income of all students at a large university?" We collect a sample from the population about which we have the question and measure the variable(s) of interest. We can then answer the question of interest for the sample. Finally, using statistical theory, we determine how close our sample answer is to what we really want to know: the true answer for the population.

9.1 Parameters, Statistics, and Statistical Inference

Most questions that we ask about large populations are translated into questions about specific summary characteristics of the group. For instance, suppose that we collect data from a sample of students at our school on how many hours they study per week. We might characterize the data by finding the mean value for the sample. Then we might use that sample summary to estimate the mean weekly hours of study for the complete population of students.

A **parameter** is a number that is a summary characteristic of a population, a random situation, or a comparison of populations. Sometimes, the phrase **population parameter** is used to make it clear that a parameter is associated with a population instead of a sample. Examples of parameters are the proportion of adults in the world who are left-handed, the probability that any baby who sleeps with a night-light will have myopia at age 20, and the difference in the mean incomes of all men and women in the same profession.

We assume that a parameter has a fixed, unchanging value. Usually, the value of a parameter is not known to us and will not be known to us because we will not be able to measure every unit in the population. Although we will not have the information necessary to find the numerical value of a population parameter, we will be able to use statistical methods to make a good guess, as you will see in the remainder of this book.

A **statistic**, also called a **sample statistic**, is a number that is computed from a sample of values taken from a larger population. It is a summary characteristic of the sample data. The sample data may be collected in a sample survey, an observational study or an experiment. The term **sample estimate** or simply **estimate** is sometimes used for a sample statistic when the statistic is used to estimate the unknown value of a population parameter.

One of the big ideas in this chapter is that possible values of sample statistics are variable. If two different samples are taken from the same population, it is likely that the sample statistics will be different for those two samples. For instance, if we were to take two different random samples of 1000 adults and find the proportion in each sample that were left-handed, the proportions would probably not be exactly the same. In this scenario, the sample proportion is an example of a sample statistic. The corresponding population parameter is the proportion of the entire population that is left-handed. Note that the population parameter remains fixed, although we do not know its numerical value. The value of a sample statistic may change from sample to sample, and we will know the value once we have measured a sample.

Statistical Inference

The results in this chapter will be used for the remainder of the book to make conclusions about population parameters on the basis of sample statistics. The procedures we will use for making these conclusions are called **statistical inference** procedures. The two most common procedures are to find *confidence intervals* and to conduct *hypothesis tests*. Here, we give just a brief introduction to these two procedures, which will be developed throughout the rest of the book.

Confidence Intervals

A **confidence interval** is an interval of values that the researcher is fairly sure will cover the true, unknown value of the population parameter. In other words, we use a confidence interval to estimate the value of a population parameter. For instance, in Example 5.3 (p. 152), we used data from a Gallup poll of $n = 1025$ adult Americans to estimate that between 53% and 59% of U.S. adults would say that religion is very important in their lives. In this example, the population parameter of interest is the percent of all U.S. adults who would say that religion is very important in their lives. We don't know the true value of this parameter, but we estimate it to be in the interval 53% to 59%. We have already encountered confidence intervals in Case Study 1.3 and in Chapter 5, when we learned about the margin of error. We will explore confidence intervals further in Chapters 10 and 11.

Hypothesis Testing

Hypothesis testing or **significance testing** uses sample data to attempt to reject a hypothesis about the population. Usually researchers want to reject the notion that chance alone can explain the sample results. We encountered this idea in Chapter 4 when we learned how to determine if the relationship between two categorical variables is statistically significant. The hypothesis that researchers set up to reject in that setting is that two categorical variables are unrelated to each other in the population, so that any observed relationship in the sample is simply due to chance. In most research settings, the *desired* conclusion is that the variables under scrutiny are related.

Hypothesis testing is applied to population parameters by specifying a **null value** for the parameter—a value that would indicate that nothing of interest is happening. For instance, a weight-loss clinic might postulate that the average weight loss for the population of clinic patrons is 0, so the null value would be 0. The clinic's goal, of course, would be to show that something more interesting is happening and that in fact the average weight loss is greater than 0.

Hypothesis testing proceeds by obtaining a sample, computing a sample statistic, and assessing how unlikely the sample statistic would be if the null parameter value were correct. In most cases, the researchers are trying to show that the null value is not correct. Achieving **statistical significance** is equivalent to rejecting the idea that the observed results are plausible if the null value is correct. For instance, if the weight-loss clinic could show that a sample of patrons had an average weight loss of 10 pounds, with minimal variability, then it would be hard to believe that a population weight loss of 0 yielded those results. We will explore hypothesis testing further in Chapters 12 and 13.

9.1 Exercises are on page 354.

IN SUMMARY **Parameters, Statistics, and Statistical Inference**

- A **parameter** is a numerical summary of a population. Its value is considered to be fixed and unchanging.
- A **statistic**, or **sample statistic**, is a numerical summary of a sample. Its value may be different for different samples. The term **sample estimate** or **estimate** is synonymous and reinforces the idea that a statistic often is used to estimate the corresponding population parameter.
- In **statistical inference**, information from a sample is used to make generalizations about a larger population. Sample statistics are used to make conclusions about population parameters. Confidence intervals and hypothesis tests are two common statistical inference techniques.
- A **confidence interval** is an interval of values that the researcher is fairly sure will cover the true, unknown value of the population parameter.
- **Hypothesis testing**, also called **significance testing**, uses sample data to attempt to reject a hypothesis about the population.

9.2 From Curiosity to Questions about Parameters

One of the most important skills that a good researcher needs is the ability to translate curiosity about something into a question that can be answered. In statistics, this process usually requires us to formulate a question about a parameter. In this chapter and Chapters 10 to 13, we will learn how to formulate and answer questions about five different types of situations, represented by five different parameters. Fortunately, the statistical methods that we use in these five situations are almost identical. The details change, but the concepts and methods remain the same.

Let's see how curiosity about something can be translated into questions about parameters. Many students think that the appropriate statistical procedure to use to answer a question depends on what type of data is available. But in fact, the process should work the other way around. The scientific question should first be expressed as a question about a parameter, and then data should be collected that are appropriate for making conclusions about the value of that parameter. Here is a diagram illustrating how the process of satisfying one's curiosity about something should work.

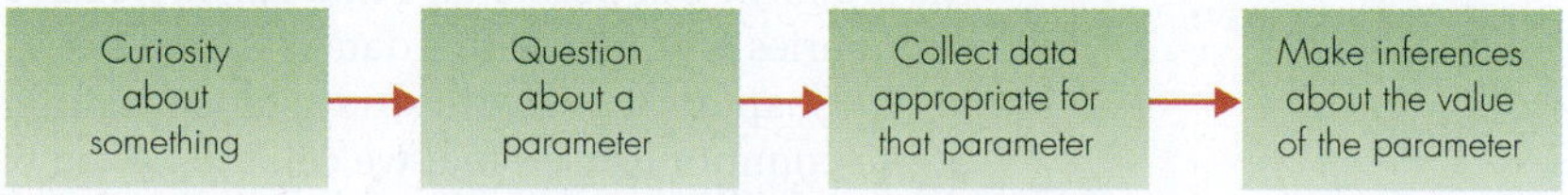

Example 9.1 illustrates how to turn curiosity into a question about a parameter. As you will see, we can ask different questions about the same issue, and each different question may lead to a different parameter of interest. To collect appropriate sample data, researchers must first decide which questions and parameters they wish to pursue.

Example 9.1 **The "Freshman 15"** Do college students really gain weight during their freshman year? The lore is that they do, and this phenomenon has been called "the freshman 15" because of speculation that students typically gain as much as 15 pounds during the first year of college. How can we turn our curiosity about the freshman 15 into a question about a parameter? There are several possibilities. We might want to know what proportion of freshmen gain weight. A related question would be whether most students gain weight. Or we might want to know what the average weight gain is across all first-year students. We might want to know whether women gain more weight than men or vice versa. Here are two ideas for satisfying our curiosity, along with the standard notation that we use for the relevant parameters:

- Parameter = p = proportion of the population of first-year college students who weigh more at the end of the year than they did at the beginning of the year.
- Parameter = μ_d = the mean (average) weight gain during the first year for the population of college students. The subscript d indicates that the raw data of interest for each student are *differences* between two values: weight at the beginning of the year and weight at the end of the year.

Most students begin college at age 17 or 18, when they might still be growing, so perhaps they would gain weight whether they went to college or not. To investigate this, we might want to compare weight gain of students during the first year of college with weight gain during the same period for people of the same age who don't go to college. In that case, the parameter would be as follows:

- Parameter = $\mu_1 - \mu_2$, where μ_1 = the mean weight gain of all college students during the first year and μ_2 = the mean weight gain of all people of the same age during that same time who do not go to college. The difference $\mu_1 - \mu_2$ measures how much more weight college students gain on average during the year than do their contemporaries of the same age. We might also want to know about the individual parameters μ_1 and μ_2—the mean weight gains for the population of first-year college students and the population of non-college students of the same age.

Researchers at Cornell University conducted a study on this topic and found that students did indeed gain more weight than others in that age group (Lang, 2003). They measured 60 students and found that the average weight gain for them during the first 12 weeks of college was 4.2 pounds. The researchers claimed that a weight gain of that magnitude is 11 times what would be expected for that age group and speculated that factors such as "all you can eat" dining facilities and more junk food were partially responsible. Note that the Cornell researchers were interested in the parameter μ_d = the mean weight gain during the first semester (12 weeks) for the *population* of college students. They could not determine this parameter's true value by measuring only 60 students. Instead, they could calculate the corresponding *sample statistic*, the sample mean weight gain of 4.2 pounds for the 60 students in the study. This sample statistic can be used to estimate the unknown value of the population parameter μ_d. In Chapter 11, we will learn how to determine how accurate this estimate is likely to be.

The Big Five Parameters

We learned in Chapter 2 that there are two basic types of variables: categorical and quantitative. We also learned that the best way to summarize categorical data is to find the *proportion* or percentage that fall into each category and that one of the most useful summaries for quantitative data is the *mean*.

In Chapters 9 to 13, we study statistical inference methods for five parameters that are so commonly used that we call them "the big five." The five parameters all involve either proportions (for categorical data) or means (for quantitative data).

It is important to remember that parameters are associated with *populations*, whereas statistics are associated with *samples*. As the parameters are introduced, the notation and an explanation of the corresponding sample statistics will be provided as well. One of the most common mistakes made by students of statistics is to confuse parameters and statistics, so make sure you understand the distinction. Table 9.1 provides a list of the five parameters along with notation for each parameter and its corresponding sample statistic. Explanations will follow.

Table 9.1 Population Parameters and Sample Statistics for the Big Five Scenarios

Parameter Name and Description	Symbol for the Population Parameter	Symbol for the Sample Statistic
For Categorical Variables		
One population proportion (or probability)	p	$\hat{p}$
Difference in two population proportions	$p_1 - p_2$	$\hat{p}_1 - \hat{p}_2$
For Quantitative Variables		
One population mean	μ	$\overline{x}$
Population mean of paired differences (dependent)	μ_d	$\overline{d}$
Difference in two population means (independent)	$\mu_1 - \mu_2$	$\overline{x}_1 - \overline{x}_2$

Reading and Pronouncing the Notation for Parameters and Statistics

Note that the Greek letter μ (spelled *mu* and pronounced "mew") is always used for population means, while the notation for sample means is $\overline{x}$ or $\overline{d}$, read as "x-bar" or "d-bar." The lowercase Roman letter p is always used for population proportions. The notation for sample proportions is $\hat{p}$, read as "p-hat." The small numbers to the lower right of these symbols are called *subscripts* and are read after the "hat" or "bar." Thus, $\hat{p}_1 - \hat{p}_2$ is read as "p-hat-one minus p-hat-two" and $\overline{x}_1 - \overline{x}_2$ is read as "x-bar-one minus x-bar-two." The notation $p_1 - p_2$ is read as "p-one minus p-two, $\mu_1 - \mu_2$ is "mu-one minus mu-two," and μ_d is "mu-d" or "mu-sub-d."

We will need to use the standard deviation as a tool for some of our methods. Remember that the notation for standard deviation is different for populations and samples as well; the lowercase Greek letter "sigma" = σ is used for the *population standard deviation* and the lowercase Roman letter s is used for the sample standard deviation. When there are two groups, we will use notation such as σ_1, which is read as "sigma-one," indicating the population standard deviation for Group 1.

Paired Differences or Independent Samples

You may have noticed that two of the parameters in Table 9.1 have similar names; they are "population mean for paired differences" and "difference in two population means." It is important to distinguish the situations for which each of these is appropriate.

We learned about *matched pairs* in Chapter 6. When we collect quantitative data on matched pairs we often are interested in the *difference* rather than in the two separate values. Data that are formed by taking the differences in matched pairs are called **paired differences**. Once we have taken differences, the parameter of interest is the **population mean for paired differences**, which is the mean that we would get if we took differences for the entire population of possible pairs. Samples taken as matched pairs are sometimes called **dependent samples** because the two values for a pair are not statistically independent.

With two **independent samples**, the individuals in one sample are not coupled in any way with the individuals in the other sample. For instance, measuring heights of brother and sister pairs would not produce independent samples, but measuring heights of a random sample of girls and boys in a city would, even if a brother and sister both happened to be chosen. Independent samples also can be assumed when one sample is divided into two groups on the basis of a categorical grouping variable, such as males and females or smokers and nonsmokers. The parameter of interest with quantitative data from independent samples is the **difference in two population means**.

Familiar Examples Translated into Questions about Parameters

Throughout the book, you have seen examples utilizing sample data that might be used to answer questions about each of the five population parameters of interest. Let's look at the types of situations and research questions that can be translated into questions about using sample statistics to estimate the big five parameters. For each of the five situations, we give two examples and then reiterate the notation and the meaning of both the population parameter and the sample estimate. Some of these examples will be familiar to you from earlier chapters, and some will become familiar in later chapters.

Situation 1. Estimating the Proportion Falling into a Category of a Categorical Variable

- *Example research questions:*

  What proportion of American adults believe there is extraterrestrial life?

  In what proportion of opposite-sex British marriages is the wife taller than her husband?

- *Population parameter:* p = proportion in the population falling into that category.
- *Sample estimate:* $\hat{p}$ = proportion in the sample falling into that category.

Situation 2. Estimating the Difference between Two Population Proportions Falling into a Category of a Categorical Variable

- *Example research questions:*

  How much difference is there between the proportions who would quit smoking if taking the antidepressant buproprion (Zyban) versus if wearing a nicotine patch?

  How much difference is there between men who snore and men who don't snore with regard to the proportion who have heart disease?

- *Population parameter:* $p_1 - p_2$, where p_1 and p_2 represent the proportions in Populations 1 and 2, respectively.
- *Sample estimate:* $\hat{p}_1 - \hat{p}_2$, the difference between the two sample proportions.

Situation 3. Estimating the Mean of a Quantitative Variable

- *Example research questions:*

  What is the mean time that college students watch TV per day?

  What is the mean pulse rate of women?

- *Population parameter:* μ = population mean for the variable.
- *Sample estimate:* $\overline{x}$ = sample mean for the variable.

Situation 4. Estimating the Mean of Paired Differences for Quantitative Variables

- *Example research questions:*

  What is the mean difference in weights for freshmen at the beginning and end of the first semester?

  What is the mean difference in age between husbands and wives in Britain?

- *Population parameter:* μ_d = population mean of the differences in the two measurements.
- *Sample estimate:* $\overline{d}$ = mean of the differences for a paired sample of the two measurements.

Situation 5. Estimating the Difference Between Two Population Means for a Quantitative Variable

- *Example research questions:*

  How much difference is there in mean weight loss for those who diet compared to those who exercise to lose weight?

  How much difference is there between the mean foot lengths of men and women?

- *Population parameter:* $\mu_1 - \mu_2$, where μ_1 and μ_2 represent the means in Populations 1 and 2, respectively.
- *Sample estimate:* $\bar{x}_1 - \bar{x}_2$, the difference between the two sample means.

9.2 Exercises are on pages 355–356.

THE ORGANIZATION OF CHAPTERS 9 TO 13

It is easier to learn something when you recognize how it relates to what you already know and when you understand recurring patterns. With that in mind, we have organized Chapters 9 to 13 to illustrate that the same basic approach is utilized in most of the statistical inference procedures covered in them. The chapters are organized around statistical inference procedures for the five parameters listed in Table 9.1.

Before we learn how to find confidence intervals and test hypotheses for each parameter, we need to investigate the relationship between the parameter of interest in a problem and the corresponding sample statistic that estimates the parameter. This relationship will be formulated into something called a *sampling distribution* for the sample statistic, which is defined in the next section. Thus, for each of the five parameters, we will study three inference topics:

Topic 1. The *sampling distribution* (SD) of the statistic that estimates the parameter

Topic 2. A *confidence interval* (CI) procedure to estimate the parameter to within an interval of possibilities

Topic 3. A *hypothesis test* (HT) procedure to test whether the parameter equals a specific value

Chapters 9 to 13 are organized and written to allow flexibility in the order of covering topics and situations. The coverage of each of the three topics above (SD, CI, and HT) begins with an *introductory module,* followed by a module for each of the big five parameters. Therefore, there are six modules for each of the three topics (SD, CI, and HT): the introduction to the basic concepts and one module for each of the five parameters.

Module locations are indicated by a color tab on the outer edge of the text page so that you can follow the three topics for a particular parameter through the chapters. The same color is used for the three topics covered for a parameter. You can study all six modules for one of the three inference topics (sampling distributions or confidence intervals or hypothesis tests) together, or you can pick one parameter of interest (such as a population mean) and learn all about methods related to it before moving on to learn about a different parameter. *Once you have studied the introductory module for a topic, you can then study the remaining ones in any order. Moreover, confidence interval modules can be covered either before or after hypothesis testing modules; the order does not matter.*

Table 9.2 shows where to find the various modules. In the module designations, SD denotes the sampling distribution topic, CI denotes the confidence interval topic, and HT denotes the hypothesis testing topic. The five parameters are numbered from 1 to 5, and an introductory topic is numbered as parameter 0. Note that all sampling distribution modules are in this chapter (Chapter 9), confidence interval modules are in Chapters 10 and 11, and hypothesis testing modules are in Chapters 12 and 13.

Table 9.2 Organization of Chapters 9 to 13

Parameter	Chapter 9: Sampling Distributions (SD)	Chapter 10: Confidence Intervals (CI)	Chapter 11: Confidence Intervals (CI)	Chapter 12: Hypothesis Tests (HT)	Chapter 13: Hypothesis Tests (HT)
0. Introductory	**SD Module 0** Overview of sampling distributions	**CI Module 0** Overview of confidence intervals		**HT Module 0** Overview of hypothesis testing	
1. Population proportion (p)	**SD Module 1** SD for one sample proportion	**CI Module 1** CI for one population proportion		**HT Module 1** HT for one population proportion	
2. Difference in two population proportions ($p_1 - p_2$)	**SD Module 2** SD for difference in two sample proportions	**CI Module 2** CI for difference in two population proportions		**HT Module 2** HT for difference in two population proportions	
3. Population mean (μ)	**SD Module 3** SD for one sample mean		**CI Module 3** CI for one population mean		**HT Module 3** HT for one population mean
4. Population mean of paired differences (μ_d)	**SD Module 4** SD for sample mean of paired differences		**CI Module 4** CI for population mean of paired differences		**HT Module 4** HT for population mean of paired differences
5. Difference in two population means ($\mu_1 - \mu_2$)	**SD Module 5** SD for difference in two sample means		**CI Module 5** CI for difference in two population means		**HT Module 5** HT for difference in two population means

9.3 SD Module 0: An Overview of Sampling Distributions

You can probably guess by now that the relationship between sample statistics and population parameters will play a crucial role in statistical inference. It is this relationship that allows us to determine the accuracy of sample statistics as estimates of population parameters and to create confidence intervals. It also allows us to determine the extent to which sample results are surprising or unusual, given that we assume a potential parameter value to be true, as we will do in hypothesis tests.

Statistics as Random Variables

Think of taking a random sample or conducting an experiment as one big random circumstance. A sample statistic is merely a number assigned to the outcome of that random circumstance. For instance, taking a random sample of 1000 adults and finding the proportion that are left-handed amounts to assigning a number (the sample proportion) to the outcome of a random circumstance (taking a random sample of 1000 adults and finding out whether they are left-handed). The number might be .12 for one sample (120 left-handers out of 1000), .09 for the next (90 left-handers), and so on. Recall from Chapter 8 that a *random variable* assigns a number to the outcome of a random circumstance. Thus, we can think of sample statistics as special cases of random variables because they fit the definition of random variables.

Remember that each random variable has a probability distribution function or probability density function, which we use to find probabilities associated with various possible outcomes for the random variable. Because a statistic is just a special case of a random variable, it too has a probability distribution, which we call a *sampling distribution*.

DEFINITION The **sampling distribution** for a statistic is the probability distribution of possible values of the statistic for repeated samples of the same size taken from the same population.

In general, we can use a random variable's probability distribution to assess possibilities and accompanying probabilities for the different values of that variable. The sampling distribution for a statistic describes the possible values the statistic might have when random samples are taken from a population. For example, the sampling distribution would tell us the probability of getting a sample proportion of more than .12 left-handers in a sample of 1000, or at most .09 left-handers, and so on. This is important information because it can be used, among other things, to find the probability that a sample statistic will be within a specified distance of the unknown population parameter. That tells us how confident we can be that an interval extending that specified distance above and below the sample statistic will reach the population parameter. This reasoning will be explained in detail when we learn how to form confidence intervals.

Statisticians have worked out what the sampling distributions look like for the big five sample statistics. You will see that the general structure of the sampling distribution is the same for each of the five situations. In Section 9.10, we describe how to find sampling distributions for other situations.

Example 9.2 **Mean Hours of Sleep for College Students** The sample of Penn State students first discussed in Chapter 2 included the question "How many hours of sleep did you get last night?" The mean of the 190 responses was 7.1 hours. Suppose that we were to ask another 190 students that same question. It is unlikely that the mean response would be exactly 7.1 hours again.

Figure 9.1 shows a histogram of possible sample means we might get from simulating this survey 500 times. A normal curve is superimposed on top of the histogram to demonstrate that the sampling distribution of possible sample means is approximately a normal distribution. We will return to this idea later in this chapter.

To understand how this histogram was developed, think about a university that has 50,000 students. Suppose the distribution of hours of sleep students had the previous night actually has a mean of 7.1 hours and standard deviation of 2 hours. You and 499 friends (you are very popular) each go out and take a random sample of 190 of those students. You ask the students in your sample how many hours they slept last night and then calculate the mean of the 190 answers. Finally, you and your friends (500 of you) get together and draw a histogram of your 500 different sample means. You would get a picture similar to Figure 9.1. Note that most of you would get a sample mean around 7.1 or slightly more or less, but occasionally, someone would get an unusual sample with a mean as high as 7.5 hours or as low as 6.7 hours.

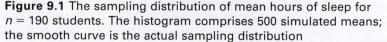

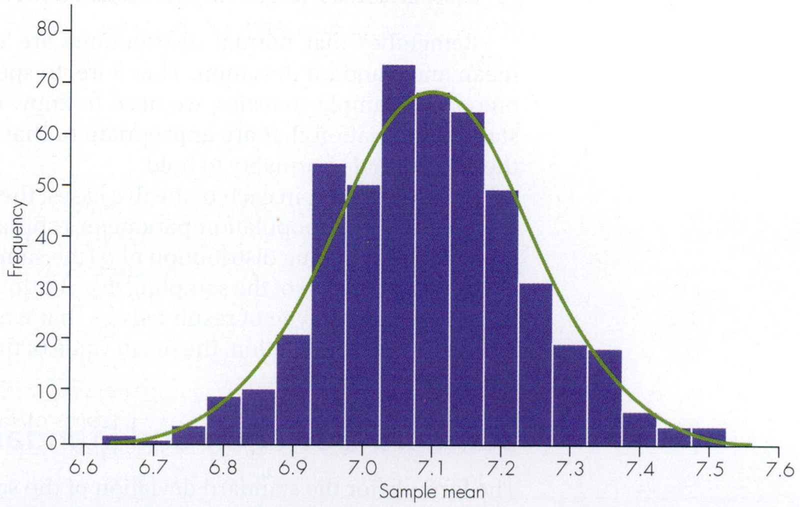

Figure 9.1 The sampling distribution of mean hours of sleep for $n = 190$ students. The histogram comprises 500 simulated means; the smooth curve is the actual sampling distribution

A sampling distribution is the probability distribution for a sample statistic. It plays the same role as probability distributions do for other random variables. We can specify the probability that a sample statistic will fall into a specific interval if we know all of the details about its sampling distribution. For instance, in Example 9.2, if we knew all of the details of the smooth curve in Figure 9.1, we could answer questions such as, "What is the probability that the mean hours of sleep (i.e., the sample mean) for a random sample of 190 students will be less than 6.9 hours?" or "What is the probability that the mean hours of sleep for a random sample of 190 students will be between 7 and 8 hours?" Note that this is *not* the same question as "What is the probability that one randomly selected student slept between 7 and 8 hours?" Sampling distributions give us information about sample statistics, not about individual values for the population.

Note that sampling distributions give us information about sample statistics, *given* that we know all about the population already. For instance, to get the sampling distribution in Figure 9.1, we assumed that we knew that the mean and standard deviation of the population of sleep times were 7.1 hours and 2 hours, respectively. In practice, knowing how to get information about possible sample statistics based on knowing values of the population parameter isn't very helpful. If we know about the population already, we don't need to find sample statistics. Therefore, why do we want to know about sampling distributions?

The answer is that we can turn the information around and use it for statistical inference. For instance, note that in Figure 9.1, the possible sample means are all within about 0.5 hours of the population mean (7.1 hours in this case). If we know how close together the sample mean and the population mean are likely to be, then once we have a sample mean, we can make a reasonably good guess about the population mean. We will learn about sampling distributions in this chapter and then learn how to turn their information around to make inferences about population parameters in Chapters 10 through 13.

A Common Format for the Five Sampling Distributions

The general format of the sampling distribution is the same for each of the five sample statistics under discussion—only the details change. Here are the common elements:

- In each case, as long as certain conditions are met, the sampling distribution is approximately normal.
- The mean of the sampling distribution is the population parameter that corresponds to the sample statistic (i.e., the parameter of interest estimated by the statistic).
- The standard deviation of a sampling distribution measures how the values of the sample statistic might vary across different samples from the same population.

Remember that normal distributions are completely specified by knowing the mean and standard deviation. Therefore, to specify the sampling distribution for any one of our sample statistics, we need to know only how to determine the mean and standard deviation that are appropriate to that case, and the conditions required for the approximate normality to hold.

As noted above, in each of the five cases, the mean of the sampling distribution for the statistic is the population parameter estimated by that statistic. For instance, the mean of the sampling distribution of $\hat{p}$ (the sample proportion) is the population proportion p. The mean of the sampling distribution of $\bar{x}$ (the sample mean) is the population mean μ. This type of result tells us that when averaged over all possible random samples from a population, the mean value of the sample statistic is equal to the population parameter.

Standard Deviation and Standard Error of a Statistic

The formula for the standard deviation of the sampling distribution differs for each of the five cases that we consider, but in each case, it depends on the sample size(s) and gets smaller for larger samples. In other words, as the sample size(s) get larger, the variability among possible values of the statistic gets smaller.

To differentiate the standard deviation of a sampling distribution from σ, the standard deviation of the individual population measurements, we need to add a qualifier to the terminology. We will do this by specifying the statistic as part of the terminology for the standard deviation. Thus, the standard deviation of the sampling distribution of a sample mean is called the **standard deviation of** $\bar{x}$. The standard deviation of the sampling distribution of a sample proportion is called **the standard deviation of** $\hat{p}$, and so on.

When we use these results for statistical inference, we often need to use sample information to estimate the standard deviation of the sampling distribution. To make it clear that the value is an estimate, we use the term **standard error** to describe the estimated standard deviation for a sampling distribution. Thus, *the standard error of the mean* or *the standard error of* $\bar{x}$ are terms used for the estimate of *the standard deviation of* $\bar{x}$, and *the standard error of* $\hat{p}$ is the term used for the estimate of *the standard deviation of* $\hat{p}$.

9.3 Exercises are on page 356.

IN SUMMARY Sampling Distribution Definition and Features

- A **sampling distribution** is the probability distribution of a sample statistic. It describes how values of a sample statistic vary across all possible random samples of a specific size that can be taken from a population. For all five scenarios considered here, the sampling distribution is approximately normal as long as the sample size(s) are large enough.
- The **mean value of a sampling distribution** is the mean value of a sample statistic over all possible random samples. For the five scenarios here, this mean equals the value of the population parameter.
- The **standard deviation of a sampling distribution** measures the variation among possible values of the sample statistic over all possible random samples. When referring to such a standard deviation, we include the name of the statistic being studied—for example, *the standard deviation of the mean*.
- The term **standard error** describes the estimated value of the standard deviation of a statistic. Because the formula is different for each statistic, we include the name of the statistic—for example, *the standard error of the mean*.

9.4 SD Module 1: Sampling Distribution for One Sample Proportion

Suppose that we conduct a binomial experiment with n trials and success probability p, and get successes on x of the trials. Or suppose that we measure a categorical variable for a representative sample of n individuals, and x of them have responses in a certain category, for which the proportion in the population that would have that response is p. The parameter of interest is the **population proportion**, p. In each case, we can compute the statistic $\hat{p}$ = the **sample proportion** $= x/n$, the proportion of trials resulting in success, or the proportion in the sample with responses in the specified category. If we repeated the binomial experiment or collected a new sample, we would probably get a different value for the sample proportion $\hat{p}$, even though the population proportion p remains fixed.

For instance, suppose that we would like to know what proportion of a large population carries the gene for a certain disease. We sample 25 people and use the sample proportion from that sample to estimate the true answer. Suppose that in truth 40% of the population carries the gene, so $p = .40$, although this fact is unknown to us.

What will happen if we randomly sample 25 people from this population? Will we always find 10 people (40%) with the gene and 15 people (60%) without? Or will the number and proportion with the gene differ for different samples of $n = 25$? You may recognize that this situation is a binomial experiment and that if X = number (out of 25) who carry the gene, then X is a *binomial random variable* with $n = 25$ and $p = .4$.

Many Possible Samples

Consider four different random samples of 25 people taken from this population. Remember that we are trying to estimate the proportion of the population with the gene, based on the sample *statistic*, which in this scenario is the *sample proportion*. We do not know that, in truth, the population proportion (the parameter) is actually .40 (40%). Here is what we would have concluded about the proportion of people who carry the gene, given four possible samples with X as specified:

Sample 1: $X = 12$, $\hat{p}$ = proportion with gene = 12/25 = .48 or 48%.
Sample 2: $X = 9$, $\hat{p}$ = proportion with gene = 9/25 = .36 or 36%.
Sample 3: $X = 10$, $\hat{p}$ = proportion with gene = 10/25 = .40 or 40%.
Sample 4: $X = 7$, $\hat{p}$ = proportion with gene = 7/25 = .28 or 28%.

Note that each one of these samples would have given a different answer and that the sample answer may or may not have matched the truth about the population, which is that .40 (40%) carry the gene.

In practice, when a researcher conducts a study similar to this one or a polling organization randomly samples a group of people to measure public opinion, only one sample is collected. There is no way to determine whether or not that particular sample is an accurate reflection of the population. However, statisticians have calculated what to expect for the vast majority of possible samples, and how much variability to expect among them.

The Sampling Distribution for a Sample Proportion

A result given in Section 8.7 is that with sufficiently large n, a binomial random variable is also approximately a normal random variable. A binomial random variable X counts the number of times an event happens in n trials, but the approximate normality also applies to the proportion, $\hat{p} = \frac{X}{n}$. Dividing each possible value of X by the sample size n does not change the shape of the distribution of possible values. In other words, for sufficiently large n, the *sampling distribution for a sample proportion is approximately a normal distribution.*

This result can be applied in two different situations in which we observe data:

Situation 1: A random sample is taken from a relatively large actual population.
Situation 2: A binomial experiment is repeated numerous times.

DEFINITION **The sampling distribution for a sample proportion** can be defined as follows:

Let p = population proportion of interest or binomial probability of success.
Let $\hat{p}$ = corresponding sample proportion or proportion of successes.

If numerous samples or repetitions of the same size n are taken, the distribution of possible values of $\hat{p}$ is approximately a normal curve distribution with

Mean = p

Standard deviation = $s.d.(\hat{p}) = \sqrt{\frac{p(1-p)}{n}}$

This approximate normal distribution is called the **sampling distribution of $\hat{p}$.**

Technical Note: The n individuals in the sample or the repetitions must be independent, equivalent to Condition 3 in a binomial experiment (p. 276).

Conditions for Which the Approximate Normality of the Sampling Distribution for a Sample Proportion Applies

In each of the two situations, three conditions must be met for the approximate normality of the sampling distribution to apply:

1. *The Physical Situation*: Either there is an actual population with a fixed proportion who have a trait or opinion of interest, or there is a repeatable situation for which an outcome of interest occurs with a fixed relative frequency probability.
2. *Data Collection*: Either a random sample is selected from an actual population, or the situation is repeated numerous times, with the outcome each time independent of outcomes all other times.
3. *Sample Size or Number of Trials*: The size of the sample or number of binomial trials must be large enough that we expect to see at least ten of each of the two possible responses or outcomes. That is, np and $n(1 - p)$ must each be at least 10.

Example 9.3 **Scratch and Win (or Lose) Lotteries** Many states sell instant lottery tickets, in which the purchaser scratches off seals to reveal whether or not the ticket is a winner. According to the California Lottery website (http://www.calottery.com), the probability that a ticket will be a winner of any prize for many of these games is 1 in 5, or $p = .2$. (The smallest prize is not monetary; it is a free ticket to play again. The probability of winning a cash prize is often closer to .1.) If you were to buy $n = 100$ tickets, what proportion of them are likely to be prize winners?

This scenario fits the conditions for the approximate normality of the sampling distribution of $\hat{p}$, with $np = 20$ and $n(1 - p) = 80$. The mean and standard deviation of the sampling distribution follow:

Mean $= p = .20$

Standard deviation $= s.d.(\hat{p}) = \sqrt{\dfrac{p(1-p)}{n}} = \sqrt{\dfrac{(.2)(.8)}{100}} = .04.$

Figure 9.2 illustrates the sampling distribution of $\hat{p}$ for this scenario. The Empirical Rule tells us that there is about a 68% chance that the proportion of winning tickets will be between .16 and .24, about a 95% chance that the proportion of winning tickets will be between .12 and .28, and about a 99.7% chance that the proportion of winning tickets will be between .08 and .32.

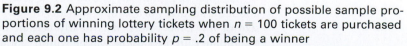

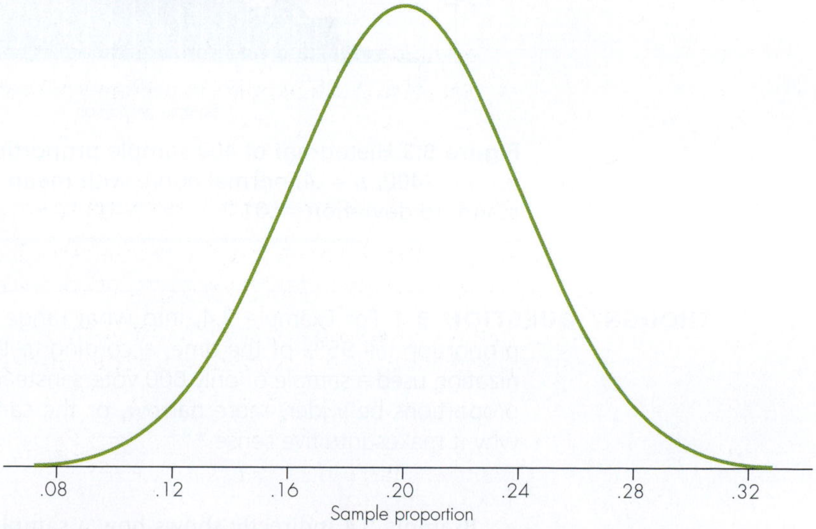

Figure 9.2 Approximate sampling distribution of possible sample proportions of winning lottery tickets when $n = 100$ tickets are purchased and each one has probability $p = .2$ of being a winner

In the next example, we determine the approximately normal sampling distribution that the theory tells us to expect for the sample proportion from a poll with a specified population proportion, and then compare it to a histogram showing sample proportions for 400 simulated polls from that population. In practice we would have only one poll and one sample proportion, but the example illustrates the idea that the one sample proportion that we observe could come from any location on the normal curve.

Example 9.4 **Possible Sample Proportions Favoring a Candidate** Suppose that of all voters in the United States, 40% are in favor of Candidate C for president. Pollsters take a sample of 2400 voters. What proportion of the sample would be expected to favor Candidate C? The proportion of the sample who favor Candidate C is a random variable that has approximately a normal distribution. The mean and standard deviation for the distribution are

Mean = p = .4 (40% expressed as a proportion)
Standard deviation =

$$\text{s.d.}(\hat{p}) = \sqrt{\frac{p(1-p)}{n}} = \sqrt{\frac{(.4)(1-.4)}{2400}} = \sqrt{\frac{.24}{2400}} = \sqrt{.0001} = .01$$

Figure 9.3 shows a histogram of the sample proportions resulting from simulating this situation 400 times. The appropriate normal curve representing the sampling distribution is superimposed on the histogram. Note that as expected from the Empirical Rule in Chapter 2, the possible values cover the range $\mu \pm 3$(s.d.), in this case, $.4 \pm 3(.01)$ or .37 to .43.

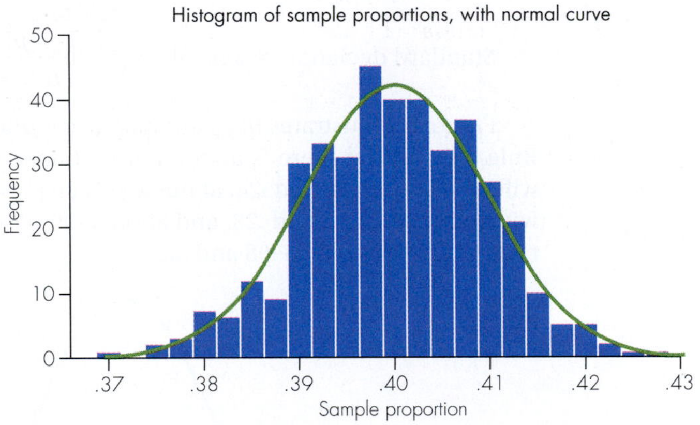

Figure 9.3 Histogram of 400 sample proportions based on $n = 2400$, $p = .4$; normal curve with mean = .4 and standard deviation = .01

THOUGHT QUESTION 9.1 For Example 9.4, into what range of possible values should the sample proportion fall 95% of the time, according to the Empirical Rule? If the polling organization used a sample of only 600 voters instead, would this range of possible sample proportions be wider, more narrow, or the same? Explain your answer, and explain why it makes intuitive sense.*

Example 9.4 indirectly shows how a sampling distribution provides information about the accuracy of a sample statistic. In that example, we learned that with a sample size of $n = 2400$ voters, it is nearly certain that the proportion of voters favoring Candidate C in the sample will be within $\pm 3(.01) = \pm .03$ of the true population proportion. In Chapters 10 and 11, we will learn how to use that kind of information to

***HINT:** Use Example 9.5 (on the next page) for guidance.

form confidence intervals that estimate population parameters. In the next example, we will see how a sampling distribution is used to judge how an observed sample result fits with a potential value for a population proportion. In Chapters 12 and 13, we will learn how to use that kind of information to test hypotheses about population parameters.

Example 9.5 **Caffeinated or Not?** Case Study 8.1 described a study in which 25 people each tried 20 times to guess whether a cola they taste was caffeinated or not caffeinated, for a total of 500 trials (25 people × 20 trials each). For some of the trials, the cola being tasted contained caffeine; in the others, it did not. Out of the 500 trials, 265 correct guesses were made about whether or not caffeine was present. The observed sample proportion is therefore $\hat{p} = 265/500 = .53$.

Suppose people cannot judge merely by taste whether caffeine is present. If this were so, then participants would actually be randomly guessing on each trial, so $p = .5$, where p is the probability of correctly guessing whether caffeine is present or not on any trial. If this is the case for all individuals, then the overall experiment can be seen as a binomial experiment with $n = 500$ trials and $p = .5$. How unlikely would a sample proportion of .53 or more be in this case if people were just guessing?

The sampling distribution for the sample proportion $\hat{p}$ provides information on potential values of $\hat{p}$, the proportion of correct guesses in the sample of 500 trials if people can't really tell whether or not the cola has caffeine in it. The sampling distribution will be approximately a normal curve with the following mean and standard deviation:

$$\text{Mean} = p = .5$$

$$\text{Standard deviation} = \text{s.d.}(\hat{p}) = \sqrt{\frac{.5(1 - .5)}{500}} = .0224$$

Figure 9.4 shows this normal curve, with the area with $\hat{p} \geq .53$ shaded. Note that the probability of observing a sample proportion that large or larger just by chance is .09, so it is quite feasible. In other words, with a sample proportion of $\hat{p} = .53$, it is possible that people were just guessing ($p = .5$), and are not able to taste whether a cola contains caffeine or not.

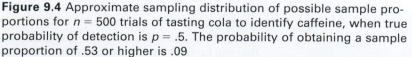

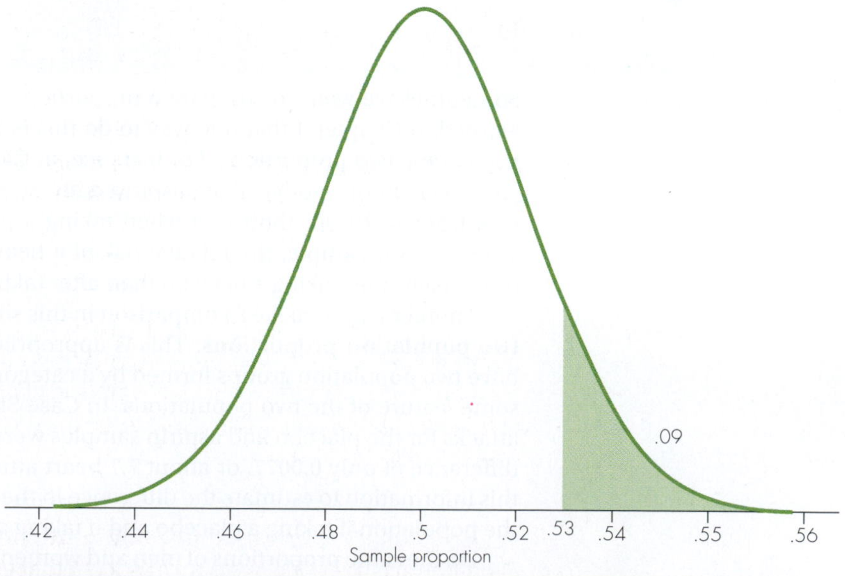

Figure 9.4 Approximate sampling distribution of possible sample proportions for $n = 500$ trials of tasting cola to identify caffeine, when true probability of detection is $p = .5$. The probability of obtaining a sample proportion of .53 or higher is .09

Estimating the Population Proportion from a Single Sample Proportion

In Example 9.4 we learned about the range of possible values to expect in polls with $n = 2400$ and $p = .4$. In practice, when we actually take a random sample of 2400 voters in a political poll, we have only one sample proportion, and we don't know the true population proportion. However, we do know how far apart the sample proportion and the true population proportion are likely to be. That information is contained in the **standard deviation of $\hat{p}$,**

$$s.d.(\hat{p}) = \sqrt{\frac{p(1-p)}{n}}$$

Note that this formula includes p, the population proportion. In many situations we will not know the value of p. Instead, we will use an observed sample proportion ($\hat{p}$) to estimate the unknown value of the parameter p. Thus, we estimate the standard deviation of $\hat{p}$ by using the sample value $\hat{p}$ in the formula. The estimated version is called the **standard error of** $\hat{p}$,

$$\text{s.e.}(\hat{p}) = \sqrt{\frac{\hat{p}(1-\hat{p})}{n}}$$

For instance, if $\hat{p} = .39$ and $n = 2400$, the standard error is $\sqrt{.39(1 - .39)/2400} = .01$. Amazingly, this value estimates the theoretical standard deviation of the sampling distribution for all possible sample proportions from this population, based on just a single sample. Because we know that the mean, which is the true proportion p, is almost surely within 3 standard deviations of the observed value $\hat{p}$, we know that p is almost surely within the range $\hat{p} \pm 3(\text{s.e.}) = .39 \pm 3(.01) = .39 \pm .03$. So we know that the true proportion who support the candidate is almost surely between .36 and .42. Note that the only numerical value we needed to determine this was the sample proportion $\hat{p}$, and of course, the sample size n. We will explore this idea in much more detail in Chapter 10.

9.4 Exercises are on pages 357–358.

9.5 SD Module 2: Sampling Distribution for the Difference in Two Sample Proportions

Sometimes we want to compare a proportion or probability for two populations. We learned in Chapter 4 that one way to do this is by using the *relative risk*, which is the *ratio* of the two proportions. For instance, in Case Study 1.6, we learned that for male volunteers in the study, taking aspirin daily reduced the chances of having a heart attack from 17.13 per thousand when taking a placebo to 9.42 per 1000 with aspirin. Thus, for this sample, the relative risk of a heart attack was 1.82, with a heart attack more likely after taking a placebo than after taking aspirin.

Another way to make a comparison in this situation is to examine the **difference in two population proportions**. This is appropriate when we have two populations or have two population groups formed by a categorical variable and we want to compare some feature of the two populations. In Case Study 1.6, the proportions having heart attacks for the placebo and aspirin samples were .01713 and .00942, respectively, for a difference of only 0.0077, or about 7.7 heart attacks per thousand people. We can use this information to estimate the difference in the probability of having a heart attack for the population if taking a placebo and if taking aspirin.

Do the same proportions of men and women support the death penalty? If not, how do the two proportions differ? Has the proportion of adults who support same-sex marriage changed in the last decade? If so, by how much has it changed? By how much does getting a flu shot reduce the probability of getting the flu? These are all questions that can be answered by looking at the difference in two proportions.

There are usually two questions of interest about a difference in two population proportions. First, we want to estimate the value of the difference. Second, often we want to test the hypothesis that the difference is 0, which would indicate that the two proportions are equal.

To use the methods described here, we need to collect *independent samples* from the two groups or populations. Remember that with independent samples, the individuals in one sample are not coupled in any way with the individuals in the other sample. For instance, measuring the voting behavior of husbands and wives would not produce independent samples, but measuring voting behavior of randomly and independently selected samples of men and women in a neighborhood would.

Another method that allows us to assume that we have independent samples is to measure two variables for everyone in a random sample, where one variable is a categorical grouping variable. As an example, we might take a random sample of adults and, for each person, record the person's sex and whether he or she supports or opposes the death penalty. The men and women constitute the independent samples, and we can then examine the difference in the proportions favoring the death penalty in these two groups.

Notation for the Difference in Two Proportions

In the notation scheme for two proportions, we will continue to use p to denote a population proportion and $\hat{p}$ to denote a sample proportion, and we will use subscripts 1 and 2 to represent the groups.

For the populations:

p_1 = *population* proportion for the first population.

p_2 = *population* proportion for the second population.

The parameter of interest is $p_1 - p_2$ = the difference in *population* proportions.

For the samples:

$\hat{p}_1$ = *sample* proportion for the sample from the first population.

$\hat{p}_2$ = *sample* proportion for the sample from the second population.

The sample statistic is $\hat{p}_1 - \hat{p}_2$ = the difference in *sample* proportions.

Conditions for the Sampling Distribution of $\hat{p}_1 - \hat{p}_2$ To Be Approximately Normal

The sampling distribution of the difference in two independent sample proportions is approximately normal when both of these conditions hold:

Condition 1: Sample proportions are available for two independent samples, randomly selected from the two populations of interest.

Condition 2: All of the quantities n_1p_1, $n_1(1 - p_1)$, n_2p_2, and $n_2(1 - p_2)$ are at least 10. These quantities represent the expected numbers of successes and failures in each of the two samples.

Mean, Standard Deviation, and Standard Error for the Sampling Distribution of $\hat{p}_1 - \hat{p}_2$

The **mean of the sampling distribution of $\hat{p}_1 - \hat{p}_2$** is:

$p_1 - p_2$ = the difference in *population* proportions

The **standard deviation of $\hat{p}_1 - \hat{p}_2$** is:

$$\text{s.d.}(\hat{p}_1 - \hat{p}_2) = \sqrt{\frac{p_1(1 - p_1)}{n_1} + \frac{p_2(1 - p_2)}{n_2}}$$

When we do not know the population proportions, we use sample proportions instead. Thus, the **standard error of $\hat{p}_1 - \hat{p}_2$** is

$$\text{s.e.}(\hat{p}_1 - \hat{p}_2) = \sqrt{\frac{\hat{p}_1(1 - \hat{p}_1)}{n_1} + \frac{\hat{p}_2(1 - \hat{p}_2)}{n_2}}$$

Remember that the mean of the sampling distribution for the scenarios we are considering is always the associated parameter. In this case, the parameter associated with the difference in two sample proportions is the difference in the corresponding population proportions.

FORMULA In general, the **standard error of the difference** between statistics from two independent samples is determined as

$$\text{s.e.(difference)} = \sqrt{[\text{s.e.(statistic 1)}]^2 + [\text{s.e.(statistic 2)}]^2}$$

The same formula applies to standard deviations, with "s.d." replacing "s.e." on both sides of the equation. The formula follows from Section 8.8 in which the result is given that the variance of the difference between two independent random variables is the sum of their variances. This result applies to the difference $\hat{p}_1 - \hat{p}_2$ because the two sample proportions $\hat{p}_1$ and $\hat{p}_2$ are independent random variables. (This is the reason for the condition requiring independent samples.) The variance of any random variable is the standard deviation squared, and we learned in Module 1 that the standard deviation of $\hat{p}$ is $\sqrt{p(1-p)/n}$. This formula applies to both $\hat{p}_1$ and $\hat{p}_2$. Putting all of these pieces together leads to the formulas given for the standard deviation and standard error of $\hat{p}_1 - \hat{p}_2$.

SD Module 2: $p_1 - p_2$

Example 9.6 **Men, Women, and the Death Penalty** Are you in favor of the death penalty, or are you opposed to it? On the basis of surveys done in the past, women are more likely to oppose the death penalty than are men. In the 2008 General Social Survey (**GSS-08** in the datasets for this book), the sample proportions opposing the death penalty were $\hat{p}_1 = 385/1017 = .360$ for women and $\hat{p}_2 = 254/885 = .285$ for men, for a difference of $\hat{p}_1 - \hat{p}_2 = .360 - .285 = .075$.

If we were to repeat this survey with a new random sample of 1017 women and 885 men, the difference $\hat{p}_1 - \hat{p}_2$ would probably not be exactly 0.075. What can we expect for the difference in repeated samples? To answer this question, we must pretend that we are omniscient and actually know the truth about the population proportions of men and women who oppose the death penalty. Let's suppose that, in truth, 37% of women and 27% of men oppose the death penalty, so $p_1 = .37$, $p_2 = .27$, and $p_1 - p_2 = .10$. Then the sampling distribution of $\hat{p}_1 - \hat{p}_2$ is approximately normal, with mean $p_1 - p_2 = .10$ and standard deviation,

$$\text{s.d.}(\hat{p}_1 - \hat{p}_2) = \sqrt{\frac{p_1(1-p_1)}{n_1} + \frac{p_2(1-p_2)}{n_2}}$$

$$= \sqrt{\frac{.37(1-.37)}{1017} + \frac{.27(1-.27)}{885}} = .021$$

Figure 9.5 illustrates this sampling distribution. It represents the possible differences in *sample* proportions of women and men (women − men) who oppose the death penalty, if in fact the difference in *population* proportions is .10, the true population proportions for women and men are .37 and .27, respectively, and the sample sizes are 1017 and 885 for women and men, respectively. Note that the difference of .075 achieved in the 2008 General Social Survey would not have been surprising in this scenario. (About 12% of the possible values fall at or below .075.) The figure shows a range spanning 3 standard deviations on either side of the mean. Lines showing 1 and 2 standard deviations on either side of the mean are illustrated. For example, using the Empirical Rule, we would expect the difference $\hat{p}_1 - \hat{p}_2$ to be between 0.079 and 0.121 in 68% of all surveys fitting this scenario.

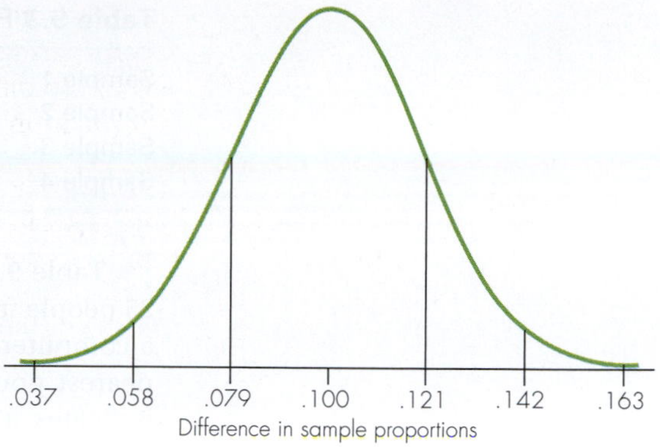

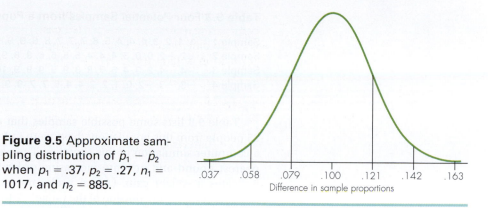

Figure 9.5 Approximate sampling distribution of $\hat{p}_1 - \hat{p}_2$ when $p_1 = .37$, $p_2 = .27$, $n_1 = 1017$, and $n_2 = 885$.

9.5 Exercises are on pages 358–359.

9.6 SD Module 3: Sampling Distribution for One Sample Mean

We now turn to the case in which the information of interest involves the mean or means of quantitative variables. For example, a company that sells oat products might want to know the mean cholesterol level that people would have if everyone had a certain amount of oat bran in their diet. To help determine financial aid levels, a large university might want to know the mean income of all students on campus who work.

Suppose that a population consists of thousands or millions of individuals, and that we are interested in estimating the mean of a quantitative variable. If we sample 25 people and compute the mean of the variable for that sample, how close will that *sample mean* be to the *population mean* we are trying to estimate? Each time we take a sample, we will get a different sample mean. Can we say anything about what we expect those means to be?

Example 9.7 **Hypothetical Mean Weight Loss** Suppose that we want to estimate the average weight loss for everyone who attends a national weight-loss clinic for 10 weeks. Suppose that, unknown to us, the distribution of weight losses for everyone in this population is approximately normal with a mean of 8 pounds and a standard deviation of 5 pounds. Figure 9.6 shows a normal curve with these characteristics. If the weight losses are approximately normal, we know from the Empirical Rule in Chapter 2 that 95% of the individuals will fall within 2 standard deviations, or 10 pounds, of the mean of 8. In other words, 95% of the individual weight losses will fall between -2 (a gain of 2 pounds) and $+18$ (a loss of 18 pounds).

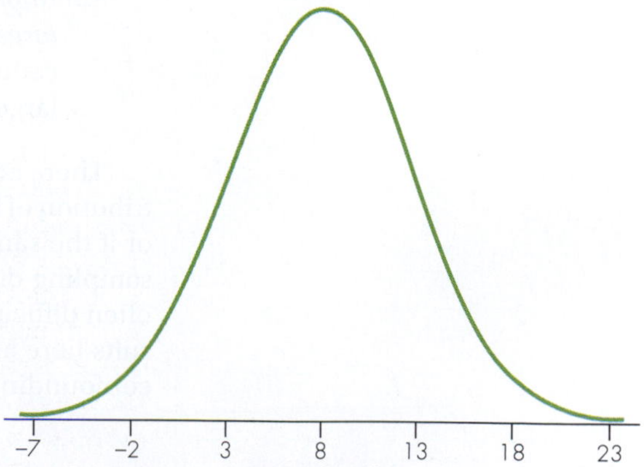

Figure 9.6 Normal curve for individual weight losses with mean $\mu = 8$ pounds and standard deviation $\sigma = 5$ pounds

SD Module 3: μ

Table 9.3 Four Potential Samples from a Population with $\mu = 8, \sigma = 5$

Sample 1	1, 1, 2, 3, 4, 4, 4, 5, 6, 7, 7, 7, 8, 8, 9, 9, 11, 11, 13, 13, 14, 14, 15, 16, 16
Sample 2	−2, −2, 0, 0, 3, 4, 4, 4, 5, 5, 6, 6, 8, 8, 9, 9, 9, 9, 9, 10, 11, 12, 13, 13, 16
Sample 3	−4, −4, 2, 3, 4, 5, 7, 8, 8, 9, 9, 9, 9, 9, 10, 10, 11, 11, 11, 12, 12, 13, 14, 16, 18
Sample 4	−3, −3, −2, 0, 1, 2, 2, 4, 4, 5, 7, 7, 9, 9, 10, 10, 10, 11, 11, 12, 12, 14, 14, 14, 19

Table 9.3 lists some possible samples that could result from randomly sampling 25 people from this population; these were indeed the first four samples produced by a computer simulation of this situation. The weight losses have been rounded to the nearest pound and put into increasing order for ease of reading. A negative value indicates a weight gain. (As an aside, note that, as expected, 5/100 = 5% of the individual values are outside of the range −2 to 18.) We are interested in the behavior of the possible sample means for samples such as these.

Here are the sample means and standard deviations for each of these four samples:

Sample 1: Mean = 8.32 pounds, standard deviation = 4.74 pounds.
Sample 2: Mean = 6.76 pounds, standard deviation = 4.73 pounds.
Sample 3: Mean = 8.48 pounds, standard deviation = 5.27 pounds.
Sample 4: Mean = 7.16 pounds, standard deviation = 5.93 pounds.

You can see that the sample means, although all different, are relatively close to the population mean of 8. You can also see that the sample standard deviations are relatively close to the population standard deviation of 5. This latter fact is important, because although we are interested in using the sample mean to estimate the population mean, we will need to use the sample standard deviation to help find a standard error for this estimate.

SD Module 3: μ

Conditions for the Sampling Distribution of the Mean To Be Approximately Normal

For the *sampling distribution of the sample mean* to be approximately normal, unlike the sampling distribution for proportions, it is not always necessary to have a large sample. If the population of measurements is bell-shaped, then the result holds for all sample sizes. The sampling distribution of the sample mean is approximately normal in both of the following types of situations:

Situation 1: The population of the measurements of interest is bell-shaped, and a random sample of any size is measured.

Situation 2: The population of measurements of interest is not bell-shaped, but a *large* random sample is measured. Thirty is usually used as an arbitrary demarcation of "large," but if there are extreme outliers, it is better to have an even larger sample size than $n = 30$.

There are actually only a limited number of situations for which the sampling distribution of the sample mean is *not* approximately normal. If the sample is not random, or if the sample size is small and the original population is not bell-shaped, then the sampling distribution will not necessarily be approximately normal. In practice, it is often difficult to get a random sample, so researchers are usually willing to use the results here as long as they can get a representative sample with no obvious sources of confounding or bias.

DEFINITION When a population of observations is bell-shaped and/or a large random sample is to be taken, the **sampling distribution for a sample mean** can be defined as follows:

- Let μ = mean for the population of interest.
- Let σ = standard deviation for the population of interest.
- Let $\overline{x}$ = sample mean (the mean for the sample).

For all possible random samples of the same size n taken from this population, the distribution of possible values of $\overline{x}$ is approximately normal, with

Mean = μ

$$\text{Standard deviation} = \text{s.d.}(\overline{x}) = \frac{\sigma}{\sqrt{n}}$$

This approximate normal distribution is called the **sampling distribution of $\overline{x}$** or the **sampling distribution of the mean.**

Technical Note: The n observations in each sample must all be independent, and they will be if random samples are used.

Standard Deviation of Measurements versus Standard Deviation of Sample Means

Be careful that you do not confuse the standard deviation for the original population of measurements, σ, with the **standard deviation of the sample means,** $\sigma/\sqrt{n}$. The parameter σ is a measure of the variability among individual measurements within the population. The parameter $\sigma/\sqrt{n}$ is a measure of the variability among sample means for the many different random samples of size n that can be taken from the population.

THOUGHT QUESTION 9.2 Construct an example of interest to you personally for which the approximate normality of the sampling distribution for the mean applies and for which a study could be done to estimate a population mean.*

Example 9.8 **Hypothetical Mean Weight Loss Revisited** For our hypothetical weight-loss clinics in Example 9.7, the population mean and standard deviation of weight losses were $\mu = 8$ pounds and $\sigma = 5$ pounds, respectively, and we were taking random samples of size 25. The definition of the sampling distribution for a sample mean tells us that the sampling distribution of $\overline{x}$ is:

- Approximately normal.
- Mean $\mu = 8$ pounds.
- Standard deviation = s.d.$(\overline{x}) = \sigma/\sqrt{n} = 5/5 = 1.0$.

From the Empirical Rule in Chapter 2, we know the following facts about possible sample means $\overline{x}$ in this situation, based on intervals extending 1, 2, and 3 standard deviations from the mean of 8:

- There is a *68%* chance that the sample mean will be between *7 and 9 pounds.*
- There is a *95%* chance that the sample mean will be between *6 and 10 pounds.*

***HINT:** Think of a quantitative variable for which you would like to estimate the mean value. For this variable, do you think that you will be in Situation 1 or Situation 2 (p. 332)?

SD Module 3: μ

- It is *almost certain* (99.7% chance) that the sample mean will be between *5 and 11 pounds*.

If you look at the hypothetical samples that we chose in Example 9.7 (Table 9.3), you will see that the four sample means range from 6.76 to 8.48 pounds, well within the range that we expect to see using these criteria. Figure 9.7 displays a histogram and superimposed normal curve for 400 simulated sample means, each based on a sample of 25 weight losses from a population with a mean of 8 pounds and a standard deviation of 5 pounds.

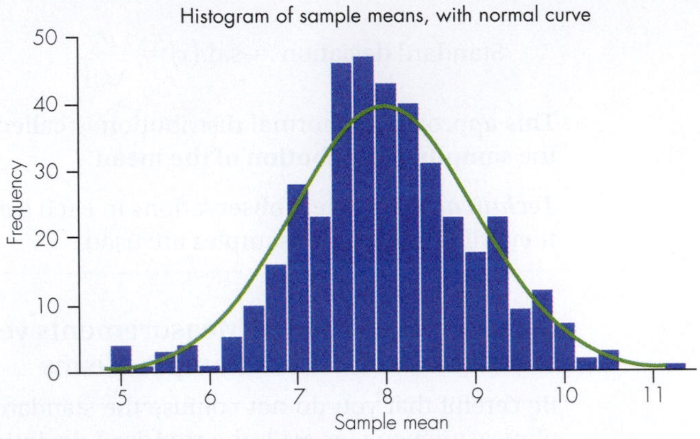

Figure 9.7 Histogram of 400 sample means for samples of $n = 25$, with $\mu = 8$ and $\sigma = 5$. Smooth curve shows the sampling distribution

Standard Error of the Mean

In practice, the population standard deviation σ is rarely known, so the sample standard deviation s is used in its place when determining the standard deviation for the sampling distribution of sample means. Consistent with what we did for sample proportions in the previous section, when making this substitution, we call the result the **standard error of the mean.** The terminology makes sense because the standard error measures roughly how much, on average, the sample mean $\bar{x}$ is in error as an estimate of the population mean μ.

We denote the standard error of the mean as s.e.($\bar{x}$), and the formula is:

$$\text{s.e.}(\bar{x}) = \frac{s}{\sqrt{n}}$$

where s is the standard deviation of the observations in the sample.

For the $n = 25$ weight losses in Sample 1 shown in Table 9.3, the sample standard deviation is $s = 4.74$ pounds. On the basis of this sample, the standard error of the mean is $s/\sqrt{n} = 4.74/\sqrt{25} = 0.948$. This value estimates $\sigma/\sqrt{n}$, the theoretical standard deviation of the sampling distribution for sample means. If we were to use a different sample of $n = 25$ observations from Table 9.3, the standard error of the mean would have a different value because s would be different. Remember that the standard deviation of $\bar{x}$, $\sigma/\sqrt{n}$, does not change because σ is a fixed population value.

Increasing the Size of the Sample

Suppose that we had considered samples of 100 people in the weight-loss example rather than samples of 25. Note that the mean of the sampling distribution of possible sample means would still be 8 pounds, but the standard deviation, s.d.($\bar{x}$), would decrease. It would now be $\sigma/\sqrt{n} = 5/\sqrt{100} = 0.5$ instead of $5/\sqrt{25} = 1.0$. Therefore, for samples of size $n = 100$, the standard deviation of potential sample means is only half of what it is for samples of size $n = 25$. In Example 9.8 we applied the Empirical Rule to see that for samples of $n = 25$, the sample means are likely to be in the range 8 ± 3 pounds, or between about 5 pounds and 11 pounds. In contrast, for samples of $n = 100$, the sample means are likely to be in the range 8 ± 1.5 pounds, or between

6.5 pounds and 9.5 pounds. Figure 9.8 compares the sampling distributions for samples of $n = 25$ and $n = 100$ and shows clearly that there is less variability among potential sample means for the larger sample size.

In general, a fourfold increase in sample size cuts the standard deviation of the distribution of possible sample means, and thus the range of likely sample means, in half. A ninefold increase in sample size cuts the standard deviation of possible means to a third of what it was, and so on. It is clear that the standard deviation of the sampling distribution quantifies what our common sense tells us: Larger samples tend to result in more accurate estimates of population values than smaller samples do.

9.6 Exercises are on pages 359–360.

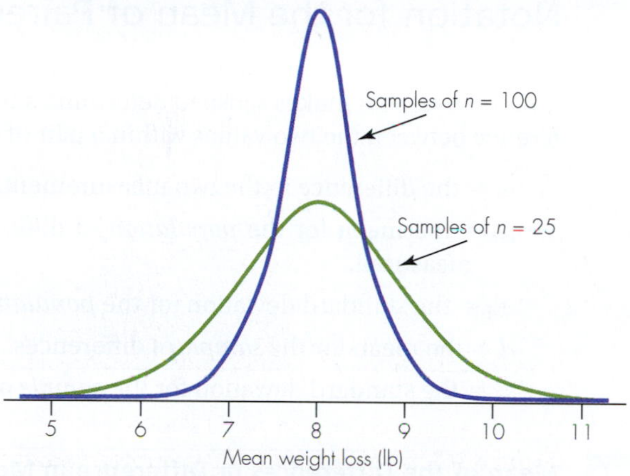

Figure 9.8 Sampling distributions of potential sample means for samples of $n = 25$ and $n = 100$ taken from a population with mean $\mu = 8$ and standard deviation $\sigma = 5$

THOUGHT QUESTION 9.3 From the weight-loss example discussed in this section, we learned that increasing the sample size fourfold would about halve the range of possible sample means. Would the range of *individual* weight losses in the sample be likely to increase, decrease, or remain about the same if the sample size were increased fourfold? Explain.*

9.7 SD Module 4: Sampling Distribution for the Sample Mean of Paired Differences

SD Module 4: μ_d

Do people tend to score higher on an IQ test after listening to Mozart than after sitting in silence? If so, by how much, on average? How much weight do college students gain, on average, during their first year, if any? Do women tend to marry men who are older or younger than themselves, and what is the average difference in age between husbands and wives? These and many more questions can be answered by measuring *paired differences* for a sample of pairs and then using them to estimate the mean difference for the population of pairs.

When two sets of observations will be compared, they may be collected as *two independent samples* in which the individuals in one sample are not coupled in any way with those in the other sample, or they may be collected as *matched pairs* in which observations are taken on two halves of a pair. A matched pair can be two measurements from the same individual measured under two conditions (IQ after listening to Mozart and after sitting in silence), or at two different times (weight at the beginning and at the end of the first year of college), and so on. Or, a pair can be two separate individuals who form a natural pair, such as twins, spouses (ages of hus-

***HINT:** Which sample size might be more likely to include some of the more extreme weight losses for individuals in the population, leading to a wider range?

bands and wives), or individuals who are matched according to one or more important criteria.

Data that are collected as matched pairs are sometimes called **dependent samples** because the two observations are not statistically independent of each other. In the next module, we consider how to compare means for *independent* samples. In this module, we are concerned with the population mean difference for *dependent* observations, taken as matched pairs. The situation is analogous to the case for one mean, except that in that case, we observed a single measurement for each individual, and in this case, we start with two measurements but reduce them to a single measurement by taking the difference.

Notation for the Mean of Paired Differences

The sample consists of *n* matched pairs of observations. The response variable(s) is quantitative, so it makes sense to determine a mean. We are interested only in the difference between the two values within a pair of observations.

d_i = the difference in the two measurements for individual *i*, where $i = 1, 2, \cdots, n$.

μ_d = the mean for the *population* of differences, if all possible pairs were to be measured.

σ_d = the standard deviation for the *population* of differences.

$\bar{d}$ = the mean for the *sample* of differences.

s_d = the standard deviation for the *sample* of differences.

TECHNICAL NOTE

Mean of the Differences or Difference in Means?

You might wonder why we make a distinction between the mean of the differences and the difference in the means, when they are in fact numerically equivalent. Why do we have to take the differences first and then find the sample statistics? Why not just find the average of the first number in all pairs, then the average of the second number and subtract the two averages? For instance, to find the average difference in ages of husbands and wives, why don't we just find the average age for the husbands and the average age for the wives and then take the difference? The answer is that if we do not have the individual difference, then we can't find the standard deviation for the differences, and we need that to estimate μ_d and test hypotheses about it. Thus, while it is true that $\mu_d = \mu_1 - \mu_2$ (the population mean of the differences = the difference in the separate population means) and $\bar{d} = \bar{x}_1 - \bar{x}_2$ (the sample mean of the differences = the difference in the sample means), there is no way to find s_d or σ_d if all we have are the separate standard deviations s_1 and s_2 or σ_1 and σ_2. We need to know how spread out the differences are, not how spread out the values in each separate set are.

Conditions for the Sampling Distribution of $\bar{d}$ To Be Approximately Normal

Remember that the mean of paired differences is analogous to the case of one mean except that the measurements are differences. Therefore, the same conditions that are required for the sampling distribution of $\bar{x}$ to be approximately normal also hold for the sampling distribution of $\bar{d}$. Either of the following two situations will work:

Situation 1: The population of differences is bell-shaped, and a random sample of any size is measured.

Situation 2: The population of differences is not bell-shaped, but a *large* random sample is measured. Thirty pairs are usually used as an arbitrary demarcation of a "large" sample, but if any of the differences are extreme outliers, it is better to have an even larger sample size than $n = 30$.

As with other situations, it often is difficult to get a random sample, and researchers usually are willing to assume that these conditions hold if they can get a representative sample with no obvious sources of confounding or bias.

Mean, Standard Deviation, and Standard Error for the Sampling Distribution of $\overline{d}$

The **mean of the sampling distribution of $\overline{d}$** is μ_d, the population mean of the differences. The **standard deviation of $\overline{d}$** is

$$\textbf{s.d.}(\overline{d}) = \frac{\sigma_d}{\sqrt{n}}$$

The **standard error of $\overline{d}$** is

$$\textbf{s.e.}(\overline{d}) = \frac{s_d}{\sqrt{n}}$$

Remember that for the situations we are considering, the mean of the sampling distribution is always the associated parameter. Also remember that the standard error is used to estimate the standard deviation when σ_d is not known.

Example 9.9 **Suppose That There Is No "Freshman 15"** In Example 9.1, we asked whether it was really true that students gain weight on average in the first year of college. We discussed a study that measured a sample of 60 students for 12 weeks and found that they gained an average of 4.2 pounds. Is it likely that the mean weight gain for a random sample of 60 students would be as high as 4.2 pounds *if* there is no average weight gain in the population of all students? To answer this question, we need to know what to expect the *sample* mean $\overline{d}$ to be for a sample of $n = 60$ measurements if the *population* mean μ_d is 0. The sampling distribution of $\overline{d}$ provides this information. To find the standard deviation for the sampling distribution, we need to know the standard deviation for the population of weight gains, σ_d. Let's be omniscient and suppose that we know that it is 7 pounds. Then the sampling distribution of $\overline{d}$ is as follows:

- Approximately normal.
- Mean $= \mu_d = 0$.
- Standard deviation $=$ s.d.$(\overline{d}) = \sigma_d/\sqrt{n} = 7/\sqrt{60} = 0.904$, which we will round off to 0.9 pounds.

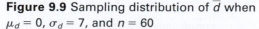

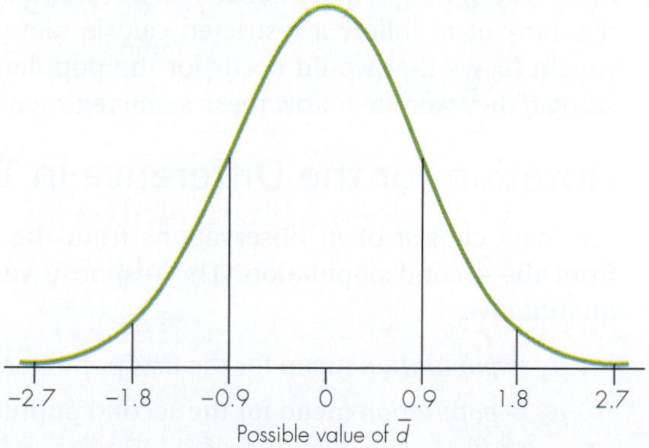

Figure 9.9 Sampling distribution of $\overline{d}$ when $\mu_d = 0$, $\sigma_d = 7$, and $n = 60$

Figure 9.9 illustrates the sampling distribution for this situation; remember that the mean is $\mu_d = 0$ pounds and the standard deviation is s.d.$(\overline{d}) = 0.9$ pounds. Values that are 1, 2, and 3 standard deviations on either side of the mean of the distribution are marked. For instance, using the Empirical Rule, we would expect the sample mean difference to fall between -1.8 pounds and $+1.8$ pounds in 95% of all samples of size 60. Again, this assumes that in fact the average weight gain in the population is 0 and the standard deviation for the population of weight gains is 7 pounds.

The mean weight gain for the 60 students in the Cornell study was $\bar{d} = 4.2$ pounds. We can see from Figure 9.9 that a sample average this large would be extremely unlikely under the assumed scenario that the true population mean weight gain is 0 and the standard deviation is 7 pounds. In Chapter 13, we will learn a more formal way to use the sample mean to determine whether it is unreasonable to assume that the population mean is a specific value—in this case, 0.

9.7 Exercises are on page 361.

9.8 SD Module 5: Sampling Distribution for the Difference in Two Sample Means

Do women who smoke during pregnancy have babies with lower average birth weight than women who don't smoke during pregnancy? If so, what is the difference? Do people who practice yoga have lower blood pressure, on average, than people who do not practice yoga? If so, by how much? Is the mean resting pulse rate for men lower than it is for women? Do men drive faster, on average, than women? If so, by how much? These are all questions that can be answered by looking at the difference in two means for independent samples.

To use the methods described here, we need to collect *independent samples* from the two groups or populations. Remember that with independent samples, the individuals in one sample are not coupled in any way with the individuals in the other sample. An equivalent procedure is to measure a categorical variable with two possible values for everyone in a random sample (along with the quantitative variable of interest) and use the categorical variable to form two groups. An example is to categorize mothers by whether or not they smoked during pregnancy and measure the birth weights of their newborn infants. The parameter of interest would be the difference in mean birth weights for children of smokers and nonsmokers.

A common application of inference for the difference in two population means is in randomized experiments, in which the two groups are formed by randomly assigning one of two treatments to each individual in the study. Volunteers generally are used for these studies rather than random samples, and the results can be extended to the population of individuals who would volunteer if given the opportunity. The population means are conceptual rather than real because they are the means that would occur if everyone in the population were to be subjected to the treatment. For example, in a study of weight loss in obese men, volunteers were randomly assigned to exercise regularly or to follow a restricted-calorie diet. The population means are the mean weight losses that would occur for the population of men similar to the ones in the study *if* they were to follow these same regimens.

Notation for the Difference in Two Means

The data consist of n_1 observations from the first population and n_2 observations from the second population. The response variable, measured in both samples, is quantitative.

μ_1 = *population* mean for the first population.

μ_2 = *population* mean for the second population.

The parameter of interest is $\mu_1 - \mu_2$ = the difference in *population* means.

$\bar{x}_1$ = *sample* mean for the sample from the first population.

$\bar{x}_2$ = *sample* mean for the sample from the second population.

The sample statistic of interest is $\bar{x}_1 - \bar{x}_2$ = the difference in *sample* means.

σ_1 = the *population* standard deviation for the first population.

σ_2 = the *population* standard deviation for the second population.

s_1 = the *sample* standard deviation for the sample from the first population.

s_2 = the *sample* standard deviation for the sample from the second population.

Conditions for the Sampling Distribution of $\bar{x}_1 - \bar{x}_2$ To Be Approximately Normal

An important condition in this situation is that the two samples must be *independent.* Three ways to obtain independent samples are as follows:

- Take separate random samples from each of two populations such as men and women.
- Take a random sample from a population and divide the sample into two groups based on a categorical variable such as smoker and nonsmoker.
- Randomly assign participants in a randomized experiment to two treatment groups such as exercise or diet.

In addition, one of the following two situations must hold:

Situation 1: The populations of measurements are both bell-shaped, and random samples of any size are measured.

Situation 2: Large random samples are measured from each population. A somewhat arbitrary definition of "large" is that both sample sizes are at least 30, but if there are extreme outliers or extreme skewness in one or both samples, it is better to have even larger samples.

Mean, Standard Deviation, and Standard Error for the Sampling Distribution of $\bar{x}_1 - \bar{x}_2$

Remember that the mean of the sampling distribution for the situations we are considering is always the associated parameter. Therefore, the **mean of the sampling distribution of $\bar{x}_1 - \bar{x}_2$** is $\mu_1 - \mu_2$ = the difference in *population* means. The **standard deviation of $\bar{x}_1 - \bar{x}_2$** is

$$\textbf{s.d.}(\bar{x}_1 - \bar{x}_2) = \sqrt{\frac{\sigma_1^2}{n_1} + \frac{\sigma_2^2}{n_2}}$$

When we don't know the values of the population standard deviations we use the sample standard deviations instead. Thus, the **standard error of $\bar{x}_1 - \bar{x}_2$** is

$$\textbf{s.e.}(\bar{x}_1 - \bar{x}_2) = \sqrt{\frac{s_1^2}{n_1} + \frac{s_2^2}{n_2}}$$

TECHNICAL NOTE

Finding the Standard Deviation of a Difference

In Section 8.8, the result was given that the variance of the difference of two independent random variables is the sum of their variances. That result applies to the difference $\bar{x}_1 - \bar{x}_2$ because $\bar{x}_1$ and $\bar{x}_2$ are independent random variables. (This is the reason for the condition requiring independent samples.) The variance of any random variable is the standard deviation squared, and we learned in Module 3 that the standard deviation of $\bar{x}$ is $\sigma/\sqrt{n}$. This formula applies to both $\bar{x}_1$ and $\bar{x}_2$, adding appropriate subscripts to σ and n. Putting these pieces together leads to the formula given for the standard deviation of $\bar{x}_1 - \bar{x}_2$.

Example 9.10 **Case Study 1.1 Revisited: Who Are the Speed Demons?** In Case Study 1.1, we examined data from a survey in which college students were asked "What's the fastest you've ever driven a car? _____ mph." For the 87 males and 102 females who responded, the sample means were $\bar{x}_1 = 107$ mph for males and $\bar{x}_2 = 88$ mph for females. Thus, the difference in means was $\bar{x}_1 - \bar{x}_2 = 107 - 88 = 19$ mph. The sample standard deviations were $s_1 = 17$ for males and $s_2 = 14$ for females.

Is the difference that we observed in this survey large enough to convince us that there is a real difference in means for the *populations* of male and female students represented by this sample, that is, that $\mu_1 - \mu_2 > 0$? To answer that question, we need to know what to expect of the difference in sample means if indeed there is no difference in the population means—if in fact $\mu_1 - \mu_2 = 0$. In that case, is a difference of 19 mph in the sample means out of the question, or is it a feasible difference? In other words, we need to find the sampling distribution of $\bar{x}_1 - \bar{x}_2$ and see if 19 mph is a plausible value in that distribution.

Let's suppose that in truth, $\mu_1 - \mu_2 = 0$. To find the standard deviation, s.d.$(\bar{x}_1 - \bar{x}_2)$, we need to know the population standard deviations as well. Let's pretend that we are all-knowing and assume that they are both 15. Then the sampling distribution of $\bar{x}_1 - \bar{x}_2$ is:

- Approximately normal.
- Mean $= \mu_1 - \mu_2 = 0$.
- Standard deviation $= s.d.(\bar{x}_1 - \bar{x}_2) = \sqrt{\dfrac{\sigma_1^2}{n_1} + \dfrac{\sigma_2^2}{n_2}} = \sqrt{\dfrac{15^2}{87} + \dfrac{15^2}{102}} = 2.2$ mph.

Figure 9.10 illustrates this sampling distribution. It represents the possible differences in *sample* means of males' and females' (males − females) reported fastest driving times if in fact the difference in *population* means is 0, the population standard deviations are both 15, and the sample sizes are 87 for males and 102 for females. Figure 9.10 shows a range spanning 3 standard deviations on both sides of the mean; remember that s.d.$(\bar{x}_1 - \bar{x}_2) = 2.2$. Note that the difference of 19 mph achieved in the sample would have been almost impossible in this scenario. Therefore, the true difference in population means is almost surely much greater than 0. We will learn how to estimate the actual difference in Chapter 11.

9.8 Exercises are on pages 361–362.

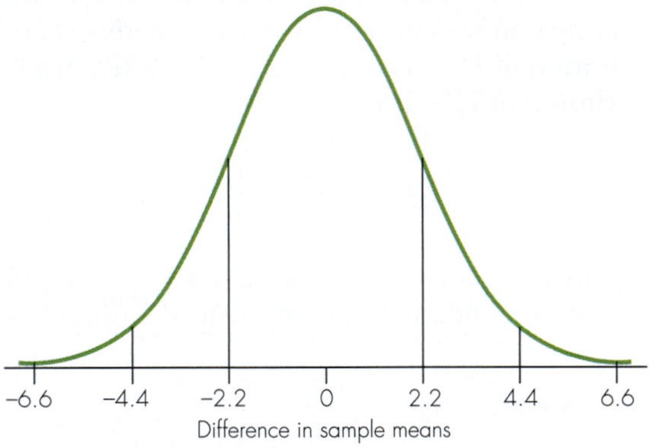

Figure 9.10 Approximate sampling distribution of $\bar{x}_1 - \bar{x}_2$ when $\mu_1 - \mu_2 = 0$, $\sigma_1 = \sigma_2 = 15$, $n_1 = 87$, and $n_2 = 102$

9.9 Preparing for Statistical Inference: Standardized Statistics

Lesson 1: Standardized Statistics for Sampling Distributions

In Chapter 2, we learned that sometimes it is useful to transform a raw score into a standardized score, or *z*-score, measuring how many standard deviations the raw score falls above or below the mean. The *z*-scores for raw scores from a normal population with mean μ and standard deviation σ are determined by the following formula:

$$z = \frac{x - \mu}{\sigma}$$

The resulting *z*-scores form a *standard normal population*, with $\mu = 0$ and $\sigma = 1$.

Similarly, sometimes it is useful to transform a raw sample statistic into its standardized version. Doing this transformation requires knowledge of the mean and standard deviation of the sampling distribution for the sample statistic. If the sampling distribution is approximately normal, then the standardized statistic has, approximately, a standard normal distribution.

As we learned earlier in this chapter, many sample statistics have approximately normal sampling distributions. Researchers often conduct studies in which they observe one value of such a statistic (such as a sample proportion or a sample mean), and they want to know how far that value falls from a hypothetical center of the sampling distribution. Sometimes it is useful to obtain that information in standardized form. In other words, it is useful to find the standardized version of the observed sample value.

We can find standardized statistics for any of the five sample statistics that we discussed previously in this chapter. In each case, as long as the conditions are satisfied for the sampling distribution to be approximately normal, the **standardized statistic** for a sample statistic is

$$z = \frac{\text{Sample statistic} - \text{Population parameter}}{\text{s.d.(sample statistic)}}$$

In this section, we show how this generic form applies to a sample proportion and a sample mean. In the exercises, you will be asked to specify the details of this generic formula for the three cases that are not explicitly covered here.

Lesson 2: Standardized Statistics for Proportions

Standardized *z*-Statistic for a Sample Proportion

Assuming that appropriate conditions are met for the sampling distribution to be approximately normal, as given in Section 9.4, the distribution of the following standardized statistic (z-statistic) for a sample proportion is approximately a standard normal distribution:

$$z = \frac{\hat{p} - p}{\text{s.d.}(\hat{p})} = \frac{\hat{p} - p}{\sqrt{\dfrac{p(1-p)}{n}}}$$

We can use this **standardized *z*-statistic** to assess the difference between an observed sample proportion ($\hat{p}$) and a possible value for the population proportion (p). This is demonstrated in the next example.

Example 9.11 **Unpopular TV Shows** It is common for television networks to cancel shows that have consistently low ratings. Ratings are based on a random sample of households, and the sample proportion of households that are tuned to the show ($\hat{p}$) is an estimate of p, the population proportion watching the show. Suppose that a network wants p to be at least .20 or the show will be canceled. In a random sample of 1600 households, 288, or a proportion of .18, are watching. Is this close enough to the desired population proportion of .20 to be attributable to random fluctuations in possible sample proportions, even if p really is .20 or higher?

The sampling distribution of $\hat{p}$ is approximately normal, with p = actual population proportion watching the show. If $p = .20$, then the mean and standard deviation of the sampling distribution are $p = .20$ and $\text{s.d.}(\hat{p}) = .01$, respectively. Therefore, the standardized version of $\hat{p} = .18$ is

$$z = \frac{\hat{p} - p}{\text{s.d.}(\hat{p})} = \frac{\hat{p} - p}{\sqrt{\dfrac{p(1-p)}{n}}} = \frac{.18 - .20}{\sqrt{\dfrac{(.2)(.8)}{1600}}} = \frac{.18 - .20}{.01} = -2.00$$

The sample proportion of .18 is about 2 standard deviations below the mean of .20. That probably isn't far enough below .20 to completely rule out sampling variability as the reason for such a low proportion, but it probably would be if the ratings were consistently that low.

Lesson 3: Standardized Statistics for Means

Standardized z-Statistic for a Sample Mean

Assuming that appropriate conditions are met for the approximate normality of the sampling distribution, as given in Section 9.6, the following standardized statistic (z-statistic) for a sample mean has approximately a standard normal distribution:

$$z = \frac{\bar{x} - \mu}{\text{s.d.}(\bar{x})} = \frac{\bar{x} - \mu}{\sigma/\sqrt{n}} = \frac{\sqrt{n}(\bar{x} - \mu)}{\sigma}$$

As an example, suppose that we think that pulse rates for adult women have mean $\mu = 72$ beats per minute and standard deviation $\sigma = 8$. Pulse rates for $n = 100$ randomly selected women are measured, and the sample mean is $\bar{x} = 75$. A z-statistic for this observed sample mean is

$$z = \frac{75 - 72}{8/\sqrt{100}} = \frac{3}{0.8} = 3.75$$

A value as large as 3.75 is unusual for a z-statistic. Either this was an odd sample or maybe our assumptions about the population mean and standard deviation are wrong.

Student's t-Distribution: Replacing σ with s

When computing standardized statistics for means, we face a dilemma because we rarely know the population standard deviations, and they are part of the formula. For instance, the denominator of the standardized statistic for $\bar{x}$ includes the population standard deviation σ, which is rarely known. The best that we can do is to approximate σ with the *sample* standard deviation s and thus approximate $\text{s.d.}(\bar{x}) = \sigma/\sqrt{n}$ with $\text{s.e.}(\bar{x}) = s/\sqrt{n}$. Unfortunately, for small sample sizes, this approximation is often off the mark, so the standardized statistic does not exactly conform to the standard normal distribution. Instead, under certain conditions (see the In Summary box on page 343), it has a probability distribution called **Student's *t*-distribution**, or just ***t*-distribution** for short.

A parameter called **degrees of freedom**, abbreviated as **df**, is associated with any *t*-distribution. In most applications, this parameter is a function of the sample size for the problem, but the specific formula for the degrees of freedom depends on the type of problem. For the standardized version of a sample mean $\bar{x}$, $\text{df} = n - 1$, where n is the sample size. A property of the *t*-distribution is that as the number of degrees of freedom increases, the distribution gets closer to the standard normal curve. The *t*-distribution with $\text{df} = \infty$ (infinity) is identical to the standard normal curve. In practice, if the number of degrees of freedom is very large, the *t*-distribution is so close to the standard normal distribution that they are used interchangeably.

Figure 9.11 shows *t*-distributions with df = 3 and df = 8 along with the standard normal curve. As can be seen in this figure, the *t*-curves are more spread out than the standard normal and the *t*-curve with 3 df is more spread out than the one with 8 df. In other words, for the *t*-distributions, there is slightly more probability in the extreme tails of the distribution than there is for the standard normal. This difference diminishes as the number of degrees of freedom increases.

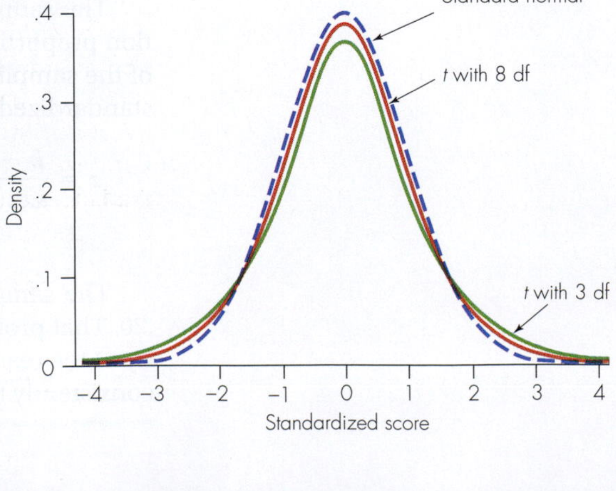

Figure 9.11 *t*-distributions with df = 3, df = 8, and a standard normal distribution

IN SUMMARY

Standardized z- and t-Statistics for $\overline{x}$

If a small random sample is taken from a normal population or a large random sample is taken from any population (in which case these results are approximate), then the standardized statistic,

$$z = \frac{\overline{x} - \mu}{\text{s.d.}(\overline{x})} = \frac{\overline{x} - \mu}{\sigma/\sqrt{n}} = \frac{\sqrt{n}(\overline{x} - \mu)}{\sigma}$$

has a standard normal distribution, and

$$t = \frac{\overline{x} - \mu}{\text{s.e.}(\overline{x})} = \frac{\overline{x} - \mu}{s/\sqrt{n}} = \frac{\sqrt{n}(\overline{x} - \mu)}{s}$$

has a t-distribution with df $= n - 1$.

Example 9.12 **Standardized Mean Weights** In Section 9.5, we considered four hypothetical samples of $n = 25$ people who were trying to lose weight at a clinic. We played the role of the all-knowing sage and assumed that we knew that $\mu = 8$ and $\sigma = 5$. If the value for μ is correct, then the standardized statistic

$$t = \frac{\overline{x} - \mu}{s/\sqrt{n}} = \frac{\overline{x} - 8}{s/\sqrt{25}}$$

has a t-distribution with df $= 25 - 1 = 24$. If we were to generate thousands of random samples of size 25 and draw a histogram of the resulting standardized t-statistics, they would adhere to this t-distribution.

In practice, we do not draw thousands of samples, and we do not know μ. Suppose we speculated that $\mu = 8$ pounds and drew one random sample, the first one given in Table 9.3, for which $\overline{x} = 8.32$ pounds and $s = 4.74$ pounds. Are the sample results consistent with the speculation that $\mu = 8$ pounds? In other words, is a sample mean of $\overline{x} = 8.32$ pounds reasonable to expect if $\mu = 8$ pounds? The standardized statistic is

$$t = \frac{\overline{x} - \mu}{s/\sqrt{n}} = \frac{\overline{x} - 8}{s/\sqrt{25}} = \frac{8.32 - 8}{4.74/5} = 0.34$$

This statistic tells us that the sample mean of 8.32 is only about one-third of a standard error above 8, which is certainly consistent with a population mean weight loss of 8 pounds.

When to Use z and When to Use t for Standardized Statistics

The need to use Student's t instead of z as the standardized statistic applies only to means. It doesn't apply to proportions. We use Student's t when we replace the population standard deviation σ with its estimate, the sample standard deviation s. Thus, when the parameter of interest is μ, μ_d, or $\mu_1 - \mu_2$, the standardized statistic is a t-statistic when the denominator is the *standard error* of the sample statistic. The standardized statistic is a z-statistic when the denominator is the *standard deviation* of the statistic. For means (but not for proportions), the basic distinction between the use of z and the use of t is as follows:

$$z = \frac{\text{Sample statistic} - \text{Population parameter}}{\text{s.d.(sample statistic)}}$$

$$t = \frac{\text{Sample statistic} - \text{Population parameter}}{\text{s.e.(sample statistic)}}$$

Note that the difference has to do with the denominator of the standardized statistic. Remember, for proportions, the standardized statistic is a *z*-statistic, whether the denominator is the standard deviation of the sample statistic or the standard error. Details on what to use as degrees of freedom for *t* in each case (one mean, mean of paired difference, and difference in two means) will be covered in Chapters 11 and 13.

Areas and Probabilities for Student's *t*-Distribution

Because Student's *t*-distribution differs for each possible df value, we cannot summarize the probability areas in one table as we could for the standard normal distribution. We would need a separate table for each possible df value. Instead, tables for the *t*-distribution are tailored to specific uses. We will explore two types of *t*-distribution tables in Chapters 11 and 13.

Many calculators and computer software programs provide probabilities (areas) for specified *t*-values and *t*-values for specified areas. For example, the Excel command TDIST(k,d,1) provides the area above the value *k* (for positive values of *k*) for a *t*-distribution with degrees of freedom *d*. (The reason for including the third value, 1, will become clear in Chapter 13.) In other words, it provides $P(t > k)$. For the situation in Example 9.12, TDIST(0.34, 24,1) = .3684. Figure 9.12 illustrates this area.

9.9 Exercises are on pages 362–364.

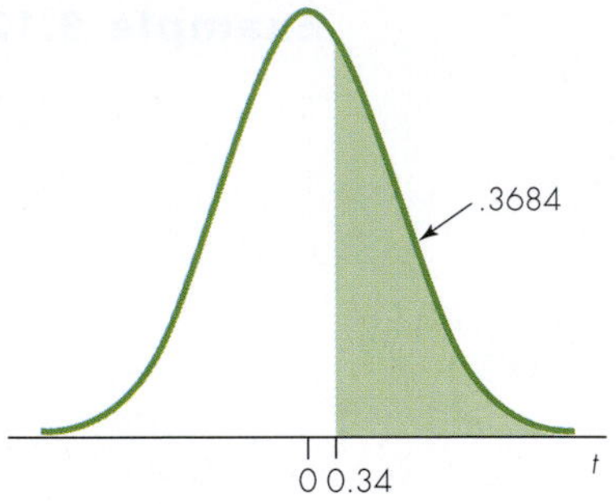

Figure 9.12 A *t*-distribution with df = 24, illustrating the location of $t = 0.34$

EXCEL TIP

To find a *t*-value corresponding to a specified area, the Excel command is a bit tricky. The command TINV(p,d) provides the value *k* from the *t*-distribution with df = *d* for which *p* is the *combined* area in the region above *k* and below $-k$. For instance, TINV(.05,24) = 2.06. This indicates that the area below −2.06 plus the area above 2.06 is .05, for a *t*-distribution with df = 24. Note that this also implies that 95% of the area is between −2.06 and +2.06, in contrast to the standard normal curve, for which 95% of the area is between −1.96 and 1.96 (about −2 and +2).

9.10 Generalizations beyond the Big Five

The five situations that we have covered so far in this chapter are all special cases of more general results. In this section, we will describe two important results in statistics, the *Law of Large Numbers* and the *Central Limit Theorem*, and show how they are related to the five situations we have discussed thus far. We also discuss sampling distributions for other sample statistics and illustrate, by example, that it is not always the case that the sampling distribution of a statistic is approximately normal.

Sampling for a Long, Long Time: The Law of Large Numbers

The sampling distribution for sample means tells us how much variability in possible sample means to expect for different sample sizes. There is a much simpler technical result in statistics called the **Law of Large Numbers**, which guarantees that the sample mean $\bar{x}$ will eventually get "close" to the population mean μ, *no matter how small a difference* you use to define *close*. The result requires that the population mean be finite (not infinite) and that the observations in the sample be independent of each other. "Eventually" could be a very long time, depending on how much variability there is in the population of measurements. In practice, the Law of Large Numbers says that for any specific population, the larger the sample size, the more you can count on $\bar{x}$ to be an accurate representation of μ.

The Law of Large Numbers holds whether the measurements are continuous or discrete. For instance, suppose a slot machine that costs \$1.00 per play pays out nothing 90% of the time and pays out \$5.00 the remaining 10% of the time. Then the mean payout per play is (.9)(0) + (.1)(\$5) = \$0.50 or 50 cents, for a net loss of 50 cents (because it costs \$1 to play). The Law of Large Numbers tells us that if you play long enough on this machine, the average amount "won" per play will be extremely close to a loss of 50 cents. That does not apply to the short run; you might lose \$15 before you ever win anything, or you might win \$10 in your first two plays. But in the long run, your average will be a loss of 50 cents a play.

The Law of Large Numbers is used by insurance companies, casinos, and other businesses to give them peace of mind about what will happen to their average profit (or loss) in the long run. They know that eventually, after enough gamblers have bet or enough customers have bought insurance and claims have been paid, they can count on the mean net profit to be close to the theoretical expected value or mean. This peace of mind comes at a price; they must make sure that they have enough cash on hand to pay claimants or gamblers through the natural fluctuations in the process.

Example 9.13 **The Long Run for the Decco Lottery Game** In Chapter 8, we computed the expected value of the California Decco lottery game and found that the state gains an average of \$0.35 for every dollar played. In other words, over all tickets sold, the state pays out an average of \$0.65 and keeps \$0.35. Obviously, this average is not achieved in the short run. The state knows that to achieve this average, it must sell hundreds of thousands of tickets. For instance, the first ten tickets sold might all be losers, or there might be a \$5000 prize winner. The state must be prepared to pay winnings in the short run. States and casinos that operate gambling games have large funds they can use to pay winning players. For this reason, even when a game is an even bet, a gambler with limited funds will still eventually lose while the "house" weathers the lows (and highs).

The Central Limit Theorem

Although it will not be immediately obvious, all of the results for the five statistics considered in this chapter follow from a result called the **Central Limit Theorem** (CLT) and the results for linear combinations of random variables given in Chapter 8. The CLT derives its name from the fact that it is indeed one of the most important, or central, results in statistics. The term *limit* in mathematics is commonly used to describe what happens when something gets arbitrarily large or, in technical terms, as it approaches infinity. The word "limit" in the CLT is used because the result describes what happens when the sample size or sizes get large.

DEFINITION The **Central Limit Theorem** states that if n is sufficiently large, the sample means of random samples from a population with mean μ and finite standard deviation σ are approximately normally distributed with mean μ and standard deviation $\sigma/\sqrt{n}$.

Technical Note: The mean and standard deviation given in the Central Limit Theorem hold for any sample size; it is only the "approximately normal" shape that requires n to be sufficiently large.

You may note that the Central Limit Theorem is nothing more than a restatement of the sampling distribution for sample means, except that in the latter, we specified the conditions that are needed to satisfy "if n is sufficiently large." The Central Limit Theorem is often used in situations in which it isn't obvious that the statistic of interest is actually a sample mean. It may be obvious that the sampling distributions of $\bar{x}$ and $\bar{d}$ follow from the Central Limit Theorem, but it may not be obvious that the sampling distribution for the sample proportion $\hat{p}$ does too.

The sampling distribution for a sample proportion follows from the Central Limit Theorem by defining each observation in the sample to be either 1 or 0. The observation is 1 if the individual sampled has the desired trait or opinion; the observation is 0 otherwise. The sample mean is simply the average of the 1s and 0s in the sample, which is the sample proportion $\hat{p}$. The population mean = p, the proportion of 1s in the population. The population standard deviation is the standard deviation of the 1s and 0s in the population, which can be shown to be $\sigma = \sqrt{p(1-p)}$, so $\sigma/\sqrt{n} = \sqrt{p(1-p)/n}$, which you should recognize as s.d.($\hat{p}$).

THOUGHT QUESTION 9.4 Verify that if the raw data for each individual in a sample is 1 when the individual has a certain trait and 0 otherwise, then the sample mean is equivalent to the sample proportion with the trait. You can do this by using a formula, explaining it in words, or constructing a numerical example.*

Example 9.14 **California Decco Losses** In Chapter 8 and in Example 9.13 of this section, we discussed the California Decco lottery game, for which the mean amount players lose per ticket over millions of tickets sold is $\mu = \$0.35$. The standard deviation is $\sigma = \$29.67$, reflecting the large variability in possible amounts won or lost, ranging from a net win of \$4999 to a net loss of \$1. Suppose that a store sells 100,000 tickets in a year. What are the possible values for the average amount players lose per ticket for that store? The Central Limit Theorem tells us the answer: Assuming that $n = 100{,}000$ is sufficiently large, the distribution of possible sample means is approximately normal, with a mean (loss) of \$0.35 and a standard deviation of $\sigma/\sqrt{n} = \$29.67/\sqrt{100{,}000} = \$0.09$. According to the Empirical Rule, the mean loss is almost surely in the range \$0.35 ± 3(\$0.09), or between 8 cents and 62 cents.

Remember that this is the average loss per ticket for a collection of 100,000 tickets. Note that the *total* loss for the 100,000 tickets is therefore likely to be between (\$0.08 × 100,000) and (\$0.62 × 100,000), or between \$8000 and \$62,000! Certainly, you can think of better ways to invest \$100,000.

Mean or Median as "Central"?

The Central Limit Theorem provides information about what to expect of the mean in large samples, but the median remains the statistic of choice for representing the "central" value in an extremely skewed set of numbers. For instance, consider two random samples of net winnings for ten Decco tickets:

Net Amount Won (n = 10)	Mean	Median
−1, −1, −1, −1, −1, −1, −1, −1, 0, 0	−\$0.80 (80 cent loss)	−\$1.00 (\$1.00 loss)
−1, −1, −1, −1, −1, −1, −1, 0, 0, 4999	\$499.20 (won)	−\$1.00 (\$1.00 loss)

Clearly, the median is a better measure of the "typical" net win in ten plays of this game. The problem is that for the occasional sample with a large outlier, the mean will be inflated, but this is not representative of the entire population. Remember that in this case, the population mean is 35 cents lost, and the population median is −\$1.00. In fact, 72.6% of all tickets (as shown in Chapter 8) result in \$1.00 lost, which is written equivalently as a −\$1.00 net win.

***HINT** : Write out a short list of 0s and 1s; then find the mean of the list, and compare that to the proportion of 1s in the list.

THOUGHT QUESTION 9.5 The Central Limit Theorem does not specify what is meant by "a sufficiently large sample." What factor(s) about the population of values do you think determine how large is large enough for the approximate normal shape to hold? Consider the California Decco example. Do you think $n = 30$ would be large enough for the distribution of possible values for the average loss to be approximately normal? Why or why not?

Now consider the handspan measurements of females. Do you think $n = 30$ would be large enough for the approximate normal shape to hold? What is different about these two examples?*

Sampling Distribution for Any Statistic

Every statistic has a sampling distribution, but the appropriate distribution may not always be normal or even be approximately bell-shaped. Statisticians have developed a theory that helps to determine the appropriate sampling distribution for many common statistics. It is also possible to construct an approximate sampling distribution for a statistic by actually taking repeated samples of the same size from a population and constructing a relative frequency histogram or table for the values of the statistic over the many samples. The next example illustrates this method.

Example 9.15 **Winning the Lottery by Betting on Birthdays** In the Pennsylvania Cash 5 lottery game, players select five numbers from the integers 1 to 39, and the grand prize is won by anyone who selects the same five numbers as the ones randomly drawn by the lottery officials. Suppose that in such a game, someone bets numbers corresponding to the days of the month on which five family members were born, which is anecdotally a common strategy. Obviously, the highest integer that someone using that strategy could select is 31, so there is no chance of winning the grand prize at all if the highest of the winning draws is bigger than 31—that is, if it is 32 to 39. What is the probability of that happening?

The statistic in question is the *highest* of five integers randomly drawn without replacement from the integers 1 to 39—let's call it H. For instance, if the numbers selected are 3, 12, 22, 36, and 37, then $H = 37$. Determining the actual sampling distribution of H is a complicated exercise in probability. However, we can approximate the sampling distribution of H using lottery data supplied by the Pennsylvania Lottery Commission for repeated plays of the Cash 5 game.

The relative frequency histogram in Figure 9.13 shows the values of H for the 1560 games between April 23, 1992, and May 18, 2000, obtained from the website http://www.palottery.com. (The game was played weekly at first, then daily.) Table 9.4 displays some of the possible values of H and the frequency with which they occurred in the 1560 games. For instance, for one game, the highest of the five numbers was only 10, while for 70 out of the 1560 games (about 4.5% of the time), the high was 30. (Examples noted in this paragraph are in boldface in the table to help you find them.) Note that the high was 31 or less for 444 of the games, which represents 28.46% of the games. In other words, all of the winning numbers are contained in the integers 1 to 31 in just over 28% of the games. For the remaining 72% of the games, the highest number is over 31. In fact, the most common outcome is $H = 39$, the highest possible number, which occurred in 211/1560, or about 13.5% of all games played.

*HINT: See Table 8.1 (p. 272) for the Decco game probability distribution. See Figure 2.5 (p. 26) in Section 2.5 for a dotplot of female handspans.

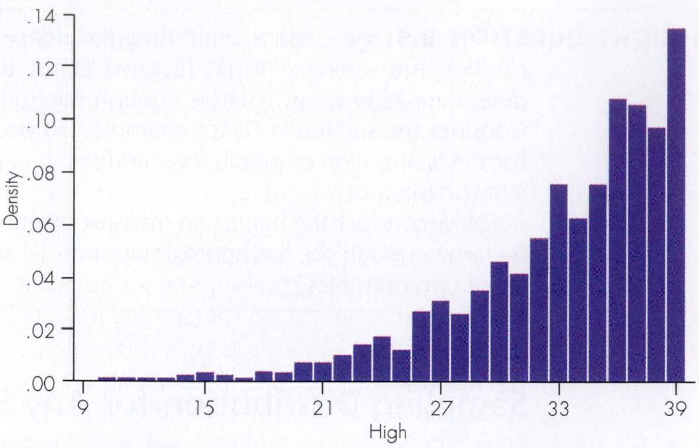

Figure 9.13 Approximate sampling distribution for highest Cash 5 number

Table 9.4 Partial Sampling Distribution for Highest Cash 5 Number

H = high	Count	Cumulative Count	Percentage	Cumulative Percentage
10	**1**	1	0.06	0.06
11	1	2	0.06	0.13
12	1	3	0.06	0.19
13	1	4	0.06	0.26
14	2	6	0.13	0.38
15	3	9	0.19	0.58
16	2	11	0.13	0.71
⋮		⋮	⋮	⋮
29	55	309	3.53	19.81
30	**70**	379	**4.49**	24.29
31	65	**444**	4.17	**28.46**
32	85	529	5.45	33.91
33	118	647	7.56	41.47
34	98	745	6.28	47.76
35	118	863	7.56	55.32
36	169	1032	10.83	66.15
37	165	1197	10.58	76.73
38	152	1349	9.74	86.47
39	**211**	1560	**13.53**	100.00

Note: *Examples from text are in boldface.*

Note that the shape of the probability distribution for H is not symmetric and is skewed to the left. This example illustrates that not all sample statistics have bell-shaped sampling distributions.

THOUGHT QUESTION 9.6 Example 9.15 described a sample statistic, H = highest number drawn, for the Cash 5 lottery game. Give another example of a sample statistic for the Cash 5 game, and describe what you think the shape of its sampling distribution would be.*

It is also possible to construct a sampling distribution for simple situations using the probability rules from Chapter 7. It requires listing all possible values of the sample statistic and finding the probability that each one will be the one achieved when a sample is taken. This can be done for discrete random variables only, because the rules of probability from Chapter 7 cannot readily be applied to continuous random variables.

***HINT:** Some possibilities are the lowest value, the median value, and the mean value.

Example 9.16 **Constructing a Simple Sampling Distribution for the Mean Movie Rating** Sean is lazy about reading, so he does not read movie reviews. Instead, he relies on numerical ratings given by two movie critics, both of whom assign ratings of from one to four stars to the movies they review. Sean decides what movies to see on the basis of the mean of the two ratings. If the mean is at least 3, Sean will go see the movie. Unknown to Sean, the reviewers are lazy too, and they both randomly choose how many stars to give to each movie as follows: The probabilities of assigning one or two stars are .2 each, and the probabilities of assigning three or four stars are .3 each. What is the sampling distribution for the mean number of stars, assuming that the two critics make these assignments independently? What percentage of movies receive a mean rating of at least three stars? (Doesn't it sometimes seem that this is how some critics assign their ratings?) The sample statistic is the mean of the two ratings. There are a total of $4 \times 4 = 16$ combinations of ratings for the two critics. The probability of each combination is found by using the multiplication rule (Rule 3b, p. 233). For instance, the probability that they both assign one star to a movie is $(.2)(.2) = .04$. The sampling distribution for $\bar{x}$ can be found by first listing all of the possible samples of two ratings along with probabilities for the 16 possible samples, as shown in Table 9.5.

Table 9.5 Possible Samples of Two Movie Ratings, Associated $\bar{x}$ and Probability

Sample	1, 1	1, 2	1, 3	1, 4	2, 1	2, 2	2, 3	2, 4	3, 1	3, 2	3, 3	3, 4	4, 1	4, 2	4, 3	4, 4
$\bar{x}$	1	1.5	2	2.5	1.5	2	2.5	3	2	2.5	3	3.5	2.5	3	3.5	4
Probability	.04	.04	.06	.06	.04	.04	.06	.06	.06	.06	.09	.09	.06	.06	.09	.09

The distinct values of $\bar{x}$ can now be listed, with their probabilities found by adding the probabilities for the samples that produced them. For example, two samples produce a mean of 1.5; they are (one star, two stars) and (two stars, one star). Their probabilities are each .04, so the probability that the mean rating is 1.5 stars is $.04 + .04 = .08$. The resulting sampling distribution for $\bar{x}$ is shown in Table 9.6.

Table 9.6 The Sampling Distribution for the Mean Movie Rating, $\bar{x}$

Value of $\bar{x}$	1	1.5	2	2.5	3	3.5	4
Probability	.04	.08	.16	.24	.21	.18	.09

Note that $.21 + .18 + .09 = .48$, or 48% of all movies will pass Sean's test of receiving a mean rating of at least 3.

9.10 Exercises are on pages 365–366.

SKILLBUILDER APPLET

9.11 Finding the Pattern in Sample Means

The **SampleMeans** *applet described in this section is available on the companion website* http://www.cengage.com/statistics/Utts4e.

The main idea for any sampling distribution is that it gives the pattern for how the potential value of a statistic may vary from sample to sample. We learned in Section 9.6 that in two common situations, a normal curve approximates the sampling distribution of the sample mean. The **SampleMeans** applet lets us see the pattern that emerges when we look at the means of many different random samples from the same population. Figure 9.14 illustrates the appearance of the applet when it is started. The histogram shown is a population of individual measurements with $\mu = 8$ and $\sigma = 5$, as they were for the weight-loss example in this chapter. Note that the distribution of individual measurements is bell-shaped, so the sampling distribution of the sample mean should be approximately normal for any sample size. At the start, the number of observations per sample is set at $n = 25$.

(continued)

SKILLBUILDER APPLET (CONTINUED)

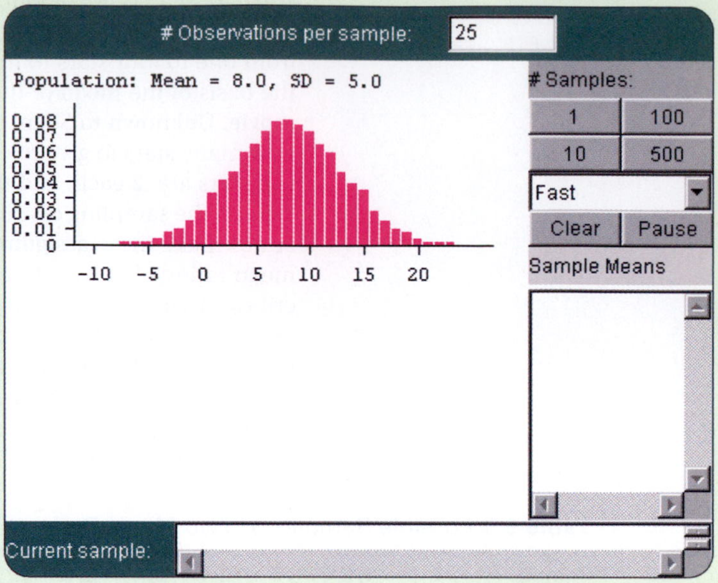

Figure 9.14 SampleMeans applet starting point

What Happens

The applet simulates choosing simple random samples from the population. For each sample selected, the mean is calculated, displayed in the box under "Sample Means" and graphed in a histogram. The individual measurements in the most recently selected sample are shown at the bottom of the display. The buttons under # ***Samples*** can be used to select either 1, 10, 100, or 500 different samples at a time. Figure 9.15 illustrates an example in which two different samples of $n = 25$ per sample have been selected, while Figure 9.16 shows an example in which 500 different samples of $n = 25$ per sample were selected. Note that in Figure 9.15 the two distinct sample means are evident in the bottom histogram, while in Figure 9.16 the histogram of the 500 sample means (the bottom histogram) looks bell-shaped.

You can control the speed of the applet with the menu under the # ***Samples*** buttons. With slow and fast sampling, you see the histogram change as samples are added. With batch sampling, the histogram of sample means is not updated until all requested samples have been selected. The "# Observations per sample" box is used to change the sample size per sample. The ***Clear*** button clears all results and should be used whenever changing the sample size ("# Observations per sample").

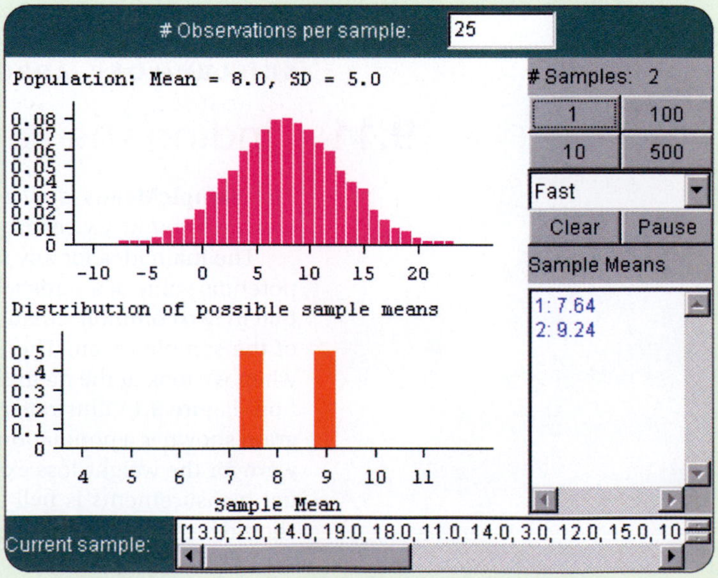

Figure 9.15 Example of SampleMeans applet after two samples of $n = 25$ have been selected

SKILLBUILDER APPLET (CONTINUED)

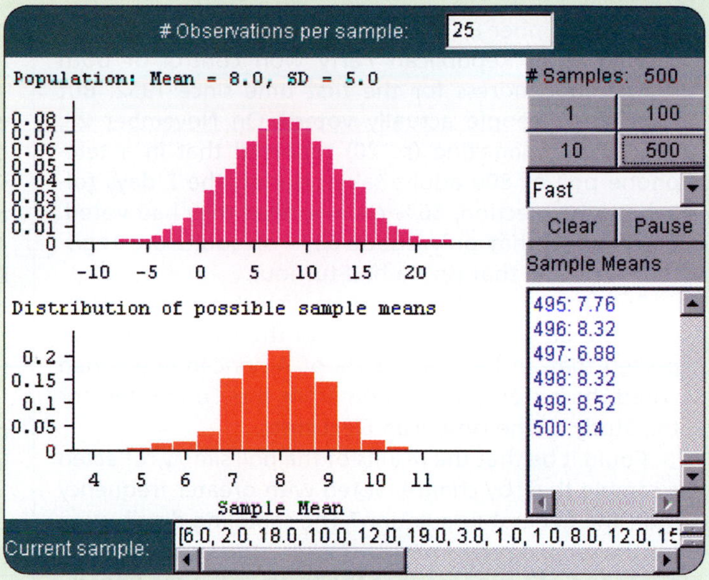

Figure 9.16 Example of SampleMeans applet after 500 samples of $n = 25$ have been selected

What to Do

With the number of observations per sample set at 25, generate many different samples. Take note of the resulting shape of the histogram of sample means. Take note of the center and range of the distribution of sample means. Use the ***Clear*** button to clear the histogram, and then enter 100 as "# Observations per sample." Generate many different samples for samples of this size. Again, take note of the shape, center, and spread of the histogram of sample means. Verify that the range of the histogram of sample means is approximately $\mu \pm 3 \times (\sigma/\sqrt{n})$. Do the same things for "# Observations per sample" = 200. Experiment with the applet to formulate answers to the following questions. What feature of the distribution of sample means is affected by changing the sample size? What features are not affected by changing the sample size? How does changing the sample size affect the chance that a sample mean will fall in the interval 7.5 to 8.5 (within 0.5 pounds of the true mean)?

Lessons Learned

This applet lets you see that a normal curve approximates the distribution of sample means for different random samples of the same size from a population. This is evident from the shape of the histogram for sample means, assuming that you select many different samples of the same size. You can also see that increasing the number of observations in a sample decreases the variability among sample means. Note that regardless of sample size, the histogram of sample means is approximately a normal curve centered at 8 pounds, the population mean. But increasing the sample size increases the likelihood that a sample mean will be close to 8 pounds, the population mean.

9.11 Exercises are on pages 366–368.

CASE STUDY 9.1 Do Americans Really Vote When They Say They Do?

On November 8, 1994, a historic election took place in which the Republican Party won control of both houses of Congress for the first time since 1952. But how many people actually voted? On November 28, 1994, *Time* magazine (p. 20) reported that in a telephone poll of 800 adults taken during the 2 days following the election, 56% claimed that they had voted. Considering that only about 68% of adults are registered to vote, that isn't a bad turnout.

But *Time* reported another disturbing fact along with the 56% turnout claimed in the survey. *Time* reported that, in fact, only 39% of American adults had voted, based on information from the Committee for the Study of the American Electorate.

Could it be that the results of the poll simply reflected a sample that, by chance, voted with greater frequency than the general population? The sampling distribution for the sample proportion in this setting can answer that question. Let's suppose that the truth about the population is, as reported by *Time*, that only 39% of American adults voted, so $p = .39$. Then the sampling distribution for a sample proportion tells us what kind of sample proportions we can expect in samples of 800 adults, the size used by the *Time* poll. The mean of the possibilities is $p = .39$, or 39%. The standard deviation of the possibilities is $\sqrt{(.39)(.61)/800} = .017$, or 1.7%.

Therefore, we are almost certain that the sample percentage based on a sample of 800 adults should fall within $3 \times 1.7\% = 5.1\%$ of the truth of 39%. In other words, if respondents were telling the truth, the sample percentage should be no higher than 44.1%—nowhere near the reported percentage of 56%!

In fact, if we combine this sampling distribution with what we learned about normal curves in Chapter 8, we can say even more about how unlikely this sample result would be. If in truth only 39% (.39) of the population voted, the standardized score for the reported value of 56% (.56) is $z = (.56 - .39)/.017 = 10.0$. We know from Chapter 8 that it is virtually impossible to obtain a standardized score of 10.

Another example of the fact that *reported* voting tends to exceed *actual* voting occurred in the 1992 U.S. presidential election. According to the *World Almanac* (1995, p. 631), 61.3% of American adults reported voting in the 1992 election. In a footnote, the *Almanac* explains:

> Total reporting voting compares with 55.9 percent of population actually voting for president, as reported by Voter News Service. Differences between data may be the result of a variety of factors, including sample size, differences in the respondents' interpretation of the questions, and the respondents' inability or unwillingness to provide correct information or recall correct information.

Unfortunately, because figures are not provided for the size of the sample, we cannot assess whether or not the difference between the actual percentage of 55.9 and the reported percentage of 61.3 can be explained by the natural variability among possible sample proportions.

IN SUMMARY Sampling Distribution Summary Table

Assuming that appropriate conditions are met, the sampling distribution of each of the statistics in this summary table is:

- Approximately normal.
- Mean = parameter.
- Standard deviation is as shown in the table.

The standard deviation is estimated using the standard error. The standardized statistic is always z when the *standard deviation* is used in the denominator. But the standardized statistic is z or t, as shown in the last column, when the standard deviation must be estimated and thus the *standard error* is used in the denominator.

	Parameter	Statistic	Standard Deviation of the Statistic	Standard Error of the Statistic	Standardized Statistic with s.e.
One Proportion	p	$\hat{p}$	$\sqrt{\frac{p(1-p)}{n}}$	$\sqrt{\frac{\hat{p}(1-\hat{p})}{n}}$	z
Difference between Proportions	$p_1 - p_2$	$\hat{p}_1 - \hat{p}_2$	$\sqrt{\frac{p_1(1-p_1)}{n_1} + \frac{p_2(1-p_2)}{n_2}}$	$\sqrt{\frac{\hat{p}_1(1-\hat{p}_1)}{n_1} + \frac{\hat{p}_2(1-\hat{p}_2)}{n_2}}$	z
One Mean	μ	$\bar{x}$	$\frac{\sigma}{\sqrt{n}}$	$\frac{s}{\sqrt{n}}$	t
Mean Difference, Paired Data	μ_d	$\bar{d}$	$\frac{\sigma_d}{\sqrt{n}}$	$\frac{s_d}{\sqrt{n}}$	t
Difference between Means	$\mu_1 - \mu_2$	$\bar{x}_1 - \bar{x}_2$	$\sqrt{\frac{\sigma_1^2}{n_1} + \frac{\sigma_2^2}{n_2}}$	$\sqrt{\frac{s_1^2}{n_1} + \frac{s_2^2}{n_2}}$	t

Key Terms

Section 9.1
parameter, 313, 315
population parameter, 313
statistic, 314, 315
sample statistic, 314, 315
sample estimate, 314, 315
estimate, 314, 315
statistical inference, 314, 315
confidence interval, 314, 315
hypothesis testing, 314, 315
significance testing, 314, 315
null value, 314
statistical significance, 315

Section 9.2
paired differences, 317
population mean for paired differences, 317
dependent samples, 317
independent samples, 317, 329
difference in two population means, 317

Section 9.3
sampling distribution, 321, 323
standard deviation of $\bar{x}$, 323
standard deviation of $\hat{p}$, 323
standard error, 323

Section 9.4
population proportion, 323
sample proportion, 323
sampling distribution for a sample proportion, 324
sampling distribution of $\hat{p}$, 324
standard deviation of $\hat{p}$, 328
standard error of $\hat{p}$, 328

Section 9.5
difference in two population proportions, 328
sampling distribution of $\hat{p}_1 - \hat{p}_2$, 329
standard deviation of $\hat{p}_1 - \hat{p}_2$, 329
standard error of $\hat{p}_1 - \hat{p}_2$, 329
standard error of the difference, 330

Section 9.6
sampling distribution for a sample mean, 333
sampling distribution of $\bar{x}$, 333
sampling distribution of the mean, 333
standard deviation of the sample mean, 333
standard error of the mean, 334

Section 9.7
dependent samples, 336
sampling distribution of $\bar{d}$, 337
standard deviation of $\bar{d}$, 337
standard error of $\bar{d}$, 337

Section 9.8
sampling distribution of $\bar{x}_1 - \bar{x}_2$, 339
standard deviation of $\bar{x}_1 - \bar{x}_2$, 339
standard error of $\bar{x}_1 - \bar{x}_2$, 339

Section 9.9
standardized statistic, 341
standardized *z*-statistic, 341, 343
Student's *t*-distribution, 342
t-distribution, 342
degrees of freedom (df), 342
standardized *t*-statistic, 343

Section 9.10
Law of Large Numbers, 344
Central Limit Theorem, 345

In Summary Boxes

Exercises

◆ Denotes that the dataset is available on the companion website, http://www.cengage.com/statistics/Utts4e, but is not required to solve the exercise.

Bold exercises have answers in the back of the text.

Section 9.1

Skillbuilder Exercises

9.1 In each situation, explain whether the value given in bold print is a statistic or a parameter:

a. A polling organization samples 1000 adults nationwide and finds that **72%** of those sampled favor tougher penalties for persons convicted of drunk driving.

b. In the year 2000 census, the U.S. Census Bureau found that the median age of all American citizens was about **35** years.

c. For a sample of 20 men and 25 women, there is a **14**-cm difference in the mean heights of the men and women.

d. A writer wants to know how many typing mistakes there are in his manuscript, so he hires a proofreader, who reads the entire manuscript and finds **15** errors.

9.2 A polling organization plans to sample 1000 adult Americans to estimate the proportion of Americans who think crime is a serious problem in this country. In the context of this poll, explain the difference between a statistic and a parameter. What is the parameter of interest to the polling organization? What will be the statistic?

9.3 As we learned in Chapter 5, a 95% confidence interval is an interval such that we are 95% confident it covers the truth. Does "the truth" in this definition refer to the truth about a sample statistic or about a population parameter? Or does it depend on the situation? Explain.

9.4 *Part (a) of this exercise is also Exercise 1.6.* Using Case Study 1.6 on pages 5–6 as an example:

a. Explain the difference between a population and a sample.

b. Use the proportion of placebo takers who had heart attacks, which was .017, to explain the difference between a population parameter and a sample statistic.

General Section Exercises

9.5 A report in *Newsweek* about the relationship among diet, genes, and disease (Underwood and Adler, 2005) discussed the E4 allele for a protein known as Apo E and said "an estimated 15 to 30 percent of the population has at least one copy of the allele." People who have the allele are at higher risk for diabetes, heart disease in smokers, and Alzheimer's disease.

a. What is the population parameter of interest in the quote?

b. Would it make more sense to find a confidence interval for the population parameter you described in part (a) or to conduct a hypothesis test for a specific null value for it? Explain. If your answer is "a hypothesis test," specify the null value that you would test.

c. The phrase "an estimated 15 to 30 percent" indicates that scientists don't know precisely what percentage of the population has the allele. Explain why they don't know.

9.6 Some stockholders want to know whether the mean salary for male employees in a large company is higher than the mean salary for female employees. The company allows them access to salary information for a random sample of 100 male and 100 female employees, and the mean salaries are $51,000 for the males and $49,500 for the females.

a. Are the mean salaries in this example statistics or parameters? Explain.

b. Based on these means, can the shareholders determine that the mean salary for males in the company is higher than the mean salary for females? Explain.

c. If they selected new random samples of 100 males and 100 females, would the average salaries of those samples be $51,000 and $49,500? Explain.

9.7 Refer to Exercise 9.6. Suppose that the 100 men and 100 women in the sample were all of the company's employees. Now answer questions a and b, repeated here:

a. Are the mean salaries in this example statistics or parameters? Explain.

b. Based on these means, can the shareholders determine that the mean salary for males in the company is higher than the mean salary for females? Explain.

9.8 In Section 9.1, read the paragraph that begins with "Hypothesis testing is applied to population parameters by specifying a null value..." on page 314. Now refer to Exercise 9.6, for which the population parameter of interest would be the difference in mean salaries for all males and all females in the company.

a. What null value for the population parameter would indicate that there is no problem with equity in salaries?

b. As stated in Exercise 9.6, the mean salaries for a random sample of 100 male and 100 female employees were $51,000 and $49,500, respectively. What numerical value could be used as a sample statistic to estimate the population parameter of interest?

c. Would statistical significance be more likely to be achieved if the sample statistic was higher than the value you specified in part (b), or if it was lower than that value? Explain.

Section 9.2

Skillbuilder Exercises

9.9 Explain whether $\hat{p}$ or p is the correct statistical notation for each proportion described:

a. The proportion that smokes in a randomly selected sample of $n = 300$ students in the eleventh and twelfth grades.
b. The proportion that smokes among all students in the eleventh and twelfth grades in the United States.
c. The proportion that is left-handed in a sample of $n = 250$ individuals.

9.10 Explain whether $\bar{x}$ or μ is the correct statistical notation for each mean described.

a. The mean hours of study per week was 15 hours for a sample of $n = 30$ students at a college.
b. A university administrator determines that the mean age of all students at a college is 20.2 years.

9.11 In a random sample of 300 parents in a school district, 104 supported a controversial new education program. The purpose of the survey was to estimate the proportion of all parents in the school district who support the new program.

a. What is the research question of interest for this survey?
b. What is the population parameter in this study? What is the appropriate statistical notation (symbol) for this parameter?
c. What is the value of the sample estimate (statistic) in this study? What is the appropriate statistical notation (symbol) for this estimate?

9.12 A large Internet provider conducted a survey of its customers. One question that it asked was how many e-mail messages the respondent had received the previous day. The mean number was 13.2.

a. What is the research question of interest for this survey?
b. What is the population parameter in this study? What is the appropriate statistical notation (symbol) for this parameter?
c. What is the value of the sample estimate (statistic) in this study? What is the appropriate statistical notation (symbol) for this estimate?

9.13 A student doing a project for a statistics class wants to estimate the difference between the mean number of music CDs owned by male and female students. He surveys 50 male students and 45 female students about the number of music CDs they own. The mean number for the males is 110, and the mean number for females is 90.

a. What is the research question of interest for this survey?
b. What is the population parameter in this study? What is the appropriate statistical notation for this parameter?
c. What is the value of the sample estimate (statistic) in this study? What is the appropriate statistical notation for this estimate?

9.14 A medical researcher wants to estimate the difference in the proportions of women with high blood pressure for women who use oral contraceptives versus women who do not use oral contraceptives. In an observational study involving a sample of 900 women, the researcher finds that .15 (15%) of the 500 women who used oral contraceptives had high blood pressure, whereas only .10 (10%) of the 400 women who did not use oral contraceptives had high blood pressure.

a. What is the research question of interest for this study?
b. What is the population parameter in this study? What is the appropriate statistical notation for this parameter?
c. What is the value of the sample estimate (statistic) in this study? What is the appropriate statistical notation for this estimate?

General Section Exercises

9.15 A study was done by randomly assigning 200 volunteers with sore throats to either drink a cup of herbal tea or use a throat lozenge to ease their pain. The percentages reporting relief for the two methods were compared. What are the research question, the population parameter, and the sample estimate for this study?

9.16 A sample of 100 students at a university was asked how many hours a week they spent studying and how many they spent socializing. The difference was computed for each student. What are the research question, the population parameter, and the sample estimate for this study?

9.17 Refer to Exercise 9.16. Does the situation described represent independent samples or paired data? Explain.

9.18 For each of the following research questions, would it make more sense to collect independent samples or paired data? Explain.

a. In the United States, on average, what is the age difference between husbands and wives?
b. What is the difference in average ages at which teachers and plumbers retire?
c. What is the difference in average salaries for high school graduates and college graduates?

9.19 Give an example of a research question for which the population parameter of interest would be $p_1 - p_2$.

9.20 Give an example of a research question for which the population parameter of interest would be $\mu_1 - \mu_2$.

9.21 For each of the following situations from earlier chapters, define the parameter of interest and the sample statistic that would be used to estimate that parameter, choosing from among the five described in Section 9.2. When defining the parameter, make sure you explain what constitutes the population.

a. In Case Study 1.3, we learned that 57% of a sample of teens who go out on dates say that they have been out with someone of another race or ethnic group.
b. In Example 2.2, we learned that 155 out of 172, or 90%, of babies who slept in darkness were free of myopia later in childhood.
c. In Example 2.2, we learned that while 90% of the babies in the sample who slept in darkness were free of myopia later in childhood, only 45% of the sample of babies who slept in full light were free of myopia later in childhood.
d. In Example 2.17, we learned that the mean height was 1602 mm for a sample of 199 married British women taken in 1980.

9.22 In Example 9.1 (p. 316), we discussed a few different parameters that might be of interest in exploring the question of whether first-year college students gain weight. In each of the following cases, explain what population(s) the sample(s) should be drawn from and what variable or variables would be measured for each individual in the sample(s).

a. The parameter of interest is p = proportion of the population of first-year college students who weigh more at the end of the year than they did at the beginning of the year.

b. The parameter of interest is μ_d = the mean (average) weight gain during the first year for the population of college students.

c. The parameter of interest is $\mu_1 - \mu_2$ = how much more weight college students gain, on average, during the year than do their contemporaries of the same age.

Section 9.3

Skillbuilder Exercises

9.23 Consider a situation in which a sample of 100 SAT scores is taken from the population of all people who took the test in a given year. Which would be more spread out: the sampling distribution of the sample mean for 100 SAT scores or a histogram of the 100 SAT scores reported in one sample? Explain.

9.24 Explain what symbol ($\hat{p}$, μ, etc.) would be used to represent each of the following:

a. The mean of the sampling distribution of the sample mean.

b. One value from the sampling distribution of the sample mean.

9.25 Explain what symbol ($\hat{p}$, μ, etc.) would be used to represent each of the following:

a. The mean of the sampling distribution of $\hat{p}$.

b. One value from the sampling distribution of $\hat{p}$.

General Section Exercises

9.26 Consider a situation in which a random sample of 1000 adults is surveyed and the proportion that primarily buys organic vegetables is found. If a new random sample of 1000 adults is taken from the same population, explain whether each of the following would change:

a. The population proportion, p.

b. The sample proportion, $\hat{p}$.

c. The standard deviation of $\hat{p}$.

d. The standard error of $\hat{p}$.

e. The sampling distribution of $\hat{p}$, including its shape, mean, and standard deviation.

9.27 Consider a situation in which researchers measure a random sample of 50 newborn babies and the average weight is calculated. If a new random sample of 50 newborns is taken from the same population, explain whether each of the following would change:

a. The population mean, μ.

b. The sample mean, $\bar{x}$

c. The standard deviation of $\bar{x}$.

d. The standard error of $\bar{x}$.

e. The sampling distribution of $\bar{x}$, including its shape, mean, and standard deviation.

9.28 Researchers would like to know the mean body temperature of runners at the completion of a marathon. They plan to take a random sample of 100 runners who complete a marathon and record their temperatures. Which of the following will the researchers know after they have taken the sample? Explain your answer in each part.

a. The value of the mean of the sampling distribution of the sample mean.

b. The approximate shape of the sampling distribution of the sample mean.

c. The standard deviation of the sampling distribution of the sample mean.

9.29 Refer to Exercise 9.28. If a different sample of 100 marathon runners was taken from the same population, which of the following would be the same as it was for the first sample, and which would be different? Explain your answer in each part.

a. The value of the mean of the sampling distribution of the sample mean.

b. The approximate shape of the sampling distribution of the sample mean.

c. The standard deviation of the sampling distribution of the sample mean.

9.30 Refer to Exercises 9.28 and 9.29. Using that context as an example, define in your own words what is meant by a sampling distribution of the sample mean.

9.31 Refer to the smooth curve in Figure 9.1 (p. 321), illustrating the sampling distribution of mean hours of sleep for a sample of 190 college students, taken from a population of sleep hours with a mean of 7.1 hours and standard deviation of 2 hours. Explain whether each of the following would be likely if you were to take a sample of 190 students from the original population.

a. The mean hours of sleep for the 190 students would be between 8 and 9 hours.

b. The mean hours of sleep for the 190 students would be between 7 and 8 hours.

c. The number of hours of sleep reported by the first student in the sample would be between 8 and 9 hours.

d. The maximum number of hours of sleep reported by any student in the sample would be less than 8 hours.

9.32 One result stated in Section 9.3 is that the standard deviation of the sampling distribution gets smaller as the sample size gets larger.

a. Using the context of Example 9.2 (p. 321), which described the sampling distribution of the mean of 190 sleep values, explain why this result makes sense.

b. Consider the extreme case in which everyone in the population is measured. In that case, what would be the standard deviation for the sampling distribution of the sample mean?

Section 9.4

Skillbuilder Exercises

9.33 Calculate the mean and the standard deviation of the sampling distribution of possible sample proportions for each combination of sample size (n) and population proportion (p).

a. $n = 400, p = .5.$
b. $n = 1600, p = .5.$

9.34 Calculate the mean and the standard deviation of the sampling distribution of possible sample proportions for each combination of sample size (n) and population proportion (p).

a. $n = 64, p = .8.$
b. $n = 256, p = .8.$

9.35 Refer to Exercises 9.33 and 9.34, and note that for both exercises the value of n in part (b) is four times what it is in part (a), while p is the same. Use the answers to Exercise 9.33 or 9.34 to describe how changing the sample size affects the mean and standard deviation of the distribution of possible sample proportions.

9.36 In a random sample of $n = 200$ drivers, 50 individuals say that they never wear a seatbelt when driving.

a. Give a numerical value of $\hat{p}$ = sample proportion that never wears a seatbelt when driving.
b. Calculate the standard error of $\hat{p}$
c. In a different survey of $n = 400$ drivers, 100 people say they never wear a seatbelt when driving. Give numerical values for $\hat{p}$ and the standard error of $\hat{p}$.

9.37 A polling organization polls $n = 100$ randomly selected registered voters to estimate the proportion of a large population that intends to vote for Candidate Y in an upcoming election. Although it is not known by the polling organization, $p = .55$ is the actual proportion of the population that prefers Candidate Y.

a. Give the numerical value of the mean of the sampling distribution of $\hat{p}$.
b. Calculate the standard deviation of the sampling distribution of $\hat{p}$.
c. Use the Empirical Rule to find values that fill in the blanks in the following sentence. In about 99.7% of all randomly selected samples of $n = 100$ from this population, the sample proportion preferring Candidate Y will be between ______ and ______.

9.38 In a random sample of 1000 adults aged 18 to 25, 590 individuals said that they drink alcohol at least once a month.

a. Calculate the value of $\hat{p}$ = sample proportion that drinks alcohol at least once a month.
b. Calculate the standard error of $\hat{p}$.

9.39 In a random sample of $n = 500$ adults, 300 individuals say that they believe in love at first sight.

a. Calculate the value of $\hat{p}$ = sample proportion that believes in love at first sight.
b. Calculate the standard error of $\hat{p}$.

General Section Exercises

9.40 Suppose the probability is $p = .2$ that a person who purchases an instant lottery ticket wins money, and this probability holds for every ticket purchased. Consider different random samples of $n = 64$ purchased tickets. Let $\hat{p}$ = sample proportion of winning tickets in a sample of 64 tickets.

a. Give the numerical value of the mean of the sampling distribution of $\hat{p}$.
b. Calculate the standard deviation of the sampling distribution of $\hat{p}$.
c. Use the Empirical Rule to find values that fill in the blanks in the following sentence: In about 68% of all randomly selected samples of $n = 64$ instant lottery tickets, the proportion of tickets that will be money winners will be between ______ and ______.
d. Use the Empirical Rule to find values that fill in the blanks in the following sentence: In about 95% of all randomly selected samples of $n = 64$ instant lottery tickets, the proportion of tickets that will be money winners will be between ______ and ______.

9.41 Suppose medical researchers think that $p = .70$ is the proportion of all teenagers with high blood pressure whose blood pressure would decrease if they took calcium supplements. To test this theory, the researchers plan a clinical trial (experiment) in which $n = 200$ teenagers with high blood pressure will take regular calcium supplements.

a. Assume that $p = .70$ actually is the population proportion that would experience a decrease in blood pressure. What are the numerical values of the mean and standard deviation of the sampling distribution of $\hat{p}$, the sample proportion, for samples of $n = 200$ teenagers?
b. Use the results of part (a) to calculate an interval that will contain the sample proportion $\hat{p}$ for about 99.7% of all samples of $n = 200$ teenagers.
c. In the clinical trial, 120 of the 200 teenagers taking calcium supplements experienced a decrease in blood pressure. What is the value of $\hat{p}$ for this sample? Is this value a parameter or a statistic?
d. Given the answers for parts (b) and (c), explain why the observed value of $\hat{p}$ could be used as evidence that the researchers might be wrong to think that $p = .70$.

9.42 An automobile club comes to the aid of stranded motorists who are members. Over the long run, about 5% of the members utilize this service in any 12-month period. A small town has 400 members of the club. Consider the 400 members to be representative of the larger club membership. Let $\hat{p}$ be the proportion of them who will utilize the service in the coming 12-month period.

a. Explain how Conditions 1 to 3 (p. 325) for the approximate normality of the sampling distribution for a sample proportion are met in this scenario. State any assumptions that you need to make to do so.
b. What are the values of n and p in this situation?
c. Describe the sampling distribution of $\hat{p}$, including values for the mean and standard deviation.

d. The towing service in town contracts with the club to come to the aid of up to 28 members in the next 12-month period. What proportion is that of the 400 members in town? What is the approximate probability that the actual proportion requiring aid will exceed that value?

9.43 Refer to Exercise 9.42, in which $\hat{p}$ is the sample proportion of 400 randomly selected club members who require aid in a 12-month period. Based on the Empirical Rule, into what interval should $\hat{p}$ fall about 95% of the time?

9.44 For each of the following, if the scenario fits the conditions for the approximate normality of the sampling distribution for a sample proportion, describe the sampling distribution of $\hat{p}$ and draw a picture of it. If the scenario doesn't fit, explain why not (in which case you do not have the tools to describe the sampling distribution of $\hat{p}$).

a. An auto insurance company has 1500 customers in a city, and over the long run, 6% of them will file a claim in any given year. Define $\hat{p}$ to be the proportion who file a claim in the next year.

b. A dean has found that over the years, about 15% of students who begin their college work in her program do not finish. This year, 90 students will begin her program; define $\hat{p}$ to be the proportion who will finish.

c. An essay contest has three winners, but the number of entrants varies each time that it is offered, so the proportion of entrants who are winners also varies. Define $\hat{p}$ to be the proportion of entrants who will be winners the next time that the contest is offered.

9.45 Recent studies have shown that about 20% of American adults fit the medical definition of being obese. A large medical clinic would like to estimate what percentage of their patients are obese, so they take a random sample of 100 patients and find that 18 are obese. Suppose that in truth, the same percentage holds for the patients of the medical clinic as for the general population, 20%. Give a numerical value for each of the following:

a. The population proportion of obese patients in the medical clinic.

b. The proportion of obese patients for the sample of 100 patients.

c. The standard error of $\hat{p}$.

d. The mean of the sampling distribution of $\hat{p}$.

e. The standard deviation of the sampling distribution of $\hat{p}$.

9.46 Refer to Exercise 9.45, in which 20% of the patients in a medical clinic are obese. If the clinic took repeated random samples of 100 observations and found the sample proportion who were obese, into what interval should those sample proportions fall about 95% of the time?

9.47 Explain the difference between the *standard error of* $\hat{p}$ and the *standard deviation of the sampling distribution of* $\hat{p}$. Write down the formula for each one. Which one is more likely to be used in practice? Why?

9.48 A student commented, "Once a poll has been taken, $\hat{p}$ (the sample proportion with a certain opinion) is a known value, so it cannot have a distribution of possible values. Therefore, I don't understand what you mean by the sampling distribution of $\hat{p}$." Write an explanation for this student.

Section 9.5

Skillbuilder Exercises

9.49 If the population proportions are equal, what will be the mean of the sampling distribution of $\hat{p}_1 - \hat{p}_2$?

9.50 For each of the following situations, specify the mean of the sampling distribution of the difference in sample proportions:

a. The populations are male and female registered voters in the United States, and it is of interest to compare the proportions of males and females who are registered with the Democratic Party. Suppose that, in truth, 30% of men and 36% of women are registered Democrats. Random samples of 500 will be taken from each population.

b. The populations are adults with college degrees and adults without college degrees. It is of interest to compare the proportions that support the legalization of marijuana. Suppose that, in truth, 30% of those with college degrees support it, and 20% of those without college degrees support it. Random samples of 100 will be taken from each population.

c. The populations are left-handed and right-handed junior high school students. It is of interest to compare the proportions that have learned to play a musical instrument. Suppose that, in truth, 40% of both populations have learned to play a musical instrument. A random sample of 800 right-handed students and 200 left-handed students will be taken.

General Section Exercises

9.51 Refer to Exercise 9.50. In each part, specify the standard deviation of the sampling distribution of the difference in sample proportions.

9.52 Refer to Exercises 9.50 and 9.51. The situations are repeated here. In each case, draw a picture of the sampling distribution of $\hat{p}_1 - \hat{p}_2$, similar to Figure 9.5 (p. 331).

a. The populations are male and female registered voters in the United States, and it is of interest to compare the proportions of males and females who are registered with the Democratic Party. Suppose that, in truth, 30% of men and 36% of women are registered Democrats. Random samples of 500 will be taken from each population.

b. The populations are adults with college degrees and adults without college degrees. It is of interest to compare the proportions that support the legalization of marijuana. Suppose that, in truth, 30% of those with college degrees support it, and 20% of those without college degrees support it. Random samples of 100 will be taken from each population.

c. The populations are left-handed and right-handed junior high school students. It is of interest to compare the proportions that have learned to play a musical instrument. Suppose that, in truth, 40% of both populations have learned to play a musical instrument. A random sample of 800 right-handed students and 200 left-handed students will be taken.

9.53 For each of the following situations, explain whether the conditions hold for the sampling distribution of $\hat{p}_1 - \hat{p}_2$ to be approximately normal:

a. A polling organization wants to compare the proportions of men and women who favor a particular candidate in an upcoming election and plans to take random samples of 1000 men and 1000 women.

b. A researcher plans to take a random sample of 500 children and compare the proportion of them who live with both parents in the household to the proportion who live with just one parent in the household.

c. A company builds an expensive component for automobiles and wants to compare two methods of manufacturing them. The company plans to take a random sample of ten of the components manufactured by using each method and compare the proportion of them that fail from each of the two methods.

9.54 There is some evidence that women tend to vote earlier in the day and men tend to vote later in the day. If men and women vote differently, then exit polls taken early in the day may not reflect the eventual results. Suppose that in a particular election, 51% of the voters who vote in the morning vote for Candidate A, while only 48% of those who vote later vote for Candidate A. An exit poll is based on 500 voters in the morning and 500 voters later in the day. Define $\hat{p}_1$ to be the proportion of the morning sample that voted for Candidate A and $\hat{p}_2$ to be the proportion of the later sample that voted for Candidate A.

a. Based on the morning voters only, describe the sampling distribution of $\hat{p}_1$. Draw a figure similar to Figure 9.2 (p. 325).

b. What is the mean of the sampling distribution of $\hat{p}_1 - \hat{p}_2$?

c. What is the standard deviation of the sampling distribution of $\hat{p}_1 - \hat{p}_2$?

d. Draw a picture of the sampling distribution of $\hat{p}_1 - \hat{p}_2$, similar to Figure 9.5 (p. 331).

e. Using the figure you drew in part (d), sketch the region of the figure that corresponds to having samples with $\hat{p}_1$ being *less* than $\hat{p}_2$.

f. Using your knowledge of how to find areas under normal curves from Chapter 8, find the probability that $\hat{p}_1 - \hat{p}_2 < 0$. In other words, find the probability that the exit poll shows a lower proportion voting for Candidate A in the morning ($\hat{p}_1$) than later in the day ($\hat{p}_2$).

9.55 When taking large random samples from two different populations and finding the sampling distribution of $\hat{p}_1 - \hat{p}_2$ explain which of the following would change if the sample sizes are changed.

a. The mean of the sampling distribution.

b. The approximate shape of the sampling distribution.

c. The standard deviation of the sampling distribution.

9.56 Researchers would like to know whether the proportions of elementary school children who are obese differ in rural and urban areas. They plan to take random samples of 900 children from urban areas and 900 children from rural areas and compare the proportions that are obese. Define Population 1 to be elementary school children in urban areas and Population 2 to be elementary school children in rural areas.

a. Define in words what the parameter p_1 represents in this situation.

b. Define in words what $\hat{p}_1$ is in this situation.

c. Suppose that, unknown to the researchers, p_1 and p_2 are both .20 (20%). Describe the sampling distribution of $\hat{p}_1 - \hat{p}_2$, including its mean, standard deviation, and shape.

d. Draw a picture of the sampling distribution that you described in part (c), similar to Figure 9.5 (p. 331).

9.57 Refer to Exercise 9.56, including the information given in part (c). Which of the following could be specified before the samples are taken? Explain.

a. The value of the mean of the sampling distribution of $\hat{p}_1 - \hat{p}_2$.

b. The approximate shape of the sampling distribution of $\hat{p}_1 - \hat{p}_2$.

c. The standard deviation of the sampling distribution of $\hat{p}_1 - \hat{p}_2$.

d. The standard error of the sampling distribution of $\hat{p}_1 - \hat{p}_2$.

9.58 Refer to Exercises 9.56 and 9.57. Which of the following could be different if two different sets of samples of 900 children are taken? Explain.

a. The value of the mean of the sampling distribution of $\hat{p}_1 - \hat{p}_2$.

b. The approximate shape of the sampling distribution of $\hat{p}_1 - \hat{p}_2$.

c. The standard deviation of the sampling distribution of $\hat{p}_1 - \hat{p}_2$.

d. The standard error of the sampling distribution of $\hat{p}_1 - \hat{p}_2$.

Section 9.6

Skillbuilder Exercises

9.59 Explain whether one (or both) of the situations for the approximate normality of the sampling distribution of the mean holds in each scenario.

a. Mean normal body temperature will be determined for a randomly selected sample of 18 individuals. In the population of all humans, normal body temperature has approximately a normal distribution with mean $\mu = 98.2$ degrees Fahrenheit and standard deviation $\sigma = 0.5$.

b. Mean number of music CDs owned will be determined for a randomly selected sample of four college students. In the population of all college students, the distribution of number of CDs owned is skewed to the right.

c. Refer to part (b). The mean number of music CDs owned will be determined for a randomly selected sample of 900 college students.

9.60 Example 9.2 (p. 321) concerned selecting random samples of $n = 190$ from a population of student responses to the question "How many hours of sleep did you get last night?" Suppose that in a large population of students, the mean amount of sleep the previous night was $\mu = 7.05$ hours and the standard deviation was $\sigma = 1.75$ hours. Consider randomly selected samples of $n = 190$ students.

a. What is the value of the mean of the sampling distribution of possible sample means?

b. Calculate the standard deviation of the sampling distribution of possible sample means.

c. Use the Empirical Rule to find values that fill in the blanks at the end of the following sentence: For 68% of all randomly selected samples of $n = 190$ students, the mean amount of sleep the previous night will be between ______ and ______ hours.

d. Use the Empirical Rule to fill in the blanks at the end of the following sentence. For 95% of all randomly selected samples of $n = 190$ students, the mean amount of sleep will be between ______ and ______ hours.

9.61 The weights of men in a particular age group have mean $\mu = 170$ pounds and standard deviation $\sigma = 24$ pounds.

a. For randomly selected samples of $n = 16$ men, what is the standard deviation of the sampling distribution of possible sample means?

b. For randomly selected samples of $n = 64$ men, what is the standard deviation of the sampling distribution of possible sample means?

c. In general, how does increasing the sample size affect the standard deviation of the sampling distribution of possible sample means? Parts (a) and (b) provide a hint.

9.62 A randomly selected sample of $n = 60$ individuals over 65 years old takes a test of memorization skills. The sample mean is $\bar{x} = 53$, and the standard deviation is $\sigma = 7.2$. Give the numerical value of the standard error of the mean.

9.63 ◆The **cholest** dataset on the companion website includes cholesterol levels measured for 28 heart attack patients 2 days after their attacks. The sample mean for this sample is 253.9, and the standard deviation is 47.7.

a. Assuming that this was a sample from a larger population of heart attack patients, what symbols should be used to represent the values given for the mean and standard deviation?

b. Calculate the standard error of the sample mean.

General Section Exercises

9.64 Vehicle speeds at a certain highway location are believed to have approximately a normal distribution with mean $\mu = 60$ mph and standard deviation $\sigma = 6$ mph. The speeds for a randomly selected sample of $n = 36$ vehicles will be recorded.

a. Give numerical values for the mean and standard deviation of the sampling distribution of possible sample means for randomly selected samples of $n = 36$ from the population of vehicle speeds.

b. Use the Empirical Rule to find values that fill in the blanks in the following sentence: For a random sample of $n = 36$ vehicles, there is about a 95% chance that the mean vehicle speed in the sample will be between ______ and ______ mph.

c. Sample speeds for a random sample of 36 vehicles are measured at this location, and the sample mean is 66 mph. Given the answer to part (b), explain whether this result is consistent with the belief that the mean speed at this location is $\mu = 60$ mph.

9.65 Small planes cannot fly well if the payload (people, luggage, and fuel) weighs too much. Suppose that an airline runs a commuter flight that holds 40 people. The airline knows that the weights of passenger plus luggage for typical customers on this flight is approximately normal with a mean of 210 pounds and a standard deviation of 25 pounds.

a. Draw a picture of this distribution.

b. Describe the sampling distribution of the mean weight of passenger plus luggage for a random sample of 40 customers.

c. Draw a picture superimposing the distributions from parts (a) and (b). (Label it clearly, and remember that the total area under each curve must equal 1.)

d. Assume that customers on any particular flight are similar to a random sample. If the *total* weight of passengers and their luggage should not exceed 8800 pounds, what is the probability that a sold-out flight (40 passengers and their luggage) will exceed the weight limit? (*Hint:* Rewrite the desired limit as an average per passenger.)

9.66 Describe the sampling distribution for the statistic of interest in each of the following stories, including numerical values for the mean and standard deviation of the sampling distribution.

a. A fisherman takes tourists out fishing and has noticed that over the long run, the weights of a certain type of fish that is typically caught is approximately normal with a mean of 10 pounds and a standard deviation of 2 pounds. During a good day, the boatload of eight people catches the limit of 2 fish each, or a total of 16 fish. The statistic of interest is the mean weight of the 16 fish caught on a good day.

b. A fisherman hangs out at the local bar in the evening and hears tourists talking about the weights of the fish they caught that day. He notices that the weights mentioned seem to have a mean of about 20 pounds and a standard deviation of about 4 pounds. He records what he hears as the weights for the next 50 "fish stories" at the bar. The statistic of interest is the mean weight for the 50 stories. Assume that they represent a random sample of all such stories.

9.67 Explain the difference between the standard deviation of the sampling distribution of the mean and the standard error of the mean. Explain which one is more likely to be used in practice and why.

9.68 A rental car company has noticed that the distribution of the number of miles customers put on rental cars per day is somewhat skewed to the right, with an occasional high outlier. The distribution has a mean of 80 miles and a standard deviation of 50 miles.

a. Draw a picture of a normal curve with this mean and standard deviation (80 miles and 50 miles), and use the information in the picture to verify that the distribution of number of miles must be skewed to the right and/or have occasional high outliers rather than being normally distributed.

b. What is the approximate distribution for the *mean* number of miles per day put on a typical one of the company's rental cars in a year (365 days)?

c. What is the approximate distribution for the *total* number of miles put on a typical one of the rental cars in a year (365 days)?

d. Do you think the necessary conditions for the approximate normality of the sampling distribution in part (b) are met in this situation? Explain.

Section 9.7

Skillbuilder exercises

9.69 Twenty sophomores are randomly selected from campus dormitories at a college, and 20 other sophomores are randomly selected from students who live off campus. The mean grade point averages of the two groups are compared. Does this comparison involve paired data or data from two independent samples?

9.70 Thirteen individuals each take a strength test both before and again after they participate in a 6-week weight-training program. The investigator wants to estimate the average gain in strength due to the training program. Does this comparison involve paired data or data from two independent samples?

9.71 Explain which parameter is being described in each of the following situations: μ, μ_d, or $\mu_1 - \mu_2$.

a. The difference in the mean number of pushups new male and female military recruits can do.

b. The mean change in the number of pushups female recruits can do at the end of basic training, compared with the number at the beginning.

9.72 Explain which parameter is being described in each of the following situations: μ, μ_d, or $\mu_1 - \mu_2$.

a. The mean grade point average for all students at your school.

b. The difference in the mean age at marriage of women who were married for the first time in 1950 and women who were married for the first time in 2000.

c. The difference in the mean height of college women and the mean height of their mothers.

9.73 Is the mean of the sampling distribution of $\bar{d}$ always 0? If so, explain why. If not, explain what it means about the population when it is 0.

General Section Exercises

9.74 In a project for a statistics class, a group of students decided to compare a name brand (Brand 1) and a store brand (Brand 2) of microwave popcorn to see which one left fewer unpopped kernels, on average, after the recommended popping time. They knew that microwave ovens can differ, so they asked each of the 50 students in the class to take one bag of each brand home and conduct the test, then report the number of unpopped kernels from each brand.

a. Define the parameter of interest in the context of this situation. Use appropriate notation.

b. Describe the sample statistic of interest in the context of this situation. Use appropriate notation.

c. Would your answers in parts (a) and (b) change if the students had randomly assigned half of the class to test one brand and the other half to test the other brand? Explain.

d. Suppose that, in truth, the store brand leaves an average of ten fewer unpopped kernels and that the standard deviation for the population of differences is five kernels. Describe the sampling distribution of the statistic you defined in part (b), giving numerical values to the extent possible.

e. Draw a picture of the sampling distribution you found in part (d).

f. Using the information provided in part (d) would you be surprised if a sample of 50 differences collected in this way resulted in a higher mean number of unpopped kernels for the store brand than for the name brand? Explain. You can explain in words or use your figure from part (e) to illustrate the situation.

9.75 Horne, Reyner, and Barrett (2003) describe a study in which 12 men took a simulated driving test after drinking alcohol at lunch, getting too little sleep the night before, neither (control condition), or both. Each of the men took the test under all four conditions. One of the measurements taken under each condition was the number of times the driver drifted out of his lane during the 2-hour test. For this problem, consider the mean difference in number of lane drifts between the condition with alcohol but enough sleep and the condition with no alcohol but too little sleep.

a. Define the parameter of interest in the context of this situation. Use appropriate notation.

b. Describe the sample statistic of interest in the context of this situation. Use appropriate notation.

c. Given that the sample size is only 12, describe what is required so that the sampling distribution for the statistic described in part (b) will be approximately normal.

d. Suppose that getting too little sleep and drinking alcohol at lunch have the same effect on the mean number of lane drifts in the 2-hour period and that the standard deviation for the population of differences is 5. Describe the sampling distribution of the statistic you gave in part (b), assuming that the situation you described in part (c) holds.

Section 9.8

Skillbuilder Exercises

9.76 Is the mean of the sampling distribution of $\bar{x}_1 - \bar{x}_2$ always 0? If so, explain why. If not, explain what condition(s) would be necessary for it to be 0.

9.77 In situations for which the parameter of interest is the difference in population means for independent samples, is it necessary for the two samples sizes n_1 and n_2 to be equal in order for the sampling distribution of the sample statistic to be approximately normal? If not, what condition is required for the sample sizes?

General Section Exercises

9.78 Volunteers who had developed a cold within the previous 24 hours were randomized to take either zinc or placebo lozenges every 2 to 3 hours until their cold symptoms were gone (Prasad et al. 2000). Twenty-five participants took zinc lozenges, and 23 participants took placebo lozenges. The mean overall duration of symptoms was found for each group.

a. What is the parameter of interest in this situation? Give notation and define it.

b. What is the sample statistic in this situation? Give notation and define it.
c. What condition(s) must be met for the sampling distribution of the statistic you defined in part (b) to be approximately normal?

9.79 Refer to Exercise 9.78, in which 25 participants took zinc lozenges and 23 participants took placebo lozenges. Suppose that, in truth, if everyone in the population took zinc lozenges, symptoms would last an average of 4.5 days with a standard deviation of 1.5 days, while if they took placebo lozenges, symptoms would last an average of 8 days with a standard deviation of 2 days.

a. What is the mean of the sampling distribution of the difference in sample means? (Use placebo − zinc.)
b. What is the standard deviation of the sampling distribution of the difference in sample means?
c. Assuming that the conditions are met to make it approximately normal, draw a picture of the sampling distribution of the difference in sample means in this situation, similar to Figure 9.10 (p. 340).
d. In the study by Prasad et al. (2000), the sample mean duration of symptoms was 4.5 days for the 25 participants who took zinc lozenges and 8.1 days for the 23 participants who took placebo lozenges, for a difference of 3.6 days. Is this difference reasonable based on the figure you drew in part (c)? You can explain in words or show where the value falls on the figure.

9.80 Refer to Exercises 9.78 and 9.79. Suppose that in truth, there is no difference in the mean duration of symptoms that would occur if everyone in the population were to take zinc lozenges and the mean duration if everyone were to take placebo lozenges. Assume that the standard deviations for duration of symptoms are as given in Exercise 9.79. Answer the same questions as in Exercise 9.79 (repeated here) for this scenario, and then answer part (e).

a. What is the mean of the sampling distribution of the difference in sample means?
b. What is the standard deviation of the sampling distribution of the difference in sample means?
c. Assuming that the conditions are met to make it approximately normal, draw a picture of the sampling distribution of the difference in sample means in this situation, similar to Figure 9.10 (p. 340).
d. In the study by Prasad et al. (2000), the sample mean duration of symptoms was 4.5 days for the 25 participants who took zinc lozenges and 8.1 days for the 23 participants who took placebo lozenges, for a difference of 3.6 days. Is this difference reasonable on the basis of the figure you drew in part (c)? You can explain in words or show where the value falls on the figure.
e. Based on the result in part (d), do you think it is reasonable to assume that taking zinc lozenges and taking placebo lozenges would result in the same mean duration of symptoms for the population? Explain.

9.81 The mean and standard deviation for the heights of adult men are about 70 inches and 3 inches, respectively; for adult women, they are about 65 inches and 2.5 inches, respectively. Heights within each sex are bell-shaped. Suppose random samples of nine men's and nine women's heights are measured and the difference in the sample means is found (men − women).

a. What is the mean of the sampling distribution of $\bar{x}_1 - \bar{x}_2$ in this situation?
b. What is the standard deviation of the sampling distribution?
c. Draw a picture of the sampling distribution of $\bar{x}_1 - \bar{x}_2$, similar to Figure 9.10 (p. 340).
d. Is it possible that the mean height for the sample of women will be greater than the mean height for the sample of men? Explain.

Section 9.9

Exercises 9.82 to 9.112 correspond to the three lessons in Section 9.9. Lesson 1 exercises are 9.82 to 9.91; Lesson 2 exercises are 9.92 to 9.96; Lesson 3 exercises are 9.97 to 9.112.

Lesson 1 Skillbuilder Exercises

9.82 In addition to the sample size n, what population value(s) must be known to find a standardized score for a sample proportion $\hat{p}$?

9.83 In addition to the sample size n, what population value(s) must be known to find a standardized score for a sample mean $\bar{x}$?

9.84 Using the formula

$$z = \frac{\text{Sample statistic} - \text{Population parameter}}{\text{s.d.(sample statistic)}}$$

give the formula for the standardized statistic in each of the following cases.

a. The sample statistic is $\bar{x}_1 - \bar{x}_2$.
b. The population parameter is $p_1 - p_2$.
c. The sample statistic is $\bar{d}$.

Lesson 1 General Section Exercises

9.85 In comparing two proportions, suppose that $p_1 = p_2 = .5$ and $n_1 = n_2 = 100$.

a. Draw a picture of the appropriate sampling distribution, and show where the value $\hat{p}_1 - \hat{p}_2 = 0.13$ falls on it.
b. Find the value of the standardized statistic corresponding to $\hat{p}_1 - \hat{p}_2 = 0.13$.
c. Draw a picture of the standard normal distribution with the standardized statistic you found in part (b) shown on it. Compare it to your picture in part (a).

9.86 In a situation for which paired differences are appropriate, suppose that $\mu_d = 0$, $\sigma_d = 10$, and $n = 25$.

a. Draw a picture of the appropriate sampling distribution, and show where the value $\bar{d} = -3$ falls on it.
b. Find the value of the standardized statistic corresponding to $\bar{d} = -3$.
c. Draw a picture of the standard normal distribution with the standardized statistic that you found in part (b) shown on it. Compare it to your picture in part (a).

9.87 In comparing means for two independent samples, suppose that $\mu_1 - \mu_2 = 0$, $\sigma_1 = \sigma_2 = 3$, and $n_1 = n_2 = 30$.

a. Draw a picture of the appropriate sampling distribution, and show where the value $\bar{x}_1 - \bar{x}_2 = 2$ falls on it.

b. Find the value of the standardized statistic corresponding to $\bar{x}_1 - \bar{x}_2 = 2$.

c. Draw a picture of the standard normal distribution with the standardized statistic you found in part (b) shown on it. Compare it to your picture in part (a).

9.88 Refer to Exercise 9.56, in which the proportions of obese children in urban and rural areas are to be compared. Random samples of 900 children from each type of area are taken. Suppose that in fact, 20% of the children are obese in both populations.

a. Give numerical values for the mean and standard deviation of the sampling distribution of $\hat{p}_1 - \hat{p}_2$.

b. Suppose that the samples are taken and the difference in sample proportions is 0.05. Find the standardized statistic corresponding to this difference.

c. Draw a picture of the sampling distribution, and illustrate where the difference of 0.05 falls on it.

d. If the population proportions really are equal, is a difference in sample proportions of 0.05 plausible? Explain.

9.89 Refer to Exercise 9.75, in which 12 drivers were tested after getting too little sleep and after drinking alcohol. In that exercise, we assumed that the mean difference in erroneous lane drifts during a 2-hour driving period would be 0 for the population, with a standard deviation of 5. Assume that the population of possible differences is bell-shaped.

a. Describe the sampling distribution of $\bar{d}$ in this situation, including its mean and standard deviation.

b. Suppose that the sample mean difference $\bar{d}$ in number of lane drifts was 8, with more lane drifts after drinking alcohol than after getting too little sleep. Would you still believe that the mean difference in the population is 0? Explain, using the standardized statistic as part of your explanation.

9.90 Suppose that a new diet barely works and that if everyone in the population of overweight adults were to follow it, the weight losses would be bell-shaped with a mean of 0.1 pound and a standard deviation of 5 pounds.

a. Describe the sampling distribution of the sample mean for $n = 100$.

b. Find the standardized statistic corresponding to a sample mean weight loss of 1 pound.

c. Using your answer in part (b), comment on whether a sample mean of 1 pound would be unusual in this situation.

d. Now suppose that the sample size is increased to $n = 900$. Repeat parts (a), (b), and (c) in that case. In particular, comment on whether a sample mean of 1 pound would be unusual.

e. Compare your answers to parts (c) and (d). Use the comparison to make a general statement about how sample size affects standardized statistics related to the sampling distribution of the sample mean.

9.91 Refer to Exercises 9.79 to 9.80, discussing research on whether zinc lozenges helped to reduce the duration of cold symptoms. In the actual experiment (Prasad et al. 2000), the difference in sample means was 3.6 days. Suppose that the population standard deviations and sample sizes are as given in Exercise 9.79, with $\sigma_1 = 1.5$, $n_1 = 25$, $\sigma_2 = 2$, and $n_2 = 23$. Assume that the populations of durations are approximately bell-shaped.

a. Describe the sampling distribution of the difference in sample means in this situation, assuming that the difference in population means is 0. (Use placebo − zinc)

b. Find the standardized statistic corresponding to the sample difference actually observed, 3.6 days.

c. Suppose that a skeptic were to assert that zinc lozenges cannot possibly reduce the duration of cold symptoms and that the difference in average duration of symptoms of 3.6 days observed in the study was just a statistical fluke. Using your result in part (b), argue for or against this assertion.

Lesson 2 Skillbuilder Exercises

9.92 On the basis of past history, a car manufacturer knows that 10% ($p = .10$) of all newly made cars have an initial defect. In a random sample of $n = 100$ recently made cars, 13% ($\hat{p} = .13$) have defects. Find the value of the standardized statistic (z-score) for this sample proportion.

9.93 In each part, give the value of the standardized statistic (z-score) for the sample proportion.

a. $\hat{p} = .60$, $p = .50$, $n = 100$.

b. $\hat{p} = .60$, $p = .50$, $n = 200$.

9.94 In each part, give the value of the standardized statistic (z-score) for the sample proportion.

a. $\hat{p} = .78$, $p = .80$, $n = 400$.

b. $\hat{p} = .82$, $p = .80$, $n = 400$.

c. $\hat{p} = .25$, $p = .40$, $n = 900$.

Lesson 2 General Section Exercises

9.95 Suppose that an ESP test consists of $n = 75$ independent tries, for which there are four possible choices on each try. Suppose that someone is just guessing, so the probability of a correct guess on each try is $p = .25$. Consider two possible sample proportions: $\hat{p} = .20$ (15 correct) and $\hat{p} = .33$ (25 correct).

a. Find the standardized z-statistics for $\hat{p} = .20$ and $\hat{p} = .33$.

b. Draw a picture of the sampling distribution of $\hat{p}$, and identify where $\hat{p} = .20$ and $\hat{p} = .33$ fall on the picture.

c. Draw a picture of the standard normal curve, and identify where the standardized statistics found in part (a) fall on the picture.

d. What is the relationship between the pictures in parts (b) and (c)?

9.96 A blind taste test is done to compare Cola 1 and Cola 2. Among $n = 75$ participants, $\hat{p} = .64$ is the proportion preferring Cola 1. If, actually, the two colas are equally preferred in a larger population of tasters, $p = .5$ is the corresponding population proportion. Assuming that $p = .5$, find the value of the standardized statistic (z-score) for the sample proportion observed in the taste test.

Lesson 3 Skillbuilder Exercises

9.97 In each part, give the value of the standardized statistic (z-score) for the sample mean:

a. $\bar{x} = 74$, $\mu = 72$, $\sigma = 10$, $n = 25$.

b. $\bar{x} = 70$, $\mu = 72$, $\sigma = 10$, $n = 25$.

9.98 In each part, give the value of the standardized statistic (z-score) for the sample mean:

a. $\bar{x} = 92$, $\mu = 100$, $\sigma = 16$, $n = 4$.
b. $\bar{x} = 92$, $\mu = 100$, $\sigma = 16$, $n = 64$.
c. $\bar{x} = 10.15$, $\mu = 10.0$, $\sigma = 5$, $n = 10{,}000$.

9.99 Home prices in a city have mean $\mu = \$200{,}000$ and standard deviation $\sigma = \$25{,}000$. For $n = 25$ randomly selected homes, the mean price is $\bar{x} = \$208{,}000$. Find the value of the standardized statistic (z-score) for this sample mean.

9.100 Explain whether or not each of the following would ever differ for two random samples of the same size from the same population.

a. The sample mean $\bar{x}$.
b. The standard deviation of the sampling distribution of $\bar{x}$.
c. The standard error of $\bar{x}$.
d. The standardized z-score for the observed value of $\bar{x}$.
e. The population mean μ.

9.101 In each situation, find the value of the t-statistic for the sample mean $\bar{x}$ and give the value of degrees of freedom (df).

a. $\bar{x} = 5$, $\mu = 10$, $\sigma = 20$, $n = 16$.
b. $\bar{x} = 15$, $\mu = 10$, $\sigma = 20$, $n = 16$.

9.102 In each situation, find the value of the t-statistic for the sample mean $\bar{x}$ and give the value of degrees of freedom (df).

a. $\bar{x} = 56$, $\mu = 50$, $\sigma = 15$, $n = 25$.
b. $\bar{x} = 50$, $\mu = 50$, $\sigma = 20$, $n = 9$.

9.103 In each situation, find the value of the standardized statistic for the sample mean, and indicate whether the standardized statistic is a t-statistic or a z-statistic.

a. $\bar{x} = 175$, $\mu = 170$, $s = 24$, $n = 4$.
b. $\bar{x} = 175$, $\mu = 170$, $\sigma = 20$, $n = 4$.
c. $\bar{x} = 161$, $\mu = 170$, $s = 18$, $n = 36$.

9.104 Draw a picture of each of the following distributions. Shade the area above the value given, and label the proportion of the distribution that falls in that region, as in Figure 9.12 (p. 344). You do not have to label the vertical axis. *Note:* For parts (b) and (c), the use of either statistical software or Excel is necessary.

a. Standard normal distribution; $z = 0.70$.
b. t-distribution with df $= 10$, $t = 0.70$.
c. t-distribution with df $= 10$, $t = -0.70$.

9.105 Explain the difference between what is represented by s and σ, the two different symbols for standard deviation.

Lesson 3 General Section Exercises

9.106 Suppose that a random sample of 36 IQ scores is drawn from a population of IQ scores with mean $= 100$ and standard deviation $= 15$.

a. Find the standardized statistic for $\bar{x} = 97$.
b. Find the standardized statistic for $\bar{x} = 105$.
c. Give numerical values for the mean and standard deviation of the sampling distribution of $\bar{x}$.
d. Draw a picture of the sampling distribution of $\bar{x}$, and identify where $\bar{x} = 97$ and $\bar{x} = 105$ fall on the picture.
e. Draw a picture of the standard normal curve, and identify where the standardized statistics found in parts (a) and (b) fall on the picture.
f. What is the relationship between the pictures in parts (d) and (e)?

9.107 In each of the following situations, which of the two distributions would be more spread out?

a. Standard normal distribution or t-distribution with df $= 10$.
b. t-distribution with df $= 5$ or t-distribution with df $= 25$.
c. t-distribution with df $= 100$ or normal distribution with standard deviation $= 100$.

9.108 The mean IQ score in a population is 100. A random sample of 25 IQ scores is taken from this population and the values for the sample mean and standard deviation are found to be $\bar{x} = 105$ and $s = 10$.

a. Compute the standardized statistic for the sample mean in this situation.
b. Draw a picture of the appropriate distribution for the standardized statistic, and find the area below the value of the standardized statistic computed in part (a). (*Note:* The use of either statistical software or Excel is necessary.)

9.109 The mean salary for a population of women who work in a traditional male job is \$80,000. A random sample of 100 women is taken from this population and the sample mean and standard deviation are found to be $\bar{x} = \$78{,}000$ and $s = \$4000$.

a. Compute the standardized statistic for the sample mean in this situation.
b. Draw a picture of the appropriate distribution for the standardized statistic, and find the area below the value of the standardized statistic computed in part (a). (*Note:* The use of either statistical software or Excel is necessary.)

9.110 Refer to Exercise 9.109. Suppose that allegations were made that the (unknown) population mean salary of women who work in that job is lower than the known population mean salary of \$80,000 for men.

a. Does the sample information for the 100 women in Exercise 9.109 support that allegation? Explain.
b. The question in part (a) is an example of hypothesis testing, the goal of which was described in Section 9.1 as "to reject the notion that chance alone can explain the sample results." Explain how the question in part (a) fits that description of hypothesis testing.

9.111 Are there any (finite) values for which the area above that value is the same for all t-distributions? If so, specify the value(s) and the area(s). If not, explain why not.

9.112 Draw a picture illustrating each of the following distributions, showing the t-values marking the specified areas and the proportion of the t-distribution that falls in that area. *Note:* The use of either statistical software or Excel is necessary.

a. t-distribution with df $= 10$, middle 95%.
b. t-distribution with df $= 20$, area above 1.725.
c. t-distribution with df $= 20$, area between -1.725 and 1.725.
d. t-distribution with df $= 10$, upper 5%.
e. t-distribution with df $= 10$, area below $-t$ and above $+t$ (same t) totals 5%.

Section 9.10

Skillbuilder Exercises

9.113 Suppose that a simple random sample of $n = 2$ numbers will be selected from the list of values 1, 3, 5, 7, 9.

a. There are ten possible equally likely samples of $n = 2$ numbers that could be selected (1 and 3, 1 and 5, etc.). List the ten possible samples, and calculate the sample mean $\bar{x}$ for each sample.

b. Summarize the results of part (a) into a table showing the sampling distribution of possible sample means. To do this, list each possible value for the sample mean along with the probability the value would occur.

9.114 Suppose that a simple random sample of $n = 2$ numbers will be selected from the list of values 1, 2, 3, 4, 5.

a. There are ten possible equally likely samples of $n = 2$ numbers that could be selected (1 and 2, 1 and 3, etc.) List the ten possible samples, and give H = highest number for each sample.

b. Summarize the results of part (a) into a table showing the sampling distribution of H. To do this, list each possible value for H along with the probability that the value would occur.

c. Which value is most likely to be the highest number in the sample?

9.115 Refer to Exercise 9.114.

a. For each of the ten possible samples listed in part (a) of Exercise 9.114, give the value of the range $R = H - L$, where H is the highest number and L is the lowest number.

b. Use the results of part (a) to find the probability that $R = 4$.

c. Summarize the results of part (a) into a table showing the sampling distribution of R. To do this, list each possible value for R along with the probability that the value would occur.

d. Which value of R is most likely to occur? Which value is least likely to occur?

General Section Exercises

9.116 A soft drink called Crash is so popular that vending machines allocate the top two selection buttons to it. Over the long run, of those who buy Crash, 60% push the top button and 40% push the lower button, and this behavior seems to be independent from one purchase to the next. Suppose that a sample of $n = 10$ people buy Crash at a vending machine one morning and X = the number who push the top button.

a. Is X a statistic? Explain.

b. Describe the sampling distribution of X.

9.117 In a large retirement community, 60% of households have no dogs, 30% have one dog, and 10% have two dogs. Following a news story about how the elderly have trouble walking their dogs in the winter weather, a magnanimous group of teenagers decides that it will randomly select two households in the retirement community each and call and offer to walk their dog(s). For each teen who participates, T is the *total* number of dogs in the two households called.

a. List all possible values for T.

b. Find the sampling distribution for T. (*Hint:* This is simply a list of possible values for T and their probabilities.)

c. If there are no dogs in the two households ($T = 0$), the teen starts over with a new sample of two households. If 1000 teens participate, for about how many of them will $T = 0$ for their first sample?

9.118 Refer to the discussion of the Cash 5 lottery game in Example 9.15 and the accompanying Table 9.4 (p. 348).

a. In what proportion of the games is the highest number 35 or more?

b. What is the median value of H for the 1560 Cash 5 games represented in Example 9.15?

c. In what proportion of the games is the highest number 15 or less?

9.119 Define L to be the *lowest* of the five numbers drawn in the Cash 5 lottery game in Example 9.15. Describe the shape of the sampling distribution of L, and explain how you know.

9.120 Use a computer to simulate the numbers drawn for the Cash 5 lottery game 1000 times, and draw a histogram for the values of H = highest number drawn. Compare your histogram to the one shown in Figure 9.13 (p. 348).

9.121 The *I Ching* is an ancient Chinese system of asking for advice. In one version, three pennies are flipped, and if the number of heads is odd (1 or 3), a "broken line" is recorded, while if it is even (0 or 2), a "solid line" is recorded. This process is repeated six times, and then the results are used to identify one of the $2^6 = 64$ possible patterns of solid and broken lines and corresponding advice. Assume that the coins are fair and the flips are independent. Define B to be the number of broken lines out of the six.

a. For each of the six lines, what is the probability that it is a broken line?

b. What are the possible values for B?

c. What is the probability that $B = 0$, so that all of the lines are solid?

d. Find the sampling distribution for B.

9.122 Are there situations for which the sampling distribution of a statistic consists of a single possible value, which has probability 1 of occurring each time? If not, explain why not. If so, explain how that could happen and give an example.

9.123 Consider the population of net gains for tickets in the California Decco lottery game described in Examples 9.13 and 9.14. The possible values are \$4999, \$49, \$4, 0, and −\$1; the mean is −\$0.35, and the standard deviation is \$29.67. Suppose that someone buys ten tickets each week.

a. Explain why the Central Limit Theorem would *not* hold for the distribution of possible mean net gains for the weekly purchases of ten tickets. In other words, explain why the sampling distribution of the sample mean is not approximately normal.

b. Although the possible means for samples of ten tickets are *not* approximately normal, you can still specify numerical values for the mean and standard deviation of the sampling distribution of the mean in this situation. What are they?

9.124 Suppose that two fair dice are rolled and H = higher of the two faces. If both dice show the same value, H is that value. Construct the sampling distribution of H, listing all possible values and their probabilities.

9.125 Suppose someone who takes a ten-question true–false test doesn't know any of the answers and randomly selects true or false for each question. Define the statistic X = number correct.

a. Describe the sampling distribution of X. (*Hint:* This distribution was introduced in Chapter 8.)
b. What is the mean of this sampling distribution?
c. What is the standard deviation of this sampling distribution?

9.126 Refer to Exercise 9.125. Now define the statistic of interest to be the *proportion* correct, $X/10$. What is the *mean* of the sampling distribution of this statistic? Explain how you found your answer.

Section 9.11: Skillbuilder Applet Exercises

For Exercises 9.127 to 9.129, use the **SampleMeans** *applet described in Section 9.11 and available on the companion website, http://www.cengage.com/statistics/Utts4e. The applet selects random samples from a bell-shaped population in which the mean $\mu = 8$ and the standard deviation $\sigma = 5$. For Exercises 9.130 and 9.131, use the related* **TVMeans** *applet.*

9.127 This exercise concerns possible sample means for random samples of $n = 36$.

a. Use the applet to generate 500 different random samples of $n = 36$ observations. What is the approximate shape of the resulting histogram of sample means?
b. Approximately, what were the smallest and largest values of the sample mean among the 500 samples you generated?
c. For random samples of $n = 36$, use the Empirical Rule to find the interval of values in which about 99.7% of all sample means will fall. Compare that interval to your answer for part (b).

9.128 This exercise concerns possible sample means for random samples of $n = 75$.

a. Use the applet to generate 500 different random samples of $n = 75$ observations. What is the approximate shape of the resulting histogram of sample means?
b. Approximately, what were the smallest and largest values among the sample means in the 500 samples you generated?
c. For random samples of $n = 75$, use the Empirical Rule to find the interval of values into which the sample mean will fall about 99.7% of the time. Compare the interval to your answer for part (b).
d. Describe any difference between the histograms created for part (a) of this exercise and part (a) of Exercise 9.127. (If you have not already generated the histogram for part (a) of Exercise 9.127, do so now.)

9.129 This exercise compares sample means for random samples of $n = 10$ and random samples of $n = 100$.

a. Use the applet to generate ten different random samples of $n = 10$. List the ten sample means, and identify the lowest and highest values in the list.
b. Use the applet to generate ten different random samples of $n = 100$. List the ten sample means, and identify the lowest and highest values in the list.
c. Compare the results of parts (a) and (b). What is indicated about the effect of sample size on the variation among sample means for different samples?

9.130 Use the **TVMeans** applet, which selects random samples from a population modeled by using responses given by college students to a question asking how many hours they watch television in a typical week. The mean for the population is $\mu = 8.352$ hours and the standard deviation is $\sigma = 7.723$ hours.

a. Generate 500 different random samples of $n = 5$ observations. Draw a sketch showing the shape of the histogram of sample means for random samples of $n = 5$.
b. Generate 500 different random samples of $n = 49$ observations. Draw a sketch showing the shape of the histogram of sample means for random samples of $n = 49$.
c. Refer to the two situations given in Section 9.6 for which the sampling distribution of the mean is approximately normal. Explain which situation is present in the **TVMeans** applet. Also, explain how parts (a) and (b) illustrate that situation.

9.131 Refer to Exercise 9.130 about the **TVMeans** applet. Assume that the sampling distribution of the sample mean is approximately normal for random samples of $n = 49$ observations.

a. Use the Empirical Rule to find the interval of values into which the sample mean will fall about 95% of the time for random samples of $n = 49$.
b. Use the Empirical Rule to find the interval of values into which the sample mean will fall about 68% of the time for random samples of $n = 49$.

Chapter Exercises

9.132 Answer Thought Question 9.1 on page 326.

9.133 Suppose you want to estimate the proportion of students at your college who are left-handed. You decide to collect a random sample of 200 students and ask them which hand is dominant. Go through the conditions for which the sampling distribution of a sample proportion is approximately normal, and explain why they would apply to this situation.

9.134 Refer to Exercise 9.133. Suppose the truth is that .12, or 12%, of the students are left-handed and you take a random sample of 200 students. Draw a picture similar to Figure 9.2 (p. 325), showing the possible sample proportions for this situation.

9.135 A recent Gallup poll found that of 800 randomly selected drivers surveyed, 70% thought that they were better-than-average drivers. It is reasonable to define "average" so that in the population, at most only 50% of all drivers can be "better than average."

a. Draw a picture of the possible sample proportions that would result from samples of 800 people from a population with a true proportion of .50.

b. Would you be unlikely to see a sample proportion of .70, based on a sample of 800 people, from a population with a proportion of .50? Explain, using your picture from part (a).

9.136 According to the *Sacramento Bee* (April 2, 1998, p. F5), "A 1997–98 survey of 1027 Americans conducted by the National Sleep Foundation found that 23% of adults say they have fallen asleep at the wheel in the last year."

a. Conditions 2 and 3 needed for the sampling distribution of $\hat{p}$ to be approximately normal are met because this result is based on a large random sample of adults. Explain how Condition 1 is also met.

b. The article also said that (based on the same survey) "37 percent of adults report being so sleepy during the day that it interferes with their daytime activities." If in truth, 40% of all adults have this problem, find the interval in which about 95% of all sample proportions should fall, based on samples of size 1027. Does the result of this survey fall into that interval?

c. Suppose that a survey based on a random sample of 1027 college students was conducted and 25% reported being so sleepy during the day that it interferes with their daytime activities. Would it be reasonable to conclude that the population proportion of college students who have this problem differs from the proportion of all adults who have the problem [using $p = .4$ from part (b)]? Explain.

9.137 *This is also Thought Question 9.4.* Verify that if the raw data for each individual in a sample is 1 when the individual has a certain trait and 0 otherwise, then the sample mean is equivalent to the sample proportion with the trait. You can do this by using a formula, explaining it in words, or constructing a numerical example.

9.138 Suppose that you are interested in estimating the average number of miles per gallon of gasoline your car can get. You calculate the miles per gallon for each of the next nine times you fill the tank. Suppose that in truth, the values for your car are bell-shaped, with a mean of 25 miles per gallon and a standard deviation of 1. Draw a picture of the possible sample means you are likely to get based on your sample of nine observations. Include the intervals into which 68%, 95%, and almost all of the potential sample means will fall, using the Empirical Rule from Chapter 2.

9.139 Refer to Exercise 9.138. Redraw the picture under the assumption that you will collect 100 measurements instead of only 9. Discuss how the picture differs from the one in the previous exercise.

9.140 Give an example of a scenario of interest to you for which the sampling distribution of $\hat{p}$ would be approximately normal. Explain why the conditions for it to be approximately normal are satisfied for your example.

9.141 Suppose the population of IQ scores in the town or city where you live is bell-shaped, with a mean of 105 and a standard deviation of 15. Describe the distribution of possible sample means that would result from random samples of 100 IQ scores.

9.142 Suppose that 35% of the students at a university favor the semester system, 60% favor the quarter system, and 5% have no preference. Would a random sample of 100 students be large enough to provide convincing evidence that the quarter system is favored? Explain.

9.143 According to *USA Today* (April 20, 1998 Snapshot), a poll of 8709 adults taken in 1976 found that 9% believed in reincarnation, while a poll of 1000 adults taken in 1997 found that 25% held that belief.

a. Assuming that a proper random sample was used, verify that the sample proportion for the 1976 poll represents the population proportion to within about 1%.

b. Based on these results, would you conclude that the proportion of all adults who believe in reincarnation was higher in 1997 than it was in 1976? Explain.

9.144 Suppose that 20% of all television viewers in the country watch a particular program.

a. For a random sample of 2500 households measured by a rating agency, describe the distribution of the possible sample proportions who watch the program.

b. The program will be cancelled if the ratings show less than 17% of households watching it in a random sample of households. Given that 2500 households are used for the ratings, is the program in danger of getting cancelled? Explain.

9.145 Use the information about the sampling distribution for a sample mean to explain why it is desirable to take as large a sample as possible when trying to estimate a population value.

9.146 According to the *Sacramento Bee* (April 2, 1998, p. F5), Americans get an average of 6 hours and 57 minutes of sleep per night. A survey of a class of 190 statistics students at a large university found that they averaged 7.1 hours of sleep the previous night, with a standard deviation of 1.95 hours.

a. Assume that the population average for adults is 6 hours and 57 minutes, or 6.95 hours of sleep per night, with a standard deviation of 2 hours. Draw a picture illustrating the sampling distribution of the sample mean for random samples of 190 adults.

b. Would the mean of 7.1 hours of sleep obtained from the statistics students be a reasonable value to expect for the sample mean of a random sample of 190 adults? Explain.

c. Can the sample taken in the statistics class be considered a representative sample of all adults? Explain.

9.147 Explain whether or not each of the following scenarios meets the conditions for which the sampling distribution of the sample proportion is approximately normal. If not, explain which condition is violated.

a. Unknown to the government, 10% of all cars in a certain city do not meet appropriate emissions standards. The government wants to estimate that percentage, so they take a random sample of 30 cars and compute the sample proportion that do not meet the standards.

b. The U.S. Census Bureau wants to estimate the proportion of households that have someone at home between 7:00 P.M. and 7:30 P.M. on weeknights to determine whether or not that would be an efficient time to collect census data. They survey a random sample of 2000 households and visit them during that time to see whether or not someone is at home.

c. You want to know what proportion of days in typical years have rain or snow in the area where you live. For the months of January and February, you record whether or not there is rain or snow each day, and then you calculate the proportion.

d. A large company wants to determine the proportion of its employees who are interested in on-site day care. For a random sample of 100 employees, the company calculates the sample proportion who are interested.

9.148 Explain whether or not you think the sampling distribution for the sample mean would be approximately normal in each of the following scenarios. If it would be, specify the population of interest and the measurement of interest. If it would not be, explain why not.

a. A researcher wants to know what the average cholesterol level would be if people restricted their fat intake to 30% of calories. He gets a group of patients who have had heart attacks to volunteer to participate, puts them on a restricted diet for a few months, and then measures their cholesterol and calculates the mean value.

b. A large corporation would like to know the average income of its workers' spouses. Rather than going to the trouble to collect a random sample, they post someone at the exit of the building at 5:00 P.M. Everyone who leaves between 5:00 P.M. and 5:30 P.M. is asked to complete a short questionnaire on the issue. There are 70 responses.

c. A university wants to know the average income of its alumni. They select a random sample of 200 alumni and mail them a questionnaire. The university follows up with a phone call to those who do not respond within 30 days.

d. An automobile manufacturer wants to know the average price for which used cars of a particular model and year are selling in a certain state. It is able to obtain a list of buyers from the state motor vehicle division, from which a random sample of 20 buyers is selected. The manufacturer makes every effort to find out what those people paid for the cars and is successful in doing so.

9.149 In Case Study 9.1 (p. 352), we learned that about 56% of American adults actually voted in the presidential election of 1992, whereas about 61% of a random sample claimed that they had voted. The size of the sample was not specified, but suppose it was based on 1600 American adults, a common size for such studies.

a. Into what interval of values should the sample proportion fall 68%, 95%, and almost all of the time?

b. Do you think that the observed value of 61% is reasonable, based on your answer to part (a)?

c. Now suppose that the sample had been of only 400 people. Compute a standardized score to correspond to the reported percentage of 61%. Comment on whether or not you believe that people in the sample could all have been telling the truth, based on your result.

9.150 Suppose that the population of grade point averages (GPAs) for students at the end of their first year at a large university has a mean of 3.1 and a standard deviation of 0.5.

a. Draw a picture of the sampling distribution of the sample mean GPA for a random sample of 100 students.

b. Find the probability that the sample mean will be 3.15 or greater.

9.151 The administration of a large university wants to take a random sample to measure student opinion of a new food service on campus. It plans to use a continuous scale from 1 to 100, on which 1 is complete dissatisfaction and 100 is complete satisfaction. The administration knows from past experience with such questions that the standard deviation for the responses is going to be about 5, but it does not know what to expect for the mean. It wants to be almost sure that the sample mean is within plus or minus 1 point of the true population mean value. How large will the random sample have to be?

Dataset Exercises

Datasets required to solve these exercises are available on the companion website, http://www.cengage.com/statistics/Utts4e.

9.152 The data for this exercise are in the **GSS-93** dataset on the companion website. The variable ***gunlaw*** is whether a respondent favors or opposes stronger gun control laws. Of the 1055 respondents, 870 are in favor. *For this exercise, assume that these respondents comprise the entire population.*

a. What is the population proportion p who favor stronger gun control laws?

b. Suppose random samples of $n = 60$ are drawn from this population. Describe and draw the approximate sampling distribution for the proportion $\hat{p}$ who would favor stronger gun control laws.

c. Simulate 200 samples of size 60 from this population, and draw a histogram of the sample proportions who favor stronger gun control laws.

d. Compare your simulated distribution in part (c) with the sampling distribution that you described in part (b).

9.153 Use the **pennstate1** dataset on the companion website for this exercise. The data for the variable ***HrsSleep*** are responses by $n = 190$ students to the question "How many hours did you sleep last night?" For this exercise, suppose that they are the whole population.

a. Simulate 200 samples of $n = 10$ responses for ***HrsSleep***. Show the commands you used, or explain how you did the simulation.

b. Draw a histogram of the 200 sample means.

c. Describe the simulated sampling distribution you found in part (b).

d. Draw a histogram of the 200 sample medians for the samples you found in part (a).

e. Compare the sampling distributions for the sample mean and the sample median.

9.154 Use the data in the **GSS-93** dataset on the companion website for this exercise.

One of the questions asked was age in years. A five-number summary for age is 18, 32, 43, 58, and 89.

a. Draw a boxplot of the ages for the entire dataset.

b. Let H = highest age in the sample for samples of size $n = 10$. Explain whether any of the rules given in this chapter provide information about the sampling distribution of *H*.

c. Simulate 200 samples of size 10, and find *H* for each one. Create a histogram for the 200 values of *H*.

d. Using the histogram in part (c) and any other numerical or graphical information you wish to provide, describe the sampling distribution of *H*.

9.155 Use the **Fantasy5** dataset on the companion website for this exercise, with data from 2318 games of California's version of the same scheme as the Pennsylvania Cash 5 game.

a. Draw a picture of the sampling distribution of *H*, the highest number each day, and compare it to the equivalent picture for the Cash 5 game, shown in Figure 9.13 (p. 348). Do you expect them to be similar? Are they?

b. Consider each game to be a random sample of size 5, drawn without replacement from the integers 1 to 39. Which column in the dataset represents the median for each of these random samples?

c. Draw a picture of the sampling distribution of the median.

d. Find the mean of each of the 2318 samples of size 5. Draw a picture of the sampling distribution of the mean.

e. Compare the sampling distributions of the median and the mean. Which one is more spread out? Why do you think that is the case?

9.156 Use the **Cash 5** dataset on the companion website for this exercise, with data from 1560 games of Pennsylvania's Cash 5 lottery game described in Section 9.10. The variables ***High*** and ***Low*** give the highest and lowest numbers picked in each game.

a. For each game, find $R =$ ***High*** $-$ ***Low***, the range between the highest and lowest numbers. Draw a histogram of *R*, and describe the shape of this histogram.

b. Find the mean for each of the variables *R*, ***High*** and ***Low***. Verify that the mean for *R* equals the difference between the means for *High* and *Low*.

c. What is the maximum possible value of the range *R* in the Cash 5 game? What values of ***High*** and ***Low*** would lead to this value?

d. Tally how often each value of *R* occurred in the 1560 games given in the dataset. How often did the maximum possible value of *R* occur?

10

Erik S. Lesser/Getty Images

Will she quit her job after this big lottery win?

See Example 10.6 *(p. 383)*

Estimating Proportions with Confidence

One of the most common types of inference procedures is to construct a *confidence interval* estimate of the unknown value of a population parameter. Here is a typical confidence interval statement: Based on this sample, we are 95% confident that somewhere between 33% and 39% of all Americans suffer from allergies.

Statistical inference procedures are used to make inferences about populations based on samples. Confidence intervals and hypothesis tests are the two most common inference methods. Confidence intervals are the topic of Chapters 10 and 11 and hypothesis tests are covered in Chapters 12 and 13.

In this chapter, we will learn what confidence intervals are, how to interpret them, and the generic format for constructing one for any of the five parameters covered in Chapter 9. We will also learn how to construct and interpret a confidence interval that estimates one population proportion and how to construct and interpret a confidence interval for the difference between two population proportions.

10.1 CI Module 0: An Overview of Confidence Intervals

This module is divided into two lessons. Lesson 1 explains the basic idea of a confidence interval. Lesson 2 gives a general format for a confidence interval that is used when we estimate any of the five parameters introduced in Chapter 9.

Lesson 1: The Basic Idea of a Confidence Interval

A confidence interval uses sample information to estimate the unknown value of a population parameter. This is done by creating an interval of numbers that we are fairly confident includes the value of the population parameter.

DEFINITION A **confidence interval** is an interval of values computed from sample data that is likely to include the unknown value of a population parameter.

Curiosity and Confidence Intervals

In Chapter 9, we explored the idea that curiosity about the world may often be translated into a question about a population parameter. The general type of question answered using a confidence interval is, "What is the value of this population parameter?" Here are some examples of problems where confidence intervals can be used to determine an answer:

- If you suffer from allergies, you might wonder how much company you have. The parameter we may wish to estimate is p = proportion of the population that suffers from allergies. This parameter is estimated in Example 10.2.

- If you won the lottery, would you quit working? What would others do? One parameter of interest is p = proportion of working adults who say that they would quit their jobs if they won a large amount in the lottery. This parameter is estimated in Example 10.6.
- Do men and women feel the same way about dating someone who has a great personality without regard for looks? To investigate this, one parameter we might estimate is $p_1 - p_2$ = the difference between population proportions of females and males who would answer "yes" if asked, "Would you date someone with a great personality even though you did not find them attractive?" This parameter is estimated in Case Study 10.3.

For all three examples just described, curiosity about something can be explored by collecting sample data and then using the data to estimate the unknown value of a population parameter.

Population Parameters and Sample Statistics

A confidence interval uses sample information to estimate the unknown value of a characteristic of a population. Let's review some key terms in this process.

- The **population** is the entire collection of units about which we would like information or is also considered to be the entire collection of measurements that we would have if we could measure the whole population.
- A **population parameter** is a fixed summary number associated with a population, and in the context of confidence intervals, it is unknown and we would like to estimate it. An example is p, the proportion of a *population* with a particular characteristic.
- The **sample** is the collection of units that we will actually measure or the collection of measurements that we will actually obtain.
- The **sample size**, denoted by the letter n, is the number of units in the sample.
- A **sample statistic** is a summary number computed from a sample that is used to estimate the corresponding population parameter. An example is $\hat{p}$, the proportion of a *sample* with a particular characteristic, found as the number in the sample with the characteristic divided by the sample size n.
- **Sample estimate,** and **point estimate** are synonyms for a *sample statistic.*

Also, remember that we would ideally like to have a *randomly selected sample* from the population, but in Chapters 5 and 6, we learned that is not always possible and that sometimes we can settle for less. Let's review the crucial criterion for statistical inference to be valid.

DEFINITION The **Fundamental Rule for Using Data for Inference** is that available data can be used to make inferences about a much larger group *if the data can be considered to be representative with regard to the question(s) of interest.*

THOUGHT QUESTION 10.1 Each day, Maria gets dozens of e-mail messages. She keeps track of what proportion of the messages are spam or other junk and what proportion are interesting. Suppose that she got 50 messages yesterday, and 20 of them were interesting. If the collection of messages on a single day is considered to be a sample of all e-mail messages she ever receives, explain the meaning of each of the following definitions in the context of this example, and give numerical values where possible: *population, sample, sample size, population parameter, sample statistic* (or *sample estimate*). Discuss whether you think the Fundamental Rule for Using Data for Inference would allow Maria to draw conclusions about the population proportion based on the sample proportion.*

***HINT:** Consider whether some days of the week are more likely than others to bring junk e-mail messages.

The Concept of a Confidence Interval as an Interval Estimate

Remember that the term **point estimate** is sometimes used as a synonym for the *sample estimate* (also called the *sample statistic*). That's because it is a single number or *point* on the number line. In contrast, the term **interval estimate** is used as a synonym for *confidence interval.* Even though it is an interval of values, an interval estimate estimates one single, fixed population value.

A confidence interval is always accompanied by a *confidence level,* which tells us how likely it is that the interval estimate actually contains the true value of the parameter we are estimating. The most common confidence level used in research and in the media is 95%. Later in this chapter, we will learn how to determine an appropriate interval for any specified confidence level.

Example 10.1

Case Study 1.3 Revisited: Teens and Interracial Dating In Case Study 1.3, we described a 1997 *USA Today*/Gallup poll of teenagers. One finding was that 57% of the 496 teens in the sample who date said that they had been out on a date with someone of another race or ethnic group. What can we say about this issue for the entire population of teenagers who date? In Case Study 1.3, we learned that we can have 95% confidence that the percentage of all teenagers who date in the United States who would say that they have dated interracially or someone of another ethnic group is between 52% and 62%. We formed this interval using the *sample estimate* of 57% and then adding and subtracting the *margin of error* of about 5%, which was reported with the poll. We will learn why this method works, as well as how to find the margin of error, in the next section of this chapter.

Figure 10.1 illustrates this situation. The population is all U.S. teens who date. The population parameter of interest is the percentage of all U.S. teens who have dated someone of another race or ethnic group. The sample used for this question is the n = 496 teens who date and the sample statistic is 57%. Using information from the sample, we estimate that the unknown value of the population parameter is between 52% and 62%.

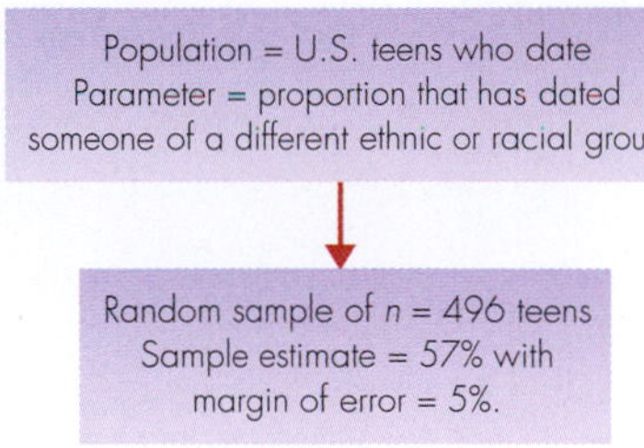
Population = U.S. teens who date
Parameter = proportion that has dated
someone of a different ethnic or racial group
Random sample of n = 496 teens
Sample estimate = 57% with
margin of error = 5%.

Figure 10.1 Population and sample results for Example 10.1

Interpreting the Confidence Level

The term **confidence level** describes the chance that an interval actually contains the true population value in the following sense. Most of the time (quantified by the confidence level), intervals that are computed in this way will capture the truth about the population, but occasionally, they will not. In any given instance, the interval either covers the truth or it does not, but we will never know which is the case. Therefore, our confidence is in the *procedure*—it works most of the time—and the "confidence level" or "level of confidence" is the percentage of the time we expect the procedure to work. For instance, a 95% confidence level means that confidence intervals computed by using the same procedure would include the true population value for 95% of all possible random samples from the population.

DEFINITION For a confidence interval, the **confidence level** is the probability that the procedure used to determine the interval will provide an interval that includes the population parameter. It is commonplace to express the confidence level as a percentage.

The confidence level has a relative frequency probability interpretation: If we consider all possible randomly selected samples of the same size from a population, the *confidence level* is the fraction or percentage of those samples for which the confidence interval would include the population parameter.

The most common confidence level that researchers use is 95%. In other words, researchers define "likely to include the true population value" to mean that they are 95% certain that the *procedure* will work. So they are willing to take a 5% risk that the interval does not actually include the true value. Sometimes researchers employ only 90% confidence, and occasionally, they use higher confidence levels such as 98% or 99%.

Be careful when giving information about a specific confidence interval computed from an observed sample. The confidence level expresses only how often the confidence interval procedure works in the long run. It does not tell us the probability that a specific interval includes the population value. For example, suppose that a 95% confidence interval estimate of the percentage of all Americans suffering from allergies is 33% to 39%. This interval either does or does not include the true unknown value of the population percentage, but we can never know which is the case because we can't know the population value (unless we ask everyone in the population).

THOUGHT QUESTION 10.2 Explain in your own words what it means to say that we have 95% confidence in the interval estimate. Then give an example of something you do in your life that illustrates the same concept: You follow the same procedure each time, and it either works (most of the time) or does not work to produce the desired result. What confidence level would you assign to the procedure in your example; that is, what percentage of the time do you think it produces your desired result?*

Lesson 2: Computing Confidence Intervals for the Big Five Parameters

The same general format applies to computing any confidence interval for any of the big five parameters of interest. Remember that the five different situations include estimating:

- One population proportion, p
- The difference in two population proportions for independent samples, $p_1 - p_2$
- One population mean, μ
- The population mean of paired differences, μ_d
- The difference in two population means for independent samples, $\mu_1 - \mu_2$

Confidence intervals for the first two parameters, involving proportions, are covered in this chapter. Confidence intervals for the remaining three parameters, involving means, are covered in Chapter 11. In each case, the goal is to estimate the *population parameter* with an interval of values that we are fairly confident covers it. In each case, we will have sample data either from one sample (for p, μ, or μ_d) or two independent samples (for $p_1 - p_2$ or $\mu_1 - \mu_2$).

We'll start with the end result: the general formula for a confidence interval. We will give specific details for each of the five parameters in the individual modules. The general format follows.

***HINT:** What is a task you frequently perform that you can't always do successfully?

FORMULA **Confidence Interval or Interval Estimate**

A **confidence interval** or **interval estimate** for any of the five parameters can be expressed as

Sample estimate ± Multiplier × Standard error

The **multiplier** is a number based on the confidence level desired and determined from the standard normal distribution (for proportions) or Student's t-distribution (for means). Details are provided in the individual modules.

Let's look at each part of the formula:

- The *sample estimate*, also called the *sample statistic* that estimates the parameter of interest. For example, if the parameter of interest is a population proportion p, then the sample estimate is the sample proportion $\hat{p}$.
- The *multiplier* determines the amount of confidence we will have in the result. For instance, for a 95% confidence interval for a proportion, the appropriate multiplier is 1.96, sometimes rounded off to 2.
- The *standard error* is the **standard error of the sample statistic**, which is an *estimate* of the standard deviation of the sample statistic. An example is the **standard error of $\hat{p}$**, which is $\sqrt{\hat{p}(1-\hat{p})/n}$.

95% Confidence Interval for One Proportion as an Illustration

To illustrate the general formula, here is how we would find a **95% confidence interval for one proportion**:

- The sample estimate is the sample proportion, denoted by $\hat{p}$.
- The appropriate multiplier is 1.96, which we will round off to 2.
- The standard error of $\hat{p}$ (from page 375 in Chapter 9) is $\sqrt{\hat{p}(1-\hat{p})/n}$.

So a 95% confidence interval for a population proportion can be written approximately as follows:

Sample estimate ± Multiplier × Standard error

$$\hat{p} \pm 2 \times \sqrt{\frac{\hat{p}(1-\hat{p})}{n}}$$

Let's look at an example of computing an interval estimate using this formula.

Example 10.2 **The Pollen Count Must Be High Today** In April 1998, the Marist Institute for Public Opinion surveyed 883 randomly selected American adults about allergies. According to a report posted at the Institute's website (http://www.mipo.marist.edu), 36% of the sample answered "yes" to the question "Are you allergic to anything?" Therefore, the sample proportion who said "yes" is $\hat{p} = .36$. We will use the sample information to calculate a 95% confidence interval estimate of the population parameter p = proportion of all American adults who are allergic to something.

Values for the parts of the formula Sample estimate ± Multiplier × Standard error are as follows:

- *Sample estimate* = $\hat{p}$ = .36
- *Multiplier* = 2 (to achieve 95% confidence)
- *Standard error* = $\sqrt{\frac{\hat{p}(1-\hat{p})}{n}} = \sqrt{\frac{.36(1-.36)}{883}} = .016$

The 95% confidence interval is .36 ± 2 × .016, which is .36 ± .032 or .328 to .392 (about 33% to 39%).

Interpretation: The confidence interval, .328 to .392, estimates the proportion of *all* American adults who have an allergy. In percentage terms, this is about 33% to 39%. The confidence level (95%) describes our confidence in the procedure used to determine the interval. In the long run, the procedure will work about 95% of the time, meaning it will provide an interval that includes the true population value.

What Determines the Width of the Interval?

The width of a confidence interval is the difference between the lower and upper values of the interval. A relatively narrow confidence interval gives a more precise estimate of the population value than does a relatively wide confidence interval. The amount that we add and subtract to the sample estimate to create the interval determines its width. Three factors affect the width of a confidence interval:

1. *The sample size, n.* Larger sample sizes tend to produce narrower intervals because the standard error of a statistic generally decreases when the sample size increases. We are more likely to obtain a sample estimate that is closer to the true population value for a large sample than for a small one.
2. *The confidence level.* The multiplier is determined by the desired confidence level. For instance, when we estimate a proportion, the multiplier is 1.96, often rounded to 2, for a 95% confidence level. Later in the chapter, we will learn that the larger the multiplier, the more confidence we have that the interval covers the truth. The price of added confidence is that the interval will be wider.
3. *The natural variability among individual units.* More natural variability results in wider intervals. When the parameter is a proportion, the sample proportion $\hat{p}$ is a measure of the natural variability. If the proportion is close to either 1 or 0, most individuals have the same trait or opinion, so there is little natural variability, and the standard error is relatively small. As $\hat{p}$ gets closer to .5 from either direction, the standard error gets larger. When the parameter involves means, natural variability is quantified by the sample standard deviation(s).

The same three factors influence the amount that is added and subtracted to the sample estimate to form a confidence interval for all five parameters. The first two factors, sample size and confidence level, can be selected by the researcher. The third factor, natural variability among the units, is an inherent property of the population being studied so it can't be selected or manipulated.

10.1 Exercises are on pages 394–396.

IN SUMMARY Confidence Interval Basics

- A **confidence interval** is an interval of values computed from sample data that is likely to include the unknown value of a population parameter.
- The **confidence level** for an interval is the likelihood that the procedure used will give an interval that contains the unknown value of the population parameter. The most common confidence level used by researchers and the media is 95%.
- The general format for a confidence interval estimate for any of the five parameters we will consider is Sample estimate ± Multiplier × Standard error.

10.2 CI Module 1: Confidence Interval for a Population Proportion

The material in this module is divided into three lessons. Lesson 1 explains how to implement the general formula to find a confidence interval for a population proportion for any confidence level. Lesson 2 explains why the formula works. In

Lesson 3, we reconcile the formulas in this chapter with the confidence interval formula in Chapter 5, in which we added and subtracted a margin of error of $1/\sqrt{n}$ to the sample proportion to find an approximate 95% confidence interval for a population proportion. We also explain why the media reports this margin of error.

Lesson 1: Details of How to Compute a Confidence Interval for a Population Proportion

The confidence interval method described in this section is applicable to a surprisingly wide range of interesting research questions. The scenarios are similar to the ones that we encountered when we studied binomial random variables in Chapter 8. There are two common settings in which we might want to estimate a population proportion or probability p:

1. *A population exists, and we are interested in knowing what proportion of it has a certain trait, opinion, characteristic, response to a treatment, and so on.* Two typical research questions for this setting follow:
   - What proportion of drivers are talking on a cell phone at any given moment?
   - What proportion of a population of smokers would quit smoking if they were to wear a nicotine patch for 8 weeks?
2. *A repeatable situation exists, and we are interested in the long-run probability of a specific outcome.* Two typical research questions for this setting follow:
   - What is the probability that a new fertility procedure will be successful for a randomly selected couple who tries it?
   - What is the probability that a randomly selected television of a certain model will fail before the warranty period is over?

We can find a confidence interval with specified confidence level to answer each of these questions by collecting appropriate sample data from the population or repeatable situation.

How to Compute a Confidence Interval for a Population Proportion

In CI Module 0, we learned that a confidence interval for any of our five situations is given by the formula,

$$\text{Sample estimate} \pm \text{Multiplier} \times \text{Standard error}$$

In the case of a confidence interval for one proportion:

$$\textit{Sample estimate} = \text{the sample proportion} = \hat{p}$$

$$\textit{Standard error} = \text{s.e.}(\hat{p}) = \sqrt{\frac{\hat{p}(1-\hat{p})}{n}}$$

Finding the Multiplier z^*

The *multiplier* is denoted by z^* and is found by using the standard normal distribution. Values of the multiplier for the most common confidence levels used by researchers are shown in Table 10.1 (next page). The multipliers are shown to three decimal places, but in practice they often are rounded to two or even fewer decimal places, especially when this rounding does not change the resulting confidence interval or confidence level by much. For example, the multiplier for a 95% confidence interval is 1.96, which is sometimes rounded to 2.0. The result of this rounding is that the actual confidence level is 95.45% instead of 95%, a minor difference.

Table 10.1 Confidence Intervals for a Population Proportion

Confidence Level	Multiplier (z^*)	Confidence Interval
90%	1.645 or 1.65	$\hat{p} \pm 1.65$ standard errors
95%	1.960, sometimes rounded to 2	$\hat{p} \pm 2$ standard errors
98%	2.326 or 2.33	$\hat{p} \pm 2.33$ standard errors
99%	2.576 or 2.58	$\hat{p} \pm 2.58$ standard errors

In general, the multiplier z^* is the standardized score such that the area between $-z^*$ and $+z^*$ under the standard normal curve corresponds to the desired confidence level. Figure 10.2 illustrates this relationship. In practice, it would be unusual for anybody to use a confidence level other than the four shown in Table 10.1. We will demonstrate how to find a z^* multiplier for any confidence level in Lesson 2 of this module.

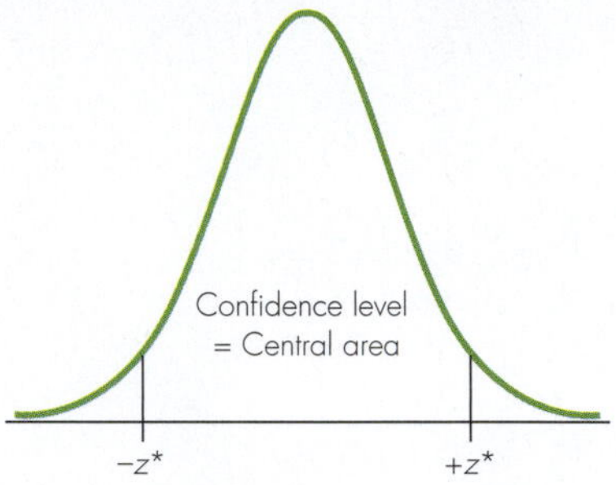

Figure 10.2 The relationship between the confidence level and the multiplier z^*. The confidence level is the probability between $-z^*$ and $+z^*$ in the standard normal curve

Example 10.3 **Is There Intelligent Life on Other Planets?** In a 2008 Scripps Howard News/Ohio University survey of 1003 randomly selected Americans, 56% of the sample said that it is either very or somewhat likely that there is intelligent life on other planets (Source: www.scrippsnews.com/node/34758). Let's use this sample information to calculate a 90% confidence interval for the proportion of all Americans who believe that it is either very or somewhat likely that there is intelligent life on other planets.

Sample estimate $= \hat{p} = .56$ (56% expressed as a proportion)

$$s.e.(\hat{p}) = \sqrt{\frac{.56(1 - .56)}{1003}} = .016$$

Multiplier = 1.645 (from Table 10.1 for .90 confidence level)

A 90% confidence interval for the population proportion is

Sample estimate ± Multiplier × Standard error

$.56 \pm 1.645 \times .016$

$.56 \pm .026$, which is .534 to .5846

Interpretation: With 90% confidence, we can say that in 2008, the proportion of Americans who believed that it is very or somewhat likely there is intelligent life on other planets was in the range .534 to .586 (or 53.4% to 58.6%).

We can increase our confidence that the interval covers the truth by increasing the confidence level. However, we pay a price because the multiplier will be larger and the interval will be wider. For instance, for 98% confidence, the z^* multiplier is 2.33 (from Table 10.1). The 98% confidence interval is

$.56 \pm 2.33 \times .016$, which is $.56 \pm .037$.

When converted to percentages, the 98% confidence interval is 56% ± 3.7%. We can be fairly sure that in 2008, between 52.3% and 59.7% of Americans believed that there is intelligent life on other planets. Because the entire interval is above 50%, we can also state with high confidence that in 2008, a majority of Americans believed that there is intelligent life on other planets.

IN SUMMARY Formula for a Confidence Interval for a Population Proportion p

The formula for a **confidence interval for a population proportion** is

$$\hat{p} \pm z^* \sqrt{\frac{\hat{p}(1-\hat{p})}{n}}$$

where

- $\hat{p}$ is the sample proportion.
- z^* denotes the multiplier.
- $\sqrt{\frac{\hat{p}(1-\hat{p})}{n}}$ is the **standard error of the sample proportion.**

Values of z^* for common confidence levels can be found in Table 10.1, or in the last row of Table A.2 in the Appendix for reasons that are explained in CI Module 3 in Chapter 11. The multiplier z^* is such that the area under the standard normal curve between $-z^*$ and z^* corresponds to the confidence level.

Conditions for Using the Formula for a Confidence Interval for a Proportion

The research questions that can be answered by using a confidence interval for a population proportion are ones that arise when data are collected in ways similar to a binomial experiment. Therefore, the conditions that must be met to use the formula for a confidence interval for one population proportion are as follows:

Condition 1. The sample is a randomly selected sample from the population.

Condition 2. Both $n\hat{p}$ and $n(1-\hat{p})$ should be at least 10 (although some authors say that these quantities need only be at least 5).

The first condition is sometimes difficult to achieve. A more lenient condition is given by the Fundamental Rule for Using Data for Inference, which we repeat here: Available data can be used to make inferences about a much larger group *if the data can be considered to be representative with regard to the question(s) of interest.* In the context of this chapter, a confidence interval based on a sample proportion can be used to estimate the population proportion if the individuals in the sample can be considered to be representative of the individuals in the population for the trait, opinion, and so on of interest.

Conditions When the Parameter Is a Long-Run Probability p

In settings for which the parameter p is a long-run probability instead of a proportion for an existing population, the first condition can be rewritten as follows:

Alternative Condition 1: The sample proportion $\hat{p}$ is the proportion of times a specified outcome occurs in n repeated independent trials with fixed probability p that the specified outcome will occur.

The sample size requirement remains as stated in Condition 2; both $n\hat{p}$ and $n(1-\hat{p})$ must be at least 10. Note that this is equivalent to having at least 10 trials with the specified outcome and at least 10 trials without it. The next example illustrates this situation.

Example 10.4 **Would You Return a Lost Wallet?** In a study that was repeated in various countries, staff from the magazine *Reader's Digest* planted "lost" wallets containing the equivalent of about $50, a name, a local address, a phone number, and family photos to find out whether people who found them would return them. For the study done in Canada, 120 wallets were "lost," and 77 of them were returned intact (Kiener, 2005), so the sample proportion of wallets returned is $\hat{p} = 77/120 = .64$. The sample size requirement is met, because there are at least 10 trials in which the wallet was returned and at least 10 in which it was not.

The parameter of interest is the probability p that a similarly "lost" wallet would be returned intact if lost in Canada. Let's find a 95% confidence interval estimate for p. The appropriate multiplier is 1.96, which we will round to 2.0, so the interval is

$$\text{Sample estimate} \pm \text{Multiplier} \times \text{Standard error}$$

$$\hat{p} \pm 2 \times \sqrt{\frac{\hat{p}(1-\hat{p})}{n}}$$

$$.64 \pm 2 \times \sqrt{\frac{(.64)(.36)}{120}}$$

$$.64 \pm 2 \times .044$$

$$.55 \text{ to } .73$$

Therefore, we can be 95% confident that the probability of a lost wallet being returned under the conditions used in the experiment is between .55 and .73.

THOUGHT QUESTION 10.3 Suppose the legislature in a particular state wanted to know what proportion of students graduating from the state university last year were permanent residents of the state. The university had information for all students showing that 3900 of the 5000 graduates were state residents. Is a confidence interval appropriate for this situation? If so, compute the appropriate interval. If not, explain why not.*

MINITAB TIP **Calculating a Confidence Interval for a Proportion**

- To compute a confidence interval for a proportion, use **Stat > Basic Statistics > 1 Proportion**. (This procedure is not in versions earlier than Version 12.)
- If the raw data are in a column of the worksheet, specify that column. If the data have already been summarized, click on "Summarized Data," and then specify the sample size and the count of how many observations have the characteristic of interest.

Note: To calculate intervals in the manner described in this chapter, use the ***Options*** button, and click on "Use test and interval based on normal distribution." Note also that the confidence level can be changed by using the ***Options*** button.

Lesson 2: Understanding the Formula

Developing the 95% Confidence Interval

The key to developing a 95% confidence interval for a proportion is the **sampling distribution** for a sample proportion described in Chapter 9. It describes the distribution of the sample proportions that would occur if we took many different random samples of the same size from a population. If the sample size is sufficiently large, the distribution of possible values of $\hat{p}$ is approximately a normal curve. For this normal distribution,

***HINT:** Do the 5000 graduates constitute a sample or a population?

- The mean is p, the population proportion.
- The standard deviation is $\sqrt{\dfrac{p(1-p)}{n}}$

Because the distribution of possible sample proportions has a normal curve, we can use the Empirical Rule from Chapter 2 to make the following statement: *In about 95% of all samples, the sample proportion $\hat{p}$ will fall within 2 standard deviations of* p, *the true population proportion. So for about 95% of all samples,*

$$-2 \text{ standard deviations} < \hat{p} - p < 2 \text{ standard deviations}$$

This means that in about 95% of all samples, the difference between the sample proportion and the true population proportion is less than 2 standard deviations.

We have one difficulty. The standard deviation formula involves p, the true population proportion, a value that we don't know. The obvious solution is to put $\hat{p}$, the sample proportion, in the place of p in the standard deviation formula. As we discussed in Chapter 9, when this is done, many researchers use the term **standard error** to describe the result. Thus the **standard error of a sample proportion** is

$$\text{s.e.}(\hat{p}) = \sqrt{\frac{\hat{p}(1-\hat{p})}{n}}$$

If we assume that the standard error of $\hat{p}$ is a reasonably close approximation to the standard deviation of $\hat{p}$, we can use the standard error in place of the standard deviation. So for approximately 95% of all random samples from a population, the difference between the sample proportion and the population proportion is described by the mathematical inequality

$$-2 \text{ standard errors} < \hat{p} - p < 2 \text{ standard errors}$$

This mathematical inequality says that in about 95% of all samples, the difference between the sample proportion and the population proportion is less than 2 standard errors. With two steps of algebra, this inequality can be changed to

$$\hat{p} - 2 \text{ standard errors} < p < \hat{p} + 2 \text{ standard errors}$$

This holds for about 95% of all random samples and thus provides the formula for the 95% confidence interval.

Developing the Interval for Other Levels of Confidence

There is nothing special about the 95% level of confidence except that the Empirical Rule provided us with the knowledge of approximately what multiplier to use (2.0). As illustrated in Figure 10.2 on page 378, for any confidence level we can find the appropriate multiplier z^* such that the desired percentage of the standard normal curve lies between $-z^*$ and $+z^*$. Call the confidence level C, expressed as a proportion. For instance, for a 95% confidence interval, $C = .95$.

To find z^*, note that the standard normal curve can be partitioned into three sections:

1. The area between $-z^*$ and z^*, which is C, the confidence level.
2. The area below $-z^*$, which is half of the area not between $-z^*$ and z^*. This area is $(1 - C)/2$.
3. The area above $+z^*$, which by symmetry equals the area below $-z^*$. This area is $(1 - C)/2$.

One strategy for determining z^* for a general confidence level is to use the fact that $-z^*$ is the z-score that has area $(1 - C)/2$ below it. This can easily be read from Table A.1 in the Appendix of Tables. Search for the value of $(1 - C)/2$ as a cumulative probability within the body of Table A.1 and then identify the corresponding value of $-z^*$.

For instance, for a 90% confidence level the area between $-z^*$ and z^* is .90, so the area below $-z^*$ is $(1 - .90)/2$ or .05. Looking up .05 in the body of Table A.1, we find that $-z^*$ is between -1.64 (area is .0505) and -1.65 (area is .0495). Therefore,

CI Module 1: *p*

we use a multiplier of 1.645 for a 90% confidence interval. Figure 10.3 illustrates the relationship between the confidence level and the values of $-z^*$ and z^* for a 90% confidence interval.

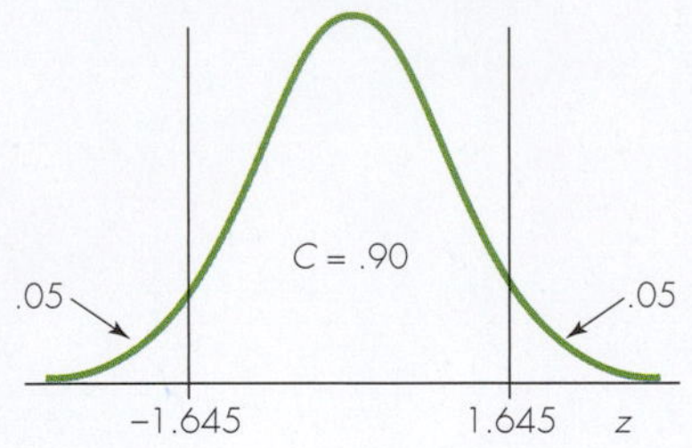

Figure 10.3 z^* for a 90% confidence interval

Example 10.5 **Example 10.3 Revisited: 50% Confidence Interval for Proportion Believing That Intelligent Life Exists Elsewhere** Example 10.3 described a 2008 survey in which the sample proportion that said it was somewhat or very likely that intelligent life exists on another planet is .56. The sample size was $n = 1003$ and the standard error of the sample proportion is .016. Suppose we want a 50% confidence interval that estimates the population proportion that believed it to be somewhat or very likely that intelligent life exists elsewhere. Figure 10.4 illustrates the problem that we have to solve to find the z^* value for a 50% confidence level.

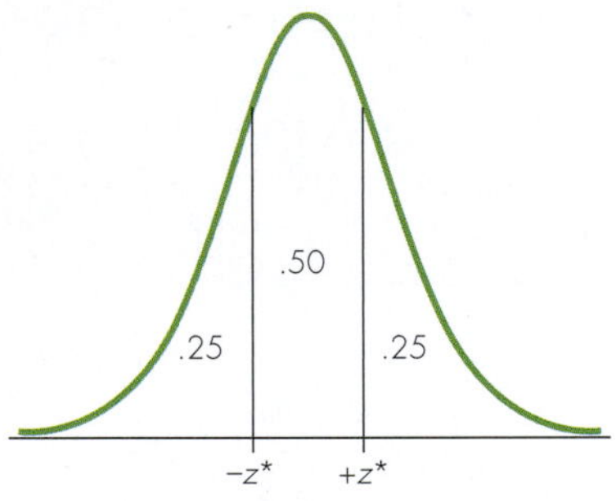

Figure 10.4 Finding z^* for the 50% confidence level

To use the Standard Normal Curve table (Table A.1 of the Appendix), we observe from Figure 10.4 that if the area between $-z^*$ and $+z^*$ is .50, the area to the left of $-z^*$ is .25, and the area to the right of z^* is also .25. From Table A.1, we learn that the z-value for which the cumulative probability is .25 is about -0.67. This is the value of $-z^*$, so a 50% confidence interval for a population proportion is *sample proportion* ± 0.67 *standard errors,* which in this example is $.56 \pm 0.67 \times .016$, or $.56 \pm .011$. This interval is narrower than the 90% and 98% confidence intervals computed previously in Example 10.3. Using a lower confidence level produces a narrower interval.

THOUGHT QUESTION 10.4 As we noted toward the end of Section 9.4, s.e.($\hat{p}$) is an estimate of s.d.($\hat{p}$), and it is generally a close estimate. Use the Empirical Rule from Chapters 2 and 8 and the sampling distribution for a sample proportion from Section 9.4 to draw a picture of possible sample proportions, using s.e.($\hat{p}$) in place of s.d.($\hat{p}$). Show what percentage of sample proportions you expect to fall into the following ranges:

Population proportion $\pm$ s.e.($\hat{p}$)
Population proportion $\pm$ 2 s.e.($\hat{p}$)
Population proportion $\pm$ 3 s.e.($\hat{p}$)

Then identify what connection margin of error has with the value "95%" in this picture.*

***HINT:** Example 9.4 (p. 326) and the Thought Question accompanying it can be used for guidance.

Lesson 3: Reconciling and Understanding Different Margin of Error Formulas

In CI Module 0, we learned that a confidence interval for any of our five situations is given by the formula,

Sample estimate ± Multiplier × Standard error

We also learned that for a 95% confidence interval for one proportion,

$$\text{Multiplier} \times \text{standard error} = 2\sqrt{\frac{\hat{p}(1-\hat{p})}{n}}$$

However, in Chapter 5, we learned that a 95% confidence interval for a population proportion is given by the formula,

Sample estimate ± Margin of error

where the margin of error was given as $1/\sqrt{n}$. How do we reconcile these two different formulas for a 95% confidence interval for a proportion?

The term **margin of error** can be generally used to describe the value of Multiplier × Standard error in a 95% confidence interval estimate of a population proportion. When estimating a proportion, it is equivalent to write either "±Multiplier × Standard error" or "±Margin of error." However, the value of the margin of error does not always equal $1/\sqrt{n}$. The formula given next describes a more precise margin of error that can be used in all circumstances.

FORMULA **Margin of Error for 95% Confidence**

For a 95% confidence level, a formula for the **margin of error for a sample proportion** is

$$\text{Margin of error} = 2\sqrt{\frac{\hat{p}(1-\hat{p})}{n}}$$

In other words, the "95% margin of error" is two standard errors, or 2 s.e.($\hat{p}$).

The two different formulas we've seen for the margin of error are equivalent when $\hat{p} = .5$. In that case, the more complicated formula for margin of error becomes

$$\text{Margin of error} = 2 \times \sqrt{\frac{.5(1-.5)}{n}} = \frac{1}{\sqrt{n}}$$

The margin of error usually reported in the news is a conservative approximation of the margin of error. It is an approximation that works best when the true population proportion p is close to .5. An important part of this story is that when you use any proportion other than .5 in the more precise formula, the answer is smaller than what you get when you use .5. This means that the conservative approximation will almost always overestimate the size of the margin of error.

Example 10.6 **Winning the Lottery and Quitting Work** What would you do for the rest of your life if you won a lot of money in a lottery? In a 1997 poll conducted by the Gallup Organization, one of the questions was "If you won 10 million dollars in the lottery would you continue to work, or would you stop working?" The results were reported at the Gallup Organization's website, http://www.gallup.com.

Surprisingly, 59% of the 616 employed respondents answered that they would continue working, 40% said that they would stop working, and 1% had no opinion. The website article also gave this information about the poll:

> The current results are based on telephone interviews with a randomly selected sample of 1014 adults, conducted August 22–25, 1997. Among this group, 616 are

employed full-time or part-time. For results based on this sample of "workers," one can say with 95% confidence that the error attributable to sampling could be plus or minus 4 percentage points.

Gallup describes the margin of error with the phrase "could be plus or minus 4 percentage points," but the phrase "could be *as large as* plus or minus 4 percentage points" would have been better. The margin of error is a likely upper limit on the possible difference between the sample and population percentages. Here, there is a 95% chance that the difference is 4 percentage points or less.

In the poll, 40% of the $n = 616$ employed respondents said that they would quit working if they won the lottery, and the margin of error is 4%. Our formulas are based on proportions, so we will translate these percentages to proportions giving a *sample estimate* of $\hat{p} = .4$ and a margin of error of .04. The reported margin of error of .04 was calculated as $1/\sqrt{n} = 1/\sqrt{616} = .0403$.

The more accurate calculation of margin of error is

$$2\sqrt{\frac{\hat{p}(1-\hat{p})}{n}} = 2\sqrt{\frac{.4(1-.4)}{616}} = 2(.01974) = .03948.$$

You can see that this value, .03948, is nearly the same as the margin of error of .04 reported by Gallup with the survey. This is because $\hat{p} = .4$ is close to .5, so the two formulas give about the same value.

Therefore, in this situation, a 95% confidence interval will be essentially the same regardless of which formula we use for the margin of error. The interval is

Sample estimate $\pm$ *margin of error*

.4 $\pm$ *.04, which is .36 to .44*

A 95% confidence interval that estimates the proportion of all working adults in the United States who would quit working if they won a large amount in the lottery is .36 to .44, or 36% to 44%.

Note that the interval resides completely below 50%. Therefore, it would be fair to conclude, with high confidence, that fewer than half of all working Americans think they would quit if they won the lottery.

Why Polling Organizations Report a Conservative Margin of Error

The formula $1/\sqrt{n}$ provides a conservative approximation of the margin of error in that it usually overestimates the actual size of the margin of error, and thus leads to an underestimate of how confident we can be that our interval estimate covers the truth. This conservative number is used by polling organizations so that they can report a single margin of error for all questions on a survey. The more precise formula involves the sample proportion, so with that formula the margin of error might change from one question to the next within the same survey. The formula $1/\sqrt{n}$ provides a margin of error that works (conservatively) for all questions based on the same sample size, even if the sample proportions differ from one question to the next.

Example 10.7 **The Gallup Poll Margin of Error for n = 1000** The Gallup Organization often uses samples of about 1000 randomly selected Americans. For $n = 1000$, the conservative estimate of margin of error is $1/\sqrt{1000} = .032$, or about 3.2%. If the proportion for which the margin of error is used is close to .5, then the reported (conservative) margin of error of .032 will be a good approximation to the real, more accurate margin of error. However, if the proportion is much smaller or larger than .5, then using .032 as the margin of error will produce an interval that is wider than necessary for the desired 95% confidence. For instance, if $\hat{p}$ is .1, then a more precise estimate of the margin of error is $2\sqrt{(.1)(.9)/1000} = .019$. You can see that using $\hat{p} \pm .032$ instead of $\hat{p} \pm .019$ would give you an interval that is much wider than needed.

The next example describes a situation in which an unnecessarily wide interval includes negative values, which are not possible for a population proportion.

Example 10.8 **Example 10.2 Revisited: Allergies and Really Bad Allergies** In the Marist Institute allergy survey, about 3% said that they experienced "severe" allergy symptoms. The sample size is $n = 883$, so a conservative approximation of the margin of error is $1/\sqrt{883} = .034$, or 3.4%. Using this conservative approximation, a 95% confidence interval estimate of the percentage of all Americans who have severe allergies is 3% ± 3.4%, or −0.4% to 6.4%. Note that the lower end of this interval is negative, an impossible value for a population percentage, so the interval would be reported as 0% to 6.4%.

Generally, when the sample proportion $\hat{p}$ is far from .5, as it is for the severe allergy symptoms group, the conservative approximation is too conservative. Using $\hat{p} = .03$ in the more precise formula produces .011, or 1.1%, as the margin of error value. The corresponding 95% confidence interval estimate of the population percentage with severe allergy symptoms is 3% ± 1.1%, which is 1.9% to 4.1%.

THOUGHT QUESTION 10.5 Suppose that we collect data on "hours of sleep last night" from 64 students who are present at an 11:00 A.M. statistics class. The sample mean is 6.8 hours, the standard deviation of the data is 1.6 hours, and the standard error of the mean is 0.2 hour. What population, if any, is represented by this sample? Assuming that there is a population represented by this sample, explain why we cannot use the formula

$$\hat{p} \pm z^* \sqrt{\frac{\hat{p}(1 - \hat{p})}{n}}$$

to compute a 95% confidence interval for the population mean. What general formula in this chapter could be used to calculate the interval? Calculate the confidence interval.*

An Intuitive Explanation of Margin of Error

The margin of error has these characteristics:

- The difference between the sample proportion and the population proportion is *less* than the margin of error *about 95% of the time*, or for about 19 of every 20 sample estimates.
- The difference between the sample proportion and the population proportion is *more* than the margin of error *about 5% of the time*, or for about 1 of every 20 sample estimates.

In other words, for most sample estimates, about 95% of them, the actual error is quite likely to be smaller than the margin of error. Occasionally, about 5% of the time, the error might be larger than the margin of error. Unfortunately, we never know the actual amount of error in a particular estimate. We know only that most of the time, the actual error is less than the reported margin of error.

If we use the margin of error to form an interval around the sample estimate, there are two things that can happen:

- If, by the luck of the draw, we have one of the sample estimates for which the actual error is *smaller* than the margin of error, then the sample estimate is close enough to the true population value that the interval "sample estimate ± margin of error" will include the true value. Because at least 95% of all sample estimates have this characteristic, our method will work at least 95% of the time.

***HINT:** Is the given formula appropriate for a mean? Find the formula box that provides a confidence interval for any of the five parameters. Table 10.1 can be used for the multiplier.

- If we happen to have one of the few sample estimates for which the actual error is *larger* than the margin of error, then the interval "sample estimate ± margin of error" will *not* include the true value. This will happen with no more than 5% of possible sample estimates.

10.2 Exercises are on pages 396–398.

10.3 CI Module 2: Confidence Intervals for the Difference in Two Population Proportions

The objective in this section is to form a **confidence interval for the difference between two population proportions.** Again, we start with the same general format that we introduced in CI Module 0 for the five parameters of interest, which is

Sample estimate ± Multiplier × Standard error

Let's review the situation that generates research questions for which the difference between two population proportions is of interest. We have two populations or groups from which independent samples are available. We are interested in comparing the two populations with respect to the proportion with a certain trait, opinion, or characteristic. For example, we might want to compare the proportions of males and females who support the death penalty. Or, we might want to compare the proportion of physicians that would have a heart attack if they were to take daily aspirin to the proportion that would have a heart attack if they were to take a daily placebo instead.

The notation is as follows:

	Population Proportion	Sample Size	Number with Trait	Sample Proportion
Population 1	p_1	n_1	X_1	$\hat{p}_1 = \dfrac{X_1}{n_1}$
Population 2	p_2	n_2	X_2	$\hat{p}_2 = \dfrac{X_2}{n_2}$

The *sample estimate* is $\hat{p}_1 - \hat{p}_2$, the difference between the two sample proportions. The *standard error* is

$$\text{s.e.}(\hat{p}_1 - \hat{p}_2) = \sqrt{\frac{\hat{p}_1(1-\hat{p}_1)}{n_1} + \frac{\hat{p}_2(1-\hat{p}_2)}{n_2}}$$

The multiplier is denoted as z^*, and it is determined by using the standard normal distribution.

FORMULA

Confidence Interval for the Difference Between Two Population Proportions

A **confidence interval for $p_1 - p_2$** is

$$(\hat{p}_1 - \hat{p}_2) \pm z^* \sqrt{\frac{\hat{p}_1(1-\hat{p}_1)}{n_1} + \frac{\hat{p}_2(1-\hat{p}_2)}{n_2}}$$

z^* is the value of the standard normal variable with the desired confidence level as the area between $-z^*$ and z^*.

Note: Values of z^* for common confidence levels can be found in Table 10.1, or in the last row of Table A.2 in the Appendix for reasons that are explained in CI Module 3 in Chapter 11.

Conditions for a Confidence Interval for the Difference in Two Proportions

As is the case for all statistical inference procedures, there are conditions that should be present in order to use the confidence interval just described. For a confidence interval for the difference between two proportions, the principal conditions have to do with the sampling method and the sample sizes for the observed samples. Both of the following conditions must hold.

Condition 1: Sample proportions are available based on independent, randomly selected samples from the two populations.

Condition 2: All of the quantities $n_1\hat{p}_1$, $n_1(1 - \hat{p}_1)$, $n_2\hat{p}_2$, and $n_2(1 - \hat{p}_2)$ are at least 10. These quantities represent the counts observed in the category of interest and not in that category, respectively, for the two samples.

Note that these are the same conditions that are required for the sampling distribution of $\hat{p}_1 - \hat{p}_2$ to be approximately normal, except that the sample size condition now applies using the sample $\hat{p}$'s instead of the population p's.

Example 10.9 **Age and Using the Internet as a News Source** One difference between younger and older people in the United States is that younger people are more likely to rely on the Internet as their main news source. In the 2008 General Social Survey, participants were asked what their main news source is. Among those surveyed in the 18- to 29-year-old age group, 92/262 = .351 (35.1%) said that the Internet was their main news source. In the 50-and-over age group, only 59/632 = .093 (9.3%) said that the Internet was their main news source (Data source: http://sda.berkeley.edu/archive.htm, accessed March 9, 2010).

We'll use this sample information to determine a 95% confidence interval that estimates the parameter $p_1 - p_2$ = difference between population proportions who use the Internet as their main source in the two age groups. With the 18- to 29-year-old age group as Population 1 and the 50-and-over group as Population 2, the sample estimate of the difference and the standard error of the difference follow:

$$\hat{p}_1 - \hat{p}_2 = .351 - .093 = .258$$

$$\text{s.e.}(\hat{p}_1 - \hat{p}_2) = \sqrt{\frac{\hat{p}_1(1 - \hat{p}_1)}{n_1} + \frac{\hat{p}_2(1 - \hat{p}_2)}{n_2}}$$

$$= \sqrt{\frac{.351(1 - .351)}{262} + \frac{.093(1 - .093)}{632}} = .0317$$

In this situation, the general format of a confidence interval estimate of $p_1 - p_2$ is

$$\hat{p}_1 - \hat{p}_2 \pm z^*\text{s.e.}(\hat{p}_1 - \hat{p}_2)$$

$$.258 \pm z^*(.0317)$$

To complete the calculation of a confidence interval, we need only to choose a confidence level in order to determine the value of z^*. The following table gives confidence intervals for three different confidence levels. For each level, we show the appropriate z^* multiplier from Table 10.1 (p. 378) and the completed answer to $.258 \pm z^*(.0317)$.

Confidence Level	z^*	Confidence Interval
90%	1.645	0.206 to 0.310
95%	1.960	0.196 to 0.320
99%	2.576	0.176 to 0.340

Note that, as always is the case, the higher the level of confidence the wider the interval. An interpretation of the 95% confidence interval is that we are 95% confident that the population proportion of 18- to 29-year-olds who use the Internet as their main source of news is between .196 and .320 higher than the proportion using the Internet as their main source of news in the population aged 50 and older. In percents, we are 95% confident that somewhere between 19.6% and 32.0% more young adults than older adults use the Internet as their main news source.

> **MINITAB TIP** **Determining the Interval for the Difference in Proportions**
>
> - To compute a confidence interval for the difference in two proportions, use **Stat > Basic Statistics > 2 Proportions**. See the following note for information about inputting data.
> - To change the confidence level, use the ***Options*** button. The default confidence level is 95%.
>
> ***Note:*** The raw data for the response may be in one column, and the raw data for group categories (Subscripts) may be in a second column. Or the raw data for the two independent groups may be in two separate columns. Or the data may already be summarized. If so, click ***Summarized Data***, and then specify the sample size (trials) and the *number* of successes for each group.

Example 10.10 **Do You Always Buckle Up When Driving?** How often do you wear a seat belt when driving a car? This was asked in the 2003 Youth Risk Behavior Surveillance System, a nationwide biennial survey of 9th through 12th graders in the United States. Possible responses to the question about seat belt use were as follows: never, rarely, sometimes, most times, and always. Example 2.1 and Table 2.2 presented data on the connection between seat belt use and course grades usually attained in school for 12th graders who said that they drive. In this example, we compare female and male 12th-grade drivers. To do so, we find a 95% confidence interval for $p_1 - p_2$ = difference between the population proportions of 12th-grade female and male drivers who would say they always wear a seat belt when driving. The Minitab output for this confidence interval follows. (The data are in the **YouthRisk03** dataset on the companion website, http://www.cengage.com/statistics/Utts4e.)

```
Session

Test and CI for Two Proportions

Event = Always

Sex       X     N  Sample p
Female  915  1467  0.623722
Male    771  1575  0.489524

Difference = p (Female) - p (Male)
Estimate for difference:  0.134198
95% CI for difference:  (0.0992116, 0.169184)
Test for difference = 0 (vs not = 0):  Z = 7.52  P-Value = 0.000
```

The sample proportions of females and males who said they always wear a seat belt are given toward the top of the output under "Sample *p*." The values under "*X*" and "*N*" are the number who said they always wear a seat belt and the sample size for each sex, respectively. Rounded to three decimal places, the sample proportion for females is $\hat{p}_1 = 915/1467 = .624$ and the sample proportion for males is $\hat{p}_2 = 771/1575 = .490$. We see that females are more likely to say that they always wear a seat belt, and the difference between the sample proportions is $\hat{p}_1 - \hat{p}_2 = .624 - .490 = .134$ (13.4%). This value is given in the line of output that begins "Estimate for..."

The 95% confidence interval for $p_1 - p_2$ rounded to three decimal places is 0.099 to 0.169. This information is given in the next-to-last line of output. The confidence interval estimates the difference in population proportions of female and male 12th-grade drivers who would say that they always wear a seat belt when driving. As always, the 95% confidence level expresses our confidence in the procedure that is used to find the interval. Because the interval does not include 0, it is reasonable to conclude that there is a difference between the population proportions. Apparently, 12th-grade female drivers are more likely than 12th-grade male drivers to say that they always buckle up when driving.

The last line of output gives results for a hypothesis test in which the null hypothesis is that there is no difference between the female and male population proportions. The details for that type of hypothesis test will be covered in Chapter 12.

TI-84 TIP **Calculating a Confidence Interval for the Difference between Two Proportions**

- Press STAT and scroll horizontally to **TESTS** (highlighted). Then, scroll vertically to **B:2-PropZInt** and press ENTER.
- Enter values for $x1$ = observed number of successes for sample 1, $n1$ = sample size for sample 1, $x2$ = observed number of successes for sample 2, $n2$ = sample size for sample 2, and *C-Level* = confidence level expressed as a decimal fraction or a percentage (e.g., enter either .90 or 90 for 90% confidence). Press ENTER.
- **CALCULATE** will be flashing. Press ENTER. The resulting display will give the confidence interval in the form (a, b) and will also give values for $\hat{p}_1$, $\hat{p}_2$, n_1, and n_2.

THOUGHT QUESTION 10.6 An environmental group is suing a manufacturer because chemicals dumped into a nearby river may be harming fish. A sample of fish from upstream (no chemicals) is compared with a sample from downstream (chemicals), and a 95% confidence interval for the difference in proportions of healthy fish is 0.01 to 0.11 (with a higher proportion of healthy fish upstream). First, interpret this interval. The statistician for the manufacturer produces a 99% confidence interval ranging from −0.01 to +0.13 and tells the judge that because it includes 0 and because it has higher accompanying confidence than the other interval, we can't conclude that there is a problem. Comment on the statistician's conclusion.*

10.3 Exercises are on pages 398–399.

10.4 Using Confidence Intervals to Guide Decisions

Two research questions that investigators often consider about population proportions are as follows:

- Is it reasonable to reject a specific hypothesized value of a population proportion?
- Is it reasonable to conclude that two population proportions are different from each other?

In Chapter 12, we will learn how to use hypothesis tests to study these questions. Confidence intervals, however, can be used as a somewhat informal method for evaluating these issues.

Principles for Using Confidence Intervals to Guide Decision Making

Three general principles are useful for examining these questions. These principles can be applied to any situation in which confidence intervals are calculated. We state the principles only in the context of proportions, but they can be used to answer questions about population means or any other population parameter.

Principle 1. A value that is not in a confidence interval can be rejected as a likely value of the population proportion. Any value that is in a confidence interval is an "acceptable" possibility for the value of a population proportion.

***HINT:** Does knowing the 99% interval change the interpretation of the 95% interval? You could also turn the argument around by noting the upper end of the 99% interval.

Principle 2. When a confidence interval for the difference in two proportions does not cover 0, it is reasonable to conclude that the two population proportions are different.

Principle 3. When the confidence intervals for proportions in two different populations do not overlap, it is reasonable to conclude that the two population proportions are different. However, if the intervals do overlap, no conclusion can be made. A confidence interval for the difference should be constructed to determine whether 0 is a plausible value.

To better understand these principles, let's look at some examples.

Example 10.11 **Which Drink Tastes Better?** Imagine a taste test done to compare two different soft drinks. In the test, 60 people taste both drinks and 55% of these 60 participants say that they like the taste of drink A better than the taste of drink B. The makers of drink A will be excited that their drink was picked by more than half of the sample and might even create commercials to brag about the results. The makers of drink B, however, can use a 95% confidence interval to show that there is not enough evidence to claim that drink A would be preferred by a majority of the population represented by the sample.

An approximate 95% confidence interval for p, the population proportion that prefers drink A, is .42 to .68. The calculation is sample proportion ± margin of error:

$$.55 \pm 2\sqrt{\frac{.55(1 - .55)}{60}}$$

$$.55 \pm .13$$

Here, the confidence interval leaves the issue unsettled. Most of the interval is above .50, so the makers of drink A may argue that it is reasonable to claim that more than half of the population prefers their drink. On the other hand, the makers of drink B can point out that it is possible that a majority of the population actually prefers their drink because the confidence interval also includes some estimates that are less than .50 (50%)—in fact as low as only 42%.

The difficulty with this hypothetical experiment is that it is too small to be definitive. In actual practice, a much larger sample size should be used to generate a more precise estimate of the proportion that prefers drink A.

10.4 Exercises are on pages 399–400.

CASE STUDY 10.1 Extrasensory Perception Works with Movies

Extrasensory perception (ESP) is the apparent ability to obtain information in ways that exclude ordinary sensory channels. Early ESP research focused on having people try to guess at simple targets, such as symbols on cards, to find out whether they could guess at a better rate than would be expected by chance. In recent years, experimenters have used more interesting targets, such as photographs, outdoor scenes, or short movie segments.

In a study of ESP reported by Bem and Honorton (1994), subjects called "receivers" described what another person, the "sender," was seeing on a television screen in another room. The receivers were shown four pictures and asked to pick which one they thought the sender had actually seen. The actual image shown to the sender had been randomly picked by the investigators from among these four choices.

Sometimes the sender was looking at a single, "static" image on the television screen and sometimes at a "dynamic" short video clip, played repeatedly. The additional three choices shown to the receiver for judging were always of the same type (static or dynamic) as the actual target to eliminate biases due to a preference for one type of picture.

Results from the experiment are shown in Table 10.2. Because the right answer is one of four possibilities, receivers who are randomly guessing will be right about 25% of the time. For the static pictures, the 27% success rate is about what would be expected from random guessing. The 41% success rate for the dynamic pictures, however, is noticeably higher than the 25% that would result from random guessing.

(continued)

Table 10.2 Results of the Bem and Honorton ESP Study

	Successful Guess?			
	Yes	No	Total	Percent Success
Static picture	45	119	164	27.4%
Dynamic picture	77	113	190	40.5%

Source: Bem and Honorton, 1994.

A skeptic might raise the possibility that the results for the dynamic pictures were the product of good luck. Is there enough evidence to safely say that the percentage of correct guesses for the dynamic pictures is significantly above 25%, the percentage that would be expected by chance? A confidence interval helps us to answer this question.

For the dynamic pictures, the 95% margin of error is

$$\text{Margin of error} = 2 \times \sqrt{\frac{.405 \times (1 - .405)}{190}} = .071$$

So an approximate 95% confidence interval for the proportion is .405 ± .071, which is .334 to .476. In percentage terms, the interval ranges from 33.4% to 47.6%, an interval that does not include 25%. This means that it is reasonable to reject the possibility that the true percentage of correct guesses is 25%. The researchers can claim that the true percentage of correct guesses is, in fact, significantly better than what would occur from random guessing.

CASE STUDY 10.2 Nicotine Patches versus Zyban®

Read the original source at the companion website, http://www.cengage.com/statistics/Utts4e.

Some of you may know from personal experience that quitting smoking is difficult. Many people who are trying to quit use nicotine replacement methods like nicotine patches or nicotine gum to ease nicotine withdrawal symptoms. As an alternative, medical researchers have investigated whether the use of an antidepressant medication might be a more effective aid to those attempting to give up cigarettes. In a study reported in the March 4, 1999, *New England Journal of Medicine* (at the companion website for this book), Dr. Douglas Jorenby and colleagues compared the effectiveness of nicotine patches to the effectiveness of the antidepressant bupropion, which is marketed with the brand name Zyban.

The 893 participants were randomly allocated to four treatment groups: placebo, nicotine patch only, Zyban only, and Zyban plus nicotine patch. To keep participants blind as to their treatments, they all used a patch (nicotine or placebo) and took a pill (Zyban or placebo). For instance, in the placebo-only group, participants used both a placebo patch and took placebo pills. The Zyban group also used placebo patches, and so on. The treatments were used for 9 weeks.

Table 10.3 displays, for each treatment group, an approximate 95% confidence interval for the proportion not smoking 6 months after the start of the experiment. Each interval is a range of estimates of the proportion that would not be smoking after 6 months in a population of individuals using that particular method. A more direct way to compare the four success rates and determine if there are differences will be given in Chapter 15. At this point, we can use the intervals shown in the table to gauge how the four treatments might differ.

Table 10.3 95% Confidence Intervals for Proportion Not Smoking after 6 Months

Treatment	Subjects	Proportion Not Smoking	Approximate 95% CI
Placebo only	160	.188	.13 to .25
Nicotine patch	244	.213	.16 to .26
Zyban	244	.348	.29 to .41
Zyban and nicotine patch	245	.388	.33 to .45

Figure 10.5 shows the 95% confidence intervals graphically.

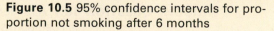

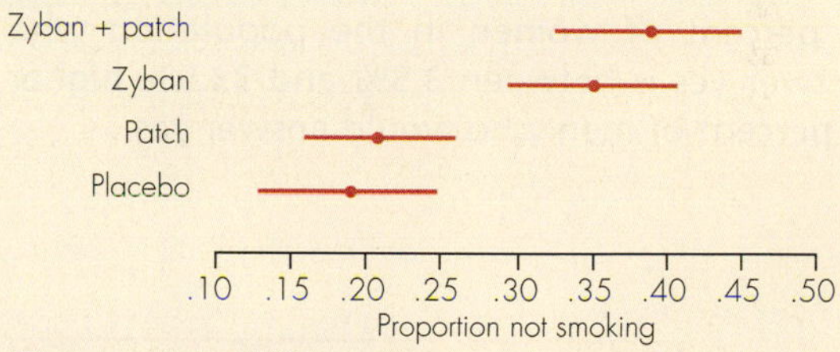

Figure 10.5 95% confidence intervals for proportion not smoking after 6 months

On the basis of the graph, we can make the following generalizations:

- Zyban appears to be effective. The Zyban groups had higher success rates, and the confidence intervals for the two groups that used Zyban do not overlap with the intervals for the two groups not using Zyban.
- There is substantial overlap in the range of estimates in the intervals for the nicotine patch and placebo groups, so we can't conclude that the nicotine patch is better than the placebo. Similarly, there is substantial overlap between the intervals for the Zyban-only group and the Zyban-plus-patch group, so we can't conclude that the nicotine patch significantly enhances the effectiveness of Zyban.

CASE STUDY 10.3 What a Great Personality

Students in a college statistics class were asked, "Would you date someone with a great personality even though you did not find them attractive?" The results were as follows:

- 61.1% of 131 women answered "yes."
- 42.6% of 61 men answered "yes."

There clearly is a difference between the men and the women in these samples. Can this difference be generalized to the populations represented by the samples? This question brings up another question: What are the populations that we are comparing? These men and women weren't randomly picked from any particular population, but for this example, we'll assume that they are like a random sample from the populations of all American college men and women.

To examine the difference between the men and women, we'll use a 95% confidence interval to estimate the parameter $p_1 - p_2 =$ difference between population proportions of women and men who would answer "yes" to this question. The calculation is

$$(.611 - .426) \pm 1.96\sqrt{\frac{.611(1 - .611)}{131} + \frac{.426(1 - .426)}{61}}$$

$.185 \pm .150$, which is .035 to .335.

The interval does not include 0 so it is reasonable to conclude that the population proportions differ. We are 95% confident that the population proportion is between 0.035 and 0.335 higher for women than it is for men. Expressed as a percent, we are 95% confident that the percent of women in the population who would answer yes is between 3.5% and 33.5% higher than the percent of men who would answer yes.

The 95% confidence intervals that separately estimate the population proportions for women and men also provide useful information.

- For women, the approximate 95% confidence interval is .527 to .694.
- For men, the approximate 95% confidence interval is .302 to .550.

The two separate intervals are compared graphically in Figure 10.6. Note that there is a slight overlap between the two intervals, so if this were our only analysis we would not be able to infer whether there is a difference in the population proportions. However, the confidence interval for the *difference* between the proportions of men and women who would answer yes did provide a conclusion.

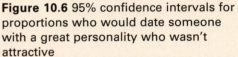

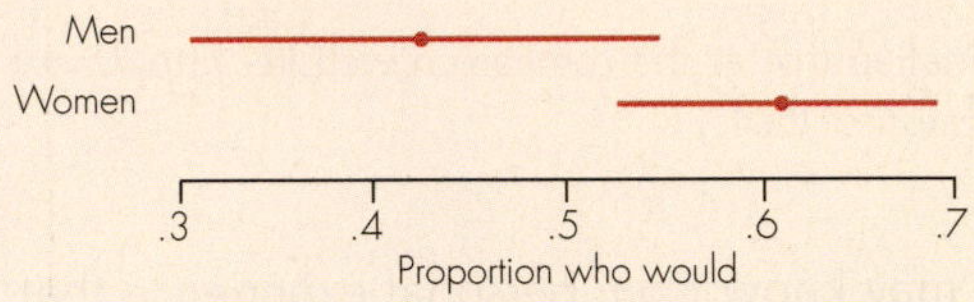

Figure 10.6 95% confidence intervals for proportions who would date someone with a great personality who wasn't attractive

We can also see the effect of sample size on interval width in Figure 10.6. The confidence interval for women is not as wide as the interval for men. There were more than twice as many women as there were men in the class, and the larger sample size for women leads to a more precise estimate of the population proportion.

THOUGHT QUESTION 10.7 A randomly selected sample of 400 students is surveyed about whether additional coed dorms should be created at their school. Of those surveyed, 57% say that there should be more coed dorms. The 95% margin of error for the survey is 5%. Compute a 95% confidence interval for the population percentage in favor of more coed dorms. On the basis of this confidence interval, can we conclude that more than 50% of all students favor more coed dorms? Explain. Can we reject the possibility that the population proportion is .60?*

***HINT:** See Example 10.11 (p. 390) and the end of Example 10.6 (pp. 383–384).

IN SUMMARY Confidence Intervals for Population Proportions and Differences in Population Proportions

In this chapter, each method covered for finding a confidence interval estimate of a population proportion (p) or a difference of two proportions is an instance of the same general format, which is

Sample estimate ± Multiplier × Standard error

The methods that were covered were as follows:

- *Confidence interval for p with exact multiplier for any confidence level:*

$$\hat{p} \pm z^* \sqrt{\frac{\hat{p}(1-\hat{p})}{n}}$$

where the symbol $\hat{p}$ represents the sample proportion, and the multiplier z^* is such that the area under the standard normal curve between $-z^*$ and z^* is the confidence level.

- *For a 95% confidence interval,* the exact multiplier is $z^* = 1.96$. This multiplier is often rounded to 2, and the term **approximate 95% confidence interval** is used in that case.
- *A 95% confidence interval is sometimes written as*

Sample estimate ± Margin of error

where the margin of error is approximately 2 × Standard error. A conservative estimate of the margin of error for 95% confidence is $1/\sqrt{n}$.

- *Confidence interval for the difference in two population proportions:*

$$(\hat{p}_1 - \hat{p}_2) \pm z^* \sqrt{\frac{\hat{p}_1(1-\hat{p}_1)}{n_1} + \frac{\hat{p}_2(1-\hat{p}_2)}{n_2}}$$

Key Terms

In Summary Boxes

Exercises

◆ Denotes that the dataset is available on the companion website, http://www.cengage.com/statistics/Utts4e, but is not required to solve the exercise.

Bold exercises have answers in the back of the text.

Section 10.1

Exercises 10.1 to 10.22 correspond to the two Lessons in Section 10.1. Lesson 1 exercises are 10.1 to 10.14; Lesson 2 exercises are 10.15 to 10.22.

Lesson 1 Skillbuilder Exercises

10.1 In each part of this question, explain whether the proportion that is described is a sample proportion or a population proportion.

a. In the 1990 U.S. Census, it was found that about one in nine Americans were at least 65 years old at that time.

b. A randomly selected group of 500 registered voters in your state is asked whom they intend to vote for in an upcoming election for governor. Based on the proportion choosing each candidate, the polling organization predicts the election winner.

c. In a clinical trial done to assess the effectiveness of a new medication for asthma, a satisfactory relief from symptoms was experienced by 55% of $n = 80$ participants.

10.2 *This is also Exercise 2.6.* For each of the following statistical summaries, explain whether it is a population parameter or a sample statistic.

a. A highway safety researcher wants to estimate the average distance at which all drivers can read a highway sign at night. She measures the distance for a sample of 50 drivers; the average distance for these 50 drivers is 495 feet.

b. The average score on the final exam is 76.8 for $n = 83$ students in a statistics class. The instructor is only interested in describing the performance of this particular class.

c. Case Study 1.3 (p. 3) reported that in a Gallup poll, 57% of $n = 496$ teens who date said they have been out with someone of another race or ethnic group.

10.3 A Gallup poll surveyed a random sample of 439 American teenagers. One question that they were asked was, "How strict are your parents compared to most of your friends' parents?" The choices were "more strict," "less strict," and "about the same." The choice "more strict" was selected by 171 of the respondents (Mazzuca, 2004).

a. What is the research question of interest?

b. What is the population parameter of interest?

c. What percentage of the sample responded with "more strict"?

d. Draw a figure like Figure 10.1 (p. 373) to illustrate this situation.

10.4 A survey is done to estimate the proportion of U.S. adults who think that cell phone use while driving should be illegal. In the survey, 54% of a randomly selected sample of 1025 individuals said that cell phone use while driving should be illegal.

a. What is the population of interest?

b. What is the population parameter of interest?

c. What is the sample?

d. What is the value of the sample statistic?

10.5 What is the likelihood that a 95% confidence interval will not cover the true population value? Explain.

10.6 Refer to Exercise 10.3 about a Gallup poll on how strict teenagers perceive their parents to be in comparison to other parents, based on a random sample of 439 teenagers. The choice "more strict" was selected by 171 of the respondents. Suppose that in fact the true population percentage of teenagers who think that their parents are more strict is 40%.

a. If a new sample of 439 teenagers from the same population were taken, which one of the following would be true? Explain your answer.

i. The number selecting the choice "more strict" would be 171 again.

ii. The number selecting the choice "more strict" could be 171 again.

iii. The number selecting the choice "more strict" could not be 171 again.

b. If a new sample of 439 teenagers from the same population were taken, which one of the following would be true? Explain your answer.

i. The true population percentage of teenagers who think that their parents are more strict would be 40%.

ii. The true population percentage of teenagers who think that their parents are more strict could be 40%.

iii. The true population percentage of teenagers who think that their parents are more strict could not be 40%.

10.7 Suppose that a survey is done to estimate the proportion of U.S. adults who think that the use of handheld cell phones while driving should be illegal. In the survey, 72% of a randomly selected national sample of $n = 836$ individuals said that the use of cell phones while driving should be illegal.

a. What is the population of interest?

b. What is the population parameter of interest?

c. What is the sample?

d. What is the value of the sample statistic?

Lesson 1 General Section Exercises

10.8 Read the examples of questions based on a "parameter of interest" in the section titled "Curiosity and Confidence Intervals" (p. 371). Create your own example of something that interests you, and translate it into a question about a population parameter.

10.9 In a 1997 survey done by the Marist College Institute for Public Opinion, 36% of a randomly selected sample of $n = 935$ American adults said that they do not get enough sleep each night (Source: http://www.mipo.marist.edu). The margin of error was reported to be 3.5%.

a. Use the survey information to create a 95% confidence interval for the percentage that feels they don't get enough sleep every night. Write a sentence that interprets this interval. Specify the population.

b. Do you think this sample could be used to estimate the percentage of college students who think they don't get enough sleep each night? Why or why not?

10.10 Answer Thought Question 10.2 on p. 374.

10.11 *Parade Magazine* reported that "nearly 3200 readers dialed a 900 number to respond to a survey in our Jan. 8 cover story on America's young people and violence" (February 19, 1995, p. 20). Of those responding, "63.3% say they have been victims or personally know a victim of violent crime." Can the methods in this chapter legitimately be used to compute a 95% confidence interval for the proportion of Americans who fit that description? Explain why or why not.

10.12 A CNN/*Time* poll conducted in the United States October 23–24, 2002, asked, "Do you favor or oppose the legalization of marijuana?" In the nationwide poll of $n = 1007$ adults, 34% said that they favored legalization. The margin of error was given as 3.1% (Source: http://www.pollingreport.com).

a. Find a 95% confidence interval estimate of the percentage of American adults who favored the legalization of marijuana at the time of the poll.

b. Write a sentence that interprets the interval computed in part (a).

10.13 Suppose 200 different researchers all randomly select samples of $n = 400$ individuals from a population. Each researcher uses his or her sample to compute a 95% confidence interval for the proportion who have blue eyes in the population. About how many of the confidence intervals will cover the population proportion? About how many of the intervals will not cover the population proportion? Briefly explain how you determined your answers.

10.14 Taking into account the purpose of a confidence interval described in Section 10.1, explain what is wrong with the following statement: "Based on the survey data, a 95% confidence interval estimate of the sample proportion is .095 to .117."

Lesson 2 Skillbuilder Exercises

10.15 Explain whether the width of a confidence interval would increase, decrease, or remain the same as a result of each of the following changes:

a. Increase the confidence level from 95% to 99%.

b. Decrease the confidence level from 95% to 90%.

10.16 Explain whether the width of a 95% confidence interval would increase, decrease, or remain the same as a result of the following changes in the design of a survey. Assume that sampling will be from the same population in all cases.

a. A planned sample size is increased from 500 to 1000.

b. A planned sample size is decreased from 1500 to 1000.

10.17 For the following examples, specify which of the five parameters stated on page 374 is appropriate. Give and define appropriate notation for the parameter of interest. If two symbols are involved, such as $\mu_1 - \mu_2$, define each symbol.

a. The parameter of interest is the proportion of the population that suffers from allergies.

b. The parameter of interest is the proportion of working adults who say that they would quit their jobs if they won a large amount in the lottery.

c. Does taking calcium help to reduce premenstrual symptoms? To investigate, we could devise a questionnaire that measured the severity of symptoms and then randomly assign women to take calcium or a placebo. The parameter of interest is the difference in what the population mean severity score would be if everyone took calcium versus the score if everyone took a placebo.

10.18 For the following examples, specify which of the five parameters stated on page 374 is appropriate. Give and define appropriate notation for the parameter of interest. If two symbols are involved, such as $\mu_1 - \mu_2$, define each symbol.

a. Do men and women feel the same away about dating someone who has a great personality without regard for looks? To investigate this, one parameter that we could use is the difference in population proportions of males and females who would answer "yes" if asked, "Would you date someone with a great personality even though you did not find them attractive?"

b. The parameter of interest is the mean number of hours that students at a college say they study per week.

c. The parameter of interest is the proportion who have type O negative blood in a population.

Lesson 2 General Section Exercises

10.19 *U.S. News and World Report* (December 19, 1994, pp. 62–71) reported on a survey of 1000 American adults, conducted by telephone December 2–4, 1994, designed to measure beliefs about apocalyptic predictions. One of the results reported was that 59% of the sample said that they believe the world will come to an end.

a. Calculate the standard error of the sample proportion who believe that the world will come to an end.

b. Calculate a 95% confidence interval for the population proportion. Interpret the interval in a way that could be understood by a statistically naïve reader.

10.20 Suppose that in a random sample of 300 employed Americans, there are 57 individuals who say that they would fire their boss if they could. Calculate a 95% confidence interval for the population proportion. Write a sentence or two that interprets this interval.

10.21 Suppose that a survey is planned to estimate the proportion of a population that is left-handed. The sample data will be used to form a confidence interval. Explain which one of the following combinations of sample size and confidence level will give the widest interval.

(i) $n = 400$, confidence level = 90%.
(ii) $n = 400$, confidence level = 95%.
(iii) $n = 1000$, confidence level = 90%.
(iv) $n = 1000$, confidence level = 95%.

10.22 Refer to Exercise 10.3, which described a Gallup poll. The poll surveyed a random sample of 439 American teenagers, and one question that they were asked was, "How strict are your parents compared to most of your friends' parents?" The choices were "more strict," "less strict," and "about the same." The choice "less strict" was selected by 30% of the respondents (Mazzuca, 2004).

a. What is the sample proportion that answered "less strict?" Use appropriate notation.
b. Calculate the standard error of the sample proportion.
c. Calculate a 95% confidence interval for the population proportion. Interpret the interval in a way that could be understood by a statistically naïve reader.

Section 10.2

Exercises 10.23 to 10.49 correspond to the three Lessons in Section 10.2. Lesson 1 exercises are 10.23 to 10.34; Lesson 2 exercises are 10.35 to 10.40; Lesson 3 exercises are 10.41 to 10.49.

Lesson 1 Skillbuilder Exercises

10.23 Suppose that a new treatment for a certain disease is given to a sample of 200 patients. The treatment was successful for 166 of the patients. Assume that these patients are representative of the population of individuals who have this disease.

a. Calculate the sample proportion successfully treated.
b. Determine a 95% confidence interval for the proportion of the population for whom the treatment would be successful.
c. Write a sentence that interprets the interval found in part (b).

10.24 ◆ In a survey of 190 college students, 134 students said that they believe there is extraterrestrial life (Data source: **pennstate3** dataset at the companion website).

a. Find $\hat{p}$ = sample proportion who believe that there is extraterrestrial life.
b. Find a 95% confidence interval estimate of the proportion of all college students who believe that extraterrestrial life exists.

10.25 Refer to Exercise 10.23. Calculate a 98% confidence interval for the proportion successfully treated. Is this interval wider or narrower than the interval computed in part (b) of Exercise 10.23?

10.26 For each confidence interval procedure, provide the confidence level.

a. Sample proportion ± 1.645 × standard error.
b. Sample proportion ± 1.96 × standard error.
c. Sample proportion ± 2.326 × standard error.
d. Sample proportion ± 2.576 × standard error.

Lesson 1 General Section Exercises

10.27 In a randomly selected sample of 400 registered voters in a community, 220 individuals say that they plan to vote for Candidate Y in the upcoming election.

a. Find the sample proportion planning to vote for Candidate Y.
b. Calculate the standard error of the sample proportion.
c. Find a 95% confidence interval for the proportion of the registered voter population who plan to vote for Candidate Y.
d. Find a 98% confidence interval for the proportion of the registered voter population who plan to vote for Candidate Y.

10.28 In a *USA Today*/Gallup poll survey carried out in 2009, 57% of 1006 randomly sampled American adults said that abortions should be legal only under certain circumstances when asked, "Do you think abortions should be legal under any circumstance, legal only under certain circumstances or illegal in all circumstances?" (Source: http://www.pollingreport.com/abortion.htm).

a. Calculate a 95% confidence interval that estimates the proportion of all Americans in 2009 who thought abortions should be legal only under certain circumstances.
b. On the basis of this confidence interval, is it reasonable to infer that in 2009 more than half of all Americans thought that abortions should be legal only under certain circumstances? Explain.
c. In the survey, 18% answered that abortions should be illegal in all circumstances. Calculate a 95% confidence interval that estimates the proportion of all Americans in 2009 who thought abortions should be illegal in all circumstances.

10.29 A question in a 2007 Gallup poll asked, "Do you think the penalties for underage drinking should be made more strict, less strict, or remain the same?" The sample proportion that responded "more strict" was .60 (60%). This result will be used to estimate the proportion of all adults in the United States who think that penalties for underage drinking should be more strict. The sample consisted of $n = 537$ randomly selected adults in the United States (Source: www.pollingreport.com/life.htm).

a. What is the population parameter of interest in this situation?
b. Calculate the standard error of the sample proportion.
c. Calculate a 95% confidence interval that estimates the population proportion who think that penalties for underage drinking should be more strict.
d. Write a sentence that interprets the confidence interval found in part (c).

10.30 In a CBS News/*New York Times* nationwide poll done in 2009, the proportion of respondents who thought that it should be illegal to use a handheld cellular telephone while driving a car was .80 (80%). The poll's sample size was 829 (Source: www.pollingreport.com/transpor.htm#Road).

a. Find the value of the standard error of the sample proportion.
b. Find a 95% confidence interval estimate of the population proportion who think that it should be illegal to use a cellular telephone while driving.
c. Find a 98% confidence interval estimate of the population proportion who think that it should be illegal to use a cellular telephone while driving.
d. Which of the intervals calculated in parts (b) and (c) is wider—the interval with 95% confidence or the interval with 98% confidence?

10.31 A randomly selected sample of 15 individuals is asked whether they are right-handed or left-handed. In the sample, only one person is left-handed. Explain why a confidence interval for the population proportion should not be computed by using the methods described in this chapter.

10.32 Refer to Example 10.3 (p. 378), in which for a sample of 1003 American adults, .56 was the proportion who think that it is somewhat or very likely that intelligent life exists elsewhere. Compute a 99% confidence interval for the population proportion.

10.33 Refer to Exercise 10.20. Calculate a 90% confidence interval for the population proportion who would fire their boss if they could.

10.34 In a Gallup poll done in 2007, a randomly selected sample of $n = 254$ U.S. parents was asked how often the television was on when they ate dinner as a family. About 33% of the sample said "always" (Source: www.pollingreport.com/life.htm).

a. Calculate the standard error of the sample proportion.
b. Calculate a 95% confidence interval that estimates the proportion of U.S. families with children who always have the television on when they eat dinner as a family.
c. Calculate a 90% confidence interval for this situation.
d. Write a sentence that interprets the confidence interval found in part (c).

Lesson 2 Skillbuilder Exercises

10.35 Determine the value of the z^* multiplier that would be used to compute an 80% confidence interval for a population proportion.

10.36 Find z^* for each of the confidence levels in parts (a) through (c), and draw a picture similar to Figure 10.4 (p. 382). Then answer part (d).

a. 60%.
b. 86%.
c. 99.8%.
d. On the basis of the values of z^* found in parts (a) through (c), explain what happens to the width of the confidence interval when the confidence level is increased.

Lesson 2 General Section Exercises

10.37 What is the confidence level for a confidence interval computed as Sample proportion $\pm$ 1.28 $\times$ Standard error?

10.38 What is the confidence level for a confidence interval computed as Sample proportion $\pm$ 3.09 $\times$ Standard error?

10.39 Refer to Exercise 10.29. Calculate a 50% confidence interval that estimates the proportion of U.S. adults who think penalties for underage drinking should be more strict.

10.40 Refer to Exercise 10.34. Calculate an 80% confidence interval that estimates the proportion of U.S. families with children that always has the television on when they eat dinner as a family.

Lesson 3 Skillbuilder Exercises

10.41 In constructing 99% confidence intervals for proportions, over the long run, about what percentage of the time would each of the following happen?

a. The difference between the sample proportion and the population proportion is less than the "Multiplier $\times$ Standard error."
b. The difference between the sample proportion and the population proportion is more than the "Multiplier $\times$ Standard error."

10.42 Suppose that a polling organization reports that the margin of error is 4% for a sample survey. Explain what this indicates about the possible difference between a percentage determined from the survey data and the population value of the percentage.

10.43 Suppose that a margin of error for a sample percentage is reported to be 3%. What is the probability that the difference between the sample percentage and the population percentage will be more than 3%? Interpret the meaning of this probability.

10.44 For each combination of sample size and sample proportion, find the margin of error for the 95% confidence level. Use the formula given on page 383.

a. $n = 400, \hat{p} = .20$.
b. $n = 400, \hat{p} = .80$.
c. $n = 1000, \hat{p} = .50$.

10.45 For each combination of sample size and sample proportion, find the margin of error for the 95% confidence level. Use the formula given on page 383.

a. $n = 100, \hat{p} = .56$.
b. $n = 400, \hat{p} = .56$.

10.46 Suppose that a polling organization is conducting a survey to estimate the proportion of Americans who regularly attend religious services. The organization plans to gather data from a randomly selected sample of 550 individuals. On the basis of the "conservative" estimate, what will be the margin of error for the survey?

Lesson 3 General Section Exercises

10.47 In Example 10.8 (p. 385), the information given was that 3% of a nationwide randomly selected sample of $n = 883$ suffered from severe allergies.

a. Determine the standard error of the sample proportion in this problem.
b. Verify that the margin of error is about 1.1% for a 95% confidence interval for the percentage that suffers from severe allergies in the population represented by this sample.

10.48 In a Gallup Youth Survey done in 2000, $n = 501$ randomly selected American teenagers were asked about how well they get along with their parents.

a. According to the Gallup Organization, the margin of error for the poll was 5%. Show how you think they determined this figure.

b. A survey result was that 54% of the sample said they get along "very well" with their parents. Using the reported margin of error, calculate a 95% confidence interval for the population proportion that gets along "very well" with their parents. Write a sentence that interprets this interval.

c. Using the more precise formula for margin of error given on page 383, calculate a 95% confidence interval. Compare the answer to the answer in part (b).

d. Another result in the same survey was that only 5% said that cheating on exams was "not at all serious." Using the reported margin of error, compute a 95% confidence interval for the population proportion who would say this. Also, compute the 95% confidence interval using the more precise formula for margin of error. Which interval is narrower?

e. You should have found that the intervals in parts (b) and (c) were quite similar to each other but that the two intervals computed in part (d) were not. Explain why the two intervals agreed more closely in parts (b) and (c) than they did in part (d).

10.49 In a survey reported in a special edition of *Newsweek* (Spring/Summer 1999, *Health for Life*), only 3% of a sample of 757 American women responded "Not at all satisfied" to the question, "How satisfied are you with your overall physical appearance?"

a. The margin of error for the poll was reported as 3.5%. Use the conservative approximation for margin of error to verify that this is roughly correct.

b. On the basis of the reported margin of error, calculate a 95% confidence interval for the percentage of all American women who are "not at all satisfied" with their physical appearance. Are all values in this interval valid estimates of the population percentage? Explain.

c. Use the more precise formula to calculate the margin of error.

d. On the basis of the margin of error in part (c), calculate a 95% confidence interval for the population percentage. Write a sentence that interprets this confidence interval for somebody who knows little about statistics.

Section 10.3

Skillbuilder Exercises

10.50 ◆ In the General Social Survey, an ongoing nationwide survey done by the National Opinion Research Center at the University of Chicago, a question asked is whether a respondent favors or opposes capital punishment (the death penalty) for persons convicted of murder. The output for this exercise compares the proportions who said that they were opposed to the death penalty in the year 2008 and the year 1993 (Data sources: **GSS-08** and **GSS-93** datasets on the companion website).

Sample	X	N	Sample p
2008	639	1902	0.336
1993	337	1488	0.226

Estimate for p(1) − p(2): 0.110
95% CI for p(1) − p(2): (0.079, 0.140)

a. What proportion of the year 2008 sample was opposed to the death penalty? What proportion of the year 1993 sample was opposed?

b. What is the estimated difference between the proportions opposed to the death penalty in the 2 years?

c. Write the 95% confidence interval given in the output. Then interpret this interval in the context of this situation.

d. Provide the formula used to calculate the interval, and substitute appropriate numerical values into the formula.

10.51 In each situation, explain whether the method covered in Section 10.3 for finding a confidence interval for the difference in two proportions should be used.

a. A survey is done to estimate the difference between the proportions of college students and high school students who smoke cigarettes. Data on cigarette smoking habits are gathered from randomly selected samples of 500 college and 500 high school students.

b. Two new treatments for a serious disease are compared. Each treatment is used for five patients. The first treatment is successful for three of the five patients who used it. The second treatment is successful for only one of the five patients who used that treatment.

c. An economist wants to estimate the difference between the mean annual income of college graduates and the mean annual income of high school graduates who did not go to college. Income data are collected from samples of 200 individuals in each educational degree group.

10.52 ◆ In Case Study 10.3 (p. 392), students in a statistics class at Penn State were asked, "Would you date someone with a great personality even though you did not find them attractive?" The results were that 61.1% of 131 women answered "yes," while 42.6% of 61 men answered "yes" (Data source: **pennstate3** dataset on the companion website).

a. A 95% confidence interval for the difference in proportions of men and women who would say "yes" to the question if asked is given as .035 to .335. Write a sentence that interprets the interval.

b. What populations do you think are represented by these samples of men and women?

10.53 Suppose that a randomly selected sample of $n = 900$ registered voters is surveyed to estimate the proportions that will vote for the two candidates in an upcoming election. Fifty-five percent of those sampled say that they will vote for one candidate, while 45% indicate a preference for the other candidate. Explain why the method in Section 10.3 for computing a confidence interval for a difference in proportions should not be used to estimate the difference in the proportions planning to vote for the two candidates.

General Section Exercises

10.54 A study was done to determine whether there is a relationship between snoring and the risk of heart disease (Norton and Dunn, 1985). Among 1105 snorers in the study, 85 had heart disease, while only 24 of 1379 nonsnorers had heart disease.

a. Determine a 95% confidence interval that estimates $p_1 - p_2$ = difference in proportions of nonsnorers and snorers who have heart disease.

b. On the basis of the confidence interval determined in part (a), can we infer that the population proportions with heart disease differ for nonsnorers and snorers? Explain.

10.55 In a study done in Maryland, investigators surveyed individuals by telephone about how often they get tension headaches (Schwartz et al., 1998). One response variable that was measured was whether or not the respondent had experienced an episodic tension-type headache (ETTH) in the prior year. A headache pattern was called "episodic" if the headaches occurred less often than 15 times a month; otherwise, the headaches were called "chronic." Of the 1600 women in the survey aged 18 to 29, a total of 653 said that they had experienced ETTHs in the last year. Of the 2122 women in the 30- to 39-year-old age group, the number having experienced episodic headaches was 995.

a. Calculate the proportion in each group who experienced an episodic headache in the prior year, and compute the difference in these two proportions.

b. Compute a 95% confidence interval for the difference between the proportions for these age groups in the population. Write a sentence that interprets this confidence interval.

10.56 Refer to the output given for Exercise 10.50.

a. Find the value of s.e.$(\hat{p}_1 - \hat{p}_2)$ = standard error of the difference between the two sample proportions.

b. Find a 90% confidence interval for the difference between the population proportions opposed to the death penalty in the years 2008 and 1993. Use Example 10.9 (p. 387) for guidance.

10.57 Refer to Exercise 10.55. In the sample, there were 4594 individuals with at least a college degree and $n = 7076$ individuals with at least a high school diploma (but not a college degree). Following is Minitab output with a 95% confidence interval for the difference in proportions experiencing episodic tension-type headaches in the two educational groups:

Sample	X	N	Sample p
1	2140	4594	0.466
2	2690	7076	0.380

Estimate for p(1) − p(2): 0.086
95% CI for p(1) − p(2): (0.068, 0.104)

a. Write two or three sentences that interpret the results.

b. Provide the formula that is used to calculate the interval, and substitute appropriate numerical values into the formula.

10.58 In a CBS News survey done in 2009, 95% of $n = 346$ randomly sampled married men said that they would marry their spouses again if they had it to do all over again. In the same survey, 85% of $n = 522$ married women said that they would marry their spouses if they had it to do all over again (Source: www.cbsnews.com/htdocs/pdf/Poll_Jan10dLove.pdf).

a. Calculate a 95% confidence interval that estimates the difference in proportions of married women and men who would marry their spouse again if they had it to do all over again.

b. On the basis of the confidence interval determined in part (a), can we infer that the population proportions who would marry their spouse again differ for men and women? Explain.

10.59 In a survey of college students, 70% (.70) of the 100 women surveyed said that they believe in love at first sight, whereas only 40% (.40) of the 80 men surveyed said that they believe in love at first sight.

a. Find the value of the difference between the sample proportions for men and women.

b. Find the standard error of the difference between the sample proportions.

c. Find an approximate 95% confidence interval for the difference between population proportions believing in love at first sight for men versus women. (Use $z^* = 2$)

d. On the basis of the confidence interval found in part (c), is it reasonable to conclude that there is a difference in the proportions of all college men and women who believe in love at first sight? Explain.

10.60 Refer to Exercise 10.52.

a. Compute a 99% confidence interval for the difference in proportions.

b. How does the width of the 99% confidence interval computed in part (a) of this problem compare to the width of the 95% confidence interval given in part (a) of Exercise 10.52?

c. Would a 90% confidence interval for the difference in proportions be wider or narrower than the 99% confidence interval computed in part (a)? Explain.

d. Does the interval computed in part (a) include the value 0? What does this tell us about whether there is a difference between college men and women for this question?

Section 10.4

Skillbuilder Exercises

10.61 A Gallup Organization poll of $n = 1004$ randomly selected American adults in July 2002 found that 55% of those surveyed felt that their weight was about right. The margin of error for the survey was given as 3% (Source: http://www.gallup.com).

a. Find a 95% confidence interval estimate of the percentage of American adults who think their weight is about right.

b. Based on the interval computed in part (a), explain whether it is reasonable to say that more than 50% of American adults think their weight is about right.

10.62 A university is contemplating switching from the quarter system to a semester system. The administration conducts a survey of a random sample of 400 students and finds that 240 of them prefer to remain on the quarter system.

a. Construct a 95% confidence interval for the true proportion of all students who would prefer to remain on the quarter system.

b. Does the interval that you computed in part (a) provide convincing evidence that more than half of all students prefer to remain on the quarter system? Explain.

10.63 Suppose that a health expert has claimed that 16% of college students smoke cigarettes. To investigate this claim, researchers survey a random sample of college students. Using this sample, a 95% confidence interval for the percentage of college students who smoke is found to be 9% to 14%. On the basis of this interval, explain whether the claim that 16% of all college students smoke is reasonable.

10.64 Suppose that a 95% confidence interval for $p_1 - p_2 =$ difference in proportions of college men and women who have ever missed a class due to drinking alcohol is .07 to .13. On the basis of this confidence interval can we infer that the population proportions ever having missed a class due to drinking alcohol differ for men and women? Explain.

General Section Exercises

10.65 As was described in Exercise 10.61, a July 2002 Gallup poll of $n = 1004$ randomly selected U.S. adults found that 55% felt that their weight was about right. In a similar poll in 1990, about 46% felt that their weight was about right. The margin of error for each poll was 3%.

a. Based on these results, is it reasonable to conclude that the population proportions thinking their weight is about right were different in 1990 and 2002?

b. Assuming that the sample size was 1004 in the 1990 poll as well, find a 95% confidence interval for the difference in the population proportions.

c. On the basis of the interval that you found in part (b), formulate a conclusion about whether the population proportion was different in 2002 than it was in 1990. Write your conclusion in words that would be understood by someone with no training in statistics.

10.66 Suppose that a 95% confidence interval for the proportion of men who experience sleep apnea (irregular breathing during sleep) is .11 to .17 and a 95% confidence interval for the proportion of women who experience sleep apnea is .04 to .08.

a. On the basis of these intervals, is it reasonable to conclude that the population proportions experiencing sleep apnea differ for men and women? Explain.

b. Find a 95% confidence interval for the difference in proportions of men and women in the population who experience sleep apnea.

c. On the basis of the interval you found in part (b), is it reasonable to conclude that the population proportions experiencing sleep apnea differ for men and women? Explain.

10.67 A survey is done to estimate the proportion of a population that favors a new tax law proposal for a city. In a randomly selected sample of $n = 400$ city residents, 43% favor the proposal.

a. Calculate a 95% confidence interval that estimates the population proportion in favor of the proposed law.

b. On the basis of the confidence interval found in part (a), can we infer that fewer than 50% of the population favors the proposed law?

10.68 *This exercise is also Exercise 1.32.* A random sample of 1001 University of California faculty members taken in December 1995 was asked, "Do you favor or oppose using race, religion, sex, color, ethnicity, or national origin as a criterion for admission to the University of California?" (Roper Center, 1996). Fifty-two percent responded "favor."

a. What is the population for this survey?

b. What is the approximate margin of error for the survey?

c. Based on the results of the survey, could it be concluded that a majority (over 50%) of *all* University of California faculty members favored using these criteria at that time? Explain.

Chapter Exercises

10.69 In each situation, explain why you think that the sample proportion should or should not be used to estimate the population proportion.

a. An Internet news organization asks visitors to its website to respond to the question, "Are you satisfied with the president's job performance?" Of $n = 3500$ respondents, 61% say that they are not satisfied with the president's performance. On the basis of this survey, the organization writes an article saying that a majority of Americans are not satisfied with the president.

b. A convenience sample of $n = 400$ college students in two classes at the same university is used to estimate the proportion that is left-handed in the nationwide population of college students.

10.70 A federal law that lowered the limit for a legal blood alcohol level for automobile drivers was signed by President Clinton in October 2000. At its website, the Gallup Organization reported the following survey results in an article dated October 26, 2000:

> 72% of Americans support lowering the drunk driving limit to 0.08% BAC....... A total 1,002 telephone interviews were conducted between July 20–Aug 3, 2000, with a representative sample of the U.S. public age 16 and older. The findings are based on 930 respondents who identified themselves as licensed drivers. The margin of error equals plus or minus three percentage points.

a. The first sentence of the information from the Gallup Organization states, "72% of Americans support lowering the drunk driving limit to 0.08% BAC." Considering the difference between a sample and a population, explain why this sentence is not necessarily correct. Rewrite the sentence so that it is correct.

b. The article provides the information that "the margin of error equals plus or minus three percentage points." Write one or two sentences interpreting this value that could be understood by someone who does not know anything about statistics.

c. Using the margin of error provided by the Gallup Organization, calculate a 95% confidence interval for the proportion that supported the lower legal blood alcohol limit at that time. Write a sentence that interprets this interval. Be sure to specify the appropriate population.

d. On the basis of the interval computed in part (c), is it reasonable to conclude that more than half of American licensed drivers supported the lower blood alcohol limit at that time?

10.71 In Chapter 2, we saw data from a statistics class activity in which 190 students were asked to randomly pick one of the numbers 1, 2, 3, 4, 5, 6, 7, 8, 9, 10. The number 7 was picked by 56 students.

a. For the sample, calculate the proportion of students who picked 7.
b. Calculate the standard error for this sample proportion.
c. Calculate a 90% confidence interval for the population proportion.
d. Calculate a 95% confidence interval for the population proportion.
e. Calculate a 98% confidence interval for the population proportion.
f. What do the results of parts (c) through (e) indicate about the effect of confidence level on the width of a confidence interval?
g. On the basis of these confidence intervals, do you think that students choose numbers "randomly"? Explain.

10.72 A study reported in the *Journal of Occupational and Environmental Medicine* (Franke et al., 1998, 40: 441–444) found that retired male Iowa policemen are more likely to have heart disease compared to other men with similar ages but different occupations. Data from the study were that about 32% of a sample of 232 retired policemen had heart disease but only 18% of a sample of 817 men with other occupations had heart disease.

a. In Chapter 6, we learned about the various types of studies. Is this an observational study or a randomized experiment? Explain.
b. Do you think that we can attribute the observed difference in heart disease percentages to job-related factors? Explain.
c. Using a bar chart, graphically compare the policemen and the other men.
d. Calculate a 95% confidence interval for the proportion of retired Iowa policemen who have heart disease.
e. Calculate a 95% confidence interval for the proportion of other retired men who have heart disease.
f. Why are the widths of the two confidence intervals different from each other?

10.73 Refer to Exercise 10.72. Find a 95% confidence interval for the difference in population proportions of male policemen and men of other occupations who have heart disease. On the basis of this confidence interval, is it safe to conclude that male police officers are more likely to experience heart disease than other men? Explain.

10.74 Use an example from this chapter to explain the difference between a sample proportion and a population proportion.

10.75 A professor wants to know whether her class of 60 students would prefer the final exam to be given as a take-home exam or an in-class exam. She surveys the class and learns that 45 of the 60 students would prefer a take-home. Explain why she should not use these data to compute a 95% confidence interval for the proportion preferring a take-home exam.

10.76 Refer to the derivation of a 95% confidence interval for p on page 381 in Lesson 2 of Section 10.2. Use a similar argument to explain why a 68% confidence interval for p is "Sample estimate ± Standard error."

10.77 The Gallup Organization used two different wordings of a question in a February 1999 poll about the death penalty. The two different questions were asked of two different random samples of Americans. Results reported at the Gallup website follow.

Question 1: Are you in favor of the death penalty for a person convicted of murder? (Based on 543 adults: margin of error plus or minus 5 percentage points)

	For	Against	No Opinion
99 Feb 8-9	71%	22%	7%

Question 2: What do you think should be the penalty for murder—the death penalty or life imprisonment with absolutely no possibility of parole? (Based on 511 adults: margin of error plus or minus 5 percentage points)

	Death Penalty	Life Imprisonment	No Opinion
99 Feb 8-9	56%	38%	6%

a. The margin of error for each question was reported as 5%. Verify that this is approximately true.
b. For the first question, calculate a 95% confidence interval for the percentage of all Americans who are "for" the death penalty. Use the margin of error reported by the Gallup Organization. Write a sentence that interprets this confidence interval.
c. For the second question, calculate a 95% confidence interval for the percentage of all Americans who would answer "death penalty" when asked whether the penalty for murder should be the death penalty or life imprisonment. Use the margin of error reported by the Gallup Organization. Write a sentence that interprets this interval.

10.78 Refer to Exercise 10.77, in which two different wordings of a question were used to estimate the proportion of Americans who are in favor of the death penalty.

a. Verify that the conditions are met to compute a confidence interval for the difference in two population proportions, where they are the proportions of the population who would support the death penalty if asked Question 1 and if asked Question 2, respectively.
b. Find a 95% confidence interval for the difference in the two population proportions.
c. On the basis of the confidence interval that you computed in part (b), formulate a conclusion about whether the wording of the question affects the proportion of the population who support the death penalty. Write a few sentences expressing your conclusion that would be understood by someone who has no training in statistics.

10.79 In 1998, two statistics classes for students in the liberal arts at a large northeastern university were asked, "Do you have a tattoo?" The responses were as follows:

	Yes	No
Women	46	227
Men	32	175

a. What proportion of the women in this sample has a tattoo?
b. What proportion of the men in this sample has a tattoo?
c. Compare the men and women, using a bar graph.
d. What population do you think is represented by this sample? In other words, to whom do you think the results can be generalized?
e. For the college women, calculate a 95% confidence interval for the proportion of the population who have a tattoo. Write a sentence that interprets this interval.
f. For the college men, calculate a 95% confidence interval for the proportion of the population who have a tattoo. Write a sentence that interprets this interval.
g. Use the results of parts (e) and (f) to compare college men and women with regard to the proportion who have a tattoo.

10.80 Refer to Exercise 10.79, which reported results for the numbers of men and women who had a tattoo.

a. Find a 90% confidence interval for the difference in population proportions of men and women with a tattoo.
b. On the basis of the interval that you computed in part (a), formulate a conclusion about whether the population proportions of men and women with tattoos differ for the populations represented in this study. Explain.

10.81 On January 30, 1995, *Time* magazine reported the results of a poll of adult Americans in which it asked, "Have you ever driven a car when you probably had too much alcohol to drive safely?" The exact results were not given, but from the information given, we can guess at what they were. Of the 300 men who answered, 189 (63%) said yes and 108 (36%) said no. The remaining three men weren't sure. Of the 300 women, 87 (29%) said yes and 210 (70%) said no; the remaining three women weren't sure.

a. Compute a 95% confidence interval for the difference in proportions of American men and women who would say yes to this question.
b. Write a sentence or two interpreting this interval.

10.82 Find the reported results of a poll in which a margin of error is also reported. Look on the Internet, in a weekly news magazine such as *Newsweek* or *Time*, or a newspaper such as the *New York Times* or *USA Today*.

a. Explain what question was asked in the poll and what margin of error was reported. Verify the margin of error with a calculation.
b. Present a 95% confidence interval for the results. Explain in words what the interval means for your example.

10.83 Refer to Exercise 10.81. Calculate a 95% confidence interval for the proportion of American men who would say that they have ever driven a car when they probably have had too much to drink, and write a sentence or two that interprets the interval. Be specific about the population that the interval describes.

10.84 ◆ A sample of college students was asked whether they would return the money if they found a wallet on the street. Of the 93 women, 84 said "yes," and of the 75 men, 53 said "yes." Assume that these students represent all college students (Data source: **UCDavis2** dataset on the companion website).

a. Find separate 95% confidence intervals for the proportions of college women and college men who would say "yes" to this question.
b. Find a 95% confidence interval for the difference in the proportions of college men and women who would say "yes" to this question.
c. Write a few sentences interpreting the intervals in parts (a) and (b).

10.85 In the table below are data collected between 1996 and 1998 on ear piercing and tattoos for male Penn State students. Assume that these men represent a random sample of male Penn State students. The results were as follows:

	Tattoo?		
Ear Piercing	**No**	**Yes**	**Total**
No	381	43	424
Yes	99	42	141
Total	480	85	565

a. For the population of men with no ears pierced, find a 90% confidence interval for the proportion with a tattoo.
b. For the population of men with an ear pierced, find a 90% confidence interval for the proportion with a tattoo.
c. Find a 99% confidence interval for the difference in proportions of men with a tattoo for the populations with and without an ear piercing. Write a sentence interpreting the interval.

10.86 For each of the following, explain whether it affects the width of a confidence interval, the center of a confidence interval, neither, or both.

a. The sample estimate, for the situations involving means.
b. The multiplier.
c. The standard error of the estimate.
d. The confidence level.
e. The sample standard deviation(s) for situations involving means.
f. The sample proportion(s) for situations involving proportions.

10.87 Fatty acids that are present in fish oil may be useful for treating some psychiatric disorders. An article reported on September 3, 1998, at the Yahoo Health website described a randomized experiment done by a Harvard Medical School researcher in which 14 bipolar patients received fish oil daily, while 16 other patients received a placebo daily. After 4 months, 9 of the 14 patients who received fish oil had responded favorably, but only 3 of the 16 placebo patients had done so.

a. Calculate the difference between the proportions showing favorable response in the two groups, and also calculate a standard error for this difference.
b. Discuss whether the method described in Section 10.3 should be used to calculate a confidence interval for the difference. If you believe that it should, calculate the confidence interval.

c. In Chapter 6, blind and double-blind procedures were discussed. Explain how those concepts would be applied to this experiment.

10.88 Answer Thought Question 10.3 on page 380.

Dataset Exercises

Datasets required to solve these exercises are available on the companion website, http://www.cengage.com/statistics/Utts4e.

10.89 For this exercise, use the **GSS-08** dataset on the companion website. The dataset contains data from the 2008 General Social Survey, and the variable ***polparty*** indicates the political party preference of the respondent.

a. Calculate a 95% confidence interval for the proportion of Americans in 2008 who indicated that they preferred the Democratic Party.

b. Write a sentence that interprets the interval found in part (a).

10.90 For this exercise, use the **GSS-08** dataset on the companion website. The dataset contains data from the 2008 General Social Survey, and the variable ***marijuan*** indicates the respondent's opinion about the legalization of marijuana.

a. Calculate a 95% confidence interval for the proportion of Americans who were opposed to the legalization of marijuana in 2008.

b. Write a sentence that interprets the interval found in part (a).

10.91 For this exercise, use the **pennstate3** dataset on the companion website. The data are from a student survey in a statistics class at Penn State. The variable ***atfirst*** gives data for whether or not a student believes in love at first sight.

a. What is the sample proportion that believes in love at first sight?

b. Find a 95% confidence interval for the population proportion that believes in love at first sight.

c. Write a sentence that interprets the interval found in part (b).

10.92 Refer to Exercise 10.91 about the variable ***atfirst*** in the **pennstate3** dataset on the companion website.

a. Create a two-way table that shows the relationship between the variables ***atfirst*** and ***Sex***. (See Sections 2.3 and 4.1 to review two-way tables.)

b. What proportion of the females in the sample believe in love at first sight? What proportion of the males believe in love at first sight?

c. For each sex separately, find a 95% confidence interval for the population proportion that believes in love at first sight.

d. Find a 95% confidence interval for the difference in the population proportions. What does the confidence interval indicate about the difference between the proportions of males and females who believe in love at first sight?

10.93 Use the dataset **GSS-08** on the companion website. The variable ***owngun*** indicates whether or not the respondent owns a gun, and the variable ***polparty*** contains the respondent's political party preference.

a. Determine a 95% confidence interval for the difference in the proportions of Republicans and Democrats who owned a gun in 2008.

b. On the basis of the interval calculated in part (a), is it reasonable to conclude that in the United States population there is a difference in the proportions of Republicans and Democrats who owned a gun in 2008? Explain.

11

Mathias Wilson 2010/Used under license from Shutterstock.com

Do men lose more weight from dieting or from exercising?

See Example 11.3 *(p. 410)*

Estimating Means with Confidence

The parameters that researchers estimate with confidence intervals are not limited to means and proportions. When more complicated procedures are needed, journal articles provide completed intervals and your job is to interpret them. Principles in this and the previous chapter can be used to understand any confidence interval.

In Chapter 10, we learned how to form and interpret a confidence interval for a population proportion and for the difference in two population proportions. Remember that a *confidence interval* is an interval of values computed from sample data that is likely to include the true population value. For example, based on a sample survey of 1000 Americans, we may be able to say with 95% confidence that the interval from .33 to .39 (or 33% to 39%) will cover the true proportion of all Americans who favor the legalization of marijuana. The *confidence level* for an interval is a proportion or percentage, such as 95%, that describes our confidence in the *procedure* that we used to determine the interval. The stated confidence level is the percentage of the confidence intervals computed using the same procedure that over the long run will actually contain the true population value.

In this chapter, we expand the use of confidence intervals to parameters involving means of populations, including the population mean of one quantitative variable, the population mean of paired differences, and the difference in means for two populations. We may, for example, wish to know the mean number of hours per day that college students watch television. As an example of the mean of paired differences, we may want to know how much the mean IQ would change if everyone in the population were to listen to classical music for 30 minutes. As an example of the difference in two population means, we may want to know the difference in effectiveness of two diets, based on comparing the mean weight that would be lost if everyone in a population were to go on one diet to the mean weight that would be lost if they were to go on the other diet. In each of these types of situations, the objective is the same. We use sample information to form a confidence interval estimate of a population value.

Note to Readers: CI Module 0 in Chapter 10 must be covered before covering Chapter 11.

11.1 Introduction to Confidence Intervals for Means

Here we will review material from Chapters 9 and 10 that is relevant to finding confidence intervals for parameters involving means. Because the material is scattered throughout those two chapters, it is useful to review it all in one place.

To begin, let's review the following two definitions, which are relevant to our discussion of the estimation of all population values:

- A **parameter** is a fixed population characteristic. The numerical value of a parameter usually cannot be determined because we cannot measure all units in the population. We have to estimate the parameter using sample information.

- A **statistic**, or **estimate**, is a characteristic of a sample. A statistic estimates a parameter.

Figure 11.1 illustrates the statistical estimation problem. We wish to estimate a characteristic of a population. In more technical terms, we want to estimate a *parameter.* To do this, we take a random sample from the population and use the sample to calculate a sample *statistic* that estimates the parameter.

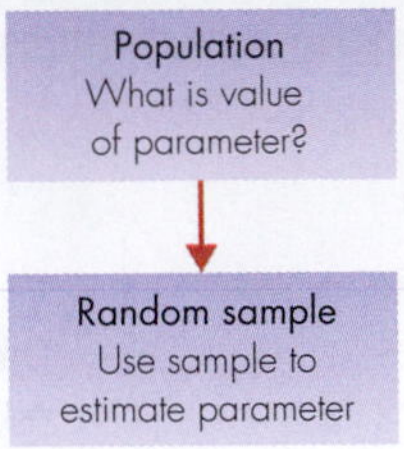

Figure 11.1 Sample information is used to estimate a population parameter

You will need to cover the CI Module 0 in Chapter 10 before the material in this chapter. From that introductory module, remember that the general format for a confidence interval for the five parameters that we are studying is

Sample estimate $\pm$ Multiplier $\times$ Standard error

In this introduction, we will discuss each of the three components of this formula for the three situations involving means. After the introduction, there is a module for each of the three parameters, expanding on the material in this section.

Following are descriptions of the three research situations that can be analyzed by using a confidence interval estimate of a population parameter involving means. The parameters in these three situations are:

1. μ = population mean.
2. μ_d = population mean of paired differences.
3. $\mu_1 - \mu_2$ = difference between two population means.

Let's revisit the research situations for these parameters that were discussed in Chapter 9 and specify the *sample estimate* used in the confidence interval formula.

Situation 1: Estimating the Mean of a Quantitative Variable

- *Example research questions:*

  What is the mean amount of time that college students spend watching TV per day?

  What is the mean pulse rate of women?

- *Population parameter:* μ (spelled *mu* and pronounced "mew") = population mean for the variable
- *Sample estimate:* $\bar{x}$ = the sample mean for the variable

Situation 2: Estimating the Population Mean of Paired Differences for a Quantitative Variable

- *Example research questions:*

  What is the mean difference in weights for freshmen at the beginning and end of the first semester?

  What is the mean difference in age between husbands and wives in Britain?

- *Population parameter:* μ_d = population mean of the differences in the two measurements.
- *Sample estimate:* $\bar{d}$ = the mean of the differences for a sample of the two measurements.

Situation 3: Estimating the Difference between Two Population Means for a Quantitative Variable (Independent Samples)

- *Example research questions:*

  How much difference is there in average weight loss for those who diet compared to those who exercise to lose weight?

  How much difference is there between the mean foot lengths of men and women?

- *Population parameter:* $\mu_1 - \mu_2$, where μ_1 and μ_2 represent the means in population 1 and 2, respectively
- *Sample estimate:* $\bar{x}_1 - \bar{x}_2$, the difference between the two sample means

THOUGHT QUESTION 11.1 Pick one of the example research questions provided in one of the three situations just described. Explain an appropriate sampling procedure, and then describe what the population parameter and the sample estimate would be for your example.*

Paired Data versus Independent Samples

It is important to understand the distinction between data collected as paired differences, for which the parameter of interest is μ_d, and data collected as independent samples, for which the parameter of interest is $\mu_1 - \mu_2$.

Paired Data

Suppose that we measure the IQ score of each person in a sample after he or she listens to a relaxing message and again after he or she listens to Mozart for the same amount of time. The difference between the two IQ measurements can be computed for each person, and we could use the sample differences to estimate the mean population difference. This would provide information about whether listening to Mozart produces higher IQ scores, on average, than listening to a relaxation tape does.

The term *paired data* (or *paired samples*) is used when pairs of variables are collected and used in this way. Generally, in such situations, we are interested only in the population (and sample) of *differences,* not in the original data. Therefore, for paired quantitative variables, it is appropriate to use methods that are designed for inferences about a single population mean. The mean in these situations is the mean for the population of differences that would result if we measured both variables on all units in the population and then found the difference. In Chapter 6, we called samples measured in this way *matched pairs.*

DEFINITION The term **paired data** means that the data have been observed in natural pairs. Some ways in which paired data can occur are as follows:

- The same measurement is taken twice on each person, under different conditions or at different times.
- Similar individuals are paired before giving the treatments in an experiment. Each member of a pair then receives a different treatment. The same response variable is measured for all individuals.
- Two different variables are measured for each individual. There is interest in the amount of difference between the two variables.

Independent Samples

Two samples are called **independent samples** when the measurements in one sample are not related to the measurements in the other sample. For example, a comparison of the mean pulse rates of men and women is a comparison of two *independent sam-*

*HINT: Sampling methods are covered in Sections 5.3 and 5.4 of Chapter 5.

ples as long as the individuals that are sampled are not related pairs, such as siblings or spouses. Independent samples are generated in a variety of ways. Here are the most common methods:

- Random samples are taken separately from two populations, and the same response variable is recorded for each individual.
- One random sample is taken, and a variable is recorded for each individual, but then units are categorized as belonging to one population or another, such as male and female.
- Participants are randomly assigned to one of two treatment conditions, such as diet or exercise, and the same response variable, such as weight loss, is recorded for each individual unit.
- Two random samples are taken from a population, and a separate variable is measured in each sample. For example, one sample may be asked about hours of internet use, whereas a separate sample is asked about hours of television viewing. Then the results are compared. This method for collecting data is not used very often.

Example 11.1 **Pet Ownership and Stress** Many studies use a combination of paired data and independent samples. The independent samples are used to compare groups or treatment conditions, and the paired data are used to measure the change in a variable of interest for each person in both groups.

For example, suppose researchers were interested in knowing if owning a pet helps reduce blood pressure for seniors who live alone. They could design an experiment as follows: Recruit a representative group of volunteers who are willing to participate in the experiment. Randomly assign half of the group to receive a pet and the other half to serve as a control group. (To be fair, they could provide a pet to the control group at the end of the study.) Measure the initial blood pressure for all volunteers at the start of the experiment. Then measure their blood pressure again after 6 months. The data of interest is the change in blood pressure.

Note that the *change in blood pressure* for each individual is an example of a *paired difference*, taken as (Initial blood pressure − Ending blood pressure) for each person. Collecting the data in this way is an example of the first method of obtaining paired data listed after the definition on page 407: *The same measurement is taken twice on each person, under different conditions or at different times.* In this example, blood pressure is measured under two different conditions: at the start of the experiment, and 6 months later, after owning a pet or not.

The *changes in blood pressure for the two groups,* those who are given a pet and the control group of those who are not given one, constitute *independent samples.* Collecting data in this way is an example of the third method of obtaining independent samples, listed above: *"Participants are randomly assigned to one of two treatment conditions, and the same response variable is recorded for each individual unit."* In this case, the two treatment conditions are being given a pet or not, and the response variable is the *change* in blood pressure from the beginning to the end of the experiment.

The parameters of interest in this experiment would be as follows:

μ_d = mean change in blood pressure for the *population* of all seniors who live alone, if they were to own a pet for 6 months.

$\mu_1 - \mu_2$ = difference in the mean change in blood pressure for the population of all seniors who live alone, *if* they were to own a pet for 6 months compared with *if* they were not to own a pet for 6 months.

Standard Errors for Sample Statistics Involving Means

Recall from Chapter 10 that the standard error of the sample statistic is a key component for computing confidence intervals for the situations that we are discussing here. Let's review what we have learned about standard errors in general and then revisit what we need to compute standard errors for statistics involving means.

THOUGHT QUESTION 11.2 Note that all of the standard error formulas in this section have the sample size(s) in the denominator. This tells us that if the sample size is increased, the standard error will decrease (assuming that the sample statistics remain the same). Refer to the rough definition of standard error, and explain why this relationship between sample size and standard error makes sense, based on that definition.*

Using Student's *t*-Distribution to Determine the Multiplier

The main difference between finding confidence intervals for proportions and confidence intervals for means is the way the appropriate *multiplier* is found. In Chapter 10, we learned that the appropriate multiplier for confidence intervals involving proportions is found from a standard normal, or *z*-distribution. The notation for the multiplier in that case is z^*. In contrast, in finding confidence intervals for parameters involving means, the appropriate multiplier is found from **Student's *t*-distribution**, introduced in Chapter 9, and the notation for the multiplier is t^*.

Remember that theoretically, the *t*-distribution arises when a standardized statistic is calculated by using the sample standard deviation in place of the population standard deviation. This almost always is what happens in estimating means, because the value of the population standard deviation σ is rarely known. The rationale for using the *t*-distribution instead of the *z*-distribution will be given in CI Module 3, where we will learn how to find a confidence interval for one mean.

A *t*-distribution has a bell shape, it is centered at 0, and it is more spread out than the standard normal curve in that there is more probability in the extreme areas than there is for the standard normal curve. This additional variability results from using s, which differs for each sample from the same population, instead of σ, which is a fixed constant for a given population of measurements.

A parameter called **degrees of freedom**, abbreviated as **df**, is associated with any *t*-distribution. For problems involving a *single mean* or the *mean of paired differences*, df $= n - 1$, where n is the sample size. We will learn more about what to use for the degrees of freedom for the difference in two means for independent samples in CI Module 5, where that situation is covered.

DEFINITION For a confidence interval for a population mean, the **t^* multiplier** is the value in a *t*-distribution with df $= n - 1$ such that the area between $-t^*$ and $+t^*$ equals the desired confidence level.

Finding t^* by Using Table A.2

Most statistical software packages will compute a confidence interval for a mean and for the other parameters involving means. If you are forced to do confidence interval calculations "by hand," it is relatively simple to complete the calculations. To find the multiplier for the desired confidence level, all you need is a table of the *t*-distribution (or an appropriate calculator).

Table A.2 in the Appendix shows t^* multipliers for seven different confidence levels and varying degrees of freedom. Figure 11.2 (on the next page) illustrates the relationship between the confidence level and the numerical value t^*. To use Table A.2, compute the degrees of freedom (df), choose a confidence level, and then look in the row and column for these values in the table. Read the value of t^* from the body of the table.

If the appropriate degrees of freedom are not in the table, find the closest values that are in Table A.2, then either average or use the larger of the two t^* values. For example if df = 149, which is not in Table A.2, the closest choices are df = 100 and df = 1000. For a 95% confidence interval, the respective t^* values are 1.98 and 1.96.

***HINT:** Roughly, a standard error measures the average difference between a sample estimate and a parameter, so it measures how closely the sample estimate approximates the parameter.

Therefore, you could use the average of 1.97 or be conservative and use the larger t^* value of 1.98.

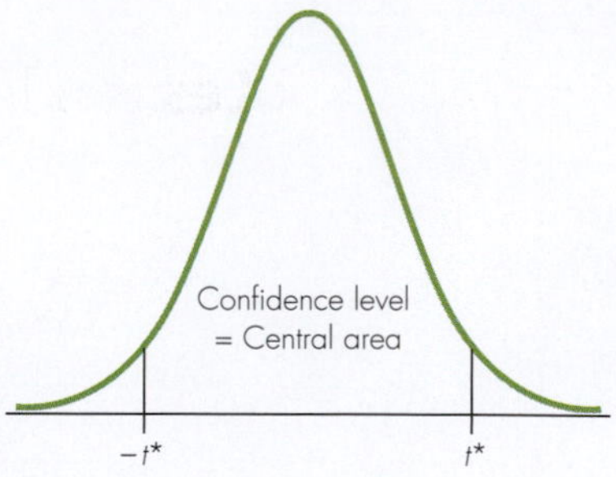

Figure 11.2 The relationship between the confidence level and the t^* multiplier

As df gets large, the t-distribution gets closer to the standard normal curve because the sample standard deviation s becomes a more accurate estimate of the population standard deviation σ. The t-distribution with df = infinity is identical to the standard normal curve. The last row of Table A.2 shows the values for that situation and can be used to find the multiplier for confidence intervals for proportions, where z^* is used instead of t^*.

Example 11.4 **Finding the t^* Values for 24 Degrees of Freedom and 95% or 99% Confidence Intervals** Suppose that you want to find a 95% confidence interval and also a 99% confidence interval for the mean of a population based on a sample of $n = 25$ values. For the 95% interval, the appropriate multiplier is the value t^* such that the area between $-t^*$ and t^* is .95, or 95%, for a Student's t-distribution with degrees of freedom $n - 1 = 24$. The appropriate value, found in Table A.2, is 2.06. Figure 11.3 illustrates the relationship between 95% confidence and $t^* = 2.06$ for this example. For a 99% confidence interval, the appropriate value is $t^* = 2.80$, found in the ".99" column of Table A.2. Note that the multiplier for 99% confidence is greater than it is for 95% confidence, which will result in a wider interval.

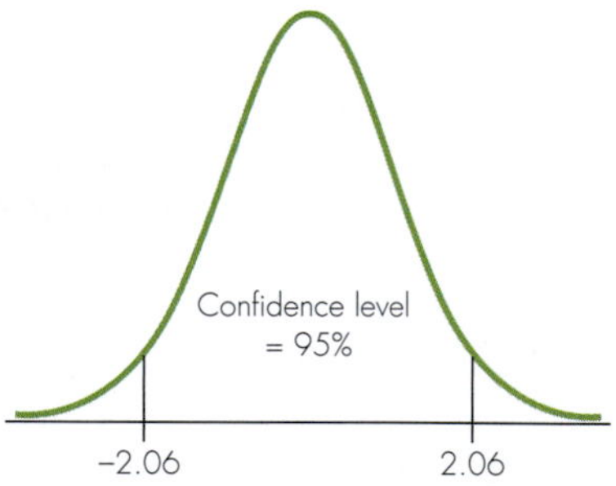

Figure 11.3 The t^* multiplier for a 95% confidence interval with df = *24*

EXCEL TIP The command TINV(p,d) provides the value t^* for which the area *between* $-t^*$ and t^* is $(1 - p)$ for a Student's t-distribution with degrees of freedom d. For example, to find t^* for a 95% confidence interval with df = 24, use TINV(.05, 24). The result is 2.06, meaning that the area between −2.06 and 2.06 is .95 (or 95%) for a t-distribution with df = 24.

11.1 Exercises are on pages 440–442.

11.2 CI Module 3: Confidence Intervals for One Population Mean

Make sure you read CI Module 0 (Section 10.1) and Section 11.1, "Introduction to Confidence Intervals for Means," before studying this section. The relevant background material is provided in those sections.

In Lesson 1 of this module, we describe how to find a confidence interval for a population mean using a sample of any size and with any level of confidence desired. In Lesson 2, we describe the special and simpler case of finding an approximate 95%

CI Module 3: μ

confidence interval when the sample is large. This is a very common situation and therefore is worth a separate discussion.

Lesson 1: Finding a Confidence Interval for a Mean for Any Sample Size and Any Confidence Level

Researchers often are interested in a very simple question about a population: If we were to measure a quantitative variable for every unit in the population, what would be the mean or average of those values? For example, what is the mean age of death for the population of left-handed adults? What is the mean body temperature for healthy human adults? What is the mean per capita income for all adults in your state? What was the mean price of a gallon of gasoline last week in the United States? What is the mean amount of sleep college students get per night?

All of these questions can be answered by taking an appropriate sample from the population of interest and computing a **confidence interval estimate of the mean** of the population, which is an interval of values computed from the sample data that we can be fairly confident covers the true population mean. How certain we can be is determined by the confidence level that we choose to use.

How to Compute the Confidence Interval Estimate

By now, you should be familiar with the generic format for a confidence interval for any of the five parameters under discussion:

Sample estimate ± Multiplier × Standard error

To find a **confidence interval for a population mean**, the ingredients of the formula are as follows:

Sample estimate = $\bar{x}$ = the mean of all of the numbers in the sample. We learned how to compute the sample mean in Chapter 2.

Multiplier = t^* = the value of Student's t-distribution with df = $n - 1$ and the desired confidence level as the area between $-t^*$ and t^*. We discussed how to find t^* using software or Table A.2 on pages 411–412 in Section 11.1.

Standard error = $s/\sqrt{n}$, where s = the sample standard deviation (available by using computer software or found by using the formula on page 48 of Chapter 2) and n = the number of individuals in the sample.

The result is sometimes called a ***t*-interval** because the multiplier is from Student's t-distribution. Putting together all of the pieces gives the t-interval formula as follows:

$$\bar{x} \pm t^* \times \frac{s}{\sqrt{n}}$$

Table 11.1 gives some examples of t^* multipliers, found using Table A.2. If you don't have access to a calculator or computer to compute confidence intervals, make sure you understand how to find t^* values.

Table 11.1 Selected t^* Values from Table A.2 and Resulting Confidence Interval Formulas

Sample Size n	df = $n - 1$	Confidence Level	Multiplier = t^*	Confidence Interval Formula
25	24	90%	1.71	$\bar{x} \pm 1.71\frac{s}{\sqrt{n}}$
25	24	95%	2.06	$\bar{x} \pm 2.06\frac{s}{\sqrt{n}}$
25	24	99%	2.80	$\bar{x} \pm 2.80\frac{s}{\sqrt{n}}$
101	100	95%	1.98	$\bar{x} \pm 1.98\frac{s}{\sqrt{n}}$

CI Module 3: μ

Note these features of t^* values, evident from Table 11.1:

- As the confidence level *increases* (for constant df), the t^* value *increases* as well. This result makes sense because to increase our confidence that the interval covers the true parameter value, we need to increase the width of the interval. Having a larger multiplier accomplishes that.
- As the degrees of freedom *increase* (for constant confidence level), the value of t^* *decreases.* This is because the uncertainty surrounding our sample estimate decreases when we have more data. A smaller value of t^* results in a more precise (narrower) interval.

Conditions Required for Using the *t* Confidence Interval

The methods that are presented in this section are derived mathematically by assuming that the sample has been randomly selected from a population in which the response variable has a normal distribution. The *t*-interval procedure, however, is a **robust procedure** because it works well over a wide range of situations. Put another way, the stated confidence level for a *t*-interval is approximately correct in many situations in which the assumption about a normally distributed response variable is not correct.

For large sample sizes, a *t*-interval is a valid estimate of the population mean, even in the presence of skewness. For small sample sizes, a *t*-interval can be used if the data are not skewed and contain no outliers.

Two Situations for Which a *t* Confidence Interval for One Mean Is Valid

Situation 1: The population of the measurements is bell-shaped, and a random sample of any size is measured. In practice, for small samples, the data should show no extreme skewness and should not contain any outliers.

Situation 2: The population of measurements is not bell-shaped, but a *large* random sample is measured. A somewhat arbitrary definition of a "large" sample is $n \geq 30$, but if there are extreme outliers, it is better to have an even larger sample size than $n = 30$.

Before calculating a confidence interval for a mean, first check that one of the situations just described holds. To determine whether the data are bell-shaped or skewed, and to check for outliers, plot the data using a histogram, dotplot, or stemplot. A boxplot can reveal outliers and will sometimes reveal skewness, but it cannot be used to determine the shape otherwise. The sample mean and median can also be compared to each other. Differences between the mean and the median usually occur if the data are skewed—that is, are much more spread out in one direction than in the other. For $n \geq 30$, the procedure in this section usually produces a valid confidence interval estimate of the population mean, but if the data are obviously skewed, you should question whether the median might be a better measure of location than the mean. (We will not cover confidence intervals for medians in this book.)

CI Module 3: μ

Example 11.5 **Are Your Sleeves Too Short? The Mean Forearm Length of Men** People are always interested in comparing themselves to others. If you are male, how does the length of your forearm, from elbow to wrist, compare to the average for the population of men? Suppose that the forearm lengths (in centimeters) for a randomly selected sample of $n = 9$ men are as follows:

25.5, 24.0, 26.5, 25.5, 28.0, 27.0, 23.0, 25.0, 25.0

Figure 11.4 shows a dotplot of these values. There is no obvious skewness, and there are no outliers. Assuming that the sample can be considered a random sample, Situation 1 holds, and the necessary conditions for computing a confidence interval for the mean are present.

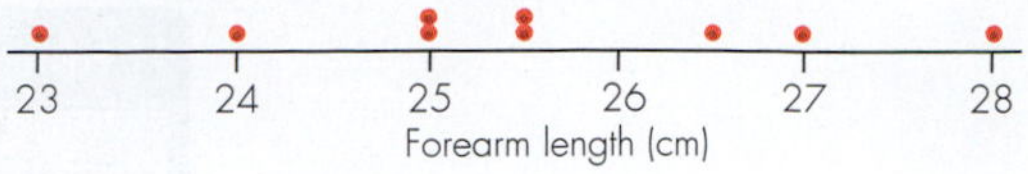

Figure 11.4 Dotplot of the forearm lengths for Example 11.5

The sample mean is $\bar{x} = 25.5$ cm and the sample standard deviation is $s = 1.52$ cm. These values can be found using methods covered in Chapter 2. Values for two of the necessary ingredients for calculating a confidence interval for the population mean are as follows.

- *Sample estimate* $= \bar{x} = 25.5$ cm
- *Standard error* $=$ s.e.$(\bar{x}) = \dfrac{s}{\sqrt{n}} = \dfrac{1.52}{\sqrt{9}} = 0.507$ cm

We'll use the 95% confidence level. For this example, degrees of freedom are df $= n - 1 = 9 - 1 = 8$. Using Table A.2, we find that the multiplier is $t^* = 2.31$ (look in df $= 8$ row and ".95" column of the table). A 95% confidence interval for μ, the population mean, is

Sample estimate $\pm$ Multiplier $\times$ Standard error

$\bar{x} \pm t^* \times$ s.e.$(\bar{x})$

$25.5 \pm 2.31 \times 0.507$

25.5 ± 1.17

24.33 to 26.67 cm

Interpretation: We can say with 95% confidence that in the population represented by the sample, the mean forearm length is between 24.33 and 26.67 cm.

IN SUMMARY Calculating a Confidence Interval for a Population Mean

A confidence interval for a population mean μ (also called a ***t*-interval**) is

$$\bar{x} \pm t^* \times \text{s.e.}(\bar{x}), \quad \text{which is } \bar{x} \pm t^* \times \frac{s}{\sqrt{n}}$$

To determine this interval:

1. Make sure the appropriate conditions apply by checking sample size and/or a shape picture. For small samples ($n < 30$) with extreme skewness or outliers, you cannot proceed.
2. Choose a confidence level.
3. Determine the sample mean and standard deviation ($\bar{x}$ and s) using software or formulas in Chapter 2.
4. Calculate the standard error of the mean: s.e.$(\bar{x}) = \dfrac{s}{\sqrt{n}}$.
5. Calculate df $= n - 1$.
6. Use Table A.2 (or statistical software) to find the multiplier t^*.

CI Module 3: μ

Example 11.6 **How Much TV Do Penn State Students Watch?** In Example 11.2, we learned the results of a survey given to 175 Penn State students, in which one question asked was "In a typical day, about how much time do you spend watching television?" We would like to estimate

μ = mean number of television viewing hours for the *population* of all Penn State students.

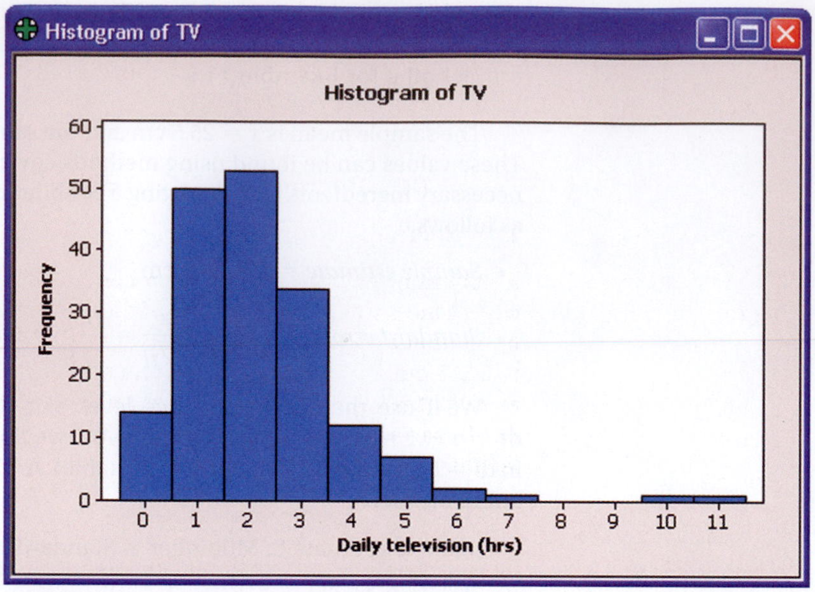

Figure 11.5 Histogram of daily hours of television for Example 11.6

Let's go through the steps for finding the t-interval:

1. A histogram of the data, created by Minitab, is shown in Figure 11.5. Note that the shape is skewed to the right and there are a few outliers. However, the sample size is large, so Situation 2 applies whether or not the values are bell-shaped.
2. Let's compute both a 95% and a 99% confidence interval.
3. The sample mean was just over 2 hours, with $\bar{x} = 2.09$ hours; the sample standard deviation was $s = 1.644$ hours.
4. The standard error of the mean was computed as $s/\sqrt{n} = 0.124$.
5. Degrees of freedom = 175 − 1 = 174.
6. Table A.2 doesn't include df = 174. Let's use the conservative approach and find the multiplier t^* corresponding to df = 100. For 95% confidence, $t^* = 1.98$. For 99% confidence, $t^* = 2.63$.

A 95% confidence interval estimate for μ is

$$\bar{x} \pm t^* \times \text{s.e.}(\bar{x}) \quad \textit{or} \quad 2.09 \pm (1.98)(0.124) \quad \textit{or} \quad 2.09 \pm 0.25 \quad \textit{or} \quad 1.84 \text{ to } 2.34 \text{ hours}$$

A 99% confidence interval estimate for μ is

$$2.09 \pm (2.63)(0.124) \textit{ or } 2.09 \pm 0.33 \textit{ or } 1.76 \text{ to } 2.42 \text{ hours}$$

Note that, as we expected, the 99% confidence interval estimate is somewhat wider than the 95% one. The added confidence that our interval does indeed cover the true population mean has cost us a small amount of precision.

Correct Interpretation of a Confidence Interval for a Mean

The interpretation of a confidence interval for a mean, and the confidence level accompanying it, is no different from the interpretation of other confidence intervals. Our confidence is in the procedure. There is no way to know whether or not an individual confidence interval estimate covers the truth. But we know that in the long run, of all confidence intervals that we compute using a 95% level of confidence, about 19 out of 20 of them (i.e., 95%) will cover the true population value. About one out of 20 of them will not. Therefore, we don't know whether or not the mean TV viewing hours for Penn State students is really between 1.84 and 2.34 hours. It is if we got one of the 95% of possible samples for which the sample mean is close enough to the true population mean for our procedure to work. It isn't if we happened to get one of the 5% of possible samples for which the sample mean is too far away from the population mean for the procedure to work.

A Common Misinterpretation of a Confidence Interval for a Mean

A common *misinterpretation* of a confidence interval for a mean is to think that it covers the specified percentage of the population. For instance, in Example 11.5, we found that a 95% confidence interval for the mean forearm length of men is 24.33 cm to 26.67 cm. We can be fairly certain that the *population mean* forearm length is covered by this interval. This does *not* give us *any* information about the range of *individual* forearm lengths. Do not fall into the trap of thinking that 95% of men's forearm lengths are in this interval.

Example 11.7 **What Type of Students Sleep More?** On the second day of classes in spring quarter 2000, a Monday, students in two statistics classes at the University of California at Davis were given a survey. One of the questions asked was "How many hours of sleep did you get last night, to the nearest half hour?" One class was Statistics 10, a statistical literacy course for liberal arts majors ($n = 25$), while the other was Statistics 13, a large introductory class in statistical methods for more technical majors ($n = 148$). Here is Minitab output that summarizes the information on this question for the two classes:*

class	N	Mean	StDev	SE Mean
Stat 10	25	7.66	1.34	0.27
Stat 13	148	6.81	1.73	0.14

Note that Minitab has computed the standard errors. For instance, for the Statistics 10 class, $\text{s.e.}(\bar{x}) = s/\sqrt{n} = 1.34/\sqrt{25} = 0.27$. The standard deviation for the Statistics 13 class is actually larger than that for the Statistics 10 class, but the standard error is much smaller because the sample size n is so much larger. Generally, the larger the sample, the more accurate is the sample mean as an estimate of the population mean.

Let's find 95% confidence intervals for the mean hours of sleep for the *populations* represented by these two classes. Note that there are some differences between the hypothetical populations. For the Statistics 10 class, students are liberal arts majors, and the course is a general education elective. The Statistics 13 students represent a wide variety of majors, and the course is required for most of them. For both classes, the night in question was a Sunday, and many of the students had just returned from the between-quarter break that day (skipping the first day of classes, a Friday).

Checking the Conditions

For each group, we must make the assumption that these students are equivalent to a random sample from the population of interest. Although this assumption is not strictly true, the students in these two classes are probably representative of all students who take these and similar classes over the years. Second, for the Statistics 10 class, the sample size is only 25, so we must assume that the population of sleep times is approximately bell-shaped. A histogram of the sample values, shown in Figure 11.6, indicates that this assumption is not unreasonable; there are no outliers and no extreme skewness.

*Note that Minitab uses N instead of n to indicate the size of a sample.

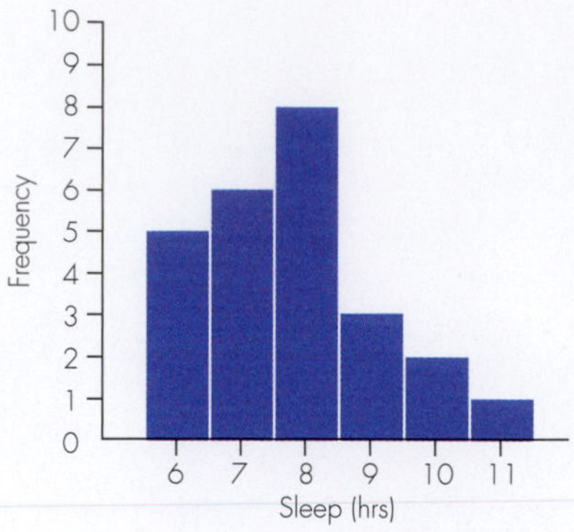

Figure 11.6 Histogram of sleep values for Statistics 10 students for Example 11.7

Calculating the Confidence Intervals

The calculations are as shown in the table below.

Class	n	df = $n - 1$	Mean	t^*	s.e.(Mean)	95% Confidence Interval
Stat 10	25	24	7.66	2.06	0.27	7.10 to 8.22 hours
Stat 13	148	147	6.81	1.98	0.14	6.53 to 7.09 hours

Interpreting the Confidence Intervals

Each interval indicates the range of values that we think covers the true *average* number of hours students in each population sleep. For instance, we can say with 95% confidence that the average hours of sleep for all students similar to those in Statistics 10 is covered by the interval from 7.10 to 8.22 hours. Note that the interval is much wider for the Statistics 10 situation than for Statistics 13 because the sample size was much smaller.

Note also that the two intervals do not overlap. Therefore, it appears that the liberal arts students' average hours of sleep really is higher than the general students' average hours of sleep. We can confirm this speculation by computing a confidence interval for the difference in the two population means. We will learn how to do this in Module 5.

Technical Explanation: Why the Confidence Interval Formula Works

In Chapter 9, we learned that the standardized statistic for a sample mean has a Student's t-distribution with df $= n - 1$. Understanding how that result is derived is beyond the scope of this text, but the result can be used to find the formula for a confidence interval for a population mean. We used a similar technique in CI Module 1 in Chapter 10 to derive the formula for a confidence interval for a population proportion.

Recall from Chapter 9 that

$$t = \frac{\bar{x} - \mu}{\text{s.e.}(\bar{x})}$$

has a t-distribution with df $= n - 1$. Suppose that we want a confidence interval with confidence level C; typically, C is .95, .90, or .99. Define the value t^* to be such that the area between $-t^*$ and t^* is C for a t-distribution with df $= n - 1$. For example, if we want a 95% confidence level, then $C = .95$, and we find t^* so that the area between $-t^*$ and t^* is .95.

Then we can write the following probability statement:

$$P\left(-t^* \leq \frac{\bar{x} - \mu}{\text{s.e.}(\bar{x})} \leq t^* \right) = C$$

$$= P(- t^* \times \text{s.e.}(\bar{x}) \leq \bar{x} - \mu \leq t^* \times \text{s.e.}(\bar{x}))$$

$$= P(- \bar{x} - t^* \times \text{s.e.}(\bar{x}) \leq - \mu \leq - \bar{x} + t^* \times \text{s.e.}(\bar{x}))$$

Multiply everything by -1, which reverses the inequality signs, to get

$$P(\bar{x} - t^* \times \text{s.e.}(\bar{x}) \leq \mu \leq \bar{x} + t^* \times \text{s.e.}(\bar{x}))$$

In other words, out of all the times we compute the interval $\bar{x} \pm t^* \times \text{s.e.}(\bar{x})$, the long-run probability that the result covers μ is given by the confidence level we chose, C.

Once we have data and compute a numerical interval, we can no longer apply the probability statement to the result. This is similar to other probability situations, such as flipping a coin. *Before* we determine the outcome (heads or tails), the probability of a head can be said to be .5. But *after* we have the outcome, the probability of getting a head is either 0 (if we got a tail) or 1 (if we got a head). It no longer makes sense to talk about the probability as .5. What we can continue to say is that the long-run probability of a head is .5.

Similarly, once we have observed data and computed a confidence interval, either it covers the true μ or it doesn't. Unlike the coin situation, which had a known outcome, we don't know whether the confidence interval covers the population mean or not. However, it still doesn't make sense to talk about the probabilities of various outcomes when we have only one observed outcome. All we can say is that we have a high degree of confidence in our procedure, determined by C, and that we are aware that there is a small chance that the confidence interval estimate does not cover the true parameter.

MINITAB TIP **Computing a Confidence Interval for a Mean**

- To compute a confidence interval for a mean, use **Stat > Basic Statistics > 1-sample t.** Specify the column that contains the raw data. To change the confidence level, use the ***Options*** button. The default confidence level is 95%.
- If the raw data are not available but the data are summarized so that you know values of the sample mean, standard deviation, and sample size, enter those values in the "Summarized data" area of the dialog box.

Note: Use the ***Graphs*** button to create visual displays of the data.

Lesson 2: Special Case: Approximate 95% Confidence Intervals for Large Samples

When we learned about Student's t-distribution, we learned that for large values of degrees of freedom, the t-distribution is very close to the standard normal distribution. Therefore, when the sample is large and thus the degrees of freedom are large, the t^* multiplier will be close to the corresponding z^* multiplier. You can see this empirically in Table A.2 in the Appendix. For 95% confidence, the z^* multiplier is 1.96 but is often rounded off to 2. The t^* multiplier for df = 1000 is 1.96 as well; for df = 100, it is 1.98; and for df = 70, 80, or 90, t^* is 1.99, which is still close to the exact z^* value of 1.96. Even with as few as 30 df, the t^* multiplier is 2.04, not far from the exact z^* of 1.96 and even closer to the rounded-off value of 2.0.

In checking the conditions for which the formula for a confidence interval for a mean applies, we see that it always applies in Situation 2, when a *large* random sample is measured, regardless of the shape of the distribution. Therefore, when we have a large sample, we can simplify matters and find an *approximate* 95% confidence interval estimate using a multiplier of 2.0 instead of t^*. The argument works for the other parameters that are covered in this chapter and in Chapter 10 as well, so let's write the result to encompass them too.

DEFINITION For sufficiently large samples, the interval

Sample estimate $\pm$ 2 $\times$ Standard error

is an **approximate 95% confidence interval** for a population parameter. This is so for a proportion, a mean, the difference between two proportions, the mean of paired differences, and the difference between two means.

Note: The 95% confidence level describes how often the procedure provides an interval that includes the population value. For about 95% of all random samples of a specific size from a population, the confidence interval covers the population parameter.

Example 11.8 **Approximate 95% Confidence Interval for TV Time** In Example 11.6, we found that a 95% confidence interval for mean number of hours Penn State students watch TV was 2.09 $\pm$ (1.98)(0.124) or 2.09 $\pm$ 0.25, or 1.84 hours to 2.34 hours. The sample mean, based on a sample $n = 175$ students, was $\bar{x} = 2.09$ hours per day. The standard error is s.e.($\bar{x}$) = 0.124, and an approximate 95% confidence interval estimate of μ, the population mean, is

Sample estimate $\pm$ 2 $\times$ standard error
$\bar{x} \pm 2 \times$ s.e.($\bar{x}$)
2.09 $\pm$ 2 $\times$ 0.124
2.09 $\pm$ 0.248
1.842 to 2.338 hours

You can see that the approximate 95% confidence interval (1.842 to 2.338) is essentially the same as the one that was computed by using the multiplier t^* (1.84 to 2.34). It's obvious that this should be the case, since the t^* multiplier was 1.98, very close to the multiplier of 2 used for the approximate interval.

In a research article, this confidence interval might be described as follows: *We are 95% confident that the mean time that Penn State students spend watching television per day is somewhere between 1.842 and 2.338 hours.* Actually, the report would probably round off the numbers and declare that, with 95% confidence, the mean is between 1.8 and 2.3 hours. Remember what is meant technically by that statement: We are 95% *confident* that the interval from 1.842 to 2.338 hours is correct in that it captures the mean television viewing hours for Penn State students. We will never know whether or not the interval actually does cover the mean in any particular instance, unless the whole population is eventually measured.

11.2 Exercises are on pages 442–443.

CI Module 3: μ

THOUGHT QUESTION 11.3

- What population do you think is represented by the sample of 175 students in Example 11.8? Do you think the Fundamental Rule for Using Data for Inference (reviewed on p. 372 of Chapter 10) holds in this case?
- Does the confidence interval in Example 11.8 tell us that *95% of the students* watch television between 1.842 and 2.338 hours per day? If not, what exactly does the interval tell us?*

***HINT:** Refer to the statement in Example 11.8 that shows how the result might be described in a research article.

11.3 CI Module 4: Confidence Interval for the Population Mean of Paired Differences

Remember that an important special case of a single mean of a population occurs when two quantitative variables are collected in pairs and we desire information about the *difference* between the two variables. For example, we may want to know the average difference between verbal SAT scores for individuals before and after taking a training course. Or we may want to know the average difference in manual dexterity for the dominant and nondominant hands. In both of these examples, we would collect two measurements from each individual in the sample.

How to Compute the Confidence Interval Estimate for the Mean of Paired Differences

To analyze paired data, we begin by calculating the difference in the two measurements for each pair in the sample. As notation for an analysis of the differences, we could simply use the notation that was already developed for a single mean, but it is preferable to use notation emphasizing that the data consist of differences.

Notation for Paired Differences

Notation that can be used for paired data is as follows:

Data: two variables for each of n individuals or pairs; use the difference $d = x_1 - x_2$.

Population parameter: μ_d = mean of differences for the population, equivalent to $\mu_1 - \mu_2$.

Sample estimate: $\bar{d}$ = sample mean of the differences, equivalent to $\bar{x}_1 - \bar{x}_2$.

Standard deviation and standard error: s_d = standard deviation of the sample of differences; s.e.$(\bar{d}) = s_d/\sqrt{n}$.

Confidence interval for μ_d: $\bar{d} \pm t^* \times$ s.e.$(\bar{d})$, where df $= n - 1$ for the multiplier t^*.

Conditions Required for Using the t Confidence Interval for the Mean of Paired Differences

Once the differences have been computed, the conditions that are required are the same as they are for a t-interval for one population mean. The distinction is that the conditions must hold for the data set of *differences*. Let's rewrite the two situations for which the t-interval formula is appropriate, using the terminology for the differences:

Situation 1: The population of *differences* is bell-shaped, and a random sample of any size is measured. In practice, for small samples, the differences in the sample should show no extreme skewness and should not contain any outliers.

Situation 2: The population of *differences* is not bell-shaped, but a *large* random sample is measured. A somewhat arbitrary definition of a "large" sample is $n \geq 30$ pairs, but if there are extreme outliers in the sample of differences, it is better to have an even larger sample size than $n = 30$.

Before calculating a **confidence interval for the population mean of paired differences** (using the formula shown above), check that there are enough pairs to satisfy Situation 2. If not, examine a plot of the differences. You can use a histogram, dotplot, or stem-and-leaf plot to examine shape, and any of those plots or a boxplot to make sure that there are no excessive outliers or extreme skewness.

Note that it is *not* equivalent to look at plots of the two original variables before taking the differences. It is quite possible to have extreme outliers or skewness in the original data but not in the differences. For example, suppose that the data represent IQ after listening to a relaxing message and after listening to Mozart for a sample of randomly selected adults. It is quite possible that there will be someone with an abnormally high or low IQ in the sample. For each of the two original IQ variables, that person's data may look like an outlier. But it is unlikely that the person's *difference* in IQ for the two conditions would stand out from the others.

Once we have the difference for each pair, we simply follow the steps for computing a confidence interval for the mean of one quantitative variable. We use the notation for differences, but everything else stays the same.

Example 11.9 **Screen Time—Computer versus TV** The 25 students in a liberal arts course in statistical literacy were given a survey that included questions on how many hours per week they watched television and how many hours a week they used a computer. The responses are shown in Table 11.2, along with the difference d for each student. From these data, let's construct a 90% confidence interval for μ_d, the average *difference* in hours spent using the computer versus watching television (Computer use − TV) for the population of students represented by this sample.

Table 11.2 Data on Weekly Hours of Computer Use and TV Viewing

Student	Computer	TV	Difference	Student	Computer	TV	Difference
1	30	2.0	28.0	14	5	6.0	−1.0
2	20	1.5	18.5	15	8	20.0	−12.0
3	10	14.0	−4.0	16	30	20.0	10.0
4	10	2.0	8.0	17	40	35.0	5.0
5	10	6.0	4.0	18	15	15.0	0.0
6	0	20.0	−20.0	19	40	5.0	35.0
7	35	14.0	21.0	20	3	13.5	−10.5
8	20	1.0	19.0	21	21	35.0	−14.0
9	2	14.0	−12.0	22	2	1.0	1.0
10	5	10.0	−5.0	23	9	4.0	5.0
11	10	15.0	−5.0	24	14	0.0	14.0
12	4	2.0	2.0	25	21	14.0	7.0
13	50	10.0	40.0				

Let's go through the steps we learned for calculating a confidence interval for one population mean, substituting notation for differences:

1. *Make sure the appropriate conditions apply by checking sample size and/or a shape picture. For small samples ($n < 30$) with extreme skewness or outliers, you cannot proceed.*

The sample size is under 30, so it is important to make sure that there are no major outliers or extreme skewness. The sample mean and median are 5.36 and 4.0 hours, respectively. In data ranging from −20 to +40, these statistics are close enough to rule out extreme skewness. The boxplot of the differences in Figure 11.7 provides further evidence that the appropriate conditions are satisfied.

CI Module 4: μ_d

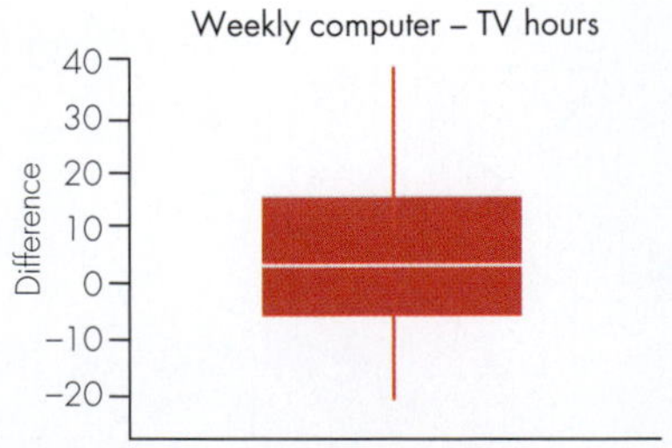

Figure 11.7 Boxplot for difference between computer and TV hours per week, $n = 25$

2. *Choose a confidence level:* Let's construct a 90% confidence interval.
3. *Determine the sample mean and standard deviation using software or formulas in Chapter 2.*

The mean of the differences is $\bar{d} = 5.36$ and the standard deviation of the differences is $s_d = 15.24$ hours.

4. *Calculate the standard error of the mean:* s.e.$(\overline{d}) = \dfrac{s_d}{\sqrt{n}}$.

With $n = 25$, the standard error is $15.24/\sqrt{25} = 3.05$.

5. *Calculate* $df = n - 1$.
6. *Use Table A.2 (or statistical software) to find the multiplier* t^*.

The df $= 25 - 1 = 24$, so from Table A.2, for a 90% confidence interval, $t^* = 1.71$. The computation for a 90% confidence interval for μ_d is

$\overline{d} \pm t^* \times$ s.e.$(\overline{d})$
$5.36 \pm (1.71)(3.05)$
5.36 ± 5.22 or 0.14 to 10.58 hours

Interpretation: We are 90% confident that the average difference between computer usage and television viewing, μ_d, for the population of students represented by this sample is covered by the interval from 0.14 to 10.58 hours per week, with more hours spent on computer usage than on television viewing.

Interpreting a Confidence Interval for the Mean of Paired Differences

In addition to the usual interpretation of a confidence interval, it is often of interest to know whether the confidence interval for the mean of paired differences covers 0. If it does cover 0, then it is possible that the population mean is 0, indicating that the population means for the two measurements could be the same. If the interval does not cover 0, then we can be fairly certain that the population means for the two variables are different. How certain we can be depends on the confidence level that is used for the computation.

In Example 11.9, we found that a 90% confidence interval for the mean difference between hours of computer use and hours of television watching is 0.14 hour to 10.58 hours per week. The interval does not cover 0, so we can be fairly certain that for the population represented by this sample, students really do spend more time on computer usage than on watching television, on average. What conclusion do you think we would have reached if we had used a 95% confidence level? (See Exercise 11.45.)

THOUGHT QUESTION 11.4 For a fixed sample, explain why it is logical that a 95% confidence interval covers a wider range of values than a 90% confidence interval. Explain this in terms of our confidence that the procedure works in any given case.*

IN SUMMARY

Calculating a Confidence Interval for the Population Mean of Paired Differences

A **confidence interval for the population mean of paired differences** μ_d (also called a **paired *t*-interval**) is

$$\overline{d} \pm t^* \times \text{s.e.}(\overline{d}), \quad \text{which is } \overline{d} \pm t^* \times \frac{s_d}{\sqrt{n}}$$

To determine this interval:

1. Make sure the appropriate conditions apply by checking sample size and/or a shape picture of the differences.

(continued)

***HINT:** Consider the extreme case. How could you construct an interval that you are 100% confident covers the truth? Would it be wider or narrower than a 90% confidence interval?

CI Module 4: μ_d

IN SUMMARY **Calculating a Confidence Interval for the Population Mean of Paired Differences (Continued)**

2. Choose a confidence level.
3. Compute the differences for the n pairs in the sample, then find the mean $\bar{d}$ and standard deviation s_d for those differences.
4. Calculate the standard error of $\bar{d}$: s.e.$(\bar{d}) = \dfrac{s_d}{\sqrt{n}}$.
5. Calculate df $= n - 1$.
6. Use Table A.2 (or statistical software) to find the multiplier t^*.

Sometimes you will be provided with the summarized data rather than the original differences, as the next example illustrates. In that case, you won't have to carry out Step 3 of the summary.

Example 11.10 **Meditation and Anxiety** Davidson et al. (2003) conducted a study in which volunteers were randomly assigned to a meditation group or a control group. The meditation group took an 8-week training course during which they were to practice meditation daily. (The control group was offered the training at the conclusion of the experiment.) The participants were given a standard test for level of anxiety, called the Spielberger State-Trait Anxiety Inventory. The "trait" part of the test, used in this experiment, is supposed to measure "trait anxiety," which is how the person generally feels. It is a characteristic that remains relatively constant. (The "state" anxiety part of the test measures how anxious someone is at the point in time when they are taking it and is used for studies of things such as test anxiety.) The test consists of 20 questions for which respondents rate themselves on a scale from 1 to 4. An example of a question is "I lack self-confidence," with choices ranging from 1 = "almost never" to 4 = "almost always." The maximum possible score on the test is 80 and the minimum is 20, but the majority of scores tend to be in the 30s and 40s.

Read the original source on the companion website, http://www.cengage.com/statistics/Utts4e.

The participants were given the test at the beginning of the study and again after 8 weeks of meditation training. The *difference* in the scores was found for each of the $n = 21$ people in the meditation group who completed the test both times, with the goal of determining whether meditation reduces anxiety. The *parameter* of interest is the mean change in the anxiety score for the population if everyone were to be given the 8-week meditation training.

The article did not provide the original data, so the results that are presented here are approximated from information that was provided. Because the original data weren't provided, there is no way to check for outliers or skewness in the differences. We will proceed under the assumption that Situation 1 is satisfied. Let's find a 95% confidence interval. The sample mean for the 21 differences was $\bar{d} = 4.5$, and the sample standard deviation was $s_d = 7.2$, so the standard error is

$$\text{s.e.}(\bar{d}) = \frac{s_d}{\sqrt{n}} = \frac{7.2}{\sqrt{21}} = 1.57$$

The df $= 21 - 1 = 20$, so from Table A.2, $t^* = 2.09$. Therefore, a 95% confidence interval estimate for the population mean reduction in anxiety if everyone were to be trained in meditation is

$$\bar{d} \pm t^* \times \text{s.e.}(\bar{d})$$
$$4.5 \pm 2.09 \times 1.57 \quad or \quad 4.5 \pm 3.3 \quad or \quad 1.2 \text{ to } 7.8$$

Note that the entire interval is above 0. Therefore, we can say with high confidence that the mean anxiety level would go down in the population if everyone were to take the meditation training course. The control group scores actually went up slightly during the 8-week period, with a mean increase of about 1 point.

11.3 Exercises are on pages 443–444.

CI Module 4: μ_d

MINITAB TIP **Computing a Confidence Interval for the Mean of Paired Differences**

These instructions are to be used if you have the raw data for the original two variables stored in two columns, with one pair in each row. There are two methods that can be used:

- First, calculate a column of differences using **Calc > Calculator**, then use **Stat > Basic Statistics > 1-sample t**. Specify the column that contains the differences. To change the confidence level, use the ***Options*** button. The default confidence level is 95%.
- Alternatively, use **Stat > Basic Statistics > Paired t**. Specify the two columns that contain the raw data for the pair of measurements.
- If the raw data are not available but you know values of the mean and standard deviation of the sample of differences as well as the sample size, enter those values in the "Summarized data" area of the dialog box.

Note: Use the ***Graphs*** menu to create visual displays of the data to check for outliers and skewness.

TI-84 TIP **Confidence Interval for a Mean or the Mean of Paired Differences**

- Press [STAT] and scroll horizontally to **TESTS**. Scroll vertically to **8:Tinterval** and press [ENTER].
- If the individual data values are stored in a list (e.g., L1) select **Data** as the input method. Enter the data list name (say, L1) and a confidence level. The entry for ***Freq:*** should be 1, the default value. Scroll to ***Calculate*** and press [ENTER].
- For paired data, enter the first value for each pair in list L1 and the second value for each pair in list L2. With a clean home screen use the keystrokes [2nd] [1] [−] [2nd] [2] [STO>] [2nd] [3] [ENTER] to create and store the differences in list L3. Then proceed as above, using L3 as the data list.
- If summary statistics are already known, select **Stats** as the input method. Enter values for the sample mean, standard deviation, sample size, and a confidence level. Scroll to ***Calculate*** and press [ENTER].

11.4 CI Module 5: Confidence Interval for the Difference in Two Population Means (Independent Samples)

Lesson 1: The General (Unpooled) Case

Let's review the situation that generates research questions for which the difference between two population means is of interest. We have two populations or groups from which independent samples are available, or we have one population for which two groups can be formed using a categorical variable. We are interested in comparing the means of a quantitative variable for the two populations or for the two groups in a population. For example, we might want to compare the mean incomes of the populations of males and females who work in a comparable job. Or we might want to compare the mean IQ for babies whose mothers smoked during pregnancy with the mean IQ for babies whose mothers did not smoke.

The notation is as follows:

	Population Mean	Sample Size	Sample Mean	Sample Standard Deviation
Population 1	μ_1	n_1	$\bar{x}_1$	s_1
Population 2	μ_2	n_2	$\bar{x}_2$	s_2

The same formula we have been using for the other parameters involving proportions and means can be applied to find a confidence interval for the difference in the two population means, $\mu_1 - \mu_2$:

Sample estimate ± Multiplier × Standard error

In this case, the sample estimate is the difference in sample means, $\bar{x}_1 - \bar{x}_2$. As is the case in confidence intervals for a single mean, the t-distribution is used to determine the multiplier in the confidence interval for the difference in means for independent samples. Because the multiplier is t^* and two samples are used, the result is sometimes called a **two-sample *t*-interval**. The general format of a confidence interval for the difference in two means is therefore the following:

Difference in sample means ± t^* × Standard error

The *standard error* of the difference in sample means is

$$\text{s.e.}(\bar{x}_1 - \bar{x}_2) = \sqrt{\frac{s_1^2}{n_1} + \frac{s_2^2}{n_2}}$$

Unfortunately, there is a muddy mathematical story underneath the calculation of a confidence interval for the difference between two population means. On the surface, however, the story appears to be easy and can be summarized as follows.

FORMULA

Confidence Interval for the Difference between Two Population Means (Independent Samples)

An approximate **confidence interval for $\mu_1 - \mu_2$** is

$$(\bar{x}_1 - \bar{x}_2) \pm t^* \times \sqrt{\frac{s_1^2}{n_1} + \frac{s_2^2}{n_2}}$$

The multiplier t^* is a t-value such that the area between $-t^*$ and $+t^*$ in the appropriate t-distribution equals the desired confidence level. Appropriate degrees of freedom are difficult to specify. Computer software will provide the approximate df, but lacking that, a conservative approximation is to use the smaller of the two sample sizes and subtract 1.

The difficulty is that for this confidence interval procedure, it's not exactly mathematically correct to use a t-distribution to determine the multiplier. It is approximately correct to do so, but the approximation involves a complicated formula for the degrees of freedom. That approximation formula (often called **Welch's approximation**) is

$$\text{df} = \frac{\left(\dfrac{s_1^2}{n_1} + \dfrac{s_2^2}{n_2}\right)^2}{\dfrac{1}{n_1 - 1}\left(\dfrac{s_1^2}{n_1}\right)^2 + \dfrac{1}{n_2 - 1}\left(\dfrac{s_2^2}{n_2}\right)^2}$$

Statistical software such as Minitab will determine this quantity and will also automate the process of finding the multiplier t^*. If software is not available, a conservative "by hand" approach is to use the lesser of $n_1 - 1$ and $n_2 - 1$ for the degrees of freedom.

TECHNICAL NOTE

Equal Variances and Pooled Standard Error

If we are willing to make the assumption that the two populations have the same (or similar) variance, then there is a procedure for which the t^* multiplier *is* mathematically correct. It involves "pooling" both samples to estimate the common variance. That procedure is discussed in Lesson 2 of this module.

Situations for Which a *t* Confidence Interval for the Difference between Two Means Is Valid

An important condition required for the formulas in this section to be valid is that the samples must be *independent* of each other. In addition, one of two different situations must hold:

Situation 1: The populations of measurements are both bell-shaped, and random samples of any size are measured. In practice, for small samples, the observed data should not show extreme skewness, and there should not be any outliers.

Situation 2: Large random samples are measured. A somewhat arbitrary definition of a "large" sample is $n \geq 30$, but if there are extreme outliers or extreme skewness in the measurements, it is better to have an even larger sample than $n = 30$.

Note: For a confidence interval for the difference in two means, one of these situations must hold for both groups.

Example 11.11 **The Effect of a Stare on Driving Behavior** Social psychologists at the University of California at Berkeley wanted to study the effect that staring at drivers would have on driver behavior (Ellsworth, Carlsmith, and Henson, 1972). In a randomized experiment, the researchers either stared or did not stare at the drivers of automobiles stopped at a campus stop sign. The researchers timed how long it took each driver to proceed from the stop sign to a mark on the other side of the intersection. Suppose that the crossing times, in seconds, were as follows:

No-Stare Group ($n = 14$): 8.3, 5.5, 6.0, 8.1, 8.8, 7.5, 7.8, 7.1, 5.7, 6.5, 4.7, 6.9, 5.2, 4.7
Stare Group ($n = 13$): 5.6, 5.0, 5.7, 6.3, 6.5, 5.8, 4.5, 6.1, 4.8, 4.9, 4.5, 7.2, 5.8

Using these data, let's determine a 95% confidence interval for $\mu_1 - \mu_2$, the difference between the mean crossing times in the populations represented by these two independent samples.

Checking the Conditions and Preliminary Graphical Analysis

The two samples are *independent* because different drivers were measured in the two different experimental conditions and measurements are not paired in any way. Boxplots of the data (see Figure 11.8) indicate that there are no outliers and that the shape of the data is more or less symmetric within each group.

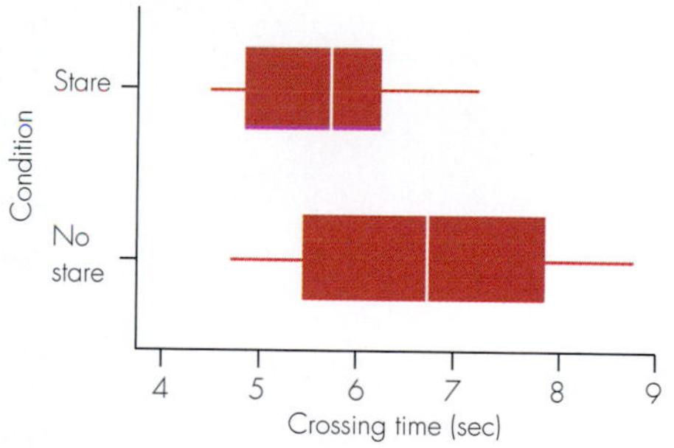

Figure 11.8 Boxplots of the intersection crossing time data for Example 11.11

The boxplots also illustrate that the crossing times in the stare sample were generally faster than crossing times in the no-stare sample. It also looks as though the data within the stare group are less variable than the data within the no-stare group, so it is probably not reasonable to assume that the populations have equal variances.

```
Two-sample T for CrossTime
Group      N    Mean    StDev    SE Mean
NoStare   14    6.63     1.36       0.36
Stare     13    5.59    0.822       0.23
95% CI for mu (NoStare) - mu (Stare): (0.14, 1.93)
T-Test mu (NoStare) = mu (Stare) (vs not: = ): T = 2.41   P = 0.025   DF = 21
```

Figure 11.9 Minitab output for Example 11.11

Calculating the Confidence Interval

Figure 11.9 displays Minitab output that includes a 95% confidence interval for the difference between population means. This output was produced by the "two-sample *t*" procedure, and it includes descriptive statistics for each group, a 95% confidence interval for the difference between means, and a significance test concerning the population means. We will examine this significance test in Chapter 13.

For the no-stare data, the sample mean is $\bar{x}_1 = 6.63$ seconds; for the stare data, $\bar{x}_2 = 5.59$ seconds; and the difference between the sample means is $6.63 - 5.59 = 1.04$ seconds. The 95% confidence interval for the difference between population means, $\mu_1 - \mu_2$, is 0.14 to 1.93 seconds.

We see from the output that df = 21 for this problem. Minitab calculated this quantity using the approximation formula for degrees of freedom. Table A.2 can be used to determine that when df = 21, $t^* = 2.08$ at the 95% level. Minitab calculated the confidence interval reported on the output as follows:

Sample estimate ± Multiplier × Standard error

$$(\bar{x}_1 - \bar{x}_2) \pm 2.08 \times \sqrt{\frac{s_1^2}{n_1} + \frac{s_2^2}{n_2}}$$

$$(6.63 - 5.59) \pm 2.08 \times \sqrt{\frac{1.36^2}{14} + \frac{0.822^2}{13}}$$

0.14 to 1.93 seconds

(Actually, the numbers that are given produce a lower endpoint of 0.15 instead of 0.14. Minitab obtained 0.14 by using more decimal places in its calculations.) If the approximate df provided by Minitab had not been available, we could have used the conservative degrees of freedom, the minimum of the two values $n_1 - 1$ and $n_2 - 1$, which is $13 - 1 = 12$. With df = 12, $t^* = 2.18$ (see Table A.2), and the resulting interval is 0.11 to 1.97 seconds, an interval that is only slightly wider than the interval provided by Minitab.

Interpreting a Confidence Interval for the Difference in Two Means

In addition to the usual interpretation of a confidence interval, we often are interested in whether the confidence interval estimate for $\mu_1 - \mu_2$ covers 0. If the interval does cover 0, then it is possible that the population means are equal. If the interval does not cover 0, then we can be fairly certain that the population means for the two groups are different. How certain we can be depends on the confidence level that is used for the computation.

In Example 11.11, a 95% confidence interval estimate for the difference in mean crossing times was 0.14 to 1.93 seconds. Because the interval does not cover 0, we can be fairly confident that the population means are not equal. For this example, we interpret this to say that for the population of drivers similar to the ones in the study, on average, they would cross the intersection somewhat faster if they were being stared at than if they were not.

Sometimes researchers provide summary statistics rather than the raw data. For the next example, the researchers provided the means and standard deviations but not a confidence interval for the difference in the population means. The example illustrates a situation in which it is of interest to know whether the confidence interval for $\mu_1 - \mu_2$ covers 0.

Example 11.12 **Parental Alcohol Problems and Child Hangover Symptoms** Slutske, Piasecki, and Hunt-Carter (2003) conducted a study on hangover symptoms in college students, using a sample of 1227 students who were enrolled in introductory psychology classes and who acknowledged that they currently use alcohol. The participants rated the severity of their experience with 13 hangover symptoms during the previous year, with possible ratings of 0 (never experience) to 4 (always experience). For the analysis in this example, the researchers simply counted how many of the symptoms received a nonzero rating for each student, so the score could range from 0 to 13. The mean and standard deviation for all of the students were 5.2 and 3.4, respectively. Thus, the average number of symptoms experienced at all during the previous year for this sample of students was 5.2 out of the 13 possible.

Read the original source on the companion website, http://www.cengage.com/statistics/Utts4e.

One concern that the researchers explored was that children whose parents had problems with alcohol might have problems of their own and therefore might experience more hangover symptoms. They asked the respondents the question, "To your knowledge, have your biological parents had any of the following problems related to their use of alcohol: fulfilling obligations at school or work, physical health or emotional problems, problems with family or friends, legal problems (e.g., DUI, DWI)?" (Slutske et al., 2003, p. 1444).

Twenty-three percent of the 1227 respondents (282) answered yes to the question, and 945 answered no. Let's find a 95% confidence interval estimate for the difference in mean hangover symptoms for the *populations* of students who would answer yes (population 1) and no (population 2). The sample statistics were as follows:

Group	Mean	Standard Deviation
Parental alcohol problems ($n_1 = 282$)	$\bar{x}_1 = 5.9$	$s_1 = 3.6$
No parental alcohol problems ($n_2 = 945$)	$\bar{x}_2 = 4.9$	$s_2 = 3.4$

Using these values, the sample estimate and the standard error are as follows:

$$\text{Sample estimate} = \bar{x}_1 - \bar{x}_2 = 5.9 - 4.9 = 1$$

$$\text{Standard error} = \text{s.e.}(\bar{x}_1 - \bar{x}_2) = \sqrt{\frac{s_1^2}{n_1} + \frac{s_2^2}{n_2}} = \sqrt{\frac{3.6^2}{282} + \frac{3.4^2}{945}} = 0.24$$

This tells us that for this sample, the students with a parental history of alcohol problems experienced an average of one more hangover symptom in the past year than did students without that history. We would like to know what we can say about the difference in the mean number of hangover symptoms for the corresponding populations.

The sample sizes are large enough to proceed without checking the shape of the data. Using the conservative df = 282 − 1 = 281, we find from Table A.2 that t^* is between 1.98 and 1.96. We will average them and use $t^* = 1.97$. The resulting confidence interval estimate for $\mu_1 - \mu_2$ is:

$$(\bar{x}_1 - \bar{x}_2) \pm t^* \times \text{s.e.}(\bar{x}_1 - \bar{x}_2)$$
$$1 \pm (1.97)(0.24)$$
$$1 \pm 0.47$$
$$0.53 \text{ to } 1.47 \text{ symptoms}$$

Therefore, we can conclude that in the population of students similar to these, students with a biological parent who experienced alcohol problems are likely to have an average number of hangover symptoms that is from 0.53 to 1.47 higher than the average number of symptoms of students without that parental history. Note that the entire interval is above 0, so we can be fairly certain that there really is a difference in the mean number of hangover symptoms in the populations of students with and without this parental history. In other words, we can be fairly certain that μ_1 really is higher than μ_2.

Approximate 95% Confidence Interval Using Multiplier 2

As with the other types of confidence intervals in Chapters 10 and 11, in certain circumstances, we can find an approximate 95% confidence interval for the difference in two means by using the multiplier 2 instead of using t^*. In finding a confidence interval for the difference in two means, the approximation will work well as long as the number of degrees of freedom is large. Remember that as degrees of freedom increase, the t^* multiplier gets closer and closer to the corresponding z^* multiplier. For 95% confidence, the z^* multiplier is 1.96 but is often rounded up to 2.0.

When both sample sizes are at least 30 the approximation works well. Note that in that case, the conservative degrees of freedom value is the minimum sample size $-1 = 30 - 1 = 29$, and from Table A.2, we see that the t^* multiplier for df = 29 is 2.05. But the df value of 29 is conservative, and in fact in this situation, the actual degrees of freedom (found by using Welch's approximation) would be quite a bit larger. Therefore, we recommend using the approximation as long as both sample sizes are at least 30.

DEFINITION When both sample sizes are at least 30, the interval

Difference in sample means $\pm$ 2 $\times$ Standard error

is an **approximate 95% confidence interval for the difference in population means**.

Example 11.13 **Confidence Interval for Difference in Mean Weight Losses by Diet or Exercise** In Example 11.3, we learned about a study comparing the weight loss for men who dieted (group 1) and men who exercised (group 2). The sample mean weight losses were 7.2 kg and 4.0 kg, for a difference of 3.2 kg. In Example 11.3, we found that the standard error was 0.81. Therefore, an approximate 95% confidence interval for the difference in population means is

Sample difference $\pm$ 2 $\times$ Standard error

$3.2 \pm 2 \times 0.81$

3.2 ± 1.62

1.58 to 4.82 kg

THOUGHT QUESTION 11.5 In Section 11.3, we learned how to find a confidence interval for the mean of paired differences, which we used in Example 11.9 to estimate the mean difference in weekly computer and TV hours for a population of liberal arts students.

- Explain why it would not have been appropriate to use the methods in this section, for two independent samples, to estimate that mean difference, even though in either case the sample estimate is the difference in sample means, 5.36 hours.
- If the methods in this section had been erroneously used by treating computer usage and TV viewing hours as independent samples, do you think the standard error of the sample estimate would have been larger or smaller than it was in Example 11.9? Explain your answer using common sense, not formulas. Think about how much natural variability there would be in the data for two independent samples, compared with measuring both sets of hours on the same individuals.*

Lesson 2: The Equal Variance Assumption and the Pooled Standard Error

In estimating the difference between two population means, it may sometimes be reasonable to assume that the populations have equal standard deviations. The term *variance* describes the squared standard deviation, so the assumption of equal standard

***HINT:** Remember that the standard error measures how much the sample estimate is likely to vary from sample to sample. Do you think the paired sample estimate $\bar{d}$ would be likely to vary more than or less than the difference in the two separate means?

deviations for the two populations is the same as assuming that the variances are equal. Using statistical notation, we can express the assumption of equal population variances as $\sigma_1^2 = \sigma_2^2 = \sigma^2$, where σ^2 represents the common value of the variance. With this **equal variance assumption**, information from both groups is combined in order to estimate the value of σ^2. The estimate of variance based on the combined or "pooled" data is called the **pooled variance**. The square root of the pooled variance is called the **pooled standard deviation**, and it is computed as follows:

$$\text{Pooled standard deviation} = s_p = \sqrt{\frac{(n_1 - 1)s_1^2 + (n_2 - 1)s_2^2}{n_1 + n_2 - 2}}$$

TECHNICAL NOTE Note that the pooled variance is a "weighted" average of the individual sample variances. The weights are $(n_1 - 1)/(n_1 + n_2 - 2)$ and $(n_2 - 1)/(n_1 + n_2 - 2)$, giving the variance of the larger sample more weight. If $n_1 = n_2$, then the weights are equal, and $s_p^2 = (s_1^2 + s_2^2)/2$.

Replacing the individual standard deviations s_1 and s_2 with the pooled version s_p in the formula for standard error leads to the **pooled standard error for the difference between two means**:

$$\text{Pooled s.e.}(\bar{x}_1 - \bar{x}_2) = \sqrt{\frac{s_p^2}{n_1} + \frac{s_p^2}{n_2}} = \sqrt{s_p^2\left(\frac{1}{n_1} + \frac{1}{n_2}\right)} = s_p\sqrt{\frac{1}{n_1} + \frac{1}{n_2}}$$

This all may seem quite complicated, but if the assumption of equal population variances is correct, this complication actually provides a cleaner mathematical solution for the determination of the multiplier t^*. In this case, the degrees of freedom are simply df $= n_1 + n_2 - 2$.

FORMULA

Pooled Confidence Interval for the Difference Between Two Population Means (Independent Samples)

If we assume that the population variances are the same, then the confidence interval for $\mu_1 - \mu_2$, the difference between the population means, is

$$(\bar{x}_1 - \bar{x}_2) \pm t^*\sqrt{s_p^2\left(\frac{1}{n_1} + \frac{1}{n_2}\right)}$$

t^* is found by using a t-distribution with df $= n_1 + n_2 - 2$, and s_p is the pooled standard deviation.

MINITAB TIP In Minitab, the default option for the two-sample t procedure is the "unpooled" version in which the population variances are not assumed to be equal. To get the pooled version, check the dialog box item that says "Assume Equal Variances."

Example 11.14 Pooled t-Interval for Difference Between Mean Female and Male Sleep Times

Students in an introductory statistics class filled out a survey on a variety of issues, including how much sleep they had the previous night. Let's use the data to estimate how much more or less male students sleep than females students, on average. The class included 83 females and 65 males. Let's assume that these students are equivalent to a random sample of all students who take introductory statistics. How much difference is there between how long female and male students represented by this sample slept the previous night? To answer that question, we'll find a 95% confidence interval for $\mu_1 - \mu_2 =$ difference in mean sleep hours for females versus males in the population. At the top of the next page is the Minitab output for the pooled procedure in which we assume equal population variances for males and females.

Two-sample T for sleep [with "Assume Equal Variance" option]				
Sex	N	Mean	StDev	SE Mean
Female	83	7.02	1.75	0.19
Male	65	6.55	1.68	0.21

Difference = mu (Female) − mu (Male)
Estimate for difference: 0.461
<u>95% CI for difference: (−0.103, 1.025)</u>
T-Test of difference = 0 (vs not =): T-Value = 1.62 P-Value = 0.108 DF = 146
Both use Pooled StDev = 1.72

Note that the sample standard deviations are very similar (1.75 for females and 1.68 for males), so it may be reasonable to assume that the population variances are similar as well. The sample mean given for females is $\bar{x}_1 = 7.02$ hours, the sample mean given for males is $\bar{x}_2 = 6.55$ hours, and the difference between these two sample means is reported on the output as "Estimate for difference: [0.461 hour]." Note that the sample means were rounded to two decimal places in the output which causes an inconsistency because the difference between 7.02 and 6.55 is 0.47. The value 0.461 is, however, the correct difference between the sample means.

The 95% confidence interval for the difference in mean hours of sleep for the populations of female and male students is −0.103 to 1.025 hours (in bold and underlined in the output display). Because the interval covers 0, we can't rule out the possibility that the population means are equal for men and women, although for this sample, the women slept an average of about half an hour more than the men did.

The pooled standard deviation given at the bottom of the output is $s_p = 1.72$. Although not given in the output, the pooled standard error is 0.285. These two values can be found as follows:

$$s_p = \sqrt{\frac{(83-1)(1.75)^2 + (65-1)(1.68)^2}{83+65-2}} = \sqrt{2.957} = 1.72$$

$$\text{Pooled s.e.}(\bar{x}_1 - \bar{x}_2) = \sqrt{\frac{(1.72^2)}{83} + \frac{(1.72^2)}{65}} = 1.72\sqrt{\frac{1}{83} + \frac{1}{65}} = 0.285$$

In the pooled procedure, the degrees of freedom are found as df $= n_1 + n_2 - 2$, which in this example equals $83 + 65 - 2 = 146$. If Table A.2 is used, a conservative estimate of the multiplier t^* is 1.98 (for df = 100 and .95 confidence). The confidence limits found by Minitab can be verified by calculating $.461 \pm (1.98)(0.285)$.

Pooled or Unpooled?

In Example 11.14, the sample standard deviations for females and males had about the same values, so it was reasonable to use the assumption that the population standard deviations were equal. The confidence interval for the difference in means, however, would have been about the same even if the assumption of equal standard deviations had not been made. With the unpooled procedure, the 95% confidence interval for the difference in population means is −0.10 to 1.03 hours, quite close to the pooled version of −0.103 to 1.025 hours. One advantage for the pooled version is that finding the value of the degrees of freedom is simpler.

Sample standard deviations in two independent samples will almost never be identical in practice. So how do we know when it might be reasonable to use the pooled version of a confidence interval for a difference between two population means? And what is the risk of using the pooled procedure when, in truth, the population standard deviations differ? We will explore that question in detail when we consider hypothesis testing for the difference between two means in Chapter 13, but here we give some preliminary guidance:

- If the larger value of the two sample standard deviations is from the group with the larger sample size, the pooled procedure will tend to give a wider confidence interval than the unpooled version and so would be a conservative estimate of the true difference, as the next example illustrates. Similar to when we used the conservative margin of error in a confidence interval for one proportion, it is acceptable to use the more conservative pooled procedure. But it is not good practice to knowingly create an interval that is wider than necessary.
- On the other hand, if the smaller of the two sample standard deviations is from the group with the larger sample size, the pooled version of the procedure may produce a misleading narrow interval.
- Generally, it is best to use the unpooled procedure unless the sample standard deviations are quite similar.

Example 11.15 **Sleep Time with and without the Equal Variance Assumption** In Example 11.7, we compared confidence intervals for a single mean for the average amount of sleep for two types of statistics students. We can also find a confidence interval for the difference in mean hours of sleep for the two types of students. Here is the Minitab output for doing so, with and without the "Assume Equal Variance" option:

```
Two-sample T-Test and Confidence Interval
Two-sample T for sleep [without "Assume Equal Variance" option]
class      N      Mean     StDev     SE Mean
10        25      7.66      1.34        0.27
13       148      6.81      1.73        0.14
95% CI for mu (10) − mu (13): (0.23, 1.46)
T-Test mu (10) = mu (13) (vs not =):   T = 2.79   P = 0.0083   DF = 38
Two-sample T for sleep [with "Assume Equal Variance" option]
class      N      Mean     StDev     SE Mean
10        25      7.66      1.34        0.27
13       148      6.81      1.73        0.14
95% CI for mu (10) − mu (13): (0.13, 1.57)
T-Test mu (10) = mu (13) (vs not =):   T = 2.33   P = 0.021   DF = 171
Both used Pooled StDev = 1.68
```

For this example, it is not reasonable to assume equal variances. Note that doing so increases the width of the confidence interval. Without the assumption, the interval is from 0.23 to 1.46 hours, but with the assumption, it is 0.13 to 1.57 hours. A comparison of the standard error of the difference for the two situations illustrates why the width increased. Using the pooled estimate increases the standard error of the mean for the sample with $n = 25$.

$$\text{Unpooled s.e.}(\bar{x}_1 - \bar{x}_2) = \sqrt{\frac{(1.34)^2}{25} + \frac{(1.73)^2}{148}} = \sqrt{0.07 + 0.02} = 0.30$$

$$\text{Pooled s.e.}(\bar{x}_1 - \bar{x}_2) = \sqrt{\frac{(1.68)^2}{25} + \frac{(1.68)^2}{148}} = \sqrt{0.11 + 0.02} = 0.36$$

This example illustrates that it is conservative to use the pooled procedure if the two sample sizes are decidedly different and the larger sample standard deviation accompanies the larger sample size.

11.4 Exercises are on pages 444–446.

MINITAB TIP **Computing a Confidence Interval for the Difference Between Two Means for Independent Samples**

- To compute a confidence interval for the difference in two means, use **Stat > Basic Statistics > 2 sample t**. Specify the location of the data. The raw data for the response (*Samples*) may be in one column, and the raw data for group categories (*Subscripts*) may be in a second column. Or the raw data for the two independent groups may be in two separate columns.
- If the raw data are not available but the data are summarized so that you know the sample mean, standard deviation, and sample size for each group, enter those values in the "Summarized data" area of the dialog box.
- To change the confidence level, use the ***Options*** button.
- To use the *pooled* standard error, click on ***Assume Equal Variances***.

Note: Use the ***Graphs*** button to create a comparative dotplot (called an "individual value plot" in the option list) or a comparative boxplot.

IN SUMMARY **Confidence Interval for the Difference in Two Population Means**

A confidence interval for the difference in two population means $\mu_1 - \mu_2$ (also called a **two-sample *t*-interval**) is

$$(\bar{x}_1 - \bar{x}_2) \pm t^* \times \text{s.e.}(\bar{x}_1 - \bar{x}_2)$$

where the degrees of freedom for t^* and the standard error depend on whether the assumption of equal variances is reasonable. To determine this interval:

1. Make sure that the appropriate conditions apply by examining both sample sizes and/or a shape picture for each sample.
2. Choose a confidence level.
3. Calculate the sample means, $\bar{x}_1$ and $\bar{x}_2$, and the sample standard deviations, s_1 and s_2.
4. Determine whether the sample standard deviations are similar enough to assume that the population standard deviations are about equal. If so, the pooled procedure can be used.
5. Calculate the standard error s.e.$(\bar{x}_1 - \bar{x}_2)$, which is

$$\sqrt{\frac{s_1^2}{n_1} + \frac{s_2^2}{n_2}} \ (\textit{unpooled}) \quad \text{or} \quad s_p\sqrt{\frac{1}{n_1} + \frac{1}{n_2}} \ (\textit{pooled})$$

where s_p is the pooled standard deviation (see p. 431).

6. Calculate the degrees of freedom. For the unpooled case, use Welch's approximation (see p. 426) or, conservatively, the smaller of $n_1 - 1$ and $n_2 - 1$. For the pooled case, use $n_1 + n_2 - 2$.
7. Use Table A.2 (or statistical software) to find the multiplier t^*.

11.5 Understanding Any Confidence Interval

The statistics and parameters that researchers examine are not limited to proportions, means, differences in means, or differences in proportions. Throughout this book, we have seen several other statistics, such as the median, the lower and upper quartiles, the correlation coefficient, and relative risk. For each of these statistics, there are formulas and procedures that allow us to compute confidence intervals for the corresponding population parameter. Some of those formulas and procedures, however, are quite complex.

In cases in which a complicated procedure is needed to compute a confidence interval, authors of journal articles usually provide the completed intervals. Your job as a reader is to be able to interpret those intervals. The principles that you have learned for understanding confidence intervals for means and proportions are directly applicable to understanding any confidence interval. As an example, let's consider the confidence intervals that were reported in one of our earlier case studies.

11.5 Exercises are on page 446.

CASE STUDY 11.1 Confidence Interval for Relative Risk: Case Study 6.4 Revisited

In Case Study 6.4, we discussed a study relating heart disease to baldness in men. A parameter of interest in that study was the relative risk of heart disease based on the degree of baldness. The investigators focused on the relative risk (RR) of myocardial infarction (heart attack) for men with baldness compared to men without any baldness. Here is how they reported some of the results:

> For mild or moderate vertex baldness, the age-adjusted RR estimates were approximately 1.3, while for extreme baldness the estimate was 3.4 (95% CI, 1.7 to 7.0).... For any vertex baldness (i.e., mild, moderate, and severe combined), the age-adjusted RR was 1.4 (95% CI, 1.2 to 1.9) (Lesko et al., 1993, p. 1000).

The 95% confidence intervals for age-adjusted relative risk are not simple to compute, and they don't take the form "sample estimate ± 2 × standard error." This is evidenced by the fact that the intervals are not symmetric about the sample relative risks given. However, these intervals have the same interpretation as any other confidence interval. For instance, with 95% confidence, we can say that the ratio of risks of a heart attack in the populations of men with extreme baldness and men with no baldness is somewhere between 1.7 and 7.0. In other words, men with extreme baldness are probably somewhere between 1.7 to 7 times as likely to experience a heart attack as men of the same age who have no baldness. Of course, these results assume that the men in the study are representative of the larger population.

THOUGHT QUESTION 11.6 Part of the quote in Case Study 11.1 said, "For any vertex baldness (i.e., mild, moderate, and severe combined), the age-adjusted RR was 1.4 (95% CI, 1.2 to 1.9)." Explain what is wrong with the following interpretation of this result, and write a correct interpretation.

Incorrect Interpretation: There is a .95 probability that the age-adjusted relative risk of a heart attack (for men with any vertex baldness compared to men without any) is between 1.2 and 1.9.*

CASE STUDY 11.2 Premenstrual Syndrome? Try Calcium

The front-page headline in the *Sacramento Bee* was "Study says calcium can help ease PMS" (Maugh, 1998). The article described a randomized, double-blind experiment in which women who suffered from premenstrual syndrome (PMS) were randomly assigned to take daily either a placebo or 1200 mg of calcium in the form of four Tums E-X tablets (Thys-Jacobs et al., 1998). The participants were 466 women with a history of PMS. In the experiment, 231 of these women were randomly assigned to the calcium treatment group and 235 women were in the placebo group.

The primary measure of interest was a composite score based on 17 PMS symptoms (insomnia, six mood-related symptoms, five water-retention symptoms, two involving food cravings, and three related to pain). Participants were asked to rate each of the 17 symptoms daily on a scale from 0 (absent) to 3 (severe). The actual "symptom complex score" was the mean rating for the 17 symptoms. Thus, a score of 0 would imply that all symptoms were absent, and a score of 3 would indicate that all symptoms were severe. The original article (Thys-Jacobs et al., 1998) presents results individually for each of the 17 symptoms in addition to the composite score.

The treatments were continued for three menstrual cycles. We report the symptom scores for the baseline premenstrual time (7 days) before treatments began and for the premenstrual time before the third cycle.

(continued)

***HINT:** The interpretation is confusing the long-run results for the procedure with the results for a specific interval. See the discussion in Chapter 10, just before Thought Question 10.2 (p. 374).

CASE STUDY 11.2 Premenstrual Syndrome? Try Calcium (Continued)

Table 11.3 shows some results given in the journal article, including the mean symptom complex scores ±1 standard deviation and the sample sizes. We see that the mean symptom scores for the two treatment groups were about equal at the baseline time but differed more noticeably before the third cycle. Note also that the sample sizes for the third cycle are slightly reduced because patients dropped out of the study.

Table 11.3 Results for Case Study 11.2

	Symptom Complex Score: mean ± s.d.	
	Placebo Group	***Calcium-Treated Group***
Baseline	0.92 ± 0.55 (n = 235)	0.90 ± 0.52 (n = 231)
Third cycle	0.60 ± 0.52 (n = 228)	0.43 ± 0.40 (n = 212)

One interesting outcome is that the mean symptom score was substantially reduced for both the placebo and calcium-treated groups. For the placebo group, the mean symptom score dropped by about a third; and for the calcium-treated group, the mean was more than cut in half. The purpose of the experiment is to find out whether taking calcium diminishes symptom severity. But we see that placebo alone can be responsible for reducing symptoms, so the appropriate comparison is between the placebo and calcium-treated groups. We should compare those two groups to each other rather than simply examining the reduction in scores for only the calcium-treated group.

We can use the results in Table 11.3 to compute a confidence interval for $\mu_1 - \mu_2$, the difference between the placebo and calcium mean symptom scores before the third cycle for the entire population of PMS sufferers. The difference in sample means (placebo − calcium) is $\bar{x}_1 - \bar{x}_2 = 0.60 - 0.43 = 0.17$. The standard error is

$$\text{s.e.}(\bar{x}_1 - \bar{x}_2) = \sqrt{\frac{0.52^2}{228} + \frac{0.40^2}{212}} = 0.044$$

An approximate 95% confidence interval for the difference in population means is

$(\bar{x}_1 - \bar{x}_2) \pm 2 \times \text{s.e.}(\bar{x}_1 - \bar{x}_2)$
$0.17 \pm 2(0.044)$
0.08 to 0.26

To put this in perspective, remember that the scores are averages over the 17 symptoms. Therefore, a reduction from a mean of 0.60 to a mean of 0.43 would, for instance, correspond to a reduction from (0.6)(17) = 10.2 or about 10 mild symptoms (rating of 1) to (0.43)(17) = 7.31, or just over 7 mild symptoms. In fact, examination of the full results of the study shows that for each one of the 17 symptoms, the reduction in mean severity score was greater in the calcium-treated group than in the placebo group. Because this is a randomized experiment and not an observational study, we can conclude that the calcium treatment actually *caused* the greater reduction in PMS symptoms.

SKILLBUILDER APPLET

11.6 The Confidence Level in Action

The Confidence Level applet described in this section is available on the companion website, http://www.cengage.com/statistics/Utts4e.

In Chapter 10, we learned that a confidence level has a relative frequency probability interpretation. If we consider all possible randomly selected samples of the same size from a population, the confidence level is the fraction or percent of these samples for which the confidence interval includes the population parameter. The **ConfidenceLevel** applet on the website for this book lets you see this concept in action.

What Happens

The applet simulates choosing random samples of $n = 16$ from a normally distributed population of measurements with mean $\mu = 170$ and standard deviation $\sigma = 20$, roughly the characteristics of weights of men aged 18 to 24 years. For each sample, a confidence interval for the population mean is found and graphed. Figure 11.10 is an example of how the display looks when the applet is first started. Note that the initial confidence level is 68% for the procedure being used to estimate the population mean. The red vertical line in the display is at the value of the population mean, $\mu = 170$. In Figure 11.10, the confidence interval shown, 167.7 to 177.7, covers the population mean.

SKILLBUILDER APPLET (CONTINUED)

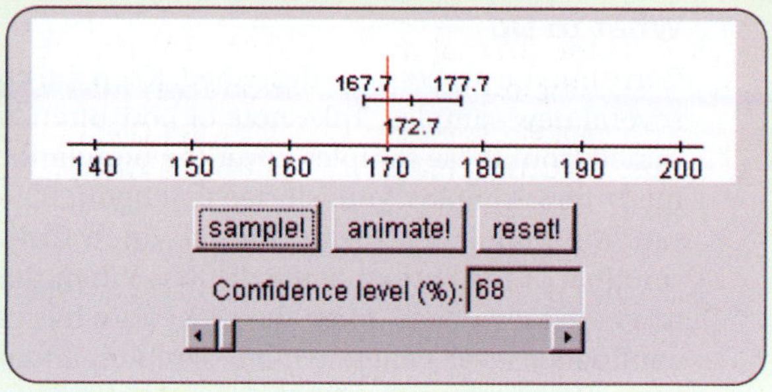

Figure 11.10 Example of the initial **ConfidenceLevel** applet display

Use the ***sample!*** button to generate a new sample. Figure 11.11 is an example of how the display might appear after this button is clicked. Note that the confidence interval that is found from this new sample also covers the population mean (located at the red line), and at the top of the display, the information "1 out of 1 (100%) captures" is given. For the first new sample that is generated, the confidence interval captured the population mean, so the procedure has been successful 100% of the time thus far in the simulation.

The applet continues to tally how often the confidence interval captures the population value (170) as new samples are generated. The ***sample!*** button is used to generate one new sample, whereas using the ***animate!*** button causes the applet to repeatedly generate samples until ***stop!*** is clicked. (The ***stop!*** button appears after ***animate!*** is clicked.)

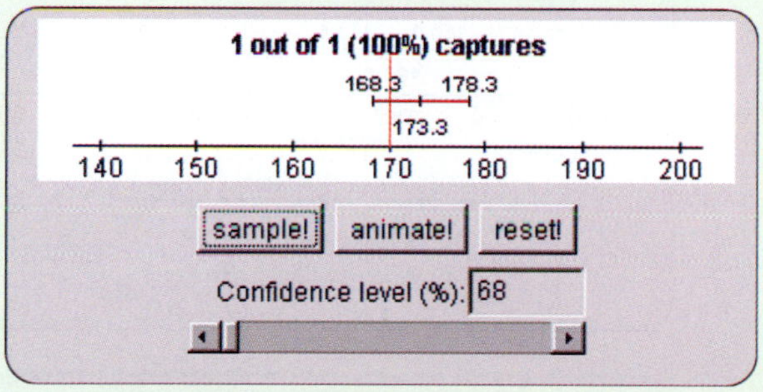

Figure 11.11 The **ConfidenceLevel** applet display after one new sample has been generated

Figure 11.12 shows how the display looked after 103 different random samples had been generated. At the top of the display, we see that 70.87% of the 103 samples gave confidence intervals that captured the population parameter. This is close to what we would expect for the 68% confidence level. Over many different samples, the confidence interval procedure captures the population mean about 68% of the time. Note, by the way, that the interval shown in Figure 11.12 did not cover the red line at 170, so that interval does not capture the population mean.

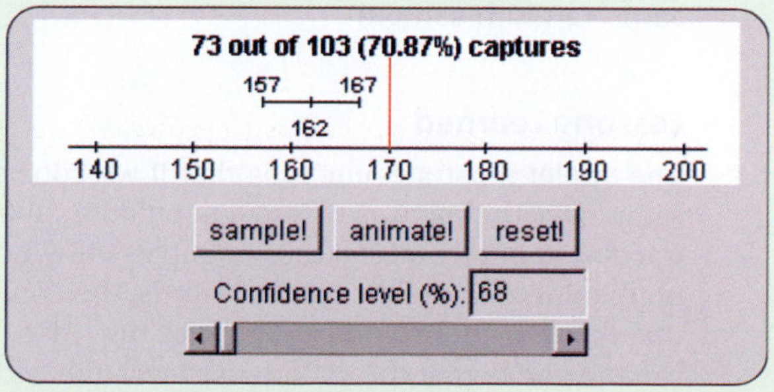

Figure 11.12 The **ConfidenceLevel** applet display after 103 new samples have been generated

SKILLBUILDER APPLET (CONTINUED)

What to Do

Carry out the process just described. Keep the confidence level set at 68%, and generate several new samples. Take note of how often the confidence intervals that are determined from these samples cover the population mean. Eventually, after you generate many new samples, you will see that about 68% of the samples you generate will capture the population mean (the red line). Change the confidence level by using the scrollbar at the bottom of the display. When this is done, the applet will automatically reset the tally of how often the procedure has captured the parameter. Using this new confidence level, generate many samples, and take note of how often the samples give intervals that capture 170, the population mean. Over many samples, the percentage shown at the top of the display will be close to the confidence level for the procedure. Examine what happens when the confidence level is set at 99%. You will see that the confidence interval will capture the population mean (170) for nearly every sample generated.

It is informative to watch what happens to the displayed confidence intervals when the scrollbar is used to change the confidence level. You will see an interval become wider as the confidence level is increased, and become narrower as the confidence level is decreased. Figure 11.13 shows the 68% and the 99% confidence intervals for the same random sample. The interval for the 99% level is much wider than the interval for the 68% level. Note that the 99% confidence interval has captured the parameter, but the 68% confidence interval has not. Generally, the wider interval is more likely to capture the parameter value.

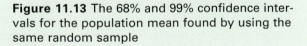

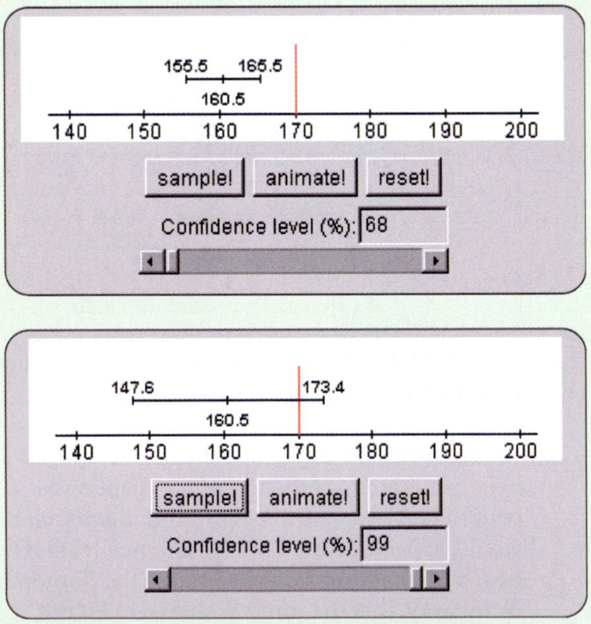

Figure 11.13 The 68% and 99% confidence intervals for the population mean found by using the same random sample

Lessons Learned

The applet illustrates the idea that if we consider many different randomly selected samples of the same size from a population, the confidence level is approximately the fraction or percentage of these samples for which the confidence interval includes the population parameter. In other words, the confidence level expresses our confidence that the procedure works in the long run. Also, you will see that for the same random sample, the higher the confidence level, the wider is the interval. These principles apply to all confidence interval procedures, regardless of the parameter being estimated.

11.6 Exercises are on pages 446–447.

IN SUMMARY Confidence Interval Procedures

The basic structure of intervals for the parameters in the following table is

Sample estimate ± Multiplier × Standard error

The table describes the specific details for each type of parameter. The *Sample estimate* in the formula is the *Statistic* in the Table.

	Parameter	Statistic	Standard Error	Multiplier	Example
One proportion	p	$\hat{p}$	$\sqrt{\frac{\hat{p}(1-\hat{p})}{n}}$	z^* (see note 4)	10.3
Difference between proportions	$p_1 - p_2$	$\hat{p}_1 - \hat{p}_2$	$\sqrt{\frac{\hat{p}_1(1-\hat{p}_1)}{n_1} + \frac{\hat{p}_2(1-\hat{p}_2)}{n_2}}$	z^* (see note 4)	10.9
One mean	μ	$\bar{x}$	$\frac{s}{\sqrt{n}}$	t^* (see note 1)	11.5
Mean difference, paired data	μ_d	$\bar{d}$	$\frac{s_d}{\sqrt{n}}$	t^* (see note 1)	11.9
Difference between means (unpooled)	$\mu_1 - \mu_2$	$\bar{x}_1 - \bar{x}_2$	$\sqrt{\frac{s_1^2}{n_1} + \frac{s_2^2}{n_2}}$	t^* (see note 2)	11.11
Difference between means (pooled)	$\mu_1 - \mu_2$	$\bar{x}_1 - \bar{x}_2$	$s_p\sqrt{\frac{1}{n_1} + \frac{1}{n_2}}$	t^* (see note 3)	11.14

[1]Use Table A.2, df = $n - 1$.

[2]Use Table A.2, df = $\dfrac{\left(\frac{s_1^2}{n_1} + \frac{s_2^2}{n_2}\right)^2}{\frac{1}{n_1 - 1}\left(\frac{s_1^2}{n_1}\right)^2 + \frac{1}{n_2 - 1}\left(\frac{s_2^2}{n_2}\right)^2}$ or use df = $\min(n_1 - 1, n_2 - 1)$.

[3]Use Table A.2, df = $n_1 + n_2 - 2$; see Section 11.4, Lesson 2 for the formula for s_p.

[4]Use last row of Table A.2, or the z^* multipliers:

Confidence Level	z^*
.90	1.645 or 1.65
.95	1.960 or 2.00
.98	2.326 or 2.33
.99	2.576 or 2.58

Key Terms

Section 11.1
parameter, 405
statistic, 406
estimate, 406
paired data, 407
independent samples, 407–408
standard error, rough definition, 409
standard error, technical definition, 409
standard error of a sample mean, 409
standard error of a sample mean of paired differences, 409
standard error of the difference, 410
standard error of the difference between two sample means, 410
Student's *t*-distribution, 411
degrees of freedom (df), 411
*t** multiplier, 411

Section 11.2
confidence interval estimate of the mean, 413
confidence interval for a population mean, 413
t-interval, 413, 415
robust procedure, 414
approximate 95% confidence interval (for large samples), 420

Section 11.3
confidence interval for the population mean of paired differences, 421, 423
paired *t*-interval, 423

Section 11.4
two-sample *t*-interval, 426, 434
confidence interval (unpooled) for the difference in two population means, 426
Welch's approximation for df, 426
approximate 95% confidence interval for the difference in two population means, 430
equal variance assumption, 431
pooled variance, 431
pooled standard deviation, 431
pooled standard error for the difference between two means, 431
pooled confidence interval for the difference between two population means, 431

In Summary Boxes

Exercises

◆ Denotes that the dataset is available on the companion website, http://www.cengage.com/statistics/Utts4e, but is not required to solve the exercise.

Bold exercises have answers in the back of the text.

Section 11.1

Skillbuilder Exercises

11.1 Suppose that you are told that the mean IQ score for children in a particular school is 105.3. Explain what additional information you would need in order to know whether the mean of 105.3 is a statistic or a parameter.

11.2 Would it be appropriate to use the ages of death of First Ladies given in Chapter 2 (Table 2.5, p. 28) to find a 95% confidence interval for the mean? If your answer is yes, what is the parameter of interest? If your answer is no, explain why not.

11.3 In each part, identify whether it would be more appropriate to use a procedure for paired data or a procedure for two independent samples to estimate the difference of interest.

a. A researcher wants to estimate the difference between the mean scores on a memorization test for people over 60 years old and people under 40 years old.

b. Women give their actual and desired weights in a survey. An estimate of the mean difference is desired.

c. Two treatments for reducing cholesterol are compared. Forty people with high cholesterol use the first treatment, and 40 other people with high cholesterol use the second treatment. An estimate of the difference between the mean decrease in cholesterol for Treatment 1 and the mean decrease in cholesterol for Treatment 2 is desired.

11.4 You have been told that for university classes, students are supposed to spend 2 hours working outside of class for each hour in class. You are curious to know the average amount of time that students at your school spend working outside of class. You contact what you think is a representative sample of 100 students at your school and ask them, "How many hours of schoolwork do you do outside of class for each hour in class?" The mean for the 100 responses is 1.2 hours.

a. What is the population of interest, and what is the research question you want to answer about it?

b. Explain in words what the population parameter is in this study, and identify the appropriate symbol for it.

c. What is the value of the sample estimate in this study, and what symbol is used for it?

11.5 In each part, use the information given to calculate the appropriate standard error.

a. One mean, $n = 100$ and standard deviation $s = 15$.

b. Mean for paired differences, $n = 49$ and standard deviation of the differences $s_d = 21$.

c. Difference in means for independent samples, $n_1 = n_2 = 50$ and $s_1 = s_2 = 10$.

11.6 In each part, use the information to calculate the appropriate standard error of the mean.

a. The mean height for a sample of $n = 25$ children is $\bar{x} = 36$ inches, and the standard deviation is $s = 1.0$ inch.

b. A sample of $n = 100$ women went on a diet and lost an average of $\bar{d} = 9$ pounds. The standard deviation for the differences in weight (before − after) was $s_d = 6$ pounds.

11.7 Use software, a calculator or Table A.2 to find the value of t^* for each of the following situations.

a. df = 15, 95% confidence level.

b. df = 30, 90% confidence level.

c. $n = 10$, 95% confidence interval for one mean.

11.8 Use software, a calculator, or Table A.2 to find the value of t^* for each of the following situations.

a. df = 18, 99% confidence level.

b. df = 100, 95% confidence level.

c. $n = 41$, 90% confidence interval for one mean.

General Section Exercises

11.9 A multinational company plans to open new offices in one of two countries where it has not operated before and is trying to decide which country to choose. One feature that the company is interested in is how much vacation time it will need to offer employees in the two countries to be competitive with companies that are already there. The company takes a representative sample of 50 workers in similar industries in each country and asks them how many days of vacation they are given per year. In Country A, the average is 22.5 days; in Country B, the average is 16.3 days.

a. Of the three parameters that we considered in Section 11.1 (μ, μ_d, and $\mu_1 - \mu_2$), two of them would be of interest in this situation. Identify which two, and explain what they would represent in this situation. Note that one of them is represented twice.

b. What is the sample estimate for the parameter $\mu_1 - \mu_2$ in this situation? Use appropriate notation, and give a numerical answer. (Assign μ_1 to Country A.)

11.10 Give an example of a research question (not covered in this chapter) for which the appropriate parameter is

a. μ.
b. μ_d.
c. $\mu_1 - \mu_2$.

11.11 Researchers at the MIND Institute at the University of California-Davis analyzed the blood of 70 autistic and 35 nonautistic children between the ages of 4 and 6 (Lau, 2005). One marker that they examined was B cells, which is a type of immune cell that produces antibodies in the blood. They found that the mean level for the autistic children was substantially higher than the mean level for the nonautistic children in the study.

a. What is the research question of interest in this part of the study? Write the question in words, not in terms of statistical notation.
b. What is the population parameter of interest? Use appropriate notation, and explain in words what it means.
c. Write the research question of interest in terms of the population parameter that you defined in part (b).

11.12 For each of the following research questions, would it make more sense to collect paired data or independent samples? Explain.

a. What is the difference in the mean price of gasoline for all gas stations in the United States last week and this week?
b. How does the mean weight of adults in college towns compare with the mean weight of adults in other towns of similar size?

11.13 For each of the following research questions, would it make more sense to collect paired data or independent samples? Explain.

a. What is the mean difference in height for the male and female in fraternal twin pairs in which there is one of each sex?
b. What is the difference in mean heights for males and females in the adult population?

11.14 In the definition of paired data on page 407, three ways in which paired data can occur were given. They are repeated in the parts of this exercise. In each case, given an example of a research question for which paired data would be collected using the method described.

a. The same measurement is taken twice on each person, under different conditions or at different times.
b. Similar individuals are paired before giving the treatments in an experiment. Each member of a pair then receives a different treatment. The same response variable is measured for all individuals.
c. Two different variables are measured for each individual. There is interest in the amount of difference between the two variables.

11.15 In the description of independent samples on page 408, four common methods were given for generating independent samples. They are repeated in the parts of this exercise. In each case, give an example of a research question for which independent samples would be collected using the method described.

a. Random samples are taken separately from two populations, and the same response variable is recorded for each individual.
b. One random sample is taken, and a variable is recorded for each individual, but then units are categorized as belonging to one population or another, such as male and female.
c. Participants are randomly assigned to one of two treatment conditions, such as diet or exercise, and the same response variable, such as weight loss, is recorded for each individual unit.
d. Two random samples are taken from a population, and a separate variable is measured in each sample. For example, one sample may be asked about hours of internet use, whereas a separate sample is asked about hours of television viewing. Then the results are compared.

11.16 In each part, use the information given to calculate the standard error of the mean.

a. Mean height for a sample of $n = 81$ women is $\bar{x} = 64.2$ inches, and the standard deviation is $s = 2.7$ inches.
b. Mean systolic blood pressure for a sample of $n = 100$ men is $\bar{x} = 123.5$, and the standard deviation is $s = 9$.
c. Mean systolic blood pressure for a sample of $n = 324$ men is $\bar{x} = 123.5$, and the standard deviation is $s = 9$.

11.17 Suppose that a randomly selected sample of $n = 64$ men has a mean foot length of $\bar{x} = 27.5$ cm, and the standard deviation of the sample is 2 cm. Calculate the standard error of the sample mean. Write one or two sentences that interpret this value.

11.18 Refer to Exercise 11.17 about the foot lengths of men. A randomly selected sample of $n = 100$ women has a mean foot length of $\bar{x} = 24.0$ cm, and the standard deviation of the sample is 2 cm.

a. Calculate the standard error of the mean. Explain why the value is smaller than the standard error for the men in Exercise 11.17.
b. Now consider the difference between the mean foot lengths of men and women. Compute the standard error for the difference between sample means for the men and women.

11.19 The pulse rates of $n = 50$ people are measured. These people then march in place for 1 minute, and the pulse rates are measured again. We are interested in estimating the mean difference between the two pulse rates for the population represented by this sample. Explain how you would compute the standard error for the sample mean difference. Be specific about what formula you would use.

11.20 If each of the following is decreased but everything else remains the same, will a confidence interval become wider, will it become narrower, or will the width remain the same? Explain.

a. The sample estimate.
b. The standard error.
c. The confidence level.

11.21 Note that the standard error formulas for means all involve sample standard deviations and sample sizes. With this knowledge, what can a researcher do, when designing a study, to try to make sure the standard error isn't so large that it produces a meaningless confidence interval?

11.22 Suppose that a teacher asks her class how many hours they studied for the midterm exam. For the 40 students who are enrolled in the class, the mean was 5.5 hours, and the standard deviation was 2 hours. Explain why the teacher should not calculate a standard error for the mean, and why she should not calculate a confidence interval to estimate the population mean.

11.23 Explain what happens to the value of t^* in each of the following cases.

a. The confidence level is increased from 90% to 95%.
b. The sample size is increased.
c. The degrees of freedom are increased.
d. The degrees of freedom are essentially infinite.
e. The standard error of the mean is decreased because the standard deviation is decreased.

11.24 Refer to Figure 11.3 (p. 412), in which the relationship between t^* and the 95% confidence level is illustrated for df = 24. Draw a similar picture for df = 24 and 90% confidence.

Section 11.2

Exercises 11.25 to 11.43 correspond to the two Lessons in Section 11.2. Lesson 1 exercises are 11.25 to 11.37, and Lesson 2 exercises are 11.38 to 11.43.

Lesson 1 Skillbuilder Exercises

11.25 In each case, use the information given to compute a confidence interval for the population mean μ. Assume that all necessary conditions are present for using the method described in Section 11.2.

a. $\bar{x} = 76$, $s = 6$, $n = 9$, 95% confidence level.
b. $\bar{x} = 76$, $s = 6$, $n = 9$, 90% confidence level.
c. $\bar{x} = 76$, $s = 6$, $n = 16$, 90% confidence level.

11.26 In each case, use the information given to compute a confidence interval for the population mean μ. Assume that all necessary conditions are present for using the method described in Section 11.2.

a. $\bar{x} = 76$, $s = 8$, $n = 16$, 95% confidence level.
b. $\bar{x} = 100$, $s = 8$, $n = 16$, 95% confidence level.
c. $\bar{x} = 100$, $s = 18$, $n = 81$, 99% confidence level.
d. $\bar{x} = 100$, $s = 18$, $n = 81$, 90% confidence level.

11.27 Use Table A.2 to find the multiplier t^* for calculating a confidence interval for a single population mean in each of the following situations.

a. $n = 22$; confidence level = .98.
b. $n = 5$; confidence level = .90.
c. $n = 5$; confidence level = .99.

11.28 Use Table A.2 to find the multiplier t^* for calculating a confidence interval for a single population mean in each of the following situations.

a. $n = 25$; confidence level = .95.
b. $n = 81$; confidence level = .95.
c. $n = 1001$, confidence level = .99.

11.29 A random sample of $n = 9$ men between 30 and 39 years old is asked to do as many sit-ups as they can in one minute. The mean number is $\bar{x} = 26.2$, and the standard deviation is $s = 6$.

a. Find the value of the standard error of the sample mean. Write a sentence that interprets this value. (Refer to Section 11.1 for guidance.)
b. Find a 95% confidence interval for the population mean.
c. Write a sentence that interprets the confidence interval in the context of this situation.

11.30 The age at first marriage is found for a random sample of $n = 81$ women who graduated from college in 1980. The mean age is $\bar{x} = 25.5$, and the standard deviation is $s = 4.5$ years.

a. Find the value of the standard error of the sample mean. Write a sentence that interprets this value. (Refer to Section 11.1 for guidance.)
b. Find a 95% confidence interval for the population mean.
c. Write a sentence that interprets the confidence interval in the context of this situation.

11.31 What three factors affect the width of a confidence interval for a population mean? For each factor, indicate how an increase in the numerical value of the factor affects the interval width.

Lesson 1 General Section Exercises

11.32 Volunteers who had developed a cold within the previous 24 hours were randomized to take either zinc or placebo lozenges every 2 to 3 hours until their cold symptoms were gone (Prasad et al., 2000). Twenty-five participants took zinc lozenges, and 23 participants took placebo lozenges. The mean overall duration of symptoms for the zinc lozenge group was 4.5 days, and the standard deviation of overall duration of symptoms was 1.6 days. For the placebo group, the mean overall duration of symptoms was 8.1 days, and the standard deviation was 1.8 days.

a. Calculate a 95% confidence interval for the mean overall duration of symptoms in a population of individuals like those who used the zinc lozenges.
b. Calculate a 95% confidence interval for the mean overall duration of symptoms in a population of individuals like those who used the placebo lozenges.
c. On the basis of the intervals computed in parts (a) and (b), is it reasonable to conclude generally that taking zinc lozenges reduces the overall duration of cold symptoms? Explain why you think this is or is not an appropriate conclusion.
d. In their paper, the researchers say that they checked whether it was reasonable to assume that the data were sampled from a normal curve population and decided that it was. How is this relevant to the calculations done in parts (a) and (b)?

11.33 Refer to Exercise 11.32. Compute a 99% confidence interval for the mean duration of symptoms for individuals using the placebo lozenges. Write one or two sentences that interpret this interval.

11.34 A randomly selected sample of $n = 12$ students at a university is asked, "How much did you spend for textbooks this semester?" The responses, in dollars, are

200, 175, 450, 300, 350, 250, 150, 200, 320, 370, 404, 250

Computer output with a 95% confidence interval for the population mean follows:

Variable	N	Mean	StDev	SE Mean	95.0% CI
spent	12	284.9	96.1	27.7	(223.9, 346.0)

a. Draw either a dotplot or a boxplot of the data. Briefly discuss whether or not the necessary conditions for doing a confidence interval for the population mean are present.

b. The standard error of the mean given in the output is 27.7 (under "SE Mean"). Verify this value by giving the correct formula and substituting appropriate values into this formula.

c. What are the degrees of freedom for the t^* multiplier? What is the value of the t^* multiplier for the given confidence interval?

d. Write a sentence that interprets the confidence interval. Be specific about the population that is being described.

e. Calculate a 90% confidence interval for the population mean.

11.35 Refer to Exercise 11.34. Use the Empirical Rule to create an interval that estimates the textbook expenses of 95% of the *individual students* at the university (not the mean).

11.36 Suppose that a highway safety researcher is studying the design of a highway sign and is interested in the mean maximum distance at which drivers are able to read the sign. The maximum distances (in feet) at which $n = 16$ drivers can read the sign are as follows:

440	490	600	540	540	600	240	440
360	600	490	400	490	540	440	490

a. Are the necessary conditions met for computing a confidence interval for the population mean distance? Assume that the sample can be considered to be a random sample from the population of interest. See page 414 for guidance.

b. Estimate the mean maximum distance at which drivers can read the sign. Use a 95% confidence interval, and write a sentence that interprets this interval. If conditions are not met, make appropriate adjustments, explain what you did, and report the results with and without the adjustment.

11.37 In Example 11.6 (p. 415), we found that a 99% confidence interval for the mean number of hours Penn State students watch television in a day is 1.76 to 2.42 hours. Explain whether each of the following statements is a correct interpretation of this result.

a. We can be fairly confident that 99% of all Penn State students watch between 1.76 hours and 2.42 hours of television a day.

b. We can be fairly confident that the mean number of hours of television Penn State students watch per day is under 3 hours.

c. The probability is .99 that the mean number of hours of TV Penn State students watch per day is between 1.76 hours and 2.42 hours.

Lesson 2 Skillbuilder Exercises

11.38 Compare the formula for an approximate 95% confidence interval given on page 420 in Lesson 2 with the general formula for a confidence interval in the formula box on page 420 in Chapter 10.

a. What is different about the two formulas?

b. In finding a confidence interval for one population mean, what condition must be satisfied in order to use the approximate 95% confidence interval formula?

11.39 The following scenario was presented in Exercise 11.16a, which asked for the standard error of the mean. Find the standard error of the mean, and then find an approximate 95% confidence interval for the population mean, using the approximation that is appropriate for large samples. *Scenario:* The mean height for a sample of $n = 81$ women is $\bar{x} = 64.2$ inches, and the standard deviation is $s = 2.7$ inches.

11.40 The following two scenarios were presented in Exercise 11.16 (b) and (c), which asked for the standard error of the mean. Find the standard error of the mean, and then find an approximate 95% confidence interval for the population mean, using the approximation that is appropriate for large samples.

a. Mean systolic blood pressure for a sample of $n = 100$ men is $\bar{x} = 123.5$, and the standard deviation is $s = 9$.

b. Mean systolic blood pressure for a sample of $n = 324$ men is $\bar{x} = 123.5$, and the standard deviation is $s = 9$.

Lesson 2 General Section Exercises

11.41 This scenario was presented in Exercise 11.17, which asked for the standard error of the mean: Suppose that a randomly selected sample of 64 men has a mean foot length of 27.5 cm, and the standard deviation of the sample is 2 cm. Calculate an approximate 95% confidence interval for the mean foot length of men. Write a sentence that interprets this interval.

11.42 Example 11.3 (p. 410) described a study that compared mean weight loss for a diet plan to mean weight loss for an exercise plan. In the study, $n = 42$ men were put on a diet. The men who dieted lost an average of 7.2 kg, with a standard deviation of 3.7 kg.

a. Compute the standard error of the mean for the men who dieted. Write a sentence or two to explain what is measured by this value.

b. Compute an approximate 95% confidence interval for the mean weight loss for the men who dieted. Write a sentence that interprets this interval.

11.43 Explain why the approximation for a 95% confidence interval for a mean that uses the value 2 as multiplier should not be used when the sample size is very small.

Section 11.3

Skillbuilder Exercises

11.44 In Example 11.10 (p. 424), we found a 95% confidence interval for the mean reduction in trait anxiety if everyone in the population were to take the meditation training offered in the study. For the 21 people in the study, the mean reduction in anxiety was 4.5, and the standard deviation was 7.2.

a. Find a 99% confidence interval for the population mean.
b. On the basis of the interval that you found in part (a), what can you conclude about the population mean reduction in anxiety?
c. Refer to the three ways that paired data can occur, listed following the definition on page 407. Which of the three best describes the situation in this exercise? Explain.

11.45 In Example 11.9 (p. 422), we compared the number of hours of computer time with the number of hours watching TV by finding a 90% confidence interval.

a. Find a 95% confidence interval for this situation, and write a sentence interpreting it.
b. On the basis of the 95% confidence interval, can we conclude that for the population represented by this sample, students really do spend more time on computer usage than on watching television, on average? Explain.
c. Refer to the three ways that paired data can occur, listed following the definition on page 407. Which of the three best describes the situation in this exercise? Explain.

General Section Exercises

11.46 A random sample of five college women was asked for their own heights and their mothers' heights. The researchers wanted to know whether college women are taller on average than their mothers. The results (in inches) follow:

Pair	1	2	3	4	5
Daughter	66	64	64	69	66
Mother	66	62	65	66	63

a. Define the parameter of interest in this situation.
b. Find a 95% confidence interval for the parameter you defined in part (a).
c. Using the interval in part (b), write a sentence or two about the relationship between women students' heights and their mothers' heights for the population. Your explanation should be written to be understood by someone with no training in statistics.
d. Explain two different things that the researchers could have done to obtain a narrower interval than the one you found in part (b).

11.47 If a confidence interval for μ_d covers 0, can we conclude that the population means for the two measurements are identical? Explain.

11.48 A researcher was interested in knowing whether the mean weight of the second baby is higher, lower, or about the same as the mean weight of the first baby for women who have at least two children. She selected a representative sample of 40 women who had at least two children and asked them for the weights of the oldest two at birth. She found that the mean of the differences (first − second) was 5 ounces, with a standard deviation of 7 ounces.

a. Why was the researcher's study design better than taking independent samples of mothers and asking the first sample about the weight of their first baby and the second sample about the weight of their second baby?
b. Define the parameter of interest. Use appropriate notation.
c. Write the mean of 5 ounces and the standard deviation of 7 ounces using appropriate statistical notation.
d. Find a 90% confidence interval for the parameter that you defined in part (b). Write a sentence or two interpreting the interval.

11.49 Refer to Exercise 11.48.

a. Find a 95% confidence interval for the mean difference in weights for the first and second baby for the population of women who have had at least two children.
b. Would it be appropriate to use the method given on page 420 for finding an approximate 95% confidence interval in this situation? If so, find the interval and compare it to the interval you found in part (a). If you do not think the method is appropriate, explain why not.

11.50 For a sample of $n = 20$ women aged 18 to 29, responses to the question "How tall would you like to be?" are recorded along with actual heights. In the sample, the mean desired height is 66.7 inches, the mean actual height is 64.9 inches, and the mean difference is $\bar{d} = 1.8$ inches. The standard deviation of the differences is $s_d = 2.1$ inches.

a. Compute a 95% confidence interval for the mean difference between desired and actual height, and write a sentence that interprets this interval. Be specific about what population is described by the interval.
b. What are the necessary conditions for the validity of the confidence interval computed in part (a)?
c. Explain why the following statement is not correct: "On the basis of the confidence interval computed in part (a), we can conclude that all women aged 18 to 29 would like to be taller."
d. Do you think the method for finding an approximate 95% confidence interval, given on page 420, is appropriate in this situation? Explain.

Section 11.4

Exercises 11.51 to 11.61 correspond to the two Lessons in Section 11.4. Lesson 1 exercises are 11.51 to 11.57; Lesson 2 exercises are 11.58 to 11.61.

Lesson 1 Skillbuilder Exercises

11.51 Suppose that you were given a 95% confidence interval for the difference in two population means. What could you conclude about the population means if

a. The confidence interval did *not* cover zero?
b. The confidence interval *did* cover zero?

11.52 ◆ Each of 63 students in a statistics class used her or his nondominant hand to print as many letters of the alphabet, in order, as they could in 15 seconds. The following output for this exercise gives results for a 95% confidence interval for the difference in population means for females and males. The "unpooled" procedure was used (*Data source:* **letters** dataset on the companion website).

Two-sample T for Exercise 11.52

Sex	N	Mean	StDev	SE Mean
Female	29	12.55	4.01	0.74
Male	34	13.65	4.46	0.77

Difference = mu (Female) − mu (Male)
Estimate for difference: −1.10
95% CI for difference: (−3.23, 1.04)
T-Test of difference = 0 (vs not =)
T-Value = −1.03 P-Value = 0.309 DF = 60

a. What was the mean number of letters printed by females in the sample? What was the mean number printed by males? What is the difference between the sample means for males and females? (Use females − males.)
b. What is the 95% confidence interval given in the output for the difference between population means? Write a sentence that interprets this interval.
c. Refer to part (b). On the basis of the 95% confidence interval, what conclusion can be made about whether the population means for males and females differ? Explain.
d. Find the value of the (unpooled) standard error of the difference between sample means.
e. Note that the value for degrees of freedom for this problem are given at the end of the output as "DF = 60." Use Table A.2 to find the value of t^* that was used in finding the interval.

11.53 Data on the testosterone levels (ng/dL in saliva) of men in different professions were given in a paper published in the *Journal of Personality and Social Psychology* (Dabbs et al., 1990). The researchers suggest that there are occupational differences in mean testosterone level. Medical doctors and university professors were two of the occupational groups for which means and standard deviations were given along with a dotplot of the raw data. Computer output for a 95% confidence interval for the difference in mean testosterone levels of medical doctors and university professors follows.

Two-sample T for MD vs Profs

	N	Mean	StDev	SE Mean
MD	16	11.60	3.39	0.85
Prof	10	10.70	2.59	0.82

Difference = mu MD − mu Profs
Estimate for difference: 0.90
95% CI for difference: (−1.54, 3.34)
DF = 22

a. The 95% confidence interval is given as −1.54 to 3.34. The unpooled procedure was used to compute this interval. Show how the interval was computed by giving the formula and identifying numerical values for all elements of the formula. Note the degrees of freedom are given as "DF = 22."
b. Interpret the confidence interval for the difference in means. Be specific about the populations that are described. What assumptions are you making in this interpretation?
c. On the basis of the 95% confidence interval given, what can be concluded generally about the difference in mean testosterone levels of medical doctors and professors?

Lesson 1 General Section Exercises

11.54 Refer to Exercise 11.32 about the effect of zinc lozenges on the duration of cold symptoms. For $n = 25$ in the zinc lozenge group, the mean overall duration of symptoms was 4.5 days, and the standard deviation was 1.6 days. For $n = 23$ in the placebo group, the mean overall duration of symptoms was 8.1 days and the standard deviation was 1.8 days.
a. Calculate $\bar{x}_1 - \bar{x}_2$ = difference in sample means (placebo − zinc) and also compute the unpooled s.e.$(\bar{x}_1 - \bar{x}_2)$ = standard error of the difference in means.
b. Compute a 95% confidence interval for the difference in mean days of overall symptoms for the placebo and zinc lozenge treatments, and write a sentence interpreting the interval. Use the unpooled standard error and use the smaller of $n_1 - 1$ and $n_2 - 1$ as a conservative estimate of degrees of freedom.
c. Is the interval computed in part (b) evidence that the population means are different? Explain.

11.55 For a sample of 36 men, the mean head circumference is 57.5 cm with a standard deviation equal to 2.4 cm. For a sample of 36 women, the mean head circumference is 55.3 cm with a standard deviation equal to 1.8 cm.
a. Find the value of the difference between the sample means for men and women. (Use men − women.)
b. Find the standard error of the difference between the sample means in this context.
c. Find an approximate 95% confidence interval for the difference between population mean head circumferences for men versus women.

11.56 In Example 11.12 (p. 428), we learned about a study of hangover symptoms in college students (Slutske et al., 2003, on the website for this book). The students answered questions about alcohol use and hangovers, including a count of how many out of a list of 13 possible hangover symptoms that they had experienced in the past year. For the 470 men, the mean number of symptoms was 5.3; for the 755 women, it was 5.1. The standard deviation was 3.4 for each of the two samples.
a. Define the parameter of interest for comparing men and women using appropriate notation.
b. Find the standard error of the difference in means.
c. Find an approximate 95% confidence interval for the difference in population means.
d. On the basis of your interval in part (c), do you think that there is a difference in the population mean number of hangover symptoms experienced by college men and women? Explain.

11.57 Refer to Exercises 11.17 and 11.18, in which the foot lengths of a random sample of 64 men had a mean of 27.5 cm and a standard deviation of 2 cm, while a random sample of 100 women had a mean foot length of 24.0 cm and a standard deviation of 2 cm. Compute an approximate 95% confidence interval for the difference between the mean foot lengths of men and women. Write a sentence that interprets this interval.

Lesson 2 Skillbuilder Exercises

11.58 Refer to the output for Exercise 11.52 about letters printed with the nondominant hand.

a. Find the value of s_p = pooled standard deviation for these data.
b. For the pooled two-sample procedure, find the (pooled) standard error of the difference in sample means.
c. What would be the degrees of freedom for the pooled procedure?

11.59 What assumption is necessary in order to use the pooled procedure for finding a confidence interval for the difference between two population means?

11.60 Following are data for 20 individuals on resting pulse rate and whether or not the individual regularly exercises.

Person	Pulse	Regularly Exercises
1	72	No
2	62	Yes
3	72	Yes
4	84	No
5	60	Yes
6	63	Yes
7	66	No
8	72	No
9	75	Yes
10	64	Yes
11	62	No
12	84	No
13	76	No
14	60	Yes
15	52	Yes
16	60	No
17	64	Yes
18	80	Yes
19	68	Yes
20	64	Yes

a. Calculate a 95% confidence interval for the difference in means for the two groups defined by whether or not the individual regularly exercises. Be sure to verify necessary conditions and state any assumptions. Use the unpooled procedure. (Use the non-exercisers as Group 1.)
b. Repeat part (a) using the pooled procedure.
c. Compare the unpooled and pooled results, and discuss which procedure is more appropriate.

11.61 Refer to Exercises 11.32 and 11.54. Use the pooled procedure to compute a 95% confidence interval for the difference in mean days of symptoms for the zinc and placebo treatments. (Use placebo − zinc.)

Section 11.5 and Case Studies 11.1 and 11.2

Skillbuilder Exercises

11.62 ◆ Responses from $n = 204$ college students to the question, "About how many CDs do you own?" are used to determine 95% confidence intervals for the median and the mean. (*Data source:* **pennstate2** dataset on the companion website)

a. A 95% confidence interval for the population median is 50 to 60. Write two or three sentences interpreting this result. Be specific about the population that the interval describes, and also describe any implied assumptions about the sample.
b. A 95% confidence interval for the mean μ is 62.9 to 82.8. Compare this result to the result in part (a), and explain why this comparison suggests that the data for music CDs owned may be skewed.

11.63 Refer to Exercise 11.62. Sample statistics summarizing the data on number of CDs are $\bar{x} = 72.84$, $s = 72$, $n = 205$. Use this information to verify the 95% confidence interval for the mean of 62.9 to 82.8 given in part (b).

11.64 Suppose that you were given a 95% confidence interval for the relative risk of disease under two different conditions. What could you conclude about the risk of disease under the two conditions if:

a. The confidence interval did not cover 1.0?
b. The confidence interval did cover 1.0?

11.65 In a 12-year study done at ten medical centers, Cole et al. (2000) investigated whether the decrease in heart rate over the first 2 minutes after stopping treadmill exercise was a useful predictor of subsequent death during the study period. Based on observations of $n = 5234$ participants, the investigators concluded: "Abnormal heart rate recovery predicted death (relative risk 2.58 [CI, 2.06 to 3.20]). After adjustment for standard risk factors, fitness, and resting and exercise heart rates, abnormal heart rate recovery remained predictive (adjusted relative risk, [1.55 CI, 1.22 to 1.98])." Explain why the given confidence intervals are evidence that abnormal heart rate recovery after treadmill exercise is a predictor of subsequent death during the study period.

General Section Exercises

11.66 Refer to Case Study 11.2 (pp. 435–436). Use the baseline data for the placebo group to find an approximate 95% confidence interval for the mean PMS complex score for the population of all women like the ones in this study. Write a sentence interpreting the interval.

11.67 Refer to Case Study 11.2 (pp. 435–436).

a. Use the baseline data to find an approximate 95% confidence interval for the difference in means for the populations represented by the placebo and calcium groups at the start of the study.
b. Explain why the researchers would want the interval computed in part (a) to cover zero.

Section 11.6: Skillbuilder Applet Exercises

For these exercises, use the **ConfidenceLevel** *applet described in Section 11.6 and available on the companion website,* http://www.cengage.com/statistics/Utts4e.

11.68 Set the confidence level at 68%.

a. Use the ***sample!*** button to generate a new random sample. Write the confidence interval displayed, and indicate whether or not it captures the population mean (170).

b. Generate 150 different samples. How many times did the confidence interval capture the population mean? What percentage of the samples gave a confidence interval that captured the population mean?
c. Refer to part (b). Compare the number and percentage of times that the confidence interval captured the population mean to what might be expected for the 68% confidence level.

11.69 Change the confidence level to 95%. Repeat the parts of Exercise 11.68 using the 95% confidence level.

11.70 Change the confidence level to 99%. Repeat the parts of Exercise 11.68 using the 99% confidence level.

11.71 For the same random sample, give the 90%, 95%, and 99% confidence intervals for the population mean. (This can be done by observing what occurs when the scrollbar is used to change the confidence level.) What is indicated about the relationship between the width of the confidence interval and the confidence level?

11.72 Write a definition of the confidence level for a confidence interval procedure.

Chapter Exercises

Be sure to include a check of the appropriate conditions as part of your work when you compute a confidence interval.

11.73 In computing a confidence interval for a population mean μ, explain whether the interval would be wider, more narrow, or neither as a result of each of the following changes. Assume that features that are not mentioned (confidence level, mean, standard deviation, sample size) remain the same.

a. The level of confidence is changed from 90% to 95%.
b. The sample size is doubled.
c. A new sample of the same size is taken, and $\bar{x}$ increases by 10.

11.74 Find an exact 95% confidence interval (use t^*) for the mean weight loss after a year of exercise for the population of men represented by those in the study in Example 11.3 (p. 410). There were 47 men in the exercise group in the sample, with mean weight loss of 4.0 kg and standard deviation of 3.9 kg.

11.75 Example 2.17 (p. 47) reported that a random sample of 199 married British women had a mean height of 1602 mm with standard deviation of 62.4 mm.

a. Find a 99% confidence interval for the mean height of all women represented by this sample. Write a sentence interpreting the interval.
b. Give a 99% confidence interval for the mean height *in inches* of all women represented by those in this sample (1 mm = .03937 inch).

11.76 Refer to Exercise 11.75. The sample included the heights of the women's husbands as well. For these men, the mean height was 1732.5 mm with standard deviation 68.8 mm.

a. Find a 90% confidence interval for the mean height of the population of men represented by this sample.
b. Given the information provided in this and the previous exercise, is it possible to compute an approximate 95% confidence interval for the average difference in height between the husband and wife in British couples? If so, compute the interval. If not, explain what additional numerical information you would need.

11.77 Example 2.16 (p. 46) gave the weights in pounds of the 18 men on the crew teams at Oxford and Cambridge universities in 1991–1992. The data were as follows:

Cambridge: 188.5, 183.0, 194.5, 185.0, 214.0, 203.5, 186.0, 178.5, 109.0
Oxford: 186.0, 184.5, 204.0, 184.5, 195.5, 202.5, 174.0, 183.0, 109.5

a. Explain why it is not appropriate to use the methods in this chapter to compute a confidence interval for the mean weight of college crew team members, even if we assume these 18 men represent a random sample of all college crew team members.
b. The first eight values listed for Cambridge represent the weights of the eight rowers on the team. Assuming these eight men represent a random sample of all Cambridge crew team rowers over the years, compute a 90% confidence interval for the mean of that population.
c. Make the same assumption as in part (b) about the weights of the first eight men in the Oxford sample. Compute a 90% confidence interval for the difference in mean weights of Cambridge and Oxford rowers. Explain whether you used the pooled or unpooled standard error, and why.
d. Write a sentence interpreting the interval in part (c).

11.78 In Stroop's Word Color Test, 100 words that are color names are shown in colors different from the word. For example, the word *red* might be displayed in blue. The task is to correctly identify the display color of each word; in the example just given the correct response would be blue. Gustafson and Kallmen (1990) recorded the time needed to complete this test for $n = 16$ individuals after they had consumed alcohol and for $n = 16$ other individuals after they had consumed a placebo drink flavored to taste as if it contained alcohol. Each group included eight men and eight women. In the alcohol group, the mean completion time was 113.75 seconds and the standard deviation was 22.64 seconds. In the placebo group, the mean completion time was 99.87 seconds and the standard deviation was 12.04 seconds.

a. Calculate the unpooled standard error for the difference in the two sample means.
b. Calculate a 95% confidence interval for the difference in population means. For the unpooled procedure, the approximate degrees of freedom are df = 22.
c. On the basis of the confidence interval computed in part (b), can we conclude that the population means for the two groups are different? Why or why not?
d. Can we verify the necessary conditions for computing the confidence interval in part (b)? If so, verify the conditions. If not, explain why it is not possible.

11.79 *This is also Thought Question 11.3.* Refer to Example 11.8, in which a confidence interval was found for the mean number of hours that Penn State students watch TV.

a. What population do you think is represented by the sample of 175 students in Example 11.8? Do you think the Fundamental Rule for Using Data for Inference (reviewed on p. 372 of Chapter 10) holds in this case?

b. Does the confidence interval in Example 11.8 tell us that 95% of the students watch television between 1.842 and 2.338 hours per day? If not, what exactly does the interval tell us?

11.80 ◆ An experiment was performed by 15 students in a statistics class at the University of California-Davis to determine whether manual dexterity was better for the dominant hand compared to the nondominant hand (left or right). Each student measured the number of beans they could place into a cup in 15 seconds, once with the dominant hand and once with the nondominant hand. The order in which the two hands were measured was randomized for each student.

a. Explain whether a comparison of beans placed using the dominant and nondominant hands is a comparison of paired data or two independent samples.

b. Referring to material earlier in the book, explain why the order of the two hands was randomized rather than, for instance, having each student test the dominant hand first.

c. The data are presented in the table below and are given also in the **beans** dataset. Assuming that these students represent a random sample from the population, compute a 90% confidence interval for the mean difference in the number of beans that can be placed into a cup in 15 seconds by the dominant and nondominant hands.

Student	1	2	3	4	5	6	7	8	9	10	11	12	13	14	15
Dominant	22	19	18	17	15	16	16	20	17	15	17	17	14	20	26
Nondominant	18	15	13	16	17	16	14	16	20	15	17	17	16	18	25
Difference	4	4	5	1	−2	0	2	4	−3	0	0	0	−2	2	1

d. Write a sentence or two using the interval in part (c) to address the question of whether manual dexterity is better, on average, for the dominant hand.

11.81 In a random sample of 199 married British couples (described in Chapter 2), there were 170 for which both ages were reported. The average difference in husband's age and wife's age was 2.24 years, with standard deviation of 4.1 years.

a. Find an approximate 95% confidence interval for the average age difference for all married British couples. Write a sentence interpreting the interval.

b. Were the samples of husband ages and wife ages collected as paired data or as two independent samples? Briefly explain.

11.82 A study was conducted on pregnant women and the subsequent development of their children (Olds et al., 1994). One of the questions of interest was whether the IQ of children would differ for mothers who smoked at least 10 cigarettes a day during pregnancy and those who did not smoke at all. The mean IQ at age 4 for the children of the 66 nonsmokers was 113.28 points, while for the children of the 47 smokers, it was 103.12 points. The standard deviations were not reported, but from other information that was provided, the pooled standard deviation was about 13.5 points.

a. Assuming that these women and children are a representative sample, find a 95% confidence interval (use t^*) for the difference in mean IQ scores at age 4 for children of mothers who do not smoke during pregnancy and those whose mothers smoke at least 10 cigarettes a day during pregnancy. Write a sentence interpreting the interval.

b. One of the statements in the article was "after control for confounding background variables . . . the average difference observed at 36 and 48 months was reduced to 4.35 points (95% CI: .02, 8.68)" (Olds et al., 1994, p. 224). Explain the purpose of this statement, and interpret the result given.

11.83 Examples 11.2, 11.6, and 11.8 in Sections 11.1 and 11.2 used a sample of $n = 175$ students to estimate the mean daily hours that Penn State students watch television. The following summary shows results for daily hours of television by sex:

Sex	N	Mean	St Dev	SE Mean
Female	116	1.95	1.51	0.14
Male	59	2.37	1.87	0.24

a. Calculate the difference between the mean values for the females and males. Is this a statistic or a parameter? Explain. (Use males − females)

b. For each sex, verify the standard error of the mean ("SE Mean" in the output) by substituting appropriate values into the correct formula.

c. An approximate 95% confidence interval for the difference between means is −0.14 to 0.98 hours. On the basis of this interval, what can we conclude, if anything, about the difference in the mean daily television watching times of men and women? Justify your answer.

d. What formula was used to compute the interval given in part (c)? Substitute appropriate values into this formula.

11.84 In Example 11.11 (p. 427), we compared the time in seconds to cross an intersection when someone was or was not staring at the driver. We did not assume equal population variances. Following is the Minitab (Version 16) output that resulted from clicking "Assume Equal Variance."

```
Two-sample T for NoStare vs Stare
Group     N   Mean    StDev   SE Mean
NoStare  14   6.63    1.36    0.36
Stare    13   5.592   0.822   0.23

Difference = mu (NoStare) − mu (Stare)
Estimate for difference: 1.036
95% CI for difference: (0.135, 1.937)
T-Test of difference = 0 (vs not =):
   T-Value = 2.37 P-Value = 0.026 DF = 25
Both use Pooled StDev = 1.1356
```

a. What is the value of s_p?

b. Compare the 95% confidence interval with the one in Example 11.11, and discuss whether or not it is reasonable to use the "equal variance" assumption for this study.

11.85 Suppose that 100 researchers each plan to independently gather data and construct a 90% confidence interval for a population mean. If X = the number of those intervals that actually cover the population means, then X is a binomial random variable.

a. What is a "success" for this binomial random variable?
b. What is the numerical value of p, the probability of a success?
c. What is the expected number of intervals that will cover their population means?
d. Will each researcher know whether he or she has a "successful" interval? Explain.
e. What is the probability that all 100 researchers actually get intervals that cover their population means of interest?

Dataset Exercises

Datasets required to solve these exercises are available on the companion website, http://www.cengage.com/statistics/Utts4e.

11.86 For this exercise, use the dataset **pennstate1** on the companion website. Case Study 1.1 presented data given by college students in response to the question, "What is the fastest you have ever driven a car? ______ mph" The variable ***fastest*** contains the responses to this question. Determine a 95% confidence interval for the difference in the mean response of college men and college women. Interpret the interval, state any assumptions about the sample, and verify necessary conditions.

11.87 For this exercise, use the dataset **pennstate1** on the companion website. The variable ***RtSpan*** has the raw data for measurements of the stretched right handspans of n = 190 Penn State students.

a. Calculate a 95% confidence interval for the mean stretched handspan in the population of men represented by this sample. Write a sentence that interprets the interval.
b. Calculate a 95% confidence interval for the mean stretched handspan in the population of women represented by this sample. Write a sentence that interprets the interval.
c. Compute a 95% confidence interval to estimate the difference in the mean handspans of college men and women. (Use men − women.) Write a sentence that interprets the interval.

11.88 For this exercise, use the dataset **UCDavis2** on the companion website. In the survey, students reported ideal height (***IdealHt***) and actual height (***height***).

a. For the men in the sample, calculate a 90% confidence interval to estimate the mean difference between ideal and actual height. Be certain to graphically analyze the data in order to verify necessary conditions.
b. Repeat part (a) for the women in the sample.
c. Based on the intervals computed in parts (a) and (b), is there a difference between men and women with regard to the mean difference between ideal and actual height? Explain.
d. Define μ_1 to be the population mean difference (ideal height − actual height) for men, and define μ_2 to be the same for women. Find a 95% confidence interval for $\mu_1 - \mu_2$. What does the result say about how men and women differ on the issue?

11.89 For this exercise, use the **cholest** dataset on the companion website. The variable ***control*** contains cholesterol levels for individuals who have not had a heart attack, while the variable ***2-day*** contains cholesterol levels for heart attack patients 2 days after the attack.

a. Calculate a 98% confidence interval for the mean cholesterol level of individuals who have not had a heart attack. Interpret the interval, state any assumptions, and verify necessary conditions.
b. Calculate a 98% confidence interval for the mean cholesterol level of heart attack patients 2 days after the attack. Interpret the interval, state any assumptions, and verify necessary conditions.
c. Calculate a 98% confidence interval for the mean difference in the cholesterol levels of the populations represented by the two samples described in this problem. (Use 2-day − control.) Interpret the interval, state any assumptions, and verify necessary conditions.

11.90 For this exercise, use the **deprived** dataset on the companion website. Students in a statistics class were asked whether they generally felt sleep deprived (variable name = ***Deprived***) and also were asked how many hours they usually slept per night (variable name = ***SleepHrs***).

a. For the sample, find the mean hours of sleep per night for students who said that they are sleep deprived and for students who said that they are not sleep deprived. What is the difference in the two sample means?
b. Find a 95% confidence interval for the difference in population mean hours of sleep for students who would say that they are not sleep deprived versus students who would say that they are sleep deprived. Interpret the interval in the context of this situation.

◆ Dataset available but not required **Bold** exercises answered in the back

12

Laurence Mouton/PhotoAlto Agency RF Collections/Getty Images

Is it their personalities or their looks?

See Example 12.15 *(p. 472)*

Testing Hypotheses about Proportions

The logic of statistical hypothesis testing is similar to the "presumed innocent until proven guilty" principle of the U.S. judicial system. In hypothesis testing, we assume that the null hypothesis is a possible truth until the sample data conclusively demonstrate otherwise.

In Chapters 10 and 11, we learned how to find confidence intervals for population parameters in five different contexts involving proportions and means. In this chapter and the next one, we will learn how to conduct tests of hypotheses for the same five population parameters.

Hypothesis testing and confidence interval estimation are related methods, and often both methods can be used to analyze the same situation. As we learned in Chapter 10, a confidence interval is a numerical answer to the question, "What is the population value?" A hypothesis test is used to answer a question about particular values for a population parameter, or a particular relationship in a population, based on information in the sample data.

Example 12.1 **Does a Majority Favor a Lower Limit for Drunk Driving?** Suppose that your state legislature is considering a proposal to lower the legal limit for the blood alcohol level that constitutes drunk driving. A legislator wants to determine whether a majority of adults in her district favor this proposal. To gather information, she surveys 200 randomly selected individuals from her district and learns that 59% of these people favor the proposal. This sample information will be used to decide whether more than 50% of the population in her district favors the proposal.

Can the legislator conclude that a majority of all adults in her district favor this proposal? Because the result is based on a sample, there is the possibility that the observed majority may have occurred just by the "luck of the draw." If a majority of the whole population actually is opposed to the proposal, how likely is it that 59% of a random sample would favor it?

In this chapter, we will learn how to use the method of *statistical hypothesis testing* to analyze the type of issue just described in Example 12.1. The **hypothesis testing** method uses data from a sample to judge whether a statement about a population may be true or not. For example, is it true that a majority of all adults in the legislator's district favor the proposal for the tougher blood alcohol standard?

We cover the basic definitions and concepts for hypothesis testing in Section 12.1. These ideas apply to any of these five population parameters: one proportion, the difference between two proportions, one mean, the mean difference between paired data values, and the difference between two means. In Sections 12.2 and 12.3 we provide details for the two situations involving population proportions, for which confidence interval methods were covered in Chapter 10. They are the following:

- Hypotheses about one population proportion. For example, full moons occur on average every 29.53 days, and there is some speculation that babies are more likely to be born during the period of a full moon than at other times. We might classify babies as to whether or not they were born during the 24-hour period surrounding a full moon. The hypothesis of interest is that the proportion of babies born during a full moon is 1/29.53, as it would be if births were uniformly distributed across days.

- Hypotheses about the difference between two population proportions. For example, we might question whether or not the proportion favoring the death penalty is the same for teenagers as it is for adults.

Chapter 13 will cover hypothesis tests for hypotheses about one population mean, the population mean difference for paired data and the difference between two population means when data are from two independents samples.

12.1 HT Module 0: An Overview of Hypothesis Testing

There are five basic steps for any hypothesis test (also called a **significance test**). The details of how those steps are implemented for our five parameters of interest are given in this chapter and Chapter 13. Hypothesis tests for some other situations are provided in Chapters 14 to 16. The same five steps are always used, although some of the details change. The five steps, first introduced in Chapter 4, are as follows:

Step 1: *Determine the null and alternative hypotheses.*

Step 2: *Verify necessary data conditions, and if they are met, summarize the data into an appropriate* test statistic.

Step 3: *Assuming that the null hypothesis is true, find the* p-*value.*

Step 4: *Decide whether the result is* statistically significant *based on the* p-value.

Step 5: *Report the conclusion in the context of the situation.*

Lessons 1 and 2 of this module will describe the basic ideas and definitions for these five steps. Lesson 3 will discuss possible errors in hypothesis testing and the factors that affect the probabilities of those errors.

Lesson 1: Formulating Hypothesis Statements

Many questions that researchers ask can be expressed as questions regarding which of two possibilities might be true for a population. Consider, for instance, the nature of these research questions:

- Does a majority of the population favor a new legal standard for the blood alcohol level that constitutes drunk driving?
- Do female students study, on average, more than male students do?
- Will side effects be experienced by fewer than 20% of people who take this new medication?

All of these questions can be answered with either a "no" or a "yes," and each possible answer corresponds to a specific hypothesis about the situation. For instance, we can view the research question about the proportion favoring a new legal standard for drunk driving as a choice between two competing *hypothesis* statements:

Hypothesis 1: The proportion favoring the new standard *is not greater than .50 and thus does not constitute a majority.*

Hypothesis 2: The proportion favoring the new standard *is greater than .50 and thus does constitute a majority.*

For the question about the comparative study habits of women and men, we can view the research question in terms of a choice between these two statements:

Hypothesis 1: On average, women *do not* study more than men do.

Hypothesis 2: On average, women *do* study more than men do.

Terminology for the Two Choices

In the language of statistics, the two possible answers to questions like the ones we just encountered are called the **null hypothesis** and the **alternative hypothesis**.

DEFINITION The **null hypothesis**, represented by the symbol $\mathbf{H_0}$, is a statement that there is nothing happening. The specific null hypothesis varies from problem to problem, but generally it can be thought of as the status quo, or no relationship, or no difference. In most situations, the researcher hopes to disprove or reject the null hypothesis.

The **alternative hypothesis**, represented by the symbol $\mathbf{H_a}$, is a statement that something is happening. In most situations, this hypothesis is what the researcher hopes to prove. It may be a statement that the assumed status quo is false, or that there is a relationship, or that there is a difference.

Each of the following statements is an example of a null hypothesis:

- There is no extrasensory perception.
- There is no difference between the mean pulse rates of men and women.
- There is no relationship between exercise intensity and the resulting aerobic benefit.

Corresponding examples of alternative hypotheses are as follows:

- There is extrasensory perception.
- Men have a lower mean pulse rate than women do.
- Increasing exercise intensity increases the resulting aerobic benefit.

Example 12.2 **Are Side Effects Experienced by Fewer Than 20% of Patients?** Suppose that a pharmaceutical company wants to be able to claim that for its newest medication, the percentage of patients who experience side effects is less than 20%. The issue can be addressed with these null and alternative hypotheses:

Null: 20% (or more) of users will experience side effects.

Alternative: Fewer than 20% of users will experience side effects.

Note that the claim that the company hopes to show is used as the alternative hypothesis. The null hypothesis is that the company's desired claim is not the case. If we convert the percentage experiencing side effects to the parameter p = proportion of all users who experience side effects, our hypotheses can be written as

H_0: $p = .20$ (or $p \geq .20$)
H_a: $p < .20$

Example 12.3 **Mean Normal Body Temperature for Men and Women** A researcher wonders whether men and women have different average normal body temperatures. To study this question, she defines the population parameters

μ_1 = mean body temperature for the population of all men
μ_2 = mean body temperature for the population of all women

If these two population parameters are equal, the difference $\mu_1 - \mu_2$ will be 0. If the two population parameters are not equal, the difference $\mu_1 - \mu_2$ will not be 0. Thus, the null and alternative hypotheses for the researcher are

H_0: $\mu_1 - \mu_2 = 0$
H_a: $\mu_1 - \mu_2 \neq 0$

One-Sided and Two-Sided Hypothesis Tests

In Example 12.2, the alternative hypothesis is that the proportion experiencing side effects is *less than* .20. This alternative specifies a single direction for the value of the population parameter. In Example 12.3, the alternative hypothesis is that the difference between mean body temperatures for men and women *is not equal to* 0. That alternative hypothesis specifies values in both directions from the null hypothesis.

DEFINITION A **one-sided hypothesis test** is one for which the alternative hypothesis specifies parameter values in a single direction from a specified "null" value. A one-sided test may also be called a **one-tailed hypothesis test**.

A **two-sided hypothesis test** is one for which the alternative hypothesis specifies parameter values in both directions from the specified null value. A two-sided test may also be called a **two-tailed hypothesis test**.

In practice, most hypothesis tests are one-sided tests because investigators usually have a particular direction in mind when they consider a question. The direction expressed in the alternative hypothesis could be a "less than" direction as in Example 12.2 or it could be a "greater than" direction. For a two-sided test, the alternative hypothesis will always involve a "not equal" type of comparison as in Example 12.3.

Notation and Null Value for the Five Situations

For each of the five parameters of interest in Chapters 12 and 13, the null hypothesis can be written as

H_0: population parameter = null value

where the **null value** is the specific number the parameter equals if the null hypothesis is true. The alternative hypothesis is written in one of the following three ways, depending on what the researchers hope to show is true about the parameter:

Two-sided alternative hypothesis:

- H_a: population parameter $\neq$ null value

One-sided alternative hypothesis (choose one):

- H_a: population parameter $>$ null value
- H_a: population parameter $<$ null value

Sometimes when the alternative hypothesis is one-sided, values in the other direction are added to the null hypothesis, as in Example 12.2 on page 453. But the specific "null value" is always used when computing the test statistic, and the alternative hypothesis never includes the equals sign.

In a specific situation, the appropriate parameter symbol and the number for the null value are used instead of the words. For example, if researchers were interested in determining whether men have a lower mean pulse rate than women, the hypotheses would be as follows:

H_0: $\mu_1 - \mu_2 = 0$

H_a: $\mu_1 - \mu_2 < 0$ (equivalent to $\mu_1 < \mu_2$)

where μ_1 and μ_2 are the mean pulse rates for the populations of all men and all women, respectively, and the null value is 0. If the researchers simply were trying to determine whether there is a difference in mean pulse rates, the alternative hypothesis would be written as

H_a: $\mu_1 - \mu_2 \neq 0$

THOUGHT QUESTION 12.1 Confidence intervals and hypothesis testing are the two major categories of statistical inference. On the basis of this information, do you think null and alternative hypotheses are generally statements about populations, samples, or both? Explain.*

Lesson 2: Test Statistic, *p*-Value, and Deciding between the Hypotheses

The logic of statistical hypothesis testing is similar to the "presumed innocent until proven guilty" principle of the U.S. judicial system. In hypothesis testing, we assume that the null hypothesis is a possible truth until the sample data conclusively demonstrate otherwise. The "something is happening" hypothesis is chosen only when the data show us that we can reject the "nothing is happening" hypothesis.

Conclusions made in hypothesis testing are based on the following question: If the null hypothesis is true about the population, what is the probability of observing sample data like that which was observed? The next example illustrates how the answer to this question is useful for deciding whether to reject the null hypothesis or not.

Example 12.4 **Stop the Pain Before It Starts** A headline on the CNN website on October 21, 1996, read, "Study: painkillers more effective before surgery than afterward." The article described a University of Pennsylvania research study that was done to test the theory that the most effective pain relief strategy is to stop the pain before it starts. In the study, men undergoing surgery for prostate cancer were randomly assigned either to an "experimental" group that began taking painkillers before the operation or to a "control" group that followed the standard practice of waiting until after the operation to begin taking painkillers. The author of the website article presented the key result in terms of the likelihood that the observed data could have occurred just by chance:

> But 9 1/2 weeks later ... only 12 members of the 60 men in the experimental group were still feeling pain. Among the 30 control group members, 18 were still feeling pain. Gottschalk said the likelihood of this difference being due to chance was only 1 in 500.

The null hypothesis, a statement of no difference, is that the effectiveness of the painkillers would be the same whether they were started before or after the operation. According to the investigator, if we assume that this null hypothesis is true, the probability is only 1 in 500 that the observed difference could have been as large as it was or larger. On the basis of this evidence, it seems reasonable to reject the null hypothesis that the two timings are equally effective.

THOUGHT QUESTION 12.2 For Example 12.4, explain what the null hypothesis and the alternative hypothesis would be. Then write a sentence answering the probability question on which hypothesis testing is based for the example. The sentence should be of the form: "If [fill in null hypothesis] is true, then the likelihood that [fill in event] would have happened is [fill in likelihood]."†

***HINT:** See the discussion of statistical inference in Section 9.1.

†HINT: The sentence might start: "If there really is no difference between the timing of starting painkillers, then the likelihood that..."

Test Statistic and *p*-Value

After the null and alternative hypotheses are formulated, the next step in any hypothesis test is to calculate a data summary called a *test statistic* that measures the difference between the sample result and the null value. Then, we compute a probability called a *p*-value, which is the probability that the test statistic would take the value it did if the null hypothesis was actually true. The *p*-value is used to make a decision between the hypotheses.

DEFINITION The **test statistic** for a hypothesis test is the data summary used to evaluate the null and alternative hypotheses.

The ***p*-value** is computed by assuming that the null hypothesis is true and then determining the probability of a test statistic as extreme as or more extreme than the observed test statistic in the direction of the alternative hypothesis.

Example 12.5 **Example 12.4 Revisited: *p*-Value for Comparing the Painkiller and Control Groups** In the study described in Example 12.4, surgery patients either took painkillers before the surgery or did not do so until afterward. Among the pre-surgery painkiller group, only 12 of 60 patients (20%) still experienced pain 9 1/2 weeks after the surgery, whereas in the other group, 18 of 30 patients (60%) still felt pain at that time. The *p*-value for this situation was reported as 1 in 500. This is the probability that a sample difference as extreme as or more extreme than the one observed would occur if there really is no difference in the two conditions (pre-surgery painkiller or not) for the population of all possible surgery patients. Because the *p*-value was so small, the researchers concluded that there really must be a difference.

Using the *p*-Value to Reach a Conclusion about the Two Hypotheses

When we perform a hypothesis test, the objective is to decide if we should reject the null hypothesis in favor of the alternative hypothesis. The decision can be made by comparing the *p*-value to a designated standard called the *level of significance* for the test.

DEFINITION The **level of significance**, denoted by the Greek letter α (alpha), is a value chosen by the researcher to be the borderline between when a *p*-value is small enough to choose the alternative hypothesis over the null hypothesis, and when it is not. When the *p*-value is less than or equal to α, we reject the null hypothesis. When the *p*-value is larger than α, we cannot reject the null hypothesis. The level of significance may also be called the **α-level** of the test.

The phrase **statistically significant** is used to describe the results when the researcher has decided that the *p*-value is small enough to decide in favor of the alternative hypothesis. This happens when the *p*-value is less than the chosen level of significance.

Most researchers use the convention that we can reject the null hypothesis when the *p*-value is less than .05 (or 5%). In other words, the *level of significance* is usually set at $\alpha = .05$, and the result is *statistically significant* when the *p*-value is less than .05. In some research disciplines, the phrase *highly significant* is used to describe the result when the *p*-value is less than .01 (1%), and the phrase *marginally significant* is used when the *p*-value is between .05 and .10. *p*-Values greater than .10 are rarely considered sufficient evidence to reject the null hypothesis. Researchers frequently criticize these somewhat arbitrary borderlines, but they are used by many scientific journals.

Note that because the *p*-value is a probability, it must be between 0 and 1. It indicates the strength of the evidence against the null hypothesis. The smaller the *p*-value is, the more conclusive is the evidence against the null hypothesis. For example, the evidence against the null hypothesis is more conclusive when the *p*-value is .001 than when it is .25. As we learned earlier in this chapter, the logic of hypothesis testing is indirect. A *p*-value *does not* tell us the probability that the null hypothesis is true. Instead, it tells us only the probability that our test statistic could be as extreme as it is, or more so, if we assume that the null hypothesis is true.

IN SUMMARY Interpreting a *p*-Value

In any statistical hypothesis test, the smaller the *p*-value is, the stronger is the evidence against the null hypothesis. It is common research practice to call a result *statistically significant* when the *p*-value is less than or equal to .05.

The Rejection Region Approach to Hypothesis Testing

Before computers and calculators were in widespread use, it was difficult or impossible to compute the *p*-value in many circumstances, so researchers used a different method, called the *rejection region approach.* The **rejection region** in a hypothesis test is the region of possible values for the test statistic that would lead to rejection of the null hypothesis. Instead of finding a *p*-value and comparing it to the desired level of significance, the computed test statistic is compared to a "rejection region." If the test statistic falls in the rejection region, the null hypothesis is rejected. If not, the null hypothesis cannot be rejected.

The rejection region and *p*-value approaches give equivalent conclusions. The *p*-value will be less than the level of significance for any test statistic that falls into the rejection region. Like *p*-values, rejection regions depend on whether the alternative hypothesis is one-sided or two-sided. Although it is less common to use this approach now, details will be given for finding the rejection region for testing each of our five parameters.

Stating the Two Possible Conclusions

The two possible conclusions of a hypothesis test are as follows:

- When the *p*-value is small, we *reject the null hypothesis* or, equivalently, we *accept the alternative hypothesis*. "Small" is defined as a *p*-value $\leq \alpha$, where $\alpha =$ level of significance (usually .05).
- When the *p*-value is not small, we conclude that we *cannot reject the null hypothesis* or, equivalently, that *there is not enough evidence to reject the null hypothesis.* "Not small" is defined as a *p*-value $> \alpha$, where $\alpha =$ level of significance (usually .05).

When the *p*-value is greater than the level of significance (usually .05), you may be tempted to say that you accept the null hypothesis. Resist that temptation because *it is almost never correct to say,* "I accept the null hypothesis." This wording incorrectly makes it seem that we are convinced that the precise value specified in the null hypothesis is the true value of the parameter. If we collect only a small amount of data, we may not see convincing evidence that the true parameter falls into the range specified for the alternative hypothesis because the sampling error is so large. That is *not* the same as accepting that the value in the null hypothesis is the true value. For instance, if we observe three births and two are boys, we would certainly not be willing to *accept* the hypothesis "2/3 of all births are boys" even though the observed data don't provide evidence that it is false.

Example 12.6 **A Jury Trial** If you are on a jury in the U.S. judicial system, you must presume that the defendant is innocent unless there is enough evidence to conclude that he or she is guilty. Therefore the two hypotheses are as follows:

Null hypothesis: The defendant is innocent.
Alternative hypothesis: The defendant is guilty.

The prosecution collects evidence in much the way that researchers collect data, in the hope that the jurors will be convinced that such evidence would be extremely unlikely if the assumption of innocence were true. Consistent with our thinking in hypothesis testing, in many cases, we would not *accept* the hypothesis that the defendant is innocent. We would simply conclude that the evidence was not strong enough to rule out the possibility of innocence. In fact, in the United States, the two conclusions juries are instructed to choose from are "guilty" and "not guilty." A jury would never conclude that "the defendant is innocent."

THOUGHT QUESTION 12.3 Suppose that an ESP test is conducted by having someone guess whether each of n coin flips will result in heads or tails. The null hypothesis is that $p = .5$, and the alternative is that $p > .5$, where p = probability of guessing correctly. Suppose one participant guesses $n = 10$ times and gets 6 right, while another guesses $n = 100$ times and gets 60 right. In each case, the percent correct was 60%. Do you think that the p-value would be lower in one case than in the other, or would it be the same? Explain.*

Computing the Test Statistic for the Five Scenarios

In general, a *test statistic* is simply a summary that compares the sample data to the null hypothesis. The chi-square statistic introduced in Chapter 4 is an example of a test statistic. For the five scenarios involving proportions and means, the test statistic is a standardized score based on the null hypothesis.

Remember that for each of the five scenarios, the null hypothesis can be written in the following format:

H_0: Parameter = null value

where the **null value** is a specific number. If the population parameter equals that number, then the null hypothesis is true.

The test statistic for each of the five scenarios is computed by determining how far the null value is from the sample statistic (sample estimate) for the parameter. If they are close, then the null value may be a plausible value for the parameter. But if they are far apart, then it is unlikely that the true parameter is equal to the null value.

There is no way to know whether the difference between a sample statistic and the null value is large without some measure of how close they would be just by chance if the null value were the correct value of the parameter. As we learned in Chapter 9, a measure of how close they should be by chance is the standard error associated with the sample statistic. In some cases, the standard error will depend on the null value. When the standard error of the sample statistic depends on the null value, we call it the **null standard error**.

FORMULA For each of our five scenarios, the **test statistic** is a *standardized statistic* computed as follows:

$$\text{Test statistic} = \text{Standardized statistic} = t \text{ or } z = \frac{\text{Sample statistic} - \text{Null value}}{\text{Null standard error}}$$

The test statistic is z for the two situations involving proportions, and it is t for the three situations involving means. Further explanation of why the standardized statistic is used as the test statistic will be given in the module on hypothesis testing for one proportion. Details for computing the test statistic for each scenario will be given in each module.

***HINT:** The p-value is the conditional probability of observing 60% or more correct, given that the participant is just guessing. Would it be harder to guess at least 6 correct in 10 tries or at least 60 correct in 100 tries?

TECHNICAL NOTE **Bayesian Statistics**

The p-value represents the *conditional probability* $P(A|B)$, where

A = observing a test statistic value as extreme as that observed, or more so, and

B = the null hypothesis is true.

We would really prefer to know $P(B|A)$, the probability that the null hypothesis is actually true, *given* the observed data. Suppose that we try using Rule 3 or 4 from Chapter 7 to find this conditional probability. We would need to find $P(A \text{ and } B)$, and to find that, we would need to know $P(B)$. But that's the probability that the null hypothesis is true. If we knew that, we wouldn't have to conduct the hypothesis test! That's why we must rely on the indirect information provided by the p-value to make our conclusion. In fact, the null hypothesis is either true or not, so even considering $P(B)$ makes sense only if we think of it as a personal probability representing our belief that the null hypothesis is true. There is a branch of statistics, called **Bayesian statistics**, that utilizes the approach of assessing $P(B)$ and then combining it with the data to find an updated $P(B)$.

Lesson 3: What Can Go Wrong: The Two Types of Errors and Their Probabilities

Whenever we use a sample of data to make a decision about a larger population, we may make a mistake. In testing statistical hypotheses, there are two potential decisions, and each decision brings the possibility of an error. Decision makers and researchers should consider the consequences of these possible errors when they create and apply their decision-making rules.

Example 12.7 **Errors in Medical Tests** Imagine that you are tested to determine whether or not you have a disease. The lab technician or physician who evaluates your results must make a choice between two hypotheses:

Null hypothesis: You do not have the disease.
Alternative hypothesis: You have the disease.

Unfortunately, many laboratory tests for diseases are not 100% accurate. There is a chance that the result is wrong. Consider the two possible errors and their consequences:

- *Possible error 1:* You are told you have the disease, but you actually don't. The test result was a **false positive**.

  Consequence: You will be unnecessarily concerned about your health, and you may receive unnecessary treatment, possible suffering adverse side effects.

- *Possible error 2:* You are told that you do not have the disease, but you actually do. The test result was a **false negative**.

  Consequence: You do not receive treatment for a disease that you have. If this is a contagious disease, you may infect others.

Which error is more serious? In most medical situations, the second possible error, a false negative, is more serious, but this could depend on the disease and the follow-up actions that are taken. For instance, in a screening test for cancer, a false negative could lead to a fatal delay in treatment. Initial test results that are "positive" for cancer are usually followed up with a retest so that a false positive may be discovered quickly.

Type 1 and Type 2 Errors in Statistical Hypothesis Testing

The situation encountered in statistical hypothesis testing is about the same as in the medical analogy. One notable difference is that in statistical hypothesis testing, random sampling error causes the errors defined in this section. When a sample is used, the "luck of the draw" could be such that the sample statistic does not properly represent the population value. This is particularly likely when the sample size is small and the corresponding standard error is large.

In the terminology of statistical hypothesis testing, the two possible errors are called **Type 1** and **Type 2** errors.

DEFINITION A **Type 1 error** can occur only when the null hypothesis is actually true. The error occurs by concluding that the alternative hypothesis is true.

A **Type 2 error** can occur only when the alternative hypothesis is actually true. The error occurs by concluding that the null hypothesis cannot be rejected.

The numbering of the error types indicates which of the two hypotheses, (1) null or (2) alternative, is actually true. For example, a Type 1 error is the error that occurs when the first hypothesis, the null, is really true but we decide in favor of the alternative.

Example 12.8 **Calcium and the Relief of Premenstrual Symptoms** In Case Study 11.2, we learned about a randomized "blind" study in which 466 women who suffered from premenstrual syndrome (PMS) were randomly assigned to take either calcium supplements or a placebo. The research group believed that the use of calcium supplements would reduce the severity of the symptoms. The hypotheses of interest to the researchers are as follows:

H_0: For reducing PMS symptoms, calcium is not better than a placebo.
H_a: For reducing PMS symptoms, calcium is better than a placebo.

A Type 1 error would occur if the researchers decided that calcium is better than a placebo when, in fact, it is not (i.e., the *null hypothesis* is true). A Type 2 error would occur if the researchers decided that calcium is not better than a placebo when, in fact, it is (i.e., the *alternative hypothesis* is true).

In the actual study, published in the *American Journal of Obstetrics and Gynecology* (Thys-Jacobs et al., 1998), the authors concluded that there was a statistically significant difference between the effectiveness of calcium supplements and placebo. Ideally, this conclusion is correct and no error was made. If future research and experience prove that calcium supplements are actually not effective, these researchers will have made a Type 1 error by choosing the alternative when the null hypothesis is true. If a Type 1 error was made in this situation, women may be taking calcium to reduce their symptoms when it does not actually work. On the other hand, if the researchers had not found a difference but calcium really is effective (a Type 2 error), then women would not know about a treatment that is actually effective.

Example 12.9 **Medical Tests Revisited** In Example 12.7, we saw the two possible errors that might occur when you are tested for the presence of disease. Let's rephrase the possible errors in statistical terminology.

- A Type 1 error occurs when the diagnosis is that you have the disease when actually you do not. In other words, *a Type 1 error is a false positive.*
- A Type 2 error occurs when the diagnosis is that you do not have the disease when actually you do. In other words, *a Type 2 error is a false negative.*

Probabilities and Consequences of the Two Types of Errors

There always is a trade-off between the chances of making the two types of errors. For instance, in the U.S. judicial system, the procedures that are designed to minimize the probability of a false conviction carry with them an increased risk of erroneous "not guilty" verdicts. But if court rules were made more lenient to decrease the risk of incorrect acquittals, the probability of false convictions would increase.

In any given situation, it is important to understand the consequences of the two types of errors and to find the right balance between making the more serious and the less serious of the two types of errors. To know how to do that, we need to know how to determine the probability of making each type of error.

Probability of a Type 1 Error and the Level of Significance

A Type 1 error can only be made when the null hypothesis is actually true (but the data leads us to decide in favor of the alternative). The probability of making a Type 1 error, when the null is actually true, is given by the designated level of significance for the test. This is the only factor under the researcher's control that affects the chance of making a Type 1 error.

DEFINITION When the null hypothesis is true, the **probability of a Type 1 error** is equal to the level of significance (the α-level). When the null hypothesis is not true, a Type 1 error cannot be made so has probability 0.

The correspondence between the probability of a Type 1 error and the level of significance means that with the standard level of significance of .05 or 5%, about 5% of all hypothesis tests done *when the null hypothesis is really true* will result in a false claim of statistical significance. Note that this is not the same thing as saying that 5% of all rejected null hypotheses should not have been rejected. In fact, some statisticians and researchers argue that in the "real world," the null hypothesis is rarely true. If this argument is correct, the overall number of Type 1 errors in practice may be quite small.

To see the correspondence between the level of significance and the probability of a Type 1 error, suppose that the null value is the correct population parameter value. This means that the null hypothesis is in fact true. In that case, the distribution of the test statistic (z or t) is known and can be used to find the p-value. Suppose that the $\alpha = .05$ level of significance is used. By chance alone, the test statistic will fall into the most extreme 5% of the appropriate z or t distribution just 5% of the time. When that happens, the p-value will be less than .05, the null hypothesis will be rejected, and a Type 1 error will have been made. The same holds for any specified value for α.

Type 2 Errors and Power

It is not simple to find the **probability of a Type 2 error**. A Type 2 error is made when the alternative hypothesis is true, but the data doesn't allow us to reject the null. However, the alternative hypothesis covers a wide range of possibilities. If the truth is close to the null value, it may be difficult for sample data to provide adequate evidence to reject the null hypothesis, so the probability of making a Type 2 error will be high. If the truth is far from the null value, then the sample data will likely provide enough evidence to reject the null value, and the probability of a Type 2 error will be low.

There are three factors that affect the probability of making a Type 2 error:

1. The sample size: A larger n reduces the probability of a Type 2 error without affecting the probability of a Type 1 error.
2. The level of significance used for the test: A larger α reduces the probability of a Type 2 error by making it easier to reject the null hypothesis.
3. The actual value of the population parameter: The further the true parameter value is from the null value (in the direction of the alternative hypothesis), the lower the probability is of a Type 2 error.

We can control the first two factors—sample size and level of significance. We cannot control the third factor, the difference between the true and null values of the parameter.

Power of a Hypothesis Test

Rather than focusing on the risk of making a mistake, many investigators prefer to focus on the chance that their sample will provide the evidence necessary to make the right choice. A common question when a study is planned is "If something is really going on in the population, what is the probability that we will be able to detect it?" This question leads to the statistical meaning of the word *power*.

DEFINITION The **power** of a hypothesis test is the probability that we decide in favor of the alternative hypothesis given a specific truth about the population. When the alternative hypothesis is actually true, power is the probability that we do not make a Type 2 error.

There are two features of power that apply to all hypothesis tests and that researchers should keep in mind when they plan a study:

- *The power increases when the sample size is increased.* This makes sense because when the sample size is increased the standard error generally is decreased, leading to the likelihood of larger values of the test statistic. Also, the sample statistic is a more accurate estimate of the population value, making it easier to detect a difference between the true population value and the null value.
- *The power increases when the difference between the true population value and the null hypothesis value increases.* This makes sense because the probability of detecting a large difference is higher than the probability of detecting a small difference. However, remember that the truth about the population is not something that the researcher can control or change.

Researchers should evaluate power before they collect data that will be used to do hypothesis tests to make sure they have sufficient power to make the study worthwhile. Occasionally, news reports, especially in science magazines, will insightfully note that a study may have failed to find a relationship between two variables because the test had low power. This is a common consequence of conducting research with samples that are too small, but it is one that is often overlooked in media reports.

Probabilities of Type 1 and Type 2 Errors are Conditional Probabilities

The probabilities of making Type 1 or Type 2 errors are *conditional probabilities.* If the null hypothesis is true, you can make a Type 1 error (by rejecting it), but you cannot make a Type 2 error. If the alternative hypothesis is true, you can make a Type 2 error (by failing to reject the null hypothesis), but you cannot make a Type 1 error. Therefore, we talk about the probability of making a Type 1 error *given* that the null hypothesis is true and the probability of making a Type 2 error *given* that the alternative hypothesis is true.

Figure 12.1 provides a tree diagram illustrating the possible errors and their probabilities. Note that the first set of branches illustrates the truth, which is either the null hypothesis H_0 or the alternative hypothesis H_a. Because one or the other of these is definitely true, the appropriate probabilities for those branches are 1 (for the one that is true) and 0 (for the one that is not true). Unfortunately, we don't know which one is true, so we cannot specify which branch should be labeled 0 and which should be labeled 1.

The second set of branches show the conditional probabilities of either making or not making a type of error given the truth specified on the first branch. There is no standard notation for *power,* but the Greek letter β is sometimes used to indicate the probability of making a Type 2 error as illustrated on the appropriate branch of the tree diagram in Figure 12.1.

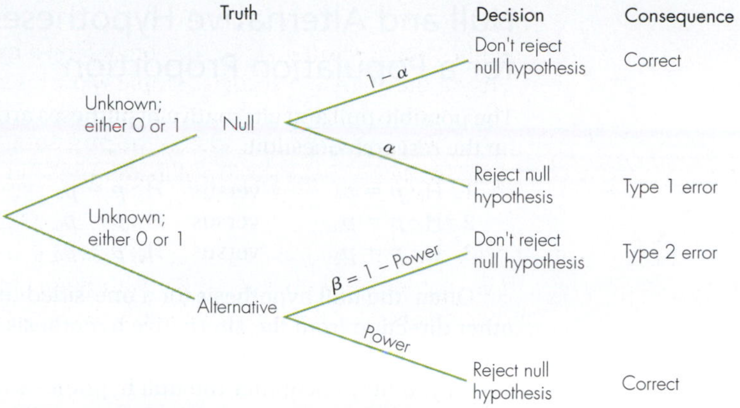

Figure 12.1 Tree diagram of possible errors and their probabilities

12.1 Exercises are on pages 487–490.

IN SUMMARY **The Two Types of Error and Power**

- A **Type 1 error** can occur only when the null hypothesis is actually true. The error occurs by concluding that the alternative hypothesis is true.
- A **Type 2 error** can occur only when the alternative hypothesis is actually true. The error occurs by concluding that the null hypothesis cannot be rejected.
- The **power** of a hypothesis test is the probability that we decide in favor of the alternative hypothesis given a specific truth about the population. When the alternative hypothesis is actually true, power is the probability that we do not make a Type 2 error.

12.2 HT Module 1: Testing Hypotheses about a Population Proportion

In this section, we describe how to apply the five steps to hypothesis testing when testing hypotheses about a population proportion. Example 12.2 was an example of how to write the null and alternative hypotheses in words and in formulas when the parameter of interest is a population proportion. In that example, about p = proportion of users of a medication who would experience side effects, the null hypothesis was H_0: $p \geq .20$ and the alternative was H_a: $p < .20$. The value specified in the hypotheses for the population proportion of interest is called the null value and is denoted by the symbol p_0. In Example 12.2 just described, the value of p_0 is .20.

Example 12.10 **Example 12.1 Revisited: Does a Majority Favor a Lower BAC Limit for Drivers?** Example 12.1 described a legislator who wondered whether more than 50% of the voters in her district favored a law that would reduce the legal blood alcohol level that defines drunk driving. We let p = proportion of all voters in the district favoring the lower limit. A majority is $p > .5$, so the null and alternative hypotheses for this situation may be written as

H_0: $p \leq .5$ (not a majority)
H_a: $p > .5$ (a majority)

The null value in this instance is $p_0 = .5$.

Null and Alternative Hypotheses for a Population Proportion

The possible null and alternative hypotheses are one of these three choices, depending on the research question:

1. H_0: $p = p_0$ versus H_a: $p \neq p_0$ (two-sided)
2. H_0: $p = p_0$ versus H_a: $p < p_0$ (one-sided)
3. H_0: $p = p_0$ versus H_a: $p > p_0$ (one-sided)

Often, the null hypothesis for a one-sided test is written as to include values in the other direction from the alternative hypothesis as well as equality with the null value. We will write it both ways in this chapter to give you practice with the two ways in which you may encounter the null hypothesis in journal articles. Remember that a p-value is computed by assuming that the null hypothesis is true, and the specific null value p_0 is what is assumed to be the truth about the population for that computation.

The z-Test for a Proportion

The general format of a test statistic for the parameters that we're covering in Chapters 9 to 13 is

$$\text{Test statistic} = \frac{\text{Sample statistic} - \text{Null value}}{\text{Null Standard error}}$$

When testing hypotheses about a proportion, the components of the formula for the test statistic are as follows:

- Sample statistic $= \hat{p}$, the sample proportion with the characteristic of interest.
- **Null value = p_0,** the value specified in the null hypothesis.
- Null standard error $= \sqrt{\dfrac{p_0(1-p_0)}{n}}$, the standard error of $\hat{p}$ if p_0 is the true value of p.

Substituting these components into the general format leads to

$$z = \frac{\hat{p} - p_0}{\sqrt{\dfrac{p_0(1-p_0)}{n}}}$$

This ***z*-test statistic** is a standardized score (z-score) that measures the difference between the sample proportion $\hat{p}$ and the null hypothesis value of the population proportion, p_0. For this reason, the test is called a ***z*-test for a proportion**. When we have a sufficiently large random sample and the null value is the true population value of p, the test statistic has approximately a standard normal distribution. This follows from the material in Chapter 9 on the approximate sampling distribution of possible values of sample proportions.

Note that the magnitude of z will be relatively large when there is a relatively large difference between the sample result ($\hat{p}$) and the null value (p_0). When the difference is large enough, in the direction of the alternative hypothesis, we will be able to reject the null hypothesis and decide in favor of the alternative hypothesis.

Computing the p-Value for the z-Test

Recall that a p-value is computed by assuming the null hypothesis is true and then determining the probability of a test statistic as extreme as or more extreme than the observed test statistic in the direction of the alternative hypothesis. For the z-test of hypotheses about a population proportion, the p-value probability is found by using the standard normal curve.

The details of how to find the p-value, i.e., the probability of a test statistic as extreme as or more extreme than the observed test statistic depend on the direction specified in the alternative hypothesis:

- For a *greater than* alternative hypothesis, find the probability that the test statistic z could have been *equal to or greater than* what it is.
- For a *less than* alternative hypothesis, find the probability that the test statistic z could have been *equal to or less than* what it is.
- For a *two-sided* alternative hypothesis, the p-value includes the probability areas in both extremes of the distribution of the test statistic z.

The correspondence between the type of alternative hypothesis and the p-value area is summarized in Table 12.1.

Table 12.1 Alternative Hypothesis Regions and p-Value Areas

Statement of H_a		p-Value Area	Normal Curve Region
$p < p_0$	(less than)	Area to the left of z (even if $z > 0$)	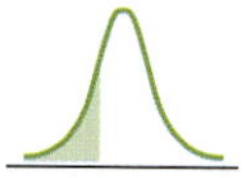
$p > p_0$	(greater than)	Area to the right of z (even if $z < 0$)	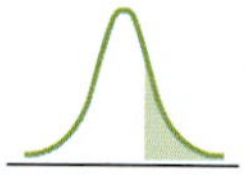
$p \neq p_0$	(not equal)	$2 \times$ area to the right of $\|z\|$	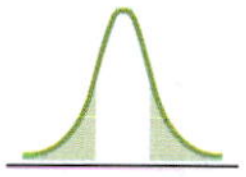

We include both extremes for a two-sided alternative hypothesis because both extremes provide equivalent support for the alternative hypothesis. Remember that "*the p-value is computed by assuming the null hypothesis is true and then determining the probability of a test statistic as extreme as or more extreme than the observed test statistic in the direction of the alternative hypothesis.*" For a two-sided test, "in the direction of the alternative hypothesis" includes values both above and below the null value.

If we use statistical software such as Minitab to do the hypothesis test, the program will automatically compute and report the p-value. To calculate the p-value "by hand," we can use Table A.1, the Standard Normal Table, or we could use a spreadsheet program like Excel (the command NORMSDIST(z) finds the area to *left* of z).

Conditions for Conducting the z-Test

Whenever we conduct a hypothesis test, we must make sure that the data meet certain conditions. These conditions are the assumptions that were made when the theory was derived for the test statistic that we are using to test our hypotheses. For testing proportions with the procedure described in this section, there are two conditions that should be true for the sample:

1. The sample should be a random sample from the population or the data should come from a binomial experiment with independent trials.
2. The quantities np_0 and $n(1 - p_0)$ should both be at least 10.

The first condition is not always practical, and most researchers are willing to use the test procedure as long as the sample is considered to be representative of the population for the question of interest. For instance, a college statistics class may be representative of all college students on questions such as which of two soft drinks they prefer or whether their right foot is longer than their left foot.

The second condition is a sample size requirement. To test the null hypothesis that $p = .5$, for example, we would need a sample of at least $n = 20$ to meet this condition. If the null hypothesis is that $p = .1$, the sample size would have to be at least $n = 100$. Our standard may be somewhat conservative; some authors suggest that np_0 and $n(1 - p_0)$ only need to be larger than 5 (e.g., Ott, 1998, p. 370).

The following example illustrates the five steps of hypothesis testing for a situation in which there is a one-sided alternative hypothesis that is a "greater than" type of statement. Remember from Table 12.1 that for this type of alternative hypothesis, the *p*-value region is the area under a standard normal curve to the right of the observed test statistic.

Example 12.11 **The Importance of Order in Voting** In a student survey, a statistics teacher asked his students to "randomly pick a letter from these two choices—*S* or *Q*." For about half of the students, the order of the letters *S* and *Q* was reversed so that the end of the instruction read "*Q* or *S*." The purpose of the activity was to determine whether there might be a preference for choosing the first letter. A tendency to pick the first choice offered has been noted in several settings, including elections.

Step 1: *Determine the null and alternative hypotheses.*

Use the letter p to represent the proportion of the population that would pick the first letter. If there is no general preference for either the first or second letter, $p = .5$ (because there are two choices). So the "null value" for p is $p_0 = .5$.

The alternative hypothesis usually states the researcher's belief or speculation. The purpose of the activity was to find out whether there is a general preference for picking the first choice. A preference for the first letter would mean that p is greater than .5. The null and alternative hypotheses can be summarized as follows:

H_0: $p = .5$
H_a: $p > .5$

- These hypotheses are statements about the larger population that this sample represents. Hypothesis tests are always used to say something about the *population* from which the sample has been taken.
- A more literal statement of the null hypothesis would be H_0: $p \leq .5$ because it is possible that less than half of the population would choose the first letter. Even if the null hypothesis is modified in this way, the value .5 is used as the assumed null value in the calculations.

Step 2: *Verify necessary data conditions, and if they are met, summarize the data into an appropriate test statistic.*

We first verify that our sample meets the necessary conditions for using the *z*-statistic. With $n = 190$ and $p_0 = .5$, both np_0 and $n(1 - p_0)$ equal 95, a quantity that is larger than 10, so the sample size condition is met. The sample, however, is not really a random sample—it is a convenience sample of students who were enrolled in this class. It does not seem that this will bias the results for this question, so we will behave as though the sample was a random sample.

In all, 114 of 190 students picked the first choice of letter on their form. Expressed as a proportion, this is $\hat{p} = 114/190 = .60$. The general format of the test statistic is

$$\text{Test statistic} = \frac{\text{Sample statistic} - \text{Null value}}{\text{Null Standard error}}$$

In this instance, the observed test statistic is

$$z = \frac{\hat{p} - p_0}{\sqrt{\dfrac{p_0(1 - p_0)}{n}}} = \frac{.6 - .5}{\sqrt{\dfrac{.5(1 - .5)}{190}}} = 2.76$$

Step 3: *Assuming that the null hypothesis is true, find the* p-*value.*

The *p*-value for this hypothesis test is the answer to this question: *If the true population proportion is .5, what is the probability that, for a sample of 190 people, the sample proportion could be as large as .60 (or larger)?*

The answer to this question is also the answer to a corresponding question about the *z*-statistic: *If the null hypothesis is true, what is the probability that the z-statistic could be as large as 2.76 (or larger)?*

To determine this *p*-value "by hand," we can use Table A.1 in the Appendix to learn that for $z = 2.76$, the area to the left is roughly .997. This is the probability that we could get a z-score smaller than 2.76, so the probability we could get a z-score greater than 2.76 is $1 - .997 = .003$. To calculate this p-value with Excel, enter = 1 − @NORMSDIST (2.76) into a cell of the spreadsheet. (The @ symbol tells Excel to invoke that function.) Figure 12.2 illustrates the connection between the p-value and the z-statistic for this example.

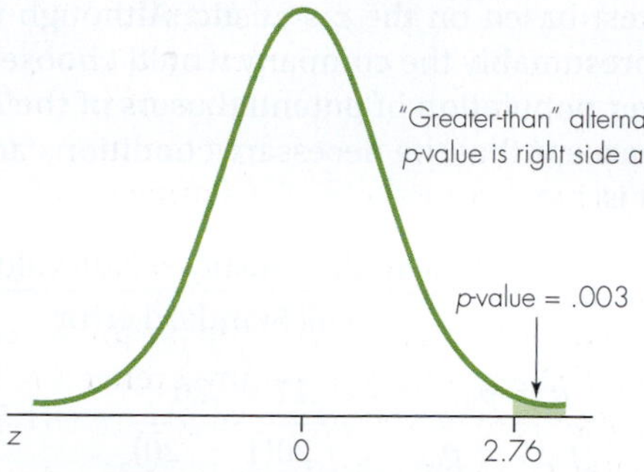

Figure 12.2 *p*-Value for the *S* or *Q* example

Note: In the specific context of hypothesis tests for proportions, it is unfortunate that the letter "p" could have two different meanings. Do not confuse the population proportion "p" with the p-value for the test, which is sometimes abbreviated to just "p" in research articles and computer output. You should be able to sort out which is which based on the context.

Step 4: *Decide whether the result is statistically significant based on the* p*-value.*

The convention used by most researchers is to declare statistical significance when the p-value is less than or equal to .05. The p-value in this example is .003, so we can reject the null hypothesis.

Step 5: *Report the conclusion in the context of the situation.*

The statistical conclusion was to reject the null hypothesis that $p = .50$. In this situation, the conclusion is that *there is statistically significant evidence that the first letter presented is preferred.* This conclusion is a generalization that applies to the population represented by this sample. In other words, we conclude that for the entire population of individuals represented by these students, there would be a preference for choosing the first letter presented.

Conclusions are strengthened when another study replicates the results. The instructor repeated this activity in another class of 327 students. In that class, 58% of the students picked the first letter, which is very close to the 60% found in the original study.

When the alternative hypothesis is a one-sided, "less than" type of statement, the p-value is found as the probability less than the observed z-statistic. The following example illustrates this type of problem.

Example 12.12 **Example 12.2 Revisited: Do Fewer Than 20% Experience Medication Side Effects?** Suppose that a pharmaceutical company wants to claim that side effects will be experienced by fewer than 20% of the patients who use a particular medication. In a clinical trial with $n = 400$ patients, they find that 68 patients experienced side effects. The sample proportion experiencing side effects is $\hat{p} = \frac{68}{400} = .17$.

Step 1: *The null and alternative hypotheses follow:*

H_0: $p \geq .20$ (company's claim is not true)
H_a: $p < .20$ (company's claim is true)

Note that the pharmaceutical company's claim is used as the alternative hypothesis. Also note that the null hypothesis is a region of values. To calculate the z-statistic, it is standard practice to use the value that separates the null and alternative hypothesis regions, in this case $p_0 = .20$.

Step 2: *Verify necessary data conditions, and if they are met, summarize the data into an appropriate test statistic.*

There were 400 patients, so both np_0 and $n(1 - p_0)$ are large enough to proceed with a test based on the *z*-statistic. Although most clinical trials use volunteer patients, presumably the company would choose volunteers who are representative of the larger population of potential users of the medication, so for practical purposes we will accept that the necessary conditions are met. The *z*-statistic for the hypothesis test is

$$\text{Test statistic} = \frac{\text{Sample statistic} - \text{Null value}}{\text{Null Standard error}}$$

$$z = \frac{\hat{p} - p_0}{\sqrt{\frac{p_0(1 - p_0)}{n}}} = \frac{.17 - .20}{\sqrt{\frac{.20(1 - .20)}{400}}} = -1.5$$

Note that the *z*-statistic is negative. This occurs because the sample proportion is less than the null value, the result that the pharmaceutical company wanted.

Step 3: *Assuming that the null hypothesis is true, find the* p*-value.*

The alternative hypothesis is that the *proportion is less than* a specified value, so the *p*-value is the *area to the left* of the observed *z*-statistic. We can use Table A.1, the Standard Normal Table, to determine this probability, or we can use the Excel instruction NORMSDIST(−1.5). For $z = -1.5$, the area to the left is about .067, and this is the *p*-value. It is the answer to this question: *If the true* p *is .2, what is the probability that, for a sample of 400 people, the sample proportion could be as small as .17 (or smaller)?*

The relationship between the *p*-value and the *z*-statistic for this problem is illustrated in Figure 12.3.

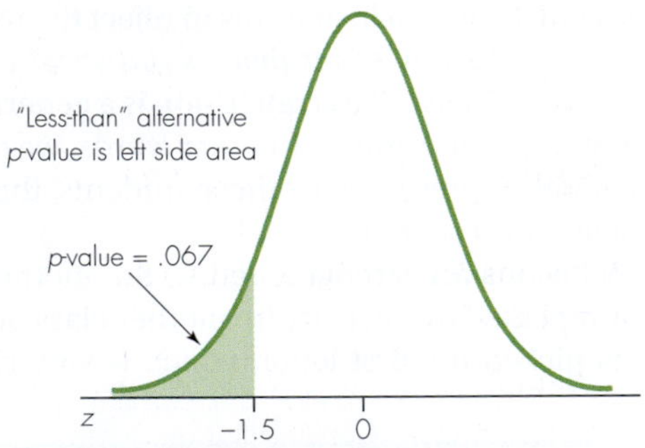

Figure 12.3 *p*-Value for medication side effects example

Step 4: *Decide whether the result is statistically significant based on the* p*-value.*

If the usual .05 standard for statistical significance is used, the *p*-value of .067 is not quite small enough to reject the null hypothesis.

Step 5: *Report the conclusion in the context of the situation.*

The sample evidence was in the desired direction for the pharmaceutical company, but it was not strong enough to conclusively reject the null hypothesis. The company cannot reject the idea that the proportion of the population who would experience side effects is .20 (or more). If this is an important issue for the company, it should consider gathering additional data. As we will learn in the next section, sample size affects statistical significance. This example also illustrates why it is not a good idea to "accept the null hypothesis." We would not be convinced that 20% or more of the population would experience side effects when only 17% of the sample did so.

The next example illustrates how to find a p-value when the alternative hypothesis is a two-sided "not equal to" statement. In these cases, the p-value is determined using both sides of the standard normal curve.

Example 12.13 **A Two-Sided Test: If Your Feet Don't Match, Is the Right One More Likely to Be Longer or Shorter?** Students in a statistics class measured the lengths of their right and left feet to the nearest millimeter. The right and left foot measurements were equal for 103 of the 215 students, but the two foot lengths were different for 112 students.

Let's consider the 112 individuals with unequal measurements and use them to represent the population of adults with unequal foot lengths. Among the population with unequal lengths, let p = proportion that has a longer right foot.

Step 1: If we have no advance belief that either foot will tend to be longer, the null and alternative hypotheses for p are as follows:

H_0: $p = .5$

H_a: $p \neq .5$

These two hypotheses are statements about the population represented by this sample of students. The null hypothesis states that among the population of people who have different foot lengths, exactly half will have a longer right foot and exactly half will have a longer left foot. The alternative states that this is not the case but does not specify a direction.

Steps 2 and 3: Both np_0 and $n(1 - p_0)$ are greater than 10, so we can proceed with the test, assuming that we think the students in the sample are representative of the larger population. In the sample, a longer right foot was reported by 63 of the 112 students who reported different foot lengths. The corresponding proportion with a longer right foot is $\hat{p} = 63/112 = .5625$. Minitab output for the hypothesis test is shown in Figure 12.5. We see that the value of the z-statistic is 1.32 and the p-value is .186.

Test of p = 0.5 vs p not = 0.5

X	N	Sample p	95.0 % CI	Z-Value	P-Value
63	112	0.5625	(0.471, 0.654)	1.32	0.186

Figure 12.4 Minitab output for foot length example

The calculation of the z-value (the answer is shown in the Minitab output as Z-value = 1.32) is

$$z = \frac{\text{Sample estimate} - \text{Null value}}{\text{Null standard error}} = \frac{.5625 - .5}{\sqrt{\dfrac{.5(1-.5)}{112}}} = \frac{.0625}{.0472} = 1.32$$

The two-tailed p-value is the total of the area to the right of $z = 1.32$ and the area to the left of $z = -1.32$, an area illustrated in Figure 12.5.

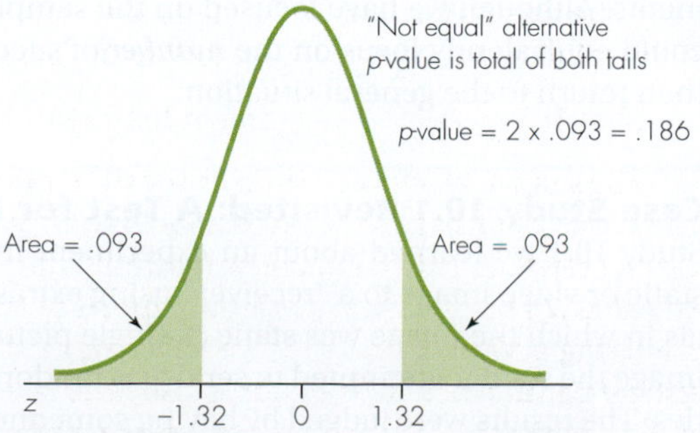

Figure 12.5 Two-tailed p-value for foot length example

Steps 4 and 5: The *p*-value (= .186) is not less than .05 so we cannot reject the null hypothesis. Although there was a tendency toward a longer right foot in the sample, there is insufficient evidence to conclude that the proportion in the population with a longer right foot is different from the proportion with a longer left foot.

IN SUMMARY

The Steps for a *z*-Test for One Proportion

Step 1: Determine the *null* and *alternative* hypotheses.

Null hypothesis: H_0: $p = p_0$ where p_0 is a specified value

Alternative hypothesis: H_a: $p \neq p_0$ *or* H_a: $p > p_0$ *or* H_a: $p < p_0$

where the format of the alternative hypothesis depends on the research question of interest. The null may include an inequality when the alternative is one-sided.

Step 2: Verify necessary data conditions, and if they are met, summarize the data into an appropriate *test statistic*.

Verify that the sample is large enough so that both np_0 and $n(1 - p_0)$ are at least 10. The sample should be a random sample from the population or the result of a binomial experiment. Compute the test statistic:

$$z = \frac{\text{Sample estimate} - \text{Null value}}{\text{Null standard error}} = \frac{\hat{p} - p_0}{\sqrt{\dfrac{p_0(1 - p_0)}{n}}}$$

Step 3: Assuming that the null hypothesis is true, find the *p*-value.

Using the *z* (standard normal) distribution, the *p*-value is the area in the tail(s) beyond the test statistic *z* as follows (refer to Table 12.1 on page 465 for illustrations):

For H_a: $p \neq p_0$, the *p*-value is 2 × area above $|z|$ (a two-tailed test).

For H_a: $p > p_0$, the *p*-value is the area above *z*, even if *z* is negative.

For H_a: $p < p_0$, the *p*-value is the area below *z*, even if *z* is positive.

For **Steps 4 and 5**, proceed as instructed on pages 456–457; see Examples 12.11 to 12.13.

Exact *p*-Values in Tests for Population Proportions

The method for finding a *p*-value shown in Table 12.1 relies on the fact that the test statistic *z* is approximately normal. However, in the special case of testing hypotheses about one population proportion, there is a method available for finding an *exact p-value*. Recall that the *p*-value is the *probability of a test statistic as extreme as or more extreme than the observed test statistic, in the direction of the alternative hypothesis, if the null hypothesis is true.*

Studies designed to test a population proportion are equivalent to binomial experiments. Although we have focused on the sample proportion of successes $\hat{p} = X/n$, we could equivalently focus on the *number* of successes *X*. Let's look at an example, and then return to the general situation.

Example 12.14 **Case Study 10.1 Revisited: A Test for Extrasensory Perception** In Case Study 10.1 we learned about an experiment in which a "sender" tried to transmit a static or video image to a "receiver" using extrasensory perception. There were 164 trials in which the image was static (a single picture instead of a video). In each trial the image the sender attempted to send was randomly selected from a set of four possibilities. The results were judged by having someone blind to the right answer compare the information produced by the receiver to the four possible target pictures. If the judge was able to pick out the right picture from the set of four, the trial was a success.

There were 45 successes in the 164 trials. Is this sufficient evidence to show that the probability of a match was higher than the value of .25 expected by random guessing? Let's carry out the five steps of hypothesis testing for this example.

Step 1: *Determine the null and alternative hypotheses.*

The parameter of interest is p = probability of a successful guess. The hypotheses follow:

H_0: $p = .25$ (random guessing)
H_a: $p > .25$ (better than random guessing)

Step 2: *Verify necessary data conditions, and if they are met, summarize the data into an appropriate test statistic.*

If all participants were randomly guessing on every trial, then this is a binomial experiment with $n = 164$ trials and $p = .25$ probability of correctly guessing each time. Our test statistic is $X = 45$, the number of successful identifications in the 164 trials.

Step 3: *Assuming that the null hypothesis is true, find the* p-*value.*

The alternative hypothesis is one-sided in the "greater than" direction, so we will find the probability that $X \geq 45$ in a binomial distribution with $n = 164$ and $p = .25$. The following Minitab output gives the p-value for the exact test in this case. The p-value = .261. Figure 12.6 illustrates the p-value region.

Test of p = 0.25 vs p > 0.25

Sample	X	N	Sample p	95% Lower Bound	Exact P-Value
1	45	164	0.274390	0.217459	0.261

In Excel, the command 1 − BINOMDIST(44, 164, .25,1) can be used to find the probability that $X \geq 45$.

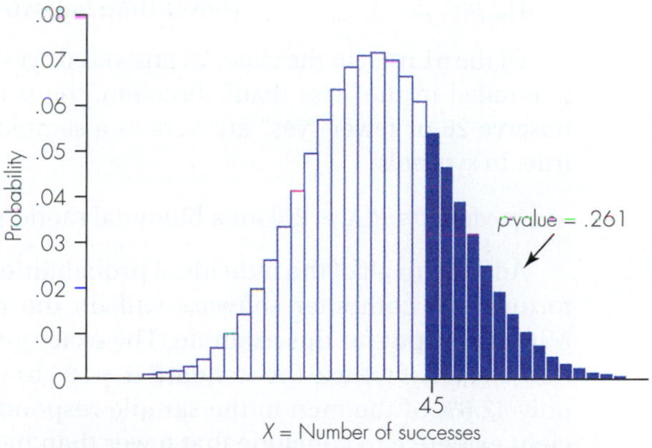

Figure 12.6 p-Value region for ESP example

Step 4: *Decide whether the result is statistically significant based on the* p-*value.*

Using the standard significance level of .05, the result is not statistically significant. The p-value of .261 is greater than .05. Therefore, we cannot reject the null hypothesis.

Step 5: *Report the conclusion in the context of the situation.*

We cannot reject the null hypothesis that $p = .25$. Therefore, we do not have sufficient evidence to conclude that the probability of a successful match based on the information obtained by receivers is greater than .25, or greater than what would be expected from random guessing.

Note that this does *not* allow us to conclude that $p = .25$. A two-sided 95% confidence interval for p is .21 to .34, an interval that contains values both above and below .25, with slightly more of the possible values above .25.

IN SUMMARY Finding the Exact p-Value for a Binomial Proportion

Define X to be a binomial random variable with n trials and probability of success p_0. Suppose that there are exactly k successes observed in the study, so $\hat{p} = k/n$. Then the **exact *p*-value** for testing the null hypothesis H_0: $p = p_0$ depends on the alternative hypothesis and is found as follows:

Statement of H_a	Exact *p*-Value for *k* Successes
$p < p_0$	$P(X \leq k)$
$p > p_0$	$P(X \geq k)$
$p \neq p_0$	$2P(X \leq k)$ if $k < np_0$; $2P(X \geq k)$ if $k > np_0$; 1.0 if $k = np_0$ } if p_0 is .5 only

For H_a: $p \neq p_0$, if p_0 is not .5 the computation is more complicated and should be done with statistical software.

Example 12.15 **What Do Men Care about in a Date?** Is it true that men care more about looks than personality? Students in a statistics class were asked, "Would you date someone with a great personality even though you did not find them attractive?" Suppose that we believed that fewer than half of men would answer "yes" to that question and wanted to test our theory. The population parameter is p = proportion in the population of men similar to those in this class who would answer "yes" to the question. The hypotheses are:

H_0: $p = .5$ (or $p \geq .5$) (half or more would answer yes)
H_a: $p < .5$ (fewer than half would answer yes)

Of the 61 men in the class, 26 answered "yes." Because the alternative hypothesis is one-tailed in the "less than" direction, the p-value is the probability that we would observe 26 or fewer "yes" answers in a sample of $n = 61$ *if* the null hypothesis were true. In symbols:

$$p\text{-value} = P(X \leq 26) \text{ for a binomial random variable with } n = 61 \text{ and } p = .5$$

Adding up all of the individual probabilities from 0 to 26 would be a tedious task; fortunately, computer software will do the task for us. Figure 12.7 provides the Minitab output for this example. The exact p-value is .153, so the null hypothesis is not rejected if we use the standard $\alpha = .05$ or even a more lenient $\alpha = .10$. Although only 42.6% of the men in the sample responded "yes," this does not provide sufficient evidence to conclude that fewer than half of the men in the *population* represented by this sample would respond this way. The p-value region is illustrated in Figure 12.8.

Test of p = 0.5 vs p < 0.5

Sample	X	N	Sample p	95.0% Upper Bound	Exact P-Value
1	26	61	0.426230	0.539523	0.153

Figure 12.7 Minitab output for Example 12.15

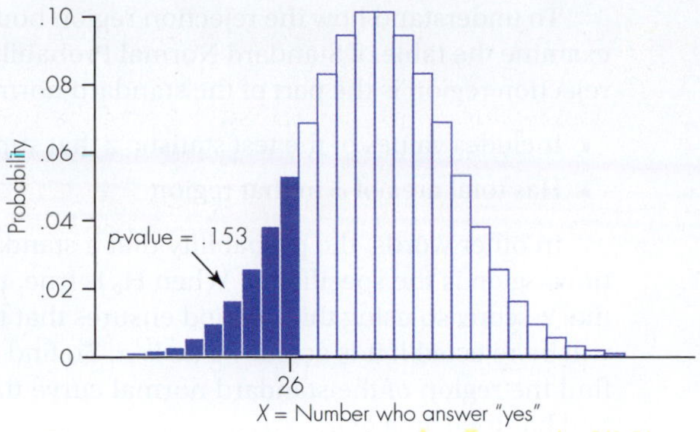

Figure 12.8 *p*-Value region for Example 12.15

The Rejection Region Approach to Hypothesis Testing for a Proportion

The **rejection region** in a hypothesis test is the region of possible values for the test statistic that would lead to rejection of the null hypothesis. *If* the null hypothesis is true, the probability that the computed test statistic will fall in the rejection region is α, the desired level of significance. There is a direct equivalence between the *p*-value approach and the rejection region approach. If an observed test statistic falls into the rejection region, the *p*-value for the test statistic will be less than or equal to the significance level.

Like *p*-values, rejection regions depend on whether the alternative hypothesis is one-sided or two-sided. Here are the rejection regions and rules for a *z*-test with levels of significance $\alpha = .05$ and .01:

	Rejection Region Rule	
Alternative Hypothesis	$\alpha = .05$	$\alpha = .01$
H_a: $p < p_0$	Reject H_0 if $z \leq -1.645$	Reject H_0 if $z \leq -2.33$
H_a: $p > p_0$	Reject H_0 if $z \geq 1.645$	Reject H_0 if $z \geq 2.33$
H_a: $p \neq p_0$	Reject H_0 if $\lvert z \rvert \geq 1.96$	Reject H_0 if $\lvert z \rvert \geq 2.58$

Note that the form of the rejection region (upper tail, lower tail, or both tails) is the same as the form of the *p*-value areas given in Table 12.1 on page 465. In both cases, the decision to reject the null hypothesis is only made if the test statistic falls into a tail that supports the alternative hypothesis. Figure 12.9 illustrates the rejection region for a test with an alternative hypothesis H_a: $p > p_0$ and level of significance $\alpha = .05$.

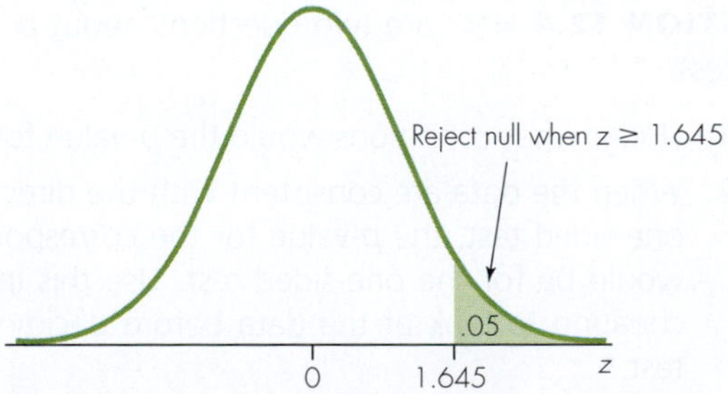

Figure 12.9 Rejection region for "greater than" alternative with $\alpha = .05$

A boundary of a rejection region is called a **critical value**. In Figure 12.9, the critical value for the test is $z = 1.645$ and the test statistic is in the rejection region whenever it is greater than or equal to this critical value.

HT Module 1: *p*

To understand how the rejection region boundaries (the critical values) are found, examine the table of Standard Normal Probabilities in the Appendix. In each case, the rejection region is the part of the standard normal curve that

- Includes values of the test statistic z that support the alternative hypothesis
- Has total area of α in that region

In other words, the probability that a standard normal z-score falls into the rejection region is the specified α. When H_0 is true, the z-statistic is simply a standard normal z-score, so using this method ensures that the probability that the null hypothesis will be rejected if it is actually true is α. To find rejection regions for other values of α, find the region of the standard normal curve that supports the alternative hypothesis and has total area of α.

Example 12.16 **Example 12.11 Revisited: Rejecting the Hypothesis of Equal Choices**

For the "choose S or Q" problem, the test statistic is $z = 2.76$ and the alternative hypothesis is that $p > .5$. On the basis of the rejection region rule for $\alpha = .05$, which is to reject H_0 if $z \geq 1.645$, we conclude that the null hypothesis can be rejected at $\alpha = .05$. Figure 12.10 illustrates the rejection region for the .05 significance level in this problem and also shows that the observed test statistic ($z = 2.76$) falls into this region. Using a more stringent $\alpha = .01$, we can also reject the null hypothesis because the test statistic $z = 2.76$ is greater than 2.33. The conclusion would be stated, "The null hypothesis can be rejected at $\alpha = .01$." In the context of the example, the result might be stated as follows: "There is statistically significant evidence that the first letter presented is preferred, using $\alpha = .01$."

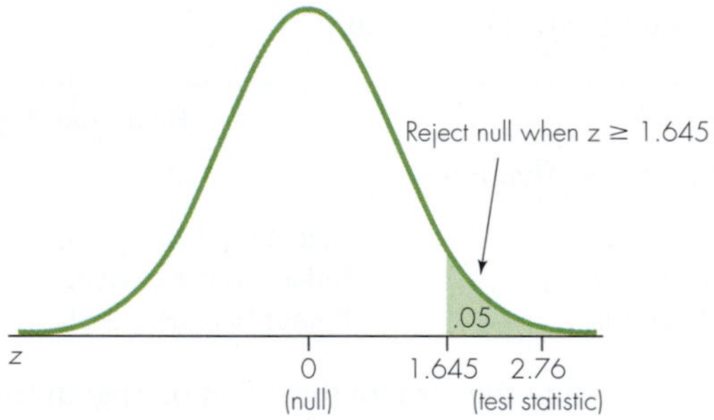

Figure 12.10 Rejection region for Example 12.10

THOUGHT QUESTION 12.4 Here are two questions about p-values and one-sided versus two-sided tests:

1. Under what conditions would the p-value for a one-sided z-test be greater than .5?
2. When the data are consistent with the direction of the alternative hypothesis for a one-sided test, the p-value for the corresponding two-sided test is double what it would be for the one-sided test. Use this information to explain why it would be cheating to look at the data before deciding whether to do a one- or two-sided test.*

***HINT:** Q1: Study Table 12.1 (p. 465). Q2: The p-value is supposed to reflect how unlikely such extreme data would be by chance. Would all possibilities for extreme data be covered if the alternative hypothesis were altered on the basis of looking at the data?

MINITAB TIP **Testing Hypotheses about a Proportion**

- To test hypotheses about a proportion, use **Stat > Basic Statistics > 1 Proportion**.
- If the raw data are in a column of the worksheet, specify that column. If the data have already been summarized, click on "Summarized Data," and then specify the sample size and the count of how many observations have the characteristic of interest.
- Select "Perform Hypothesis Test" and specify the null value.
- Use the ***Options*** button to select the type of alternative hypothesis, and also click on "Use test and interval based on normal distribution."
- If the "Use ... normal distribution" option is not selected, then the reported p-value will be exact, based on adding the relevant binomial probabilities. The confidence interval also will be exact. See the Minitab Help feature for details on how it is computed.
- If the alternative hypothesis selected is one-sided, the reported confidence interval also will be one-sided. For instance, if the alternative hypothesis is "greater than p_0," a lower bound will be given for the confidence interval. The interval extends from the lower bound to 1.0. To produce a two-sided confidence interval, go back to the default option of "not equal" as the alternative hypothesis.

TI-84 TIP **Testing Hypotheses about One Proportion**

- Press STAT, scroll across to **TESTS** (highlighted) and then scroll down to **5:1-PropZTest**. Press ENTER.
- Enter values for p_0, the observed number of successes (x) and the sample size (n).
- On the line that begins with *prop*, scroll to the desired alternative hypothesis and press ENTER.
- Scroll to **Calculate** and press ENTER. The display shows, in order, values for the test statistic z, the p-value, the sample estimate $\hat{p}$, and n. Selecting **Draw** rather than **Calculate** will produce a normal curve with the p-value shown as a shaded area.

12.2 Exercises are on pages 490–492.

12.3 HT Module 2: Testing Hypotheses about the Difference in Two Population Proportions

Researchers often wish to examine the difference between two populations with regard to the proportions that fall into a particular category of a response variable. It is almost always of interest to test whether or not the two population proportions are equal, and that is the only case we will consider. The two populations may be represented by two categories of a categorical variable, such as when we want to compare the proportions of Republicans and Democrats who support a certain political issue. Or the two populations may be hypothetical, represented by different treatments in an experiment. For instance, we may want to examine the proportions of smokers who would quit smoking if wearing a nicotine patch compared to if wearing a placebo patch.

To use the methods in this section we need measurements on the categorical variable of interest for *independent samples* from the two groups or populations. Common methods for accomplishing this are as follows:

- Choose separate random samples from the two populations and measure the same categorical response variable for the individuals in them. For example, separate random samples of adults may be asked the same question in two different years, to see if the proportion with a certain opinion has changed from one year to the next.

- Choose one random sample and group the individuals in it based on a categorical grouping variable. For instance, to compare the proportions of Republicans and Democrats who support a certain issue, a random sample of adults could be asked their opinion on the issue as well as their party affiliation. The groups would be formed based on party affiliation, and the proportions with a certain opinion would be compared for the two groups.
- Randomly assign participants in a randomized experiment to two treatment groups such as wearing a nicotine patch or a placebo patch, and record the outcome of interest such as whether or not they quit smoking.

The notation for comparing two proportions is as follows:

	Population Proportion	Sample Size	Number with Trait	Sample Proportion
Population 1	p_1	n_1	X_1	$\hat{p}_1 = \frac{X_1}{n_1}$
Population 2	p_2	n_2	X_2	$\hat{p}_2 = \frac{X_2}{n_2}$

Parameter of interest $= p_1 - p_2$

Sample statistic $= \hat{p}_1 - \hat{p}_2$

$$\textit{Standard deviation of } \hat{p}_1 - \hat{p}_2 = \sqrt{\frac{p_1(1-p_1)}{n_1} + \frac{p_2(1-p_2)}{n_2}}$$

Remember that the appropriate standard error is found by replacing unknown values in the standard deviation with estimates from the data.

If the null hypothesis is true, the proportions p_1 and p_2 are equal. This fact is used to determine the *null standard error* for the test statistic, by combining data from both samples to create a common estimate of p_1 and p_2 (given on p. 477). Thus the null standard error formula that is used for the test differs from the standard error formula used in Chapter 11 to find confidence intervals for the difference in two proportions. It is appropriate to use the phrase *null standard error* rather than simply *standard error* because the calculation is connected to assuming the null hypothesis to be true.

Example 12.17 **The Prevention of Ear Infections** On the basis of its biochemical properties, Finnish researchers hypothesized that regular use of the sweetener xylitol might be useful for preventing ear infections in preschool children and carried out a study to test this hypothesis (Uhari, 1998). In a randomized experiment, $n_1 = 165$ children took five daily doses of a placebo syrup, and 68 of these children got an ear infection during the 3 months of the study. Another $n_2 = 159$ children took five daily doses of xylitol, and 46 of these children got an ear infection during the study. The sample proportions getting an ear infection are $\hat{p}_1 = 68/165 = .412$ for the placebo group and $\hat{p}_2 = 46/159 = .289$ for the xylitol group, and the difference between these two proportions is .123 (12.3%). Is this observed difference in proportions large enough to conclude in general that using xylitol reduces the risk of ear infection? We will return to this example after summarizing the steps for the **hypothesis test for comparing two population proportions.**

Read the original source on the companion website, http://www.cengage.com/statistics/Utts4e.

(Reproduced from Pediatrics, *Vol. 102, pp. 879–884. "A novel use of Xylitol sugar in preventing acute otitis media," Uhari, M., T. Kenteerkar, and M. Niemala. Copyright © 1998 by the AAP.)*

Step 1: Determine the Null and Alternative Hypotheses

The *null* hypothesis is that there is no difference in population proportions:

$$H_0: p_1 - p_2 = 0 \quad (\text{or } p_1 = p_2)$$

The *alternative* hypothesis may be either two-sided or one-sided, depending on the research question of interest:

$$H_a: p_1 - p_2 \neq 0 \quad \textit{or} \quad H_a: p_1 - p_2 > 0 \quad \textit{or} \quad H_a: p_1 - p_2 < 0$$

The alternative hypothesis may also be written as

$$H_a: p_1 \neq p_2 \quad or \quad H_a: p_1 > p_2 \quad or \quad H_a: p_1 < p_2$$

The direction of the alternative hypothesis for a one-sided test depends on which group is defined as Population 1 and which is defined as Population 2. It is often easier to clarify the correct direction by writing the alternative hypothesis in the second format.

Step 2: Verify Necessary Data Conditions, and If They Are Met, Summarize the Data into an Appropriate Test Statistic

Conditions Necessary for a z-Test of the Difference in Two Proportions to Be Valid

- Independent samples are available from the two populations.
- The number with the trait or response of interest and the number without the trait or response of interest is at least 10 in each sample.

Make sure the data collection was done so as to ensure independent samples, and then simply make sure there are at least 10 with and without the trait or response of interest in each of the two samples.

Continuing Step 2: Computing the Test Statistic for a *z*-Test for Two Proportions

The sample statistic $\hat{p}_1 - \hat{p}_2$ (difference between sample proportions) estimates the parameter $p_1 - p_2$ (difference between population proportions). To find the null standard error, assume that the null hypothesis is true, so that $p_1 = p_2 = p$. Estimate the common population proportion p using all of the data:

$$\hat{p} = \frac{n_1\hat{p}_1 + n_2\hat{p}_2}{n_1 + n_2} = \frac{X_1 + X_2}{n_1 + n_2}$$

This combined estimate is simply the sample proportion for both samples combined. It is used instead of separate estimates of p_1 and p_2 to find the null standard error. The rationale for this substitution is that the test is based on the distribution of the test statistic *assuming that the null hypothesis is true.* If $p_1 = p_2$, then $\hat{p}$ is the best estimate for each of the population proportions because it makes use of all available sample data. The standardized test statistic follows:

$$z = \frac{\text{Sample statistic} - \text{Null value}}{\text{Null standard error}} = \frac{\hat{p}_1 - \hat{p}_2 - 0}{\sqrt{\frac{\hat{p}(1-\hat{p})}{n_1} + \frac{\hat{p}(1-\hat{p})}{n_2}}}$$

$$= \frac{\hat{p}_1 - \hat{p}_2 - 0}{\sqrt{\hat{p}(1-\hat{p})\left(\frac{1}{n_1} + \frac{1}{n_2}\right)}}$$

If the null hypothesis is true, the sampling distribution of this *z*-statistic is approximately the standard normal curve.

Step 3: Assuming That the Null Hypothesis Is True, Find the *p*-Value

The standard normal curve is used to find the *p*-value:

- For $H_a: p_1 - p_2 > 0$, the *p*-value is the area above z (a one-tailed test), even if z is negative.
- For $H_a: p_1 - p_2 < 0$, the *p*-value is the area below z (a one-tailed test), even if z is positive.
- For $H_a: p_1 - p_2 \neq 0$, the *p*-value is $2 \times$ area above $|z|$ (a two-tailed test).

See Table 12.1 (p. 465) for guidance. Do not confuse the *p-value* with the population or sample *proportions,* which also use the letter *p*.

Steps 4 and 5: Decide Whether the Result Is Statistically Significant Based on the *p*-Value and Make a Conclusion in the Context of the Situation

Choose a significance level ("alpha"); standard practice is $\alpha = .05$. The result is statistically significant if the *p*-value $\leq \alpha$. Interpret the conclusion in the context of the situation. Whenever two populations are compared, the manner in which the data were collected should also be considered. Remember from Chapter 6 that an experiment is generally a stronger proof of causation than an observational study is.

Example 12.17 *(cont.)* **The Prevention of Ear Infections** **Step 1:** *Determine the null and alternative hypotheses.* The researchers hoped to show that xylitol reduces ear infections. The parameter p_1 is the proportion who would get an ear infection in the population of children similar to those in the study if taking a placebo. The parameter p_2 is the proportion who would get an ear infection in that population if taking xylitol. Therefore, the null and alternative hypotheses of interest are as follows:

$$H_0: p_1 - p_2 = 0 \quad (\text{or } p_1 = p_2)$$
$$H_a: p_1 - p_2 > 0 \quad (\text{or } p_1 > p_2)$$

A one-sided alternative is used because the researchers want to show that the proportion getting an ear infection is significantly lower in the xylitol group than in the placebo group.

Step 2: *Verify necessary data conditions, and if they are met, summarize the data into an appropriate test statistic.* There are at least ten children in each sample who did and did not get ear infections, so the conditions are met.

- $\hat{p}_1 = \dfrac{68}{165} = .412$ and $\hat{p}_2 = \dfrac{46}{159} = .269.$
- The sample statistic is $\hat{p}_1 - \hat{p}_2 = .412 - .289 = .123.$
- The combined proportion is $\hat{p} = \dfrac{68 + 46}{165 + 159} = \dfrac{114}{324} = .35.$
- $\text{null s.e.}(\hat{p}_1 - \hat{p}_2) = \sqrt{\hat{p}(1 - \hat{p})\left(\dfrac{1}{n_1} + \dfrac{1}{n_2}\right)}$

$$= \sqrt{(.35)(.65)\left(\frac{1}{165} + \frac{1}{159}\right)} = .053.$$

- $z = \dfrac{\text{Sample statistic} - \text{Null value}}{\text{Null standard error}} = \dfrac{.123}{.053} = 2.32.$

Steps 3, 4, and 5: Because the alternative is $H_a: p_1 - p_2 > 0$, the *p*-value is the area above $z = 2.32$. Figure 12.11 illustrates this probability. From the normal probability table (Appendix Table A.1), we learn that the probability is .9898 that z is less than 2.32, so *p*-value $= 1 - .9898 = .0102$. We can reject the null hypothesis and attach the label "statistically significant" to the result because the *p*-value is small. On the basis of this experiment, we can conclude that taking xylitol would reduce the proportion of ear infections in the population of similar preschool children in comparison to taking a placebo.

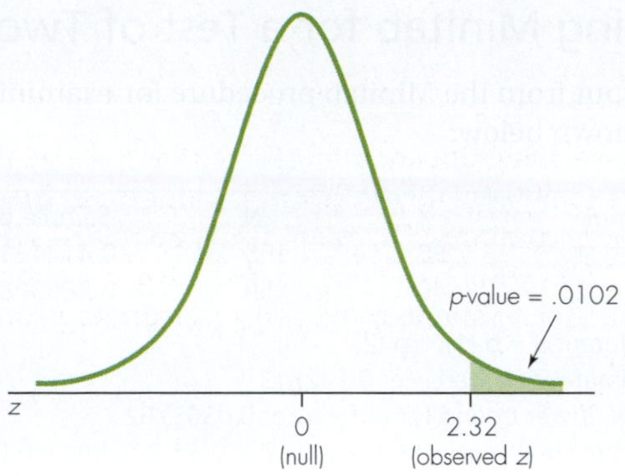

Figure 12.11 *p*-Value region for Example 12.17

IN SUMMARY

The Steps for a *z*-Test for the Difference in Two Proportions

Step 1: Determine the *null* and *alternative* hypotheses.

Null hypothesis: $H_0: p_1 - p_2 = 0$

Alternative hypothesis: $H_a: p_1 - p_2 \neq 0$ *or* $H_a: p_1 - p_2 > 0$ *or* $H_a: p_1 - p_2 < 0$

where the format of the alternative hypothesis depends on the research question of interest and the order in which Population 1 and Population 2 are defined.

Step 2: Verify necessary data conditions, and if they are met, summarize the data into an appropriate *test statistic.*

Verify that the samples are large enough so that for each sample, $n\hat{p}$ and $n(1 - \hat{p})$ are at least 10. The samples also must be independent. Compute the test statistic:

$$z = \frac{\text{Sample statistic} - \text{Null value}}{\text{Null standard error}} = \frac{(\hat{p}_1 - \hat{p}_2) - 0}{\sqrt{\hat{p}(1 - \hat{p})\left(\frac{1}{n_1} + \frac{1}{n_2}\right)}}$$

where

$$\hat{p}_1 = \frac{X_1}{n_1},\ \hat{p}_2 = \frac{X_2}{n_2} \text{ and } \hat{p} = \frac{X_1 + X_2}{n_1 + n_2}$$

Step 3: Assuming that the null hypothesis is true, find the *p*-value.

Using the *z* (standard normal) distribution, the *p*-value is the area in the tail(s) beyond the test statistic *z* as follows (refer to Table 12.1 on page 465 for illustrations):

For $H_a: p_1 - p_2 \neq 0$, the *p*-value is 2 × area above $|z|$ (a two-tailed test).

For $H_a: p_1 - p_2 > 0$, the *p*-value is the area above *z*, even if *z* is negative.

For $H_a: p_1 - p_2 < 0$, the *p*-value is the area below *z*, even if *z* is positive.

For **Steps 4 and 5**, proceed as instructed on page 478.

HT Module 2: $p_1 - p_2$

Using Minitab for a Test of Two Proportions

Output from the Minitab procedure for examining two proportions for Example 12.17 is shown below:

```
Sample        X        N        Sample p
  1          68       165       0.412121
  2          46       159       0.289308

Difference = p (1) − p (2)
Estimate for difference: 0.122813
95% lower bound for difference: 0.0363762
Test for difference = 0 (vs > 0): Z = 2.31 P-Value = 0.010
```

The default in Minitab is *not* to use the combined sample proportion in the formula for the standard error, because the Minitab procedure for two proportions is used to compute a confidence interval as well as to carry out a test of hypotheses. In other words, Minitab is using the standard error formula in Chapter 10 in the denominator of the test statistic, rather than the null standard error formula given in this chapter. However, if you use "Options" and click "Use pooled estimate of *p* for test," the appropriate null standard error will be used.

MINITAB TIP **Computing a Two-Sample *z*-Test for the Difference in Proportions**

- To test hypotheses about the difference in two proportions, use **Stat > Basic Statistics > 2 Proportions**. See the note below for information about inputting data.
- To specify the alternative hypothesis, use the ***Options*** button.
- To compute the *z*-statistic described in this section, use ***Options***, and then click ***Use pooled estimate of* p *for test***.

Note: There are three possibilities for inputting data. The raw data for the response (*Samples*) may be in one column and the raw data for group categories (*Subscripts*) may be in a second column. Or, the raw data for the two independent groups may be in two separate columns. Or, the data may already be summarized. If so, click ***Summarized Data***, and then specify the sample size and the *number* of successes for each group.

THOUGHT QUESTION 12.5 The raw data for Example 12.17 (p. 476) can be thought of as two categorical variables (with two possible categories each) for each participant in the experiment. What are the two variables and their categories? Construct a contingency table in the format learned in Chapter 4, with the explanatory variable as rows. In Chapter 4, you learned how to look for relationships by finding a chi-square statistic and associated *p*-value. In the context of this example, how do you think the null and alternative hypotheses would be stated if the problem is viewed from that perspective? (You will learn how to do this in Chapter 15.)*

12.3 Exercises are on pages 492–493.

12.4 Sample Size, Statistical Significance, and Practical Importance

There is a big difference between the following two beginnings to an advertisement for a toothpaste:

- 2 out of 3 dentists recommend...
- 2/3 of the 500 dentists surveyed recommend...

***HINT:** One variable is treatment (xylitol or placebo). What else did the researchers measure for each child?

We should be suspicious that the first claim may be based literally on the recommendations of only three dentists. If this is the case, we know that the claim is meaningless. A sample of just three individuals is not large enough to be conclusive. On the other hand, the second claim seems stronger. A sample of 500 dentists seems like a pretty big sample, although we should ask how the 500 dentists were selected for whatever survey was done.

The size of the sample affects our ability to make firm conclusions based on that sample. With a small sample, we may not be able to conclude anything. With larger samples, there is more chance that we can make firm conclusions in hypothesis testing situations, and in fact, with large samples we may find statistical significance even though the effect is minor and unimportant. The results of hypothesis testing should always be interpreted with the size of the sample in mind. We will learn more about this idea in the remainder of this section and in Chapter 13.

Cautions about Sample Size and Statistical Significance

- When there is a small to moderate difference between the null value and the true population value, a small sample has little chance of providing statistically significant support for the alternative hypothesis. The power of the test will be low.
- With a large sample, even a small and unimportant difference between the null value and the true population value may lead to a conclusion of statistical significance.

Example 12.18 **How the Same Sample Proportion Can Produce Different Conclusions** In Example 10.11 of Chapter 10, we used a confidence interval to analyze a taste test in which 55% of the 60 participants liked the taste of drink A better than the taste of drink B. On the basis of the 95% confidence interval for the "true" proportion that prefers drink A, we were unable to conclude whether a majority of the population prefers the taste of drink A.

We can also examine these data with a hypothesis test in which the null hypothesis is that there is no general preference for either drink. The null and alternative hypotheses are as follows:

H_0: $p = .5$ (no preference)
H_a: $p \neq .5$ (preference for one or the other)

where p represents the proportion in the population that would prefer drink A.

Now, suppose that a much larger sample size of 960 was used and that the result was still that 55% of the sample prefers drink A. What effect will the larger sample size have on the statistical significance of the data?

Figure 12.12 displays hypothesis test and confidence interval results (from using Minitab) for both sample sizes, assuming that the sample proportion remains at 55%. When the sample size is 60, the p-value is .439, so we cannot reject the null hypothesis as a possible truth. When the sample size is 960, the p-value is .002, so the result is *statistically significant*. In that case, we can reject the null hypothesis and conclude that a majority of the population would prefer drink A.

Test of p = 0.5 vs p ≠ 0.5

Results when n = 60 and sample p = 0.55

X	N	Sample p	95.0% CI	Z-Value	P-Value
33	60	0.55	(0.42, 0.68)	0.77	0.439

Results when n = 960 and sample p = 0.55

X	N	Sample p	95.0% CI	Z-Value	P-Value
528	960	0.55	(0.52, 0.58)	3.10	0.002

Figure 12.12 Taste test results for $n = 60$ and $n = 960$

The sample size also influences the width of the 95% confidence intervals for the true p. When the sample size is only 60, the 95% confidence interval for the true p is .42 to .68, an imprecise, wide range of estimates. When the sample size is 960, the 95% confidence interval is .52 to .58, a relatively precise range of estimates and an interval range that is entirely in the region defined by the alternative hypothesis.

Example 12.18 shows us again why it is not advisable to use the phrase *accept the null hypothesis.* The sample of only 60 participants was too small to be conclusive about whether we should reject the statement that the population proportion is .5. The larger sample size of 960 participants allowed a stronger conclusion, even though the sample proportion was the same in both samples.

Why is the same sample proportion more significant for a larger sample size? The conceptual, or intuitive, answer to this question is that with more data, we have more accurate sample estimates, and we reduce our uncertainty about what population values are likely. The technical answer is that increasing the sample size decreases the standard error of the sample proportion.

The z-value for the hypothesis test is

$$z = \frac{\text{Sample estimate} - \text{Null value}}{\text{Null standard error}}$$

For both of our hypothesis tests, the sample proportion is .55 and the null hypothesis value is .5, so the top portion of the calculation is the same for both cases. The z-value changes because the sample size affects the standard error.

- When $n = 60$, the null standard error $= \sqrt{\frac{.5(1 - .5)}{60}} \approx .065.$
- When $n = 960$, the null standard error $= \sqrt{\frac{.5(1 - .5)}{960}} \approx .016.$

Note that increasing the sample size decreases the null standard error. Remember that a standard error roughly measures the average difference between a statistic and a population parameter. If the null hypothesis that $p = .5$ were actually true, the consequence of taking a larger sample is that the sample proportion would be likely to fall closer to the null value (.5) than it would for a smaller sample. As a result, the difference between the sample proportion of .55 and the null value of .5 is relatively greater when the sample size is large than when it is small.

One more comment is in order about the example just presented. If in fact the null hypothesis were true, and $p = .50$, then a sample proportion as large as .55 would probably not have occurred in the larger sample. It is quite feasible for a sample of $n = 60$ to produce a sample proportion that differs from the truth by as much as .55 differs from .50. But it is quite unlikely for a sample of $n = 960$ to produce a sample proportion this far from the true proportion.

In summary, it is the size of the standard error that changes based on the size of the sample, and increasing the sample size gives a smaller standard error. As a result, a particular absolute difference between the sample proportion and null value is more significant with a large sample than with a small sample. No matter how large the sample, the sample proportion is likely to fall within two or so standard errors of the true population proportion. If the difference between the sample proportion and the null value is more than this, it is significant.

THOUGHT QUESTION 12.6 If the null hypothesis is true, the correct conclusion is "cannot reject the null hypothesis." If the alternative hypothesis is true, the correct conclusion is "reject the null hypothesis." Increasing the sample size increases the probability of making the correct conclusion in one of these cases, but not in the other. Your job is to figure out which is which. Here is a hint: The conclusion "cannot reject the null hypothesis" is made as long as the sample proportion falls within a reasonable number of standard errors of the null value. When the null hypothesis is true, does the likelihood of this happening change as the sample size increases? Explain, and use this explanation to determine which conclusion has an increased probability of being made correctly with increasing sample size.*

***HINT:** Think about the probability that the sample proportion falls within 2 standard errors of the null value p_0 when it is the true population value. Does that probability depend on the sample size?

Practical Importance versus Statistical Significance

The phrase *statistically significant* means only that the data are strong enough to reject a null hypothesis. A result that is significant in the statistical meaning of the word is not necessarily significant in the more common meaning of the word. One reason for this, of course, is that the research study might not address an important real-world issue. Even when the research issue is important, however, statistical significance does not guarantee practical significance.

A *p*-value provides information about the conclusiveness of the evidence against the null hypothesis, but it does *not* provide information about the *magnitude* of the effect. In some instances, the magnitude of a statistically significant effect can be so small that the practical effect is not important. If the sample size is large enough, almost any null hypothesis can be rejected because there is almost always at least a slight relationship between two variables, a slight difference between two groups, or a slight deviation from the status quo.

Example 12.19 **Birth Month and Height** The headline of a Reuters news article posted on the Yahoo Health News website on February 18, 1998, was "Spring Birthday Confers Height Advantage." The article describes an Austrian study of the heights of 507,125 military recruits. In an article in the journal *Nature*, the researchers reported their finding that men born in the spring were, on average, about 0.6 centimeter taller than men born in the fall (Weber et al., *Nature*, 1998, 391: 754–755). This is a small difference; 0.6 centimeter is only about 1/4 of an inch. The sample size for the study is so large that even a very small difference will earn the title *statistically significant*. Do you think the practical importance of this difference warranted the headline?

Example 12.20 **Case Study 1.7 Revisited: The Internet and Loneliness** We encountered the concept that a statistically significant result may not have practical importance when we read Case Study 1.7. That case study examined research done at Carnegie Mellon University in which the principal finding was that Internet usage may lead to feelings of loneliness and depression. This finding received a lot of media attention, but a close look at the study shows that the actual effects were quite small. For example, according to the *New York Times*, "one hour a week on the Internet was associated, on average, with an increase of 0.03, or 1 percent on the depression scale" (Harman, August 30, 1998, p. A3).

THOUGHT QUESTION 12.7 In this section, we noted that a research finding of "statistical significance" does not necessarily indicate that the finding is of practical importance. In situations in which the hypothesis test is about a specified value for a population parameter, explain why it would be helpful to present a confidence interval for the parameter along with the finding of statistical significance.*

Power and Sample Size for Testing One Proportion

Remember that the *power* of a statistical test is the probability that we decide in favor of the alternative hypothesis given a specific true population parameter value. As an example, suppose that we are testing hypotheses about the value of a population proportion. In Section 12.2 on page 473, we learned that in a one-sided test with a "greater than" alternative (and $\alpha = .05$), we can reject the null hypothesis if the test statistic (z) is 1.645 or larger. With this rejection region, the power of a test for a specific as-

***HINT:** Think about what information a confidence interval for a proportion provides.

sumed true value of the population proportion is the probability that the test statistic (z) will be 1.645 or greater.

For a specified true value and level of significance, we can either calculate the power for a given sample size or we can compute the sample size required to achieve a desired power. When researchers are planning a study, they might plan to take a sample large enough to achieve power of .80 or better for a reasonable guess of what the true population parameter value is.

Example 12.21 **Power and Sample Size for a Survey of Students** A university is considering a plan to offer regular classes year-round by making the summer session a regular term but will do so only if there is sufficient student interest. A university administrator is planning a student survey to determine whether a majority of all students at the school would attend summer session under that structure. The plan will be implemented only if a majority would attend, so the status quo is to assume that the proportion who would attend is not a majority. With p representing the proportion of all students who would attend in the summer, a hypothesis testing structure is as follows:

H_0: $p \leq .50$ (the proportion who would attend is not a majority)
H_a: $p > .50$ (a majority would attend)

Note that the alternative hypothesis includes a broad range of possibilities and does not specify an exact value. If the true population proportion that would attend is any value greater than .50, the correct decision is to pick the alternative hypothesis.

Table 12.2 shows the *power* for three different "true" proportions and for three different sample sizes, for a test with a significance level = .05. As you look at the table, keep in mind that the power is the probability that the sample evidence will lead us to conclude that a majority of the student population would attend a regular summer term. For example, suppose that in truth, 60% of students think they would attend a regular summer term and $n = 100$ students are surveyed. The probability that the sample proportion would be large enough to conclude that the population proportion $p > .50$ is given in Table 12.2 as .64. That means the probability is $1 - .64 = .36$ that a Type 2 error would be made, and the null hypothesis would not be rejected. Similarly, if the truth is that 52% think they would attend and a sample of 400 is chosen, the power is only .20, and the probability of a Type 2 error is .80.

Table 12.2 Power for Selected Sample Sizes and True Proportions, $\alpha = .05$

		True Population Proportion		
		.52	*.60*	*.65*
Sample Size	***n* = 50**	.09	.41	.69
	***n* = 100**	.11	.64	.92
	***n* = 400**	.20	.99	Nearly 1

Figure 12.13 shows **power curves** for the three sample sizes in this example. Each curve is for a different sample size. The horizontal axis gives possible true values of the population proportion and the vertical axis gives the power. These power values were determined for tests done with a .05 level of significance for tests of H_0: $p = .5$ versus H_a: $p > .5$. Don't worry about how to do the calculations. (Minitab was used to get the numbers for Table 12.2 and Figure 12.13.)

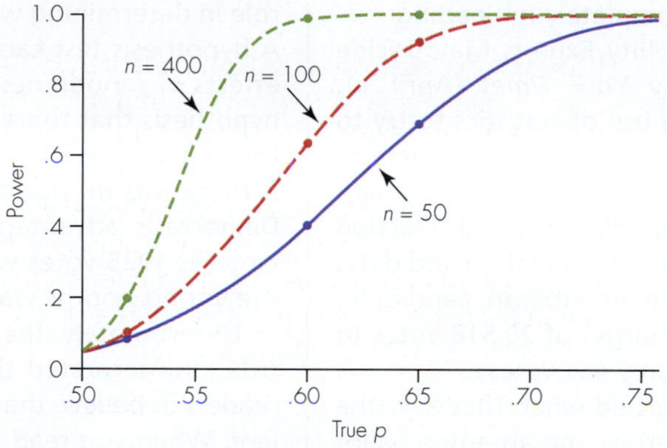

Figure 12.13 Power for testing H_0: $p = .5$ versus H_a: $p > .5$ as a function of true population value and sample size (level of significance = .05.)

Table 12.2 and Figure 12.13 illustrate what we learned about power in HT Module 0, Lesson 3:

- The power increases when the sample size is increased. We can see this by looking down any column of Table 12.2 and by comparing the three curves in Figure 12.10.
- The power increases when the difference between the true population value and the null hypothesis value increases. We can see this by looking across any row of Table 12.2 or by looking at how the power curves change when we move horizontally in Figure 12.10; remember that the null value is .50.

Researchers should evaluate power before they collect data that will be used to do hypothesis tests to make sure they have sufficient power to make the study worthwhile. In Example 12.21, we see that a sample of 50 students would have low power even if the true proportion who would attend in the summer was as high as .60 or even .65, a definite majority. If this hypothesis test is important to the administrator for making a decision, he or she should sample more than 50 students. We also see that if the true population proportion is .52, only a slight majority, there is little chance that the administrator will be able to decide in favor of the alternative, even if 400 students are surveyed.

12.4 Exercises are on pages 493–495.

CASE STUDY 12.1 An Interpretation of a *p*-Value Not Fit to Print

One of the most common mistakes that researchers and journalists make has to do with the interpretation of the *p*-value. We learned that a *p*-value does not tell us the probability that a hypothesis is true. When a *p*-value is erroneously interpreted in this manner, the resulting statements can be completely misleading.

In an article titled "Probability Experts May Decide Pennsylvania Vote," the *New York Times* (April 11, 1994, p. A15) reported on the use of statistics to try to decide if there was fraud in a special election held in Philadelphia. When a state senator from Pennsylvania's Second Senate District died in 1993, a special election was held to fill the seat. The Democratic candidate, William Stinson, defeated the Republican candidate, Bruce Marks, by the narrow margin of 20,518 votes to 20,057 votes, a difference of only 461 votes.

The Republicans were disturbed when they saw the comparison of voting booth results and absentee ballot results, shown in Table 12.3. The Republican Marks beat Stinson in the voting booths, but Stinson, the election winner, had a huge advantage in the absentee ballots. The Republicans charged that the election was fraudulent and asked the courts to disallow the absentee ballot votes on the basis of suspicion of fraud. In February 1994, 3 months after the election, Philadelphia Federal District Court Judge Clarence Newcomer disqualified all absentee ballots and ruled that the Marks should be seated. The Democrats appealed this ruling, and both sides hired statisticians to help sort out what might have happened.

Table 12.3 Election Results by Type of Vote

	Democrat Stinson	Republican Marks	Difference
Voting Booth	19,127	19,691	−564
Absentee	1391	366	+1025
Total	20,518	20,057	+461

One statistical expert, Orley Ashenfelter, examined 22 previous senatorial elections in Philadelphia to determine the relationship between votes cast in the voting booth and those cast by absentee ballot. Using the previous data, he calculated a regression equation to predict the difference in absentee ballot votes for the two parties based on a given amount of difference between the parties in the voting booth. (See Case Study 14.1 on pages 571–572 for details of the regression analysis.)

In the voting booth, there was a difference of 564 votes in favor of the Republicans. Using his equation, Ashenfelter estimated that when there was this much difference in favor of the Republicans in the voting booths, there would be a difference of 133 votes in favor of the Republicans in the absentee ballots. Instead, the difference for absentee ballots in the disputed election was 1025 votes in favor of the *Democrats*.

Of course, everyone knows that chance events play a role in determining what happens in any given election. A hypothesis test can be used to evaluate the possible effects of randomness. Ashenfelter considered the null hypothesis that there was no fraud and calculated that the *p*-value was 6%. In other words, he determined that *if there was no fraud,* the probability was 6% that the Democratic advantage in absentee votes would be as large as 1025 votes when the Republican advantage in the voting booths was 564 votes.

Unfortunately, the author of the *New York Times* article misinterpreted this probability in a way that leads readers to believe that the election was probably fraudulent. When you read the following quote from the article, see if you can detect the mistake in interpretation:

> More to the point, there is some larger probability that chance alone would lead to a sufficiently large Democratic edge on the absentee ballots to overcome the Republican margin on the machine balloting. And the probability of such a swing ... Professor Ashenfelter calculates, was about 6 percent. Putting it another way, if past elections are a reliable guide to current voting behavior, *there is a 94 percent chance that irregularities in the absentee ballots, not chance alone,* swung the election to the Democrat, Professor Ashenfelter concludes (Passell, 1994, p. A15, italics added).

The author's statement that there is a 94% chance that irregularities swung the election is wrong. He mistakenly interpreted the *p*-value to be the probability that the null hypothesis is true, so he reported what he thought to be the probability that the election was fraudulent. In this case, the *p*-value tells us only that if we assume that there was no fraud, there is a 6% chance that we would see results like these. The *p*-value cannot be used to find the probability that the election was fraudulent.

The case went through two appeals, but the original decision made by Judge Newcomer to disallow the absentee ballots was upheld each time. The statistical evidence was not a major factor in these decisions. The appeal judges found that there was enough evidence of irregular absentee voting procedures to justify Judge Newcomer's decision.

(Source: "Probability Experts May Decide Pennsylvania Vote," by P. Passell, The New York Times, *April 11, 1994, p. A15. [See p. 701 for complete credit.])*

Key Terms

Section 12.1
hypothesis testing, 451
significance test, 452
null hypothesis (H_0), 453
alternative hypothesis (H_a), 453
one-sided hypothesis test, 454
one-tailed hypothesis test, 454
two-sided hypothesis test, 454
two-tailed hypothesis test, 454
null value, 454, 458
test statistic, 456, 458
p-value, 456, 458
level of significance, 456
α-level, 456
statistically significant, 456
rejection region, 457
null standard error, 458
Bayesian statistics, 459
false positive, 459
false negative, 459
Type 1 error, 460
Type 2 error, 460
probability of a Type 1 error, 461
probability of a Type 2 error, 461
power, 462

Section 12.2
null value p_0, 464
z-test statistic for proportion, 464
z-test for a proportion, 464
p-value, test for proportion, 464–465
exact *p*-value, test for proportion, 470–471
critical value, 473
rejection region, test for a proportion, 473

Section 12.3
hypothesis test for comparing two population proportions, 476
z-test (hypothesis test) for difference in two proportions, 477
conditions for *z*-test for difference in two proportions, 477

Section 12.4
sample size and statistical significance, 481
practical importance vs. statistical significance, 483
power and sample size, 483–484
power curve, 485

In Summary Boxes

Interpreting a *p*-Value, 457
The Two Types of Error and Power, 463
The Steps for a *z*-Test for One Proportion, 470
Finding the Exact *p*-Value for a Binomial Proportion, 472
The Steps for a *z*-Test for the Difference in Two Proportions, 479

Exercises

◆ Denotes that the dataset is available on the companion website, http://www.cengage.com/statistics/Utts4e, but is not required to solve the exercise.

Bold exercises have answers in the back of the text.

Section 12.1

Exercises 12.1 to 12.40 correspond to the three Lessons in Section 12.1. Lesson 1 exercises are 12.1 to 12.18; Lesson 2 exercises are 12.19 to 12.28; Lesson 3 exercises are 12.29 to 12.40.

Lesson 1 Skillbuilder Exercises

12.1 One example of a possible hypothesis of interest given on pp. 451–452 is based on the fact that full moons occur on average every 29.53 days, and we might classify babies as to whether or not they were born during the 24-hour period surrounding a full moon. The hypothesis of interest is that the proportion of babies born during a full moon is 1/29.53, as it would be if births were uniformly distributed across days.

a. What is the population of interest in this situation?
b. Specify the population parameter of interest.
c. Write the hypothesis of interest using appropriate notation (introduced in Chapter 9).

12.2 One example of a possible hypothesis of interest given on p. 452 is whether or not the proportion favoring the death penalty is the same for teenagers as it is for adults.

a. What are the populations of interest in this situation?
b. Specify the population parameter of interest.
c. Write the hypothesis of interest using appropriate notation (introduced in Chapter 9).

12.3 Define the parameter $p_1 - p_2$ to be the difference in the proportions of 21-year-old men and women in the United States who have a high school diploma. Researchers are interested in the hypothesis that $p_1 - p_2 = 0$.

a. What are the populations of interest in this situation?
b. Write the hypothesis in words, being careful to specify the populations and the parameter of interest, by completing this sentence: "Researchers are interested in the hypothesis that..."

12.4 Retired professional tennis players Martina Navratilova, Monica Seles, John McEnroe, and Jimmy Connors are all left-handed. Define the parameter *p* to be the proportion of professional tennis players (current and retired) who are left-handed. Researchers are interested in the hypothesis that $p > .10$. (*Note:* About 10% of the *general* population is left-handed.)

a. What is the population of interest in this situation?
b. Write the hypothesis in words, being careful to specify the population and the parameter of interest, by completing this sentence: "Researchers are interested in the hypothesis that. . ."

12.5 Determine whether each of these statements is an example of a null hypothesis or an example of an alternative hypothesis.

a. The average weight of Canadian geese is the same as the average weight of Canadian warblers.

b. The proportion of books in the local public library that are novels is higher than the proportion of books in the university library that are novels.

c. The average price of wool jackets in New York City is lower in the summer than in the winter.

d. The proportion of students who receive A grades from Professor Harrington is the same as or higher than the proportion of students who receive A grades from Professor Cantor.

12.6 For each of the following situations, write the alternative hypothesis.

a. The null hypothesis is H_0: $p = .30$, and it is a two-sided hypothesis test.

b. The null hypothesis is H_0: $p \leq .45$.

c. The null hypothesis is H_0: $p \geq .60$.

12.7 In each part, is the statement a valid null hypothesis? If not, explain why not.

a. In a sample of students, the mean pulse rate for the men is equal to the mean pulse rate for the women.

b. The average weight of newborn boys is the same as the average weight of newborn girls.

c. The proportion of cars in California that are white is higher than the proportion of cars in Maine that are white.

12.8 State the null and alternative hypotheses for each of the following potential research questions. In each situation, also indicate whether the hypothesis test will be one-sided or two-sided.

a. Do female college students study more, on average, than male college students do?

b. Compared to men who are not bald, are bald men more likely to have heart disease?

c. Is there a correlation between height and head circumference?

12.9 State the null and alternative hypotheses for each of the following potential research questions. In each situation, also indicate whether the hypothesis test will be one-sided or two-sided.

a. Will an increase in the speed limits on interstate highways lead to an increase in the highway fatality rate?

b. When a coin is spun on its edge, is the probability that it will land heads up equal to .5?

Lesson 1 General Section Exercises

12.10 Suppose that the present success rate in the treatment of a particular psychiatric disorder is .65 (65%). A research group hopes to demonstrate that the success rate of a new treatment will be better than this standard. Use the letter p to represent the success rate of the new treatment. Write null and alternative hypotheses for p.

12.11 Suppose that a statistics teacher asks his students to each randomly pick one of the numbers 0, 1, 2, 3, 4. His general theory is that the proportion who pick the number 0 will be less than what would be expected from random selection. Use p to represent the population proportion that would pick the number 0.

a. What is the specific value of p that corresponds to true random selection?

b. In terms of p, write null and alternative hypotheses for this situation.

12.12 Answer Thought Question 12.1 on page 455.

12.13 Do you think researchers should determine whether to use a one-sided or two-sided hypothesis test before they look at the sample data or after they look at it? Explain.

12.14 For each of the following situations, write the null and alternative hypotheses in words and in symbols (as in Example 12.2, p. 453).

a. An anthropologist is trying to determine if the people in a certain region are descended from the same ancestors as those in another region she has studied. She knows that in the region she has studied, 15% of the people have a certain unique genetic trait. She plans to take a random sample of people in the new region and test them for that trait.

b. Paul likes a certain candy that comes in a bag with mixed colors, each with its own flavor. The candy company's website claims that 30% of these candies are red, which is Paul's favorite. He doubts their claim and thinks that fewer than 30% are red. He is willing to assume that the candy is randomly placed into bags, and he plans to buy several bags of candy and count the proportion of red pieces.

12.15 In the last census, it was determined that 6% of school-aged children in a certain state lived with their grandparent(s). Suppose that to support a bill on tax breaks for seniors, a congressional member plans to take a random sample of school-aged children to determine if that percentage has increased.

a. For this situation, write the null and alternative hypotheses in words and in symbols (as in Example 12.2, p. 453).

b. Explain what the population is and what the proportion of interest is in this situation.

12.16 Refer to Exercise 12.14. In each of the two situations, the hypotheses are about a population proportion p. Explain what the population is and what the proportion of interest is in each case.

12.17 Suppose that a null hypothesis, in words, is that the mean weight for the population of newborn babies is the same in the United States as it is in England.

a. Write the null hypothesis in symbols. Use Example 12.3 on p. 453 for guidance.

b. What is the null value in this situation?

12.18 Suppose that the sticker on a new car states that the car gets 32 miles per gallon for highway driving. Marisa wants to test that hypothesis.

a. Write the null hypothesis in symbols. Remember from Chapter 9 that the symbol for one population mean is μ.

b. What is the null value in this situation?

Lesson 2 Skillbuilder Exercises

12.19 State the conclusion that would be made in each of the following situations.

a. Level of significance = .05, p-value = .10.
b. Level of significance = .05, p-value = .01.
c. Level of significance = .01, p-value = .99.
d. Level of significance = .01, p-value = .002.

12.20 Consider this quote: "In a recent survey, 61 out of 100 consumers reported that they preferred plastic bags instead of paper bags for their groceries. If there is no difference in preference in the population, the chance of such extreme results in a sample of this size is about .03. Because .03 is less than .05, we can conclude that there is a statistically significant difference in preference." Give a numerical value for each of the following.

a. The p-value.
b. The level of significance, α.
c. The sample proportion.
d. The sample size.
e. The null value.

12.21 Consider testing the null hypothesis that there is no relationship between smoking and getting a certain disease versus the alternative hypothesis that smokers are more likely than nonsmokers to get the disease. Explain in words what would be concluded about the relationship in the population if

a. The p-value is .33, and the level of significance is .05.
b. The p-value is .03, and the level of significance is .05.

12.22 Given the convention of declaring that a result is "statistically significant" if the p-value is .05 or less, what decision would be made concerning the null and alternative hypotheses in each of the following cases? Be explicit about the wording of the decision.

a. p-value = .35.
b. p-value = .001.
c. p-value = .04.

12.23 Suppose that the value of a test statistic falls into the rejection region for the test. Will the p-value be less than the specified level of significance? Explain.

Lesson 2 General Section Exercises

12.24 Suppose that a woman thinks she might be pregnant, so she takes a pregnancy test. Considering this situation to be analogous to hypothesis testing, write each of the following in words.

a. The null hypothesis. (*Hint:* Remember that the null hypothesis generally states that there is nothing going on.)
b. The alternative hypothesis.
c. The conclusion if there is not enough evidence to reject the null hypothesis.
d. The conclusion if the null hypothesis is rejected.

12.25 About 10% of the human population is left-handed. Suppose that a researcher speculates that artists are more likely to be left-handed than are other people in the general population. The researcher surveys 150 artists and finds that 18 of them are left-handed.

a. Define the parameter of interest and give the null value.
b. State the researcher's null and alternative hypotheses.
c. What proportion of the sample of artists is left-handed?
d. To calculate a p-value for the hypothesis test, what probability should the researcher calculate? Make your answer specific to this situation.
e. Write the formula for the test statistic in words and substitute the appropriate values in the numerator.

12.26 An article in *USA Today* (Elias, June 3, 1999, p. D1) describes a study done by Georgia State University psychologist James Dabbs in which he found that women who are trial lawyers (litigators) are more likely to have male children than are women lawyers who are not trial lawyers. According to the article, "58% of litigators' kids were boys vs. 44% [boys] for the others. The odds of this happening by chance are less than 5%, he says."

a. How much difference was there between the percentages of male children for the two types of lawyers?
b. Consider the second sentence in the quote. What statistical term describes this result?
c. The quote from the article tells us that a p-value is less than 5%. Write one sentence that describes exactly what the p-value measures in this setting.
d. Dabbs believes that the result can be attributed to the women's testosterone levels because in a previous study, he found that, on average, women litigators have higher testosterone levels than women lawyers who aren't litigators. Do you think this conclusion is justified? Explain why or why not.

12.27 Two researchers are testing the null hypothesis that a population proportion p is .25 and the alternative hypothesis that $p > .25$. Both take a sample of 100 observations. Researcher A finds a sample proportion of .29, and Researcher B finds a sample proportion of .33. For which researcher will the p-value of the test be smaller? Explain without actually doing any computations.

12.28 A physician claims that as soon as his patients have a positive pregnancy test, he is generally able to predict the sex of the baby and that his probability of being right is greater than the one-half that would be expected if he were just guessing. A skeptic challenges his claim, and the physician decides to collect data for the next ten pregnancies to try to support his claim.

a. From the perspective of the physician, what are the null and alternative hypotheses for this situation?
b. Out of the ten pregnancies, he is correct six times. Write in words what probability needs to be computed to find the p-value.
c. The p-value for this test is .377. Does that prove that the skeptic is correct? Explain.
d. State the appropriate conclusion, using $\alpha = .05$.
e. Assuming that the physician really does have some ability to make these predictions, what would you recommend that he do to increase the likelihood that he can convince the skeptic?

Lesson 3 Skillbuilder Exercises

12.29 Explain whether each of the following statements is true or false.

a. The *p*-value is the probability that the null hypothesis is true.
b. If the null hypothesis is true, then the level of significance is the probability of making a Type 1 error.
c. A Type 2 error can only occur when the null hypothesis is true.
d. The probability of making a Type 1 error plus the probability of making a Type 2 error is always 1.

12.30 Can the two types of error both be made in the same hypothesis test? Explain.

12.31 Define events as follows: A = Null hypothesis is true. A^C = Alternative hypothesis is true. B = Null hypothesis is not rejected. B^C = Null hypothesis is rejected. For each of the following outcomes, explain whether a Type 1 error, a Type 2 error, or a correct decision has been made.

a. A and B.
b. A^C and B.
c. A and B^C.
d. A^C and B^C.

12.32 *This is also Exercise 1.20.* Explain what is meant by a "false positive" in the context of conclusions in statistical studies.

Lesson 3 General Section Exercises

12.33 A medical insurance company wants to know whether the proportion of its customers requiring a hospital stay during a year will decrease if it provides coverage for certain types of alternative medicine. The company conducts a 1-year study in which it gives insurance coverage for alternative medicine to 5000 randomly selected customers. Using the data from this study, the company tests the following hypotheses about the effect of offering the alternative medicine coverage to its customers:

H_0: proportion requiring a hospital stay will not decrease

H_a: proportion requiring a hospital stay will decrease

If the null hypothesis is rejected, the company will offer this coverage to all of its customers in the future.

a. Explain what a Type 1 error would be in this situation.
b. Explain what a Type 2 error would be in this situation.
c. Explain which type of error would be more serious for the insurance company.
d. Explain which type of error would be more serious for the customers.

12.34 Medical researchers now believe there may be a link between baldness and heart attacks in men.

a. State the null hypothesis and the alternative hypothesis for a study used to investigate whether or not there is such a relationship.
b. Discuss what would constitute a Type 1 error in this study.
c. Discuss what would constitute a Type 2 error in this study.

12.35 Consider medical tests, in which the null hypothesis is that the patient does not have the disease and the alternative hypothesis is that the patient has the disease.

a. Give an example of a medical situation in which a Type 1 error would be more serious.
b. Give an example of a medical situation in which a Type 2 error would be more serious.

12.36 Explain which type of error, Type 1 or Type 2, could be made in each of the following cases.

a. The null hypothesis is true.
b. The alternative hypothesis is true.
c. The null hypothesis is not rejected.
d. The null hypothesis is rejected.

12.37 A politician is trying to decide whether to vote for a new tax bill that calls for substantial reforms. A random sample of voters in his district leads him to believe the alternative hypothesis, H_a: $p > .5$, where p is the proportion of all voters in his district who support the bill. As a consequence, he decides to vote for the bill.

a. What would a Type 1 error be in this situation, and what would be the consequences for the politician?
b. What would a Type 2 error be in this situation, and what would be the consequences for the politician?
c. Explain which error would be more serious, if either, for the politician in this example.
d. Given the situation described, could the politician have made a Type 1 error or a Type 2 error?

12.38 Refer to Exercise 12.24, in which a woman thinks that she might be pregnant so she takes a pregnancy test. This situation is analogous to hypothesis testing.

a. Explain in words what a Type 1 error would be in this situation.
b. Explain in words what a Type 2 error would be in this situation.
c. Which type of error do you think is more serious in this situation? Explain.

12.39 A researcher is deciding whether to use a sample size of 100 or whether to increase the sample size to 200. Explain how this choice will affect the power of any hypothesis tests conducted using data from the resulting sample.

12.40 A researcher is deciding whether to use a level of significance equal to .05 or a level of significance equal to .01. Explain how this choice will affect the power of the hypothesis test.

Section 12.2

Skillbuilder Exercises

12.41 Find the *p*-value for each of these situations, taking into account whether the test is one-sided or two-sided.

a. z-statistic = 2.10, H_0: $p = .10$, H_a: $p \neq .10$.
b. z-statistic = 2.00, H_0: $p = .6$, H_a: $p < .6$.
c. z-statistic = −1.09, H_0: $p = .5$, H_a: $p < .5$.
d. z-statistic = 4.25, H_0: $p = .25$, H_a: $p > .25$.

12.42 Refer to Exercise 12.41. For each of parts (a) to (d), specify the rejection region for $\alpha = .05$, and then reach a conclusion for the test using the rejection region rule.

12.43 In each of the following, determine whether the conditions for conducting a z-test for a proportion are met. If not, explain why not.

a. Twenty students are randomly selected from the list of all sorority and fraternity members at a university to determine if a majority of sorority and fraternity students favor a new policy on alcohol on campus. The hypotheses are as follows:

H_0: $p = .50$

H_a: $p > .50$

b. Twenty employees of a large company are randomly selected to determine whether the proportion of company employees who are left-handed exceeds the national proportion of 10% who are left-handed. The hypotheses are as follows:

H_0: $p = .10$

H_a: $p > .10$

12.44 In each of the following, determine whether the conditions for conducting a z-test for a proportion are met. If not, explain why not.

a. A company employs 500 stockbrokers. They are all surveyed to find out whether a majority believes the market will go up in the next year.

H_0: $p = .50$

H_a: $p > .50$

b. A market research firm wants to know whether more than 30% of the people who visit a mall actually buy something. A researcher stands by the exit door starting at noon and asks 50 people as they are leaving whether they bought anything.

H_0: $p = .30$

H_a: $p > .30$

12.45 Refer to Exercise 12.44. In each of the parts, define in words the population parameter p.

12.46 For each of the following, calculate the z-statistic:

a. Sample size $n = 30$; sample proportion $\hat{p} = .60$.

H_0: $p = .50$

H_a: $p \neq .50$

b. Sample size $n = 60$; sample proportion $\hat{p} = .10$.

H_0: $p = .25$

H_a: $p < .25$

12.47 For each of the following, calculate the z-statistic:

a. Sample size $n = 500$; sample proportion $\hat{p} = .30$.

H_0: $p = .20$

H_a: $p > .20$

b. Sample size $n = 200$; sample proportion $\hat{p} = .50$.

H_0: $p = .80$

H_a: $p < .80$

12.48 Refer to Exercise 12.47. In each case, calculate the p-value for the test.

General Section Exercises

12.49 Explain the difference between the *null standard error* used in the denominator of a z-statistic for a test for a proportion and the *standard error* used in finding a confidence interval for a proportion.

12.50 In Exercise 12.25, we described a survey done to determine whether artists are more likely to be left-handed than others in the general population. If we use p to represent the proportion of all artists who are left-handed, the hypotheses are H_0: $p = .10$ and H_a: $p > .10$. The sample result was that 18 artists of 150 surveyed (or 12%) are left-handed.

a. From the given information, do we know if the conditions necessary to use a z-statistic for this test are met? Explain.

b. Conduct a hypothesis test using level of significance $\alpha = .05$. Clearly give the details of the five steps. Be sure to write a conclusion.

12.51 *Time* magazine reported that in a 1994 survey of 507 randomly selected adult American Catholics, 59% answered yes to the question "Do you favor allowing women to be priests?" (*Time*, 26 December–2 January 1995, pp. 74–76).

a. Set up the null and alternative hypotheses for deciding whether more than half of American Catholics in 1994 favored allowing women to be priests.

b. What conditions are necessary for using a z-statistic to test the hypotheses in part (a)? Are those conditions met here? Explain.

c. Compute the test statistic for this situation.

d. Calculate the p-value for the test.

e. On the basis of the p-value, make a conclusion for this situation. Use level of significance $\alpha = .05$. Write the conclusion in both statistical language and in words that someone with no training in statistics would understand.

f. Calculate a two-sided 95% confidence interval for the proportion of American Catholics in 1994 who favored allowing women to be priests. Does the interval support the conclusion that a majority of American Catholics in 1994 favored allowing women priests? Explain.

12.52 Suppose that a one-sided test for a proportion resulted in a p-value of .04. What would the p-value be if the test was two-sided instead?

12.53 Suppose that a two-sided test for a proportion resulted in a p-value of .07.

a. Using the usual $\alpha = .05$ criterion for hypothesis testing, would we conclude that the population proportion was different from the null hypothesis value? Explain.

b. Suppose that the test had been constructed as a one-sided test instead and that the sample proportion was in the direction to support the alternative hypothesis. Using the usual $\alpha = .05$ criterion for hypothesis testing, would we be able to decide in favor of the alternative hypothesis? Explain.

12.54 In Example 12.13 (p. 469), results were given for a survey of students in a statistics class who measured the lengths of their right and left feet to the nearest millimeter. The right and left foot measurements were equal for 103 of the 215 students. Assuming that this class is representative of all college students, is there evidence that the proportion of college students whose feet are the *same* length differs from one-half? Go through the five steps of hypothesis testing for this situation. Use $\alpha = .05$.

12.55 A Gallup poll released on October 13, 2000 (Chambers, 2000) found that 47% of the 1052 U.S. adults surveyed classified themselves as "very happy" when given the choices of "very happy," "fairly happy," or "not too happy." Suppose that a journalist who is a pessimist took advantage of this poll to write the headline "Poll finds that U.S. adults who are very happy are in the minority." If p = proportion of all U.S. adults who were very happy in 2000, go through the five steps of hypothesis testing and determine if the headline is justified. Use level of significance $\alpha = .05$. Be sure to comment on the headline in your conclusion in Step 5.

12.56 Describe how to find the exact p-value in each of the following situations. What probability distribution should be used? What is the interval of possible X values for which a probability must be found?

a. H_0: $p = .25$; H_a: $p < .25$; $n = 50$; $X = 17$ successes.
b. H_0: $p = .5$; H_a: $p \neq .5$; $n = 50$; $X = 17$ successes.

12.57 Describe how to find the exact p-value in a test of H_0: $p = .25$ versus H_a: $p > .25$ when the sample size is $n = 50$ and we have observed $X = 17$ successes. What probability distribution should be used? What is the interval of possible X values for which a probability must be found?

12.58 (*Computer software or calculator is required.*) A multiple-choice test consists of 15 questions with four choices each. The teacher wants to test the hypothesis that a student is just guessing versus the hypothesis that the probability of a correct answer on each question is higher than it would be if the student were guessing.

a. Specify the parameter of interest.
b. Write the null and alternative hypotheses.
c. If a student chooses the correct answer on eight of the 15 questions, what is the exact p-value for the hypothesis test?
d. What conclusion can be made about whether someone who got eight correct answers was guessing? Explain. Indicate the level of significance that you used in determining your conclusion.

12.59 (*Computer software or calculator is required.*) Refer to Exercise 12.57.

a. What is the exact p-value for that situation?
b. What conclusion can be made about the hypotheses? Explain.

Section 12.3

Skillbuilder Exercises

12.60 Find the p-value and draw a sketch of the p-value area for each of the following situations in which the value of z is the test statistic for the hypotheses given:

a. H_0: $p_1 - p_2 = 0$, H_a: $p_1 - p_2 \neq 0$; $z = 1.75$.
b. H_0: $p_1 - p_2 = 0$, H_a: $p_1 - p_2 \neq 0$; $z = -1.75$.

12.61 Find the p-value and draw a sketch of the p-value area for each of the following situations in which the value of z is the test statistic for the hypotheses given:

a. H_0: $p_1 - p_2 = 0$, H_a: $p_1 - p_2 > 0$; $z = 1.75$.
b. H_0: $p_1 - p_2 = 0$, H_a: $p_1 - p_2 > 0$; $z = -1.75$.

12.62 State a conclusion for each of the following situations, using $\alpha = .05$:

a. H_0: $p_1 - p_2 = 0$, H_a: $p_1 - p_2 < 0$; $z = 0.33$.
b. H_0: $p_1 - p_2 = 0$, H_a: $p_1 - p_2 > 0$; $z = 0.33$.

12.63 State a conclusion for each of the following situations, using $\alpha = .05$:

a. H_0: $p_1 - p_2 = 0$, H_a: $p_1 - p_2 < 0$; $z = -1.99$.
b. H_0: $p_1 - p_2 = 0$, H_a: $p_1 - p_2 > 0$; $z = 1.78$.

12.64 Refer to Exercise 12.62 and to the discussion of the rejection region approach to hypothesis testing (for z-tests) in Section 12.2. In each part (a and b), find the rejection region that would be used for $\alpha = .05$, and then state a conclusion for the z-statistic given.

12.65 Is the proportion of men who write a shopping list before going shopping less than the proportion of women who do so? Let p_1 and p_2 represent the population proportions of men and women, respectively, who use shopping lists. Suppose that a random sample of 50 men and 50 women were asked if they use a shopping list when they have more than ten items to buy. Of the 50 men, 20 said yes and of the 50 women, 30 said yes.

a. Write the null and alternative hypotheses in terms of p_1 and p_2, taking into account the initial question asked at the beginning of this exercise.
b. Provide numerical values for $\hat{p}_1$ and $\hat{p}_2$.
c. To find the null standard error, the combined sample proportion $\hat{p}$ is required. Compute $\hat{p}$ for this situation.
d. Find the test statistic for this situation.
e. Form a conclusion using the .05 level of significance.

12.66 Exercises 4.10 and 4.69 presented data on grumpy old men and heart disease. In Chapter 4, the goal was to determine if there was a relationship between the two variables. A one-tailed test may also make sense. We might hypothesize that if there is a difference in the probability of developing heart disease for men with no anger and men with lots of anger, the latter group would have a higher probability. Minitab output for testing this claim follows. Note that p_1 is the probability of developing heart disease for men with no anger and p_2 is the probability of developing heart disease for men with the most anger.

Sample	X	N	Sample p
1	8	199	0.040201
2	59	559	0.105546

Estimate for p(1) − p(2): −0.0653446
95% CI for p(1) − p(2): (−0.102675, −0.0280137)
Test for p(1) − p(2) = 0 (vs < 0): Z = −2.79, P-Value = 0.003

a. Write the null and alternative hypotheses in terms of p_1 and p_2, taking into account the claim being tested.

b. What is the value of the test statistic?

c. What conclusion would you reach using level of significance $\alpha = .01$? Write your conclusion in statistical terms and in the context of the problem.

General Section Exercises

12.67 ◆ Exercise 10.84 presented results of a survey asking college students whether they would return the money if they found a wallet on the street. Of the 93 women, 84 said they would, and of the 75 men, 53 said they would. Assume that these students represent all college students. Test the null hypothesis that equal proportions of college men and women would say that they would return the money versus the alternative hypothesis that a higher proportion of women would do so. Use $\alpha = .05$. (Data source: **UCDavis2** dataset on the companion website).

12.68 Exercise 10.85 presented a table in which men were classified according to whether or not they had a pierced ear and a tattoo. Of the 424 men with no pierced ear, 43 had a tattoo. Of the 141 men with at least one pierced ear, 42 had a tattoo. Assuming that these men are a random sample of college men, test the hypothesis that college men with at least one ear pierced are more likely to have a tattoo than college men with no ears pierced. Show all five steps of the hypothesis testing procedure. Be sure to specify the value of α that you chose to use.

12.69 A Gallup poll taken in May 2000 asked the question, "In general, do you feel that the laws covering the sale of firearms should be made: more strict, less strict, or kept as they are now?" Of the $n = 493$ men who responded, 52% said "more strict," while of the $n = 538$ women who responded, 72% said "more strict." Assuming that these respondents constitute random samples of U.S. men and women, is there sufficient evidence to conclude that a higher proportion of women than men in the population think that these laws should be made more strict? Show all five steps of the hypothesis testing procedure. Be sure to specify the value of α that you chose to use.

12.70 Refer to Exercise 12.69. The same poll asked the question, "Which of the following do you think is the primary cause of gun violence in America—the availability of guns, the way parents raise their children, or the influences of popular culture such as movies, television, and the Internet?" Fifty-one percent of the $n = 493$ men and 38% of the $n = 538$ women responded, "Way parents raise kids."

a. Using level of significance $\alpha = .05$, is there sufficient evidence to conclude that a higher proportion of men than women in the population think the "way parents raise kids" is the primary cause? Justify your answer.

b. Refer to the data in this exercise as well as the previous one. Note that 52% of the men thought that laws for firearm sales should be more strict, and 51% of men thought that the way parents raise kids is the primary cause of gun violence. Can the two-sample test for proportions covered in Section 12.3 be used with these data to test whether the corresponding population proportions differ? If so, carry out the test. If not, explain why not.

12.71 Case Study 10.3 (p. 392) reported on a survey of Penn State students asking the question, "Would you date someone with a great personality even though you did not find them attractive?" Of the $n = 131$ women, 61.1% said yes, and of the $n = 61$ men, 42.6% said yes. Is there sufficient evidence to conclude that for the populations represented by these students, a higher proportion of women than men would answer yes to this question? Use $\alpha = .05$.

12.72 The headline for an article in the *Sacramento Bee* read, "Women appear to be better investors than men in study" (Jack Sirard, April 24, 2005, p. D1). The conclusion was based on a telephone poll of 500 men and 500 women. One quote in the article was "men are much more likely to stick with a losing investment than women (47 percent to 35 percent)."

a. Assuming that there was no preconceived idea of whether men or women would be more likely to stick with a losing investment, state the null and alternative hypotheses for testing whether the population proportions of men and women who would stick with a losing investment differ.

b. Carry out the remaining steps to test the hypotheses that you stated in part (a).

12.73 Exercise 4.11 described a case-control study comparing short to not-short English secondary school students. Of $n = 92$ short students, 42 said they had been bullied in school. Of $n = 117$ not-short students, 30 said they had been bullied in school. Here is Minitab output for comparing two proportions:

Sample	Bullied	N	Sample p
Short	42	92	0.456522
Not Short	30	117	0.256410

Estimate for p(1) − p(2): 0.200111
95% lower bound for p(1) − p(2): 0.0919200
Test for p(1) − p(2) = 0 (vs > 0): Z = 3.02
P-Value = 0.001

a. The researchers wanted to know whether short students are bullied more often. Write the null and alternative hypotheses for this question, defining and using appropriate notation.

b. Use the Minitab output to carry out the remaining steps of the hypothesis test. Use $\alpha = .05$.

c. The Minitab procedure found a one-sided 95% confidence interval for the difference in the two proportions, and the lower end of the interval is given in the output. What is the upper end of the interval? Write the complete interval.

d. Use the one-sided 95% confidence interval to test the hypotheses stated in part (a).

Section 12.4

Skillbuilder Exercises

12.74 Suppose that the null and alternative hypotheses in a test are as follows:

H_0: $p = .70$

H_a: $p \neq .70$

For each of the following sample sizes, calculate the null standard error.

a. $n = 40$.
b. $n = 100$.
c. $n = 500$.
d. $n = 1000$.

12.75 Refer to Exercise 12.74. In each case, suppose the sample proportion is $\hat{p} = .75$ and compute the test statistic z and the p-value. Then make a conclusion using a .05 level of significance. Comment on the relationship between the sample size and the conclusions.

12.76 Sometimes a result that is statistically significant does not have practical significance. Is this more likely to happen when the sample size is very large or when it is very small? Explain.

12.77 Researchers know that 20% of a certain ethnic group has a distinctive trait. They want to know whether the proportion with the trait is higher in a second ethnic group. They take a random sample of n members of the second group and find the sample proportion $\hat{p}$ with the trait. The parameter of interest is p, the population proportion with the trait for the second group. In each of the following scenarios, explain which situation is more likely to lead to a statistically significant finding, assuming that all other conditions remain the same and it's true that $p > .20$.

a. $n = 100$ or $n = 1000$.
b. $p = .25$ or $p = .45$.
c. $\hat{p} = .25$ or $\hat{p} = .45$.

12.78 One possible problem in hypothesis testing is that researchers fail to find a statistically significant result even though the null hypothesis is false. Is this problem more likely to occur with a very small sample or with a very large sample? Explain.

General Section Exercises

12.79 Which do you think is more informative when you are given the results of a hypothesis test — the p-value or the decision about whether to reject the null hypothesis? Explain.

12.80 In reporting the results of a study, explain why a distinction should be made between "statistical significance" and "significance" as the term is used in ordinary language.

12.81 Suppose you were to read that a new study based on 100 men had found that there was *no difference* in heart attack rates for men who exercised regularly and men who did not. Do you think the study found *exactly* the same rate of heart attacks for the two groups of men? In the context of the material in this chapter, what would you suspect was the reason for that reported finding?

12.82 Which do you think is more informative when you are given the results of a study, a confidence interval or a p-value? Explain.

12.83 An advertisement for Claritin, a drug for seasonal nasal allergies, made this claim: "Clear relief without drowsiness. In studies, the incidence of drowsiness was similar to placebo" (*Time*, February 6, 1995, p. 43). The advertisement also reported that 8% of the 1926 Claritin takers and 6% of the 2545 placebo takers reported drowsiness as a side effect. A one-sided test of whether a higher proportion of Claritin takers than placebo takers would experience drowsiness in the population results in a p-value of about .005.

a. From this information, would you conclude that the incidence of drowsiness for the Claritin takers is statistically significantly higher than for the placebo takers?
b. Does the answer to part (a) contradict the statement in the advertisement that the "incidence of drowsiness was similar to placebo"? Explain.
c. Use this example to discuss the importance of making the distinction between the common use and the statistical use of the word *significant*.

12.84 Suppose that a study is designed to choose between the following hypotheses:

Null hypothesis: Population proportion is .20.

Alternative hypothesis: Population proportion is higher than .20.

On the basis of a random sample of size 400, the sample proportion is .25.

a. Compute the z-score corresponding to the sample proportion of .25, assuming the null hypothesis is true.
b. What is the p-value for the standardized score computed in part (a)?
c. On the basis of the result from part (b), form a conclusion. Be explicit about the wording of your conclusion and justify your answer. Indicate the α that you chose to use.
d. Suppose that the sample size had only been 100, rather than 400, but the sample proportion was again .25. Using this smaller sample size, repeat parts (a) to (c).
e. Explain what this problem illustrates about the effect of sample size on statistical significance.

12.85 Minitab can provide power for a variety of situations. Suppose that a test for ESP has four choices, and that the probability of a correct guess by chance on each trial is .25. A researcher believes that the true probability of a correct guess is .33. The following output shows the power of the one-sided test for this situation for three possible sample sizes:

```
Test for One Proportion
Testing proportion = 0.25 (versus > 0.25)
Calculating power for proportion = 0.33
Alpha = 0.05 Difference = 0.08

Sample
Size        Power
  50        0.3776
 100        0.5740
 400        0.9705
```

a. What is the power of the test if $n = 50$ trials are used?
b. Write a sentence providing the power of the test for $n = 100$ and explaining its meaning.
c. If the researcher wants to have at least a .95 probability of detecting ESP in the study, and is correct that the true probability of a success is .33, would a sample of size 400 be sufficient? Explain.

d. If the true probability of success is actually .40 on each trial, would the power for each sample size be higher or lower than that shown in the output? Explain.

12.86 Refer to Case Study 1.6, comparing heart attack rates for men who had taken aspirin or a placebo. Suppose that the observed proportions of .017 and .0094 are actually the correct population proportions who would have heart attacks with a placebo and with aspirin. The following Minitab output shows the power of a one-sided test for two proportions for this situation for three sample sizes. The samples are the number of participants in *each* group (aspirin and placebo). Suppose you are the statistician advising a research team about conducting a new study to confirm the results of the old study. The researchers comment that samples of size 500 in each condition should be sufficient, since the effect is obviously so strong, based on the small *p*-value for the previous study. What would you advise? Explain.

```
Testing proportion 1 = proportion 2 (versus >)
Calculating power for proportion 1 = 0.017 and proportion 2 = 0.0094
Alpha = .05 Difference = 0.0076

Sample
Size       Power
  500      0.2768
 1000      0.4380
 3000      0.8250
```

Chapter Exercises

12.87 Refer to the five steps of any hypothesis test, given in Section 12.1. Which step(s) are different if the rejection region approach is used instead of the *p*-value approach?

12.88 Consider the two conditions necessary for conducting a *z*-test for a proportion. One is a sample size condition, and the other is a condition about how the sample was collected. For each of the two conditions, explain why it is required in order for the test and the results to be valid.

12.89 *This is also Exercise 1.17.* A randomized experiment was done in which overweight men were randomly assigned to either exercise or go on a diet for a year. At the end of the study there was a statistically significant difference in average weight loss for the two groups. What additional information would you need in order to determine if the difference in average weight loss had *practical* importance?

12.90 A student has been accused of cheating on an examination by copying another student's paper, and you have been asked to serve on the panel that must decide the student's fate. If the student is found guilty, he will fail the course and will have to write an essay on honesty for the school paper.

a. What are the null and alternative hypotheses for this situation?
b. What is a Type 1 error in this situation, and what are the consequences?
c. What is a Type 2 error in this situation, and what are the consequences?
d. As a member of the panel making the decision, you must evaluate the seriousness of the two types of error. Explain which type you think is more serious.

12.91 In a survey of 240 students in an elementary statistics class at the University of California-Davis (UCD), 20 said that they were left-handed and 220 said that they were right-handed. Assume that the students are representative of all students at the school. Does this provide evidence that the proportion of UCD students who are left-handed differs from the national proportion of .10? Carry out the five steps of the hypothesis test for this situation.

12.92 Refer to the five steps in any hypothesis test listed in Section 12.1. Only one of the steps can be performed before the data are collected. Which step is it?

12.93 Give an example of a possible hypothesis of interest about the difference between two population proportions. Be sure to specify the populations and the population parameter the hypothesis is about.

12.94 Give an example of a possible hypothesis of interest about a population proportion. Be sure to specify the population and the population parameter that the hypothesis is about.

12.95 Max likes to keep track of birthdays of people he meets. He has 170 birthdays listed on his birthday calendar. One cold January night, he comes up with the theory that people are more likely to be born in October than they would be if all 365 days were equally likely. He consults his birthday calendar and finds that 22 of the 170 birthdays are in October.

a. Write down the null and alternative hypotheses for Max's test. Be sure to carefully specify what *p* represents.
b. Carry out the remaining four steps of the hypothesis test for this situation.

12.96 Specify the null and alternative hypotheses in words for each of the following research questions.

a. Does listening to Mozart increase performance on an intelligence test?
b. Does talking to plants result in better growth for the plants?
c. Does drinking fluoridated water lead to increased bone fractures in elderly people?

12.97 For the following two situations, define the population parameter in words. Then specify the null and alternative hypotheses in symbols.

a. For people suffering from a certain type of chronic pain, 70% experience temporary relief when they take a standard medication. If a new medication appears to be more effective, it will replace the standard. Does the new medication provide pain relief to a larger proportion than the standard medication?
b. A student at a large university is thinking about starting a new service in which students can order their textbooks from his website and he will buy them at the bookstore and deliver them for a small service charge. He plans to survey a random sample of students and ask them whether or not they would be willing to pay for such a service. If he is convinced that more than 5% of students would use his service, he will go forward with it.

12.98 Refer to Exercise 12.97. For each of the two situations described, explain the consequences of a Type 1 error and the consequences of a Type 2 error. For part (a), explain which type would be more serious for the patients, and for part (b) explain which type would be more serious for the student who is planning the new service.

12.99 For a one-sided hypothesis test for a proportion in which the alternative hypothesis is H_a: $p < p_0$, for what values of the test statistic z will the p-value be *greater* than .5? Explain your answer.

12.100 For a test of H_0: $p = .25$ versus H_a: $p > .25$, for what range of values of the sample proportion $\hat{p}$ will the p-value for the test be *greater* than .5? Explain your answer.

12.101 ◆ In Chapter 2, we saw data from a statistics class activity in which students were asked to "randomly" pick a number (integer) from 1 to 10. Of the 190 students, 56 picked the number 7. Carry out the five steps of hypothesis testing to determine if people similar to these students are more likely to pick the number 7 than would be expected by chance if all numbers were equally likely (Data source: **pennstate1** dataset on the companion website).

12.102 A professor planned to give an examination in a large class on the Monday before Thanksgiving vacation. Some students asked whether he could change the date because so many of their classmates had at least one other exam on that date. They speculated that at least 40% of the class had this problem. The professor agreed to poll the class, and if there was convincing evidence that the proportion with at least one other exam on that date was greater than .40, he would change the date. Of the 250 students in the class, 109 reported that they had another exam on that date.

a. What proportion of the class reported that they had another exam on that date?
b. Is the proportion you found in part (a) a sample proportion or a population proportion?
c. The professor conducted a z-test of H_0: $p = .40$ versus H_a: $p > .40$ and found $z = 1.16$ and p-value $= .123$. He said that he would not move the exam because the null hypothesis cannot be rejected, and there is not convincing evidence that the population proportion is greater than .40. What is wrong with his reasoning? [*Hint:* Refer to part (b).]

12.103 In a survey of students at the University of California, Davis (UCD), students were asked which of two popular soft drinks they preferred—let's call them brand C and brand P. Of the 159 respondents, 80 preferred brand P and 79 preferred brand C. Assuming that these students represent all students at UCD, test whether there is a preference for one drink over the other for students at UCD.

12.104 According to *USA Today* (Snapshot, April 20, 1998, also referenced in Exercise 9.143), a random sample of 8709 adults taken in 1976 found that 9% believed in reincarnation. A poll of 1000 adults in 1997 found that 25% believed in reincarnation. Test the hypothesis that the proportion of adults believing in reincarnation in 1997 was higher than the proportion in 1976.

12.105 Refer to Exercise 12.103. Some students were asked if they preferred C or P, while other students were asked if they preferred P or C. Of the 159 respondents, 86 responded with the first drink presented. Test the hypothesis that the first drink presented is more likely to be selected than the second drink presented.

12.106 Refer to Example 12.11 (p. 466), in which convincing evidence was found for the importance of order in a "random" selection between two letters. Suppose that a student in your class ran for president of the campus student association and lost. After reading Example 12.11, she realized that because her name is at the end of the alphabet, she was listed after all of her opponents on the ballot and the election was unfair to her. She demands a reelection with names listed in random order.

a. Which type of error, Type 1 or Type 2, could have been made in Example 12.11?
b. If the type of error that you specified in part (a) was made, is the reelection justified? Explain.
c. Now suppose the other type of error had been made, and the student read the results of the study in that case instead. Explain what the consequences would be for the student and the student government.
d. Which type of error do you think is more serious for the student who ran for president? Explain.

12.107 A Gallup poll taken on a random sample of Canadian adults in February 2000 asked the question, "Do you favour or oppose marriages between people of the same sex?" A similar poll was taken in April 1999 (Edwards and Mazzuca, 2000). The Minitab output for the "two proportions" procedure (with the "pooled estimate for the test") is as follows, where Sample 1 is the February 2000 poll and Sample 2 is the April 1999 poll. X is the number who answered "favour":

```
Sample        X       N     Sample p
1           431    1003     0.429711
2           360    1000     0.360000
Estimate for p(1) - p(2): 0.0697109
95% CI for p(1) - p(2): (0.0270067, 0.112415)
Test for p(1) - p(2) = 0 (vs not = 0): Z = 3.19
P-Value = 0.001
```

a. Define appropriate notation and write the null and alternative hypotheses to test whether the proportion of adult Canadians who favor marriages between people of the same sex was different in April 1999 and February 2000.
b. Using the Minitab output, go through the remaining four steps of hypothesis testing to test the hypotheses that you defined in part (a).
c. Use the confidence interval given in the Minitab output to test the hypotheses in part (a).

(Source: Copyright © The Gallup Organization)

12.108 Refer to Example 12.17 (pp. 476 and 478), comparing the proportion of children with ear infections while taking xylitol or a placebo. Read the discussion of Type 1 and Type 2 errors, and explain the consequences of each type of error if it were to have been made in this study.

12.109 Case Study 1.6 presented data on 22,071 physicians who were randomly assigned to take aspirin or a placebo every other day for five years. Of the 11,037 taking aspirin, 104 had a heart attack, while of the 11,034 taking placebo, 189 had a heart attack.

Testing Hypotheses about Means

To evaluate the quality of a research study, consider such features as sampling procedure, methods used to measure individuals, wordings of questions, and possible lurking variables. When conclusions are based on hypothesis tests, additional issues and cautions discussed in this chapter should be considered.

In this chapter, we continue our discussion of hypothesis testing, initiated in Chapter 12. Make sure that you are familiar with HT Module 0, "An Overview of Hypothesis Testing" (Section 12.1), before covering this chapter.

Before we delve into the details, it is worth reviewing three cautions we have already encountered concerning the use of data to make inferences beyond the sample:

1. Inference is only valid if the sample is representative of the population for the question of interest.
2. Hypotheses and conclusions apply to the larger population(s) represented by the sample(s).
3. If the distribution of a quantitative variable is highly skewed, we should consider analyzing the median rather than the mean. Methods for testing hypotheses about medians are a special case of **nonparametric methods**, covered as Supplemental Topic 2 (on the companion website, http:/www.cengage.com/statistics/Utts4e) and briefly in Section 16.3.

13.1 Introduction to Hypothesis Tests for Means

In Chapter 12, we learned the logic and basic steps of hypothesis testing and applied them to *z*-tests for one proportion and the difference in two proportions. In this chapter, we will look at the details of hypothesis tests for the same three situations for which we developed confidence intervals in Chapter 11, all of which involve population means:

- Hypotheses about one population mean, μ. For example, we might wish to determine whether the mean body temperature for humans is 98.6 degrees Fahrenheit.
- Hypotheses about the population mean difference for paired data, μ_d. For example, we might ask whether the mean difference in height between college men and their fathers is 0. The data for each pair is "son's height − father's height," and we are interested in the mean of those differences for the population of father-son pairs in which "son" is a college student.
- Hypotheses about the difference between the means of two populations, $\mu_1 - \mu_2$. For example, we might ask whether the mean time of television watching per week is the same for men and women.

Tests to answer these kinds of questions are sometimes called **significance tests**.

Note to Readers: HT Module 0 in Chapter 12 must be covered before covering Chapter 13.

The terms **hypothesis testing** and **significance testing** are synonymous. The term *significance testing* arises because the conclusion about whether to reject a null hypothesis is based on whether a sample statistic is "significantly" far from what would be expected if the null hypothesis were true in the population. We declare **statistical significance** and reject the null hypothesis if there is a relatively small probability that an observed difference or relationship in the sample would have occurred if the null hypothesis holds in the population.

The significance tests for the three situations that we examine in this chapter all have the same general format. This basic format was introduced in Chapter 12 and is as follows:

- *The null hypothesis defines a specific value of a population parameter, called the null value.* In Example 13.1, for instance, the null hypothesis is that the mean normal body temperature of young adults is $\mu = 98.6$ degrees Fahrenheit.
- *A relevant statistic is calculated from sample information and summarized into a "test statistic."* In each case, we begin by determining the sample equivalent of the parameter of interest. For instance, we use $\bar{x}$ = mean of a sample of temperature measurements to estimate μ = population mean temperature. We then measure the difference between the sample statistic and the null value using the **standardized statistic**, which for hypotheses involving means is:

$$t = \frac{\text{Sample statistic} - \text{Null value}}{\text{Null standard error}}$$

  For hypotheses about means, the *standardized statistic* is called a ***t*-statistic**, and the *t*-distribution is used to find the *p*-value, whereas for hypotheses about proportions, the *standardized statistic* is called a *z*-statistic, and the standard normal distribution is used to find the *p*-value.
- *A p-value is computed on the basis of the standardized "test statistic."* The *p*-value is calculated by temporarily assuming the null hypothesis to be true and then calculating the probability that the test statistic could be as large in magnitude as it is (or larger) in the direction(s) specified by the alternative hypothesis.
- *On the basis of the p-value, we either reject or fail to reject the null hypothesis.* The most commonly used criterion (level of significance) is that we reject the null hypothesis when the *p*-value is less than .05. In many research articles, *p*-values are simply reported, and readers are left to draw their own conclusions. Remember that a *p*-value measures the strength of the evidence *against* the null hypothesis, and the smaller the *p*-value, the *stronger* the evidence against the null hypothesis and *for* the alternative hypothesis.

These general ideas are consistent with the five steps outlined in Chapters 4 and 12 that are used for *any* hypothesis test. Let's review those five steps:

Step 1: Determine the *null* and *alternative* hypotheses.

Step 2: Verify necessary data conditions, and if they are met, summarize the data into an appropriate *test statistic.*

Step 3: Assuming that the null hypothesis is true, find the *p-value.*

Step 4: Decide whether the result is *statistically significant* based on the *p-value.*

Step 5: Report the conclusion in the context of the situation.

13.1 Exercises are on pages 535–536.

13.2 HT Module 3: Testing Hypotheses about One Population Mean

For questions about **the mean** of a quantitative variable **for one population**, the null hypothesis typically has the form

$H_0: \mu = \mu_0$ (population mean μ is a specified value called μ_0)

The alternative may either be one-sided (H$_a$: $\mu < \mu_0$ or H$_a$: $\mu > \mu_0$) or two-sided (H$_a$: $\mu \neq \mu_0$), depending on the research question of interest. The usual procedure for testing these hypotheses is called a **one-sample *t*-test** because the *t*-distribution is used to determine the *p*-value. Let's look at an example of a situation in which a one-sample *t*-test would be used.

Example 13.1 **Normal Body Temperature for Young Adults** What is normal body temperature? A paper published in the *Journal of the American Medical Association* presented evidence that normal body temperature may be less than 98.6 degrees Fahrenheit, the long-held standard (Mackowiak et al., 1992). The value 98.6 degrees seems to have come from determining the mean in degrees Celsius, rounding up to the nearest whole degree (37 degrees), and then converting that number to Fahrenheit using $32 + (1.8)(37) = 98.6$. Rounding up may have produced a result higher than the actual average, which may therefore be lower than 98.6 degrees. To test this, the null hypothesis is $\mu = 98.6$, and we are only interested in rejecting in favor of lower values, so the alternative is $\mu < 98.6$. We can write these hypotheses as follows:

H$_0$: $\mu = 98.6$
H$_a$: $\mu < 98.6$

The dataset ***bodytemp*** on the companion website, http://www.cengage.com/statistics/Utts4e, includes body temperatures for healthy adults, taken at a blood bank when they were donating blood. Because body temperature drops with age, let's consider only the young adults in the sample, that is, those under age 30. There were $n = 16$ such individuals, and their body temperatures were:

98.4, 98.6, 98.8, 98.8, 98.0, 97.9, 98.5, 97.6, 98.4, 98.3, 98.9, 98.1, 97.3, 97.8, 98.4, 97.4

For these data, the sample mean is $\bar{x} = 98.20$ degrees Fahrenheit. This is lower than 98.6, but to determine whether this is a statistically significant difference from 98.6, the question we must answer is: What is the probability that the sample mean for 16 individuals could be 98.20 or less if the population mean is actually 98.6 (the null value)?

The answer is the *p*-value that we will use to decide between the two hypotheses. We will return to this example after we provide some details for the five steps of a one-sample *t*-test, but we won't keep you in suspense. The *p*-value for this example is .003, so the data strongly support the alternative hypothesis.

The Five Steps for a One-Sample *t*-Test

For a one-sample *t*-test, details of the five steps in the hypothesis test are as follows:

Step 1: Determine the Null and Alternative Hypotheses

Remember that in general, the null hypothesis states the status quo and the alternative hypothesis states the research question of interest. For a one-sample *t*-test, the possible null and alternative hypotheses are one of these three choices, depending on the research question:

(1) H$_0$: $\mu = \mu_0$ versus H$_a$: $\mu \neq \mu_0$ (two-sided)
(2) H$_0$: $\mu = \mu_0$ versus H$_a$: $\mu < \mu_0$ (one-sided)
(3) H$_0$: $\mu = \mu_0$ versus H$_a$: $\mu > \mu_0$ (one-sided)

Sometimes, when the alternative hypothesis is one-sided, the null hypothesis is written to include values in the other direction from the alternative hypothesis as well as equality, for instance, H$_0$: $\mu \leq \mu_0$. But the "null value" used when computing the test statistic is always μ_0, and the alternative hypothesis never includes the equals sign.

Step 2: Verify Necessary Data Conditions, and If They Are Met, Summarize the Data into an Appropriate Test Statistic

Recall from the previous three chapters that whenever we carry out a statistical inference procedure, we must check to see whether certain data conditions are met. These conditions are connected to the assumptions that were made when the theory was

HT Module 3: μ

derived for the particular situation. Carrying out a t-test for one population mean requires that one of the same two situations holds as for constructing a confidence interval for a mean, given in Chapter 11. Here are those situations, slightly modified from the way in which they were stated in Chapter 11.

Situations for Which a One-Sample *t*-Test Is Valid

Situation 1: The population of the measurements of interest is approximately normal, and a random sample of any size is measured. In practice, the method is used as long as there is no evidence that the shape of the population is notably skewed or that there are extreme outliers.

Situation 2: The population of measurements of interest is not approximately normal, but a *large* random sample is measured. Thirty is usually used as an arbitrary demarcation of "large," but if there are extreme outliers or extreme skewness, it is better to have an even larger sample size than $n = 30$.

Use a histogram, dotplot, or stem-and-leaf plot to examine the shape of the sample data and to check for notable skewness or extreme outliers. A boxplot can reveal outliers and will sometimes reveal skewness, but it cannot be used to determine the shape otherwise. Checking the relative sizes of the sample mean and sample median might also be useful. Skewness and outliers both can cause these two statistics to differ from each other.

If either skewness or outliers are present in a small or moderate-sized sample, it could mean that these characteristics are also present in the population, and a one-sample t-test should not be used unless a large sample can be obtained. In this event, a statistical test called the *sign test* can be used to analyze hypotheses about the *median*. We provide the details of the sign test in Supplemental Topic 2 on the companion website, http://www.cengage.com/statistics/Utts4e. The procedure is available in most statistical software.

For a large sample size, the t-test will work well for testing hypotheses about the mean even if there is some skewness. However, when the data are obviously skewed, you should question whether the mean is the appropriate parameter to analyze. For skewed data, the median may be a better measure of location, and test procedures for analyzing the median may be more appropriate than the t-test.

Continuing Step 2: The Test Statistic for a One-Sample *t*-Test

The next part of step 2 is to compute a test statistic. The statistic $\bar{x}$, the sample mean, estimates the population mean μ. The "null standard error" for the denominator is the usual standard error of $\bar{x}$, s.e.$(\bar{x}) = s/\sqrt{n}$. Because this standard error does not depend on the null value μ_0, the word *null* can be omitted from the description of the denominator standard error. The test statistic, called the t-statistic, is a standardized score for measuring the difference between the sample mean and the null hypothesis value of the population mean and is written as

$$t = \frac{\text{Sample mean} - \text{Null value}}{\text{Standard error}} = \frac{\bar{x} - \mu_0}{s/\sqrt{n}} = \frac{\sqrt{n}(\bar{x} - \mu_0)}{s}$$

This particular t-statistic has approximately a t-distribution with df $= n - 1$.

Step 3: Assuming That the Null Hypothesis Is True, Find the *p*-Value

Using the t-distribution with df $= n - 1$, the p-value is the area in the tail(s) beyond the test statistic t, as follows:

- For H_a: $\mu < \mu_0$ (a one-sided test), the p-value is the area below t, even if t is positive.
- For H_a: $\mu > \mu_0$ (a one-sided test), the p-value is the area above t, even if t is negative.
- For H_a: $\mu \neq \mu_0$ (a two-sided test), the p-value is $2 \times$ area above $|t|$.

Table 13.1 summarizes and illustrates this information. The p-value is found in exactly the same way for the other hypothesis-testing situations considered in this chapter, including tests for the population mean difference (paired data) and the difference in population means (independent samples).

Table 13.1 Alternative Hypothesis Regions and *p*-Value Areas

Statement of H_a	*p*-Value Area	*t*-Curve Region
$\mu < \mu_0$ (less than)	Area to the left of t (even if $t > 0$)	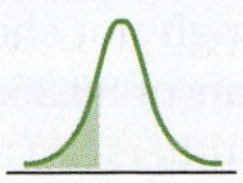
$\mu > \mu_0$ (greater than)	Area to the right of t (even if $t < 0$)	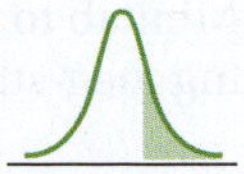
$\mu \neq \mu_0$ (not equal)	$2 \times$ area to the right of $\|t\|$	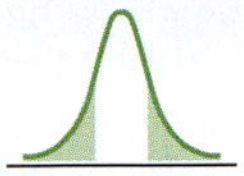

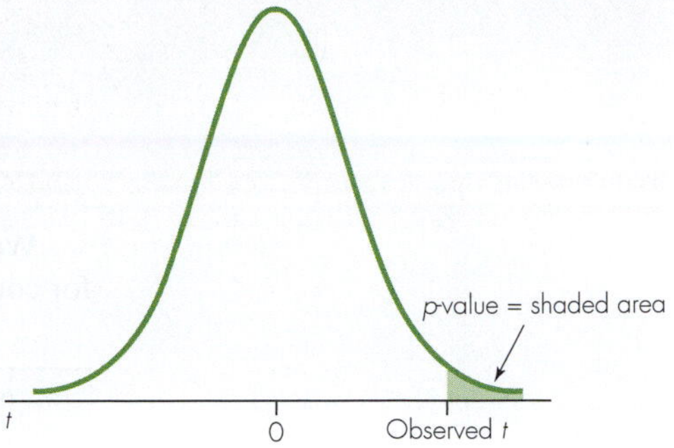

Figure 13.1 The *p*-value for a one-tailed *t*-test with H_a: $\mu > \mu_0$

Most software programs, including Minitab and Excel, will do the work of finding the *p*-value for you. You can also find a range for the *p*-value using tables of the Student *t*-distribution. Table A.3 in the Appendix gives the areas to the right of eight different values of *t* for various values of degrees of freedom. The type of area given in Table A.3 is shown in Figure 13.1, which shows the area comprising the *p*-value for a one-tailed *t*-test with H_a: $\mu > \mu_0$. For instance, in Example 13.2 later in this chapter (p. 508), the *p*-value is the area to the right of $t = 2.68$, and the degrees of freedom are df = 9. In Table A.3, the *t*-value $t = 2.68$ is between the column headings 2.5 and 3.0, so the *p*-value is somewhere between .015 and .007. A *t*-distribution is symmetric, so Table A.3 can also be used to estimate *p*-values for a "less than" alternative hypothesis. This will be demonstrated when we complete Example 13.1.

Steps 4 and 5: Decide Whether the Result Is Statistically Significant Based on the *p*-Value and Report the Conclusion in the Context of the Situation

These two steps remain the same for all of the hypothesis tests considered in this book. Choose a level of significance α, and reject the null hypothesis if the *p*-value is less than α. Otherwise, conclude that there is not enough evidence to support the alternative hypothesis. It is standard to use $\alpha = .05$.

Example 13.1 (cont.) **The Five Steps for Testing the Hypotheses about Normal Body Temperature for Young Adults** Step 1: The appropriate null and alternative hypotheses for this situation are as follows:

H_0: $\mu = 98.6$ degrees

H_a: $\mu < 98.6$ degrees

The parameter μ is the mean body temperature for the population of young adults. The research question is whether μ is smaller than the long-held standard, so the alternative hypothesis displays a one-sided test.

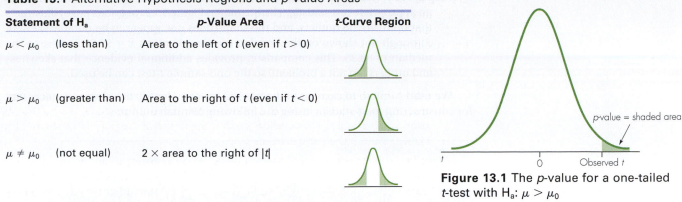

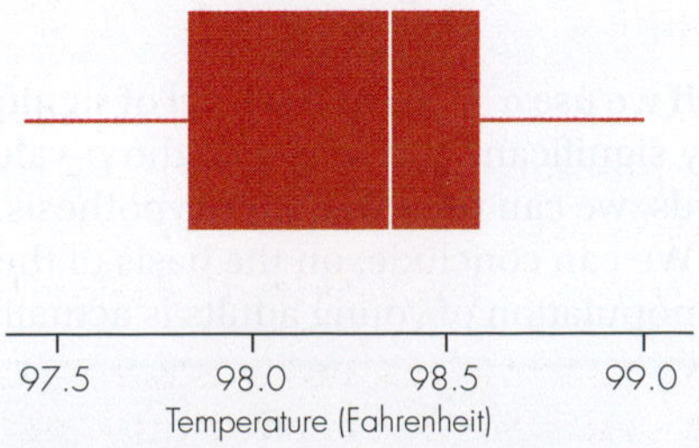

Figure 13.2 Boxplot of the data for Example 13.1

Step 2: Verify necessary data conditions, and if they are met, summarize the data into an appropriate test statistic. The boxplot in Figure 13.2 (previous page) illustrates that there are no outliers, and the shape does not appear to be notably skewed. Although not shown on the boxplot, the sample mean $\bar{x} = 98.20$ is close to the median of 98.35. This comparison provides additional evidence that skewness and outliers are not a problem, so the one-sample *t*-test can be used.

We used Minitab to carry out the test, and we can identify the elements necessary for constructing the *t*-statistic using the following Minitab output:

Test of mu = 98.6 vs < 98.6

Variable	N	Mean	StDev	SE Mean	95% Upper Bound	T	P
Temperature	16	98.200	0.497	0.124	98.418	−3.22	0.003

The key elements are the following:

- The sample statistic is $\bar{x} = 98.20$ (under "Mean").
- The standard error is $\text{s.e.}(\bar{x}) = \frac{s}{\sqrt{n}} = \frac{0.497}{\sqrt{16}} = 0.12425$, rounded to 0.124 (under "SE Mean").
- $t = \frac{\text{Sample statistic} - \text{Null value}}{\text{Standard error}} = \frac{98.20 - 98.6}{0.12425} = \frac{-0.40}{0.12425} = -3.22$ (under "T")

Step 3: The *p*-value is given as .003 in the last column of output. It was calculated as the area to the left of $t = -3.22$ in a *t*-distribution with $n - 1 = 16 - 1 = 15$ degrees of freedom, and Figure 13.3 illustrates this (barely visible) area. If we use Table A.3 in the Appendix, we can determine that the *p*-value is less than .004. Because the *t*-distribution is symmetric, the area to the left of $t = -3.22$ equals the area to the right of $t = +3.22$. The value $t = 3.22$ is bigger than the largest heading in the table, 3.00, and for 15 degrees of freedom, the corresponding one-sided *p*-value shown in the table is .004. We know that the area above 3.22 is even smaller than the area above 3.00, so the *p*-value is less than .004.

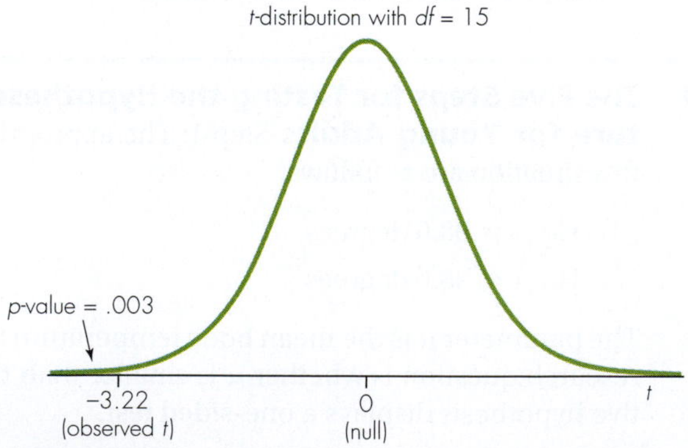

Figure 13.3 *p*-Value region for Example 13.1

Step 4: If we use $\alpha = .05$ as the level of significance criterion, the results are statistically significant because .003, the *p*-value of the test, is less than .05. In other words, we can reject the null hypothesis.

Step 5: We can conclude, on the basis of these data, that the mean temperature in the population of young adults is actually less than 98.6 degrees.

EXCEL TIP **Finding the *p*-Value for a *t*-Test**

In Example 13.1, the *p*-value can also be found in Excel by typing @TDIST(3.22,15,1) in any cell. Note that Excel requires you to give the absolute value of the test statistic. The "1" indicates that you want the area in one tail only, since this is a one-tailed test. (An = sign can also be used, in place of the @ symbol.)

Caution about using Excel for one-sided alternative hypothesis: If the sample mean happens to be in the direction opposite to that specified by the alternative hypothesis, then *p*-value > .5 and Excel gives 1 − *p*-value. For instance, in Example 13.1, if the sample mean had been above 98.6 and had resulted in $t = +3.22$, then the appropriate *p*-value would *still* be the area to the *left* of *t*, or to the *left* of 3.22, which is 1 − .003 = .997. Excel gives only areas to the *right* of the specified value of *t* for one-tailed tests and allows only positive values of *t* to be specified.

IN SUMMARY

The Steps for Testing a Single Mean with a "One-Sample *t*-Test"

Step 1: Determine the *null* and *alternative* hypotheses.

Null hypothesis: $H_0: \mu = \mu_0$
Alternative hypothesis: $H_a: \mu \neq \mu_0$ *or* $H_a: \mu > \mu_0$ *or* $H_a: \mu < \mu_0$

where the format of the alternative hypothesis depends on the research question of interest and must be decided before looking at the data.

Step 2: Verify necessary data conditions, and if they are met, summarize the data into an appropriate *test statistic.*

If *n* is large, or if there are no extreme outliers or skewness, compute

$$t = \frac{\text{Sample mean} - \text{Null value}}{\text{Standard error}} = \frac{\bar{x} - \mu_0}{s/\sqrt{n}} = \frac{\sqrt{n}(\bar{x} - \mu_0)}{s}$$

Step 3: Assuming that the null hypothesis is true, find the *p-value.*

Using the *t*-distribution with df = $n - 1$, the *p*-value is the area in the tail(s) beyond the test statistic *t*, as follows:

For $H_a: \mu \neq \mu_0$, the *p*-value is 2 × area above $|t|$ (a two-tailed test).
For $H_a: \mu > \mu_0$, the *p*-value is the area above *t* (a one-tailed test) even if *t* is negative.
For $H_a: \mu < \mu_0$, the *p*-value is the area below *t* (a one-tailed test) even if *t* is positive.

Step 4: Decide whether the result is *statistically significant* based on the *p-value.*

Choose a significance level ("alpha"); the standard is $\alpha = .05$. The result is statistically significant if the *p*-value $\leq \alpha$.

Step 5: Report the conclusion in the context of the situation.

THOUGHT QUESTION 13.1 Consider the two possible one-sided alternative hypotheses in a one-sample t-test:

H_a: $\mu > \mu_0$ and H_a: $\mu < \mu_0$

In each case, what range of values of the t-statistic would result in a p-value $> .5$? What range of values of the sample mean would result in a p-value $> .5$? Write a general rule for one-sided tests explaining conditions under which the p-value $> .5$.*

The Rejection Region Approach for *t*-Tests

In Chapter 12 we learned that the decision in hypothesis testing can be made by finding a **rejection region**, which is the region of possible values for the test statistic that would lead to rejection of the null hypothesis. The method is based on finding the **critical value**, which is a value such that if the test statistic is beyond it in one or both tails (depending on H_a), the result is statistically significant. When this approach is used, the null hypothesis is rejected if the test statistic falls into the rejection region.

This method consists of replacing Steps 3 and 4 with the following:

Substitute Step 3: Find the critical value and rejection region for the test.

Substitute Step 4: If the test statistic is in the rejection region, conclude that the result is *statistically significant* and reject the null hypothesis. Otherwise, do not reject the null hypothesis.

In Chapter 12, we learned what values are in the rejection region for z-tests with a level of significance of .05 or .01. This approach can be used for t-tests as well, but the rejection region changes depending on the degrees of freedom accompanying the test. Table A.2 can be used to find the critical value and rejection region.

Steps for Finding the Critical Value and Rejection Region for a *t*-Test

1. Determine the degrees of freedom for the test.
2. Decide on α, the level of significance for the test; .05 is the most common, but sometimes .10, .01, or some other value is used.
3. Determine whether the alternative hypothesis is for a one-tailed test or a two-tailed test.
4. In Table A.2, find the *row* with the appropriate degrees of freedom, listed on the left-hand side of the table. Find the *column* with the appropriate one-tailed or two-tailed level of significance α, listed across the bottom of the table. Read the critical value t^* from the body of the table in the appropriate row and column. Compare the test statistic t to the critical value t^*. The rejection region depends on the direction of the alternative hypothesis, as shown in Table 13.2.

Table 13.2 How to Find the Rejection Region for a *t*-Test

Alternative Hypothesis	Column (bottom of Table A.2)	Rejection Region	Example: Rejection Region for df = 10, $\alpha = .05$
H_a: Parameter $\neq$ null value	Two-tailed α	$\lvert t\rvert \geq t^*$	$\lvert t\rvert \geq 2.23$
H_a: Parameter $>$ null value	One-tailed α	$t \geq t^*$	$t \geq 1.81$
H_a: Parameter $<$ null value	One-tailed α	$t \leq t^*$	$t \leq -1.81$

***HINT:** Remember that for H_a: $\mu < \mu_0$, the p-value is the area below the test statistic t. What values of t have more than half of the area below them? What relationship between the sample mean and μ_0 would result in those values of t?

HT Module 3: μ

IN SUMMARY Conducting a *t*-Test Using the Rejection Region Approach

Substitute Step 3: Choose α and then find the critical value t^* and rejection region from Table A.2, using the instructions in Table 13.2.

Substitute Step 4: Reject the null hypothesis and accept the alternative hypothesis if the test statistic t falls in the rejection region. Otherwise, do not reject the null hypothesis.

Example 13.1 (cont.) **The Critical Value and Rejection Region for Testing Normal Body Temperature for Young Adults** In Example 13.1, the alternative hypothesis was H_a: $\mu < 98.6$ degrees, the degrees of freedom = 15, and $\alpha = .05$. From Table A.2, the critical value is found to be $t^* = -1.75$, using the df = 15 row and the one-tailed $\alpha = .05$ column. The rejection region is $t \le -1.75$. The test statistic $t = -3.22$ is in the rejection region, so the null hypothesis is rejected. This is the same conclusion reached by finding the *p*-value and comparing it to .05.

Comparing the *p*-Value Approach to the Rejection Region Approach

Once the level of significance α has been decided, the rejection region method and the *p*-value method will always arrive at the same conclusion about statistical significance. It is always true that if the *p*-value $\le \alpha$, then the test statistic falls into the rejection region and vice versa. However, the *p*-value method provides additional information because it allows us to determine what decision would be made for every possible value of α. The rejection region method provides a decision only for the specified α. For that reason, and because statistical software programs and research journals generally report *p*-values, we will continue to emphasize the *p*-value approach.

MINITAB TIP **Computing a One-Sample *t*-Test**

- To carry out a one-sample *t*-test, use **Stat > Basic Statistics > 1-Sample t.** Specify the column that contains the raw data, check "Perform hypothesis test" and specify the null value in the "hypothesized value" box. (For Version 14 and lower, check "Test mean:" and specify the null value.) To specify the *alternative hypothesis,* use the ***Options*** button.
- If the raw data are not available, but the data are summarized so that you know values of the sample mean, standard deviation, and sample size, enter those values in the "Summarized data" area of the dialog box.

Note: Use the ***Graphs*** button to create visual displays of the data.

13.2 Exercises are on pages 536–538.

13.3 HT Module 4: Testing Hypotheses about the Population Mean of Paired Differences

Recall from Chapter 11 that the term **paired data** is used to describe data that are collected in natural pairs. In Chapter 6, we learned about *matched-pair* designs in which units are paired and one unit in each pair receives each treatment. Data are collected in pairs in other types of studies as well, such as when cases and controls are matched in case-control studies. Often, paired data occur when the researcher collects two measurements from each observational unit. For instance, if we record weights both before and after a diet program for each person in a sample, we have paired data. Or we might record performance on a college entrance exam for each individual before and after a training program designed to boost performance on that type of exam.

In most cases, paired data are collected because the researchers want to know about the differences and not about the original observations. In particular, it is often of interest to know whether the mean difference in the population is different from 0. A one-sample t-test can be used on the sample of differences to examine whether the sample mean difference is significantly different from 0. When this is done, the test is called a **paired t-test**. It is nothing more than a one-sample t-test conducted on the n differences. To emphasize that the data used in the test are differences, it is commonplace to use d_i instead of x_i to denote the original data values, $\bar{d}$ instead of $\bar{x}$ for the sample mean of the differences, s_d instead of s for the sample standard deviation of the differences, and μ_d instead of μ for the population mean of the differences. Other than those notational differences, once the difference value has been computed for each pair, the test proceeds exactly like a one-sample t-test.

IN SUMMARY The Steps for a Paired *t*-Test

Step 1: Determine the *null* and *alternative* hypotheses.

Null hypothesis: $H_0: \mu_d = 0$

Alternative hypothesis: $H_a: \mu_d \neq 0$ *or* $H_a: \mu_d > 0$ *or* $H_a: \mu_d < 0$

where the format of the alternative hypothesis depends on the research question of interest and the order in which the differences are taken.

Step 2: Verify necessary data conditions, and if they are met, summarize the data into an appropriate *test statistic. Before* proceeding with the details of this step, calculate the difference d between the two observations for each pair. Then verify that n is large or that there are no extreme outliers or skewness in the *differences.* Compute the sample mean and standard deviation of the differences, $\bar{d}$ and s_d, and then compute the test statistic:

$$t = \frac{\text{Sample mean} - \text{Null value}}{\text{Standard error}} = \frac{\bar{d} - 0}{s_d/\sqrt{n}} = \frac{\sqrt{n}(\bar{d} - 0)}{s_d}$$

Steps 3, 4, and 5: Proceed exactly as instructed in the In Summary box, "The Steps for Testing a Single Mean with a 'One-Sample t-Test'" on page 505, except that in the instructions for Step 3, replace μ with μ_d and replace μ_0 with 0.

HT Module 4: μ_d

Example 13.2 **Why Can't the Pilot Have a Drink?** Ten pilots performed tasks at a simulated altitude of 25,000 feet. Each pilot performed the tasks in a completely sober condition and, three days later, after drinking alcohol. The response variable is the time (in seconds) of useful performance of the tasks for each condition. The longer a pilot spends on useful performance, the better. The research hypothesis is that useful performance time decreases with alcohol use, so the data of interest are the decrease (or increase) in performance with alcohol compared to when sober. Note that it would have been a better study design if the order of the sober and alcohol conditions had been randomized for each pilot, in case there was a learning effect. But any learning effect would benefit the pilots in the alcohol condition, so if anything, the results in this case would be conservative. The data (Devore and Peck, 1993, p. 575) are as follows:

Pilot	No Alcohol	Alcohol	Difference = No Alcohol − Alcohol
1	261	185	76
2	565	375	190
3	900	310	590
4	630	240	390
5	280	215	65
6	365	420	−55
7	400	405	−5
8	735	205	530
9	430	255	175
10	900	900	0

Step 1: This is a paired-data design. Let μ_d = population mean difference between no alcohol and alcohol measurements if all pilots were to take these tests. Null and alternative hypotheses about μ_d follow:

H_0: $\mu_d = 0$ seconds

H_a: $\mu_d > 0$ seconds (i.e., no alcohol > alcohol)

The alternative hypothesis is one-sided because we hope to show that if performance does change, there is longer useful performance in the "No Alcohol" condition.

Step 2: Figure 13.4 displays a boxplot of the differences for the ten participants. A check of the necessary conditions for doing a *t*-test reveals that while the sample size is small and the dataset of differences for the sample does have some skewness, outliers and extreme skewness do not appear to be serious problems. So a paired *t*-test will be used to examine this question.

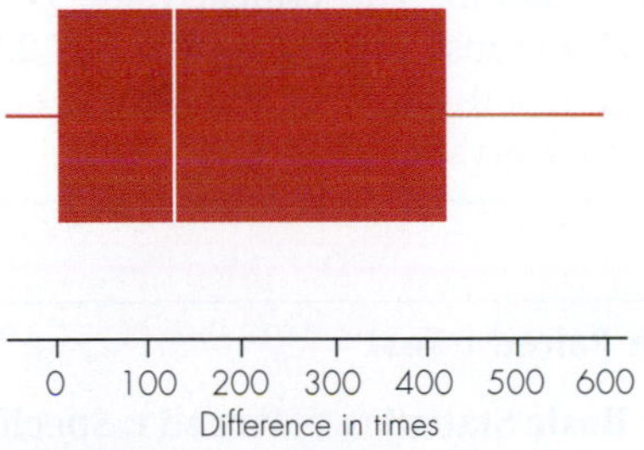

Figure 13.4 Boxplot of differences for Example 13.2

Minitab output for analyzing the mean difference follows. Note that the analysis involves only the sample of $n = 10$ differences and does not otherwise utilize the original performance scores.

```
Test of mu = 0 vs > 0

                                              95% Lower
Variable   N   Mean   StDev   SE Mean            Bound      T       P
Diff      10  195.6   230.5      72.9             62.0   2.68   0.013
```

The *t*-statistic shown in the output is $t = 2.68$ (under "T" in the output). The calculation of the test statistic is as follows:

- The sample statistic is the observed mean difference, $\bar{d} = 195.6$.
- The standard error is $\text{s.e.}(\bar{d}) = \dfrac{s_d}{\sqrt{n}} = \dfrac{230.5}{\sqrt{10}} = 72.9$
- $t = \dfrac{\text{Sample statistic} - \text{Null value}}{\text{Standard error}} = \dfrac{195.6 - 0}{72.9} = 2.68$

Step 3: At the right side of the output, we see that the p-value is .013. Because the alternative hypothesis was "greater than," this p-value was computed as the area to the right of 2.68 in a t-distribution with df $= 10 - 1 = 9$, an area that is illustrated in Figure 13.5. If we had used Table A.3, we would have found that $.007 < p\text{-value} < .015$. Finding the exact p-value of .013 requires a computer or calculator that gives areas for the t-distribution. For instance, in Excel, if you type @TDIST(2.68,9,1) into any cell, the result (rounded to three decimal places) will be .013.

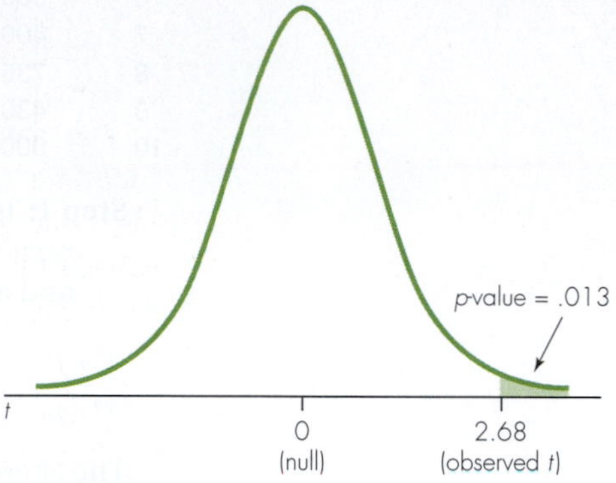

Figure 13.5 p-Value region for Example 13.2

Steps 4 and 5: The p-value is .013, so we can reject the null hypothesis using the standard significance level of .05. Even with such a small experiment, we can declare that alcohol has a statistically significant effect and decreases useful performance time.

Substitute Steps 3 and 4: To find the critical value, use Table A.2 with df = 9. Because the alternative hypothesis is H_a: $\mu_d > 0$, for $\alpha = .05$ use the one-tailed $\alpha = .05$ column. The critical value is $t^* = 1.83$ and the rejection region is $t \geq 1.83$. Because the test statistic $t = 2.68$ is in the rejection region, we reject the null hypothesis. This is the same conclusion we reached in Step 4 on the basis of the p-value.

MINITAB TIP **Computing a Paired t-Test**

- Use **Stat > Basic Statistics > Paired t.** Specify the two columns that contain the raw data for the pair of measurements.
- If the raw data are not available, but you know values of the mean and standard deviation of the sample of differences, and the sample size, enter those values in the "Summarized data" area of the dialog box. To specify the *alternative hypothesis,* use the ***Options*** button. If you want to test a null value other than 0, you can specify that using the ***Options*** button as well.
- Alternatively, you can first calculate a column of differences using **Calc > Calculator**, and then use the ***1-sample t*** procedure, entering 0 as the null value.

Note: Use the ***Graphs*** button to create visual displays of the data.

THOUGHT QUESTION 13.2 Suppose that in Example 13.2, the purpose of the study was to determine whether pilots should be allowed to consume alcohol the evening prior to their flights, and that in the alcohol consumption condition of the study, the consumption occurred 12 hours before the measurement of time of useful performance. Refer to the discussion of Type 1 and Type 2 errors in Chapter 12. Explain what the consequences of each type of error would be in this example. Which would be more serious? Given the data and results of the study, which type of error could have been made?*

***HINT:** A Type 1 error can be made only when the null hypothesis is true. In this case, the null hypothesis is that, on average, alcohol has no effect on pilots' performance.

Example 13.3 **Do You Know How Tall You Really Are?** Height and weight are important confounding variables in many health studies, and often these measurements are reported by participants rather than being measured by experimenters. A research team in Britain wondered how accurate these "self-reported" height and weight measurements tend to be (Spencer et al., 2002). They obtained data on self-reported and measured height and weight for a sample of 5140 middle-aged men and women. Height was reported and measured to the nearest centimeter and weight to the nearest 0.1 kg. Let's look at the women in the youngest age group, ages 35 to 49, to see if they were accurate in reporting their heights, or if they over- or underreported them, on average.

The data pairs were used to find differences (self-reported height − measured height) for the $n = 1168$ women in this age group. The mean difference was $\bar{d} = 0.15$ cm with a standard deviation of $s_d = 2.703$ cm. Is this sufficient evidence to reject the hypothesis that the average difference for the population is 0? Let's carry out the five steps to test this hypothesis.

Step 1: Determine the *null* and *alternative* hypotheses. There is no reason to hypothesize a particular direction for the difference so we will use a two-sided alternative hypothesis. So the null and alternative hypotheses are:

$H_0: \mu_d = 0 \qquad H_a: \mu_d \neq 0$

Step 2: Verify data conditions and if met, summarize data into an appropriate test statistic. The sample size is large, so we can proceed. The test statistic is

$$t = \frac{\bar{d} - 0}{s_d/\sqrt{n}} = \frac{0.15 - 0}{2.703/\sqrt{1168}} = \frac{0.15}{0.079} = 1.90$$

Step 3 and Substitute Step 3: Find the *p*-value and the rejection region. Because this is a two-sided test, the *p*-value is the area to the left of −1.90 plus the area to the right of +1.90, for a *t* with df = 1167. This is equivalent to finding 2 × area to the right of 1.90. Statistical software gives this area as .029, so the *p*-value is 2 × .029 = .058. Figure 13.6 (next page) illustrates this situation.

Using Table A.3, we would only learn that the area to the right of 1.90 is between the areas to the right of 1.80 and 2.00, which is between .023 and .036. That would tell us that the *p*-value is between .046 and .072, which isn't helpful in deciding whether we can reject the null hypothesis. Using the rejection region method, we find the critical value from Table A.2 to be $t^* = 1.962$ (use df = 1000 and two-tailed $\alpha = .05$). Therefore, we can reject the null hypothesis if $|t| \geq 1.962$.

Step 4 and Substitute Step 4: Decide whether the result is statistically significant.

Using $\alpha = .05$ we cannot reject the null hypothesis, because the computed *p*-value of .058 > .05. Using the rejection region we reach the same conclusion because the computed test statistic of $t = 1.90$ is not in the rejection region. The rejection region only included values with $|t| \geq 1.962$.

Step 5: Make a conclusion in context. We cannot reject the null hypothesis. We do not have sufficient evidence to conclude that for the population of British women in the age group 35 to 49 the average self-reported height is different from actual height as measured.

It is informative to compute a confidence interval. A 95% confidence interval for the population mean difference μ_d is −0.01 cm to +0.30 cm. Note that most of the interval is above 0, but even if there is a mean difference between self-reported and measured height for women in this age group, it is quite small. As an aside, the same is not true for weight—for which the average underreporting across all age groups for men was 1.85 kg (about 4 pounds) and for women was 1.40 kg (about 3 pounds).

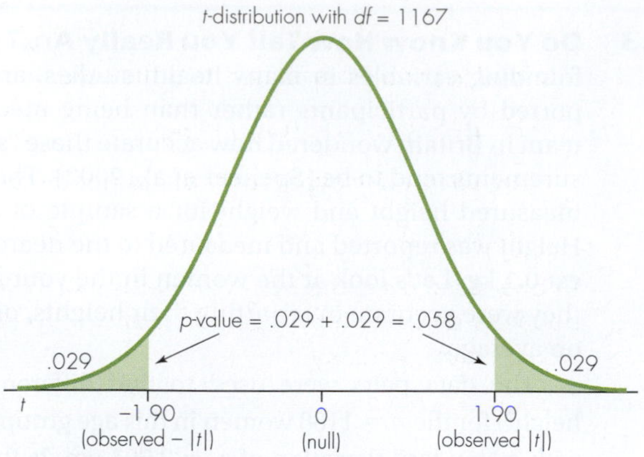

13.3 Exercises are on pages 538–539.

Figure 13.6 *p*-Value region for Example 13.3

13.4 HT Module 5: Testing Hypotheses about the Difference in Two Population Means (Independent Samples)

Lesson 1: The General (Unpooled) Case

It is often of interest to determine whether the means of populations represented by two independent samples of a quantitative variable differ. The two populations may be represented by two categories of a categorical variable, such as males and females, or they may be two hypothetical populations represented by different treatment groups in an experiment. In most cases, when comparing two means, the null hypothesis is that they are equal:

$$H_0: \mu_1 - \mu_2 = 0 \quad (\text{or } \mu_1 = \mu_2)$$

The procedure for testing this null hypothesis is called the **two-sample *t*-test** or ***t*-test for the difference in two means**. Let's look at an example of a situation where a two-sample *t*-test would be used.

Example 13.4 **The Effect of a Stare on Driving Behavior** In Example 11.11, we discussed an experiment done by social psychologists at the University of California at Berkeley. The researchers either did not stare or did stare at automobile drivers stopped at a campus stop sign. In the experiment, the response variable was the time (in seconds) it took the drivers to drive from the stop sign to a mark on the other side of the intersection. The two populations represented by the observed times are the hypothetical ones that would consist of the times drivers like these would take to move through a similar intersection, either under normal conditions (no stare) or the experimental condition (stare). The hypothesis the researchers wished to test was that the stare would speed up the crossing times, so the mean crossing time would be greater (slower) for those who did not experience the stare than it would be for those who did. Thus, the null and alternative hypotheses are

$$H_0: \mu_1 - \mu_2 = 0 \quad (\text{or } \mu_1 = \mu_2)$$
$$H_a: \mu_1 - \mu_2 > 0 \quad (\text{or } \mu_1 > \mu_2)$$

In these hypotheses, the subscript 1 is used to denote the No Stare population, and the subscript 2 is used to denote the Stare population. For the data given in Chapter 11, the mean crossing time was $\bar{x}_1 = 6.63$ seconds for $n = 14$ drivers who crossed under normal conditions (No Stare) and $\bar{x}_2 = 5.59$ seconds for $n = 13$ drivers who crossed in the experimental condition (Stare). The difference between the sample means is 1.04 seconds. Is this difference large enough to be statistically significant evidence against the null hypothesis? We will answer this question after describing the details of the five hypothesis-testing steps for a two-sample *t*-test.

Step 1: Determine the Null and Alternative Hypotheses

The *null* hypothesis is usually that the difference in means is 0:

$$H_0: \mu_1 - \mu_2 = 0 \quad (\text{or } \mu_1 = \mu_2)$$

The *alternative* hypothesis may either be one-sided or two-sided:

$$H_a: \mu_1 - \mu_2 > 0 \quad or \quad H_a: \mu_1 - \mu_2 < 0 \quad or \quad H_a: \mu_1 - \mu_2 \neq 0$$

In one-sided tests, the null hypothesis may be written to include an inequality in the opposite direction from the alternative hypothesis, but the alternative hypotheses never include equality.

Step 2: Verify Necessary Data Conditions, and If They Are Met, Summarize the Data into an Appropriate Test Statistic

The situations in which the t-test for the difference in two means is valid are similar to those given in Chapter 11 for constructing a confidence interval for the difference in two means.

Situations for Which a t-Test for the Difference in Two Population Means Is Valid

Situation 1: Both populations represented by the measurements of interest are approximately normal, and random samples of any size are measured. In practice, the method is used as long as there is no evidence that the shapes of the populations are notably skewed or that there are extreme outliers.

Situation 2: The populations of measurements of interest are not approximately normal, but a *large* random sample is measured from each one. Thirty is usually used as an arbitrary demarcation of "large," but if there are extreme outliers or extreme skewness, it is better to have even larger sample sizes than $n = 30$.

Independence: In either situation, the samples must be independent. In other words, they must not have been measured as paired or blocked data. A detailed description of methods for generating independent samples is given on page 408 in Section 11.1.

To verify the necessary conditions, examine histograms, stem-and-leaf plots, boxplots, or dotplots for each sample. Within each sample, a comparison of the mean and median can also provide information about possible skewness and outliers. If both of the two independent samples are large, we are in "Situation 2," and the two-sample t-test generally is valid for comparing the means unless there are remarkably extreme outliers. Keep in mind, however, that for skewed data, the median may be the better measure of location.

Continuing Step 2: The Test Statistic for a Two-Sample t-Test

The relevant sample statistic is $\bar{x}_1 - \bar{x}_2$ = the difference between sample means. As with a one-sample t-test, the standard error of $\bar{x}_1 - \bar{x}_2$ does not depend on whether the null hypothesis is true, so the term *null* is dropped from the denominator standard error. The "standardized" test statistic is

$$t = \frac{\text{Sample statistic} - \text{Null value}}{\text{Standard error}} = \frac{(\bar{x}_1 - \bar{x}_2) - 0}{\sqrt{\dfrac{s_1^2}{n_1} + \dfrac{s_2^2}{n_2}}}$$

Step 3: Assuming That the Null Hypothesis Is True, Find the p-Value

To find the p-value, use the t-distribution with appropriate degrees of freedom:

- For $H_a: \mu_1 - \mu_2 < 0$, the p-value is the area below t, even if t is positive.
- For $H_a: \mu_1 - \mu_2 > 0$, the p-value is the area above t, even if t is negative.
- For $H_a: \mu_1 - \mu_2 \neq 0$, the p-value is $2 \times$ area above $|t|$.

As discussed in Chapter 11, the t-distribution is only an approximation for the distribution of the t-statistic, and appropriate degrees of freedom are found by a complicated formula called *Welch's approximation.* The formula is given in Section 11.4 and will not be repeated here. In practice, if computer software is used, it will provide the numerical value of degrees of freedom found from the approximation. If software is not available, a conservative approach is to use the smaller of $n_1 - 1$ and $n_2 - 1$ as the degrees of freedom.

TECHNICAL NOTE **Equal Variances and Pooled Standard Error**

If we are willing to make the assumption that the two populations we are comparing have the same (or similar) variance, then there is a procedure for which the t-distribution *is* the correct distribution of the t-statistic when the null hypothesis is true. It involves pooling both samples to estimate the common variance. The procedure is called a *pooled two-sample* t-*test*, and it is discussed in Lesson 2 of this module.

Steps 4 and 5: Decide Whether the Result Is Statistically Significant Based on the p-Value and Report the Conclusion in the Context of the Situation

These steps proceed as with any hypothesis test. Choose a significance level ("alpha"), usually $\alpha = .05$, and declare the difference in means to be statistically significant if the p-value $\leq \alpha$. Report a conclusion appropriate for the situation. In drawing a conclusion about differences between two populations, it also is important to consider the way in which the data were gathered. Remember from Chapter 6 that data collected in an experiment can provide stronger evidence of causation than data from an observational study.

Example 13.4 (cont.) **The Effect of a Stare on Driving Behavior**

Step 1: *State the hypotheses.* The subscript 1 is used to represent the No Stare population, and the subscript 2 is used to represent the Stare population. As was stated above, the appropriate null and alternative hypotheses are

$$H_0: \mu_1 - \mu_2 = 0 \quad (\text{or } \mu_1 = \mu_2)$$
$$H_a: \mu_1 - \mu_2 > 0 \quad (\text{or } \mu_1 > \mu_2)$$

The alternative hypothesis reflects the fact that the researchers thought staring would cause drivers to speed up, so the mean time should be slower (larger values) for the No Stare group and faster (smaller values) for the Stare group.

Step 2: *Verify necessary data conditions, and if they are met, determine the test statistic.* The data conditions were checked in Example 11.11. We saw that the conditions for conducting the t-test appeared to be valid, as there were no extreme outliers and no extreme skewness in the data. There were $n_1 = 14$ observations in the No Stare group and $n_2 = 13$ observations in the Stare group. The following Minitab output provides the results for a one-tailed test:

```
Two-sample T for No Stare vs Stare

            N    Mean   StDev   SE Mean
No Stare   14    6.63    1.36      0.36
Stare      13   5.592   0.822      0.23

Difference = mu (No Stare) − mu (Stare)
Estimate for difference: 1.036
95% lower bound for difference: 0.297
T-Test of difference = 0 (vs >): T-Value = 2.41   P-Value = 0.013   DF = 21
```

The last line of output shows us that the t-statistic is "T = 2.41." The elements of the calculation are as follows:

- The sample statistic is $\bar{x}_1 - \bar{x}_2 = 6.63 - 5.59 = 1.04$ seconds (shown as 1.036 in the output).
- $\text{s.e.}(\bar{x}_1 - \bar{x}_2) = \sqrt{\frac{s_1^2}{n_1} + \frac{s_2^2}{n_2}} = \sqrt{\frac{1.36^2}{14} + \frac{0.822^2}{13}} = 0.43$

- $t = \dfrac{\text{Sample statistic} - \text{Null value}}{\text{Standard error}} = \dfrac{\bar{x}_1 - \bar{x}_2 - 0}{\text{s.e.}(\bar{x}_1 - \bar{x}_2)} = \dfrac{1.036 - 0}{0.43} = 2.41$

Steps 3, 4, and 5: *Determine the* p*-value and form a conclusion in context.* Note from the output that the p-value of the test is .013, so we can reject the null hypothesis and attach the label "statistically significant" to the results. The p-value is the probability that a sample difference could be as far or farther from 0 as 1.04 seconds is *in the positive direction,* if the population difference is actually 0. This probability is determined by using a t-distribution with df = 21 (the approximate degrees of freedom computed by Minitab using the Welch approximation formula described in Chapter 11) and finding the area to the right of the test statistic value, $t = 2.41$. From Table A.3, we would learn that the p-value is between .009 and .015. We can conclude that if all drivers were stared at, the mean crossing times at an intersection would be faster than under normal conditions. It can also be noted that these data were collected in a randomized experiment, so it is reasonable to conclude that the staring or not staring caused the difference between the mean crossing times.

Substitute Steps 3 and 4: As is always the case, the same conclusion is reached by using the *rejection region* approach. To find the *critical value* for this example, use Table A.2 with df = 21. (If Minitab had not provided the degrees of freedom, we would use the conservative value of smaller sample size − 1, which is 13 − 1 = 12, but we will assume that we know df = 21.) Because the alternative hypothesis is one-sided, H_a: $\mu_1 - \mu_2 > 0$, for $\alpha = .05$ use the one-tailed $\alpha = .05$ column. The critical value is $t^* = 1.72$, and the rejection region is $t \geq 1.72$. Figure 13.7 illustrates this rejection region. Because the test statistic $t = 2.41$ is in the rejection region, we reject the null hypothesis. This is the same conclusion that we reached on the basis of the p-value of .013, and because the p-value is more informative than the rejection region approach, it is always better to report it if it is known.

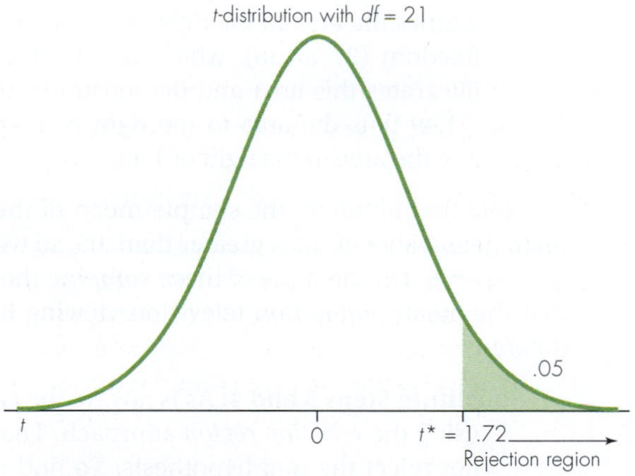

Figure 13.7 Rejection region for Example 13.4

Example 13.5 **A Two-Tailed Test of Television Watching for Men and Women** In Example 11.2 we used information from $n = 175$ Penn State students to estimate the mean amount of time that college students watch television in a typical day. Do you think that the mean time watching television is *different* for the population of college men than it is for the population of college women?

Step 1: A two-sided alternative hypothesis is appropriate because there is no prior knowledge that one specific sex might have a higher mean. Therefore, if μ_1 and μ_2 are the mean population daily television-viewing hours for men and women, respectively, for the populations represented by these students, then the appropriate hypotheses are

$$H_0: \mu_1 - \mu_2 = 0$$
$$H_a: \mu_1 - \mu_2 \neq 0$$

Step 2: The conditions are met for a two-sample t-test to be valid because television viewing hours are available for large independent samples of men and women. Minitab provides the test statistic and furnishes the information necessary to show how to compute it by hand:

Sex	N	Mean	StDev	SE Mean
Male	59	2.37	1.87	0.24
Female	116	1.95	1.51	0.14

Difference = mu (Male) − mu (Female)
Estimate for difference: 0.418
95% CI for difference: (−0.139, 0.974)
T-Test of difference = 0 (vs not =): T-Value = 1.49 P-Value = 0.140 DF = 97

In the last line of output, we see that the test statistic is "T = 1.49." The calculation is as follows:

$$\bar{x}_1 - \bar{x}_2 = 2.37 - 1.95 = 0.42 \text{ hour (shown as 0.418 in the output)}$$

$$\text{s.e.}(\bar{x}_1 - \bar{x}_2) = \sqrt{\frac{s_1^2}{n_1} + \frac{s_2^2}{n_2}} = \sqrt{\frac{1.87^2}{59} + \frac{1.51^2}{116}} = 0.281$$

$$t = \frac{\text{Sample statistic} - \text{Null value}}{\text{Standard error}} = \frac{0.42 - 0}{0.281} = 1.49$$

Steps 3, 4, and 5: *p-value and conclusion.* The p-value is the probability that a sample mean difference could be as far or farther from 0 as 0.42 hour is, *in either direction,* if the population mean difference is actually 0. This is determined by finding the probability that the value of a t-distribution is above 1.49 or below −1.49. For this test, use approximate df = 97 (see the Minitab output above) or, if necessary, use df = 59 − 1 = 58 (smaller sample size −1). In either case, the *p-value* is 2 times the area to the right of 1.49 in a t-distribution with the stated degrees of freedom (97 or 58), which is .14 for either degrees of freedom. Figure 13.8 illustrates this area and demonstrates that it is actually the area to the left of −1.49 plus the area to the right of 1.49. It is simpler computationally to find 2 × the area to the right of 1.49.

Note that although the sample mean of the daily viewing hours is higher for the men, the p-value of .14 is greater than .05, so we cannot reject the null hypothesis that $\mu_1 - \mu_2 = 0$. On the basis of these *samples,* there is insufficient evidence to conclude that the mean *population* television viewing hours for college men and women are different.

Substitute Steps 3 and 4: As is always the case, the same conclusion is reached by using the *rejection region* approach. The conclusion in this case is that we cannot reject the null hypothesis. To find the critical value for this example, use Table A.2 with df = 58. (If Minitab had not provided the degrees of freedom, we would have to use the conservative value of smaller sample size −1, which is 59 − 1 = 58, so we will use it here to illustrate that method.) Because the alternative hypothesis is two-sided, H_a: $\mu_1 - \mu_2 \neq 0$, for $\alpha = .05$ use the two-tailed $\alpha = .05$ column. The critical value falls between 2.01 (for df = 50) and 2.00 (for df = 60); we will use the more conservative 2.01 as the critical value. The rejection region is $t \leq -2.01$ and $t \geq 2.01$. Because the test statistic $t = 1.49$ is *not* in the rejection region, we *cannot* reject the null hypothesis. This is the same conclusion we reached based on the p-value of .14.

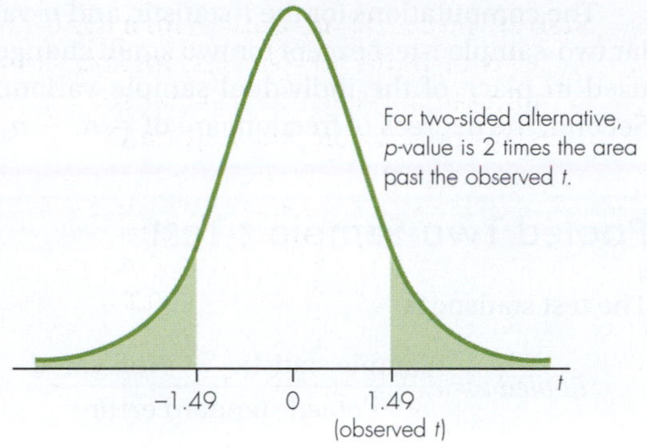
For two-sided alternative,
p-value is 2 times the area
past the observed t.
t
−1.49
0
1.49
(observed t)

Figure 13.8 p-Value region for Example 13.5

IN SUMMARY **The Steps for a Two-Sample t-Test (Unpooled)**

Step 1: Determine the *null* and *alternative* hypotheses.

Null hypothesis: $H_0: \mu_1 - \mu_2 = 0$

Alternative hypothesis: $H_a: \mu_1 - \mu_2 \neq 0$ *or* $H_a: \mu_1 - \mu_2 > 0$ *or* $H_a: \mu_1 - \mu_2 < 0$

where the format of the alternative hypothesis depends on the research question of interest and the order in which Population 1 and Population 2 are defined.

Step 2: Verify necessary data conditions, and if they are met, summarize the data into an appropriate *test statistic.*

Verify that both n's are large *or* that there are no extreme outliers or skewness in either sample. The samples must also be independent. Compute the test statistic:

$$t = \frac{\text{Sample statistic} - \text{Null value}}{\text{Standard error}} = \frac{(\bar{x}_1 - \bar{x}_2) - 0}{\sqrt{\dfrac{s_1^2}{n_1} + \dfrac{s_2^2}{n_2}}}$$

Steps 3, 4, and 5: Proceed exactly as instructed in the In Summary box, "The Steps for Testing a Single Mean with a 'One-Sample t-Test'" on page 505, except in the instructions for Step 3, use df = (smaller of $n_1 - 1$, $n_2 - 1$) or df as determined by the Welch approximation formula on page 426 in Chapter 11. Also, replace μ with $\mu_1 - \mu_2$ and μ_0 with 0.

Lesson 2: The Pooled Two-Sample t-Test

The use of the t-distribution for finding p-values for the two-sample t-test is an approximation. However, if the population standard deviations are equal, then a more precise method is available. The **pooled two-sample t-test** is based on the assumption that $\sigma_1 = \sigma_2 = \sigma$, the common standard deviation for both populations. The sample variances are combined to provide a **pooled sample variance, s_p^2,** and **pooled standard deviation, s_p.** The formula for the pooled sample variance is

$$\text{Pooled sample variance} = s_p^2 = \frac{(n_1 - 1)s_1^2 + (n_2 - 1)s_2^2}{n_1 + n_2 - 2}$$

The computations for the t-statistic and p-value remain the same as with the regular two-sample t-test except for two small changes. First, the pooled sample variance is used in place of the individual sample variances in computing the standard error. Second, the degrees of freedom are $df = n_1 + n_2 - 2$.

IN SUMMARY

Pooled Two-Sample t-Test

The test statistic is

$$\textit{Pooled } t = \frac{\text{Sample statistic} - \text{Null value}}{\text{Pooled standard error}} = \frac{(\bar{x}_1 - \bar{x}_2) - 0}{\text{Pooled standard error}}$$

where

$$\text{Pooled standard error} = \text{Pooled s.e.}(\bar{x}_1 - \bar{x}_2) = \sqrt{\frac{s_p^2}{n_1} + \frac{s_p^2}{n_2}} = s_p\sqrt{\frac{1}{n_1} + \frac{1}{n_2}}$$

Under the null hypothesis, the pooled t-statistic has a t-distribution with $df = n_1 + n_2 - 2$. All other steps are the same as for the unpooled two-sample t-test.

Although the pooled t-test provides an exact solution and the unpooled version provides only an approximate solution, we do not recommend that the pooled t-test be used except in very specific circumstances. It is better to have an approximate solution that is known to work well than to use an exact solution based on an assumption that cannot be verified. Let's look at some guidelines for when to use the pooled test.

Guidelines for Using the Pooled Two-Sample t-Test

- When $n_1 = n_2$, the pooled and unpooled standard errors are equal, so the t-statistic also is the same for both the unpooled and pooled procedures. The pooled procedure, however, is not exact unless the population standard deviations are equal, so it is generally preferable to use Welch's approximate df. If this value is not available, when the two sample sizes are equal, it is an acceptable substitute to use $df = n_1 + n_2 - 2$. This recommendation can be extended to include situations in which the sample sizes are close.
- When n_1 and n_2 are very different, the pooled test can be quite misleading unless the sample standard deviations are similar. Some books recommend using the pooled test as long as the ratio of sample standard deviations (larger s/smaller s) is ≤ 2, but this cutoff is completely arbitrary, and we do not recommend it.
- It can be shown by using algebra that

$$\text{Pooled s.e.}(\bar{x}_1 - \bar{x}_2) \approx \sqrt{\frac{s_1^2}{n_2} + \frac{s_2^2}{n_1}}$$

  where the symbol "≈" means "is approximately equal to." Therefore, using the pooled standard error is essentially equivalent to reversing the roles of n_1 and n_2 in the standard error. If the larger sample produces the smaller sample standard deviation, the standard error will tend to be underestimated, the t-statistic will tend to be too large, and the null hypothesis will be rejected too easily. Therefore, if the sample sizes are very different and the smaller standard deviation accompanies the larger sample size, we do not recommend using the pooled procedure.
- If the sample sizes are very different, the standard deviations are similar, and the larger sample size produced the larger standard deviation, the pooled t-test is acceptable because it will be conservative.

Example 13.6 **Misleading Pooled *t*-Test for Television Watching for Men and Women** In Example 13.5 (p. 515), we compared television viewing hours for male and female college students and could not reject the hypothesis that they had the same population means. The sample size for males was only about half of what it was for females, but the standard deviation was larger for males. The summary statistics follow:

	Sample Size	Mean	Standard Deviation	Variance
Male	$n_1 = 59$	2.37	1.87	3.50
Female	$n_2 = 116$	1.95	1.51	2.28

The difference in means is 0.42 hour, the unpooled standard error is 0.281, the test statistic is 1.49, and the resulting *p*-value is .14. Here are the computations for the *pooled* t*-test:*

$$s_p^2 = \frac{(n_1 - 1)s_1^2 + (n_2 - 1)s_2^2}{n_1 + n_2 - 2} = 2.69, \quad s_p = \sqrt{2.69} = 1.64$$

$$\text{Pooled s.e.}(\bar{x}_1 - \bar{x}_2) = \sqrt{\frac{s_p^2}{n_1} + \frac{s_p^2}{n_2}} = s_p\sqrt{\frac{1}{n_1} + \frac{1}{n_2}} = 1.64\sqrt{\frac{1}{59} + \frac{1}{116}} = 0.262$$

$$t = \frac{\text{Sample statistic} - \text{Null value}}{\text{Pooled standard error}} = 0.42/0.262 = 1.60$$

$$\text{df} = 59 + 116 - 2 = 173; \quad p\text{-value} = .11$$

Although the *p*-value for the pooled test still would not lead to rejection of the null hypothesis, it is noticeably smaller than the *p*-value of .14 for the unpooled test. In similar situations, in which sample sizes are very different and the larger sample has the smaller variance, the pooled *t*-test and inflated test statistic could provide a misleading conclusion to reject the null hypothesis.

Example 13.7 **Legitimate Pooled *t*-Test for Comparing Male and Female Sleep Time** In Example 11.14 (p. 431) we compared the mean sleep time for male and female college students by finding a confidence interval for the difference. The data were collected in statistics classes at the University of California, Davis, at the beginning of the spring 2000 quarter. Here is a summary of the results from Minitab, displaying both the unpooled and pooled versions:

```
Two Sample T-Test and Confidence Interval
Two sample T for sleep [without "Assume Equal Variance" option]

sex     N    Mean    StDev   SE Mean
f       83   7.02    1.75       0.19
m       65   6.55    1.68       0.21

95% CI for mu (f) − mu (m): (−0.10, 1.02)
T-Test mu (f) = mu (m) (vs. not =): T = 1.62   P = 0.11   DF = 140
Two sample T for sleep [with "Assume Equal Variance" option]

sex     N    Mean    StDev   SE Mean
f       83   7.02    1.75       0.19
m       65   6.55    1.68       0.21

95% CI for mu (f) − mu (m): (− 0.10, 1.03)
T-Test mu (f) = mu (m) (vs. not =): T = 1.62   P = 0.11   DF = 146
Both use Pooled StDev = 1.72
```

Note that in this case, the sample sizes are similar, as are the sample standard deviations. Consequently, the pooled standard deviation of 1.72 is similar to the separate standard deviations of 1.75 and 1.68. The test statistic of 1.62 is identical under both procedures, and the p-value of .11 is the same. The use of the pooled procedure was warranted in this case.

MINITAB TIP **Computing a Two-sample *t*-Test to Compare Means**

- To carry out a two-sample t-test, use **Stat > Basic Statistics > 2-sample t.** Specify the location of the data. The raw data for the response (*Samples*) may be in one column, and the raw data for group categories (*Subscripts*) may be in a second column. Or, the raw data for responses in the two independent groups may be in two separate columns.
- If the raw data are not available, but the data are summarized so that you know the sample mean, standard deviation, and sample size for each group, enter those values in the "Summarized data" area of the dialog box.
- To use the pooled standard error, click on ***Assume Equal Variances***.
- To specify the alternative hypothesis, use the ***Options*** button.

SPSS TIP **Computing a Two-Sample *t*-Test**

- To compare two population means using data from two independent samples, use **Analyze > Compare Means > Independent-Samples T Test.** Results are given for both the unpooled and pooled procedures.
- The p-values given are for a two-sided alternative hypothesis. For a one-sided alternative, divide a given p-value by 2 (assuming that the ordering of the sample means is consistent with the alternative hypothesis).

THOUGHT QUESTION 13.3 The paired t-test introduced in Section 13.3 and the two-sample t-test introduced in this section are both used to compare two sets of measurements, and the null hypothesis in both cases is usually that the mean population difference is 0. Explain the difference in the situations for which they are used. Suppose that researchers wanted to know if college students spend more time watching TV or exercising. Explain how they could collect data appropriate for a paired t-test and how they could collect data appropriate for a two-sample t-test.*

13.4 Exercises are on pages 539–542.

Note to Readers: The remainder of this chapter applies to all of the material covered in Chapters 10 to 13, not just to hypothesis tests for means.

13.5 The Relationship Between Significance Tests and Confidence Intervals

In the situations covered in Chapters 10 through 13, you may have noticed a direct correspondence between the values covered by a confidence interval and the results of two-sided hypothesis tests. The correspondence is precise for tests involving one or two population means and a two-sided alternative.

***HINT:** If you were in the study and they asked you (and other participants) about both time watching TV and time exercising, would the data be paired or independent samples?

Confidence Intervals and Tests with Two-Sided Alternatives

In testing one population mean or the difference in two population means with the following hypotheses:

$$H_0\text{: parameter} = \text{null value} \quad \textit{and} \quad H_a\text{: parameter} \neq \text{null value}$$

- If the null value is covered by a $(1 - \alpha)100\%$ confidence interval, the null hypothesis is not rejected, and the test is not statistically significant at level α.
- If the null value is not covered by a $(1 - \alpha)100\%$ confidence interval, the null hypothesis is rejected, and the test is statistically significant at level α.

Don't be confused by the cumbersome notation $(1 - \alpha)100\%$. The correspondence is between a 95% confidence interval and a significance level of $\alpha = .05$, between a 99% confidence interval and a significance level of $\alpha = .01$, and so on.

Example 13.8 **Mean Daily Television Hours of Men and Women** In Example 13.5 (p. 515), we tested whether the population mean daily television viewing hours differed for male and female college students. The Minitab output is reprinted here:

```
Sex       N   Mean   StDev   SE Mean
Male     59   2.37    1.87      0.24
Female  116   1.95    1.51      0.14

Difference = mu (Male) − mu (Female)
Estimate for difference: 0.418
95% CI for difference: (−0.139, 0.974)
T-Test of difference = 0 (vs not =): T-Value = 1.49  P-Value = 0.140  DF = 97
```

Note that a 95% confidence interval for the difference in population means is −0.139 to +0.974 hour. In other words, with 95% confidence, we can say that $-0.139 \leq \mu_1 - \mu_2 \leq 0.974$ hour. The null value for the test was 0 hours, which is covered by this interval. Therefore, the difference in sample means, which is 0.42 hour, is not significantly different from the population null value of 0.

The correspondence also holds for one or two proportions, with one caveat. In those situations, the standard error that is used for the confidence interval is slightly different from the null standard error used for the hypothesis test. Therefore, the correspondence may not hold if the null value is close to one of the boundaries of the confidence interval. In most cases, however, this minor technical detail will not interfere with the correspondence.

This correspondence means that a confidence interval can be used as another way to conduct a two-sided significance test. Some researchers think that whenever possible, a confidence interval should be used to test hypotheses because an interval provides information about the value of a parameter as well.

Confidence Intervals and One-Sided Tests

It is possible to construct one-sided confidence intervals by allowing one end of the interval to extend to the minimum or maximum possible value of the parameter and then adjusting the confidence level appropriately. In that case, the same correspondence would hold between one-sided confidence intervals and one-sided tests. Starting with Version 14, by default Minitab provides a one-sided confidence interval to accompany one-tailed hypothesis tests.

It is also possible to use a standard two-sided confidence interval to test a one-sided alternative by appropriately adjusting the significance level of the test or the confidence level for the interval. We present the general result, but the caveat given immediately after Example 13.8 for the two-sided case for one or two proportions holds in this case as well.

When testing the hypotheses H_0: parameter = null value versus a one-sided alternative, compare the null value to a two-sided $(1 - 2\alpha)100\%$ confidence interval:

- If the null value is covered by the interval, the test is not statistically significant at level α.
- For the alternative H_a: parameter > null value, the test is statistically significant at level α if the entire interval falls *above* the null value.
- For the alternative H_a: parameter < null value, the test is statistically significant at level α if the entire interval falls *below* the null value.

For instance, if a two-sided 90% confidence interval falls completely above the null value for a test and the alternative hypothesis is H_a: parameter > null value, then the null hypothesis can be rejected at a significance level of $\alpha = .05$. As an illustration, a 90% confidence interval for $\mu_1 - \mu_2$ that contained only positive values would allow researchers to accept the alternative hypothesis H_a: $\mu_1 - \mu_2 > 0$ using $\alpha = .05$.

Example 13.9 **Ear Infections and Xylitol** For Example 12.17, children were randomly assigned to take xylitol or a placebo in a clinical trial done to see if xylitol might reduce the risk of getting ear infections. The Minitab output on page 480 of Chapter 12 provides 0.036 as a lower bound for a one-sided 95% confidence interval for the difference between population proportions getting ear infections for placebo versus xylitol. On the basis of that lower bound, we can reject the null hypothesis that the difference is 0 using a .05 level of significance because the interval is entirely above 0.

Read the original source on the companion website, http://www.cengage.com/statistics/Utts4e.

A two-sided 95% confidence interval for $p_1 - p_2$, the difference in proportions of children who would get ear infections with placebo and xylitol, is 0.020 to 0.226. This interval also tells us that we can reject H_0: $p_1 - p_2 = 0$ and accept H_a: $p_1 - p_2 > 0$ because the entire two-sided 95% confidence interval falls above the null value of 0. But in this case, we can use $\alpha = .025$ because we used a 95% confidence level in our two-sided interval, which is equivalent to a 97.5% one-sided interval. The *p*-value for the test was .01, and since $.01 < .025$, this conclusion is confirmed by using the standard approach as well. Note that if the alternative hypothesis had been in the opposite direction, that is, $p_1 - p_2 < 0$, the null hypothesis would *not* be rejected even though the null value of 0 is not contained in the interval. This conclusion fits with our common sense. The interval indicates that the possible values for the difference are all *above* 0 (0.02 to 0.226), so the alternative hypothesis that the true difference is *less* than 0 is certainly not supported by the interval.

(Source: Reproduced from Pediatrics, *Vol. 102, pp. 879–884. [See p. 701 for complete credit.])*

13.5 Exercises are on page 542.

THOUGHT QUESTION 13.4 Refer to Example 13.9. A 95% confidence interval for the difference in proportions of children who would get ear infections with placebo compared to with xylitol was 0.02 to 0.226. On the basis of this information, specify a one-sided 97.5% confidence interval and explain how you would use it to test H_0: $p_1 - p_2 = 0$ versus H_a: $p_1 - p_2 > 0$ with $\alpha = .025$.*

13.6 Choosing an Appropriate Inference Procedure

In this chapter and the previous three, we have learned how to construct confidence intervals and conduct hypothesis tests to make inferences for several different population parameters. The focus has been on how to carry out and interpret these inference procedures after the procedure and population parameter of interest have already been identified. In this section, we provide guidance on how to identify the appropriate

***HINT:** The maximum possible value for $p_1 - p_2$ is 1.0, so that is the upper end of the interval.

procedure based on the research question(s) of interest. Necessary decisions include what parameter to investigate and whether to construct a confidence interval, conduct a hypothesis test, or do both.

Confidence Interval or Hypothesis Test?

When using sample data to answer a question about a larger population, how can you decide whether to construct a confidence interval, conduct a hypothesis test, or do both? The key point to consider is whether the main purpose is to estimate the numerical value of a parameter or to make a "maybe not" or "maybe yes" type of conclusion about a specific hypothesized value. If the purpose is to estimate a numerical value, a confidence interval is appropriate. If the purpose is to make a conclusion about a specific value, hypothesis testing is appropriate. Because these procedures provide different types of information about the population, it may be appropriate to use both in many situations.

As an example, think about the difference between the types of information asked for by the following two questions:

Question 1: What is the mean number of hours that students study each week?

Question 2: Is there a difference between mean hours of study per week for females versus males?

Note that Question 1 asks for a numerical value. The answer might be something like "about 18 hours per week." A confidence interval can be used to give an estimate that reflects the uncertainty due to using a sample to estimate the population parameter. For example, we might answer the question about mean number of hours studying with a statement like "With 95% confidence, we estimate that the mean hours of study per week for all students at our school is between 17.2 and 18.8 hours." In general, a confidence interval is used if the main purpose is to estimate the value of a parameter.

How would you answer Question 2? It doesn't ask us to estimate numerical values of mean hours of study for men and women but instead simply asks whether there is a difference. Given the question wording, we only have to answer with something like "Yes, there appears to be a difference" or "We cannot say there is a difference." We can create an answer by conducting a hypothesis test (also called a significance test) of the null hypothesis that the difference in means for men and women is 0 versus the alternative hypothesis that the difference in means is not 0. In general, a hypothesis test is used if the main purpose is to provide a "maybe not" or "maybe yes" type of answer for a research question.

In many situations, it may be informative to conduct a significance test and construct a confidence interval as well. For Question 2 above, for instance, it would not only be useful to know whether there was a difference, it would also be informative to know the magnitude of a possible difference. A confidence interval for the difference in means for the populations of men and women would give us that information. It is not always the case that both procedures should be used. When the primary purpose is to estimate the numerical value of a single parameter, it usually doesn't make sense to conduct a hypothesis test because there is no specific value of interest. As an example, there is no specific or hypothesized value for the parameter of interest for Question 1 above.

Let's summarize what can be learned from constructing a confidence interval or conducting a hypothesis test:

- A confidence interval estimate provides a range of values that is likely to contain the true value of the population parameter, so it gives information about the *magnitude* of the parameter. The confidence level, typically 95%, gives information about how often in the long run the procedure will result in an interval that successfully covers the parameter value.
- The *p*-value and conclusion from hypothesis testing provide a "maybe not" or "maybe yes" answer to the question "Is the parameter equal to (or different from) a certain value?" They do not give information about the magnitude of a parameter.
- In many cases in which hypothesis testing is the primary goal, it makes sense to construct a confidence interval to accompany the results of the hypothesis test.

- In most cases in which a confidence interval estimate is the primary goal, it does not make sense to conduct a hypothesis test to accompany the interval because there is no specific parameter value to be tested.

Determining the Appropriate Parameter

In Chapters 10 through 13, we learned inference procedures for five specific population parameters:

p: One proportion = population proportion in a category of a categorical variable, or long-run probability of a particular outcome.

$p_1 - p_2$: Difference in two population proportions or long-run probabilities; used when independent samples are available.

μ: One mean = mean for a quantitative variable in a population, or a quantitative outcome in a random circumstance.

μ_d: Paired difference mean = population mean for the difference in two paired variables or quantitative outcomes; used when paired samples are available.

$\mu_1 - \mu_2$: Difference in two means = difference in population means for two populations; used when independent samples are available.

Determining which of these is the parameter of interest depends on the answers to these questions:

- Is there one sample or two? If there are two, are they independent samples or are they paired?
- For one sample, is the variable that is measured on each unit quantitative or categorical?
- For two independent samples, one variable identifies the two samples and is categorical. Is the other variable also categorical?

On the basis of the answers to these questions, you can identify which of the five parameters is appropriate. Table 13.3 illustrates the process.

Table 13.3 Parameters and Types of Data Used for Inferences about Them

Variable Type (Parameter Type)	One Sample (No Pairing)	Paired Data	Two Independent Samples
Categorical (Proportions)	p	None	$p_1 - p_2$
Quantitative (Means)	μ	μ_d	$\mu_1 - \mu_2$

When you are first learning to distinguish among the situations, it helps to have some clues. Here are tips for helping you to determine which parameter to use, assuming that the choices are from the five covered in Chapters 10 through 13.

- *Are two different sample sizes given, and/or is it clear that two independent samples were taken?* If so, the parameter of interest must be either $p_1 - p_2$ or $\mu_1 - \mu_2$.
- *Are one or more sample means given?* If so, the parameter is probably a mean or difference in two means.
- *Are sample counts or proportions given?* If so, the parameter is probably a proportion or a difference in two proportions.

After you determine the parameter of interest and whether to conduct a hypothesis test and/or construct a confidence interval, you can use the summary tables at the end of this chapter (p. 534) and at the end of Chapter 11 (p. 439) to review the computational details of the analysis.

Examples of Choosing an Inference Procedure

The following examples utilize studies from other chapters in this textbook. In each case, think about what type of inference is appropriate.

Example 13.10 **Kids and Weight Lifting** In Case Study 6.2 (p. 195), a randomized experiment was presented in which children were randomly assigned to one of two weight-lifting conditions or a control condition. Muscular strength and endurance were measured after eight weeks. One of the results reported in Case Study 6.2 was "leg extension strength significantly increased in both exercise groups compared with that in the control subjects" (Faigenbaum et al., 1999, p. 85).

Read the original source on the companion website, www.cengage.com/statistics/Utts4e.

Note that the researchers are comparing each of the two weight-lifting conditions to the control condition. Let's focus on comparing only the heavy-load group to the control group for the reported response variable, leg extension strength.

Decision 1: Confidence interval, hypothesis test, or both? Based on the statement about the results, the researchers were clearly interested in knowing *if* there is a significant difference in leg extension strength after heavy lifting compared with no weight lifting. They were also most likely interested in the *magnitude* of the difference. Therefore, both inference procedures would be appropriate.

Decision 2: What is the appropriate parameter to investigate? Let's investigate the three questions that are used to determine which parameter is appropriate.

- *Is there one sample or two? If there are two, are they independent samples, or are they paired?* It is clear that there are two samples and that they are independent. One group of children did heavy lifting, and the other group did no weight training. The children in the two groups were not paired.
- *For one sample, is the variable that is measured on each unit quantitative or categorical?* This question is not relevant because there are two samples.
- *For two independent samples, one variable identifies the two samples and is categorical. Is the other variable also categorical?* The training (heavy or none) identifies the two samples. The other variable that is measured is the leg extension strength, which is not categorical. Presumably, a quantitative measurement was made for each child.

On the basis of the answers to these questions, the appropriate column of Table 13.3 is "Two Independent Samples" and the appropriate row is "Quantitative (Means)." Therefore, the parameter of interest is $\mu_1 - \mu_2$, the difference in population means for leg extension strength if all similar children were to participate in the heavy-lifting or control conditions.

(Source: Reproduced with permission from Pediatrics, *Vol. 104(1), p. e5. [See p. 701 for complete credit.])*

Example 13.11 **Loss of Cognitive Functioning** Example 6.5 (p. 206) reported on an observational study of 6000 older people, which found that 70% of the participants did not lose functioning over time. What inference procedure can be used for this result?

Decision 1: Confidence interval, hypothesis test, or both? There is no information provided about the study that would lead to a natural hypothesis testing question. The reported result, that 70% of the *sample* did not lose cognitive functioning over time, leads to a consideration of the margin of error and confidence interval estimate for the proportion of the *population* in this age group that does not lose cognitive functioning over time. Therefore, hypothesis testing is not appropriate, but a confidence interval is.

Decision 2: What is the appropriate parameter to investigate? The sample statistic reported is a sample proportion for a single sample. People in the study were categorized by whether or not they lost cognitive functioning. Therefore, the parameter of interest is p = proportion that does not lose cognitive functioning over time for the population of older people represented by this sample.

13.6 Exercises are on pages 542–543.

13.7 Effect Size

In most of the hypothesis-testing situations in Chapters 12 and 13, we are interested in comparing a population mean or proportion to a specific null value or to another population mean or proportion. In many research situations, we would like to know something about the magnitude of the comparison. The test statistic and *p*-value for a test are not useful for this purpose because they depend on the size of the sample. A confidence interval may not be useful because it depends on the units of measurement, and we want a measure that is independent of the units of measurement. This is particularly true if we want to compare research results across studies in which different measurement units may have been used or slightly different treatments may have been assigned.

In general, the **effect size** for a research question is a measure of how much the truth differs from chance or from a control condition. The most common effect-size measure used for a single mean is

$$d = \frac{\mu_1 - \mu_0}{\sigma}$$

where μ_1 is the true population mean, μ_0 is the null value, and σ is the population standard deviation. Note that d measures the distance between the true mean and the null mean value in terms of number of population standard deviations.

As an example, suppose it is known that the mean IQ in a certain population under usual conditions is 110, with a standard deviation of 15. Researchers speculate that listening to classical music temporarily boosts IQ. If indeed the mean IQ after listening to classical music is 115, then the effect size is $(115 - 110)/15 = 1/3$. In other words, listening to classical music boosts the mean IQ by one third of a standard deviation.

The most common effect-size measure for comparing two means is

$$d = \frac{\mu_1 - \mu_2}{\sigma}$$

where μ_1 and μ_2 are the two population means and σ is the population standard deviation, assumed to be the same for both populations. If this assumption is not reasonable and if one population corresponds to a control group condition, then the standard deviation for the control condition is used. If that is not possible either, then a pooled standard deviation is used. Note that d measures the distance between μ_1 and μ_2 in terms of number of standard deviations, using the standard deviation for the individual populations as the yardstick. In most situations, the order of the difference is not important to the magnitude of the effect, so the absolute value of d is reported.

As an example, suppose that resting pulse rates for those who don't exercise have a mean of 76 and standard deviation of 8 beats per minute, while resting pulse rates for those who do exercise have a mean of 68 with the same standard deviation. Then the effect size for the difference is $(76 - 68)/8 = 1$. In other words, average resting pulse rates for those who exercise and those who don't exercise differ by 1 standard deviation.

There are many uses for effect sizes, some of which require d to be estimated from data. There are various suggestions for estimating d based on complicated statistical principles (see, e.g., Hedges and Olkin, 1985), but the simplest method is to substitute sample values for the population parameters in d.

FORMULA

Estimating Effect Sizes for Means for One and Two Samples

For a single sample, the estimated effect size is $\hat{d} = \frac{\bar{x} - \mu_0}{s}$, where $\bar{x}$ and s are the sample mean and standard deviation, respectively.

For two samples, the estimated effect size is $\hat{d} = \frac{\bar{x}_1 - \bar{x}_2}{s}$, where $\bar{x}_1$ and $\bar{x}_2$ are the two sample means and s is either the standard deviation of the control condition measurements (if there is one) or the pooled sample standard deviation.

Note that there is a relationship between the test statistic t and the estimated effect size. The formula for the estimated effect size includes everything in the formula for the t statistic *except* the influence of the sample size(s).

FORMULA **Relationship between Test Statistic t and Estimated Effect Size**

For one sample: $$t = \frac{\bar{x} - \mu_0}{\frac{s}{\sqrt{n}}} = \frac{\sqrt{n}(\bar{x} - \mu_0)}{s} = \sqrt{n}\hat{d}, \text{ so } \hat{d} = \frac{t}{\sqrt{n}}$$

For two samples: $$\text{Pooled } t = \frac{\bar{x}_1 - \bar{x}_2}{s\sqrt{\frac{1}{n_1} + \frac{1}{n_2}}} = \frac{\hat{d}}{\sqrt{\frac{1}{n_1} + \frac{1}{n_2}}}$$

Although we will not cover other effect-size measures, the relationship for t-tests holds in most hypothesis-testing situations for which effect sizes are used. Rosenthal (1991) provides a summary of many of these procedures, and notes that in most cases,

Test statistic = Size of effect × Size of study

as we have seen in the tests for means, where "Size of study" is a function of the sample size(s) that increases as the sample size(s) increase.

Interpreting an Effect Size

Effect sizes provide information about how strong a difference or effect is in the population, relative to another population or a hypothesized value. The interpretation is similar to the interpretation of a standardized score for a specific individual value. In fact, you should note that the effect size d for a single mean of a normal population is simply the standardized score for μ_1 relative to the normal distribution with mean μ_0 and standard deviation σ.

Cohen (1988) defined small, medium, and large effect sizes for $|d|$ somewhat arbitrarily to be 0.2, 0.5, and 0.8, respectively. In other words, if the true mean μ_1 is half of a standard deviation away from the null mean value ($|d| = 0.5$), Cohen called this a medium effect. If the true mean μ_1 is only one-fifth of a standard deviation away from the null mean ($|d| = 0.2$), Cohen called it a small effect; similarly for 0.8 as a large effect. These are arbitrary, but he said that a small effect should be detectable only through statistics, a medium effect should be obvious to a careful observer, and a large effect should be obvious to any observer.

Example 13.12 **Could Aliens Tell That Women Are Shorter?** Suppose that an alien were to visit Earth and observe humans. Would it be obvious that men are taller, on average, than women? The mean heights of adult men and women are about 70 inches and 65 inches, respectively, with standard deviations of about 2.5 inches within each population. Therefore, the effect size for the difference in heights is approximately:

$$d = (70 - 65)/2.5 = 2.0.$$

Such a large difference should be immediately obvious to anyone.

However, suppose that the difference in men's and women's heights was smaller. The following table shows what differences correspond to small, medium, and large effect sizes. Think about whether you would notice the difference in mean heights for two groups whose means differed by these amounts.

Effect Size Magnitude	Difference in Heights when $\sigma = 2.5$	Interpretation
Small (d = 0.2)	2.5 × 0.2 = 0.50 inch	Not obvious without statistics
Medium (d = 0.5)	2.5 × 0.5 = 1.25 inches	Obvious to careful observer
Large (d = 0.8)	2.5 × 0.8 = 2.00 inches	Obvious to most observers

Example 13.13 **Normal Body Temperature** Have you noticed that many people comment on how their normal body temperature seems to be lower than the previously held standard of 98.6? If so, then you have detected a large effect size, which should be evident to most observers with access to sufficient data. In Example 13.1, data were presented for normal body temperature for 16 young adults, taken from a larger data set of 100 healthy adults. Because temperatures differ substantially for men and women, let's look at the effect size for the 46 women in the sample. The sample mean and standard deviation were 98.10 and 0.65, respectively. The estimated effect size for comparing normal body temperature of women to the previously held standard of 98.6 is

$$\hat{d} = \frac{\bar{x} - \mu_0}{s} = \frac{98.1 - 98.6}{0.65} = \frac{-0.5}{0.65} = -0.77$$

This is a large effect size according to Cohen's categorization, which should be evident to an observer such as a health care worker who routinely measures the temperature of healthy women.

Comparing Effect Sizes across Studies

Effect sizes have become increasingly important in the past few decades with the increase in the use of **meta-analysis**, which is the statistical combination of results on the same topic across studies. Studies are all somewhat different and may use different units of measurement, different sample sizes, and so on. It is useful to have a method of comparison across studies that is not influenced by sample size or units of measurement, and effect sizes serve that purpose.

In early attempts to review many studies on the same topic, researchers made the mistake of using p-values. In fact, they often used a **vote count**, in which they simply counted the proportion of studies that achieved statistical significance. As we have learned, the problem with this method is that statistical significance is heavily dependent on sample size. To illustrate the problem, consider the scenario in the following example.

Example 13.14 **The Hypothesis-Testing Paradox** Suppose that a researcher conducts a one-sample t-test in a study with $n = 100$ and finds that the test statistic is $t = 2.5$, so with a two-tailed test the p-value is .014. The results of the study are clearly statistically significant. Just to be sure, the researcher decides to repeat the study, but this is just for confirmation, so a smaller size of $n = 25$ is used. The researcher is disappointed to find that in the replication study, $t = 1.25$, so the p-value is only .22, and the result is not statistically significant. Perplexed about why the effect disappeared, the researcher decides to combine the data. For the combined data, the researcher is very surprised to find that $t = 2.795$ with df $= 124$, and the p-value is .006! The paradox is that when the researcher considered the second study alone, it seemed to detract from the evidence for an effect because it was not statistically significant. When the researcher considered the second study in conjunction with the first study, it seemed to strengthen the evidence for an effect.

Now let's look at the effect size for each study and for the combined studies. Here is a summary of the results:

	Sample Size	Test Statistic t	p-Value	Effect Size $\hat{d} = t/\sqrt{n}$
Study #1	100	2.50	.014	0.25
Study #2	25	1.25	.22	0.25
Combined	125	2.795	.006	0.25

Note that the effect size is exactly the same for all three studies! The problem is that the researcher was using statistical significance as the basis for the conclusion about whether the first study replicated the second one. The different sample sizes rendered that comparison meaningless.

Thus, effect sizes are quite useful for comparing the results of studies on the same topic. Replication of results requires that similar effect sizes be obtained across studies and not that statistical significance or the lack of it be obtained. Comparing *p*-values or statistical significance across studies is valid only if the sample sizes are similar.

Effect Size and Power

In Chapter 12, Example 12.21 (p. 484), we learned that the *power* for a test for a population proportion depends on the difference between the null value and the true value of the parameter. However, in tests for population means, the difference between the null value and the true population mean is not sufficient because the test depends on the standard deviation as well. For means, the relevant difference is the effect size, *d*, which is the difference expressed as number of standard deviations.

Suppose that a researcher wants to find a statistically significant result as long as the truth is a certain size effect, say *D*. All that he or she needs to do in addition to specifying *D* is to choose a significance level, usually .05 or .01. There is then a trade-off between sample size(s) and power that can be quantified. Many researchers aim for power of .80, meaning that there is an 80% chance of achieving statistical significance if the true effect size is *D*. Many software packages will compute the sample size(s) needed to achieve the specified power, or the power that will result from a specified sample size.

Example 13.15 **Planning a Weight-Loss Study** Suppose that you were convinced that your new weight-loss plan worked, and you wanted to conduct a study to verify that. How would you decide how many participants to include? You need to specify four things: (1) the effect size you hope to detect, (2) the planned level of significance α, (3) whether the test is one- or two-tailed, and (4) the desired power. Let's suppose that if there is at least a medium effect size (i.e. *d* is at least 0.5), you hope to detect the effect. You plan to use $\alpha = .05$ for a one-tailed test, and you want power = .80. Using Minitab, we find that a sample of $n = 27$ participants is required. To detect a small effect size of 0.2 with power of .80, a sample of $n = 156$ is required.

The interpretation of an effect size of 0.2 or 0.5 in terms of weight lost depends on the standard deviation of the weight losses. If the standard deviation is 10 pounds, then effect sizes of 0.2 and 0.5 correspond to average weight losses of 2 pounds and 5 pounds, respectively.

Table 13.4 shows some additional power and sample size results for this situation. Note that if indeed the effect is very small ($d = 0.1$), it will be very difficult to achieve statistical significance. Even with $n = 100$ participants, the probability of detecting a statistically significant effect is only .26. On the other hand, if the true effect is a medium effect size, then even with only 20 participants there is a fairly high probability (.695) of achieving a statistically significant result.

Figure 13.9, created by Minitab Version 16, displays **power curves** for the sample sizes given in Table 13.4. These curves show how the power increases as the true effect size increases, for a fixed sample size. Comparing curves illustrates that the power increases as the sample size increases, for a fixed effect size. The dots on the curves illustrate the values shown in Table 13.4. For example, the curve for a sample size of $n = 20$ shows that the power is about .7 (shown as .695 in Table 13.4) if the true effect size is 0.5. This means that there is about a 70% chance (.7 probability) of correctly rejecting the null hypothesis for a medium effect size (0.5) with a sample of $n = 20$ in a one-sided t-test using $\alpha = .05$.

Table 13.4 Power Based on True Effect Size Values for One-Sided, One-Sample t-Test, $\alpha = .05$

	True Effect Size			
Sample Size	**0.1**	**0.2**	**0.5**	**0.8**
$n = 20$	.11	.22	.695	.964
$n = 50$	.17	.40	.967	nearly 1
$n = 100$	.26	.63	.9996	nearly 1

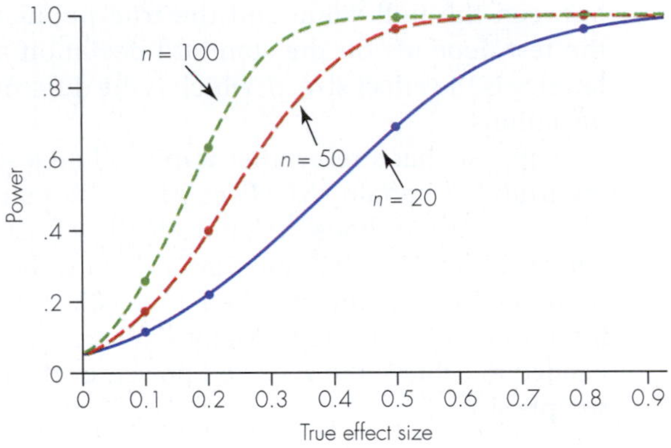

Figure 13.9 Power curves for Example 13.15

MINITAB TIP **Computing Sample Size for Specified Power or Power for Specified Sample Size**

1. To compute power for a one-sample t-test, use **Stat > Power and Sample Size >1-Sample t.** Enter "1" in the "Standard deviation" box. As "Differences" enter the effect size(s) you hope to detect.
2. There are two choices for what to compute:
   - Calculate the sample size for each power value. If you want to use this option, fill in a list of power values.
   - Calculate power for each sample size. If you want to use this option, fill in the sample sizes for which you want to find the power.
3. Use ***Options*** to specify the form of the alternative hypothesis (not equal to, greater than, less than), the level of significance, and optional storage locations.

13.7 Exercises are on pages 543–544.

13.8 Evaluating Significance in Research Reports

To evaluate the quality of a research study, you should always consider features like the sampling procedure, the methods used to measure individuals, the wording of questions, and the possibility of lurking variables. Remember that cause and effect cannot generally be concluded unless the study is a randomized experiment, and inferences to a larger population cannot be made unless the sample is representative of that population for the question of interest.

When the conclusions are based on hypothesis tests, several additional items should be considered. We learned in Chapter 12 and in the previous section that the test statistic and p-value are a function of both the magnitude of the difference between the truth and the null value, and the size of the sample. A small difference will be easily detected with a large sample, but even a moderate difference may not be strong enough to be detected with a small sample. So it's important to consider the sample size when you interpret the results of hypothesis tests.

Multiple Testing and Multiple Comparisons

There is one more problem with statistical inference that is important to consider when interpreting the results of a research study. Many studies include multiple hypothesis tests and/or confidence intervals. With each test and interval there is a chance that an erroneous conclusion will be made, so it stands to reason that if multiple tests and intervals are considered, the chance of an erroneous conclusion will increase. This problem of **multiple testing** or **multiple comparisons** should be acknowledged and accounted for when considering p-values and confidence levels.

As a simple example, suppose that 20 independent 95% confidence intervals are to be computed. How many of them will *not* cover the truth? Because they are independent and the probability of not covering the truth is .05 for each interval, the answer is a binomial random variable with $n = 20$ and $p = .05$. Therefore, the expected value for the number that miss the truth is $20(.05) = 1$. If 100 intervals are computed, the expected number that miss the truth is $100(.05) = 5$, and so on. The same reasoning holds for hypothesis tests. Remember that when the null hypothesis is true the level of significance α is also the probability of making a Type 1 error. If 20 independent tests are conducted using a significance level of .05, and if *all* of the null hypotheses are true, on average we should expect one **false positive** in which a Type 1 error is made, and the null hypothesis is rejected even though it is true.

The situation is rarely as simple as the illustration just given, because the tests and confidence intervals done in one study generally use the same individuals and thus are not independent. So it is almost impossible to ascertain the probability of a false positive. However, statisticians have developed methods for handling multiple comparisons. The simplest (and most conservative) is the **Bonferroni method**. The method proceeds by dividing up the significance level (or confidence level) and apportioning it across tests (or confidence intervals). For example, if a total Type 1 error probability of .05 is desired and five tests are to be done, each test is done using $\alpha = .05/5 = .01$. This method controls the overall probability of making at least one Type 1 error to be the sum of the individual significance levels used for each test.

Unfortunately, the media rarely mention how many tests and confidence intervals are done in one study, and the focus is often placed on surprising outcomes. It is thus very easy for false positive results to gain wide media attention. This is one reason that replication of studies is important in science. A false positive is unlikely to be replicated in multiple studies.

Case Study 1.8 (page 7) provides an example of multiple testing, in which much media attention was given to a study purporting to show that eating breakfast cereal before becoming pregnant could increase the chances of having a boy. But the study measured 133 different foods, and thus many, many hypotheses were considered. If the truth is that none of the foods increases the likelihood of having a boy, it is still quite likely that several will appear to do so. These findings, though, will actually be false positives.

A Checklist for Evaluating Significance in Research Reports

Here is a list of issues you should consider when you read the results of a statistical study:

1. Is the *p-value* reported? If you know the p-value, you can make your own decision, based on the severity of a Type 1 error and the size of the p-value.

2. If the word *significant* is used to describe a result, determine whether the word is being used in the everyday sense or in the statistical sense only. The phrase *statistically significant* just means that a null hypothesis has been rejected, which is no guarantee that the result has real-world importance, as illustrated in Case Study 1.7 on page 6.
3 If you read that "no difference" or "no relationship" has been found in a study, try to determine whether the sample size was small. The hypothesis test may have had very low power because not enough data were collected to be able to make a firm conclusion, as illustrated in Example 12.18 on page 481.
4. Think carefully about conclusions based on extremely large samples. If a study is based on a very large sample size, even a weak relationship or a small difference can be statistically significant, as illustrated by Example 12.19 on page 483.
5. If possible, determine what confidence interval should accompany a hypothesis test. This interval will provide information about the magnitude of the effect as well as information about the margin of error in the sample estimate.
6. Determine how many hypothesis tests were conducted in the study. Sometimes researchers perform a multitude of tests, but only a few of the tests achieve statistical significance. Remember that if all of the null hypotheses tested are true, then about 1 in 20 tests will achieve statistical significance just by chance at the .05 level of significance. The media often fails to report this multiple-testing phenomenon, focusing instead on the one or two tests that were statistically significant out of many that were conducted in a study, as illustrated by Case Study 1.8 on page 7.

13.8 Exercises are on pages 544–546.

CASE STUDY 13.1 Beat the Heat with a Frozen Treat

If you are an athlete who exercises in hot weather, you probably realize that your endurance suffers from the heat. Researchers in Australia have some advice for you: drink a fruit-flavored slushie before hitting the track. In a study published in the journal *Medicine and Science in Sports and Exercise* (Siegel et al., 2010),* researchers asked volunteers to run on a treadmill in a hot room (34 degrees Centigrade, about 93 degrees Fahrenheit) until exhaustion. On average, the participants lasted almost 10 minutes longer before reaching exhaustion when they drank a fruit slushie before beginning, compared to when they drank fruit-flavored cold water.

Ten healthy males with an average age of 28 and who routinely did moderate exercise were recruited to participate. This was a matched-pair design because all 10 men were given both treatments. The two treatments in the study were to drink a fruit-flavored ice slushie, with a temperature just below freezing, and to drink a glass of cold water with the same fruit-flavored syrup as the slushie, but with a temperature equivalent to water taken from a refrigerator. The order of the treatments was randomized for each participant, and they were administered a few weeks apart. To avoid a learning effect, each participant also did a practice run under the same conditions a few weeks before the first experimental run.

In each condition, the participants were told to run on a treadmill in a hot room until they couldn't run any longer. The response variable was time until exhaustion. The parameter of interest is μ_d, the population mean difference in times that men similar to the ones in the study would be able to run until exhaustion, under each of the two conditions. We might also want to estimate the population mean times for each condition separately. Let's define the difference for participant i as d_i = time until exhaustion after slushie − time until exhaustion after cold water for $i = 1, 2, \ldots 10$. Then the hypotheses of interest are as follows:

H_0: $\mu_d = 0$ Ha: $\mu_d > 0$

Because the sample size is small ($n = 10$), it is important to check for outliers. Although the authors did not provide the original data, they did provide plots, which showed that there were no outliers. The sample mean and standard deviation for the differences were $\overline{d} = 9.5$ minutes and $s_d = 3.6$ minutes. Therefore, the test statistic is

*Read the original source on the companion website, www.cengage.com/statistics/Utts4e.

$$t = \frac{\overline{d} - 0}{s_d/\sqrt{n}} = \frac{(9.5 - 0)}{3.6/\sqrt{10}} = \frac{9.5}{1.14} = 8.3$$

Even with very low df = 10 − 1 = 9, the *p*-value is essentially 0. Therefore, we can reject the null hypothesis, and conclude that there is a statistically significant increase in running time until exhaustion after drinking an ice slushie, compared to drinking fruit-flavored cold water.

One of the disadvantages of a paired *t*-test and corresponding confidence interval is that information about the individual means is not provided. Often they are not of interest, but in this example it is informative to know how long the men lasted until exhaustion under the two conditions. Here are the sample data and separate confidence intervals for the two conditions:

Condition	$\overline{x}$	s	95% Confidence Interval
Ice slushie	50.2	8.5	44.1 to 56.3 minutes
Cold water	40.7	7.2	35.6 to 45.9 minutes

But wait, these confidence intervals overlap, yet we clearly rejected the null hypothesis that the population means were equal! What's going on? Remember that the data were collected as matched pairs. The standard error of the mean difference (1.14) is much smaller than the standard error to accompany a two-sample *t*-test (3.5). In fact, a two-sample test, which would be incorrect here, yields a t of 2.70 and a *p*-value of .008, compared with the correct t of 8.3 and *p*-value of essentially 0. You can see that it is very important to use the appropriate procedure.

Let's evaluate this study using the checklist given on pages 531–532.

1. The *p*-value is reported to be essentially 0, so it is unlikely that a Type 1 error has been made.
2. The difference in the sample of almost 10 minutes, an increase from 40.7 minutes to 50.2 minutes, indicates that the result is of practical importance as well as statistical significance.
3. This study did not report that there was "no difference," so this item does not apply.
4. This study used a very small sample size, so this item does not apply.
5. You can easily compute a confidence interval in this situation, and are asked to do so in Exercise 13.109. The confidence interval gives important information about how small or how large the population difference might be.
6. There were indeed multiple tests done in this study. In addition to the one reported here, the researchers reported the results of tests on body and skin temperature, heart rate, sweating rate, thermal sensation, and perceived exertion, for a total of eight hypothesis tests. Using the Bonferroni method, maintaining an overall level of significance of .05 would mean using .05/8 = .00625 for each test. Fortunately, the main hypothesis of interest here still achieves statistical significance, even using $\alpha = .00625$.

As a side note, this study was reported in the *New York Times* (Kolata, 2010), and a few of the other tests were mentioned, but the total number of tests done was not. As is often the case, the only way to learn the details of a study is to find the original journal article and assess it for yourself.

(Source: Siegel, R., J. Maté, M.B. Brearley, G. Watson, K. Nosaka, and P. B. Laursen (2010). "Ice slurry ingestion increases core temperature capacity and running time in the heat," Medicine & Science in Sports & Exercise, *Vol. 42, No. 4, pp. 717–725.)*

IN SUMMARY Hypothesis Testing Procedures

The basic structure of hypothesis testing for the parameters in the table below is as follows:

1. Null hypothesis: H_0: parameter = null value

   Alternative hypothesis: $\left\{\begin{array}{l} H_a\text{: parameter} \neq \text{null value} \\ H_a\text{: parameter} > \text{null value} \\ H_a\text{: parameter} < \text{null value} \end{array}\right\}$ Choose one.

2. Test statistic: t or $z = \dfrac{\text{Sample statistic} - \text{Null value}}{\text{Null standard error}}$

3. p-value: For H_a: parameter ≠ null value, p-value is 2 × area above $|t|$ or $|z|$.
   For H_a: parameter > null value, p-value is the area *above* t or z, even if t or z is negative.
   For H_a: parameter < null value, p-value is the area *below* t or z, even if t or z is positive.

4. Decision: If p-value > significance level, do not reject H_0.
   If p-value ≤ significance level, reject H_0, accept H_a, and conclude that the result is statistically significant.

5. Conclusion: Report the conclusion in the context of the situation.

	Parameter	Statistic	Null Value	Null Standard Error	t or z; df for t	Example
One Proportion	p	$\hat{p}$	p_0	$\sqrt{\dfrac{p_0(1-p_0)}{n}}$	z	12.12
Difference between Proportions	$p_1 - p_2$	$\hat{p}_1 - \hat{p}_2$	Usually 0	$\sqrt{\hat{p}(1-\hat{p})\left(\dfrac{1}{n_1}+\dfrac{1}{n_2}\right)}$	z	12.17
One Mean	μ	$\bar{x}$	μ_0	$\dfrac{s}{\sqrt{n}}$	t df = $n - 1$	13.1
Mean Difference, Paired Data	μ_d	$\bar{d}$	Usually 0	$\dfrac{s_d}{\sqrt{n}}$	t df = $n - 1$	13.2
Difference between Means (unpooled)	$\mu_1 - \mu_2$	$\bar{x}_1 - \bar{x}_2$	Usually 0	$\sqrt{\dfrac{s_1^2}{n_1}+\dfrac{s_2^2}{n_2}}$	t df = Welch's approximation or smaller of $n_1 - 1$, $n_2 - 1$	13.4
Difference between Means (pooled)	$\mu_1 - \mu_2$	$\bar{x}_1 - \bar{x}_2$	Usually 0	$s_p\sqrt{\dfrac{1}{n_1}+\dfrac{1}{n_2}}$	t df = $n_1 + n_2 - 2$	13.7

Key Terms

In Summary Boxes

Exercises

◆ Denotes that the dataset is available on the companion website, http://www.cengage.com/statistics/Utts4e, but is not required to solve the exercise.

Bold exercises have answers in the back of the text.

For all exercises requiring hypothesis tests, make sure that you check conditions, define all parameters, and state your conclusion in words in the context of the problem.

Section 13.1

Skillbuilder Exercises

13.1 Explain whether each of the following statements is true.

a. One of the two possible conclusions in hypothesis testing is to accept the null hypothesis.

b. The statements "reject the null hypothesis" and "accept the alternative hypothesis" are equivalent.

13.2 Explain whether each of the following statements is true.

a. Hypotheses and conclusions from hypothesis testing apply only to the samples on which they are based.

b. The *p*-value is calculated with the assumption that the null hypothesis is true.

13.3 For each of the following research questions, specify whether the parameter of interest is one population mean, the population mean of paired differences, or the difference between the means of two populations.

a. Nutrition trends have changed over the years, and this may affect growth. Researchers want to know whether the mean height of 25-year-old women is the same as the mean height of 45-year-old women.

b. You plan to fly from New York to Chicago and have a choice of two flights. You are able to find out how many minutes late each flight was for a random sample of 25 days over the past few years. (You have data for both flights on the same 25 days.) You want to know whether one flight has a higher average delay (i.e., a higher population mean for the number of minutes the flight is late).

13.4 For each of the following research questions, specify whether the parameter of interest is one population mean, the population mean of paired differences, or the difference between the means of two populations.

a. Residents of a neighborhood have been complaining about speeding cars. The speed limit in the area is

25 miles per hour. The local police monitor the situation by recording the speed of 100 randomly selected cars. They want to know whether the mean speed of all cars that drive through the area is higher than 25 miles per hour.

b. Refer to part (a). After collecting the data described in part (a), the police decide to install an electronic roadside sign that shows cars how fast they are driving. They then record the driving speed of another 100 randomly selected cars. They want to know whether the mean speed is lower after installation of the sign than it was beforehand.

13.5 Refer to Exercise 13.3. For each part, write and define the notation that would be used for the population parameter of interest. Make sure that you specify the population(s) to which the parameter applies.

13.6 Refer to Exercise 13.4. For each part, write and define the notation that would be used for the population parameter of interest. Make sure that you specify the population(s) to which the parameter applies.

13.7 Refer to Exercises 13.3 and 13.5. Specify the null value that the researchers are interested in testing, and write the null hypothesis using the appropriate symbol(s) for the parameter of interest. If you haven't already done so in Exercise 13.5, make sure that you specify the population(s) to which the parameter applies.

13.8 Refer to Exercises 13.4 and 13.6. Specify the null value that the researchers are interested in testing, and write the null hypothesis using the appropriate symbol(s) for the parameter of interest. If you haven't already done so in Exercise 13.6, make sure that you specify the population(s) to which the parameter applies.

General Section Exercises

13.9 Explain why a "standardized statistic" is used in hypothesis testing, instead of simply using the difference between the sample statistic and the null value as the test statistic.

13.10 Explain why the null hypothesis for a significance test is rejected when the *p*-value is small rather than when it is large.

13.11 Explain why each of the following terms is used for the procedure outlined in Section 13.1.

a. Hypothesis testing.
b. Significance testing.

Section 13.2

Skillbuilder Exercises

13.12 If you were given a set of data for which the question of interest was about the population mean, explain how you would determine whether it would be valid to do a one-sample *t*-test using that set of data.

13.13 Review the five steps for any hypothesis test. Can any of the five steps be done *before* collecting the data? Explain.

13.14 Define the parameter of interest, and then use it to write the null and alternative hypotheses (in symbols) for each of the following research questions.

a. Many cars have a recommended tire pressure of 32 psi (pounds per square inch). At a roadside vehicle safety checkpoint, officials plan to randomly select 50 cars for which this is the recommended tire pressure and measure the actual tire pressure in the front left tire. They want to know whether drivers on average have too little pressure in their tires.

b. The box of Yvette's favorite cereal states that the net contents weigh 12 ounces. Yvette is suspicious of this claim because the package never seems full to her. She plans to measure the weight of the contents of the next 20 boxes she buys and find out whether she is being short-changed.

13.15 Define the parameter of interest, and then use it to write the null and alternative hypotheses (in symbols) for the following situation. Nonprofit hospitals must provide a certain amount of charity care to maintain nonprofit, tax-exempt status. Although the amount differs by area, let's suppose that all such hospitals are required to provide charity services equivalent to 4% of net patient revenue. A legislator is concerned that there is inadequate enforcement of this regulation. He plans to audit a random sample of 30 nonprofit hospitals and assess the percentage of net patient revenue they spend on charity care. He will then test whether the population mean is at least 4%.

13.16 Give the value of the test statistic *t* in each of the following situations.

a. H_0: $\mu = 50$, $\bar{x} = 60$, $s = 90$, $n = 100$.
b. Null value $= 100$, sample mean $= 98$, $s = 15$, sample size $= 40$.

13.17. Give the value of the test statistic *t* for a test with H_0: $\mu = 250$, $\bar{x} = 270$, standard error $= 5$ and $n = 100$.

13.18 Find the *p*-value and draw a sketch showing the *p*-value area for each of the following situations in which the value of *t* is the test statistic for the hypotheses given:

a. H_0: $\mu = \mu_0$, H_a: $\mu > \mu_0$, $n = 28$, $t = 2.00$.
b. H_0: $\mu = \mu_0$, H_a: $\mu > \mu_0$, $n = 28$, $t = -2.00$.

13.19 Find the *p*-value and draw a sketch showing the *p*-value area for each of the following situations in which the value of *t* is the test statistic for the hypotheses given:

a. H_0: $\mu = \mu_0$, H_a: $\mu \neq \mu_0$, $n = 81$, $t = 2.00$.
b. H_0: $\mu = \mu_0$, H_a: $\mu \neq \mu_0$, $n = 81$, $t = -2.00$.

13.20 Use Table A.2 to find the critical value and rejection region in each of the following situations. Then determine whether the null hypothesis would be rejected. In each case, the null hypothesis is H_0: $\mu = 100$.

a. H_a: $\mu > 100$, $n = 21$, $\alpha = .05$, test statistic $t = 2.30$.
b. H_a: $\mu > 100$, $n = 21$, $\alpha = .01$, test statistic $t = 2.30$.
c. H_a: $\mu \neq 100$, $n = 21$, $\alpha = .05$, test statistic $t = 2.30$.
d. H_a: $\mu \neq 100$, $n = 21$, $\alpha = .01$, test statistic $t = 2.30$.

13.21 Use Table A.2 to find the critical value and rejection region in each of the following situations. Then determine whether the null hypothesis would be rejected. In each case the null hypothesis is $H_0: \mu = 50$.

a. $H_a: \mu > 50$, $n = 10$, $\alpha = .05$, test statistic $t = 1.95$.
b. $H_a: \mu < 50$, $n = 10$, $\alpha = .05$, test statistic $t = -1.95$.
c. $H_a: \mu < 50$, $n = 10$, $\alpha = .05$, test statistic $t = 1.95$.
d. $H_a: \mu \neq 50$, $n = 10$, $\alpha = .05$, test statistic $t = 1.95$.

General Section Exercises

13.22 ◆ The dataset **cholest** on the companion website includes cholesterol levels for heart attack patients and for a group of control patients. It is recommended that people try to keep their cholesterol level below 200. The following Minitab output is for the control patients:

Test of mu = 200.00 vs. mu < 200.00

Variable	N	Mean	StDev	SE Mean	T	P
control	30	193.13	22.30	4.07	−1.69	0.051

a. What are the null and alternative hypotheses being tested? Write them in symbols.
b. What is the mean cholesterol level for the sample of control patients?
c. How many patients were in the sample?
d. Use the formula for the standard error of the mean to show how to compute the value of 4.07 reported by Minitab.
e. What values does Minitab report for the test statistic and the p-value?
f. Identify the numbers that were used to compute the t-statistic, and verify that the reported value is correct.
g. What conclusion would be made in this situation, using a .05 level of significance?

13.23 Suppose that a study is done to test the null hypothesis $H_0: \mu = 100$. A random sample of $n = 50$ observations results in $\bar{x} = 102$ and $s = 15$.

a. What is the null standard error in this case?
b. Plug numbers into the formula

$$t = \frac{\text{Sample statistic} - \text{Null value}}{\text{Null standard error}}$$

c. On the basis of the information given, can the p-value for this test be found? If so, find it. If not, explain what additional information would be needed.

13.24 It has been hypothesized that the mean pulse rate for college students is about 72 beats per minute. A sample of Penn State students recorded their sex and pulse rate. Assume that the sample is representative of all Penn State men and women for pulse rate measurements. The summary statistics follow:

Sex	n	Mean	Standard Deviation
Women	35	76.9	11.6
Men	57	70.42	9.95

a. Test whether or not the pulse rates of all Penn State men have a mean of 72.
b. Test whether or not the pulse rates of all Penn State women have a mean of 72.
c. Write a sentence or two summarizing the results of parts (a) and (b) in words that would be understood by someone with no training in statistics.

13.25 A cell phone company knows that the mean length of calls for all of its customers in a certain city is 9.2 minutes. The company is thinking about offering a senior discount to attract new customers, but first wants to know whether or not the mean length of calls for current customers who are seniors (65 and over) is the same as it is for the general customer pool. The only way to identify seniors is to conduct a survey and ask people whether they are over age 65. Using this method, the company contacts a random sample of 200 seniors and records the length of their last call. The sample mean and standard deviation for the 200 calls are 8 minutes and 10 minutes, respectively.

a. Do you think that the data collected on the 200 seniors are approximately bell-shaped? Explain.
b. Is it valid to conduct a one-sample t-test in this situation? Explain.
c. In spite of how you may have answered part (b), carry out the five steps to test the hypotheses of interest in this situation using $\alpha = .05$. (Use a p-value, not a rejection region.)

13.26 Refer to Exercise 13.14, part (a), which posed the following research question: "Many cars have a recommended tire pressure of 32 psi (pounds per square inch). At a roadside vehicle safety checkpoint, officials plan to randomly select 50 cars for which this is the recommended tire pressure and measure the actual tire pressure in the front left tire. They want to know whether drivers on average have too little pressure in their tires." Suppose that the experiment is conducted, and the mean and standard deviation for the 50 cars tested are 30.1 psi and 3 psi, respectively. Carry out the five steps to test the appropriate hypotheses using $\alpha = .05$. (Use a p-value, not a rejection region.)

13.27 Refer to Exercise 13.25. Carry out the hypothesis test in part (c) using the rejection region method, that is, using the Substitute Steps 3 and 4 described on pages 506–507 instead of finding the p-value. Use $\alpha = .05$.

13.28 Refer to Exercise 13.26. Carry out the hypothesis test using the rejection region method, that is, using the Substitute Steps 3 and 4 described on pages 506–507 instead of finding the p-value. Use $\alpha = .05$.

13.29 A university is concerned that it is taking students too long to complete their requirements and graduate; the average time for all students is 4.7 years. The dean of the campus honors program claims that students who participated in that program in their first year have had a lower mean time to graduation. Unfortunately, there is no automatic way to pull the records of all of the thousands of students who have participated in the program; they must be pulled individually and checked. A random sample of 30 students who had participated is taken, and the mean and standard deviation for the time to completion for those students are 4.5 years and 0.5 year, respectively. Carry out the five steps to test the hypotheses of interest in this situation. Use $\alpha = .05$.

13.30 ◆ The survey in the **UCDavis2** dataset on the companion website asked students if they preferred to sit in the front, middle, or back of the class and also asked them their heights. The following data are the heights for 15 female students who said they prefer to sit in the back of the class.

68, 62, 65, 69, 68, 69, 64, 66, 69, 68, 62, 64, 67, 68, 65

The mean height for the population of college females is 65 inches. Carry out the five steps to test the claim that the mean height for females who prefer to sit in the back of the room is higher than it is for the general population. In other words, test whether females who prefer to sit in the back of the room are taller than average. Use $\alpha = .05$.

13.31 Refer to Exercise 13.30. The following data are the heights for the 38 females who said they prefer to sit in the front of the classroom

66, 63, 63, 66, 65.5, 63, 60, 64, 63, 68, 68, 66, 62.5, 65, 64, 63, 66, 63, 63, 67, 66, 66, 62, 65, 63.5, 60, 61, 62, 63, 60, 65, 62, 63, 63, 62, 65, 63, 66

The mean height for the population of college females is 65 inches. Carry out the five steps to test the claim that the mean height for females who prefer to sit in the front of the room is lower than it is for the general population. In other words, test whether females who prefer to sit in the front of the room are shorter than average. Use $\alpha = .05$.

13.32 *This exercise is part of Thought Question 13.1.* Consider a one-sample t-test with the one-sided alternative hypothesis H_a: $\mu < \mu_0$.

a. What range of values of the test statistic t would result in a p-value $> .5$?
b. What range of values of the sample mean would result in a p-value $> .5$?
c. Write a general rule for one-sided tests explaining the conditions under which the p-value would be greater than .5.

Section 13.3

Skillbuilder Exercises

13.33 Explain how a paired t-test and a one-sample t-test are different and how they are the same.

13.34 Suppose you were given a data set consisting of pairs of observations for which the question of interest was whether or not the population mean of the differences was 0. Explain the steps you would take to determine whether it is valid to use a paired t-test.

13.35 Give the value of the test statistic t and the p-value or p-value range for this situation: H_0: $\mu_d = 0$, H_a: $\mu_d \neq 0$, $\bar{d} = -4$, $s_d = 15$, $n = 50$.

13.36 Give the value of the test statistic t in each of the following situations, and then find the p-value or p-value range for a two-tailed test.

a. H_0: $\mu_d = 0$, $\bar{d} = 4$, $s_d = 15$, $n = 50$.
b. H_0: $\mu_d = 0$, $\bar{d} = 0$, $s_d = 15$, $n = 50$.

General Section Exercises

13.37 ◆ Data from the dataset **UCDavis1** on the companion website included information on height (***height***) and mother's height (***momheight***) for 93 female students. Here is the output from the Minitab paired t procedure comparing these heights:

Paired T for height – momheight

	N	Mean	StDev	SE Mean
height	93	64.4495	2.5226	0.2616
momheight	93	63.1645	2.6284	0.2726
Difference	93	1.28495	2.64719	0.27450

95% lower bound for mean difference: 0.82884
T-Test of mean difference = 0 (vs. > 0): T-Value = 4.68
P-Value = 0.000

a. It has been hypothesized that college students are taller than they were a generation ago, and therefore that college women should be significantly taller than their mothers. State the null and alternative hypotheses to test this claim. Be sure to define any parameters you use.
b. Using the information in the Minitab output, the test statistic is $t = 4.68$. Identify the numbers that were used to compute the t-statistic, and verify that the stated value is correct.
c. What are the degrees of freedom for the test statistic?
d. Carry out the remaining steps of the hypothesis test.
e. Draw a sketch that illustrates the connection between the t-statistic and the p-value in this problem.

13.38 ◆ In Exercise 11.80 a study was reported in which students were asked to place as many dried beans into a cup as possible in 15 seconds with their dominant hand, and again with their nondominant hand (in randomized order). The differences in number of beans (dominant hand–nondominant hand) for 15 students were as follows:

4, 4, 5, 1, −2, 0, 2, 4, −3, 0, 0, 0, −2, 2, 1

The data also are given in the dataset **beans** on the companion website.

a. The research question was whether students have better manual dexterity with their dominant hand than with their nondominant hand. Write the null and alternative hypotheses.
b. Check the necessary conditions for doing a one sample t-test.
c. Carry out the test using $\alpha = .05$. (Find the p-value, not the rejection region.)
d. Carry out the test using $\alpha = .10$. (Find the p-value, not the rejection region.)
e. Write a conclusion about this situation that would be understood by other students of statistics.

13.39 Most people complain that they gain weight during the December holidays, and Yanovski et al. (2000) wanted to determine if that is the case. They sampled the weights of 195 adults in mid-November and again in early to mid-January. The mean weight change for the sample was a gain of 0.37 kg, with a standard deviation of 1.52 kg. State and test the appropriate hypotheses. Use $\alpha = .01$ instead of .05. Be sure to carefully define the population parameter(s) you are testing. *(Source: From* New England Journal of Medicine, *2000, Vol. 342 (12): 861–867. [See p. 701 for complete credit.])*

13.40 Refer to Exercise 13.38. Carry out all parts of the Exercise, but in parts (c) and (d) use the rejection region method, that is, use the Substitute Steps 3 and 4 described on pages 506–507 instead of finding the p-value.

13.41. Refer to Exercise 13.39. Carry out the hypothesis test using the rejection region method, that is, using the Substitute Steps 3 and 4 described on pages 506–507 instead of finding the p-value. Use $\alpha = .01$ instead of .05.

13.42 In Case Study 3.1 (p. 97), results were presented for a sample of 63 men who were asked to report their actual weight and their ideal weight. The mean difference between actual and ideal weight was 2.48 pounds, and the standard deviation of the differences was 13.77 pounds. Is there sufficient evidence to conclude that for the population of men represented by this sample, the actual and ideal weights differ, on average? Use $\alpha = .10$. Justify your answer by showing all steps of a hypothesis test.

13.43 Although we have not emphasized it, the paired t-test can be used to test hypotheses in which the null value is something other than 0. For example, suppose that the proponents of a diet plan claim that the mean amount of weight lost in the first 3 weeks of following the plan is 10 pounds. A consumer advocacy group is skeptical and measures the beginning and ending weights for a random sample of 20 people who follow the plan for 3 weeks. The mean and standard deviation for the difference in weight at the two times are 8 pounds and 4 pounds, respectively.

a. What is the parameter of interest? Be sure to specify the appropriate population.
b. What are the null and alternative hypotheses?
c. What is the value of the test statistic?
d. What is the p-value for the test?
e. What conclusion can the consumer advocacy group make? Use $\alpha = .05$.

13.44 A company manufactures a homeopathic drug that it claims can reduce the time it takes to overcome jet lag after long-distance flights. A researcher would like to test that claim. She recruits nine people who take frequent trips from San Francisco to London and assigns them to take a placebo for one of their trips and the drug for the other trip, in random order. She then asks them how many days it took to recover from jet lag under each condition. The results are as follows:

	Person								
	1	*2*	*3*	*4*	*5*	*6*	*7*	*8*	*9*
Placebo	7	8	5	6	5	3	7	8	4
Drug	4	4	4	6	6	2	8	6	2

Carry out the five steps to test the appropriate hypotheses. Use $\alpha = .05$.

13.45 Many people have high anxiety about visiting the dentist. Researchers want to know if this affects blood pressure in such a way that the mean blood pressure while waiting to see the dentist is higher than it is an hour after the visit. Ten individuals have their systolic blood pressures measured while they are in the dentist's waiting room and again an hour after the conclusion of the visit to the dentist. The data are as follows:

	Person									
	1	*2*	*3*	*4*	*5*	*6*	*7*	*8*	*9*	*10*
B.P. Before	132	135	149	133	119	121	128	132	119	110
B.P. After	118	137	140	139	107	116	122	124	115	103

a. Write the parameter of interest in this situation.
b. Write the null and alternative hypotheses of interest.
c. Carry out the remaining steps to test the hypotheses you specified in part (b). Use $\alpha = .05$.

13.46 *This is Thought Question 13.2.* Read Example 13.2 on page 508. Suppose that the purpose of the study was to determine whether pilots should be allowed to consume alcohol the evening prior to their flights, and that in this study the alcohol consumption occurred 12 hours before the measurement of the time of useful performance. Refer to the discussion of Type 1 and Type 2 errors in Chapter 12.

a. Explain what the consequences of each type of error would be in this situation.
b. Which type of error would be more serious? Explain.
c. Given the data and results of the study, which type of error could have been made?

Section 13.4

Exercises 13.47 to 13.62 correspond to the two Lessons in Section 13.4. Exercises 13.47 to 13.56 require Lesson 1; Exercises 13.57 to 13.62 require both Lesson 1 and Lesson 2.

Skillbuilder Exercises

13.47 In each of the following situations, determine whether the alternative hypothesis was $H_a: \mu_1 - \mu_2 > 0$, $H_a: \mu_1 - \mu_2 < 0$, or $H_a: \mu_1 - \mu_2 \neq 0$.

a. $H_0: \mu_1 - \mu_2 = 0$, $t = -2.33$, df = 8, p-value = .024.
b. $H_0: \mu_1 - \mu_2 = 0$, $t = 2.33$, df = 8, p-value = .976.

13.48 In each of the following situations, determine whether the alternative hypothesis was $H_a: \mu_1 - \mu_2 > 0$, $H_a: \mu_1 - \mu_2 < 0$, or $H_a: \mu_1 - \mu_2 \neq 0$.

a. $H_0: \mu_1 - \mu_2 = 0$, $t = 2.33$, df = 8, p-value = .048.
b. $H_0: \mu_1 - \mu_2 = 0$, $t = -2.33$, df = 8, p-value = .976.

13.49 For each of the following situations, identify whether the appropriate test is a paired t-test or a two-sample t-test:

a. The weights of a sample of 15 marathon runners were taken before and after a training run to test whether marathon runners lose dangerous levels of fluids during a run.
b. Random samples of 200 new freshmen and 200 new transfer students at a university were given a 50-question

test on current events to test whether the level of knowledge of current events differs for new freshmen and transfer students.

13.50 For each of the following situations, identify whether the appropriate test is a paired t-test or a two-sample t-test:

a. Sixty students were matched by initial pulse rate, with the two with the highest pulse forming a pair, and so on. Within each pair, one student was randomly chosen to drink a caffeinated beverage, while the other one drank an equivalent amount of water. Their pulse rates were measured 10 minutes later, to test whether caffeine consumption elevates pulse rates.

b. To determine whether lack of sleep increases appetite, researchers recruited 50 volunteers and randomly assigned 25 of them to sleep at least 8 hours a night and the other 25 to sleep at most 5 hours a night, for 3 days. Calorie intake was recorded for all 50 volunteers for the 3-day period.

13.51 Calculate the value of the unpooled test statistic t in each of the following situations. In each case, assume the null hypothesis is $H_0: \mu_1 - \mu_2 = 0$.

a. $\bar{x}_1 = 35, s_1 = 10, n_1 = 100; \bar{x}_2 = 33, s_2 = 9, n_2 = 81$.

b. The difference in sample means is 48, $s.e.(\bar{x}_1 - \bar{x}_2) = 22$.

13.52. Using the following Minitab output, calculate the value of the unpooled test statistic t for the null hypothesis $H_0: \mu_1 - \mu_2 = 0$.

	N	Mean	StDev	SE Mean
Sample 1	68	80.58	4.22	0.51
Sample 2	68	78.55	3.31	0.40

13.53 ◆ The data in Example 13.1, on body temperature for young adults, was part of a larger set of body temperatures taken on 100 adults at a blood donor center. The data are in the file ***bodytemp*** on the website for this book. The Minitab output below shows the results of the "2-sample t" procedure for comparing the body temperatures of males and females.

Sex	N	Mean	StDev	SE Mean
Female	46	98.102	0.651	0.096
Male	54	97.709	0.762	0.10

```
Difference = mu (Female) − mu (Male)
Estimate for difference: 0.393
95% CI for difference: (0.112, 0.673)
T-Test of difference = 0 (vs not =): T-Value = 2.78
P-Value = 0.007   DF = 97
```

a. Give the null and alternative hypotheses using symbols.

b. What is the value of the test statistic t?

c. Identify the numbers that were used to compute the t-statistic, and verify that the reported value is correct.

d. What conclusion would be made using a .05 level of significance? Write the conclusion in statistical terms and in the context of the problem.

13.54 ◆ Do hardcover and softcover books likely to be found on a professor's shelf have the same average number of pages? Data on the number of pages for ten hardcover and eight softcover books from a professor's shelf were presented in Example 6.2 (p. 192) and are in the file **ProfBooks** on the companion website. The Minitab output from the "2-sample t" procedure follows. Assume that the books are equivalent to a random sample.

Type	N	Mean	StDev	SE Mean
Hard	10	439	158	50
Soft	8	710	163	58

```
Difference = mu (Hard) − mu (Soft)
Estimate for difference: −270.7
95% CI for difference: (−434.2, −107.1)
T-Test of difference = 0 (vs not =): T-Value = −3.55
P-Value = 0.003   DF = 14
```

a. Give the null and alternative hypotheses using symbols.

b. What is the value of the test statistic t?

c. Identify the numbers that were used to compute the t-statistic, and verify that the reported value is correct.

d. What conclusion would be made using a .05 level of significance? Write the conclusion in statistical terms and in the context of the problem.

General Section Exercises

13.55 Example 2.16 (p. 46) gave the weights of eight rowers on each of the Cambridge and Oxford crew teams. The weights are shown again here. Assume that these men represent appropriate random samples from the population of members of these crew teams over all time. Test the hypothesis that the mean weights of the populations of rowers on the Cambridge and Oxford crew teams over all time are equal versus the alternative that they are not equal. Specify the value of α that you chose to use.

Cambridge: 188.5, 183.0, 194.5, 185.0, 214.0, 203.5, 186.0, 178.5

Oxford: 186.0, 184.5, 204.0, 184.5, 195.5, 202.5, 174.0, 183.0

13.56 Case Study 1.1 presented data given in response to the question "What is the fastest you have ever driven a car (mph)?" The summary statistics are:

Females: $n = 102$, mean $= 88.4$, standard deviation $= 14.4$

Males: $n = 87$, mean $= 107.4$, standard deviation $= 17.4$

Assuming that these students represent a random sample of college students, test the null hypothesis that the mean fastest speed driven by college men and college women is equal versus the alternative that it is higher for men. Specify the value of α that you chose to use.

13.57 Example 11.3 (p. 410) presented data from a study in which sedentary men were randomly assigned to be placed on a diet or exercise for a year to lose weight. Forty-two men were placed on a diet, while the remaining 47 were put on an exercise routine. The group on a diet lost an average of 7.2 kg, with a standard deviation of 3.7 kg. The men who exercised lost an average of 4.0 kg, with a standard deviation of 3.9 kg.

a. State and test appropriate null and alternative hypotheses to determine whether the mean weight loss would be different under the two routines for the population of men similar to those in this study. Specify the value of α that you chose to use.

b. Explain how you decided to do a pooled or unpooled test in part (a).

13.58 Example 11.7 (p. 417) presented results for the number of hours slept the previous night from a survey given in two statistics classes. One class was a liberal arts class; the other class was a general introductory class. The survey was given following a Sunday night after classes had started. For simplicity, let's assume that these classes represent a random sample of sleep hours for college students in liberal arts and non-liberal arts majors. The data are as follows:

	n	Mean	Standard Deviation
Liberal arts	25	7.66	1.34
Non-liberal arts	148	6.81	1.73

a. Test the hypothesis that the mean number of hours of sleep for the two populations of students are equal versus the alternative that they are not equal. Use the unpooled *t*-test. Specify the value of α that you chose to use. (*Note:* The approximate df $= 38$ for the unpooled test.)

b. The figure below displays a dotplot of the data. Briefly explain what is indicated about the necessary conditions for doing a two-sample *t*-test.

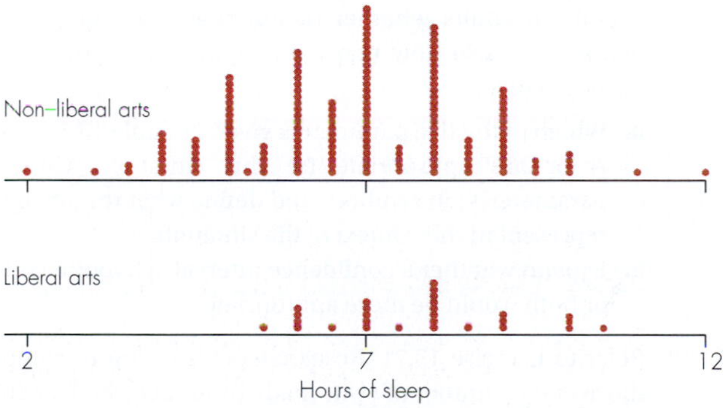

c. Repeat the hypothesis test using the pooled procedure. Compare the results to those in part (a), and discuss which procedure you think is more appropriate in this situation.

13.59 Do students sleep more in Pennsylvania or in California? Data from surveys in elementary statistics classes at Penn State University and the University of California at Davis resulted in the following summary statistics for the number of hours students sleep:

	n	Mean	Standard Deviation	Standard Error of Mean
UC-Davis	173	6.93	1.71	0.13
Penn State	190	7.11	1.95	0.14

Assume that these students are representative of all students at those two schools. Is there sufficient evidence to conclude that the mean hours of sleep are different at the two schools? Carry out all steps of the hypothesis test, and define all parameters. Specify the value of α that you chose to use.

13.60 Students in a statistics class at Penn State were asked, "About how many minutes do you typically exercise in a week?" Responses from the *women* in the class were as follows:

60, 240, 0, 360, 450, 200, 100, 70, 240, 0, 60, 360, 180, 300, 0, 270

Responses from the *men* in the class were as follows:

180, 300, 60, 480, 0, 90, 300, 14, 600, 360, 120, 0, 240

a. Draw appropriate graphs to check whether the conditions for conducting a two-sample *t*-test are met. Discuss the results of your graphs.

b. What additional assumption or condition is required if conclusions are to be made about amount of exercise for the population of all Penn State students on the basis of these sample results?

c. Assume that the conditions are met, and conduct a test to determine whether the mean amount of exercise differs for men and women. Specify the value of α that you chose to use.

13.61 Researchers speculate that drivers who do not wear a seatbelt are more likely to speed than drivers who do wear one. The following data were collected on a random sample of 20 drivers who were clocked to see how fast they were driving (miles per hour), and then were stopped to see whether they were wearing a seat belt (Y = yes, N = no).

Driver	1	2	3	4	5	6	7	8	9	10	11	12	13	14	15	16	17	18	19	20
Speed	62	60	72	85	68	64	72	72	75	63	62	84	76	60	66	63	64	80	52	64
Seatbelt	Y	Y	N	N	Y	Y	Y	N	Y	Y	N	N	N	Y	N	N	Y	Y	Y	Y

Do these results support the claim that the mean speed is higher for the population of drivers who do not wear seatbelts than for the population of drivers who do?

a. Carry out the five steps of hypotheses testing using the unpooled procedure.

b. Repeat part (a) using the pooled procedure.

c. Compare the unpooled and pooled results and discuss which procedure is more appropriate.

d. Carry out the unpooled test using the rejection region approach with $\alpha = .05$.

13.62 In Example 11.12 (p. 428) a study by Slutske, Piasecki, and Hunt-Carter (2003; and on the website for this book) was presented, in which the mean number of hangover symptoms was compared for students whose parents have alcohol problems and students whose parents do not. Researchers are interested in knowing if the mean number of hangover symptoms is higher for the population of students whose parents have alcohol problems than for the population whose parents do not. The sample statistics are as follows:

Group	Mean	Standard Deviation
Parental alcohol problems ($n_1 = 282$)	$\bar{x}_1 = 5.9$	$s_1 = 3.6$
No parental alcohol problems ($n_2 = 945$)	$\bar{x}_2 = 4.9$	$s_2 = 3.4$

a. Carry out the five steps of hypothesis testing using the unpooled procedure.
b. Repeat part (a) using the pooled procedure.
c. Compare the unpooled and pooled results and discuss which procedure is more appropriate.

Section 13.5

Skillbuilder Exercises

13.63 In each of the following cases, explain whether the null hypothesis H_0: $\mu = 25$ can be rejected. Use $\alpha = .05$.

a. 95% confidence interval for μ is (10 to 30), H_a: $\mu \neq 25$.
b. 90% confidence interval for μ is (10 to 30), H_a: $\mu > 25$.
c. 90% confidence interval for μ is (26 to 50), H_a: $\mu > 25$.

13.64 In each of the following cases, explain whether the null hypothesis H_0: $\mu = 25$ can be rejected. Use $\alpha = .05$.

a. 95% confidence interval for μ is (26 to 50), H_a: $\mu \neq 25$.
b. 90% confidence interval for μ is (10 to 30), H_a: $\mu < 25$.
c. 90% confidence interval for μ is (26 to 50), H_a: $\mu < 25$.

13.65 Refer to the rules for the relationship between confidence intervals and tests with two-sided alternatives, given in the two bullets on page 521.

a. Rewrite the rules specifically for $\alpha = .05$.
b. Rewrite the rules specifically for $\alpha = .01$.

13.66 Refer to the rules for the relationship between confidence intervals and one-sided tests given in the three bullets and sentence preceding them on page 522.

a. Rewrite the rules specifically for $\alpha = .05$.
b. Rewrite the rules specifically for $\alpha = .01$.

General Section Exercises

13.67 Each of the following presents a two-sided 95% confidence interval and the alternative hypothesis of a corresponding hypothesis test. In each case, state a conclusion for the test, including the level of significance you are using.

a. C.I. for μ is (101 to 105), H_a: $\mu \neq 100$.
b. C.I. for p is (.12 to .28), H_a: $p < .10$.
c. C.I. for $\mu_1 - \mu_2$ is (3 to 15), H_a: $\mu_1 - \mu_2 > 0$.
d. C.I. for $p_1 - p_2$ is (−0.15 to 0.07), H_a: $p_1 - p_2 \neq 0$.

13.68 As was stated in Section 13.5, "a confidence interval can be used as another way to conduct a two-sided significance test." If a test were conducted by using this method, would the p-value for the test be available? Explain.

13.69 For each of the following situations, can you conclude whether a two-sided 90% confidence interval for μ would include the value 10? If so, make the conclusion. If not, explain why you can't tell.

a. H_0: $\mu = 10$, H_a: $\mu < 10$, do not reject the null hypothesis for $\alpha = .05$.
b. H_0: $\mu = 10$, H_a: $\mu < 10$, reject the null hypothesis for $\alpha = .05$.
c. H_0: $\mu = 10$, H_a: $\mu \neq 10$, do not reject the null hypothesis for $\alpha = .10$.
d. H_0: $\mu = 10$, H_a: $\mu \neq 10$, reject the null hypothesis for $\alpha = .10$.

Section 13.6

Skillbuilder Exercises

13.70 Researchers want to know what proportion of a certain type of tree growing in a national forest suffers from a disease. They test a representative sample of 200 of the trees from around the forest and find that 15 of them have the disease.

a. Which of the five parameters given in Table 13.3 (p. 524) is or are appropriate for this situation? Give the parameter(s) in symbols, and define what the symbol(s) represent in the context of the situation.
b. Explain whether a confidence interval, a hypothesis test, or both would be more appropriate.

13.71 Researchers want to know what percentage of adults have a fear of going to the dentist. They also want to know the average number of visits made to a dentist in the past 10 years for adults who have that fear. They ask a random sample of adults whether or not they fear going to the dentist and also how many times they have gone in the past 10 years.

a. Which of the five parameters given in Table 13.3 (p. 524) is or are appropriate for this situation? Give the parameter(s) in symbols, and define what the symbol(s) represent in the context of the situation.
b. Explain whether a confidence interval, a hypothesis test, or both would be more appropriate.

13.72 Refer to Exercise 13.71. Researchers also want to compare the average number of visits made by adults who fear going to the dentist with the average number of visits for those who don't have the fear.

a. Which of the five parameters given in Table 13.3 (p. 524) is or are appropriate for this situation? Give the parameter(s) in symbols, and define what the symbol(s) represent in the context of the situation.
b. Explain whether a confidence interval, a hypothesis test, or both would be more appropriate.

13.73 Researchers want to compare the mean running times for the 50-yard dash for first-grade boys and girls. They select a random sample of 35 boys and 32 girls in first grade and time them running the 50-yard dash.

a. Which of the five parameters given in Table 13.3 (p. 524) is or are appropriate for this situation? Give the parameter(s) in symbols, and define what the symbol(s) represent in the context of the situation.

b. Explain whether a confidence interval, a hypothesis test, or both would be more appropriate.

General Section Exercises

13.74 Give an example of a situation for which the appropriate inference procedure would be each of the following:

a. A hypothesis test for one proportion.
b. A hypothesis test and confidence interval for a paired difference mean.
c. A confidence interval for one mean.
d. A confidence interval for the difference in two means for independent samples.

13.75 Refer to the following two scenarios from exercises in various chapters of this book. In each case, determine the most appropriate inference procedure(s), including the appropriate parameter. If you think that inference about more than one parameter may be of interest, answer the question for all parameters of interest. Explain your choices.

a. A sample of college students was asked whether they would return the money if they found a wallet on the street. Of the 93 women, 84 said "yes," and of the 75 men, 53 said "yes." Assume that these students represent all college students.
b. A study was conducted on pregnant women and subsequent development of their children (Olds et al., 1994). One of the questions of interest was whether the IQs of children are lower for mothers who smoke at least ten cigarettes a day during pregnancy than for children of mothers who do not smoke at all.

13.76 Refer to the following two scenarios from exercises in various chapters of this book. In each case, determine the most appropriate inference procedure(s), including the appropriate parameter. If you think that inference about more than one parameter may be of interest, answer the question for all parameters of interest. Explain your choices.

a. Max likes to keep track of birthdays of people he meets. He has 170 birthdays on his birthday calendar. One cold January night, he comes up with the theory that people are more likely to be born in October than they would be if all 365 days were equally likely. He consults his birthday calendar and finds that 22 of the 170 birthdays are in October.
b. The dataset **cholest** reports cholesterol levels of heart attack patients 2, 4, and 14 days after the heart attack. Data for days 2 and 4 were available for 28 patients. The mean difference (2 day − 4 day) was 23.29, and the standard deviation of the differences was 38.28.

Section 13.7

Skillbuilder Exercises

13.77 Compute the effect size for each of the following situations, and state whether it would be considered closer to a small, medium, or large effect:

a. In a one-sample test with $n = 100$, the test statistic is $t = 2.24$.
b. In a one-sample test with $n = 50$, the test statistic is $t = -2.83$.

13.78 Compute the effect size for each of the following situations, and state whether it would be considered closer to a small, medium, or large effect:

a. In a paired-difference test with $n = 30$ pairs, the test statistic is $t = 1.48$.
b. In a test for the difference in two means with independent samples, with $n_1 = 40$ and $n_2 = 50$, the test statistic is $t = -2.33$.

13.79 Refer to Table 13.4 (p. 530), which presents power for a one-sided, one-sample t-test. In a test of H_0: $\mu_d = 0$ versus H_a: $\mu_d > 0$, suppose that the truth is that the population of differences is a normal distribution with mean $\mu_d = 2$ and standard deviation $\sigma = 4$.

a. Recalling the Empirical Rule from Chapter 2, draw a picture of this distribution, showing the ranges into which 68% and 95% of the differences fall.
b. On your picture, indicate where the null value of 0 falls.
c. If a sample of size 20 is taken, what is the power for the test, assuming that a .05 level of significance will be used?
d. Explain in words what probability the power of the test represents.

13.80 In Table 13.4 (p. 530), it is shown that the power is .40 for a one-sided, one-sample t-test with .05 level of significance, $n = 50$, and true effect size of 0.2. Would the power be higher or lower for each of the following changes?

a. The true effect size is 0.4.
b. The sample size used is $n = 75$.
c. The level of significance used is .01.

General Section Exercises

13.81 (Computer software is required.) Find the power for the following one-sample t-test situations. In each case, assume that a .05 level of significance will be used.

a. Effect size = 0.3, sample size = 45, H_a: $\mu > \mu_0$.
b. Sample size = 30, H_a: $\mu \neq 10$, true mean = 13, $\sigma = 4$.
c. Effect size = −1.0, sample size = 15, H_a: $\mu < \mu_0$.

13.82 (Computer software is required.) In parts (a) to (d), find the sample size necessary to achieve power of .80 for a one-sample t-test with H_a: $\mu > \mu_0$ and level of significance of .05 for each of the following effect sizes:

a. 0.2.
b. 0.4.
c. 0.6.
d. 0.8.
e. Make a scatterplot of the sample size (vertical axis) versus the effect size for the effect sizes in parts (a) to (d). Does the relationship between the effect size and the sample size required to achieve 80% power appear to be linear? If not, what is the nature of the relationship?

13.83 Refer to Figure 13.9 on page 530, showing power curves for a one-sided t-test. Use the Figure to give the approximate power in each of the following situations, and write a sentence explaining what probability is represented by the power.

a. $n = 20$, true effect size $= 0.4$.
b. $n = 50$, true effect size $= 0.4$.
c. $n = 100$, true effect size $= 0.4$.
d. $n = 20$, true effect size $= 0.7$.

13.84 Refer to Figure 13.9 on page 530, showing power curves for a one-sided t-test. Use the Figure to answer the following questions for a one-sided t-test with $\alpha = .05$. Explain your answers.

a. If the true effect size is 0.3, would a sample of size $n = 20$ be large enough to achieve power of at least .4?
b. If the true effect size is 0.4 and the sample size is 50, what is the approximate probability that the null hypothesis will be rejected?
c. Of the three sample sizes shown, which (one or more) of them is large enough to be sure that the power is at least .8 for a true effect size of 0.4?

13.85 Explain why it is more useful to compare effect sizes than p-values in trying to determine whether many studies about the same topic have found similar results.

13.86 For a z-test for one proportion, a possible effect size measure is $(p - p_0)/\sqrt{p_0(1 - p_0)}$ where p_0 is the null value and p is the true population proportion, which can be estimated using the sample proportion.

a. What is the relationship between this effect size and the z-test statistic for this situation?
b. Does the effect size fit the relationship "Test statistic = Size of effect × Size of study"? If not, explain why not. If so, show how it fits.
c. What would be a reasonable way to estimate this effect size?

13.87 Refer to the effect-size measure in Exercise 13.86. For parts (a) to (c), compute the effect size.

a. $p = .35$, $p_0 = .25$.
b. $p = .15$, $p_0 = .05$.
c. $p = .95$, $p_0 = .85$.
d. On the basis of the results in parts (a) to (c), does this effect size stay the same when $p - p_0$ stays the same? Explain.
e. In statistical software for computing power for a test for a single proportion, unlike for a single mean, both p and p_0 must be specified, rather than just the difference between them. Explain why it is not enough to specify the difference, using the results of the previous parts of this exercise.

Section 13.8 and Case Study 13.1

Skillbuilder Exercises

13.88 Refer to the checklist of issues on pp. 531–532. Explain which ones should be of concern if the sample size(s) for a test are small.

13.89 Refer to the checklist of issues on pp. 531–532. Explain which ones should be of concern if the sample size(s) for a test are large.

General Section Exercises

13.90 Refer to the statement in Item 6 in the checklist of issues on pp. 531–532 that begins with "Remember that if all of the null hypotheses tested...." Is that statement the same thing as saying that the null hypothesis is likely to be true in about 1 out of 20 tests that have achieved statistical significance? Explain.

13.91 Refer to the checklist of issues on pp. 531–532.

a. For which of the concerns would the p-value for a test be useful to have? Explain why in each case.
b. For which of the concerns would a confidence interval estimate for the parameter be useful to have? Explain why in each case.
c. For which of the concerns would the sample size(s) be useful to know? Explain why in each case.

13.92 *This is also Exercise 1.42b.* Suppose that you were to read the following news story: "Researchers compared a new drug to a placebo for treating high blood pressure, and it seemed to work. But the researchers were concerned because they found that significantly more people got headaches when taking the new drug than when taking the placebo. Headaches were the only problem out of the 20 possible side effects the researchers tested." Do you think the researchers are justified in thinking the new drug would cause more headaches in the population than the placebo would? Explain.

13.93 *This is also Exercise 1.19.* A (hypothetical) study of what people do in their spare time found that people born under the astrological sign of Aries were significantly more likely to be regular swimmers than people born under other signs. What additional information would you want to know to help you determine if this result is a false positive?

13.94 Using the Bonferroni method, what significance level should be used for each individual test in each of the following situations?

a. Overall significance level is .10 and five tests are to be conducted.
b. Overall significance level is .05 and two tests are to be conducted.

13.95 Answer the following questions about the study reported as Case Study 13.1. Explain your answers.

a. Was the study a randomized experiment or an observational study?
b. Based on your answer in part (a), can it be concluded that drinking an ice slushie before exercising in the heat *causes* time to exhaustion to be longer?
c. Could this study have been done as a double-blind study, a single-blind study, or neither? Explain whether this could have affected the results of the study.

Chapter Exercises

13.96 For each of the following research questions, specify the parameter and the value that constitute the null hypothesis of "parameter = null value." In other words, define the population parameter of interest and specify the null value that is being tested.

a. Do a majority of Americans between the ages of 18 and 30 think that the use of marijuana should be legalized?

b. Is the mean of the Math SAT scores in California in a given year different from the target mean of 500 set by the test developers?

c. Is the mean age of death for left-handed people lower than that for right-handed people?

d. Is there a difference in the proportions of male and female college students who smoke cigarettes?

13.97 Refer to Exercise 13.96. In each case, specify whether the alternative hypothesis would be one-sided or two-sided.

13.98 A study is conducted to find out whether results of an IQ test are significantly higher after listening to Mozart than after sitting in silence. Explain what has happened in each of the following scenarios:

a. A Type 1 error was committed.

b. A Type 2 error was committed.

c. The power of the test was too low to detect the difference that actually exists.

d. The power of the test was so high that a very small difference resulted in a statistically significant finding.

13.99 Suppose that a one-sample *t*-test of H_0: $\mu = 0$ versus H_a: $\mu \neq 0$ results in a test statistic of $t = 0.65$ with df = 14. Suppose that a new study is done with $n = 150$, and the sample mean and standard deviation turn out to be exactly the same as in the first study.

a. What conclusion would you reach in the original study, using $\alpha = .05$?

b. What conclusion would you reach in the new study, using $\alpha = .05$?

c. Compare your results in parts (a) and (b) and comment.

13.100 *This exercise is Thought Question 13.3.* The paired *t*-test introduced in Section 13.3 and the two-sample *t*-test introduced in Section 13.4 are both used to compare two sets of measurements, and the null hypothesis in both cases is usually that the mean population difference is 0.

a. Explain the difference in the situations for which they are used.

b. Suppose that researchers wanted to know if college students spend more time watching TV or exercising. Explain how they could collect data appropriate for a paired *t*-test and how they could collect data appropriate for a two-sample *t*-test.

13.101 In a random sample of 170 married British couples, the difference between the husband's and wife's ages had a mean of 2.24 years and a standard deviation of 4.1 years.

a. Test the hypothesis that British men are significantly older than their wives, on average. Specify the value of α that you chose to use.

b. Explain what is meant by the use of the term *significant* in part (a), and discuss how it compares with the everyday use of the word.

13.102 Suppose that a highway safety researcher makes modifications to the design of a highway sign. The researcher believes that the modifications will make the mean maximum distance at which drivers are able to read the sign greater than 450 feet. The maximum distances (in feet) at which $n = 16$ drivers can read the sign are as follows:

440, 490, 600, 540, 540, 600, 240, 440, 360, 600, 490, 400, 490, 540, 440, 490

a. Check whether the conditions are met for using a *t*-test to determine if the mean maximum sign-reading distance at which drivers can read the sign is greater than 450. If they are not met, explain why not.

b. Show all five steps of a hypothesis test, and be sure to state a conclusion. If you determined in part (a) that conditions are not met, make appropriate adjustments before doing the test. Describe any assumptions that you made when carrying out the test. Use $\alpha = .05$.

13.103 *This is Thought Question 13.4.* Refer to Example 13.9. A 95% confidence interval for the difference in proportions of children who would get ear infections with placebo compared to with xylitol was 0.02 to 0.226. On the basis of this information, specify a one-sided 97.5% confidence interval and explain how you would use it to test H_0: $p_1 - p_2 = 0$ versus H_a: $p_1 - p_2 > 0$ with $\alpha = .025$.

13.104 In Exercise 11.82, a study is described in which the mean IQs at age 4 for children of smokers (at least ten cigarettes a day) and nonsmokers were compared. The mean for the children of the 66 nonsmokers was 113.28 points, while for the children of the 47 smokers it was 103.12 points. Assume that the pooled standard deviation for the two samples is 13.5 points and that a pooled *t*-test is appropriate. Test the hypothesis that the population mean IQ at age 4 is the same for children of smokers and of nonsmokers versus the alternative that it is higher for children of nonsmokers. Use $\alpha = .01$.

13.105 Refer to Exercise 13.104. One of the statements made in the research article and reported in Exercise 11.82 was "After control for confounding background variables ... the average difference observed at 36 and 48 months was reduced to 4.35 points (95% CI: 0.02, 8.68)." Use this statement to test the null hypothesis that the difference in the mean IQ scores for the two populations is 0 versus the alternative that it is greater than 0. (Note that the results have now been adjusted for confounding variables such as parents' IQ, whereas the data given in Exercise 13.104 had not.) Specify the value of α that you must use given the available information.

13.106 It is believed that regular physical exercise leads to a lower resting pulse rate. Following are data for $n = 20$ individuals on resting pulse rate and whether the individual regularly exercises or not. Assuming that this is a random sample from a larger population, use this sample to determine if the mean pulse is lower for those who exercise.

Clearly show all five steps of the hypothesis test. Use $\alpha = .05$ and specify whether you are using the pooled or unpooled procedure.

Person	Pulse	Regularly Exercises	Person	Pulse	Regularly Exercises
1	72	No	11	62	No
2	62	Yes	12	84	No
3	72	Yes	13	76	No
4	84	No	14	60	Yes
5	60	Yes	15	52	Yes
6	63	Yes	16	60	No
7	66	No	17	64	Yes
8	72	No	18	80	Yes
9	75	Yes	19	68	Yes
10	64	Yes	20	64	Yes

13.107 In an experiment conducted by one of the authors, ten students in a graduate-level statistics course were given this question about the population of Canada: "The population of the U.S. is about 270 million. To the nearest million, what do you think is the population of Canada?" (The population of Canada at the time was slightly over 30 million.) The responses (in millions) were as follows:

20, 90, 1.5, 100, 132, 150, 130, 40, 200, 20

Eleven other students in the same class were given the same question with different introductory information: "The population of Australia is about 18 million. To the nearest million, what do you think is the population of Canada?" The responses (in millions) were as follows:

12, 20, 10, 81, 15, 20, 30, 20, 9, 10, 20

The experiment was done to demonstrate the **anchoring effect**, which is that responses to a survey question may be "anchored" to information provided to introduce the question. In this experiment, the research hypothesis was that the individuals who saw the U.S. population figure would generally give higher estimates of Canada's population than would the individuals who saw the Australian population figure.

a. Write null and alternative hypotheses for this experiment. Use proper notation.

b. Test the hypotheses stated in part (a). Be sure to state a conclusion in the context of the experiment. Specify the value of α that you chose to use.

c. As a step in part (b), you should have created a graphical summary to verify necessary conditions. Do you think that any possible violations of the necessary conditions have affected the results of part (b) in a way that produced a misleading conclusion? Explain why or why not.

13.108 Refer to Exercise 13.107. The experiment was repeated in 2010 using updated population figures of 308 million for the United States and 22 million for Australia when the questions were asked. At the time, the population of Canada was about 34 million. Here are the summary statistics for the results:

Reference country	n	Mean	Standard Deviation	Standard Error of Mean
United States	54	172	100	14
Australia	34	39.4	35.4	6.1

a. Do you think that a pooled t-test would be appropriate for these results? Explain.

b. Write null and alternative hypotheses for this experiment. Use proper notation. (Refer to Exercise 13.107 for details of the hypotheses of interest.)

c. Test the hypotheses stated in part (b). Be sure to state a conclusion in the context of the experiment.

13.109 Refer to Case Study 13.1. The mean and standard deviation for the 10 differences in times to exhaustion after drinking a slushie and drinking cold water were $\bar{d} = 9.5$ minutes and $s_d = 3.6$ minutes.

a. Compute a two-sided 95% confidence interval for the population mean difference.

b. Use the interval you computed in part (a) to test the hypothesis that the population mean difference is 0 versus the hypothesis that it is not equal to 0.

c. Use the interval in part (a) to compute a 97.5% one-sided confidence interval for the mean difference, that could be used to test the alternative hypothesis that the mean difference is greater than 0.

d. Use the interval you computed in part (c) to test the alternative hypothesis that the population mean difference is greater than 0. What significance level accompanies this test?

13.110 ◆ The dataset **cholest** on the companion website reports cholesterol levels of heart attack patients 2, 4, and 14 days after the heart attack. Data for days 2 and 4 were available for 28 patients. The mean difference (2 day − 4 day) was 23.29, and the standard deviation of the differences was 38.28. A histogram of the differences was approximately bell-shaped. Is there sufficient evidence to indicate that the mean cholesterol level of heart attack patients decreases from the second to fourth days after the heart attack? Use $\alpha = .01$.

13.111 Refer to Exercise 13.110. Suppose that physicians will use the answer to that question to decide whether to retest patients' cholesterol levels on day 4. If there is no conclusive evidence that the cholesterol level goes down, they will use the day 2 level to decide whether to prescribe drugs for high cholesterol. If there is evidence that cholesterol goes down between days 2 and 4, patients will all be retested on day 4, and the prescription drug decision will be made then.

a. What are the consequences of Type 1 and Type 2 errors for this setting?

b. Which type of error do you think is more serious?

Dataset Exercises

Datasets required to solve these exercises are available on the companion website, http://www.cengage.com/statistics/Utts4e.

13.112 Refer to Exercises 13.110 and 13.111. Using the dataset **cholest,** determine whether there is sufficient evidence to conclude that the cholesterol level drops, on average, from day 2 to day 14 after a heart attack.

13.113 The dataset **UCDavis2** includes information on ***Sex***, ***Height***, and ***Dadheight***. Use the data to test the hypothesis that college men are taller, on average, than their fathers. Assume that the male students in the survey represent a random sample of college men.

13.114 The dataset **UCDavis2** includes grade point average ***GPA*** and answers to the question: "Where do you typically sit in a classroom (circle one): Front, Middle, Back." The answer to this question is coded as F, M, B for the variable ***Seat***. Assuming these students represent all college students, test whether there is a difference in mean GPA for students who sit in the front versus the back of the classroom.

13.115 The dataset **deprived** includes information on self-reported amount of sleep per night and whether a person feels sleep-deprived for $n = 86$ college students (Source: Laura Simon, Pennsylvania State University). Assume that the students represent a random sample of college students. Is the mean number of hours of sleep per night lower for the population of students who feel sleep-deprived than it is for the population of students who do not? Use an unpooled test.

14

How is her handspan related to her height?

See Figure 14.1 *(p. 550)*

Inference about Simple Regression

In Chapter 3, we used regression to describe a relationship in a sample. Now we make inferences about the population represented by the sample. What is the relationship between handspan and height in the population? What is the mean handspan for people who are 65 inches tall? What interval covers the handspans of most individuals of that height?

We learned in Chapter 3 that a straight line often describes the pattern of a relationship between two quantitative variables. For instance, in Example 3.6 we explored the relationship between the handspans (centimeters) and heights (inches) of 167 college students, and we found that the pattern of the relationship in this sample could be described by the equation

Average handspan = −3 + 0.35 (Height)

An equation like the one relating handspan to height is called a *regression equation*, and the term **simple regression** is sometimes used to describe the analysis of a straight-line relationship (*linear relationship*) between a response variable (y variable) and an explanatory variable (x variable).

In Chapter 3, we used regression methods only to describe a sample and did not make statistical inferences about the larger population. Now we consider how to make inferences about a relationship in the population represented by the sample. Some questions involving the population that we might ask when analyzing a relationship are the following:

1. Does the observed relationship also occur in the population? For example, is the observed relationship between handspan and height strong enough to conclude that a similar relationship also holds in the population?
2. For a linear relationship, what is the slope of the regression line in the population? For example, in the larger population, what is the slope of the regression line that connects handspans to heights?
3. What is the mean value of the response variable (y) for individuals with a specific value of the explanatory variable (x)? For example, what is the mean handspan in a population of people who are 65 inches tall?
4. What interval of values predicts the value of the response variable (y) for an individual with a specific value of the explanatory variable (x)? For example, what interval predicts the handspan of an individual who is 65 inches tall?

14.1 Sample and Population Regression Models

A **regression model** describes the relationship between a quantitative response variable (the y variable) and one or more explanatory variables (x variables). The y variable is sometimes called the **dependent variable**, and because regression models may be used to make predictions, the x variables may be called the **predictor variables**. The labels *response variable* and *explanatory variable* may be used for the variables on the y axis and x axis, respectively, even if there is no obvious way to assign these labels in the usual sense.

Any regression model has two important components. The most obvious component is the equation that describes how the mean value of the y variable is connected to specific values of the x variable. The equation for the connection between handspan and height, Average handspan = −3 + 0.35 (Height), is an example. In this chapter, we focus on *linear relationships*, so a straight-line equation will be used, but it is important to note that some relationships are *curvilinear*.

The second component of a regression model describes how individuals vary from the regression line. Figure 14.1, which is identical to Figure 3.6, displays the raw data for the sample of $n = 167$ handspans and heights along with the regression line that estimates how the mean handspan is connected to specific heights. Note that most individuals vary from the line. When we examine sample data, we will find it useful to estimate the general size of the deviations from the line. When we consider a model for the relationship within the population represented by a sample, we will state assumptions about the distribution of deviations from the line.

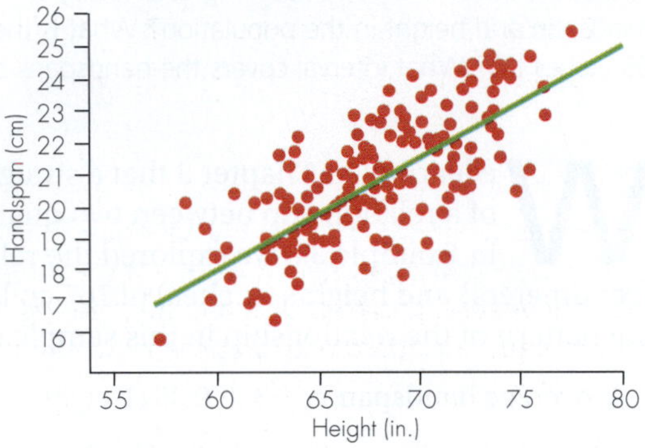

Figure 14.1 Regression line linking handspan and height for a sample of college students

If the sample represents a larger population, we need to distinguish between the **regression line for the sample** and the **regression line for the population**. The observed data can be used to determine the regression line for the sample, but the regression line for the population can only be imagined. Because we do not observe the whole population, we will not know numerical values for the intercept and slope of the regression line in the population. As in nearly every statistical problem, the statistics from a sample are used to estimate the unknown population parameters, which in this case are the slope and intercept of the regression line.

The Regression Line for the Sample

In Chapter 3, we introduced this notation for the regression line that describes sample data:

$$\hat{y} = b_0 + b_1x$$

In any given situation, the sample is used to determine values for b_0 and b_1.

- $\hat{y}$ is pronounced "y-hat" and is also referred to as *predicted y* or *estimated mean of y*.
- b_0 is the **intercept** of the straight line, also called the **y-intercept**. The *intercept* is the value of $\hat{y}$ when $x = 0$.
- b_1 is the **slope** of the straight line. The *slope* tells us how much of an increase (or decrease) there is for $\hat{y}$ when the x variable increases by one unit. The sign of the slope tells us whether $\hat{y}$ increases or decreases when x increases. If the slope is 0, there is no linear relationship between x and y because $\hat{y}$ is the same for all values of x.

The equation describing the relationship between handspan and height for the sample of college students can be written as follows:

$$\hat{y} = -3 + 0.35x$$

In this equation,

- $\hat{y}$ estimates the average handspan for any specific height x. If height = 70 inches, for instance, $\hat{y} = -3 + 0.35(70) = 21.5$ cm.
- The *intercept* is $b_0 = -3$. While necessary for the line, this value does not have a useful statistical interpretation in this example. It estimates the average handspan for individuals who have height = 0 inches, an impossible height, far from the range of the observed heights. It also is an impossible handspan.
- The *slope* is $b_1 = 0.35$. This value tells us that the *average increase* in handspan is 0.35 cm for every 1-inch increase in height.

We learned in Chapter 3 that a regression equation can be used to predict y for a given value of x, or to estimate the mean value of y for all individuals with a given value of x. Therefore, in this instance the equation can be written in words in the following ways:

Estimated mean handspan = −3 + 0.35(height)

Predicted handspan = −3 + 0.35(height)

Reminder: The Least-Squares Criterion

In Chapter 3, we described the least-squares criterion. This mathematical criterion is used to determine numerical values of the intercept and slope of a sample regression line. The **least-squares line** is the line, among all possible lines, that has the smallest sum of squared differences between the sample values of y and the corresponding values of $\hat{y}$.

Deviations from the Regression Line in the Sample

The terms *random error*, *residual variation*, and *residual error* are used as synonyms for the term **deviation**. Most commonly, the word **residual** is used to describe the deviation of an observed y value from the sample regression line. A *residual* is easy to compute. It is simply the difference between the observed y value for an individual and the value of $\hat{y}$ determined from the x value for that individual.

Example 14.1 **Residuals in the Handspan and Height Regression** Consider a person who is 70 inches tall whose handspan is 23 cm. The sample regression line is $\hat{y} = -3 + 0.35x$, so predicted handspan $= \hat{y} = -3 + 0.35(70) = 21.5$ cm for this person. The *residual* = observed y − predicted $y = y - \hat{y} = 23 - 21.5 = 1.5$ cm. Figure 14.2 illustrates this residual.

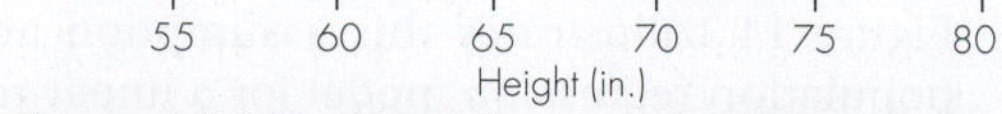

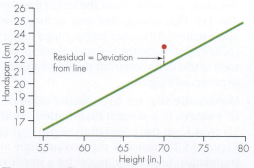

Figure 14.2 Residual for a person 70 inches tall with a handspan = 23 cm. The residual is the difference between observed $y = 23$ and $\hat{y} = 21.5$, the predicted value for a person 70 inches tall.

DEFINITION For an observation y_i in the sample, the **residual** is

$$e_i = y_i - \hat{y}_i$$

$y_i =$ the value of the response variable for the observation.

$\hat{y}_i = b_0 + b_1 x_i$, where x_i is the value of the explanatory variable for the observation.

Note that the residual is the *vertical* distance from the data point to the regression line.

TECHNICAL NOTE The sum of the residuals is 0 for any least-squares regression line. The least-squares formulas for determining the equation always result in $\sum y_i = \sum \hat{y}_i$, so $\sum e_i = 0$.

The Regression Line for the Population

The regression equation for a simple linear relationship in a population can be written as follows:

$$E(Y) = \beta_0 + \beta_1 x$$

- $E(Y)$ represents the mean or expected value of y for individuals in the population who all have the same particular value of x. Note that $\hat{y}$ is an estimate of $E(Y)$.
- β_0 is the **intercept** of the straight line in the **population**. It is estimated by b_0.
- β_1 is the **slope** of the straight line in the **population**. It is estimated by b_1. Note that if the slope $\beta_1 = 0$, there is no linear relationship in the population.

Unless we measure the entire population, we cannot know the numerical values of β_0 and β_1. These are population parameters that we estimate using the corresponding sample statistics. In the handspan and height example, $b_1 = 0.35$ is a sample statistic that estimates the population parameter β_1, and $b_0 = -3$ is a sample statistic that estimates the population parameter β_0.

Assumptions about Deviations from the Regression Line in the Population

To make statistical inferences about the population, two assumptions about how the y values vary from the population regression line are necessary:

- We assume that there is a **constant variance**, meaning that the general size of the deviation of y values from the line is the same for all values of the explanatory variable (x). This assumption may or may not be correct in any particular situation, and a scatterplot should be examined to see whether it is reasonable. In Figure 14.1 (p. 550), the constant variance assumption looks reasonable because the magnitude of the deviation from the line appears to be about the same across the range of observed heights.
- We assume that for any specific value of x, the distribution of the population of y values is a normal distribution. Equivalently, this assumption is that deviations from the population regression line have a normal curve distribution. Figure 14.3 illustrates this assumption along with the other elements of the population regression model for a linear relationship. The line $E(Y) = \beta_0 + \beta_1 x$ describes the mean of y, and the normal curves describe deviations from the mean.

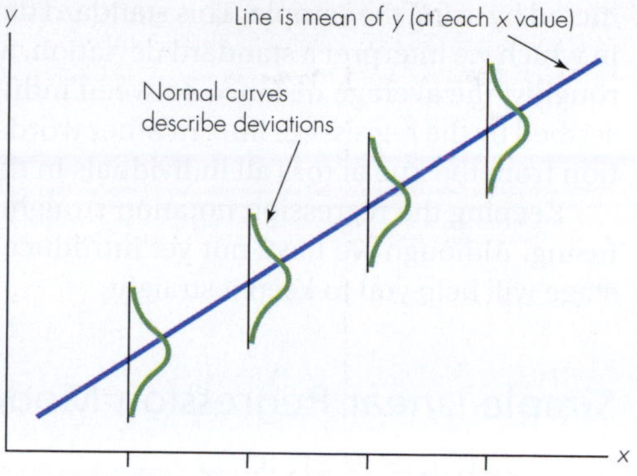

Figure 14.3 Regression model for population

The Simple Regression Model for a Population

A useful format for expressing the components of the population regression model is

y = Mean + Deviation = Part of y explained by knowing x + Unexplained part of y.

This conceptual equation states that for any individual, the value of the response variable (y) can be constructed by combining two components:

1. The *mean* for all individuals in the population who have the same value of x, which in the population is the line $E(Y) = \beta_0 + \beta_1 x$, assuming the relationship is linear. This part of y is determined, or *explained*, by knowing the value of x for the individual.
2. The individual's *deviation* = *y* − *mean*, which is what is left *unexplained* after accounting for the mean y value at that individual's x value.

This format also applies to the sample, but we use the term *estimated mean* when referring to the sample regression line and *residual* when referring to the deviation.

TECHNICAL NOTE There are possible relationships other than linear, such as curvilinear, a special case of which is a quadratic relationship, $E(Y) = \beta_0 + \beta_1 x + \beta_2 x^2$. Relationships that are not linear will not be discussed further in this book.

Example 14.2 **Mean and Deviation for Height and Handspan Regression** Recall from Example 14.1 that the sample regression line for handspans (y) and heights (x) is $\hat{y} = -3 + 0.35x$. Although it is not likely to be true, let's assume for convenience that this equation also holds in the population, so $E(Y) = -3 + 0.35x$. If your height is $x =$ 70 inches and your handspan is $y = 23$ cm, then,

$$\text{Mean} = E(Y) = -3 + 0.35(70) = 21.5$$
$$\text{Deviation} = y - \text{Mean} = 23 - 21.5 = 1.5$$
$$y = 23 = \text{Mean} + \text{Deviation} = 21.5 + 1.5$$

In other words, the mean handspan for people with your height is 21.5 cm, and your handspan is 1.5 cm above that mean.

In the theoretical development of procedures for making statistical inferences for a regression model, the collection of all *deviations* in the population is assumed to have a normal distribution with mean 0 and standard deviation σ (so the variance is σ^2). The value of the standard deviation σ is an unknown population parameter that is estimated by using the sample. This standard deviation can be interpreted in the usual way in which we interpret a standard deviation, applied to the deviations from the line. It is, roughly, the average distance between individual values of y and the mean of y as described by the regression line. In other words, it is roughly the size of the average deviation from the line across all individuals in the range of x values.

Keeping the regression notation straight for populations and samples can be confusing. Although we have not yet introduced all relevant notation, a summary at this stage will help you to keep it straight.

IN SUMMARY

Simple Linear Regression Model: Population and Sample Versions

For $(x_1, y_1), (x_2, y_2), \dots, (x_n, y_n)$, a sample of n observations of the explanatory variable x and the response variable y from a large population, the **simple linear regression model** describing the relationship is as follows.

Population Version

Mean: $E(Y) = \beta_0 + \beta_1 x$

Individual: $y_i = \beta_0 + \beta_1 x_i + \varepsilon_i = E(Y_i) + \varepsilon_i$

The deviations ε_i are assumed to follow a normal distribution with mean 0 and standard deviation σ.

Sample Version

Mean: $\hat{y} = b_0 + b_1 x$

Individual: $y_i = b_0 + b_1 x_i + e_i = \hat{y}_i + e_i$

where e_i is the *residual* for individual i. The sample statistics b_0 and b_1 *estimate* the population parameters β_0 and β_1. The mean of the residuals is 0, and the residuals are used to estimate the population standard deviation σ.

Multiple Regression

In **multiple regression**, the mean of the response variable is a function of two or more explanatory variables. Put another way, in multiple regression we use the values of more than one explanatory (predictor) variable to predict the value of a response variable. For example, a college admissions committee might predict college GPA for an applicant based on using Verbal SAT, Math SAT, high school GPA, and class rank. The general structure of an equation for doing this might be

$$\text{College GPA} = \beta_0 + \beta_1 \text{ Verbal SAT} + \beta_2 \text{ Math SAT} + \beta_3 \text{ HS GPA} + \beta_4 \text{ Class Rank}$$

As in simple regression, numerical estimates of the parameters β_0, β_1, β_2, β_3, and β_4 would be determined from a sample.

As another example, when a girl is born can we use her parents' heights to predict how tall she will be as an adult? The dataset **ucdwomht** on the companion website, http://www.cengage.com/statistics/Utts4e, can be used for this purpose. It includes heights for 90 female students, along with the heights of the student's mother and father. Using this dataset, we find the regression relationship:

$$\text{Predicted height} = 24.5 + 0.313 \text{ momheight} + 0.291 \text{ dadheight}$$

Thus, if a girl's mother is 62 inches tall and her father is 67 inches tall, we predict her height to be:

$$24.5 + 0.313 \text{ momheight} + 0.291 \text{ dadheight}$$
$$= 24.5 + 0.313(62) + 0.291(67) = 63.4 \text{ inches.}$$

As in simple linear regression, not all women whose parents have these heights will be exactly 63.4 inches tall. There is substantial natural variability in heights, but knowing parents' heights partially explains why a woman reaches the height she does.

Multiple regression is covered in detail as Supplemental Topic 3 on the companion website and will not be covered further in this chapter.

14.1 Exercises are on pages 573–574.

14.2 Estimating the Standard Deviation for Regression

Recall that the standard deviation in the regression model measures, roughly, the average deviation of y values from the mean (the regression line). Expressed another way, the **standard deviation for regression** measures the general size of the residuals. This is an important and useful statistic for describing individual variation in a regression problem, and it also provides information about how accurately the regression equation might predict y values for individuals. A relatively small standard deviation from the regression line indicates that individual data points generally fall close to the line, so predictions based on the line will be close to the actual values.

The calculation of the estimate of standard deviation is based on the sum of the squared residuals for the sample. This quantity is called the **sum of squared errors** and is denoted by **SSE**. Synonyms for "sum of squared errors" are **residual sum of squares** or **sum of squared residuals.** To find the SSE, residuals are calculated for all observations, and then the residuals are squared and summed. This sum is used to find the *standard deviation from the regression line*:

$$s = \sqrt{\frac{\text{Sum of squared residuals}}{n - 2}} = \sqrt{\frac{\text{SSE}}{n - 2}}$$

and this sample statistic estimates the population standard deviation σ.

FORMULA

Estimating the Standard Deviation for a Simple Regression Model

The standard deviation σ for a simple regression model is estimated as follows:

$$\text{SSE} = \Sigma(y_i - \hat{y}_i)^2 = \Sigma e_i^2$$

$$s = \sqrt{\frac{\text{SSE}}{n - 2}} = \sqrt{\frac{\Sigma(y_i - \hat{y}_i)^2}{n - 2}}$$

The statistic s is an estimate of the population standard deviation σ.

TECHNICAL NOTE

Note how the estimate of σ in the regression situation differs from what it would be if we simply had a random sample of the y_i's without information about the x_i's:

Sample of y's only: $s = \sqrt{\frac{\Sigma(y_i - \bar{y})^2}{n - 1}}$

Sample of (x, y) pairs, linear regression: $s = \sqrt{\frac{\Sigma(y_i - \hat{y}_i)^2}{n - 2}}$

Remember that in the regression context, σ is the standard deviation of the population of y values at *each* x, not the standard deviation of the whole population of y values.

Example 14.3 **Relationship between Height and Weight for College Men** Figure 14.4 displays regression results from the Minitab program and a scatterplot for the relationship between y = weight (pounds) and x = height (inches) in a sample of $n = 43$ men in a statistics class. The regression line for the sample is $\hat{y} = -318 + 7x$, and this line is drawn onto the plot. We see from the plot that there is considerable variation from the line at any given height. The standard deviation, shown in the last row of the computer output to the left of the plot, is "S = 24.00." This value roughly measures, for any specific height, the general size of the deviations of individual weights from the mean weight for that particular height.

The standard deviation from the regression line can be interpreted in conjunction with the Empirical Rule for bell-shaped data stated in Section 2.7. Recall, for instance, that about 95% of individuals will fall within 2 standard deviations of the mean. As an example, consider men who are 72 inches tall. For men with this height, the estimated average weight determined from the regression equation is $-318 + 7.00(72) = 186$ pounds. The estimated standard deviation from the regression line is $s = 24$ pounds, so we can estimate that about 95% of men 72 inches tall have weights within $2 \times 24 = 48$ pounds of 186 pounds, which is 186 ± 48, or 138 to 234 pounds. Think about whether this makes sense for all the men you know who are 72 inches (6 feet) tall.

The regression equation is
Weight = −318 + 7.00 Height

Predictor	Coef	SE Coef	T	P
Constant	−317.9	110.9	−2.87	0.007
Height	6.996	1.581	4.42	0.000

S = 24.00   R-Sq = 32.3%   R-Sq(adj) = 30.7%

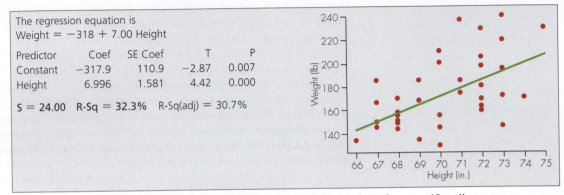

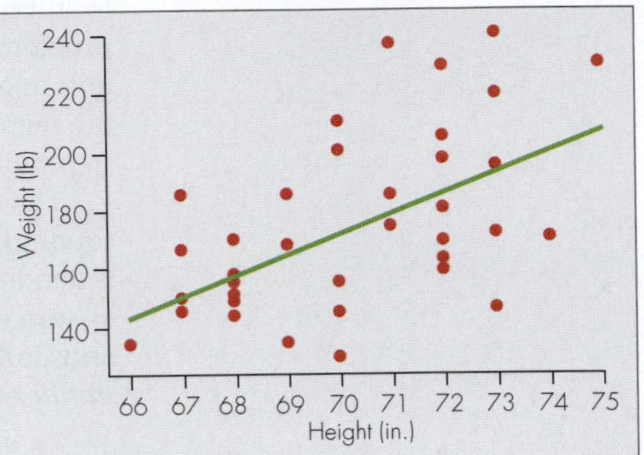

Figure 14.4 The relationship between weight and height for $n = 43$ college men

IN SUMMARY Interpreting the Standard Deviation for Regression

The standard deviation for regression estimates the standard deviation of the differences between values of y and the regression equation that relates the mean value of y to x. In other words, it measures the general size of the differences between actual and predicted values of y.

THOUGHT QUESTION 14.1 Regression equations can be used to predict the value of a response variable for an individual. What is the connection between the accuracy of predictions based on a particular regression line and the value of the standard deviation from the line? If you were deciding between two different regression models for predicting the same response variable, how would your decision be affected by the relative values of the standard deviations for the two models?*

The Proportion of Variation Explained by *x*

In Chapter 3, we learned that the squared correlation r^2 is a useful statistic. It is used to measure how well the explanatory variable explains the variation in the response variable. This statistic is also denoted as R^2 (rather than r^2), and the value is commonly expressed as a percentage. Researchers typically use the phrase "**proportion of variation explained by *x***" in conjunction with the value of r^2. For example, if $r^2 = .60$ (or 60%), the researcher may write that the explanatory variable explains 60% of the variation in the response variable.

The formula for r^2 presented in Chapter 3 is

$$r^2 = \frac{\text{SSTO} - \text{SSE}}{\text{SSTO}}$$

***HINT:** Read the first paragraph of this section (p. 555).

Recall from Chapter 3 that the quantity **SSTO** is the sum of squared differences between observed *y* values and the sample mean $\bar{y}$. It measures the size of the deviations of the *y* values from the overall mean of *y*, whereas **SSE** measures the size of the deviations of the *y* values from the predicted values of *y*, denoted by $\hat{y}$.

Example 14.4 **R^2 for Heights and Weights of College Men** In Figure 14.4 for Example 14.3 (p. 556), we can find the information "R-sq = 32.3%" for the relationship between weight and height. A researcher might write "the variable height explains 32.3% of the variation in the weights of college men." This isn't a particularly impressive statistic. As we noted before, there is substantial deviation of individual weights from the regression line, so a prediction of a college man's weight based on height may not be particularly accurate.

THOUGHT QUESTION 14.2 Look at the formula for SSE, and explain in words under what condition SSE = 0. Now explain what happens to r^2 when SSE = 0, and explain whether that makes sense according to the definition of r^2 as "proportion of variation in y explained by x."*

Example 14.5 **Driver Age and Highway Sign-Reading Distance** In Example 3.2 (p. 71), we examined data for the relationship between y = maximum distance (feet) at which a driver can read a highway sign and x = the age of the driver. There were n = 30 observations in the dataset. Figure 14.5 displays Minitab regression output for these data. The equation describing the linear relationship in the sample is

$$\text{Average distance} = 577 - 3.01 \times \text{Age}$$

From the output, we learn that the standard deviation from the regression line is s = 49.76 (feet) and R-sq = 64.2%. Roughly, the average deviation from the regression line is about 50 feet, and the proportion of variation in sign-reading distances explained by age is .642 or 64.2%.

The regression equation is
Distance = 577 − 3.01 Age

Predictor	Coef	SE Coef	T	P
Constant	576.68	23.47	24.57	0.000
Age	**−3.0068**	**0.4243**	**−7.09**	**0.000**

S = 49.76 R-Sq = 64.2% R-Sq (adj) = 62.9%

Analysis of Variance

Source	DF	SS	MS	F	P
Regression	1	124333	124333	50.21	0.000
Residual Error	28	69334	2476		
Total	29	193667			

Unusual Observations

Obs	Age	Distance	Fit	SE Fit	Residual	St Resid
27	75.0	460.00	351.17	13.65	108.83	2.27R

R denotes an observation with a large standardized residual

Figure 14.5 Minitab output: Sign-reading distance and driver age

***HINT:** Remember that SSE stands for "sum of squared errors." The formula for r^2 is given just before Example 14.4.

The **analysis of variance table** provides information about the different sources of variability needed to compute r^2 and s:

SSE = 69,334, found in the "SS" column and the "Residual Error" row.

$$s = \sqrt{\frac{\text{SSE}}{n-2}} = \sqrt{\frac{69{,}334}{28}} = 49.76$$

SSTO = 193,667, found in the "SS" column and the "Total" row.

$$\text{SSTO} - \text{SSE} = 193{,}667 - 69{,}334 = 124{,}333$$

$$r^2 = \frac{\text{SSTO} - \text{SSE}}{\text{SSTO}} = \frac{124{,}333}{193{,}667} = .642, \text{ or } 64.2\%$$

14.2 Exercises are on pages 574–575.

14.3 Inference about the Slope of a Linear Regression

In this section, we will learn how to carry out a hypothesis test to determine if we can infer that two variables are linearly related in the larger population represented by a sample. We will also learn how to use sample regression results to calculate a confidence interval estimate of a population slope.

Hypothesis Test for a Population Slope

The **statistical significance of a linear relationship** can be evaluated by testing whether or not the population slope is 0. If the slope is 0 in a simple linear regression model, the two variables are not related because changes in the x variable will not lead to changes in the y variable. The usual null hypothesis and alternative hypothesis about β_1, the slope of the population regression line $E(Y) = \beta_0 + \beta_1 x$, are

H_0: $\beta_1 = 0$ (the population slope is 0, so y and x are *not linearly related*)

H_a: $\beta_1 \neq 0$ (the population slope is not 0, so y and x *are linearly related*)

As usual the alternative hypothesis may be one-sided or two-sided depending on the research question of interest, but most statistical software uses the two-sided alternative.

The test statistic used to conduct the hypothesis test is a t-statistic with the same general format for a test statistic that we used in Chapters 12 and 13. That format, and its application to this situation, is

$$t = \frac{\text{Sample statistic} - \text{Null value}}{\text{Standard error}} = \frac{b_1 - 0}{\text{s.e.}(b_1)}$$

This is a standardized statistic for the difference between the sample slope and 0, the null value. Note that a large value of the sample slope (either positive or negative) relative to its standard error will give a large absolute value of t. If the mathematical assumptions about the population model described in Section 14.1 are correct, the statistic has a t-distribution with $n - 2$ degrees of freedom. The p-value for the test is determined using that distribution.

It is important to be sure that the necessary conditions are met when using any statistical inference procedure. The necessary conditions for using this test, and how to check them, will be discussed in Section 14.5.

"By hand" calculations of the sample slope and its standard error are cumbersome. Fortunately, the regression analysis of most statistical software includes a t-statistic and a p-value for this significance test.

FORMULA **Formula for the Sample Slope and Its Standard Error**

In case you ever need to compute the values by hand, here are the formulas for the sample slope and its standard error:

$$b_1 = r\frac{s_y}{s_x}$$

$$\text{s.e.}(b_1) = \frac{s}{\sqrt{\sum(x_1 - \bar{x})^2}} \quad \text{where } s = \sqrt{\frac{\text{SSE}}{n-2}}$$

In the formula for the sample slope, s_x and s_y are the sample standard deviations of the x and y values, respectively, and r is the correlation between x and y.

Example 14.6 **Hypothesis Test for Driver Age and Sign-Reading Distance** Figure 14.5 (p. 557) for Example 14.5 presents Minitab output for the regression of sign-reading distance (y) and driver age. The part of the output that is used to test the statistical significance of the observed relationship is shown in bold. This line of output gives values for the sample slope, the standard error of the sample slope, the t-statistic, and the p-value for the test of

H_0: $\beta_1 = 0$ (the population slope is 0, so y and x are *not* linearly related)

H_a: $\beta_1 \neq 0$ (the population slope is not 0, so y and x *are* linearly related)

The test statistic is

$$t = \frac{\text{Sample statistic} - \text{Null value}}{\text{Standard error}} = \frac{b_1 - 0}{\text{s.e.}(b_1)} = \frac{-3.0068 - 0}{0.4243} = -7.09$$

The p-value (underlined in the output) is given to three decimal places as .000. This means that the probability is virtually 0 that the sample slope could be as far as it is from 0 or farther if the population slope really is 0. Because the p-value is so small, we can reject the null hypothesis and infer that the linear relationship observed between the two variables in the sample represents a real relationship in the population.

TECHNICAL NOTE Most statistical software reports a p-value for a two-sided alternative hypothesis when doing a test for whether or not the slope in the population is 0. It may sometimes make sense to use a one-sided alternative hypothesis instead. In that case, the p-value for the one-sided alternative is (reported p)/2 if the sign of b_1 is consistent with H_a, but is 1 − (reported p)/2 if it is not.

Confidence Interval for the Population Slope

The significance test of whether or not the population slope is 0 tells us only whether we can declare the relationship to be statistically significant. If we decide that the true slope is not 0, we might ask, "What is the value of the slope?" We can answer this question with a **confidence interval for β_1, the population slope.**

The format for this confidence interval is the same as the general format used in Chapters 10 and 11, which is

Sample statistic ± Multiplier × Standard error

The sample statistic is b_1, the slope of the least-squares regression line for the sample. As has been shown already, the standard error formula is complicated, and we will usually rely on statistical software to determine this value. The "multiplier" will be labeled t^* and is determined by using a t-distribution with df $= n - 2$. Table A.2 can be used to find the multiplier for the desired confidence level.

FORMULA **Formula for Confidence Interval for β_1, the Population Slope**

A confidence interval for β_1 is

$$b_1 \pm t^* \times \text{s.e.}(b_1)$$

The multiplier t^* is found by using a t-distribution with $n - 2$ degrees of freedom and is such that the probability between $-t^*$ and t^* equals the confidence level for the interval.

Example 14.7 **95% Confidence Interval for Slope between Age and Sign-Reading Distance** In Figure 14.5 (p. 557), we see that the sample slope is $b_1 = -3.01$ and s.e.$(b_1) = 0.4243$. There are $n = 30$ observations, so df = 28 for finding t^*. For a 95% confidence level, $t^* = 2.05$ (see Table A.2). The 95% confidence interval for the population slope is

$$-3.01 \pm 2.05 \times 0.4243$$
$$-3.01 \pm 0.87$$
$$-3.88 \text{ to } -2.14$$

With 95% confidence, we can estimate that in the population of drivers represented by this sample, the *mean* sign-reading distance decreases somewhere between 2.14 and 3.88 feet for each 1-year increase in age.

THOUGHT QUESTION 14.3 In previous chapters, we learned that a confidence interval can be used to determine if a hypothesized value for a parameter can be rejected. How would you use a confidence interval for the population slope to determine whether there is a statistically significant relationship between x and y? For example, why is the interval that we just computed for the sign-reading example evidence that sign-reading distance and age are related?*

SPSS TIP **Calculating a 95% Confidence Interval for the Slope**

- Use **Analyze > Regression > Linear Regression**. Specify the *y* variable in the *Dependent* box and specify the *x* variable in the *Independent(s)* box.
- Click ***Statistics*** and then select "Confidence intervals" under "Regression Coefficients."

Testing Hypotheses about the Correlation Coefficient

In Chapter 3, we learned that the correlation coefficient is 0 when the regression line is horizontal. In other words, if the slope of the regression line is 0, the correlation is 0. This means that the results of a hypothesis test for the population slope can also be interpreted as applying to equivalent hypotheses about the correlation between x and y in the population.

We use different notation to distinguish between a correlation computed for a sample and a correlation within a population. It is commonplace to use the Greek letter ρ (pronounced "rho") to represent the correlation between two variables within a population. Using this notation, null and alternative hypotheses of interest are as follows:

H_0: $\rho = 0$ (x and y are not correlated)

H_a: $\rho \neq 0$ (x and y are correlated)

The results of the hypothesis test described on page 558 for the population slope β_1 can be used for these hypotheses as well. If we reject H_0: $\beta_1 = 0$, we also reject H_0: $\rho = 0$. If we decide in favor of H_a: $\beta_1 \neq 0$, we also decide in favor of H_a: $\rho \neq 0$.

***HINT:** What is the null value for the slope? Section 13.5 discusses the connection between confidence intervals and significance tests.

Many statistical software programs will give a p-value for testing whether or not the population correlation is 0. This p-value will be the same as the p-value given for testing whether or not the population slope is 0.

Example 14.8 **Is Pulse Rate Related to Weight?** The following Minitab output is for the relationship between pulse rate and weight in a sample of 35 college women. Note that .292 is given as the p-value for testing that the slope is 0 (look under "P" in the regression results) and the same p-value is given for testing that the correlation is 0. Because this is not a small p-value, we cannot reject the null hypotheses for the slope and the correlation. Even though the slope of the sample regression line is 0.159 and the correlation in the sample is 0.183, there is insufficient evidence to conclude that pulse rate and weight are linearly related in the population of college women.

Regression Analysis: Pulse Versus Weight

The regression equation is
Pulse = 57.2 + 0.159 Weight

Predictor	Coef	SE Coef	T	P
Constant	57.17	18.51	3.09	0.004
Weight	0.1591	0.1487	1.07	0.292

Correlations: Pulse, Weight

Pearson correlation of Pulse and Weight = 0.183
P-Value = 0.292

The Effect of Sample Size on Significance

The size of a sample always affects whether a specific observed result achieves statistical significance. For example, $r = 0.183$ is not a statistically significant correlation for a sample size of $n = 35$, as shown in Example 14.8, but it would be statistically significant if $n = 1000$. With very large sample sizes, weak relationships with low correlation values can be statistically significant. The "moral of the story" here is that with a large sample size, it may not be saying much to say that two variables are significantly related. This means only that we think that the correlation is not precisely 0. To assess the practical significance of the result, we should carefully examine the observed strength of the relationship.

14.3 Exercises are on pages 575–577.

TECHNICAL NOTE

The usual t-statistic for testing whether or not the population slope is 0 in a linear regression could also be found by using a formula that involves only n = sample size and r = correlation between x and y. The algebraic equivalence is

$$t = \frac{b_1}{\text{s.e.}(b_1)} = \sqrt{n-2}\frac{r}{\sqrt{1-r^2}}$$

In the output for Example 14.8, notice that the t-statistic for testing whether or not the slope $\beta_1 = 0$ is $t = 1.07$. This was calculated as

$$t = \frac{b_1}{\text{s.e.}(b_1)} = \frac{0.1591}{0.1487} = 1.07$$

The sample size is $n = 35$, and the correlation is $r = 0.183$, so an equivalent calculation of the t-statistic is

$$t = \sqrt{n-2}\frac{r}{\sqrt{1-r^2}} = \sqrt{35-2}\frac{0.183}{\sqrt{1-0.183^2}} = 1.07$$

This second method for calculating the t-statistic illustrates two ideas. First, there is a direct link between the correlation value and the t-statistic that is used to test whether or not the slope is 0. Second, notice that for any fixed value of r, increasing the sample size n will increase the size of the t-statistic. And the larger the value of the t-statistic, the stronger is the evidence against the null hypothesis.

14.4 Predicting *y* and Estimating Mean *y* at a Specific *x*

In this section, we cover two different types of intervals that are used to make inferences about the response variable (y). The first type of interval *predicts the value of* y for an individual with a specific value of x. For example, we may want to predict the freshman year GPA of a college applicant who has a 3.6 high school GPA. The second type of interval *estimates the mean value of* y for a population of individuals who all have the same specific value of x. As an example, we may want to estimate the mean (average) freshman year GPA of all college applicants who have a 3.6 high school GPA.

Predicting the Value of *y* for an Individual

An important use of a regression equation is to estimate or predict the unknown value of a response variable for an *individual* with a known specific value of the explanatory variable. Using the data described in Example 14.5 (p. 557), for instance, we can predict the maximum distance at which an individual can read a highway sign by substituting his or her age for x in the sample regression equation. Consider a person who is 21 years old. The predicted distance for such a person is approximately $\hat{y} = 577 - 3.01(21) = 513.79$, or about 514 feet.

There will be variation among 21-year-olds with regard to the sign-reading distance, so the predicted distance of 513.79 feet is not likely to be the exact distance for the next 21-year-old who views the sign. Rather than predicting that the distance will be exactly 513.79 feet, we should instead predict that the distance will be within a particular interval of values.

A **95% prediction interval** describes the values of the response variable (y) for 95% of all individuals with a particular value of x. This interval can be interpreted in two equivalent ways:

1. The 95% prediction interval estimates the central 95% of the values of y for members of the population with a specified value of x.
2. The probability is .95 that the value of y for a randomly selected individual from the population with a specified value of x falls into the corresponding 95% prediction interval.

We don't always have to use a 95% prediction interval. A prediction interval for the value of the response variable (y) can be found for any specified central percentage of a population with a specified value of x. For example, a 75% prediction interval describes the central 75% of a population of individuals with a particular value of the explanatory variable (x).

DEFINITION A **prediction interval** estimates the value of y for an individual with a particular value of x, or equivalently, the range of values of the response variable for a specified central percentage of a population with a particular value of x.

Note that a prediction interval differs conceptually from a confidence interval. A confidence interval estimates an unknown population parameter, which is a numerical characteristic or summary of the population. An example in this chapter is a confidence interval for the slope of the population line. A prediction interval, however, does not estimate a parameter; instead, it estimates the potential data value for an individual. Equivalently, it describes an interval into which a specified percentage of the population may fall.

As with most regression calculations, the "by hand" formulas for prediction intervals are formidable. Statistical software can be used to create the interval.

Example 14.9 **Predicting When Someone Can Read a Sign** In Example 14.5, we found the regression relationship between x = age and y = maximum distance at which someone can read a highway sign. Figure 14.6 shows Minitab output that includes the 95%

prediction intervals for three different ages (21, 30, and 45) for this example. The intervals are toward the bottom-right side of the display in a column labeled "95% PI" and are highlighted with bold type. The ages for which the intervals were computed are shown at the bottom of the output. (***Note:*** The term **fit** is a synonym for $\hat{y}$, the estimate of the average response at the specific x value.) From Figure 14.6, here is what we can conclude:

- The probability is .95 that a randomly selected 21-year-old will read the sign at somewhere between 406.69 and 620.39 feet.
- The probability is .95 that a randomly selected 30-year-old will read the sign at somewhere between 381.26 and 591.69 feet.
- The probability is .95 that a randomly selected 45-year-old will read the sign at somewhere between 337.63 and 545.12 feet.

```
The regression equation is
Distance = 577 − 3.01 Age

Predictor      Coef      SE Coef       T          P
Constant     576.68      23.47       24.57      0.000
Age          −3.0068     0.4243      −7.09      0.000

S = 49.76       R-Sq = 64.2%       R-Sq(adj) = 62.9%

Analysis of Variance
Source          DF       SS          MS          F          P
Regression       1     124333      124333      50.21      0.000
Residual Error  28      69334        2476
Total           29     193667

Unusual Observations
Obs     Age     Distance      Fit      SE Fit     Residual     St Resid
27      75.0    460.00       351.17    13.65      108.83       2.27R

R denotes an observation with a large standardized residual
Predicted Values for New Observations

New Obs     Fit      SE Fit      95.0% CI              95.0% PI
1         513.54     15.64     (481.50, 545.57)     (406.69, 620.39)
2         486.48     12.73     (460.41, 512.54)     (381.26, 591.69)
3         441.37      9.44     (422.05, 460.70)     (337.63, 545.12)

Values of Predictors for New Observations
New Obs      Age
1            21.0
2            30.0
3            45.0
```

Figure 14.6 Minitab output showing prediction intervals of distance

We can also interpret each interval as an estimate of the sign-reading distances for the central 95% of a population of drivers with a specified age. For instance, about 95% of all drivers 21 years old will be able to read the sign at a distance somewhere between roughly 407 and 620 feet.

With most statistical software, we can describe any central percentage of the population that we wish. For example, here is Minitab output showing 50% prediction intervals for the sign-reading distance at the three specific ages we considered in Example 14.9.

Age	Fit	50.0% PI
21	513.54	(477.89, 549.18)
30	486.48	(451.38, 521.58)
45	441.37	(406.76, 475.98)

For each specific age, the 50% prediction interval estimates the central 50% of the maximum sign-reading distances in a population of drivers with that age. For example, we can estimate that 50% of drivers 21 years old would have a maximum sign-reading distance somewhere between about 478 feet and 549 feet. The distances for the other 50% of 21-year-old drivers would be predicted to be outside this range, with 25% above about 549 feet and 25% below about 478 feet.

TECHNICAL NOTE The formula for the prediction interval for y at a specific x (denoted as x_j) is

$$\hat{y} \pm t^* \times \sqrt{s^2 + [\text{s.e.(fit)}]^2}$$

where

$$\text{s.e.(fit)} = s\sqrt{\frac{1}{n} + \frac{(x_j - \bar{x})^2}{\sum(x_i - \bar{x})^2}}$$

The multiplier t^* is found by using a t-distribution with $n - 2$ degrees of freedom and is such that the probability between $-t^*$ and t^* equals the desired level for the interval.

Note:

- The s.e.(fit), and thus the width of the interval, depends on how far the specified x value is from $\bar{x}$. The further the specific x is from the mean, the wider is the interval. This is evident from the numerator of the relevant part of s.e.(fit).
- When n is large, s.e.(fit) will be small, and the prediction interval will be approximately $\hat{y} \pm t^*s$. This is evident from the fact that the denominators of both pieces under the square root sign of s.e.(fit) get large as n gets large.

THOUGHT QUESTION 14.4 If we knew the population parameters β_0, β_1, and σ, under the usual regression assumptions, we would know that the population of y values at a specific x value was normal with mean $\beta_0 + \beta_1 x$ and standard deviation σ. In that case, what interval would cover the central 95% of the y values for that x value? Use your answer to explain why a prediction interval would not have zero width even with complete population details (as long as σ is not 0).*

Estimating the Mean *y* for a Specified *x*

Thus far in this section, we have focused on the estimation of the values of the response variable for individuals. A researcher may instead want to estimate the *mean* value of the response variable for individuals with a particular value of the explanatory variable. We might ask, "What is the mean weight for college men who are 6 feet tall?" This question asks only about the mean weight in a group with a common height; it is not concerned with the deviations of individuals from that mean.

In technical terms, we wish to estimate the population mean $E(Y)$ for a specific value of x that is of interest to us. To make this estimate, we use a confidence interval. The format for this confidence interval is the same one that we used in Chapters 10 and 11:

***HINT:** Remember the Empirical Rule, and also recall that the regression equation gives the mean y for a specific x.

FORMULA **Estimating the Mean y for a Specified x**

The formula for estimating the mean of the of y values for everyone in the population with a specific x value is

Sample statistic ± Multiplier × Standard error

where

- The *sample statistic* is the value of $\hat{y}$ that is determined by substituting the x value of interest into $\hat{y} = b_0 + b_1x$, the least-squares regression line for the sample.
- The *standard error* of $\hat{y}$ is the s.e.(fit) shown in the technical note on page 564, and its value is usually provided by statistical software.
- The *multiplier* is found by using a t-distribution with df $= n - 2$, and Appendix A.2 can be used to determine its value.

Example 14.10 **Estimating Mean Weight of College Men at Various Heights** Based on the sample of $n = 43$ college men in Example 14.3 (p. 555), let's estimate the mean weight in the population of college men for each of three different heights: 68 inches, 70 inches, and 72 inches. Figure 14.7 shows Minitab output that includes the three different confidence intervals for these three different heights. These intervals are in bold toward the bottom of the display in a column labeled "95% CI." The heights used to find the confidence intervals are listed under the heading "Values of Predictors for New Observations," at the end of the output. Thus:

- For college men who are 68 inches tall, we can say with 95% confidence that their mean weight is between 147.78 and 167.81 pounds.
- For college men who are 70 inches tall, we can say with 95% confidence that their mean weight is between 164.39 and 179.19 pounds.
- For college men who are 72 inches tall, we can say with 95% confidence that their mean weight is between 176.25 and 195.31 pounds.

The regression equation is
Weight = −318 + 7.00 Height

Predictor	Coef	SE Coef	T	P
Constant	−3.179	110.9	−2.87	0.007
Height	6.996	1.581	4.42	0.000

S = 24.00 R-Sq = 32.3% R-Sq(adj) = 30.7%

— Some Output Omitted —

Predicted Values for New Observations

New Obs	Fit	SE Fit	95.0% CI	95.0% PI
1	157.80	4.96	**(147.78, 167.81)**	(108.31, 207.29)
2	171.79	3.66	**(164.39, 179.19)**	(122.76, 220.82)
3	185.78	4.72	**(176.25, 195.31)**	(136.38, 235.18)

Values of Predictors for New Observations

New Obs	Height
1	68.0
2	70.0
3	72.0

Figure 14.7 Minitab output for Example 14.10, with confidence intervals for mean weight

Each of these confidence intervals estimates $E(Y)$, the *mean* weight for individuals with a particular height (e.g., 68 inches). These intervals do not describe the variation among *individual* weights. The prediction intervals for individual responses describe the variation among individuals.

You may have noticed that 95% prediction intervals, labeled "95% PI," are next to the confidence intervals in the output. Among men who are 70 inches tall, for instance, we would estimate that 95% of the *individual* weights would be in the interval from about 123 to about 221 pounds. Note that this prediction interval is much wider than the corresponding confidence interval for the mean weight of men of this height, which is 164.39 to 179.19 pounds. That's because the prediction interval incorporates the wide range of individual differences in weights for college men of the same height.

14.4 Exercises are on page 557.

IN SUMMARY Prediction Intervals for *y* and Confidence Intervals for *E*(*Y*)

In regression, the difference between a prediction interval for *y* and a confidence interval for *E*(*Y*) is as follows:

- A **prediction interval for *y*** predicts the value of the response variable (*y*) for an *individual* with a particular value of *x*. This interval also estimates the range of values for a specified central percentage of a population of individuals with the same particular value of *x*.
- A **confidence interval for *E*(*Y*)** estimates the *mean* value of *y* for a population of individuals who all have the same particular value of *x*.

THOUGHT QUESTION 14.5 Draw a picture similar to the one in Figure 14.3 (p. 553), illustrating the regression line and the normal curves for the *y* values at several values of *x*. Use it to illustrate the difference between a prediction interval for *y* and a confidence interval for the mean of the *y*'s at a specific value of *x*.*

MINITAB TIP Creating Prediction Intervals for *y* and Confidence Intervals for *E*(*Y*)

- Use **Stat > Regression > Regression**, then use the ***Options*** button of the dialog box.
- In the box labeled "Prediction intervals for new observations," either specify the numerical value of a single *x* value of interest or specify a worksheet column that contains two or more different *x* values of interest. The confidence level can be specified in the box labeled "Confidence Level."

Note: Minitab computes both a prediction interval for *y* and a confidence interval for *E*(*Y*) for each specified value of *x*.

14.5 Checking Conditions for Using Regression Models for Inference

There are a few conditions that should be at least approximately true when we use a regression model to make an inference about a population. You already have encountered some of these conditions in Section 3.4, where we discussed the "difficulties and disasters" that could result from outliers in regression, or from assuming a relationship is linear when it is not. Of the five conditions that follow, the first two are particularly crucial when using sample data to make inferences about a population.

Conditions for Linear Regression

1. The form of the equation that links the mean value of *y* to *x* must be correct. For instance, we will not make proper inferences if we use a straight line to describe a curved relationship. This is called the **linear relationship condition**.

***HINT:** Each bell curve in Figure 14.3 describes individual variation at a specific *x*-value. How does this relate to the purpose of a prediction interval?

2. There should not be any extreme outliers that influence the results unduly. This is called the **no outlier condition**.
3. The standard deviation (and thus the variance) of the values of y at each value of x is the same regardless of the value of x. In other words, y values are similarly spread out around the line at all values of x. This is called the **constant variance condition**.
4. For individuals in the population with the same particular value of x, the distribution of the values of y is a normal distribution. Equivalently, the distribution of deviations from the mean value of y is a normal distribution. This is called the **normal distribution condition**. This condition can be relaxed if the sample size is large.
5. Observations in the sample are independent of each other. This is called the **independence condition**.

Figure 14.3 on page 553 illustrates all of these conditions except the independence condition, which cannot be illustrated with this type of picture.

Checking the Conditions with Plots

A scatterplot of the raw data and **plots of the residuals**, also called **residual plots**, provide information about the validity of the assumptions. Remember that a residual is the difference between an observed value and the predicted value for that observation and that some assumptions made for a linear regression model have to do with how y values deviate from the regression line. If the properties of the residuals for the sample appear to be consistent with the mathematical assumptions that are made about deviations within the population, we can use the model to make statistical inferences.

Conditions 1, 2, and 3 can be checked by using two useful plots:

- A scatterplot of y versus x for the sample
- A scatterplot of the residuals versus x for the sample

If Condition 1 holds for a linear relationship, then:

- The plot of y versus x should show points randomly scattered around an imaginary straight line.
- The plot of residuals versus x should show points randomly scattered around a horizontal line at residual $= 0$.

If Condition 2 holds, extreme outliers should not be evident in either plot.

If Condition 3 holds, neither plot should show increasing or decreasing spread in the points as x increases.

THOUGHT QUESTION 14.6 A residual is the difference between an observed value of y and the predicted value of y for that observation. Based on the size of a residual for an observation, how would you decide if the observation is an outlier? Is it enough to know the value of the residual, or do you need to know other information to make this judgment? How could you apply the methods for detecting outliers described in Chapter 2?*

Example 14.11 **Checking Conditions 1 to 3 for the Weight and Height Problem** Figure 14.4 (p. 556) displays a scatterplot of the weights and heights of $n = 43$ college men. In that plot, it appears that a straight line is a suitable model for how mean weight is linked to height, so the linear relationship condition is met. Figure 14.8 shows a plot of the residuals (e_i) versus the corresponding values of height for these 43 men. This plot is further evidence that the appropriate model has been used. If the appropriate model has been used, the way in which individuals deviate from the line (residuals) will not be affected by the value of the explanatory variable. The somewhat random-looking blob of

***HINT:** We must know the general size of the residuals to judge the size of a single residual. Section 2.5 discusses how to use the five-number summary to identify outliers.

points in Figure 14.8 is the way that a plot of residuals versus x should look if the appropriate equation for the mean has been used. Both plots (Figures 14.4 and 14.8) also show that there are no extreme outliers and that the weights have approximately the same variance across the range of heights in the sample, and so that the constant variance condition appears to be met. Therefore, Conditions 1 to 3 appear to be met.

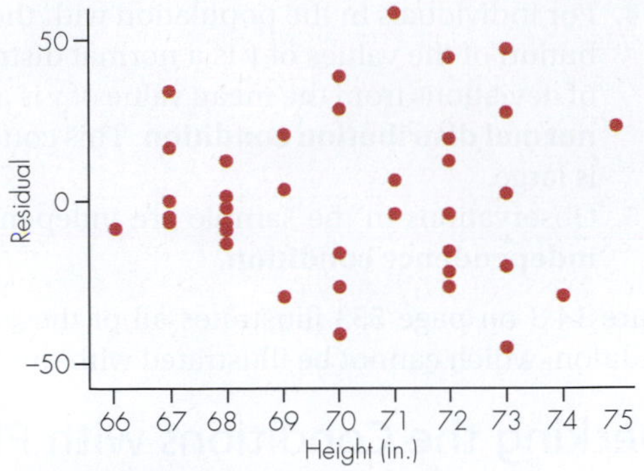

Figure 14.8 Plot of residuals versus x for Example 14.3; the absence of a pattern indicates the appropriate model has been used

Condition 4, which is that deviations from the regression line are normally distributed, is difficult to verify, but it is also the least important of the conditions because the inference procedures for regression are **robust**. This means that if there are no major outliers or extreme skewness, the inference procedures work well even if the distribution of y values is not a normal distribution. In Chapters 11 and 13, we learned that confidence intervals and hypothesis tests for a mean, for a mean of paired differences, and for a difference between two means also are robust.

To examine the distribution of the deviations from the line, a *histogram of the residuals* is useful, although for small samples, a histogram may not be informative. A more advanced plot called a *normal probability plot* can also be used to check whether the residuals are normally distributed, but we do not provide the details in this text. Figure 14.9 displays a histogram of the residuals for the data from Examples 14.3 (p. 555) and 14.11 (p. 567). It appears that the residuals are approximately normally distributed, so Condition 4 is met.

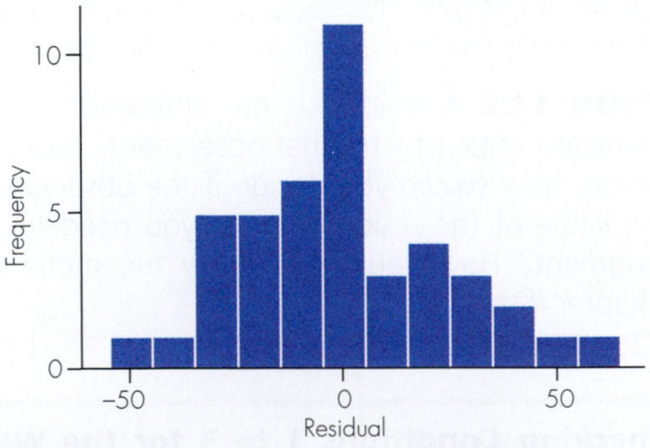

Figure 14.9 Histogram of residuals for Example 14.3

Condition 5, the independence condition, follows from the data collection process. It is met as long as the units are measured independently. It would not be met if the same individuals were measured across the range of x values, such as if x = average speed and y = gas mileage were to be measured for multiple tanks of gas on the same cars. More complicated diagnostic plots and models are needed for *dependent* observations; these plots and models will not be discussed in this book.

Corrections When Conditions Are Not Met

There are some steps that can be taken if Condition 1, 2, or 3 is not met. If Condition 1 is not met, more complicated models can be used. For instance, Figure 14.10 shows a typical plot of residuals that occurs when a straight-line model is used to describe data that are curvilinear. It may help to think of the residuals as prediction errors that would occur if we used the regression line to predict the value of y for the individuals in the sample. In the plot shown in Figure 14.10, the "prediction errors" are all negative in the central region of x and nearly all positive for outer values of x. This occurs because the wrong model is being used to make the predictions. A curvilinear model, such as the quadratic model in the Technical Note on page 553, may be more appropriate.

Condition 2, that there are no influential outliers, can be checked graphically with the scatterplot of y versus x and the plot of residuals versus x. The appropriate correction if there are outliers depends on the reason for the outliers. The same considerations and corrective action that we discussed in Chapter 2 would be taken, depending on the cause of the outlier.

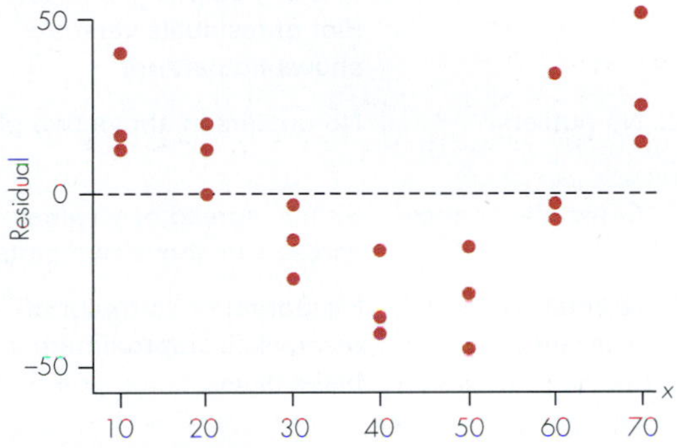

Figure 14.10 A residual plot indicating the wrong model has been used

Example 14.12 **Chug-Time and Weight** Figure 14.11 shows a scatterplot and a residual plot for the data of Exercise 3.92. In this example, the x variable is weight, and the y variable is time to chug a beverage. A potential outlier is seen in both plots. The outlier probably represents a legitimate data value. The relationship appears to be linear for weights ranging up to about 210 pounds, but then it appears to change. It could either become quadratic or level off. We do not have enough data to determine what happens for higher weights. The solution in this case would be to remove the outlier and use the linear regression relationship only for body weights under about 210 pounds. Determining the relationship for higher body weights would require a larger sample of individuals in that range.

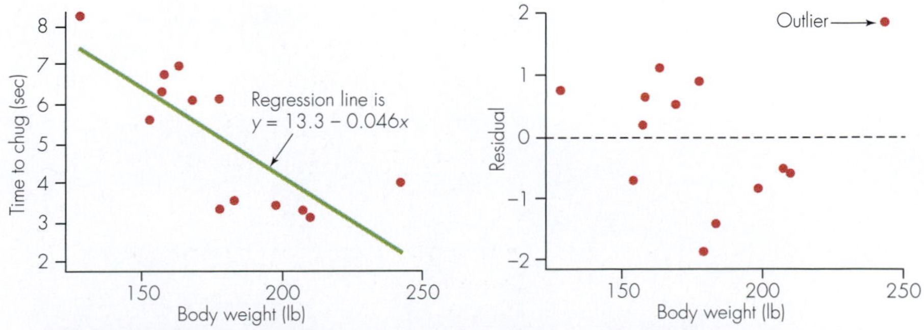

Figure 14.11 Scatterplot and corresponding residual plot with an outlier

If either Condition 1 or Condition 3 is not met, a **transformation** of the x values, the y values, or both may be required. This is equivalent to using a different model. Fortunately, often the same transformation will correct problems with Conditions 1, 3, and 4. For instance, when the response variable is monetary, such as salaries, it is often more appropriate to use the following relationship:

$$\ln(y) = b_0 + b_1x + e$$

In other words, we assume that there is a linear relationship between the natural log of y and the x values. This is called a **log transformation on the y's**. We will not pursue transformations further in this book.

14.5 Exercises are on page 578.

IN SUMMARY Conditions for Linear Regression and How to Check Them

Here are the five conditions required for inference in linear regression to be valid, how to check them, and what corrective action to take:

Condition	How to Check	Corrective Action if Not Met
1. Linear relationship	Plot of y versus x is linear. Plot of residuals versus x shows no pattern.	Use more complex model. Transform x and/or y.
2. No outliers	No outliers in above two plots.	See Section 2.6 on how to handle outliers.
3. Constant variance	Similar spread of y values across x in above two plots.	Transform y values.
4. Normal distribution	Histogram or stemplot of residuals is approximately bell-shaped.	No correction needed if not extreme and/or n is large. Otherwise, transform x and/or y.
5. Independence of observations	Make sure units were measured independently.	Use complex models that allow dependent observations.

MINITAB TIP **Plotting and Storing Residuals**

- Use **Stat > Regression > Regression**. The ***Graphs*** button of the dialog box offers options for several possible residual plots for the regression.
- To store residuals in a column of the worksheet, use the ***Storage*** button of the dialog box for regression.

CASE STUDY 14.1 A Contested Election

Case Study 12.1 (p. 486), discussed allegations of fraud in a 1993 Philadelphia special election for state senator. The Democratic candidate, William Stinson, beat the Republican candidate, Bruce Marks, by a difference of 461 votes, but Judge Clarence Newcomer overturned the results in a decision issued on April 24, 1994. Judge Newcomer wrote, "The court will order the certification of Bruce Marks as the winner of the 1993 Special Election in the Second Senatorial District, because this court finds that the record of evidence overwhelmingly supports the finding that Bruce Marks would have won the election but for wrongdoing" (Newcomer, 1994).

Judge Newcomer based his decision partially on statistical evidence presented by expert witnesses. Other evidence suggested that a number of absentee ballots had been filed illegally, and a statistical analysis indicated that the absentee count was indeed unusual when compared to the machine vote count.

The data used for the statistical analysis consisted of two variables for each of the 21 senatorial elections held in the previous 11 years (1982–1992) in the Philadelphia area. The variables follow:

y = Diff absentee = (Democrat candidate − Republican candidate) votes by absentee ballot

x = Diff machine = (Democrat candidate − Republican candidate) votes by machine ballot

The data for the 21 prior elections were given by economist and expert witness Orley Ashenfelter in a report to Judge Newcomer (Ashenfelter, 1994), and they are presented in Table 14.1 along with the contested 1993 results.

Table 14.1 Differences in Democratic and Republican Votes in Philadelphia Elections

Year	District	Diff Absentee	Diff Machine
82	2	346	26,427
82	4	282	15,904
82	8	223	42,448
84	1	593	19,444
84	3	572	71,797
84	5	−229	−1,017
84	7	671	63,406
86	2	293	15,671
86	4	360	36,276
86	8	306	36,710
88	1	401	21,848
88	3	378	65,862
88	5	−829	−13,194
88	7	394	56,100
90	2	151	700
90	4	−349	11,529
90	8	160	26,047
92	1	1,329	44,425
92	3	368	45,512
92	5	−434	−5,700
92	7	391	51,206
(Contested) 93	2	1,025	−564

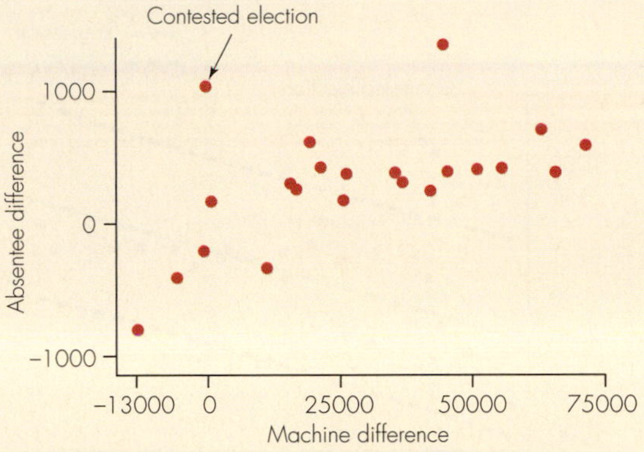

Figure 14.12 Plot of absentee difference versus voting machine difference

Figure 14.12 shows a plot of the data with the contested election indicated. In the contested election, the absentee difference (Democrat − Republican) appears well above what might be expected based on the overall connection between absentee and machine votes. Note also that there is an additional outlier—for the election held in District 1 in 1992. No explanation or discussion of that outlier occurred in the court records.

Is the combination of absentee and machine differences in the contested election at all reasonable, if all other factors remain similar to circumstances of the other elections? The output shown at the bottom of this column is for a simple linear regression computed using data from the 21 prior elections. Included in the results are 95% confidence and prediction intervals for the difference in absentee ballots when the machine ballot difference is −564, the value observed in the contested election. In particular, notice that the 95% prediction interval for the absentee ballot difference for this machine ballot difference is −854.7 to +588.6. The actual absentee difference in the contested election was +1025, a difference far outside this range. Figure 14.13 (next page) displays this result graphically, giving "prediction limits" for the entire range of possible values of machine differences covered by the data.

The analysis reported in the *New York Times* and quoted in Case Study 12.1 used the linear relationship that we have just used. However, an examination of the plot of the data in Figure 14.12 indicates that the

```
The regression equation is
Diff Absentee = -126 + 0.0127 Diff Machine

Predictor       Coef       StDev        T        P
Constant      -125.9       114.3    -1.10    0.284
Diff Mac    0.012703    0.002980     4.26    0.000

S = 324.8      R-Sq = 48.9%      R-Sq(adj) = 46.2%

Predicted Values

   Fit    StDev Fit         95.0% CI              95.0% PI
-133.1       115.6    (-375.0, 108.8)    (-854.7, 588.6)
```

(continued)

CASE STUDY 14.1 A Contested Election (continued)

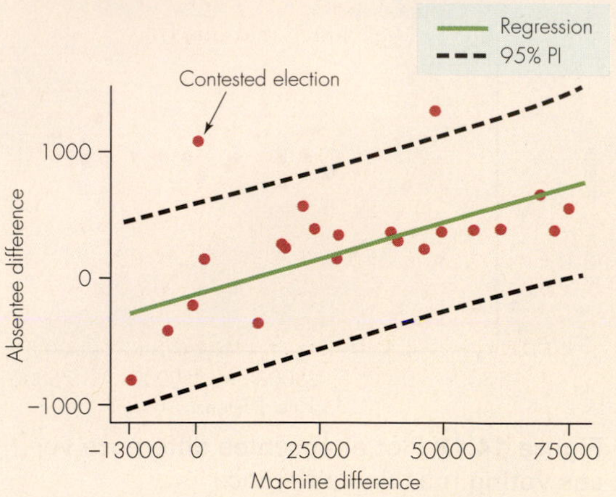

Figure 14.13 Prediction limits for Diff Absentee, given machine difference

relationship may actually be curvilinear. A plot of the residuals versus x (not shown) confirms this effect. Therefore, another analysis was done by including a "quadratic" term, (Diff Machine)^2, in the regression equation. Results are shown in the output below, including 95% confidence and prediction intervals for the contested election. Note that the prediction interval is even more disparate with the observed result of 1025 than it was with the simple linear regression equation.

```
The regression equation is
Diff Absentee = -219 + 0.0297 Diff Machine -0.000000
Quadratic

Predictor        Coef        StDev       T       P
Constant       -219.0        105.4   -2.08   0.052
Diff Mac     0.029709     0.006887    4.31   0.000
Quadrati  -0.00000028   0.00000011   -2.67   0.016

S = 282.6       R-Sq = 63.3%       R-Sq(adj) = 59.3%

Predicted Values

    Fit   StDev Fit       95.0% CI             95.0% PI
 -235.9     107.7    (-462.1, -9.6)     (-871.2, 399.5)
```

The *p*-value reported for the "quadratic term" is .016, and it is for a test of the null hypothesis that the quadratic term is not needed. Therefore, it is clear that the new model, including the quadratic term, is better than the simple linear model. Figure 14.14 shows the prediction limits based on the quadratic model over the range of machine differences for the model including the quadratic term.

The regression analysis and accompanying figures cast doubt on the results of the contested election, assuming that all other factors were the same as in the past. There may have been other factors that led more people who cast votes for the Democratic candidate to do so by absentee ballot in this election, but that issue cannot be addressed using the statistical evidence available. Judge Newcomer was convinced that other potential differences could not account for the statistical anomaly displayed by the regression analysis.

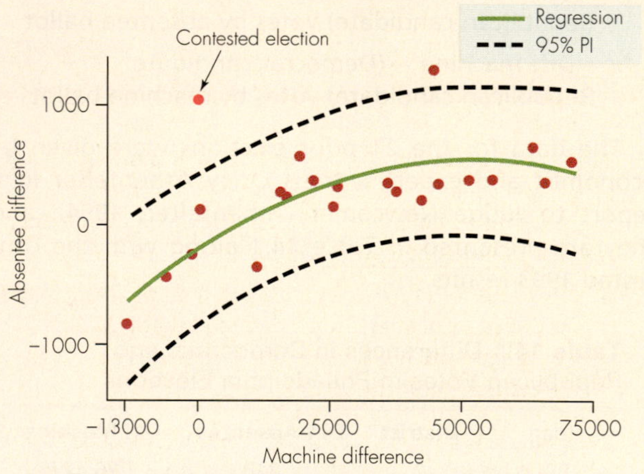

Figure 14.14 Prediction limits for quadratic model

Key Terms

Section 14.1
simple regression, 549
regression model, 549
dependent variable, 549
predictor variables, 549
regression line for the sample, 550
regression line for the population, 550, 552
intercept, 550
y-intercept, 550
slope, 550
least-squares line, 551
deviation in regression, 551
residual, 551, 552
population intercept, β_0, 552
population slope, β_1, 552
constant variance in regression, 552
regression model, for population, 553
simple linear regression model, 554
multiple regression, 554

Section 14.2
standard deviation for regression, 555
sum of squared errors (SSE), 555
residual sum of squares, 555
sum of squared residuals, 555
proportion of variation explained by *x*, R^2, 556
SSTO, 557
analysis of variance table, 558

Section 14.3
hypothesis test for population slope, 558
statistical significance of a linear relationship, 558
confidence interval for population slope, 559–560

In Summary Boxes

Exercises

◆ Denotes that the dataset is available on the companion website, http://www.cengage.com/statistics/Utts4e, but is not required to solve the exercise.

Bold exercises have answers in the back of the text.

Section 14.1

Skillbuilder Exercises

14.1 ◆ The **temperature** dataset on the companion website gives y = mean April temperature (Fahrenheit) and x = geographic latitude for 20 U.S. cities. The simple linear regression equation for the sample is $\hat{y} = 119 - 1.64x$.

a. The value of latitude for Pittsburgh is 40. What is the predicted April temperature for Pittsburgh?

b. The mean April temperature for Pittsburgh is 50. What is the residual for Pittsburgh?

14.2 For each of the following, specify whether the notation or word applies to the population or the sample. If it applies to the population, give the notation or word for the corresponding feature in the sample, and if it applies to the sample, give the notation or word for the corresponding feature in the population.

a. b_0.

b. $E(Y)$.

c. ε_i

d. Residual.

14.3 The population model for simple linear regression can be defined as

$$y_i = \beta_0 + \beta_1 x_i + \varepsilon_i$$

a. Explain what the parameters β_0 and β_1 represent.

b. Write the equation that represents how the mean of the y variable relates to the x variable.

14.4 Ages of husbands and wives are recorded for a randomly selected sample of $n = 200$ opposite-sex married couples in the United States. The data will be used to do a simple linear regression with y = husband's age and x = wife's age.

a. Explain the distinction between the regression line for the sample and the regression line for the population in this situation. Include descriptions of the sample and the population.

b. Write the population model for the simple linear regression that describes how the age of a husband is related to the age of a wife. Write any necessary assumptions about deviations from the line.

14.5 Suppose that medical researchers use sample data to find a regression line describing the relationship between y = systolic blood pressure and x = age for men between 40 and 60 years old and that the sample regression line is

Estimated mean systolic blood pressure = 85 + 0.9(Age)

a. Give the value of the slope, and write a sentence that interprets it in the context of this situation.

b. Estimate the mean systolic blood pressure of men who are 50 years old.

c. Suppose that the systolic blood pressure of a 50-year-old man is 125. What is the value of the residual for this man?

d. Explain in words what the residual you found in part (c) indicates about the relationship between the systolic blood pressure for that individual and the estimated mean systolic blood pressure for 50-year-old men.

14.6 The linear relationship between the number of bedrooms a house has and the selling price of the house (in dollars) is analyzed for a sample of $n = 250$ recently sold houses in a large metropolitan area. The sample regression line is

Predicted selling price = 144000 + 50000 (Bedrooms)

a. Give the value of the slope of the regression line, and write a sentence that interprets it in the context of this situation.

b. Predict the selling price for a house with three bedrooms.

c. Suppose that a three-bedroom house sold for $300,000. What is the residual for this house?

d. Explain in words what the residual you found in part (c) indicates about the relationship between the selling price for this house and the estimated mean selling price for three-bedroom houses.

General Section Exercises

14.7 Heights and answers to the question, "How much would you like to weigh?" were recorded for $n = 126$ women in a statistics class. A regression equation for y = desired weight (in pounds) and x = height (in inches) is

$$\text{Estimated mean desired weight} = -65.4 + 2.90\,(\text{Height})$$

(*Data source:* William Harkness.)

a. What is the value of the slope of this regression equation? Write a sentence that explains what this value shows about the linear relationship between desired weight and height for these women.

b. Is the proper notation for the slope of this equation b_1 or β_1? Explain.

c. What is the estimated mean desired weight for college women who are 65 inches tall?

14.8 Refer to Exercise 14.7. Suppose a woman is 63 inches tall and her desired weight is 115 pounds. Give numerical values for each of the following for this individual:

a. x.

b. y.

c. $\hat{y}$.

d. The residual.

e. The component of y that is explained by knowing x.

f. The unexplained part of y.

14.9 Suppose that the *population* regression equation relating y = weight (pounds) and x = height (inches) for men aged 18 to 29 years old is

$$\text{Mean weight} = -250 + 6(\text{Height}).$$

An individual in this population is 70 inches tall and weighs 180 pounds. Give numerical values for each of the following for this individual:

a. x.

b. y.

c. $E(Y)$ for this individual's height.

d. The deviation for this individual.

e. The component of y that is explained by knowing x.

f. The unexplained part of y.

14.10 In the simple linear regression model for a population, the relationship between the x variable and the mean of the y variable is $E(Y) = \beta_0 + \beta_1 x$.

a. Suppose $\beta_1 = 0$. What would be the appearance of this line? Draw a sketch illustrating your answer.

b. Explain how the answer to part (a) illustrates the lack of a linear relationship between x and y.

Section 14.2

Skillbuilder Exercises

14.11 Find r^2 in each of the following situations:

a. SSTO = 500, SSE = 300.

b. SSTO = 200, SSE = 40.

c. SSTO = 80, SSE = 0.

d. SSTO = 100, SSE = 95.

14.12 ◆ An example in Section 14.1 described the relationship between y = handspan (centimeters) and x = height (inches). The sample consisted of measurements of both variables for $n = 167$ college students, available in the dataset **handheight** on the companion website. Computer output for this example is as follows:

The regression equation is
HandSpan = −3.00 + 0.351 Height

Predictor	Coef	SE Coef	T	P
Constant	−3.002	1.694	−1.77	0.078
Height	0.35057	0.02484	14.11	0.000

S = 1.301 R-Sq = 54.7%

a. What is the value of the standard deviation from the regression line? Write a sentence that interprets this value.

b. What is the value of R^2 given in the output? Write a sentence that interprets this value.

14.13 Refer to Exercise 14.1 about mean April temperatures and geographic latitudes for U.S. cities. Minitab output for the linear regression is as follows:

The regression equation is
AprTemp = 119 − 1.64 latitude

Predictor	Coef	SE Coef	T	P
Constant	118.776	4.467	26.59	0.000
latitude	−1.6436	0.1165	−14.11	0.000

S = 2.837 R-Sq = 91.7%

a. In the output, what value is given for the standard deviation from the regression line? Write a sentence that interprets this value.

b. In the output, the value of R^2 is given as 91.7%. Write a sentence that interprets this value.

14.14 ◆ The dataset **bodytemp** on the companion website gives age in years and body temperature in degrees Fahrenheit for 100 blood donors ranging in age from 17 to 84 years old. Minitab output for the linear regression is as follows:

The regression equation is
Temperature = 98.6 − 0.0138 Age

Predictor	Coef	SE Coef	T	P
Constant	98.6012	0.2290	430.60	0.000
Age	−0.013835	0.004239	−3.26	0.002

S = 0.703089 R-Sq = 9.8%

a. In the output, what value is given for the standard deviation from the regression line? Write a sentence that interprets this value.
b. In the output, the value of R^2 is given as 9.8%. Write a sentence that interprets this value.

General Section Exercises

14.15 ◆ Example 14.5 (p. 557) gave the regression equation $\hat{y} = 577 - 3.01x$ for the relationship between y = maximum distance at which a driver can read a highway sign and x = driver age. The data are in the dataset **signdist** on the companion website.

a. What is the predicted value of the maximum sign-reading distance for a driver who is 21 years old?
b. Compute the residual for a 21-year-old driver who can read the sign at a maximum distance of 525 feet.
c. The standard deviation from the regression line in this example is approximately $s = 50$ feet. Use the Empirical Rule to give an interval that describes the maximum sign-reading distances for about 95% of all drivers who are 21 years old.
d. Would it be unusual for a 21-year-old driver to be able to read the sign from 650 feet? Justify your answer.

14.16 Refer to Exercise 14.14, in which the regression relationship between y = body temperature and x = age was given in Minitab output as $\hat{y} = 98.6 - 0.0138(\text{Age})$.

a. What is the predicted body temperature for someone who is 21 years old?
b. There were two individuals in the sample who were 21 years old. Their body temperatures were 98.3 and 98.4 degrees. Compute the residual for each of these individuals.
c. The standard deviation from the regression line in this example is approximately $s = 0.70$ degrees. Use the Empirical Rule to give an interval that describes body temperature for about 95% of the population of people who are 21 years old.
d. Would it be unusual for a 21-year-old to have a body temperature of 98.6 degrees? Justify your answer.

14.17 The least-squares regression line is $\hat{y} = 9 + 2x$ for these six observations of x and y:

x	1	1	3	3	5	5
y	10	12	13	17	17	21

a. For each observation, calculate the residual. Verify that the sum of the residuals is 0.
b. Compute SSE, the sum of squared errors (residuals).
c. For the sample, calculate the standard deviation from the regression line.

14.18 Sketch sample data for which the standard deviation from the regression line is $s = 0$.

14.19 Refer to Exercises 14.1 and 14.13 in which results are given for the relationship between y = mean April temperature and x = geographic latitude for 20 U.S. cities. The analysis of variance table for this relationship is as follows:

Analysis of Variance					
Source	DF	SS	MS	F	P
Regression	1	1601.9	1601.9	198.99	0.000
Residual Error	18	144.9	8.1		
Total	19	1746.8			

Find or compute the following values based on this information. Refer to Example 14.5 (p. 557) for guidance.

a. SSTO.
b. SSE.
c. r^2.

14.20 Refer to Exercise 14.14 in which results are given for the relationship between y = body temperature and x = age. The analysis of variance table for this relationship is as follows:

Analysis of Variance					
Source	DF	SS	MS	F	P
Regression	1	5.2653	5.2653	10.65	0.002
Residual Error	98	48.4447	0.4943		
Total	99	53.7100			

Find or compute the following values based on this information. Refer to Example 14.5 (p. 557) for guidance.

a. SSTO.
b. SSE.
c. r^2.

Section 14.3

Skillbuilder Exercises

14.21 ◆ The **UCDavis1** dataset on the companion website includes grade point averages and self-reported hours per week of watching television for students in a statistics class. In a simple linear regression for the relationship between y = grade point average and x = hours per week of television watching, the p-value is .552 for testing H_0: $\beta_1 = 0$ versus H_a: $\beta_1 \neq 0$.

a. Is the observed relationship statistically significant? Explain.
b. What is the p-value for testing the null hypothesis that the correlation between the variables is 0 in the population represented by the sample? Assume a two-sided alternative hypothesis.

14.22 For men over the age of 40, a linear regression is done to examine the relationship between y = systolic blood pressure and x = age. A 95% confidence interval for the population slope β_1 is 0.3 to 0.7. In the context of this situation, write a sentence that interprets this confidence interval.

14.23 Refer to the output given in Exercise 14.13 for the simple linear regression with y = average April temperature and x = geographic latitude for 20 U.S. cities.

a. Write null and alternative hypotheses for testing whether there is a linear relationship between mean April temperature and latitude. Use proper statistical notation.
b. Use information given in the output to test the hypotheses written in part (a). State a conclusion and give a reason for the conclusion.
c. Using values given in the output, show how to calculate the *t*-statistic for testing the hypotheses stated in part (a).
d. What is the *p*-value for testing $H_0: \rho = 0$ versus $H_a: \rho \neq 0$?

14.24 Refer to the output given in Exercise 14.14 for the simple linear regression with y = body temperature and x = age.

a. Write null and alternative hypotheses for testing whether there is a linear relationship between body temperature and age. Use proper statistical notation.
b. Use information given in the output to test the hypotheses written in part (a). State a conclusion and give a reason for the conclusion.
c. Using values given in the output, show how to calculate the *t*-statistic for testing the hypotheses stated in part (a).
d. What is the *p*-value for testing $H_0: \rho = 0$ versus $H_a: \rho \neq 0$?

General Section Exercises

14.25 ◆ This output is for a linear regression relating y = GPA and x = hours per week spent using the computer for $n = 162$ college students. The data are in the **UCDavis1** dataset on the companion website. Note that 11 of the 173 students surveyed did not provide complete data.

The regression equation is
GPA = 3.01 − 0.00555 computer

162 cases used, 11 cases contain missing values

Predictor	Coef	SE Coef	T	P
Constant	3.00588	0.07335	40.98	0.000
computer	−0.005548	0.003979	−1.39	0.165

a. Is there a statistically significant relationship between the two variables? Justify your answer using information from the output. Be specific about the null and alternative hypotheses that you are evaluating.
b. Suppose that ρ denotes the correlation between the two variables in the population represented by the sample. What is the *p*-value for testing $H_0: \rho = 0$ versus $H_a: \rho \neq 0$?

14.26 The relationship between y = hours of watching television in a typical day and x = age was examined in Example 3.14. The data were gathered in the 2008 General Social Survey done by the National Opinion Research Center at the University of Chicago, and there were $n = 1288$ observations. The correlation between the variables is $r = 0.171$, a value that is indicative of a weak connection between hours of watching television and age.

The regression equation is
tvhours = 1.77 + 0.0225 age

Predictor	Coef	SE Coef	T	P
Constant	1.7695	0.1784	9.92	0.000
age	0.022474	0.003619	6.21	0.000

S = 2.11802   R-Sq = 2.9%   R-Sq(adj) = 2.8%

a. What is the value of the slope of the sample regression equation? Write a sentence that interprets this value in the context of this situation.
b. Using proper statistical notation, write null and alternative hypotheses for testing the statistical significance of the relationship between the two variables.
c. Use information in the output to determine if there is a statistically significant relationship between age and daily hours of watching television. Include relevant information in your answer.
d. Do you think that the observed relationship has practical significance? Explain.

14.27 In Example 14.3 (p. 555), a sample of $n = 43$ college men was used to determine the regression equation, Weight = −318 + 7(Height). Weights were measured in pounds and heights were measured in inches.

a. The slope of the line is 7. Is the proper notation for this slope b_1 or β_1? Explain.
b. The standard error for the slope was given as 1.581 in Figure 14.4. Calculate an approximate 95% confidence interval for the slope in the population. Write a sentence that interprets this interval.
c. It has been stated that for each 1-inch increase in height, average weight increases by 5 pounds. Does the interval computed in part (b) support this statement? Explain.

14.28 In Exercise 14.14 the regression relationship between age and body temperature based on 100 blood donors was given as Temperature = 98.6 − 0.0138(Age).

a. The slope of the line is −0.0138. Explain whether the proper notation for this slope is b_1 or β_1.
b. The standard error for the slope was given in the output accompanying Exercise 14.14 as 0.0042. Calculate an approximate 95% confidence interval for the slope in the population. Write a sentence that interprets this interval.
c. Use your answer in part (b) to find an approximate 95% confidence interval for the predicted decrease in body temperature as someone ages by 10 years.

14.29 A sample of 837 ninth-grade girls was asked how much time they spent watching music videos each week. The girls also were asked to rate how concerned they were about their weight on a scale of 0 to 100 (100 = extremely concerned) as well as to rate their perception of how important appearance is on a scale of 1 to 6 (6 = very important). The researchers (Borzekowski et al., 2000) wrote the following about the results: "When media use was separated into distinct media genres, only hours of watching music videos was related to perceived importance of appearance and weight concerns ($r = .12$, $p < .001$, and $r = .08$, $p < .05$, respectively)."

a. What are the null and alternative hypotheses for which *p*-values are given in the researchers' statement?
b. What conclusions have the researchers made about the hypotheses you gave in part (a)?

14.30 Refer to Exercise 14.29, in which correlations were given for the relationship between hours of watching music videos and both perceived appearance and weight concerns. The correlations were based on data from 837 ninth-grade girls.

a. Do the correlation values given by the researchers indicate weak or strong relationships between the variables? Support your answer with a sketch illustrating the appearance of a scatterplot for the given correlation values.
b. Read the paragraph on page 561 about the effect of sample size on significance. Discuss the issue addressed in that paragraph in the context of this example.

Section 14.4

Skillbuilder Exercises

14.31 For each of the following two situations, explain whether it would be more appropriate to use a confidence interval for the mean or a prediction interval for a value of y.

a. Estimate the first-year college GPA for a student whose high school GPA was 3.5.
b. Estimate the mean first-year college GPA for students whose high school GPA was 3.5.

14.32 For each of the following two situations, explain whether it would be more appropriate to use a confidence interval for the mean or a prediction interval for a value of y.

a. Estimate the average income for individuals who have 20 years of experience within a particular occupation.
b. Estimate the adult height of a daughter of a woman who is 64 inches tall.

14.33 Suppose a regression model is used to analyze the relationship between y = grade point average and x = number of classes missed in a typical week for college students.

a. In the context of this situation, explain what would be predicted by a prediction interval for y when $x = 2$ classes missed per week.
b. In the context of this situation, explain what would be estimated by a confidence interval for $E(Y)$ when $x = 2$ classes missed per week.

14.34 A college finds that among students who had a total SAT score of approximately 1200, about 90% have a first-year GPA in the range 2.7 to 3.7. Is this interval a prediction interval or a confidence interval for the mean? Explain.

14.35 ◆ Forty students measure their resting pulse rates, then march in place for 1 minute and measure their pulse rates after the marching. The regression line for the sample is $\hat{y} = 17.79 + 0.894x$, where y = pulse after marching and x = pulse before marching. The following Minitab output gives a confidence interval and a prediction interval for pulse after marching when pulse before marching is 70. (*Data source:* **pulsemarch** dataset on the companion website.)

Fit	SE Fit	95.0% CI	95.0% PI
80.35	1.15	(78.02, 82.67)	(65.50, 95.19)

a. Write the 95% confidence interval for $E(Y)$. In the context of this situation, write a sentence that interprets this interval.
b. Write the 95% prediction interval for y. In the context of this situation, write a sentence that interprets this interval.

General Section Exercises

14.36 ◆ The dataset **oldfaithful** on the companion website provides data collected on 299 eruptions of the Old Faithful geyser in Yellowstone Park. The variables are x = duration of the eruption (in minutes) and y = time until the next time the geyser erupts (in minutes). The following Minitab output shows the regression relationship, and a confidence interval and prediction interval when $x = 4$ minutes.

The regression equation is
TimeNext = 35.0 + 10.7 Duration

New Obs	Fit	SE Fit	95% CI	95% PI
1 [x = 4]	77.620	0.482	(76.670, 78.571)	(64.307, 90.934)

a. Carried out to more decimal places than shown in the output, the regression equation is TimeNext = 34.98 + 10.66(Duration). Use this regression equation to verify that $\hat{y}$ when $x = 4$ is the number shown in the column labeled "Fit."
b. Write the 95% confidence interval for $E(Y)$. In the context of this situation, write a sentence that interprets this interval.
c. Write the 95% prediction interval for y. In the context of this situation, write a sentence that interprets this interval.
d. If you were a tourist waiting to see Old Faithful erupt, and were provided with the duration of the last eruption, would you be more interested in knowing the confidence interval for $E(Y)$, or the prediction interval for y? Explain.

14.37 Example 2.17 (p. 47) described the heights of a sample of married British women. The ages of the women and their husbands were available in the same dataset for $n = 170$ couples. For the sample, the linear regression line relating y = husband's age and x = wife's age is $\hat{y} = 3.59 + 0.9667x$. The following results computed with Minitab give a confidence interval and prediction interval for husband's age when wife's age is 40:

Fit	SE Fit	95.0% CI	95.0% PI
42.258	0.313	(41.641, 42.875)	(34.200, 50.316)

a. Verify that the "Fit" is consistent with the predicted value that would be given by the regression equation.
b. Explain why the "95% PI" is much wider than the "95% CI."

14.38 Refer to Exercise 14.37, giving the relationship between husbands' and wives' ages for a sample of British couples.

a. Interpret the "95% CI" given by Minitab. Be specific about what the interval estimates.
b. Interpret the "95% PI" given by Minitab. Be specific about what the interval estimates.

14.39 Refer to Exercise 14.37. Using the general format for a 95% confidence interval, verify the confidence interval for the mean given by Minitab. Note that the standard error of the "Fit" is given.

14.40 Suppose that a linear regression analysis of the relationship between y = systolic blood pressure and x = age is done for women between 40 and 60 years old. For women who are 45 years old, a 90% confidence interval for $E(Y)$ is determined to be 128.2 to 131.3. Explain why it is incorrect to conclude that about 90% of women who are 45 years old have a systolic blood pressure in this range. Write a sentence that correctly interprets the interval.

Section 14.5

Skillbuilder Exercises

14.41 There are five conditions listed at the beginning of Section 14.5 that should be at least approximately true for linear regression. Which of the conditions can be checked by using each of the following methods? In each case, list all of the conditions that can be checked.

a. Drawing a histogram of the residuals.
b. Drawing a scatterplot of the residuals versus the x values.
c. Learning how the data were collected.
d. Drawing a scatterplot of the raw data, y versus x.

14.42 Refer to Exercise 14.35 about a linear regression for y = pulse after marching in place and x = pulse before marching in place. The figure for this exercise is a plot of residuals versus the pulse before marching for a sample of 40 students. Discuss what the plot indicates about Conditions 1, 2, and 3 for linear regression listed at the beginning of Section 14.5.

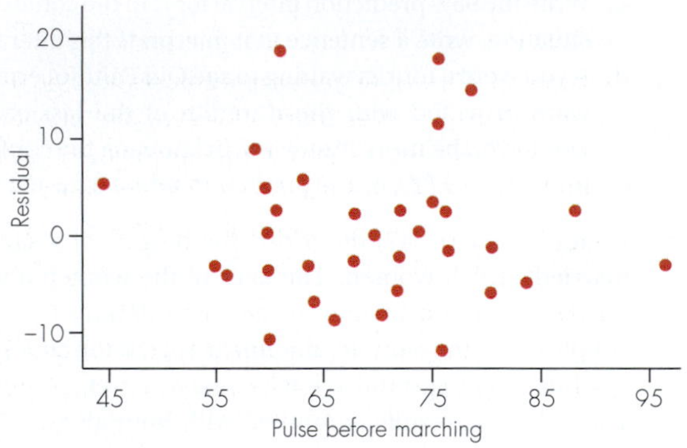

14.43 ◆ The figure for this exercise is a histogram of the residuals for a linear regression relating y = height (inches) and x = foot length (centimeter) for a sample of college men. Discuss what the histogram indicates about Conditions 2 (no outliers) and 4 (normality) for linear regression listed at the beginning of Section 14.5. (*Data source:* **heightfoot** dataset on the companion website.)

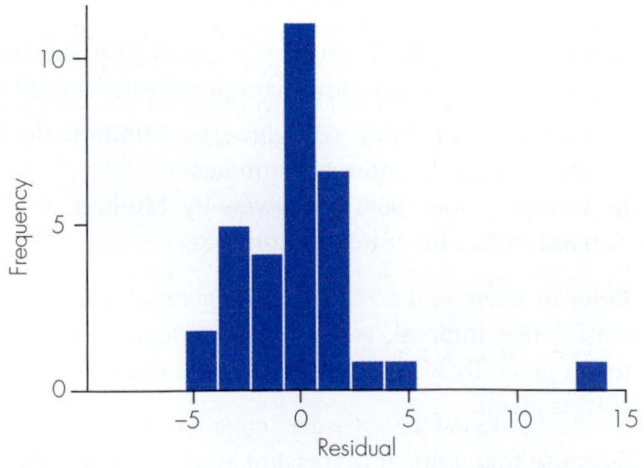

General Section Exercises

14.44 The figure accompanying this exercise is a histogram of the residuals for a simple linear regression. What does this plot indicate about the necessary conditions for conducting a linear regression? Be specific about which of the five necessary conditions are verified in this figure.

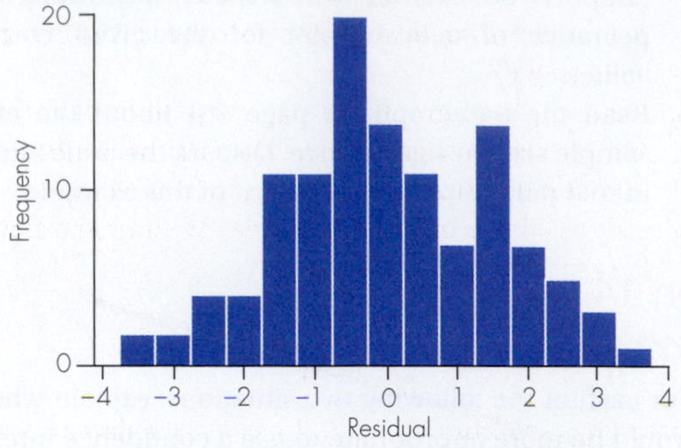

14.45 ◆ Observed data along with the sample regression line for the relationship between body weight (pounds) and neck girth (inches) for 19 female bears of various ages are shown in the figure below. (*Note:* The data are in the dataset **bears-female** on the companion website.)

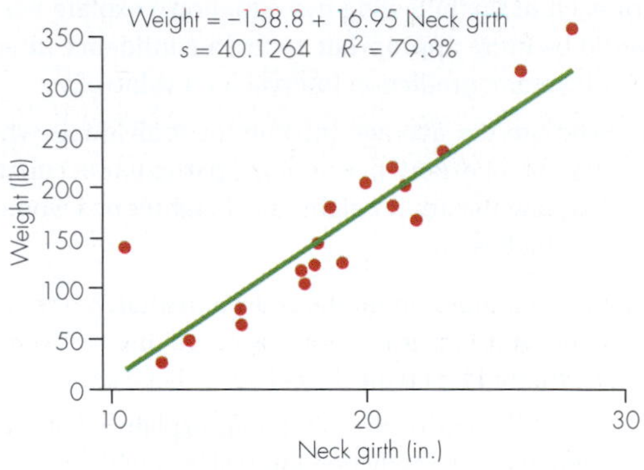

a. Which one of the necessary conditions for linear regression appears to be violated in this dataset?
b. What corrective action(s) would you consider in order to properly estimate the relationship between body weight and neck girth?
c. Sketch a histogram that illustrates the pattern of the distribution of the residuals for this problem. Your sketch does not have to be accurate in the numerical details but should correctly show the pattern.

14.46 The figure for this exercise shows data for the relationship between the average stopping distance (feet) of a car when the brakes are applied and vehicle speed (miles per hour). The regression line for these data is also shown on the plot. (*Note:* The raw data were given in Exercise 3.7.)

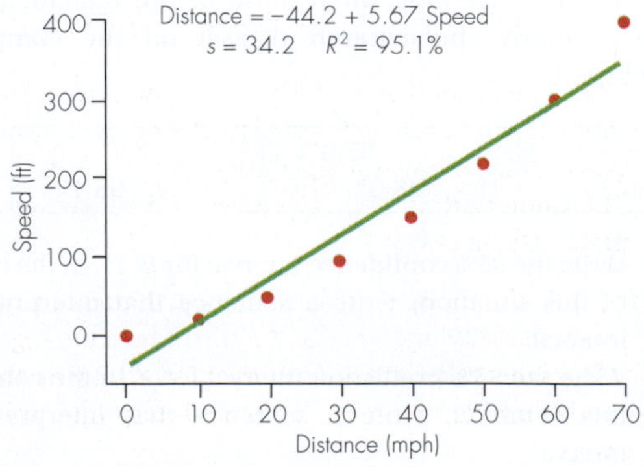

a. Which one of the five necessary conditions for linear regression appears to be violated in this situation?
b. What corrective action would you take to correctly estimate the connection between stopping distance and vehicle speed?
c. Make a sketch that illustrates the pattern of a plot of residuals versus speed for the data shown in the figure in this exercise. Your sketch does not need to be numerically accurate but should correctly show the pattern of the plot.

Chapter Exercises

14.47 Data for y = hours of sleep the previous day and x hours of studying the previous day for $n = 116$ college students were shown in Figure 3.14 (p. 85) and described in Example 3.15. Some regression results for those data are as follows:

The regression equation is
Sleep = 7.56 − 0.269 Study

Predictor	Coef	SE Coef	T	P
Constant	7.5555	0.2239	33.74	0.000
study	−0.26917	0.06616	—	0.000

S = 1.509 R-Sq = 12.7% R-Sq(adj) = 11.9%

a. What is the estimated mean decrease in hours of sleep per 1-hour increase in hours of studying? What notation is used for this value?
b. Calculate an approximate 95% confidence interval for β_1. Write a sentence that interprets this interval.
c. Using proper statistical notation, write null and alternative hypotheses for assessing the statistical significance of the relationship.
d. We omitted from the output the t-statistic for testing the hypotheses of part (c). Compute the value of this t-statistic using other information shown in the output. What are the degrees of freedom for this t-statistic?
e. Is there a statistically significant relationship between hours of sleep and hours of studying? Justify your answer on the basis of information shown in the output.

14.48 Refer to Exercise 14.47 about hours of sleep and hours of study.
a. What is the value of the standard deviation from the regression line? Write a sentence that interprets this value.
b. Calculate the predicted value of hours of sleep the previous day for a student who studied 4 hours the previous day.
c. Using the Empirical Rule, determine an interval that describes hours of sleep for approximately 95% of students who studied 4 hours the previous day.
d. The value of R^2 is given as 12.7%. Write a sentence that interprets this value.

14.49 Refer to Exercises 14.47 and 14.48 about hours of sleep and hours spent studying. What is the intercept of the regression line? Does this value have a useful interpretation in the context of this problem? If so, what is the interpretation? If not, why not?

14.50 Regression results for the relationship between y = hours of sleep the previous day and x = hours spent studying the previous day were given in Exercise 14.47. The figure for this exercise is a plot of residuals versus hours spent studying. What does the plot indicate about the necessary conditions for conducting a linear regression? Be specific about which of the five conditions given in Section 14.5 are verified by this plot.

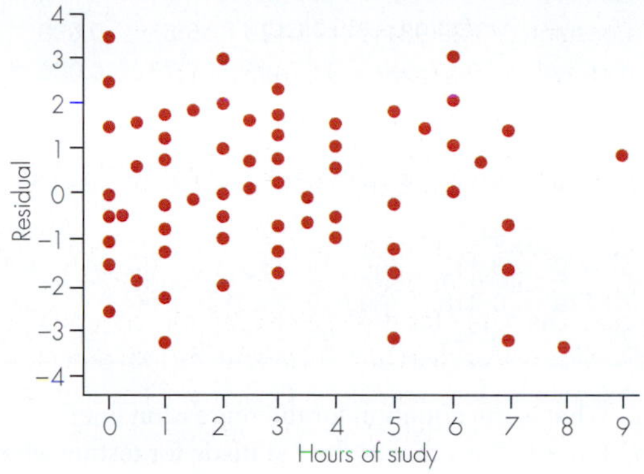

14.51 Exercise 14.47 gave linear regression results for the relationship between y = hours of sleep the previous day and x = hours spent studying the previous day. Following is Minitab output showing a confidence interval and a prediction interval for hours of sleep when hours of studying = 3 hours:

Fit	SE Fit	95.0% CI	95.0% PI
6.748	0.142	(6.466, 7.029)	(3.746, 9.750)

Write down the interval given for predicting the hours of sleep when hours of studying is 3 hours. Give two different interpretations of this interval.

14.52 Refer to the output given in Exercise 14.12 about handspan and height. Compute a 90% confidence interval for the population slope. Write a sentence that interprets this interval.

14.53 The following data are x = average on five quizzes before the midterm exam and y = score on the midterm exam for $n = 11$ students randomly selected from a multiple-section statistics class of about 950 students:

x = Quizzes	80	68	94	72	74	83	56	68	65	75	88
y = Exam	72	71	96	77	82	72	58	83	78	80	92

a. Plot the data, and describe the important features of this plot.
b. Using statistical software, calculate the regression line for this sample.
c. What is the predicted midterm exam score for a student with a quiz average equal to 75?
d. With statistical software, determine a 50% prediction interval for the midterm exam score of a student whose quiz average was 75.

14.54 ◆ This exercise refers to the following Minitab output, relating y = son's height to x = father's height for a sample of $n = 76$ college males. (*Note:* The data are in the dataset **UCDavis1** on the companion website.)

```
The regression equation is
Height = 30.0 + 0.576 dadheight

76 cases used, 3 cases contain missing values

Predictor      Coef     SE Coef      T       P
Constant     29.981       5.129   5.85   0.000
dadheigh    0.57568     0.07445   7.73   0.000

S = 2.657       R-Sq = 44.7%

Predicted Values [for dad's heights of 65, 70, and 74]

Fit       SE Fit        95.0% CI              95.0% PI
67.400    0.415    (66.574, 68.226)    (62.041, 72.759)
70.279    0.318    (69.645, 70.913)    (64.946, 75.612)
75.581    0.494    (71.596, 73.566)    (67.195, 77.967)
```

a. What is the equation for the regression line?

b. Identify the value of the t-statistic for testing whether or not the slope is 0. Verify that the value is correct using the formula for the t-statistic and the information provided by Minitab for the parts that go into the formula.

c. State and test the hypotheses about whether or not the population slope is 0. Use relevant information provided in the output.

d. Compute a 95% confidence interval for β_1, the slope of the relationship in the population. Write a sentence that interprets this interval.

14.55 Refer to Exercise 14.54.

a. What is the value of R^2 for the observed linear relationship between height and father's height? Write a sentence that interprets this value.

b. What is the value of the correlation coefficient r?

14.56 Refer to Exercises 14.54 and 14.55. The output provides prediction intervals and confidence intervals for father's heights of 65, 70, and 74 inches.

a. Verify that the "Fit" given by Minitab for father's height of 65 inches is consistent with the predicted height that would be given by the regression equation.

b. Write down the interval Minitab provided for predicting an individual son's height if his father's height is 70 inches. Provide two different interpretations for the interval.

c. Write a sentence that interprets the "95% CI" given for a son's height when the father's height is 74 inches.

d. Explain why the prediction interval is much wider than the corresponding confidence interval for each father's height provided.

e. The accompanying figure is a plot of the residuals versus father's height. Which one of the five necessary conditions for regression listed in Section 14.5 appears to be violated? What corrective action(s) should you consider?

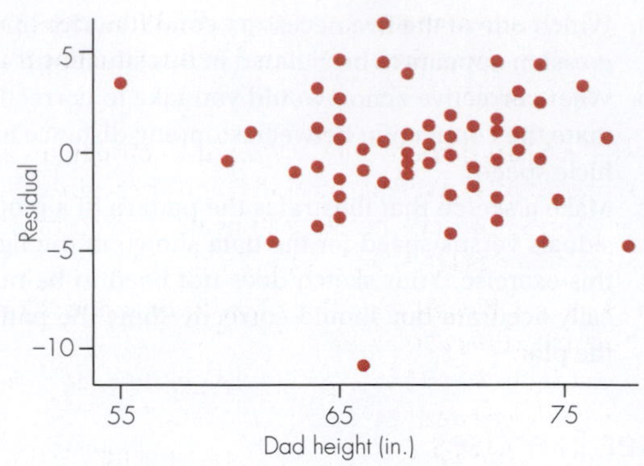

14.57 The five steps for hypothesis testing were given in Chapters 12 and 13. Describe those steps as they apply to testing whether there is a relationship between two variables in the simple linear regression model.

14.58 Explain why rejecting H_0: $\beta_1 = 0$ in a simple linear regression model does not prove that the relationship is linear. To answer this question, you might find it helpful to consider the figure in Exercise 14.46, which shows stopping distance and vehicle speed for automobiles.

Dataset Exercises

Datasets required to solve these exercises are available on the companion website, http://www.cengage.com/statistics/Utts4e.

14.59 Use the dataset **letters** from the companion website for this exercise. A sample of 63 students wrote as many capital letters of the alphabet in order as they could in 15 seconds using their dominant hand, and then they repeated this task using their nondominant hand. The variables ***dom*** and ***nondom*** contain the raw data for the results.

a. Plot y = ***dom*** versus x = ***nondom***. Describe the important features of the plot.

b. Compute the simple regression equation for the relationship. What is the equation?

c. What are the values of the standard deviation from the regression line and R^2? Interpret these values in the context of this problem.

d. Determine a 95% confidence interval for the population slope. Write a sentence that interprets this interval.

e. Consider the statement, "On average, a student can write about 23 more letters in 15 seconds with the dominant hand than with the nondominant hand." What regression equation would accompany this statement? Based on your answers to parts (b) and (d), explain why this statement is reasonable.

14.60 Refer to the previous exercise about letters written with the dominant (y) and nondominant (x) hands.

a. Plot residuals versus x = ***nondom***. What does this plot indicate about conditions for using the linear regression model?

b. Create a histogram of the residuals. What does this plot indicate about the conditions for using the linear regression model?

14.61 Use the **bears-female** dataset from the companion website for this exercise. Weights (pounds) and chest girths (inches) are given for $n = 19$ female wild bears. The corresponding variable names are ***Weight*** and ***Chest***.

a. Plot $y =$ ***Weight*** versus $x =$ ***Chest***. Describe the important features of the plot.
b. Compute a simple linear regression equation for $y =$ ***Weight*** and $x =$ ***Chest***.
c. What is the value of R^2 for this relationship? Write a sentence interpreting this value.
d. What is the predicted weight for a bear with a chest girth of 40 inches?
e. Compute a 95% prediction interval for the weight of a bear with a chest girth of 40 inches. Write a sentence interpreting this interval.
f. Compute a 95% confidence interval for the mean weight of bears with a chest girth of 40 inches. Write a sentence interpreting this interval.

14.62 Use the dataset **heightfoot** from the companion website for this exercise. Heights (inches) and foot lengths (centimeters) are given for 33 men.

a. Plot $y =$ height (***height***) versus $x =$ foot length (***foot***). What important features are evident in the plot?
b. Omit any outliers evident in the plot in part (a), and compute the linear regression line for predicting $y =$ height from $x =$ foot length.
c. Determine a 90% prediction interval for the height of a man whose foot length is 28 cm.
d. Discuss whether height can be accurately predicted from foot length. Use regression results to justify your answer.
e. For the data used for part (b), plot residuals versus $x =$ foot length. What does this plot indicate about the necessary conditions for conducting a linear regression?

14.63 Refer to Exercise 14.62 about the relationship between height (***height***) and foot length (***foot***) for the dataset **heightfoot**.

a. Do not omit any outliers. Use the complete dataset to determine a 90% prediction interval for the height of a man whose foot length is 28 cm.
b. Explain why the interval computed in part (a) is wider than the interval computed for part (c) of the previous exercise.
c. For the regression based on all data points, plot residuals versus $x =$ foot length. What does this plot indicate about the necessary conditions for conducting a linear regression?

15

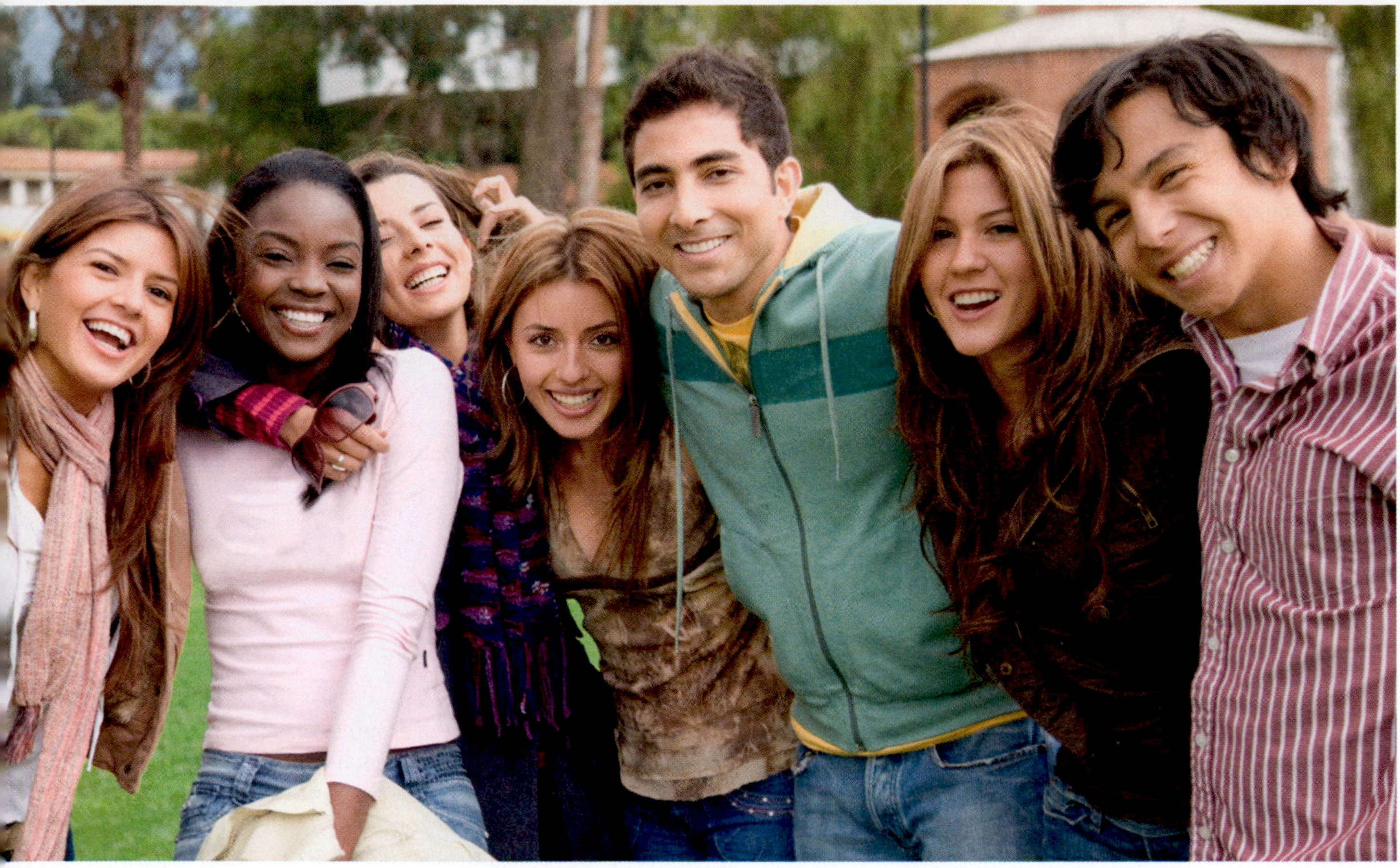

Andresr 2010/Used under license from Shutterstock.com

With whom is it easiest to make friends?

See Example 15.2 *(p. 585)*

More about Inference for Categorical Variables

Are the digits that are drawn in lotteries actually equally likely, as claimed? Are belief and performance on ESP tests related? Do males and females have the same opinions about whether it's easier to make friends with the same sex or the opposite sex? Chi-square tests can be used to answer these kinds of questions about categorical variables.

In Chapter 4, we learned techniques for analyzing a fundamental question that researchers often ask about two categorical variables, which is

Is there a relationship between the variables so that the chance that an individual falls into a particular category for one variable depends upon the particular category they fall into for the other variable?

For instance, in Example 4.2, we saw that the likelihood of divorce is greater for smokers than it is for nonsmokers. The chi-square test is a procedure for assessing the statistical significance of a relationship between categorical variables. We introduced the basic ideas of this statistical method in Chapter 4; in this chapter, we examine the chi-square test more fully. We will use the chi-square procedure to test hypotheses about two-way tables, and we will also learn how a different chi-square procedure is used to test hypotheses about a single categorical variable.

15.1 The Chi-Square Test for Two-Way Tables

In Chapter 4, we learned these terms:

- The raw data from a **categorical variable** consist of group or category names that don't necessarily have any logical ordering. Your sex (male or female) and your handedness (left, right, ambidextrous), for instance, are both categorical variables.
- A **two-way table** displays the counts of how many individuals fall into each possible combination of categories of two categorical variables. A two-way table may also be called a **contingency table**.
- Each combination of a row variable category and a column variable category is referred to as a **cell** of the contingency table.
- **Row percentages** are the percentages within a row of a contingency table. They are based on the total number of observations in the row.
- **Column percentages** are the percentages within a column of a contingency table. They are based on the total number of observations in the column.

When we consider two categorical variables, a question of interest is whether a relationship that is observed in a sample can be used to infer that there is a relationship in the population(s) from which the sample(s) was drawn. This question can be answered by using a procedure called a chi-square test. Section 4.4 gave a brief introduction to the chi-square test. In this section and the next, we expand that discussion. As usual, five steps are required to test for statistical significance.

Step 1: The Null and Alternative Hypotheses

The **null** and **alternative hypotheses** about the categorical variables that form a two-way table are as follows:

H_0: The two variables are not related.

H_a: The two variables are related.

It is commonplace to use the word *associated* in place of the word *related.* Within a specific context, we may be able to write more informative versions of the general statements about the relationship. The next two examples illustrate different variations on how we might express null and alternative hypotheses for a two-way table.

Example 15.1

Read the original source on the companion website, http://www.cengage.com/statistics/Utts4e.

Ear Infections and Xylitol Sweetener Xylitol is a food sweetener that may also have antibacterial properties. In an experiment conducted in Finland, researchers investigated whether the regular use of chewing gum containing xylitol could reduce the risk of a middle ear infection for children in daycare centers (Uhari et al., 1998). The investigators randomly divided 533 children in daycare centers into three groups. One group regularly chewed gum that contained xylitol, another group regularly took xylitol lozenges, and the third group regularly chewed gum that did not contain xylitol. The experiment lasted for 3 months and for each child the researchers recorded whether the child had an ear infection during that period. (This experiment was also discussed in Example 12.17, on p. 476, where the data concerned xylitol administered in a syrup form rather than in gum.)

Table 15.1 displays the observed counts along with row percentages. The explanatory variable, which is the type of gum or lozenge each child took, is displayed as rows. The response variable, which is whether the child had an ear infection, is displayed as columns. Note that only 16.2% of the children in the xylitol gum group got an ear infection during the experiment, compared to 22.2% of the xylitol lozenge group and 27.5% of the placebo gum group.

Table 15.1 Ear Infections and the Use of Xylitol

	Ear Infection in 3 Months?		
Group	**No**	**Yes**	**Total**
Placebo gum	129 (72.5%)	49 (27.5%)	178
Xylitol gum	150 (83.8%)	29 (16.2%)	179
Xylitol lozenge	137 (77.8%)	39 (22.2%)	176
Total	416 (78%)	117 (22%)	533

The purpose of the experiment was to compare the three groups with regard to the proportion of children who experienced an ear infection. For this purpose, there are three relevant population parameters:

p_1 = proportion who would get an ear infection in a population given placebo gum

p_2 = proportion who would get an ear infection in a population given xylitol gum

p_3 = proportion who would get an ear infection in a population given xylitol lozenges

If the chance of falling into the ear infection category is not related to treatment method, the risk of an ear infection would be the same for the three treatments. So null and alternative hypotheses about the three parameters are as follows:

H_0: $p_1 = p_2 = p_3$ (No relationship between treatment and outcome.)

H_a: p_1, p_2, and p_3 are not all the same. (There is a relationship.)

Note that the alternative hypothesis simply states that the three proportions are not all the same. A limitation of chi-square tests is that no particular direction for the relationship can be stated in the alternative hypothesis. We will revisit this ear infection study in several examples as we learn how to do a chi-square test.

(Source: reproduced from Pediatrics, *Vol. 102, pp. 879–844. [See p. 701 for complete credit.])*

Example 15.2 **With Whom Do You Find It Easiest to Make Friends?** Students in a statistics class were asked, "With whom do you find it easiest to make friends?" Possible responses were "opposite sex," "same sex," and "no difference." Students were also categorized as male or female. We would like to determine whether there is a relationship between the sex of the respondent and response. To emphasize the comparison of men and women, we might express the null and alternative hypotheses as follows:

H_0: There would be no difference in the distribution of responses of men and women if the populations of them were asked this question. (No relationship between sex of respondent and response.)

Ha: There would be a difference in the distribution of responses of men and women if the populations of them were asked this question. (There is a relationship between sex of respondent and response.)

As is always the case in hypothesis testing, these hypotheses are statements about the larger populations represented by the samples of men and women. Again, note that the alternative hypothesis does not specify any particular way in which the men and women might differ.

Table 15.2 compares the responses of males and females, and Figure 15.1 displays the relevant conditional percentage distributions for females and males. We see that there appears to be a relationship. For instance, a decidedly higher percentage of females than males responded "opposite sex." We will continue the analysis of these data in Example 15.8 (p. 591).

Table 15.2 Data for Example 15.2

	With Whom Is It Easiest to Make Friends?			
	Opposite Sex	Same Sex	No Difference	Total
Females	58 (42%)	16 (12%)	63 (46%)	137
Males	15 (22%)	13 (19%)	40 (59%)	68
Total	73 (36%)	29 (14%)	103 (50%)	205

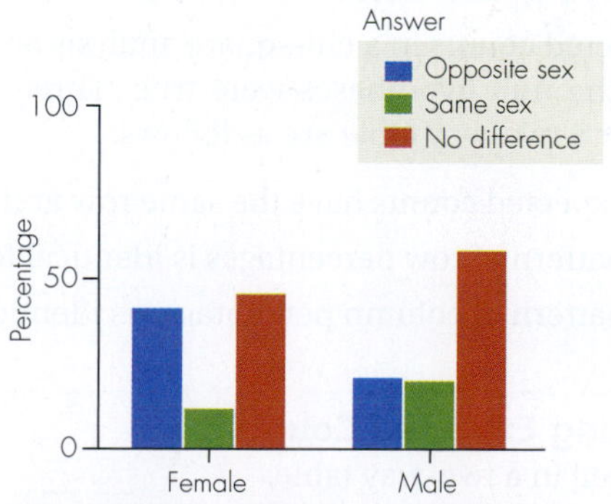

Figure 15.1 Conditional distributions by sex for "With whom is it easiest to make friends?"

TECHNICAL NOTE **Homogeneity and Independence**

It is generally sufficient to state the null and alternative hypotheses as H_0: No relationship, and H_a: There is a relationship. There are two specific variations of these general statements, and the method of sampling dictates which variation is appropriate in a particular situation. If samples have been taken from separate populations, the null hypothesis statement of no relationship is a statement of **homogeneity** (sameness) among the populations. Example 15.1 is such an instance. If a sample has been taken from a single population and two categorical variables are measured for each individual, the null hypothesis statement of no relationship is a statement of **independence** between the two variables. For instance, a null hypothesis may be that hair color and eye color are independent characteristics. (***Note:*** Independence was discussed in Chapter 7.) For sufficiently large samples, the same chi-square test is appropriate for both situations, so we will not be concerned about whether the null hypothesis is about homogeneity or about independence.

Step 2: The Chi-Square Statistic for Two-Way Tables and Necessary Conditions

The basics of computing the chi-square statistic used for two-way tables are as follows:

- The counts in the cells of a two-way table of the sample data are called the **observed counts**.
- The chi-square statistic measures the difference between the observed counts and corresponding **expected counts**. The *expected counts* are hypothetical counts that would occur if the null hypothesis were true. The expected counts can be calculated as

$$\text{Expected} = \frac{\text{Row total} \times \text{Column total}}{\text{Total } n}$$

- The **chi-square statistic** is

$$\chi^2 = \sum_{\text{all cells}} \frac{(\text{Observed} - \text{Expected})^2}{\text{Expected}}$$

The symbol Σ stands for "sum," and the sum is over all cells of the two-way table (not including cells with the row and column totals). A large chi-square value occurs when there is a large difference between the observed and expected counts.

Interpreting and Calculating Expected Counts

The expected counts in a chi-square analysis are the counts that would be expected to occur if the null hypotheses were true. Three important properties of the expected counts for a two-way table are as follows:

- The expected counts have the same row and column totals as the observed counts.
- The pattern of row percentages is identical for all rows of expected counts.
- The pattern of column percentages is identical for all columns of expected counts.

FORMULA **Calculating Expected Counts**

For any cell in a two-way table,

$$\text{Expected count} = \frac{\text{Row total} \times \text{Column total}}{\text{Total } n}$$

"Row total" is the total sample size in the row, "Column total" is the total sample size in the column, and "Total n" is the total sample size for the entire table.

Example 15.3 **Calculation of Expected Counts and Chi-Square for the Xylitol and Ear Infection Data** In the experiment discussed in Example 15.1 about the use of xylitol to prevent ear infections, the observed counts were as follows:

Read the original source on the companion website, http://www.cengage.com/statistics/Utts4e.

	Ear Infection in 3 Months		
Group	**No**	**Yes**	**Total**
Placebo gum	129	49	178
Xylitol gum	150	29	179
Xylitol lozenge	137	39	176
Total	416	117	533

The calculations of expected counts use only the total counts for rows and the total counts for columns. As an example, consider the "Placebo gum, Yes ear infection" cell of the table. In the "Placebo gum" row, the total count is 178, and in the "Yes" column, the total count is 117. For this cell, the expected count is

$$\frac{\text{Row total} \times \text{Column total}}{\text{Total } n} = \frac{178 \times 117}{533} = 39.07$$

The calculation of expected counts for all cells in the table is shown in Table 15.3.

Table 15.3 Calculation of Expected Counts for Example 15.3

	No Ear Infection	Yes Ear Infection	Row Total
Placebo gum	$\frac{178 \times 416}{533} = 138.93$	$\frac{178 \times 117}{533} = 39.07$	178
Xylitol gum	$\frac{179 \times 416}{533} = 139.71$	$\frac{179 \times 117}{533} = 39.29$	179
Xylitol lozenge	$\frac{176 \times 416}{533} = 137.37$	$\frac{176 \times 117}{533} = 38.63$	176
Column total	416	117	533

The formula for expected counts was used for all cells in Table 15.3, but this wasn't necessary. For a table of this size, it is actually only necessary to use the formula for two cells that don't constitute a whole row (or column). The remaining expected counts can be determined because the expected counts must have the same row and column totals as the observed data.

The calculation of the chi-square statistic is

$$\chi^2 = \sum_{\text{all cells}} \frac{(\text{Observed} - \text{Expected})^2}{\text{Expected}}$$

$$= \frac{(129 - 138.93)^2}{138.93} + \frac{(49 - 39.07)^2}{39.07} + \frac{(150 - 139.71)^2}{139.71} + \frac{(29 - 39.29)^2}{39.29}$$

$$+ \frac{(137 - 137.37)^2}{137.37} + \frac{(39 - 38.63)^2}{38.63} = 6.69$$

Technical Note Note that the expected counts are *not* rounded off to whole numbers. Much like the expected values we computed for random variables in Chapter 8, these are hypothetical average expected counts over the long run, not what we expect in a single experiment. Thus, they are not necessarily whole numbers.

Necessary Conditions

The chi-square test described in this section can be used to make inferences about two-way tables with any numbers of rows and columns. This procedure is an approximate method that requires a "large" sample; the larger the sample the better the approximation. Commonly used guidelines for the term *large sample* are as follows:

1. All expected counts should be greater than 1.
2. At least 80% of the cells should have an expected count greater than 5.

When these conditions are violated, the results may not be valid.

TECHNICAL NOTE

Notation for Chi-Square Statistic for Two-Way Tables

The following notation can be used for the chi-square test:

O_{ij} = observed count for the cell in row i and column j of the table

E_{ij} = expected count for the cell in row i and column j

R_i = total count of observations in row i

C_j = total count of observations in column j

n = total count for the entire table

$$E_{ij} = \frac{R_i C_j}{n} = \text{expected count for the cell in row } i \text{ and column } j$$

$$\chi^2 = \sum_{i,j} \frac{(O_{ij} - E_{ij})^2}{E_{ij}}, \text{ summed over all combinations of } i = 1, \ldots, c \text{ and } j = 1, \ldots, r.$$

r = number of row variable categories; c = number of column variable categories

Step 3: The *p*-Value for the Chi-Square Test

The ***p*-value** for the chi-square test is the probability that the chi-square statistic, χ^2, could be as large as it is or larger than it is if the null hypothesis is true. We use the upper tail only because small values of the chi-square statistic would mean that the observed and expected values are close together, thus supporting the null hypothesis. A chi-square probability distribution is used to determine this probability.

A chi-square distribution has a parameter called *degrees of freedom.* The **degrees of freedom** are given by df = (Rows − 1)(Columns − 1), where "Rows" indicates the number of rows in the table (excluding the row of totals), and "Columns" indicates the number of columns in the table (excluding the column of totals).

FORMULA

For the chi-square statistic for a two-way table,

$$\text{Degrees of freedom} = \text{df} = (r - 1)(c - 1)$$

r = number of row variable categories; c = number of column variable categories.

The Chi-Square Family of Distributions

A **chi-square distribution** is a skewed probability distribution that stretches to the right. The minimum value is 0 for a variable with a chi-square distribution. A single parameter, called degrees of freedom (df), completely determines the precise characteristics of a specific chi-square distribution. Figure 15.2 shows a chi-square distribution with 2 degrees of freedom and also shows a chi-square distribution with 5 degrees of freedom.

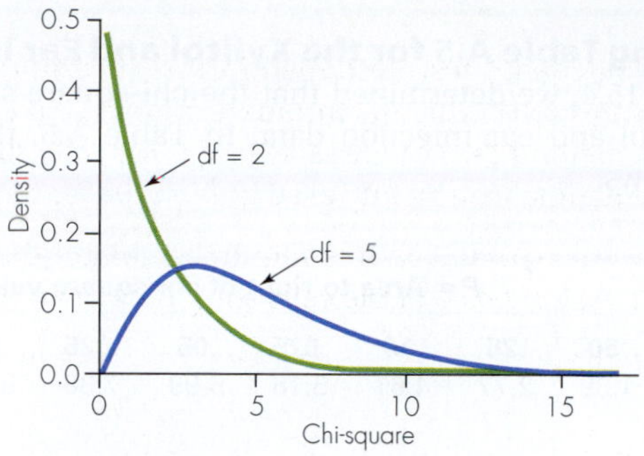

Figure 15.2 Two different chi-square distributions

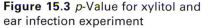

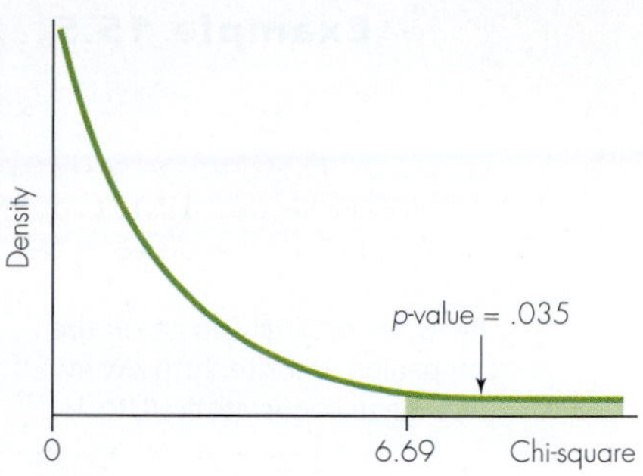

Figure 15.3 *p*-Value for xylitol and ear infection experiment

Example 15.4 ***p*-Value Area for the Xylitol Example** Figure 15.3 illustrates the *p*-value area for the xylitol and ear infection problem. In Example 15.3, we found that the chi-square statistic equals 6.69. There are three treatment groups (rows) and two ear infection possibilities (columns). Thus, degrees of freedom = $(3 - 1)(2 - 1) = 2$. The *p*-value is the probability that a chi-square statistic with df = 2 would be 6.69 or larger. This is the area to the right of 6.69 under a chi-square probability density curve with df = 2. This area equals .035. Next, we will learn the specifics for finding this area.

Read the original source on the companion website, http://www.cengage.com/statistics/Utts4e.

Three Ways to Determine the *p*-Value

The chi-square test for two-way tables is used so frequently that any statistical software program will include this procedure and provide the *p*-value necessary for deciding between the null and alternative hypotheses about the relationship in the population represented by the table. Some of you might not be using statistical software, however, so we consider three different ways to determine the *p*-value:

- Use statistical software to do the chi-square test. The *p*-value will be part of the output. We used Minitab to find the *p*-value shown in Figure 15.3 for the xylitol and ear infection data.
- Use the Excel command CHIDIST(chi-square value, df). This command will return the area to the right of the value you specify. For instance, CHIDIST(6.69,2) returns .035, the *p*-value for the xylitol and ear infection example.
- Use Table A.5 in the Appendix to give a range for the *p*-value.

Using Table A.5 to Approximate the *p*-Value

First, determine the degrees of freedom for your two-way table. Then look in the corresponding "df" row of Table A.5. Scan across that row until you locate approximately where the calculated chi-square statistic falls.

- If the value of the chi-square statistic falls between two table entries, the *p*-value is between the values of *p* (column headings) for these two entries.
- If the value of the chi-square statistic is larger than the entry in the rightmost column (labeled .001), the *p*-value is less than .001 (written as $p < .001$).
- If the value of the chi-square statistic is smaller than the entry in the leftmost column (labeled .50), the *p*-value is greater than .50 ($p > .50$).

The next three examples demonstrate how to use Table A.5 to approximate the *p*-value for a chi-square test.

Example 15.5 **Using Table A.5 for the Xylitol and Ear Infection Problem** In Examples 15.3 and 15.4, we determined that the chi-square statistic equals 6.69 and df = 2 for the xylitol and ear infection data. In Table A.5, the information given for df = 2 is as follows:

Read the original source on the companion website, http://www.cengage.com/statistics/Utts4e.

	P = Area to right of chi-square value								
df	**.50**	**.25**	**.10**	**.075**	**.05**	**.025**	**.01**	**.005**	**.001**
2	1.39	2.77	4.61	5.18	5.99	7.38	9.21	10.60	13.82

Scanning to the right, we see that 6.69 falls between 5.99 and 7.38. The corresponding column headings are .05 and .025. This means that the *p*-value is less than .05 but greater than .025. We write this as .025 < *p*-value < .05, or simply $.025 < p < .05$.

Example 15.6 **A Moderate *p*-Value** Suppose that a two-way table has three rows and three columns and the chi-square statistic is $\chi^2 = 8.12$. The degrees of freedom are df = (3 − 1)(3 − 1) = 4. To find the *p*-value, we scan the df = 4 row in Table A.5 and learn that 8.12 is between the entries 7.78 ($p = .10$) and 8.50 ($p = .075$). Thus, the *p*-value is between .075 and .10. We would write this as .075 < *p*-value < .10 or $.075 < p < .10$. In Excel, the command CHIDIST(8.12,4) tells us that the *p*-value is .087.

Example 15.7 **A Tiny *p*-Value** Suppose that a two-way table has four rows and two columns and the chi-square statistic is $\chi^2 = 17.67$. The degrees of freedom are (4 − 1)(2 − 1) = 3. In Table A.5, we see that 17.67 is larger than the rightmost entry in the df = 3 row, so the *p*-value is less than .001 ($p < .001$). In Excel, the command CHIDIST(17.67,3) tells us that the *p*-value is .00051.

THOUGHT QUESTION 15.1 Consider Example 15.2 (p. 585) about with whom it's easiest to make friends. What are the degrees of freedom for the chi-square statistic for these data? To be statistically significant at the .05 level, how large would the calculated chi-square statistic have to be?*

Steps 4 and 5: Making a Decision and Reporting a Conclusion

An approximate *p*-value, or even a range for the *p*-value, is enough information to make a decision about the null hypothesis. If we use the .05 significance level, we simply have to determine whether the *p*-value is less than .05. In Example 15.6, for instance, we learned from Table A.5 that the *p*-value is between .075 and .10. Thus, we know that it is not less than .05, and we would fail to reject the null hypothesis.

The value in the .05 column of Table A.5 is referred to as the **critical value** for the .05 significance level. The region of chi-square values greater than the critical value is called the **rejection region**. The *p*-value will be less than .05 when the chi-square statistic is greater than this critical value—in other words, if the chi-square statistic is in the rejection region. Consequently, if the level of significance is $\alpha = .05$, two equivalent rules for declaring statistical significance based on a chi-square statistic are

- Reject the null hypothesis when the *p*-value is less than or equal to .05.
- Reject the null hypothesis when the chi-square statistic is greater than or equal to the entry in the .05 column of Table A.5 (the critical value), in other words, when the chi-square statistic is in the rejection region.

***HINT:** You should use Table A.5 to determine this.

Reporting a Conclusion in Context

Remember that the two possible conclusions for hypothesis tests are "Do not reject the null hypothesis" or "Reject the null hypothesis." The latter conclusion is equivalent to "accept the alternative hypothesis" or "conclude that the results are statistically significant."

In chi-square tests for two-way tables, it is important to write these conclusions in context. Suppose that we were testing whether there is a relationship between smoking (yes or no) and drinking alcohol (never, occasionally, often). Here are various ways to write the possible conclusions.

Ways to write the conclusion "do not reject the null hypothesis"

- The relationship between smoking and drinking alcohol is not statistically significant.
- The proportions of smokers who never drink, drink occasionally, and drink often are not significantly different from the proportions of nonsmokers who do so.
- There is insufficient evidence to conclude that there is a relationship in the population between smoking and drinking alcohol.

Ways to write the conclusion "reject the null hypothesis"

- There is a statistically significant relationship between smoking and drinking alcohol.
- The proportions of smokers in the population who never drink, drink occasionally, and drink often are not all the same as the proportions of nonsmokers who do so.
- Smokers have significantly different drinking behavior than nonsmokers.

Example 15.8 **Making Friends** Figure 15.4 shows partial Minitab output for a chi-square test of the relationship between sex and response to the question, "With whom is it easiest to make friends?" (opposite sex, same sex, no difference). The data were first given in Table 15.2 (p. 585) as part of Example 15.2. The null hypothesis is that the response to the question is not related to the sex of the respondent. The alternative hypothesis is that these two variables are related.

Expected counts are printed below observed counts

	Opposite	Same	No Diff	Total
Female	58	16	63	137
	48.79	19.38	68.83	
Male	15	13	40	68
	24.21	9.62	34.17	
Total	73	29	103	205

Chi-Sq = 1.740 + 0.590 + 0.494 +
3.507 + 1.188 + 0.996 = 8.515

Figure 15.4 Minitab output for Example 15.8

In the last line of output, we see that the chi-square statistic is 8.515. Although Minitab provides the p-value, this value has been omitted to provide practice using Table A.5, the chi-square distribution table. The contingency table has two rows and three columns, so df $= (2 - 1)(3 - 1) = 2$. In the df $= 2$ row of Table A.5, the computed chi-square statistic falls between the entries in the .025 column (7.38) and the .01 column (9.21). The p-value is between .01 and .025, so the result is statistically significant if we use the standard .05 significance level.

Equivalently, we can compare the value of the chi-square statistic to a critical value from Table A.5. Using the df $= 2$ row and the .05 column of Table A.5, we find that the critical value is 5.99. Reject the null hypothesis because the computed chi-square statistic of 8.515 is greater than the critical value of 5.99.

In the context of this situation, the conclusion is that there is a relationship between the respondent's sex and response to the question asked. This conclusion applies to the larger population represented by the sample. Equivalently, we can conclude that there would be a difference in the distribution of responses of men and women if the populations of them were asked this question. Using the bar chart of row percentages shown in Figure 15.1 (p. 585), we can describe the nature of the difference between the females and males. Compared to men, women are more likely to say "opposite sex" and less likely to say "same sex." These conclusions are about the populations of men and women represented by the sample data.

Supporting Analyses

We don't learn anything about the specific nature of a relationship from a chi-square test. The results of a statistically significant chi-square test should be accompanied by some other descriptive or inferential statistics that elaborate on the connection between the variables. Some or all of the following procedures may be useful for this purpose:

- Description of row (or column) percentages.
- Bar chart of counts or percentages.
- Examination of $\frac{(\text{Observed} - \text{Expected})^2}{\text{Expected}}$ within each cell. These values are called **contributions to chi-square**. The cells with the largest values have contributed the most to the statistical significance of the relationship, so they deserve attention in any description of the relationship.
- Confidence intervals for important proportions or for differences between proportions. For instance, confidence intervals for the proportions getting an ear infection in the three treatment groups of the xylitol experiment would be informative.

15.1 Exercises are on pages 605–608.

IN SUMMARY

The Five Steps in a Chi-Square Test of a Relationship between Two Variables

Step 1: *Determine the null and alternative hypotheses.*

- *Null hypothesis* H_0: The two variables are not related.
- *Alternative hypothesis:* H_a: The two variables are related.

Step 2: *Verify necessary data conditions, and if they are met, summarize the data into an appropriate test statistic.* If at least 80% of the expected counts are greater than 5 and none are less than 1, compute

$$\chi^2 = \sum_{\text{all cells}} \frac{(\text{Observed} - \text{Expected})^2}{\text{Expected}}$$

The *expected count* for a cell is $\frac{\text{Row total} \times \text{Column total}}{\text{Total } n}$

Step 3: *Assuming that the null hypothesis is true, find the p-value.* Using the χ^2 distribution with df = $(r - 1)(c - 1)$, where r = number of row variable categories and c = number of column variable categories, the *p*-value is the area in the tail to the right of the calculated test statistic χ^2.

Step 4: *Decide whether the result is statistically significant based on the* p-*value.* Choose a significance level ("alpha"); the standard is $\alpha = .05$. The result is statistically significant if the *p*-value $\leq \alpha$.

Step 5: *Report the conclusion in the context of the situation.*

MINITAB TIP **Computing a Chi-Square Test for a Two-Way Table**

- If the raw data are stored in columns of the worksheet, use **Stat > Tables > Cross Tabulation and Chi-Square**. In the dialog box, specify a variable in the "For rows" box and a variable in the "For columns" box. Click ***Chi-Square***, and select "Chi-Square analysis" along with any other desired options such as "Expected cell counts."
- If the data are already summarized into counts, enter the table of counts into columns of the worksheet, and then use **Stat > Tables > Chi-Square Test (Two-Way Table in Worksheet)**. In the dialog box, specify the worksheet columns that contain the counts.

15.2 Analyzing 2 × 2 Tables

There are a number of special circumstances that apply to analyzing 2 × 2 tables. First, there is a handy computational shortcut for computing the chi-square statistic for a 2 × 2 table that does not require computing expected counts. It can be accomplished with a calculator, without needing to write down any intermediate results.

Shortcut Formula for a 2 × 2 Table

Suppose that we label a 2 × 2 table as follows:

	Column 1	Column 2	Total
Row 1	A	B	R_1
Row 2	C	D	R_2
Total	C_1	C_2	N

Then the test statistic is

$$\chi^2 = \frac{N(AD - BC)^2}{R_1R_2C_1C_2}$$

with df = 1.

Example 15.9 **Sex of Driver and Drinking before Driving** Case Study 4.2 (p. 133) described an Oklahoma roadside survey that was done to estimate the proportions of male and female drivers under 20 years old who had been drinking alcohol within the previous 2 hours. Table 15.4 displays observed counts for a two-way table summarizing sex of driver and whether he or she had consumed alcohol.

Table 15.4 Results of Roadside Survey for Young Drivers

	Drank Alcohol in Last 2 Hours?		
	Yes	**No**	**Total**
Males	77	404	481
Females	16	122	138
Total	93	526	619

The table has two rows (males and females) and two columns (yes and no), so the shortcut formula for the chi-square statistic can be used. The calculation is

$$\frac{N(AD - BC)^2}{R_1R_2C_1C_2} = \frac{619(77 \times 122 - 404 \times 16)^2}{481 \times 138 \times 93 \times 526} = 1.637$$

This matches the value of chi-square given in Figure 4.5 for Case Study 4.2. The remaining steps of the hypothesis test are as usual. In Case Study 4.2, the *p*-value was reported to be .201. This is not sufficiently small to allow us to conclude that there is a relationship in the population of young drivers represented by this sample.

Chi-Square Test or *z*-Test for Difference in Two Proportions?

In Section 12.3, we learned that a *z*-test can be used to compare two proportions. In such situations, we also can create a 2×2 table of the observed counts. The two groups would constitute the two rows, and whether or not the trait of interest is present would constitute the two columns. When the data are presented as a contingency table, we may naturally consider using a chi-square test. Does it make a difference whether we choose to use the *z*-test of Section 12.3 or instead use the chi-square test to compare two proportions? The answer depends on the nature of the alternative hypothesis. If the desired alternative hypothesis has no specific direction (is two-sided), the two tests give exactly the same *p*-value. If the desired alternative hypothesis has a direction (is one-sided), the *z*-test should be used because in a chi-square test, a specific direction cannot be specified in the alternative hypothesis.

The only essential difference between the two tests is the ability to specify a one-sided alternative hypothesis. In fact, there is an algebraic connection between the values of the test statistics for the two procedures. For any 2×2 table, the squared value of the *z*-statistic equals the chi-square statistic ($z^2 = \chi^2$). Or, put another way, $|z| = \sqrt{\chi^2}$. This connection between the two test statistics is also the connection between the theoretical standard normal distribution and the theoretical chi-square distribution with df = 1.

THOUGHT QUESTION 15.2 Suppose that you read that men are more likely to be left-handed than women are. To investigate this claim, you survey your class and find that 11 of 84 men and 7 of 78 women are left-handed. Should you compare the men and women using a *z*-test or a chi-square test? Or does it matter?*

Example 15.10 **Age and Tension Headaches** Exercise 10.55 was about a study for which the response variable was whether or not the respondent had experienced an episodic tension-type headache during the previous year (Schwartz et al., 1998). Table 15.5 is a contingency table that compares the incidence of tension headaches for women 18 to 29 years old and women 30 to 39 years old.

Table 15.5 Age Group and the Incidence of Tension Headaches

	Episodic Tension Headaches		
Age	**Yes**	**No**	**Total**
18 to 29	653 (40.8%)	947 (59.2%)	1600
30 to 39	995 (46.9%)	1127 (53.1%)	2122
Total	1648 (44.3%)	2074 (57.7%)	3722

Let p_1 and p_2 represent the proportions experiencing tension headaches in the populations of women aged 30 to 39 and women aged 18 to 29, respectively. Because

*HINT: Is the claim one-sided or two-sided?

the researchers did not speculate about the possible relative sizes of p_1 and p_2, the alternative hypothesis is two-sided. Null and alternative hypotheses are as follows:

H_0: $p_1 = p_2$
H_a: $p_1 \neq p_2$

The null hypothesis is equivalent to stating that there isn't a relationship between age group and the risk of episodic tension headaches. In this example, we can use either a chi-square test or a two-sample z-test because no specific direction is specified in the alternative hypothesis. Using statistical software (Minitab) to do a chi-square test, we found that $\chi^2 = 13.66$, and the p-value was reported to be .000 (only three decimal places were given). The observed relationship between age group and the incidence of tension headaches is statistically significant, and the row percentages show us that the older women are more likely to experience tension headaches. If we do a z-test for comparing two proportions, the value of z is $z = \sqrt{13.66} \approx 3.70$, and the p-value is identical to the p-value for the chi-square test.

The next example describes a situation in which it would be better to do a z-test for comparing two proportions rather than a chi-square test. The example involves a one-sided alternative hypothesis.

Example 15.11 **Sheep, Goats, and ESP** In parapsychology, the Sheep–Goat effect is that people who believe in extrasensory perception (Sheep) tend to be more successful in ESP experiments than people who do not (Goats). The effect is slight, but it has been observed in many forced-choice ESP experiments (Lawrence, 1993; Schmeidler, 1988, pp. 56–58). In a forced-choice experiment, the participant "guesses" which of several known possibilities has been (or will be) selected by the experimenter.

Students in a statistics class classified themselves as Sheep (believers) or Goats (nonbelievers). Among 192 students, there were 112 Sheep and 80 Goats. Each student was paired with another student in the same belief group. One member of each pair flipped a coin ten times while the other member guessed the outcome of each flip. Students then reversed their roles and repeated the flipping, so each person was able to make ten guesses. The guessing process was uncontrolled, so each student could decide whether he or she would guess in advance of the flip, during the flip, or after the flip. Table 15.6 displays data for the numbers of successful and unsuccessful guesses in each group. The Sheep did slightly better than the Goats, guessing correctly about 52% of the time compared to 49% for the Goats.

Table 15.6 ESP Belief and Success at Guessing Coin Flips

	Successful Guess		
Group	**Yes**	**No**	**Total**
Sheep (Believers)	582 (52%)	538 (48%)	1120
Goats (Nonbelievers)	389 (49%)	411 (51%)	800
Total	971 (51%)	949 (49%)	1920

The activity was done to determine if Sheep would do better than Goats, so a one-sided alternative hypothesis is appropriate. Define p_1 and p_2 to be the population proportions of successful guesses for Sheep and Goats, respectively. The null and alternative hypotheses of interest are as follows:

H_0: $p_1 \leq p_2$
H_a: $p_1 > p_2$

A z-test for comparing two proportions should be done because the alternative hypothesis describes a relationship in a specific direction. Figure 15.5 (next page) displays Minitab output for this test. From the bottom line of output, we learn that the p-value is .075. Thus, the null hypothesis cannot be rejected at the .05 significance level.

Sample	X	N	Sample p
1	582	1120	0.519643
2	389	800	0.486250

Estimate for p(1) − p(2): 0.0333929
Test for p(1) − p(2) = 0 (vs > 0): Z = 1.44 P-Value = 0.075

Figure 15.5 Minitab output for Example 15.11

Fisher's Exact Test for 2 × 2 Tables

For 2 × 2 tables, **Fisher's Exact Test** is another test that can be used to analyze the statistical significance of the relationship. In theory, this test can be used for any 2 × 2 table, but most commonly, it is used when necessary sample size conditions for using the z-test or the chi-square test are violated.

Although the computations are cumbersome when done by hand, the idea behind Fisher's Exact Test is easy. Suppose, for example, that a randomized experiment is done to determine whether taking the herb echinacea reduces the risk of getting a cold, and the observed data are that 1 of 10 people taking echinacea get a cold during the study while 4 of 10 people taking a placebo get a cold. In all, 5 of the 20 participants got a cold, but only 1 was in the echinacea group. The p-value for a one-sided Fisher's Exact Test is the answer to the following question:

> Given that 5 of 20 participants get a cold regardless of treatment method, what is the probability that only 1 or fewer would be in the echinacea group just by chance?

Note that the test statistic is simply the count of how many got a cold in the echinacea group. As always, the p-value question addresses how likely it is that the test statistic would be as extreme as or more extreme than it is in the direction of the alternative hypothesis if the null hypothesis is true. The answer (it is .152) is determined by using a probability distribution called the *hypergeometric distribution,* which is discussed further in Supplemental Topic 1 on the companion website for this book, http://www.cengage.com/statistics/Utts4e.

Some, but not all statistical software programs include the Fisher's Exact Test. The Statistical Package for the Social Sciences (SPSS) does, as does Minitab in Version 14 and above. Several websites provide a calculator that is designed to give the p-value for Fisher's Exact Test.

Example 15.12 **Butterfly Ballots** The winner of the 2000 U.S. presidential election (George W. Bush) was not determined until about a month after election day. There were several court hearings in Florida about the legality of vote counting and voting procedures in that state. One of the Florida controversies concerned a ballot format called the "butterfly ballot," used in Palm Beach County. Candidates were listed in two columns. To indicate a choice, a voter punched a hole in the appropriate place in an area between the two columns of names. The first three presidential candidates were listed as follows:

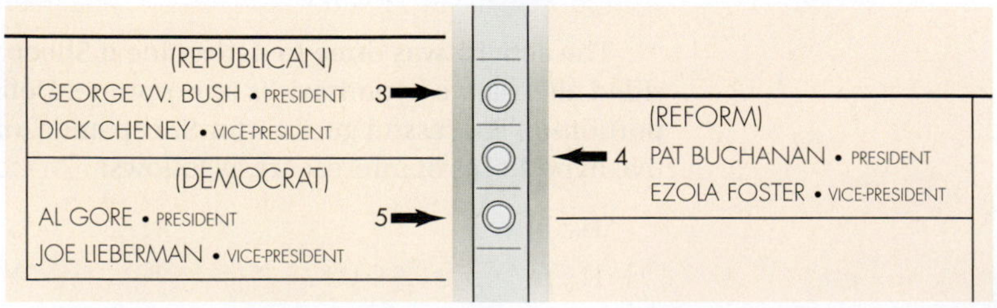

After the election, several Palm Beach County voters said that they meant to vote for Al Gore but mistakenly voted for Pat Buchanan, the Reform Party candidate. These voters said that they punched a hole in the second punch location because they saw

Al Gore's name listed second in the left column. That location, however, was for Pat Buchanan, who was listed in the right column. Perhaps not coincidentally, Buchanan's vote total in Palm Beach County was decidedly higher than expected.

A brief communication published in *Nature* (Sinclair et al., 2000) gave results from an experiment done in Canada to study whether the butterfly ballot increases the likelihood of voter mistakes. Shortly after the U.S. election in 2000, the researchers recruited people in an Edmonton, Alberta, shopping mall to vote in a mock election for the prime minister of Canada. Participants were randomly assigned to use either the double-column butterfly format or the more conventional single-column format. After they voted, the participants were asked for whom they had voted. Of 55 voters using the butterfly format, 4 voted for a different candidate than they intended. Of 52 voters using the single-column format, none made a mistake.

Is the observed result statistically significant? Define p_1 = population proportion that makes a mistake with the butterfly ballot and p_2 = population proportion that makes a mistake with the single-column ballot. The null and alternative hypotheses of interest are as follows:

H_0: $p_1 \leq p_2$
H_a: $p_1 > p_2$

The observed counts don't satisfy the necessary conditions for doing a z-test, which are that at least 10 observations with and without the trait of interest should be observed in each group (see Section 12.3). Fisher's Exact Test is more appropriate for these data, and the p-value can be determined by answering the following question:

> Given that 4 of 107 participants will make a mistake, what is the probability that all 4 are among the 55 voters randomly selected to use the butterfly ballot?

The answer is .0661, a p-value that is not quite below the usual .05 standard for statistical significance. It is important to note, however, that given that four participants will make a mistake, the evidence for the alternative is as strong as possible since all four were using the butterfly ballot format, so it may be reasonable to use a higher level of significance and reject the null hypothesis. Also, the researchers asked all participants to rate the confusion level of the ballot they used on a 7-point scale, with 7 representing the most confusing. The mean confusion score for the butterfly ballot format was significantly higher than that for the single-column format.

MINITAB TIP **Fisher's Exact Test (Version 14 and Higher)**

- If the raw data are stored in columns of the worksheet, use **Stat > Tables > Cross Tabulation and Chi-Square**. Specify a categorical variable in the "For rows" box and a second categorical variable in the "For columns" box. Then click ***Other Stats*** and select "Fisher's exact test for 2 × 2 tables."
- If the data are already summarized into counts, enter the summarized information into three columns of the worksheet. For each cell in the 2 × 2 table, enter the row variable category (1 or 2) into one column, the column variable category (1 or 2) into a second column, and the observed count into a third column. Use **Stat > Tables > Cross Tabulation and Chi-Square** as above, but also specify the column with observed counts in the "Frequencies are in" box.

Note: The p-value given by the procedures above will be a two-sided test. In Version 15 and above, Fisher's Exact Test is also part of the output for **Stat > Basic Statistics > 2 Proportions**. This procedure can be used for either one-sided or two-sided tests.

Statistical Significance and Sample Size

Although we have used the common standard of rejecting the null hypothesis when the p-value $\leq$ .05, there is nothing magic about .05. In Examples 15.11 and 15.12, the p-values were .075 and .0661, respectively. In these and similar cases, it is up to you as the reader of the research to decide whether to reject the null hypothesis. Remember that the level of significance used corresponds to the conditional probability of rejecting a true null hypothesis. Sometimes there may be reasons to set that probability at a higher (or lower) level than .05. Where to set the cutoff is something that you as a reader should always decide, which is why it is important for researchers to report *p*-values rather than simply to report a decision about whether they have rejected the null hypothesis.

As we learned in Chapters 12 and 13, sample size affects the power of a hypothesis test. With a small sample size, the power of a test may be low, which means that we may not be able to detect a relationship or difference that actually exists in the population. The following example demonstrates that the same magnitude of observed sample difference can be nonsignificant with a relatively small sample size, but can be statistically significant with a larger sample size.

Example 15.13 **Sex of Student and Car Accidents** In a class activity, students in a statistics class were asked to report their sex (male, female) and whether they had ever been in a car accident. Table 15.7 displays the observed data.

Table 15.7 Sex and Likelihood of Having Been in a Car Accident

	Ever in a Car Accident?		
Sex	**No**	**Yes**	**Total**
Male	7	16	23
Female	16	18	34
Total	23	34	57

Source: Jessica Utts.

To analyze the relationship between the two variables, we can use a chi-square test or, equivalently, a *z*-test to analyze the difference between the population proportions of men and women who have been in a car accident. For a chi-square test, $\chi^2 = 1.575$, with df $= (2 - 1)(2 - 1) = 1$. The *p*-value is .209, computed as the probability that a chi-square value with df $= 1$ could be greater than or equal to 1.575. With this *p*-value, we are not able to conclude that sex and the likelihood of having been in a car accident are related.

In the sample, the relative risk of men and women reporting they have been in a car accident is about 1.32, and the difference in proportions is almost .17 (or 17%). For men, the proportion is $16/23 = .696$ compared to $18/34 = .529$ for the women. This disparity is consistent with figures reported by the U.S. Department of Transportation (Cerrelli, 1998), showing that the accident rate per driver for females aged 19 to 24 is only about 70% of what the rate is for males in that age group, for a relative risk (men to women) of about 1.43. Yet, despite this large difference, we are not able to reject the hypothesis of no difference. What if this same difference had been observed for a sample that was four times as large as the observed sample? Table 15.8 is a contingency table in which all values in Table 15.7 are multiplied by four. The proportions having been in a car accident are the same for men and women as they were in Table 15.7.

Table 15.8 Counts in Table 15.7 Multiplied by 4

	Ever in a Car Accident?		
Sex	**No**	**Yes**	**Total**
Male	28	64	92
Female	64	72	136
Total	92	136	228

For the larger sample, shown in Table 15.8, chi-square test results are that $\chi^2 = 6.301$ (df $= 1$) and the *p*-value $= .012$. This is a statistically significant result, in contrast to the result for the smaller sample size. Of course, it's not likely that the difference would have remained exactly the same if data actually had been collected from a larger group, but the idea is that the statistical significance of a fixed difference is affected by the sample size. Be careful about the interpretation of nonsignificant results when the sample size is small. It's possible that the sample size was too small to detect a real difference.

An interesting side note is that rates of accidents for male and female drivers are about the same when they are computed based on miles driven, rather than on number of drivers. Males drive more than females, so even though the rate of accidents per mile driven is about the same, the per-driver accident rate is higher for males (Cerrelli, 1998).

McNemar's Test for Change or Difference in Paired Data

In Chapters 9, 11, and 13 we learned what to do with quantitative data measured as matched pairs (use inference for μ_d), but what about when categorical data are measured as matched pairs? When *paired data* are measured with a categorical response variable that has two categories, **McNemar's Test** can be used to analyze the difference in proportions for the paired data. The following situations are examples of questions that can be analyzed using McNemar's Test.

- The day before a televised debate between two political candidates, a sample of registered voters is asked which candidate they prefer. The day after the debate the same voters are again asked which candidate they prefer. Is the proportion that prefers Candidate X higher after the debate than before?
- Husbands and wives in 200 opposite-sex married couples are asked if they are satisfied with their marriage. Is there a difference between the proportions of husbands and wives who are satisfied?

McNemar's test uses only the data pairs for which the two paired responses differ (called **discordant pairs**). Consider, for example, the study of marital satisfaction just described. Suppose that the data are as follows:

	Number of Couples
Both husband and wife satisfied	160
Neither husband nor wife satisfied	20
Husband satisfied, wife is not	15
Wife is satisfied, husband is not	5
Total	200

The discordant pairs are the 15 + 5 = 20 couples for which there is disagreement about satisfaction. The difference between the two types of discordant pairs is a measure of the difference in the proportion of husbands and the proportion of wives who are satisfied. In this sample, more husbands are satisfied because there are 15 couples in which the husband is satisfied but the wife is not, compared to 5 couples in which the wife is satisfied but the husband is not. McNemar's test analyzes whether this imbalance is statistically significant. The basic steps are as follows:

1. Determine the number of pairs of both possible types of disagreement (the discordant pairs).
2. Define p = proportion of one specific type of disagreement, given that there is disagreement, that would occur in the population represented by the sample if everyone was measured.
3. Write the null hypothesis H_0: $p = .5$. The alternative hypothesis may be either one-sided or two-sided depending upon the research purpose.
4. For an exact test, use the binomial distribution to find the p-value. The test statistic is the number of discordant pairs of one specific type. The "sample size" for the test is the total number of discordant pairs. The null value is $p_0 = .5$. An approximate test can be done using the z-statistic described in Section 12.2.

Note: For a two-sided test with a sufficiently large sample size, it is common to use a chi-square goodness of fit test, described in Section 15.3, to analyze the balance or imbalance between the two types of discordant pairs.

Example 15.14 **Asthma Prevalence over Time** Suppose that a researcher records whether $n = 500$ children have asthma or not at age 13, and then again records whether these same 500 individuals have asthma 7 years later. Table 15.9 displays (made up) data for such a study.

Table 15.9 Asthma Prevalence at Ages 13 and 20

	Asthma at Age 20?		
Asthma at Age 13?	**Yes**	**No**	**Total**
Yes	50	22	72
No	8	420	428
Total	58	442	500

Suppose that the researcher is interested in the difference in asthma rates at the two ages. To focus on the difference in asthma rates at ages 13 and 20, we need only consider cases with different responses at the two ages. There were 30 such cases, 22 with asthma at age 13 but not at age 20, and 8 that did not have asthma at age 13 but did at age 20.

Suppose that the purpose is to see if asthma is generally more prevalent at the younger age. To test statistical significance, we define p = population proportion with asthma at age 13 but not at age 20, among people not in the same asthma category at both ages. The hypotheses are H_0: $p = .5$ versus H_a: $p > .5$. The test statistic is $X = 22$, the number with asthma at age 13 but not at age 20. A binomial distribution with $n = 30$ and $p = .5$ can be used to find the p-value, the probability that $X \geq 22$. Minitab gives this answer as .008, so we have observed a statistically significant difference and can conclude, for the population, that asthma is more prevalent at age 13 than at age 20.

In the example just completed, we had a single sample of $n = 500$ individuals observed at two different times, not two independent samples of 500 different individuals at each age. In the example, $72/500 = .144$ was the asthma rate at age 13 and $58/500 = .116$ was the asthma rate at age 20. If we had erroneously treated this information as statistics for two independent samples, and used a z-test to compare two population proportions, the p-value would have been .094, a nonsignificant result. When we correctly considered the data as paired data, we were able to determine that there was, in fact, a significant difference between the asthma rates for the two ages.

15.2 Exercises are on pages 608–609.

IN SUMMARY Methods for Analyzing 2 × 2 Tables

- For a 2 × 2 table, a two-sided z-test for the difference in two population proportions for independent samples and a **chi-square test of homogeneity** give equivalent results.
- A one-sided z-test should be used when the alternative hypothesis is a one-sided statement about the difference in two population proportions for independent samples.
- For small sample sizes, **Fisher's Exact Test** can be used to analyze the relationship between two categorical variables.
- For paired data with a categorical response variable, **McNemar's Test** should be used to analyze the difference between the pairs.

15.3 Testing Hypotheses about One Categorical Variable: Goodness of Fit

THOUGHT QUESTION 15.3 Imagine tossing a six-sided die 60 times. How many times would you expect each side to occur? What did you assume when you calculated these *expected counts*? How would you measure the difference between the observed and expected counts of how often each side occurs in the 60 tosses?*

We sometimes wish to test whether the probabilities for being in the possible categories of a single categorical variable are given by a specified set of values. For example, if a lottery drawing involves randomly drawing digits between 0 and 9, we may wish to analyze observed data to confirm that the probability over the long run is 1/10 for drawing each of the ten digits from 0 to 9. With two modifications from how it is applied to two-way tables, the chi-square statistic can be used to test hypotheses about the probability distribution of a single categorical variable. In such instances, the significance test is called the **chi-square goodness-of-fit test**.

The first modification of the chi-square procedure necessary for the goodness-of-fit situation involves the calculation of the expected counts. This calculation is straightforward and intuitively obvious. Suppose that we hypothesize, for example, that 45% of a population has brown eyes, 35% has blue eyes, and 20% has an eye color other than brown or blue. If we randomly select $n = 200$ people from this population, it probably makes sense to you that we would expect about 45% of 200 = $.45 \times 200 = 90$ people to have brown eyes if our hypothesis is correct. We also would expect about $.35 \times 200 = 70$ people to have blue eyes and $.20 \times 200 = 40$ people to have an eye color other than brown or blue.

The other modification of the chi-square procedure concerns the degrees of freedom. For a goodness-of-fit test, the degrees of freedom are computed as $\text{df} = k - 1$, where k is the number of categories for the variable of interest. In the eye color example of the previous paragraph, there were three eye color categories. Consequently, the degrees of freedom for a chi-square goodness-of-fit test would be $\text{df} = 3 - 1 = 2$.

THOUGHT QUESTION 15.4 The "degrees of freedom" for the chi-square test for a two-way table represent the largest number of cells for which you are "free" to find expected counts. The remaining expected counts can be determined because the row and column totals have to be the same as they are for the observed counts. Explain how the same principle applies in specifying the degrees of freedom for a chi-square goodness-of-fit test, which are $k - 1$ when there are k categories.†

***HINT:** How have the differences between observed and expected counts been measured in other chi-square tests? Would that work here?

†HINT: The total expected count must equal the total observed count. If you know $k - 1$ expected counts, what must the remaining one be?

IN SUMMARY The Five Steps in a Chi-Square Goodness-of-Fit Test

Step 1: *Determine the null and alternative hypotheses.*

- *Null hypothesis:* H_0: The probabilities for the k categories of a categorical variable are given by $p_1, p_2, \ldots, p_k$.
- *Alternative hypothesis:* H_a: Not all probabilities specified in H_0 are correct.

Note: The probabilities specified in the null hypothesis must sum to 1.

Step 2: *Verify necessary data conditions, and if they are met, summarize the data into an appropriate test statistic.* If at least 80% of the expected counts are greater than 5 and none are less than 1, compute

$$\chi^2 = \sum_{\text{all categories}} \frac{(\text{Observed} - \text{Expected})^2}{\text{Expected}}$$

where the expected count for the i-th category is computed as np_i.

Step 3: *Assuming that the null hypothesis is true, find the* p*-value.* Using the χ^2 distribution with df $= k - 1$, the p-value is the area to the right of the test statistic χ^2.

Step 4: *Decide whether the result is statistically significant based on the* p*-value.* Choose a significance level ("alpha"); the standard is $\alpha = .05$. The result is statistically significant if the p-value $\le \alpha$. Equivalently, the result is statistically significant if the test statistic χ^2 is as large as or larger than the value in the α column and df $= k - 1$ row of Table A.5.

Step 5: *Report the conclusion in the context of the situation.*

Example 15.15 **The Pennsylvania Daily Number** The Pennsylvania Daily Number is a state lottery game in which the state constructs a three-digit number by drawing a digit between 0 and 9 from each of three different containers. If the digits that were drawn, in order, were 3, 6, and 3, for example, then the daily number would be 363. In this example, we focus only on draws from the first container. If numbers are randomly selected, each value between 0 and 9 would be equally likely to occur. This leads to the following null hypothesis:

H_0: $p = 1/10$ for each of the 10 possible digits in first container.

Simply stated, the alternative hypothesis is that the null hypothesis is false. In this setting, that would mean that the probability of selection for some digits is different from 1/10.

Figure 15.6 shows the frequency distribution of draws from the first container for the $n = 500$ days between July 19, 1999, and November 29, 2000. The observed data were obtained from the Pennsylvania lottery website. The expected count for each of the 10 possible outcomes is $500 \times 1/10 = 50$, and the red dashed line in the figure marks this value. The greatest difference between observed and expected occurs for the number 5, which was drawn from the first container only 39 times over these 500 days.

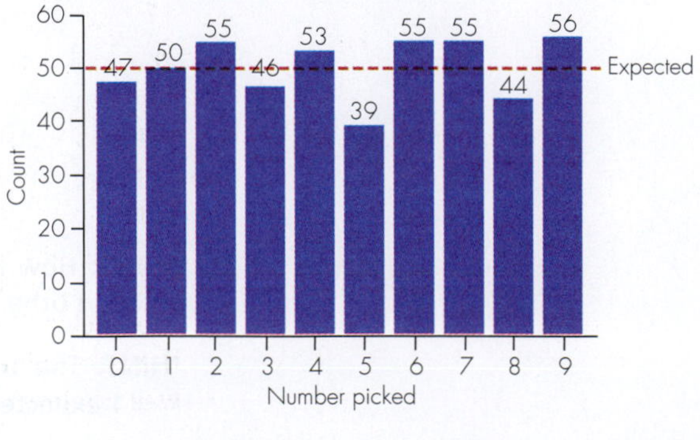

Figure 15.6 Observed and expected counts for first digit drawn in the Pennsylvania Daily Number Game, July 19, 1999, through November 29, 2000

The chi-square goodness-of-fit statistic for these data follows:

$$\chi^2 = \sum_{\text{categories}} \frac{(\text{Observed} - \text{Expected})^2}{\text{Expected}}$$

$$= \frac{(47-50)^2}{50} + \frac{(50-50)^2}{50} + \frac{(55-50)^2}{50} + \frac{(46-50)^2}{50} + \frac{(53-50)^2}{50}$$

$$+ \frac{(39-50)^2}{50} + \frac{(55-50)^2}{50} + \frac{(55-50)^2}{50} + \frac{(44-50)^2}{50} + \frac{(56-50)^2}{50} = 6.04$$

There are 10 possible outcomes for the number drawn, so df = 10 − 1 = 9. To find the p-value, we find the probability that a chi-square statistic would be as large as or larger than 6.04 using a chi-square distribution with df = 9. From Table A.5, the chi-square distribution table, we learn that the p-value is greater than .50 because 6.04 is smaller than (to the left of) the entry of 8.34 under .50 in the row for df = 9. The Excel command CHIDIST(6.04,9) provides the information that the p-value is .736. This p-value is illustrated in Figure 15.7. The result is not statistically significant; the null hypothesis is not rejected.

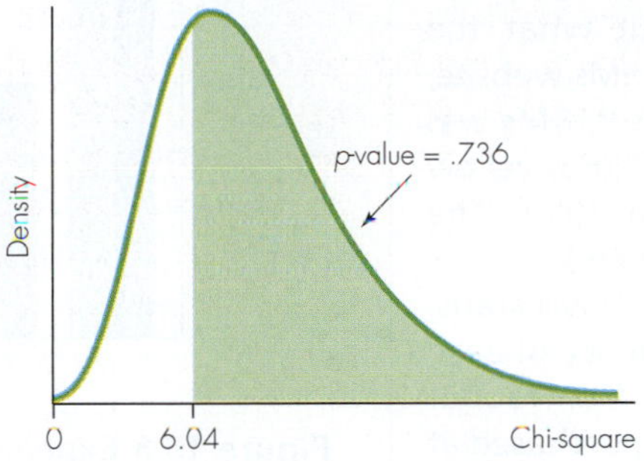

Figure 15.7 p-Value for Example 15.15 is area to right of 6.04 in chi-square distribution with df = 9

TECHNICAL NOTE

Mathematical Notation for Chi-Square Goodness-of-Fit Statistic

The following notation can be used for the chi-square goodness-of-fit test:

O_i = observed count in category i

p_i = hypothesized probability for category i

$E_i = np_i$ = expected count in category i

$$\chi^2 = \sum_i \frac{(O_i - E_i)^2}{E_i}$$

k = number of categories; df = $k - 1$

Sample Size and the Goodness-of-Fit Test

The null hypothesis often is the desired hypothesis for the investigators when they do a goodness-of-fit test. In Example 15.15 about the Pennsylvania Daily Number, lottery officials would hope to find that all numbers have an equal chance to be selected. This creates a technical and logical difficulty because with a small sample, a null hypothesis might be retained simply because there are not enough data to reject it. This was the basis for the advice given in Chapter 12 to use the phrase "cannot reject the null" rather than "accept the null." If the goal is to not reject the null hypothesis, the most scientifically valid approach is to use as large a sample as possible. This prevents the possible criticism that the null hypothesis was retained only because the sample was not large enough to provide conclusive evidence. On the other hand, the risk of a large sample size is that small, relatively unimportant departures from the null hypothesis can achieve statistical significance. It is often useful to include confidence intervals for the parameters, if possible, to determine the magnitude of departures from the null hypothesis.

15.3 Exercises are on pages 610–611.

MINITAB TIP **Calculating the Chi-Square Statistic for Goodness-of-Fit**

- Use **Stat > Tables > Chi-Square Goodness-of-Fit Test (One Variable)**. Enter counts for all categories into the "Observed Counts" box.
- To test that all categories are equally likely, select "Equal proportions" under "Test." Otherwise, select "Specific proportions" and enter the null hypothesis proportions in the same order as the counts were entered.

CASE STUDY 15.1 Do You Mind If I Eat the Blue Ones?

You'll find six different colors of M&Ms in a bag of milk chocolate M&Ms: brown, red, yellow, blue, orange, and green. If you're like some people, you may have a color preference (or dislike) when it comes to eating M&Ms and might have fleeting thoughts about what the color distribution is or should be. On the M&Ms website, the distribution of colors for milk chocolate M&Ms was given as follows in 2000: brown 30%, red 20%, yellow 20%, blue 10%, orange 10%, and green 10%. (The color distribution has been changed since then.)

Students in a University of California at Davis statistics class (fall 2000 term) counted the frequency of each color among 6918 milk chocolate M&Ms from about 130 small (1.69-ounce) bags. All bags were purchased at the same store in California. The observed distribution of colors in this sample was as follows:

	Color					
	Brown	*Red*	*Yellow*	*Blue*	*Orange*	*Green*
Count	1911	1072	1308	804	821	1002
Percent	27.6%	15.5%	18.9%	11.6%	11.9%	14.5%
Expected %	30%	20%	20%	10%	10%	10%

Generally, the observed percentages in the UC Davis data are not far from the percentages advertised by the manufacturer, but some differences are evident. Most notably (especially for those who like red M&Ms), there were fewer red and more green candies than expected.

We can use a chi-square goodness-of-fit test to determine whether the difference between the expected and observed distributions is statistically significant. If we assign numbers to colors using the order in which the colors are listed above (1 = brown, 2 = red, and so on), the null hypothesis of interest is

$$H_0: p_1 = .3, p_2 = .2, p_3 = .2, p_4 = .1, p_5 = .1, p_6 = .1$$

The first step in computing the chi-square statistic is to calculate the expected counts for each color. To do so, multiply the null hypothesis proportion for a color by the total sample size, $n = 6918$. For example, the expected count for brown is $.3 \times 6918 = 2075.4$. Figure 15.8 displays expected counts and observed counts for each color. As we noted before, the largest differences between observed and expected counts occur for red and green.

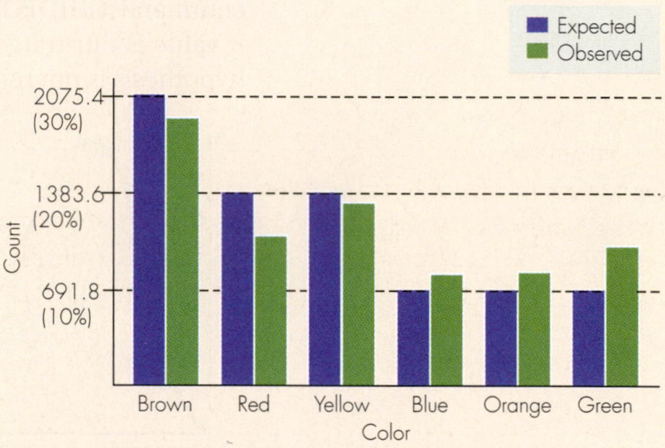

Figure 15.8 Expected and observed counts for a sample of 6918 plain M&Ms

The chi-square goodness-of-fit statistic for these data is $\chi^2 = 268.75$, with degrees of freedom = number of categories − 1 = 6 − 1 = 5. In Table A.5, the largest entry (under .001) for df = 5 is 20.51. The computed chi-square statistic, 268.75, is so much larger than 20.51 that it is reasonable to say that the *p*-value is essentially 0. The result is highly significant in that it would be nearly impossible for a random sample of 6918 M&Ms to have a color distribution that differed from the null hypothesis distribution by as much as or more than the observed sample does.

After obtaining these results, one of the authors posted a message about them to an Internet e-mail list for teachers of Advanced Placement statistics courses. She received several replies from other instructors who had done this activity. Nearly all described statistically significant departures from the color distribution given on the M&Ms website, but the nature of the difference varied considerably. In some cases, for example, the number of red candies was much more than expected;

(continued)

while in other cases, the number of red candies was much less than expected.

Why do so many different samples provide statistically significant evidence against the distribution given by the manufacturer? The answer probably has to do with the manufacturing and distribution process. A bag of M&Ms, or a group of bags purchased at the same store, may not be a *random sample* from a larger population. The color composition of the bags produced in a manufacturing plant on a given day could be affected by the available quantities of the various colors. Also, the process that is used to put candies into the bags may not produce a well-mixed random sampling of the colors. In other words, students counting M&M colors most likely are not using random or representative samples of the larger population.

Key Terms

In Summary Boxes

Exercises

◆ Denotes that the dataset is available on the companion website, http://www.cengage.com/statistics/Utts4e, but is not required to solve the exercise.

Bold exercises have answers in the back of the text.

Section 15.1

Skillbuilder Exercises

15.1 For each pair of variables, indicate whether a two-way table would be appropriate for summarizing the relationship. In each case, briefly explain why or why not.

a. Satisfaction with quality of local K through 12 schools (satisfied or not satisfied) and political party (Republican, Democrat, etc.).

b. Height (centimeters) and foot length (centimeters).

15.2 For each pair of variables, indicate whether a two-way table would be appropriate for summarizing the relationship. In each case, briefly explain why or why not.

a. Sex (female, male) and amount willing to spend on a home theater system.

b. Age group (under 20, 21–29, etc.) and handedness (right-handed, left-handed, or ambidextrous).

15.3 In a nationwide survey, college students are asked how important religion is in their life (very, fairly, or not very) and whether they have ever cheated on a college exam (no or yes).

a. Write null and alternative hypotheses about the two variables in this situation. Make your hypotheses specific to the context of this situation.

b. Suppose that the *p*-value is .127 for a chi-square test of the hypotheses you wrote for part (a). Using a .05 level of significance, write a conclusion about the two variables in this exercise.

c. What is the value of the degrees of freedom for the chi-square statistic in this situation?

15.4 Sex (female or male) and handedness (right-handed or left-handed) are recorded for a randomly selected sample of adults. Of the 100 women in the sample, 92 women are right-handed. Of the 80 men in the sample, 70 men are right-handed.

a. Write a two-way table of observed counts.

b. Determine expected counts for all combinations of sex and handedness.

c. Calculate the value of the chi-square test statistic.

15.5 In each of the following situations, give the *p*-value for the given chi-square statistic. Either use the information in Table A.5 to provide a range for the *p*-value or use software to determine an exact value.

a. $\chi^2 = 3.84$, df = 1.
b. $\chi^2 = 6.7$ for a table with 3 rows and 3 columns.
c. $\chi^2 = 26.23$ for a table with 2 rows and 3 columns.
d. $\chi^2 = 2.28$, df = 9.

15.6 For each of the following situations, determine whether the result is statistically significant at the .05 level of significance.

a. $\chi^2 = 2.89$, df = 1.
b. $\chi^2 = 5.00$, df = 1.
c. $\chi^2 = 23.60$, df = 4.
d. $\chi^2 = 23.60$, df = 15.

15.7 Recall that the critical value for a chi-square test is the chi-square value for which the area to its right equals the level of significance. Use Table A.5 to determine the critical value in each of the following situations.

a. Level of significance is $\alpha = .05$; df = 1.
b. Level of significance is $\alpha = .01$; table has 3 rows and 4 columns.
c. Level of significance is $\alpha = .05$; df = 6.

15.8 Refer to Exercise 15.6. For each part, determine whether the result is statistically significant at the .01 level of significance.

General Section Exercises

15.9 ◆ Students from two different statistics classes at UC Davis reported their sex (male, female) and recorded which class they were taking. One class is for liberal arts students, and the other is for non–liberal arts students. The results for the 173 students are given below. In each cell, the observed counts are printed first, with the expected counts below them. Assume that these students are a representative sample of all UC Davis students who would take either class. (Raw data are in **UCDavis1** dataset on the companion website.)

	NonLib	LibArt	All
Female	83 80.42	11 13.58	94 94.00
Male	65 67.58	14 11.42	79 79.00
All	148 148.00	25 25.00	173 173.00

a. State the null and alternative hypotheses that could be tested using these data.
b. What are the degrees of freedom for the chi-square test statistic?
c. Compute the value of the chi-square test statistic.
d. Find the *p*-value or *p*-value range for the chi-square statistic found in part (c).
e. Make a conclusion and write it in the context of the situation. Be sure to state the level of significance that you are using.

15.10 Suppose that investigators conduct a study on the relationship between birth order (first or only child, not first or only child) and activity preference (indoor or outdoor).

a. Write null and alternative hypotheses for the two variables in this situation.
b. Suppose that the *p*-value of the study is not small enough to reject the null hypothesis. Write this conclusion in the context of the situation.
c. Now suppose that the *p*-value of the study is small enough to reject the null hypothesis. In the context of the situation, express the conclusion in two different ways.

15.11 In the 2008 General Social Survey conducted by the National Opinion Research Center at the University of Chicago, participants were asked:

Do you favor or oppose the death penalty for persons convicted of murder?

Do you think the use of marijuana should be made legal or not?

A two-way table of counts for the responses to these two questions is as follows:

		Marijuana?		
		Legal	**Not Legal**	**Total**
	Favor	313	461	774
Death Penalty?	**Oppose**	165	247	412
	Total	478	708	1186

Data source: http://sda.berkeley.edu/archive.htm.

a. Is there an obvious choice for which variable should be the explanatory variable in this situation? Explain.
b. Calculate row and/or column percentages for the table. Use these percentages to describe the relationship (or absence of a relationship).
c. State null and alternative hypotheses about the two variables that have been used to create the contingency table.
d. Carry out a chi-square test of the hypotheses stated in part (c). State a conclusion, and support your conclusion with a *p*-value. Use level of significance $\alpha = .05$.

15.12 The data for this exercise are from a sample of twelfth-graders, collected as part of the 2001 Youth Risk Behavior Surveillance System. The students were asked how often they wear a seatbelt while driving, and for this exercise, we combine the responses for "never" and "rarely" and for "most times" and "always" and ignore the "sometimes" responses. A contingency table of the results for males and females follows.

	Seatbelt Use		
	Most Times or Always	**Rarely or Never**	**Total**
Female	964	97	1061
Male	924	254	1178
Total	1888	351	2239

Data source: http://www.cdc.gov/nccdphp/dash/yrbs/.

a. Calculate the row percentages for females and the row percentages for males. Compare them.
b. Specify the null and alternative hypotheses that could be tested by using a chi-square statistic in this situation.

c. Calculate the expected count for the cell "Male, Most times or Always."
d. Using the expected count you calculated in part (c), obtain the expected counts for the remaining cells by subtraction. Remember that row and column totals are the same for expected counts as they are for observed counts.
e. Compute the chi-square statistic.
f. Use the chi-square statistic computed in part (e) to find a p-value or p-value range.
g. Form a conclusion, and write it in the context of the situation. Be sure to state the level of significance that you are using.

15.13 For the expected counts shown in Table 15.3 (p. 587) for xylitol and ear infection data, verify that the null hypothesis "expected" proportion getting an ear infection is the same for the three treatment groups.

15.14 An article on the Gallup website titled "SOCIAL AUDIT, Gambling in America" included the following comparison of the responses of teenagers and adults to the question, "Generally speaking, do you approve or disapprove of legal gambling or betting?" The survey was conducted in April 1999.

	Opinion of Legal Gambling			
	Approve	**Disapprove**	**No Opinion**	**Total**
Adults	959 (63%)	488 (32%)	76 (5%)	1523
Teens	261 (52%)	235 (47%)	5 (1%)	501
Total	1220	723	81	2024

Source: Copyright © 1999 The Gallup Organization.

a. State null and alternative hypotheses for this contingency table.
b. Here is Minitab output for a chi-square test. What conclusion can we reach about the hypotheses stated in part (a)? Justify this conclusion using information from the output.

```
Expected Counts are shown below observed counts
Rows: Age Group      Columns: Opinion

          Approve    Disappro      No Opini       All
Adults        959         488            76      1523
           918.01      544.04         60.95   1523.00

Teens         261         235             5       501
           301.99      178.96         20.05    501.00

All          1220         723            81      2024
          1220.00      723.00         81.00   2024.00

Chi-Square = 45.723,     DF = 2,     P-Value = 0.000
```

c. What is the expected count for the "Teens, Approve" cell? Show how to calculate this value.

15.15 In Exercise 2.30 data were presented on seatbelt use and grades for a sample of 2530 twelfth-graders, collected as part of the 2001 Youth Risk Behavior Surveillance System. Students were asked how often they wear a seatbelt while driving; possible choices were never, rarely, sometimes, most times, and always. For this exercise, we combine the responses for "never" and "rarely" and for "most times" and "always" and ignore the "sometimes" responses. They also were asked what grades they typically get. The following contingency table summarizes the results.

	Seat Belt Use		
Typical Grades	**Most Times or Always**	**Rarely or Never**	**Total**
A or B	1354	180	1534
C	428	125	553
D or F	65	40	105
Total	1847	345	2192

Data source: http://www.cdc.gov/nccdphp/dash/yrbs/.

Carry out a chi-square test for the relationship between seatbelt use and typical grades for twelfth-graders. Be sure to cover all five steps.

15.16 The following drawing illustrates the water-level task. Several developmental psychologists have investigated performance on this task. The figure on the left shows the water level in a glass of water that is half full (or is it half empty?). The figure on the right shows the same glass tipped to the right, but the water level is not shown. The task is to correctly draw the water line in this glass.

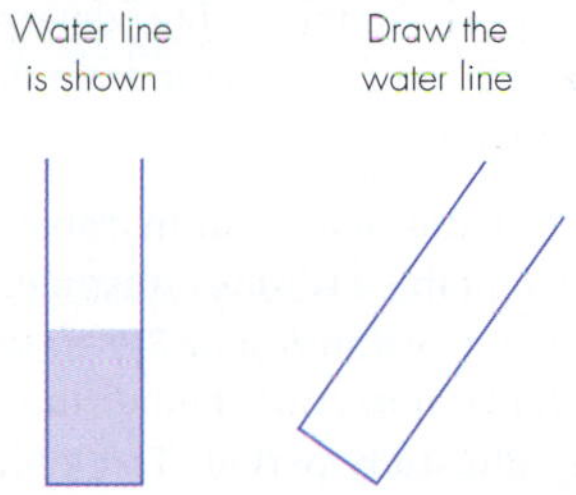

A University of South Carolina psychologist examined the water-level task success rate of 50 students from each of five different colleges within the university (Kalichman, 1986). Within each college, the sample included 25 men and 25 women. The investigators showed each participant eight different drawings of tipped glasses. If the participant correctly drew the water line on four or more of the drawings, he or she was given a "passing" result. Here is a contingency table of the results:

	Result on Task		
College	**Pass**	**Fail**	**Total**
Business	33	17	50
Language Arts	32	18	50
Social Sciences	25	25	50
Natural Sciences	38	12	50
Engineering	43	7	50
Total	171	79	250

a. Compare the colleges using row percentages. What differences are apparent from these percentages?
b. Write null and alternative hypotheses for this table.
c. What are the degrees of freedom for a chi-square test?
d. The chi-square statistic for this table is 16.915. Determine a p-value, and state a conclusion about the null and alternative hypotheses. Use level of significance $\alpha = .05$.

e. For each college, what are the expected counts for "pass" and "fail" outcomes?
f. (Obviously optional) What is the correct water line in the glass on the right?

15.17 The following table was first given in Exercise 4.4 in Chapter 4. It contains counts and row percentages for data on age group and frequency of reading newspapers for respondents in the 2008 General Social Survey.

	Frequency of Reading Newspapers				
Age Group	Every Day	A Few Times a Week	Once a Week	Less Than Once a Week	Total
18–29	45	68	38	83	234
30–49	118	125	100	175	518
50+	260	99	69	126	554
Total	423	292	207	384	1306

Data source: http://sda.berkeley.edu/archive.htm, accessed March, 2010.

a. Compare the age groups using row percentages. What differences are apparent from these percentages?
b. Write null and alternative hypotheses for this table.
c. Carry out a chi-square test for the hypotheses written in part (b). Calculate the chi-square value, determine a *p*-value and state a conclusion. Be sure to state the level of significance that you are using.

15.18 The data for this exercise were first given in Table 4.3 for Example 4.2. The following table classifies Australian couples who were married at the beginning of a 3-year study by the smoking habits of the couple and whether the couple separated or not during the study period. The table gives counts and row percentages. (Source: Butterworth et al., 2008. Read the original source on the companion website.)

	Marital Status		
Smoking	Separated	Not Separated	Total
Neither smoked	41 (4.2%)	931 (95.8%)	972 (100%)
One smoked	41 (12.4%)	290 (87.6%)	331 (100%)
Both smoked	32 (16.4%)	163 (83.6%)	195 (100%)
Total	114 (7.6%)	1384 (92.4%)	1498 (100%)

a. Write null and alternative hypotheses for this table.
b. Carry out a chi-square test for the hypotheses written in part (a). Calculate the chi-square value, determine a *p*-value, and state a conclusion. Use $\alpha = .05$.

Section 15.2

Skillbuilder Exercises

15.19 Refer to the data given for Exercise 15.11 about opinion on the death penalty and opinion on the legalization of marijuana. Use the shortcut formula to calculate the value of the chi-square statistic for those data.

15.20 Refer to the data given for Exercise 15.9 about sex (male, female) and type of statistics class students were taking. Use the shortcut formula to calculate the value of the chi-square statistic for those data.

15.21 The data in the table below first appeared in Exercise 4.11. The variables are height (short or not) and whether or not the student had ever been bullied in school for 209 secondary school students in England. The researchers gathered the data to test their hypothesis that short students are more likely to be bullied in school than are other students.

Height and Bullying in School

	Ever Bullied		
Height	Yes	No	Total
Short	42	50	92
Not Short	30	87	117
Total	72	137	209

a. Should these data be analyzed by using a chi-square test or a *z*-test for comparing two proportions? Explain.
b. Explain why it makes sense to designate height as the explanatory variable.
c. Carry out an appropriate hypothesis test. Be sure to state null and alternative hypotheses, and clearly indicate your conclusion about these hypotheses.

15.22 Refer to the data given for Exercise 15.21 about height and the risk of having been bullied for secondary school students. Use the shortcut formula to calculate the value of the chi-square statistic for those data.

General Section Exercises

15.23 Refer to Exercise 15.14 about age and opinion on gambling. Would a *z*-test for the difference in two proportions have been appropriate instead of a chi-square test? If your answer is yes, explain whether the results would have been identical to the results of the chi-square test. If your answer is no, explain why not.

15.24 Refer to Exercise 15.12, in which the relationship between sex (male, female) and seatbelt use was examined.

a. Compute the chi-square statistic using the shortcut formula given in Section 15.2. Show your work.
b. What are the degrees of freedom for this situation?
c. Find the *p*-value or *p*-value range for the chi-square statistic found in part (a).
d. Form a conclusion, and write it in the context of the situation. Be sure to state the level of significance that you are using.

15.25 The use of magnets has been proposed as a cure for various illnesses. Suppose that researchers conduct a study with ten participants to determine whether using magnets as therapy reduces pain from migraine headaches. Five participants are randomly assigned to receive the magnet treatment, and of those, two report that their pain is reduced. The remaining five participants are given an indistinguishable sham treatment, and of those, none report that their pain is reduced. Fisher's Exact Test would be appropriate in this situation.

a. Write in words the question that the *p*-value would answer in this situation.
b. (Appropriate software is required.) Carry out Fisher's Exact Test.

15.26 In each of the following situations, explain which test would be most appropriate: a chi-square test, a one-sided z-test for the difference in two proportions, or a Fisher's Exact Test.

a. The manufacturer of a safety seal that is used in cars wants to know whether the safety seals perform as well in extreme cold temperatures as they do in normal temperatures. Testing the seals is expensive, so the manufacturer tests only six seals at an extreme cold temperature and six other seals at a normal temperature. Two of the six seals tested in the extreme cold fail and none of the six seals tested at a normal temperature fails.

b. In June, before an upcoming November presidential election, a random sample of 1000 voters was asked which candidate they would vote for if the election were held then. In September, a new random sample of 1000 voters is asked the same question. The polling agency wants to know if the population proportions planning to vote for each candidate are different in September than they were in June.

c. Frequent computer use may be a cause of carpal tunnel syndrome, a painful condition that affects people's hands and wrists. Researchers hypothesized that people who took typing classes in school would be less likely to suffer from carpal tunnel syndrome than those who did not take typing classes. To examine this hypothesis, the researchers asked a random sample of 3000 people who frequently used computers in their jobs whether they had ever taken typing classes and whether they had ever suffered from carpal tunnel syndrome.

15.27 Refer to Exercise 15.26. In each case, specify the null and alternative hypotheses.

15.28 Weindling et al. (1986; also in Hand et al., 1994, p. 15) were interested in the health of juvenile delinquents. They classified 16 boys who failed a vision test by whether or not they wore glasses and whether or not they were a juvenile delinquent. They were interested in knowing whether the likelihood of wearing glasses differed for juvenile delinquents compared to nondelinquents. The results are shown in the table below.

Data for Exercise 15.28

	Wears Glasses	Does Not Wear Glasses	Total
Delinquent	1	8	9
Nondelinquent	5	2	7
Total	6	10	16

a. Specify the appropriate null and alternative hypotheses.

b. Explain why a chi-square test is not appropriate in this situation.

c. Find statistical software or a website that will conduct a Fisher Exact Test, and carry out the test. Use level of significance $\alpha = .05$.

d. Explain in words what the p-value for this test represents.

15.29 Explain how sample size affects the statistical significance of a fixed amount of difference between two sample proportions.

15.30 In an activity in a statistics class, students were asked if they had ever been pulled over by a police officer while driving. The following table summarizes results, as counts, classified by sex.

Sex	Ever Pulled Over by an Officer? No	Yes	Total
Male	7	15	22
Female	16	18	34
Total	23	136	56

a. Determine the (separate) proportions of males and females who had ever been pulled over by an officer.

b. Carry out a chi-square test to determine if there is a relationship between the sex of a student and whether he or she has ever been pulled over by a police officer. Determine the chi-square value, degrees of freedom, a range for the p-value, and state a conclusion.

c. Multiply all counts in the table by 5. Then, repeat part (b) for this new table of data.

d. Compare the results of parts (b) and (c). How has sample size affected the statistical significance of the observed relationship?

15.31 Suppose that 400 registered voters are surveyed about which of two candidates (X and Y) for a political office they prefer both before and after a televised debate between the candidates. The following table summarizes preferences both before and after the debate.

	Number of Voters
Preferred X before and after	190
Preferred Y before and after	170
Preferred X before, Y after	26
Preferred Y before, X after	14

a. What percentage of the 400 registered voters surveyed preferred X before the debate? What percentage preferred X after the debate?

b. How many total discordant pairs of before and after preferences were there?

c. Suppose that it was believed before the debate that X was not as good a debater as Y, so it was hypothesized that the proportion preferring X would decrease after the debate. Let p = population proportion preferring X after the debate among those whose opinions were changed by the debate. Write null and alternative hypotheses about p.

d. Use McNemar's Test to test the hypothesis written in part (c). Give a p-value and state a conclusion. (*Hint:* The p-value should be found as the probability that $X \leq 14$ in a binomial distribution with $n = 40$ and $p = .5$.) Be sure to state the level of significance that you are using.

15.32 Refer to Exercise 15.31. Suppose that 600 registered voters had been surveyed both before and after the debate and that the observed data were

	Number of Voters
Preferred X before and after	300
Preferred Y before and after	265
Preferred X before, Y after	25
Preferred Y before, X after	10

Use McNemar's Test to determine if the proportion of all registered voters who prefer X decreased after the debate. Give a p-value and state a conclusion. Use level of significance $\alpha = .05$.

Section 15.3

Skillbuilder Exercises

15.33 In the following situations, give the expected count for each of the k categories.

a. $k = 3$, H_0: $p_1 = p_2 = p_3 = 1/3$, and $n = 300$.
b. $k = 3$, H_0: $p_1 = 1/4$, $p_2 = 1/4$, $p_3 = 1/2$, and $n = 1000$.

15.34 In the following situations, give the expected count for each of the k categories.

a. $k = 4$, H_0: $p_1 = .2$, $p_2 = .4$, $p_3 = .1$, $p_4 = .3$, and $n = 2000$.
b. $k = 5$, H_0: all p_i are the same, and $n = 2500$.

15.35 Explain whether each of these is possible in a chi-square goodness-of-fit test.

a. The chi-square statistic is negative.
b. The chi-square statistic is 0.
c. The expected counts are not whole numbers.
d. The observed counts are not whole numbers.
e. The probabilities specified in the null hypothesis sum to less than 1.
f. The degrees of freedom are larger than the number of categories.

15.36 In a chi-square goodness-of-fit test, is it possible for all of the expected counts to be larger than the corresponding observed counts? Explain.

15.37 ◆ The following table shows the results from 190 students "randomly" choosing among the integers from 1 to 10. For example, there were two students who chose the digit 1 and nine students who chose the digit 2. (Raw data are in the **pennstate1** dataset on the companion website.)

Digit	1	2	3	4	5	6	7	8	9	10
Frequency	2	9	22	21	18	23	56	19	14	6

Assuming that the students were actually choosing digits at random, the expected number of students choosing any digit is 19 ($19 = 190/10$), so all expected counts are 19.

a. State the null and alternative hypotheses. (*Hint:* Refer to Example 15.15, p. 602.)
b. Calculate the chi-square goodness-of-fit test statistic.
c. What are the degrees of freedom for this test statistic?
d. Find the p-value or p-value range.
e. Form a conclusion, and write it in the context of the situation. Be sure to state the level of significance that you are using.

15.38 Suppose that on a typical day, the proportion of students who drive to campus is .30 (30%), the proportion who bike is .60 (60%), and the remaining .10 (10%) come to campus in some other way (e.g., walk, take the bus, get a ride). The campus sponsors a "spare the air" day to encourage people not to drive to campus on that day. They want to know whether the proportions using each mode of transportation on that day differ from the norm. To test this hypothesis, a random sample of 300 students that day was asked how they got to campus, with the following results:

Method of Transportation	Drive	Bike	Other	Total
Frequency	80	200	20	300

a. State the null and alternative hypotheses.
b. Find the expected counts for the three modes of transportation.
c. Calculate the chi-square goodness-of-fit statistic.
d. Find the p-value or p-value range.
e. Form a conclusion, and write it in the context of the situation. Be sure to state the level of significance that you are using.

General Section Exercises

15.39 Suppose that a statistics teacher assigns her class of 60 students the homework task of flipping a coin two times, counting the number of heads, and submitting that number to her during the next class period.

a. List the possible sequences of results for two flips of a coin. Using this list, determine the probability distribution for X = number of heads in two flips if the coin is fair.
b. Using proper notation, express the probabilities determined in part (a) as a null hypothesis about the number of heads that may occur in two flips of a fair coin.
c. The distribution of the number of heads reported by the 60 students in the class was as follows:

Number of Heads	0	1	2
Number of Students	8	40	12

Using these data, carry out a chi-square goodness-of-fit test to test the null hypothesis stated in part (b). Clearly show all steps, and state a conclusion about the null hypothesis. Use level of significance $\alpha = .05$.
d. If you were the teacher, what would you conclude about how the class may have done this assignment? Briefly justify your answer.

15.40 In a class survey done in a statistics class, students were asked, "Suppose that you are buying a new car and the model you are buying is available in three colors: silver, blue, or green. Which color would you pick?" Of the $n = 111$ students who responded, 59 picked silver, 27 picked green, and 25 picked blue. Is there sufficient evidence to conclude that the colors are not equally preferred? Carry out a significance test. Be sure to state the null hypothesis, and specify the population to which your conclusion applies.

15.41 Refer to Exercise 15.40. Suppose that a car manufacturer had hypothesized that 50% would prefer silver, 30% would prefer blue, and 20% would prefer green. Test the manufacturer's hypothesis. Use $\alpha = .05$.

15.42 The following table contains the observed distribution of the last digit of the forecasted high temperature on December 10, 2000, for $n = 150$ U.S. and international cities. (*Source: New York Times*, December 10, 2000, p. 47.)

Last Digit	0	1	2	3	4	5	6	7	8	9
Count	11	21	11	23	10	17	11	15	13	18

a. Compute expected counts for the null hypothesis that all digits 0, 1, ... , 9 are equally likely to be the last digit of the forecasted high temperature.
b. Calculate the chi-square goodness-of-fit statistic for these data. What are the degrees of freedom for this statistic?
c. State a conclusion about the null hypothesis that all digits are equally likely to be the last digit in the forecasted high temperature. Explain.

15.43 One of the authors of this book purchased four 1-pound bags of plain M&Ms at different stores in Pennsylvania to compare the color distribution to the one stated on the manufacturer's website in 2000. The observed results for the combined bags and the proportions alleged by the manufacturer are given in the following table. State and test the appropriate hypotheses to determine whether it is reasonable to assume that the observed colors are a random sample from a population with the manufacturer's alleged proportions.

	Brown	Red	Yellow	Blue	Orange	Green	Total
Observed	602 (.289)	396 (.190)	379 (.182)	227 (.109)	242 (.116)	235 (.113)	2081
Alleged proportions	.30	.20	.20	.10	.10	.10	

15.44 The California Daily 3 lottery game is identical to the Pennsylvania Daily Number game described in Example 15.15 (p. 602). The following table contains the observed distribution of all digits drawn in the California Daily 3 on the 200 days between May 14, 2000, and November 29, 2000. Three digits are drawn each day, so the sample size is $n = 600$ for this table.

Digit	0	1	2	3	4	5	6	7	8	9
Count	49	61	64	62	50	64	59	65	63	63

Carry out a significance test of the null hypothesis that all digits 0, 1, ... , 9 are equally likely to be drawn. Be sure to state the level of significance that you are using.

Chapter Exercises

15.45 In a survey reported in a special issue of *Newsweek* magazine (Special Edition: Health for Life, Spring/Summer 1999), $n = 747$ randomly selected women were asked, "How satisfied are you with your overall appearance?" There were four possible responses to this question, and the following table shows the distribution of counts for the possible responses for each of three age groups. (*Note:* The counts were estimated from percentages given in *Newsweek*.)

	How Satisfied Are You with Your Overall Appearance?				
Age	*Very*	*Somewhat*	*Not Too*	*Not at All*	*Total*
Under 30	45	82	10	4	141
30–49	73	168	47	6	294
50+	106	153	41	12	312
Total	224	403	98	22	747

a. Determine conditional percentages within each age group. Summarize these data using a bar chart, and describe what is shown about the relationship between age and satisfaction with appearance.
b. Minitab output for a chi-square test of association between age group and satisfaction with appearance follows part (e). Verify that the degrees of freedom shown are correct.
c. Is there a statistically significant association between the two variables for this problem? Justify your answer with information from the output. Use level of significance $\alpha = .05$.
d. Refer to the output below. Draw a sketch that illustrates the p-value for this problem, and write a sentence interpreting the p-value.
e. Refer to the output below. What is the expected count for the "50+, Not Too Satisfied" cell? Show how to calculate this expected count.

```
Expected counts are printed below observed counts

         Very  Somewhat  Not Too  Not at All  Total
18–29      45        82       10           4    141
        42.28     76.07    18.50        4.15

30–49      73       168       47           6    294
        88.16    158.61    38.57        8.66

50+       106       153       41          12    312
        93.56    168.32    40.93        9.19

Total     224       403       98          22    747

Chi-Sq = 0.175 + 0.463 + 3.904 + 0.006 +
         2.607 + 0.556 + 1.842 + 0.816 +
         1.655 + 1.395 + 0.000 + 0.860 = 14.278
DF = 6,      P-Value = 0.027
1 cells with expected counts less than 5.0
```

15.46 Answer Thought Question 15.1 on page 590.

15.47 Exercise 4.61 gave data on ear pierces and tattoos for a sample of 1375 college women. The data are presented again in the following table. The ear-pierce response is the total number of ear pierces for a woman, and this has been categorized.

Ear Pierces and Tattoos, 1375 College Women

Pierces	**No Tattoo**	**Have Tattoo**
2 or less	498	40
3 or 4	374	58
5 or 6	202	77
7 or more	73	53

Data source: One of the authors.

a. Describe the relationship with relevant percentages.
b. Write null and alternative hypotheses for the relationship between ear pierces and tattoos.
c. Test the null and alternative hypotheses written for part (b). Use $\alpha = .05$ for the level of significance. State a conclusion.

15.48 Refer to Exercise 15.45. Note that the "contributions to chi-square" are given in the output for all cells. For instance, the "contribution to chi-square" for the row "18–29" and the column "Very" is 0.175, and for the row "18–29" and the column "Not at All" it is 0.006.

a. Identify the two cells with the highest "contributions to chi-square." Specify the numerical value of the "contribution" and the row and column categories for each of the two cells.
b. For each of the two cells identified in part (a), determine whether the expected count is higher or lower than the observed count.
c. Using the information in parts (a) and (b), explain how the women in those category combinations contribute to the overall conclusion for this study.

15.49 Example 15.11 (p. 595) described an experiment in which students were classified as "Sheep" who believe in ESP or as "Goats" who do not. Each student then guessed the results of ten coin tosses. We classify students as "Stars" if they guessed five or more correctly and as "Duds" if they did not. The results are shown in the following table:

	Stars	Duds	Total
Sheep	79	33	112
Goats	48	32	80
Total	127	65	192

a. Describe the relationship between belief and performance with relevant percentages.
b. Write null and alternative hypotheses for the relationship between belief and performance. Be sure to state the population to which your hypotheses apply.
c. Test the null and alternative hypotheses written for part (b). Use $\alpha = .05$ for the level of significance. State a conclusion.

15.50 In a study reported in the *Annals of Internal Medicine* (Lotufo et al., 2000), the investigators examined the possible relationship in men between baldness and the risk of coronary heart disease. Other researchers have reported a possible link between these two variables (see Case Study 6.4, for example). In this study, 19,112 men aged 40 to 84 years old were followed for 11 years. All men were free of coronary heart disease at the beginning of the study. During the study, the men were asked about any baldness pattern that they may have had at the age of 45. The following table shows counts for the cross-classification of hair loss pattern at age 45 and whether a participant developed coronary heart disease during the study period or not. (*Note:* Vertex baldness occurs on top in the area of the crown or peak of the head.)

Hair Loss Pattern and Coronary Heart Disease

	Coronary Heart Disease		
Hair Pattern	**Yes**	**No**	**Total**
No baldness	548	7611	8159
Baldness			
Frontal	333	4,075	4,408
Mild vertex	275	3,148	3,423
Moderate vertex	163	1,608	1,771
Severe vertex	127	1,224	1,351
Total	1,446	17,666	19,112

a. Analyze the relationship between hair loss pattern and the risk of coronary heart disease. Write a short report in which you provide relevant descriptive statistics as well the results of a test for statistical significance. If there is an association, describe the nature of the association.
b. Is this an observational study or an experiment? Briefly explain how the answer affects the interpretation of any observed link between hair loss pattern and coronary heart disease in men.

15.51 The data in this exercise were first presented in Exercise 4.8. In the 2008 General Social Survey, religious preference and opinion about when premarital sex is wrong were among the measured variables. The contingency table of counts for these variables is shown in Exercise 4.8 and again here:

Religious Preference and Opinion about Premarital Sex

	When Is Premarital Sex Wrong?				
Religion	**Always**	**Almost Always**	**Sometimes**	**Never**	**Total**
Protestant	221	54	98	288	661
Catholic	45	17	54	179	295
Jewish	2	1	8	18	29
None	15	10	32	164	221
Other	20	7	12	41	80
Total	303	89	204	690	1286

Data source: http://sda.berkeley.edu/archive.htm, accessed May 13, 2010.

a. Are the conditions necessary for carrying out a chi-square test met? Explain.
b. Test whether there is a statistically significant relationship between these two variables. Show all five steps for the hypothesis test. Be sure to state the level of significance that you are using.

15.52 Refer to Exercise 15.49, in which each student guessed the results of ten coin flips. If all students are just guessing, and if the coins are fair, then the number of correct guesses for each student should follow a binomial distribution.

a. What are the parameters n and p for the binomial distribution, assuming that the coins are fair and students were just guessing?
b. Specify the probabilities of getting 0 correct, 1 correct, ... , 10 correct for this experiment if students were just guessing. (*Hint:* These are the probabilities in the pdf for a binomial distribution with parameters specified in part (a).)
c. The following table shows how many students got two or less right, three right, four right, and so on, separately, for students classified as Sheep (believe in ESP) and classified as Goats (don't believe in ESP). Using your results from part (b), fill in the null probabilities that correspond to the hypothesis that students are just guessing.

Number Correct	Sheep	Goats	Null Probabilities
≤2	6	5	
3	11	12	
4	16	15	
5	29	19	
6	28	16	
7	14	10	
≥8	8	3	
Total	112	80	

d. Test the hypothesis that the distribution of correct guesses for the population of Sheep represented by these students is given by the probabilities specified in part (c). Give a *p*-value, and state a conclusion. Use $\alpha = .05$ for the level of significance.

e. Repeat part (d) for the population of Goats represented by these students.

15.53 Household sizes for the households participating in the 1996 General Social Survey conducted by the National Opinion Research Center at the University of Chicago are shown in the following table. The table also shows the proportion of households of each size in the survey and the corresponding proportions of all U.S. households of each size.

Household Size	Number of Cases	Sample Proportion	U.S. Proportion
1	744	.256	.257
2	988	.340	.322
3	454	.156	.169
4	453	.156	.149
5 or more	265	.081	.103
Total	2904		

Sources: Sample data: SDA data archive, http://csa.berkeley.edu:7502; *Population data:* www.census.gov/populatio/socdemo/hh-fam/98ppla.txt (Table A. Households by Type and Selected Characteristics, 1998).

a. Carefully state the null and alternative hypotheses for testing whether the sizes of households in the General Social Survey reflect the size of households for the United States as a whole.

b. Carry out the appropriate test for your hypotheses in part (a). State your conclusion in the context of the situation.

15.54 Case Study 10.3 (p. 392) described a survey in which students in a statistics class were asked, "Would you date someone with a great personality even though you did not find them attractive?" The results were that 80 of 131 women answered "yes" and 26 of 61 men answered "yes."

a. Construct a contingency table for this situation, being careful to identify the explanatory and response variables.

b. State the appropriate hypotheses for this situation.

c. Given the hypotheses that you specified in part (b), would it be more appropriate to conduct a chi-square test or a *z*-test, or does it matter?

d. Conduct the appropriate test of the hypotheses you specified in part (b). Carry out all five steps. Be sure to state the level of significance that you are using.

15.55 Refer to Exercise 15.54. Use the shortcut formula for 2×2 tables to compute the chi-square statistic. Show the formula with all numbers entered.

15.56 Wilding and Cook (2000) asked 352 males and 376 females to listen to a male voice and a female voice. One week later, they attempted to identify the voices they had heard in line-ups of six male and six female voices. The results are shown in the following table. Using appropriate subsets of the data, conduct a test of the null hypotheses in parts (a) to (d). For all parts, use $\alpha = .05$ for the level of significance.

	Identified Male Voice?		Identified Female Voice?	
Listener's Sex	*Yes*	*No*	*Yes*	*No*
Male ($n = 352$)	145	207	132	220
Female ($n = 376$)	162	214	191	185
Total ($n = 728$)	307	421	323	405

a. H_0: Sex of listener and ability to identify a male voice are not related.

b. H_0: Sex of listener and ability to identify a female voice are not related.

c. H_0: Sex of listener and ability to identify a voice of the same sex are not related.

d. H_0: Sex of listener and ability to identify a voice of the opposite sex are not related.

e. Using the results from parts (a) to (d), write a short paragraph about the relationship between listener's sex, speaker's sex, and ability to subsequently identify a voice.

15.57 In a class survey, statistics students were asked, "Which one of these choices describes your perception of your weight: about right, overweight, or underweight?" The table in Exercise 4.12 displayed the results by sex. Displayed below are those results along with the Minitab output for a chi-square test. (*Source:* The authors.)

```
Expected counts are printed below observed counts

        About right    Overwt    Underwt    Total
F             87         39          3        129
           91.88      25.56      11.56

M             64          3         16         83
           59.12      16.44       7.44

Total        151         42         19        212

Chi-Sq = 0.259 + 7.072 + 6.340 +
         0.403 + 10.991 + 9.853 = 34.918
DF = 2,        P-Value = 0.000
```

a. State the null and alternative hypotheses being tested in this situation.

b. State a conclusion for the test, using $\alpha = .05$.

c. Using the results shown in the Minitab output, identify the cells with large "contributions to chi-square." Explain in words that would be understood by someone with no training in statistics why the contributions are so large for those cells.

15.58 In the 2008 General Social Survey, participants were asked, "Should divorce in this country be easier or more difficult to obtain than it is now?" The results are shown in the following table and Minitab output.

Sex	Easier	More Difficult	Stay Same	Total
Male	165	278	159	602
Female	199	331	140	670
Total	364	609	299	1272

Data source: http://sda.berkeley.edu/archive.htm, accessed May 13, 2010.

Expected counts are printed below observed counts

	Easier	More Diff	Stay Same	Total
Male	165	278	159	602
	172.27	288.22	141.51	
Female	199	331	140	670
	191.73	320.78	157.49	
Total	364	609	299	1272

Chi-Sq = 5.376, DF = 2, P-Value = 0.068

a. Carry out the five steps for a chi-square test for this situation.

b. Draw a picture of the appropriate chi-square distribution, and shade the region for the p-value for this test.

15.59 Refer to Exercise 15.58. Suppose that the same question were to be asked in many independent surveys. Assuming that the null hypothesis is true, into what range should the test statistic fall about 95% of the time, where 0 is at the lower end of the range?

15.60 Explain why a chi-square test statistic cannot be negative.

15.61 Students (n = 183) were asked to identify their own eye color as well as the eye color to which they are most attracted. The results are shown here:

Eye Color and Attraction

	Eyes Attracted To				
Own Eyes	*Brown*	*Blue*	*Hazel*	*Green*	*Total*
Brown	30	22	6	13	71
Blue	15	37	3	11	66
Hazel	4	12	7	7	30
Green	4	8	1	3	16
Total	53	79	17	34	183

a. The Minitab results of a chi-square analysis follow. On the basis of these results, explain why a chi-square analysis is not appropriate.

	Brown	Blue	Hazel	Green	Total
Brown	30	22	6	13	71
	20.56	30.65	6.60	13.19	
Blue	15	37	3	11	66
	19.11	28.49	6.13	12.26	
Hazel	4	12	7	7	30
	8.69	12.95	2.79	5.57	
Green	4	8	1	3	16
	4.63	6.91	1.49	2.97	
Total	53	79	17	34	183

Chi-Sq = 4.331 + 2.441 + 0.054 + 0.003 + 0.886 + 2.541 + 1.599 + 0.130 + 2.530 + 0.070 + 6.369 + 0.365 + 0.087 + 0.173 + 0.159 + 0.000 = 21.738

DF = 9, P-Value = 0.010

4 cells with expected counts less than 5.0

b. Combine the categories for "hazel" and "green," and rerun the analysis. Carry out all five steps of the hypothesis test.

c. Refer to the results from part (b). Identify the cells that have the largest "contributions to chi-square," and explain why the results in those cells support the alternative hypothesis.

15.62 Refer to Exercise 15.61. Using the original data (not combining green and hazel), construct a contingency table for the two variables:

Explanatory variable: Eye color.

Response variable: Finds own eye color most attractive, yes or no.

a. Conduct a chi-square test to determine if these two variables are related. Specify all five steps of the test.

b. Determine which eye color (as respondent's own color) contributes the most to the chi-square statistic. Explain whether that eye color is more attractive to people who have it as their own eye color than would be expected under the null hypothesis, or less so.

15.63 Gillespie (1999) and Chambers (2000) reported on two Gallup polls, taken in August 1999 and August 2000 using independent samples, which asked parents the question, "How satisfied are you with the quality of education your oldest child is receiving? Would you say completely satisfied, somewhat satisfied, somewhat dissatisfied or completely dissatisfied?" The results of the two polls are shown in the following table:

Results of Gallup Polls Taken in 1999 and 2000

	August 24–26, 1999	**August 24–27, 2000**
Completely satisfied	125	87
Somewhat satisfied	155	133
Somewhat dissatisfied	41	34
Completely dissatisfied	7	17
Just starting school (Volunteered answer)	7	11
No opinion	3	0
Total	338	282

a. Refer to the technical note on page 586 about the difference between tests for homogeneity and tests for independence. Which is more appropriate for this situation, for comparing the results of the survey for the 2 years? Specify the appropriate null and alternative hypotheses.

b. Ignore the last two categories (Just starting school; No opinion). Carry out the five steps for a chi-square test to determine if opinions differed in 1999 and 2000. Be sure to identify the appropriate populations represented by these data.

Dataset Exercises

Datasets required to solve these exercises are available on the companion website, http://www.cengage.com/statistics/Utts4e.

15.64 For this exercise, use the **GSS-08** dataset on the companion website. The variable ***owngun*** indicates whether or not the respondent owns a gun, and the variable ***polparty*** contains the respondent's political party preference. Is there a significant relationship between ***owngun*** and ***polparty***?

a. Determine the chi-square statistic and p-value, and then state a conclusion.
b. Provide percentages that describe the relationship between the two variables.

15.65 For this exercise, use the **UCDavis2** dataset. Respondents were categorized as male or female (***Sex***) and were asked whether they typically sit in the front, middle, or back of the classroom (***Seat***).

a. Carry out the five steps to determine whether there is a significant relationship between these two variables.
b. The p-value for your test in part (a) should be about .02. Explain in words which cells contribute the most to the chi-square statistic and how those cells support the alternative hypothesis.

15.66 For this exercise, use the **UCDavis2** dataset. Two variables that were measured were whether the respondent was left- or right-handed (***Hand***) and whether the respondent finds it easier to make friends with people of the same or opposite sex (***Friends***). Carry out the five steps to determine whether there is a significant relationship between those variables.

15.67 For this exercise, use the **UCDavis2** dataset. Identify two variables for which it would be of interest to you to test whether there is a relationship. Carry out the five steps of the chi-square test. If one or both of the variables are quantitative, create reasonable categories. For instance, you could classify students as nondrinkers, moderate drinkers, or heavy drinkers using the variable ***Alcohol***. Make sure you explain what variables you used and any recoding you did.

15.68 *This is also Exercise 4.80.* Use the **Student2010** dataset. The variable ***UseCell*** gives student responses to a question that asked students how they mainly used a cell phone (to talk, to text).

a. Create a two-way table that summarizes the relationship between ***Sex*** and ***UseCell***.
b. Fill in the following table with row percentages.

	UseCell		
Sex	**To talk**	**To text**	**Total**
Female			
Male			

c. Explain why the percentages found in part (b) are evidence that ***Sex*** and ***UseCell*** are related in the sample.
d. Write null and alternative hypotheses about the possible relationship between ***Sex*** and ***UseCell***.
e. What is the p-value of a chi-square test for the relationship between ***Sex*** and ***UseCell***? Is the observed relationship statistically significant?

15.69 Use the **Student2010** dataset. The variable ***LiveWhere*** indicates whether a student lives in a dorm or off-campus. The variable ***AlcMissClass*** indicates whether a student has ever missed a class due to drinking alcohol.

a. Create a two-way table that summarizes the relationship between ***LiveWhere*** and ***AlcMissClass***.
b. Fill in the following table with row percentages.

	AlcMissClass		
LiveWhere	**No**	**Yes**	**Total**
Dorm			
Off-campus			

c. Explain why the percentages found in part (b) are evidence that the two variables are related in the sample.
d. Write null and alternative hypotheses about the possible relationship between the two variables.
e. What is the p-value of a chi-square test for the relationship between the variables? Is the observed relationship statistically significant?

16

Nick Daly / Photonica / Getty Images

Is GPA related to where a student likes to sit?

See Example 16.1 *(p. 618)*

Analysis of Variance

Analysis of variance is a versatile tool for analyzing how the mean value of a quantitative response variable is affected by one or more categorical explanatory factors. For instance, it can be used to compare the mean weight loss for three different weight-loss programs or the mean testosterone levels of men in seven different occupations.

Suppose that a researcher wants to compare the mean weight loss for three different weight-loss programs or that another researcher wants to compare the mean testosterone levels of men in seven different occupations. What statistical methods can be used to make the desired comparisons? In the data analysis, the researchers might, at some point, use confidence intervals and significance tests to compare two means at a time (e.g., compare mean testosterone levels of occupations 1 and 2, occupations 1 and 3, and so on). Usually, however, an important first step in the analysis of more than two means is to do a significance test to determine if there are any differences at all among the population means being compared. The significance test for doing this is part of a procedure called the **analysis of variance**, which is also sometimes referred to as **ANOVA**.

In Chapters 11 and 13, we learned how to compare the means of two populations. In this chapter, we focus on the comparison of the means of more than two populations. When different values or levels of a single categorical explanatory variable (weight-loss programs, for instance) define the populations being compared, the ANOVA procedure is called **one-way analysis of variance**. It is sometimes called **one-factor analysis of variance** because a categorical explanatory variable is sometimes called a **factor**. In general, *analysis of variance* is a versatile tool for analyzing how the mean value of a quantitative response variable is related to one or more categorical explanatory factors. While most of the chapter is about one-way analysis of variance, the final section gives an overview of the concepts of *two-way* analysis of variance, which is used to examine the effects of two categorical explanatory variables on the mean value of a quantitative response variable.

16.1 Comparing Means with an ANOVA *F*-Test

When we compare the means of populations represented by independent samples of a quantitative response variable, a null hypothesis of interest is that all means have the same value. An alternative hypothesis is that the means are not all equal. Note that this alternative hypothesis does not require that all means must differ from each other. The alternative would be true, for example, if only one of the means were different from the others. If $k =$ the number of populations, the null and alternative hypotheses for comparing population means can be written as

H_0: $\mu_1 = \mu_2 = \ldots = \mu_k$

H_a: The population means are not all equal.

To test these hypotheses we follow the same five steps for hypothesis testing that we have used in other chapters. The appropriate test statistic is called an ***F*-statistic**, the general procedure is called *one-way analysis of variance*, and the significance test is called an ***F*-test**. The *F*-statistic is sensitive to differences among a set of sample means. The greater the variation among the sample means, the larger is the value of the test statistic. The smaller the variation among the observed means, the smaller the value of the test statistic. In this section, we are concerned with the general ideas of the *F*-test. The specific details and formulas are given in the next section.

Conceptually, the *F*-statistic can be viewed as follows:

$$F = \frac{\text{Variation between sample means}}{\text{Natural variation within groups}}$$

The variation between sample means is 0 if all k of the sample means are exactly equal and gets larger the more spread out they are. If that variation is large enough, it is evidence that at least one of the k population means is different from the other means. In that event, the null hypothesis should be rejected. It is this comparison of the variation between sample means for the groups to the natural variation within groups that gives "analysis of variance" its name. We are comparing two sources of variability to test whether means are equal, so the procedure might more naturally be called "analysis of means."

The denominator of the *F*-statistic, the natural variation within groups, provides a yardstick for determining whether the numerator is large enough to reject the null hypothesis. Much like the standard error in a z-statistic, it standardizes the numerator so that the p-value can be found from common tables. In fact, the denominator of the *F*-statistic is simply a pooled estimate of the variance within each group, a fact that will be discussed in more detail in the next section.

To find the p-value, which is the probability that the computed *F*-statistic would be as large as it is (or larger) if the null hypothesis is true, a probability distribution called the ***F*-distribution** is used. As with all other significance tests, the null hypothesis is rejected if the p-value is as small as or smaller than the desired level of significance (usually $\alpha = .05$). When the null hypothesis is rejected, the conclusion is that not all of the population means have the same value.

Example 16.1 **Classroom Seat Location and Grade Point Average** Is it true that the best students sit in the front of a classroom, or is that a false stereotype? In surveys done in two statistics classes at the University of California at Davis, students reported their grade point averages and also answered the question, "Where do you typically sit in a classroom (front, middle, back)?" In all, 384 students gave valid responses to both questions, and among these students, 88 said that they typically sit in the front, 218 said they typically sit in the middle, and 78 said they typically sit in the back.

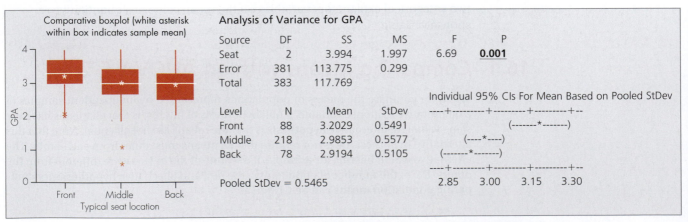

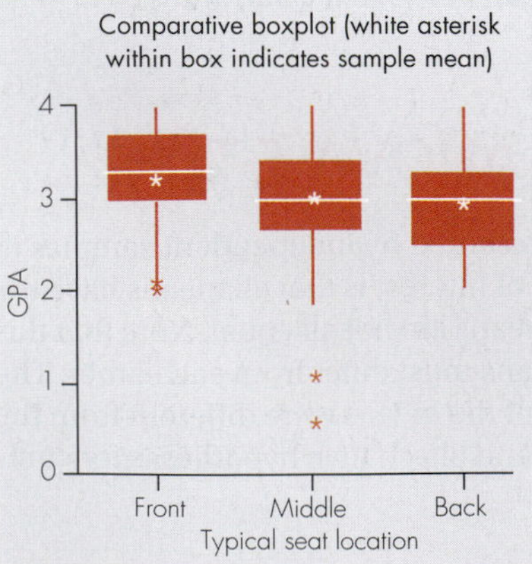


```
Analysis of Variance for GPA
Source     DF        SS       MS       F        P
Seat        2     3.994    1.997    6.69    0.001
Error     381   113.775    0.299
Total     383   117.769
                                 Individual 95% CIs For Mean Based on Pooled StDev
Level       N      Mean    StDev  ----+---------+---------+---------+--
Front      88    3.2029   0.5491                  (-------*-------)
Middle    218    2.9853   0.5577          (----*----)
Back       78    2.9194   0.5105     (-------*-------)
                                  ----+---------+---------+---------+--
Pooled StDev = 0.5465                 2.85      3.00      3.15      3.30
```

Figure 16.1 Comparison of GPA for three classroom seat locations

Figure 16.1 shows a boxplot comparing the GPAs for the three seat locations and Minitab analysis of variance results that can be used to test whether or not the mean GPAs are the same for the three locations. In the boxplot, we see that students sitting in the front generally have slightly higher GPAs than the others; toward the bottom of the computer output, the sample means are given as Front = 3.2029, Middle = 2.9853, Back = 2.9194.

The analysis of variance results in Figure 16.1 can be used to test the following:

H_0: $\mu_1 = \mu_2 = \mu_3$
H_a: The three means are not all equal.

where μ_1, μ_2, and μ_3 are the *population* mean GPAs for the populations of students who typically sit in the front, middle, and back of the classroom, respectively. The *p*-value for the *F*-test is under "P" in the rightmost column of the table titled "Analysis of Variance for GPA," and its value is 0.001 (bold, underlined). With a *p*-value this small, we can reject the null hypothesis and thus conclude that there are differences among the means in the populations represented by the samples.

Note that the output in Figure 16.1 also shows individual 95% confidence intervals for the three population means. The location of the interval for the "Front" mean does not overlap with the other two intervals, which indicates a significant difference between the mean GPA for the front-row sitters and the mean GPA for the other students. It is not clear whether there is a significant difference between the "Middle" and "Back" groups, since their confidence intervals overlap. We will further analyze the seat location differences in Example 16.6 (p. 624).

SPSS TIP **Calculating a One-Way Analysis of Variance**

- Use **Analyze > Compare Means > One-Way ANOVA**. Specify the *y* variable in the *Dependent List* box and specify the variable that defines the groups in the *Factor* box.

Notation for Summary Statistics

Useful notation for summarizing statistics from the observed samples is

k = number of groups.

$\bar{x}_i$, s_i, and n_i are the mean, standard deviation, and sample size for the *i*-th sample group.

N = total sample size = $n_1 + n_2 + \ldots + n_k$.

Example 16.2 **Application of Notation to the GPA and Classroom Seat Sample** In Example 16.1, the three seat locations (1 = front, 2 = middle, 3 = back) are compared, so $k = 3$. The group sample sizes are $n_1 = 88$, $n_2 = 218$, $n_3 = 78$. The total sample size is $N = 88 + 218 + 78 = 384$. Toward the bottom of the output shown in Figure 16.1, the sample means given are $\bar{x}_1 = 3.2029$, $\bar{x}_2 = 2.9853$, $\bar{x}_3 = 2.9194$. The sample standard deviations given in Figure 16.1 are $s_1 = 0.5491$, $s_2 = 0.5577$, and $s_3 = 0.5105$.

Assumptions and Necessary Conditions for the *F*-Test

In the derivation of the *F*-statistic, the assumptions are similar to those for the pooled two-sample *t*-procedures described in Chapters 11 and 13. The assumptions about the populations and samples representing them follow:

- The samples are independent random samples.
- The distribution of the response variable is a normal curve within each population.

- The different populations may have different means under the alternative hypothesis; the null hypothesis is that the means are equal.
- All populations have the same standard deviation, σ.

These assumptions make up a model for the probability distributions in the populations being compared. Figure 16.2 illustrates, for three populations, how this model might look if the alternative hypothesis were true and how it would look if the null hypothesis were true. Note that in the alternative hypothesis case, the only difference among the populations is the difference among their means, μ_1, μ_2, and μ_3. The spread of the normal curve is assumed to be the same for each population. Therefore, when the null hypothesis is true the probability distributions for the populations are all identical.

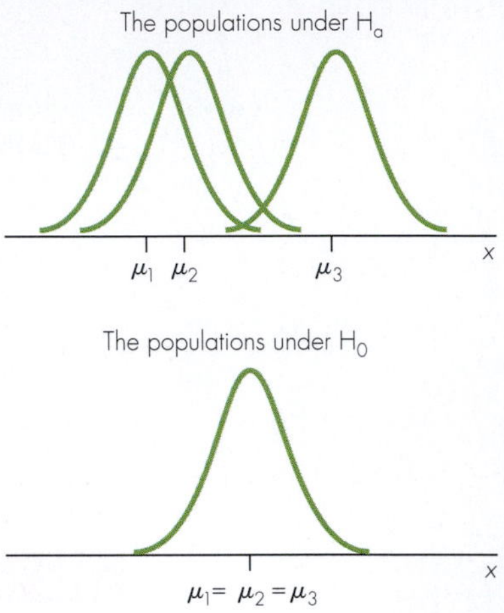

Figure 16.2 The model for one-way analysis of variance under the two different hypotheses

You may recall from Chapter 13 that when we compare two means, it is not imperative to assume equal population standard deviations because an "unpooled" procedure is available that does not require this assumption. Unfortunately, when we do a one-way analysis of variance, the equal standard deviations assumption is *necessary*. In practice, equality of the observed standard deviations does not have to hold exactly for the *F*-test to provide a valid significance test of differences among population means. One commonly used criterion is that the *F*-test can be used if the largest of the sample standard deviations is not more than twice as large as the smallest of the sample standard deviations.

There are significance tests for the equality of several standard deviations, but those tests tend not to work well when the data are not normally distributed. They can also have low power for small or moderate sample sizes, leading to false acceptance of equal variances when they are actually quite different. Therefore, we do not recommend using those tests.

The assumption of normally distributed data is not strictly necessary in practice either. Unless there are extreme outliers or extreme skewness in the data, the *F*-test works well even for relatively small sample sizes. When the raw data are available, a visual display should be created and examined to evaluate whether skewness or outliers are present.

IN SUMMARY Necessary Conditions for Using the *F*-Statistic to Compare Means

- The *F*-statistic can be used if the data are not extremely skewed, there are no extreme outliers, and the group standard deviations are not markedly different.
- As with the *t*-tests and confidence intervals in Chapters 11 and 13, tests based on the *F*-statistic are valid for data with skewness or outliers if the sample sizes are large.
- A rough criterion for standard deviations is that the largest of the sample standard deviations should not be more than twice as large as the smallest of the sample standard deviations.

Example 16.3 **Assessing the Necessary Conditions for the GPA and Seat Location Data** The boxplot in Figure 16.1 (p. 618) shows two outliers in the group of students who typically sit in the middle of a classroom, but there are 218 students in that group so these outliers don't have much influence on the results. The standard deviations for the three groups are nearly the same, and the data do not appear to be skewed. The necessary conditions for doing an *F*-test are satisfied in this example.

THOUGHT QUESTION 16.1 To what populations do the conclusions of Example 16.1 apply? Do you think it matters that the data were collected at a single university? Does it matter that the surveys were done only in statistics classes?*

Example 16.4 **Occupational Choice and Testosterone Level** Many research studies have been done to examine how testosterone affects human behavior; conversely, some studies have considered how behavior affects testosterone level. In a study done at Georgia State University, psychologists compared the salivary testosterone concentrations of men who were ministers, salesmen, firemen, professors, physicians, professional football players, and actors. There were 66 men in the overall sample, and the sample sizes in the seven occupational groups ranged from 6 (actors) to 16 (physicians).

The researchers reported in the *Journal of Personality and Social Psychology* (Dabbs et al., 1990, p. 1262) that "analysis of variance showed an overall difference among the groups, $F(6,59) = 2.50$, $p < 0.05$, and a Newman-Keuls test indicated that actors and football players were significantly higher than ministers." In this quote, the information given is that the *p*-value for an *F*-test was less than .05, so the null hypothesis that all population means are the same was rejected.

Additional information is given about the pattern of the difference, which is that actors and football players had higher testosterone levels than ministers. Presumably, all other differences involving the other occupational groups were not statistically significant. The researchers focused on the difference between actors and ministers and reported on two additional studies of theirs that replicated the finding. In their discussion, they related the observed testosterone differences to the social, competitive, and dominance behaviors of ministers and actors. This example will be continued in Example 16.5 (p. 622).

The Family of *F*-Distributions

An *F*-distribution is used to find the *p*-value for an ANOVA *F*-test of the null hypothesis that several population means are equal. The family of *F*-distributions is a family of skewed distributions, each with a minimum value of 0. A specific *F*-distribution is indicated by two parameters called *degrees of freedom*. The first of the two parameters is called the **numerator degrees of freedom**; the second is called the **denominator degrees of freedom**. The values of the two degrees of freedom parameters always are given in the order *numerator df, denominator df.* In one-way ANOVA, the numerator df $= k - 1$ (number of groups $- 1$), and the denominator df $= N - k$ (total sample size $-$ number of groups).

Figure 16.3 (next page) shows an *F*-distribution with 2 and 30 degrees of freedom along with an *F*-distribution with 6 and 30 degrees of freedom. Note that both curves are skewed and the minimum possible value is 0. These are properties of all *F*-distributions, regardless of the values for the numerator and denominator degrees of freedom.

***HINT:** How universal do you think are the characteristics of students selecting the three different classroom seat locations?

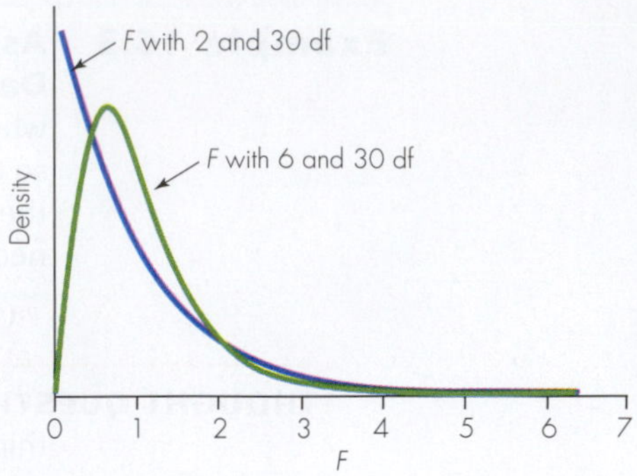

Figure 16.3 An *F*-distribution with 2 and 30 df and an *F*-distribution with 6 and 30 df

Determining the *p*-Value

The *p*-value for the *F*-test is the area to the right of the value of the *F*-statistic, under an *F*-distribution with $k - 1$ and $N - k$ degrees of freedom. All statistical software programs incorporate analysis of variance procedures, and the *p*-value will be reported as part of the output. In situations where an *F*-statistic has been computed without statistical software, either a statistical calculator or the Excel command FDIST (*value, numerator df, denominator df*) can be used to find the *p*-value. It is also possible to find a range for the *p*-value using tabled values for the appropriate *F*-distribution.

Table A.4 in the Appendix gives **critical values** for two different levels of significance, $\alpha = .05$ and $\alpha = .01$. If the observed *F*-statistic is greater than or equal to the critical value for a particular level of significance, the result is statistically significant at that level. The region of *F*-statistic values greater than the critical value is called the **rejection region**.

Using Table A.4 to Judge Statistical Significance

First, determine values for the numerator degrees of freedom and the denominator degrees of freedom. Possible values for the numerator degrees of freedom are given as column headings across the top of Table A.4. Possible values for the denominator degrees of freedom are given as row labels in the leftmost column.

- If the value of the *F*-statistic is greater than or equal to the critical value in the $\alpha = .05$ portion of the table, the result is statistically significant at the .05 significance level. In this case, the *p*-value is less than .05, which may be written $p < .05$.
- If the value of the *F*-statistic is greater than or equal to the critical value in the $\alpha = .01$ portion of the table, the result is statistically significant at the .01 significance level. In this case, the *p*-value is less than .01, which may be written $p < .01$.
- If the value of the *F*-statistic is between the critical values given for $\alpha = .05$ and $\alpha = .01$, the *p*-value is between .05 and .01 ($.01 < p < .05$). In this case, the result is statistically significant for $\alpha = .05$ but not for $\alpha = .01$.
- If the value of the *F*-statistic is less than the critical value in the $\alpha = .05$ portion of the table, the result is *not* statistically significant at the .05 significance level. In this case, the *p*-value is greater than .05, and as in other hypothesis tests, we conclude that we cannot reject the null hypothesis.

As an example, suppose the *F*-statistic for comparing four means is $F = 4.26$ and numerator df = 3 and denominator df = 20. In Table A.4, for $\alpha = .05$, the critical value is given as 3.10 (in the column labeled "3" and row labeled "20"). Because the observed $F = 4.26$ is greater than the critical value = 3.10, the result is statistically significant at the $\alpha = .05$ level. The critical value for the $\alpha = .01$ level is given as 4.94 in Table A.4. Because 4.26 is not greater than this critical value, the result is not significant at the .01 level. Thus, the *p*-value is between .01 and .05.

Example 16.5 **The *p*-Value for the Testosterone and Occupational Choice Example** The authors of the study described in Example 16.4 (p. 621) reported that for comparing the mean testosterone levels of seven occupations, the value of the *F*-statistic was $F = 2.5$. There were $k = 7$ occupational groups and $N = 66$ total observations. Thus, for the

F-distribution that is used to find the p-value, numerator df $= k - 1 = 7 - 1 = 6$ and denominator df $= N - k = 66 - 7 = 59$.

Figure 16.4 illustrates the p-value, which is the area to the right of 2.5 under an F-distribution with 6 and 59 df. In Excel, FDIST(*value, numerator df, denominator df*) can be used to find the probability of an F-statistic as large as or larger than the specified value. In this case, FDIST(2.5,6,59) = .032, the p-value for this situation. Using Table A.4, we learn that the approximate critical value is 2.25 for $\alpha = .05$. This is the value given for numerator df = 6 and denominator df = 60 (df value closest to 59 in the table). Because 2.5 is larger than this critical value, the result is statistically significant for the .05 significance level, and the p-value is smaller than .05.

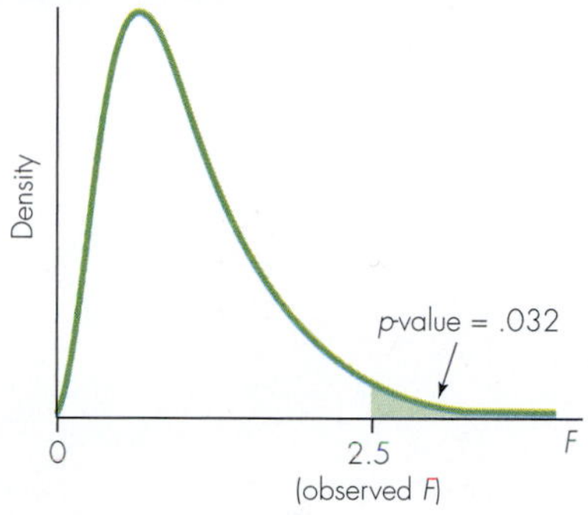

Figure 16.4 The p-value for Example 16.4 is the area to the right of $F = 2.5$ under an F-distribution with 6 and 59 degrees of freedom

Multiple Comparisons

The term **multiple comparisons** is used when two or more comparisons are made to examine the specific pattern of differences among means. The most commonly analyzed set of multiple comparisons is the set of all **pairwise comparisons** among population means. In Example 16.1 (p. 618) about GPA and classroom seat location, the possible pairwise comparisons are front versus middle, middle versus back, and front versus back. In Example 16.4 (p. 621), about occupation and testosterone level, the mean testosterone levels for 21 different pairs of occupations can be compared (actor versus minister, football player versus professor, and so on).

Either of two equivalent approaches can be used to make inferences about pairs of means. For each pair, a significance test could be done to determine whether the two means significantly differ. Or a confidence interval for the difference in each pair of means could be computed and then interpreted in terms of statistical significance. If a confidence interval for a difference does not include the value 0, there is a statistically significant difference.

Remember that when the $\alpha = .05$ significance level is used, about 1 in 20 independent statistical tests will achieve statistical significance even if all null hypotheses are true. In other words, when many statistical tests are done, there is an increased risk of making at least one Type 1 error (erroneously rejecting a null hypothesis). Consequently, several procedures have been developed to control the overall Type 1 error rate or the overall confidence level when inferences for a set (family) of multiple comparisons are done.

DEFINITION

A **family Type 1 error rate** for a set of significance tests is the probability of making one or more Type 1 errors (erroneously rejecting a null hypothesis) when more than one significance test is done.

A **family confidence level** for a procedure used to create a set of confidence intervals is the long-run proportion of times that all intervals in the set would capture their true parameter values if the procedure were to be repeated numerous times.

Tukey's procedure is one of several available procedures that control the family Type 1 error rate (and family confidence level) for multiple comparisons between pairs of population means. If the family error rate is not a concern, Fisher's procedure may be used. In **Fisher's procedure** as it is used for confidence intervals, a confidence level (say, 95%) is specified for each separate comparison of two means.

Example 16.6 **Pairwise Comparisons of GPAs Based on Seat Locations** Example 16.1 (p. 618) was about GPA and preferred classroom seat location for college students. Figure 16.5 shows Minitab output for pairwise comparisons of the mean GPA by seat location. The family confidence level used for the Tukey procedure was .95, while for Fisher's procedure a .95 confidence level was used for each individual interval. For both methods, the output gives a confidence interval for the difference between population means for each of the three possible differences between two locations (back versus middle, back versus front, and middle versus back). The numbers given under "Lower" and "Upper" are the lower and upper ends of a confidence interval for the difference between population means for the two locations. (The value under "Center" is the difference between sample means.) For example, in the results for the Tukey procedure, the "Lower" and "Upper" values for "Back subtracted from Front" are 0.0846 and 0.4824, respectively. Thus, a confidence interval for the difference in means ($\mu_{\text{Front}} - \mu_{\text{Back}}$) is 0.0846 to 0.4824. In the results given for the Fisher procedure, the confidence interval for ($\mu_{\text{Front}} - \mu_{\text{Back}}$) is 0.1164 to 0.4506.

Tukey 95% Simultaneous Confidence Intervals

Seat = Back subtracted from:

Seat	Lower	Center	Upper
Middle	−0.1028	0.0659	0.2347
Front	0.0846	0.2835	0.4824

Seat = Middle subtracted from:

Seat	Lower	Center	Upper
Front	0.0561	0.2176	0.3791

Fisher 95% Individual Confidence Intervals

Seat = Back subtracted from:

Seat	Lower	Center	Upper
Middle	−0.0759	0.0659	0.2077
Front	0.1164	0.2835	0.4506

Seat = Middle subtracted from:

Seat	Lower	Center	Upper
Front	0.0819	0.2176	0.3533

Figure 16.5 Confidence intervals for pairwise differences in the GPA and seat location example

In general, two means are significantly different if the confidence interval for the difference does not cover 0. Note that the two procedures lead to the same conclusions about statistical significance in this example. In both procedures, the only confidence interval that covers 0 is the one for ($\mu_{\text{Middle}} - \mu_{\text{Back}}$). We conclude that for students who sit in the front, the population mean GPA is different from the population mean for students who sit in the middle and the back of the room. We are not able to say that there is a difference between the population means of those who sit in the middle and those who sit in the back.

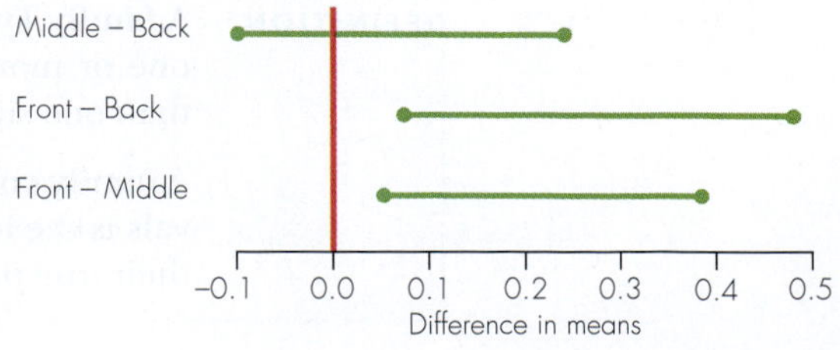

Figure 16.6 Confidence intervals for differences between pairs of means in the GPA and seat location example (Tukey procedure)

Figure 16.6 is a graph of the confidence intervals found using the Tukey method. Each interval estimates the difference between the population mean GPAs for two different seat locations. A vertical line is drawn at 0, the value that would indicate no difference between two means. Note that the confidence interval for $(\mu_{\text{Middle}} - \mu_{\text{Back}})$ covers the line drawn at 0, whereas confidence intervals for $(\mu_{\text{Front}} - \mu_{\text{Back}})$ and $(\mu_{\text{Front}} - \mu_{\text{Middle}})$ do not.

TECHNICAL NOTE

The Bonferroni Method for Multiple Comparisons

As first described in Section 13.8, the simplest (and most conservative) method for controlling the family error rate for multiple comparisons is the **Bonferroni method.** The method proceeds by dividing up the significance level (or confidence level) across tests (or confidence intervals). In Example 16.6, three pairwise comparisons of seat locations are made: middle versus back, middle versus front, and back versus front. If a total Type 1 error probability of .05 is desired and 3 tests are to be done, each test uses level of significance $\alpha = .05/3 = .0167$. This method controls the overall probability of making at least one Type 1 error to be at most the sum of the individual significance levels used for each test. Or, if we were to compute confidence intervals for the three possible pairwise differences in population means using the Bonferroni method, using a confidence level $= 1 - (.05/3) = .9833$ for each interval would provide a family confidence of at least .95.

The Tukey method is applicable only for comparing pairs of means. The Bonferroni method is more general, and can be used for any set of comparisons. For instance, in Example 16.4 we might want to compare the mean testosterone level for various subsets of occupations to those for other subsets (such as professors and physicians compared to firemen and football players, actors and salesmen compared to ministers, and so on). If we wanted to do five such comparisons and control the family error rate at .05, using the Bonferroni method we could use $.05/5 = .01$ for each comparison.

16.1 Exercises are on pages 639–641.

IN SUMMARY

Concepts for Comparing More Than Two Population Means

- **One-way analysis of variance** is used to compare more than two population means.
- The *null and alternative hypotheses* used for comparing *k* population means can be written as follows:

  $H_0: \mu_1 = \mu_2 = \ldots = \mu_k$

  H_a: The means are not all equal.

- The *test statistic* is an ***F*-statistic** that can be expressed conceptually as follows:

$$F = \frac{\text{Variation between sample means}}{\text{Natural variation within groups}}$$

- An ***F*-distribution** is used to determine the *p*-value that is used as the basis for reaching a conclusion about the hypotheses.

MINITAB TIP One-Way Analysis of Variance

- Use **Stat > ANOVA > One-way** if values of the response variable are in one column (*Response*) and values or names for groups are in a second column (*Factor*). In the dialog box, the ***Comparisons*** button is used for pairwise comparisons.
- Use **Stat > ANOVA > One-way (unstacked)** if the data for different groups are stored in separate columns. Specify the separate columns for the groups in the dialog box. There is not a pairwise comparisons capability in this procedure in Versions 13 or lower, but there is in Versions 14 and higher.

16.2 Details of One-Way Analysis of Variance

In this section, we provide some computational details for the *F*-statistic and give the necessary conditions for using the statistic. In practice, statistical software is nearly always used to do a one-way analysis of variance, so it is not necessary to learn formulas to do the computations "by hand." The formulas, however, provide useful insights into the concepts of analysis of variance and how the *F*-test is constructed.

The Analysis of Variance Table

A fundamental concept in one-way analysis of variance is that the variation among the data values in the overall sample can be separated into (1) differences between group means, and (2) natural variation among observations within a group. The calculations and theory for analysis of variance stem from the fact that for a particular way of measuring variation, the sum of the variation between group means and the variation among observations within groups equals the total variation. Expressed as an equation, the relationship is

Total variation = Variation between group means + Variation within groups

An **analysis of variance table**, like the one shown in Figure 16.1 for Example 16.1 (p. 618), is used to display information about the sources of variation in the response variable. The *F*-statistic that is used to compare population means measures the relative size of the variation between group means and the natural variation within groups.

Example 16.7 **Comparison of Weight-Loss Programs** Suppose that a researcher does a randomized experiment to compare the mean weight loss for three different programs for losing weight, and the observed weight losses after 3 months are as follows:

Program 1	Program 2	Program 3
7	9	15
9	11	12
5	7	18
7		

A look at the dotplot of the data shown in Figure 16.7 shows that over the total dataset, the weight losses ranged from 5 to 18 pounds. The sample means for the three programs are $\bar{x}_1 = 7$, $\bar{x}_2 = 9$, $\bar{x}_3 = 15$. While there is some variation among individuals within each program, generally it appears that differences *between* weight-loss programs account for much of the variation in the total dataset. The weight losses for Program 3 in particular are noticeably greater than the weight losses for the other two programs.

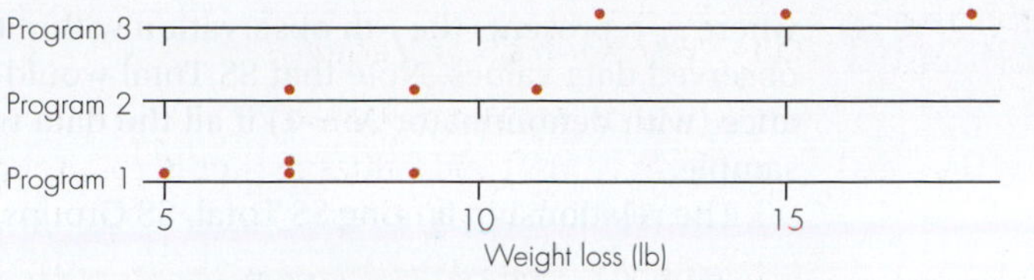

Figure 16.7 Dotplot for Example 16.7

Measuring Variation between Group Means

The variation between group means is measured with a weighted sum of squared differences between the sample means (the $\bar{x}_i$) and $\bar{x}$, the overall mean of all data. Each squared difference is multiplied by the appropriate group sample size, n_i, in this sum. This quantity is called **sum of squares for groups** or **SS Groups**. A formula is

$$\text{SS Groups} = n_1(\bar{x}_1 - \bar{x})^2 + n_2(\bar{x}_2 - \bar{x})^2 + \cdots + n_k(\bar{x}_k - \bar{x})^2 = \sum_{\text{groups}} n_i(\bar{x}_i - \bar{x})^2$$

The numerator of the F-statistic for comparing means is called the **mean square for groups** or **MS Groups**, and it is calculated as follows:

$$\text{MS Groups} = \frac{\text{SS Groups}}{k - 1}$$

Measuring Variation within Groups

To measure the variation among individuals within groups, find the sum of squared deviations between data values and the sample mean in each group, and then add these quantities. This is called the **sum of squared errors** or **SSE**.

A formula in terms of the sample standard deviations (denoted by s_i for Group i) is

$$\text{SS Error} = (n_1 - 1)s_1^2 + (n_2 - 1)s_2^2 + \cdots + (n_k - 1)s_k^2 = \sum_{\text{groups}} (n_i - 1)s_i^2$$

The denominator of the F-statistic is called the **mean square error** or **MSE**, and it is calculated as follows:

$$\text{MSE} = \frac{\text{SSE}}{N - k} = \frac{(n_1 - 1)s_1^2 + (n_2 - 1)s_2^2 + \cdots + (n_k - 1)s_k^2}{n_1 + n_2 + \cdots + n_k - k}$$

Note that MSE is just a weighted average of the sample variances for the k groups. In fact, if all n_i are equal, MSE is simply the average of the k sample variances. The square root of MSE, denoted by s_p and called the **pooled standard deviation**, estimates the population standard deviation of the response variable. Recall that the populations being compared are assumed to have equal standard deviations.

DEFINITION The **pooled standard deviation** $s_p = \sqrt{\text{MSE}}$ estimates the population standard deviation, σ. This assumes that each population has the same value of σ.

Measuring Total Variation

The total variation in the data from all samples combined is measured by computing the sum of squared deviations between data values and the mean of all the data. This quantity is referred to as the **total sum of squares** or **SS Total**. The total sum of squares may also be referred to as SSTO. A formula for the sum of squared differences from the overall mean is

$$\text{SS Total} = \sum_{\text{values}} (x_{ij} - \bar{x})^2$$

where x_{ij} represents the j-th observation within the i-th group, and $\bar{x}$ is the mean of all observed data values. Note that SS Total would be the numerator of the sample variance (with denominator $N - 1$) if all the data were combined and treated as a single sample.

The relationship among SS Total, SS Groups, and SS Error follows:

$$\text{SS Total} = \text{SS Groups} + \text{SS Error}$$

This additive relationship is useful computationally. For example, if the values of SS Total and SS Groups are known, then SS Error can be calculated as

$$\text{SS Error} = \text{SS Total} - \text{SS Groups}$$

Example 16.8 **Analysis of Variation among Weight Losses** For Example 16.7, the $k = 3$ sample means are $\bar{x}_1 = 7$, $\bar{x}_2 = 9$, and $\bar{x}_3 = 15$, and the mean of all N = 10 observations is $\bar{x} = 10$.

$$\begin{aligned}\text{SS Groups} &= n_1(\bar{x}_1 - \bar{x})^2 + n_2(\bar{x}_2 - \bar{x})^2 + n_3(\bar{x}_3 - \bar{x})^2 \\ &= 4(7 - 10)^2 + 3(9 - 10)^2 + 3(15 - 10)^2 = 114\end{aligned}$$

$$\text{MS Groups} = \frac{\text{SS Groups}}{k - 1} = \frac{114}{3 - 1} = 57$$

$$\begin{aligned}\text{SS Total} &= (7 - 10)^2 + (9 - 10)^2 + (5 - 10)^2 + (7 - 10)^2 \\ &\quad + (9 - 10)^2 + (11 - 10)^2 + (7 - 10)^2 \\ &\quad + (15 - 10)^2 + (12 - 10)^2 + (18 - 10)^2 = 148\end{aligned}$$

$$\text{SS Error} = \text{SS Total} - \text{SS Groups} = 148 - 114 = 34$$

$$\text{MSE} = \frac{\text{SS Error}}{N - k} = \frac{34}{10 - 3} = 4.857$$

$$F = \frac{\text{MS Groups}}{\text{MSE}} = \frac{57}{4.857} = 11.74 \quad \text{with } k - 1 = 2 \text{ and } N - k = 7 \text{ df}$$

Figure 16.8 shows Minitab output for this example. The analysis of variance table format used is standard for most software. The line labeled "Programs" gives information about SS Groups and MS Groups and includes the F-statistic and the p-value. The label for this line is the name the user gives the explanatory variable when the dataset is created. In some instances, "Factor" may be used as a generic label for the explanatory variable. The line labeled "Error" gives information about SS Error and MSE. At the bottom of the output, the pooled standard deviation is given as 2.204. This can be computed as $\sqrt{\text{MSE}} = \sqrt{4.86}$.

```
Analysis of Variance
Source      DF       SS       MS        F       P
Programs     2   114.00    57.00    11.74   0.006
Error        7    34.00     4.86
Total        9   148.00
                                  Individual 95% CIs For Mean
                                  Based on Pooled StDev
Level        N     Mean    StDev  ----------+----------+----------+-----
Program1     4    7.000    1.633   (------*------)
Program2     3    9.000    2.000      (-------*-------)
Program3     3   15.000     3.00                 (-------*-------)
                                  ----------+----------+----------+-----
Pooled StDev =    2.204                    8.0       12.0       16.0
```

Figure 16.8 Minitab output for Example 16.8

Example 16.9 **Top Speeds of Supercars** Kitchens (1998, p. 783) gives data gathered by *Car and Driver* magazine on the top speeds of five supercars from five different countries: Acura NSX-T from Japan, Ferrari F355 from Italy, Lotus Esprit S4S from Great Britain, Porsche 911 Turbo from Germany, and Dodge Viper RT/10 from the United States. The data represent the top speeds for six runs on each car, using as much distance as necessary without exceeding the engine's redline. To cancel grade or wind effects, there were three runs in each direction on the test facility. Figure 16.9 includes a dotplot of the data, as well as output for a one-way analysis of variance that compares the five cars. Some important features of the results shown in the figure follow:

- The p-value is .000, so we can reject the null hypothesis that the population mean speeds are the same for all five cars.
- $F = 25.15$. The calculation was $F = \text{MS Cars}/\text{MS Error} = 364.1/14.5$. An F-distribution with 4 and 25 df was used to determine the p-value.
- The necessary conditions for doing an ANOVA F-test are present. The data are not skewed and there are no extreme outliers. The largest sample standard deviation (5.02 for the Viper) is not more than twice as large as the smallest standard deviation (2.92 for the Acura).
- MS Error = 14.5 is an estimate of the variance of the top speed for the hypothetical distribution of all possible runs with one car. The estimated standard deviation for each car is $\sqrt{\text{MS Error}} = \sqrt{14.5} = 3.81$. This value is given as the pooled standard deviation at the bottom of the output.
- On the basis of the sample means and corresponding confidence intervals for population means, the Porsche and Ferrari seem to be faster on average than the other three cars.

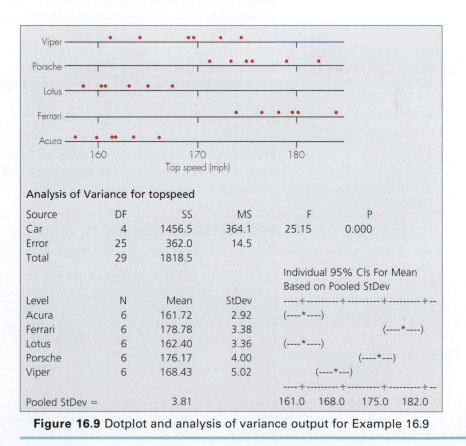

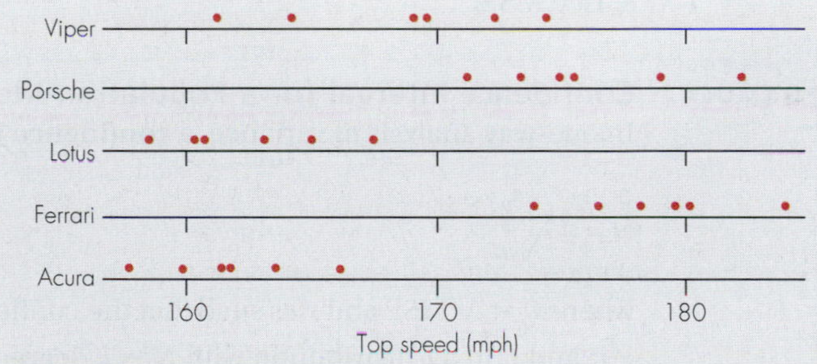

Analysis of Variance for topspeed

Source	DF	SS	MS	F	P
Car	4	1456.5	364.1	25.15	0.000
Error	25	362.0	14.5		
Total	29	1818.5			

```
                                 Individual 95% CIs For Mean
                                 Based on Pooled StDev
Level      N     Mean    StDev   ----+---------+---------+---------+--
Acura      6   161.72     2.92   (----*----)
Ferrari    6   178.78     3.38                          (----*----)
Lotus      6   162.40     3.36   (----*----)
Porsche    6   176.17     4.00                     (----*---)
Viper      6   168.43     5.02             (----*---)
                                 ----+---------+---------+---------+--
Pooled StDev =     3.81          161.0     168.0     175.0     182.0
```

Figure 16.9 Dotplot and analysis of variance output for Example 16.9

IN SUMMARY General Format of a One-Way ANOVA Table

The general format of the one-way analysis of variance table is

Source	df	SS	MS	F
Between groups (due to factor)	$k-1$	SS Groups $= \sum_{\text{groups}} n_i(\bar{x}_i - \bar{x})^2$	$\dfrac{\text{SS Groups}}{k-1}$	$F = \dfrac{\text{MS Groups}}{\text{MSE}}$
Error (within groups)	$N-k$	SSE $= \sum_{\text{groups}} (n_i - 1)s_i^2$	$\dfrac{\text{SSE}}{N-k}$	
Total	$N-1$	SSTO $= \sum_{\text{values}} (x_{ij} - \bar{x})^2$		

Computation of 95% Confidence Intervals for the Population Means

It is informative to examine and compare confidence intervals for the population means. In Figure 16.9 (p. 629) we see that the Minitab output for one-way analysis of variance includes a graph showing a 95% confidence interval for each population mean. These intervals are computed by using the same general format we learned in Chapter 11 for a confidence interval for a mean:

Sample mean ± Multiplier × Standard error

The sample mean for the i-th group is $\bar{x}_i$. For the i-th sample mean, the standard error is $\text{s.e.}(\bar{x}_i) = s_p/\sqrt{n_i}$. Note that the pooled standard deviation is used regardless of group, but the sample size is specific to the group. The "multiplier" is determined by using a t-distribution as described in Section 11.1. But in this situation df $= N - k$, the degrees of freedom that accompany MSE, because the estimated standard deviation s_p is $\sqrt{\text{MSE}}$.

FORMULA

Confidence Interval for a Population Mean

In one-way analysis of variance, a **confidence interval for a population mean** μ_i is

$$\bar{x}_i \pm t^* \frac{s_p}{\sqrt{n_i}}$$

where $s_p = \sqrt{\text{MSE}}$ and t^* is such that the confidence level is the probability between $-t^*$ and t^* in a t-distribution with $N - k$ degrees of freedom.

Example 16.10 **95% Confidence Intervals for Mean Car Speeds** The pooled standard deviation for Example 16.9 (p. 629) is $s_p = 3.81$ mph, a value displayed at the bottom of Figure 16.9. For each car, $n_i = 6$, so the standard error of the mean for any car is $\text{s.e.}(\bar{x}_i) = 3.81/\sqrt{6}$. The df for the t^* multiplier are $N - k = 30 - 5 = 25$. From the table of t^* multipliers (Table A.2), we learn that for a .95 confidence level and df $= 25$, the multiplier is $t^* = 2.06$. A 95% confidence interval for any mean is $\bar{x}_i \pm 2.06(3.81/\sqrt{6})$, or $\bar{x}_i \pm 3.02$. For example, the 95% confidence interval for the mean speed for the Acura is 161.72 ± 3.02 mph, for the Ferrari it is 178.78 ± 3.02 mph, and so on.

16.2 Exercises are on pages 642–643.

THOUGHT QUESTION 16.2 In Example 16.10, each 95% confidence interval had the same width. Why did this happen? When would the 95% confidence intervals have different widths?*

***HINT:** In the formula given just before Example 16.10, which of the quantities affecting interval width may or may not differ from group to group?

IN SUMMARY Steps for the *F*-Test for Comparing Several Means

Step 1: *Determine the* null *and* alternative *hypotheses.*

- *Null hypothesis:* H_0: $\mu_1 = \mu_2 = \ldots = \mu_k$
- *Alternative hypothesis:* H_a: The μ's are not all equal.

Step 2: *Verify necessary data conditions, and if met, summarize the data into an appropriate* test statistic. The *F*-statistic can be used if the data are not extremely skewed, there are no extreme outliers, and the group standard deviations are not markedly different. A criterion for standard deviations is that the largest of the sample standard deviations should not be more than twice as large as the smallest of the sample standard deviations. The test statistic is

$$F = \frac{\text{MS Groups}}{\text{MS Error}} = \frac{\dfrac{\sum n_i(\bar{x}_i - \bar{x})^2}{k-1}}{\dfrac{\sum (n_i - 1)s_i^2}{N-k}}$$

Step 3: *Assuming that the null hypothesis is true, find the* p-*value.* Using the *F*-distribution with numerator df $= k - 1$ and denominator df $= N - k$, the *p*-value is the area in the tail to the right of the test statistic *F*.

Step 4: *Decide whether the result is* statistically significant *based on the* p-*value.* Choose a significance level ("alpha"); standard is $\alpha = .05$. The result is statistically significant if the *p*-value $\leq \alpha$.

Step 5: *Report the conclusion in the context of the situation.*

Note: In analysis of variance, when the null hypothesis is rejected these steps are usually followed by a multiple comparison procedure to determine which group means are significantly different from each other.

16.3 Other Methods for Comparing Populations

You probably will not be surprised to learn that the necessary conditions for using an analysis of variance *F*-test do not hold for all datasets. In this section, we discuss methods that can be used when one or both of the assumptions about equal population standard deviations and normal distributions are violated. It is important to remember that no inference method is appropriate if the observed data do not represent the population for the question of interest.

Example 16.11 **Drinks per Week and Seat Location** In the two surveys described for Example 16.1 (p. 618), students were asked, "How many alcoholic beverages do you consume each week?" Let's compare responses for the same three groups compared in Example 16.1: students who typically sit in the front, middle, or back of a classroom. Table 16.1 contains summary statistics for the three seat locations and Figure 16.10 displays a boxplot of the data. The sample sizes are a bit different here than in Example 16.1 because some students did not give a response to every question.

Table 16.1 Summary Statistic by Seat Location for Number of Alcoholic Beverages per Week

Location	*n*	Mean	Median	St. Dev.
Front	87	1.6	0	3.4
Middle	207	3.9	1	6.5
Back	79	8.5	5	10.5

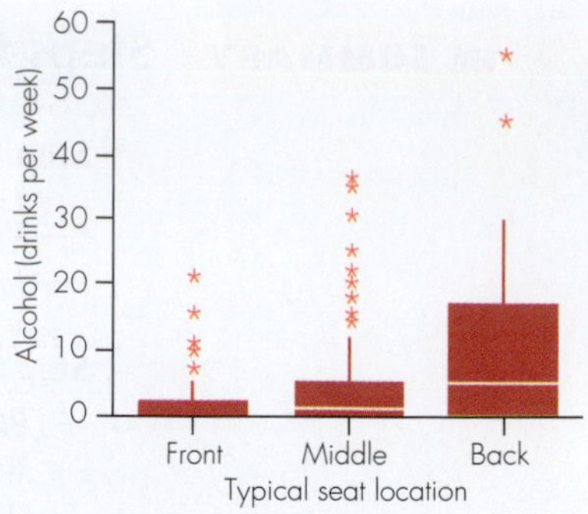

Figure 16.10 Boxplot by seat location for number of alcoholic beverages per week

The information in Figure 16.10 and Table 16.1 reveals that the students who typically sit in the back reported drinking more alcohol than the other students did. The information in the figure and table also shows that the group standard deviations differ, and the data appear to be skewed. The boxplot shows much more variation in the responses of the back of the classroom group than in the other two groups. In Table 16.1, we see that the standard deviation in the "back" group (10.5) is more than twice as large as the standard deviation in the "front" group (3.4). Also, the mean is greater than the median in each group, which is evidence of skewness in the data. The necessary conditions for doing an analysis of variance are violated in this dataset. We will use other methods to analyze these data in Examples 16.12 and 16.13.

Hypotheses about Medians

When the observed data are skewed or when extreme outliers are present, it usually is better to analyze the median rather than the mean. This is also what should usually be done if the response variable is an ordinal variable. When several population medians are compared, null and alternative hypotheses of interest are the following:

H_0: Population medians are equal.

H_a: Population medians are not all equal.

The notation used for a population median varies from author to author. The most commonly used symbol is η (pronounced "eta"). With this notation, the null hypothesis about the medians of k populations is H_0: $\eta_1 = \eta_2 = \ldots = \eta_k$. The letter M is generally used to denote a sample median. In considering several samples, the median of the i-th sample would be written M_i.

Kruskal-Wallis Test for Comparing Medians

One test for comparing medians is the **Kruskal-Wallis test**. It is based on a comparison of the relative rankings (sizes) of the data in the observed samples, and for this reason is called a **rank test**. The term **nonparametric test** also is used to describe this test because there are no assumptions made about a specific distribution for the population of measurements.

The general idea is that values in the total dataset of N observations are ranked from lowest to highest (lowest = 1, highest = N). The ranks of the values are averaged for each group, and the test statistic measures the variation among the average ranks of the groups. If most of the small data values were in one particular group, for example, that group would have a lower average rank than the other groups. A p-value can be determined by finding the probability that the variation among the set of rank averages for the groups would be as large (or larger) as it is if the null hypothesis is true. More details about the Kruskal-Wallis test as well as other nonparametric methods are given in Supplemental Topic 2 on the companion website, http://www.cengage.com/statistics/Utts4e.

Example 16.12 **Kruskal-Wallis Test for Alcoholic Beverages per Week by Seat Location** Figure 16.11 contains output from Minitab for a Kruskal-Wallis test of the null hypothesis that the *median* number of alcoholic beverages consumed per week is the same for the three classroom seat locations. The *p*-value shown at the bottom of the output is P = 0.000, which indicates strong evidence against the null hypothesis. You'll see that two *p*-values are shown; the second is followed by the phrase "adjusted for ties." There is a "tie" in the rankings when two or more observations have the same value. For example, all students who reported 0 drinks per week are tied with each other. It is possible to determine an adjustment to the *p*-value based on the number of ties that occur. In practice, the adjusted and unadjusted values rarely differ by much.

```
Kruskal-Wallis test on alcohol

Seat        N      Median      Ave Rank       Z
Front      87    0.00E + 00      143.7    -4.27
Middle    207    1.00E + 00      186.6    -0.68
Back       79    5.00E + 00      243.6     5.25

Overall   373                    187.0
H = 35.98   DF = 2     P = 0.000
H = 40.18   DF = 2     P = 0.000 (adjusted for ties)
```

Figure 16.11 Kruskal-Wallis test for comparing median number of alcoholic beverages per week in three classroom locations

The output in Figure 16.11 includes the sample median and the average rank within each group. Oddly, the medians are given by Minitab in scientific notation. Note that the median number of alcoholic beverages per week for the "back" group is 5, which is clearly higher than the medians for the other two groups. Because the reported data values tended to be higher in the back group, the average rank is highest for this group. The data values tended to be lowest for the front group (median = 0), so the average rank is lowest for that group.

The Kruskal-Wallis statistic measures the variation among the average ranks shown for the three groups much like the *F*-statistic measures the variation among the sample means in analysis of variance.

Mood's Median Test for Comparing Medians

Another *nonparametric* test used to compare population medians is **Mood's median test**. The idea is easy to grasp. First, the median (M) of the total dataset with all groups combined is determined. Then, within each group, the number of observations less than or equal to M and the number of observations greater than M are counted. These counts can be displayed in the two-way table shown in Table 16.2. A chi-square statistic for two-way tables (Chapter 15) is used to test the null hypothesis that the population medians are the same. The test is equivalent to testing that the two variables creating the rows and columns in Table 16.2 are not related. A statistically significant result indicates that the medians are not all the same.

Table 16.2 Format of Two-Way Table for Mood's Median Test

Group	Number ≤ M	Number > M
1		
2		
⋮		
k		

M = median of all data in overall dataset.

THOUGHT QUESTION 16.3 Show why the null hypothesis of equal population medians is equivalent to the null hypothesis that the two variables in Table 16.2 are not related, as follows. Assume for simplicity that no values are tied with the sample median M. In that case, fill in the marginal totals for Table 16.2 using the notation for group sizes of $n_1, n_2, \ldots, n_k$; $N = n_1 + n_2 \ldots + n_k$. What would be the expected cell counts for the test of the null hypothesis that the two variables are unrelated? Explain why those are the same cell counts that would be expected under the null hypothesis that all population medians are equal.*

Example 16.13 **Mood's Median Test for the Alcoholic Beverages and Seat Location Example** Output from Minitab for Mood's median test in the alcoholic beverages consumed per week by classroom seat location example is shown in Figure 16.12. The first line shows that the *p*-value for the chi-square test is .000, indicating that the null hypothesis of equal population medians can be rejected. Sample medians for "front," "middle," and "back" groups are given as 0, 1, and 5 drinks per week, respectively. Confidence intervals for the three population medians are given graphically. It looks reasonable to say that the population median is greatest for those who prefer to sit in the back.

```
Mood Median Test for alcohol

Chi-Square = 26.54        DF = 2        P = 0.000

                                                      Individual 95.0% CIs
Seat        N<=      N>     Median     Q3–Q1    +---------+---------+---------+------
Front        62      25       0.00      2.00    +
Middle      115      92       1.00      5.00    (--*---)
Back         25      54       5.00     12.00              (------*---------------)
                                                +---------+---------+---------+------
                                                0.0      3.0       6.0       9.0

Overall median = 1.00
```

Figure 16.12 Mood's median test for the alcoholic beverages per week by seat location

The overall median of all responses is shown at the bottom of the output as "Overall median = 1.00." In the central portion of the output, information is given about how many responses in each seat location group were less than or equal to this value and how many were greater than this value. In the "front" group, for instance, 62 students reported consuming one or fewer alcoholic beverages per week and 25 reported drinking more than this. A much different pattern is evident in the "back" group. In that group, 25 reported consuming one or fewer drinks per week compared to 54 who reported drinking more than one drink per week.

16.3 Exercises are on pages 643–644.

MINITAB TIP **Nonparametric Procedures for Comparing Several Medians**

- The Kruskal-Wallis and Mood's median tests are listed in the menu that results from using **Stat > Nonparametrics**.

***HINT:** If the null were true, what fraction of each group would be on either side of the overall median? See Section 15.1 to review expected counts for a two-way table.

TECHNICAL NOTE

Transforming the Response Variable for Analysis of Variance

When the group standard deviations and the group means are related, it may be possible to transform the values of the response variable so that the necessary conditions for ANOVA are satisfied on the transformed scale. An analysis of variance is then done on the transformed data. Three transformations that frequently work for this purpose are square root, logarithm, and reciprocal. For instance, if the ratio $s_i / \bar{x}_i$ is about the same for all groups, then the values of log(response) tend to have equal standard deviations for all groups. Using this "log transformation" frequently works for financial data, such as salaries, for which the variability increases as the mean increases.

The type of pattern for which a transformation might work is present in Example 16.11. Note that sample means and sample standard deviations are ordered in the same way over the three groups. In this example, if x = alcoholic beverages per week, the group standard deviations are nearly equal for the transformed response $1/(0.5 + x)$. (Adding 0.5 to x avoids dividing by 0.) It is difficult, however, to interpret the transformed responses, so a comparison of medians may be preferable.

16.4 Two-Way Analysis of Variance

A **two-way analysis of variance** is used to examine how two categorical explanatory variables affect the mean of a quantitative response variable. For example, an industrial psychologist may want to see how type of background music and loudness of background music affect productivity in a workplace. Or an economist may be interested to learn the effects of sex (male, female) and race on mean income. In problems like these, there is interest in the effect of each separate explanatory factor, and there is interest in the combined effect of the two explanatory factors.

When there is an **interaction** between two explanatory variables, the effect on the response variable of one explanatory variable depends on the specific value or level present for the other explanatory variable. A statement like "being overweight caused greater increases in blood pressure for men than for women" is a statement describing an interaction between weight and sex. In this statement, the way that weight (overweight or not) affects blood pressure depends upon sex (male or female).

Suppose that an instructor wants to see if students take longer to finish a final exam if it's open book or if it's closed book. She gives her final exam as an open-book exam one semester and as a closed-book exam the next semester. She also categorizes students by whether or not they have an A going into the final. Table 16.3 gives a hypothetical set of means for minutes needed to finish the exam, classified by type of exam and grade before the final.

Table 16.3 Mean Time (in Minutes) to Complete Exam by Grade and Type of Exam

	Exam type	
Grade before Final	**Open-book**	**Closed-book**
A	80	100
Not A	110	90

The average of the two grade-level groups is 95 minutes for both types of exam, so if we compared only the average times we might conclude that students complete the two types of exams in equal time. But that wouldn't tell the full story. There is an interaction between exam type and grade before the final. The A students took 20 minutes less, on average, to finish the open-book exam compared to the closed-book exam. Conversely, the non-A students took 20 minutes more to finish the open-book exam compared to the closed-book exam. Possibly the A students were able to make efficient use of the book, but took longer to work things out without the book. And, the non-A students spent time trying to learn things with the open-book exam, but finished the closed-book exam more quickly because they didn't know how to do some of the problems.

The term **main effect** describes the overall effect of a single explanatory variable on the response variable. A main effect has to do with the differences in the mean responses for the categories of a single explanatory variable. In the "type of exam" and "grade before final" example just described, the main effect of the explanatory variable "type of exam" is the difference between overall mean completion times for the closed book versus open book exams. The main effect of "grade before final" is the difference between overall mean completion times for A students versus non-A students.

Note that the main effect for "type of exam" is 0, because on average, the means are the same for the two types of exams (95 minutes for each). The main effect for "grade before final" is not 0, because the A students completed the exam in an average of 90 minutes, while the average for non-A students was 100 minutes. However, a main effect may not be meaningful if an interaction is present. Here, it would not be useful to report that the mean time to complete the exam was the same for the two types of exams. Instead we should report that the difference between the mean completion times for the two types of exams was different for the A students than for the non-A students.

Example 16.14 **Happy Faces and Restaurant Tips** When a restaurant server writes a friendly note or draws a "happy face" on your restaurant check, is this just a friendly act, or is there a financial incentive? Temple University psychologists conducted a randomized experiment to investigate whether drawing a happy face on the back of a restaurant bill increased the average tip given to the server (Rind and Bordia, 1996). One female server and one male server in a Philadelphia restaurant either did or did not draw a happy face on checks during the experiment. In all, they drew happy faces on 45 checks (22 for the female, 23 for the male) and did not draw happy faces on 44 other checks (23 for the female and 21 for the male). The sequence of drawing the happy face or not was randomized in advance.

Figure 16.13 is a graph of the mean tip percentage for each of the four combinations of sex of the server (female or male) and check message (none or happy face drawing). The type of graph shown is called an **interaction plot**. The mean response (tip percentage) is graphed on the vertical axis and message type (none or happy face) is graphed on the horizontal axis. The higher line in the figure is for the female server (she got better tips than the male). We see that her mean tip percentage was about 5% higher for the "happy face" checks. The lower line in the figure shows the results for the male server, and for him, drawing a happy face decreased the mean tip percentage by about 4%. The effect of drawing a happy face depended on whether the server was female or male, so there is an interaction between sex of server and message type. The researchers speculated that customers might have felt that drawing a happy face was not gender appropriate for males.

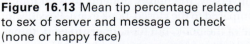

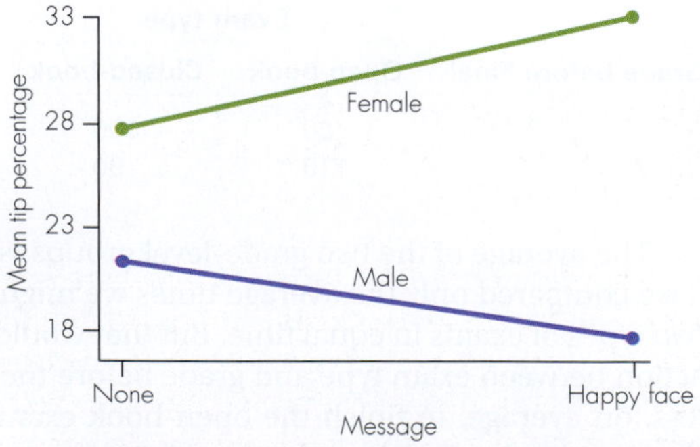

Figure 16.13 Mean tip percentage related to sex of server and message on check (none or happy face)

THOUGHT QUESTION 16.4 In Example 16.14, there was only one server of each sex. What problem does this cause in the interpretation and generalization of the results? How would you have designed the experiment to better examine the interaction between sex of the server and drawing a happy face (or not)?*

Example 16.15 **You've Got to Have Heart** Psychologist Lee Salk (May 1973) observed 287 mothers within 4 days after giving birth and found that 83% of the right-handed mothers and 78% of the left-handed mothers held their babies on the left side. When asked why they chose the left side, the right-handed mothers said that it was so their right hand would be free. The left-handed mothers said that it was because they could hold the baby better with their dominant hand. Salk speculated that "it is not in the nature of nature to provide living organisms with biological tendencies unless such tendencies have survival value." He surmised that there must be survival value to placing a newborn infant close to the sound of the mother's heart.

To test his conjecture, Salk arranged for a baby nursery at a New York City hospital to have the continuous sound of a human heartbeat played over a loudspeaker. At the end of 4 days, he measured how much weight $n = 102$ babies in the nursery had gained or lost. Later, with a new group of $n = 112$ babies in the nursery, no sound was played, and weight gains (or losses) were again measured after 4 days. Because initial birth weight affects weight gain or loss in the early stages of life, Salk also categorized the babies into three birth-weight categories.

In all, there are six possible combinations of experimental group (heartbeat or not) and birth weight (low, medium, high). Figure 16.14 is an interaction plot of the sample mean weight gain or loss (in grams) for these combinations. (Salk did not give means; we have estimated them from the data shown in a dotplot in his *Scientific American* report.) Considering the purpose of the investigation, the most important finding revealed by Figure 16.14 is that the weight gain was generally greater for the heartbeat group. This confirmed what Salk had suspected. Although they did not eat more than the control group, the infants who were treated to the sound of the heartbeat gained more weight (or lost less). Furthermore, they spent much less time crying. Salk's conclusion was that "newborn infants are soothed by the sound of the normal adult heartbeat." Somehow, mothers intuitively know that it is important to hold their babies on the left side.

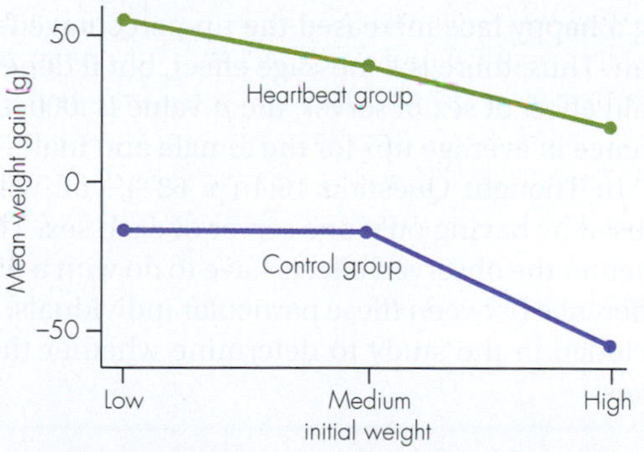

Figure 16.14 Mean weight gain (or loss) by initial weight and experimental conditions (heartbeat or not)

The difference in weight gain evident in Figure 16.14 for all three birth-weight groups indicates that there is a *main effect* for the explanatory variable of "heartbeat or control group." The difference between the mean weight gain or loss in the heartbeat and control groups is roughly the same in each of the birth-weight groups. This is evident from the approximately equal distance between the two lines (heartbeat and

***HINT:** Do the results reflect a general difference between females and males or simply the difference between two servers?

control) shown in Figure 16.14 regardless of initial weight. This similarity across groups indicates that there may not be an *interaction* between the experimental condition variable and initial birth-weight group. In this example, it is sensible to report the main effect of heartbeat versus control averaged over birth-weight groups.

Two-Way Analysis of Variance and *F*-Tests

In a two-way analysis of variance, three *F*-statistics are constructed. One is used to test the statistical significance of the interaction, while the other two are used to test the significance of the two separate main effects. The computational details and a more extensive discussion of two-way ANOVA are in Supplemental Topic 4 on the companion website, http://www.cengage.com/statistics/Utts4e.

Example 16.16 **Two-Way Analysis of Variance for Happy Face Example** The Minitab output in Figure 16.15 is for a two-way analysis of variance of the data for the happy face and restaurant tip example. (Professor Bruce Rind of Temple University provided the raw data.) Consider only the *p*-values shown in the last column of the table, and don't worry about the other details.

Source	DF	Adj SS	Adj MS	F	P
Message	1	14.7	14.7	0.13	0.715
Sex	1	2602.0	2602.0	23.69	0.000
Interaction	1	438.7	438.7	3.99	0.049
Error	85	9335.5	109.8		
Total	88	12407.9			

Figure 16.15 Two-way analysis of variance of restaurant tipping data

The significance test of the effect of "Message" (happy face or none) has a *p*-value of .715, indicating a nonsignificant effect of message type. That result, however, must be interpreted carefully due to the nature of the interaction between message type and sex of server. The interaction, which is significant with a *p*-value of .049, was that drawing a happy face increased the tip percentage for the woman but decreased it for the man. Thus, there is a message effect, but it depends upon the sex of the server. For the main effect of sex of server, the *p*-value is .000, indicating a statistically significant difference in average tips for the female and male.

In Thought Question 16.4 (p. 637), you were asked to think about the problem caused by having only one server of each sex. The answer is that we cannot be certain whether the observed effects have to do with a difference between the sexes or with the difference between these particular individuals. Multiple servers of each sex should be included in the study to determine whether there really is a difference between the sexes.

16.4 Exercises are on pages 644–646.

MINITAB TIP **Drawing an Interaction Plot**

- To create an interaction plot of means, use **Stat > ANOVA > Interactions Plot**. Enter the column name for the quantitative response variable in the "Responses" box. Enter the names of the two explanatory variables in the "Factors" box.

Key Terms

Section 16.1

Section 16.2

Section 16.3

Section 16.4

In Summary Boxes

Exercises

◆ Denotes that the dataset is available on the companion website, http://www.cengage.com/statistics/Utts4e, but is not required to solve the exercise.

Bold exercises have answers in the back of the text.

Section 16.1

Skillbuilder Exercises

16.1 In each situation, determine whether one-way analysis of variance could be an appropriate method for analyzing the data described. Briefly explain why or why not for each part.

a. A researcher compares the mean blood pressures of men over 50 years old for three different ethnic groups. He samples 100 men in each ethnic group, and measures their blood pressures.

b. Fifty individuals all listen to five songs, and rate each song on a scale of 0 to 100. The mean scores for the five songs are compared.

16.2 In each situation, determine whether one-way analysis of variance could be an appropriate method for analyzing the data described. Briefly explain why or why not for each part.

a. A psychiatrist compares four treatment programs for clinical depression. The response variable is whether or not a patient shows improvement after 2 months of treatment. A randomized experiment is done in which a different group of 20 patients is randomly allocated to each treatment.

b. Three methods for memorizing information are compared. Forty-five participants are randomly divided into three groups, and each group memorizes information with a different method. All participants then take a test on the memorized information, and the scores are used to compare the methods.

16.3 Researchers studying the connection between body weight and age report the following finding: "The *p*-value was .003 for a one-way analysis of variance done to compare body mass index values for the three age groups."

a. What null hypothesis did the researchers test? Write the null hypothesis in words, and also write it using proper statistical notation.

b. Explain what conclusion can be made about the comparison of body mass index values for the three age groups.

16.4 ◆ The output for this exercise is an analysis of variance comparing hours spent watching television in a typical day for five categories of highest educational degree attained for

$n = 1317$ respondents in the 2008 General Social Survey. Educational degree groups are graduate, bachelor's, junior college, high school, and less than high school (*Data source:* **GSS-08** dataset on the companion website).

Analysis of Variance for tvhours

Source	DF	SS	MS	F	P
degree	4	562.32	140.58	29.27	0.000
Error	1312	6301.68	4.80		
Total	1316	6863.99			

Level	N	Mean
NotHS	184	4.065
HS	671	3.066
JunColl	117	2.547
Bachelor	225	2.120
Graduate	120	1.808

a. Write null and alternative hypotheses for this situation. Write the hypotheses in words, and also write the null hypothesis using proper notation.
b. What are the values of the F-statistic and the p-value for testing the hypotheses written in part (a)? What conclusion can be reached about the hypotheses?
c. Compare the sample means given in the output below the analysis of variance table. Describe how mean hours of watching television per day relates to highest degree attained.

16.5 ◆ For $n = 153$ female students in the **UCDavis2** dataset on the companion website, mean height (inches) by student's preferred seat location in a classroom is shown in the table below.

Location	Mean
Front ($n = 38$)	63.86
Middle ($n = 93$)	64.80
Back ($n = 22$)	66.55

a. The p-value is .001 for an F-test that compares the mean heights of female students in the three seating locations. In the context of this situation, what conclusion can be made?
b. Describe how mean height of female students in the dataset relates to preferred seating location.

16.6 In each situation, use Table A.4 to find a critical value and then state a conclusion for an F-test of the null hypothesis of equal population means:

a. F-statistic $= 3.27$ with 3 and 20 degrees of freedom; $\alpha = .05$.
b. F-statistic $= 3.27$ with 3 and 20 degrees of freedom; $\alpha = .01$.

16.7 In each situation, use Table A.4 to find a critical value and then state a conclusion for an F-test of the null hypothesis of equal population means:

a. F-statistic $= 6.27$ with 2 and 12 degrees of freedom; $\alpha = .05$.
b. F-statistic $= 1.63$ with 5 and 70 degrees of freedom; $\alpha = .05$.

General Section Exercises

16.8 ◆ In the 2002 General Social Survey, a randomly selected sample of U.S. adults was asked what they think is the ideal number of children for a couple to have. The output for this exercise is for a one-way analysis of variance that compares the mean response in four age groups: 18–29, 30–44, 45–59, and 60+ (*Data source:* **GSS-02** dataset on the companion website).

```
Analysis of variance for chldidel
Source   DF       SS      MS     F      P
agegrp    3   17.848   5.949  7.41  0.000
Error   805  646.350   0.803
Total   808  664.198
                                Individual 95% CIs For Mean
                                Based on Pooled StDev
Level    N     Mean   StDev  +---------+---------+---------+---------
18-29  178   2.6067  0.8909                  (--------*--------)
30-44  241   2.3568  0.8198  (------*------)
45-59  207   2.3671  0.8979  (-------*-------)
60-89  183   2.6995  0.9903                        (--------*--------)
                             +---------+---------+---------+---------
                                    2.25      2.40      2.55      2.70
Pooled StDev = 0.8961
```

a. Write null and alternative hypotheses for this problem. State the hypotheses in words and also state them using statistical notation. To what populations do these hypotheses apply?
b. What values are given in the output for the F-statistic and the p-value? What conclusion can be reached about the hypotheses you wrote in part (a)?
c. Based on the 95% confidence intervals given for the population means, describe the differences among the age groups.
d. Is the assumption of equal population standard deviations reasonable for these data? Briefly explain.

16.9 Refer to Exercise 16.8 about the ideal number of children for a couple to have. The following output is for a Tukey procedure done to examine all pairwise comparisons of the four age groups. A 95% family confidence level was used.

Tukey 95% Simultaneous Confidence Intervals
All Pairwise Comparisons among Levels of agegroup

agegroup = 18–29 subtracted from:

agegroup	Lower	Center	Upper
30–44	−0.4772	−0.2499	−0.0226
45–59	−0.4747	−0.2396	−0.0045
60+	−0.1494	0.0927	0.3348

agegroup = 30–44 subtracted from:

Agegroup	Lower	Center	Upper
40–59	−0.2077	0.0103	0.2283
60+	0.1171	0.3426	0.5681

agegroup = 45–59 subtracted from:

Agegroup	Lower	Center	Upper
60+	0.0989	0.3323	0.5657

a. Which pairs of age groups exhibit statistically significant differences? In one or two sentences, summarize the differences in mean response for the four age groups.

b. Briefly explain what it means to say that a 95% family confidence level was used for the six confidence intervals shown in the output for this problem.

16.10 Refer to Exercise 16.5 about female height and preferred classroom seating location. The following output from Minitab gives confidence intervals for all pairwise comparisons of seating locations. A Tukey procedure with 95% family confidence level was used.

```
Tukey 95% Simultaneous Confidence Intervals
All Pairwise Comparisons among Levels of Seat
Seat = Front subtracted from:
Seat        Lower      Center     Upper
Middle     −0.261       0.940     2.142
Back        1.018       2.690     4.362
Seat = Middle subtracted from:
Seat        Lower      Center     Upper
Back        0.270       1.750     3.230
```

a. What confidence interval is given for the difference between the mean heights of female students who prefer to sit in the back versus those who prefer to sit in the front? Explain whether the interval gives evidence that the population means differ for these two locations.

b. What confidence interval is given for the difference between the mean heights of female students who prefer to sit in the back versus those who prefer to sit in the middle? Explain whether the interval gives evidence that the population means differ for these two locations.

c. What confidence interval is given for the difference between the mean heights of female students who prefer to sit in the front versus those who prefer to sit in the middle? Explain whether the interval gives evidence that the population means differ for these two locations.

16.11 To evaluate a standard test of the flammability of fabric for children's sleepwear, the American Society for Testing Materials had five laboratories each test 11 pieces of the same type of fabric. The response variable is the length of the char mark made when the fabric sample is held over a flame for a specified time period. Ideally, all labs using the standardized test to test the same fabric should observe about the same value for this response variable. Ryan and Joiner (2001, p. 269) give the following data for the observed char lengths. (The measurement unit was not specified by Ryan and Joiner.)

Lab 1	2.9	3.1	3.1	3.7	3.1	4.2	3.7	3.9	3.1	3.0	2.9
Lab 2	2.7	3.4	3.6	3.2	4.0	4.1	3.8	3.8	4.3	3.4	3.3
Lab 3	3.3	3.3	3.5	3.5	2.8	2.8	3.2	2.8	3.8	3.5	3.8
Lab 4	3.3	3.2	3.4	2.7	2.7	3.3	2.9	3.2	2.9	2.6	2.8
Lab 5	4.1	4.1	3.7	4.2	3.1	3.5	2.8	3.5	3.7	3.5	3.9

a. Graph the data in a way that is useful for comparing the laboratories. Describe any differences among laboratories that can be identified in your graph.

b. Use the graph drawn for part (a) to assess whether the necessary conditions are present for using one-way analysis of variance to compare the five laboratories.

16.12 Refer to Exercise 16.11 about testing the flammability of children's sleepwear. Output for a one-way analysis of variance of the data is the following:

```
Analysis of Variance for Charlng
Source  DF      SS      MS     F      P
Lab      4   2.987   0.747  4.53  0.003
Error   50   8.233   0.165
Total   54  11.219
                              Individual 95% CIs For Mean
                              Based on Pooled StDev
Level    N    Mean   StDev  --------+---------+---------+---------+-
1       11  3.3364  0.4523             (-------*-------)
2       11  3.6000  0.4604                     (-------*-------)
3       11  3.3000  0.3715            (-------*-------)
4       11  3.0000  0.2864  (-------*-------)
5       11  3.6455  0.4321                      (-------*-------)
                            --------+---------+---------+---------+-
Pooled StDev = 0.4058            3.00      3.30      3.60      3.90
```

a. Write null and alternative hypotheses for comparing the mean char lengths for the five labs. State the hypotheses in words, and also state them using statistical notation. To what populations do these hypotheses apply?

b. Are there statistically significant differences among the labs? Justify your answer using the *F*-statistic and *p*-value found in the output.

c. Draw a sketch illustrating how the *p*-value was determined for this problem. See Figure 16.4 (p. 623) for guidance.

d. Based on the sample means and the 95% confidence intervals given for the population means, describe the differences among the five laboratories. Be specific about which laboratories appear to have different mean values from the others.

16.13 Koopmans (1987, p. 93) gave data on the testosterone levels (measured as milligrams per 100 milliliters of blood samples) of 46 women classified into three occupational groups: (1) not employed, (2) employed in job not requiring an advanced degree, and (3) employed in job requiring advanced degree. The data are displayed in the following dotplot.

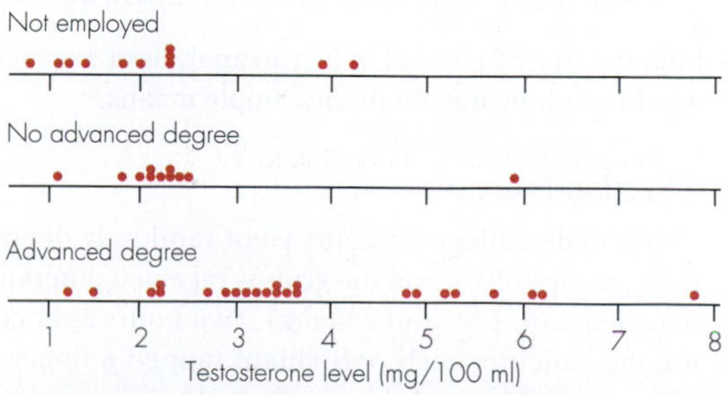

a. Write null and alternative hypotheses for comparing the mean testosterone levels in the three groups. Use proper notation.

b. What does the dotplot indicate about the necessary conditions for doing a one-way analysis of variance of the data? Do any conditions appear to be violated? If so, which condition(s)?

Section 16.2

Skillbuilder Exercises

16.14 Give a value for each of the missing elements in the following analysis of variance table.

Source	df	SS	MS	F
Between groups	2	10	—	—
Error	—	—	—	
Total	30	300		

16.15 Give a value for each of the missing elements in the following analysis of variance table.

Source	df	SS	MS	F
Between groups	5	40	—	—
Error	10	60	—	
Total	—	—		

16.16 Refer to Exercises 16.5 and 16.10 about female student height and preferred classroom seating location. Sample sizes, means, and standard deviations by seating location follow.

Location	N	Mean	Std. Dev.
Front	38	63.86	2.09
Middle	93	64.80	2.80
Back	22	66.55	2.76
All	153	64.81	

a. Show calculations verifying that SS Groups is approximately 101.

b. Use the formula SS Error $= \sum_{\text{groups}} (n_i - 1)s_i^2$ to verify that SS Error is approximately 1043.

c. The value of df for error is 150. Find the value of MSE, and find the value of the pooled standard deviation s_p.

16.17 Suppose that SS Groups = 0 in an analysis of variance. What would this indicate about the sample means?

General Section Exercises

16.18 Thirty male college students were randomly divided into three groups of 10, and the groups received different doses of caffeine (0, 100, and 200 mg). Two hours after consuming the caffeine, each participant tapped a finger as rapidly as possible, and the number of taps per minute was recorded. The data are displayed in a comparative boxplot in the following figure (*Data source:* Hand et al., 1994, dataset 50).

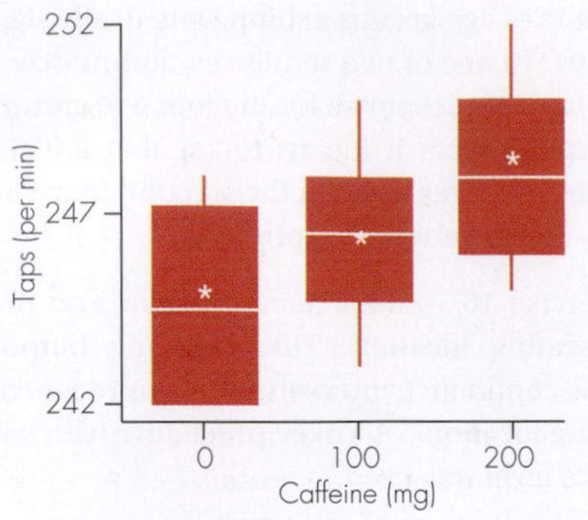

a. Write null and alternative hypotheses for comparing the mean taps per minute in the three caffeine amount groups. Use proper notation.

b. Using the figure, evaluate the necessary conditions for doing a one-way analysis of variance to compare the mean taps per minute for the three caffeine amount groups.

16.19 Refer to Exercise 16.18 about finger tapping and caffeine amount. Here is a partial analysis of variance table for the problem with some elements intentionally omitted:

Source	df	SS	MS	F	p
Caffeine	—	61.40	—	—	.006
Error	—	—	—		
Total	29	195.50			

a. Write out the completed analysis of variance table. Indicate how you determined a value for each element that is not shown in the table.

b. Calculate the pooled standard deviation, s_p. Describe what parameter is estimated by this statistic.

c. Based on the p-value given in the ANOVA table, what conclusion can be reached about the mean taps per minute in the population for the three caffeine amounts?

16.20 Refer to Exercise 16.8 about the ideal number of children for a couple to have. Assuming that the equal standard deviations assumption is valid, calculate 95% confidence intervals for each of μ_1, μ_2, μ_3, and μ_4. The output for Exercise 16.8 shows the sample means and the pooled standard deviation.

16.21 Refer to Exercises 16.11 and 16.12 about testing the flammability of children's sleepwear. Assuming that the equal standard deviations assumption is valid, calculate 95% confidence intervals for each of μ_1, μ_2, μ_3, μ_4, and μ_5. The output for Exercise 16.12 shows the sample means and the pooled standard deviation.

16.22 Suppose that four soil treatments that might be useful for improving the yield of alfalfa grown in fields are compared in an experiment. Each treatment is used in six fields, and the response variable is the crop yield per acre. Suppose further that SS Groups = 150 and SS Error = 200.

a. Write null and alternative hypotheses for comparing the mean crop yields in the four groups. Use proper notation.

b. Calculate the F-statistic for comparing the mean yield of the four treatments.
c. What are the degrees of freedom for the F-statistic?
d. The correct p-value is about .01. Draw a sketch illustrating how the p-value is found.
e. Based on the p-value given in part (d), what conclusion can be reached about the population mean yields for the four soil treatments?

16.23 Suppose that three drugs used to reduce cholesterol are compared in a randomized experiment in which three people use each drug for a month. The data for the reductions in cholesterol level for the $N = 9$ participants follow.

Drug 1	Drug 2	Drug 3
6	10	9
4	14	12
2	9	6

a. Calculate the overall mean, $\bar{x}$, and the group means, $\bar{x}_1$, $\bar{x}_2$, and $\bar{x}_3$.
b. Calculate SS Groups, the sum of squares for groups.
c. Calculate SS Total, the total sum of squares.
d. Determine SS Error, the sum of squared errors.
e. Calculate the F-statistic for comparing the mean cholesterol reductions for the three drugs. What are the degrees of freedom for this statistic?

16.24 Suppose that three drugs used to reduce systolic blood pressure are compared in a randomized experiment in which four people use each drug for a month. The data for the reductions in systolic blood pressure for the $N = 12$ participants follow.

Drug 1	Drug 2	Drug 3
7	8	16
2	12	9
5	0	5
6	4	10

a. Calculate the overall mean, $\bar{x}$, and the group means, $\bar{x}_1$, $\bar{x}_2$, and $\bar{x}_3$.
b. Calculate SS Groups, the sum of squares for groups.
c. Calculate SS Total, the total sum of squares.
d. Determine SS Error, the sum of squared errors.
e. Calculate the F-statistic for comparing the mean cholesterol reductions for the three drugs. What are the degrees of freedom for this statistic?

16.25 Refer to Exercise 16.22. Use the information given or determined in that exercise to write out a completed analysis of variance table.

Section 16.3

Skillbuilder Exercises

16.26 ◆ Refer to Exercise 16.4 about hours of watching television in a typical day and the highest educational degree attained for respondents in the 2008 General Social Survey. Minitab output for a Mood's median test of the same data follows.

Mood median test for tvhours

Chi-Square = 77.86 DF = 4 P = 0.000

degree	N<=	N>	Median
0	62	122	3.00
1	338	333	2.00
2	69	48	2.00
3	153	72	2.00
4	92	28	2.00

Overall median = 2.00

a. In the context of this situation, write the null and alternative hypotheses that are tested by Mood's median test. (Note that degree 0 is No HS, 1 is HS, etc.)
b. What is the p-value for testing the hypotheses written in part (a)? What conclusion can be reached about the hypotheses?
c. Which group had the highest median hours of television watched in a typical day? What was the median value for that group?
d. The overall median for the sample is 2 hours of watching television in a typical day. What percentage of respondents in the high school degree group (degree = 1) said that they watch television more than 2 hours in a typical day? What percentage of those who did not get a high school degree (degree = 0) watch more than 2 hours of television in a typical day?

16.27 A sample of $n = 729$ college students is asked to rate how much they like various types of music on a scale of 1 to 6 on which 1 = don't like at all and 6 = like a lot. The students also were asked whether their hometown was a big city, rural, a small town, or suburban. Results for a Kruskal-Wallis test comparing ratings of reggae music for the four types of hometown are given in the output for this exercise.

Kruskal-Wallis Test on Reggae

Hometown	N	Median	Ave Rank	Z
Big city	89	5.000	476.2	5.32
Rural	96	3.000	335.2	−1.49
Small town	176	3.000	360.3	−0.34
Suburban	368	3.000	348.2	−2.18
Overall	729		365.0	

H = 29.17 DF = 3 P = 0.000

a. Write null and alternative hypotheses for comparing the four hometown types with regard to the rating of reggae music.
b. What p-value is given in the output? Explain what conclusion can be made based on this p-value.
c. For which type of hometown was the median rating of reggae the highest? What was the median rating in that type?

16.28 In the survey that was described in Exercise 16.27, students also were asked how many times they pray per week and what they felt was the importance of religion in their life (very important, fairly important, or not very). Descriptive statistics comparing number of times praying per week for the three religious importance groups follow.

Importance	n	Mean	Median	St. Dev.
Very	169	9.57	7	8.64
Fairly	312	4.76	3	2.15
Not very	252	0.77	0	1.83

(Total sample size differs from that in Exercise 16.27 because not all students answered all questions.) Why would it be preferable to compare the prayer frequency for the three religious importance groups using a nonparametric technique rather than analysis of variance? (*Hint:* Which of the necessary conditions for using an *F*-statistic to compare means are violated by these data?)

General Section Exercises

16.29 A sample of $n = 235$ college students was asked, "Consider how important a person's personality and looks (attractiveness) are to you. Rate this importance on a scale of 1 (personality is most important) to 25 (looks are most important)." Minitab results for a Mood's median test that compares the responses of men and women follow.

```
Chi-Square = 15.05      DF = 1       P = 0.000

Sex     N<=      N>     Median
F       104      46      10.00
M        37      48      13.00

Overall median = 12.00
A 95.0% CI for median(F) – median(M): (–4.19,–2.00)
```

a. Explain whether the response is a quantitative variable, a categorical variable, or an ordinal variable.

b. What are the null and alternative hypotheses of the Mood's median test in this instance? Make your answer specific to this situation.

c. What is the overall median for all responses? What percentage of the women gave an answer that was less than or equal to the overall median? What percentage of the men gave an answer that was less than or equal to the overall median?

d. What conclusion can be reached about the difference between men and women with regard to the relative importance of looks versus personality? Explain how you made this decision, and be specific about the populations to which you think the conclusion applies.

16.30 Refer to Exercise 16.29. The *p*-value is given as P = 0.000, and the value of the chi-square statistic is given as Chi-Square = 15.05. Explain the connection between the *p*-value and the chi-square statistic. (A picture may help.)

16.31 Refer to Exercise 16.13 about women's testosterone levels. Results are shown in the accompanying output for a Kruskal-Wallis test done to compare the testosterone levels of the three groups of women.

```
Group       N     Median    Ave Rank        Z
Adv Degr   24      3.400        30.1     3.46
No Adv D   11      2.200        17.6    –1.67
Not Empl   11      2.000        15.1    –2.38
Overall    46                   23.5

H = 12.19   DF = 2   P = 0.002
H = 12.21   DF = 2   P = 0.002 (adjusted for ties)
```

a. Write null and alternative hypotheses for the Kruskal-Wallis test. Make your answer specific to this situation.

b. Based on the output, what conclusion can be reached about testosterone levels in the three groups? Justify your answer using information given in the output.

c. What are the sample medians for the three groups? Based on these medians, and the appearance of the dotplot in the figure for Exercise 16.13, describe the differences among the three groups.

Section 16.4

Skillbuilder Exercises

16.32 ◆ The following table shows mean age for participants in the 2008 General Social Survey classified by sex of participant and whether a participant had either a high school or a bachelor's degree (*Data source*: **GSS-08** dataset on the website for this book).

	Degree	
Sex	**High School**	**Bachelors**
Women	47.09 ($n = 356$)	44.18 ($n = 110$)
Men	44.47 ($n = 302$)	47.08 ($n = 109$)

a. For women, which degree group had the greater mean? How much greater was the mean age for that degree group compared to the other group?

b. For men, which degree group had the greater mean? How much greater was the mean age for that degree group compared to the other group?

c. Compare the answers to parts (a) and (b). Explain why this comparison indicates a possible interaction between sex and degree with regard to age.

16.33 For each scenario, explain whether an interaction is described or not. If there is an interaction, what are the two variables that interact?

a. For women, there was no difference between the mean hours spent studying per week for members and nonmembers of Greek organizations. For men, there was a 5-hour difference between mean hours spent studying per week for members of Greek organizations versus nonmembers.

b. The difference between the mean survival times for the two treatments was about 6 months in each patient age group.

16.34 Explain whether the following statement is an example of a main effect or an interaction: The mean number of classes missed per week was significantly lower for students who think religion is very important than it was for students who think religion is either fairly or not very important.

16.35 Exercise 16.27 described a survey in which college students were asked to rate various types of music on a 1 to 6 scale on which 1 = don't like and 6 = like a lot. The figure for this exercise is an interaction plot showing how the mean rating of rock music relates to type of hometown and sex of the student.

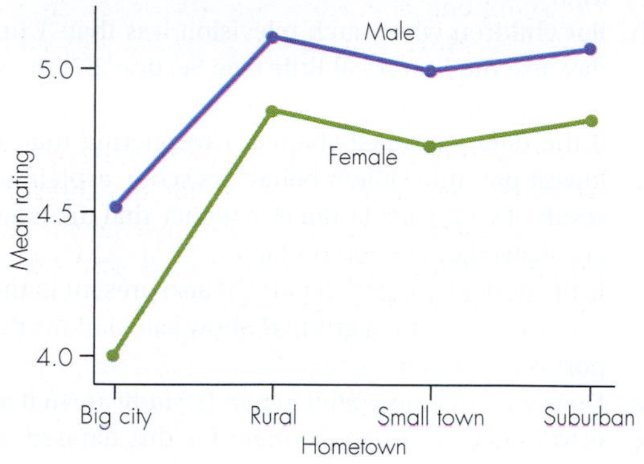

a. Which sex liked rock music better?
b. What is the evidence of a possible main effect of type of hometown?
c. Discuss whether you think there is evidence of a possible interaction between sex of the student and type of hometown.

General Section Exercises

16.36 ◆ The figure for this exercise shows how mean hours slept the previous night is related to the respondent's sex and preferred classroom seat location for a sample of $n = 173$ college students (*Data source:* **UCDavis1** dataset on the companion website).

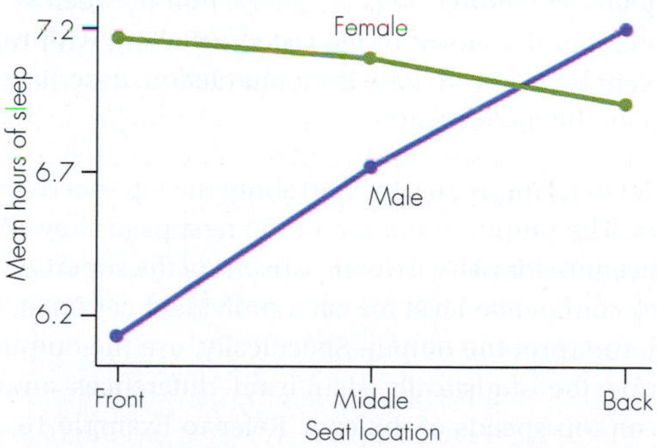

a. For males, describe how mean hours slept the previous night is related to preferred classroom seat location.
b. For females, describe how mean hours slept the previous night is related to preferred classroom seat location.
c. Does the plot give evidence of a possible interaction between respondent's sex and seat location?

16.37 In Example 16.1 about mean GPA and typical classroom seat location, the finding was that there were differences among the mean grade point averages for the three locations. Suppose that the sex of the students is also considered along with the seat location.

a. What would be indicated about the difference in the mean GPA of men and women if there were an interaction between sex of the student and seat location?
b. What would be indicated about the difference in the mean GPA of men and women if there were not an interaction between sex of the student and seat location?

16.38 Give an example not given in this chapter in which two-way analysis of variance would be used to make a comparison of means. Be specific about what the response variable of interest is, what the explanatory variables are, and what specific levels or categories of the explanatory variables are used.

16.39 Ryan and Joiner (2001) give data collected in an experiment done to examine the effects of storage temperature on the rot that occurs in stored potatoes. Potatoes were injected with bacteria known to cause potato rot, and three different bacteria amounts (low, medium, and high) were used. For each bacteria amount, half of the injected potatoes were stored at 10 °C, while the others were stored at 16 °C. The response variable was the diameter (millimeters) of the rot in a potato after being stored. The mean rot diameters for all combinations of bacteria amount and temperature are shown in the following table. For each combination, the sample size is 9.

Mean Diameter of Potato Rot by Storage Temperature and Bacteria Amount

	Temperature	
Bacteria	**10°C**	**16°C**
Low	3.6	7.0
Medium	4.8	13.6
High	8.0	19.6

a. Draw an interaction plot of the means. Figures 16.13 (p. 636) and 16.14 (p. 637) may provide guidance for how to do this.
b. What is the evidence that there may be an interaction between bacteria amount and storage temperature with regard to amount of potato rot? Briefly describe the pattern of this possible interaction.

16.40 Students in a statistics class gave information on their height and weight and also reported their perception of their weight (about right, underweight, overweight). The mean body mass index (BMI) is shown in the following table, along with sample size, for each combination of sex (female or male) and perception of weight category. (BMI is calculated as weight in kilograms/height in meters^2.)

Mean Body Mass Index by Sex and Perception of Weight

	Perception of Weight		
	Underweight	**About Right**	**Overweight**
Females	19.3 ($n = 7$)	21.1 ($n = 107$)	24.4 ($n = 31$)
Males	20.1 ($n = 14$)	23.9 ($n = 56$)	27.8 ($n = 13$)

a. Draw an interaction plot of the means. Figures 16.13 (p. 636) and 16.14 (p. 637) may provide guidance for how to do this.

b. Based on the given data, do you think the respondent's sex and perception of weight are interacting variables with regard to body mass index? Briefly explain.

Chapter Exercises

16.41 Refer to Figure 16.1 (p. 618) for Example 16.1 about GPA by classroom seat location. Explain why the 95% confidence intervals for the mean GPA in the three seat locations have unequal widths.

16.42 Give an example not already given anywhere in this chapter in which one-way analysis of variance could be used to make a comparison of means. Be specific about what the response variable of interest is and what groups are compared.

16.43 A sample of students in grades 3 to 8 in 11 public schools in Ohio were asked how often they engage in violent behaviors and how often they watch television (Singer et al., 1998). A violent behaviors score was calculated by adding frequency scores (0 = never, 3 = almost every day) for five different violent acts: threatening, hitting before being hit, hitting after being hit, beating up, and attacking with knife. For five groups determined by self-reported hours of television, the sample size, sample mean, and sample standard deviation for the violent behaviors scores follow.

Daily Hours of Television	n_i	$\bar{x}_i$	s_i
<1	227	2.94	2.73
1–2	526	2.59	2.44
3–4	666	2.87	2.36
5–6	310	3.10	2.45
>6	488	4.03	2.81

a. Write null and alternative hypotheses for comparing the mean violent behaviors score for the five television-watching groups. State the hypotheses in words, and also state them using statistical notation.

b. What do you think are the populations to which the results of an analysis of variance F-test would apply?

c. The researchers did a one-way analysis of variance and reported the results as $F = 22.95$, $p < .001$. What does this result indicate about the null and alternative hypotheses for this problem?

d. Note that the highest mean violent behaviors score is in the group that watches the most television. Explain why this cannot necessarily be interpreted to mean that watching a lot of television causes violent behavior in children. What is another explanation for the observed result?

16.44 Refer to Exercise 16.43.

a. On the basis of the information given, which of the necessary conditions for doing a one-way analysis of variance appear to be satisfied? What is the evidence of this?

b. For children who watch television less than 1 hour per day, use the Empirical Rule (see Section 2.7) to estimate the interval in which the middle 95% of scores would fall if the data were bell-shaped. Considering that 0 is the lowest possible violent behaviors score, explain why the result of your calculation is evidence that the data either are skewed or contain outliers.

c. Is the difficulty noted in part (b) also present in the other television-watching groups? Show calculations that support your answer.

d. Does the difficulty noted in part (b) indicate that analysis of variance is not appropriate for this dataset? Explain why or why not.

16.45 Refer to Exercises 16.43 and 16.44. The investigators reported results separately for boys and girls in the five television-watching groups. In this two-way classification, the approximate observed mean violent behaviors scores were as follows:

	Hours of Daily Television Watching				
	<1	1–2	3–4	5–6	6+
Boys	3.7	3.0	3.4	3.4	4.1
Girls	2.3	2.3	2.4	2.8	3.9

Explain whether there is an interaction between sex of the student and amount of television watching with regard to violent behavior. If there is an interaction, describe the pattern of the interaction.

16.46 Refer to Example 16.9 (p. 629) about the top speeds of supercars. The output at the top of the next page shows Fisher's procedure for pairwise comparisons of the supercars, with a 95% confidence level for each individual confidence interval. Interpret the output. Specifically, use the output to describe the statistically significant differences among the mean top speeds of the cars. Refer to Example 16.9 to see which were the fastest and slowest cars.

Fisher 95% Individual Confidence Intervals			
All Pairwise Comparisons			
Acura subtracted from:			
	Lower	Center	Upper
Ferrari	12.542	17.067	21.592
Lotus	−3.842	0.683	5.208
Porsche	9.925	14.450	18.975
Viper	2.192	6.717	11.242
Ferrari subtracted from:			
	Lower	Center	Upper
Lotus	−20.908	−16.383	−11.858
Porsche	−7.142	−2.617	1.908
Viper	−14.875	−10.350	−5.825
Lotus subtracted from:			
	Lower	Center	Upper
Porsche	9.242	13.767	18.292
Viper	1.508	6.033	10.558
Porsche subtracted from:			
	Lower	Center	Upper
Viper	−12.258	−7.733	−3.208

16.47 Refer to Exercise 16.18 about finger tapping and amount of caffeine consumed. Note that caffeine amount is a quantity, so it could be analyzed as a quantitative variable. What statistical technique other than one-way analysis of variance could be used to analyze the relationship between taps per minute and caffeine amount? What would be an advantage of using that technique to analyze this relationship? (*Tip:* It might help to look again at the boxplot for Exercise 16.18.)

16.48 At a private university in southern California, 334 freshman (197 males and 137 females) reported their frequency of drinking alcoholic beverages and also completed a test of personality and psychological characteristics (Ichimaya and Kruse, 1998). One response variable was the score on a 12-item test of conscientiousness for which a high score indicates dependability. Summary statistics for three groups defined by frequency of binge drinking (five or more drinks in one session) follow.

Drinking Group	n_i	$\bar{x}_i$	s_i
Nonbinge	162	32.59	7.15
Occasional binge	67	29.93	6.43
Frequent binge	105	30.10	6.45

a. Write null and alternative hypotheses for comparing the mean conscientiousness score for the three groups. Write the hypotheses in words, and also write the null hypothesis using proper statistical notation.

b. What do you think are the populations to which the results of an analysis of variance *F*-test would apply?

c. The researchers did a one-way analysis of variance and reported the results as $F = 5.95$, $p < .0005$. What does this result indicate about the null and alternative hypotheses for this problem? (*Note:* The *p*-value reported by the researchers is incorrect. The correct *p*-value is $p = .003$.)

d. Using the reported sample means, describe the differences among the mean conscientiousness scores for the three groups. Explain whether you think any observed differences have practical significance.

e. Is it reasonable to assume that population standard deviations are equal? Why or why not?

16.49 Refer to Exercise 16.48. Use formulas in Section 16.2 to compute SS Error, MSE, and the pooled standard deviation, s_p.

16.50 Refer to the example given on page 635 about comparing time to finish an open-book exam and a closed-book exam.

a. Use the means given for the four groups to draw an interaction plot. See Figures 16.13 and 16.14 for guidance.

b. Explain why the figure you drew in part (a) indicates that there is an interaction between the two variables of grade before final and exam type, with regard to time taken to finish the exam.

c. Explain why the figure you drew in part (a) indicates that there is no overall main effect for exam type. Do this by indicating on the figure where the overall mean is for each exam type.

16.51 Refer to Example 16.6 on page 624, in which Tukey and Fisher confidence intervals are given for comparing mean GPAs for students who typically sit in the front, middle and back of a classroom.

a. What is the Tukey confidence interval for the difference in means ($\mu_{\text{Front}} - \mu_{\text{Back}}$)?

b. What is the Fisher confidence interval for the difference in means ($\mu_{\text{Front}} - \mu_{\text{Back}}$)?

c. Suppose that you wanted to use the Bonferroni method to compute confidence intervals for all pairs of differences in means for this example, using an overall family confidence level of .95. What confidence level would you need to use for each confidence interval?

d. The Bonferroni confidence interval for the difference in means ($\mu_{\text{Front}} - \mu_{\text{Back}}$) using a family confidence level of .95 is 0.0791 to 0.4878. Compare the Bonferroni interval to the Tukey and Fisher intervals you reported in parts (a) and (b). Comment on which of the three methods gives the widest and the most narrow interval, and include a discussion of whether the confidence level of 95% in each case applies to the set of intervals for all pairs of means, or if it applies to each interval individually.

Dataset Exercises

Datasets required to solve these exercises are available on the companion website, http://www.cengage.com/statistics/Utts4e.

16.52 Use the dataset **wineratings** on the companion website. The variable ***Quality*** gives ratings of overall quality for 38 samples of Pinot Noir wine made in three different wine-producing regions. The variable ***Region*** gives the region (1, 2, or 3).

a. Create a comparative boxplot to compare overall quality ratings in the three different regions. Describe any regional differences.
b. Determine and compare sample means of overall quality ratings for the three regions. Find sample standard deviations for the three regions.
c. Use software to do a one-way analysis of variance to compare overall quality ratings for the three regions. State null and alternative hypotheses, give the *p*-value, and clearly state a conclusion about the regions.

16.53 Use the dataset **wineratings** on the companion website. The variable ***Aroma*** gives ratings of aroma for 38 samples of Pinot Noir wine made in three different wine-producing regions. The variable ***Region*** gives the region (1, 2, or 3) for a sample.

a. Determine and compare sample mean aroma ratings for the three regions.
b. Create a comparative boxplot to compare aroma ratings in the three different regions. Describe any regional differences.
c. Use software to do a one-way analysis of variance to compare aroma ratings for the three regions. State null and alternative hypotheses, give the *p*-value, and clearly state a conclusion about the regions.

16.54 Use the dataset **Student0405** on the companion website, which contains data that were gathered in surveys of statistics classes at a large university in the United States during 2004 and 2005. The variable ***MissClass*** gives student estimates of how many classes they typically miss in a week. The variable ***Seat*** gives responses to the question, "Where do you prefer to sit in a classroom (front, middle, back)?"

a. Determine and compare sample means of missed classes per week for the three seat locations.
b. Use a comparative boxplot to compare missed classes for the three seat locations. Describe any possible differences.
c. Do a one-way analysis of variance *F*-test to compare mean missed classes for the seat locations. State null and alternative hypothesis, give the *p*-value, and clearly state a conclusion about the locations.
d. Explain whether the data satisfy the necessary conditions for using an *F*-test to compare population means.

16.55 Use the dataset **Student0405** on the companion website, which contains data that were gathered in surveys of statistics classes at a large university in the United States during 2004 and 2005. The variable ***StudyHrs*** gives student estimates of how many hours they typically study in a week. The variable ***ReligImp*** gives responses to the question, "How important is religion in your own life (very, fairly, not very)?"

a. Graphically compare study hours for the three religious importance groups. Describe the results.
b. Use software to do a one-way analysis of variance to compare mean weekly hours of study for the three religious importance groups. State null and alternative hypotheses, give the *p*-value for the *F*-test, and clearly state a conclusion about whether population means differ.
c. Explain whether the data satisfy the necessary conditions for using an *F*-test to compare population means.

16.56 Refer to Exercise 16.55. Compare study hours for the three religious importance groups using either a Mood's median test or a Kruskal-Wallis test. Give the output as part of your answer, and state a clear conclusion in the context of this situation.

16.57 Use the **GSS-02** dataset on the companion website. The variable ***degree*** gives the highest educational degree achieved (five categories) for respondents in the 2002 General Social Survey, and ***tvhours*** gives responses for hours of watching television in a typical day.

a. Find mean and median weekly television watching amounts for each educational degree group. Describe any differences among the groups.
b. Use either a Kruskal-Wallis test or a Mood's median test to compare the amount of television watching in the five educational degree groups. Give null and alternative

hypotheses, give a p-value for the test you chose to do, and state a conclusion in the context of this situation.

c. Compare the amount of television watching in the five educational degree groups using a one-way analysis of variance. Give null and alternative hypotheses, give a p-value for the test, and state a conclusion in the context of this situation.

16.58 Use the **GSS-02** dataset on the companion website. The variable ***degree*** gives the highest educational degree achieved (there are five categories) for respondents in the 2002 General Social Survey, ***sex*** is the sex of the respondent, and ***tvhours*** gives responses for hours of watching television in a typical day.

a. Create an interaction plot to examine how the variables ***degree*** and ***sex*** affect daily television watching. Explain whether there is an interaction between the variables ***degree*** and ***sex***.

b. Describe the main effects of ***degree*** and ***sex***.

16.59 Use the dataset **Student0405** on the companion website, which contains data that were gathered in surveys of statistics classes at a large university in the United States during 2004 and 2005. ***Sex*** is sex of the respondent, and ***Seat*** gives responses to the question, "Where do you prefer to sit in a classroom (front, middle, back)?" The variable ***MissClass*** gives student estimates of how many classes they typically miss in a week.

a. Create an interaction plot to examine how the variables ***Sex*** and ***Seat*** relate to mean missed classes per week. Explain whether or not there is an interaction in this situation.

b. Describe the main effects of ***Sex*** and ***Seat***.

17

Jeff Greenberg/PhotoEdit

Is it the wine or the cheese that protects French hearts?

See Example 17.6 *(p. 659)*

Turning Information into Wisdom

In the opening paragraph of Chapter 1, we said, "We hope you will agree that learning about statistics may be interesting and useful." We hope that by the time you have read this far in the book, you have seen some interesting examples and information. But are the methods in this book useful? Why would anyone care about what can be done with statistics?

In this chapter, we speculate on the wisdom to be gained from statistical studies and why people are interested in such studies. Certainly, you can live without them. But information developed through the use of statistics has enhanced our understanding of how life works, helped us to learn about each other, allowed control over some societal issues, and helped individuals to make informed decisions. There is almost no area of knowledge that has not been advanced by statistical studies. As individuals, we learn from our own and others' experiences. But as a society, we learn through scientific studies—many of them based on statistical methods.

This chapter contains a number of examples that illustrate the wisdom to be gained from statistical studies. Before delving into examples, let's review the basic situations that allow us to turn data into information and information into wisdom.

17.1 Beyond the Data

Much of the appeal of statistical methods is that they allow us to describe, predict, and sometimes control the world around us. This is possible because with the methods of statistical inference, we can use relatively small amounts of data to provide information and sometimes reach conclusions about immense populations. This process is one of the truly remarkable features of modern statistics.

However, it is important to remember that the extent to which data descriptions can be generalized to something more rests on two issues: *what group of individuals was measured* and, if relevant, *whether randomization was used to assign conditions.* These are two different aspects of study design, and they lead to two different possible extensions of simple descriptive measures.

Random, Representative, or Restrictive Sample?

If pressed to be completely intellectually honest, most statisticians would have to agree with the statement made in *The Statistical Sleuth* (Ramsey and Schafer, 2002, p. 7): "Inferences to **populations** can be drawn from random sampling studies, but not otherwise." The mathematics supporting the material on statistical inference in this book demands simple random samples. However, most researchers, including statisticians, realize that true random samples are almost impossible to obtain. Methods of statistical inference would be of little use if they were restricted to situations in which true random samples were available. Instead, in practice, most researchers follow the Fundamental Rule for Using Data for Inference, which we introduced in Chapter 5.

DEFINITION The **Fundamental Rule for Using Data for Inference** is that available data can be used to make inferences about a much larger group *if the data can be considered to be representative with regard to the question(s) of interest.*

One of the key steps in interpreting a statistical study is the capacity to determine if this rule holds for the questions of interest in the study. We will look at some examples in this chapter.

Randomized Experiments, Observational Studies, and Causal Conclusions

As we discussed in Chapter 6, one of the most common errors made by the media in interpreting statistical studies is to conclude that a causal relationship has been established when that conclusion is not warranted by the way the study was conducted. If a statistically significant relationship between the explanatory and response variables is demonstrated, it is tempting to assume that the differences in the explanatory variable *caused* the differences in the response variable. But it is important to remember that even if a cause-and-effect conclusion is sensible, it is not justified by the results of an observational study. Let's revisit the Rule for Concluding Cause and Effect, which we introduced in Chapter 6.

DEFINITION The **Rule for Concluding Cause and Effect** is that cause-and-effect relationships can be inferred from randomized experiments but not from observational studies.

The reasoning is simple. When individuals are randomized into different treatment conditions, the effect of confounding variables should be similar in all treatment groups. Therefore, differences in the response variable(s) can be attributed to the one thing that is known to differ across groups—the explanatory variable, which is the set of treatment conditions. In observational studies, the groups that define the explanatory variable occur naturally, and they may have other natural differences between them in addition to the explanatory variable. These other possible differences produce confounding variables, the effect of which cannot be separated from the effect of the explanatory variable. Therefore, differences in the response variable cannot be attributed to differences in the explanatory variable.

Here are some examples of headlines from various news services on the Internet, all based on observational studies. In each instance, the headline implies a cause-and-effect relationship. We are sure you can think of confounding variables that may have led to the observed relationship:

- *Emotional Support Helps Breast Cancer Survival* (Reuters News Service, http://dailynews.yahoo.com/headlines/hl/, December 2, 2000)
- *Study: Even a Little* [of the drug] *Ecstasy Can Lower IQ* (USA Today Health News, http://www.usatoday.com/life/health/archive.htm, May 16, 2000)
- *Study: Junk Food Raises Teens' Risk of Heart Disease* (CNN Health News, http://www.cnn.com/2000/HEALTH/03/14/teen.arteries/index.html, March 14, 2000)
- *Lack of Folic Acid Causes Defects* (Associated Press, http://dailynews.yahoo.com/htx /ap/20001129/hl/folic_acid_1.html, November 29, 2000)
- *City Living Increases Men's Death Risk* (ABC News, http://www.abcnews.go.com/sections/living/DailyNews/cityrisk001130.html, November 30, 2000)

In most cases, neither the original researchers nor the authors of the news articles made the mistake that is made in the headline. The reporters who write the articles usually do not write the headlines, and the reporters may be better trained in how to interpret studies than are headline writers. Unfortunately, the public may not read beyond the headline.

Using Nonstatistical Considerations to Assess Cause and Effect

You might wonder why researchers conduct observational studies at all, given that the goal is probably to make a causal conclusion. There are some nonstatistical considerations that can be used to help assess whether a causal link is reasonable based on observational studies. Here are some features that lend evidence to a causal connection.

1. *There is a reasonable explanation of cause and effect.* A potential causal connection will be more believable if an explanation exists for how the cause and effect occurs. For instance, in Example 6.2 (p. 192), we established that for hardcover books the number of pages is correlated with the price. We would probably not contend that higher prices cause more pages but could reasonably argue that more pages cause higher prices. We can imagine that publishers set the price of a book based on the cost of producing it and that the more pages there are, the higher is the cost of production. Thus, we have a reasonable explanation for how an increase in the length of a book could cause an increase in the price.
2. *The connection happens under varying conditions.* If many observational studies conducted under different conditions all find the same link between two variables, that strengthens the evidence for a causal connection. This is especially true if the studies are not likely to have the same confounding variables. The evidence is also strengthened if the same type of relationship holds when the explanatory variable falls into different ranges. For example, numerous observational studies have related cigarette smoking and lung cancer. Furthermore, the studies have shown that the higher the number of cigarettes smoked, the greater are the chances of developing lung cancer; similarly, a connection has been established with the age at which smoking began. These facts make it more plausible that smoking actually causes lung cancer.
3. *Potential confounding variables are ruled out.* When a relationship appears in an observational study, potential confounding variables may immediately come to mind. For example, the researchers in Case Study 6.1 (p. 193), which showed a relationship between lead exposure and tooth decay in children, did consider that income level, carbohydrates, calcium, and lack of dental care may have been confounding variables. However, they were able to measure those variables and examine them in the analysis, and the relationship between lead exposure and tooth decay persisted. The greater the number of confounding factors that can be ruled out, the more convincing is the evidence for a causal connection.

Chapter exercises are on pages 664–666.

IN SUMMARY

Guidelines about Extending Results to a Population and Concluding Cause and Effect

Results can be extended to a larger population if a representative sample was chosen, ideally a *random sample*, but not otherwise. Cause-and-effect conclusions can be made if *randomization* (random assignment) was used, but generally not if an observational study was used. Do not confuse *random sample* with *randomization*. The combination of representative sample (yes, no) and randomization (yes, no) leads to four possible outcomes, as shown in the following table:

	Sample represents population for question of interest	Sample does not represent population
Randomization used (randomized experiment)	Cause-and-effect inferences about population possible	Cause-and-effect inferences possible for sample, but not for population
Randomization not used (observational study)	Generalization to population possible, but not with cause-and-effect explanations	No inference possible; can describe sample results only

17.2 Transforming Uncertainty into Wisdom

We live in a world that is filled with uncertainty and variability. The field of statistics exists because this is so. Statistical methods allow us to gain information and control because we can use these methods to measure variability and to find predictable patterns in the presence of uncertainty. Here are some reasons that statistical methods are popular and useful:

- As individuals, we need to make personal decisions about things like:

  What to eat to remain healthy
  How to enhance our children's well-being
  Which treatment to undergo in a medical situation
  How to invest our time to enhance our enjoyment of life

- As a society, we want to have some control over things like:

  Reducing rates of cancer, heart attacks, and other diseases
  Growing better plants (pest-resistant, high-yield, and so on)
  Determining what public education strategies help to reduce risky behavior for HIV infection
  Deciding what new laws to enact to enhance driver safety

- As intelligent and curious beings, we want to understand things like:

  Are psychic abilities real? If so, under what conditions?
  Do left-handed people die at a younger average age?
  Does listening to classical music improve scores on intelligence tests?
  Are there differences in how men's and women's brains work?

- As social and curious beings, we want to know about other people:

  Who are they planning to vote for in the next election?
  What television programs are they watching?
  How much alcohol do they drink?
  How healthy are they?

17.3 Making Personal Decisions

Results of statistical studies and other statistical information can sometimes help when you are required to make a decision involving uncertainty. Such a decision usually involves both factual and emotional reasoning, so statistical information alone is often not sufficient. It is sometimes helpful to think about the decision in the framework of hypothesis testing and to consider the consequences of the two types of error we discussed in Chapter 12. In many cases, the decision can be formulated as follows:

H_0: I will be better off if I take no action.

H_a: I will be better off if I do take action.

In such cases a Type 1 error would correspond to taking action when you would have been better off not doing so. A Type 2 error would correspond to taking no action when you would have been better off taking action.

Example 17.1 **Playing the Lottery** Suppose that there is a lottery in your area and the prize money is many millions of dollars (or the equivalent in your currency). You have picked the numbers you would like to play but you cannot decide whether to buy a ticket. For simplicity, let's forget about the smaller possible prizes and consider only the decision about whether you should spend money for a ticket based on the big prize. Here are your hypotheses:

H_0: The numbers I have chosen will not win the big prize.
H_a: The numbers I have chosen will win the big prize.

Let's consider the possible consequences of choosing each of these hypotheses and then analyze the choices.

	Likelihood	You Choose H_0 (Don't Buy)	You Choose H_a (Buy Ticket)
H_0 Is True (Won't Win)	Very high	No loss	Lose cost of ticket
H_a Is True (Would Win)	Very low	Forgo millions, agony of seeing your numbers win	Win the big prize

The likelihood that the null hypothesis is true is very high, yet many people behave as if the alternative hypothesis is true. The reason becomes obvious by examining the consequences of each choice in the last two columns of the table. Although there is no loss for correctly choosing H_0, the possible error associated with choosing H_0 is devastating. The possible error associated with choosing H_a is minor for many people (lose the cost of the ticket) and the possible payoff is very high.

To summarize: if you choose H_a the likelihood that you are making a Type 1 error is very high but the consequences of it are minor. If you choose H_0 the likelihood that you are making a Type 2 error is extremely small, but the consequences of it are not minor. By the way, the long-run average loss of 30 to 60 cents per ticket in most lotteries is apparently insignificant to most individual ticket buyers, who make their decision (consciously or not) based on the consequences shown in the table.

Example 17.2 **Surgery or Uncertainty?** Sometimes life-or-death decisions must be made based on little information. A somewhat common situation is as follows. Suppose that your doctor discovers a lump and cannot tell whether it is a malignant or benign growth. You have two choices: You can have the affected organ removed with surgery, or you can wait and see if the lump continues to grow or to spread. What would you do? The hypotheses you must choose between are as follows:

H_0: The lump is benign.
H_a: The lump is malignant.

Let's examine the choices and their consequences:

	Likelihood	You Choose H_0 (No Surgery)	You Choose H_a (Surgery)
H_0 Is True (Benign)	??	No loss	Lose organ needlessly
H_a Is True (Malignant)	??	Possibly life-threatening	Stop possible spread

Statistical information may be available to assess the likelihood of each hypothesis, and presumably, the medical practitioner could provide reasonable probabilities. You must then weigh the possible consequences of each choice on the basis of nonstatistical issues. For instance, depending on which organ is involved, its loss may be minor, or it may be extremely serious and require lifelong medical intervention.

Example 17.3 **Fish Oil and Psychiatric Disorders** A friend's child was recently diagnosed with a possible psychiatric disorder, and the child's physician recommended that the parents give the child fish oil. In an Internet search for more information, the parents found the following news story:

Thursday September 3, 1998

Fish Oil May Fight Psychiatric Disorders

NEW YORK, Sep 03 (Reuters)—The consumption of omega-3 polyunsaturated fatty acids found in fish and fish oil may reduce the symptoms of a variety of psychiatric illnesses, including schizophrenia, bipolar disorder, and depression, researchers report.

"Research suggests that (fatty acids) may have a role in psychiatric disorders," said Dr. Joseph Hibbeln of the National Institute on Alcohol Abuse and Alcoholism, part of the National Institutes of Health (NIH) in Bethesda, Maryland.

Hibbeln is one of a number of researchers attending an NIH-sponsored workshop on the issue in Bethesda this week.

The workshop was prompted in part by the results of three recent studies.

Findings from one study, conducted by Dr. Andrew Stoll of the Harvard Medical School in Boston, Massachusetts, suggest that fish oil supplementation could help alleviate the symptoms of bipolar (manic-depressive) disorder.

For a 4-month period, Stoll gave bipolar patients daily supplements of either fish oil or a "dummy pill," or placebo. He found that "overall, 9 of 14 patients responded favorably to the addition of omega-3 fatty acids (to their diet), compared to only 3 of 16 patients receiving placebo."

Another study focused on the effects of one fish-oil fatty acid, eicosapentaenoic (EPA), in the treatment of schizophrenia. A 3-month trial conducted by Dr. Malcolm Peet of Northern General Hospital in Sheffield, England, concluded that [there was] "a 25% improvement (in schizophrenic symptoms) in the EPA treated group," compared with patients receiving either docosahexaenoic acid (DHA, another omega-3 fatty acid) or placebo.

A third study, conducted by Hibbeln, focused on levels of omega-3 fatty acids in the blood of 50 patients hospitalized after attempting suicide.

Hibbeln found that, among nondepressive (but not depressive) patients, high blood concentrations of EPA "predicted strikingly lower (better) scores in 6 different psychological rating scales which are related to suicidal risk." The NIH researcher says these findings suggest that "some subgroups of suicidal patients may reduce their suicidal risk with the consumption of EPA."

Hibbeln also noted that another study showed that dietary intake of EPA and DHA may influence serotonin function in the brain. "Such an alteration in serotinergic function may possibly reduce depressive, suicidal and violent behavior, but these changes have not yet been demonstrated in ... clinical trials," he said in a statement.

Hibbeln explained that the brain's synaptic membranes, where much of the brain's neurological signaling takes place, "have a large proportion of essential fatty acids in them—fatty acids which are derived entirely from the diet."

He points out that "in the last century, (Western) diets have radically changed and we eat grossly fewer omega-3 fatty acids now. We also know that rates of depression have radically increased by perhaps a hundred-fold" over the same period of time.

Links between fish consumption and neurological health may be supported by the results of global studies. According to Hibbeln, those findings suggest that "rates of major depression are markedly different across countries, depending upon how much fish is consumed in those countries."

Source: Mundell, E. J. (1998). "Fish oil may fight psychiatric disorders," Reuters News press release, Sept. 3, 1998.

In a more recent study (DeNoon, 2010), researchers randomly assigned 81 teens at high risk for psychosis to receive either fish oil or a placebo for 12 weeks. A year later, 11 of the 40 teens who received the placebo had developed a psychotic disorder, while only two of the 41 teens who received the fish oil had developed a disorder. The p-value for a Fisher's Exact Test for this study is .006.

What would you do if you were the parents of a child at high risk of a psychiatric disorder? Would you start giving fish oil to your child? The two hypotheses are as follows:

H_0: Fish oil will not help my child.
H_a: Fish oil will help my child.

The possible decisions and consequences are as follows:

	Likelihood	You Choose H_0 (No Fish Oil)	You Choose H_a (Fish Oil)
H_0 Is True (Fish Oil Won't Help)	??	No loss or gain	Child takes fish oil needlessly; presumably no harm
H_a Is True (Fish Oil Will Help)	??	Child does not receive a treatment that would help	Child is helped by fish oil

In this case, parents must assess the likelihood that fish oil will help their child and the consequences of making each decision. Some of the research is based on randomized experiments, so in those studies, there appears to be a causal benefit from fish oil.

The participants in those studies may or may not be representative of the same group as any particular child, and parents need to assess that issue. Finally, the consequences of choosing H_a erroneously are minor, so even with a moderate possibility that H_a is true, the parents might be wise to decide in favor of giving fish oil to the child.

Chapter exercises are on pages 664–666.

17.4 Control of Societal Risks

Sometimes statistical studies can be used to guide policy decisions. In these cases, lawmakers, government regulatory agencies, and other decision makers must weigh decisions and their consequences.

Example 17.4 **Go, Granny, Go or Stop, Granny, Stop?** Efforts to improve driving safety are often based on results of statistical studies. For instance, studies have shown that wearing a seatbelt is likely to reduce the chances of death or serious injury in an accident, that driving after consuming alcohol increases the risk of an accident, and so on. Laws based on these findings have been passed to help protect drivers and passengers. But there is clearly a trade-off between protection and personal freedom. If you were a lawmaker concerned about driver safety, would the following article convince you to take action?

> Tuesday April 7, 1998 6:46 PM EDT Yahoo 1998
>
> *Visual Field Loss Ups Elderly Car Crashes*
>
> NEW YORK (Reuters)—A 40% loss in range of vision among older drivers more than doubles their risk for a car accident, experts say. They believe around one third of all seniors suffer from such vision impairment.
>
> "Visual dysfunction and eye disease deserve further examination as causes of motor vehicle crashes and injury," say ophthalmology researchers at the University of Alabama at Birmingham, and the Western Kentucky University in Bowling Green, Kentucky.
>
> Their study, published in the April 8th issue of The Journal of the American Medical Association (JAMA), focused on the 3-year motor vehicle accident rates of 294 Alabama drivers between 55 and 87 years of age. Each of the individuals received various vision tests, including field of vision examinations.
>
> The researchers discovered that, of all the vision-related factors they studied, "impaired useful field of view was the only one that demonstrated a marked elevation (in risk). Older drivers with a 40% or greater reduction in the useful field of view were 2.1 times more likely to have incurred a crash ... compared with those with less than 40% reduction."
>
> They found that an older driver's crash risk increased by 16% with every 10-point drop in their field of vision.
>
> Since previous research has found that one in three older adults has a greater than 40% reduction in their field of vision, the study authors speculate that field of vision tests might "be a good candidate for inclusion in a functional test battery" aimed at determining driver safety. At the present time, most of the nation's drivers are only obligated to pass visual acuity tests that assess their sharpness of vision in order to obtain license renewals.
>
> However, JAMA contributing editor Dr. Thomas Cole believes field of view tests are "probably too time-consuming and expensive to be used as a screening test in driver licensing offices, and should instead be considered a diagnostic test for drivers with suspected subtle (vision) impairment."
>
> *(Source:* Journal of American Medical Association, *1998, Vol. 279(14), pp. 1083–1088 [See p. 701 for complete credit.])*

Should additional vision screening be required for older drivers? There are statistical and nonstatistical issues involved in that decision. First, the research was based on an observational study; drivers cannot be randomly assigned to have vision impairment or not. There are obviously potential confounding factors that may affect accident rates. Also, the researchers mentioned that they studied many aspects of vision but that "impaired useful field of view" was the only one that was related to accident rates. It is possible that they made a Type 1 error, which as we learned in Section 13.8, can easily happen for one or more hypotheses when a large number are tested. Finally, it is not clear how representative the participants in this study are of all older drivers, and the results may not apply to all older drivers.

Even if the relationship is real and causal, it may not be appropriate to restrict driving privileges based on visual field impairment. In addition to the statistical issues, there are issues of personal freedom and of the cost of vision testing. Policymakers must weigh all of these factors when making decisions based on statistical studies.

Example 17.5 **When Smokers Butt Out Does Society Benefit?** The debate over the health risks of breathing smoke from other people's cigarettes has been largely statistical in nature, and new information released at the end of 2000 added fuel to the fire. After more than 10 years of active antismoking measures, the lung cancer rate in California was down substantially compared to the rest of the United States. The situation was summarized in the following news article:

Thursday November 30, 2000 8:13 PM ET

Anti-Tobacco Measures Lessen Cancer

By JENNIFER COLEMAN, Associated Press Writer

SACRAMENTO, Calif. (AP)—California's tough anti-smoking measures and public health campaigns have resulted in a 14 percent decrease in lung cancer over the past 10 years, the government reported Thursday.

Other regions of the country reported only a 2.7 percent decrease over the same period, the Centers for Disease Control and Prevention said.

"Based on the California experience, we would hope to see similar effects in other states using similar programs," said Dr. Terry Pechacek, CDC associate director for science and public health.

Lung cancer develops slowly and the full benefits of quitting can take up to 15 years to be realized. However, Pechacek said, researchers can start seeing some results within five years.

Smoking rates in California began dropping in the late 1980s, helped in part by Proposition 99 in 1988. The voter-approved measure added a 25-cent-per-pack tax on tobacco products that paid for anti-smoking and education programs. Local governments also began restricting smoking in public buildings and workplaces.

Two years ago, voters bumped the price of cigarettes an additional 50 cents per pack, money also earmarked for education. And this year alone, the state will spend $136 million on smoking prevention, cessation and research—some $45 million of it on anti-tobacco advertising.

"California has the most comprehensive program for protecting nonsmokers from secondhand smoke," said Ken August, spokesman for the state health department. "Restaurants, bars and almost all indoor workplaces are smoke-free."

The effect of the anti-tobacco efforts has been fewer smokers and fewer deadly cases of cancer related to smoking, health officials said. August and Pechacek both said they expect the trend to continue.

August said that means there will be up to 4,000 fewer lung cancer cases in California this year and about 2,000 fewer deaths.

In its report, the CDC compared cancer registries in California, Connecticut, Hawaii, Iowa, New Mexico and Utah, as well as Seattle, Atlanta and Detroit.

In 1988, the lung cancer rate in California was 72 cases per 100,000 people, slightly higher than that of the other regions studied. By 1997, California's rate had dropped to about 60 per 100,000.

The CDC averages the statistics for the first two years and the last two years studied to arrive at an accurate representation, health officials said. The numbers the CDC used were 71.9 for 1988 and 70.3 for 1989, averaging to 71.1; and 62.2 for 1996 and 60.1 for 1997, averaging to 61.15. That computes to a 14 percent drop in lung cancer cases.

While lung cancer rates for women in the other regions rose 13 percent, the rate for California women dropped 4.8 percent. Among California men, lung cancer rates dropped 23 percent, compared with a 13 percent drop among men elsewhere.

Dr. David Burns, a volunteer with the American Lung Association in California, said: "This is an accomplishment of Proposition 99 money being invested wisely by the state to help people change their smoking behavior."

Source: http://dailynews.yahoo.com/htx/ap/20001130/hl/cancer_study_2.html.

Suppose that you were a lawmaker in another state and read these results. Would you attempt to enact legislation similar to that in California? The evidence is certainly tempting, but there are some statistical issues that need to be considered. First, these results are based on observational data. California was not randomly chosen to apply tough antismoking laws. People in California may be more health conscious than people in other states. California has other strict air pollution legislation that was enacted in the same time period, such as tough maintenance and inspection laws for vehicle emissions. California industry has changed over the years, favoring high-tech industries that are likely to employ younger, healthier workers.

If you were a lawmaker, you would need to weigh consequences of possible decisions. For instance, as mentioned, California has had an active antismoking education program. The consequences of launching such a program would presumably not be harmful, so a decision to do so would have positive or neutral consequences. On the other hand, a decision to ban smoking in bars and restaurants would involve more controversial consequences.

Chapter exercises are on pages 664–666.

17.5 Understanding Our World

Sometimes statistical studies are done to help us understand ourselves and our world, without involving any decisions. There are thousands of academic journals containing the results of statistical studies, most of which will never be used to change personal or societal decisions or behavior. Scan any major news source for a few weeks and you will find reports of many interesting studies that are done to help us understand the world. Many of these studies are exploratory in nature and have results that are controversial. That's part of why they make interesting news.

Example 17.6 **Is It Wining or Dining That Helps French Hearts?** The "French paradox" refers to the fact that French citizens have lower rates of heart disease than people in neighboring countries, yet seem to eat more saturated fats. Speculation in the early 1990s focused on red wine as a partial cause. The following article discusses some other possible explanations. Although the article is over 10 years old, the mystery of the French paradox continues, as does research trying to explain it. Note that no one is likely to change laws or behavior as a result of this research, it is simply interesting information.

Friday May 28 1:24 PM ET (1999)

Wine May Not Explain France's Lower Heart Disease Rate

NEW YORK, May 28 (Reuters Health)—Previous research has suggested that red wine consumption may explain why the French have a much lower rate of heart disease than other nations. But in a report published this week in the *British Medical Journal*, two UK experts dispute this theory, suggesting instead that the reduced risk lies in the fact that the French diet traditionally contains less animal fat.

And because the levels of animal fat and cholesterol in the French diet have risen in the last 15 years, their explanation suggests that it is only a matter of time before the heart disease rate in France catches up with the rate in Britain, where heart disease mortality is about four times higher.

Dr. Malcolm Law and Professor Nicholas Wald from the Wolfson Institute of Preventive Medicine in London, UK, suggest that the tendency of French doctors to attribute heart disease deaths to other causes "could account for about 20% of the difference," according to the report.

Law and Wald do give red wine some credit as a preventive factor, but say that this effect is small. The high consumption of red wine in France, they write, "explains (less than 5%) of the difference."

On the other hand, they provide considerable statistical evidence to support their "time lag" theory. "Mortality from ... heart disease," the authors write, "was strongly associated with past animal fat consumption ... and past (serum cholesterol) values, but not with recent values."

"Animal fat consumption and serum cholesterol concentration have been similar in France and Britain for a relatively short time—about 15 years," write Law and Wald. "Serum cholesterol concentration in 1970 was ... lower in France than in Britain ... and this explains most of its lower mortality from heart disease," they suggest.

Several related editorials question Law and Wald's hypothesis. Meir Stampfer and Eric Rimm from Harvard University, Boston, Massachusetts, write, "Obviously, other factors must play a role.... We think it more likely that the difference in coronary mortality rests on behavioral (especially dietary) differences that have not received adequate attention."

In another commentary, D.J.P. Barker from the University of Southampton in Southampton, UK, writes that "recent trends in coronary heart disease are only weakly related to trends in serum cholesterol" and argues that "coronary heart disease originates in utero, through adaptations that the fetus makes to undernutrition."

Finally, Johan Mackenbach and Anton Kunst from Erasmus University in Rotterdam, the Netherlands, contend that "heterogeneity of populations should be taken into account."

But Law and Wald dismiss these arguments. "We believe," they write in response, "that the time lag explanation is the major reason and that the alternative explanations offered in the commentaries are quantitatively unimportant."

Source: British Medical Journal (1999), 318, 1471–1480.

Example 17.7 **Give Her the Car Keys** Memory is a fascinating ability, and most of us wish we had more of that ability. Studies about memory may eventually help us find ways to improve it. In the meantime, the following study is simply interesting. If it leads to any decision at all, it would be related to who thinks they should get the car keys and who actually should. We'll let you figure it out.

Thursday March 26, 1998 1:30 PM EST Yahoo Health News

Women Remember Item Location Better

NEW YORK (Reuters)—Women are better at remembering where objects are than men, but show less confidence in their memory, according to a study.

"When it comes to memory, women have more skill than confidence, and men have more confidence than skill," concludes Dr. Robin West, a University of Florida (UF) psychology professor and researcher.

West, along with UF psychology graduate student Duana Welch, compared the spatial memories of over 300 healthy men and women of various ages in a study funded by the National Institutes of Mental Health.

They used specially-designed computer tests that had each participant place 20 common household "objects" into one of 12 "rooms" (e.g., kitchen, bedroom, bathroom) depicted on the screen. After a 40-minute interval, each participant was asked to remember the location of each object.

According to Welch, the test focused on "something that has very important day-to-day meaning for people; that is, the ability to find where you put something." The results? West and Welch say they found "younger adults performing better than older adults, and women performing better than men." In fact, women scored an average of 14.4 points on the test, compared with the men's average score of 13.5.

Those scores didn't meet male expectations of their own memory skills, the researchers say. In a statement issued by the university, West and Welch say that when questioned, "men tended to overestimate their ability to remember object locations," while women tended to underestimate their powers of recall. The study authors note these findings "are in keeping with other literature suggesting that women are not as confident about their cognitive (intellectual) abilities as they should be."

Those types of gender-based insecurities may be ending, however. In a previous study involving married couples, West discovered that gender-specific differences in memory-confidence disappeared among younger (ages 35 to 45) couples. West credits those results on "historical changes in the beliefs and attitudes of our culture. The older women grew up in a time when men were thought to have more mental ability than women. Such beliefs—learned in childhood and reinforced over time—are hard to change."

A detailed meta-analysis of 86 effect sizes derived from studies of gender differences in memory of object locations was published in 2007 (Voyer et al.). They found that females persisted in having better memory of where most types of objects were located, for all age groups except children under age 13, where the difference between males and females was negligible. The average effect size for the difference was a somewhat small value of 0.269. (See Section 13.7 for an explanation of effect size.) One exception to the female advantage was for objects classified as "masculine," for which males displayed better memory.

Chapter exercises are on pages 664–666.

17.6 Getting to Know You

As social beings, we are curious about how others think and behave. What do they do with their time? Are we in the majority with our opinions on controversial issues? At what age do people typically get married? What proportion of the population is left-handed? What proportion is gay or lesbian? Are people basically honest? Many of these questions can be answered by surveying random or representative samples.

Most national governments have agencies that collect samples to answer some of these questions on a routine basis. The U.S. government collects diverse and numerous statistics on the behavior and opinions of its people. The website http://www.fedstats.gov, which is headlined "Celebrating over 10 years of making statistics from more than 100 agencies available to citizens everywhere," provides links to all kinds of information about the U.S. population.

Example 17.8 **Lifestyle Statistics from the Census Bureau** The U.S. Census Bureau collects voluminous data on many aspects of American life. The Current Population Survey and American Community Survey poll a random sample of U.S. households on a wide variety of topics on an ongoing basis. From these surveys, trends in lifestyle decisions can be tracked. Here are a few examples of interesting statistics from resources available at http://search.census.gov:

- The median age at first marriage in 2009 was 28.1 years for men and 25.9 years for women, the highest ages since reporting began in 1890. The lowest ages from 1890 to 2009 were reported for 1956, with medians of 22.5 years for men and 20.1 years for women. The biggest difference in median ages for men and women occurred in 1890, when men had a median age of 26.1, fully 4.1 years older than the median age of 22.0 for women. Since 1948 the gap in ages has always been under 3 years, but the median has always been higher for men than for women (Source: http://www.census.gov/population/socdemo/hh-fam/ms2.xls).
- In 1970 the most common type of household in the United States was married couples with children, representing 40.3% of all households. But in 2007 only 22.5% of households were married couples with children. The most common type of household in 2007 was unmarried couples without children, comprising 28.3% of households (Source: http://www.census.gov/population/socdemo/hh-fam/p20-561.pdf).
- The mean income for all households in the United States in 2006 was $66,570, but this included single-person households. For households that included couples, income differed by status of the couple. For married opposite-sex couples the mean was $86,747, for unmarried opposite-sex couples it was $61,440, for unmarried male–male couples it was the highest, at $105,289, and for unmarried female–female couples it was still higher than for married couples, at $89,528. One possible explanation is that opposite-sex couples are more likely to have children and to have one partner who is not working. Also, remember that very high incomes raise the mean income. The median income for all households was only $48,201; medians were not available for different household types (Source: www.census.gov/population/www/socdemo/files/ssex-tables-2006.xls).

Note that these statistics are presented as if they were known for the population, but in fact, they are based on random samples of various sizes. Consequently, they have a margin of error that could be computed from knowing the sample size. Some of the reports gave the margin of error, while others did not.

Example 17.9 **In Whom Do We Trust?** If you want to know about the opinions of Americans on almost any topic, visit the website for the Gallup Organization (http://www.gallup.com) and type in the keyword of your choice. (Unfortunately, Gallup now charges for the results of some of its polls.) For instance, a news release on December 9, 2009, was headlined "U.S. Clergy, Bankers See New Lows in Honesty/Ethics Ratings." The accompanying story (Jeffrey M. Jones, http://www.gallup.com/poll/124628/Clergy-Bankers-New-Lows-Ethics-Ratings.aspx) reported on a poll that has been taken yearly since 1977. Respondents were asked, "Please tell me how you would rate the honesty and ethical standards of people in these different fields—very high, high, average, low, or very low," and were then given a list of several dozen occupations, in random order. Some of the results highlighted in the news release follow.

- "The percentage of Americans rating the honesty and ethics of clergy as very high or high is down to 50% in 2009, the lowest percentage it has been in the 32 years Gallup has measured it. In last year's Honesty and Ethics update, 56% of Americans rated the clergy's honesty and ethics very high or high. Still, ratings of the clergy remain high on a relative basis, ranking 8th of the 22 professions tested this year."
- "In addition to the clergy and bankers, ratings of stockbrokers have hit a new low, and ratings of business executives, members of Congress, and lawyers have tied their previous lows."
- "The 63% very high/high ratings for police officers are their best since 2001—shortly after the Sept. 11 terrorist attacks—and the second highest in the 30+ years Gallup has asked about this profession."
- "The new poll also documents significant decreases in the evaluated honesty of dentists [57% very high/high] and psychiatrists [33% very high/high] since 2006. Additionally, the four-point decline in ratings of senators over this time period leaves them with a new low rating [11% very high/high]."
- "Nurses continue to rate as the most highly regarded profession in terms of honesty and ethics [83% very high/high]."

Of course, to interpret the results correctly, you need to know that "[r]esults are based on telephone interviews with 1017 national adults, aged 18 and older, conducted Nov. 20–22, 2009. For results based on the total sample of national adults, one can say with 95% confidence that the maximum margin of sampling error is ±4 percentage points."

Chapter exercises are on pages 664–666.

17.7 Words to the Wise

We hope that by now you realize that you are likely to encounter statistical studies in your personal and professional life and that they can provide useful information. We also hope you realize that results from such studies can be misleading if you don't interpret them wisely.

Throughout this book we have presented warnings and examples of how statistical studies are often misinterpreted. The ten guiding principles presented here are a synthesis of those warnings. Keeping these principles in mind when you read statistical studies will help you gain wisdom about the world around you while maintaining a healthy dose of skepticism about legitimate conclusions.

IN SUMMARY Ten Guiding Principles

1. A representative sample can be used to make inferences about a larger population, but descriptive statistics are the only useful results for an unrepresentative sample.
2. Cause and effect can be inferred from randomized experiments but generally not from observational studies, in which confounding variables are likely to cloud the interpretation.
3. A conservative estimate of *sampling* error in a survey is the margin of error, $1/\sqrt{n}$. This value provides a bound on the difference between the true proportion and the sample proportion that holds for at least 95% of properly conducted surveys.
4. The margin of error does *not* include nonsampling error, such as errors due to biased wording, nonresponse, and so on.
5. When the individuals measured constitute the whole population, there is no need for statistical inference because the truth is known.
6. A significance test based on a very large sample is likely to produce a statistically significant result even if the true value is close to the null value. In such cases, it is wise to examine the magnitude of the parameter with a confidence interval to determine whether the result has practical importance.
7. A significance test based on a small sample may not produce a statistically significant result even if the true value differs substantially from the null value. Because this is so, it is important that the null hypothesis not be "accepted," even when we don't have enough evidence to reject it.
8. When deciding how readily to reject the null hypothesis (what significance level to use), it is important to consider the consequences of Type 1 and Type 2 errors. If a Type 1 error has serious consequences, the level of significance should be small. If a Type 2 error is more serious, a higher level of significance should be used.
9. A study that examines many hypotheses could find one or more statistically significant results just by chance, so you should try to find out how many tests were conducted when you read about a significant result. For instance, with a level of significance of .05, about 1 test in 20 will result in statistical significance when all of the null hypotheses are correct and the tests are independent. It is common in large studies to find that one test attracts media attention, so it is important to know whether that test was the only one out of many conducted that achieved statistical significance.
10. You will sometimes read that researchers were surprised to find "no effect" and that a study "failed to replicate" an earlier finding of statistical significance. In that case, consider two possible explanations. One is that the sample size was too small and the test had low power. The other possibility is that the result in the first study was based on a Type 1 error. This explanation is particularly likely if the effect in that study was moderate and was part of a larger study that covered multiple hypotheses.

We hope that we have convinced you that statistics is interesting and useful to you in your daily life. If not, check to make sure you are still breathing, your heart is beating, and your mind has not turned off!

In Summary Boxes

Note to the Reader: Exercises for all sections are combined for this chapter.

Exercises

◆ Denotes that the dataset is available on the companion website, http://www.cengage.com/statistics/Utts4e, but it is not required to solve the exercise.

Bold exercises have answers in the back of the text.

Skillbuilder Exercises

17.1 In each of the following cases, explain whether the results can be extended to a larger population. If so, explain what the population is. If not, explain why not.

a. A company measured the salaries of all of its employees and found that the average salary for men was 10% higher than the average salary for women.

b. A geologist at Yellowstone National Park measured the length of time for each eruption and the waiting time until the next eruption for the Old Faithful geyser from August 1–8 and 16–23, 1978 (data from Weisberg, 1985, p. 231). The correlation between the two measurements was 0.877, indicating that waiting time until the next eruption is strongly correlated with the duration of the previous eruption.

17.2 In each of the following cases, explain whether the results can be extended to a larger population. If so, explain what the population is. If not, explain why not.

a. A poll following presidential debates in an election year was conducted by using random-digit dialing. When people answered the phone, they were asked whether they had watched the debates. If the answer was yes, they were asked who they thought won. If the answer was no, the call was terminated. The parameters of interest were the proportions of U.S. adults who thought each of the two candidates had won the debate.

b. A website conducts a poll by placing a box in the lower right corner asking "Is this website your major source for entertainment news?" Viewers can click a "yes" button or a "no" button to answer the question.

17.3 A study of twins adopted and raised apart found that in situations for which one was raised by parents with college degrees and the other was not, there was an average 5-point difference in IQs, with the adopted child of the college-educated parents having the higher IQ. Can we conclude that having parents with college degrees causes adopted children to have higher IQs? Explain.

17.4 Elderly patients were randomly assigned to receive either acupuncture or massage to help reduce lower back pain. It was found that those who received massage reported less pain on the day following treatment than did those receiving acupuncture. Can we conclude that massage caused the pain to be lessened, compared with acupuncture?

17.5 Explain what condition(s) need to be met in a study in order for a cause-and-effect conclusion to be made.

17.6 Explain what condition(s) need to be met to allow the results of a study to be extended to a larger population rather than simply applied to the sample data.

17.7 Suppose that it is morning and you need to be outside for half an hour later in the day. Your favorite weather website says there is a 40% chance of rain, and you must decide whether to take an umbrella with you.

a. Create a table for this situation, similar to the ones shown in the examples of "making personal decisions" in Section 17.3.

b. What decision would you be likely to make? Explain your reasoning.

17.8 Suppose that you have some food in your refrigerator that doesn't seem to be spoiled, but the "use by" date on it is two weeks ago. You must decide whether to eat it or to throw it away.

a. Create a table for this situation, similar to the ones shown in the examples of "making personal decisions" in Section 17.3.

b. What decision would you be likely to make? Explain your reasoning.

17.9 Refer to Case Study 13.1, in which men were able to exercise longer in a warm room after ingesting an ice slushie than after ingesting cold water.

a. Can we conclude that ingesting the slushie was the *cause* of the increased stamina? Explain.

b. Ten healthy males with an average age of 28 and who routinely did moderate exercise were recruited to participate in this study. Do you think the results can be extended to a population beyond those ten men? If so, to what population? Explain.

17.10 Refer to the study published in 2010 by DeNoon, described in Example 17.3, in which teens at risk of psychosis were given either fish oil or a placebo.

a. Can a cause-and-effect conclusion be made? Explain.

b. Do you think the results can be extended to a population beyond the 81 teens included in the study? If so, to what population? Explain.

Chapter Exercises

17.11 Refer to the headlines on page 652, in which observational studies were done, but the headlines implied cause-and-effect conclusions. Find an example of this phenomenon on a news website. State the headline, briefly explain how the study was done, and provide the web address in your answer.

17.12 Find an example of an observational study in the news and answer these questions about it:

a. Explain why it fits the definition of an observational study rather than a randomized experiment.

b. Describe an explanatory and response variable considered in the study.

c. Using the variables in part (b), give an example of a possible confounding variable and explain why it fits the definition of a confounding variable.

d. Write two headlines to accompany the study, one that is correct and one that implies a conclusion that is not justified. Explain which is which and why. (You may use the actual headline accompanying the story as one of these.)
e. Can the results of the study be extended to a larger population? Explain.

17.13 Find an example of a randomized experiment in the news and answer these questions about it:

a. Explain why it fits the definition of a randomized experiment rather than an observational study.
b. Describe an explanatory and response variable considered in the study.
c. Using the variables in part (b), give an example of a possible interacting variable, and explain why it fits the definition of an interacting variable. You may use one that is actually measured in the study and discussed in the news report.
d. Give the headline that was provided with the news article, and discuss whether it is justified based on the study.
e. Can the results of the study be extended to a larger population? Explain.

17.14 Example 17.5 (p. 658) describes California's antismoking measures and subsequent reduction in lung cancer rates. Read the example and the material that follows it. Using the "nonstatistical considerations to assess cause and effect" given in Section 17.1, discuss the extent to which you think a cause-and-effect conclusion can be made about the relationship between the antismoking measures and reduction in lung cancer rates.

17.15 Refer to Example 17.5 (p. 658), which describes California's antismoking measures and reduction in lung cancer. An article in *The New England Journal of Medicine* (Fichtenberg and Glantz, 2000) studied mortality from heart disease in California during the same period. The conclusion the authors reached was that "[a] large and aggressive tobacco-control program is associated with a reduction in deaths from heart disease in the short run."

a. The conclusion is based on results obtained through methods similar to those in Example 17.5. Is the wording of the conclusion justified? Explain.
b. Write two headlines to accompany a news report on this study, one that presents the conclusion accurately and one that implies a conclusion that is not justified. Explain your reasoning.

17.16 Find a statistical study in the news that is of interest to you. Refer to the four reasons given in Section 17.2 for why statistical studies are popular and useful. Describe the study, and explain which one (or more) of the four reasons are relevant to it.

Exercises 17 to 34 are based on the following three reports taken from news organization websites. In each case, the headline and a few sentences from the news story are provided. The original source for Studies #2 and #3 are on the companion website.

Study #1: "*Calcium Lowers Blood Pressure in Black Teens:* In the new study, researchers followed 116 African-American teens who were enrolled in grades 10 through 12 at three high schools in Los Angeles. Over the course of eight weeks, the study participants were given either a placebo or a calcium supplement. They were also asked to fill out food questionnaires. Researchers measured the teens' blood pressure at the outset of the study and two, four and eight weeks later" (Linda Carroll, Medical Tribune News Service, September 7, 1998; study published in the *American Journal of Clinical Nutrition* (1998), 68, 648–655).

Study #2: "*Daily Two-Mile Walk Halves Death Risk:* Between 1980 and 1982, multicenter researchers in the Honolulu Heart Program studied 707 nonsmoking, retired men, aged 61 to 81 years, and collected mortality data on these men over the following 12 years. During the study, 208 of the men died. The study results show that while 43.1% of men who walked less than one mile per day died, only half this figure—21.5%—of the men who walked more than two miles per day died" (Reuters News, January 8, 1998; study published in *The New England Journal of Medicine* (1998), 338, 94–99).

Study #3: "*Tea Doubles Chance of Conception:* Tea for two might make three, according to a new study. The investigators asked 187 women, each of whom said they were trying to conceive, to record information concerning their daily dietary intake over a one-year period. According to the researchers, an analysis of the data suggests drinking one-half cup or more of tea daily approximately doubled the odds of conception per cycle, compared with non-tea drinkers. The mechanism behind tea's apparent influence on fertility remains unclear. But the investigators point out that, on average, tea drinking is associated with a 'preventive or healthier' lifestyle, which might encourage conception" (Reuters News, February 27, 1998; study published in the *American Journal of Public Health* (1998), 88, 2, 270–274).

17.17 Read the headlines and description for Study #1.

a. What are the explanatory and response variables for this study?
b. Does the headline imply a cause-and-effect relationship?

17.18 Read the headlines and description for Study #2.

a. What are the explanatory and response variables for this study?
b. Does the headline imply a cause-and-effect relationship?

17.19 Read the headlines and description for Study #3.

a. What are the explanatory and response variables for this study?
b. Does the headline imply a cause-and-effect relationship?

17.20 Do you think Study #1 was a randomized experiment or an observational study? Explain. Based on your answer, explain whether the headline is justified.

17.21 Do you think Study #2 was a randomized experiment or an observational study? Explain. Based on your answer, explain whether the headline is justified.

17.22 Give an example of a possible confounding variable for Study #2. Explain why you think it fits the definition of a confounding variable (see p. 191).

17.23 Do you think Study #3 was a randomized experiment or an observational study? Explain. Based on your answer, explain whether the headline is justified.

17.24 Give an example of a possible confounding variable for Study #3. Explain why you think it fits the definition of a confounding variable (see p. 191).

17.25 Can the results of Study #1 be applied to a larger population? If so, to what population? If not, why not?

17.26 Can the results of Study #2 be applied to a larger population? If so, to what population? If not, why not?

17.27 Can the results of Study #3 be applied to a larger population? If so, to what population? If not, why not?

17.28 In Study #1, participants were also asked to fill out a food questionnaire. Another quote from the news article was: "Overall, diastolic blood pressure—the second of the two numbers given when blood pressure is measured—dropped about two points in the group given calcium supplements. But the drop was almost five points in teens with a particularly low intake of calcium-containing foods."

a. Why do you think the researchers wanted the information provided by the food questionnaire?

b. Is "intake of calcium-containing foods" a confounding variable or an interacting variable in this study? Explain.

17.29 The article for Study #3 also noted: "The study results also show that the women who consumed the most tea 'had significantly higher energy and fat intakes' than women who drank less tea."

a. Do you think the word *significantly* in this quote refers to statistical significance? If not, why not? If so, describe in words the null and alternative hypotheses you think the researchers tested to reach that conclusion.

b. Are energy and fat intake confounding variables or interacting variables in this study? Explain.

17.30 Here are some additional quotes from the article for Study #2. Comment on each quote using issues covered in this chapter.

a. "The investigators also report that 'when the distance walked is increased by one mile per day ... the risk of death can be reduced by 19%.'"

b. "Cancer was the most common cause of death in the men studied. The research team found that 13.4% of the men who walked less than one mile per day died of cancer, but only 5.3% of those who walked more than two miles per day died of cancer. Even when age was included in the analysis, this result held."

c. "Walking also appeared to protect against dying from heart disease and stroke, but the findings were not statistically significant, meaning that they could have occurred by chance."

17.31 The article for Study #3 also said: "The California researchers say they found no significant association between coffee intake and fertility, but 'in later cycles ... there was a suggestion that both total caffeine and coffee were associated with a nonsignificant reduction in fertility,' a finding in keeping with previous research which pointed to the possibility that coffee or caffeine might lower conception rates."

a. The phrase "*no* significant association between coffee intake and fertility" seems to contradict the "suggestion that both total caffeine and coffee were associated with a nonsignificant reduction in fertility." Explain this apparent contradiction.

b. Refer to Item 10 in Section 17.7, and discuss it in the context of this quote.

17.32 The two main variables measured in Study #1 were the treatment given to the participant and the participant's blood pressure.

a. For each of the two variables, determine whether it was recorded as a categorical variable or a quantitative variable. If it was a categorical variable, specify the categories.

b. Researchers measured the participants' blood pressure at the beginning of the study and 2, 4, and 8 weeks later. If they wanted to determine whether blood pressure had dropped significantly from the beginning of the study to eight weeks later for the participants in the calcium group, would they use a paired *t*-test or a test for the difference in means for independent samples? Explain.

c. Refer to part (b). Write the null and alternative hypotheses that the researchers would use.

d. If the researchers wanted to examine the change in blood pressure after eight weeks of taking calcium compared to 8 weeks of taking a placebo, what hypothesis-testing method would they use?

e. Refer to part (d). Write the null and alternative hypotheses that the researchers would use.

17.33 Refer to the headline in Study #2 and to the statement "while 43.1% of men who walked less than one mile per day died, only half this figure—21.5%—of the men who walked more than two miles per day died."

a. What two categorical variables measured on each participant are used in making that statement? What are the categories for each one?

b. If the researchers wanted to determine whether the result was statistically significant, what inference procedure would be appropriate? Be specific.

17.34 The article for Study #3 reported that "[a]t the end of each calendar month, each woman in the study completed a food-frequency questionnaire covering the past month's intake of common foods, including the frequency and serving size of regular and decaffeinated coffees, teas, and sodas" (p. 271). Which one of the issues discussed in the "Ten Guiding Principles" in Section 17.7 is likely to be a concern, given this information about the study? Explain.

(Source for Exercises 17.17 to 17.34, Study #3: Caan B, Quesenberry CP, Coates AO [1998]. "Differences in fertility associated with caffeinated beverage consumption." American Journal of Public Health, *88(2):270-274 Reprinted with permission from APHA Books.)*

(Source for Exercises 17.18, 17.21, 17.22, 17.26, 17.30, 17.33: From New England Journal of Medicine, *1998, Vol. 338 (2), pp. 94–99. [See p. 701 for complete credit.])*

Appendix of Tables

Table A.1 Standard Normal Probabilities (for $z \le 0$)

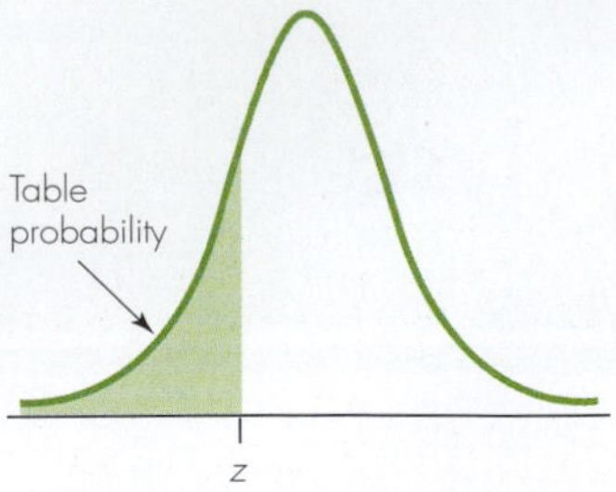

z	.00	.01	.02	.03	.04	.05	.06	.07	.08	.09
−3.4	.0003	.0003	.0003	.0003	.0003	.0003	.0003	.0003	.0003	.0002
−3.3	.0005	.0005	.0005	.0004	.0004	.0004	.0004	.0004	.0004	.0003
−3.2	.0007	.0007	.0006	.0006	.0006	.0006	.0006	.0005	.0005	.0005
−3.1	.0010	.0009	.0009	.0009	.0008	.0008	.0008	.0008	.0007	.0007
−3.0	.0013	.0013	.0013	.0012	.0012	.0011	.0011	.0011	.0010	.0010
−2.9	.0019	.0018	.0018	.0017	.0016	.0016	.0015	.0015	.0014	.0014
−2.8	.0026	.0025	.0024	.0023	.0023	.0022	.0021	.0021	.0020	.0019
−2.7	.0035	.0034	.0033	.0032	.0031	.0030	.0029	.0028	.0027	.0026
−2.6	.0047	.0045	.0044	.0043	.0041	.0040	.0039	.0038	.0037	.0036
−2.5	.0062	.0060	.0059	.0057	.0055	.0054	.0052	.0051	.0049	.0048
−2.4	.0082	.0080	.0078	.0075	.0073	.0071	.0069	.0068	.0066	.0064
−2.3	.0107	.0104	.0102	.0099	.0096	.0094	.0091	.0089	.0087	.0084
−2.2	.0139	.0136	.0132	.0129	.0125	.0122	.0119	.0116	.0113	.0110
−2.1	.0179	.0174	.0170	.0166	.0162	.0158	.0154	.0150	.0146	.0143
−2.0	.0228	.0222	.0217	.0212	.0207	.0202	.0197	.0192	.0188	.0183
−1.9	.0287	.0281	.0274	.0268	.0262	.0256	.0250	.0244	.0239	.0233
−1.8	.0359	.0351	.0344	.0336	.0329	.0322	.0314	.0307	.0301	.0294
−1.7	.0446	.0436	.0427	.0418	.0409	.0401	.0392	.0384	.0375	.0367
−1.6	.0548	.0537	.0526	.0516	.0505	.0495	.0485	.0475	.0465	.0455
−1.5	.0668	.0655	.0643	.0630	.0618	.0606	.0594	.0582	.0571	.0559
−1.4	.0808	.0793	.0778	.0764	.0749	.0735	.0721	.0708	.0694	.0681
−1.3	.0968	.0951	.0934	.0918	.0901	.0885	.0869	.0853	.0838	.0823
−1.2	.1151	.1131	.1112	.1093	.1075	.1056	.1038	.1020	.1003	.0985
−1.1	.1357	.1335	.1314	.1292	.1271	.1251	.1230	.1210	.1190	.1170
−1.0	.1587	.1562	.1539	.1515	.1492	.1469	.1446	.1423	.1401	.1379
−0.9	.1841	.1814	.1788	.1762	.1736	.1711	.1685	.1660	.1635	.1611
−0.8	.2119	.2090	.2061	.2033	.2005	.1977	.1949	.1922	.1894	.1867
−0.7	.2420	.2389	.2358	.2327	.2296	.2266	.2236	.2206	.2177	.2148
−0.6	.2743	.2709	.2676	.2643	.2611	.2578	.2546	.2514	.2483	.2451
−0.5	.3085	.3050	.3015	.2981	.2946	.2912	.2877	.2843	.2810	.2776
−0.4	.3446	.3409	.3372	.3336	.3300	.3264	.3228	.3192	.3156	.3121
−0.3	.3821	.3783	.3745	.3707	.3669	.3632	.3594	.3557	.3520	.3483
−0.2	.4207	.4168	.4129	.4090	.4052	.4013	.3974	.3936	.3897	.3859
−0.1	.4602	.4562	.4522	.4483	.4443	.4404	.4364	.4325	.4286	.4247
−0.0	.5000	.4960	.4920	.4880	.4840	.4801	.4761	.4721	.4681	.4641

In the Extreme (for $z < 0$)

z	−3.09	−3.72	−4.26	−4.75	−5.20	−5.61	−6.00
Probability	.001	.0001	.00001	.000001	.0000001	.00000001	.000000001

S-PLUS was used to determine information for the "In the Extreme" portion of the table.

Table A.1 Standard Normal Probabilities (for $z \geq 0$)

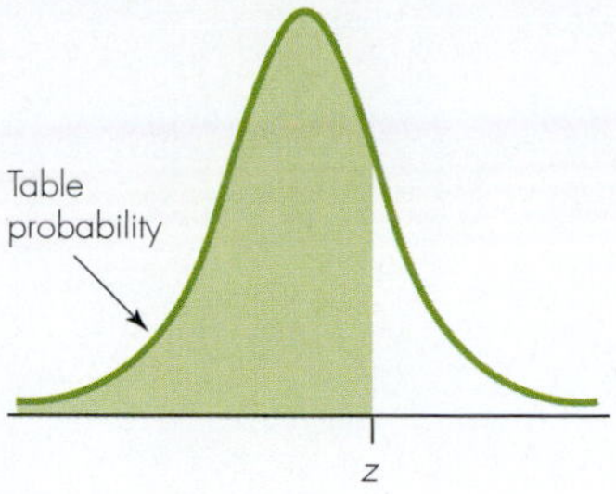

z	.00	.01	.02	.03	.04	.05	.06	.07	.08	.09
0.0	.5000	.5040	.5080	.5120	.5160	.5199	.5239	.5279	.5319	.5359
0.1	.5398	.5438	.5478	.5517	.5557	.5596	.5636	.5675	.5714	.5753
0.2	.5793	.5832	.5871	.5910	.5948	.5987	.6026	.6064	.6103	.6141
0.3	.6179	.6217	.6255	.6293	.6331	.6368	.6406	.6443	.6480	.6517
0.4	.6554	.6591	.6628	.6664	.6700	.6736	.6772	.6808	.6844	.6879
0.5	.6915	.6950	.6985	.7019	.7054	.7088	.7123	.7157	.7190	.7224
0.6	.7257	.7291	.7324	.7357	.7389	.7422	.7454	.7486	.7517	.7549
0.7	.7580	.7611	.7642	.7673	.7704	.7734	.7764	.7794	.7823	.7852
0.8	.7881	.7910	.7939	.7967	.7995	.8023	.8051	.8078	.8106	.8133
0.9	.8159	.8186	.8212	.8238	.8264	.8289	.8315	.8340	.8365	.8389
1.0	.8413	.8438	.8461	.8485	.8508	.8531	.8554	.8577	.8599	.8621
1.1	.8643	.8665	.8686	.8708	.8729	.8749	.8770	.8790	.8810	.8830
1.2	.8849	.8869	.8888	.8907	.8925	.8944	.8962	.8980	.8997	.9015
1.3	.9032	.9049	.9066	.9082	.9099	.9115	.9131	.9147	.9162	.9177
1.4	.9192	.9207	.9222	.9236	.9251	.9265	.9279	.9292	.9306	.9319
1.5	.9332	.9345	.9357	.9370	.9382	.9394	.9406	.9418	.9429	.9441
1.6	.9452	.9463	.9474	.9484	.9495	.9505	.9515	.9525	.9535	.9545
1.7	.9554	.9564	.9573	.9582	.9591	.9599	.9608	.9616	.9625	.9633
1.8	.9641	.9649	.9656	.9664	.9671	.9678	.9686	.9693	.9699	.9706
1.9	.9713	.9719	.9726	.9732	.9738	.9744	.9750	.9756	.9761	.9767
2.0	.9772	.9778	.9783	.9788	.9793	.9798	.9803	.9808	.9812	.9817
2.1	.9821	.9826	.9830	.9834	.9838	.9842	.9846	.9850	.9854	.9857
2.2	.9861	.9864	.9868	.9871	.9875	.9878	.9881	.9884	.9887	.9890
2.3	.9893	.9896	.9898	.9901	.9904	.9906	.9909	.9911	.9913	.9916
2.4	.9918	.9920	.9922	.9925	.9927	.9929	.9931	.9932	.9934	.9936
2.5	.9938	.9940	.9941	.9943	.9945	.9946	.9948	.9949	.9951	.9952
2.6	.9953	.9955	.9956	.9957	.9959	.9960	.9961	.9962	.9963	.9964
2.7	.9965	.9966	.9967	.9968	.9969	.9970	.9971	.9972	.9973	.9974
2.8	.9974	.9975	.9976	.9977	.9977	.9978	.9979	.9979	.9980	.9981
2.9	.9981	.9982	.9982	.9983	.9984	.9984	.9985	.9985	.9986	.9986
3.0	.9987	.9987	.9987	.9988	.9988	.9989	.9989	.9989	.9990	.9990
3.1	.9990	.9991	.9991	.9991	.9992	.9992	.9992	.9992	.9993	.9993
3.2	.9993	.9993	.9994	.9994	.9994	.9994	.9994	.9995	.9995	.9995
3.3	.9995	.9995	.9995	.9996	.9996	.9996	.9996	.9996	.9996	.9997
3.4	.9997	.9997	.9997	.9997	.9997	.9997	.9997	.9997	.9997	.9998

In the Extreme (for $z > 0$)

z	3.09	3.72	4.26	4.75	5.20	5.61	6.00
Probability	.999	.9999	.99999	.999999	.9999999	.99999999	.999999999

S-PLUS was used to determine information for the "In the Extreme" portion of the table.

Table A.2 *t** Multipliers for Confidence Intervals and Rejection Region Critical Values

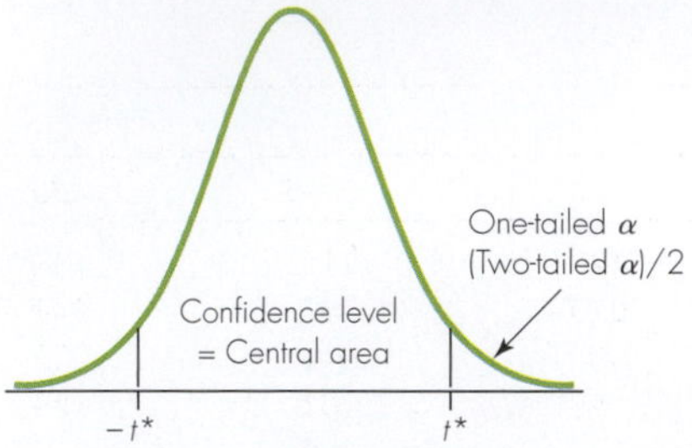

	Confidence Level						
df	*.80*	*.90*	*.95*	*.98*	*.99*	*.998*	*.999*
1	3.08	6.31	12.71	31.82	63.66	318.31	636.62
2	1.89	2.92	4.30	6.96	9.92	22.33	31.60
3	1.64	2.35	3.18	4.54	5.84	10.21	12.92
4	1.53	2.13	2.78	3.75	4.60	7.17	8.61
5	1.48	2.02	2.57	3.36	4.03	5.89	6.87
6	1.44	1.94	2.45	3.14	3.71	5.21	5.96
7	1.41	1.89	2.36	3.00	3.50	4.79	5.41
8	1.40	1.86	2.31	2.90	3.36	4.50	5.04
9	1.38	1.83	2.26	2.82	3.25	4.30	4.78
10	1.37	1.81	2.23	2.76	3.17	4.14	4.59
11	1.36	1.80	2.20	2.72	3.11	4.02	4.44
12	1.36	1.78	2.18	2.68	3.05	3.93	4.32
13	1.35	1.77	2.16	2.65	3.01	3.85	4.22
14	1.35	1.76	2.14	2.62	2.98	3.79	4.14
15	1.34	1.75	2.13	2.60	2.95	3.73	4.07
16	1.34	1.75	2.12	2.58	2.92	3.69	4.01
17	1.33	1.74	2.11	2.57	2.90	3.65	3.97
18	1.33	1.73	2.10	2.55	2.88	3.61	3.92
19	1.33	1.73	2.09	2.54	2.86	3.58	3.88
20	1.33	1.72	2.09	2.53	2.85	3.55	3.85
21	1.32	1.72	2.08	2.52	2.83	3.53	3.82
22	1.32	1.72	2.07	2.51	2.82	3.50	3.79
23	1.32	1.71	2.07	2.50	2.81	3.48	3.77
24	1.32	1.71	2.06	2.49	2.80	3.47	3.75
25	1.32	1.71	2.06	2.49	2.79	3.45	3.73
26	1.31	1.71	2.06	2.48	2.78	3.43	3.71
27	1.31	1.70	2.05	2.47	2.77	3.42	3.69
28	1.31	1.70	2.05	2.47	2.76	3.41	3.67
29	1.31	1.70	2.05	2.46	2.76	3.40	3.66
30	1.31	1.70	2.04	2.46	2.75	3.39	3.65
40	1.30	1.68	2.02	2.42	2.70	3.31	3.55
50	1.30	1.68	2.01	2.40	2.68	3.26	3.50
60	1.30	1.67	2.00	2.39	2.66	3.23	3.46
70	1.29	1.67	1.99	2.38	2.65	3.21	3.44
80	1.29	1.66	1.99	2.37	2.64	3.20	3.42
90	1.29	1.66	1.99	2.37	2.63	3.18	3.40
100	1.29	1.66	1.98	2.36	2.63	3.17	3.39
1000	1.282	1.646	1.962	2.330	2.581	3.098	3.300
Infinite	1.282	1.645	1.960	2.326	2.576	3.090	3.291
Two-tailed α	.20	.10	.05	.02	.01	.002	.001
One-tailed α	.10	.05	.025	.01	.005	.001	.0005

Note that the *t*-distribution with infinite df is the standard normal distribution.

Table A.3 One-Sided *p*-Values for Significance Tests Based on a *t*-Statistic

- A *p*-value in the table is the area to the right of *t*.
- Double the value if the alternative hypothesis is two-sided (not equal).

	Absolute Value of *t*-Statistic							
df	*1.28*	*1.50*	*1.65*	*1.80*	*2.00*	*2.33*	*2.58*	*3.00*
1	.211	.187	.173	.161	.148	.129	.118	.102
2	.164	.136	.120	.107	.092	.073	.062	.048
3	.145	.115	.099	.085	.070	.051	.041	.029
4	.135	.104	.087	.073	.058	.040	.031	.020
5	.128	.097	.080	.066	.051	.034	.025	.015
6	.124	.092	.075	.061	.046	.029	.021	.012
7	.121	.089	.071	.057	.043	.026	.018	.010
8	.118	.086	.069	.055	.040	.024	.016	.009
9	.116	.084	.067	.053	.038	.022	.015	.007
10	.115	.082	.065	.051	.037	.021	.014	.007
11	.113	.081	.064	.050	.035	.020	.013	.006
12	.112	.080	.062	.049	.034	.019	.012	.006
13	.111	.079	.061	.048	.033	.018	.011	.005
14	.111	.078	.061	.047	.033	.018	.011	.005
15	.110	.077	.060	.046	.032	.017	.010	.004
16	.109	.077	.059	.045	.031	.017	.010	.004
17	.109	.076	.059	.045	.031	.016	.010	.004
18	.108	.075	.058	.044	.030	.016	.009	.004
19	.108	.075	.058	.044	.030	.015	.009	.004
20	.108	.075	.057	.043	.030	.015	.009	.004
21	.107	.074	.057	.043	.029	.015	.009	.003
22	.107	.074	.057	.043	.029	.015	.009	.003
23	.107	.074	.056	.042	.029	.014	.008	.003
24	.106	.073	.056	.042	.028	.014	.008	.003
25	.106	.073	.056	.042	.028	.014	.008	.003
26	.106	.073	.055	.042	.028	.014	.008	.003
27	.106	.073	.055	.042	.028	.014	.008	.003
28	.106	.072	.055	.041	.028	.014	.008	.003
29	.105	.072	.055	.041	.027	.013	.008	.003
30	.105	.072	.055	.041	.027	.013	.008	.003
40	.104	.071	.053	.040	.026	.012	.007	.002
50	.103	.070	.053	.039	.025	.012	.006	.002
60	.103	.069	.052	.038	.025	.012	.006	.002
70	.102	.069	.052	.038	.025	.011	.006	.002
80	.102	.069	.051	.038	.024	.011	.006	.002
90	.102	.069	.051	.038	.024	.011	.006	.002
100	.102	.068	.051	.037	.024	.011	.006	.002
1000	.100	.067	.050	.036	.023	.010	.005	.001
Infinite	.1003	.0668	.0495	.0359	.0228	.0099	.0049	.0013

Note that the *t*-distribution with infinite df is the standard normal distribution.

Table A.4 Critical Values for *F*-Test (α = .05)

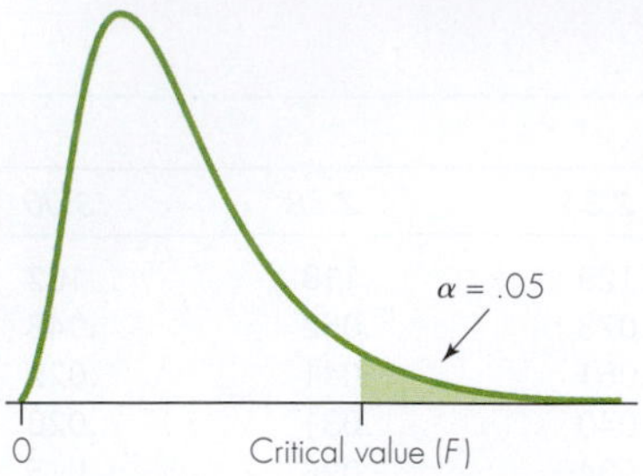

	Numerator df									
Denom df	*1*	*2*	*3*	*4*	*5*	*6*	*7*	*8*	*9*	*10*
1	161.45	199.50	215.71	224.58	230.16	233.99	236.77	238.88	240.54	241.88
2	18.51	19.00	19.16	19.25	19.30	19.33	19.35	19.37	19.38	19.40
3	10.13	9.55	9.28	9.12	9.01	8.94	8.89	8.85	8.81	8.79
4	7.71	6.94	6.59	6.39	6.26	6.16	6.09	6.04	6.00	5.96
5	6.61	5.79	5.41	5.19	5.05	4.95	4.88	4.82	4.77	4.74
6	5.99	5.14	4.76	4.53	4.39	4.28	4.21	4.15	4.10	4.06
7	5.59	4.74	4.35	4.12	3.97	3.87	3.79	3.73	3.68	3.64
8	5.32	4.46	4.07	3.84	3.69	3.58	3.50	3.44	3.39	3.35
9	5.12	4.26	3.86	3.63	3.48	3.37	3.29	3.23	3.18	3.14
10	4.96	4.10	3.71	3.48	3.33	3.22	3.14	3.07	3.02	2.98
11	4.84	3.98	3.59	3.36	3.20	3.09	3.01	2.95	2.90	2.85
12	4.75	3.89	3.49	3.26	3.11	3.00	2.91	2.85	2.80	2.75
13	4.67	3.81	3.41	3.18	3.03	2.92	2.83	2.77	2.71	2.67
14	4.60	3.74	3.34	3.11	2.96	2.85	2.76	2.70	2.65	2.60
15	4.54	3.68	3.29	3.06	2.90	2.79	2.71	2.64	2.59	2.54
16	4.49	3.63	3.24	3.01	2.85	2.74	2.66	2.59	2.54	2.49
17	4.45	3.59	3.20	2.96	2.81	2.70	2.61	2.55	2.49	2.45
18	4.41	3.55	3.16	2.93	2.77	2.66	2.58	2.51	2.46	2.41
19	4.38	3.52	3.13	2.90	2.74	2.63	2.54	2.48	2.42	2.38
20	4.35	3.49	3.10	2.87	2.71	2.60	2.51	2.45	2.39	2.35
21	4.32	3.47	3.07	2.84	2.68	2.57	2.49	2.42	2.37	2.32
22	4.30	3.44	3.05	2.82	2.66	2.55	2.46	2.40	2.34	2.30
23	4.28	3.42	3.03	2.80	2.64	2.53	2.44	2.37	2.32	2.27
24	4.26	3.40	3.01	2.78	2.62	2.51	2.42	2.36	2.30	2.25
25	4.24	3.39	2.99	2.76	2.60	2.49	2.40	2.34	2.28	2.24
26	4.23	3.37	2.98	2.74	2.59	2.47	2.39	2.32	2.27	2.22
27	4.21	3.35	2.96	2.73	2.57	2.46	2.37	2.31	2.25	2.20
28	4.20	3.34	2.95	2.71	2.56	2.45	2.36	2.29	2.24	2.19
29	4.18	3.33	2.93	2.70	2.55	2.43	2.35	2.28	2.22	2.18
30	4.17	3.32	2.92	2.69	2.53	2.42	2.33	2.27	2.21	2.16
40	4.08	3.23	2.84	2.61	2.45	2.34	2.25	2.18	2.12	2.08
50	4.03	3.18	2.79	2.56	2.40	2.29	2.20	2.13	2.07	2.03
60	4.00	3.15	2.76	2.53	2.37	2.25	2.17	2.10	2.04	1.99
70	3.98	3.13	2.74	2.50	2.35	2.23	2.14	2.07	2.02	1.97
80	3.96	3.11	2.72	2.49	2.33	2.21	2.13	2.06	2.00	1.95
90	3.95	3.10	2.71	2.47	2.32	2.20	2.11	2.04	1.99	1.94
100	3.94	3.09	2.70	2.46	2.31	2.19	2.10	2.03	1.97	1.93
200	3.89	3.04	2.65	2.42	2.26	2.14	2.06	1.98	1.93	1.88
500	3.86	3.01	2.62	2.39	2.23	2.12	2.03	1.96	1.90	1.85
1000	3.85	3.00	2.61	2.38	2.22	2.11	2.02	1.95	1.89	1.84

Table A.4 Critical Values for F-Test ($\alpha = .01$)

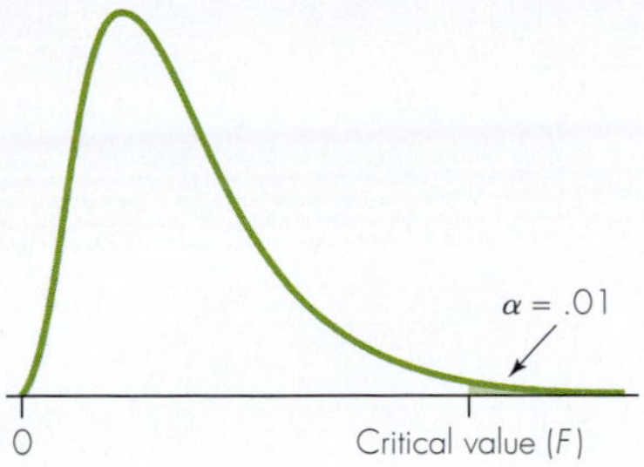

Denom df	Numerator df 1	2	3	4	5	6	7	8	9	10
1	4052	4999	5403	5625	5764	5859	5928	5981	6022	6056
2	98.50	99.00	99.17	99.25	99.30	99.33	99.36	99.37	99.39	99.40
3	34.12	30.82	29.46	28.71	28.24	27.91	27.67	27.49	27.35	27.23
4	21.20	18.00	16.69	15.98	15.52	15.21	14.98	14.80	14.66	14.55
5	16.26	13.27	12.06	11.39	10.97	10.67	10.46	10.29	10.16	10.05
6	13.75	10.92	9.78	9.15	8.75	8.47	8.26	8.10	7.98	7.87
7	12.25	9.55	8.45	7.85	7.46	7.19	6.99	6.84	6.72	6.62
8	11.26	8.65	7.59	7.01	6.63	6.37	6.18	6.03	5.91	5.81
9	10.56	8.02	6.99	6.42	6.06	5.80	5.61	5.47	5.35	5.26
10	10.04	7.56	6.55	5.99	5.64	5.39	5.20	5.06	4.94	4.85
11	9.65	7.21	6.22	5.67	5.32	5.07	4.89	4.74	4.63	4.54
12	9.33	6.93	5.95	5.41	5.06	4.82	4.64	4.50	4.39	4.30
13	9.07	6.70	5.74	5.21	4.86	4.62	4.44	4.30	4.19	4.10
14	8.86	6.51	5.56	5.04	4.69	4.46	4.28	4.14	4.03	3.94
15	8.68	6.36	5.42	4.89	4.56	4.32	4.14	4.00	3.89	3.80
16	8.53	6.23	5.29	4.77	4.44	4.20	4.03	3.89	3.78	3.69
17	8.40	6.11	5.19	4.67	4.34	4.10	3.93	3.79	3.68	3.59
18	8.29	6.01	5.09	4.58	4.25	4.01	3.84	3.71	3.60	3.51
19	8.18	5.93	5.01	4.50	4.17	3.94	3.77	3.63	3.52	3.43
20	8.10	5.85	4.94	4.43	4.10	3.87	3.70	3.56	3.46	3.37
21	8.02	5.78	4.87	4.37	4.04	3.81	3.64	3.51	3.40	3.31
22	7.95	5.72	4.82	4.31	3.99	3.76	3.59	3.45	3.35	3.26
23	7.88	5.66	4.76	4.26	3.94	3.71	3.54	3.41	3.30	3.21
24	7.82	5.61	4.72	4.22	3.90	3.67	3.50	3.36	3.26	3.17
25	7.77	5.57	4.68	4.18	3.85	3.63	3.46	3.32	3.22	3.13
26	7.72	5.53	4.64	4.14	3.82	3.59	3.42	3.29	3.18	3.09
27	7.68	5.49	4.60	4.11	3.78	3.56	3.39	3.26	3.15	3.06
28	7.64	5.45	4.57	4.07	3.75	3.53	3.36	3.23	3.12	3.03
29	7.60	5.42	4.54	4.04	3.73	3.50	3.33	3.20	3.09	3.00
30	7.56	5.39	4.51	4.02	3.70	3.47	3.30	3.17	3.07	2.98
40	7.31	5.18	4.31	3.83	3.51	3.29	3.12	2.99	2.89	2.80
50	7.17	5.06	4.20	3.72	3.41	3.19	3.02	2.89	2.78	2.70
60	7.08	4.98	4.13	3.65	3.34	3.12	2.95	2.82	2.72	2.63
70	7.01	4.92	4.07	3.60	3.29	3.07	2.91	2.78	2.67	2.59
80	6.96	4.88	4.04	3.56	3.26	3.04	2.87	2.74	2.64	2.55
90	6.93	4.85	4.01	3.53	3.23	3.01	2.84	2.72	2.61	2.52
100	6.90	4.82	3.98	3.51	3.21	2.99	2.82	2.69	2.59	2.50
200	6.76	4.71	3.88	3.41	3.11	2.89	2.73	2.60	2.50	2.41
500	6.69	4.65	3.82	3.36	3.05	2.84	2.68	2.55	2.44	2.36
1000	6.66	4.63	3.80	3.34	3.04	2.82	2.66	2.53	2.43	2.34

Table A.5 Chi-Square Distribution

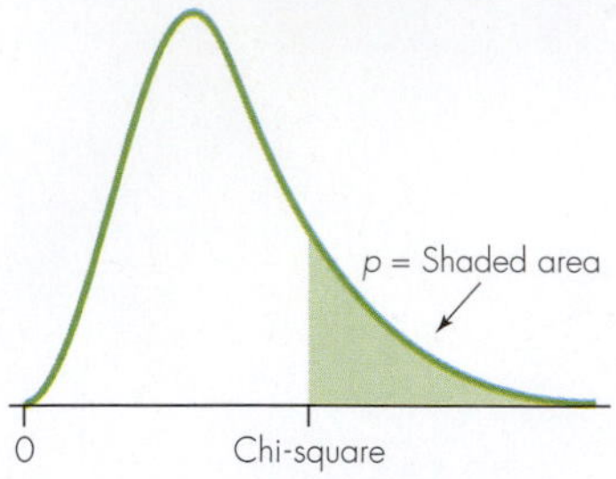

	p = Area to Right of Chi-Square Value								
df	*.50*	*.25*	*.10*	*.075*	*.05*	*.025*	*.01*	*.005*	*.001*
1	0.45	1.32	2.71	3.17	3.84	5.02	6.63	7.88	10.83
2	1.39	2.77	4.61	5.18	5.99	7.38	9.21	10.60	13.82
3	2.37	4.11	6.25	6.90	7.81	9.35	11.34	12.84	16.27
4	3.36	5.39	7.78	8.50	9.49	11.14	13.28	14.86	18.47
5	4.35	6.63	9.24	10.01	11.07	12.83	15.09	16.75	20.52
6	5.35	7.84	10.64	11.47	12.59	14.45	16.81	18.55	22.46
7	6.35	9.04	12.02	12.88	14.07	16.01	18.48	20.28	24.32
8	7.34	10.22	13.36	14.27	15.51	17.53	20.09	21.95	26.12
9	8.34	11.39	14.68	15.63	16.92	19.02	21.67	23.59	27.88
10	9.34	12.55	15.99	16.97	18.31	20.48	23.21	25.19	29.59
11	10.34	13.70	17.28	18.29	19.68	21.92	24.72	26.76	31.26
12	11.34	14.85	18.55	19.60	21.03	23.34	26.22	28.30	32.91
13	12.34	15.98	19.81	20.90	22.36	24.74	27.69	29.82	34.53
14	13.34	17.12	21.06	22.18	23.68	26.12	29.14	31.32	36.12
15	14.34	18.25	22.31	23.45	25.00	27.49	30.58	32.80	37.70
16	15.34	19.37	23.54	24.72	26.30	28.85	32.00	34.27	39.25
17	16.34	20.49	24.77	25.97	27.59	30.19	33.41	35.72	40.79
18	17.34	21.60	25.99	27.22	28.87	31.53	34.81	37.16	42.31
19	18.34	22.72	27.20	28.46	30.14	32.85	36.19	38.58	43.82
20	19.34	23.83	28.41	29.69	31.41	34.17	37.57	40.00	45.31
21	20.34	24.93	29.62	30.92	32.67	35.48	38.93	41.40	46.80
22	21.34	26.04	30.81	32.14	33.92	36.78	40.29	42.80	48.27
23	22.34	27.14	32.01	33.36	35.17	38.08	41.64	44.18	49.73
24	23.34	28.24	33.20	34.57	36.42	39.36	42.98	45.56	51.18
25	24.34	29.34	34.38	35.78	37.65	40.65	44.31	46.93	52.62
26	25.34	30.43	35.56	36.98	38.89	41.92	45.64	48.29	54.05
27	26.34	31.53	36.74	38.18	40.11	43.19	46.96	49.64	55.48
28	27.34	32.62	37.92	39.38	41.34	44.46	48.28	50.99	56.89
29	28.34	33.71	39.09	40.57	42.56	45.72	49.59	52.34	58.30
30	29.34	34.80	40.26	41.76	43.77	46.98	50.89	53.67	59.70

References

Abelson, R. P., E. F. Loftus, and A. G. Greenwald (1992). "Attempts to improve the accuracy of self-reports of voting," in *Questions about Questions,* J. M. Tanur (ed.), New York: Russell Sage Foundation, pp. 138–153.

American Statistical Association (1999). Ethical guidelines for statistical practice; available at http://www.amstat.org/about/ethicalguidelines.cfm.

Anastasi, Anne (1988). *Psychological Testing,* 6th ed., New York: Macmillan.

Anderson, M. J., and S. E. Fienberg (2001). *Who Counts? The Politics of Census-Taking in Contemporary America,* New York: Russell Sage Foundation.

Araf, J. (1994). "Leave Earth or perish: Sagan," *China Post,* December 14, 1994, p. 4.

Ashenfelter, Orley (1994). "Report on expected absentee ballot," revised March 29, 1994. Unpublished report submitted to Judge Clarence Newcomer, U.S. District Court, Eastern District of Pennsylvania, March 30, 1994.

Baird, D. D., and A. J. Wilcox (1985). "Cigarette smoking associated with delayed conception," *Journal of the American Medical Association,* Vol. 253, pp. 2979–2983.

Bem, D., and C. Honorton (1994). "Does psi exist? Replicable evidence for an anomalous process of information transfer," *Psychological Bulletin,* Vol. 115, No. 1, pp. 4–18.

Bickel, P. J., E. A. Hammel, and J. W. O'Connell (1975). "Sex bias in graduate admissions: Data from Berkeley," *Science,* Vol. 187, pp. 298–304.

Bloom B., and R. A. Cohen (2007). Summary Health Statistics for U.S. Children: National Health Interview Survey, 2006. National Center for Health Statistics. *Vital Health Stat,* Vol. 10, No. 234, pp. 1–79.

Borzekowski, D. L. G., T. N. Robinson, and J. D. Killen (2000). "Does the camera add 10 pounds? Media use, perceived importance of appearance, and weight concerns among teenage girls," *Journal of Adolescent Health,* Vol. 26, pp. 36–41.

Bryson, M. C. (1976). "The *Literary Digest* poll: Making of a statistical myth," *American Statistician,* Vol. 30, pp. 184–185.

Butterworth, P., T. Oz, B. Rodgers, and H. Berry (2008). Factors associated with relationship dissolution of Australian families with children. Social Policy Research Paper No. 37, Australian Government, Dept. of Families, Housing, Community Services and Indigenous Affairs. Accessed March 2010 at http://www.facsia.gov.au.

Capron, C., and M. Duyme (1991). "Children's IQs and SES of biological and adoptive parents in a balanced crossfostering study," *European Bulletin of Cognitive Psychology,* Vol. 11, No. 3, pp. 323–348.

Carpenter, J., J. Hagemaster, and B. Joiner (1998). Letter to the editor, *Journal of the American Medical Association,* Vol. 280, No. 22, p. 1905.

Carter, C. L., D. Y. Jones, A. Schatzkin, and L. A. Brinton (1989). "A prospective study of reproductive, familial, and socioeconomic risk factors for breast cancer using NHANES I data," *Public Health Reports,* Vol. 104, January–February 1989, pp. 45–49.

Casella, G., and R. L. Berger (2002). *Statistical Inference,* 2nd ed., Belmont, CA: Brooks/Cole.

Cerrelli, E. C. (1998). "Crash data and rates for age-sex groups of drivers, 1996" *Research Note, U.S. Department of Transportation.* Posted January 1998, Accessed May 14, 2010 at http://www.nhtsa.gov/people/ncsa/agesex.html.

Chambers, C. (2000). "Americans are overwhelmingly happy and optimistic about the future of the United States," Gallup Poll Press Release, October 13, 2000, www.gallup.com/poll/releases/pr001013.asp.

——— (2000). "Americans dissatisfied with U.S. education in general, but parents satisfied with their kids' schools," Gallup Poll Release, Sept. 5, 2000 (www.gallup.com).

Citro, C. F., and J. L. Norwood (eds.) (1997). *The Bureau of Transportation Statistics: Priorities for the Future,* Washington, D.C.: National Academy Press.

Clark, H. H., and M. F. Schober (1992). "Asking questions and influencing answers," in *Questions about Questions,* J. M. Tanur (ed.), New York: Russell Sage Foundation, pp. 15–48.

Cohen, J. (1988). *Statistical Power Analysis for the Behavioral Sciences,* 2nd ed., Hillsdale, NJ: Lawrence Erlbaum Associates.

Cohn, Victor, and Lewis Cope (2001). *News and Numbers: A Guide to Reporting Statistical Claims and Controversies in Health and Other Fields,* 2nd ed., Ames, IA: Iowa State University Press.

Cole, C. R., J. M. Foody, E. H. Blackstone, and M. S. Lauer (2000). "Heart rate recovery after submaximal exercise as a predictor of mortality in a cardiovascularly healthy cohort," *Annals of Internal Medicine,* Vol. 132, pp. 552–555.

Coren, S., and D. Halpern (1991). "Left-handedness: A marker for decreased survival fitness," *Psychological Bulletin,* Vol. 109, No. 1, pp. 90–106.

Crossen, Cynthia (1994). *Tainted Truth,* New York: Simon and Schuster.

Dabbs, J. M., D. de LaRue, and P. Williams (1990). "Testosterone and occupational choice: Actors, ministers, and other men," *Journal of Personality and Social Psychology,* Vol. 59, No. 6, pp. 1261–1265.

Davidson, R. J., J. Kabat-Zinn, J. Schumacher, M. Rosenkranz, D. Muller, S. F. Santorelli, F. Urbanowski, A. Harrington, K. Bonus, and J. F. Sheridan (2003). "Alterations in brain and immune function produced by mindfulness meditation," *Psychosomatic Medicine,* Vol. 65, pp. 564–570.

Davis, R. (1998). "Prayer can lower blood pressure," *USA Today,* August 11, 1998, p. 1D.

DeNoon, D. J. (2010). "Fish oil may fight psychosis: Fish oil prevents psychosis in high-risk teens," *WebMDHealthNews,* Posted February 1, 2010, Accessed July 28, 2010 at http://www.webmd.com/schizophrenia/news/20100201/fish-oil-vs-psychosis.

Devore, J., and R. Peck (1993). *Statistics: The Exploration and Analysis of Data,* 2nd ed., Belmont, CA: Duxbury Press.

Diaconis, P., and F. Mosteller (1989). "Methods for studying coincidences," *Journal of the American Statistical Association,* Vol. 84, pp. 853–861.

Eddy, D. M. (1982). "Probabilistic reasoning in clinical medicine: Problems and opportunities," Chapter 18 in D. Kahneman, P. Slovic, and A. Tversky (eds.), *Judgment under Uncertainty: Heuristics and Biases,* Cambridge, U.K.: Cambridge University Press.

Edwards, Gary, and Josephine Mazzuca (2000). "About four in ten Canadians accepting of same sex marriages, adoption," Gallup Poll Release, March 7, 2000 (www.gallup.com).

Elias, Marilyn (2001). "Web use not always a downer. Study disputes link to depression," *USA Today,* July 23, 2001, p. 1D.

Ellsworth, P. C., J. M. Carlsmith, and A. Henson (1972). "The stare as a stimulus to flight in human subjects: A series of field experiments," *Journal of Personality and Social Psychology,* Vol. 21, pp. 302–311.

Faigenbaum, A. D., W. L. Westcott, R. LaRosa Loud, and C. Long (1999). "The effects of different resistance training protocols on muscular strength and endurance development in children," *Pediatrics Electronic Article,* Vol. 104, No. 1 (July), p. e5.

Fichtenberg, C. M., and S. A. Glantz (2000). "Association of the California Tobacco Control program with declines in cigarette consumption and mortality from heart disease," *New England Journal of Medicine,* Vol. 343, No. 24, 1772–1777.

Fletcher, S. W., B. Black, R. Harris, B. K. Rimer, and S. Shapiro (1993). "Report of the international workshop on screening for breast cancer," *Journal of the National Cancer Institute,* Vol. 85, No. 20, pp. 1644–1656.

Freedman, D., R. Pisani, R. Purves, and A. Adhikari (1991). *Statistics,* 2nd ed., New York: W. W. Norton.

Freinkel, A. (1998). Letter to the editor, *Journal of the American Medical Association,* Vol. 280, No. 22, p. 1905.

French, J. R. P. (1953). "Experiments in field settings," in L. Festinger and D. Katz (eds.), *Research Method in the Behavioral Sciences,* New York: Holt, pp. 98–135.

Gastwirth, J. L. (1988). *Statistical Reasoning in Law and Public Policy,* New York: Academic Press.

Gastwirth, J. L. (1988). *Statistical Reasoning in Law and Public Policy, Vol. 2: Tort Law, Evidence and Health,* Boston: Academic Press.

Gehring, W. J., and A. R. Willoughby (2002). "The medial frontal cortex and the rapid processing of monetary gains and losses," *Science,* Vol. 295, pp. 2279–2284.

Gillespie, M. (1999). "Local schools get passing grades: Americans support higher pay for teachers," Gallup Poll Release, Sept. 8, 1999.

Griffiths, R. R., and E. M. Vernotica (2000). "Is caffeine a flavoring agent in cola soft drinks?" *Archives of Family Medicine,* Vol. 9, pp. 727–734.

Gustafson, R., and H. Kallmen (1990). "Effects of alcohol on cognitive performance measured with Stroop's Color Word Test," *Perceptual and Motor Skills,* Vol. 71, pp. 99–105.

Gwiazda, E., E. Ong, R. Held, R. N., and F. Thorn (2000). "Myopia and ambient night-time lighting." *Nature,* Vol. 404, No. 6744, p. 144.

Hand, D. J., F. Daly, A. D. Lunn, K. J. McConway, and E. Ostrowski (1994). *A Handbook of Small Data Sets,* London: Chapman and Hall.

Harmon, Amy (1998). "Sad, lonely world discovered in cyberspace," *New York Times,* August 30, 1998, p. A3.

Hedges, L. V., and I. Olkin (1985). *Statistical Methods for MetaAnalysis,* New York: Academic Press.

Hite, S. (1987). *Women and Love,* New York: Knopf.

Holbrook, M. B., and R. M. Schindler (1989). "Some exploratory findings on the development of musical tastes," *Journal of Consumer Research,* Vol. 16, pp. 119–124.

Horne, J. A., L. A. Reyner, and P. R. Barrett (2003). "Driving impairment due to sleepiness is exacerbated by low alcohol intake," *Journal of Occupational and Environmental Medicine,* Vol. 60, pp. 689–692.

Horner M. J., L. A. G. Ries, M. Krapcho, N. Neyman, R. Aminou, N. Howlander, S. F. Altekruse, E. J. Feuer, L. Huang, A. Mariotto, B. A. Miller, D. R. Lewis, M. P. Eisner, D. G. Stinchcomb, and B. K. Edwards (eds). SEER Cancer Statistics Review, 1975–2006, National Cancer Institute. Bethesda, MD, Accessed July 27, 2010 at http://seer.cancer.gov/csr/1975_2006.

Howell, David C. (1997). *Statistical Methods for Psychology,* 4th ed., Belmont, CA: Duxbury.

Hurt, R., L. Dale, P. Fredrickson, C. Caldwell, G. Lee, K. Offord, G. Lauger, Z. Maruisic, L. Neese, and T. Lundberg (1994). "Nicotine patch therapy for smoking cessation combined with physician advice and nurse follow-up," *Journal of the American Medical Association,* Vol. 271, No. 8, pp. 595–600.

Ichimaya, M. A., and M. I. Kruse (1998). "The social contexts of binge drinking among university freshmen," *Journal of Alcohol and Drug Education,* Vol. 44, pp. 18–33.

Iman, R. L. (1994). *A Data-Based Approach to Statistics,* Belmont, CA: Wadsworth.

Jorenby, D. E., S. J. Leischow, M. A. Nides, S. I. Rennard, J. A. Johnston, A. R. Hughes, S. S. Smith, M. L. Muramoto, D. M. Daughton, K. Doan, M. C. Fiore, and T. B. Baker (1999). "A controlled trial of sustained-release bupropion, a nicotine patch, or both for smoking cessation," *New England Journal of Medicine,* Vol. 340, No. 9, pp. 685–691.

Kalichman, S. C. (1986). "Horizontality as a function of sex and academic major," *Perceptual and Motor Skills,* Vol. 63, pp. 903–908.

Kawachi, I., D. Sparrow, A. Spiro III, P. Vokonas, and S. T. Weiss (1994). "A prospective study of anger and coronary heart disease: The normative aging study," *Circulation,* Vol. 94, pp. 2090–2095.

Kiener, R. (1997). "How honest are we?" *Reader's Digest Canada,* Accessed June 3, 2005, at http://www.readersdigest.ca/mag/1997/03/think_01.html.

Kitchens, L. J. (1998). *Exploring Statistics: A Modern Introduction to Data Analysis and Inference,* 2nd ed., Boston: Duxbury Press.

Kohlmeier, L., G. Arminger, S. Bartolomeycik, B. Bellach, J. Rehm, and M. Thamm (1992). "Pet birds as an independent risk factor for lung cancer: Case-control study," *British Medical Journal,* Vol. 305, pp. 986–989.

Kolata, G. (2010). "To beat the heat, drink a slushie first," *New York Times,* April 27, 2010.

Koopmans, L. H. (1987). *An Introduction to Contemporary Statistical Methods,* 2nd ed., Boston: Duxbury Press.

Krantz, Les (1992). *What the Odds Are,* New York: Harper-Perennial.

Kraut, R., V. Lundmark, M. Patterson, S. Kiesler, T. Mukopadhyay, and W. Scherlis (1998). "Internet paradox: A social technology that reduces social involvement and psychological well-being?" *American Psychologist,* Vol. 53, No. 9, pp. 1017–1031.

Lang, S. (2003). "CU nutritionists: Junk food, all-you-can-eat make 'freshman 15' a reality," *Cornell Chronicle,* Posted August 28, 2003, Accessed July 27, 2010, at http://www.news.cornell.edu/Chronicle/03/8.28.03/freshman_15.html.

Langer, G. (2008). Poll: Four in 10 would fly in space; just knock $198,000 off the ticket. Posted February 8, 2008, Accessed July 27, 2010, at http://abcnews.go.com/PollingUnit/story?id=4255841.

Langer, G. and M. Mokrzycki (2010). Parents on kids: Pierced ears, sure; but a late night out–not so fast: Poll finds significant parental constraints on children's activities. Posted February 18, 2010, Accessed July 27, 2010, at http://abcnews.go.com/PollingUnit/parents-kids-pierced-ears-late-night-fast/story?id=9874817.

Lau, E. (2005). "Autism study finds key markers in blood," *The Sacramento Bee,* May 7, 2005, pp. A1 and A26.

Lawrence, T. R. (1993). "Gathering in the sheep and goats. A Meta-analysis of forced choice sheep-goat ESP studies, 1947–1993," *Proceedings of Presented Papers: The 36th Annual Convention of the Parapsychological Association,* pp. 75–86.

Lesko, S. M., L. Rosenberg, and S. Shapiro (1993). "A case-control study of baldness in relation to myocardial infarction in men,"

Journal of the American Medical Association, Vol. 269, No. 8, pp. 998–1003.

Levin, A. (1999). “Planes get closer in midair as traffic control errors rise,” *USA Today,* February 24, 1999, p. 1A.

Levy, S. (2005). “Does your iPod play favorites?” *Newsweek,* January 31, 2005, p. 10.

Liss, S. (1997). “The Effects of Survey Methodology Changes in the NPTS,” Technical Report, Office of Highway Policy Information, Federal Highway Administration.

Loftus, E. F., and J. C. Palmer (1974). “Reconstruction of automobile destruction: An example of the interaction between language and memory,” *Journal of Verbal Learning and Verbal Behavior,* Vol. 13, pp. 585–589.

Lotufo, P. A., C. U. Chae, U. A. Ajani, C. H. Henneken, and J. E. Manson (2000). “Male pattern baldness and coronary heart disease: The physicians’ health study,” *Archives of Internal Medicine,* Vol. 160, pp. 165–171.

Mackowiak, P. A., S. S. Wasserman, and M. M. Levine (1992). “A critical appraisal of 98.6 degrees F, the upper limit of the normal body temperature, and other legacies of Carl Reinhold August Wunderlich,” *Journal of the American Medical Association,* Vol. 268, No. 12, pp. 1578–1580.

Martin, J.A, et al. (2009). “Births: Final Data for 2006,” *National Vital Statistics Reports*, Vol. 57, No. 7. Hyattsville, MD: National Center for Health Statistics.

Martin, M. E., M. L. Straf, and C. F. Citro (eds.) (2001). *Principles and Practices for a Federal Statistical Agency,* 2nd ed., Washington, D.C.: National Academy Press.

Mathews, F., P. J. Johnson, and A. Neil (2008). “You are what your mother eats: evidence for maternal preconception diet influencing foetal sex in humans,” *Proceedings of The Royal Society B*, Vol 275, pp. 1661–1668.

Mathews, F., P. Johnson, and A. Neil (2009). “Reply to Comment by Young et al,” *Proceedings of The Royal Society B*, Vol 276, pp. 1213–1214.

Maugh, T. H., II. (1998). “Study says calcium can help ease PMS,” *Sacramento Bee,* August 26, 1998, pp. A1 and A9.

Mazzuca, J. (2004). “Do U.S. parents rule with an iron fist?” Press release at http://www.gallup.com on Dec. 7, 2004.

Mervis, J. (1998). “Report deplores science-media gap,” *Science,* Vol. 279, p. 2036.

Milgram, S. (1983). *Obedience to Authority: An Experimental View,* New York: Harper/Collins.

Moore, D. S. (1997). *Statistics: Concepts and Controversies,* 4th ed., New York: W. H. Freeman.

Morin, R. (1995). “What informed public opinion?” *The Washington Post National Weekly Edition,* April 10–16, 1995, p. 36.

Morrison, C. M., and H. Gore (2010). “The relationship between excessive internet use and depression: A questionnaire-based study of 1,319 young people and adults,” *Psychopathology*, Vol 43, pp. 121–126.

Newcomer, C. (1994). Judgment in the Civil Action No. 93–6157, *Bruce S. Marks et al.* v. *William Stinson et al.*, United States District Court for the Eastern District Court of Pennsylvania, entered April 26, 1994.

Norton, P. G., and E. V. Dunn (1985). “Snoring as a risk factor for disease,” *British Medical Journal,* Vol. 291, pp. 630–632.

Olds, D. L., C. R. Henderson, Jr., and R. Tatelbaum (1994). “Intellectual impairment in children of women who smoke cigarettes during pregnancy,” *Pediatrics,* Vol. 93, No. 2, pp. 221–227.

Ott, R. L. (1998). *An Introduction to Statistical Methods and Data Analysis,* 4th ed., Belmont, CA: Duxbury Press.

Pagano, M., and K. Gauvreau (1993). *Principles of Biostatistics,* Belmont, CA: Duxbury Press.

Passell, P. (1994). “Probability experts may decide Pennsylvania vote,” *New York Times,* April 11, 1994, p. A15.

Paulos, J. A. (1995). *A Mathematician Reads the Newspaper,* New York: Basic Books.

Perry, D. F., J. DiPetro, and K. Costigan (1999). “Are women carrying ‘basketballs’ really having boys? Testing pregnancy folklore,” *Birth,* Vol. 26, pp. 172–177.

Peterson, Karen S. (1997). “Interracial dates common among today’s teenagers,” *Sacramento Bee,* November 3, 1997, p. A6.

Plewes, Thomas J. (1994). “Federal agencies introduce redesigned Current Population Survey,” *Chance,* Vol. 7, No. 1, pp. 35–41.

Plous, Scott (1993). *The Psychology of Judgment and Decision Making,* New York: McGraw-Hill.

Plous, S., and H. Herzog (2001, July 27). Reliability of protocol reviews for animal research. *Science,* Vol. 293, pp. 608–609; available at http://www.socialpsychology.org/articles/scipress.htm.

Prasad, A. S., J. T. Fitzgerald, B. Bao, F. W. J. Beck, and P. H. Chandrasekar (2000). “Duration of symptoms and plasma cytokine levels in patients with the common cold treated with zinc acetate,” *Annals of Internal Medicine,* Vol. 133, pp. 245–252.

Prescott, E., M. Hippe, H. O. H. Schnor, and J. Vestbo (1998). “Smoking and risk of myocardial infarction in women and men: Longitudinal population study,” *British Medical Journal,* Vol. 316, pp. 1043–1047.

Prevention Magazine’s Giant Book of Health Facts (1991). John Feltman (ed.), New York: Wings Books.

Ramsey, F. L., and D. W. Schafer (1997). *The Statistical Sleuth: A Course in Methods of Data Analysis,* Belmont, CA: Duxbury.

——— (2002). *The Statistical Sleuth: A Course in Methods of Data Analysis,* 2nd edition. Belmont, CA: Duxbury Press.

Rind, B., and P. Bordia (1996). “Effect on restaurant tipping of male and female servers drawing a happy face on the backs of customers’ checks,” *Journal of Social Psychology,* Vol. 26 (3), pp. 215–225.

Roper Center (1996). “University of California Faculty Opinion Survey on Affirmative Action Policies,” Roper Center for Public Opinion Research, Press Release, January 15, 1996.

The Roper Organization (1992). *Unusual Personal Experiences: An Analysis of the Data from Three National Surveys,* Las Vegas: Bigelow Holding Corp.

Rosa, L., E. Rosa, L. Sarner, and S. Barrett (1998). “A close look at therapeutic touch,” *Journal of the American Medical Association,* Vol. 279, No. 13, pp. 1005–1010.

Rosenthal, R. (1991). *Meta-Analytic Procedures for Social Research,* revised ed., Newbury Park, CA: Sage Publications.

Rosenthal, R., and K. L. Fode (1963). “The effect of experimenter bias on the performance of the albino rat,” *Behavioral Science,* Vol. 8, pp. 183–189.

Ryan, B., and B. L. Joiner (2001). *MINITAB Handbook,* Fourth ed., Belmont, CA: Duxbury Press.

Salk, L. (1973). “The role of the heartbeat in the relations between mother and infant,” *Scientific American,* May, pp. 262–264.

SAMSHA (1998). “Driving After Drug and Alcohol Use: Findings from the 1996 National Household Survey on Drug Abuse.” United States Department of Health and Human Services.

Schmeidler, G. R. (1988). *Parapsychology and Psychology: Matches and Mismatches,* Jefferson, NC: McFarland.

Schmidt, S. M. (1998). Letter to the editor, *Journal of the American Medical Association,* Vol. 280, No. 22, p. 1906.

Schuman, H., and J. Scott (1987). “Problems in the use of survey questions to measure public opinion,” *Science,* Vol. 236, pp. 957–959.

Schwartz, B. S., W. F. Stewart, D. Simon, and R. B. Lipton (1998). “Epidemiology of tension-type headache,” *Journal of the American Medical Association,* Vol. 279, pp. 381–383.

Service, R. F. (1994). “Dropping cholesterol—safely,” *Science,* Vol. 266, No. 5189, p. 1323.

Siegel, R., J. Maté, M. B. Brearley, G. Watson, K. Nosaka and P. B. Laursen (2010). Ice slurry ingestion increases core temperature capacity and running time in the heat," *Medicine & Science in Sports & Exercise,* Vol. 42, No. 4, pp. 717–725.

Sinclair, R. C., M. M. Mark, S. E. Moore, L. A. Lavis, and A. S. Soldat (2000). "An electoral butterfly effect," *Nature,* Vol. 408, pp. 665–666.

Singer, M. J., K. Slovak, T. Frierson, and P. York (1998). "Viewing preferences, symptoms of psychological trauma, and violent behaviors among children who watch television," *Journal of the American Academy of Child and Adolescent Psychiatry,* Vol. 37, No. 10, pp. 1041–1049.

Slutske, W. S., T. M. Piasecki, and E. E. Hunt-Carter (2003). "Development and initial validation of the Hangover Symptoms Scale: Prevalence and correlates of hangover symptoms in college students," *Alcoholism: Clinical and Experimental Research,* Vol. 27, pp. 1442–1450.

Spencer, E. A., P. N. Appleby, G. K. Davey, and T. J. Key (2002). "Validity of self-reported height and weight in 4808 EPIC-Oxford participants," *Public Health Nutrition,* Vol. 5, No. 4, pp. 561–565.

Spilich, G. J., L. June, and J. Renner (1992). "Cigarette smoking and cognitive performance," *British Journal of Addiction,* Vol. 87, 1313–1326.

The Steering Committee of the Physicians' Health Study Research Group (1988). "Preliminary report: Findings from the aspirin component of the ongoing Physicians' Health Study," *New England Journal of Medicine,* Vol. 318, No. 4, pp. 262–264.

Stephens, R., J. Atkins, and A. Kingston (2009). "Swearing as a response to pain," *NeuroReport,* Vol. 20, pp. 1056–1060.

Stigler, S. M. (1986). *The History of Statistics: The Measurement of Uncertainty Before 1900.* Cambridge, MA: Belknap Press.

——— (1989). "Francis Galton's account of the invention of correlation," *Statistical Science,* Vol. 4, pp. 73–79.

Stinchfield, R. (2001). "A comparison of gambling by Minnesota public school students in 1992, 1995 and 1998," *Journal of Gambling Studies,* Vol. 17, No. 4, pp. 273–296.

Taubes, G. (1993). "Claim of higher risk for women smokers attacked," *Science,* Vol. 262, p. 1375.

Thys-Jacobs, S., P. Starkey, D. Bernstein, J. Tian, and the Premenstrual Syndrome Study Group (1998). "Calcium carbonate and the premenstrual syndrome: Effects on premenstrual and menstrual symptoms," *American Journal of Obstetrics and Gynecology,* Vol. 179, No. 2, pp. 444–452.

Tuna, C. (2009). "When combined data reveal the flaw of averages," *Wall Street Journal,* Dec. 2, 2009. Accessed March 2010 at http://online.wsj.com/article/SB125970744553071829.html.

Turner, C. F., L. Ku, S. M. Rogers, L. D. Lindberg, J. H. Pleck, and F. L. Sonenstein (1998). "Adolescent sexual behavior, drug use, and violence: Increased reporting with computer survey technology," *Science,* Vol. 280, pp. 867–873.

Tversky, A., and D. Kahneman (1982). "Judgment under uncertainty—Heuristics and biases," Chapter 1 in D. Kahneman, P. Slovic, and A. Tversky (eds.), *Judgment under Uncertainty: Heuristics and Biases,* Cambridge, U.K.: Cambridge University Press.

U.S. Department of Labor, Bureau of Labor Statistics (1992). *BLS Handbook of Methods,* Sept. 1992, Bulletin 2414.

Uhari, M., T. Kenteerkar, and M. Niemala (1998). "A novel use of Xylitol sugar in preventing acute otitis media," *Pediatrics,* Vol. 102, pp. 879–884.

Underwood, Anne, and Jerry Adler (2005). "Health for life: Diet and genes," *Newsweek,* January 17, 2005, pp. 40–48.

University of California, Berkeley (1991). *The Wellness Encyclopedia,* Boston: Houghton Mifflin.

Villenueve, P. J., and Y. Mao (1994). "Lifetime probability of developing lung cancer, by smoking status, Canada," *Canadian Journal of Public Health,* Vol. 85, No. 6, pp. 385–388.

Voss, L. D., and J. Mulligan (2000). "Bullying in school: Are short pupils at risk? Questionnaire study in a cohort," *British Medical Journal,* Vol. 320, pp. 612–613.

Voyer, D., A. Postma, B. Brake, and J. Imperato-McGinley (2007). "Gender differences in object location memory: A meta-analysis," *Psychonomic Bulletin & Review,* Vol. 14, No. 1, pp. 23–38.

Wagner, C. H. (1982). "Simpson's Paradox in real life," *The American Statistician,* Vol. 36, No. 1, pp. 46–48.

Wald, M. L. (2007). "Fatal airplane crashes drop 65%," *New York Times,* October 1, 2007, online.

Weaver, W. (1963). *Lady Luck: The Theory of Probability,* Garden City, NY: Doubleday.

Weber, G. W., H. Prossinger, and H. Seidler (1998). "Height depends on month of birth," *Nature,* 391, pp. 754–755.

Wechsler, H., and M. Kuo (2000). "College students define binge drinking and estimate its prevalence: Results of a national survey," *Journal of American College Health,* Vol. 49, pp. 57–64.

Weiden, C. R., and B. C. Gladen (1986). "The beta-geometric distribution applied to comparative fecundability studies," *Biometrics,* Vol. 42, pp. 547–560.

Weindling, A. M., F. N. Bamford, and R. A. Whittall (1986). "Health of juvenile delinquents," *British Medical Journal,* Vol. 292, p. 447.

Weisberg, S. (1985). *Applied Linear Regression,* 2nd ed. New York: Wiley and Sons.

Wilding, J., and S. Cook (2000). "Sex differences and individual consistency in voice identification," *Perceptual and Motor Skills,* Vol. 91, pp. 535–538.

Wood, R. D., M. L. Stefanick, D. M. Dreon, B. Frey-Hewitt, S. C. Garay, P. T. Williams, H. R. Superko, S. P. Fortmann, J. J. Albers, K. M. Vranizan, N. M. Ellsworth, R. B. Terry, and W. L. Haskell (1988). "Changes in plasma lipids and lipoproteins in overweight men during weight loss through dieting as compared with exercise," *New England Journal of Medicine,* Vol. 319, No. 18, pp. 1173–1179.

World Almanac and Book of Facts (1993). Mark S. Hoffman (ed.), New York: Pharos Books.

——— (1995). Mahwah, NJ: Funk and Wagnalls.

Writing Group for the Women's Health Initiative Investigators (2002). "Risks and benefits of estrogen plus progestin in healthy postmenopausal women: Principal results from the Women's Health Initiative randomized controlled trial," *Journal of the American Medical Association,* Vol. 288, No. 3, pp. 321–333.

Yang, J., S. W. Marshall, M. J. Bowling, C. Runyan, F. O. Mueller, and M. A. Lewis (2005). "Use of discretionary protective equipment and rate of lower extremity injury in high school athletes," *American Journal of Epidemiology,* Vol. 161, No. 6, pp. 511–519.

Yanovski, J. A., S. Z. Yanovski, K. N. Sovik, T. T. Nguyen, P. M. O'Neil, and N. G. Sebring (2000). "A prospective study of holiday weight gain," *The New England Journal of Medicine,* Vol. 342, No. 12, pp. 861–867.

Young, S.S., H. Bang, and K. Oktay (2009). "Cereal-induced gender selection? Most likely a multiple testing false positive," *Proceedings of The Royal Society B,* Vol. 276, pp. 1211–1212.

Zadnik, K., L. A. Jones, B. C. Irvin, R. N. Kleinstein, R. E. Manny, J. A. Shin, and D. O. Mutti (2000). "Myopia and ambient night-time lighting," *Nature,* Vol. 404, No. 6744, pp. 143–144.

Answers to Selected Odd-Numbered Exercises

*The following are partial or complete answers to the exercises numbered in **bold** in the text.*

Chapter 1

1.1 a. 150 mph. b. 55 mph. c. 95 mph. d. 1/2. e. 51.

1.3 a. .00043. b. .00043. c. Rate is based on past data; risk uses past data to predict an individual's likelihood of developing cervical cancer.

1.5 a. All teens in the U.S. at the time the poll was taken. b. All teens in the U.S. who had dated at the time the poll was taken.

1.7 a. All adults in the U.S. at the time the poll was taken. b. $\frac{1}{\sqrt{1048}} = .031$ or 3.1%. c. 30.9% to 37.1%.

1.9 a. 400.

1.11 a. Self-selected or volunteer sample. b. No; readers with strong opinions will respond.

1.13 a. Randomized experiment. b. Observational study. c. Observational study.

1.15 Answers will vary, but one possibility is general level of activity.

1.17 How large the difference in weight loss was for the two groups.

1.19 How many different relationships were examined.

1.21 189/11,034, or about 17/1000, based on placebo group.

1.23 a.

	Minutes of exercise per week	
Median	180	
Quartiles	37	330
Extremes	0	600

1.25 a. Observational study. b. No. c. An example is amount of exercise.

1.27 Base rate or baseline risk.

1.29 For Caution 1: 46% who quit with nicotine versus 20% who quit with placebo patch is a large difference and has practical importance. For Caution 2: It is not possible that whether someone quit or not influenced the type of patch they were assigned.

1.31 Neither caution applies.

1.33 a. .026. b. .113 to .165.

1.35 a. Self-selected or volunteer sample. b. Probably higher; those who saw a ghost would be more likely to call in.

1.37 Looking at the data in a variety of ways until something interesting to report emerges.

1.39 It may not be possible or ethical to do a randomized experiment.

1.41 Randomly assign volunteers to either eat lots of chocolate or not eat any chocolate for a period of time, and give them a questionnaire about depression at the beginning and the end of the time period. Then compare the change in depression scores for the two groups.

1.43 They made a cause-and-effect conclusion, which was not justified because this was an observational study.

Chapter 2

2.1 a. 4. b. State in the United States. c. $n = 50$.

2.3 a. Whole population. b. Sample.

2.5 a. Population parameter. b. Sample statistic. c. Sample statistic.

2.7 Sex and self-reported fastest ever driven speed. b. Students in a statistics class. c. Answer depends on whether interest is in this class only or in a larger group represented by this class.

2.9 Population summary if we restrict interest to fiscal year 1998. Sample summary if 1998 value is used to represent errors in other years.

2.11 a. Categorical. b. Quantitative. c. Quantitative. d. Categorical.

2.13 a. Categorical. b. Ordinal. c. Quantitative

2.15 a. Explanatory is score on the final exam; response is final course grade. b. Explanatory is gender; response is opinion about the death penalty.

2.17 a. Not continuous. b. Continuous. c. Continuous.

2.19 a. Support ban or not; categorical. b. Gain on verbal and math SATs after program; quantitative.

2.21 a. Pulse rate and gender. b. Gender is categorical; pulse rate is quantitative. c. Is there a difference between the mean pulse rates of men and women? Need mean pulse for each sex.

2.23 Example: Letter grades (A, B, etc.) converted to GPA.

2.25 a. Dominant hand is a categorical variable and IQ is a quantitative variable. Explanatory is dominant hand and response is IQ. b. Eventual divorce status and pet ownership are both categorical variables. Explanatory is pet ownership and response is eventual divorce status.

2.27 a. 1427/2530 = .564, or 56.4%. b. 100% − 56.4% = 43.6%. c. Never: 105/2530 = .042; Rarely: 248/2530 = .098; Sometimes: 286/2530 = .113. Most times: 464/2530 = .183; Always: 1427/2530 = .564.

2.29 a.

	Main Use		
	To Talk	To Text	Total
Women	22	84	106
Men	34	49	83

b. To talk = 20.8%; to text = 79.2%. c. To talk = 41.0%; to text = 59.0%. d. Women more likely than men to respond "to text."

2.31 a. Explanatory is smoked or not; response is developed Alzheimer's disease or not. b. Explanatory is political party; response is whether a person voted or not. c. Explanatory is income level; response is whether a person has been subjected to a tax audit or not.

2.33 a. Explanatory is gender and response is how they feel about their weight.

b.

	Feelings about Weight			
Gender	Over-weight	About Right	Under-weight	Total
Female	38 (26.6%)	99 (69.2%)	6 (4.2%)	143
Male	18 (23.1%)	35 (44.9%)	25 (32.1%)	78

c. Overweight = 26.6%; about right = 69.3%; underweight = 4.2%.
d. Overweight = 23.1%; about right = 44.9%; underweight = 32.1%. e. Males more likely than females to say they are underweight; females more likely than males to say their weight is about right.

2.35 a.

	Picked S	Picked Q	Total
S listed first	61	31	92
Q listed first	45	53	98
Total	106	84	190

b. Picked S = 66.3%; picked Q = 33.7%. c. Picked S = 45.9%; picked Q = 54.1%.

2.37 a. 150 mph. b. 55 mph. c. 95 mph. d. 1/2. e. 51.

2.39 a. Center for females is at a greater percentage than for males. b. Data are more spread for females. c. The greatest two female percentages are set apart from the bulk of the data. Values are about 65% and 72%.

2.41 a. Median = 65 in. b. Overall spread is from 59 to 71 in; middle 50% of data is between 63.5 and 67.5 in.

2.43 a. Approximately symmetric and bell-shaped.
b. No noticeable outliers. c. 7 hr. d. About 14.

2.45 a. Roughly symmetric. b. Highest = 92.
c. Lowest = 64. d. 25% (5/20).

2.47 a.

5	6
6	2
6	8
7	55
8	02344
8	77
9	13
9	5

2.49 a.

0	689
1	11111122233444
1	566667777779
2	011
2	5555778889
3	0011
3	7

c. Skewed slightly to the right.

2.51 Yes. Values inconsistent with the bulk of the data will be obvious.

2.53 Skewed to the left.

2.55 a. Shape is better evaluated by using a histogram.
b. Boxplot is useful for identifying outliers, evaluating spread, and comparing groups.

2.57 Females tended to have higher tip percentages. The median is clearly greater for females. Female data shows greater spread than male data.

2.59 a. Mean = 74.33; median = (72 + 76)/2 = 74.
b. Mean = 25; median = 7. c. Mean = 27.5; median = 30.

2.61 a. Range = 225 − 123 = 102. b. IQR = 35. c. 50%.

2.63 a. Low = 109, Q_1 = 180.75, median = 186, Q_3 = 199, high = 214. b. 109 is an outlier. c. The crew member who weighed 109.

2.65 The fastest speeds ever driven tend to be higher for males than for females. Ignoring outliers, the spread is greater for males than for females.

2.67 a. Low = 0, Q_1 = 37, median = 180, Q_3 = 330, high = 600. b. Amounts vary between 0 and 600 min/wk. Median is 180 min. Middle half of the data are between 37 and 330 min/wk.

2.69 The median is 16.72 in. The data values vary from 6.14 to 37.42 in. The middle one-half of the data are between 12.05 and 25.37 in, so "typical" annual rainfall covers quite a wide range.

2.71 Low = 6.14 in; Q_1 = 12.05 in; median = 16.72 in; Q_3 = 25.37 in; high = 37.42 in.

2.73 Median is located between 21 and 25 words. Q_1 is in the interval 16–20 words and Q_3 is in the interval 31–35, so IQR may be about 15 or so. Low is in the interval 1–5 words and high is in the interval 56–60, so the range may be about 55 or so.

2.75 Low = 42, Q_1 = 52, median = 57.5, Q_3 = 62, high = 78.

2.77 Mean = 57.84 yr. The median is 57.5 yr. Mean and the median are similar because the data are more or less symmetric in shape.

2.79 Example: Age of a person 80 years old would be an outlier at a traditional college but not at a retirement home.

2.81 The 1982 rainfall total is high (37.42 in.), but not high enough to be an outlier.

2.83 Most likely a data entry mistake was made. The instructor should correct the value if possible, or delete the value from the dataset.

2.85 a. Between 5.3 and 8.7 hr. b. Between 3.6 and 10.4 hr.
c. Between 1.9 and 12.1 hr.

2.87 a. Mean = 25, s = 4.24. b. Mean = 30, s = 9.13.

2.89 Draw a bell curve centered at 7 with nearly all of the curve over the interval 1.9 to 12.1. Then indicate the intervals found in Exercise 2.85.

2.91 a. −0.5. b. 1.1. c. 3. d. −1.4.

2.93 50, 50, 50, 50, 50, 50, 50; no.

2.95 a. Range = 98 − 41 = 57. b. $s \approx 57/6 = 9.5$.

2.97 The Empirical Rule predicts about 68%, 95%, 99.7% within 1, 2, and 3 standard deviations of the mean, respectively. Data show 72%, 97%, 98%, so the set of measurements fits well.

2.99 a. Somewhat unusual. Only about 2.5% of adult males will have a smaller measurement because 52 cm is 2 standard deviations below the mean. b. Rare. Only about 0.15% (about 3 in 2,000 men) will have a larger circumference because 62 cm is 3 standard deviations above the mean.

2.101 a. Draw a bell curve centered at 540 with middle 68% between 490 and 590, middle 95% between 440 and 640, and middle 99.7% between 390 and 690. b. 2,500.

2.103 a. 590. b. 640. c. 490.

2.105 No. A variable must be quantitative to have a distribution with any particular shape.

2.107 a. ***Sleep, Dad's Height, Ideal Height*** for females, and ***Handspan*** for males are well described by the Empirical Rule, as is ***Handspan*** for females if outliers are ignored. ***TV, Exercise,*** and ***Alcohol*** are not well described by the Empirical Rule. b. Variables well described have roughly a bell shape. Variables not well described have a skewed shape.

2.109 a. Shape is skewed, making it hard to judge outliers because extreme points may be part of the skewed pattern. b. The Empirical Rule will not hold. c. Interval is −3.863 to 12.08, and 90.42% of the data values are in this interval, more than the expected 68% if the Empirical Rule applies. Also, the interval contains negative values, which are impossible for this variable. d. Interval is −11.83 to 20.05; 97.01% of the data values are in this interval, about the 95% expected if Empirical Rule applies. But, the interval contains negative values, which are impossible for this variable.

2.111 a. Outlier at 100 hr. The remaining data is skewed so the Empirical Rule does not hold. b. Outliers occur at 13 and 14 cm. Remaining data is roughly bell-shaped so the Empirical Rule may hold. c. Outlier at 55 in. Remaining data is roughly bell-shaped so the Empirical Rule may hold.

2.113 a. $Q_1 = 0$ hours, and 25% of the values are at or below the lower quartile. b. 0 to 2 hr. c. 2 to 70 hr. d. Yes, 70 hr would be marked as an outlier. e. Range/6 = (70-0)/6 = 11.67. The outlier (70 hours) and skewness in the data cause Range/6 to differ from the standard deviation. f. Mean is greater than the median. The outlier and skewness to the right are responsible.

2.115 a. Categorical. b. Quantitative. c. Quantitative. d. Categorical.

2.117 a. Yes; night light use, for example. b. No. c. Yes. d. Yes.

2.119 a. Mean will be larger. b. Mean will be larger. c. Mean will be larger. d. Mean is larger. Mean is 10.33 cents and median is 5 cents.

2.121 Interval is −6.1 to 22.7 hr. It includes negative values, which are impossible times, so a bell-shaped curve would not reflect reality.

2.123 a. Population. b. Population s.d. = 15.13.

2.125 a. Explanatory is amount of beer consumed per unit of time (week, etc.); response is systolic blood pressure. b. Explanatory is calories of protein consumed per day on average; response is whether they had colon cancer.

2.127 a. Ignoring 2 outliers, Empirical Rule may hold. b. Yes, range is 10.75 cm, close to 6 standard deviations. Expect between 4 and 6 standard deviations.

2.129 a. Mean = 57.84 yr; standard deviation = 7.00 yr (population standard deviation = 6.93). b. Range is 36 yr, 5.14 standard deviations, so it holds. c. z for youngest CEO is −2.26, z for oldest CEO is 2.88; about as expected from the Empirical Rule.

2.131 What percentage of kindergarten children live with their mother only? Their father only? One or both grandparents?

2.133 Is the average amount of coffee consumed per day the same for married people as it is for single people?

2.135 a. Low = 0, $Q_1 = 22.5$, median = 55, $Q_3 = 175$, high = 450. b. 450 will be marked as an outlier.

2.137 Box covers the interval 180.75 to 199 with median marked at 186. Lines extend to 178.5 and 214, and 109 is marked as an outlier.

2.139 Outliers affect the standard deviation, but not the IQR.

2.141 a. The boxplot does not show the bimodal nature of the distribution. b. The distribution is bimodal.

2.143 a. Generally, the distribution is bell-shaped, although there are two outliers at 16 hr of sleep. Disregarding the outliers, the center of the distribution is somewhere near 7 hr. b. Low = 3, $Q_1 = 6$, median = 7, $Q_3 = 8$, high = 16. c. Range = 16 − 3 = 13 hours. $IQR = Q_3 - Q_1 = 8 - 6 = 2$ hr.

2.145 a. 66.56%.

b.

	Favor	Oppose	All
Democrat	372 (53.9%)	318 (46.1%)	690
Independent	457 (67.3%)	222 (32.7%)	679
Other	22 (64.8%)	14 (35.2%)	36
Republican	406 (83.2%)	82 (16.8%)	488
All	1257 (66.4%)	636 (33.6%)	1893

c. Yes. The percentage in favor differs for the different political party affiliations.

2.147 b. It's difficult to describe the shape of the histogram, but it's not bell-shaped, and it's not skewed. Theoretically, it is bimodal, with one peak near the mean height of women and the other near the mean height for men. d. The histogram is more informative. We're able to see the two distinct peaks in the distribution.

Chapter 3

3.1 a. Negative association. b. No association. c. Negative association.

3.3 a. Negative association. b. Roughly linear. c. Highest average math SAT is about 600; fewer than 5% took the test. d. Lowest average math SAT is about 475; about 60% took the test.

3.5 a. Yes, both variables are quantitative. b. No, both variables are categorical.

3.7 a. The speed of the car is the explanatory variable, and the stopping distance is the response variable. b. Positive association, appears to be curvilinear.

3.9 b. The pattern appears to be linear and the direction is negative. c. Phoenix may be an outlier because its August temperature (92°F) is high for its latitude (33). San Francisco may be an outlier because its August temperature (64°F) is low for its latitude (38).

3.11 a. Asking price is the response variable and square footage is the explanatory variable. c. "The individual in question belongs to a different group than the bulk of individuals measured." Probably the other houses are in suburban neighborhoods, while the outlier is a beachfront mansion. d. Yes. The outlier is a different type of house than the rest of them, which are suburban residences.

3.13 a. Average weight = −250 + 6(70) = 170 lb. b. On average, weight increases 6 pounds per each 1-inch increase in height.

3.15 a. Average math SAT decreases 1.11 points per each 1-percent increase in the percentage of high school graduates taking the test. b. Predicted average math SAT = 575 − 1.11(8) = 566.12. c. Residual = 573 − 566.12 = 6.88.

3.17 Deterministic relationship.

3.19 a. Height increases an average of 0.7 inch for each centimeter increase in handspan. b. 65.1 inches. c. Residual = 66.5 − 65.1 = 1.4 inches.

3.21 a. Success rate = 76.5 − 3.95(10) = 37%. b. The slope of −3.95 shows when the distance is increased by a foot, the success rate decreases by about 3.95%, on average.

3.23 a. $\hat{y} = 17.8 + 0.894(70) = 80.38$. b. Residual = Actual y − Predicted y = 76 − 80.38 = −4.38.

3.25 a. For line 1, SSE = 10; for line 2, SSE = 4. b. Line 2 is better because SSE is smaller.

3.27 −1.7 and 2.5. They are not between −1 and +1.

3.29 The two variables do not have a linear association.

3.31 a. Chest girth increases as length increases. b. The value of r is high, so the relationship apparently is strong. c. $r = +0.82$; changing units does not change the correlation.

3.33 It would still be 0.95. Changing the units of measurement does not change the correlation.

3.35 $r = 1.0$

3.37 Graph 2 is the strongest; Graph 3 is the weakest.

3.39 The correlation, $r = +0.12$, is low. In the sample, age and hours of watching television per day have a very weak, positive association.

3.41 a. There is no reason to expect any correlation, but you could argue for a positive correlation if students with little social life both study more and watch more television, or a negative correlation, if students who study more have less time to watch television. b. Positive, because larger cities have more of both. c. Negative, because strength decreases with age.

3.43 a. $r^2 = (0.4)^2 = 0.16$. This means that height explains 16% of the variation in weight. b. The correlation would still be 0.40.

3.45 $r^2 = (-0.36)^2 = 0.1296$, or about 13%. So, hours of study explains about 13% of the observed variation in hours of sleep.

3.47 $r^2 = \dfrac{SSTO - SSE}{SSTO} = \dfrac{800 - 200}{800} = 0.75.$

3.49 We know there is a positive association, but we do not know the strength of the association.

3.51 Answers may vary, but in general it is not wise to extrapolate with the assumption that a present trend will continue indefinitely.

3.53 Your sketch should place the outlier away from the bulk of the data, but not in line with the other points.

3.55 a. Estimated stopping distance when speed is 80 miles per hour is $-44.2 + 5.7(80) = 411.8$ ft. No, this is probably not an accurate estimate. The relationship is likely not linear. b. From the plot, the relationship looks curvilinear. With the curvilinear model, estimated stopping distance is about 500 feet. c. No. Weather conditions and weight of the car are two examples of variables that would affect stopping distance.

3.57 Men and women should not be combined. Women probably have higher scores but lower heights.

3.59 The answer will vary. An example is predicting the median price of a house in your town or city in the year 2080, based on data from 1950 to 2010.

3.61 a. At 2 feet, the predicted success rate is $76.5 - 3.95(2) = 68.6\%$, well below the observed success rate of 93.3%. At 20 feet, the predicted success rate is $76.5 - 3.95(20) = -2.5\%$, much different than the observed rate of 15.8% and also an impossible value because a success rate cannot be negative. b. The equation predicts success rates that are much too low in both cases, indicating that the true relationship is probably not linear much beyond the range of 5 feet to 15 feet. c. The relationship for the entire range from 2 feet to 20 feet is likely to be curvilinear.

3.63 Both variables will have relatively high values in the winter and relatively low values in the summer.

3.65 Gender differences create the observed negative correlation. Females tend to be shorter and to have more ear pierces.

3.67 a. While smoking might actually cause people to die at a younger age, there are possible confounding variables that are related to both, such as alcohol consumption and lack of exercise. b. Both variables are related to the number of people at the ski resort on a given day.

3.69 Answers will vary, but a randomized experiment gives the strongest evidence of a cause and effect relationship, so the answer should describe a randomized experiment.

3.71 Answers will vary. As an example, yearly amount of travel on highways in a country and yearly sales of shoes in the country are likely to be increasing over time due to increasing population size.

3.73 No, correlation does not prove causation. Possible confounding variables include alcohol use, higher sugar content in the diet, lack of exercise and so on. Any variable that is systematically different between countries with high-fat and low-fat diets could be contributing to the observed relationship.

3.75 Answer will vary, but an example is a plot that looks approximately like Figure 3.12 (p. 84) in the text.

3.77 Answers will vary, but the best line through the points will be flat. Examples are the two plots in Figure 3.20 (p. 92).

3.79 Try putting a rectangular grid of points in the upper left corner of the plot. Put the outlier in the lower right corner.

3.81 a. Body temperature is the response variable and age is the explanatory variable.
b. $\hat{y} = 98.6 - 0.0138x = 98.6 - 0.0138(50) = 97.91$ degrees.
c. Residual = actual − predicted = $97.6 - 97.91 = -0.31$. His body temperature is 0.31 degrees lower than would be predicted based on his age.

3.83 a. 19.42, no meaningful interpretation because height cannot be 0. b. 0.658, indicating that if one woman's father was 1 inch taller than another woman's father, the one with the taller father would be predicted to be 0.658 inches taller.
c. $\hat{y} = 19.42 + 0.658x = 19.42 + 0.658(70) = 65.48$ inches.
d. Residual = actual − predicted = $67 - 65.48 = 1.52$ inches. The woman is 1.52 inches taller than would be predicted based on her father's height. e. No. The relationship is based on female students. Male students are taller in general, and a different relationship would hold for relating their heights with their fathers' heights.

3.85 a. For a man, $53 + 0.7(140) = 151$ pounds. For a woman, $44 + 0.6(140) = 128$ pounds. The man wants to weigh 11 pounds more; the woman wants to weigh 12 pounds less. b. No, there are not actual weights close to 0.
c. Yes. For each increase of 1 pound in actual weight, the slope provides the estimated average increase in ideal weight, which is 0.6 pounds for women and 0.7 pounds for men.

3.87 a. Average foot length increases 0.384 centimeters for each one-inch increase in height. This is the slope of the line.
b. The predicted difference in foot lengths is $(10)(0.384) = 3.84$ cm.
c. The predicted foot length is $\hat{y} = 0.25 + 0.384(70) = 27.13$cm., and the prediction error (residual) is $28.5 - 27.13 = 1.37$cm.

3.89 b. The correlation between price and pages is $r = -0.309$.
c. For hardcover books, the correlation is $r = 0.497$. For softcover books, the correlation is $r = 0.626$. d. Inappropriately combining groups.

3.91 a. ***Perhouse*** = 44.4 − (0.0209)(Year). The estimated number of persons per household in 2010 is $44.4 - (0.0209)(2010) = 2.391$ persons. b. Slope = −0.0209, indicating that the average number of people per household decreases by about 0.02 per year, or about 0.2 every 10 years. c. For 2200, the estimated value is $44.4 - 0.0209(2200) = -1.58$, which is an impossible value because 1 is the lowest possible number of persons per household. d. In the future, the pattern will flatten out and perhaps even increase.

3.93 Answers will vary, but one example is shoe size and head circumference. Neither causes the other to be larger, but they are both related to the overall size of a person.

3.95 a. Determine the nature of the relationship. b. Predict college grade point average in the future, based on SAT score.
c. Predict future height at age 21, based on height at age 4.
d. Determine the nature of the relationship.

3.97 a. There is a positive linear association between the two variables. There is a moderately strong association. There are no notable outliers. b. $\hat{y} = 4.267 + 1.373x$, where y = birth rate for females 15 to 17 years old, and x = poverty rate (as a percent).
c. The slope is 1.373. On average, the birth rate (per 1000 persons) for females 15 to 17 years old increases by 1.373 for each 1% increase in the poverty rate. d. Predicted birth rate = $\hat{y} = 4.267 + 1.373(15) = 24.86$.

3.99 a. There is a moderately strong positive association between the variables. There are no data points that have a combination of values that are not consistent with the overall pattern. Some may note the gap between the bulk of the data and the individual with the highest cholesterol values.
b. $\hat{y} = 62.37 + 0.6627x$, where y = 4-day measurement and x = 2-day measurement. c. Slope = 0.6627. Average 4-day measurement increases 0.6627 per each one-unit increase in the 2-day measurement. d. With 2-day = 200, predicted 4-day = $62.37 + 0.6627(200) = 194.90$. With 2-day = 250, predicted 4-day = $62.37 + 0.6627(250) = 228.04$. With 2-day = 300, predicted 4-day = $62.37 + 0.6627(300) = 261.17$. e. The difference between the 2-day and the 4-day measurements becomes greater as the 2-day measurement increases.

3.101 a. There is a strong positive association with a linear pattern.
b. Hawaii may be an outlier because its average verbal SAT (483) is low relative to its math average (513). (Additional suggestion: Calculate the difference between the two scores for all states and then draw either a dotplot or boxplot of those differences. Hawaii will be an obvious outlier.)

3.103 a. There is a positive association with a linear pattern. The difference increases, on average, as actual weight increases. Note that a positive difference occurs when actual weight is more than ideal weight. b. Diff = −52.5 + 0.312 (Actual). For men who weigh 150 pounds, the estimated average difference is $-52.5 + 0.312(150) = -5.7$ pounds. This is actual − ideal, so they want to weigh more. c. The estimated average difference is $-52.5 + 0.312(200) = 9.9$ pounds. On average, 200-pound men want to weigh less than they do. d. $r^2 = 0.353$ or 35.3%.

3.105 a. The regression equation is *height* = 24.7 + 0.6 *midparent*. b. $\hat{y} = 24.7 + 0.6(68) = 65.5$ inches. c. The mid-parent height is $(70+62)/2 = 66$ inches, so $\hat{y} = 24.7 + 0.6(66) = 64.3$ inches. d. A scatterplot and a correlation (or r^2).

Chapter 4

4.1 a. 1322/2057 = 64.3%; row percent. b. 83/548 = 15.1%; column percent. c. 285/2858 = 10.0%. d. Partial answer: As and Bs row percentages are 64.3%, 21.9%, 13.9%. e. There appears to be a relationship.

4.3 a. 170/270 = 63.0%. Among freshmen, 63.0% are females. b. 170/300 = 56.7%. Among females, 56.7% are freshmen. c. Juniors. d. Males.

4.5 a. Yes, both variables are categorical. b. No, both variables are quantitative.

4.7 No. Totals are given for categories of each variable, but counts for combinations are not provided.

4.9 b. The response variable is preference for type of final exam. The explanatory variable is grade on the midterm. c. Among students who got an A on the midterm, 40% prefer a take-home final exam (and 60% prefer an in-class exam). Among students who did not get an A on the midterm, 60% prefer a take-home final exam (and 40% prefer an in-class exam). There is a relationship between the two variables.

4.11 a. 45.65%. b. 25.64%. c. Yes, there is a relationship.

4.13 Percentage distributions are about the same for men and women, so the variables do not appear to be related.

4.15 a. 1. b. 1. c. 0%.

4.17 a. Drug 1: 10/100 = .10; drug 2: 5/100 = .05. b. .10/.05 = 2. c. 100%. d. (10/90)/(5/95) = 2.11.

4.19 5% (or .05 if expressed as a proportion).

4.21 a. 53%. b. 2.40.

4.23 No, we would need to know the baseline risks for males and females.

4.25 a. Relative risk is .106/.04 = 2.65. b. Percent increase in risk is $100\% \times (10.6 - 4.0)/4.0 = 165\%$. c. 2.82.

4.27 a. Aged 18 to 29, 13.9%; aged 30 or over, 10.6%. b. Relative risk is .139/.106 = 1.31. Adults under 30 are 1.31 times as likely to report seeing a ghost as are adults 30 years old or older. c. Increased risk = $(1.31 - 1) \times 100\% = 31\%$. Adults under 30 years old are 31% more likely to report seeing a ghost than adults over 30 years old. d. 1 to 8.4

4.29 There may be more homes in your town now than in 1990. The *rate* of burglaries per number of homes now might be the same as or lower than in 1990.

4.31 The principal question to ask is, "What percentage usually fails?"

4.33 Probably not. The relationship between blood pressure and religious activity would need to be reversed when separate categories of a third variable are considered.

4.35 Occupation; men may be more likely to have jobs that require long-distance driving.

4.37 a. Of the male applicants 45% were admitted. Of the female applicants 35% were admitted. Overall, men were more successful at gaining admission. b. Program A admission rates: men, 61.5%; women, 67%. Program B admission rates: men, 14.3%; women, 29.4%. In each program a higher percentage of women was admitted. c. In both programs, the percentage of women applicants admitted was higher than the percentage of men applicants admitted. But, in the overall combined data, the percentage of women applicants admitted was lower than the percentage of men applicants admitted.

4.39 a. Response variable is outcome (successful or not). c. Compare treatments separately within the two initial severity groups.

4.41 b.

Sex	Had Injury	No Injury	Total
Girls	$\frac{999 \times 227}{2666} = 85.1$	$\frac{999 \times 2439}{2666} = 913.9$	999
Boys	$\frac{1667 \times 227}{2666} = 141.9$	$\frac{1667 \times 2439}{2666} = 1525.1$	1667
Total	227	2439	2666

c. Chi-square =

$$\frac{(74 - 85.1)^2}{85.1} + \frac{(925 - 913.9)^2}{913.9} + \frac{(153 - 141.9)^2}{141.9} + \frac{(1514 - 1525.1)^2}{1525.1} = 1.44 + 0.13 + 0.86 + 0.08 = 2.51$$

4.43 a. Null: Sex and opinion on death penalty are not related. Alternative: Sex and opinion on death penalty are related. b. Conclude that sex and opinion are related because p-value < .05. c. Males: 28.7% opposed; females: 37.9% opposed. The difference in percentages is evidence of a relationship.

4.45 Null: No relationship between opinion about banning texting while driving and whether or not individual has texted while driving. Alternative: There is a relationship between opinion about banning texting while driving and whether or not individual has texted while driving.

4.47 a. Reject the null hypothesis. b. Cannot reject the null hypothesis. c. Reject the null hypothesis. d. Cannot reject the null hypothesis.

4.49 No.

4.51 a. Yes. b. Yes. c. No.

4.53 $(212 - 174.93)^2/174.93$

4.55 a. Null hypothesis: No relationship between sex and opinion about capital punishment. Alternative hypothesis: There is a relationship between sex and opinion about capital punishment. b. The relationship is not statistically significant because .19 > .05. c. Chi-square statistic = 1.714.

4.57 A small sample size limits our ability to make conclusions about a population. Failure to declare statistical significance is not absolute proof that there is no relationship in the population.

4.59 a. Null hypothesis: There is not a relationship between sex and opinion about legalization of marijuana. Alternative hypothesis: There is a relationship between sex and opinion about legalization of marijuana. b. Sex and opinion about legalization of marijuana are related variables. c. Males: 56.7% opposed; females: 63.1% opposed. The difference is evidence of a possible relationship.

4.61 a. Yes, there appears to be a relationship based on the differences among the following percents:

Piercings	% with Tattoo
2 or less	40/538 = 7.4%
3 or 4	58/432 = 13.4%
5 or 6	77/279 = 27.6%
7 or more	53/126 = 42.1%

b. The graph shows that women who have a tattoo are more likely to have a large number of ear piercings. c. (77 + 53)/228 = 57% of women with a tattoo have five or more ear piercings. (202 + 73)/1147 = 24% of women with no tattoo have five or more ear piercings. d. 228/1375 = 16.6%. e. 498/1375 = 36.2%.

4.63 a.

	Cancer	Control	Total
Bird	98	101	199
No Bird	141	328	469
Total	239	429	668

b. Bird owners: 98/199 = .4925, or 49.25%. Nonowners: 141/469 = .301, or 30.1%. c. No. The researchers purposely sampled more lung cancer patients than would naturally occur in a group of 668 people. d. .4925/.301 = 1.64. e. The "baseline" risk of lung cancer for people like yourself. f. No. This is an observational study, and there may be confounding factors that explain the results. g. Compare the smoking habits of the bird owners and nonowners and see whether or not the bird owners generally smoke more.

4.65 Chi-square = 4.817, p-value = .028; relationship is statistically significant.

4.67 a. 1.28 b. 1.53% (or .0153 as a proportion). c. From the relative risk we would know only that the risk is 1.28 times greater for the hormone therapy group, but would not know the actual risk.

4.69 There is a statistically significant relationship.

4.71 If we use the approximate odds given in Exercise 4.70, the odds ratio is 24/8.5 = 2.82. The precise odds ratio is found by taking (191 / 8) / (500 / 59) = 2.817. The odds of remaining free of heart disease versus getting heart disease for men with no anger are about 2.8 times the odds of those events for men with the most anger.

4.73 a. For men, the odds for Program A are 400 to 250, or about 1.6 to 1. For women, the odds are 50 to 25, or 2 to 1. The odds ratio for women compared to men is 2/1.6, or 1.25. Women have better odds. b. For men, the odds for the combined programs are 450 to 550, or about .8 to 1. For women, the odds are 175 to 325, or about .54 to 1. The odds ratio (women to men) is .54/.8 = .675. Women have worse odds. c. The last sentence in the solutions for parts (a) and (b) can be used for this purpose.

4.75 b. Of the men, 15.48% think they are underweight. Of the women, only 4.73% think they are underweight. c. The relationship is statistically significant.

4.77 a. Null hypothesis: There is not a relationship between gun ownership (or not) and opinion about required police permits for guns. Alternative hypothesis: There is a relationship between gun ownership (or not) and opinion about required police permits for guns. b. 34.75%. c. Gun in home: 65.7% favor permits. No gun in home: 86.4% favor permits. The difference is evidence of a possible relationship between gun ownership and opinion about permits. d. p-value $\approx$ 0. The observed relationship is statistically significant. e. An observed relationship is statistically significant if it is unlikely that a relationship as strong or stronger would be observed in a sample if there were no relationship in the population.

4.79 c. Students who prefer to sit in the back are more likely than others to say that religion is not important in their lives and less likely to say that religion is very important.

Chapter 5

5.1 When the data can be considered to be representative of a much larger group with regard to the question(s) of interest.

5.3 a. Yes. The heights of women in the class probably are similar to the heights of all women at the college. b. No. The sample of daycare parents is likely to be more supportive than the general population.

5.5 a. All registered voters in the community. b. The 400 randomly selected individuals.

5.7 Selection bias occurs if the method for selecting the participants produces a sample that does not represent the population of interest.

5.9 For example, testing manufactured parts and the test damages the products.

5.11 A sample survey measures a subgroup of a population in order to learn something about that larger population, whereas a census measures everyone in the population.

5.13 No. All possible sets of four songs do not have the same probability to be a sample. For example, a sample consisting of the first four songs on the first CD is impossible.

5.15 a. Response bias. b. Selection bias. c. Nonparticipation bias.

5.17 a. Selection bias. The sample represents only those who own automobiles. b. Nonparticipation bias (nonresponse bias). People who feel strongly about the mayor's performance are more likely to respond.

5.19 a. Does not hold. Professional basketball players do not represent any larger population. b. Probably holds. Students taking statistics most likely have representative pulse rates.

5.21 Answers will vary. One type of example is a survey mailed to a random sample of a population, but designed so that only those with a strong opinion about the issue are likely to respond.

5.23 .014 or 1.4%.

5.25 a. .105 or 10.5%. b. .695 to .905.

5.27 .022 or 2.2%.

5.29 a. .032 or 3.2%. b. .40 ± .032 or .368 to .432; 40% ± 3.2% or 36.8% to 43.2%. c. We have 95% confidence that the interval includes the percentage of all adult Americans thinking Internet shopping poses more of a threat. d. Yes. The entire interval is below 50%.

5.31 a. 95. b. No.

5.33 .03 or 3%.

5.35 a. .49 ± .03 = .46 to .52. b. .47 ± .03 = .44 to .50. c. No, we cannot conclude that a clear majority of the population has either opinion.

5.37 a. About 5%. b. 2500.

5.39 Sample B.

5.41 At least n = 400

5.43 a. Any sampling method for which it is possible to specify the chance that a particular individual in the population will be selected for the sample. b. Every conceivable group of units of the required size has the same chance of being the selected sample.

5.45 a. The 366 possible dates from January 1 to December 31. b. Answers will vary, but an example is that they could write each date on a piece of paper, put them in a large paper bag, mix well, and select one from the bag.

5.47 a. The whole numbers from 1 to 49. c. It doesn't matter. Chances of winning are the same.

5.49 d. It's possible, but unlikely. All 10 selected values would have to be 20.0 cm.

5.51 Simple random sample.

5.53 a. Stratified sample: Use the three types of schools as strata. Create a list of all students for each of the three strata; draw a simple random sample from each of the three lists. b. Cluster sample: Use individual schools or individual classes as clusters. Take a random sample of clusters; measure all students in those clusters. c. Simple random sample: Obtain a list of all students in the classes at all schools; take a simple random sample from that combined list.

5.55 Cluster sample because a sample of exchanges is found and then only numbers within those exchanges are sampled.

5.57 a. All taxpayers. b. Parents of school children in the local schools. c. They are missing taxpayers who do not have children in the schools, and their opinions on supporting schools may differ.

5.59 a. Convenience sample. b. Self-selected sample.

5.61 You would probably start with your friends. After that, you would probably ask people who seem friendly and approachable, rather than people who seem mean or in a rush. You may feel most comfortable asking people close to your own age, or people you perceive to be compatible with you in some way (perhaps even in their opinions). The polls failed to predict the winner because quota sampling is not likely to provide a sample representative of the population of interest.

5.63 a. Dentists who subscribe to one or both of the two dental magazines. Yes, because not all dentists subscribe to those two magazines. b. Nonresponse. c. Send a reminder or call those who didn't respond.

5.65 a. Self-selected or volunteer sample. b. No. Readers with strong opinions will respond.

5.67 a. Self-selected or volunteer sample. b. Probably higher given the nature of late-night radio talk shows and the fact that people with an interesting experience are more likely to call.

5.69 Answers will vary. One example is "Should former drug dealers be allowed to work in hospitals after they are released from prison, or not?"

5.71 Anonymous testing.

5.73 Desire to please; confidentiality and anonymity.

5.75 Any two questions in which one question changes the way respondents would think about the other question. An example: "Are you aware that over 30% of homeless people in this city are mothers with children?" and "Do you think more public money should be used to help homeless people?"

5.77 a. Door-to-door interview. b. Mail survey. c. Traditionally, it is thought that bias due to a perceived lack of confidentiality would most likely occur with a door-to-door interview and would least likely occur with a mail survey. However, many students who have answered this question thought that a perceived lack of confidentiality would most likely occur with a telephone interview, perhaps because of caller ID.

5.79 a. Closed-form. b. No. Many very popular movies were not included in the list.

5.81 a. Open-form question. b. Yes, because of all the publicity he had just received. c. Probably lower.

5.83 a. .10 or 10%. b. The answer will vary, but will be between 0 and 5 for about 99% of all students.

5.85 a. Answers will vary. It is likely to be in the vicinity of 68 inches. b. Answers will vary. It is likely to be about 4 or 5 inches (from about 66 to 70).

5.87 Any list that contains the two-digit strings in any order: 00, 07, 15, 19, 24, 33, 44, 51, 65, 99. For instance, 24190 03351 99076 54415.

5.91 a. All adults in the U.S. at the time the poll was taken. b. $\frac{1}{\sqrt{1048}} = .031$ or 3.1% c. 30.9% to 37.1%.

5.93 a. Put an ad in the local paper asking people to fill out the survey. b. Ask "Don't you agree that there is too much trash in our streets and that more public trash containers are needed?" c. Ask "Do you think there is too much trash on the streets of our city?" Then follow with "Do you think there should be more trash containers available?" d. Send interviewers out for a door-to-door survey.

5.95 Use of a self-selected sample.

5.97 a. .022 or 2.2%. b. .148 to .192; 14.8% to 19.2%.

5.99 a. 1111. b. 52% to 58%. c. Answers will vary. An example is "In a survey of about 1000 adults, 55% of those asked favor gun control. From this result, we can conclude that somewhere between 52% and 58% of the population are likely to favor gun control."

5.101 a. All students at the university. b. All students enrolled in statistics classes that term. c. The 500 students to whom the survey was mailed. d. The extent to which the sample represents the population of interest depends on what types of students are enrolled in statistics classes that term.

5.103 The sample sizes or margins of error. The 20% versus 25% may be within the margins of error for the surveys.

5.105 The "quickie poll" would probably be most representative.

5.107 a. .027 or 2.7%. b. 17.3% to 22.7%.

5.109 Nonresponse.

5.111 a. Send the survey only to coffee bar owners. b. Send the survey to a legitimate random sample, but make the questions so outrageous that only those who support the position would respond; others would not take it seriously. c. Word the question to elicit the desired response.

5.113 Answers will vary, but some websites have such polls, such as http://www.cnn.com's "Quick vote."

5.115 No, only those who are likely to vote should be used.

5.117 Larger; the sample size for Republicans only would be smaller.

5.119 a. .078; .66 ± .078 or 58.2% to 73.8%. b. .066; .38 ± .066 or 31.4% to 44.6%. c. Yes, intervals do not overlap.

Chapter 6

6.1 a. Observational study. b. Randomized experiment. c. Randomized experiment. d. Observational study.

6.3 a. Explanatory variable = sorority membership (or not); response variable = grade point average. b. Explanatory variable = medication used; response variable = extent of allergy relief for patient. c. Explanatory variable = server introducing self or not; response variable = tip amount (or percentage). d. Explanatory variable = television watching amount; response variable = bullying frequency.

6.5 a. Math skills and shoe size will both increase as children get older. b. Both the explanatory variable and the response variable may be related to time spent socializing. More social people may be more likely to get colds because they sleep less and have many contacts with other people. More social people may also procrastinate in doing assignments.

6.7 a. Randomized experiment. b. Observational study. c. Observational study.

6.9 a. Unit = college student. Variables are procrastination habits and illness frequency. b. Unit = an SUV. Variables are manufacturer and damage sustained in crash test.

6.11 a. Probably not, because long-term meditation is a matter of choice, not easily randomly assigned.
b. Yes, volunteers could be randomly assigned to attend the program or not.
c. For part (a), the explanatory variable is long-term practice of meditation; the response is blood pressure. For part (b), the explanatory variable is whether or not a person took a special training program; the response variable is their score on a standard college admissions test.
d. For part (a), amount of salt and sodium in the diet or other diet-related factors. For part (b), intelligence.

6.13 a. Randomized experiment. b. The explanatory variable is what treatment was received (xylitol in the different forms or placebo in the different forms); the response variable is whether or not a child got an ear infection. c. No.

6.15 The researchers could not assign individuals either to attend a religious service once a week and pray regularly or to not engage in these practices.

6.17 Randomly divide the 60 participants into two groups of 30. Assign one group to diet and the other to exercise.

6.19 a. Yes. b. Yes. c. No. d. No. e. No. f. Yes.

6.21 a. Randomly divide stores into two groups of 10, and then assign a different method to each group. b. Pair the stores based on weekly sales or store size. In each pair, randomly assign one store to each method.

6.23 a. Form pairs of participants by age. Members of a pair would have similar ages. Randomly assign each member of a pair to use a different method. b. A matched-pair design is a good idea because we have the problem of memorization decreasing with age and we have such an age variation.

6.25 One possibility is to number the volunteers from 1 to 100, and then use the software to randomly permute the integers from 1 to 100. Assign volunteers with the first 25 numbers in the list to the first treatment, volunteers with the next 25 numbers to the second treatment, and so on.

6.27 a. Single-blind; customers knew what plan they had, but the technician did not. b. Double-blind; the participants did not know which tea they were drinking, and the psychologists did not know who was drinking the herbal tea. c. Neither single-blind nor double-blind; everyone knew which of the three packaging designs was being used in each store.

6.29 a. The explanatory variable is the rate plan; the response is usage during peak hours.
b. The explanatory variable is whether or not the herbal ingredient was taken; the response is the change in mood or level of depression after 1 month of drinking the tea.
c. The explanatory variable is the packaging design used; the response is how much was sold in 2 months.

6.31 a. A control group was used (matched with the treatment group); a placebo was not.
b. A control group and a placebo treatment were used.
c. Neither a control group nor a placebo was used.

6.33 a. The teacher might record attendance information throughout the term. b. At the end of the term, ask students to remember and report their attendance for the term. c. The disadvantage of the prospective study is that student behavior might be altered if they know that the teacher is keeping attendance records. The disadvantage of the retrospective study is that students may not accurately remember or report their attendance habits.

6.35 a. Yes. b. No. c. No.

6.37 Gather a sample of people over 50 who have skin cancer (cases) and a sample of people over 50 who do not have skin cancer (controls). Ask each individual to recall his or her lifetime sun exposure.

6.39 Observational study because you most likely could not randomly assign students to live either off-campus or in a dorm.

6.41 a. Observational study. b. The explanatory variable was whether or not the couple owned a pet, and the response variables were marriage satisfaction and stress levels. c. An example is the amount of business travel they do. Couples who travel frequently may be less likely to own pets and may also have more stress.

6.43 A randomized experiment provides stronger evidence of a cause-and-effect relationship.

6.45 a. Observational study. b. No. c. Answers will vary, but the variable should be likely to differ for vegetarians and non-vegetarians, and should affect death from cancer and/or heart attacks. An example is whether or not someone smokes.

6.47 Interacting variable.

6.49 Hawthorne effect.

6.51 The elephants are in captivity, so their behavior will differ from what it would be in the wild.

6.53 a. Yes. b. Yes. c. No.

6.57 Confounding variables (e.g., current diet). Relying on memory; it would be hard for people to remember how high in fat their childhood diet was.

6.59 The term "effect modifier" makes sense because the presence of other smokers in the home (or not) modified the effect of the nicotine patch.

6.61 Observational study.

6.63 The explanatory variable is the hormone therapy taken, and this is a categorical variable. The categories are the possible hormone therapies, which are (1) estrogen, (2) estrogen and progestin, and (3) no hormones. The response variable is whether or not a woman developed breast cancer. This is a categorical variable; the categories are (1) had breast cancer, and (2) did not have breast cancer.

6.65 Interacting variables.

6.67 Extending results inappropriately. If women today take lower levels of hormones than the women in the study, these results might not apply to women currently on hormone therapy.

6.69 Randomized experiment.

6.71 Could not have been double-blind. Could have been single-blind, but not enough information is given to know.

6.73 This is an example of an interacting variable, because the effect on self-esteem of thinking about their bad hair is different for men than for women.

6.75 There are situations in which it is impossible to randomly assign treatments. For instance, we could not assign people to be either a smoker or a non-smoker.

6.77 Completely randomized design.

6.79 a. Yes. b. Yes. c. No.

6.81 a. Individual unit is a tomato plant. The two variables were the number of tomatoes produced and whether the tomato plant was raised in full sunlight or partial shade.
b. Individual unit is an automobile. The two variables were the gas mileage and whether the tires were under-inflated or inflated to their maximum possible pressure.
c. Individual unit is a classroom. The two variables measured were the number of children who did better than average on standardized tests and whether or not the classroom took a morning fruit snack break.

6.83 This is an observational study so a figure similar to Figure 6.2 is appropriate.

6.85 Yes.

6.87 a. Telling the experimenters they had maze-bright rats and telling the experimenters they had maze-dull rats. b. 12 individual experimenters. c. Answers may vary, but one idea is to give each experimenter two sets of rats, one called "maze bright" and the other called "maze dull." Randomly assign the order in which the two conditions are presented.

6.89 a. Yes, a variable can be both a confounding variable and a lurking variable. b. No, a variable cannot be both a response variable and a confounding variable. c. No, a variable cannot be both an explanatory variable and a dependent variable.

6.91 When participants are randomly assigned to treatments groups, they are just as likely to be in one treatment group as another. Any potential confounding variables are likely to be balanced over the treatment groups.

6.93 a. Students were measured two different times. b. The order of either swearing or not swearing should be randomized.
c. 64 students participated.

Chapter 7

7.1 Random circumstance: Domestic flight is on time or not. The probability is .787 that flight is on time.

7.3 1/16 = .0625.

7.5 Random Circumstance 1: Song on the radio when first turned on
- Robin's favorite song is playing
- Robin's favorite song is not playing

Random Circumstance 2: Color of traffic light when Robin approaches the main intersection
- Traffic light is green when Robin arrives
- Traffic light is red or yellow when Robin arrives

Random Circumstance 3: Nearest available parking space
- Robin finds an empty parking space in front of the building
- Robin does not find an empty parking space in front of the building

7.7 a. Answers will vary, but one example is whether a ticket for a lottery is going to be a winner for a lottery drawing that will occur in the future. b. Answers will vary, but one example is whether a ticket for a "scratcher" lottery game is a winner; in these games the player scratches off a covering to reveal whether the ticket is a winner or not.

7.9 1/125

7.11 a. Relative frequency probability. b. Personal probability.
c. Relative frequency probability.

7.13 Random circumstance: Traffic light is green or not. Determine probability by observing the proportion of time it's green over many months of driving to work.

7.15 Play over and over again, and record how many games he or she won out of the total number of games played.
7.17 No.
7.19 17/172
7.21 {0, 1, 2, 3, 4, 5, 6, 7}
7.23 a. Yes. b. No.
7.25 a. Yes. b. No. c. Yes.
7.27 a. Yes, B and C are independent. b. Now B and C are not independent.
7.29 a. Mutually exclusive (red die cannot be both 3 and 6), but not independent. b. Independent, but not mutually exclusive (red die can be 3 in the same toss that green die is 6).
7.31 Prior to any draws, the probability is 3/50 that Alicia will be picked. On any given day, 3 of the 50 students are chosen to answer questions. If the drawing is fair, each of the 50 students has the same probability to be picked for any one of the three questions.
7.33 a. P(A and B) = 0, because they can't both happen for the same student. b. No; $P(A) \approx 18/190$, but $P(A|B) = 0$. c. Yes; The same student cannot pick both 5 and an even number.
7.35 No; knowing the woman's age changes the probability that she is fertile.
7.37 a. 1/50 b. 0 c. No, C_1 and C_3 are not independent events. If one happens, the other cannot happen.
7.39 a. Getting at least one tail. b. 7/8
7.41 a. {Monday, Tuesday, Wednesday, Thursday, Friday, Saturday, Sunday} b. 2/7
7.43 a. A and B. b. A and C, or B and C. c. Yes, A and B are complements.
7.45 a. No. b. .8
7.47 a. With replacement. b. Without replacement. c. Without replacement.
7.49 a. 11/12 b. 11/12 c. 121/144 d. 23/144
7.51 .4998
7.53 a. (1/4)(1/4) = .0625. b. (10/40)(9/39) = .0577. c. (3/4)(3/4) = .5625. d. (3/4)(29/39) = .5577.
7.55 a.

Magazine Type	International	National	Total
News	20	10	30
Sports	5	15	20
Total	25	25	50

b. 2/3 c. 1/2
7.57 b. 24% of the students are seniors.
7.59 a. P(A and B) = .25. b. P(B|A) = .3125. c. $P(B^c|A) = .6875$.
7.61 (1/13)(48/51) = .0724.
7.63 P(calculus student) = .4; P(Junior|calculus student) = .15/.4 = .375.
7.65 The first set of branches represent whether the first student drawn is left-handed or not, and have probabilities 3/30 (left-handed) and 27/30 (not). The next set of branches represent whether the second student drawn is left-handed or not, and has conditional probabilities of 2/29 and 27/29 for the branches connected to the first student being left-handed, and 3/29 and 26/29 for the other set. The requested probability is found as (3/30)(2/29) = 1/145.
7.67 45/10000 = .0045.
7.69 a. 8/50 b. 7/31 c. 1/50
7.71 Use only two digits, such as 8 and 9, to represent correct guesses.
7.73 a. 73/1000 = .073. b. This will differ for each student, but should be close to 73/1000. c. Probably not. This answer is a conjecture, but the explanation should refer to the fact that people do not choose "randomly" when asked to do so.
7.75 No.
7.77 They are equally likely.
7.79 a. 1/11 = .0909 b. Although the probability of testing positive for those with the disease is high, the reverse is not true. If there are a very large number of people who do not have the disease, then even if only a small percentage of those test positive, the result will be a large number of positive tests in healthy people.
7.81 a. 1/365 b. $1 - \left(\frac{364}{365}\right)^5 = .0136$. c. No, that specific event has low probability, but it is not unlikely that someone in the class would have a birthday matching a teammate's family member's birthday.
7.83 No, birth outcomes are independent.
7.85 This will differ for each student, but it should not be surprising to encounter it at least once during the day. Of course it will depend on how one spends the day.
7.87 Not surprising; P(match) = 1 − P(no matches) = .1829.
7.89 The relative frequency interpretation.
7.91 a. .27 b. .86 c. .0225 d. .0064 e. .01445
7.93 a. .9999 b. .9996 c. .0004
7.95 a. 40% b. 10% c. 31%
7.97 The table is as follows.

Attendance	A	Not A	Total
Regular	28,000	42,000	70,000
Not regular	3,000	27,000	30,000
Total	31,000	69,000	100,000

7.99 a. .5 (problem stated that no responses equal median). b. .0625. c. No, they would not be independent.
7.101 This figure is based on a combination of personal probability, long run relative frequency and physical assumptions about the world.
7.103 a. (.35)(.35) = .1225. b. No, their statuses are not independent. c. Relative frequency. d. Statement must be about a separate randomly selected foreign-born person in each of the two years.
7.105 485/1669.
7.107 238/612.
7.109 .4123. (If selected from the sample only, it is .4125.)
7.111 .2246 (use Rule 4).

Chapter 8

8.1 a. Discrete. b. Continuous. c. Discrete. d. Discrete.
8.3 a. Discrete. b. Continuous. c. Continuous. d. Discrete.
8.5 Continuous example: amount of rainfall during the event; S includes the range from 0 to the maximum possible rain in 1 day. Discrete example: number of emergency vehicles with sirens driving by during event; $S = \{0,1,2, \ldots, k\}$, where k is the logical maximum for the situation.
8.7 Probabilities for exact outcomes can be found for discrete variables, but not for continuous variables.
8.9 a. 3/6. b. .1.
8.11 a.

Meals, X	1	2	3	4
Probability	.10	.42	.98	1

b. .42.
8.13

k	0	1	2	3
$P(X = k)$	1/8	3/8	3/8	1/8

8.15 a. .09.
b. .27.
c.

k	0	1	2	3	4
$P(X \leq k)$	.73	.89	.95	.98	1

8.17 1/2.
8.19 a. Condition 1: Sum = 1; Condition 2: Each of the four probabilities is between 0 and 1.

c.

k	0	1	2	3
P (X ≤ k)	.1	.2	.5	1

8.21 a. Sample space events = G, BG, BBG, BBBG, BBBB.
b. Probability of G = .5, probability of BG = .25, probability of BBG = .125, probability of BBBG = .0625, and probability of BBBB = .0625.
c.

k	1	2	3	4
P (X = k)	.5	.25	.125	.125

8.23 a. Cumulative probability distribution function. b. Probability distribution function. But cumulative distribution function may be of interest to someone who has had *k* exposures and wants to know the probability of infection so far.

8.25 a.

X	+$4	−$2
Probability	.3	.7

b. −$0.20. c. Over a large number of plays, player will lose an average of 20 cents per play.

8.27 $E(X) = (\$100)(.01) + (-\$5)(.99) - \$3.95$.

8.29 a. 0.6. b. The mean number of wins is 0.6 over (infinitely) many repeats of playing 3 times.

8.31 a. 0. b. 2. c. 1.414.

8.33 $E(X) = 1.5$.

8.35 Standard deviation = $29.67.

8.37 $E(X) = (\$500)(.10) + (\$0)(.90) = \$50$ per person (in the long run).

8.39 a. 1.66. b. No, not as an average number of parents a child was living with.

8.41 a. Yes, $n = 100$, $p = .5$. b. 100.

8.43 a. $n = 30$, $p = 1/6$. b. $n = 10$ and $p = 1/100$. c. $n = 20$ and $p = 3/10$.

8.45 a. .0264. b. .3020. c. .2684. d. .000004.

8.47 a. $\mu = 5$ and $\sigma = 1.5811$. b. $\mu = 25$ and $\sigma = 4.33$. c. $\mu = 500$ and $\sigma = 20$. d. $\mu = .1$ and $\sigma = .3$. e. $\mu = 12$ and $\sigma = 2.683$.

8.49 a. The probability of success does not remain the same from one trial to the next. b. There are not a prespecified number of trials. c. Outcomes are not independent from one trial to the next; probability of success changes.

8.51 a. .3281. b. .9185. c. 1 − .9185 = .0815.

8.53 a. .2051. b. .3504.

8.55 a. The desired probability is $P(X \geq 250)$, $n = 1000$, and the value of p is not known. b. The desired probability is $P(X \geq 110)$, n is 500, and p is .20. c. The desired probability is $P(X \geq 14)$, $n = 20$, and $p = .50$.

8.57 a. .3. b. .4. c. .5.

8.59 a. X is a uniform random variable because the spinner is equally likely to fall anywhere in the interval from 0 to 100. b. The density function is 1/100 for any x between 0 and 100; and 0 for any x not between 0 and 100. c. $P(X \leq 15 \text{ seconds}) = (15 - 0)/100 = .15$. d. $P(X \geq 40 \text{ seconds}) = (60)(1/100) = .60$. f. 50.

8.61 This may differ for each student, but the answer must be something that is equally likely to fall anywhere in an interval. A simple example is the position of the second-hand on a clock when you glance at it.

8.63 a. $z = 1.5$. b. $z = -1.0$. c. $z = -2$. d. $z = -1$.

8.65 a. .5000. b. .3632. c. .6368.

8.67 a. .8413. b. .2266. c. .7734.

8.69 a. .0808. b. .9192. c. .8384. d. .0808.

8.71 a. $P(Z \leq -0.5) = .3085$. b. $P(Z \geq 1.25) = .1016$. c. $P(-2 \leq Z \leq 2.5) = .9710$.

8.73 64.76, or about 65.

8.75 a. .0918 (for $z = -1.33$). b. .0228. c. .3830. d. .0023.

8.77 a. −1.96. b. 1.96. c. 1.96.

8.79 $P(Z \leq -0.675)$ is about .25, so about 25% of the women are shorter than $-0.675 \times 2.7 + 65 = 63.2$ inches.

8.81 a. .0571. b. .0749. c. .0743. d. Part (c) result is most accurate; part (a) result is least accurate.

8.83 a. .9015. b. .9074.

8.85 a. $\mu = 140$, $\sigma = 6.481$. b. .0320, or .0380 using continuity correction.

8.87 a. 400 seconds. b. 20 seconds. c. .0228.

8.89 a. .25. b. $n = 7$; $p = .25$. c. $n = 8$; $p = .25$. d. $n = 15$; $p = .25$.

8.91 a. The name of the distribution cannot be given. Mean = 13, variance = 7.3, standard deviation = 2.702. b. $X + Y$ is normal with mean = 150, variance = 325, standard deviation = 18.03. c. The name of the distribution cannot be given. Mean = 150, variance = 262.5, standard deviation = 16.202.

8.93 a. For $n = 10$, mean = 2.5; for $n = 20$, mean = 5; for $n = 50$, mean = 12.5. b. 20. c. $n = 10$, $P(X \geq 4) = .2241$; $n = 20$, $P(X \geq 8) = .1018$; $n = 50$, $P(X \geq 20) = .0139$. d. .0010. e. The combined data.

8.95 $P(Z < -1.21) = .1131$.

8.97 Probably not. It will be a mixture of two bell-shaped distributions with different means.

8.99 a. X is −$2, or $70; $P(X = -\$2) = 37/38$, $P(X = \$70) = 1/38$. b. $E(X) = -\$4/38 = -\$0.1053$; players lose about 10.5 cents per $2 bet over the long run.

8.101 5.01 courses per student.

8.103 a. Probability of success is not the same for all trials. b. Outcomes are not independent from one trial to the next and the probability of success changes from one trial to the next.

8.105 a. .0021. b. .25.

8.107 a. Binomial distribution with $n = 4$ and $p = .5$. b. The distribution for Karen will be the same as it is for Kim. c. Binomial distribution with $n = 8$ and $p = .5$.

8.109 a. .6915. b. .1056. c. .3345.

8.111 If 1000 repetitions are done, the simulated answer is likely to be in the range .06 to .10. The theoretical probability is .079.

8.113 .0918 (for $Z = -1.33$).

Chapter 9

9.1 a. Statistic. b. Parameter. c. Statistic. d. Parameter.

9.3 Population parameter.

9.5 a. The proportion of the population that has as least one copy of the E4 allele for Apo E. b. A confidence interval would make more sense. c. The scientists could not possibly have measured everyone in the population.

9.7 a. Parameters; they represent the population values. b. Yes, they know the means exactly, and the mean for males is $1500 higher.

9.9 a. $\hat{p}$. b. p. c. $\hat{p}$.

9.11 a. What proportion of parents in the school district support the new program? b. Parameter is p = proportion of parents in school district who support the new program. c. Sample estimate is $\hat{p} = 104/300 = .347$.

9.13 a. How much difference is there between the mean number of CDs owned in the populations of male and female students? b. Parameter = $\mu_1 - \mu_2$, where μ_1 = population mean for males and μ_2 = population mean for females. c. $\bar{x}_1 - \bar{x}_2 = 110 - 90 = 20$.

9.15 *Research question:* How much difference is there between the proportions getting relief from sore throat symptoms in the population if herbal tea is used versus if throat lozenges are used? *Population parameter:* $p_1 - p_2$ = difference in proportions reporting relief if everyone in the population were to use herbal tea compared with if everyone in the population were to use throat lozenges. *Sample estimate:* $\hat{p}_1 - \hat{p}_2$ = difference in observed proportions reporting relief for the two different methods in the study of 200 volunteers.

9.17 Paired data. Two variables are measured on each individual.

9.19 Answers will vary, but should involve proportions for populations from which independent samples can be taken. See Exercise 9.50 on page 358 for examples.

9.21 a. Parameter is p = proportion of all teens who go out on dates who would say that they have been out with someone of another race or ethnic group. Sample statistic is $\hat{p} = .57$. b. The population of interest is people who slept in darkness as babies and the parameter of interest is p = the proportion of that population who were free of myopia later in childhood. The sample statistic is $\hat{p} = .90$. c. The parameter of interest is $p_1 - p_2$ = the difference in proportions of myopia in later childhood for the populations who slept in darkness as babies and who slept in full light as babies, respectively. The sample statistic is $\hat{p}_1 - \hat{p}_2 = .90 - .45 = .45$.
d. The parameter of interest is μ = the mean height for the population of married British women in 1980. The sample statistic is $\bar{x} = 1602$ millimeters.

9.23 Histogram of the 100 SAT scores.

9.25 a. p. b. $\hat{p}$.

9.27 a. No. b. Yes. c. No. d. Yes. e. No.

9.29 a. The same; it would be μ. b. The same, approximately normal. c. The same; it would be $\sigma/\sqrt{n}$.

9.31 a. Not likely. b. Possible. As shown in Figure 9.1, the mean could be as high as 7.4 hours, so there are some likely values in the interval 7 to 8 hours. c. Likely. d. Very unlikely.

9.33 a. Mean = .5; s.d. = $\sqrt{\frac{(.5)(.5)}{400}} = .025$. b. Mean = .5; s.d. = .0125.

9.35 A change in the sample size does not affect the value of the mean of the sampling distribution. Increasing the sample size decreases the value of the standard deviation. For a four-fold increase in sample size the standard deviation is cut in half.

9.37 a. .55. b. $\sqrt{\frac{(.55)(.45)}{100}} = .0497$, or about .05. c. $.55 \pm 3(.05)$, or between about .40 and .70.

9.39 a. $\hat{p} = 300/500 = .60$. b. s.e.($\hat{p}$) = .022.

9.41 a. Mean = .70; s.d.($\hat{p}$) = .0324. b. .6028 to .7972.
c. .60. Statistic. d. The value .60 is slightly below the interval of possible sample proportions for 99.7% of all random samples of 200 from a population where $p = .70$.

9.43 $.05 \pm 2\sqrt{\frac{(.05)(.95)}{400}}$, or .03 to .07.

9.45 a. .20. b. .18. c. .0384. d. .20. e. .04.

9.47 The standard error will be used more often.

9.49 $p_1 - p_2 = 0$.

9.51 a. .0297. b. .0608. c. .0387.

9.53 a. Condition 1 is clearly met because sample proportions will be available for independent random samples. Condition 2 will be met as long as the proportions who favor the candidate are between .01 and .99 for each population, thus ensuring that all of the quantities $n_1 p_1$, $n_1(1 - p_1)$, $n_2 p_2$, $n_2(1 - p_2)$ are at least 10.
b. Condition 1 is not met; the researcher is not taking separate random samples from two populations. The two sample proportions described will not be independent. c. No. The sample size condition is not met. With a total of 10 from each method, there could not be at least 10 that succeed and 10 that fail.

9.55 a. Would not change. b. Would not change as long as the samples were large. c. Would change.

9.57 a. Yes, it is $p_1 - p_2 = 0$. b. Yes; approximately normal.
c. Yes. d. No.

9.59 a. Yes, this falls into Situation 1 described in Section 9.6.
b. No. For a skewed population, the sample size is too small.
c. Yes, this falls into Situation 2.

9.61 a. 6. b. 3. c. s.d.($\bar{x}$) decreases when sample size is increased.

9.63 a. $\bar{x}$ and s. b. s.e.($\bar{x}$) = $47.7/\sqrt{28} = 9.014$.

9.65 b. Approximately normal with a mean of 210 pounds and a standard deviation of 3.95 pounds. d. $P(z > 2.53) = .0057$.

9.67 Standard deviation of the sampling distribution is $\sigma/\sqrt{n}$. Standard error of the mean replaces population σ with the sample version s, and it is more commonly used because σ is rarely known.

9.69 Two independent samples.

9.71 a. The difference in population means for independent samples, $\mu_1 - \mu_2$. b. The mean of paired differences, μ_d.

9.73 No. The difference will be 0 only if the population means for the two measurements are the same.

9.75 a. μ_d = population mean difference in number of drift times if men similar to the ones in this experiment were to drive two hours after consuming alcohol and two hours after getting too little sleep. b. $\bar{d}$, the mean of the differences in number of lane shifts under the two conditions for the 12 men in the sample.
c. The population of differences must be bell-shaped (at least approximately). d. Approximately normal, mean = 0,
$$s.d.(\bar{d}) = \frac{5}{\sqrt{12}} = 1.44.$$

9.77 No. Both populations must be approximately bell-shaped, or both sample sizes must be large.

9.79 a. 3.5 days. b. 0.514. d. Yes, it corresponds to a z-score of $(3.6 - 3.5)/0.514 = 0.19$, so it is reasonable.

9.81 a. 5 inches. b. 1.302 inches. d. Yes. The probability that the mean for women is greater than the mean for men is $P(\bar{x}_1 - \bar{x}_2 < 0) = P(Z < -3.85) \approx .00006$.

9.83 Mean μ and standard deviation σ.

9.85 b. 1.84

9.87 b. 2.58

9.89 a. Normal, mean = 0, s.d. = 1.44. b. No, it is not a reasonable value. The standardized statistic is $(8 - 0)/1.44 = 5.56$.

9.91 a. Approximately normal with mean = 0 and standard deviation = 0.5137. b. 7.01. c. A standardized score of 7.01 is too large to be considered a statistical fluke.

9.93 a. $z = 2$. b. $z = 2.828$.

9.95 a. −1.0 and +1.6. d. The pictures are the same except for the labels on the horizontal axis.

9.97 a. $z = 1$. b. $z = -1$.

9.99 $z = 1.6$.

9.101 a. $t = -1$; df = 16 − 1 = 15. b. $t = 1$; df = 15.

9.103 a. $t = 0.417$. b. $z = 0.5$. c. $t = -3$.

9.105 s = standard deviation of a sample; σ = standard deviation of the population.

9.107 a. t-distribution. b. df = 5. c. Normal distribution.

9.109 a. $t = -5$.

9.111 Yes, the value 0. The area above it is always .5.

9.113 a.

Sample	$\bar{x}$	Sample	$\bar{x}$
1,3	2	3,7	5
1,5	3	3,9	6
1,7	4	5,7	6
1,9	5	5,9	7
3,5	4	7,9	8

b.

$\bar{x}$	2	3	4	5	6	7	8
Probability	1/10	1/10	2/10	2/10	2/10	1/10	1/10

9.115 b. 1/10.
c.

R	1	2	3	4
Probability	4/10	3/10	2/10	1/10

d. $R = 1$ is most likely, and $R = 4$ is least likely.

9.117 a. 0, 1, 2, 3, 4.
b.

t	0	1	2	3	4
$P(T = t)$	.36	.36	.21	.06	.01

c. About (1000)(.36) = 360 teens will have T = 0 on their first try.

9.119 Because of the symmetry of the situation, it should be a mirror image of the sampling distribution of H, the highest number, given in Figure 9.13, so it should be highly skewed to the right.

9.121 a. 1/2. b. Integers from 0 to 6. c. $(1/2)^6 = 1/64$. d. B is binomial with $n = 6$ and $p = ½$; Probabilities for $B = 0, 1, 2, 3, 4, 5, 6$ are 1/64, 6/64, 15/64, 20/64, 15/64, 6/64, and 1/64, respectively.

9.123 a. The population of possible net gains is highly skewed (not bell-shaped), and the sample size is small. b. −$0.35; $9.38.

9.125 a. Binomial; $n = 10, p = .5$. b. 5. c. 1.58.

9.127 a. Approximately a normal curve. b. Answer will vary, but should be close to 5.5 and 10.5. c. The interval is 5.5 to 10.5; comparison to part (b) will vary.

9.129 c. There is less variation among sample means for the larger sample size.

9.131 a. 6.146 to 10.558 hours or about 6.1 to 10.6 hours. b. 7.249 to 9.455 hours or about 7.2 to 9.5 hours.

9.133 Actual population exists (students at your college); a fixed proportion are left-handed; $n = 200$ is large enough for number of left-handers to be sufficient.

9.135 b. Yes; the sample proportion is almost surely in the range .447 to .553, so .70 is unlikely.

9.137 Mean $= \bar{x} = \left(\sum_{w/\,trait} 1 + \sum_{w/out\,trait} 0 \right)/n =$ (number with trait)/n = proportion with trait.

9.139 Mean is still 25 mpg, but standard deviation is now .1 mpg.

9.141 Approximately normal with a mean of 105 and a standard deviation of 1.5.

9.143 a. The standard error in 1976 is .003 so the difference between the sample and population proportions is almost surely less than 3(.003) = .009, or 0.9% or about 1%. b. Yes, population proportion was higher in 1997 than in 1976.

9.145 Increasing sample size decreases standard deviation.

9.147 a. No, sample size is too small. b. Yes. c. No, days in January and February only do not constitute a random sample. d. Yes.

9.149 a. 68% : .5476 to .5724; 95%: .535 to .585; almost all: .523 to .597. b. Not reasonable based on "almost all" interval in part (a). c. $z = 2$. If everybody told the truth, the sample result is unusual, but not nearly as unusual as with the sample size of 1600.

9.151 At least 225.

9.153 c. Distribution skewed somewhat to the right. e. The outliers do not affect the median, so the sample median is almost never higher than 8 hours.

9.155 a. They should be similar. b. The median is in the column labeled *Third*. e. Both histograms are centered at 20, but the histogram for the median is more spread out.

Chapter 10

10.1 a. Population proportion. b. Sample proportion. c. Sample proportion, .55.

10.3 a. How do American teenagers rate their parents on strictness, compared to their friends' parents? b. Proportion of American teenagers who think their parents are stricter than their friends' parents. c. 39%.

10.5 100% − 95% = 5%.

10.7 a. All U.S. adults. b. Proportion of all U.S. adults who think that the use of handheld cell phones while driving should be illegal. c. The $n = 836$ who were surveyed. d. .72 (or 72%).

10.9 a. 36% ± 3.5%, or 32.5% to 39.5%. With 95% confidence, we can say that the percentage of American adults who feel that they don't get enough sleep is between 32.5% and 39.5%. b. The answer depends on whether you think that college students generally differ from the general population with regard to feeling they don't get enough sleep.

10.11 No. This is not a randomly selected sample. It is a self-selected sample, and this type of sample often is biased toward a particular response. In this situation, those who had experienced violence were probably more likely to respond.

10.13 About .95 × 200 = 190 intervals will cover the population proportion, so about 10 intervals will not.

10.15 a. Increase. b. Decrease

10.17 a. p = the proportion of the population that suffers from allergies. b. p = the proportion of the population of working adults who say that they would quit their jobs if they won a large amount in the lottery. c. μ_1 = mean severity score for premenstrual symptoms in the population if all women with premenstrual symptoms were to take calcium. μ_2 = mean severity score if all women with premenstrual symptoms were to take a placebo. The parameter of interest is the difference $\mu_1 - \mu_2$.

10.19 a. .016. b. .559 to .621. We are 95% confident that between 55.9% and 62.1% of all Americans think the world will come to an end.

10.21 $n = 400$, confidence level = 95%. This combination has smallest sample size and greatest confidence level.

10.23 a. $\hat{p} = \frac{166}{200} = .83$. b. $.83 \pm 1.96\sqrt{\frac{(.83)(1 - .83)}{200}}$, or .778 to .882. c. With 95% confidence, we can say that in the population of patients with this disease, the proportion who would be cured is between .778 and .882.

10.25 $.83 \pm 2.33\sqrt{\frac{(.83)(.17)}{200}}$, or .83 ± .062. The 98% confidence interval is wider.

10.27 a. $\hat{p} = \frac{220}{400} = .55$. b. $\text{s.e.}(\hat{p}) = \sqrt{\frac{.55(1 - .55)}{400}} = .025$. c. .55 ± (1.96)(.025), or .501 to .599. d. .55 ± (2.33)(.025), or .492 to .608.

10.29 a. p = proportion of all adults in the U.S. who think that penalties for underage drinking should be stricter. b. .021. c. .60 ± (1.96)(.021), or .559 to .641. d. With 95% confidence, we estimate that between .559 and .641 of all adults in the U.S. think that penalties for underage drinking should be stricter.

10.31 The number in each category should be at least 10, but there is only one left-handed person in the sample.

10.33 $.19 \pm 1.645\sqrt{\frac{.19(1 - .19)}{300}}$, which is .153 to .227.

10.35 1.28

10.37 80% (or .80).

10.39 .60 ± .014, or .586 to .614.

10.41 a. 99%. b. 1%.

10.43 1 − .95 = .05. This is the relative frequency of random samples of this size for which the difference between sample and population percentages will be more than 3%.

10.45 a. .0993. b. .0496.

10.47 a. $\sqrt{\frac{(.03)(.97)}{883}} = .00574$. b. $2\sqrt{\frac{(.03)(.97)}{883}} = .011$.

10.49 a. $\frac{1}{\sqrt{757}} \times 100\% = 3.6\%$ (roughly 3.5%). b. −0.5% to 6.5%. c. .006. d. 1.8% to 4.2%.

10.51 a. Yes, there are two independent samples with sufficiently large sample sizes. b. No, the sample sizes are too small. c. No, the situation is about the difference in two means, not two proportions.

10.53 This is a difference between the proportions in two different categories of a variable in the same sample. The method in Section 10.3 is for a difference between proportions with the same trait in two independent samples.

10.55 a. Age 30–39: .469; age 18–29 = .408; difference is 0.061. b. 0.028 to 0.094. With 95% confidence, we estimate that the difference in proportions of women in the two age groups who experience episodic tension-type headaches is between 0.028 and 0.094.

10.57 b. $(.466 - .380) \pm 2\sqrt{\frac{.466(1-.466)}{4594} + \frac{.380(1-.380)}{7076}}$.

10.59 a. 0.30. b. .0714. c. 0.157 to 0.443. d. Yes. The interval does not include 0.

10.61 a. 55% ± 3%, or 52% to 58%. b. The confidence interval is entirely above 50%, so it is reasonable to say that more than 50% of the population thinks their weight is about right.

10.63 The 16% claim is not reasonable because 16% is not within the confidence interval.

10.65 a. Ninety-five percent (95%) confidence intervals for population percentages in the 2 years do not overlap. It is reasonable to conclude that the population percent is higher in 2002. b. 0.046 to 0.134.

10.67 a. .381 to .479. b. Yes, because the interval is entirely below .50 (50%).

10.69 a. Should not. People dissatisfied with the president may be more likely to respond to the question. b. Should. It's doubtful that the handedness characteristics of this sample should be biased toward having a different proportion that is left-handed than in the population of all college students.

10.71 a. .295. b. .033. c. .241 to .349. d. .229 to .361. e. .218 to .372. f. Width of interval increases.

10.73 Yes. The interval does not include 0.

10.75 A confidence interval is unnecessary because the population value is observed and does not have to be estimated.

10.77 b. 71% ± 5% or 66% to 76%. We can be 95% confident that somewhere between 66% and 76% of all adult Americans would answer that they are for the death penalty when asked Question 1. c. 56% ± 5%, which is 51% to 61%. We are 95% confident that between 51% and 61% of all adult Americans would respond "death penalty" when asked Question 2.

10.79 a. .168. b. .155. d. All college students with liberal arts majors in the northeast. e. .122 to .214 f. .105 to .205.

10.81 a. $(.63 - .29) \pm 2\sqrt{\frac{(.63)(1-.63)}{300} + \frac{(.29)(1-.29)}{300}}$, which is .263 to .417. b. With 95% confidence, we estimate that the difference in proportions of American men and women who have driven after drinking too much is between .265 and .415. (Men were more likely to have done so.)

10.83 .574 to .686.

10.85 a. .077 to .126. b. .235 to .361. c. A 99% confidence interval is about 0.090 to 0.302, computed as 0.1965 ± (2.576)(.0412).

10.87 a. Difference = 0.4554; s.e. = .161. b. No, the sample sizes are too small. c. In a blind experiment, the patient would not know which treatment he or she was given. In double blind, neither the patient nor the evaluator would know the treatment assignment.

10.89 a. .337 to .379. b. With 95% confidence, we estimate that in 2008 the proportion of Americans who preferred the Democratic Party was between .337 and .379.

10.91 a. .619. b. .546 to .693.

10.93 a. 0.227 to 0.354. b. Yes. The interval does not include 0.

Chapter 11

11.1 Are you asking a question about children in this school only, or are you using them to represent a larger population?

11.3 a. Two independent samples. b. Paired data. c. Two independent samples.

11.5 a. 1.5. b. 3. c. 2.

11.7 a. 2.13. b. 1.70. c. 2.26.

11.9 a. μ for Country A, μ for Country B and $\mu_1 - \mu_2$. b. $\bar{x}_1 - \bar{x}_2 = 22.5 - 16.3 = 6.2$ days.

11.11 a. Are the mean levels of B cells the same for the populations of autistic and nonautistic children? If not, by how much do the mean levels differ? b. $\mu_1 - \mu_2$. c. Does $\mu_1 - \mu_2 = 0$? If not, what is its value?

11.13 a. Paired data. b. Independent samples.

11.15 a. Answers will vary, but an example is that the two populations are women who got married in 1979 and women who got married in 2009, and a random sample is taken at the end of each of those years. The response variable is the woman's age when she got married. The question of interest is whether the average age at marriage for women changed between 1979 and 2009. b. Answers will vary, but an example is that the two populations are women with a college education and women without one. A random sample of married women is taken and each woman is asked whether she has a college degree and the age at which she got married. The question of interest is whether the average age at marriage is the same for women with and without a college education. c. Answers will vary, but an example is that the two treatments are two different training programs for salespeople in a company. The question of interest is which one is more effective in increasing average sales. d. Answers will vary, but an example is that a large company wants to know if it is more efficient to encourage customers to phone or send email when they have a request. The company measures a random sample of each type of request and measures how much time its employees spend dealing with the request.

11.17 s.e.$(\bar{x}) = \frac{2}{\sqrt{64}} = 0.25$ cm. Roughly, for all possible samples of this size, the average difference between the sample and population means is 0.25 cm.

11.19 Calculate the difference for each person. The standard error of the sample mean difference is calculated using the formula $s_d/\sqrt{n}$, where s_d = standard deviation of the sample of differences and $n = 50$.

11.21 Take a large enough sample or samples.

11.23 a. Increases. b. Decreases. c. Decreases. d. Found from the standard normal distribution. e. No effect.

11.25 a. $76 \pm 2.31\frac{6}{\sqrt{9}}$. b. $76 \pm 1.86\frac{6}{\sqrt{9}}$. c. $76 \pm 1.75\frac{6}{\sqrt{16}}$.

11.27 a. $t^* = 2.52$ (df = 21). b. $t^* = 2.13$ (df = 4). c. $t^* = 4.60$

11.29 a. s.e. $(\bar{x}) = \frac{s}{\sqrt{n}} = \frac{6}{\sqrt{9}} = 2$. b. 21.6 to 30.8 sit-ups. c. With 95% confidence, we can estimate that in the population of men represented by this sample the mean number of sit-ups in a minute is between 21.6 and 30.8.

11.31 (1) confidence level, (2) sample size, and (3) standard deviation.

11.33 The confidence interval is about 7.04 to 9.16 days, computed as $8.1 \pm 2.82 \times \frac{1.8}{\sqrt{23}}$.

11.35 The calculation is $\bar{x} \pm 2 \times s$. Rounding to the nearest dollar, about 95% of the students had textbook expenses in the interval $285 ± (2)($96), which is $93 to $477.

11.37 a. Not a correct interpretation. b. Correct interpretation. c. Not a correct interpretation.

11.39 s.e.$(\bar{x}) = \frac{s}{\sqrt{n}} = \frac{2.7}{\sqrt{81}} = 0.3$ inch. Approximate 95% confidence interval is 64.2 ± (2 × 0.3), or 63.6 to 64.8 inches.

11.41 The interval is 27 to 28 cm, computed as 27.5 ± 2 × 0.25. With approximate 95% confidence, we can say that in the population of men represented by this sample, the mean foot length is between 27 and 28 cm.

11.43 Using the value of 2 as an approximation for the t^* multiplier will work well only if the t^* multiplier for 95% confidence is in fact close to 2. For small df, t^* is much greater than 2.

11.45 a. $5.36 \pm (2.06)(3.05)$ or -0.92 to 11.64. b. No, the interval covers 0. c. Two different variables are measured for each individual.

11.47 No. Any value in the interval is a plausible value for the difference.

11.49 a. $5 \pm (2.03 \times 1.107)$, or 5 ± 2.25, or 2.75 to 7.25 ounces. (df = 39, use $t^* = 2.03$ from Table A.2, the average of the values for df = 30 and df = 40.) b. Yes because $n = 40$ is large enough. Interval is $5 \pm (2.0 \times 1.107)$, or 5 ± 2.21, or 2.79 to 7.21, almost the same as the answer in part (a).

11.51 a. The population means differ. b. The difference in the population means plausibly could be any of the values covered by the interval. In particular, because the interval includes 0, we cannot conclude that the means differ.

11.53 a. $(11.6 - 10.7) \pm 2.07\sqrt{\frac{3.39^2}{16} + \frac{2.59^2}{10}}$. b. With 95% confidence, we can say that in the populations of medical doctors and professors, the difference in mean testosterone levels is between -1.54 and 3.34. We are assuming that the samples are representative of all male medical doctors and university professors and that neither sample showed outliers or extreme skewness. c. Anything in the interval is a plausible difference in the two population means. Because the interval covers 0, we cannot conclude that the means differ.

11.55 a. 2.2 cm. b. $\sqrt{\frac{2.4^2}{36} + \frac{1.8^2}{36}} = 0.5$. c. $2.2 \pm (2 \times 0.5)$, or 1.2 to 3.2 cm.

11.57 The interval is 2.86 to 3.14 cm. *Interpretation*: With 95% confidence, we can say that in the population(s) represented by these sample(s), the difference in mean foot lengths of men and women is between 2.86 cm and 3.14 cm.

11.59 The variances (or standard deviations) in the two populations are equal.

11.61 $s_p = 1.7$. The 95% confidence interval is
$(8.1 - 4.5) \pm 2.01\sqrt{\frac{1.7^2}{23} + \frac{1.7^2}{25}}$, or about 2.6 to 4.6 days.

11.63 The interval is approximately $72.84 \pm 1.97 \times \frac{72}{\sqrt{205}}$, or 72.84 ± 9.91.

11.65 Both intervals are entirely above 1, so it is reasonable to conclude that risk of death is greater in the abnormal heart rate recovery group than in the normal heart rate group.

11.67 a. Approximate 95% confidence interval is -0.08 to 0.12. b. If the interval did not cover 0, it would show that the means of the placebo and calcium groups were different at the beginning of the study.

11.69 a. Answer will vary but for most students (about 95%) the interval will cover 170. b. Answer will vary. In most instances it will be between 91% and 99%. c. Answer will vary. Expected number is $150(.95) \approx 142$ or so.

11.71 Specific intervals will vary. In all instances, the width of the interval will increase as the confidence level is increased.

11.73 a. The interval will be wider for 95% confidence. b. The interval will become more narrow because the standard error will decrease. c. Assuming that the standard deviation stays the same, the interval width stays the same.

11.75 a. A 99% confidence interval is about 1590.5 to 1613.5 mm.
Parameter is μ = mean height in the population.
Confidence interval formula is $\bar{x} \pm t^* s.e.(\bar{x})$.
$\bar{x} = 1602$ mm, $s.e.(\bar{x}) = \frac{s}{\sqrt{n}} = \frac{62.4}{\sqrt{199}} = 4.4234$,
and $t^* \approx 2.60$ (df = 199 − 1 = 198).
In Excel, TINV(.01,198) gives $t^* = 2.60$. In Table A.2, use the entry for df = 100 as an approximation.
Check necessary conditions: The sample size is sufficiently large. Assume the sample represents a random sample from the population.
Interpretation: We are 99% confident that the mean height of all women represented by this sample is between 1590.5 and 1613.5 mm.
b. About 62.6 inches to 63.5 inches.

11.77 a. On each team, the ninth weight is the weight of the coxswain, who gives instructions about the rowing cadence but does not row. A coxswain's weight is much less than the weights of the rowers, so there are two substantial outliers. b. 183.7 to 199.6 pounds. c. The 90% confidence interval is about -7.50 to 12.25 pounds if the unpooled standard error is used. With a pooled standard error, the interval is about -7.45 to 12.20 pounds. Therefore, the two methods are essentially the same for this example. d. *Interpretation*: With 90% confidence, we can say that the difference between the mean weights in the two populations represented by the Cambridge and Oxford rowers is between -7.5 and 12.2 pounds. Based on this interval, we cannot determine which population of rowers, if either, has the greater mean weight.

11.79 a. It is important to recognize that the appropriate population depends on what was measured, which in this case is TV viewing hours. The TV viewing habits of the students in a statistics class at a particular university at a particular time may represent the TV viewing habits of all students at that school at that time, or the subset who will take a statistics class, or perhaps students at all similar schools who take statistics at some time in college. b. No. The interval tells us that the *average* of TV viewing hours for all students in the population is between 1.842 and 2.338 hours. It doesn't tell us anything about individual students.

11.81 a. A 95% confidence interval is about 1.60 to 2.88 years. *Interpretation*: With 95% confidence, we can say that in the population of British married couples, the mean difference (husband age − wife age) is between 1.60 years and 2.88 years. b. Paired data.

11.83 a. 0.42 hours. This is a statistic. b. Females: $1.51/\sqrt{116} = 0.14$. Males: $1.87/\sqrt{59} = 0.24$. c. We cannot conclude that there is a difference in the population represented by the sample. d. The formula for a confidence interval for the difference between two independent means. The calculation is $(2.37 - 1.95) \pm 2\sqrt{\frac{1.87^2}{59} + \frac{1.51^2}{116}}$.

11.85 a. A "success" is an interval that captures the population mean. b. .90. c. 90. d. No. e. $.90^{100}$. (About 2.6561×10^{-5}.)

11.87 a. 22.247 to 22.868 cm. *Interpretation*: With 95% confidence, we can say that in the population of men represented by the sample, the mean stretched right handspan is between 22.247 and 22.868 cm. b. 19.672 to 20.362 cm. *Interpretation*: With 95% confidence, we can say that in the population of women represented by the sample, the mean stretched right handspan is between 19.672 and 20.362 cm. c. Unpooled is 2.079 to 3.002 cm; pooled is 2.072 to 3.009 cm. *Interpretation*: With 95% confidence, we can say that of men and women in the populations represented by the samples, the difference in the mean stretched handspans is between approximately 2.1 and 3 cm.

11.89 a. 183.11 to 203.16. *Interpretation*: With 98% confidence, we can say that in the population of individuals who have not had a heart attack, the mean cholesterol level is between 183.11 and 203.16. b. 231.63 to 276.22. *Interpretation*: With 98% confidence, we can say that in the population of individuals who have had a heart attack, the mean cholesterol level 2 days after the attack is between 231.63 and 276.22. c. Unpooled procedure result is 36.74 to 84.85. *Interpretation*: With 98% confidence we can say that in the population of people who have suffered a heart attack, the mean cholesterol (measured 2 days after the attack) is between 36.74 and 84.85 points higher than the mean cholesterol in a population of people who have not had a heart attack.

Chapter 12

12.1 a. All babies. b. p = proportion of babies in the population born during the 24-hour period surrounding a full moon. c. H_0: $p = \frac{1}{29.53} = .03386$.

12.3 a. Twenty-one-year-old men and women in the United States. b. Researchers are interested in the hypothesis that equal proportions of 21-year-old men and women in the United States have high school diplomas.

12.5 a. Null hypothesis. b. Alternative hypothesis. c. Alternative hypothesis. d. Null hypothesis.

12.7 a. No. (Hypotheses are about populations, not samples.) b. Yes. c. No.

12.9 a. H_0: Increasing speed limits on interstate highways does not increase the fatality rate; H_a: Increasing speed limits on interstate highways increases the fatality rate. The hypothesis test is one-sided. b. H_0: Probability = .5 that a coin spun on its edge lands heads up; H_a: Probability ≠ .5 that a coin spun on its edge lands heads up. The hypothesis test is two-sided.

12.11 a. There are five choices, so $p = 1/5$, or .2. b. H_0: $p = .2$, H_a: $p < .2$.

12.13 Before they look at the data.

12.15 a. H_0: $p = .06$, H_a: $p > .06$. In words:
Null: The percent of children in the state who live with their grandparent(s) remains at 6% (or possibly even less).
Alternative: The percent of children in the state who live with their grandparent(s) has increased since the last census and is now greater than 6%.
b. The population is all school-aged children in the state; p is the proportion of them living with one or more grandparents.

12.17 a. H_0: $\mu_1 - \mu_2 = 0$, where μ_1 is the mean weight of all newborn babies in England, and μ_2 is the mean weight of all newborn babies in the United States. b. Null value is 0.

12.19 a. Do not reject the null hypothesis. b. Reject the null hypothesis. c. Do not reject the null hypothesis. d. Reject the null hypothesis.

12.21 a. There is not sufficient evidence to conclude that smokers are more likely to get the disease. b. There is enough evidence in the sample to conclude that smokers are more likely to get the disease in the population.

12.23 The p-value will be less than the level of significance. For a test statistic falling exactly at the boundary of the critical region, the p-value will equal the level of significance. The p-value will be smaller for more extreme values of the test statistic.

12.25 a. The parameter of interest is p = population proportion of artists who are left-handed. The null value is .10. b. H_0: $p = .10$, H_a: $p > .10$. c. $\hat{p} = .12$. d. The p-value is the probability that the sample proportion would be .12 or larger (for a sample of $n = 150$) if the population proportion actually is .10. e. The test statistic is $z = \frac{\text{Sample estimate} - \text{Null value}}{\text{Null standard error}}$. The numerator is (.12 − .10).

12.27 The p-value will be smaller for Researcher B.

12.29 a. False. b. True. c. False. d. False.

12.31 a. Correct decision. b. Type 2 error. c. Type 1 error. d. Correct decision.

12.33 a. The company believes that the proportion requiring a hospital stay will decrease, but in fact it will not decrease. b. The company believes the proportion requiring a hospital stay will not decrease, but in fact it will. c. Type 1. d. Type 2.

12.35 a. Example: a diagnosis that leads to major surgery. b. Example: an infection that could be cured by antibiotics but gets very serious if untreated.

12.37 a. A Type 1 error occurs if the politician believes that more than one-half of voters in his district support the new tax bill when the proportion really is not more than one-half. The consequence is that he would vote for a bill that is not supported by a majority of the voters in his district. b. A Type 2 error occurs if the politician believes the proportion of voters in the district supporting the tax bill is not a majority when really it is a majority. The consequence is that that he would vote against a bill that is supported by a majority of the voters. c. Type 1 is probably more serious: voting for the bill that implements substantial change when actually a majority of voters does not support it. d. Type 1.

12.39 The power will be greater with the larger sample size.

12.41 a. .0358. b. .0228. c. .1379. d. .00001.

12.43 a. Yes. b. No.

12.45 a. The proportion of all stockbrokers believing the market will go up next year. b. The proportion of all mall visitors that buy something.

12.47 a. $z = 5.59$. b. $z = -10.61$.

12.49 Null standard error uses the null value p_0, whereas standard error uses the sample proportion.

12.51 a. H_0: $p \leq .50$, H_a: $p > .50$, p = proportion of all adult American Catholics who favor allowing women to be priests. b. The sample was randomly selected and the sample size is large enough. c. $z = 4.05$. d. p-value ≈ .00003. e. Reject the null hypothesis and decide in favor of the alternative hypothesis. f. .55 to .63. The conclusion is supported because .5 is not covered by the interval.

12.53 a. No, .07 > .05. b. Yes, .035 < .05.

12.55 *Step 1*: H_0: $p \geq .5$, H_a: $p < .5$, p = proportion of adults in the U.S. population who classify themselves as "very happy."
Step 2: The sample was randomly selected and the values of np and $n(1 - p)$ are large enough for a z-test.

Sample estimate = $\hat{p} = .47$; $z = -1.95$

Step 3: p-value = .0256.
Step 4: Reject the null hypothesis. *Step 5*: The data support the journalist's headline that a minority (less than .5) of all U.S. adults classify themselves as "very happy" but note that $\hat{p}$ was only slightly less than .5.

12.57 Probability of 17 or more successes for a binomial distribution with $n = 50$ and $p = .25$.

12.59 a. $P(X \geq 17) = .0983$. b. We cannot reject the null hypothesis using $\alpha = .05$ (but could reject it using $\alpha = .10$).

12.61 a. p-value = .0401, the area to the right of $z = 1.75$ under a standard normal curve. b. p-value = .9599, the area to the right of $z = -1.75$ under a standard normal curve.

12.63 a. p-value = .0233; reject the null hypothesis. b. p-value = .0375; reject the null hypothesis.

12.65 a. H_0: $p_1 - p_2 = 0$, or equivalently, $p_1 = p_2$, H_a: $p_1 - p_2 < 0$, or equivalently, $p_1 < p_2$. b. $\hat{p}_1 = .4$, $\hat{p}_2 = .6$. c. $\hat{p} = .5$. d. Test statistic is $z = -2.00$. Null standard error = .10. e. p-value = .0228 (in Table A.1); reject the null hypothesis.

12.67 $z = 3.26$, $p = .0006$; reject the null hypothesis and conclude that a higher proportion of women in the population would claim that they would return the money.

12.69 $z = 6.26$, p-value < .000000001. We can reject the null hypothesis. We can conclude that in the population(s) represented by the sample(s), the proportion of women who would say there should be "more strict" laws covering the sale of firearms is higher than the proportion of men who would say this.

12.71 $z = 2.40$, p-value ≈ .008. Reject the null hypothesis. The conclusion is that in the population(s) represented by the sample(s), a higher proportion of women than men would claim that they would date someone with a great personality even if they did not find the person physically attractive.

12.73 a. H_0: $p_1 - p_2 = 0$, or equivalently, $p_1 = p_2$.
H_a: $p_1 - p_2 > 0$, or equivalently, $p_1 > p_2$.
p_1 = proportion ever bullied in the population of short students.
p_2 = proportion ever bullied in the population of students who are not short.
b. *Step 2*: Sample sizes are sufficiently large so that observed counts in both categories (bullied or not) are greater than 10 for both groups.
Test statistic is $z = 3.02$.
Steps 3, 4, and 5: *p*-value = .001. Reject the null hypothesis.
c. The upper limit is 1. The complete interval is .09192 to 1.
d. The interval is entirely above 0, so it supports the alternative hypothesis.

12.75 a. $z = 0.690$; $p = .4902$, so fail to reject null hypothesis.
b. $z = 1.091$; $p = .2758$, so fail to reject null hypothesis.
c. $z = 2.439$; $p = .0146$, so reject the null hypothesis.
d. $z = 3.448$; $p = .0006$, so reject the null hypothesis.

12.77 a. 1000. b. .45. c. .45.

12.79 *p*-value.

12.81 Probably the finding was that there was no *significant* difference, in the statistical meaning, not the ordinary-language meaning. It could have been because there really was no difference or because the sample was too small to detect it and a Type 2 error was made.

12.83 a. Yes, the difference is statistically significant. b. This is not a contradiction. There is not much (if any) practical importance to the observed difference in incidence of drowsiness (6% vs 8%), but the large sample sizes led to a *statistically* significant difference. c. A statistically significant difference indicates that the difference in the population is not zero, but does not indicate that it has any practical significance.

12.85 a. .3776. b. With a sample of $n = 100$, the probability is .5740 that the null hypothesis will be rejected if the true probability of a correct guess is $p = .33$. c. Yes. d. Higher.

12.87 Steps 3 and 4 are different.

12.89 How large the difference in weight loss was for the two groups.

12.91 Step 1: H_0: $p = .10$, H_a: $p \neq .10$.
Step 2: Sample conditions are met. Sample proportion is = .0833; $z = -0.86$.
Step 3: Exact *p*-value = .395; *p*-value for *z* is .3898.
Step 4: Cannot reject the null hypothesis; result is not statistically significant.
Step 5: Cannot conclude that proportion of UCD students who are left-handed differs from the national proportion.

12.93 The answers will vary. An example is the alternative hypothesis that different proportions of men and women are unemployed. The populations are men and women, and the parameter of interest is the difference in proportions who are unemployed.

12.95 a. H_0: $p = .0849$, H_a: $p > .0849$. *p* is the proportion of the population born in October. b. *Step 2*: $n = 170$ and $p_0 = 31/365 = .0849$, so sample is large enough.
Sample estimate = $\hat{p} = .1294$, $z = 2.08$.
Step 3: *p*-value for z = .0188. Exact *p*-value = .031.
Step 4: Reject the null hypothesis.
Step 5: Conclude that people are more likely to be born in October than they would be if all 365 days were equally likely.

12.97 a. Population parameter = proportion of all people who suffer from this chronic pain who would experience temporary relief if taking the new medication. Hypotheses are H_0: $p = .7$, H_a: $p > .7$. b. Population parameter = proportion of all students at the university who would pay for the textbook ordering service. Hypotheses are H_0: $p \leq .05$, H_a: $p > .05$.

12.99 $z > 0$.

12.101 H_0: $p = .1$, H_a: $p > .1$; $\hat{p} = .2947$, $z = 8.95$, *p*-value ≈ 0; reject the null hypothesis; conclude that the proportion of the population who would pick 7 is greater than .1.

12.103 *Step 1:* H_0: $p = .5$, H_a; $p \neq .5$. p = proportion of population who prefer brand P.
Step 2: $n = 159$ and $p_0 = .5$, so the sample is large enough.
$\hat{p} = .503$. Null standard error = .03965. $z = 0.08$.
Step 3: *p*-value ≈ .94.
Step 4: We cannot reject the null hypothesis.
Step 5: We cannot conclude that the proportion within the population that prefers either brand is different from .5.

12.105 H_0: $p = .5$, H_a: $p > .5$, $z = 1.03$, *p*-value = .151. Cannot conclude that the first drink presented is more likely to be selected.

12.107 a. H_0: $p_1 - p_2 = 0$, or equivalently, $p_1 = p_2$, H_a: $p_1 - p_2 \neq 0$, or equivalently, $p_1 \neq p_2$.
p_1 = proportion of adult Canadians in February 2000 who favor marriages between people of the same sex, and
p_2 = proportion of adult Canadians in April 1999 who favor marriages between people of the same sex.
b. $z = 3.19$; samples are large enough and independent, *p*-value = .001, conclude the population proportions were significantly different. c. The confidence interval is entirely above 0, so it is evidence in favor of the alternative hypothesis.

12.109 a. *Step 1:* H_0: $p_1 - p_2 = 0$, or equivalently, $p_1 = p_2$, H_a: $p_1 - p_2 < 0$, or equivalently, $p_1 < p_2 \cdot p_1$ = proportion that would have heart attack in population of men if they regularly take aspirin. p_2 = proportion that would have heart attack in population of men if they regularly take placebo.
Step 2: $z = -5.00$. Combined $\hat{p} = .01328$. Null standard error = .001541.
Step 3: *p*-value < .000001 (or ≈ 0).
Steps 4 and *5*: We can reject the null hypothesis. In the population represented by the sample, the proportion that would have a heart attack is less if men were to take aspirin daily than if men were to take a placebo daily.
b. Type 1, people would take aspirin although it doesn't help; Type 2, people would not take aspirin and heart attacks that could be prevented would not be; Type 2 is more serious.

12.111 $z = 0.98$, *p*-value = .1635 (exact *p*-value = .189); do not reject H_0; cannot guess at significantly better than chance level.

12.113 H_0: $p \leq .50$, H_a: $p > .50$ (a majority were dissatisfied); $\hat{p} = .6104$; $z = 7.05$, *p*-value ≈ 0; reject the null hypothesis; conclude that in the population of U.S. adults in August 2000, a majority were dissatisfied with the quality of K-12 education.

12.115 a. $z = 0.62$, *p*-value = .266; do not reject H_0 for $\alpha = .05$.
b. $z = 4.59$; yes.

12.117 $z = 2.31$, *p*-value = .021; using $\alpha = .05$, conclude that population proportions are significantly different.

Chapter 13

13.1 a. Not true. The correct conclusion if the null hypothesis cannot be rejected is to "fail to reject the null hypothesis."
b. True.

13.3 a. Difference between the means of two populations.
b. Population mean of paired differences.

13.5 a. $\mu_1 - \mu_2$, where μ_1 is the mean height of the population of 25-year-old women, and μ_2 is the mean height of the population of 45-year-old women. b. The parameter of interest is μ_d = the mean difference in number of minutes late the two flights of interest are, for the population of all days over several years.

13.7 a. H_0: $\mu_1 - \mu_2 = 0$. b. H_0: $\mu_d = 0$.

13.9 We need some way to know if the difference between the sample statistic and the null value is larger than what would be expected by chance. Just knowing the difference without any reference to how large it would be by chance doesn't tell us much.

13.11 a. Because we state two hypotheses about the truth in the population and use data to assess them. b. Because we assess whether an observed difference between sample data and our null hypothesis value is substantial enough, or significant enough, to lead us to reject the null hypothesis value as the truth.

13.13 Yes. The hypotheses can (and should) be specified.

13.15 μ = mean percentage of net patient revenue spent on charity care for the population of all nonprofit hospitals. H_0: $\mu \geq 4\%$ and H_a: $\mu < 4\%$.

13.17 4.00.

13.19 a. .048. The figure should show shaded areas below −2.00 and above +2.00. b. .048. The figure should be the same as the figure for part (a).

13.21 a. Rejection region is $t \geq 1.83$; reject H_0. b. Rejection region is $t \leq -1.83$; reject H_0. c. Rejection region is $t \leq -1.83$; do not reject H_0. d. Rejection region is $|t| \geq 2.26$; do not reject H_0.

13.23 a. 2.121. b. 0.943. c. No. We need to know if the alternative is one-sided or two-sided, and the direction if one-sided.

13.25 a. No. Even one standard deviation below the mean is negative. b. Yes, n is large. c. Step 1: H_0: $\mu = 9.2$ versus H_a: $\mu \neq 9.2$, where μ = mean length of calls for seniors. Step 2: Sample size (200) is sufficiently large to proceed; $t = -1.70$. Steps 3, 4, and 5: df = 199; p-value = .09; we cannot reject the null hypothesis.

13.27 Step 1: H_0: $\mu = 9.2$ versus H_a: $\mu \neq 9.2$, where μ = mean length of calls for seniors. Step 2: Sample size (200) is sufficiently large to proceed; $t = -1.70$. Steps 3, 4, and 5: df = 199. Rejection region is $|t| \geq 1.98$. We cannot reject the null hypothesis.

13.29 Step 1: H_0: $\mu = 4.7$, H_a: $\mu < 4.7$, where μ = mean time to graduate (in years) for students who participated in the honors program in their first year of college. Step 2: Sample size (30) is sufficiently large to proceed; $t = -2.19$. Steps 3, 4, and 5: df = 29; p-value = .018. We can reject the null hypothesis and conclude that the mean time to graduate is lower for the population of students who participated in the honors program than it is for the general population of students.

13.31 $\bar{x} = 63.86$, $s = 2.086$, $t = -3.38$, df = 37, p-value = .001, reject H_0.

13.33 Different because the paired t-test is concerned with comparing two means, whereas a one-sample t-test is concerned with comparing one mean to a fixed value. Once the sample differences have been computed for paired data, the two tests are computationally exactly the same.

13.35 $t = -1.89$, df = 49, $.05 < p < .078$ (using df = 50). Exact $p = .0646$.

13.37 a. H_0: $\mu_d = 0$.
H_a: $\mu_d > 0$ (on average, student height > mom's height).
μ_d = mean "student height − mom's height" difference for the population of college students represented by the sample.
b. $t = 4.68$.
Sample statistic is observed mean difference in heights, $\bar{d} = 1.285$ inches. Null value is $\mu_d = 0$. Null standard error 0.2745.
c. df = 92. d. Reported p-value = 0.000. Reject the null hypothesis. e. p-value is essentially 0.

13.39 $t = 3.39$; reject the null hypothesis.

13.41 Rejection region is $t \geq 2.35$ (or 2.36 from Table A.2); reject the null hypothesis.

13.43 a. μ_d = mean weight loss (in pounds) in the first 3 weeks of following a diet plan for the population of all people who would follow the plan. b. H_0: $\mu_d = 10$, H_a: $\mu_d < 10$.
c. $t = \dfrac{8 - 10}{4/\sqrt{20}} = -2.236$. d. df = 19, $.015 < p < .03$.
Exact $p = .0188$. e. Reject the null hypothesis.

13.45 a. μ_d = mean difference in blood pressure (bp) while waiting to see a dentist and an hour after visiting the dentist (waiting bp − later bp). b. H_0: $\mu_d = 0$. H_a: $\mu_d > 0$ (Blood pressure is higher on average while waiting for the dentist than an hour after.) c. Step 2: Stem-and-leaf plot shows no outliers or skewness; $t = 2.99$. Step 3: p-value = .008. Steps 4 and 5: Reject the null hypothesis because $.008 < .05$. Conclude that on average, blood pressure is higher while waiting to see a dentist than it is an hour after the visit.

13.47 a. H_a: $\mu_1 - \mu_2 < 0$. b. H_a: $\mu_1 - \mu_2 < 0$.

13.49 a. Paired. b. Two-sample.

13.51 a. 1.414. b. 2.18.

13.53 a. H_0: $\mu_1 - \mu_2 = 0$, H_a: $\mu_1 - \mu_2 \neq 0$. b. $t = 2.78$ (as reported by Minitab). c. $t = \dfrac{(98.102 - 97.709) - 0}{\sqrt{\dfrac{(0.651)^2}{46} + \dfrac{(0.762)^2}{54}}} = \dfrac{0.393}{0.1413} = 2.78.$
d. Reject the null hypothesis. Conclude that men and women do not have equal mean body temperatures.

13.55 Plots show necessary conditions are satisfied; $t = 0.43$. Don't reject the null hypothesis for any reasonable value of α.

13.57 a. Unpooled $t = 3.97$, pooled $t = 3.96$; reject the null hypothesis for any reasonable value of α. b. Can use pooled because sample sizes and standard deviations are close.

13.59 Step 1: H_0: $\mu_1 - \mu_2 = 0$.
H_a: $\mu_1 - \mu_2 \neq 0$.
μ_1 = mean sleep hours for population of UC Davis students.
μ_2 = mean sleep hours for population of Penn State students.
Step 2: The sample sizes are sufficiently large to proceed.
For unpooled procedure, $t = -0.94$. Sample statistic = −0.18 hours. Standard error = 0.192. For pooled procedure, $t = -0.93$. Pooled standard deviation = 1.834, and pooled standard error is 0.193.
Step 3: p-value ≈ .35 for either procedure.
Steps 4 and 5: Do not reject the null hypothesis for any reasonable value of α.

13.61 a. From Minitab: $t = 1.89$, df = 13, p-value = .041, reject the null hypothesis. b. From Minitab: $t = 1.96$, df = 18, p-value = .033, reject the null hypothesis. c. Unpooled procedure is better because the smaller s is from the larger sample, but pooled is not too bad because sample standard deviations are similar. d. Null hypothesis can be rejected.

13.63 a. No. b. No. c. Yes.

13.65 a. If the null value is covered by a 95% confidence interval, the null hypothesis is not rejected and the test is not statistically significant at level .05. If the null value is not covered by a 95% confidence interval, the null hypothesis is rejected and the test is statistically significant at level .05. b. If the null value is covered by a 99% confidence interval, the null hypothesis is not rejected and the test is not statistically significant at level .01. If the null value is not covered by a 99% confidence interval, the null hypothesis is rejected and the test is statistically significant at level .01.

13.67 a. Reject H_0 for $\alpha = .05$. b. Do not reject H_0 for $\alpha = .025$. c. Reject H_0 for $\alpha = .025$. d. Do not reject H_0 for $\alpha = .05$.

13.69 a. Can't tell. The entire interval could be greater than 10, or it could include 10. b. The interval would not include the value 10. If it did, the null hypothesis would not have been rejected. c. The interval would include the value 10. Otherwise, the null hypothesis would have been rejected. d. The interval would not include the value 10. If it did, the null hypothesis would not have been rejected.

13.71 a. There are two separate questions of interest. For the first one, p = proportion of all adults who have a fear of going to the dentist. (The proportion can be converted to a percent.) For the second one, μ = mean number of visits made to a dentist in the past 10 years for the population of adults who fear going to the dentist. b. Confidence interval for each parameter.

13.73 a. $\mu_1 - \mu_2$, where μ_1 and μ_2 are the mean running times for the 50-yard dash for the populations of first-grade boys and girls, respectively. b. Both.

13.75 a. Estimate the difference in proportions of college men and women who would answer yes, $p_1 - p_2$, and test whether or not $p_1 - p_2 = 0$. May also want to estimate p_1 and p_2 separately with confidence intervals. b. Test H_0: $\mu_1 - \mu_2 = 0$ versus H_a: $\mu_1 - \mu_2 < 0$, where μ_1 and μ_2 are the mean IQs for the populations of children whose mothers smoke at least 10 cigarettes a day during pregnancy and for those whose mothers don't smoke during pregnancy.

13.77 a. 0.224; small. b. −0.4; closer to medium.

13.79 a. 68% of the differences fall in the range −2 to +6 and 95% fall in the range −6 to +10. b. It falls halfway between the mean and one standard deviation below the mean. c. .695. (The effect size is 0.5.) d. The probability that the null hypothesis H_0: $\mu_d = 0$ will be rejected, when in fact μ_d is 2 and the standard deviation of the differences is 4, for a one-sided test with $\alpha = .05$.

13.81 a. .632. b. .978. c. .979.

13.83 a. .53. b. .87. c. .99. d. .91.

13.85 It is more useful to compare effect sizes because they reflect the magnitude of the true effect. The p-value is a function of the effect size and the sample size.

13.87 a. 0.23. b. 0.459. c. 0.280. d. No. e. The effect size changes when p_0 changes even if the difference $p - p_0$ stays fixed, because the denominator depends on p_0.

13.89 Issues 2, 4, and 5.

13.91 a. All except issue 5 rely on knowing the p-value. For instance, for #2 it would be useful to know if the result was just barely statistically significant, or if the p-value was extremely small. b. Issues 2, 3, and 4. For instance, for #2 and #4, a confidence interval would allow us to see if the magnitude of the effect has real world importance. c. Issues 2, 3 and 4. For instance, #3 is only a concern if the sample size is small.

13.93 How many different relationships were examined.

13.95 a. Randomized experiment. b. Yes, because this was a randomized experiment. c. It could be single-blind, but not double-blind. The volunteers would have to know which drink they had before each session. The experimenter recording the length of time to exhaustion would not need to know which drink was consumed. This could be a problem because the volunteers may have altered their behavior to create the outcome they or the researchers wanted to see.

13.97 a. One-sided. b. Two-sided. c. One-sided. d. Two-sided.

13.99 a. Do not reject the null hypothesis. b. Reject the null hypothesis. c. The effect size was the same for both studies, but the conclusion was different because of the different sample sizes.

13.101 a. H_0: $\mu_d = 0$, H_a: $\mu_d > 0$, $t = 7.12$, df $= 169$, p-value ≈ 0. Reject H_0 for any reasonable value of α. b. "Significance" in part (a) refers to statistical significance. We conclude that the mean difference in ages is not 0. However, the observed magnitude of the mean difference in ages, 2.24 years, may not have much practical importance, as it is not a large difference.

13.103 The maximum possible value of $p_1 - p_2$ is 1.0, so a one-sided 97.5% confidence interval is 0.02 to 1.0. Because the interval does not cover 0, we can reject H_0 using $\alpha = .025$.

13.105 Reject the null hypothesis using $\alpha = .025$.

13.107 a. H_0: $\mu_1 - \mu_2 = 0$. H_a: $\mu_1 - \mu_2 < 0$.
μ_1 = mean estimate for population of participants who would be told Australia's population.
μ_2 = mean estimate for population of participants who would be told the U.S. population.
b. $t = -3.03$, p-value is .006. Reject the null hypothesis and conclude that in the population of students represented by this sample, the mean estimate of the population of Canada is lower if information about the population of Australia is provided than if information about the population of the United States is provided. c. The outlier within the Australia group has not affected the results.

13.109 a. 6.9 to 12.0 minutes. b. Reject the null hypothesis (using $\alpha = .05$) because 0 is not in the interval. c. 6.9 to infinity. d. Reject the null hypothesis using $\alpha = .025$, because 0 is not in the interval.

13.111 a. Type 1, patients are retested perhaps without cause. Type 2, patients need to be retested but are not, so the cholesterol readings may be too high. b. Type 2 seems more serious because a medical decision is made without using the most accurate data.

13.113 Step 1: H_0: $\mu_d \le 0$; H_a: $\mu_d > 0$; μ_d = mean "own height − dad height" difference in population of college men. Step 2: We assume that the sample represents a random sample from a larger population of college men. Delete unreasonable outlier, $d = 37$. Step 2 continued and Steps 3, 4, and 5: The test statistic is $t = 3.12$ and the p-value is .001. Reject the null hypothesis and conclude that in the population of college men, students' heights are greater, on average, than their fathers' heights.

13.115 Step 1: H_0: $\mu_1 - \mu_2 = 0$, or equivalently, $\mu_1 = \mu_2$.
H_a: $\mu_1 - \mu_2 > 0$, or equivalently, $\mu_1 > \mu_2$.
μ_1 = mean hours of sleep for population of students who do not feel sleep deprived.
μ_2 = mean hours of sleep for population of students who do feel sleep deprived.
Step 2: The sample sizes are sufficiently large to proceed.
Steps 3, 4, and 5: There is an outlier at 0 that it may make sense to remove (response was supposed to be typical sleep per night). In either case, p-value $\approx$ .000. Reject the null hypothesis and conclude that in the population of students represented by this sample, students who feel sleep deprived do in fact get less sleep on average than does the general population.

Chapter 14

14.1 a. $\hat{y} = 119 - 1.64(40) = 53.4$. b. $y - \hat{y} = 50 - 53.4 = -3.4$.

14.3 a. β_0 = intercept; β_1 = slope. b. $E(Y_i) = \beta_0 + \beta_1 x_i$.

14.5 a. Slope = 0.9. Mean blood pressure increases 0.9 per 1-year increase in age. b. Estimated systolic blood pressure = 130. c. Residual = −5. d. The man's blood pressure is 5 points lower than the predicted blood pressure for someone his age.

14.7 a. Slope = 2.90. Mean desired weight increases 2.90 pounds per each 1-inch increase in height. b. b_1. c. 123.1 pounds.

14.9 a. $x = 70$. b. $y = 180$. c. $E(Y) = 170$. d. $180 - 170 = 10$. e. 170. f. 10.

14.11 a. $r^2 = \dfrac{\text{SSTO} - \text{SSE}}{\text{SSTO}} = \dfrac{500 - 300}{500} = .40$.
b. $r^2 = \dfrac{200 - 40}{200} = .80$.
c. $r^2 = \dfrac{80 - 0}{80} = 1.00$.
d. $r^2 = \dfrac{100 - 95}{100} = .05$.

14.13 a. $s = 2.837$ is roughly the average deviation of y-values (in degrees Fahrenheit) from the sample regression line.
b. Geographic latitude explains 91.7% of the variation in mean April temperatures.

14.15 a. $\hat{y}_i = 513.79$, or about 514 feet.
b. $e_i = 11.21$, or about 11 feet.
c. Approximately 514 ± 100 feet.
d. 650 feet is more than 2 standard deviations from the mean distance for drivers 21 years old, so it would be unusual.

14.17 a.

$e = y - \hat{y}$	−1	1	−2	2	−2	2

The sum of residuals is $\sum e_i = -1 + 1 - 2 + 2 - 2 + 2 = 0$.
b. SSE = 18.
c. $s = \sqrt{\dfrac{18}{6 - 2}} = 2.12$.

14.19 a. SSTO = 1746.8
b. SSE = 144.9
c. $r^2 = \dfrac{1746.8 - 144.9}{1746.8} = 0.917$.

14.21 a. The p-value is not less than .05, so the result is not statistically significant.
b. .552.

14.23 a. H_0: $\beta_1 = 0$ versus H_a: $\beta_1 \neq 0$, where β_1 = slope of the regression equation.
b. Reject the null hypothesis; conclude that there is a linear relationship. The value of the t-statistic is -14.11 and the p-value is 0.000.
c. $t = \dfrac{-1.6436 - 0}{0.1165} = -14.11.$
d. The same as for the test in part (a), so 0.000.

14.25 a. No, p-value = .165. Do not reject H_0: $\beta_1 = 0$.
b. .165.

14.27 a. b_1.
b. 3.84 to 10.16. With 95% confidence, we can say that in the population of college men, the mean increase in weight per 1-inch increase in height is somewhere between about 3.84 and 10.16 pounds.
c. The confidence interval for the slope contains 5, so 5 is a possible value, but there are many other possible values supported by the interval as well.

14.29 a. H_0: $\rho = 0$ versus H_a: $\rho \neq 0$. (It may be that the researchers' alternative hypothesis was $\rho > 0$.)
b. The researchers have decided to reject the null hypotheses, quoting p-values less than .001 and less than .05.

14.31 a. Prediction interval for a value of y.
b. Confidence interval for the mean.

14.33 a. Grade point average of an individual who misses two classes per week.
b. Mean grade point average of all students who miss two classes per week.

14.35 a. 78.02 to 82.67. With 95% confidence, we can say that the mean pulse rate after marching is between 78.02 and 82.67 for those in the population whose pulse is 70 before marching.
b. 65.50 to 95.19. In a population of individuals with a pulse rate of 70 before marching, about 95% of the individuals will have a pulse between 65.50 and 95.19 after marching.

14.37 a. $\hat{y}_i = 3.59 + (.9667)(40) = 42.258$.
b. The width of the prediction interval reflects the variation among the individual ages of husbands married to women 40 years old. The confidence interval is narrow because, with $n = 170$, the mean age can be estimated with good precision.

14.39 $42.258 \pm (2 \times 0.313)$, which gives approximately the interval provided by Minitab. A more exact interval can be computed using $t^* = 1.974$, the multiplier for df $= n - 2 = 168$.

14.41 a. Conditions 2 (no outliers) and 4 (normality).
b. Conditions 1, 2, and 3.
c. Condition 5.
d. Conditions 1, 2, and 3.

14.43 There is an outlier; aside from the outlier, the distribution might be normal.

14.45 a. There appears to be an outlier, so Condition 2 is violated.
b. The outlier is so inconsistent with the remainder of the data set that it is almost certainly a mistake. If possible, check the source of the data to see if a mistake has been made.

14.47 a. $b_1 = -0.269$, so the estimated *decrease* is 0.269 hours.
b. $-0.26917 \pm (2 \times 0.06616)$, or about -0.401 to -0.137.
c. H_0: $\beta_1 = 0$ versus H_a: $\beta_1 \neq 0$.
d. $t = -4.07$, df = 114.
e. The relationship is statistically significant (p-value = 0.000).

14.49 $b_0 = 7.56$. This is the estimated mean hours of sleep for students who studied 0 hours.

14.51 (3.746, 9.750). Of students in the population who studied 3 hours, about 95% slept between 3.746 and 9.75 hours. Or the probability is .95 that the hours of sleep will be in the given interval for a student randomly selected from the population of students who studied 3 hours.

14.53 a. There is a positive association and the relationship might be linear although it's difficult to be certain because there are so few observations and there may be outliers.
b. $\hat{y} = 25.38 + 0.7069x$.
c. $\hat{y} = 78.4$.
d. 72.97 to 83.83.

14.55 a. $R^2 = 44.7\%$. Knowing fathers' heights explains about 44.7% of the variation in sons' heights.
b. $r = \sqrt{0.447} = 0.67$. The sign is positive because the slope is positive.

14.57 Step 1: The hypotheses are H_0: $\beta_1 = 0$ versus H_a: $\beta_1 \neq 0$.
Step 2: Examine residual plots as well as a scatter plot of y versus x. The test statistic is $t = \dfrac{b_1}{s.e.(b_1)}$, and statistical software will provide the value.
Step 3: Statistical software will provide the p-value. A t-distribution with df $= n - 2$ is used to find this value.
Step 4: Reject the null hypothesis if the p-value is less than the specified significance level (which most often is $\alpha = .05$).
Step 5: Write a conclusion that describes whether or not the relationship between the explanatory and response variables is statistically significant.

14.59 a. There is a positive association with a linear pattern.
b. $\hat{y} = 22.693 + 0.876x$.
c. $s = 5.051$; $r^2 = 35.7\%$.
d. An approximate 95% interval is about 0.575 to 1.177. With 95% confidence, we can say that the slope of the regression line in the population is somewhere in this interval.
e. *dom* = 23 + *nondom*; the intercept is 22.7 for the sample regression line and the value 1 is contained within the interval determined in part (d) for the population slope. Thus, the statement is reasonable.

14.61 a. There is a strong, positive association between weight and chest girth. The two observations with the greatest chest girths could be outliers, or there might be a gently curving pattern.
b. $\hat{y} = -206.9 + 10.76x$. c. $R^2 = 95.1\%$. d. $\hat{y}_i = 223.5$.
e. 95% prediction interval is (180.47, 266.49). f. 95% confidence interval is (211.41, 235.55).

14.63 a. Minitab gives the 90% prediction interval as 67.050 inches to 77.731 inches. b. The interval from part (c) of the previous exercise was 68.517 inches to 75.512 inches. The interval computed in part (a) of this problem is wider because the estimated standard deviation from the regression line is larger than it is in the previous exercise due to the inclusion of the outlier. The inflated standard deviation creates a wider interval describing the variation among individual heights. This makes sense, because with the outlier included, there is indeed more variability among the heights in the sample. c. The residual plot clearly shows that there is an outlier in the data set.

Chapter 15

15.1 a. Appropriate. Both variables are categorical.
b. Not appropriate. Both variables are quantitative.

15.3 a. Null: Exam cheating is not related to feelings about importance of religion. Alternative: Exam cheating is related to feelings about importance of religion. b. We cannot conclude that the two variables are related. c. 2.

15.5 a. p-value = .05. b. $.10 < p\text{-value} < .25$, exact is .153.
c. $p\text{-value} < .001$. d. $p\text{-value} > .50$.

15.7 a. 3.84. b. 16.81 (df = 6). c. 12.59.

15.9 a. H_0: Sex of student and type of class taken are not related for the population of students. H_a: Sex of student and type of class taken are related for the population of students.
b. df = 1. c. 1.258. d. p-value = .262. e. Do not reject the null hypothesis for $\alpha = .05$. The relationship between sex of student and type of class taken is not statistically significant.

15.11 a. No. b. Favor death penalty: 40.4% favor marijuana legalization; oppose death penalty: 40.0% favor marijuana legalization. c. H_0: Opinion about death penalty and opinion about marijuana legalization are not related. H_a: Opinion about death penalty and opinion about marijuana legalization are related. d. $\chi^2 = 0.017$, df = 1, *p*-value = .896; cannot reject the null hypothesis.

15.13 For expected counts, the proportion with an ear infection is .22 within each treatment.

15.15 Step 1: H_0: Typical grades and seatbelt use are not related for the population of twelfth-graders; H_a: Typical grades and seatbelt use are related for the population of twelfth-graders.
Step 2: Expected counts are all greater than 5 so proceed with the chi-square test. Test statistic is $\chi^2 = 77.776$, df = 2.
Steps 3, 4, and 5: *p*-value = .000.
Reject the null hypothesis. Infer that typical grades and seatbelt use are related for the population of twelfth-graders.

15.17 a. Older age group reads newspapers more frequently than younger groups do. b. H_0: Age group and frequency of reading newspapers are not related. H_a: Age group and frequency of reading newspapers are related. c. $\chi^2 = 96.488$, df = 6, *p*-value ≈ 0; reject the null and conclude that the age and frequency of reading newspapers are related.

15.19 $\chi^2 = 1186(313 \times 247 - 461 \times 165)^2 / (774 \times 412 \times 478 \times 708) = 0.017$.

15.21 a. The researcher probably thought that short students would be more likely to be bullied. That is a one-sided hypothesis, so a *z*-test is appropriate. b. The researchers want to know if proportions bullied were different for short and not short students.
c. Step 1: H_0: $p_1 - p_2 = 0$ versus Ha: $p_1 - p_2 > 0$
p_1 = proportion ever bullied in population of short students
p_2 = proportion ever bullied in population of students not short
Step 2: The test statistic is $z = 3.02$. Steps 3, 4, and 5: *p*-value = .001. Reject the null hypothesis, and conclude that a higher proportion of short students than non-short students have been bullied.

15.23 No. The response variable has three categories.

15.25 a. Given that 2 out of 10 participants have reduced pain, what is the probability that both of them would be in the magnet-treated group?

15.27 a. H_0: Seal performance and temperature are unrelated; H_a: Seal performance depends on temperature.
b. H_0: Population proportions planning to vote for each candidate were the same in September and June;
H_a: Population proportions planning to vote for each candidate are differed in September and June. c. H_0: People who take a typing class do not have a reduced chance of getting carpal tunnel syndrome; H_a: People who take a typing class have a reduced chance of getting carpal tunnel syndrome.

15.29 The larger the sample size, the more likely it is that a sample difference will be statistically significant.

15.31 a. Before: (190 + 26)/400 = .54, or 54%; after: (190 + 14)/400 = .51 , or 51%. b. 26 + 14 = 40.
c. H_0: $p \geq .5$ versus H_a: $p < .5$. d. *p*-value = .0403; reject the null hypothesis.

15.33 a. 100 for each of the three categories. b. 250, 250, and 500, respectively.

15.35 a. No. b. Yes. c. Yes. d. No. e. No. f. No.

15.37 a. H_0: $p = \frac{1}{10}$ 1/10 for each of the 10 numbers, where p = the probability of a student choosing that number. The alternative hypothesis is that not all of the probabilities are 1/10$\frac{1}{10}$.
b. 104.32. c. 9. d. The *p*-value is essentially 0. e. Using any reasonable level of significance the conclusion is the same. At least two of the probabilities are not 1/10. Students are not equally likely to choose each of the digits.

15.39 a. HH, TH, HT, TT, $X = 0$, $p = 1/4$; $X = 1$, $p = 1/2$; $X = 2$, $p = 1/4$. b. H_0: $p_0 = .25$, $p_1 = .50$, $p_2 = .25$. c. $\chi^2 = 7.2$, df = 2, .025 < *p*-value < .05 (.027). d. Exactly one head occurred more often than expected. Perhaps some students made up their data and may have had a tendency to claim they got one head in the two flips.

15.41 $\chi^2 = 1.766$, df = 2, .25 < *p*-value < .50 (.413).

15.43 $\chi^2 = 15.8$, df = 5, .005 < *p*-value < .01. Reject H_0: proportions are as had been stated at M&M website, and conclude at least two proportions are not as had been stated.

15.45 b. df = (3 − 1)(4 − 1) = 6. c. Yes, the relationship is statistically significant (*p*-value is reported to be .027). e. Expected count = 40.93.

15.47 a. There appears to be a relationship. As the number of ear pierces increases, the percentage with a tattoo also increases.

Pierces	% with tattoo
2 or less	7.4% (40/538)
3 or 4	13.4% (58/432)
5 or 6	27.6% (77/279)
7 or more	42.1% (53/126)

b. H_0: Number of ear pierces and likelihood of having a tattoo are not related. H_a: Number of ear pierces and likelihood of having a tattoo are related. c. $\chi^2 = 119.279$, df = 3, *p*-value < .001. Decide H_a.

15.49 a. Sheep performed slightly better. For the Sheep, 70.5% were Stars and 29.5% were Duds. For the Goats, 60% were Stars and 40% were Duds. An appropriate graph is a bar graph displaying these percentages. b. The hypotheses are H_0: $p_1 - p_2 = 0$ versus H_a: $p_1 - p_2 > 0$. p_1 = proportion of Stars in the population of students who are Sheep; p_2 = proportion of Stars in the population of students who are Goats
c. Step 2: The test statistic is $z = 1.52$. Steps 3, 4, and 5: *p*-value = .064 (reported in output). Do not reject the null hypothesis. Based on these data it cannot be concluded that the proportions of Stars differ in the populations of Sheep and Goats.

15.51 a. The conditions are met. b. $\chi^2 = 108.1$, df = 12.

15.53 a. H_0: $p_1 = .257$, $p_2 = .322$, $p_3 = .169$, $p_4 = .149$, $p_5 = .103$ (U.S. proportions). H_a: probabilities are not all as specified in the null hypothesis. b. Step 1: Hypotheses given in part (a). Step 2: $\chi^2 = 10.6$, df = 4.
Expected counts follow:

Household size	1	2	3	4	5
Expected	746.328	935.088	490.776	432.696	299.112

Steps 3, 4, and 5: *p*-value = .031. Reject the null hypothesis. Conclude that the observed distribution of household sizes in the GSS survey is inconsistent with the U.S. distribution.

15.55 5.704.

15.57 a. H_0: Sex of student and perception of weight are not related; H_a: Sex of student and perception of weight are related.
b. Reject the null hypothesis, because *p*-value = .000.There is a statistically significant relationship between sex of student and weight perception. c. There are large contributions in the underweight and overweight categories for each sex. This reflects a large difference between the sexes in those categories.

15.59 Between 0 and 5.99.

15.61 a. Too many cells have expected counts less than 5.
b. $\chi^2 = 15.5$, df = 4, .001 < *p*-value < .005. There is a statistically significant relationship. c. The largest contributions occur in cells where student eye color and the eye color the student finds attractive are both blue or both brown.

15.63 a. Homogeneity. b. Step 2: Test statistic is $\chi^2 = 7.96$, df = 3. Steps 3, 4, and 5: *p*-value = .047. Reject the null hypothesis.

Conclude that opinion differed in the 2 years. Examination of conditional percentages for the 2 years shows that a higher percentage were "completely satisfied" in 1999 than in 2000, while a higher percentage were "completely dissatisfied" in 2000 than in 1999.

15.65 a. Step 1: H_0: Sex of student and typical seat location are not related; H_a: Sex of student and typical seat location are related. Step 2: $\chi^2 = 7.112$; df = 2. Step 3: *p*-value = .029 (given in Minitab output). Steps 4 and 5: Reject the null hypothesis. Conclude that typical seat location is related to sex of student. The conditional percentages by sex show that males are more likely than females to sit in the back and females are more likely than males to sit in the front. b. The calculated chi-square value is 7.112. The contributions to chi-square are largest for the "back" location for males (3.489) and females (1.939). This reflects a large difference between the males and females for the likelihood of sitting in the back.

15.69 c. Students who live off campus are more likely to have missed class due to drinking alcohol. d. H_0: Where students live and whether they have missed class due to drinking alcohol are not related. H_a: Where students live and whether they have missed class due to drinking alcohol are related. e. $\chi^2 = 12.702$, df = 1, *p*-value ≈ 0. The observed relationship is statistically significant.

Chapter 16

16.1 a. Appropriate. b. Not appropriate. Not a comparison of independent groups.

16.3 a. Mean body mass index is equal for the populations in the three age groups; H_0: $\mu_1 = \mu_2 = \mu_3$. b. Mean body mass index is not equal for the populations in the three age groups.

16.5 a. Mean height is not the same in the three seating location populations of female students. b. Mean height increases from front to back.

16.7 a. Reject the null; *F*-statistic is greater than 3.89, the critical value. b. Cannot reject the null. Critical value is 2.35.

16.9 a. 18 − 29 and 60+ age groups differ from the other two age groups. b. 95% confident that all six intervals capture the corresponding population parameters.

16.11 a. The data could be graphed using a comparative dotplot or a side-by-side boxplot. Both graphs show that measurements tend to be larger in Labs 2 and 5 and tend to be smaller in Lab 4. The data for Labs 1 and 3 fall in between. b. Variations is about the same for each lab and there are no outliers.

16.13 a. H_0: $\mu_1 = \mu_2 = \mu_3$ versus H_a: not all μ_i are the same, where μ_i is the population mean testosterone level for all women in occupation group *i*. b. There is an extreme outlier in one group and the variation substantially differs among the groups.

16.15 The completed table follows:

Source	df	SS	MS	F
Between groups	5	40	8	1.333
Error	10	60	6	
Total	15	100		

16.17 The sample means would all equal the same value.

16.19 a.

Source	df	SS	MS	F	p
Caffeine	2	61.40	30.70	6.18	.006
Error	27	134.10	4.97		
Total	29	195.50			

b. $s_p = \sqrt{4.97} = 2.23$. c. Infer that the population means are not the same for the three caffeine amounts.

16.21 Each interval has the form $\bar{x}_i \pm 2.01\frac{0.4058}{\sqrt{11}}$, or $\bar{x}_i \pm 0.246$.

16.23 a. $\bar{x} = 8, \bar{x}_1 = 4, \bar{x}_2 = 11, \bar{x}_3 = 9$. b. SS Groups = 78. c. SS Total = 118. d. SSE = 118 − 78 = 40. e. *F* = 5.85, df = 2, 6.

16.25

Source	df	SS	MS	F	p
Treatment	3	150	50	5	0.01
Error	20	200	10		
Total	23	350			

16.27 a. Null is that median ratings would be equal for populations of students from the four types of hometowns; alternative is that median ratings would not be equal for the four types of hometowns. b. *p*-value = 0.000; conclude that median ratings would not be equal. c. Big city, median = 5.

16.29 a. Ordinal. b. The null hypothesis is that the median response is the same for the populations of college men and women represented by this sample. The alternative hypothesis is that the population medians are not equal. c. 12. Among women, 69.3%; among men, 43.5%. d. The null hypothesis defined in part (b) can be rejected.

16.31 a. H_0: Median testosterone levels are the same for the three populations of occupational groups. H_a: Population medians are not all the same. b. Decide in favor of H_a (*p*-value = .002). c. The sample medians are 3.4 for the "advanced degree" group, 2.2 for the "no advanced degree" group, and 2.0 for the "not employed" group.

16.33 a. Interaction. Amount of difference in study time between members and nonmembers of Greek organizations depends on whether students are male or female. b. Not an interaction. The statement describes the main effect of treatment and notes that it is the same for each age group.

16.35 a. Males. b. Students from big cities rate rock music lower than students from other types of hometowns. c. Pattern of relationship between rating and hometown is about the same for males and females, so the interaction is weak or nonexistent.

16.37 a. The amount of difference between mean GPA of men and women would depend on seat location. b. The amount of difference between men's and women's mean GPAs would be the same in each seat location.

16.39 b. As the bacteria amount increases, the difference in mean rot for 10°C and 16°C increases.

16.41 The sample sizes differ for each group. Smaller sample sizes produce wider confidence intervals.

16.43 a. H_0: $\mu_1 = \mu_2 = \mu_3 = \mu_4 = \mu_5$ versus H_a: Not all μ_i the same. b. Population might be all public school children of this age in Ohio. We cannot be certain whether Ohio is representative of the nation as a whole. c. Decide in favor of H_a. Not all the means are the same. d. Perhaps children who engage in more violent behaviors also like to watch more television.

16.45 There may be interaction. The difference between boys and girls is smaller in the 6+ hours of television group than it is in other television watching groups.

16.47 Regression methods could be used. Both variables are quantitative.

16.49 SSE = 15,286, $s_p = \sqrt{46.18} = 6.796$.

16.51 a. 0.0846 to 0.4824. b. 0.1164 to 0.4506. c. Confidence level = 1 − (.05/3) = .9833 for each interval. d. Fisher gives narrowest interval, and Bonferroni gives widest. Fisher gives 95% confidence for each interval, whereas Bonferroni and Tukey give 95% family confidence level (so higher confidence for each individual interval).

16.53 a. The 3 means are 4.359, 4.278 and 5.967, so the mean for Region 3 is much higher than the other two. b. The aroma ratings are generally higher for wines from Region 3 than for Regions 1 and 2. The boxplot for Region 3 shows more spread than the boxplots for the other regions. c. Null: Population mean aroma ratings are the same for the three regions. Alternative: At least one of the population mean aroma ratings is not the same as the others. $F = 18.05$, p-value ≈ 0. Reject the null hypothesis.

16.55 a. Students who say that religion is very important in their lives tend to study more. The median and the third quartile are greater for this group than for the other two groups. b. Null: Population mean study hours is the same for the three "importance of religion" groups. Alternative: At least one of the population means is not the same as the others. p-value ≈ 0. Reject the null hypothesis. c. Yes, they satisfy necessary conditions.

16.57 a. The means and medians decrease as the years of education increase.

	Mean	Median
Not HS	3.970	3
High school	3.156	3
Junior college	2.436	2
Bachelor's	2.200	2
Graduate	1.843	1

b. Null: Population median television watching amount is the same for the five degree groups. Alternative: At least one of the population medians is not the same as the others. p-value ≈ 0. c. Null: Population mean television watching amount is the same for the five degree groups. Alternative: At least one of the population means is not the same as the others. $F = 16.22$ and the p-value ≈ 0.

16.59 a. There may possibly be a weak interaction. b. On average, males miss more classes than females. On average, students who prefer to sit in the front miss fewer classes than the others and students who prefer to sit in the back miss more.

Chapter 17

17.1 a. No, the entire population was already measured. b. Yes, assuming that the mechanism causing geyser eruptions stays the same.

17.3 No. This is an observational study. Children cannot be randomly assigned to adoptive parents.

17.5 The study needs to be a randomized experiment.

17.7 a. The null hypothesis is that it will not rain and the alternative is that it will rain. Type 1 error is that it does not rain, but you take an umbrella. Type 2 error is that it does rain but you do not have an umbrella. b. Reasoning is what counts here, but Type 2 error is probably more serious because you would get wet. So you probably would want to act as if the alternative hypothesis will be true and carry an umbrella.

17.9 a. Yes, because the study was a randomized experiment. b. Yes, those men are probably representative of healthy men in the same age group, and possibly a larger population.

17.11 Answers will vary, but make sure the study is an observational study rather than a study in which random assignments are used.

17.13 Answers will differ for each student, but make sure that the study either used random assignment of the treatments to participants, or randomized the order of the treatments if each individual received all treatments.

17.15 a. Yes, the wording is justified. It does not imply that there is a cause-and-effect relationship. b. Answers will differ for each student. An example of a conclusion that is not justified is "Aggressive anti-smoking campaign reduces deaths from heart disease."

17.17 a. Explanatory variable: Calcium supplement or placebo. Response variable: Blood pressure. b. Yes.

17.19 a. Explanatory variable: Tea consumption. Response variable: Successful conception or not. b. Yes.

17.21 Observational study; the headline is not justified.

17.23 Observational study; the headline is not justified.

17.25 The results can probably be applied to the larger population of African-American teens in the United States.

17.27 Probably. We are not told who was in the sample, but there is no indication that it was an unrepresentative group with regard to diet and conception.

17.29 a. Statistical significance; the null hypothesis is that energy and fat intake are not related to tea consumption. b. Confounding variables.

17.31 a. There was an association, but it was not strong enough to be *statistically* significant for the sample size used. b. Two possibilities are suggested in item 10 in Section 17.7, and both possibilities exist here. We are not given enough information about the details of the prior research to be able to judge which reason is more likely.

17.33 a. Amount walked per day (less than 1 mile, between 1 and 2 miles, more than 2 miles) and whether the man died during the study (yes, no). b. A test for the difference in two proportions would be appropriate.

Text Credits

This page constitutes an extension of the copyright page. We have made every effort to trace the ownership of all copyrighted material and to secure permission from copyright holders. In the event of any question arising as to the use of any material, we will be pleased to make the necessary corrections in future printings. Thanks are due to the following authors, publishers, and agents for permission to use the material indicated.

Chapter 1 **5** Case Study 1.5: *International Journal of Psychiatry in Medicine* by Koenig, H. G., L. K. George, J. C. Hays, and D. B. Larson. Copyright 1998 by Baywood Pubg. Co. Inc. Reproduced with permission of Baywood Pubg. Co. Inc. **5** Case Study 1.6 and Table 1.5: From *The New England Journal of Medicine,* 1989, Vol. 321(3), pp. 129–135. Final report on the aspirin component of the ongoing Physicians' Health Study: Steering Committee of the Physicians' Health Study Research Group. Copyright © 1989 Massachusetts Medical Society. All Rights Reserved. **6** Case Study 1.7: "Sad, Lonely World Discovered in Cyberspace," by A. Harmon, *The New York Times,* August 30, 1998, p. A3. Reprinted with permission of the New York Times Company.

Chapter 2 **22** Table 2.3: From *Nature* 1999, Vol. 399, pp. 113–114. Myopia and ambient lighting at night. Quinn, G. E., C. H. Shin, M. G. Maguire, et al. Reprinted by permission of Macmillan Publishers Ltd. Copyright © 1999.

Chapter 3 **72** Figure 3.3: *The Journal of Consumer Research,* Vol. 16, (1), pp. 119–124. Some exploratory findings on the development of musical tastes. Holbrook, M. B., and R. M. Schindler. Copyright © 1989 by The University of Chicago Press. Reprinted with permission.

Chapter 6 **193** Case Study 6.1: "Lead Exposure Linked to Bad Teeth in Children," by S. Sternberg, *USA Today*. June 23, 1999. Internet. Reprinted with permission.

Chapter 12 **476** Example 12.17: Reproduced from *Pediatrics,* Vol. 102, pp. 879–884. A novel use of Xylitol sugar in preventing acute otitis media, Uhari, M., T. Kenteerkar and M. Niemala. Copyright © 1998 by the AAP. **486** Case Study 12.1: Excerpts from "Probability experts may decide Pennsylvania vote," by P. Passell, *The New York Times,* April 11, 1994, p. A15. Copyright © 1994 *The New York Times.* All rights reserved. Used by permission and protected by the Copyright Laws of the United States. The printing, copying, redistribution, or retransmission of the material without express written permission is prohibited. **496** Exercise 12.107: Copyright © 2000 The Gallup Organization.

Chapter 13 **522** Example 13.9: Reproduced from *Pediatrics,* Vol. 102, pp. 879–884. A novel use of Xylitol sugar in preventing acute otitis media, Uhari, M., T. Kenteerkar and M. Niemala. Copyright © 1998 by the AAP. **525** Example 13.10: Reproduced with permission from *Pediatrics,* Vol. 104(1), p. e5, "The effects of different resistance training protocols on muscular strength and endurance development in children." Faigenbaum A. D., W. L. Westcott, R. L. Loud, and C. Long. Copyright © 1999 by the AAP. **532** Case Study 13.1: Siegel, R., J. Maté, M.B. Brearley, G. Watson, K. Nosaka and P. B. Laursen (2010). "Ice slurry ingestion increases core temperature capacity and running time in the heat," *Medicine & Science in Sports & Exercise,* Vol. 42, No. 4, pp. 717–725. **538** Exercise 13.39: From *New England Journal of Medicine,* 2000, Vol. 342(12): 861–867. "A prospective study of holiday weight gain." Yanovski, J. A., S. Z. Yanovski, K. N. Sovik, T. T. Nguyen, P. M. O'Neil, and N. G. Sebring. Copyright © 2000 Massachusetts Medical Society. All rights reserved.

Chapter 15 **584** Example 15.1: Reproduced from *Pediatrics,* Vol. 102, pp. 879–884. A novel use of Xylitol sugar in preventing acute otitis media, Uhari, M., T. Kenteerkar and M. Niemala. Copyright © 1998 by the AAP.

Chapter 17 **657** Example 17.4: *Journal of the American Medical Association,* 1998, Vol. 279(14), pp. 1083–1088. "Visual Processing Impairment and Risk of Motor Vehicle Crash Among Older Adults." Owsley, C., K. Ball, G. McGwin, Jr, M. E. Sloane, D. L. Roenker, M. F. White, and E. T. Overley. Reprinted with permission American Medical Association. **665–666** Exercises 17.18, 17.21, 17.22, 17.26, 17.30, 17.33: From *New England Journal of Medicine,* 1998, Vol. 338(2), pp. 94–99, "Effects of Walking on Mortality among Nonsmoking Retired Men," Hakim, A. A., H. Petrovitch, C. M. Burchfiel, G. W. Ross, B. L. Rodriguez, L. R. White, K. Yano, J. D. Curb, and R. D. Abbott. Copyright © 1998 Massachusetts Medical Society. All rights reserved. **665–666** Exercises 17.17 to 17.34, Study #3: Caan B, Quesenberry CP, Coates AO (1998). Differences in fertility associated with caffeinated beverage consumption. *American Journal of Public Health,* 88(2):270–274 Reprinted with permission from APHA Books.

Index

Note: Index entries referring to content in the Supplemental Chapters on the companion website (http://www.cengage.com/statistics/Utts4e use the format *S#-#* [e.g., *S2-12* would refer to Supplemental Chapter 2, page 12].

D

E

Q

R

U

V

W

X

Y

Z